Integrated Art:
Weaving illustrations into the text

Flowering Plant Reproduction

34.1 Angiosperm Reproduction

Although reproduction varies greatly among the members of the plant kingdom, we focus in this chapter on reproduction among flowering plants. While the evolution of their unique sexual reproductive features, flowers and fruits, have contributed to their success, angiosperms also reproduce asexually.

Asexual Reproduction

In **asexual reproduction**, an individual inherits all of its chromosomes from a single parent and is, therefore, genetically identical to that parent. Asexual reproduction produces a "clone" of the parent.

In a stable environment, asexual reproduction may prove more advantageous than sexual reproduction because it allows individuals to reproduce with a lower investment of energy and maintain successful traits. A common type of asexual reproduction, called *vegetative reproduction*, results when new individuals are simply cloned from parts of the parent.

Vegetative reproduction in plants varies:

Runners. Some plants reproduce by means of runners—long, slender stems that grow along the surface of the soil. The strawberry plant shown here reproduces by runners. Notice that at node regions on the stem, adventitious roots form, extending into the soil. Leaves and flowers form, and a new shoot is sent out, continuing the runner.

Rhizomes. Rhizomes are underground horizontal stems that create a network underground. As in runners, nodes give rise to new flowering shoots. The noxious character of many weeds results from this type of growth pattern, but so do grasses and many garden plants such as irises. Other specialized stems, called tubers, function in food storage and reproduction. White potatoes are specialized underground stems whose "eyes" give rise to new plants.

Suckers. The roots of some plants produce "suckers," or sprouts, that give rise to new plants, such as found in cherry, apple, raspberry, and blackberry plants. When the root of a dandelion is broken, which may occur if one tries to pull it from the ground, each root fragment may give rise to a new plant.

Adventitious Plantlets. In a few species, even the leaves are reproductive. One example is the house plant *Kalanchoë daigremontiana*, familiar to many people as the "maternity plant," or "mother of thousands." The common names of this plant are based on the fact that numerous plantlets arise from meristematic tissue located in notches along the leaves, as you can see in the photo shown here. The maternity plant is ordinarily propagated by means of these small plants, which drop to the soil and take root when they mature.

Sexual Reproduction

Plant sexual life cycles are characterized by an alternation of generations, in which a diploid *sporophyte generation* gives rise to a haploid *gametophyte generation*, as described in chapter 32. In angiosperms, the developing gametophyte generation is completely enclosed within the tissues of the parent sporophyte. The male gametophytes are **pollen grains**, and they develop from *microspores*. Microspores and megaspores are discussed further on the following pages. The female gametophyte is the **embryo sac**, which develops from a *megaspore*. Pollen grains and the embryo sac both are produced in separate, specialized structures of the angiosperm flower. The pollen grains form on filaments that extend out from the flower, like this lily, and the embryo sac sits at the base of the flower.

Female gametophyte (embryo sac) is inside base of flower

Parent sporophyte tissue (plant and flowers)

Male gametophytes (pollen grains) are on these filaments

Key Learning Outcome 34.1 Reproduction

Connections:
Where biology impacts society

Biology and *Staying Healthy*

Why Don't Men Get Breast Cancer?

In 2007, an estimated 240,510 new cases of ... cancer were expected among women in the United S... 2,0... new cases among men—over 99% of the new b... victims were women. Similarly, of 40,460 breast ... deaths anticipated that year, all but 450 of the... women. It is impossible not to wonder, why ... An obvious answer would be that men do... but men do have breast tissue, it just isn't... woman's. So why so few men?

While 5% of breast cancers are due t... genetic mutations (*BRCA1* and *BRCA2*), the ... 95%—the overwhelming majority—remains a ... Hormone differences seem the most promising pla... to start, as the female sex hormone estrogen controls breast development in women; men, by contrast, lack physiologically significant amounts of estrogen. A logical suggestion is that in breast cancer patients, the effect ...ls is being altered by exposure ... disrupter. Endocrine disrupters ...ls that mimic hormones. By ...ecules are perfectly shaped to fit ...eptors. In this case, the culprit ...mic of estrogen that promotes ...ast cells.

...bisphenol A (BPA), a molecule that ... estrogen, with carbon rings at ...H groups. Used to form the plastic ...ds and drinks, as well as the clear ...ood and beverage cans, BPA is a ... us are exposed daily. Six billion ... ...rldwide each year.

... since 1938 that BPA promotes ...on in rats. Alarmingly, in 1993 ... the same effects ...r cells growing ...arming was ...measured at ...s 2 parts per ... the levels to ...outinely exposed. ...nduce breast ...bility, Dr. ...rsity School of ...rch team in 2006 ...which get breast ... ...e way as humans ...ant female rats to a range of ...d after 50 days sacrificed them ...r breast tissue. The researchers ...aberrant cell growth patterns in ...al hyperplasias, which in both ...sidered to be the precursors of

...h higher doses. ... control, no BPA was administered.

The histogram above shows what the researchers found. Amounts of BPA administered to the rats are reported in micrograms of BPA per kilogram of body weight per day. The incidence of breast cancer detected is presented as the percent of examined breast tissue ducts which were hyperplastic (that is, precancerous). Vertical black lines (called error bars) represent 5% confidence intervals (a measure of scatter among the data—95% of replicate experiments would be expected to fall within the ranges of the error bars).

What are we to make of the data the researchers obtained? Did any of the four doses of BPA they administered result in a percent of hyperplastic ducts significantly higher than that seen in the BPA control group? (Hint: If their error bars overlap with the control's, the difference is not significant.) The answer—yes, all four doses are significant. Do higher doses of BPA increase the incidence of hyperplastic ducts? Do the error bars of the four doses fail to overlap? No, all doses yielded similar results. It is difficult to avoid the conclusion that exposure to even low levels of bisphenol A induces ductal hyperplasias in laboratory rats.

These results suggest the rather alarming conclusion that BPA, a chemical to which we are all exposed every day, may cause breast cancer. The study is small, and a rat is not a human, but the possibility that BPA is a carcinogen certainly merits further investigation. Most government-funded breast cancer research has focused on the search for more effective breast cancer treatments; far less money is spent on searching for the causes of breast cancer. The most encouraging aspect of this study is its suggestion that such a search, if it became a priority, might prove fruitful.

Bisphenol A

Estrogen

Experiments:
Learning the process of science

INQUIRY & ANALYSIS

Does Clear-Cutting Forests Do Permanent Damage?

The lumber industry practice called "clear-cutting" has been common in many states. Loggers find it more efficient to simply remove all trees from a watershed, and sort the logs out l...r, than to selectively cut only the most desirable ma... trees. While the open cuts seem a desolation to ...al observer, the loggers claim that ...fores... ...some established more readily in the ...ght now more easily reaches seedlings ... Ecologists counter that clear-cutting ...nges the forest in ways that cannot

...he most direct way to find out is to ... watch it very carefully. Just this ...est was carried out in a now-classic ...bbard Brook Experimental Forest ... Hubbard Brook is the central stream ...ed that drains a region of temperate ...est in northern New Hampshire. The ...am, led by then-Dartmouth College professors ...ert Bormann and Gene Likens, first gathered a great deal of information about the forest watershed. Starting in 1963, they censused the trees, measured the flow of water through the watershed, and carefully documented the levels of minerals and other nutrients in the water leaving the ecosystem via Hubbard Brook. To keep track, they constructed concrete dams across each of the six streams that drain the forest and monitored the runoff, chemically analyzing samples. The undisturbed forest proved very efficient at retaining nitrogen and other nutrients. The small amounts of nutrients that entered the ecosystem in rain and snow were approximately equal to the amounts of nutrients that ran out of the valleys into Hubbard Brook.

Now came the test. In the winter of 1965 the investigators felled all the trees and shrubs in 48 acres drained by one stream (as shown in the photo), and examined the water running off. The immediate effect was dramatic: The amount of water running out of the valley increased by 40%. Water that otherwise would have been taken up by vegetation and released into the atmosphere through evaporation was now simply running off.

The Hubbard Brook Deforestation Experiment

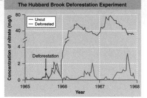

It was clear that the forest was not retaining water as well, but what about the soil nutrients, the key to future forest fertility?

The red line in the graph above shows nitrogen minerals leaving the ecosystem in the runoff water of the stream draining the clear-cut area; the blue line shows the nitrogen runoff in a neighboring stream draining an adjacent uncut portion of the forest.

1. **Applying Concepts**
   a. Variable. In the graph, what is the dependent variable?
   b. Scale. What is the significance of the break in the vertical axis between 4 and 40?

2. **Interpreting Data**
   a. What is the approximate concentration of nitrogen in the runoff of the uncut valley before cutting? of the cut valley before cutting?
   b. What is the approximate concentration of nitrogen in the runoff of the uncut valley one year after cutting? of the clear-cut valley one year after cutting?

3. **Making Inferences**
   a. Is there any yearly pattern to the nitrogen runoff in the uncut forest? Can you explain it?
   b. How does the loss of nitrogen from the ecosystem in the clear-cut forest compare with nitrogen loss from the uncut forest?

4. **Drawing Conclusions**
   a. What is the impact of this forest's trees upon its ability to retain nitrogen?
   b. Has clear-cutting harmed this ecosystem? Explain.

The Living World

SEVENTH EDITION

George B. Johnson

Washington University
St. Louis, Missouri

THE LIVING WORLD, SEVENTH EDITION

Published by McGraw-Hill, a business unit of The McGraw-Hill Companies, Inc., 1221 Avenue of the Americas, New York, NY 10020. Copyright © 2012 by The McGraw-Hill Companies, Inc. All rights reserved. Previous editions © 2010, 2008, and 2006. No part of this publication may be reproduced or distributed in any form or by any means, or stored in a database or retrieval system, without the prior written consent of The McGraw-Hill Companies, Inc., including, but not limited to, in any network or other electronic storage or transmission, or broadcast for distance learning.

Some ancillaries, including electronic and print components, may not be available to customers outside the United States.

This book is printed on acid-free paper.

1 2 3 4 5 6 7 8 9 0 DOW/DOW 1 0 9 8 7 6 5 4 3 2 1

ISBN 978–0–07–802417–7
MHID 0–07–802417–X

Vice President, Editor-in-Chief: *Marty Lange*
Vice President, EDP: *Kimberly Meriwether David*
Senior Director of Development: *Kristine Tibbetts*
Publisher: *Michael S. Hackett*
Executive Editor: *Eric Weber*
Senior Developmental Editor: *Anne L. Winch*
Senior Marketing Manager: *Tamara Maury*
Lead Project Manager: *Sheila M. Frank*
Senior Buyer: *Kara Kudronowicz*
Senior Media Project Manager: *Tammy Juran*
Senior Designer: *Laurie B. Janssen*
Cover Image: *©Anup Shah/Gettyimages*
Senior Photo Research Coordinator: *Lori Hancock*
Photo Research: *Emily Tietz/Editorial Image, LLC*
Compositor: *Electronic Publishing Services Inc., NYC*
Typeface: *10.5/12 Times Roman*
Printer: *R. R. Donnelley*

All credits appearing on page or at the end of the book are considered to be an extension of the copyright page.

Library of Congress Cataloging-in-Publication Data

Johnson, George B. (George Brooks), 1942-
    The living world / George B. Johnson. — 7th ed.
        p. cm.
    Includes index.
    ISBN 978–0–07–802417–7 — ISBN 0–07–802417–X (hard copy : alk. paper) 1. Biology. I. Title.
    QH308.2.J62 2012
    570—dc22
                                2010032045

www.mhhe.com

Brief Contents

Preface

No one who teaches biology today can fail to appreciate how important a subject it has become for our modern world. From global warming to stem cell initiatives to teaching intelligent design in classrooms, biology permeates the news, and in large measure will define our students' futures. As a teacher, I have stood in front of classrooms for over 30 years and attempted to explain biology to puzzled and sometimes uninterested students, an experience that has been both fun and frustrating: fun because biology is a joy to teach, rich in ideas and interesting concepts, and increasingly key to many important public issues; frustrating because in every biology class there are always some students who do not do well, who not only miss out on the fun but also fail to acquire a tool that will be essential to their futures.

This text, *The Living World*, is my attempt to address this problem. It is short enough to use in one semester, without a lot of technical details to intimidate wary students. I have tried to write it in an informal, friendly way, to engage as well as to teach. The focus of the book is on the biology each student ought to know to live as an informed citizen in the twenty-first century. I have at every stage addressed ideas and concepts, rather than detailed information, trying to teach *how* things work and *why* things happen the way they do rather than merely naming parts or giving definitions.

Focusing on the Essential Concepts

More than most subjects, biology is at its core a set of ideas, and if students can master these basic ideas, the rest comes easy. Unfortunately, while most of today's students are very interested in biology, they are put off by the terminology. When you don't know what the words mean, it's easy to slip into thinking that the matter is difficult, when actually the ideas are simple, easy to grasp, and fun to consider. It's the terms that get in the way, that stand as a wall between students and science. With this text I have tried to turn those walls into windows, so that readers can peer in and join the fun.

Analogies Analogies have been my tool. In writing *The Living World* I have searched for simple analogies that relate the matter at hand to things we all know. As science, analogies are not exact, but I do not count myself compromised. Analogies trade precision for clarity. If I do my job right, the key idea is not compromised by the analogy I use to explain it, but rather revealed.

Key Biological Process Boxes There is no way to avoid the fact that some of the important ideas of biology are complex. No student encountering photosynthesis for the first time gets it all on the first pass. To aid in learning the more difficult material, I have given special attention to key concepts and processes like photosynthesis and osmosis that form the core of biology. The essential processes of biology are not optional learning. A student must come to understand every one of them if he or she is to master biology as a science. A student's learning goal should not be simply to memorize a list of terms, but rather to be able to visualize and understand what's going on. With this goal in mind, I have prepared some four dozen "this is how it works" Key Biological Process boxes explaining the important concepts and processes that students encounter in introductory biology. Each of these Key Biological Process boxes walks the student through a complex process, one step at a time, so that the central idea is not lost in the details.

Teaching Biology as an Evolutionary Journey

This text, and its companion text *Essentials of The Living World*, were the first texts to combine Evolution and Diversity into one continuous narrative. Traditionally, students had been exposed to weeks of evolution before being dragged through a detailed tour of the animal phyla, the two areas presented as if unrelated to each other. I chose instead to combine these two areas, presenting biological diversity as an evolutionary journey. This has proven a very powerful way to teach evolution's role in biology, and today you would be hard pressed to find a text that does not organize the material in this way.

Evolutionary Explanations Evolution not only organizes biology, it explains it. It is not enough to say that a frog is an amphibian, transitional between fish and reptiles. This correctly organizes frogs on the evolutionary spectrum, but fails to explain *why* frogs are they way they are, with a tadpole life stage and wet skin. Only when the student is taught that amphibians evolved as highly successful land animals, often as big as ponies and armor plated, can students get the point: of 37 families of amphibians, all but the two that lived in water (frogs and salamanders) were driven extinct with the advent of reptiles. A frog has evolved to *invade* water, not escape it. It is in this way that evolution explains biology, and that is how I have tried to use evolution in this text, to explain.

Confronting Evolution's Critics As evolution continues to be a controversial subject to the general public, I have provided students with an explicit consideration of the objections raised by evolution's critics, focusing in detail on the claims of so-called "intelligent design." I feel strongly that no student's education in biology is complete without a frank discussion of this contentious issue.

Linking Biology to Everyday Life

One of the principal roles of nonmajors biology courses is to create educated citizens. In writing *The Living World* I have endeavored wherever possible to connect what the students are learning to their own everyday lives.

Connections Throughout *The Living World* are full-page features written by me that make connections between a chapter's contents and the everyday world: *Biology and Staying Healthy* discusses health issues that impact each student; *Today's Biology* examines advances that importantly affect society; *A Closer Look* examines interesting points in more detail; and *Author's Corner* takes a more personal view (mine) of how science relates to our personal lives. See examples on pages ix and x.

Using Art That Teaches

Art has always been a core component of this text, as today's students are visual learners. To help students learn, *The Living World* has a clean and simple art style that focuses on concepts and minimizes detail. In this edition I have sought to amplify the power of illustrations to teach concepts by linking the interior content of illustrations directly to the text that describes that part of the illustration:

Integrated Art Many topics, like the introduction to photosynthesis (treated on pages 118-121), involve several different interacting processes, each explained with its own illustration. In these instances bouncing back and forth between illustration and text makes it difficult for a student to gain or retain perspective, and so I have chosen in these instances to integrate the illustrations directly into the text, providing a single narrative. You will find this "integrated art" throughout the book—working art into the text narrative has been one of my chief tasks in this revision.

Bubble Numbers In complex diagrams where there is a lot going on, I have placed numbers (set off in colored balls) at key positions, and the same "bubble numbers" at those locations in the text where that element of the illustration is being described. This makes it much easier for a student to walk through the complex illustration and see how the parts are related.

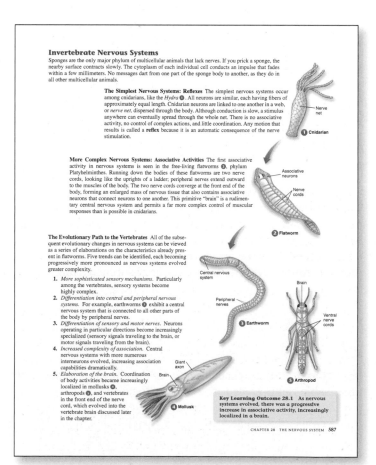

Helping Students Learn

Chapter Zero In over thirty years of teaching, I have seen students do well and others do poorly, and one of the best predictors of who would do well has been how well a student is prepared to learn. Entering a large freshman course, does a student know how to take notes? Does a student know how to use these notes effectively with the textbook? Can a student even read a graph? In this edition I have decided to tackle this problem head on, and have added a Chapter 0 at the beginning of the text to help students with these very basic but essential learning tools. Learning to take effective notes during lecture, to recopy these notes promptly, and to key them to the text for efficient review when studying, are skills that will improve students' performance not only in this course but throughout their college careers. Learning how to read a graph is a skill that will stand students in good stead throughout their lives.

Learning Objectives and Outcomes Students learn best when they are given a clear idea of what it is they are supposed to learn. With this in mind, I have in this edition begun each chapter with a list of specific learning objectives, keyed to that chapter's numbered segments. A **learning objective** is the intended result of study, a specific statement of what a student should be able to do when he or she successfully learns the material covered in that chapter segment. Learning objectives are typically discrete and quite specific—the small building blocks with which a student constructs understanding.

At the end of each numbered chapter segment, I have placed one or more Key Learning Outcomes. A **learning outcome** is a student's achieved result after studying, a concrete statement of what a student is expected to know or be able to do after successfully mastering the chapter segment. Said simply, a learning outcome is a realized learning objective.

Learning objectives and outcomes together provide a student with a powerful learning tool. Learning objectives help the student focus on the important points as they read a chapter section, and review of key learning outcomes provides the student a very pointed indicator of how well he or she has mastered the objectives that were the target of their learning.

Assessing Progress At the end of every chapter a student is provided with a powerful assessment tool to see how well the learning objectives at the beginning of the chapter have been transformed into learning outcomes as a result of the student's study. I have grouped questions according to Bloom's taxonomy of learning categories into three levels of increasing sophistication: *Test Your Understanding* questions evaluate knowledge and comprehension, *Apply Your Understanding* questions challenge the student to carry out application and analysis using illustrations from the text, and *Synthesize What You Have Learned* questions require synthesis and evaluation.

CONNECT Each chapter in *The Living World* is paired with assignable, auto-graded assessments focused on the learning objectives I outline at the beginning of each chapter. Instructors can use this content to deliver assignments, quizzes, and tests online, allowing students to practice important skills at their own pace and on their own schedule. The Connect Plus

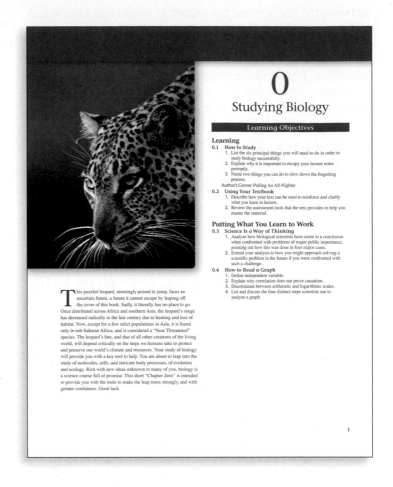

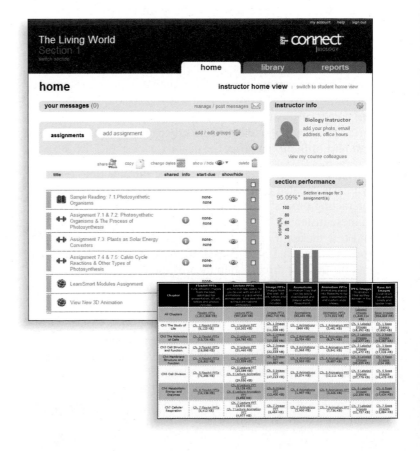

version of Connect™ provides students with access to the full textbook as a dynamic, media-rich eBook to aid them in successfully completing their work, wherever and whenever they choose. Additional details about this exciting new facet of *The Living World* can be found on pages xiii–xiv of the preface.

Teaching Science As a Process

Inquiry and Analysis One of the most useful things a student can take away from his or her biology class is the ability to judge scientific claims that they encounter as citizens, long after college is over. As a way of teaching that important skill, I have greatly expanded the *Inquiry and Analysis* features I introduced in the previous edition. Every chapter now ends with a full-page presentation of an actual scientific investigation that requires the student to analyze the data and reach conclusions. Few pages in this text provide more bang for the buck in learning that lasts.

New to This Edition

Many chapters of this revision of *The Living World* have been updated to reflect scientific advances and to improve the text as a learning tool.

Unit One
- Moved "How Scientists Analyze and Present Experimental Results" to chapter 0, where it is integrated into the discussion of how to study biology.
- Moved the content of chapter 2 "Evolution and Ecology" into the appropriate evolution and ecology chapters. The placement of this chapter in the sixth edition was an attempt to give students an early introduction to the topics of evolution and ecology, providing students with a "macro" view to begin their study, but was seen as confusing when placed out of context.
- Chapter 1 now contains a full-page "Inquiry & Analysis" feature.

Unit Two
- Added several pieces of new Integrated Art: chemical bonds, protein structure, the plasma membrane, the endomembrane system, and using electrons to make ATP.
- Added a new "Biology and Staying Healthy" feature to chapter 3: "Anabolic Steroids in Sports."
- Added a new "Author's Corner" feature to chapter 3: "My Battle with Cholesterol."
- Added a new "A Closer Look" feature to chapter 4: "How Water Crosses the Plasma Membrane."
- Added a new discussion on the endomembrane system and health to chapter 4.
- Revised the discussions in chapter 4 of diffusion, facilitated diffusion, and osmosis, adding two new figures.
- Added a new "A Closer Look" feature to chapter 7: "Beer and Wine—Products of Fermentation."

Unit Three
- Added a new discussion and illustration of growth factors and cell signaling pathways to chapter 8.

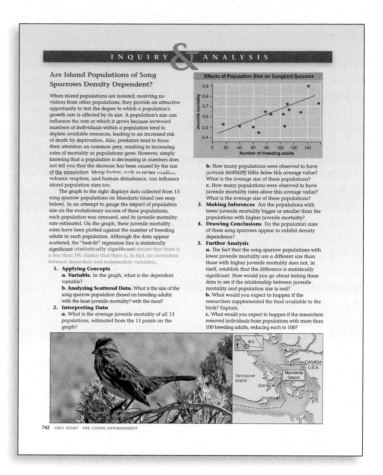

- Added a new discussion and illustration of telomeres and the Hayflick limit to chapter 8.
- Expanded the treatment of human chromosomes in chapter 10.
- Added a new discussion with illustrations to chapter 10, on how geneticists study human pedigrees, including examples such as albinism and color blindness.
- Added new Integrated Art: DNA replication.
- Added two new "Today's Biology" features to chapter 11: "DNA Fingerprinting" and "Tracing the DNA of Irish Kings."
- Expanded the overview of gene expression and discussion of transcription in chapter 12, including the addition of two new illustrations.
- Added a new illustration showing the eukaryotic transcription complex to chapter 12.
- Combined the Genomics, Gene Technology, and Cell Technology chapters of the previous edition into one chapter, chapter 13: The New Biology.
- Added two new "Today's Biology" features to chapter 13: "DNA and the Innocence Project" and "DNA Timeline."
- New developments in therapeutic cancer vaccines added to chapter 13.
- Revised the gene therapy discussion in chapter 13 to include recent studies that investigate new vectors.

Unit Four

- Darwin's history and his theory of evolution by natural selection is now covered in chapter 14.
- Added Integrated Art to chapter 14: arguments of Darwin's critics, agents of evolution, and types of selection.
- Added two new figures to chapter 14: the bottleneck effect and selection for coat color in mice.
- Added a discussion of how selection acts on peppered moths to chapter 14, including coverage of melanic mice.
- Revised the illustration on classification of organisms in chapter 15.
- Added a new discussion with illustrations to chapter 15 on "How do you read a family tree?"
- Added a new "Today's Biology" feature to chapter 15: "DNA Bar Codes."
- Added new photos of prokaryotes to chapter 16.
- Added "Biology and Staying Healthy" feature to chapter 16: "Bird and Swine Flu."
- Added Integrated Art to chapter 16: gram-negative and gram-positive bacteria and bacterial flagellum.
- Added a new "Inquiry & Analysis" feature: "Does HIV Infect All White Blood Cells?" to chapter 16.
- Chapter 18 now contains a full-page "Inquiry & Analysis" feature: "Are chytrids killing the frogs?"

Unit Five

- In chapter 19, the section on the key transitions in the animal body plan has been revised to include molting versus growth by incremental increase.
- Added a new "A Closer Look" feature to chapter 20: "Are Birds Dinosaurs?"
- Added a new "Today's Biology" feature to chapter 21: "Race and Medicine."
- Revised the Java Man discussion in chapter 21, and discussed new discoveries of *Homo* fossils.
- Discussed the significance of the newly-sequenced Neanderthal genome in chapter 21.
- Added a new full-page "Inquiry & Analysis" to chapter 21: "Increasing Brain Size in Hominids."

Unit Six

- In chapter 22, added molting to text and table discussing innovations in animal body design.
- Deleted the unnecessary space-consuming figure showing the body systems of humans.
- Added a new figure to chapter 22 on osteoporosis.
- Added new "Biology and Staying Healthy" feature to chapter 22: "Losing Bone: Osteoporosis."
- Added several pieces of Integrated Art: types of circulatory systems, types of respiratory systems, nephron and osmoregulation in fish, and invertebrate nervous systems.
- Added a discussion of lung cancer among nonsmokers to chapter 24.
- Revised the discussion of the mammalian kidney in chapter 26 to be more precise.

- Updated the AIDS data presented in chapter 27 and incorporated new information on AIDS vaccines.
- Added a new "Biology and Staying Healthy" feature to chapter 27: "AIDS Drugs Target Different Phases of the HIV Infection Cycle."
- Added a new discussion with illustrations of the evolution of the vertebrate brain to chapter 28.
- Placed the coverage of sensory systems in its own chapter now, chapter 29.
- Added a new discussion of "Other Vertebrate Senses" to chapter 29.
- Added a new "A Closer Look" feature to chapter 29: "How the Platypus Sees With Its Eyes Shut."
- Added six figures to section 31.2, "Evolution of Vertebrate Reproduction," expanding the section to four pages from two pages.
- Added two other new figures to chapter 31: fertilization and layers of the egg cell and stages of childbirth.
- Updated the "Biology and Staying Healthy" feature "Why Don't Men Get Breast Cancer?"

Unit Seven

- Added several new pieces of Integrated Art: moss and pine gametophytes, diagram of a flower, ground tissue, dermal tissue, vascular tissue, roots, and pollen and egg formation.
- Added many new figures to chapter 34: sexually reproducing plant, structure of flower and monoecious plants, different kinds of flowers, incompatibility in pollination, pollination and fertilization, and identifying nutritional requirements of plants.
- Expanded the discussion explaining the structure of flowers in chapter 34.
- Added a new section to chapter 34, "Growth and Nutrition," which covers macronutrients and micronutrients.
- Added a new "Inquiry & Analysis" feature to chapter 34: "Are Pollinators Responsible for the Evolution of Flower Color?"

Unit Eight

- Added a new section to chapter 35: "What Is Ecology?", which includes a treatment of "The Environmental Challenge."
- Added a new discussion of Population Range to chapter 35, including range expansions and contractions.
- Revised the discussion of Population Distribution in chapter 35.
- Streamlined the sections covering Symbioses and Predation in chapter 35.
- Added Integrated Art: trophic levels, atmospheric circulation, and the rain shadow effect.
- Added a new figure to chapter 37: examples of migration.
- Added a table summarizing classes of animal behaviors.
- Added a new "Inquiry & Analysis" feature to chapter 38: "How Real Is Global Warming?"

Giving Biology a Personal Perspective

The author has written full-page boxed readings to help students make connections to the everyday world.

Author's Corner essays take a more personal view of how science relates to our everyday lives.

Today's Biology closely examines important advances in science and medicine.

Author's Corner

The Author Works Out

No one seeing the ring of fat decorating my middle would take me for a runner. Only in my memory do I get up with the robins, lace up my running shoes, bounce out the front door, and run the streets around Washington University before going to work. Now my 5-K runs are 30-year-old memories. Any mention I make of my running in a race only evokes screams of laughter from my daughters, and an arch look from my wife. Memory is cruelest when it is accurate.

I remember clearly the day I stopped running. It was a cool fall morning in 1978, and I was part of a mob running a 5-K (that's 5 kilometers for the uninitiated) race, winding around the hills near the university. I started to get flashes of pain in my legs below the knees—like shin splints, but much worse. Imagine fire pouring on your bones. Did I stop running? No. Like a bonehead I kept going, "working through the pain," and finished the race. I have never run a race since.

I had pulled a muscle in my thigh, which caused part of the pain. But that wasn't all. The pain in my lower legs wasn't shin splints, and didn't go away. A trip to the doctor revealed multiple stress fractures in both legs. The X rays of my legs looked like tiny threads had been wrapped around the shaft of each bone, like the red stripe on a barber's pole. It was summer before I could walk without pain.

What went wrong? Isn't running supposed to be GOOD for you? Not if you run improperly. In my enthusiasm to be healthy, I ignored some simple rules and paid the price. The biology lesson I ignored had to do

is flexible but weak, but like the epoxy of fiberglass, it acts to spread any stress over many crystals, making bone resistant to fracture. As a result, bone is both strong and flexible.

When you subject a bone in your body to stress—say, by running—the bone grows so as to withstand the greater workload. How does the bone "know" just where to add more material? When stress deforms the collagen fibers of a leg bone, the interior of the collagen fibers becomes exposed, like opening your jacket and exposing your shirt. The fiber interior produces a minute electrical charge. Cells called fibroblasts are attracted to the electricity like bugs to night lights, and secrete more collagen there. As a result, new collagen fibers are laid down on a bone along the lines of stress. Slowly, over months, calcium phosphate crystals convert the new collagen to new bone. In your legs, the new bone forms along the long stress lines that curve down along the shank of the bone.

Now go back 30 years, and visualize me pounding happily down the concrete pavement each morning. I had only recently begun to run on the sidewalk, and for an hour or more at a stretch. Every stride I took those mornings was a blow to my shinbones, a stress to which my bones no doubt began to respond by forming collagen along the spiral lines of stress. Had I run on a softer surface, the daily stress would have been far less severe. Had I gradually increased my running, new bone would have had time to form properly in response to the added stress. I gave my leg bones a lot of stress, and no time to respond to it. I pushed them too hard, too fast, and they gave way.

Nor was my improper running limited to overstressed

IT MOVES   491

Today's Biology

Meet Our Hobbit Cousin

For the last several years, the field of paleoanthropology, the study of human evolution, has been stood on its head by a tiny fossil from Indonesia.

This story begins in 1996, when Australian paleoanthropologist Michael Morwood reported that his team had excavated stone tools from a site on Flores Island 300 miles south of Borneo in Indonesia. The site was almost a million years old, long before the period when scientists thought sophisticated humans arrived there. Even more to the point, Flores is isolated by a deep water trench from the rest of Asia. This deep-water barrier is called the Wallace Line after Darwin's contemporary Alfred Russel Wallace who first noticed the marked differences between animals living on either side of it. On one side they resemble Asian species but on the other side they resemble Australian species. Might early humans have somehow managed to cross?

Morwood and Indonesian collaborators started looking for fossils on both sides of the Wallace Line. In 2001 they began to examine caves that had been explored years earlier by other archaeologists, and in 2003 they paydirt at Liang Bua ("cool cave"). Six meters under the floor of the cave, in 18,000-year-old sediment, Morwood's team found the nearly complete fossil skeleton of a hominid, an early human. The teeth are worn, and the skull bones are knitted together in an adult way, so the fossilized individual was an adult—but an adult hominid only 1 meter tall, with a brain of only 380 cc (cubic centimeters)! This is the stature and brain size of a chimpanzee. The fossil foot, essentially complete, is extremely large. Workers at the site nicknamed the newly discovered hominid "the hobbit" after the J.R.R. Tolkien characters in the book *The Lord of the Rings*.

A firestorm of controversy broke out among the scientific community. This *Homo floresiensis*, as Morwood dubbed the fossil, makes no sense when viewed against the detailed picture paleoanthropologists have

Homo floresiensis (far right) compared with a modern human female. *H. floresiensis* preyed upon a dwarf species of elephant, which also occurred on the island of Flores.

pieced together of human evolution. For one thing, it is too young. In a 3-million-year-old prehuman ancestor, the Flores fossil's tiny dimensions would not be surprising, but the fossil is only 18,000 years old. That means it was alive when *Homo sapiens* (modern human) had evolved and even crossed the Wallace Line to Australia.

For another thing, the Flores individual is too short. Early African human ancestors called australopithecines were this size, 20 kg and about a meter tall, but every adult human fossil ever found is much bigger and taller. The humans alive 18,000 years ago were not tiny but the same stature as humans today.

And yet, despite all these objections that the Flores fossil could not exist, there it is, hard data. What are we to make of it? Morwood's suggestion is that the Flores fossil is a member of a human species that evolved in isolation on Flores island from ancient humans that reached there long ago. The ancestor of *Homo sapiens*, called *Homo erectus*, shares certain characteristics with the Flores fossil and so is the most likely candidate. Morwood goes on to point out that large mammals colonizing remote small islands often tend to evolve into isolated dwarf species. Many examples are known, including pygmy hippos, ground sloths, deer, and even dinosaurs. Indeed, pygmy elephants no bigger than cattle lived on Flores island at that very time; fossils of hunted specimens are found in the Liang Bua cave! Why do dwarf species evolve? The theory is that on islands with few or no predators, there is no advantage to being big, and with limited food there is great advantage to being small. On Flores island, Morwood suggests, the human evolutionary story has played out differently than it did in Africa, evolution favoring the small.

"Wait a minute," cry researchers from the Field Museum in Chicago, who in 2006 challenged the classification of the Flores fossil as a new species. Its brain is simply too small for a dwarf species, they claim—judging from other instances in which pygmy species have developed in isolation, the Flores brain size of 380 cc would indicate a creature only 1 foot tall! They propose that instead it may be a modern human suffering from microcephaly, a genetic disorder that results in small brain size.

Morwood's team is not ready to concede the argument. In subsequent digs at the Liang Bua cave in 2004 they have found a total of eight more tiny individuals, all of them even smaller than the complete fossil unearthed the previous year. It stretches the imagination to believe that all suffer from microcephaly. Strengthening their claim that the Flores fossils are not modern humans was the discovery of two complete lower jaws, neither of which have chins. Chins are found on the lower jaws of all modern human fossils.

The Floresians of the Liang Bua region seem to have perished after an eruption of one of the island's many volcanoes about 13,000 years ago. But they may have survived much longer elsewhere on Flores, Dr. Morwood believes. The search for more fossils goes on.

CHAPTER 21 HOW HUMANS EVOLVED   467

ix

Giving Biology a Personal Perspective

In **A Closer Look** the author provides a more detailed examination of an interesting point from the chapter.

Biology and Staying Healthy essays
focus on issues affecting health.

A Closer Look

How the Platypus Sees With Its Eyes Shut

The duck-billed platypus (*Ornithorhynchus anatinus*) is abundant in freshwater streams of eastern Australia. These mammals have a unique mixture of traits— in 1799, British scientists were convinced that the platypus skin they received from Australia was a hoax. The platypus is covered in soft fur and has mammary glands, but in other ways it seems very reptilian. Females lay eggs as reptiles do, and like reptilian eggs, the yolk of the fertilized egg does not divide. In addition, the platypus has a tail not unlike that of a beaver, a bill not unlike that of a duck, and webbed feet!

It turns out that platypuses also have some very unique behaviors. Until recently, few scientists had studied the platypus in its natural habitat— it is elusive, spending its days in burrows it constructs on the banks of waterways. Also, a platypus is active mostly at night, diving in streams and lagoons to capture bottom-dwelling invertebrates such as shrimps and insect larvae. Interestingly, unlike whales and other marine mammals, a platypus cannot stay under water long. Its dives typically last a minute and a half. (Try holding your breath that long!)

When scientists began to study the platypus' diving behavior, they soon observed a curious fact: The eyes and ears of a platypus are located within a muscular groove, and when a platypus dives, the sides of these grooves close over tightly. Imagine pulling your eyebrows down to your cheeks—effectively blindfolded, you wouldn't be able to see a thing! To complete its isolation, the nostrils at the end of the snout also close. So how in the world does the animal find its prey?

For over a century biologists have known that the soft surface of the platypus bill is pierced by hundreds of tiny openings. In recent years Australian neuroscientists (scientists that study the brain and nervous system) have learned that these pores contain sensitive nerve endings. Nestled in an interior cavity, the nerves are protected

streamwater via the pore. The nerve endings act as sensory receptors, communicating to the brain information about the animal's surroundings. These pores in the platypus bill are its diving "eyes."

Platypuses have two types of sensory cells in these pores. Clustered in the front are so-called mechanoreceptors, which act like tiny pushrods. Anything pushing against them triggers a signal. Your ears work the same way, sound waves pushing against tiny mechanoreceptors within your ears. These pushrods evoke a response over a much larger area of the platypus brain than does stimulation from the eyes and ears—for the diving platypus, the bill is the primary sense organ. What responses do the pushrod receptors evoke? Touching the bill with a fine glass probe reveals the answer—a lightning-fast, snapping movement of its jaws. When the platypus contacts its prey, the pushrod receptors are stimulated, and the jaws rapidly snap and seize the prey.

But how does the platypus locate its prey at a distance, in murky water with its eyes shut? That is where the other sort of sensory receptor comes in. When a platypus feeds, it swims along steadily wagging its bill from side to side, two or three sweeps per second, until it detects and homes in on prey. How does the platypus detect the prey individual and orient itself to it? The platypus does not emit sounds like a bat, which rules out the possibility of sonar as an explanation. Instead, electroreceptors in its bill sense the tiny electrical currents generated by the muscle movements of its prey as the shrimp or insect larva moves to evade the approaching platypus!

It is easy to demonstrate this, once you know what is going on. Just drop a small 1.5-volt battery into the stream. A platypus will immediately orient to it and attack it, from as far away as 30 centimeters. Some sharks and fishes have the same sort of sensory system. In muddy murky waters, sensing the muscle movements of a prey individual is far superior to trying to see its body or hear it move—which is why the platypus that you see in

ENSES **617**

Biology and *Staying Healthy*

Bird and Swine Flu

The influenza virus has been one of the most lethal viruses in human history. Flu viruses are animal RNA viruses containing 11 genes. An individual flu virus resembles a sphere studded with spikes composed of two kinds of protein. Different strains of flu virus, called subtypes, differ in their protein spikes. One of these proteins, hemagglutinin (H), aids the virus in gaining access to the cell interior. The other, neuraminidase (N), helps the daughter virus break free of the host cell once virus replication has been completed. Flu viruses are currently classified into 13 distinct H subtypes and 9 distinct N subtypes, each of which requires a different vaccine to protect against infection. Thus, the virus that caused the Hong Kong flu epidemic of 1968 has type 3 H molecules and type 2 N molecules, and is called H3N2.

Recombination within humans
A person infected with a flu virus can become infected with another type of flu virus by direct contact with birds. The two viruses can undergo genetic recombination to produce a third type of virus, which can spread from human to human.

Recombination within pigs
Pigs can contract flu viruses from both birds and humans. The flu viruses can undergo genetic recombination in the pig, to produce a new kind of flu virus, which can spread from pigs to humans.

How New Flu Strains Arise
Worldwide epidemics of the flu in the last century have been caused by shifts in flu virus H-N combinations. The "killer flu" of 1918, H1N1, thought to have passed directly from birds to humans, killed between 40 and 100 million people worldwide. The Asian flu of 1957, H2N2, killed over 100,000 Americans, and the Hong Kong flu of 1968, H3N2, killed 70,000 Americans.

It is no accident that new strains of flu usually originate in the Far East. The most common hosts of influenza virus are ducks, chickens, and pigs, which in Asia often live in close proximity to each other and to humans. Pigs are subject to infection by both bird and human strains of the virus, and individual animals are often simultaneously infected with multiple strains. This creates conditions favoring genetic recombination between strains, as illustrated above, sometimes putting together novel combinations of H and N spikes unrecognizable by human immune defenses specific for the old configuration. The Hong Kong flu, for example, arose from recombination between H3N8 from ducks and H2N2 from humans. The new strain of influenza, in this case H3N2, then passed back to humans, creating an epidemic because the human population had never experienced that H-N combination before.

Conditions for a Pandemic
Not every new strain of influenza creates a worldwide flu epidemic. Three conditions are necessary: (1) The new strain must contain a novel combination of H and N spikes, so that the human population has no significant immunity from infection; (2) the new strain must be able to replicate in humans and cause death—many bird influenza viruses are harmless to people because they cannot multiply in human cells; (3) the new strain must be efficiently transmitted between humans. The H1N1 killer flu of 1918

spread in water droplets exhaled from infected individuals and subsequently inhaled into the lower respiratory tract of nearby people.

The new strain need not be deadly to every infected person in order to produce a pandemic—the H1N1 flu of 1918 had an overall mortality rate of only 2%, and yet killed 40 to 100 million people. Why did so many die? Because so much of the world's population was infected.

Bird Flu
A potentially deadly new strain of flu virus emerged in Hong Kong in 1997, H5N1. Like the 1918 pandemic strain, H5N1 passes to humans directly from infected birds, usually chickens or ducks, and for this reason has been dubbed "bird flu." Bird flu satisfies the first two conditions of a pandemic: H5N1 is a novel combination of H and N spikes for which humans have little immunity, and the resulting strain is particularly deadly, with a mortality of 59% (much higher than the 2% mortality of the 1918 H1N1 strain). Fortunately, the third condition for a pandemic is not yet met: The H5N1 strain of flu virus does not spread easily from person to person, and the number of human infections remains small.

Swine Flu
A second potentially pandemic form of flu virus, H1N1, emerged in Mexico in 2009, passing to humans from infected pigs. It seems to have arisen by multiple genetic recombination events between humans, birds, and pigs. Like the 1918 H1N1 virus, this flu (dubbed "swine flu") passes easily from person to person, and within a year it spread around the world. Also like the 1918 virus, the initial wave of swine flu infection triggered only mild symptoms in most people. The 1918 H1N1 virus only became deadly in a second wave of infection, after the virus had better adapted to living in the human body. Like bird flu, public health officials continue to watch swine flu carefully, for fear that a subsequent wave of infection may also become more lethal.

Acknowledgements

I have for several years enjoyed and profited from collaborations with Jonathan Losos of Harvard University, a friend of long standing and a delight to work with. My coauthor on the previous edition, Jonathan had to relinquish that role on this edition. A coauthor on *Biology,* classroom teacher, and active researcher, this text has proven one load too many. I miss the fun of working with him.

Every author knows that he or she labors on the shoulders of many others; the text you see is the result of hard work by an army of "behind-the-scenes" editors, spelling and grammar checkers, photo researchers, and artists that perform their magic on our manuscript; and an even larger army of production managers and staff that then transform this manuscript into a bound book. I cannot thank them all. Anne Winch was my editor, with whom I worked every day. She provided valuable advice and support to a sometimes querulous and always anxious author. Publisher Michael Hackett aggressively backed this major revision; I am awfully glad he did! Marty Lange, Vice President, Editor-in-Chief, oversaw all the problems an over-eager author inadvertently creates with humor and consistent support. Sheila Frank spearheaded our production team, which worked miracles with a strange new "integrated art" program that violated all the rules. The photo program was carried out by Lori Hancock, who as always did a super job (just look at the cover image!). Laurie Janssen did a great job with the design, and was unbelievably tolerant of the author's many "creative" changes. The book was produced by Electronic Publishing Services Inc.

My long-time, off-site developmental editor and right arm Megan Berdelman has again played an invaluable role in overseeing every detail of a complex revision. Her intelligence and perseverance have, over seven editions, made a significant contribution to the quality of this book. For the first six editions she was one of two off-site editors I used to help me with the complex task of writing and revising *The Living World.* The other, Liz Sievers, was Megan's twin in importance to the book and in the sheer pleasure of working with her. Liz has gone on to work for *The Living World*'s publisher rather than its author; both Megan and I miss her a lot.

Over its last three successful editions, the marketing of *The Living World* has been planned and supervised by Tamara Maury, a battle-wise general not afraid to fight in the trenches alongside the many able sales reps that present my books to instructors. I am very pleased to be able to report that she will have this edition under her wing as well. The seventh edition will also benefit from the tactical assistance of Michelle Watnick, Director of Marketing for Life Science.

No text goes through seven editions without the strong support of its editors, past and present. I would like to extend my special thanks to Pat Reidy, Executive Marketing Manager–Life Sciences, who got me over many rough bumps in early editions, and particularly to Michael Lange, Vice President, Editorial Director, for his early and continued strong support of this project.

George Johnson

Reviewers

I have authored other texts, and all of my writing efforts have taught me the great value of reviews in improving my texts. Scientific colleagues from around the country have provided numerous suggestions on how to improve the content of the seventh edition, and many instructors and students using the sixth edition have suggested ways to clarify explanations, improve presentations, and expand on important topics. The instructors listed below provided detailed comments. I have tried to listen carefully to all of you. Everyone of you has my thanks!

Chander Arora
Los Angeles Valley College

David Boehmer
Volunteer State Community College

Bob Boykin
Bossier Parish Community College

Elizabeth Cantwell
University of Tennessee

Peter Chen
College of DuPage

Thomas Chen
Santa Monica College

Elizabeth Collins
Iowa Central Community College

Julie Ehresmann
Iowa Central Community College

April Ann Fong
Portland Community College, Sylvania Campus

Bernard Frye
University of Texas at Arlington

Julie Gibbs
College of DuPage

Shashuna J. Gray
Germanna Community College

Zafer Hatahet
Northwestern State University

Rebecca K. Hoffman
Drexel University

Carina Howell
Lock Haven University

Carrie McMahon Hughes
San Jacinto College – Central Campus

Barbara Hunnicutt
Seminole State College

Roishene Johnson
Bossier Parish Community College

Alan Journet
Southeast Missouri State University

Anthony J. Julis
Wilmington University

John Kell
Radford University

David T. Kidwell
Southeast Community & Technical College

Linda Fergusson-Kolmes
Portland Community College

Anne Macek Lachelt
Iowa Central Community College

Mary V. McDonald
University of Central Arkansas

Tiffany McFalls
Elizabethtown Community and Technical College

Jerry Mimms
University of Central Arkansas

Sheila Miracle
Southeast Kentucky Community and Technical College

Milwood Motley
Columbus State University

Joseph R. Newhouse
Lock Haven University

Kumkum Prabhakar
Nassau Community College

Karen Raines
Colorado State University

Michael Rutledge
Middle Tennessee State University

Kim Sadler
Middle Tennessee State University

Methea Sapp
Spokane Community College

Brian W. Schwartz
Columbus State University

Ramona Smith
Brevard Community College, Palm Bay Campus

Patricia Steinke
San Jacinto College Central

Season Thomson
Germanna Community College

John Topinka
American River College

Mary Elizabeth Torrano
American River College

Ryan Wagner
Millersville University

Carol Wake
South Dakota State University

Robert Wise
University of Wisconsin–Oshkosh

David Wolfe
American River College

Maury Wrightson
Germanna Community College

Kerry Yurewicz
Plymouth State University

Focus Group Participants
Introductory Biology

Caroline Ballard
Rock Valley College

Peggy Brickman
University of Georgia

Lori Buckley
Oxnard College

Jane Caldwell
Washington and Jefferson College

James Claiborne
Georgia Southern University

Scott Cooper
University of Wisconsin – La Crosse

David Cox
Lincoln Land Community College

Bruce Fink
Kaskaskia College

Brandon Foster
Wake Tech Community College

Douglas Gaffin
University of Oklahoma

Carlos Garcia
Texas A&M University – Kingsville

John P. Gibson
University of Oklahoma

Kelly Hogan
University of North Carolina at Chapel Hill

Jessica Hopkins
University of Akron

Tim Hoving
Grand Rapids Community College

Dianne Jennings
Virginia Commonwealth University

Leslie Jones
Valdosta State University

Hinrich Kaiser
Victor Valley Community College

David Lemke
Southwest Texas State University

Mark Lyford
University of Wyoming

Richard Musser
Western Illinois University

Ikemefuna Nwosu
Lake Land College

Murad Odeh
South Texas College

Karen Plucinski
Missouri Southern State University

Maretha Roberts
Olive-Harvey College

Frank Romano
Jacksonville State University

Felicia Scott
Macomb Community College – Clinton Twp

Cara Shillington
Eastern Michigan University

Brian Shmaefsky
Lone Star College - Kingwood

Wendy Stankovich
University of Wisconsin – Platteville

Bridget Stuckey
Olive-Harvey College

Sharon Thoma
University of Wisconsin - Madison

Kip Thompson
Ozarks Technical Community College

Sue Trammell
John A. Logan School

Valerie Vander Vliet
Lewis University

Mark E. Walvoord
University of Oklahoma

Daniel Ward
Waubonsee Community College

Scott Wells
Missouri Southern State University

Stephen White
Ozarks Technical Community College

Sonia Williams
Oklahoma City Community College

Jo Wu
Fullerton College

Connect Offers...
Everything You and Your Students Need In One Place!

McGraw-Hill Connect™ interactive learning platform provides auto-graded assessments and powerful reporting against learning outcomes and level of difficulty—all in an **easy-to-use interface**.

A Customizable, Interactive ebook

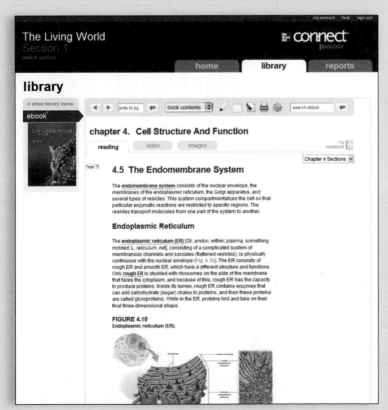

The McGraw-Hill ConnectPlus™ version of Connect includes the full textbook as an integrated, dynamic **eBook**. ConnectPlus provides all of the Connect features plus the following:

- An integrated, printable eBook, allowing for anytime, anywhere access to the textbook.
- **Dynamic links** between the problems or questions you assign to your students and the location in the ebook where those problems or questions are covered.
- You can assign fully integrated, self-study questions.
- Pagination that **exactly matches** the printed text, allowing students to rely on ConnectPlus as the complete resource for your course.
- Embedded media, including **animations and videos**.
- Customize the text for your students by **adding and sharing your own notes and highlights**.

www.mcgrawhillconnect.com

Save Time...
with Auto-Graded **Assessments** and Impressive **Reporting** Solutions

Fully editable, customizable, **auto-graded assignments** using high-quality art from the textbook take you way beyond multiple choice questions.

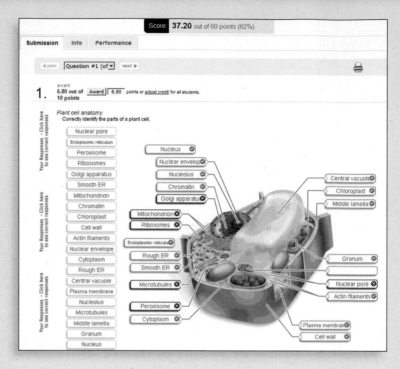

With Connect's detailed reporting, instructors can **quickly assess** how students are doing in regards to overall class performance, specific objectives, individual assignments, and more.

Graded reports can be **easily integrated** with a Learning Management System, such as Blackboard.

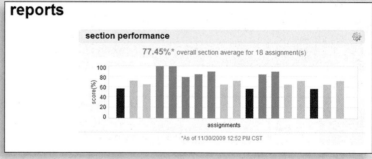

Instructors can easily modify a lecture if students aren't ready to move on.

Quickly identify content in your assignments that is causing your students trouble.

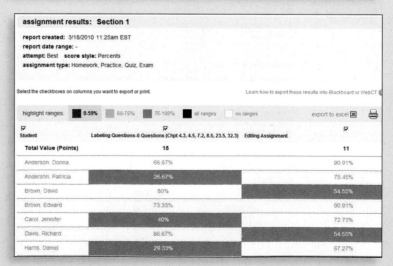

www.mcgrawhillconnect.com

Create...
Teaching Resources to Match the Way *You* Teach!

Craft your teaching resources to match the way you teach!

With McGraw-Hill Create™, **www.mcgrawhillcreate.com**, you can easily rearrange chapters, combine material from other content sources, and quickly upload content you have written like your course syllabus or teaching notes. Find the content you need in Create by searching through thousands of leading McGraw-Hill textbooks. Arrange your book to fit your teaching style. Create even allows you to personalize your book's appearance by selecting the cover and adding your name, school, and course information. Order a Create book and you'll receive a complimentary print review copy in 3–5 business days or a complimentary electronic review copy (eComp) via email in minutes.

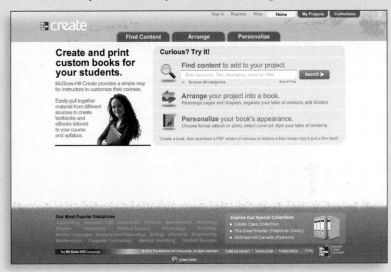

Go to **www.mcgrawhillcreate.com** today and register to experience how McGraw-Hill Create empowers you to teach your students your way.

Do More

McGraw-Hill Higher Education and Blackboard® have teamed up.

Blackboard, the Web-based course-management system, has partnered with McGraw-Hill to better allow students and faculty to use online materials and activities to complement face-to-face teaching. Blackboard features exciting social learning and teaching tools that foster more logical, visually impactful and active learning opportunities for students. You'll transform your closed-door classrooms into communities where students remain connected to their educational experience 24 hours a day.

This partnership allows you and your students access to McGraw-Hill's Connect and Create right from within your Blackboard course—all with one single sign-on.

Not only do you get single sign-on with Connect and Create, you also get deep integration of McGraw-Hill content and content engines right in Blackboard. Whether you're choosing a book for your course or building Connect™ assignments, all the tools you need are right where you want them—inside of Blackboard.

Gradebooks are now seamless. When a student completes an integrated Connect™ assignment, the grade for that assignment automatically (and instantly) feeds your Blackboard grade center.

McGraw-Hill and Blackboard can now offer you easy access to industry leading technology and content, whether your campus hosts it, or we do. Be sure to ask your local McGraw-Hill representative for details.

Animations for a new generation

Dynamic, 3D animations of key biological processes bring an unprecedented level of control to the classroom. Innovative features keep the emphasis on teaching rather than entertaining.

- An options menu lets you control the animation's level of detail, speed, length, and appearance, so you can create the experience you want.
- Draw on the animation using the white board pen to highlight important areas.
- The scroll bar lets you fast forward and rewind while seeing what happens in the animation, so you can start at the exact moment you want.
- A scene menu lets you instantly jump to a specific point in the animation.
- Pop ups add detail at important points and help students relate the animation back to concepts from lecture and the textbook.
- A complete visual summary at the end of the animation reminds students of the big picture.

Animation topics include: Cellular Respiration, Photosynthesis, Molecular Biology of the Gene, DNA Replication, Cell Cycle and Mitosis, Membrane Transport, and Plant Transport.

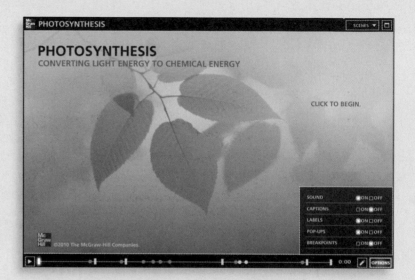

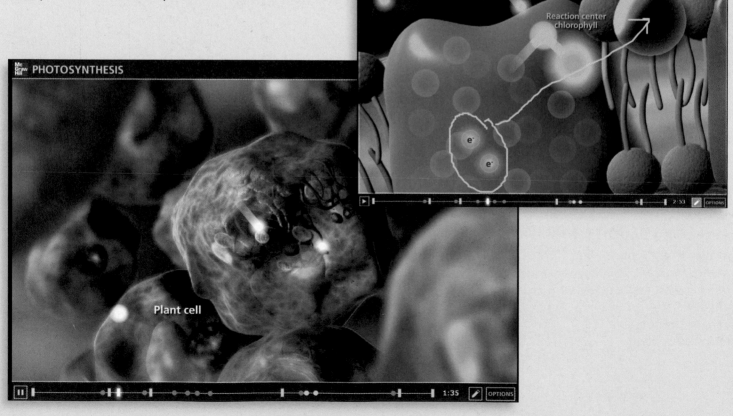

FlexArt

- **FlexArt Image PowerPoints:** Every piece of art is sized and cropped specifically for superior presentations, and labels can be edited. Key figures can be deconstructed and reshaped to fit your needs. Tables, photographs, and unlabeled art pieces are also included.
- **Lecture PowerPoints with Animations:** Animations illustrating important processes are embedded in the lecture material.
- **Animation PowerPoints:** Animations are embedded in PowerPoint slides to easily paste into your existing lecture.
- **Labeled JPEG Images:** Download folders with full-color digital files of all illustrations and most photos from the book that can be readily incorporated into presentations, exams, or custom-made classroom materials.
- **Base Art Image Files:** Download unlabeled digital files of all illustrations.

www.mhhe.com/tlw7

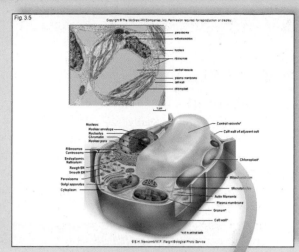

Move figure components.

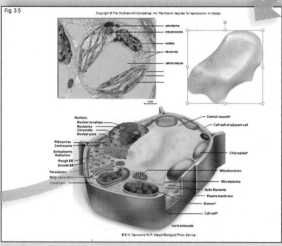

Delete elements you don't need and increase the size of key structures.

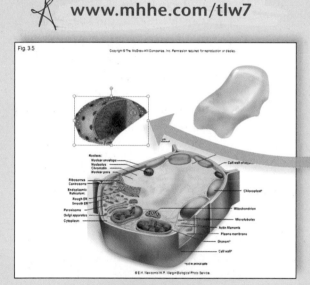

Presentation Center
Access the depth of McGraw-Hill content

Available to adopters, this online digital library contains photos, artwork, animations and other media from McGraw-Hill textbooks covering a broad array of subject areas. Create customized lectures and assessment or enhance your online course materials with our peer-reviewed content.

Make your lectures available anytime/anywhere

McGraw Hill **tegrity campus** Tegrity Campus™ records and distributes your class lecture, with just a click of a button. Students can view anytime and anywhere via computer, mp3 player, or smartphone. It indexes as it records your PowerPoint presentations and anything shown on your computer so students can use keywords to find exactly what they need to review. Tegrity is available as an integrated feature of McGraw-Hill Connect or standalone.

Contents

UNIT SEVEN   PLANT LIFE

Applications Index

0
Studying Biology

This puzzled leopard, seemingly poised to jump, faces an uncertain future, a future it cannot escape by leaping off the cover of this book. Sadly, it literally has no place to go. Once distributed across Africa and southern Asia, the leopard's range has decreased radically in the last century due to hunting and loss of habitat. Now, except for a few relict populations in Asia, it is found only in sub-Saharan Africa, and is considered a "Near Threatened" species. The leopard's fate, and that of all other creatures of the living world, will depend critically on the steps we humans take to protect and preserve our world's climate and resources. Your study of biology will provide you with a key tool to help. You are about to leap into the study of molecules, cells, and intricate body processes, of evolution and ecology. Rich with new ideas unknown to many of you, biology is a science course full of promise. This short "Chapter Zero" is intended to provide you with the tools to make the leap more strongly, and with greater confidence. Good luck.

0.1 How to Study

Some students will do well in this course, others poorly. One of the best predictors of how well you will do is how well you are prepared to learn. Entering an introductory science course like this one, do you know how to take lecture notes? Do you know how to use these notes effectively with your textbook? Can you read a graph? This edition of *The Living World* tackles this problem head-on by providing you with this "Chapter Zero" at the beginning of the text. It is intended to help you master these very basic but essential learning tools.

Taking Notes

Listening to lectures and reading the text are only the first steps in learning enough to do well in a biology course. The key to mastering the mountain of information and concepts you are about to encounter is to take careful notes. Studying from poor-quality notes that are sparse, disorganized, and barely intelligible is not a productive way to approach preparing for an exam.

There are three simple ways to improve the quality of your notes:

1. **Take many notes.** Always attempt to take the most complete notes possible during class. If you miss class, take notes yourself from a tape of the lecture, if at all possible. It is the process of taking notes that promotes learning. Using someone else's notes is but a poor substitute. When someone else takes the notes, that person tends to do most of the learning as well.

2. **Take paraphrased notes.** Develop a legible style of abbreviated note taking. Obviously, there are some things that cannot be easily paraphrased (referred to in a simpler way), but using abbreviations and paraphrasing will permit more comprehensive notes. Attempting to write complete organized sentences in note taking is frustrating and too time consuming—people just talk too fast!

3. **Revise your notes.** As soon as possible after lecture, you should decipher and revise your notes. Nothing else in the learning process is more important, because this is where most of your learning will take place. By revising your notes, you meld the information together and put it into a context that is understandable to you. As you revise your notes, organize the material into major blocks of information with simple "heads" to identify each block. Add ideas from your reading of the text and note links to material in other lectures. Clarify terms and concepts that might be confusing with short notes and definitions. Thinking through the ideas of the lecture in this organized way will crystallize them for you, which is the key step in learning.

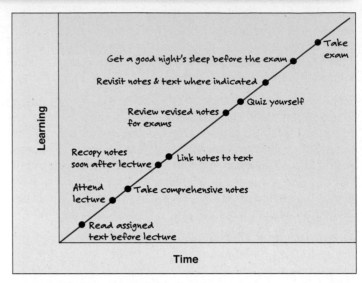

Figure 0.1   A learning timeline.

Remembering and Forgetting

Learning is the process of placing information in your memory. Just as in your computer, there are two sorts of memory. The first, *short-term memory*, is analogous to the RAM (random access memory) of a computer, holding information for only a short period of time. Just as in your computer, this memory is constantly being "written over" as new information comes in. The second kind of memory, *long-term memory*, consists of information that you have stored in your memory banks for future retrieval, like storing files on your computer's hard drive. In its simplest context, learning is the process of transferring information to your hard drive.

Forgetting is the loss of information stored in memory. Most of what we forget when taking exams is the natural consequence of short-term memories not being effectively transferred to long-term memory. Forgetting occurs very rapidly, dropping to below 50% retention within one hour after learning and leveling off at about 20% retention after 24 hours.

There are many things you can do to slow down the forgetting process (**figure 0.1**). Here are two important ones:

1. **Recopy your notes as soon as possible after lecture.** Remember, there is about a 50% memory loss in the first hours. Optimally, you should use your textbook as well while recopying your notes.

2. **Establish a purpose for reading.** When you sit down to study your textbook, have a definite goal to learn a particular concept. Each chapter begins with a preview of its key concepts—let them be your guides. Do not try and learn the entire contents of a chapter in one session; break it up into small pieces that are "easily digested."

Learning

Learning may be viewed as the efficient transfer of information from your short-term memory to your long-term memory. This transfer is referred to as *rehearsal* by learning strategists. As its name implies, rehearsal always involves some form of repetition. There are three general means of rehearsal in the jargon of education called "critical thinking skills" (**figure 0.2**).

Repeating The most obvious form of rehearsal is repetition. To learn facts, the sequence of events in a process, or the names of a group of things, you write them down, say them aloud, and mentally repeat them over and over until you have "memorized" them. This often is a first step on the road to learning. Many students mistake this as the only step. It is not, as it involves only rote memory instead of understanding. If all you do in this course is memorize facts, you will not succeed.

Organizing It is important to organize the information you are attempting to learn because the process of sorting and ordering increases retention. For example, if you place a sequence of events in order, like the stages of mitosis, the entire sequence can be recalled if you can remember what gets the sequence started.

Connecting You will learn biology much more effectively if you relate what you are learning to the world about you. The many challenges of living in today's world are often related to the information presented in this course, and understanding these relationships will help you learn. In each chapter of this textbook you will encounter full-page Connection essays that allow you to briefly explore a "real-world" topic related to what you are learning. One appears on page 5. Read these essays. You may not be tested on these essays, but reading them will provide you with another "hook" to help you learn the material on which you will be tested.

Studying to Learn

If I have heard it once, I have heard it a thousand times, "Gee, Professor Johnson, I studied for 20 hours straight and I still got a D." By now, you should be getting the idea that just throwing time at the material does not necessarily ensure a favorable outcome.

Studying, said simply, is putting your learning skills to work. It should come as no surprise to you that how you set about doing this matters. Three simple strategies can make your study sessions more effective:

1. **Study at intervals.** The length of time you spend studying and the spacing between study or reading sessions directly affects how much you learn. If you had 10 hours to spend studying, you would be better off if you broke it up into 10 one-hour sessions than to spend it all in one or two sessions. There are two reasons for this:

    First, we know from formal cognition research (as well as from our everyday life experiences) that we remember "beginnings" and "endings" but tend to forget "middles." Thus, the learning process can benefit from many "beginnings" and "endings."

Figure 0.2 Learning requires work.
Learning is something you do, not something that happens to you.

Second, unless you are unusual, after 30 minutes or an hour your ability to concentrate is diminished. Concentration is a critical component of studying to learn. Many short, topic-focused study sessions maximize your ability to concentrate effectively.

2. **Avoid distractions.** It makes a surprising amount of difference *where* you study. Why? Because effective studying requires concentration. For most of us, effective concentration requires a comfortable, quiet environment with no outside distractions like loud music or conversations.

    It is for this reason that studying in front of a loudly playing television or stereo, or at a table in a busy cafeteria, is a recipe for failure. A quiet room, a desk in the library, outside on a sunny day—all these study locations are quiet, offering few distractions and allowing you to focus your concentration on what you are trying to learn. Keep your mobile phone off— texting while studying is as distracting as it is while driving, and as much to be avoided.

3. **Reward yourself.** At the end of every study interval, schedule something fun, if only to get away from studying for a bit. This "carrot and stick" approach tends to make the next study interval more palatable.

Learning Is an Active Process

It is important to realize that learning biology is not something you can do passively. Many students think that simply possessing a lecture video or a set of class notes will get them through. In and of themselves, videos and notes are no more important than the Nautilus machine an athlete works out on. It is not the machine per se, but what happens when you use it effectively that is of importance.

Common sense will have a great deal to do with your success in learning biology, as it does in most of life's endeavors. Your success in this biology course will depend on doing some simple, obvious things (**figure 0.3**):

- *Attend class.* Go to all the lectures and be on time.
- *Read the assigned readings before lecture.* If you have done so, you will hear things in lecture that will be familiar to you, a recognition that is a vital form of learning reinforcement. Later you can go back to the text to check details.
- *Take comprehensive notes.* Recognizing and writing down lecture points is another form of recognition and reinforcement. Later, studying for an exam, you will have already forgotten lecture material you did not record, and so even if you study hard, you will miss exam questions on this material.
- *Revise your notes soon after lecture.* Actively interacting with your class notes while you still hold much of the lecture in short-term memory provides perhaps the most powerful form of reinforcement, and will be a key to your success.

The process of revising your lecture notes can and should be a powerful learning tool. For the best results, don't simply transcribe more legibly what you scribbled down so rapidly in class. Instead, focus on how the lecture was organized, and use that framework to organize your revised notes. Most lectures are organized much like each chapter is in this textbook, with three or four main topics, each covered in a series of steps. To revise your class lecture notes most effectively, you should try to *outline* what was said in lecture: First write down the three or four main headings, and then under each heading place the block of lecture material that addressed that topic.

Perhaps more than you have realized, a lecture in a biology course is a network of ideas. Going through your class lecture notes and identifying the main topics is a powerful first step in sorting these ideas out in your mind. The second step, laying out the material devoted to each topic in a logical order (which is, hopefully, the order in which it was presented), will make clearer to you the ideas that link the material together—and this is, in the final analysis, much of what you are trying to learn.

As you proceed through this textbook, you will encounter a blizzard of terms and concepts. Biology is a field

Figure 0.3 Critical learning occurs in the classroom. Learning occurs in at least four distinct stages: attending class; doing assigned textbook readings before lecture; listening and taking notes during lecture; and recopying notes shortly after lecture. If you are diligent in these steps, then studying lecture notes and text assignments before exams is much more effective. Skipping any of these stages makes successful learning far less likely.

rich with ideas and the technical jargon needed to describe them. What you discover reading this textbook is intended to support the lectures that provide the core of your biology course. Integrating what you learn here with what you learn in lecture will provide you with the strongest possible tool for successfully mastering the basics of biology. The rest is just hard work.

> **Key Learning Outcome 0.1** Studying biology successfully is an active process. To do well, you should attend lectures, do assigned readings before lecture, take complete class notes, rewrite those notes soon after class, and study for exams in short, focused sessions.

Pulling an All-Nighter

At some point in the next months you will face that scary rite, the first exam in this course. As a university professor, I get to give the exams rather than take them, but I can remember with crystal clarity when the shoe was on the other foot. I didn't like exams a bit as a student—what student does? But in my case I was often practically paralyzed with fear. What scared me about exams was the possibility of unanticipated questions. No matter how much I learned, there was always something I didn't know, some direction from which my teacher could lob a question that I had no chance of answering.

I lived and died by the all-nighter. Black coffee was my closest friend in final exam week, and sleep seemed a luxury I couldn't afford. My parents urged me to sleep more, but I was trying to cram enough in to meet any possible question, and couldn't waste time sleeping.

Now I find I did it all wrong. In work published over the last few years, researchers at Harvard Medical School have demonstrated that our memory of newly learned information improves only after sleeping at least six hours. If I wanted to do well on final exams, I could not have chosen a poorer way to prepare. The gods must look after the ignorant, as I usually passed.

Learning is, in its most basic sense, a matter of forming memories. The Harvard researchers' experiments showed that a person trying to learn something does not improve his or her knowledge until after he or she has had more than six hours of sleep (preferably eight). It seems the brain needs time to file new information and skills away in the proper slots so they can be retrieved later. Without enough sleep to do all this filing, new information does not get properly encoded into the brain's memory circuits.

To sort out the role of sleep in learning, the Harvard Medical School researchers used Harvard undergrads as guinea pigs. The undergraduates were trained to look for particular visual targets on a computer screen, and to push a button as soon as they were sure they had seen one. At first, responses were relatively sluggish—it typically took 400 milliseconds for a target to reach a student's conscious awareness. With an hour's training, however, many students were hitting the button correctly in 75 milliseconds.

How well had they learned? When retested from 3 to 12 hours later on the same day, there was no further improvement past a student's best time in the training session. If the researchers let a student get a little sleep, but less than six hours, then retested the next day, the student still showed no improvement. For students who slept more than six hours, the story was very different. Sleep greatly improved performance. Students who achieved 75 milliseconds in the training session would reliably perform the target identification in 62 milliseconds after a good night's sleep! After several nights of ample sleep, they often got even more proficient.

Why six or eight hours, and not four or five? The sort of sleeping you do at the beginning of a night's sleep and the sort you do at the end are different, and both, it appears, are required for efficient learning.

The first two hours of sleeping are spent in deep sleep, what psychiatrists call slow wave sleep. During this time, certain brain chemicals become used up, which allows information that has been gathered during the day to flow out of the memory center of the brain, the hippocampus, and into the cortex, the outer covering of the brain, where long-term memories are stored. Like moving information in a computer from active memory to the hard drive, this process preserves experience for future reference. Without it, long-term learning cannot occur.

Over the next hours, the cortex sorts through the information it has received, distributing it to various locations and networks. Particular connections between nerve cells become strengthened as memories are preserved, a process that is thought to require the time-consuming manufacture of new proteins.

If you halt this process before it is complete, the day's memories do not get fully "transcribed," and you don't remember all that you would have, had you allowed the process to continue to completion. A few hours are just not enough time to get the job done. Four hours, the Harvard researchers estimate, is a minimum requirement.

The last two hours of a night's uninterrupted sleep are spent in rapid-eye-movement (rem) sleep. This is when dreams occur. The brain shuts down the connection to the hippocampus and runs through the data it has stored over the previous hours. This process is also important to learning, as it reinforces and strengthens the many connections between nerve cells that make up the new memory. Like a child repeating a refrain to memorize it, the brain goes over things until practice makes perfect.

That's why my college system of getting by on three or four hours of sleep during exam week and crashing for 12 hours on weekends didn't work. After a few days, all of the facts I had memorized during one of my "all-nighters" faded away. Of course they did. I had never given them a chance to integrate properly into my memory circuits.

As I look back, I see now that how well I did on my exams probably had far less to do with how hard I studied than with how much I slept. It doesn't seem fair.

Using Your Textbook

A Textbook Is a Tool

A student enrolled in an introductory biology course as you are almost never learns everything from the textbook. Your text is a tool to explain and amplify what you learn in lecture. No textbook is a substitute for attending lectures, taking notes, and studying them. Success in your biology course is like a stool with three legs: lectures, class notes, and text reading—all three are necessary. Used together, they will take you a long way toward success in the course.

When to Use Your Text While you can glance at your text at any time to refresh your memory or answer a question that pops into your mind, your use of your text as a learning tool should focus on providing support for the other two "legs" of course success: lectures and class notes.

Do the Assigned Reading. Many instructors assign reading from the text, reading that is supposed to be done before lecture. The timing here is very important: If you already have a general idea of what is being discussed in lecture, it is much easier to follow the discussion and take better notes.

Link the Text to Your Lecture Notes. Few lectures cover exactly what is in the text, and much of what is in the text may not be covered in lecture. That said, much of what you will hear in lecture *is* covered in your text. This coverage provides you with a powerful tool to reinforce ideas and information you encounter in lecture. Text illustrations and detailed explanations can pound home an idea quickly grasped in lecture, and answer any questions that might occur to you as you sort through the logic of an argument. Thus it is absolutely essential that you follow along with your text as you recopy your lecture notes, keying your notes to the textbook as you go. Annotating your notes in this way will make them far better learning tools as you study for exams later.

Review for Exams. It goes without saying that you should review your recopied lecture notes to prepare for an exam. But that is not enough. What is often missed in gearing up for an exam is the need to also review that part of the text that covers the same material. Reading the chapter again, one last time, helps place your lecture notes in perspective, so that it will be easier to remember key points when a topic explodes at you off the page of your exam.

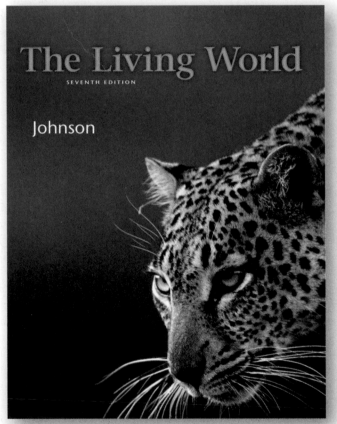

The Living World
SEVENTH EDITION

Johnson

How to Use Your Text The single most important way to use your text is to read it. As your biology course proceeds and you move through the text, read each assigned chapter all the way through at one sitting. This will give you valuable perspective. Then, guided by your lecture notes, go back through the chapter one topic at a time, and focus on learning that one topic as you recopy your notes. As discussed earlier, building a bridge between text and lecture notes is a very powerful way to learn. Remember, your notes don't take the exam, and neither does the textbook; you do, and the learning that occurs as you integrate text pages and lecture notes in your mind will go a long way toward you taking it well.

Learning Tools at Your Disposal

Learning Objectives. Every chapter begins by telling you precisely what each section of the chapter is attempting to teach you. Called "Learning Objectives," these items describe what you are intended to know after studying that section. Use them. They are a road map to success in the course.

Quiz Yourself. When you have finished studying a chapter of your text, it will be very important for you to be able to assess how good a job you have done. Waiting until a class exam to find out if you have mastered the key points of a chapter is neither necessary nor wise. To give you some handle on how you are doing, three sorts of questions appear at the end of every chapter.

Test Your Understanding. At the top of the last page of each chapter, you will find a "Test Your Understanding" section composed of ten multiple choice questions. These questions are not difficult and are intended as a quick check to see if you have understood the key ideas and identified essential information.

Apply Your Understanding. Beneath the Test Your Understanding you will find two or more "Apply Your Understanding" questions. One of the easiest mistakes to make in studying a chapter is to slide over its figures as if they were simply decoration. In fact, they often illustrate key ideas and processes. These questions will help you apply what you have learned to what the figures are trying to teach you.

Synthesize What You Have Learned. At the bottom of the page, you will find a series of "Synthesize What You Have Learned" questions. These questions do not test your memory, but rather your understanding. None of them are easy; all of them make you think.

Let the Illustrations Teach You. All introductory biology texts are rich with colorful photographs and diagrams. They are not there to decorate but to aid your comprehension of ideas and concepts. When the text refers you to a specific figure, look at it—the visual link will help you remember the idea much better than restricting yourself to cold words on a page.

Three sorts of illustrations offer particularly strong reinforcement:

Key Biological Processes. While you will be asked to learn many technical terms in this course, learning the names of things is not your key goal. Your goal is to master a small set of concepts. A few dozen key biological processes explain how organisms work the way they do. When you have understood these processes, much of the heavy lifting in learning biology is done. Every time you encounter one of these key biological processes in the text, you will be provided with an illustration to help you better understand it. These illustrations break the process down into easily understood stages, so that you can grasp how the overall process works without being lost in a forest of details (**figure 0.4***a*).

Bubble Links. Illustrations teach best when they are simple. Unfortunately, some of the structures and processes being illustrated just aren't simple. Every time you encounter a complex diagram in the text, it will be "predigested" for you, the individual components of the diagram each identified with a number in a colored circle, or bubble. This same number is also placed in the text narrative right where that component is discussed. These bubble links allow the text to step you through the illustration, explaining what is going on at each stage—the illustration is a feast you devour one bite at a time.

Phylum Facts. Not all of what you will learn are concepts. Sometimes you will need to soak up a lot of information, painting a picture with facts. Nowhere is this more true than when you study animal diversity. In chapter 19 you will encounter a train of animal phyla (a phylum is a major category of organisms) with which you must become familiar. In such a sea of information, what should you learn? Every time you encounter a phylum in chapter 19, you will be provided with a *Phylum Facts* illustration that selects the key bits of information about the body and lifestyle of that kind of animal (**figure 0.4***b*). If you learned and understood only the items highlighted there, you would have mastered much of what you need to know.

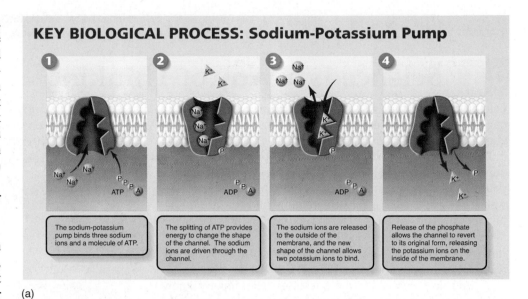

(a)

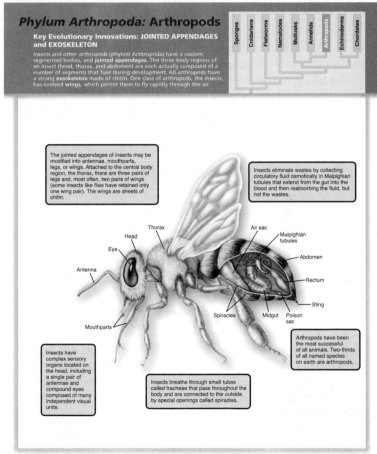

(b)

Figure 0.4 Visual learning tools.

(*a*) An example of a Key Biological Process illustration. (*b*) An example of a Phylum Facts illustration.

Key Learning Outcome 0.2   Your text is a tool to reinforce and clarify what you learn in lecture. Your use of it will only be effective if coordinated with your development of recopied lecture notes.

0.3 Science Is a Way of Thinking

One of the most important things you can learn in a biology course is how to evaluate scientific claims. Long after this class is completed, you will be making decisions that involve biology, and they will be better, more informed decisions if you have acquired the skill to evaluate scientific claims for yourself. Because it is printed in the newspaper or cited on a website doesn't make a scientific claim valid. **Figure 0.5** illustrates the sorts of biology topics you encounter today in the news—and this is just a small sample. They are, all of them, important issues that will affect your own life. How do you reach informed opinions about them?

You do it by asking the question, "How do we know this?" In this text you will encounter a great deal of information and explanation, and will be asked to accept that what you are being taught is "true," that it accurately reflects reality. In fact, what it accurately reflects is what we know about reality. Some of what we now know will be altered by future scientists as they learn more. How do they know they are right? Science is a way of thinking that demands to see the evidence, that challenges the validity of every claim. If you can learn to do this, to apply this skill in the future to personal decisions about biology as it impacts your life, you will have taken from this course a valuable lesson.

How Do We Know What We Know?

A useful way to learn how scientists think, how they constantly check and question what they know, is to look at real cases. What follows are four instances where biologists have come to a conclusion. These conclusions will be taught in this textbook, reflecting the world about us as best as science can determine. All four of these cases will be treated at length in later chapters—here they serve only to introduce the process of scientific questioning.

Does Cigarette Smoking Cause Lung Cancer? The American Cancer Society estimates that 562,340 Americans died of cancer in 2009. Fully one in four of the students using this textbook can be expected to die of it. Twenty-nine percent of these cancer victims, almost a third, die of lung cancer.

As you might imagine, something that kills so many of us has been the subject of much research. The first step biologists took was to ask a simple question: "Who gets lung cancer?" The answer came back loud and clear: Fully 87% of lung cancer deaths are cigarette smokers. Delving into this more closely, researchers looked to see if the incidence of lung cancer (that is, how many people contract it per 100,000 people) can be predicted by how many cigarettes a person smokes each day. As you can see in the graph in **figure 0.6** (which is shown again on page 634), it can. The more cigarettes smoked, the higher the occurrence of lung cancer. Based on this study, and lots of others like it, examined in more detail on pages 230 and 524, biologists concluded that smoking cigarettes causes lung cancer.

Figure 0.5 Biology in the news.

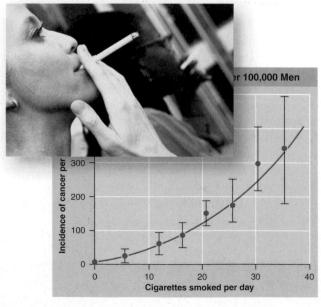

Figure 0.6 Does smoking cause lung cancer?

Does Carbon Dioxide Cause Global Warming? Our world is getting warmer. Looking for the cause, atmospheric scientists soon began to suspect what might at first seem an unlikely culprit: carbon dioxide (CO_2), a gas that is a minor component (0.03%) of the air we breathe. As you will learn in this course, burning coal and other fossil fuels releases CO_2 into the atmosphere. Problems arise because CO_2 traps heat. As the modern world industrializes, more and more CO_2 is released. Does this lead to a hotter earth? To find out, researchers looked to see if the rise in global temperature reflected a rise in the atmosphere's CO_2. As you can see in the graph in **figure 0.7** (which is shown again on page 796), it does. After these and other careful studies we will explore in detail on pages 796 and 812, scientists concluded that rising CO_2 levels are indeed the cause of global warming.

Does Obesity Lead to Type 2 Diabetes? The United States is in the midst of an obesity epidemic. Over the last 16 years, the percentage of Americans who are obese has almost tripled, from 12% to over 34%. Coincidentally (or is it a coincidence?), the number of Americans suffering from type 2 diabetes (a disorder in which the body loses its ability to regulate glucose levels in the blood, often leading to blindness and amputation of limbs) has more than tripled over the same 16-year period, from 7 million to more than 23 million (that's one in every 14 Americans!).

What is going on here? When researchers compared obesity levels with type 2 diabetes levels, they found a marked correlation, clearly visible in the graph in **figure 0.8** (which is shown again on page 629). Investigating more closely, the researchers found that an estimated 80% of people who develop type 2 diabetes are obese. Detailed investigations described on page 629 have now confirmed the relationship that these early studies hinted at: Overeating triggers changes in the body that lead to type 2 diabetes.

What Causes the Ozone Hole? Twenty-five years ago, atmospheric scientists first reported a loss of UV-absorbing ozone (O_3) gas high in the atmosphere over Antarctica. Trying to understand the reason for this "ozone hole," researchers began to suspect chlorofluorocarbons, or CFCs. CFCs are supposedly inert chemicals that are widely used as heat exchangers in air conditioners. However, further studies, detailed on pages 22–23, indicated that CFCs are not inert after all—in the intense cold temperatures high over Antarctica, they cause O_3 to be converted to O_2. Scientists concluded that CFCs were indeed causing the ozone hole over Antarctica. As you can see in the graph in **figure 0.9**, the size of the ozone hole soon stopped expanding after CFC production was restricted.

Looking at the Evidence

One thing these four cases have in common is that in each, scientists reached their conclusion not by applying established rules but rather by looking in detail at what was going on and then testing possible explanations. In short, they gathered data and analyzed it. If you are going to think independently about scientific issues in the future, then you will need to learn how to analyze data and understand what it is telling you. In each case above, the data is presented in the form of a graph. Said simply, you will need to learn to read a graph.

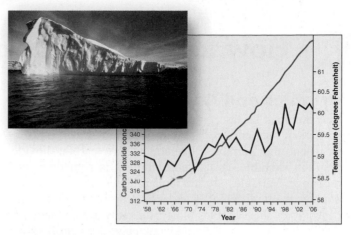

Figure 0.7   Does carbon dioxide cause global warming?

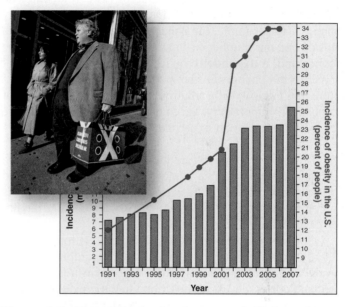

Figure 0.8   Does obesity lead to type 2 diabetes?

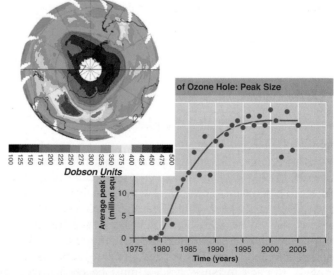

Figure 0.9   What causes the ozone hole?

> **Key Learning Outcome 0.3**   Scientists investigate by gathering data and analyzing it to form possible explanations they can test.

0.4 How to Read a Graph

Variables and Graphs

In the previous section, you encountered four graphs illustrating what happened to variables, such as global temperature, when other variables, like atmospheric carbon dioxide, change. A **variable**, as its name implies, is something that can change. Variables are the tools of science, and you will encounter many different kinds as you proceed through this text. Many of the variables biologists study are examined in graphs like you saw on the previous pages. A **graph** shows what happens to one variable when another one changes.

There are two types of variables. The first kind, an **independent variable,** is one that a researcher deliberately changes—for example, the concentration of a chemical in a solution, or the number of cigarettes smoked per day. The second kind, a **dependent variable,** is what happens in response to the changes in the independent variable—for example, the intensity of a solution's color, or the incidence of lung cancer. Importantly, the change in a dependent variable that is measured in an experiment is not predetermined by the investigator.

In science, all graphs are presented in a consistent way. The independent variable is always presented and labeled across the bottom, called the *x axis*. The dependent variable is always presented and labeled along the side (usually the left side), called the *y axis* (**figure 0.10**).

Some research involves examining correlations between sets of variables, rather than the deliberate manipulation of a variable. For example, a researcher who measures both diabetes and obesity levels (as described in the previous section) is actually comparing two dependent variables. While such a comparison can reveal correlations and so suggest potential relationships, *correlation does not prove causation*. What is happening to one variable may actually have nothing to do with what happens to the other variable. Only by manipulating a variable (making it an independent variable) can you test for causality. Just because people that are obese tend to also have diabetes does not establish that obesity *causes* diabetes. Other experiments are needed to determine causation.

Using the Appropriate Scale and Units

A key aspect of presenting data in a graph is the selection of proper scale. Data presented in a table can utilize many scales, from seconds to centuries, with no problems. A graph, however, typically has a single scale on the *x* axis and a single scale on the *y* axis, which might consist of microscopic units (for example, nanometers, microliters, micrograms) or macroscopic units (for example, feet, inches, liters, days, milligrams). In each instance, a scale must be chosen that fits what is being measured. Changes in centimeters would not be obvious in a graph scaled in kilometers. Also, if a variable

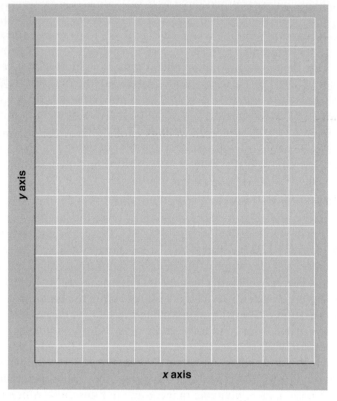

Figure 0.10   The two axes of a graph.
The independent variable is almost always presented along the *x* axis, and the dependent variable is usually shown along the *y* axis.

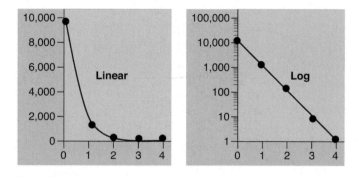

Figure 0.11   Linear and log scale: two ways of presenting the same data.

changes a great deal over the course of the experiment, it is often useful to use an expanding scale. A **log** or **logarithmic scale** is a series of numbers plotted as powers of 10 (1, 10, 100, 1,000,...) rather than in the linear progression seen on most graphs (2,000, 4,000, 6,000...). Consider the two graphs in **figure 0.11**, where the *y* axis is plotted on a linear scale on the left and on a log scale on the right. You can see that the log scale more clearly displays changes in the dependent variable (the y axis) for the upper values of the independent variable (the x axis, values 2, 3, and 4). Notice that the interval *between* each y axis number is not linear either—the interval between each number is itself subdivided on a log scale. Thus, 50 (the fourth tick mark between 10 and 100) is plotted much closer to 100 than to 10.

Individual graphs use different units of measurement, each chosen to best display the experimental data. By international convention, scientific data are presented in **metric units,** a system of units expressed as powers of 10. For example, weight is expressed in units called *grams*. Ten grams make up a decagram, and 1,000 grams is a kilogram. Smaller weights are expressed as a portion of a gram—for example, a centigram is a hundredth of a gram, and a milligram is a thousandth of a gram. The units of measurement employed in a graph are by convention indicated in parentheses next to the independent variable label on the x axis and the dependent variable label on the y axis.

Drawing a Line

Most of the graphs that you will find in this text are **line graphs,** which are graphs composed of data points and one or more lines. Line graphs are typically used to present *continuous data*—that is, data that are discrete samples of a continuous process. An example might be data measuring how quickly the ozone hole develops over Antarctica in August and September of each year. You could in principle measure the area of the ozone hole every day, but to make the project more manageable in time and resources, you might actually take a measurement only once a week. Measurements reveal that the ozone hole increases in area rapidly for about six weeks before shrinking, yielding six data points during its expansion. These six data points are like individual frames from a movie—frozen moments in time. The six data points might indicate a very consistent pattern, or they might not.

Consider the hypothetical data in the graphs of **figure 0.12.** The data points on the left graph are changing in a very consistent way, with little variation from what a straight line (drawn in red) would predict. The graph in the middle shows more experimental variation, but a straight line still does a good job of revealing the overall pattern of how the

data are changing. Such a straight "best-fit line" is called a **regression line** and is calculated by estimating the distance of each point to possible lines, adding the values, and selecting the line with the lowest sum. The data points in the graph on the right are randomly distributed and show no overall pattern, indicating that there is no relationship between the dependent and the independent variables.

Other Graphical Presentations of Data

Sometimes the independent variable for a data set is not continuous but rather represents discrete sets of data. A line graph, with its assumption of continuity, cannot accurately represent the variation occurring in discrete sets of data, where the data sets are being compared with one another. In these cases, the preferred presentation is that of a **histogram,** a kind of bar graph. For example, if you were surveying the heights of pine trees in a park, you might group their heights (the independent variable) into discrete "categories" such as 0 to 5 meters tall, 5 to 10 meters, and so on. These categories are placed on the x axis. You would then count the number of trees in each category and present that dependent variable on the y axis, as shown in **figure 0.13.**

Some data represent proportions of a whole data set, for example the different types of trees in the park as a percentage of all the trees. This type of data is often presented in a **pie chart** (**figure 0.14**).

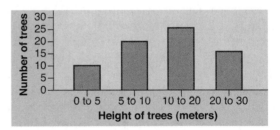

Figure 0.13 Histogram: the frequency of tall trees.

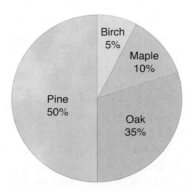

Figure 0.14 Pie chart: the composition of a forest.

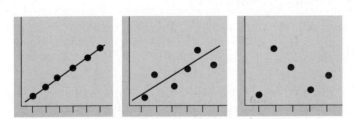

Figure 0.12 Line graphs: hypothetical growth in size of the ozone hole.

Putting Your Graph-Reading Skills to Work: Inquiry & Analysis

The sorts of graphs you have encountered here in this brief introduction are all used frequently by scientists in analyzing and presenting their experimental results, and you will encounter them often as you proceed through this text.

Learning to read a graph and understanding what it does and does not tell you is one of the most important things you can take away from a biology course. To help you develop this skill, every chapter of this text ends with a full-page *Inquiry & Analysis* feature. Each of these end-of-chapter features describes a real scientific investigation. You will be introduced to a question, and then given a hypothesis posed by a researcher (a hypothesis is a kind of explanation) to answer that question. The feature will then tell you how the researcher set about evaluating his or her hypothesis with an experiment and will present a graph of the data the researcher obtained. You are then challenged to analyze the data and reach a conclusion about the validity of the hypothesis.

As an example, consider research on the ozone hole, also discussed further in chapter 1. What sort of graph might you expect to see? A *line graph* can be used to present data on how the size of the ozone hole changes over the course of one year. Because the dependent variable is the size of the ozone hole measured continuously over a single season, a smooth curve accurately portrays what is actually going on. In this case, the regression line is not a straight line, but rather a curve (**figure 0.15**).

Can line graphs and histograms be used to present the same data? In some cases, yes. The mode of its presentation does not alter the data; it only serves to emphasize the point being investigated. The *histogram* in **figure 0.16** presents data on how the peak size of the ozone hole has changed in two-year intervals over the past 26 years. However, this same data could also have been presented as a line graph—as it was in **figure 0.9**.

Presented with a graph of the data obtained in the investigation of the *Inquiry & Analysis*, it will be your job to analyze it. Every analysis of a graph involves four distinct steps, some more complex than others but all essential to the process.

Applying Concepts. Your first task is to make sure you understand the nature of the variables, and the scale at which they are being presented in the graph. As a self-test, it is always a good idea to ask yourself to identify the dependent variable.

Interpreting Data. Look at the graph. What is changing? How much? How quickly? Is the change continuous? Progressive? What in fact has happened?

Making Inferences. Looking at what has happened, can you logically infer that the independent variable has caused the change you see in the dependent variable?

Drawing Conclusions. Does the inference you were able to make support the hypothesis that the experiment set out to test?

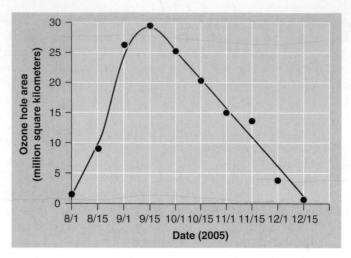

Figure 0.15   Changes in the size of the ozone hole over one year.
This graph shows how the size of the ozone hole changes over time, first expanding in size and then getting smaller.

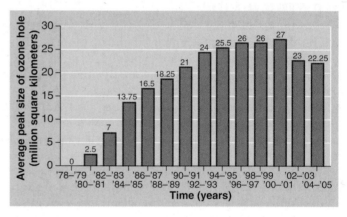

Figure 0.16   The peak sizes of the ozone hole.
This histogram shows how the size of the ozone hole increased in size (determined by its peak size) for 20 years before it then started to decrease.

This process of inquiry and analysis is the nuts and bolts of science, and by mastering it, you will go a long way toward learning how a scientist thinks. Long after this course is completed, you will find yourself compelled to make judgements about competing scientific claims you encounter as a citizen: Should the Government ban the chemical BPA from plastic bottles? Should you vote in favor of caps on factory carbon emissions? Are swine flu immunizations better delivered as nasal squirts (which use live flu viruses) or shots (which use dead flu viruses)? Few pages in this text provide more bang for the buck in learning that lasts.

Key Learning Outcome 0.4   Scientists often present data in standardized graphs, which portray how a dependent variable changes when an independent variable is changed.

1
The Science of Biology

These Antarctic Gentoo penguins share many properties with you and all living things. Their bodies are made up of cells, just as yours is. They have families, with children that resemble their parents, just as your parents did. They grow by eating, as you do, although their diet is limited to fish and krill they catch in the cold Antarctic waters. The sky above them shields them from the sun's harmful UV radiation, just as the sky above you shields you. Not in the summer, however. In the Antarctic summer an "ozone hole" appears, depleting the ozone above these penguins and exposing them to the danger of UV radiation. Scientists are analyzing this situation by a process of observation and experimentation, rejecting ideas that do not match their data. Proceeding in this way they are learning more and more about what is going on. The study of biology is a matter of observing carefully and asking the right questions. When a possible answer—what a scientist calls a *hypothesis*—was proposed, that destruction of Antarctic ozone is the result of leakage of industrial chemicals containing chlorine into the world's atmosphere, scientists carried out experiments and further observations in an attempt to prove this hypothesis wrong. Nothing they have learned so far leads them to reject the hypothesis. It appears human activities far to the north are having a serious impact on the environment of these penguins. This chapter begins your study of biology, the science of life, of penguins and people. Its study helps us to better understand ourselves, our world, and our impact on it.

Learning Objectives

Biology and the Living World
1.1 The Diversity of Life
1. List the six kingdoms of life.
2. Identify the kingdom to which you belong.

1.2 Properties of Life
1. Name the five basic properties shared by all living things.
2. Explain why complexity, movement, and response to stimulation are not properties that define life.

1.3 Organization of Life
1. List the 13 hierarchal levels of the organization of life.
2. Factor them into three general levels of complexity.
3. Define emergent property, and describe one for each of the three general levels of life's complexity.

1.4 Biological Themes
1. List the five general themes that unify biology as a science.
2. Describe the flow of energy among living organisms.
3. Define symbiosis.
4. Identify the key ability that allows a complex body like yours to maintain homeostasis.

The Scientific Process
1.5 How Scientists Think
1. Differentiate between deductive and inductive reasoning.
2. Identify the form of reasoning used in most scientific studies.

1.6 Science in Action: A Case Study
1. Describe the mechanism producing the ozone "hole" over Antarctica.
2. Explain why this ozone depletion is dangerous to humans.

1.7 Stages of a Scientific Investigation
1. State the six stages of a scientific investigation.

1.8 Theory and Certainty
1. Define hypothesis.
2. Relate hypothesis to theory.
3. Contrast how scientists and the public use the word theory.
4. Appraise the so-called "scientific method."
Author's Corner: Where Are All My Socks Going?

Core Ideas of Biology
1.9 Four Theories Unify Biology as a Science
1. Identify the four major theories that unite biology as a science.
2. Describe the cell theory.
3. Define gene.
4. Identify in what way the chromosomal theory of inheritance extends Mendel's ideas.
5. Explain how Darwin's theory of evolution is related to the gene theory.

Inquiry & Analysis: Does the Presence of One Species Limit the Population Size of Others?

1.1 The Diversity of Life

In its broadest sense, biology is the study of living things—the science of life. The living world is teeming with a breathtaking variety of creatures—whales, butterflies, mushrooms, and mosquitoes—all of which can be categorized into six groups, or **kingdoms,** of organisms. Representatives from each kingdom can be seen in **figure 1.1**. All organisms that are placed into a kingdom possess similar characteristics with all other organisms in that same kingdom and are very different from organisms in the other kingdoms.

Biologists study the diversity of life in many different ways. They live with gorillas, collect fossils, and listen to whales. They isolate bacteria, grow mushrooms, and examine the structure of fruit flies. They read the messages encoded in the long molecules of heredity and count how many times a hummingbird's wings beat each second. In the midst of all this diversity, it is easy to lose sight of the key lesson of biology, which is that all living things have much in common.

Key Learning Outcome 1.1   The living world is very diverse, but all things share many key properties.

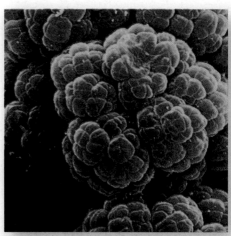

Archaea. This kingdom of prokaryotes (the simplest of cells that do not have nuclei) includes this methanogen, which manufactures methane as a result of its metabolic activity.

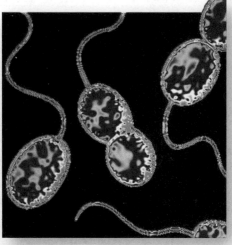

Bacteria. This group is the second of the two prokaryotic kingdoms. Shown here are purple sulfur bacteria, which are able to convert light energy into chemical energy.

Protista. Most of the unicellular eukaryotes (those whose cells contain a nucleus) are grouped into this kingdom, and so are the multicellular algae pictured here.

Fungi. This kingdom contains nonphotosynthetic organisms, mostly multicellular, that digest their food externally, such as these mushrooms.

Plantae. This kingdom contains photosynthetic multicellular organisms that are terrestrial, such as the flowering plant pictured here.

Animalia. Organisms in this kingdom are nonphotosynthetic multicellular organisms that digest their food internally, such as this ram.

Figure 1.1   The six kingdoms of life.
Biologists assign all living things to six major categories called *kingdoms*. Each kingdom is profoundly different from the others.

1.2 Properties of Life

Biology is the study of life—but what does it mean to be alive? What are the properties that define a living organism? This is not as simple a question as it seems because some of the most obvious properties of living organisms are also properties of many nonliving things—for example, *complexity* (a computer is complex), *movement* (clouds move in the sky), and *response to stimulation* (a soap bubble pops if you touch it). To appreciate why these three properties, so common among living things, do not help us to define life, imagine a mushroom standing next to a television: The television seems more complex than the mushroom, the picture on the television screen is moving while the mushroom just stands there, and the television responds to a remote control device while the mushroom continues to just stand there—yet it is the mushroom that is alive.

All living things share five basic properties, passed down over millions of years from the first organisms to evolve on earth: *cellular organization, metabolism, homeostasis, growth and reproduction,* and *heredity.*

1. **Cellular organization.** All living things are composed of one or more cells. A cell is a tiny compartment with a thin covering called a *membrane.* Some cells have simple interiors, while others are complexly organized, but all are able to grow and reproduce. Many organisms possess only a single cell, like the paramecia in **figure 1.2**; your body contains about 10–100 trillion cells (depending on how big you are)—that's how many centimeters long a string would be wrapped around the world 1,600 times!

2. **Metabolism.** All living things use energy. Moving, growing, thinking—everything you do requires energy. Where does all this energy come from? It is captured from sunlight by plants and algae through photosynthesis. To get the energy that powers our lives, we extract it from plants or from plant-eating animals. That's what the kingfisher is doing in **figure 1.3**, eating a fish that ate algae. The transfer of energy from one form to another in cells is an example of *metabolism.* All organisms require energy to grow,

Figure 1.3 Metabolism.
This kingfisher obtains the energy it needs to move, grow, and carry out its body processes by eating fish. It metabolizes this food using chemical processes that occur within cells.

and all organisms transfer this energy from one place to another within cells using special energy-carrying molecules called ATP molecules.

3. **Homeostasis.** All living things maintain stable internal conditions so that their complex processes can be better coordinated. While the environment often varies a lot, organisms act to keep their interior conditions relatively constant; a process called *homeostasis.* Your body acts to maintain an internal temperature of 37°C (98.6°F), however hot or cold the weather might be.

4. **Growth and reproduction.** All living things grow and reproduce. Bacteria increase in size and simply split in two as often as every 15 minutes, while more complex organisms grow by increasing the number of cells and reproduce sexually (some, like the bristlecone pine of California, have reproduced after 4,600 years).

5. **Heredity.** All organisms possess a genetic system that is based on the replication and duplication of a long molecule called *DNA (deoxyribonucleic acid).* The information that determines what an individual organism will be like is contained in a code that is dictated by the order of the subunits making up the DNA molecule, just as the order of letters on this page determines the sense of what you are reading. Each set of instructions within the DNA is called a *gene.* Together, the genes determine what the organism will be like. Because DNA is faithfully copied from one generation to the next, any change in a gene is also preserved and passed on to future generations. The transmission of characteristics from parent to offspring is a process called *heredity.*

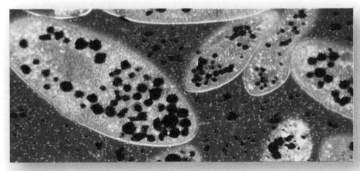

Figure 1.2 Cellular organization.
These paramecia are complex single-celled protists that have just ingested several yeast cells. Like these paramecia, many organisms consist of just a single cell, while others are composed of trillions of cells.

Key Learning Outcome 1.2 All living things possess cells that carry out metabolism, maintain stable internal conditions, reproduce themselves, and use DNA to transmit hereditary information to offspring.

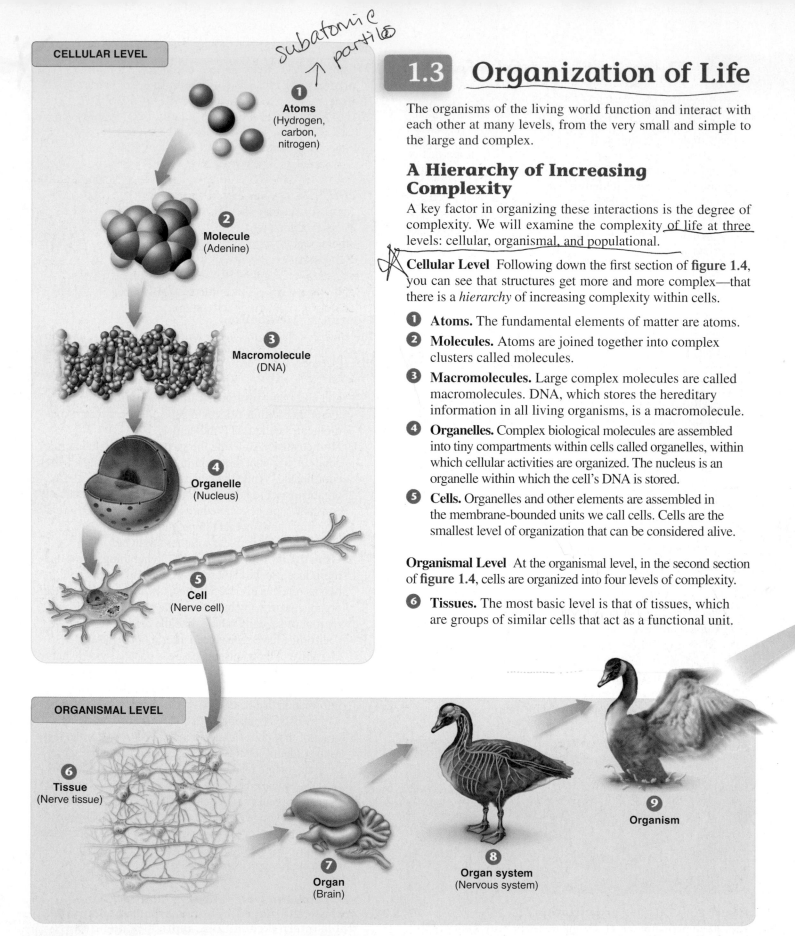

CELLULAR LEVEL

subatomic ↑ particles

1 **Atoms**
(Hydrogen, carbon, nitrogen)

2 **Molecule**
(Adenine)

3 **Macromolecule**
(DNA)

4 **Organelle**
(Nucleus)

5 **Cell**
(Nerve cell)

ORGANISMAL LEVEL

6 **Tissue**
(Nerve tissue)

7 **Organ**
(Brain)

8 **Organ system**
(Nervous system)

9 **Organism**

1.3 Organization of Life

The organisms of the living world function and interact with each other at many levels, from the very small and simple to the large and complex.

A Hierarchy of Increasing Complexity

A key factor in organizing these interactions is the degree of complexity. We will examine the complexity of life at three levels: cellular, organismal, and populational.

Cellular Level Following down the first section of **figure 1.4**, you can see that structures get more and more complex—that there is a *hierarchy* of increasing complexity within cells.

1 **Atoms.** The fundamental elements of matter are atoms.

2 **Molecules.** Atoms are joined together into complex clusters called molecules.

3 **Macromolecules.** Large complex molecules are called macromolecules. DNA, which stores the hereditary information in all living organisms, is a macromolecule.

4 **Organelles.** Complex biological molecules are assembled into tiny compartments within cells called organelles, within which cellular activities are organized. The nucleus is an organelle within which the cell's DNA is stored.

5 **Cells.** Organelles and other elements are assembled in the membrane-bounded units we call cells. Cells are the smallest level of organization that can be considered alive.

Organismal Level At the organismal level, in the second section of **figure 1.4**, cells are organized into four levels of complexity.

6 **Tissues.** The most basic level is that of tissues, which are groups of similar cells that act as a functional unit.

Figure 1.4 Levels of organization.
A traditional and very useful way to sort through the many ways in which the organisms of the living world interact is to organize them in terms of levels of organization, proceeding from the very small and simple to the very large and complex. Here we examine organization within the cellular, organismal, and populational levels.

Nerve tissue is one kind of tissue, composed of cells called neurons that are specialized to carry electrical signals from one place to another in the body.

7 **Organs.** Tissues, in turn, are grouped into organs, which are body structures composed of several different tissues grouped together in a structural and functional unit. Your brain is an organ composed of nerve cells and a variety of connective tissues that form protective coverings and distribute blood.

8 **Organ systems.** At the third level of organization, organs are grouped into organ systems. The nervous system, for example, consists of sensory organs, the brain and spinal cord, neurons that convey signals to and from them, and supporting cells.

9 **Organism.** Separate organ systems function together to form an organism.

Populational Level Organisms are further organized into several hierarchical levels within the living world, as you can see in the third section of **figure 1.4**.

10 **Population.** The most basic of these is the population, which is a group of organisms of the same species living in the same place. A flock of geese living together on a pond is a population.

11 **Species.** All the populations of a particular kind of organism together form a species, its members similar in appearance and able to interbreed. All Canada geese, whether found in Canada, Minnesota, or Missouri, are basically the same, members of the species *Branta canadensis*. Sandhill cranes are a different species.

12 **Community.** At a higher level of biological organization, a community consists of all the populations of different species living together in one place. Geese, for example, may share their pond with ducks, fish, grasses, and many kinds of insects. All interact in a single pond community.

13 **Ecosystem.** At the highest tier of biological organization, a biological community and the soil and water within which it lives together constitute an ecological system, or ecosystem.

Emergent Properties

At each higher level in the living hierarchy, novel properties emerge, properties that were not present at the simpler level of organization. These **emergent properties** result from the way in which components interact, and often cannot be guessed just by looking at the parts themselves. You have the same array of cell types as a giraffe, for example. Yet, examining a collection of its individual cells gives little clue of what your body is like.

The emergent properties of life are not magical or supernatural. They are the natural consequence of the hierarchy, or structural organization, that is the hallmark of life. Water, which makes up 50%–75% of your body's weight, and ice are both made of H_2O molecules, but one is liquid and the other solid because the H_2O molecules in ice are more organized.

Functional properties emerge from more complex organization. Metabolism is an emergent property of life. The chemical reactions within a cell arise from interactions between molecules that are orchestrated by the orderly environment of the cell's interior. Consciousness is an emergent property of the brain that results from the interactions of many neurons in different parts of the brain.

> **Key Learning Outcome 1.3**
> Cells, multicellular organisms, and ecological systems each are organized in a hierarchy of increased complexity. Life's hierarchical organization is responsible for the emergent properties that characterize so many aspects of the living world.

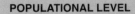

POPULATIONAL LEVEL

10 Population

11 Species

12 Community

13 Ecosystem

Biological Themes

Just as every house is organized into thematic areas such as bedroom, kitchen, and bathroom, so the living world is organized by major *themes,* such as how energy flows within the living world from one part to another. As you study biology in this text, five general themes will emerge repeatedly, themes that serve to both unify and explain biology as a science (**table 1.1**):

1. evolution;
2. the flow of energy;
3. cooperation;
4. structure determines function;
5. homeostasis.

Evolution

Evolution is genetic change in a species over time. Charles Darwin was an English naturalist who, in 1859, proposed the idea that this change is a result of a process called **natural selection.** Simply stated, those organisms whose characteristics make them better able to survive the challenges of their environment live to reproduce, passing their favorable characteristics on to their offspring. Darwin was thoroughly familiar with variation in domesticated animals (in addition to many nondomesticated organisms), and he knew that varieties of pigeons could be selected by breeders to exhibit exaggerated characteristics, a process called **artificial selection.** You can see some of these extreme-looking pigeons pictured in **table 1.1** under the heading "evolution." We now know that the characteristics selected are passed on through generations because DNA is transmitted from parent to offspring. Darwin visualized how selection in nature could be similar to that which had produced the different varieties of pigeons. Thus, the many forms of life we see about us on earth today, and the way we ourselves are constructed and function, reflect a long history of natural selection. Evolution will be explored in more detail in chapter 14.

The Flow of Energy

All organisms require energy to carry out the activities of living—to build bodies and do work and think thoughts. All of the energy used by most organisms comes from the sun and is passed in one direction through ecosystems. The simplest way to understand the flow of energy through the living world is to look at who uses it. The first stage of energy's journey is its capture by green plants, algae, and some bacteria by the process of photosynthesis. This process uses energy from the sun to synthesize sugars that photosynthetic organisms like plants store in their bodies. Plants then serve as a source of life-driving energy for animals that eat them. Other animals, like the eagle in **table 1.1**, may then eat the plant eaters. At each stage, some energy is used for the processes of living, some is transferred, and much is lost, primarily as heat. The flow of energy is a key factor in shaping ecosystems, affecting how many and what kinds of animals live in a community.

Cooperation

The ants cooperating in the upper right photo in **table 1.1** protect the plant on which they live from predators and from shading by other plants, while this plant returns the favor by providing the ants with nutrients (the yellow structures at the tips of the leaves). This type of cooperation between different kinds of organisms has played a critical role in the evolution of life on earth. For example, organisms of two different species that live in direct contact, like the ants and the plant on which they live, form a type of relationship called **symbiosis.** Animal cells possess organelles that are the descendants of symbiotic bacteria, and symbiotic fungi helped plants first invade land from the sea. The coevolution of flowering plants and insects—where changes in flowers influenced insect evolution and, in turn, changes in insects influenced flower evolution—has been responsible for much of life's great diversity.

Structure Determines Function

One of the most obvious lessons of biology is that biological structures are very well suited to their functions. You will see this at every level of organization: Within cells, the shape of the proteins called enzymes that cells use to carry out chemical reactions are precisely suited to match the chemicals the enzymes must manipulate. Within the many kinds of organisms in the living world, body structures seem carefully designed to carry out their functions—the long tongue with which the moth in **table 1.1** sucks nectar from deep within a flower is one example. The superb fit of structure to function in the living world is no accident. Life has existed on earth for over 2 billion years, a long time for evolution to favor changes that better suit organisms to meet the challenges of living. It should come as no surprise to you that after all this honing and adjustment, biological structures carry out their functions well.

Homeostasis   balance—set point

The high degree of specialization we see among complex organisms is only possible because these organisms act to maintain a relatively stable internal environment, a process introduced earlier called homeostasis. Without this constancy, many of the complex interactions that need to take place within organisms would be impossible, just as a city cannot function without rules to maintain order. Maintaining homeostasis in a body as complex as yours or the hippo's in **table 1.1** requires a great deal of signaling back-and-forth between cells.

As already stated, you will encounter these biological themes repeatedly in this text. But just as a budding architect must learn more than the parts of buildings, so your study of biology should teach you more than a list of themes, concepts, and parts of organisms. Biology is a dynamic science that will affect your life in many ways, and that lesson is one of the most important you will learn. It is also a great deal of fun.

> **Key Learning Outcome 1.4**   The five general themes of biology are (1) evolution, (2) the flow of energy, (3) cooperation, (4) structure determines function, and (5) homeostasis.

TABLE 1.1 | **BIOLOGICAL THEMES**

Cooperation Latin American ants live within the hollow thorns of certain species of acacia trees. The nectar at the bases of the leaves and at the tips of the leaflets provide food. The ants supply the trees with organic nutrients and protection.

Evolution Charles Darwin's studies of artificial selection in pigeons provided key evidence that selection could produce the sorts of changes predicted by his theory of evolution. The differences that have been obtained by artificial selection of the wild European rock pigeon (*top*) and such domestic races as the red fantail (*middle*) and the fairy swallow (*bottom*), with its fantastic tufts of feathers around its feet, are indeed so great that the birds probably would, if wild, be classified in different major groups.

The Flow of Energy Energy passes from the sun to plants to plant-eating animals to animal-eating animals, such as this eagle.

Homeostasis Homeostasis often involves water balance to maintain proper blood chemistry. All complex organisms need water—some, like this hippo, luxuriate in it. Others, like the kangaroo rat that lives in arid conditions where water is scarce, obtain water from food and never actually drink.

Structure Determines Function With its long tongue, this moth is able to reach the nectar deep within these flowers.

1.5 How Scientists Think

Deductive Reasoning

Science is a process of investigation, using observation, experimentation, and reasoning. Not all investigations are scientific. For example, when you want to know how to get to Chicago from St. Louis, you do not conduct a scientific investigation—instead, you look at a map to determine a route. In other investigations, you make individual decisions by applying a "guide" of accepted general principles. This is called **deductive reasoning.** Deductive reasoning, using general principles to explain specific observations, is the reasoning of mathematics, philosophy, politics, and ethics; deductive reasoning is also the way a computer works. All of us rely on deductive reasoning to make everyday decisions—like whether you need to slow down while driving along a city street, as in **figure 1.5.** We use general principles as the basis for examining and evaluating these decisions.

Inductive Reasoning

Where do general principles come from? Religious and ethical principles often have a religious foundation; political principles reflect social systems. Some general principles, however, are not derived from religion or politics but from observation of the physical world around us. If you drop an apple, it will fall whether or not you wish it to and despite any laws you may pass forbidding it to do so. Science is devoted to discovering the general principles that govern the operation of the physical world.

How do scientists discover such general principles? Scientists are, above all, observers: They look at the world to understand how it works. It is from observations that scientists determine the principles that govern our physical world.

This way of discovering general principles by careful examination of specific cases is called **inductive reasoning.** Inductive reasoning first became popular about 400 years ago, when Isaac Newton, Francis Bacon, and others began to conduct experiments and from the results infer general principles about how the world operates. The experiments were sometimes quite simple. Newton's consisted simply of releasing an apple from his hand and watching it fall to the ground. This simple observation is the stuff of science. From a host of particular observations, each no more complicated than the falling of an apple, Newton inferred a general principle—that all objects fall toward the center of the earth. This principle was a possible explanation, or *hypothesis,* about how the world works. You also make observations and formulate general principles based on your observations, like forming a general principle about the timing of traffic lights in **figure 1.5.** Like Newton, scientists work by forming and testing hypotheses, and observations are the materials on which they build them.

> **Key Learning Outcome 1.5** Science uses inductive reasoning to infer general principles from detailed observation.

DEDUCTIVE REASONING

An Accepted General Principle

When traffic lights along city streets are "timed" to change at the time interval it takes traffic to pass between them, the result will be a smooth flow of traffic.

DEDUCTIVE REASONING

Using a General Principle to Make Everyday Decisions

Traveling at the speed limit, you approach each intersection anticipating that the red light will turn green as you reach the intersection.

INDUCTIVE REASONING

Observations of Specific Events

Driving down the street at the speed limit, you observe that the red traffic light turns green just as you approach the intersection.

Maintaining the same speed, you observe the same event at the next several intersections: the traffic lights turn green just as you approach the intersections. When you speed up, however, the light doesn't change until after you reach the intersection.

INDUCTIVE REASONING

Formation of a General Principle

You conclude that the traffic lights along this street are "timed" to change in the time it takes your car, traveling at the speed limit, to traverse the distance between them.

Figure 1.5 Deductive and inductive reasoning.
A driver who assumes that the traffic signals are timed can use deductive reasoning to expect that the traffic lights will change predictably at intersections. In contrast, a driver who is not aware of the general control and programming of traffic signals can use inductive reasoning to determine that the traffic lights are timed as the driver encounters similar timing of signals at several intersections.

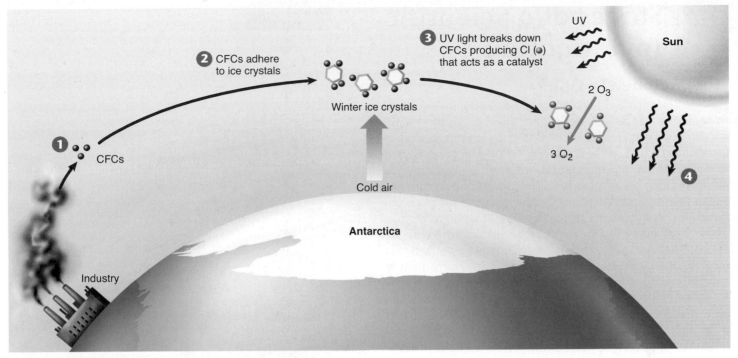

Figure 1.6 How CFCs attack and destroy ozone.
CFCs are stable chemicals that accumulate in the atmosphere as a by-product of industrial society ❶. In the intense cold of the Antarctic, these CFCs adhere to tiny ice crystals in the upper atmosphere ❷. UV light causes the breakdown of CFCs, producing chlorine (Cl). Cl acts as a catalyst, converting O_3 into O_2 ❸. As a result, more harmful UV radiation reaches the earth's surface ❹.

1.6 Science in Action: A Case Study

In 1985 Joseph Farman, a British earth scientist working in Antarctica, made an unexpected discovery. Analyzing the Antarctic sky, he found far less ozone (O_3, a form of oxygen gas) than should be there—a 30% drop from a reading recorded five years earlier in the Antarctic!

At first it was argued that this thinning of the ozone (soon dubbed the "ozone hole") was an as-yet-unexplained weather phenomenon. Evidence soon mounted, however, implicating synthetic chemicals as the culprit. Detailed analysis of chemicals in the Antarctic atmosphere revealed a surprisingly high concentration of chlorine, a chemical known to destroy ozone. The source of the chlorine was a class of chemicals called **chlorofluorocarbons (CFCs).** CFCs (the purple balls ❶ in figure 1.6) have been manufactured in large amounts since they were invented in the 1920s, largely for use as coolants in air conditioners, propellants in aerosols, and foaming agents in making Styrofoam. CFCs were widely regarded as harmless because they are chemically unreactive under normal conditions. But in the atmosphere over Antarctica, CFCs condense onto tiny ice crystals ❷; in the spring, the CFCs break down and produce chlorine, which acts as a catalyst, attacking and destroying ozone, turning it into oxygen gas without the chlorine being used up ❸.

The thinning of the ozone layer in the upper atmosphere 25 to 40 kilometers above the surface of the earth is a serious matter. The ozone layer protects life from the harmful ultraviolet (UV) rays from the sun that bombard the earth continuously. Like invisible sunglasses, the ozone layer filters out these dangerous rays. So when ozone is converted to oxygen gas, the UV rays are able to pass through to the earth ❹. When UV rays damage the DNA in skin cells, it can lead to skin cancer. It is estimated that every 1% drop in the atmospheric ozone concentration leads to a 6% increase in skin cancers.

The world currently produces less than 200,000 tons of CFCs annually, down from 1986 levels of 1.1 million tons. As scientific observations have become widely known, governments have rushed to correct the situation. By 1990, worldwide agreements to phase out production of CFCs by the end of the century had been signed. Production of CFCs declined by 86% in the following 10 years.

Nonetheless, most of the CFCs manufactured since they were invented are still in use in air conditioners and aerosols and have not yet reached the atmosphere. As these CFCs move slowly upward through the atmosphere, the problem can be expected to continue. Ozone depletion is still producing major ozone holes over the Antarctic.

But the worldwide reduction in CFC production is having a major impact. The period of maximum ozone depletion will peak in the next few years, and researchers' models predict that after that the situation should gradually improve, and that the ozone layer will recover by the middle of the 21st century. Clearly, global environmental problems can be solved by concerted action.

Key Learning Outcome 1.6 Industrially produced CFCs catalytically destroy ozone in the upper atmosphere.

1.7 Stages of a Scientific Investigation

How Science Is Done

How do scientists establish which general principles are true from among the many that might be? They do this by systematically testing alternative proposals. If these proposals prove inconsistent with experimental observations, they are rejected as untrue. After making careful observations concerning a particular area of science, scientists construct a hypothesis, which is a suggested explanation that accounts for those observations. A hypothesis is a proposition that might be true. Those hypotheses that have not yet been disproved are retained. They are useful because they fit the known facts, but they are always subject to future rejection if—in the light of new information—they are found to be incorrect.

We call the test of a hypothesis an experiment. Suppose that a room appears dark to you. To understand why it appears dark, you propose several hypotheses. The first might be, "The room appears dark because the light switch is turned off." An alternative hypothesis might be, "The room appears dark because the light bulb is burned out." And yet another alternative hypothesis might be, "I am going blind." To evaluate these hypotheses, you would conduct an experiment designed to eliminate one or more of the hypotheses. For example, you might reverse the position of the light switch. If you do so and the light does not come on, you have disproved the first hypothesis. Something other than the setting of the light switch must be the reason for the darkness. Note that a test such as this does not prove that any of the other hypotheses are true; it merely demonstrates that one of them is not. A successful experiment is one in which one or more hypotheses is demonstrated to be inconsistent with the results and is thus rejected.

As you proceed through this text, you will encounter a great deal of information, often accompanied by explanations. These explanations are hypotheses that have withstood the test of experiment. Many will continue to do so; others will be revised as new observations are made. Biology, like all science, is in a constant state of change, with new ideas appearing and replacing old ones.

The Scientific Process

Joseph Farman, who first reported the ozone hole, is a practicing scientist, and what he was doing in Antarctica was science. Science is a particular way of investigating the world, of forming general rules about why things happen by observing particular situations. A scientist like Farman is an observer, someone who looks at the world in order to understand how it works.

Scientific investigations can be said to have six stages as illustrated in **figure 1.7**: ❶ observing what is going on; ❷ forming a set of hypotheses; ❸ making predictions; ❹ testing them and ❺ carrying out controls, until one or more of the hypotheses have been eliminated; and ❻ forming conclusions based on the remaining hypothesis.

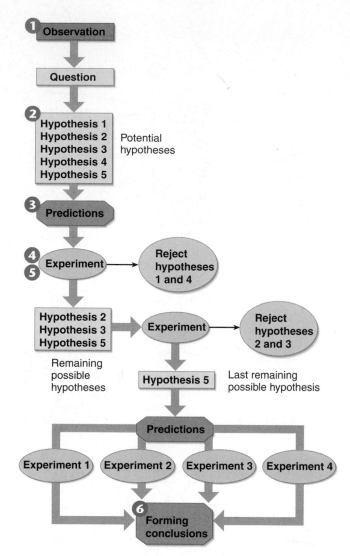

Figure 1.7 The scientific process.
This diagram illustrates the stages of a scientific investigation. First, observations are made that raise a particular question. Then a number of potential explanations (hypotheses) are suggested to answer the question. Next, predictions are made based on the hypotheses, and several rounds of experiments (including control experiments) are carried out in an attempt to eliminate one or more of the hypotheses. Finally, any hypothesis that is not eliminated is retained. Further predictions can be made based on the accepted hypothesis and tested with experiments. If it is validated by numerous experiments and stands the test of time, a hypothesis may eventually become a theory.

1. **Observation.** The key to any successful scientific investigation is careful **observation.** Farman and other scientists had studied the skies over the Antarctic for many years, noting a thousand details about temperature, light, and levels of chemicals. You can see an example in **figure 1.8**, where the purple colors represent the lowest levels of ozone that the scientists recorded. Had these scientists not kept careful records of what they observed, Farman might not have noticed that ozone levels were dropping. Observations usually generate questions, such as: Why were ozone levels dropping?

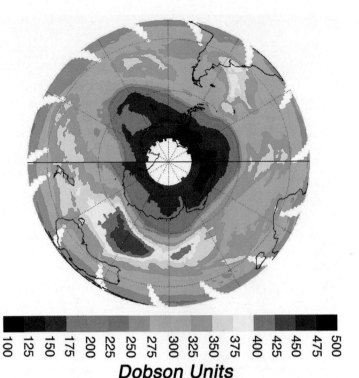

Dobson Units

Figure 1.8   The ozone hole.
The swirling colors represent different concentrations of ozone over the South Pole as viewed from a satellite on September 15, 2001. As you can easily see, there is an "ozone hole" (the *purple* areas) over Antarctica covering an area about the size of the United States. (The color *white* indicates areas where no data were available.)

2. **Hypothesis.** When the unexpected drop in ozone was reported and questioned, environmental scientists made a guess to answer their questions—perhaps something was destroying the ozone; maybe the culprit was CFCs. Of course, this was not a guess in the true sense; scientists had some working knowledge of CFCs and what they might be doing in the upper atmosphere. We call such a guess a **hypothesis.** A hypothesis is a guess that might be true. What the scientists guessed was that chlorine from CFCs was reacting chemically with ozone over the Antarctic, converting ozone (O_3) into oxygen gas (O_2) and in the process removing the ozone shield from our earth's atmosphere. Often, scientists will form **alternative hypotheses** if they have more than one guess about what they observe. In this case, there were several other hypotheses advanced to explain the ozone hole. One suggestion explained it as the result of convection: the ozone spun away from the polar regions much as water spins away from the center as a clothes washer moves through its spin cycle. Another hypothesis was that the ozone hole was a transient phenomenon, due perhaps to sunspots, and would soon disappear.

3. **Predictions.** If the CFC hypothesis is correct, then several consequences can reasonably be expected. We call these expected consequences **predictions.** A prediction is what you expect to happen if a hypothesis

is true. The CFC hypothesis predicts that if CFCs are responsible for producing the ozone hole, then it should be possible to detect CFCs in the upper Antarctic atmosphere as well as the chlorine released from CFCs that attack the ozone.

4. **Testing.** Scientists set out to test the CFC hypothesis by attempting to verify some of its predictions. We call the test of a hypothesis an **experiment.** To test the hypothesis, atmospheric samples were collected from the stratosphere over 6 miles up by a high-altitude balloon. Analysis of the samples revealed CFCs, as predicted. Were the CFCs interacting with the ozone? The samples contained free chlorine and fluorine, confirming the breakdown of CFC molecules. The results of the experiment thus support the hypothesis.

5. **Controls.** Events in the upper atmosphere can be influenced by many factors. We call each factor that might influence a process a **variable.** To evaluate alternative hypotheses about one variable, all the other variables must be kept constant so that we do not get misled or confused by these other influences. This is done by carrying out two experiments in parallel: In the first experimental test, we alter one variable in a known way to test a particular hypothesis; in the second, called a **control experiment,** we do *not* alter that variable. In all other respects, the two experiments are the same. To further test the CFC hypothesis, scientists carried out control experiments in which the key variable was the amount of CFCs in the atmosphere. Working in laboratories, scientists reconstructed the atmospheric conditions, solar bombardment, and extreme temperatures found in the sky far above the Antarctic. If the ozone levels fell without addition of CFCs to the chamber, then CFCs could not be what was attacking the ozone Carefully monitoring the chamber, however, scientists detected no drop in ozone levels in the absence of CFCs.

6. **Conclusion.** A hypothesis that has been tested and not rejected is tentatively accepted. The hypothesis that CFCs released into the atmosphere are destroying the earth's protective ozone shield is now supported by a great deal of experimental evidence and is widely accepted. While other factors have also been implicated in ozone depletion, destruction by CFCs is clearly the dominant phenomenon. A collection of related hypotheses that have been tested many times and not rejected is called a **theory.** A theory indicates a higher degree of certainty; however, in science, nothing is "certain." The theory of the ozone shield—that ozone in the upper atmosphere shields the earth's surface from harmful UV rays by absorbing them—is supported by a wealth of observation and experimentation and is widely accepted. The explanation for the destruction of this shield is still at the hypothesis stage.

Key Learning Outcome 1.7   Science progresses by systematically eliminating potential hypotheses that are not consistent with observation.

Theory and Certainty

A theory is a unifying explanation for a broad range of observations. Thus we speak of the theory of gravity, the theory of evolution, and the theory of the atom. Theories are the solid ground of science, that of which we are the most certain. There is no absolute truth in science, however, only varying degrees of uncertainty. The possibility always remains that future evidence will cause a theory to be revised. A scientist's acceptance of a theory is always provisional. For example, in another scientist's experiment, evidence that is inconsistent with a theory may be revealed. As information is shared throughout the scientific community, previous hypotheses and theories may be modified, and scientists may formulate new ideas.

Very active areas of science are often alive with controversy, as scientists grope with new and challenging ideas. This uncertainty is not a sign of poor science but rather of the push and pull that is the heart of the scientific process. The hypothesis that the world's climate is growing warmer due to humanity's excessive production of carbon dioxide (CO_2), for example, has been quite controversial, although the weight of evidence has increasingly supported the hypothesis.

The word theory is thus used very differently by scientists than by the general public. To a scientist, a theory represents that for which he or she is most certain; to the general public, the word theory implies a *lack* of knowledge or a guess. How often have you heard someone say, "It's only a theory!"? As you can imagine, confusion often results. In this text the word theory will always be used in its scientific sense, in reference to a generally accepted scientific principle.

The Scientific "Method"

It was once fashionable to claim that scientific progress is the result of applying a series of steps called the **scientific method;** that is, a series of logical "either/or" predictions tested by experiments to reject one alternative. The assumption was that trial-and-error testing would inevitably lead one through the maze of uncertainty that always slows scientific progress. If this were indeed true, a computer would make a good scientist—but science is not done this way! If you ask successful scientists like Farman how they do their work, you will discover that without exception they design their experiments with a pretty fair idea of how they will come out. Environmental scientists understood the chemistry of chlorine and ozone when they formulated the CFC hypothesis, and they could imagine how the chlorine in CFCs would attack ozone molecules. A hypothesis that a successful scientist tests is not just any hypothesis. Rather, it is a "hunch" or educated guess in which the scientist integrates all that he or she knows. The scientist also allows his or her imagination full play, in an attempt to get a sense of what *might* be true. It is because insight and imagination play such a large role in

Figure 1.9 Nobel Prize winner.
Sherwood Rowland, along with Mario Molina and Paul Crutzen, won the 1995 Nobel Prize in Chemistry for discovering how CFCs act to catalytically break down atmospheric ozone in the stratosphere, the chemistry responsible for the "ozone hole" over the Antarctic.

scientific progress that some scientists are so much better at science than others (**figure 1.9**)—just as Beethoven and Mozart stand out among composers.

The Limitations of Science

Scientific study is limited to organisms and processes that we are able to observe and measure. Supernatural and religious phenomena are beyond the realm of scientific analysis because they cannot be scientifically studied, analyzed, or explained. Supernatural explanations can be used to explain any result, and cannot be disproven by experiment or observation. Scientists in their work are limited to objective interpretations of observable phenomena.

It is also important to recognize that there are practical limits to what science can accomplish. While scientific study has revolutionized our world, it cannot be relied upon to solve all problems. For example, we cannot pollute the environment and squander its resources today, in the blind hope that somehow science will make it all right sometime in the future. Nor can science restore an extinct species. Science identifies solutions to problems when solutions exist, but it cannot invent solutions when they don't.

> **Key Learning Outcome 1.8** A scientist does not follow a fixed method to form hypotheses but relies also on judgement and intuition.

Where Are All My Socks Going?

All my life, for as far back as I can remember, I have been losing socks. Not pairs of socks, mind you, but single socks. I first became aware of this peculiar phenomenon when, as a young man, I went away to college. When Thanksgiving rolled around that first year, I brought an enormous duffle bag of laundry home. My mother, instead of braining me, dumped the lot into the washer and dryer, and so discovered what I had not noticed—that few of my socks matched anymore.

That was over 40 years ago, but it might as well have been yesterday. All my life, I have continued to lose socks. This last Christmas I threw out a sock drawer full of socks that didn't match, and took advantage of sales to buy a dozen pairs of brand-new ones. Last week, when I did a body count, three of the new pairs had lost a sock!

Enough. I set out to solve the mystery of the missing socks. How? The way Sherlock Holmes would have, scientifically. Holmes worked by eliminating those possibilities that he found not to be true. A scientist calls possibilities "hypotheses" and, like Sherlock, rejects those that do not fit the facts. Sherlock tells us that when only one possibility remains unrejected, then—however unlikely—it must be true.

Hypothesis 1: It's the socks. I have four pairs of socks bought as Christmas gifts but forgotten until recently. Deep in my sock drawer, they have remained undisturbed for five months. If socks disappear because of some intrinsic property (say the manufacturer has somehow designed them to disappear to generate new sales), then I could expect at least one of these undisturbed ones to have left the scene by now. However, when I looked, all four pairs were complete. Undisturbed socks don't disappear. Thus I reject the hypothesis that the problem is caused by the socks themselves.

Hypothesis 2: Transformation, a fanciful suggestion by science fiction writer Avram Davidson in his 1958 story "Or All the Seas with Oysters" that I cannot get out of the quirky corner of my mind. I discard the socks I have worn each evening in a laundry basket in my closet. Over many years, I have noticed a tendency for socks I have placed in the closet to disappear. Over that same long period, as my socks are disappearing, there is something in my closet that seems to multiply—COAT HANGERS! Socks are larval coat hangers! To test this outlandish hypothesis, I had only to move the laundry basket out of the closet. Several months later, I was still losing socks, so this hypothesis is rejected.

Hypothesis 3: Static cling. The missing single socks may have been hiding within the sleeves of sweatshirts or jackets, inside trouser legs, or curled up within seldom-worn garments. Rubbing around in the dryer, socks can garner quite a bit of static electricity, easily enough to cause them to cling to other garments. Socks adhering to the outside of a shirt or pant leg are soon dislodged, but ones that find themselves within a sleeve, leg, or fold may simply stay there, not "lost" so much as misplaced. However, after a diligent search, I did not run across any previously lost

© Eric Lewis/The New Yorker Collection/www.cartoonbank.com

socks hiding in the sleeves of my winter garments or other seldom-worn items, so I reject this hypothesis.

Hypothesis 4: I lose my socks going to or from the laundry. Perhaps in handling the socks from laundry basket to the washer/dryer and back to my sock drawer, a sock is occasionally lost. To test this hypothesis, I have pawed through the laundry coming into the washer. No single socks. Perhaps the socks are lost after doing the laundry, during folding or transport from laundry to sock drawer. If so, there should be no single socks coming out of the dryer. But there are! The singletons are first detected among the dry laundry, before folding. Thus I eliminate the hypothesis that the problem arises from mishandling the laundry. It seems the problem is in the laundry room.

Hypothesis 5: I lose them during washing. Perhaps the washing machine is somehow "eating" my socks. I looked in the washing machine to see if a sock could get trapped inside, or chewed up by the machine, but I can see no possibility. The clothes slosh around in a closed metal container with water passing in and out through little holes no wider than a pencil. No sock could slip through such a hole. There is a thin gap between the rotating cylinder and the top of the washer through which an errant sock might escape, but my socks are too bulky for this route. So I eliminate the hypothesis that the washing machine is the culprit.

Hypothesis 6: I lose them during drying. Perhaps somewhere in the drying process socks are being lost. I stuck my head in our clothes dryer to see if I could see any socks, and I couldn't. However, as I look, I can see a place a sock could go—behind the drying wheel! A clothes dryer is basically a great big turning cylinder with dry air blowing through the middle. The edges of the turning cylinder don't push hard against the side of the machine. Just maybe, every once in a while, a sock might get pulled through, sucked into the back of the machine.

To test this hypothesis, I should take the back of the dryer off and look inside to see if it is stuffed with my missing socks. My wife, knowing my mechanical abilities, is not in favor of this test. Thus, until our dryer dies and I can take it apart, I shall not be able to reject hypothesis 6. Lacking any other likely hypothesis, I take Sherlock Holmes' advice and tentatively conclude that the dryer is the culprit.

1.9 Four Theories Unify Biology as a Science

The Cell Theory: Organization of Life

As was stated at the beginning of this chapter, all organisms are composed of cells, life's basic units. Cells were discovered by Robert Hooke in England in 1665. Hooke was using one of the first microscopes, one that magnified 30 times. Looking through a thin slice of cork, he observed many tiny chambers that reminded him of monks' cells in a monastery. Not long after that, the Dutch scientist Anton van Leeuwenhoek used microscopes capable of magnifying 300 times, and discovered an amazing world of single-celled life in a drop of pond water like you see in **figure 1.10**. He called the bacterial and protist cells he saw "wee animalcules." However, it took almost two centuries before biologists fully understood their significance. In 1839, the German biologists Matthias Schleiden and Theodor Schwann, summarizing a large number of observations by themselves and others, concluded that all living organisms consist of cells. Their conclusion forms the basis of what has come to be known as the **cell theory.** Later, biologists added the idea that all cells come from other cells. The cell theory, one of the basic ideas in biology, is the foundation for understanding the reproduction and growth of all organisms. The nature of cells and how they function is discussed in detail in chapter 4.

The Gene Theory: Molecular Basis of Inheritance

Even the simplest cell is incredibly complex, more intricate than a computer. The information that specifies what a cell is like—its detailed plan—is encoded in a long cablelike molecule called **DNA (deoxyribonucleic acid).** Researchers James Watson and Francis Crick discovered in 1953 that each DNA molecule is formed from two long chains of building blocks, called nucleotides, wound around each other. You can see in **figure 1.11** that the two chains face each other, like two lines of people holding hands. The chains contain information in the same way this sentence does—as a sequence of letters. There are four different nucleotides in DNA (symbolized as A, T, C, and G in the figure), and the sequence in which they occur encodes the information. Specific sequences of several hundred to many thousand nucleotides make up a *gene*, a discrete unit of hereditary information. A gene might encode a particular protein, or a different kind of unique molecule called RNA, or a gene might act to regulate other genes. All organisms on earth encode their genes in strands of DNA. This prevalence of

Figure 1.10   Life in a drop of pond water.
All organisms are composed of cells. Some organisms, including these protists, are single-celled, while others, such as plants, animals, and fungi, consist of many cells.

DNA led to the development of the **gene theory.** Illustrated in **figure 1.12**, the gene theory states that the proteins and RNA molecules encoded by an organism's genes determine what it will be like. The entire set of DNA instructions that specifies a cell is called its **genome.** The sequence of the human genome, 3 billion nucleotides long, was decoded in 2001, a triumph of scientific investigation. How genes function is the subject of chapter 12. In chapter 13 we explore how detailed knowledge of genes is revolutionizing biology and having an impact on the lives of all of us.

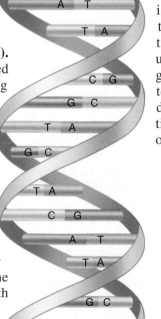

Figure 1.11   Genes are made of DNA.

Winding around each other like the rails of a spiral staircase, the two strands of a DNA molecule make a double helix. Because of its size and shape, the nucleotide represented by the letter A can only pair with the nucleotide represented by the letter T, and likewise G can only pair with C.

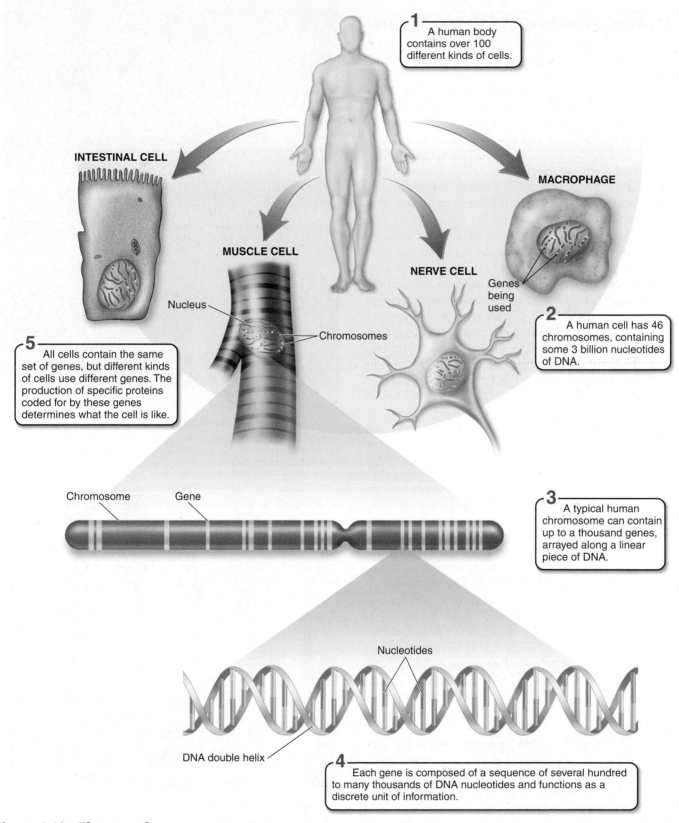

1 A human body contains over 100 different kinds of cells.

INTESTINAL CELL

MACROPHAGE

MUSCLE CELL

NERVE CELL

Genes being used

Nucleus

Chromosomes

2 A human cell has 46 chromosomes, containing some 3 billion nucleotides of DNA.

5 All cells contain the same set of genes, but different kinds of cells use different genes. The production of specific proteins coded for by these genes determines what the cell is like.

Chromosome   Gene

3 A typical human chromosome can contain up to a thousand genes, arrayed along a linear piece of DNA.

Nucleotides

DNA double helix

4 Each gene is composed of a sequence of several hundred to many thousands of DNA nucleotides and functions as a discrete unit of information.

Figure 1.12 The gene theory.

The gene theory states that what an organism is like is determined in large measure by its genes. Here you see how the many kinds of cells in the body of each of us are determined by which genes are used in making each particular kind of cell.

The Theory of Heredity: Unity of Life

The storage of hereditary information in genes composed of DNA is common to all living things. The **theory of heredity,** first advanced by Gregor Mendel in 1865, states that the genes of an organism are inherited as discrete units. A triumph of experimental science developed long before genes and DNA were understood, Mendel's theory of heredity is the subject of chapter 10. Soon after Mendel's theory gave rise to the field of genetics, other biologists proposed what has come to be called the **chromosomal theory of inheritance,** which in its simplest form states that the genes of Mendel's theory are physically located on chromosomes, and that it is because chromosomes are parceled out in a regular manner during reproduction that Mendel's regular patterns of inheritance are seen. In modern terms, the two theories state that genes are a component of a cell's chromosomes (like the 23 pairs of human chromosomes you see in **figure 1.13**), and that the regular duplication of these chromosomes during sexual reproduction is responsible for the pattern of inheritance we call Mendelian segregation. Sometimes a character is conserved essentially unchanged in a long line of descent, reflecting a fundamental role in the biology of the organism, one not easily changed once adopted. Other characters might be modified due to changes in DNA.

The Theory of Evolution: Diversity of Life

The unity of life, which we see in the retention of certain key characteristics among many related life-forms, contrasts with the incredible diversity of living things that have evolved to fill the varied environments of earth. These diverse organisms are sorted by biologists into six kingdoms, as you learned in section 1.1. Organisms placed in the same kingdom have in common some general characteristics. In recent years, biologists have added a classification level above kingdoms, based on fundamental differences in cell structure. The six kingdoms are each now assigned into one of three great groups called *domains:* Bacteria, Archaea, and Eukarya (**figure 1.14**).

The **theory of evolution,** advanced by Charles Darwin in 1859, attributes the diversity of the living world to natural selection. Those organisms best able to respond to the challenges of living will leave more offspring, he argued, and thus their traits become more common in the population. It is because the world offers diverse opportunities that it contains so many different life-forms.

Today scientists can decipher many of the thousands of genes (the genome) of an organism. One of the great triumphs of science in the century and a half since Darwin is the detailed understanding of how Darwin's theory of evolution is related to the gene theory—of how changes in life's diversity can result from changes in individual genes (**figure 1.15**).

> **Key Learning Outcome 1.9** The theories uniting biology state that cellular organisms store hereditary information in DNA. Sometimes DNA alterations occur, which when preserved result in evolutionary change. Today's biological diversity is the product of a long evolutionary journey.

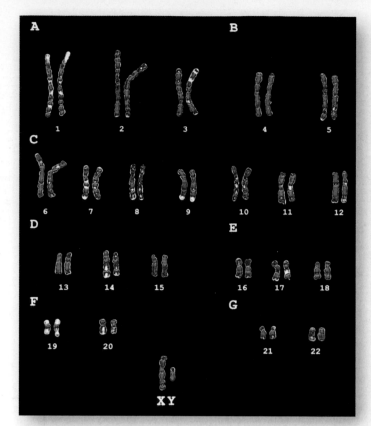

Figure 1.13 Human chromosomes.

The chromosomal theory of inheritance states that genes are located on chromosomes. This human karyotype (an ordering of chromosomes) shows banding patterns on chromosomes that represent clusters of genes.

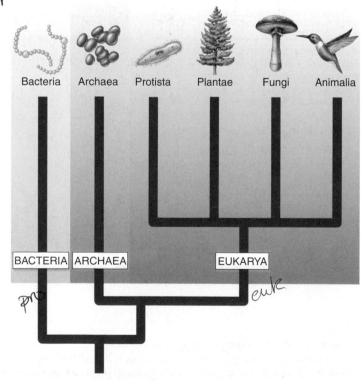

Figure 1.14 The three domains of life.

Biologists categorize all living things into three overarching groups called domains: Bacteria, Archaea, and Eukarya. Domain Bacteria contains the kingdom Bacteria, and domain Archaea contains the kingdom Archaea. Domain Eukarya is composed of four more kingdoms: Protista, Plantae, Fungi, and Animalia.

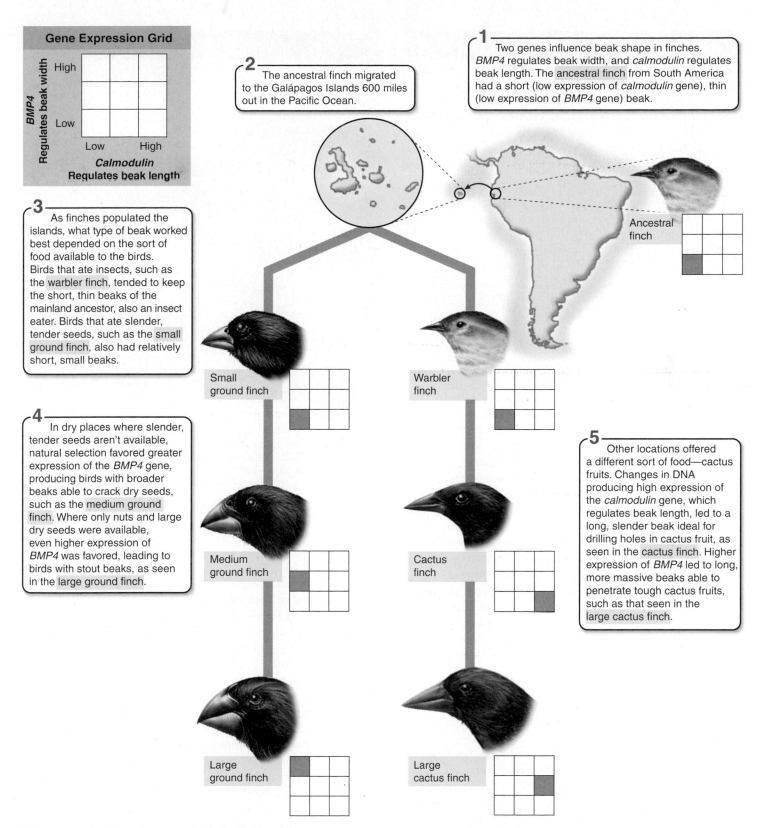

Gene Expression Grid

BMP4 — Regulates beak width: High / Low

Calmodulin — Regulates beak length: Low / High

1 Two genes influence beak shape in finches. *BMP4* regulates beak width, and *calmodulin* regulates beak length. The ancestral finch from South America had a short (low expression of *calmodulin* gene), thin (low expression of *BMP4* gene) beak.

2 The ancestral finch migrated to the Galápagos Islands 600 miles out in the Pacific Ocean.

3 As finches populated the islands, what type of beak worked best depended on the sort of food available to the birds. Birds that ate insects, such as the warbler finch, tended to keep the short, thin beaks of the mainland ancestor, also an insect eater. Birds that ate slender, tender seeds, such as the small ground finch, also had relatively short, small beaks.

4 In dry places where slender, tender seeds aren't available, natural selection favored greater expression of the *BMP4* gene, producing birds with broader beaks able to crack dry seeds, such as the medium ground finch. Where only nuts and large dry seeds were available, even higher expression of *BMP4* was favored, leading to birds with stout beaks, as seen in the large ground finch.

5 Other locations offered a different sort of food—cactus fruits. Changes in DNA producing high expression of the *calmodulin* gene, which regulates beak length, led to a long, slender beak ideal for drilling holes in cactus fruit, as seen in the cactus finch. Higher expression of *BMP4* led to long, more massive beaks able to penetrate tough cactus fruits, such as that seen in the large cactus finch.

Ancestral finch

Small ground finch

Warbler finch

Medium ground finch

Cactus finch

Large ground finch

Large cactus finch

Figure 1.15 The theory of evolution.

Darwin's theory of evolution proposes that many forms of a gene may exist among members of a population, and that those members with a form better suited to their particular habitat will tend to reproduce more successfully, and so their traits become more common in the population, a process Darwin dubbed "natural selection." Here you see how this process is thought to have worked on two pivotal genes that helped generate the diversity of finches on the Galápagos Islands, visited by Darwin in 1831 on his round-the-world voyage on HMS *Beagle*.

Does the Presence of One Species Limit the Population Size of Others?

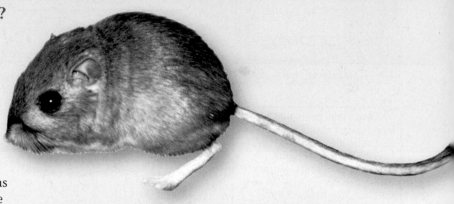

Implicit in Darwin's theory of evolution is the idea that species in nature compete for limiting resources. Does this really happen? Some of the best evidence of competition between species comes from experimental field studies, studies conducted not in the laboratory but out in natural populations. By setting up experiments in which two species occur either alone or together, scientists can determine whether the presence of one species has a negative impact on the size of the population of the other species. The experiment discussed here concerns a variety of seed-eating rodents that occur in North American deserts. In 1988, researchers set up a series of 50-meter × 50-meter enclosures to investigate the effect of kangaroo rats on smaller seed-eating rodents. Kangaroo rats were removed from half of the enclosures, but not from the other enclosures. The walls of all the enclosures had holes that allowed rodents to come and go, but in plots without kangaroo rats the holes were too small to allow the kangaroo rats to enter.

The graph to the right displays data collected over the course of the next three years as researchers monitored the number of the smaller rodents present in the enclosures. To estimate the population sizes, researchers determined how many small rodents could be captured in a fixed interval. Data were collected for each enclosure immediately after the kangaroo rats were removed in 1988, and at three-month intervals thereafter. The graph presents the relative population size—that is, the total number of captures averaged over the number of enclosures (an **average** is the numerical mean value, calculated by adding a list of values and then dividing this sum by the number of items in the list. For example, if a total of 30 rats were captured from 3 enclosures, the average would be 10 rats). As you can see, the two kinds of enclosures do not contain the same number of small rodents.

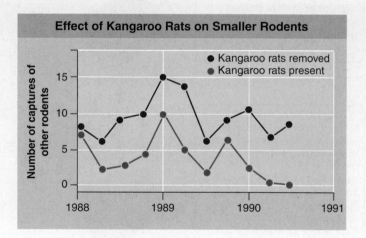

Effect of Kangaroo Rats on Smaller Rodents

● Kangaroo rats removed
● Kangaroo rats present

1. **Applying Concepts**
   a. Variable. In the graph, what is the dependent variable?
   b. Relative Magnitude. Which of the two kinds of enclosures maintains the highest population of small rodents? Does it have kangaroo rats or have they been removed?
2. **Interpreting Data**
   a. What is the average number of small rodents in each of the two plots immediately after kangaroo rats were removed? After one year? After two?

b. At what point is the difference between the two kinds of enclosures the greatest?
3. **Making Inferences**
   a. What precisely is the observed impact of kangaroo rats on the population size of small rodents?
   b. Examine the magnitude of the difference between the number of small rodents in the two plots. Is there a trend?
4. **Drawing Conclusions**
   Do these results support the hypothesis that kangaroo rats compete with other small rodents to limit their population sizes?
5. **Further Analysis**
   a. Can you think of any cause other than competition that would explain these results? Suggest an experiment that could potentially eliminate or confirm this alternative.
   b. Do the populations of the two kinds of enclosures change in synchrony (that is, grow and shrink at the same times) over the course of a year? If so, why might this happen? How would you test this hypothesis?

Biology and the Living World

1.1 The Diversity of Life

- Biology is the study of life. All living organisms share common characteristics, but they are also diverse and are categorized into six groups called kingdoms. The six kingdoms are Bacteria, Archaea, Protista, Fungi, Plantae, and Animalia (**figure 1.1**).

1.2 Properties of Life

- All living organisms share five basic properties: cellular organization, metabolism, homeostasis, growth and reproduction, and heredity. Cellular organization indicates that all living organisms are composed of cells. Metabolism means that all living organisms, like the kingfisher shown here from **figure 1.3**, use energy. Homeostasis is the process whereby all living organisms maintain stable internal conditions. The properties of growth and reproduction indicate that all living organisms grow in size and reproduce. The property of heredity describes how all living organisms possess genetic information in DNA that determines how each organism looks and functions, and this information is passed on to future generations.

1.3 Organization of Life

- Living organisms exhibit increasing levels of complexity within their cells a(cellular level), within their bodies (organismal level), and within ecosystems (populational level) (**figure 1.4**).

- Novel properties that appear in each level of the hierarchy of life are called emergent properties.

1.4 Biological Themes

- Five themes emerge from the study of biology: evolution, the flow of energy, cooperation, structure determines function, and homeostasis. These themes are used to examine the similarities and differences among organisms (**table 1.1**).

The Scientific Process

1.5 How Scientists Think

- Scientists use reasoning. Deductive reasoning is the process of using general principles to explain individual observations. Inductive reasoning is the process of using specific observations to formulate general principles (**figure 1.5**).

1.6 Science in Action: A Case Study

- Scientists observed a thinning of the ozone layer over Antarctica. Their scientific investigation of the "ozone hole" revealed that industrially produced CFCs were responsible for the thinning of the ozone layer in the earth's atmosphere (**figure 1.6**).

Winter ice crystals

Cold air

Antarctica

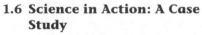

1.7 Stages of a Scientific Investigation

- Scientists use observations to formulate hypotheses. Hypotheses are possible explanations that are used to form predictions. These predictions are tested experimentally. Some hypotheses are rejected based on experimental results, while others are tentatively accepted.

- Scientific investigations often use a series of stages, called the scientific process, to study a scientific question. These stages are observations, forming hypotheses, making predictions, testing, establishing controls, and drawing conclusions (**figure 1.7**).

- The discovery of the hole in the ozone required careful observations of data collected from the atmosphere. Scientists proposed a hypothesis to explain what caused a decrease in the levels of ozone over the Antarctic. They then formed predictions and tested the hypothesis against controls.

1.8 Theory and Certainty

- Hypotheses that hold up to testing over time are combined into statements called theories. Theories carry a higher degree of certainty, although no theory in science is absolute.

- Science can only study what can be tested experimentally. A hypothesis can only be established through science if it can be tested and potentially disproven.

Core Ideas of Biology

1.9 Four Theories Unify Biology as a Science

- There are four unifying theories in biology: cell theory, gene theory, the theory of heredity, and the theory of evolution.

- The cell theory states that all living organisms are composed of cells, which grow and reproduce to form other cells (**figure 1.10**).

- The gene theory states that long molecules of DNA carry instructions for producing cellular components. These instructions are encoded in the nucleotide sequences in the strands of DNA, like this section of DNA from **figure 1.11**. The nucleotides are organized into discrete units called genes, and the genes determine how an organism looks and functions (**figure 1.12**).

- The theory of heredity states that the genes of an organism are passed as discrete units from parent to offspring (**figure 1.13**).

- Organisms are organized into kingdoms based on similar characteristics. The kingdoms are further organized into three major groups called domains based on their cellular characteristics. The three domains are Bacteria, Archaea, and Eukarya (**figure 1.14**).

- The theory of evolution states that modifications in genes that are passed from parent to offspring result in changes in future generations. These changes lead to greater diversity among organisms over time (**figure 1.15**).

1. Biologists categorize all living things based on related characteristics into large groups, called
   a. kingdoms.   c. populations.
   b. species.   d. ecosystems.

2. Living things can be distinguished from nonliving things because they have
   a. complexity.   c. cellular organization.
   b. movement.   d. response to a stimulus.

3. Living things are organized. Choose the answer that illustrates this organization and that is arranged from smallest to largest.
   a. cell, atom, molecule, tissue, organelle, organ, organ system, organism, population, species, community, ecosystem
   b. atom, molecule, organelle, cell, tissue, organ, organ system, organism, population, species, community, ecosystem
   c. atom, molecule, organelle, cell, tissue, organ, organ system, organism, community, population, species, ecosystem
   d. atom, molecule, cell wall, cell, organ, organelle, organism, species, population, community, ecosystem

4. At each level in the hierarchy of living things, properties occur that were not present at the simpler levels. These properties are referred to as
   a. novelistic properties.   c. incremental properties.
   b. complex properties.   d. emergent properties.

5. The five general biological themes include
   a. evolution, energy flow, competition, structure determines function, and homeostasis.
   b. evolution, energy flow, cooperation, structure determines function, and homeostasis.
   c. evolution, growth, competition, structure determines function, and homeostasis.
   d. evolution, growth, cooperation, structure determines function, and homeostasis.

6. When you are trying to understand something new, you begin by observation, and then put the observations together in a logical fashion to form a general principle. This method is called
   a. inductive reasoning.   c. theory production.
   b. rule enhancement.   d. deductive reasoning.

7. When trying to figure out explanations for observations, you usually construct a series of possible hypotheses. Then you make predictions of what will happen if each hypothesis is true, and
   a. test each hypothesis, using appropriate controls, to determine which hypothesis is true.
   b. test each hypothesis, using appropriate controls, to rule out as many as possible.
   c. use logic to determine which hypothesis is most likely true.
   d. reject those that seem unlikely.

8. Which of the following statements is correct regarding a hypothesis?
   a. After sufficient testing, you can conclude that it is true.
   b. If it explains the observations, it doesn't need to be tested.
   c. After sufficient testing, you can accept it as probable, being aware that it may be revised or rejected in the future.
   d. You never have any degree of certainty that it is true; there are too many variables.

9. Cell theory states that
   a. all organisms have cell walls and all cell walls come from other cells.
   b. all cellular organisms undergo sexual reproduction.
   c. all living organisms use cells for energy, either their own or they ingest cells of other organisms.
   d. all living organisms consist of cells, and all cells come from other cells.

10. The gene theory states that all the information that specifies what a cell is and what it does
    a. is different for each cell type in the organism.
    b. is passed down, unchanged, from parents to offspring.
    c. is contained in a long molecule called DNA.
    d. All of the above.

Apply Your Understanding

1. For over two centuries global temperatures have been warming, and over this same period of time the number of pirate ship attacks has steadily decreased. Does this graph support the conclusion that the number of pirate attacks has decreased because of warmer temperatures? Explain.

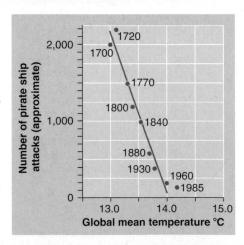

2. **Figure 1.5** You notice that on cloudy days people often carry umbrellas, folded or in a case. You also note that when umbrellas are open there are many car accidents. You conclude that open umbrellas cause car accidents. Referring back to figure 1.5, explain the type of reasoning used to reach this conclusion, and why it can sometimes be a problem.

Synthesize What You Have Learned

1. You are the biologist in a group of scientists who have traveled to a distant star system and landed on a planet. You see an astounding array of shapes and forms. You have three days to take samples of living things before returning to earth. How do you decide what is alive?

2. St. John's wort is an herb that has been used for hundreds of years as a remedy for mild depression. How might a modern-day scientist research its effectiveness?

2

The Chemistry of Life

These trees have been seriously damaged by acid rain. The death of this forest must have seemed a calamity to the animals that lived there. A porcupine knows no chemistry, has no way to comprehend what has happened, or why. Later in this chapter, you will explore what causes acid rain and snow, and how the acid has killed forests like this one. A famous conservation saying is that "you cannot save what you don't understand." In order to understand acid rain, you must first come to understand some simpler things, the nuts and bolts that underlie what happens in nature. All living things—in fact, everything you can see in the picture above—are made of tiny particles called atoms, linked together in assemblies called molecules. This is where we will have to start, if we want to understand things like what happened to this forest. Then, with molecules under our belt, we will need to get more specific and consider the nature of rain. What are rain and snow made of? Water. We will need to take a very careful look at water. When we do, we will see that when some chemicals are added to water, a chemically active mixture called an acid results. Acid rain is water containing such chemicals. Understanding this gives us the mental tool we need to attack the problem of what happened to this forest and determine how to stop it. In just this way, chemistry underlies much of what you will learn in biology.

2.1 Atoms

Biology is the science of life, and all life, in fact even all nonlife, is made of substances. **Chemistry** is the study of the properties of these substances. So, while it may seem tedious or unrelated to examine chemistry in a biology text, it is essential. Organisms are chemical machines (**figure 2.1**), and to understand them we must learn a little chemistry.

Any substance in the universe that has mass and occupies space is defined as **matter.** All matter is composed of extremely small particles called **atoms.** An atom is the smallest particle into which a substance can be divided and still retain its chemical properties.

Every atom has the same basic structure you see in **figure 2.2.** At the center of every atom is a small, very dense nucleus formed of two types of subatomic particles, **protons** (illustrated by purple balls) and **neutrons** (the pink balls). Whizzing around the core is an orbiting cloud of a third kind of subatomic particle, the **electron** (depicted by yellow balls on concentric rings). Neutrons have no electrical charge, whereas protons have a positive charge and electrons have a negative one. In each atom, there is an orbiting electron for every proton in the nucleus. The electron's negative charge balances the proton's positive charge. The atom is said to be electrically neutral.

An atom is typically described by the number of protons in its nucleus or by the overall mass of the atom. The terms *mass* and *weight* are often used interchangeably, but they have slightly different meanings. Mass refers to the amount of a substance, whereas weight refers to the force gravity exerts on a substance. Hence, an object has the same mass whether it is on the earth or the moon, but its weight will be greater on the earth, because the earth's gravitational force is greater than the moon's. For example, an astronaut weighing 180 pounds on earth will weigh about 30 pounds on the moon. He didn't lose any significant mass during his flight to the moon, there is just less gravitational pull on his mass.

The number of protons in the nucleus of an atom is called the **atomic number.** For example, the atomic number of carbon is 6 because it has six protons. Atoms with the same atomic number (that is, the same number of protons) have the same chemical properties and are said to belong to the same **element.** Formally speaking, an element is any substance that cannot be broken down into any other substance by ordinary chemical means.

Neutrons are similar to protons in mass, and the number of protons and neutrons in the nucleus of an atom is called the **mass number.** A carbon atom that has six protons and six neutrons has a mass number of 12. An electron's contribution to the overall mass of an atom is negligible. The atomic numbers and mass numbers of some of the most common elements on earth are shown in **table 2.1.**

Figure 2.1 Replacing electrolytes.
During extreme exercise, athletes will often consume drinks that contain "electrolytes," chemicals such as calcium, potassium, and sodium that play an important role in muscle contraction. Electrolytes can also be depleted in other types of dehydration.

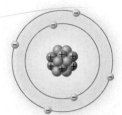

Hydrogen
Nucleus contains
1 proton

1 electron in orbit
around nucleus

Carbon
Nucleus contains
6 protons
6 neutrons

6 electrons in orbit
around nucleus

Proton ⊕ (Positive charge)   Neutron ⬤ (No charge)   Electron ⊝ (Negative charge)

Figure 2.2 Basic structure of atoms.
All atoms have a nucleus consisting of protons and neutrons, except hydrogen, the smallest atom, which has only one proton and no neutrons in its nucleus. Carbon, for example, has six protons and six neutrons in its nucleus. Electrons spin around the nucleus in orbitals a far distance away from the nucleus. The electrons determine how atoms react with each other.

TABLE 2.1	ELEMENTS COMMON IN LIVING ORGANISMS		
Element	**Symbol**	**Atomic Number**	**Mass Number**
Hydrogen	H	1	1.008
Carbon	C	6	12.011
Nitrogen	N	7	14.007
Oxygen	O	8	15.999
Sodium	Na	11	22.989
Phosphorus	P	15	30.974
Sulfur	S	16	32.064
Chlorine	Cl	17	35.453
Potassium	K	19	39.098
Calcium	Ca	20	40.080
Iron	Fe	26	55.847

Electrons Determine What Atoms Are Like

Electrons have very little mass (only about 1/1,840 the mass of a proton). Of all the mass contributing to your weight, the portion that is contributed by electrons is less than the mass of your eyelashes. And yet electrons determine the chemical behavior of atoms because they are the parts of atoms that come close enough to each other in nature to interact. Almost all the volume of an atom is empty space. Protons and neutrons lie at the core of this space, whereas orbiting electrons are very far from the nucleus. If the nucleus of an atom were the size of an apple, the orbit of the nearest electron would be more than a mile out!

Electrons Carry Energy

Because electrons are negatively charged, they are attracted to the positively charged nucleus, but they also repel the negative charges of each other. It takes work to keep them in orbit, just as it takes work to hold an apple in your hand when gravity is pulling the apple down toward the ground. The apple in your hand is said to possess **energy,** the ability to do work, because of its position—if you were to release it, the apple would fall. Similarly, electrons have energy of position, called *potential energy*. It takes work to oppose the attraction of the nucleus, so moving the electron farther out from the nucleus, as shown by the set of arrows on the right side of **figure 2.3**, requires an input of energy and results in an electron with greater potential energy. Moving an electron in toward the nucleus has the opposite effect (the set of arrows on the left side); energy is released, and the electron has less potential energy. Consider again an apple held in your hand. If you carry the apple up to a second-story window, it has a greater potential energy when you drop it, compared to when it is dropped at ground level. Similarly, if you lower the apple until it is 6 inches from the ground, it has less potential energy. Cells use the potential energy of atoms to drive chemical reactions, as we will discuss in chapter 5.

While the energy levels of an atom are often visualized as well-defined circular orbits around a central nucleus as was shown in **figure 2.2**, such a simple picture is not accurate. These energy levels, called *electron shells*, often consist of complex three-dimensional shapes, and the exact location of an individual electron at any given time is impossible to specify. However, some locations are more probable than others, and it is often possible to say where an electron is *most likely* to be located. The volume of space around a nucleus where an electron is most likely to be found is called the **orbital** of that electron.

Each electron shell has a specific number of orbitals, and each orbital can hold up to two electrons. The first shell in any atom contains one orbital. Helium, shown in **figure 2.4a**, has one electron shell with one orbital that corresponds to the lowest energy level. The orbital contains two electrons, shown above and below the nucleus. In atoms with more than one electron shell, the second shell contains four orbitals and holds up to eight electrons. Nitrogen, shown in **figure 2.4b**, has two electron shells; the first one is completely filled with two electrons, but three of the four orbitals in the second electron shell are not filled because nitrogen's second shell contains only five electrons (openings in orbitals are indicated with dotted circles). In atoms with more than two electron shells, subse-

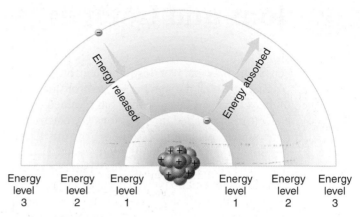

| Energy level 3 | Energy level 2 | Energy level 1 | Energy level 1 | Energy level 2 | Energy level 3 |

Figure 2.3   The electrons of atoms possess potential energy.

Electrons that circulate rapidly around the nucleus contain energy, and depending on their distance from the nucleus, they may contain more or less energy. Energy level 1 is the lowest potential energy level because it is closest to the nucleus. When an electron absorbs energy, it moves from level 1 to the next higher energy level (level 2). When an electron loses energy, it falls to a lower energy level closer to the nucleus.

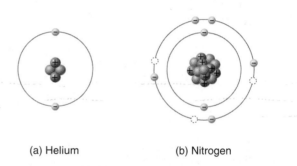

(a) Helium   (b) Nitrogen

Figure 2.4   Electrons in electron shells.

(*a*) An atom of helium has two protons, two neutrons, and two electrons. The electrons fill the one orbital in its one electron shell, the lowest energy level. (*b*) An atom of nitrogen has seven protons, seven neutrons, and seven electrons. Two electrons fill the orbital in the innermost electron shell, and five electrons occupy orbitals in the second electron shell (the second energy level). The orbitals in the second electron shell can hold up to eight electrons; therefore there are three vacancies in the outer electron shell of a nitrogen atom.

quent shells also contain up to four orbitals and a maximum of eight electrons. Atoms with unfilled electron orbitals tend to be more reactive because they lose, gain, or share electrons in order to fill their outermost electron shell. Losing, gaining, or sharing electrons is the basis for chemical reactions in which chemical bonds form between atoms. Chemical bonds will be discussed later in this chapter.

> **Key Learning Outcome 2.1**   Atoms, the smallest particles into which a substance can be divided, are composed of electrons orbiting a nucleus that contains protons and neutrons. Electrons determine the chemical behavior of atoms.

2.2 Ions and Isotopes

Ions

Sometimes an atom may gain or lose an electron from its outer shell. Atoms in which the number of electrons does not equal the number of protons because they have gained or lost one or more electrons are called **ions.** All ions are electrically charged. For example, an atom of sodium (on the left in **figure 2.5**) becomes a positively charged ion, called a *cation* (on the right), when it loses an electron, such that one proton in the nucleus is left with an unbalanced charge (11 positively charged protons and only 10 negatively charged electrons). Negatively charged ions, called *anions*, also form when an atom gains an electron from another atom.

Isotopes

The number of neutrons in an atom of a particular element can vary without changing the chemical properties of the element. Atoms that have the same number of protons but different numbers of neutrons are called **isotopes.** Isotopes of an

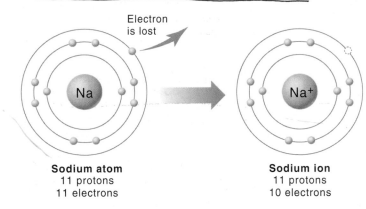

Sodium atom
11 protons
11 electrons

Sodium ion
11 protons
10 electrons

Figure 2.5 Making a sodium ion.
An electrically neutral sodium atom has 11 protons and 11 electrons. Sodium ions bear one positive charge when they ionize and lose one electron. Sodium ions have 11 protons and only 10 electrons.

atom have the same atomic number but differ in their mass number. Most elements in nature exist as mixtures of different isotopes. For example, there are three isotopes of the element carbon, all of which possess six protons (the purple balls in **figure 2.6**). The most common isotope of carbon (99% of all carbon) has six neutrons (the pink balls). Because its mass number is 12 (six protons plus six neutrons), it is referred to as carbon-12. The isotope carbon-14 (on the right) is rare (1 in 1 trillion atoms of carbon) and unstable, such that its nucleus tends to break up into particles with lower atomic numbers, a process called **radioactive decay.** Radioactive isotopes are used in medicine and in dating fossils.

Medical Uses of Radioactive Isotopes

When most people hear the word "radioactive" they picture atomic bombs exploding into mushroom clouds and the devastation that results. While it is true that the radiation emitted from radioactive isotopes can damage cells of the human body, it is also true that isotopes can be used in many medical procedures. Short-lived isotopes, those that decay fairly rapidly and produce harmless products, are commonly used as tracers in the body. A **tracer** is a radioactive substance that is taken up and used by the body. Emissions from the radioactive isotope tracer are detected using special laboratory equipment, and can reveal key diagnostic information about the functioning of the body. For example, PET and PET/CT (positron emission tomography/computerized tomography) imaging procedures can be used to identify a cancerous area in the body. First, a radioactive tracer is injected into the body. This tracer is taken up by all cells, but it is taken up in larger amounts in cells with higher metabolic activities, such as cancer cells. Images are then taken of the body, and areas emitting greater amounts of the tracer can be seen. For example, in the images in **figure 2.7**, the radioactive-emitting cancer site appears as a black area on the left and a yellow glowing area on the right. There are many other uses of radioactive isotopes in medicine, both in detection and treatment of disorders.

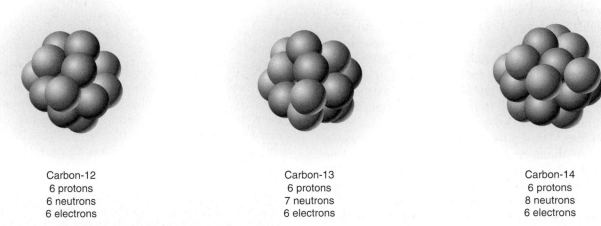

Carbon-12
6 protons
6 neutrons
6 electrons

Carbon-13
6 protons
7 neutrons
6 electrons

Carbon-14
6 protons
8 neutrons
6 electrons

Figure 2.6 Isotopes of the element carbon.
The three most abundant isotopes of carbon are carbon-12, carbon-13, and carbon-14. The yellow "clouds" in the diagrams represent the orbiting electrons, whose numbers are the same for all three isotopes. Protons are shown in purple, and neutrons are shown in pink.

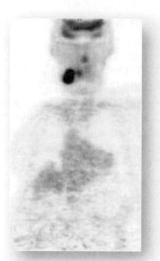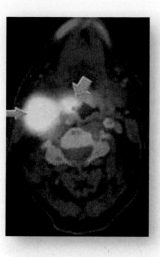

Figure 2.7   Using a radioactive tracer to identify cancer.

In certain medical imaging procedures, the patient ingests or is injected intravenously with a radioactive tracer that is absorbed in greater amounts by cancer cells. The tracer emits radioactivity that is detected using PET and PET/CT equipment. A cancerous area in the neck is seen in these two images, as a dark black area on the left and a bright yellow glowing area on the right.

Dating Fossils

Fossils are created when the remains, footprints, or other traces of organisms become buried in sand or sediment. Over time, the calcium in bone and other hard tissues becomes mineralized as the sediment is converted to rock. A fossil is any record of prehistoric life—generally taken to mean older than 10,000 years. By dating the rocks in which fossils occur, biologists can get a very good idea of how old the fossils are. Rocks are usually dated by measuring the degree of radioactive decay of certain radioactive isotopes among rock-forming minerals. A radioactive isotope is one whose nucleus is unstable and eventually flies apart, creating more stable atoms of another element. Because the rate of decay of a radioactive element (the percent of isotopes that undergo decay in a minute) is constant, scientists can use the amount of radioactive decay to date fossils. The older the fossil, the greater the fraction of its radioactive isotopes that have decayed.

A widely employed method of dating fossils less than 50,000 years old is the carbon-14 (^{14}C) **radioisotopic dating** method illustrated in **figure 2.8**. Most carbon atoms have a mass number of 12 (^{12}C). However, a tiny but fixed proportion of the carbon atoms in the atmosphere consists of carbon atoms with a mass number of 14 (^{14}C). This isotope of carbon is created by the bombardment of nitrogen-14 atoms with cosmic rays. This proportion of ^{14}C (designated A in the figure) is captured by plants in photosynthesis, and is the proportion present in the carbon molecules of the animal's body that eats the plants, in this case a rabbit. After the plant or animal dies, it no longer accumulates carbon, and the ^{14}C present at the time of death gradually decays over time back to nitrogen-14 (^{14}N). The amount of ^{14}C (A) decreases while the amount of ^{12}C stays the same. Scientists can determine how long ago an organism died by measuring the ratio of ^{14}C to ^{12}C

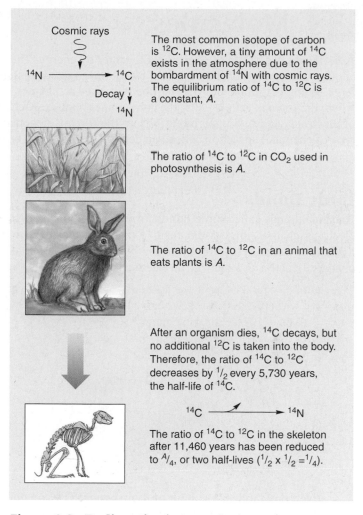

Cosmic rays

^{14}N $\longrightarrow$ ^{14}C

Decay

^{14}N

The most common isotope of carbon is ^{12}C. However, a tiny amount of ^{14}C exists in the atmosphere due to the bombardment of ^{14}N with cosmic rays. The equilibrium ratio of ^{14}C to ^{12}C is a constant, A.

The ratio of ^{14}C to ^{12}C in CO_2 used in photosynthesis is A.

The ratio of ^{14}C to ^{12}C in an animal that eats plants is A.

After an organism dies, ^{14}C decays, but no additional ^{12}C is taken into the body. Therefore, the ratio of ^{14}C to ^{12}C decreases by $^1/_2$ every 5,730 years, the half-life of ^{14}C.

^{14}C $\longrightarrow$ ^{14}N

The ratio of ^{14}C to ^{12}C in the skeleton after 11,460 years has been reduced to $^A/_4$, or two half-lives ($^1/_2 \times {}^1/_2 = {}^1/_4$).

Figure 2.8   Radioactive isotope dating.

This diagram illustrates radioactive dating using carbon-14, a short-lived isotope.

in its remains or in the fossilized rock. Over time, the ratio of ^{14}C to ^{12}C decreases. It takes 5,730 years for half of the ^{14}C ($1/2A$ or $A/2$) present in the sample to be converted to ^{14}N by this process. This length of time is called the **half-life** of the isotope. Because the half-life of an isotope is a constant that never changes, the extent of radioactive decay allows you to date a sample. Thus a sample that had a quarter of its original proportion of ^{14}C remaining ($1/4A$ or $A/4$) would be approximately 11,460 years old (two half-lives—5,730 years for half of the ^{14}C to decay to a level of $A/2$ and another 5,730 years for the remaining ^{14}C to decay to a level of $A/4$).

For fossils older than 50,000 years, there is too little ^{14}C remaining to measure precisely, and scientists instead examine the decay of potassium-40 (^{40}K) into argon-40 (^{40}Ar), which has a half-life of 1.3 billion years.

Key Learning Outcome 2.2   When an atom gains or loses one or more electrons, it is called an ion. Isotopes of an element differ in the number of neutrons they contain, but all have the same chemical properties.

2.3 Molecules

A **molecule** is a group of atoms held together by energy. The energy acts as "glue," ensuring that the various atoms stick to one another. The energy or force holding two atoms together is called a **chemical bond.** Chemical bonds determine the shapes of the large biological molecules that will be discussed in chapter 3. There are three principal kinds of chemical bonds: ionic bonds, where the force is generated by the attraction of oppositely charged ions; covalent bonds, where the force results from the sharing of electrons; and hydrogen bonds, where the force is generated by the attraction of opposite partial electrical charges. Another type of chemical attraction called van der Waals forces will be discussed later, but keep in mind that this type of interaction is not considered a chemical bond.

Ionic Bonds

Chemical bonds called **ionic bonds** form when atoms are attracted to each other by opposite electrical charges. Just as the positive pole of a magnet is attracted to the negative pole of another, so an atom can form a strong link with another atom if they have opposite electrical charges. Because an atom with an electrical charge is an ion, these bonds are called ionic bonds.

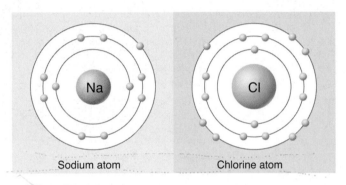
Sodium atom   Chlorine atom

Everyday table salt is built of ionic bonds. The sodium and chlorine atoms of table salt are ions. The sodium you see in the yellow panels gives up the sole electron in its outermost shell (the shell underneath has eight) and chlorine, in the light green panels, gains an electron to complete its outermost shell. Recall from section 2.1 that an atom is more stable when its outermost electron shell is filled (with two electrons in the innermost shell or eight electrons in shells that are farther out from the nucleus).

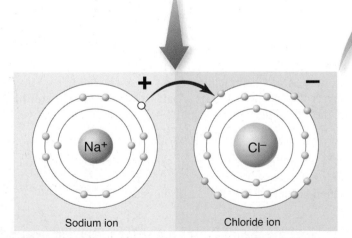
Sodium ion   Chloride ion

To achieve this stability, an atom will give up or accept electrons from another atom. As a result of this electron hopping, sodium atoms in table salt are positive sodium ions and chlorine atoms are negative chloride ions.

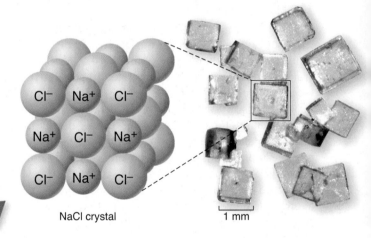
NaCl crystal   1 mm

Because each ion is attracted electrically to all surrounding ions of opposite charge, this causes the formation of an elaborate matrix of sodium and chloride ionic bonds—a crystal. The sodium chloride crystal shown above reveals an organized structure of alternating sodium (yellow) and chloride (light green) ions. That is why table salt is composed of tiny crystals and is not a powder.

The two key properties of ionic bonds that make them form crystals are that they are strong (although not as strong as covalent bonds) and that they are *not* directional. A charged atom is attracted to the electrical field contributed by all nearby atoms of opposite charge. Ionic bonds do not play an important part in most biological molecules because of this lack of directionality. Complex, stable shapes require the more specific associations made possible by directional bonds.

Covalent Bonds

Strong chemical bonds called **covalent bonds** form between two atoms when they share electrons. Most of the atoms in your body are linked to other atoms by covalent bonds. Why do atoms in molecules share electrons? Remember, all atoms seek to fill up their outermost shell of orbiting electrons, which in all atoms (except tiny hydrogen and helium) takes eight electrons.

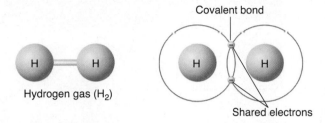

Covalent bond

Shared electrons

Hydrogen gas (H_2)

A covalent bond is formed when electrons are shared between atoms. The atoms sharing the electrons may be of the same element or different elements. Some atoms, like hydrogen (H), can form only one covalent bond, because hydrogen needs only one more electron to fill its outermost shell.

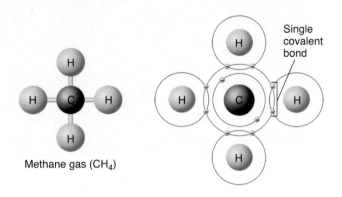

Single covalent bond

Methane gas (CH_4)

Other atoms, such as carbon (C), nitrogen (N), or oxygen (O), can form more than one covalent bond, depending on the space available in their outermost electron shells. The carbon atom has four electrons in its outermost shell, and carbon can form as many as four covalent bonds in its attempt to fully populate its outermost shell of electrons. Because there are many ways four covalent bonds can form, carbon atoms participate in many different kinds of molecules.

Double covalent bond

Oxygen gas (O_2)

Most covalent bonds are *single bonds,* which involve the sharing of two electrons, but *double bonds* (in which four electrons are shared) are also common. *Triple bonds* (in which six electrons are shared) are much less frequent in nature, but are found in some common compounds, like nitrogen gas (N_2).

Energy is often released when covalent bonds are broken. The *Hindenberg* dirigible was filled with hydrogen gas when it exploded and burned in 1937; the energy of the inferno came from the breaking of H_2 covalent bonds.

Polar and Nonpolar Covalent Bonds When a covalent bond forms between two atoms, one nucleus may be much better at attracting the shared electrons than the other, an aspect of the atom called its *electronegativity*. In water, for example, the shared electrons are much more strongly attracted to the oxygen atom than to the hydrogen atoms; oxygen has a higher electronegativity. When this happens, shared electrons spend more time in the vicinity of the oxygen atom, which as a result becomes somewhat negative in charge; they spend less time in the vicinity of the hydrogens, and these become somewhat positive in charge. The charges are not full electrical charges like in ions but rather tiny *partial charges* (see next page), signified by the Greek letter delta (δ). What you end up with is a sort of molecular magnet, with positive and negative ends, or "poles." Molecules like this are said to be **polar molecules,** and the bonds between the atoms are called *polar covalent bonds.* Molecules that don't exhibit a large difference in electronegativities of its atoms, like the carbon-hydrogen bonds of methane, are called **nonpolar molecules** and contain *nonpolar covalent bonds.*

The two key properties of covalent bonds that make them ideal for their molecule-building role in living systems are that (1) they are strong, involving the sharing of lots of energy; and (2) they are very directional—bonds form between two specific atoms, rather than a generalized attraction of one atom for its neighbors.

Hydrogen Bonds

Polar molecules like water are attracted to one another, a special type of weak chemical bond called a **hydrogen bond.** Hydrogen bonds occur when the positive end of one polar molecule is attracted to the negative end of another, like two magnets drawn to each other.

In a hydrogen bond, an electropositive hydrogen from one polar molecule is attracted to an electronegative atom, often oxygen (O) or nitrogen (N), from another polar molecule.

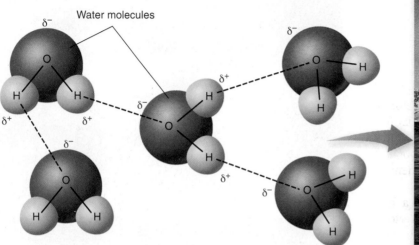

Because the oxygen atoms in water molecules are more electronegative than the hydrogen atoms, water molecules are polar. Water molecules form strong hydrogen bonds with each other, giving liquid water many unique properties. Each oxygen has a partial negative charge (δ^-), and each hydrogen has a partial positive charge (δ^+). Hydrogen bonds (shown as dashed lines) form between the positive end of one polar molecule and the negative end of another polar molecule. This attraction of partial charges attracts water molecules to one another.

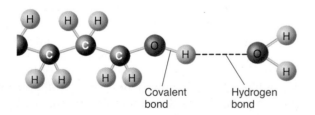

Two key properties of hydrogen bonds cause them to play an important role in the molecules found in organisms. First, they are weak and so are not effective over long distances like more powerful covalent and ionic bonds. Hydrogen bonds are too weak to actually form stable molecules by themselves. Instead, they act like Velcro, forming a tight bond by the additive effects of many weak interactions. Second, hydrogen bonds are highly directional. In chapter 3, we will discuss the role of hydrogen bonding in maintaining the structures of large biological molecules such as proteins and DNA.

Van der Waals Forces

Another important kind of weak chemical attraction is a non-directional attractive force called *van der Waals forces* (or van der Waals interactions). These chemical forces come into play only when two atoms are very close to one another. The attraction is very weak, and disappears if the atoms move even a little apart. It becomes significant when numerous atoms in one molecule simultaneously come close to numerous atoms of another molecule—that is, when the shapes of the molecules match precisely. For example, this interaction is important when antibodies in your blood recognize the shape of an invading virus as foreign.

Key Learning Outcome 2.3 Molecules are atoms linked together by chemical bonds. Ionic bonds, covalent bonds, and hydrogen bonds are the three principal types of bonds, and van der Waals forces are weaker interactions.

2.4 Hydrogen Bonds Give Water Unique Properties

Three-fourths of the earth's surface is covered by liquid water. About two-thirds of your body is water, and you cannot exist long without it. All other organisms also require water. It is no accident that tropical rain forests are bursting with life, whereas dry deserts seem almost lifeless except after rain. The chemistry of life, then, is water chemistry.

Water has a simple atomic structure, an oxygen atom linked to two hydrogen atoms by single covalent bonds. The chemical formula for water is thus H_2O. It is because the oxygen atom attracts the shared electrons more strongly than the hydrogen atoms that water is a *polar molecule* and so can form *hydrogen bonds*. Water's ability to form hydrogen bonds is responsible for much of the organization of living chemistry, from membrane structure to how proteins fold.

The weak hydrogen bonds that form between a hydrogen atom of one water molecule and the oxygen atom of another produce a lattice of hydrogen bonds within liquid water. Each of these bonds is individually very weak and short-lived—a single bond lasts only 1/100,000,000,000 of a second. However, like the grains of sand on a beach, the cumulative effect of large numbers of these bonds is enormous and is responsible for many of the important physical properties of water (**table 2.2**).

Heat Storage

The temperature of any substance is a measure of how rapidly its individual molecules are moving. Because of the many hydrogen bonds that water molecules form with one another, a large input of thermal energy is required to disrupt the organization of liquid water and raise its temperature. Because of this, water heats up more slowly than almost any other compound and holds its temperature longer. That is a major reason why your body is able to maintain a relatively constant internal temperature.

Ice Formation

If the temperature is low enough, very few hydrogen bonds break in water. Instead, the lattice of these bonds assumes a crystal-like structure, forming a solid we call ice. Interestingly, ice is less dense than water—that is why icebergs and ice cubes float. Why is ice less dense? This is best understood by comparing the molecular structures of water and ice that you see in **figure 2.9**. At temperatures above freezing (0°C or 32°F), water molecules in **figure 2.9a** move around each other with hydrogen bonds breaking and forming. As temperatures drop, the movement of water molecules decreases, allowing hydrogen bonds to

TABLE 2.2	THE PROPERTIES OF WATER
Property	**Explanation**
Heat storage	Hydrogen bonds require considerable heat before they break, minimizing temperature changes.
Ice formation	Water molecules in an ice crystal are spaced relatively far apart because of hydrogen bonding.
High heat of vaporization	Many hydrogen bonds must be broken for water to evaporate.
Cohesion	Hydrogen bonds hold molecules of water together.
High polarity	Water molecules are attracted to ions and polar compounds.

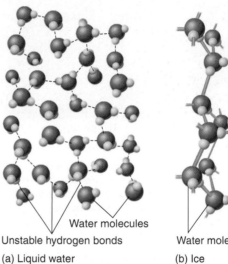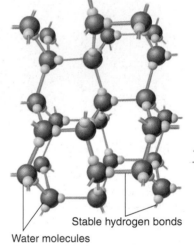

Water molecules

Stable hydrogen bonds

Unstable hydrogen bonds

Water molecules

(a) Liquid water

(b) Ice

Figure 2.9 Ice formation.
When water (*a*) cools below 0°C, it forms a regular crystal structure (*b*) that floats. The individual water molecules are spaced apart and held in position by hydrogen bonds.

stabilize, holding individual molecules farther apart, as in **figure 2.9b**, making the ice structure less dense.

High Heat of Vaporization

If the temperature is high enough, many hydrogen bonds break in water, with the result that the liquid is changed into vapor. A considerable amount of heat energy is required to do this—every gram of water that evaporates from your skin removes 2,452 joules of heat from your body, which is equal to the energy released by lowering the temperature of 586 grams of water 1°C. That is why sweating cools you off; as the sweat evaporates (vaporizes) it takes energy with it, in the form of heat, cooling the body.

(a)

(b)

Figure 2.10 Cohesion.

(a) Cohesion allows water molecules to stick together and form droplets. (b) Surface tension is a property derived from cohesion—that is, water has a "strong" surface due to the force of its hydrogen bonds. Some insects, such as this water strider, literally walk on water.

Cohesion

Because water molecules are very polar, they are attracted to other polar molecules—hydrogen bonds bind polar molecules to each other. When the other polar molecule is another water molecule, the attraction is called **cohesion.** The surface tension of water is created by cohesion. Surface tension is the force that causes water to bead, like on the spider web in **figure 2.10,** or supports the weight of the water strider. When the other polar molecule is a different substance, the attraction is called **adhesion.** Capillary action—such as water moving up into a paper towel—is created by adhesion. Water clings to any substance, such as paper fibers, with which it can form hydrogen bonds. Adhesion is why things get "wet" when they are dipped in water and why waxy substances do not—they are composed of nonpolar molecules that don't form hydrogen bonds with water molecules.

High Polarity

Water molecules in solution always tend to form the maximum number of hydrogen bonds possible. Polar molecules form hydrogen bonds and are attracted to water molecules. Polar molecules are called **hydrophilic** (from the Greek *hydros,*

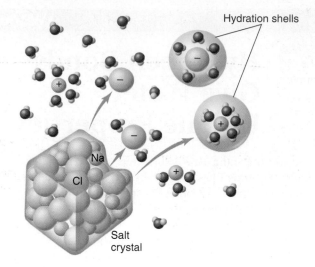
Hydration shells

Na

Cl

Salt crystal

Figure 2.11 How salt dissolves in water.

Salt is soluble in water because the partial charges on water molecules are attracted to the charged sodium and chloride ions. The water molecules surround the ions, forming what are called hydration shells. When all of the ions have been separated from the crystal, the salt is said to be dissolved.

water, and *philic,* loving), or water-loving, molecules. Water molecules gather closely around any molecule that exhibits an electrical charge, whether a full charge (ion) or partial charge (polar molecule). When a salt crystal dissolves in water as you see happening in **figure 2.11,** what really happens is that individual ions break off from the crystal and become surrounded by water molecules. The blue hydrogen atoms of water are attracted to the negative charge of the chloride ions and the red oxygen atoms are attracted to the positive charge of the sodium ions. Water molecules orient around each ion like a swarm of bees attracted to honey, and this shell of water molecules, called a *hydration shell,* prevents the ions from reassociating with the crystal. Similar shells of water form around all polar molecules, and polar molecules that dissolve in water in this way are said to be **soluble** in water.

Nonpolar molecules like oil do not form hydrogen bonds and are not water-soluble. When nonpolar molecules are placed in water, the water molecules shy away, instead forming hydrogen bonds with other water molecules. The nonpolar molecules are forced into association with one another, crowded together to minimize their disruption of the hydrogen bonding of water. It seems almost as if the nonpolar compounds shrink from contact with water, and for this reason they are called **hydrophobic** (from the Greek *hydros,* water, and *phobos,* fearing). Many biological structures are shaped by such hydrophobic forces, as will be discussed in chapter 3.

> **Key Learning Outcome 2.4** Water molecules form a network of hydrogen bonds in liquid and dissolve other polar molecules. Many of the key properties of water arise because it takes considerable energy to break liquid water's many hydrogen bonds.

2.5 | Water Ionizes

The covalent bonds within a water molecule sometimes break spontaneously. When it happens, one of the protons (hydrogen atom nuclei) dissociates from the molecule. Because the dissociated proton lacks the negatively charged electron it was sharing in the covalent bond with oxygen, its own positive charge is no longer counterbalanced, and it becomes a positively charged ion, **hydrogen ion** (H^+). The rest of the dissociated water molecule, which has retained the shared electron from the covalent bond, is negatively charged and forms a **hydroxide ion** (OH^-). This process of spontaneous ion formation is called **ionization.** It can be represented by a simple chemical equation, in which the chemical formulas for water and the two ions are written down, with an arrow showing the direction of the dissociation:

$$H_2O \longleftrightarrow OH^- + H^+$$
$$\text{water} \quad\quad \text{hydroxide} \quad\quad \text{hydrogen}$$
$$\text{ion} \quad\quad\quad \text{ion}$$

Because covalent bonds are strong, spontaneous ionization is not common. In a liter of water, only roughly 1 molecule out of each 550 million is ionized at any instant in time, corresponding to 1/10,000,000 (that is, 10^{-7}) of a mole of hydrogen ions. (A mole is a measurement of weight. One mole of any object is the weight of 6.022×10^{23} units of that object.) The concentration of H^+ in water can be written more easily by simply counting the number of decimal places after the digit "1" in the denominator:

$$[H^+] = \frac{1}{10,000,000}$$

pH

A more convenient way to express the hydrogen ion concentration of a solution is to use the pH scale (**figure 2.12**). This scale defines pH as the negative logarithm of the hydrogen ion concentration in the solution:

$$pH = -\log [H^+]$$

Since the logarithm of the hydrogen ion concentration is simply the exponent of the molar concentration of H^+, the pH equals the exponent times –1. Thus, pure water, with an $[H^+]$ of 10^{-7} mole/liter, has a pH of 7. Recall that for every hydrogen ion formed when water dissociates, a hydroxide ion is also formed, meaning that the dissociation of water produces H^+ and OH^- in equal amounts. Therefore, a pH value of 7 indicates neutrality—a balance between H^+ and OH^-—on the pH scale.

Note that the pH scale is *logarithmic,* which means that a difference of 1 on the pH scale represents a 10-fold change in hydrogen ion concentration. This means that a solution with a pH of 4 has *10 times* the concentration of H^+ present in one with a pH of 5.

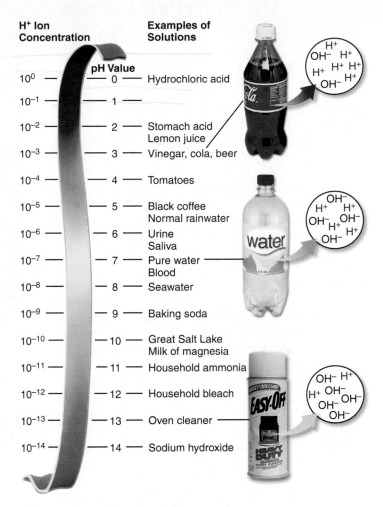

Figure 2.12 The pH scale.
A fluid is assigned a value according to the number of hydrogen ions present in a liter of that fluid. The scale is logarithmic, so that a change of only 1 means a 10-fold change in the concentration of hydrogen ions; thus lemon juice with a pH of 2 is 100 times more acidic than tomatoes with a pH of 4, and seawater is 10 times more basic than pure water.

Acids Any substance that dissociates in water to increase the concentration of H^+ is called an **acid.** Acidic solutions have pH values below 7. The stronger an acid is, the more H^+ it produces and the lower its pH. For example, hydrochloric acid (HCl), which is abundant in your stomach, ionizes completely in water. This means that a dilution of 10^{-1} mole/liter of HCl will dissociate to form 10^{-1} mole/liter of H^+, giving the solution a pH of 1. The pH of champagne, which bubbles because of the carbonic acid dissolved in it, is about 3.

Bases A substance that combines with H^+ when dissolved in water is called a **base.** By combining with H^+, a base lowers the H^+ concentration in the solution. Basic (or alkaline) solutions, therefore, have pH values above 7. Very strong bases, such as sodium hydroxide (NaOH), have pH values of 12 or more.

Today's *Biology*

Acid Rain

As you study biology, you will learn that hydrogen ions play many roles in the chemistry of life. When conditions become overly acidic—too many hydrogen ions—serious damage to organisms often results. One important example of this is acid precipitation, more informally called **acid rain.** Acid precipitation is just what it sounds like, the presence of acid in rain or snow. Where does the acid come from? Tall smokestacks from coal-burning power plants send smoke high into the atmosphere through these stacks, each of which is over 65 meters tall. The smoke the stacks belch out contains high concentrations of sulfur dioxide (SO_2), because the coal that the plants burn is rich in sulfur. The sulfur-rich smoke is dispersed and diluted by winds and air currents. Since the 1950s, such tall stacks have become popular in the United States and Europe—there are now over 800 of them in the United States alone.

In the 1970s, 20 years after the stacks were introduced, ecologists began to report evidence that the tall stacks were not eliminating the problems associated with the sulfur, just exporting the ill effects elsewhere. The lakes and forests of the Northeast suffered drastic drops in biodiversity, forests dying and lakes becoming devoid of life. It turned out that the SO_2 introduced into the upper atmosphere by high smokestacks combines with water vapor to produce sulfuric acid (H_2SO_4). When this water later falls back to earth as rain or snow, it carries the sulfuric acid with it. When schoolchildren measured the pH of natural rainwater as part of a nationwide project in 1989, rain and snow in the Northeast often had a pH as low as 2 or 3—more acidic than vinegar.

After accumulating in soils for over 50 years, the effects of acid rain are now only too evident. The impact of acid rain on forests first became evident in the Northeast. Some 15% of the lakes in New England have become chronically acidic and are dying biologically as their pH levels fall to below 5.0. Many of the forests of the northeastern United States and Canada have also been seriously damaged. The trees in this photo and on the first page of this chapter show the ill effects of acid precipitation. In the last decades, acid added to forest soils has caused the loss from these soils of over half the essential plant nutrients calcium and magnesium. Researchers blame excess acids for dissolving Ca^{++} and Mg^{++} ions into drainage waters much faster than weathering rocks can replenish them. Without them, trees stop growing and die.

Now, some 30 years later, acid rain effects are becoming apparent in the Southeast as well. Researchers suggest the reason for the delay is that southern soils are generally thicker than northern ones and thus able to

sponge up far more acid. But now that southern forest soils are becoming saturated, they too are beginning to die. In a third of the southeastern streams studied, fish are declining or already gone.

The solution is straightforward: Capture and remove the emissions instead of releasing them into the atmosphere. Progressively tougher pollution laws over the past three decades have reduced U.S. emissions of sulfur dioxide by about 40% from its 1973 peak of 28.8 metric tons a year. Despite this significant progress, much remains to be done. Unless levels are cut further, researchers predict forests may not recover for centuries.

An informed public will be essential. While textbook treatments have in the past tended to minimize the impact of this issue on students ("the vast majority of North American forests are not suffering substantially from acid precipitation"), it is important that we face the issue squarely and support continued efforts to address this serious problem.

Buffers

The pH inside almost all living cells, and in the fluid surrounding cells in multicellular organisms, is fairly close to 7. The many proteins that govern metabolism are all extremely sensitive to pH, and slight alterations in pH can cause the molecules to take on different shapes that disrupt their activities. For this reason, it is important that a cell maintain a constant pH level. The pH of your blood, for example, is 7.4, and you would survive only a few minutes if it were to fall to 7.0 or rise to 7.8.

Yet the chemical reactions of life constantly produce acids and bases within cells. Furthermore, many animals eat substances that are acidic or basic; Coca-Cola, for example, is acidic, and egg white is basic. What keeps an organism's pH constant? Cells contain chemical substances called buffers that minimize changes in concentrations of H^+ and OH^-.

A **buffer** is a substance that takes up or releases hydrogen ions into solution as the hydrogen ion concentration of the solution changes. Hydrogen ions are donated to the solution when their concentration falls and taken from the solution when their concentration rises. The graph in **figure 2.13** shows how buffers work. The blue line indicates changes in pH. As a base is added to the solution, the H^+ concentration falls and the pH should rise sharply, but by contributing H^+ to the solution the buffer works to keep the pH within a range, called the buffering range (the darker blue bar). Only when the buffering capacity is exceeded does the pH begin to rise. What sort of substance will act in this way? Within organisms, most buffers consist of pairs of substances, one an acid and the other a base.

The key buffer in human blood is an acid-base pair consisting of *carbonic acid* (acid) and *bicarbonate* (base). These two substances interact in a pair of reversible reactions. First, carbon dioxide (CO_2) and H_2O join to form carbonic acid (H_2CO_3) (step ❷ in **figure 2.14**), which in a second reaction dissociates to yield bicarbonate ion (HCO_3^-) and H^+ (step ❸). If some acid or other substance adds H^+ to the blood, the

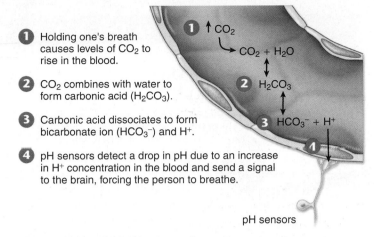

❶ Holding one's breath causes levels of CO_2 to rise in the blood.

❷ CO_2 combines with water to form carbonic acid (H_2CO_3).

❸ Carbonic acid dissociates to form bicarbonate ion (HCO_3^-) and H^+.

❹ pH sensors detect a drop in pH due to an increase in H^+ concentration in the blood and send a signal to the brain, forcing the person to breathe.

pH sensors

Figure 2.14 Holding your breath.
CO_2 accumulates in the blood when a person holds his/her breath. CO_2 combines with water, forming carbonic acid. Carbonic acid dissociates into bicarbonate and H^+, which acts to lower the pH. The drop in pH is detected by sensors that stimulate the brain, causing the person to breathe.

HCO_3^- acts as a base and removes the excess H^+ by forming H_2CO_3. Similarly, if a basic substance removes H^+ from the blood, H_2CO_3 dissociates, releasing more H^+ into the blood. The forward and reverse reactions that interconvert H_2CO_3 and HCO_3^- thus stabilize the blood's pH.

For example, when you breathe in, your body takes up oxygen from the air and when you breathe out, your body releases carbon dioxide. When you hold your breath, CO_2 accumulates in your blood and drives the chemical reactions in **figure 2.14**, producing carbonic acid. Can you hold your breath indefinitely? No, but not for the reason you might think. It is not lack of oxygen that forces you to breathe, but too much carbon dioxide. If you try and hold your breath for very long, CO_2 accumulates in the blood, as shown in step ❶ in **figure 2.14**, triggering the formation of carbonic acid (step ❷) that dissociates into bicarbonate ion and H^+ (step ❸). This increase in H^+ causes the blood to become more acidic. If the pH in the blood drops too low, pH sensors that are located in some of the major blood vessels of the body detect the change (step ❹) and send signals to the brain. These signals, along with other sensory processes, stimulate the area of the brain that controls respiration, causing it to increase the rate of breathing. Hyperventilating, breathing very quickly, has the opposite effect, lowering the levels of CO_2 in the blood. That is why you are told to breathe into a paper bag when you are hyperventilating, to increase your intake of CO_2.

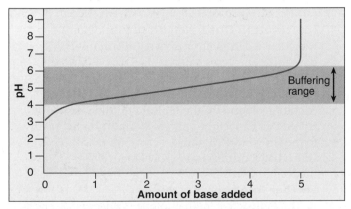

Figure 2.13 Buffers minimize changes in pH.
Adding a base to a solution neutralizes some of the acid present and so raises the pH. Thus, as the curve moves to the right, reflecting more and more base, it also rises to higher pH values. What a buffer does is to make the curve rise or fall very slowly over a portion of the pH scale, called the "buffering range" of that buffer.

Key Learning Outcome 2.5 A tiny fraction of water molecules spontaneously ionize at any moment, forming H^+ and OH^-. The pH of a solution is a measure of its H^+ concentration. Low pH values indicate high H^+ concentrations (acidic solutions), and high pH values indicate low H^+ concentrations (basic solutions).

Using Radioactive Decay to Date the Iceman

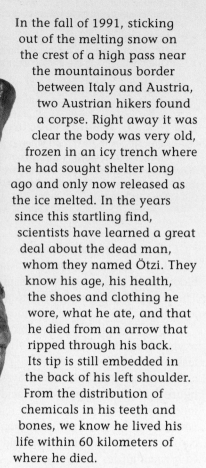

In the fall of 1991, sticking out of the melting snow on the crest of a high pass near the mountainous border between Italy and Austria, two Austrian hikers found a corpse. Right away it was clear the body was very old, frozen in an icy trench where he had sought shelter long ago and only now released as the ice melted. In the years since this startling find, scientists have learned a great deal about the dead man, whom they named Ötzi. They know his age, his health, the shoes and clothing he wore, what he ate, and that he died from an arrow that ripped through his back. Its tip is still embedded in the back of his left shoulder. From the distribution of chemicals in his teeth and bones, we know he lived his life within 60 kilometers of where he died.

How long ago did this Iceman die? Scientists answered this key question by measuring the degree of decay of the short-lived carbon isotope ^{14}C in Ötzi's body. This procedure is discussed earlier in this chapter (see **figure 2.8**). The graph to the right displays the radioactive decay curve of the carbon isotope carbon-14 (^{14}C); it takes 5,730 years for half of the ^{14}C present in a sample to decay to nitrogen-14 (^{14}N). When Ötzi's carbon isotopes were analyzed, researchers determined that the ratio of ^{14}C to ^{12}C (a **ratio** is the size of one variable relative to another), also written as the fraction $^{14}C/^{12}C$, in Ötzi's body is 0.435 of the fraction found in tissues of a person who has recently died.

1. **Applying Concepts**
   a. Variable. In the graph, what is the dependent variable?
   b. Proportion. What proportion (a **proportion** is the size of a variable relative to the whole) of the ^{14}C present in Ötzi's body when he died is still there today?

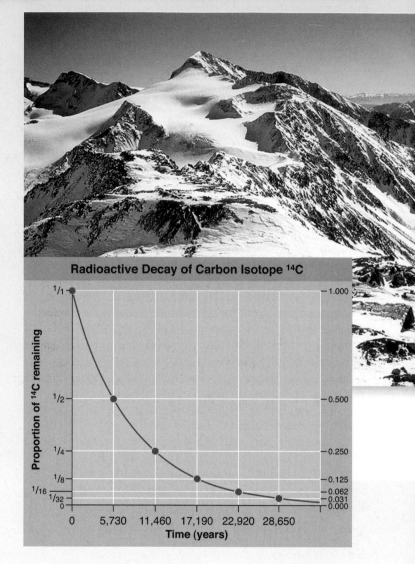

Radioactive Decay of Carbon Isotope ^{14}C

2. **Interpreting Data** Plot this proportion on the ^{14}C radioactive decay curve above. How many half-lives does this point represent?
3. **Making Inferences** If Ötzi were indeed a recent corpse, made to look old by the harsh weather conditions found on the high mountain pass, what would you expect the ratio of ^{14}C to ^{12}C to be, relative to that in your own body?
4. **Drawing Conclusions** How old are the remains of the Iceman Ötzi?
5. **Further Analysis**
   a. The radioactive iodine isotope ^{131}I decays at a half-life of eight days. Plotted on the graph above, would its radioactive decay curve be above or below that of ^{14}C?
   b. Scientists often employ the radioactive decay of isotope potassium-40 (^{40}K) into argon-40 (^{40}Ar) to date old material. ^{40}K has a half-life of 1.3 billion years. Would it be a better or poorer isotope than ^{14}C to use in dating Ötzi?

Some Simple Chemistry

2.1 Atoms

- An atom is the smallest particle that retains the chemical properties of its substance. Atoms, like the carbon atom shown here from **figure 2.2**, contain a core nucleus of protons and neutrons; electrons spin around the nucleus. The number of electrons equals the number of protons in an atom.

- The number of protons in an atom is called its atomic number. The mass that is contributed by the protons and neutrons is called the atom's mass number. All atoms that have the same atomic number are said to be the same element.

- Protons are positively charged particles and neutron particles carry no charge. Electrons are negatively charged particles that orbit around the nucleus at different energy levels. Electrons determine the chemical behavior of an atom because they are the subatomic particles that interact with other atoms.

- It takes energy to hold the electrons in their orbits; this energy of position is called potential energy. The amount of potential energy of an electron is based on its distance from the nucleus (**figure 2.3**).

- Most electron shells hold up to eight electrons and atoms will undergo chemical reactions in order to fill the outermost electron shell, either by gaining, losing, or sharing electrons (**figure 2.4**).

2.2 Ions and Isotopes

- Ions are atoms that have either gained one or more electrons (negative ions called anions) or lost one or more electrons (positive ions called cations) (**figure 2.5**).

- Isotopes are atoms that have the same number of protons but differing numbers of neutrons (**figure 2.6**). Isotopes tend to be unstable and break up into other elements through a process called radioactive decay. Some isotopes have applications in medicine (**figure 2.7**) and dating fossils (**figure 2.8**).

2.3 Molecules

- Molecules form when atoms are held together with energy. The force holding atoms together is called a chemical bond. There are three main types of chemical bonds.

- Ionic bonds form when ions of opposite charge are attracted to each other. Table salt is formed by ionic bonds between positive sodium ions and negative chloride ions (**integrated art, page 38**).

- Covalent bonds form when two atoms share electrons, attempting to fill empty electron orbitals (**integrated art, page 39**). Covalent bonds are stronger when more electrons are shared.

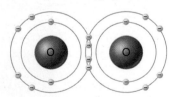

- The atoms in a polar molecule are held together by covalent bonds in which the shared electrons are unevenly distributed around their nuclei, giving the molecule a slightly positive end and a slightly negative end. Hydrogen bonds form when the positive end of one polar molecule is attracted to the negative end of another (**integrated art, page 40**).

Water: Cradle of Life

2.4 Hydrogen Bonds Give Water Unique Properties

- Water molecules are polar molecules that form hydrogen bonds with each other and with other polar molecules. Many of the physical properties of water are attributed to hydrogen bonding.

- Water molecules held together through hydrogen bonding are more difficult to separate, and as a result, a significant amount of heat energy is needed to pull the molecules apart. For this reason, water heats up slowly and holds it temperature longer.

- The hydrogen bonds that hold water molecules together become more stable at lower temperatures and as a result, they lock water molecules into place in solid crystal structures called ice, such as that shown here from **figure 2.9**.

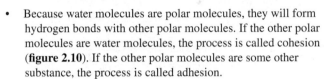

- In order for water to vaporize into a gas, a significant input of heat energy is needed to break the hydrogen bonds. This high heat of vaporization is a property of water used by our bodies in thermoregulation.

- Because water molecules are polar molecules, they will form hydrogen bonds with other polar molecules. If the other polar molecules are water molecules, the process is called cohesion (**figure 2.10**). If the other polar molecules are some other substance, the process is called adhesion.

- When water molecules form hydrogen bonds with other polar molecules, water molecules will tend to surround other polar molecules, forming a barrier around them called a hydration shell. Polar molecules are said to be hydrophilic and are water-soluble (**figure 2.11**). Nonpolar molecules do not form hydrogen bonds and will cluster together when placed in water. They are said to be hydrophobic and are water-insoluble.

2.5 Water Ionizes

- Water molecules dissociate forming negatively charged hydroxide ions (OH^-) and positively charged hydrogen ions (H^+). This property of water is significant because the concentration of hydrogen ions in a solution determines its pH.

- A solution with a higher hydrogen ion concentration is an acid with specific chemical properties, and a solution with a lower hydrogen ion concentration is a base with different chemical properties (**figure 2.12**).

- Substances called buffers control changes in pH by taking up or releasing H^+ into the solution and control pH within a range called the buffering range (**figure 2.13**).

- A buffer that functions in the human body is an acid-base pair consisting of carbonic acid and bicarbonate. This buffering process involves a series of two reversible reactions, one that generates hydrogen ions, reducing the pH, and the reverse reaction that takes up hydrogen ions from solution, increasing pH (**figure 2.14**).

- This carbonic acid-bicarbonate buffering system works in the blood to regulate blood pH (**figure 2.14**).

Test Your Understanding

1. The smallest particle into which a substance can be divided and still retain all of its chemical properties is
   a. matter.
   c. a molecule.
   b. an atom.
   d. mass.

2. An atom that has gained or lost one or more electrons is
   a. an isotope.
   c. an ion.
   b. a neutron.
   d. radioactive.

3. Atoms are held together by a force called a bond. The three types of bonds are
   a. positive, negative, and neutral.
   b. hydrophobic, hydrophilic, and van der Waals interactions.
   c. magnetic, electric, and radioactive.
   d. ionic, covalent, and hydrogen.

4. Carbon has four electrons in its outer electron shell, therefore
   a. it has a completely filled outer electron shell.
   b. it can form four single covalent bonds.
   c. it does not react with any other atom.
   d. it has a positive charge.

5. The partial separation of charge in the water molecule
   a. results from the electrons' greater attraction to the oxygen atom.
   b. means the molecule has a positive end and a negative end.
   c. indicates that the water molecule is a polar molecule.
   d. All of the above.

6. Water has some very unusual properties. These properties occur because of the
   a. hydrogen bonds between the individual water molecules.
   b. covalent bonds between the individual water molecules.
   c. hydrogen bonds within each individual water molecule.
   d. ionic bonds between the individual water molecules.

7. Which of the following properties are somehow related to the need for significant heat energy to break hydrogen bonds?
   a. cohesion and adhesion
   b. hydrophobic and hydrophilic
   c. heat storage and heat of vaporization
   d. ice formation and high polarity

8. The attraction of water molecules to other water molecules is called
   a. cohesion.
   b. capillary action.
   c. solubility.
   d. adhesion.

9. Water sometimes ionizes, a single molecule breaking apart into a hydrogen ion and a hydroxide ion. Other materials may dissociate in water, resulting in either (1) an increase of hydrogen ions or (2) a decrease of hydrogen ions in the solution. We call the results
   a. (1) acids and (2) bases.
   b. (1) bases and (2) acids.
   c. (1) neutral solutions and (2) neutronic solutions.
   d. (1) hydrogen solutions and (2) hydroxide solutions.

10. Which of the following is not true about buffers?
    a. A buffer takes up H^+ from the solution.
    b. A buffer keeps the pH relatively constant.
    c. A buffer stops water from ionizing.
    d. A buffer releases H^+ into the solution.

Apply Your Understanding

1. **Figure 2.8** Based on the figure, explain why it is difficult to use carbon-14 dating on 100-million-year-old dinosaurs.

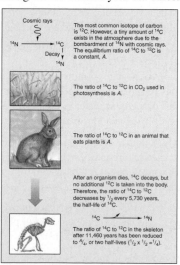

2. This figure shows an oxygen atom forming covalent bonds with two hydrogen atoms. A carbon atom, like oxygen, has two electrons in its innermost shell, but only four electrons in its outermost shell. Using this water molecule and page 39 as guides, draw a diagram showing how carbon forms covalent bonds with two oxygen atoms in a carbon dioxide (CO_2) molecule.

3. **Figure 2.10** This insect is walking on water. Why doesn't it sink?

Synthesize What You Have Learned

1. Imagine you were to find a fossilized animal bone containing one-eighth the amount of carbon-14 present in earth's atmosphere today. How long ago did the animal die?

2. Explain why it is possible for a bacterium to break the triple covalent bond of N_2 gas, whereas you and other animals cannot.

3. You are on a 10-day backpacking trip with a small group of friends. Yvonne has washed out a set of water bottles with bleach. Before she can rinse them, Carlos, hot and thirsty, picks one up and drinks it, drops it, and begins to choke. What is Carlos' problem, and what can you do for him?

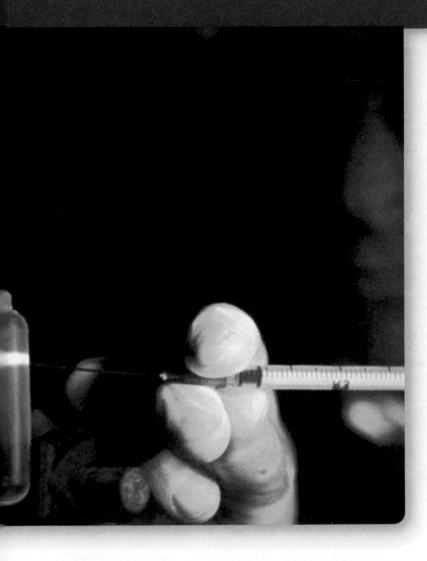

3
Molecules of Life

Using a syringe, this researcher is gently removing a glowing band of DNA, to be used in an experiment studying heredity. DNA, the carrier of an organism's genes, is one of several kinds of very large molecules found in all organisms. A molecule is a collection of tiny atoms linked together. Atoms are the basic chemical elements. Only a few are found in any significant numbers in living things. An essential atom for life is carbon, which can assemble into DNA and other very large molecules. Interacting with water, these long carbon chains twist about each other, or fold up into compact masses. Much of the chemistry that goes on in organisms, determining what each individual is like, depends on the actions of large folded molecules called proteins. By promoting particular chemical reactions, proteins trigger the production of structural materials like carbohydrates and energy storage molecules like lipids. Because DNA encodes the information needed to assemble each protein present in an organism, it is the library of life.

3.1 Polymers Are Built of Monomers

The bodies of organisms contain thousands of different kinds of molecules and atoms. Organisms obtain many of these molecules from their surroundings and from what they consume. You might be familiar with some of the substances listed on nutritional labels, such as the one shown in **figure 3.1**. But what do the words on these labels mean? Some of them are names of minerals (atoms, see chapter 2), such as calcium and iron (also discussed in chapters 22, 24, 33, and others). Others are vitamins, which are discussed in later chapters (see chapter 25). Still others are the subject of this chapter: large molecules that make up the bodies of organisms and are found in our food, such as proteins, carbohydrates (including sugars), and lipids (including, fats, trans fats, saturated fats, and cholesterol). These molecules, called **organic molecules,** are formed by living organisms and consist of a carbon-based core with special groups attached. These groups of atoms have special chemical properties and are referred to as *functional groups*. Functional groups tend to act as units during chemical reactions and to confer specific chemical properties on the molecules that possess them. Five principal functional groups are listed in **figure 3.2**; the last column indicates the types of organic molecules that contain these functional groups.

The bodies of organisms contain thousands of different kinds of organic molecules, but much of the body is made of just four kinds: *proteins, nucleic acids, carbohydrates,* and *lipids*. Called **macromolecules** because they can be very large, these four are the building materials of cells, the "bricks and mortar" that make up the bodies of cells and the machinery that runs within them.

The body's macromolecules are assembled by sticking smaller bits, called **monomers,** together, much as a train is built by linking railcars together. A molecule built up of long chains of similar subunits is called a **polymer.** Table 3.1 lists the monomers in the first column that make up the polymers that are the basis for many cellular structures.

Figure 3.1 What's in a nutritional label?

Fats, cholesterol, carbohydrates, sugars, and proteins are just some of the molecules found in popcorn and discussed in this chapter.

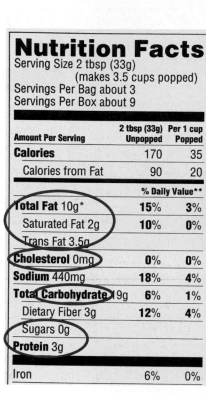

Nutrition Facts

Serving Size 2 tbsp (33g)
(makes 3.5 cups popped)
Servings Per Bag about 3
Servings Per Box about 9

Amount Per Serving	2 tbsp (33g) Unpopped	Per 1 cup Popped
Calories	170	35
Calories from Fat	90	20
	% Daily Value**	
Total Fat 10g*	15%	3%
Saturated Fat 2g	10%	0%
Trans Fat 3.5g		
Cholesterol 0mg	0%	0%
Sodium 440mg	18%	4%
Total Carbohydrate 9g	6%	1%
Dietary Fiber 3g	12%	4%
Sugars 0g		
Protein 3g		
Iron	6%	0%

Group	Structural Formula	Ball-and-Stick Model	Found In
Hydroxyl	—OH		Carbohydrates
Carbonyl	C=O		Lipids
Carboxyl	—C O OH		Proteins
Amino	—N H H		Proteins
Phosphate	O⁻ —O—P—O⁻ O		DNA, ATP

Figure 3.2 Five principal functional groups.

These functional groups can be transferred from one molecule to another and are common in organic molecules.

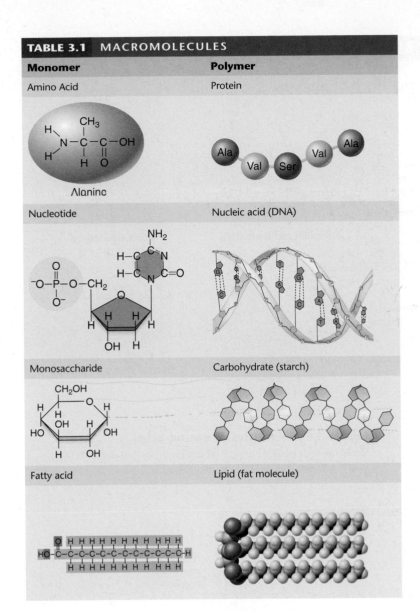

TABLE 3.1 MACROMOLECULES

Monomer	Polymer
Amino Acid	Protein
Alanine	Ala, Val, Ser, Val, Ala
Nucleotide	Nucleic acid (DNA)
Monosaccharide	Carbohydrate (starch)
Fatty acid	Lipid (fat molecule)

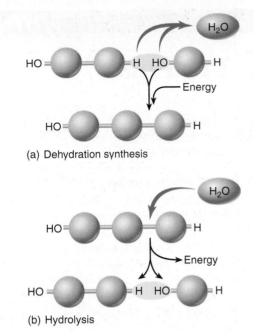

(a) Dehydration synthesis

(b) Hydrolysis

Figure 3.3 Dehydration and hydrolysis.

(*a*) Biological molecules are formed by linking subunits. The covalent bond between subunits is formed in dehydration synthesis, a process during which a water molecule is eliminated. (*b*) Breaking such a bond requires the addition of a water molecule, a reaction called hydrolysis.

Making (and Breaking) Macromolecules

The four different kinds of macromolecules (proteins, nucleic acids, carbohydrates, and lipids) are all put together in the same way: a covalent bond is formed between two subunits in which a hydroxyl group (OH) is removed from one subunit and a hydrogen (H) is removed from the other. This process (illustrated in **figure 3.3***a*) is called **dehydration synthesis** because, in effect, the removal of the OH and H groups (highlighted by the blue oval) constitutes removal of a molecule of water—the word *dehydration* means "taking away water." This process requires the help of a special class of proteins called **enzymes** to facilitate the positioning of the molecules so that the correct chemical bonds are stressed and broken. The process of tearing down a molecule, such as the protein or fat contained in food that is consumed, is essentially the reverse of dehydration synthesis: Instead of removing a water molecule, one is added. When a water molecule comes in, as shown in **figure 3.3***b*, a hydrogen becomes attached to one subunit and a hydroxyl to another, and the covalent bond is broken. The breaking up of a polymer in this way is called **hydrolysis.**

Key Learning Outcome 3.1 Macromolecules are formed by linking subunits together into long chains, removing a water molecule as each link is formed. In contrast, macromolecules are broken down into their subunits by hydrolysis reactions, the addition of water molecules.

3.2 Proteins

Complex macromolecules called **proteins** are a major group of biological macromolecules within the bodies of organisms. Perhaps the most important proteins are *enzymes,* which have the key role in cells of lowering the energy required to initiate particular chemical reactions. Other proteins play structural roles. Cartilage, bones, and tendons all contain a structural protein called collagen. Keratin, another structural protein, forms hair, the horns of a rhinoceros, and feathers. Still other proteins act as chemical messengers within the brain and throughout the body. **Figure 3.4** presents an overview of the wide-ranging functions of proteins.

Amino Acids

Despite their diverse functions, all proteins have the same basic structure: a long polymer chain made of subunits called amino acids. **Amino acids** are small molecules with a simple basic structure: a central carbon atom to which an amino group (—NH$_2$), a carboxyl group (—COOH), a hydrogen atom (H), and a functional group, designated "R," are bonded.

There are 20 common kinds of amino acids that differ from one another by the identity of their functional R group. The 20 amino acids are classified into four general groups, with representative amino acids shown in **figure 3.5** (their R groups are highlighted in white). Six of the amino acids are nonpolar, differing chiefly in size—the most bulky contain ring structures (like phenylalanine in the upper left), and amino acids containing them are called *aromatic.* Another six are polar but uncharged (like asparagine in the upper right), and these differ from one another in the strength of their polarity. Five more are polar and are capable of ionizing to a charged form (like aspartic acid in the lower left). The remaining three possess special chemical groups (like the white highlighted area of proline in the lower right) that are important in forming links between protein chains or in forming kinks in their shapes. The polarity of the R groups is important to the proper folding of the protein into its functional shape, which is discussed later.

Linking Amino Acids

An individual protein is made by linking specific amino acids together in a particular order, just as a word is made by linking specific letters of the alphabet together in a particular order. The covalent bond linking two amino acids together is called a **peptide bond** and forms by dehydration synthesis. Recall from section 3.1 that in dehydration synthesis, water is formed as a by-product of the reaction. You can see in **figure 3.6** that a water molecule is released as the peptide bond forms. Long chains of amino acids linked by peptide bonds are called **polypeptides.** Functional polypeptides are more commonly called proteins.

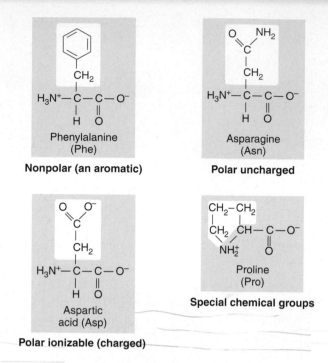

Figure 3.5 Examples of amino acids.
There are four general groups of amino acids that differ in their functional groups (highlighted in *white*).

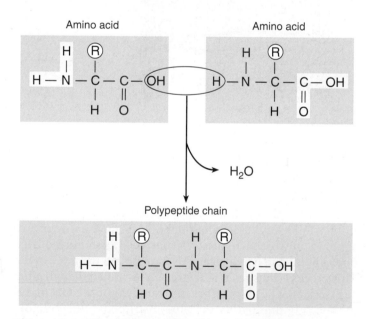

Figure 3.6 The formation of a peptide bond.
Every amino acid has the same basic structure, with an amino group (–NH$_2$) at one end and a carboxyl group (–COOH) at the other. The only variable is the functional, or "R," group. Amino acids are linked by dehydration synthesis to form peptide bonds. Chains of amino acids linked in this way are called polypeptides and are the basic structural components of proteins.

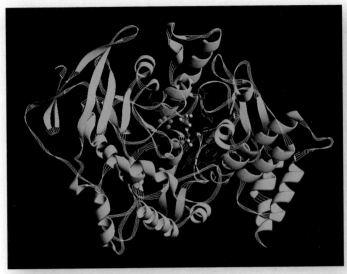

(a) **Enzymes:** Globular proteins called enzymes play a key role in many chemical reactions.

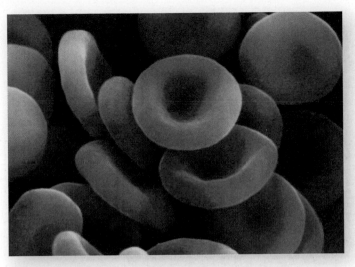

(b) **Transport proteins:** Red blood cells contain the protein hemoglobin, which transports oxygen and carbon dioxide in the body.

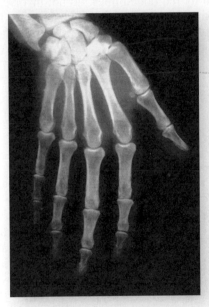

(c) **Structural proteins (collagen):** Collagen is present in bones, tendons, and cartilage.

(d) **Structural proteins (keratin):** Keratin forms hair, nails, feathers, and components of horns.

Figure 3.4 Some of the different types of proteins.

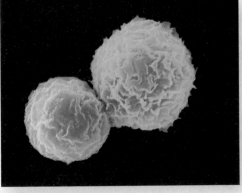

(e) **Defensive proteins:** White blood cells destroy cells without the proper identity proteins and make antibody proteins that attack invaders.

(f) **Contractile proteins:** Proteins called actin and myosin are present in muscles.

Protein Structure

Some proteins form long, thin fibers, whereas others are globular, their strands coiled up and folded back on themselves. The shape of a protein is very important because it determines the protein's function. There are four general levels of protein structure: primary, secondary, tertiary, and quaternary; all are ultimately determined by the sequence of amino acids.

Primary Structure The sequence of amino acids of a polypeptide chain is termed the polypeptide's **primary structure.** The amino acids are linked together by peptide bonds, forming long chains like a "beaded strand." The primary structure of a protein, the sequence of its amino acids, determines all other levels of protein structure. Because amino acids can be assembled in any sequence, a great diversity of proteins is possible.

Primary structure

Amino acids

Secondary structure

β-pleated sheet

α-helix

Secondary Structure Hydrogen bonds forming between different parts of the polypeptide chain stabilize the folding of the polypeptide. As you can see, these stabilizing hydrogen bonds, indicated by red dotted lines, do not involve the R groups themselves, but rather the polypeptide backbone. This initial folding is called the **secondary structure** of a protein. Hydrogen bonding within this secondary structure can fold the polypeptide into coils, called α-helices, and sheets, called β-pleated sheets.

Tertiary structure

Can see coils + sheets

Tertiary Structure Because some of the amino acids are nonpolar, a polypeptide chain folds up in water, which is very polar, pushing nonpolar amino acid functional groups from the watery environment. The final three-dimensional shape, or **tertiary structure,** of the protein, folded and twisted in the case of a globular molecule, is determined by exactly where in a polypeptide chain the nonpolar amino acids occur.

Quaternary structure

more complex 3D

Quaternary Structure When a protein is composed of more than one polypeptide chain, the spatial arrangement of the several component chains is called the **quaternary structure** of the protein. For example, four subunits make up the quaternary structure of the protein hemoglobin.

How Proteins Fold into Their Functional Shape

The polar nature of the watery environment in the cell influences how the polypeptide folds into the functional protein. A protein is folded in such a way that allows it to carry out its function.

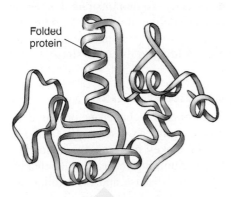

Folded protein

If the polar nature of the protein's environment changes by either increasing temperature or lowering pH, both of which alter hydrogen bonding, the protein may unfold, as in the lower right of the figure. When this happens the protein is said to be **denatured.**

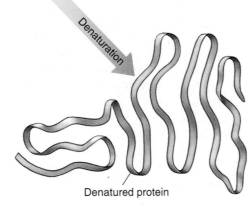

Denaturation

Denatured protein

When the polar nature of the solvent is reestablished, some proteins may spontaneously refold. When proteins are denatured, they usually lose their ability to function properly. That is the rationale behind traditional methods of salt-curing and pickling food. Prior to the ready availability of refrigerators and freezers, the only practical way to keep microorganisms from growing in food was to keep the food in a solution containing a high concentration of salt or vinegar, which denatured proteins in the microorganisms and kept them from growing on the food.

Protein Structure Determines Function

The structure of a protein determines its function, and because the primary structure of a protein, its sequence of amino acids, determines how the protein folds into its functional shape, a change in the identity of even one amino acid can have profound effects on a protein's ability to function properly.

Enzymes are globular proteins that have three-dimensional shapes. For enzymes to function properly, they need to fold correctly. Enzymes have grooves or depressions that precisely fit a particular sugar or other chemical (like the red molecule binding to the enzyme to the right); once in the groove, the chemical is encouraged to undergo a reaction—often, one of its chemical bonds is stressed as the chemical is bent by the enzyme, like a foot in a flexing shoe. This process of enhancing chemical reactions is called **catalysis,** and proteins are the catalytic agents of cells, determining what chemical processes take place and where and when.

Many structural proteins form long cables that have architectural roles in cells, providing strength and determining shape. As you will discover in chapter 4, cells contain a network of protein cables that maintain the shape of the cell and function in transporting materials throughout the cell (**figure 3.7**). Contractile proteins function in muscle contraction, which is the shortening of a muscle. A muscle shortens when two proteins that are anchored on opposite ends of a muscle fiber slide past each other, bringing the ends of the fiber closer together (discussed in more detail in chapter 22).

Active-site cleft

Figure 3.7 Protein structure determines function.
Fluorescently-labeled structural proteins within a cell.

Chaperone Proteins

How does a protein fold into a specific shape? As just discussed, nonpolar amino acids play a key role. Until recently, investigators thought that newly made proteins fold spontaneously as hydrophobic interactions with water shove nonpolar amino acids into the protein interior. We now know this is too simple a view. Proteins can fold in so many different ways that trial and error would simply take too long. In addition, as the open chain folds its way toward its final form, nonpolar "sticky" interior portions are exposed during intermediate stages. If these intermediate forms are placed in a test tube in the same protein environment that occurs in a cell, they stick to other unwanted protein partners, forming a gluey mess.

How do cells avoid this? A vital clue came in studies of unusual mutations (changes in DNA) that prevented viruses from replicating in bacterial cells—it turned out the virus proteins could not fold properly! Further study revealed that normal cells contain special proteins called **chaperone proteins** that help new proteins fold correctly. When the bacterial gene encoding its chaperone protein is disabled by mutation, the bacteria die, clogged with lumps of incorrectly folded proteins. Fully 30% of the bacteria's proteins fail to fold into the right shape.

Molecular biologists have now identified more than 17 kinds of proteins that act as molecular chaperones. Many are heat shock proteins, produced in greater amounts if a cell is exposed to elevated temperature; high temperatures cause proteins to unfold, and heat shock chaperone proteins help the cell's proteins refold.

To understand how a chaperone works, examine **figure 3.8** closely. The misfolded protein enters inside the chaperone. There, in a way not clearly understood, the visiting protein is induced to unfold, and then refold again, before it leaves. You can see in the third panel of the diagram the protein has unfolded into a long polypeptide chain. In the fourth panel, the polypeptide chain has then refolded into a different shape. The chaperone protein has in this way "rescued" a protein that was caught in a wrongly folded state, and given it another chance to fold correctly. To demonstrate this rescue capability, investigators "fed" a deliberately misfolded protein malate dehydrogenase to chaperone proteins; the malate dehydrogenase was rescued, refolding to its active shape.

Protein Folding and Disease

There are tantalizing suggestions that chaperone protein deficiencies may play a role in Alzheimer's disease. By failing to facilitate the intricate folding of key proteins, the deficiency leads to the amyloid protein clumping in brain cells characteristic of the disease. Mad cow disease and the similar human disorder called variant Creutzfeldt-Jacob disease are both caused by misfolded brain proteins called **prions.** The misfolded prions induce other brain prion proteins to misfold in turn, creating a chain reaction of misfolding that kills ever more brain cells, leading to progressive loss of brain function and eventual death.

> **Key Learning Outcome 3.2** Proteins are made up of chains of amino acids that fold into complex shapes. The sequence of its amino acids determines a protein's function. Chaperone proteins help newly produced proteins to fold properly.

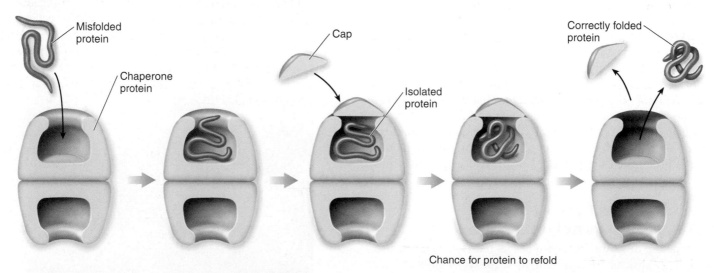

Chance for protein to refold

Figure 3.8 How one type of chaperone protein works.

This barrel-shaped chaperone protein is a heat shock protein, produced in elevated amounts at high temperatures. An incorrectly folded protein enters one chamber of the barrel, and a cap seals the chamber and confines the protein. The isolated protein is now prevented from aggregating with other misfolded proteins, and it has a chance to refold properly. After a short time, the protein is ejected, folded or unfolded, and the cycle can repeat itself.

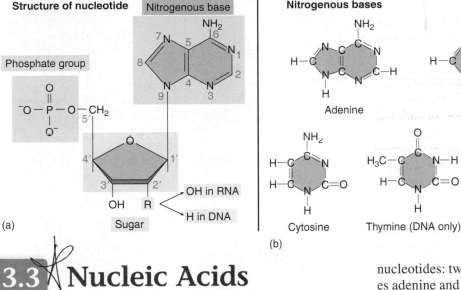

Structure of nucleotide

Nitrogenous base

Phosphate group

Sugar

OH in RNA
H in DNA

(a)

Nitrogenous bases

Adenine

Guanine

Cytosine

Thymine (DNA only)

Uracil (RNA only)

(b)

Figure 3.9
The structure of a nucleotide.

(a) Nucleotides are composed of three parts: a five-carbon sugar, a phosphate group, and an organic nitrogenous base. (b) The nitrogenous base can be one of five.

3.3 Nucleic Acids

Very long polymers called **nucleic acids** serve as the information storage devices of cells, just as CDs or hard drives store the information that computers use. Nucleic acids are long polymers of repeating subunits called **nucleotides.** Each nucleotide is a complex organic molecule composed of three parts shown in **figure 3.9a:** a five-carbon sugar (in blue), a phosphate group (in yellow, PO$_4$), and an organic nitrogen-containing base (in orange). In the formation of a nucleic acid, the individual sugars with their attached nitrogenous bases are linked in a line by the phosphate groups in very long **polynucleotide chains** (shown to the right).

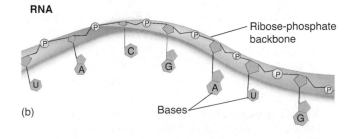

How does the long, chainlike structure of a nucleic acid permit it to store the information necessary to specify what a human being is like? If nucleic acids were simply a monotonous repeating polymer, it could not encode the message of life. Imagine trying to write a story using only the letter *E* and no spaces or punctuation. All you could ever say is "EEEEEEE. . . ." You need more than one letter to communicate—the English alphabet uses 26 letters. Nucleic acids can encode information because they contain more than one kind of nucleotide. There are five different kinds of

nucleotides: two larger ones that contain the nitrogenous bases adenine and guanine (shown in the top row of **figure 3.9b**), and three smaller ones that contain the nitrogenous bases cytosine, thymine, and uracil (in the bottom row). Nucleic acids encode information by varying the identity of the nucleotide at each position in the polymer.

DNA and RNA

Nucleic acids come in two varieties, **deoxyribonucleic acid (DNA)** and **ribonucleic acid (RNA),** both polymers of nucleotides with some differences. RNA is similar to DNA, but with two major chemical differences. First, RNA molecules contain the sugar ribose, in which the 2' carbon (this is the carbon labeled 2' in **figure 3.9a**) is bonded to a hydroxyl group (—OH). In DNA, this hydroxyl group is replaced with a hydrogen atom. Second, RNA molecules do not contain the thymine nucleotide; they contain uracil instead. Structurally, RNA is also different. RNA is a long, single strand of nucleotides and is used by cells in making proteins using genetic instructions encoded within DNA. The sequence of nucleotides in DNA determines the order of amino acids in the primary structure of the protein. DNA consists of *two* polynucleotide chains wound around each other in a **double helix,** like strands of a pearl necklace twisted together. You can see this difference in structure by comparing the blue double-stranded DNA molecule in **figure 3.10** with the green single-stranded RNA molecule.

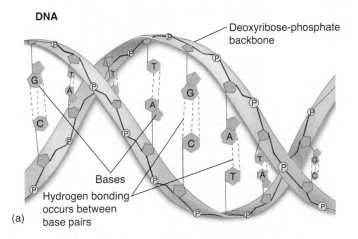

DNA

Deoxyribose-phosphate backbone

Bases
Hydrogen bonding occurs between base pairs

(a)

RNA

Ribose-phosphate backbone

Bases

(b)

Figure 3.10 How DNA structure differs from RNA.

(a) DNA contains two polynucleotide strands wrapped around each other, while (b) RNA is single-stranded.

The Double Helix

Why is DNA a *double* helix? When scientists looked carefully at the structure of the DNA double helix, they found that the bases of each chain point inward toward the other (like the DNA strands shown in **figure 3.11**). The bases of the two chains are linked in the middle of the molecule by hydrogen bonds (the dotted lines between the two strands), like two columns of people holding hands across. The key to understanding why DNA is a double helix is revealed by looking at the bases: *only two base pairs are possible.* Because the distance between the two strands is consistent, this suggests that two big bases cannot pair together—the combination is simply too bulky to fit; similarly, two little ones cannot pair, as they pinch the helix inward too much. To form a double helix, it is necessary to pair a big base with a little one. *In every DNA double helix, adenine (A) pairs with thymine (T) and guanine (G) pairs with cytosine (C).* The reason A doesn't pair with C and G doesn't pair with T is that these base pairs cannot form proper hydrogen bonds—the electron-sharing atoms are not aligned with each other.

A and C cannot properly align to form hydrogen bonds.

G and T cannot properly align to form hydrogen bonds.

A and T can align to form two hydrogen bonds.

G and C can align to form three hydrogen bonds.

The simple A–T, G–C base pairs within the DNA double helix allow the cell to copy the information in a very simple way. It just unzips the helix and adds the complementary bases to each new strand! That is the great advantage of a double helix—it actually contains two copies of the information, one the mirror image of the other. If the sequence of one chain is ATTGCAT, the sequence of its partner in the double helix *must* be TAACGTA. The fidelity with which hereditary information is passed from one generation to the next is a direct result of this simple double-entry bookkeeping, which makes accurate copying of the genetic message possible.

> **Key Learning Outcome 3.3** Nucleic acids like DNA are composed of long chains of nucleotides. The sequence of the nucleotides specifies the amino acid sequence of proteins.

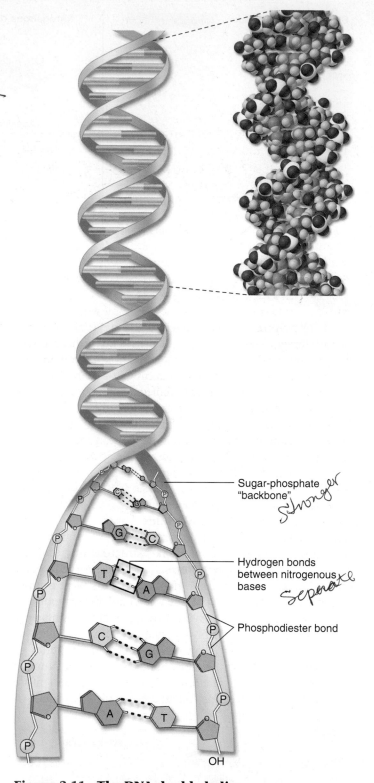

Figure 3.11 The DNA double helix.
The DNA molecule is composed of two polynucleotide chains twisted together to form a double helix. The two chains of the double helix are joined by hydrogen bonds between the A–T and G–C base pairs. The section of DNA on the upper right is a space-filling model of DNA, where atoms are indicated by colored balls.

Sugar-phosphate "backbone"

Hydrogen bonds between nitrogenous bases

Phosphodiester bond

A **Closer** Look

Discovering the Structure of DNA

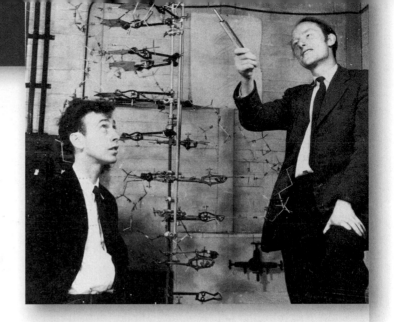

By the middle of the last century, biologists were increasingly sure that DNA was the molecule that stored the hereditary information, but investigators were puzzled over how such a seemingly simple molecule could carry out such a complex function.

A key observation was made by chemist Erwin Chargaff shortly after the end of the Second World War. He noted that in DNA molecules, the amount of adenine, A, always equals the amount of thymine, T, and the amount of guanine, G, always equals the amount of cytosine, C. This observation (A=T, G=C), known as Chargaff's rule, strongly suggested that DNA has a regular structure, but did not reveal what it was.

The significance of the regularities pointed out by Chargaff were not immediately obvious, but they became clear when a young British chemist, Rosalind Franklin, carried out an X-ray diffraction analysis of DNA. In X-ray diffraction, a molecule is bombarded with a beam of X rays. When individual rays encounter atoms, their path is bent or diffracted, and the diffraction pattern is recorded on photographic film. The pattern that resulted using DNA resembled the ripples created by tossing a rock into a smooth lake. When carefully analyzed, a molecule's pattern can reveal information about the three-dimensional structure of the molecule.

X-ray diffraction works best on substances that can be prepared as perfectly regular crystalline arrays. However at the time of Franklin's analysis, it was impossible to obtain true crystals of natural DNA, so she had to use DNA in the form of fibers. Franklin worked in the same laboratory as Oxford biochemist Maurice Wilkins, who was able to prepare more uniformly oriented DNA fibers than anyone had previously. Using these fibers, Franklin succeeded in obtaining crude diffraction information on natural DNA. The diffraction patterns she obtained seemed to suggest that the DNA molecule had the shape of a coiled spring or corkscrew, a form called a helix.

Learning informally of Franklin's results before they were published in 1953, James Watson and Francis Crick, two young investigators at Cambridge University, quickly worked out a likely structure for the DNA molecule, which we now know was substantially correct. The key to their understanding the structure of DNA was Watson and Crick's insight that each DNA molecule is actually made up of two chains of nucleotides that are intertwined—a double helix.

In Watson and Crick's historic 1953 model (in the photograph, Watson is peering at the model as Crick points), each DNA molecule is composed of two complementary polynucleotide strands that form a double helix, with the bases extending into the interior of the helix. An analogy that is often made is to a spiral staircase where the two strands of the double helix are the handrails on the staircase.

What holds the two strands together? Watson and Crick proposed that the bases from opposite strands can form hydrogen bonds with each other to join the two complementary strands. Although each individual base pair is of low energy, the sum of many base pairs has enough energy that the molecule is very stable. To return to our spiral staircase analogy, where the backbone is the handrails, the base pairs are the stairs themselves.

Because of differences in size and position of particular atoms, only two hydrogen bonding pairs are possible in such a double helix: adenine (A) can form hydrogen bonds with thymine (T), and guanine (G) can form hydrogen bonds with cytosine (C). The Watson-Crick model thus, in a very direct and simple way, explained what had until then been one of the great mysteries of DNA, Chargaff's observation that adenine and thymine always occur in the same proportions in any DNA molecule, as do guanine and cytosine.

At the heart of the Watson-Crick model of DNA is a seemingly simple concept with some profound implications. Because only two base pairs are possible, if we know the sequence of one strand, we automatically know the sequence of the other strand; wherever there is an A in one strand, there must be a T in the other, and wherever there is a G in one strand, there must be a C in the other. This concept of a mirror-image relationship is called *complementarity*. You see the importance: If more than two base pairs were possible, then the sequence of one DNA strand would not allow us to know for sure the sequence of the other. It is this fundamental insight that has made Watson and Crick's discovery one of the most profound of the 20th century.

Watson and Crick continued on in the area of DNA research, but Franklin's career was cut short by her untimely death due to cancer at the age of 37.

3.4 Carbohydrates

Polymers called **carbohydrates** make up the structural framework of cells and play a critical role in energy storage. A carbohydrate is any molecule that contains carbon, hydrogen, and oxygen in the ratio 1:2:1. Some carbohydrates are small monomers or dimers and are called **simple carbohydrates.** Others are long polymers and are called **complex carbohydrates.** Because they contain many carbon-hydrogen (C–H) bonds, carbohydrates are well-suited for energy storage. Such C–H bonds are the ones most often broken by organisms to obtain energy. **Table 3.2** on the facing page shows some examples of carbohydrates.

Simple Carbohydrates

The simplest carbohydrates are the *simple sugars* or **monosaccharides** (from the Greek *monos,* single, and *saccharon,* sweet). These molecules consist of one subunit. For example, glucose, the sugar that carries energy to the cells of your body, is made of six carbons and has the chemical formula $C_6H_{12}O_6$. A molecule of glucose is pictured in several ways in **figure 3.12**. The long chain of carbon atoms at the top of the figure is its formal chemical structure. When placed in water, the chain folds into the ring structure shown on the lower right. The individual atoms are depicted in the "3-D" space-filling model you see in the lower left. Another type of simple carbohydrate is a **disaccharide,** which forms when two monosaccharides link together through a dehydration reaction. In **figure 3.13** you can see how the disaccharide sucrose (table sugar) is made by linking two six-carbon sugars together, a glucose (orange) and a fructose (green).

Complex Carbohydrates

Organisms store their metabolic energy by converting sugars, which are soluble, into insoluble forms that can be deposited in specific storage areas in the body. This trick is achieved by linking the sugars together into long polymer chains called **polysaccharides.** Plants and animals store energy in polysaccharides formed from glucose. The glucose polysaccharide that plants use to store energy is called **starch**—that is why potatoes are referred to as "starchy" food. In animals, energy is stored in **glycogen,** a highly insoluble macromolecule formed of glucose polysaccharides that are very long and highly branched. Plants and animals also use glucose chains as building materials, linking the subunits together in different orientations not recognized by most enzymes. These structural polysaccharides are **chitin,** in animals, and **cellulose,** in plants. The cellulose deposited in the cell walls of plant cells, like the cellulose strand shown in **figure 3.14**, cannot be digested by humans and makes up the fiber in our diets.

> **Key Learning Outcome 3.4** Carbohydrates are molecules made of C, H, and O atoms. As sugars they store energy in C–H bonds, and as long polysaccharide chains they can provide structural support.

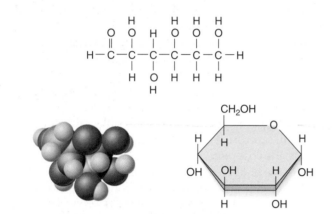

Figure 3.12 The structure of glucose.
Glucose is a monosaccharide and consists of a linear six-carbon molecule that forms a ring when added to water. This illustration shows three ways glucose can be represented diagrammatically.

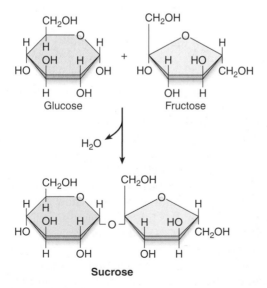

Figure 3.13 Formation of sucrose.
The disaccharide sucrose is formed from glucose and fructose in a dehydration reaction.

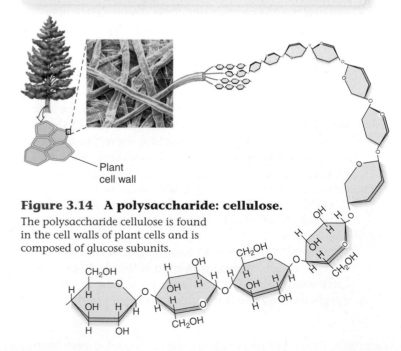

Figure 3.14 A polysaccharide: cellulose.
The polysaccharide cellulose is found in the cell walls of plant cells and is composed of glucose subunits.

TABLE 3.2   CARBOHYDRATES AND THEIR FUNCTIONS

Carbohydrate	Example	Description
Transport Disaccharides		
	Lactose	Glucose is transported within some organisms as a disaccharide. In this form, it is less readily metabolized because the normal glucose-utilizing enzymes of the organism cannot break the bond linking the two monosaccharide subunits. One type of disaccharide is called lactose. Many mammals supply energy to their young in the form of lactose, which is found in milk.
	Sucrose	Another transport disaccharide is sucrose. Many plants transport glucose throughout the plant in the form of sucrose, which is harvested from sugarcane to make sugar.
Storage Polysaccharides		
	Starch	Organisms store energy in long chains of glucose molecules called polysaccharides. The chains tend to coil up in water, making them insoluble and ideal for storage. The storage polysaccharides found in plants are called starches, which can be branched or unbranched. Starch is found in potatoes and in grains, such as corn and wheat.
	Glycogen	In animals, glucose is stored as glycogen. Glycogen is similar to starch in that it consists of long chains of glucose that coil up in water and are insoluble. But glycogen chains are much longer and highly branched. Glycogen can be stored in muscles and the liver.
Structural Polysaccharides		
	Cellulose	Cellulose is a structural polysaccharide found in the cell walls of plants; its glucose subunits are joined in a way that cannot be broken down readily. Cleavage of the links between the glucose subunits in cellulose requires an enzyme most organisms lack. Some animals, such as cows, are able to digest cellulose by means of bacteria and protists they harbor in their digestive tract, which provide the necessary enzymes.
	Chitin	Chitin is a type of structural polysaccharide found in the external skeletons of many invertebrates, including insects and crustaceans, and in the cell walls of fungi. Chitin is a modified form of cellulose with a nitrogen group added to the glucose units. When cross-linked by proteins, it forms a tough, resistant surface material.

3.5 Lipids

For long-term energy storage, organisms usually convert glucose into fats, another kind of storage molecule that contains more energy-rich C–H bonds than carbohydrates. Fats and all other biological molecules that are not soluble in water but soluble in oil are called **lipids**. Lipids are insoluble in water not because they are long chains like starches but rather because they are nonpolar. In water, fat molecules cluster together because they cannot form hydrogen bonds with water molecules. This is why oil forms into a layer on top of water when the two substances are mixed.

Fats

Fat molecules are lipids composed of two kinds of subunits: fatty acids (the gray boxed structures in **figure 3.15a**) and glycerol (the orange boxed structure). A **fatty acid** is a long chain of carbon and hydrogen atoms (called a hydrocarbon) ending in a carboxyl (—COOH) group. The three carbons of glycerol form the backbone to which three fatty acids are attached in the dehydration reaction that forms the fat molecule. That is why the carboxyl groups of the fatty acids in **figure 3.15a** are not apparent, because they are involved in bonds with glycerol. Because it contains three fatty acids, the resulting fat molecule is sometimes called a *triacylglycerol*, or **triglyceride.**

Fatty acids with all internal carbon atoms forming covalent bonds with two hydrogen atoms contain the maximum number of hydrogen atoms. Fats composed of these fatty acids are said to be **saturated** (**figure 3.15b**). Saturated fats are solid at room temperature. On the other hand, fats composed of fatty acids that have double bonds between one or more pairs of carbon atoms contain fewer than the maximum number of hydrogen atoms and are called **unsaturated** (**figure 3.15c**). The double bonds create kinks in the fatty acid tails, which usually makes the unsaturated fats liquid at room temperature. Many plant fats are unsaturated and occur as oils. Animal fats, in contrast, are often saturated and occur as hard fats. In some cases, unsaturated fats in food products may be artificially *hydrogenated* (industrial addition of hydrogens) to make them more saturated, extending the shelf life of these products. In some cases, the hydrogenation creates *trans fats,* a type of unsaturated fat in which some of the double bonds are less kinked than those in naturally occurring unsaturated fats. Eating trans fats and saturated fats may increase the risk of heart disease.

Other Types of Lipids

Organisms also contain other types of lipids that play many roles in cells in addition to energy storage. The male and female sex hormones testosterone and estradiol are lipids called *steroids*. Unlike the structure of fats shown in **figure 3.15**, the structure of steroids consists of multiple ring structures and looks somewhat like a section of chicken wire. Other important biological lipids include phospholipids, cholesterol (also a steroid), rubber, waxes, and pigments, such as the chlorophyll that makes plants green and the retinal that your eyes use to detect light (**figure 3.16**).

If you look at **figure 3.17**, you will see two types of lipids that play key roles in the membranes that encase the cells of your body: *phospholipid* molecules and *cholesterol.* Phospholipids are modified triacylglycerol molecules, where one of the carbons in the glycerol backbone bonds to a

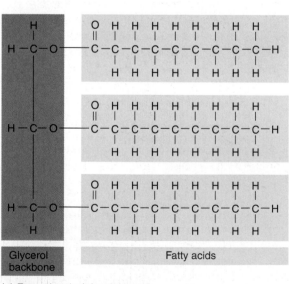

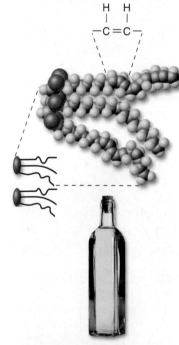

(a) Fat molecule (triacylglycerol)

(b) Hard fat (saturated): Fatty acids with single bonds between all carbon pairs

(c) Oil (unsaturated): Fatty acids that contain double bonds between one or more pairs of carbon atoms

Figure 3.15 Saturated and unsaturated fats.

(a) Fat molecules each contain a three-carbon glycerol to which is attached three fatty acid tails. (b) Most animal fats are "saturated" (every carbon atom carries the maximum load of hydrogens). Their fatty acid chains fit closely together, and these triacylglycerols form immobile arrays called hard fats. (c) Most plant fats are unsaturated, which prevents close association between triacylglycerols and produces oils.

Biology and *Staying Healthy*

Anabolic Steroids in Sports

Among the most notorious of lipids in recent years has been the class of synthetic hormones known as anabolic steroids. Since the 1950s some athletes have been taking these chemicals to build muscle and so boost athletic performance. Both because of the intrinsic unfairness of this and because of health risks, the use of anabolic steroids has been banned in sports for decades. Controversy over their use in professional baseball has recently returned anabolic steroids to the nation's front pages.

Anabolic steroids were developed in the 1930s to treat hypogonadism, a condition in which the male testes do not produce sufficient amounts of the hormone testosterone for normal growth and sexual development. Scientists soon discovered that by slightly altering the chemical structure of testosterone, they could produce synthetic versions that facilitated the growth of skeletal muscle in laboratory animals. The word "anabolic" means growing or building. Further tweaking reduced the added impact of these new chemicals on sexual development. More than 100 different anabolic steroids have been developed, most of which have to be injected to be effective. All require a prescription to be used legally in the United States, and all are banned in professional, college, and high school sports.

Another way to increase the body's level of testosterone is to use a chemical that is not itself anabolic but one that the body converts to testosterone. One such chemical is 4-androstenedione, more commonly called "andro." It was first developed in the 1970s by East German scientists to try to enhance their athletes' Olympic performances. Because andro does not have the same side effects as anabolic steroids, it was legally available until 2004. It was used by Mark McGwire, but it is now banned in all sports, and possession of andro is a federal crime.

Anabolic steroids work by signalling muscle cells to make more protein. They bind to special "androgenic receptor" proteins within the cells of muscle tissue. Like jabbing these proteins with a poker, the binding prods the receptors into action, causing them to activate genes on the cell's chromosomes that produce muscle tissue proteins, triggering an increase in protein synthesis. At the same time, the anabolic steroid molecules bind to so-called "cortisol receptor" proteins in the cell, preventing these receptors from doing their job of causing protein breakdown, the muscle cell's way of suppressing inflammation and promoting the use of proteins for fuel during exercise. By increasing protein production and inhibiting the breakdown of proteins in muscle cells after workouts, anabolic steroids significantly increase the mass of an athlete's muscle tissue.

Home run slugger Barry Bonds was involved in a steroid controversy in 2006.

If the only effect of anabolic steroids on your body was to enhance your athletic performance by increasing your muscle mass, using them would still be wrong, for one very simple and important reason: fairness. To gain advantage in competition by concealed use of anabolic steroids—"doping"—is simply cheating. That is why these drugs are banned in sports.

The use of anabolic steroids by athletes and others is not only wrong, but also illegal, because increased muscle mass is not the only effect of using these chemicals. Among adolescents, anabolic steroids can also lead to premature termination of the adolescent growth spurt, so that for the rest of their lives, users remain shorter than they would have been without the drugs. Adolescents and adults are also affected by steroids in the following ways. Anabolic steroids can lead to potentially fatal liver cysts and liver cancer (the liver is the organ of the body that attempts to detoxify the blood), cholesterol changes and hypertension (both of which can promote heart attack and stroke), and acne. Other signs of steroid use in men include reduced size of testicles, balding, and development of breasts. In women, signs include the growth of facial hair, lowering of the voice, and cessation of menstruation.

In the fall of 2003, athletic organizations learned that some athletes were using a new performance-enhancing anabolic steroid undetectable by standard antidoping tests, tetrahydrogestrinone (THG). The use of THG was only discovered because an anonymous coach sent a spent syringe to U.S. antidoping officials. THG's chemical structure is similar to gestrinone, a drug used to treat a form of pelvic inflammation, and can be made from it by simply adding four hydrogen atoms, an easy chemical task. THG tends to break down when prepared for analysis by standard means, which explains why antidoping tests had failed to detect it. New urine tests for THG that were developed in 2004 have been used to catch several well-known sports figures. Olympic athlete Marion Jones and baseball sluggers Rafael Palmeiro, Barry Bonds, and Mark McGwire have all been involved in steroid use.

Author's Corner

My Battle with Cholesterol

By a cruel twist of fate I was born loving steak: a 1 1/2 inch Porterhouse is my idea of culinary perfection. I love French fries, too, and more than anything else, Big Macs. What is cruel about my love affair with Big Macs is that a dozen years ago my doctors informed me I have high levels of cholesterol in my blood, well over the upper recommended level of 200. Because all this cholesterol in my blood tends to deposit itself along the insides of arteries, this puts me at high risk of heart attack if I don't do something about it.

Thus began my decade-long battle with cholesterol. My old friends the Big Mac and French fries pretty much became history, and steak a much more casual acquaintance. My wife and three daughters, knowing my weakness of character when it comes to things bovine, started to do more of the grocery shopping, buying lots of chicken.

Didn't do any good. After two years of Big Mac-less days and steakless nights, my cholesterol levels rose higher than before. When measured in October of 1999, my total cholesterol was 274, the highest it had ever been. When you consider that every 1% increase in level over 200 increases my risk of heart disease 2%, this report was frightening.

I called my brothers to report the bad news, only to find that they too wage the same battle. The problem, it appears, is that I and my brothers inherited a defective steak-hating gene. We suffer from hypercholesterolemia. This mouthful of a word simply means "inherited high cholesterol." People inheriting a copy of the defective gene from one of their parents have elevated levels of cholesterol in their blood serum that dietary restriction (taking away my steaks) cannot reduce.

My frustrating on-going battle with cholesterol is not unique. Over half a million Americans suffer from hypercholesterolemia, the most frequent of all gene disorders. None of my hundreds of thousands of brothers and sisters in this war, fellow victims of this genetic quirk, are able to reduce cholesterol levels by diet and exercise alone. They, and I, must rely on modern medicine to defeat our Mendelian enemy.

Even a quick look at the biology of cholesterol suggests a likely line of attack. Cholesterol is a special kind of fat that your body uses to control the flexibility of its cell membranes and to insulate nerves. Your liver makes about 80% of the cholesterol in your body, up to 800 mg a day. You take in the rest when you eat steak and other fatty foods.

You can see why my attempting to lower total cholesterol by reducing dietary input of fat (steaks, peanut butter, fried food, chocolate, ice cream—all the good stuff)

didn't get the job done. Dietary cholesterol is only 20% of my body's total. Most is manufactured by my liver, and because of hypercholesterolemia, my liver is churning out cholesterol twice as fast as a normal liver does. To solve the problem, I need to put the brakes on my liver's cholesterol-making frenzy.

In the 1980s researchers determined that the rate-limiting step in the liver's manufacture of cholesterol occurs early in the process, when a six-carbon molecule called mevalonate is converted to something called hydroxy methyl glutaryl CoA (HMG-CoA, for short). This reaction is carried out by an enzyme called HMG-CoA reductase. If we want to slow cholesterol production, that's our target.

In the last 20 years a series of drugs called statins have been developed that reduce levels of cholesterol by inhibiting HMG-CoA reductase. Among them are fluvastatin (Lescol), lovastatin (Mevacor), simvastatin (Zocor), and pravastatin (Pravachol).

Atorvastatin (Lipitor), introduced in 1997, is a particularly potent inhibitor of HMG-CoA reductase with fewer side effects than other statins. Lipitor is a blockbuster drug. It generated $4.4 billion in sales its first year, and became the top-selling global drug by 2005, with sales in excess of $10 billion.

It is upon Lipitor that I have banked my future. I started taking it in 1999, a little white pill every morning before breakfast. It is likely I shall be taking it the rest of my life.

How well did it work? Very well indeed. Within three months of starting daily Lipitor tablets, my cholesterol level has fallen to below 200. In the years since then, it has never risen above 200. Not once. When last measured at the start of 2010, my cholesterol level was 158.

In *The Naming of Cats,* the T.S. Eliot poem on which the Broadway musical *Cats* was based, "a cat must have three different names." I think of my anticholesterol drug as being like T.S. Eliot's cat. Like his cat, my drug has a "sensible everyday name", Lipitor, a name that anyone can use. Then Eliot says a cat needs a second name, "a name that's particular," "a name that never belonged to more than one cat." Atorvastatin is such a name, one that drug manufacturers and scientists use.

According to T.S. Eliot, every cat also needs a third very private name, a "deep and inscrutable singular name." For Lipitor, this name, its chemical soul, is [R-(R*,R*)]-2-(4-fluorophenyl)-β,δ-dihydroxy-5-(1-methylethyl)-3-phenyl-4-[(phenylamino)carbonyl]-1H-pyrrole-1-heptanoic acid.

To someone looking at a cholesterol level 116 points below 274, that's beautiful.

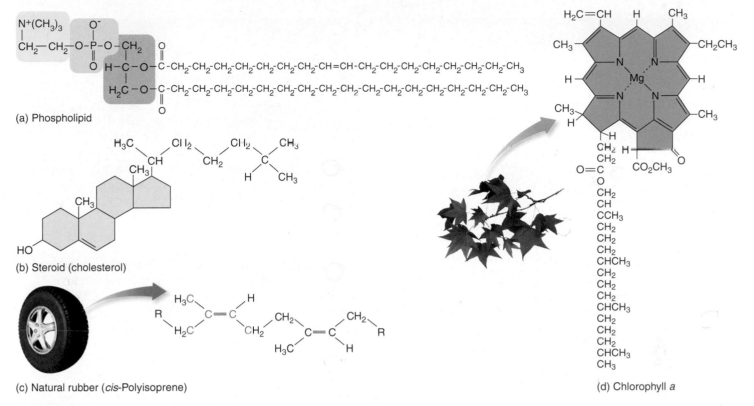

(a) Phospholipid

(b) Steroid (cholesterol)

(c) Natural rubber (*cis*-Polyisoprene)

(d) Chlorophyll *a*

Figure 3.16 Types of lipid molecules.

(*a*) A phospholipid is similar to a fat molecule with the exception that one of the fatty acid tails is replaced with a polar group linked to the glycerol by a phosphate group. (*b*) Steroids like cholesterol are lipids with complex ring structures. (*c*) Natural rubber is a linear polymer of a 5-carbon unit called isoprene (two isoprene units are shown here; one unit has its carbons in *green* and the other in *red*). Rubber, harvested from the rubber tree, is used in many products such as the tires on your car. (*d*) Chlorophyll *a* has a multiringed area and a long hydrocarbon tail. Chlorophyll is the primary pigment used in photosynthesis, and it is what makes leaves green, as you will see in chapter 6.

phosphate group rather than to a third fatty acid (compare **figures 3.15***a* and **3.16***a*). This gives the molecule an overall shape of a polar head and two nonpolar tails. In water, the nonpolar ends of the phospholipids group together through hydrophobic interactions, leaving the polar heads to interact with water molecules. These interactions result in the formation of two layers of molecules with the nonpolar tails pointed inside—a lipid bilayer. All biological membranes, those that surround the cell and those found inside the cell, have this arrangement. Most animal cell membranes also contain the steroid cholesterol, a lipid composed of four carbon rings (see **figure 3.16***b*). Cholesterol helps membranes stay flexible. However, excess saturated fat intake can cause plugs of cholesterol to form in the blood vessels, which may lead to blockage, high blood pressure, stroke, or heart attack. Membranes are discussed in more detail in chapter 4.

Key Learning Outcome 3.5 Lipids are not water-soluble. Fats contain chains of fatty acid subunits that store energy. Other lipids include phospholipids, steroids, rubber, and pigment molecules.

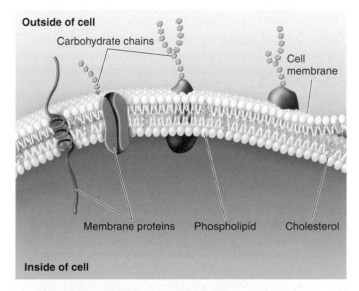

Figure 3.17 Lipids are a key component of biological membranes.

Lipids are one of the most common molecules in the human body, because the membranes of all the body's 10 trillion cells are composed largely of lipids called phospholipids. Membranes also contain cholesterol, another type of lipid.

How Does pH Affect a Protein's Function?

The red blood cells you see to the right carry oxygen to all parts of your body. These cells are red because they are chock full of a large iron-rich protein called *hemoglobin*. The iron atoms in each hemoglobin molecule provide a place for oxygen gas molecules to stick to the protein. When oxygen levels are highest (in the lungs), oxygen atoms bind to hemoglobin tightly, and a large percent of the hemoglobin molecules in a cell possess bound oxygen atoms. When oxygen levels are lower (in the tissues of the body), hemoglobin doesn't bind oxygen atoms as tightly, and as a consequence hemoglobin releases its oxygen to the tissues. What causes this difference between lungs and tissues in how hemoglobin loads and unloads oxygen? Oxygen concentration is not the only factor that might be responsible. Blood pH, for example, also differs between lungs and body tissues (**pH** is a measure of how many H^+ ions a solution contains). Tissues are slightly more acid (that is, they have more H^+ ions and a lower pH) because their metabolic activities release CO_2 into the blood, which you will recall from chapter 2, quickly becomes converted to carbonic acid.

The graph to the right displays so-called "oxygen loading curves" that reveal the effectiveness with which hemoglobin binds oxygen. The more effective the binding, the less oxygen required before hemoglobin becomes fully loaded, and the farther to the left a loading curve is shifted. To assess the impact of pH on this process, O_2 loading curves were carried out at three different blood pH values. In the graph, oxygen levels in the blood are presented on the *x* axis, and for each data point the corresponding % hemoglobin saturation (a **%**, or **percent**, is the numerator [top part] of a fraction whose denominator [bottom part] is 100—in this case, a measure of the fraction of the hemoglobin that is bound to oxygen) is presented on the *y* axis. The oxygen-loading curve was repeated at pH values of 7.6, 7.4, and 7.2, corresponding to the blood pH that might be expected in resting, exercising, and very active muscle tissue, respectively.

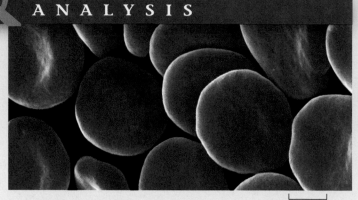

0.8 µm

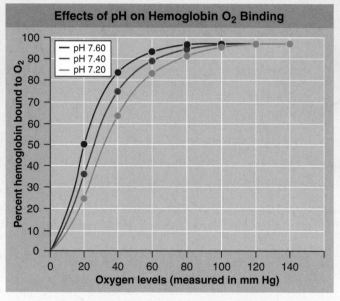

1. **Applying Concepts**
   a. Variable. In the graph, what is the dependent variable?
   b. Concentration. Which of the three pH values represents the highest concentration of hydrogen ions? (The **concentration** of a substance is the amount of that substance present in a given volume.) Is this value more acidic or more basic than the other two?
2. **Interpreting Data**
   a. What is the percent hemoglobin bound to O_2 for each of the three pH concentrations at saturation?

At an oxygen level of 20 mm Hg? At 40 mm Hg? at 60 mm Hg?
   b. What general statement can be made regarding the effect of the oxygen levels in the blood (as measured by partial pressure of oxygen, measured in mm Hg) on the binding of oxygen to hemoglobin?
   c. Are there any significant differences in the hemoglobin saturation values for the three pHs at high oxygen levels?
3. **Making Inferences** At an oxygen level of 40 mm Hg, would hemoglobin bind oxygen more tightly at a pH of 7.8 or 7.0?
4. **Drawing Conclusions** How does pH affect the release of oxygen from hemoglobin?
5. **Further Analysis**
   a. Carbon dioxide acts to lower the pH of the blood. Predict what would happen to hemoglobin loading when carbon dioxide enters the blood as the blood circulates through the tissues of the body.
   b. In the lungs, oxygen levels are high and carbon dioxide leaves the blood and is exhaled, leading to lower carbon dioxide levels in the blood. Predict what happens to hemoglobin's oxygen loading under these conditions.

Forming Macromolecules

3.1 Polymers Are Built of Monomers

- Living organisms produce organic macromolecules, which are large carbon-based molecules. The chemical properties of these molecules are due to unique functional groups that are attached to the carbon core (**figure 3.2**).

- Large molecules called macromolecules are formed by dehydration reactions, like the reaction shown here from **figure 3.3a**, in which molecular subunits called monomers are linked together through covalent bonding. The reaction is called a dehydration reaction because a water molecule is a product of the reaction.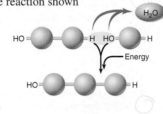

- The breakdown of a macromolecule involves a hydrolysis reaction, where a water molecule is broken down into H⁺ and OH⁻. These ions disrupt the covalent bond that holds two monomers together, causing the bond to break (**figure 3.3b**).

- Amino acids are the subunits that link together to form polypeptides. Nucleotide monomers link together to form nucleic acids. Monosaccharide monomers link together to form carbohydrates. Fatty acids are the monomers that link together to form a type of lipid called fats (**table 3.1**).

Types of Macromolecules

3.2 Proteins

- Proteins are macromolecules that carry out many functions in the cell (**figure 3.4**). They are produced by the linking together of amino acid subunits that form a chain called a polypeptide.

- There are 20 different amino acids found in proteins. All amino acids have the same basic core structure. They differ in the type of functional group attached to the core. The functional groups are referred to as R groups. Some functional groups are polar, some are nonpolar, and still others give the amino acid unique chemical properties (**figure 3.5**).

- Amino acids are linked together with covalent bonds referred to as peptide bonds (**figure 3.6**).

- The sequence of amino acids within the polypeptide is the primary structure of the protein (**integrated art, page 54**). The chain of amino acids can twist into a secondary structure, where hydrogen bonding holds portions of the polypeptide in a coiled shape, called an α-helix, or in sheets called β-pleated sheets. Further bending and folding of the polypeptide results in its tertiary structure. When two or more polypeptides are present in a protein, the interaction of these polypeptide subunits is its quaternary structure.

- Changes in environmental conditions that disrupt hydrogen bonding can cause a protein to unfold, a process called denaturation (**integrated art, page 55**). Globular proteins cannot function if they are denatured. Some denatured proteins can refold back into their functional shapes. The structure of the protein determines its function (**figure 3.7**).

- Chaperone proteins aid in the folding of the amino acid chain into the correct shape (**figure 3.8**). Some diseases may be caused by misfolded, and therefore nonfunctional, proteins.

3.3 Nucleic Acids

- Nucleic acids, such as DNA and RNA, are long chains of nucleotides. Nucleotides contain three parts: a five-carbon sugar, a phosphate group, and a nitrogenous base (**figure 3.9**). DNA and RNA function as information storage molecules in the cell, storing the information needed to build proteins. The information is stored as different sequences of nucleotides.

- DNA and RNA differ chemically in that the sugar found in DNA is deoxyribose and in RNA is ribose. The nitrogenous bases in DNA are cytosine, adenine, guanine, and thymine, and the same are found in RNA except for thymine, which is substituted in RNA with uracil.

- DNA and RNA also differ structurally. DNA contains two strands of nucleotides wound around each other, called a double helix. RNA, as shown here from **figure 3.10**, is a single strand of nucleotides.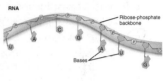

- The two nucleotide strands of the DNA double helix are held together through hydrogen bonding between nitrogenous bases: adenine (A) pairs with thymine (T) and cytosine (C) pairs with guanine (G) (**figure 3.11**).

3.4 Carbohydrates

- Carbohydrates, also referred to as sugars, are macromolecules that serve two primary functions in the cell: structural framework and energy storage.

- Carbohydrates that consist of only one or two monomers are called simple carbohydrates (**figures 3.12** and **3.13**). Carbohydrates that consist of long chains of monomers are called complex carbohydrates or polysaccharides (**figure 3.14**).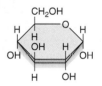

- Polysaccharides such as starch and glycogen provide a means of storing energy in the cell. They are broken down in the cells when energy is needed. Carbohydrates such as cellulose and chitin provide structural integrity (**table 3.2**).

3.5 Lipids

- Lipids are large nonpolar molecules that are insoluble in water. Fats that function in long-term energy storage include saturated fats, as shown here from **figure 3.15**, and unsaturated fats. Saturated fats are solid at room temperature and are found in animals, while unsaturated fats are liquid (oils) at room temperature and found in plants.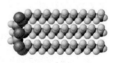

- Other lipids include the steroids (including the sex steroids and cholesterol), rubber, and pigments such as chlorophyll. All have very different structures and very different functions (**figure 3.16**).

- All cells are encased in a plasma membrane, called the lipid bilayer, that is composed of two layers of modified fats called phospholipids (**figure 3.17**).

Test Your Understanding

1. The four kinds of organic macromolecules are
   a. hydroxyls, carboxyls, aminos, and phosphates.
   b. proteins, carbohydrates, lipids, and nucleic acids.
   c. DNA, RNA, simple sugars, and amino acids.
   d. carbon, hydrogen, oxygen, and nitrogen.

2. Organic molecules are made up of monomers. Which of the following is *not* considered a monomer of organic molecules?
   a. amino acids
   b. simple sugars
   c. polypeptides
   d. nucleotides

3. Your body is filled with many types of proteins. Each type has a distinctive sequence of amino acids that determines both its specialized _____ and its specialized _____.
   a. number, weight
   b. length, mass
   c. structure, function
   d. charge, pH

4. A peptide bond forms
   a. by the removal of a water molecule.
   b. by a dehydration reaction.
   c. between two amino acids.
   d. All of the above.

5. Nucleic acids
   a. are the energy source for our bodies.
   b. act on other molecules, breaking them apart or building new ones to help us function.
   c. are only found in a few, specialized locations within the body.
   d. are information storage devices found in body cells.

6. The two strands of a DNA molecule are held together through hydrogen bonds between nucleotide bases. Which of the following best describes this base pairing in DNA?
   a. Adenine forms hydrogen bonds with thymine.
   b. Adenine forms hydrogen bonds with cytosine.
   c. Cytosine forms hydrogen bonds with thymine.
   d. Guanine forms hydrogen bonds with adenine.

7. Carbohydrates are used for
   a. structure and energy.
   b. information storage.
   c. fat storage and hair.
   d. hormones and enzymes.

8. Which of the following carbohydrates is *not* found in plants?
   a. glycogen
   b. cellulose
   c. starch
   d. All are found in plants.

9. A characteristic common to fat molecules is
   a. that they contain long chains of C–H bonds.
   b. that they are insoluble in water.
   c. that they have a glycerol backbone.
   d. All of these are characteristics of fats.

10. Lipids are used for
    a. motion and defense.
    b. information storage.
    c. energy storage and for some hormones.
    d. enzymes and for some hormones.

Apply Your Understanding

1. **Figures 3.6 and 3.13** The molecule below, on the left, is a peptide made from monomers of amino acids. The molecule below, on the right, is a disaccharide made from monomers of simple sugars. Both molecules were synthesized using a common chemical reaction. What is the chemical reaction that formed these molecules and what is the common by-product of both these reactions?

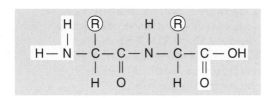

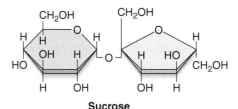

Sucrose

2. **Figure 3.15** The following are two lipid molecules. The lipid on the left is a saturated fat and the one on the right is an unsaturated fat. What is the difference in the chemical structure of their fatty acid tails and how does this affect their physical properties?

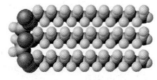 

Synthesize What You Have Learned

1. How many molecules of water are used up in the breakdown of a polypeptide that is 15 amino acids in length?

2. The amylase enzyme present in the cells of your body can break the bonds between the glucose monomers in starch but it cannot break the bonds between the glucose monomers in cellulose. You learned that enzymes are very specific in what molecules they bind to. After examining the chemical structures of starch and cellulose in table 3.2, explain why the same enzyme that breaks down starch cannot break down cellulose.

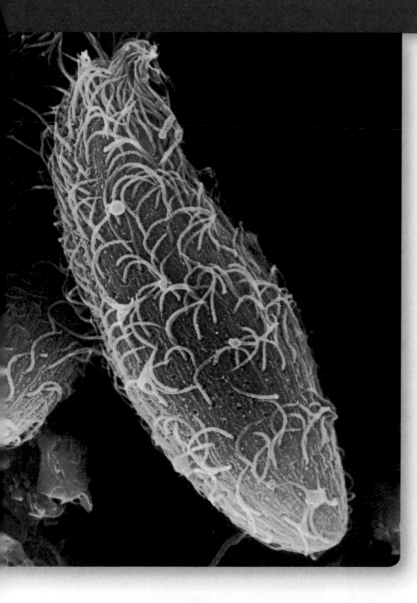

4
Cells

This alarming-looking creature is the single-celled organism *Dileptus,* magnified a thousand times. Too small to see with the unaided eye, *Dileptus* is one of hundreds of inhabitants of a drop of pond water. Everything that a living organism does to survive and prosper, *Dileptus* must do with only the equipment this tiny cell provides. Just as you move about using legs to walk, so *Dileptus* uses the hairlike projections (called cilia) that cover its surface to propel itself through the water. Just as your brain is the control center of your body, so the compartment called the nucleus, deep within the interior of *Dileptus,* controls the many activities of this complex and very active cell. *Dileptus* has no mouth, but it takes in food particles and other molecules through its surface. This versatile protist is capable of leading a complex life because its interior is subdivided into compartments, in each of which it carries out different activities. Functional specialization is the hallmark of this cell's interior, a powerful approach to cellular organization that is shared by all eukaryotes.

4.1 | Cells

Hold your finger up and look at it closely. What do you see? Skin. It looks solid and smooth, creased with lines and flexible to the touch. But if you were able to remove a bit and examine it under a microscope, it would look very different—a sheet of tiny, irregularly shaped bodies crammed together like tiles on a floor. **Figure 4.1** takes you on a journey into your fingertip. The crammed bodies you see in panels ❸ and ❹ are skin cells, laid out like a tiled floor. As your journey inward continues, you travel inside one of the cells and see organelles, structures in the cell that perform specific functions. Proceeding even farther inward, you encounter the molecules of which the structures are made, and finally the atoms shown in panels ❽ and ❾. While some organisms are composed of a single cell, your body is composed of many cells. A single

human being has as many cells as the stars in a galaxy, between 10 and 100 trillion (depending on how big you are). All cells, however, are small. In this chapter we look more closely at cells and learn something of their internal structure and how they communicate with their environment.

The Cell Theory

Because cells are so small, no one observed them until microscopes were invented in the mid-seventeenth century. Robert Hooke first described cells in 1665, when he used a microscope he had built to examine a thin slice of nonliving plant tissue called cork. Hooke observed a honeycomb of tiny, empty (because the cells were dead) compartments. He called the compartments in the cork *cellulae* (Latin, small rooms), and the term has come down to us as **cells.** For another century and a half, however, biologists failed to recognize the importance

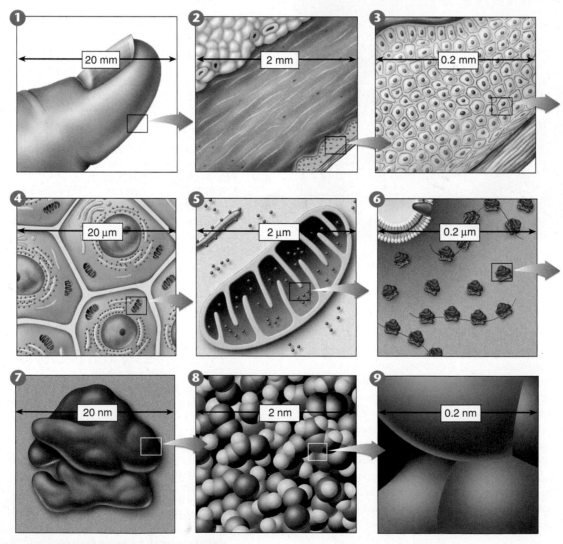

Figure 4.1 The size of cells and their contents.

This diagram shows the size of human skin cells, organelles, and molecules. In general, the diameter of a human skin cell ❹ is a little less than 20 micrometers (μm), of a mitochondrion ❺ is 2 μm, of a ribosome ❼ is 20 nanometers (nm), of a protein molecule ❽ is 2 nm, and of an atom ❾ is 0.2 nm.

of cells. In 1838, botanist Matthias Schleiden made a careful study of plant tissues and developed the first statement of the cell theory. He stated that all plants "are aggregates of fully individualized, independent, separate beings, namely the cells themselves." In 1839, Theodor Schwann reported that all animal tissues also consist of individual cells.

The idea that all organisms are composed of cells is called the **cell theory.** In its modern form, the cell theory includes three principles:

1. All organisms are composed of one or more cells, within which the processes of life occur.
2. Cells are the smallest living things. Nothing smaller than a cell is considered alive.
3. Cells arise only by division of a previously existing cell. Although life likely evolved spontaneously in the environment of the early earth, biologists have concluded that no additional cells are originating spontaneously at present. Rather, life on earth represents a continuous line of descent from those early cells.

Most Cells Are Very Small

Most cells are relatively small, but not all are the same size. The cells of your body are typically from 5 to 20 micrometers (a micrometer, μm, is one-millionth of a meter) in diameter, too small to see with the naked eye. Bacteria cells are even smaller than yours, only a few micrometers thick. However, there are some cells that are larger; individual marine alga cells, for example, can be up to 5 centimeters long—as long as your little finger.

Why Aren't Cells Larger?

Why are most cells so tiny? Most cells are small because larger cells do not function as efficiently. In the center of every cell is a command center that must issue orders to all parts of the cell, directing the synthesis of certain enzymes, the entry of ions and molecules from the exterior, and the assembly of new cell parts. These orders must pass from the core to all parts of the cell, and it takes them a very long time to reach the periphery of a large cell. For this reason, an organism made up of relatively small cells is at an advantage over one composed of larger cells.

Another reason cells are not larger is the advantage of having a greater **surface-to-volume ratio.** As cell size increases, volume grows much more rapidly than surface area. For a round cell, surface area increases as the square of the diameter, whereas volume increases as the cube. To visualize this, consider the two single cells in **figure 4.2.** The large cell to the right is 10 times bigger than the small cell, but while its surface area is 100 times greater (10^2), its volume is 1,000 times (10^3) the volume of the smaller cell. A cell's surface provides the interior's only opportunity to interact with the environment, with substances passing into and out of the cell across its surface. Large cells have far less surface for each unit of volume than do small ones.

Some larger cells, however, function quite efficiently in part because they have structural features that increase surface area. For example, cells in the nervous system called neurons

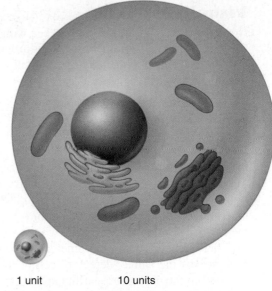

Cell radius (r)	1 unit	10 units
Surface area ($4\pi r^2$)	12.57 units^2	1,257 units^2
Volume ($\frac{4}{3}\pi r^3$)	4.189 units^3	4,189 units^3

Figure 4.2 Surface-to-volume ratio.
As a cell gets larger, its volume increases at a faster rate than its surface area. If the cell radius increases by 10 times, the surface area increases by 100 times, but the volume increases by 1,000 times. A cell's surface area must be large enough to meet the needs of its volume.

are long, slender cells, some extending more than a meter in length. These cells efficiently interact with their environment because although they are long, they are thin, some less than 1 micrometer in diameter, and so their interior regions are not far from the surface at any given point.

Another structural feature that increases the surface area of a cell are small "fingerlike" projections called microvilli. The cells that line the small intestines of the human digestive system are covered with microvilli that dramatically increase the surface area of the cells.

With few exceptions however, cells don't usually grow much larger than 50 micrometers. For organisms to get much larger, they are usually composed of many cells. By grouping together many smaller cells, these multicellular organisms vastly increase their total surface-to-volume ratio.

An Overview of Cell Structure

All cells are surrounded by a delicate membrane, called a *plasma membrane,* that controls the permeability of the cell to water and dissolved substances. A semifluid matrix called *cytoplasm* fills the interior of the cell. It used to be thought that the cytoplasm was uniform, like Jell-O™, but we now know that it is highly organized. Your cells, for example, have an internal framework that both gives the cell its shape and positions components and materials within its interior. In the following sections, we explore the membranes that encase all living cells and then examine in detail their interiors.

Visualizing Cells

How many cells are big enough to see with the unaided eye? Other than egg cells, not many are (**figure 4.3**). Most are less than 50 micrometers in diameter, far smaller than the period at the end of this sentence.

The Resolution Problem How do we study cells if they are too small to see? The key is to understand why we can't see them. The reason we can't see such small objects is the limited resolution of the human eye. **Resolution** is defined as the minimum distance two points can be apart and still be distinguished as two separated points. On the visibility scale in **figure 4.3** below, you can see that the limit of resolution of the human eye (the blue bar at the bottom) is about 100 micrometers. This limit occurs because when two objects are closer together than about 100 micrometers, the light reflected from each strikes the same "detector" cell at the rear of the eye. Only when the objects are farther apart than 100 micrometers will the light from each strike different cells, allowing your eye to resolve them as two objects rather than one.

Microscopes One way to increase resolution is to increase magnification so that small objects appear larger. Robert Hooke and Anton van Leeuwenhoek used glass lenses to magnify small cells and cause them to appear larger than the 100-micrometer limit imposed by the human eye. The glass lens adds additional focusing power. Because the glass lens makes the object appear closer, the image on the back of the eye is bigger than it would be without the lens.

Modern *light microscopes* use two magnifying lenses (and a variety of correcting lenses) to achieve very high magnification and clarity. The first lens focuses the image of the object on the second lens, which magnifies it again and focuses it on the back of the eye. Microscopes that magnify in stages using several lenses are called **compound microscopes.** They can resolve structures that are separated by more than 200 nanometers (nm). The six entries in the upper portion of **table 4.1** are images viewed through various types of light microscopes.

Increasing Resolution Light microscopes, even compound ones, are not powerful enough to resolve many structures within cells. For example, a membrane is only 5 nanometers thick. Why not just add another magnifying stage to the microscope and so increase its resolving power? Because when two objects are closer than a few hundred nanometers, the light beams reflecting from the two images start to overlap. The only way two light beams can get closer together and still be resolved is if their wavelengths are shorter.

One way to avoid overlap is by using a beam of electrons rather than a beam of light. Electrons have a much shorter wavelength, and a microscope employing electron beams has 1,000 times the resolving power of a light microscope. A **transmission electron microscope (TEM),** so called because the electrons used to visualize the specimens are transmitted through the material, is capable of resolving objects only 0.2 nanometer apart—just twice the diameter of a hydrogen atom! The entry on the left under electron microscopes in **table 4.1** is an example of an image captured using a TEM.

A second kind of electron microscope, the **scanning electron microscope (SEM),** beams the electrons onto the surface of the specimen. The electrons reflected back from the surface of the specimen, together with other electrons that the specimen itself emits as a result of the bombardment, are amplified and transmitted to a screen, where the image can be viewed and photographed. Scanning electron microscopy yields striking three-dimensional images and has improved our understanding of many biological and physical phenomena. The entry on the right in **table 4.1** under electron microscopes is an SEM image.

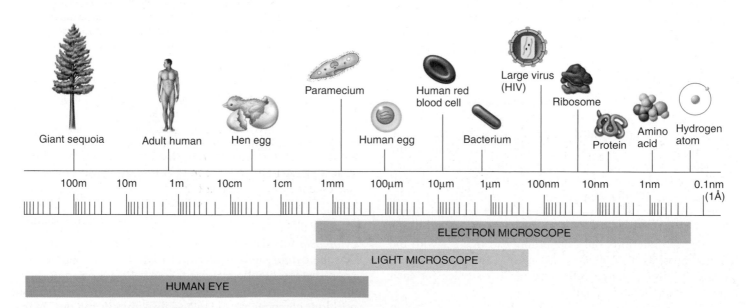

Figure 4.3 A scale of visibility.

Most cells are microscopic in size, although vertebrate eggs are typically large enough to be seen with the unaided eye. Prokaryotic cells are generally 1 to 2 micrometers (μm) across.

TABLE 4.1	TYPES OF MICROSCOPES

Light Microscopes

28.36 µm

Bright-field microscope: Light is simply transmitted through a specimen in culture, giving little contrast. Staining specimens improves contrast but requires that cells be fixed (not alive), which can cause distortion or alteration of components.

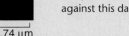

67.74 µm

Dark-field microscope: Light is directed at an angle toward the specimen; a condenser lens transmits only light reflected off the specimen. The field is dark, and the specimen is light against this dark background.

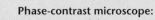

32.81 µm

Phase-contrast microscope: Components of the microscope bring light waves out of phase, which produces differences in contrast and brightness when the light waves recombine.

26.6 µm

Differential-interference-contrast microscope: Out-of-phase light waves that produce differences in contrast are combined with two beams of light travelling close together, which creates even more contrast, especially at the edges of structures.

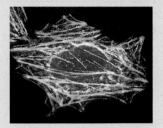

Fluorescence microscope: A set of filters transmits only light that is emitted by fluorescently stained molecules or tissues.

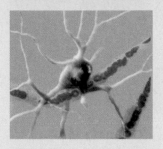

Confocal microscope: Light from a laser is focused to a point and scanned across the specimen in two directions. Clear images of one plane of the specimen are produced, while other planes of the specimen are excluded and do not blur the image. Fluorescent dyes and false coloring enhance the image.

Electron Microscopes

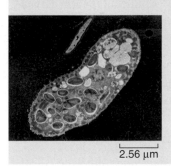

2.56 µm

Transmission electron microscope: A beam of electrons is passed through the specimen. Electrons that pass through are used to form an image. Areas of the specimen that scatter electrons appear dark. False coloring has been added to this image.

6.76 µm

Scanning electron microscope: An electron beam is scanned across the surface of the specimen, and electrons are knocked off the surface. Thus, the surface topography of the specimen determines the contrast and the content of the image. False coloring enhances the image.

Visualizing Cell Structure by Staining Specific Molecules

A powerful tool for the analysis of cell structure has been the use of stains that bind to specific molecular targets. This approach has been used in the analysis of tissue samples, or histology, for many years and has been improved dramatically with the use of antibodies that bind to very specific molecular structures. This process, called immunocytochemistry, uses antibodies generated in animals such as rabbits or mice. When these animals are injected with specific proteins, they will produce antibodies that specifically bind to the injected protein, which can be purified from their blood. These purified antibodies can then be chemically bonded to enzymes, stains, or fluorescent molecules that glow when exposed to specific wavelengths of light. When cells are washed in a solution containing the antibodies, the antibodies bind to cellular structures that contain the target molecule and can be seen with light microscopy. The image produced using fluorescence microscopy in **table 4.1** shows the cytoskeleton made of cablelike structures inside the cell. This approach has been used extensively in the analysis of cell structure and function.

Key Learning Outcome 4.1 All living things are composed of one or more cells, each a small volume of cytoplasm surrounded by a plasma membrane. Most cells and their components can only be viewed using microscopes.

4.2 The Plasma Membrane

Encasing all living cells is a delicate sheet of molecules called the **plasma membrane.** It would take more than 10,000 of these molecular sheets, which are about 5 nanometers thick, piled on top of one another to equal the thickness of this sheet of paper. However, the sheets are not simple in structure, like a soap bubble's skin. Rather, they are made up of a diverse collection of proteins floating within a lipid framework like small boats bobbing on the surface of a pond. Regardless of the kind of cell they enclose, all plasma membranes have the same basic structure of proteins embedded in a sheet of lipids, called the **fluid mosaic model.**

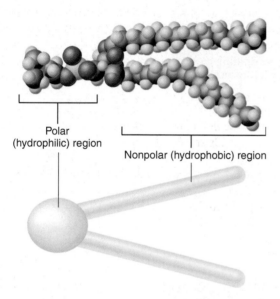

Polar (hydrophilic) region

Nonpolar (hydrophobic) region

The lipid layer that forms the foundation of a plasma membrane is composed of modified fat molecules called **phospholipids.** A phospholipid molecule can be thought of as a polar head with two nonpolar tails attached to it, as shown above. The head of a phospholipid molecule has a phosphate chemical group linked to it, making it extremely polar (and thus water-soluble). The other end of the phospholipid molecule is composed of two long fatty acid chains. Recall from chapter 3 that fatty acids are long chains of carbon atoms with attached hydrogen atoms. The carbon atoms are the gray spheres you see above. The fatty acid tails are strongly nonpolar and thus water-insoluble. The phospholipid is often depicted diagrammatically as a ball with two tails.

Imagine what happens when a collection of phospholipid molecules is placed in water. A structure called a **lipid bilayer** forms spontaneously. How can this happen? The long nonpolar tails of the phospholipid molecules are pushed away by the water molecules that surround them, shouldered aside as the water molecules seek partners that

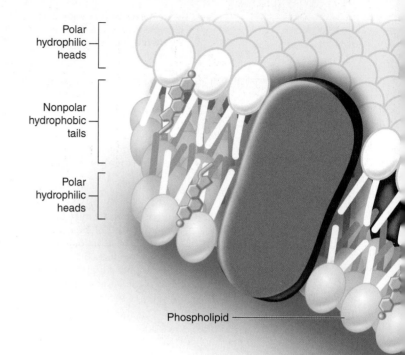

Polar hydrophilic heads

Nonpolar hydrophobic tails

Polar hydrophilic heads

Phospholipid

can form hydrogen bonds. After much shoving and jostling, every phospholipid molecule ends up with its polar head facing water and its nonpolar tail facing away from water. The phospholipid molecules form a *double* layer, called a bilayer. As you can see in the figure above, the watery environments inside and outside the plasma membrane push the nonpolar tails to the interior of the bilayer. Because there are two layers with the tails facing each other, no tails are ever in contact with water. Thus, the interior of a lipid bilayer is completely nonpolar, and it repels any water-soluble molecules that attempt to pass through it, just as a layer of oil stops the passage of a drop of water (that's why ducks do not get wet).

Cholesterol, another nonpolar lipid molecule, resides in the interior portion of the bilayer. Cholesterol is a multi-ringed molecule that affects the fluid nature of the membrane. Although cholesterol is important in maintaining the integrity of the plasma membrane, it can accumulate in blood vessels, forming plaques that lead to cardiovascular disease.

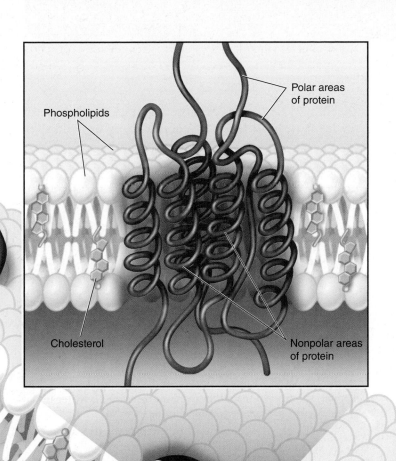

Phospholipids

Polar areas
of protein

Cholesterol

Nonpolar areas
of protein

Proteins Within the Membrane

The second major component of every biological membrane is a collection of **membrane proteins** that float within the lipid bilayer. Membrane proteins function as channels, receptors, and cell surface markers. As you can see here in the figure, some proteins pass through the lipid bilayer, providing channels through which molecules and information pass. While some membrane proteins are fixed into position, others move about freely.

Many membrane proteins project up from the surface of the plasma membrane like buoys, often with carbohydrate chains or lipids attached to their tips like flags. These **cell surface proteins** act as markers to identify particular types of cells, or as beacons to bind specific hormones or proteins to the cell.

Protein channels that extend all the way across the bilayer provide passageways for ions and polar molecules like water so they can pass into and out of the cell. How do these **transmembrane proteins** manage to span the membrane, rather than just floating on the surface in the way that a drop of water floats on oil? The part of the protein that actually traverses the lipid bilayer is a specially constructed spiral helix of nonpolar amino acids—the red coiled areas of the transmembrane protein you see in the inset. Water responds to these nonpolar amino acids much as it does to nonpolar lipid chains, and as a result the helical spiral is held within the lipid interior of the bilayer, anchored there by the strong tendency of water to avoid contact with these nonpolar amino acids.

Protein channel

Cholesterol

Receptor protein

Key Learning Outcome 4.2 All cells are encased within a delicate lipid bilayer sheet, the plasma membrane, within which are embedded a variety of proteins that act as markers or channels through the membrane.

Cell identity
marker

How Water Crosses the Plasma Membrane

One of the enduring mysteries of cellular biology has been the free movement of water into and out of cells. As early as the middle of the nineteenth century, biologists understood that there must be a way for water to pass across the plasma membrane. With the understanding in the middle of the 1950s that the plasma membrane was composed of a lipid bilayer, the problem posed by the free movement of water became even more puzzling. How could very polar water molecules traverse the very nonpolar environment found in the lipid core of the bilayer?

While some proposed that water leaked into cells through tiny imperfections in the bilayer, or through gaps that open up in the bilayer when the hydrocarbon tails flex and bend, these hypotheses were soon rejected, as they failed to explain how cell membranes managed to prevent the diffusion of protons (hydrogen ions), which are smaller than water. This ability is crucial to the life of a cell, as the difference in proton concentration between the inside and outside of cell organelles is the basis of energy metabolism, as described in chapters 6 and 7.

Clearly the answer to this puzzle lay with the proteins associated with the lipid bilayer. For 30 years researchers searched for the protein machinery that prevented ions from passing across the membrane while allowing water molecules to pass freely. With the acceptance of the fluid mosaic model of membrane structure proposed in 1972, the search focused on proteins that bridge the bilayer. In the mid-1980s, a Johns Hopkins University researcher named Peter Agre, studying red blood cell proteins, identified a previously unknown protein. "No one had seen it before, but we found that it was the fifth most abundant protein in the cell," said Agre. "That's like coming across a big town that's not on the map. It gets your attention."

Agre determined the amino acid sequence of the mysterious protein, and saw that it had long nonpolar segments that would allow it to repeatedly traverse the lipid bilayer, just as a cellular water channel would have to do. Perhaps this was the protein so many had sought.

To test this hypothesis, Agre carried out a simple experiment. He compared normal red blood cells that contained the protein with mutant red blood cells that lacked it. When he placed the cells in distilled lab water, cells with the protein in their membrane absorbed water and began to swell, while cells lacking the protein did nothing—they failed to absorb water and did not swell.

To be sure that some other undiscovered membrane protein could not be the cause of the water movement across the red blood cell plasma membrane, Agre repeated the experiment with liposomes, which are artificial cells made of pure lipid bilayer with no proteins—in essence, soap bubbles. As you might expect, water could not cross into liposomes, and he found

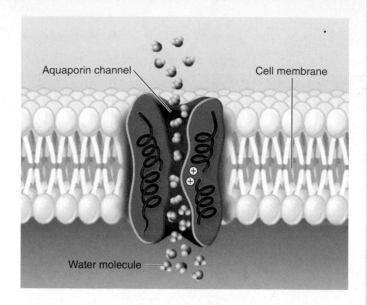

they did not swell when immersed in distilled lab water. However, the liposomes did become permeable to water if Agre planted the protein in their bilayer.

As a final test, Agre knew that mercury ions poison cells by keeping them from taking up and releasing water, and he showed that water transport through liposomes with his protein in their membranes was prevented by mercury too. Agre concluded that the protein he had discovered was indeed the water channel, and named the protein *aquaporin*, "water pore." Peter Agre was awarded the Nobel Prize in Chemistry in 2003 for his discovery.

Working with other research teams on this exciting discovery, Agre reported in 2000 the results of X-ray diffraction studies that revealed the three-dimensional structure of aquaporin in atomic detail. Now it was possible to see in detail how the water channel functions.

The aquaporin channel Agre described is illustrated above. Individual water molecules pass through the narrow open channel single file, snaking their way along by orienting themselves in the local electrical field formed by the atoms of the channel wall. And now we see the key feature that allows the channel to exclude protons while passing larger water molecules. Look in the center of the pore. A cluster of positively-charged amino acids line the pore there. While this has no impact on the passage of water molecules because of their partial negative charges, the positive charges repel protons, which are also positively charged. The positive charges in the interior of the aquaporin channel act as a filter to prevent protons leaking through the passage.

In the last 10 years researchers have identified aquaporins in many kinds of bacteria, plants, and animals. There are at least 11 kinds in the human body alone. One kind, called AQP1, acts in your kidney to recover water that would otherwise be lost in your urine. Over 24 hours, AQP1 channels in your kidneys recover about 120 liters of water!

4.3 Prokaryotic Cells

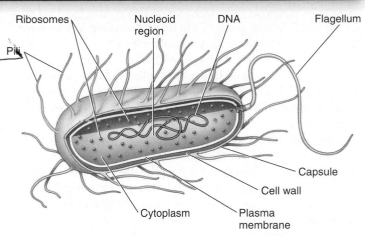

There are two major kinds of cells: prokaryotes and eukaryotes. **Prokaryotes** have a relatively uniform cytoplasm that is not subdivided by interior membranes into separate compartments. They do not, for example, have special membrane-bounded compartments, called *organelles,* or a *nucleus* (a membrane-bounded compartment that holds the hereditary information). As discussed in chapter 1, **figure 1.1,** the two main groups of prokaryotes are *bacteria* and *archaea;* all other organisms are eukaryotes.

Prokaryotes are the simplest cellular organisms. Over 5,000 species are recognized, but doubtless many times that number actually exist and have not yet been described. Although these species are diverse in form, their organization is fundamentally similar: They are single-celled organisms; the cells are small (typically about 1 to 10 micrometers thick); the cells are enclosed by a plasma membrane; and there are no distinct interior compartments (**figure 4.4**). Outside of almost all bacteria and archaea is a *cell wall,* composed of different molecules in different groups (see **table 15.1** and section 16.3). In some bacteria another layer called the *capsule* encloses the cell wall. Archaea are an extremely diverse group that inhabit diverse environments (**figure 4.5a**). Bacteria are abundant and play critical roles in many biological processes. Bacteria assume many shapes, like the sausage or spiral shapes shown in **figure 4.5b, c.** They can also adhere in chains and masses like the spherical cells in **figure 4.5d,** but in these cases the individual cells remain functionally separate from one another.

The interior of a prokaryotic cell has little or no structural support (the cell wall supports the cell's shape), but scattered throughout the cytoplasm are small structures called *ribosomes.* Ribosomes are the sites where proteins are made, but they are not considered organelles because they lack a membrane boundary. Prokaryotic DNA is found in a region of the cytoplasm called the *nucleoid region,* which is not enclosed within an internal membrane. Some prokaryotes use a **flagellum** (plural, **flagella**) to move. Flagella are long, threadlike structures, made of protein fibers that project from the surface of a cell. They are used in locomotion and feeding. There may be none, one, or more per cell depending on the species. Bacteria can swim at speeds of up to 20 cell diameters per second, rotating their flagella like screws. **Pili** (singular, **pilus**) are short flagella (only several micrometers long, and about 7.5 to 10 nanometers thick) that occur on the cells of some prokaryotes. Pili help the prokaryotic cells attach to appropriate substrates and aid in the exchange of genetic information between cells.

Figure 4.4 Organization of a prokaryotic cell.
Prokaryotic cells lack internal compartments. Not all prokaryotic cells have a flagellum or a capsule like the one illustrated here, but all have a nucleoid region, ribosomes, a plasma membrane, cytoplasm, and a cell wall.

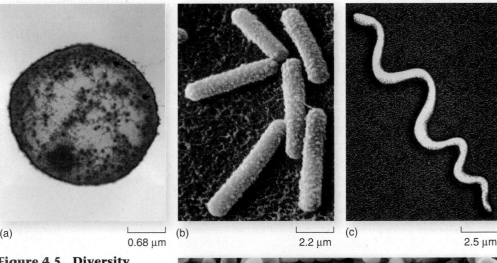

(a)   0.68 µm   (b)   2.2 µm   (c)   2.5 µm

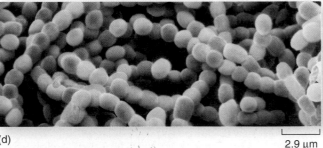

(d)   2.9 µm

Figure 4.5 Diversity of prokaryotes.
(a) *Methanococcus* is only able to survive in environments with no oxygen. (b) *Bacillus* is a rod-shaped bacterium. (c) *Treponema* is a coil-shaped bacterium; rotation of internal filaments produces a corkscrew movement. (d) *Streptomyces* is a more or less spherical bacterium in which the individuals adhere in chains.

Key Learning Outcome 4.3 Prokaryotic cells lack a nucleus and do not have an extensive system of interior membranes.

4.4 Eukaryotic Cells

For the first 1 billion years of life on earth, all organisms were prokaryotes, cells with very simple interiors. Then, about 1.5 billion years ago, a new kind of cell appeared for the first time, the eukaryotic cell. Eukaryotic cells are much larger and profoundly different from prokaryotic cells, with a complex interior organization. All cells alive today except bacteria and archaea are of this new kind.

Figures 4.6 and 4.7 present cross-sectional diagrams of idealized animal and plant cells. As you can see, the interior of a eukaryotic cell is much more complex than the prokaryotic cell you encountered in figure 4.4. The **plasma membrane** ❶ encases a semifluid matrix called the **cytoplasm** ❷, which contains within it the nucleus and various cell structures called organelles. An **organelle** is a specialized structure within which particular cell processes occur. Each organelle, such as a **mitochondrion** ❸, has a specific function in the eukaryotic cell. The organelles are anchored at specific locations in the cytoplasm by an interior scaffold of protein fibers, the **cytoskeleton** ❹.

One of the organelles is very visible when these cells are examined with a microscope, filling the center of the cell like the pit of a peach. Seeing it, the English botanist Robert Brown in 1831 called it the **nucleus** ❺ (plural, *nuclei*), from the Latin word for "kernel." Inside the nucleus, the DNA is wound tightly around proteins and packaged into compact units called chromosomes. It is the nucleus that gives **eukaryotes** their name, from the Greek words *eu*, true, and *karyon*, nut; by way of contrast, the earlier-evolving bacteria and archaea are called prokaryotes ("before the nut").

If you examine the organelles in **figures 4.6** and **4.7**, you can see that most of them form separate compartments within the cytoplasm, bounded by their own membranes. *The hallmark of the eukaryotic cell is this compartmentalization.* This internal compartmentalization is achieved by an extensive **endomembrane system** ❻ that weaves through the cell interior, providing extensive surface area for many membrane-associated cell processes to occur.

Vesicles ❼ (small membrane-bounded sacs that store and transport materials) form in the cell either by budding off of the endomembrane system or by the incorporation of lipids and protein in the cytoplasm. These many closed-off compartments allow different processes to proceed simultaneously without interfering with one another, just as rooms do in a house. Thus the

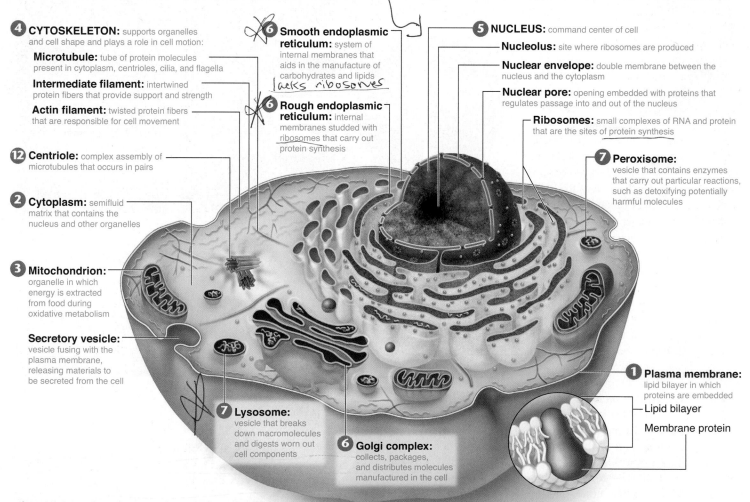

❹ **CYTOSKELETON:** supports organelles and cell shape and plays a role in cell motion:

Microtubule: tube of protein molecules present in cytoplasm, centrioles, cilia, and flagella

Intermediate filament: intertwined protein fibers that provide support and strength

Actin filament: twisted protein fibers that are responsible for cell movement

⓬ **Centriole:** complex assembly of microtubules that occurs in pairs

❷ **Cytoplasm:** semifluid matrix that contains the nucleus and other organelles

❸ **Mitochondrion:** organelle in which energy is extracted from food during oxidative metabolism

Secretory vesicle: vesicle fusing with the plasma membrane, releasing materials to be secreted from the cell

❻ **Smooth endoplasmic reticulum:** system of internal membranes that aids in the manufacture of carbohydrates and lipids *lacks ribosomes*

❻ **Rough endoplasmic reticulum:** internal membranes studded with ribosomes that carry out protein synthesis

❼ **Lysosome:** vesicle that breaks down macromolecules and digests worn out cell components

❻ **Golgi complex:** collects, packages, and distributes molecules manufactured in the cell

❺ **NUCLEUS:** command center of cell

Nucleolus: site where ribosomes are produced

Nuclear envelope: double membrane between the nucleus and the cytoplasm

Nuclear pore: opening embedded with proteins that regulates passage into and out of the nucleus

Ribosomes: small complexes of RNA and protein that are the sites of protein synthesis

❼ **Peroxisome:** vesicle that contains enzymes that carry out particular reactions, such as detoxifying potentially harmful molecules

❶ **Plasma membrane:** lipid bilayer in which proteins are embedded

Lipid bilayer

Membrane protein

Figure 4.6 Structure of an animal cell.

In this generalized diagram of an animal cell, the plasma membrane encases the cell, which contains the cytoskeleton and various cell organelles and interior structures suspended in a semifluid matrix called the cytoplasm. Some kinds of animal cells possess fingerlike projections called microvilli. Other types of eukaryotic cells, for example many protist cells, may possess flagella, which aid in movement, or cilia, which can have many different functions.

organelles called *lysosomes* are recycling centers. Their very acid interiors break down old organelles, and the component molecules are recycled. This acid would be very destructive if released into the cytoplasm. Similarly, chemical isolation is essential to the function of the organelles called *peroxisomes*. Toxic chemicals are degraded and food molecules are processed within peroxisomes by enzymes that act by removing electrons and associated hydrogen atoms. If not isolated within the peroxisomes, these enzymes would tend to short-circuit chemical reactions occurring in the cytoplasm, which often involves adding hydrogen atoms to molecules.

If you compare **figure 4.6** with **figure 4.7**, you will see the same set of organelles, with a few interesting exceptions. For example, the cells of plants, fungi, and many protists have strong thick exterior **cell walls ❽** composed of cellulose or chitin fibers, while the cells of animals lack cell walls. All plants and many kinds of protists have **chloroplasts ❾**, within which photosynthesis occurs. No animal or fungal cells contain chloroplasts. Plant cells also contain a large **central vacuole ❿** that stores water, and cytoplasmic connections through openings in the cell wall called **plasmodesmata ⓫**. **Centrioles ⓬** are present in animal cells but absent in plant and fungal cells. Some kinds of animal cells possess fingerlike projections called *microvilli*. Many animal and protist cells possess *flagella*, which aid in movement, or *cilia*, which have many different functions. Flagella occur in sperm of a few plant species, but are otherwise absent in plant and fungal cells.

We will now journey into the interior of a typical eukaryotic cell and explore it in more detail, using diagrams with the particular organelle you are examining highlighted. While the various organelles are color-coded for easier identification, remember that most are actually colorless.

> **Key Learning Outcome 4.4** Eukaryotic cells have a system of interior membranes and organelles that subdivide the interior into functional compartments.

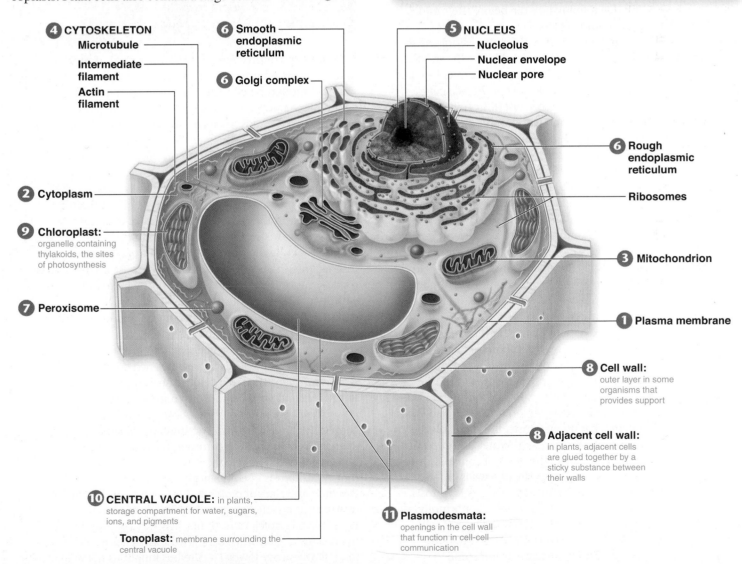

Figure 4.7 Structure of a plant cell.

Most mature plant cells contain large central vacuoles that occupy a major portion of the internal volume of the cell and organelles called chloroplasts, within which photosynthesis takes place. The cells of plants, fungi, and some protists have cell walls, although the composition of the walls varies among the groups. Plant cells have cytoplasmic connections through openings in the cell wall called plasmodesmata. Flagella occur in sperm of a few plant species, but are otherwise absent in plant and fungal cells. Centrioles are also absent in plant and fungal cells.

Today's *Biology*

Membrane Defects Can Cause Disease

The year 1993 marked an important milestone in the treatment of human disease. That year the first attempt was made to cure **cystic fibrosis (CF),** a deadly genetic disorder, by transferring healthy genes into sick individuals. Cystic fibrosis is a fatal disease in which the body cells of affected individuals secrete a thick mucus that clogs the airways of the lungs. The cystic fibrosis patient in the photograph is breathing into a Vitalograph, a device that measures lung function. These same secretions block the ducts of the pancreas and liver so that the few patients who do not die of lung disease die of liver failure. Cystic fibrosis is usually thought of as a children's disease because until recently few affected individuals lived long enough to become adults. Even today half die before their mid-twenties. There is no known cure.

Cystic fibrosis results from a defect in a single gene that is passed down from parent to child. It is the most common fatal genetic disease of Caucasians. One in 20 individuals possesses at least one copy of the defective gene. Most of these individuals are not afflicted with the disease; only those children who inherit a copy of the defective gene from each parent succumb to cystic fibrosis—about 1 in 2,500 infants.

Cystic fibrosis has proven difficult to study. Many organs are affected, and until recently it was impossible to identify the nature of the defective gene responsible for the disease. In 1985 the first clear clue was obtained. An investigator, Paul Quinton, seized on a commonly observed characteristic of cystic fibrosis patients, that their sweat is abnormally salty, and performed the following experiment. He isolated a sweat duct from a small piece of skin and placed it in a solution of salt (NaCl) that was three times as concentrated as the NaCl inside the duct. He then monitored the movement of ions. Diffusion tends to drive both the sodium (Na^+) and the chloride (Cl^-) ions into the duct because of the higher outer ion concentrations. In skin isolated from normal individuals, Na^+ and Cl^- both entered the duct, as expected. In skin isolated from cystic fibrosis individuals, however, only Na^+ entered the duct—no Cl^- entered. For the first time, the molecular nature of cystic fibrosis became clear. Water accompanies chloride, and was not entering the ducts because chloride was not, creating thick mucus. Cystic fibrosis is a defect in a plasma membrane protein called CFTR (cystic fibrosis *transmembrane conductance regulator*) that normally regulates passage of Cl^- into and out of the body's cells.

The defective *cf* gene was isolated in 1987, and its position on a particular human chromosome (chromosome 7) was pinpointed in 1989. Interestingly, many cystic fibrosis patients produce a CFTR protein with a normal amino acid sequence. The *cf* mutation in these

cases appears to interfere with how the CFTR protein folds, preventing it from folding into a functional shape.

Soon after the *cf* gene was isolated, experiments were begun to see if it would be possible to cure cystic fibrosis by gene therapy—that is, by transferring healthy *cf* genes into the cells with defective ones. In 1990 a working *cf* gene was successfully transferred into human lung cells growing in tissue culture, using adenovirus, a cold virus, to carry the gene into the cells. The CFTR-defective cells were "cured," becoming able to transport chloride ions across their plasma membranes. Then in 1991 a team of researchers successfully transferred a normal human *cf* gene into the lung cells of a living animal—a rat. The *cf* gene was first inserted into the adenovirus genome because adenovirus is a cold virus and easily infects lung cells. The treated virus was then inhaled by the rat. Carried piggyback, the *cf* gene entered the rat lung cells and began producing the normal human CFTR protein within these cells!

These results were very encouraging, and at first the future for all cystic fibrosis patients seemed bright. Clinical tests using adenovirus to introduce healthy *cf* genes into cystic fibrosis patients were begun with much fanfare in 1993.

They were not successful. As described further in chapter 13, there were insurmountable problems with the adenovirus being used to transport the *cf* gene into cystic fibrosis patients. The difficult and frustrating challenge that cystic fibrosis researchers had faced was not over. Research into clinical problems is often a time-consuming and frustrating enterprise, never more so than in this case. Recently, as chapter 13 recounts, new ways of introducing the healthy *cf* gene have been tried with better results. The long, slow journey toward a cure has taught us not to leap to the assumption that a cure is now at hand, but the steady persistence of researchers has taken us a long way, and again the future for cystic fibrosis patients seems bright.

4.5 The Nucleus: The Cell's Control Center

Comparing the animal and plant cells on the previous pages, you cannot help but notice that many parts of the cells are remarkably similar. In paramecia, petunias, and primates, cell organelles look similar and carry out similar functions (see **table 4.2** on pages 88–89).

If you were to journey far into the interior of one of your cells, you would eventually reach the center of the cell. There you would find, cradled within a network of fine filaments like a ball in a basket, the **nucleus (figure 4.8)**. The nucleus is the command and control center of the cell, directing all of its activities. It is also the genetic library where the hereditary information is stored.

Nuclear Membrane

The surface of the nucleus is bounded by a special kind of membrane called the **nuclear envelope.** The nuclear envelope is actually *two* membranes, one outside the other, like a sweater over a shirt. The nuclear envelope acts as a barrier between the nucleus and the cytoplasm, but substances need to pass through the envelope. The exchange of materials occurs through openings scattered over the surface of this envelope. Called **nuclear pores,** these openings form when the two membrane layers of the nuclear envelope pinch together. A nuclear pore is not an empty opening, however; rather, it has many proteins embedded within it that permit proteins and RNA to pass into and out of the nucleus.

Chromosomes

In both prokaryotes and eukaryotes, all hereditary information specifying cell structure and function is encoded in DNA.

However, unlike the circular prokaryotic DNA, the DNA of eukaryotes is divided into several segments and associated with protein, forming **chromosomes.** The proteins in the chromosome permit the DNA to wind tightly and condense during cell division. Under a light microscope, these condensed chromosomes are readily seen in dividing cells as densely staining rods. After cell division, eukaryotic chromosomes uncoil and fully extend into threadlike strands called **chromatin** that can no longer be distinguished individually with a light microscope within the nucleoplasm. Once uncoiled, the chromatin is available for protein synthesis. RNA copies of genes are made from the DNA in the nucleus. The RNA molecules leave the nucleus through the nuclear pores and enter the cytoplasm where proteins are synthesized.

Nucleolus

To make its many proteins, the cell employs a special structure called a **ribosome,** a kind of platform on which the proteins are built. Ribosomes read the RNA copy of a gene and use that information to direct the construction of a protein. Ribosomes are made up of several special forms of RNA called *ribosomal RNA,* or *rRNA,* bound up within a complex of several dozen different proteins.

You will notice in **figure 4.8** that one region of the nucleus appears darker than the rest; this darker region is called the **nucleolus.** There a cluster of several hundred genes encode rRNA where the ribosome subunits assemble. These subunits leave the nucleus through the nuclear pores and enter the cytoplasm, where final assembly of ribosomes takes place.

> **Key Learning Outcome 4.5** The nucleus is the command center of the cell, issuing instructions that control cell activities. It also stores the cell's hereditary information.

**Figure 4.8
The nucleus.**

The nucleus is composed of a double membrane, called a nuclear envelope, enclosing a fluid-filled interior containing the chromosomes. In cross section, the individual nuclear pores are seen to extend through the two membrane layers of the envelope. The pore is lined with protein, which acts to control access through the pore.

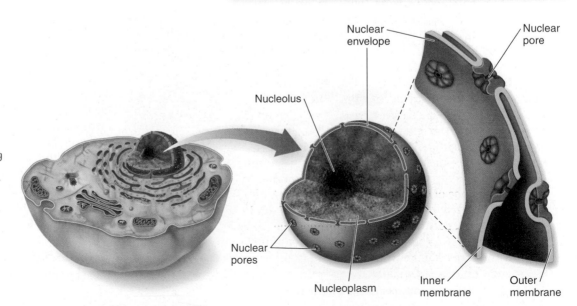

The Endomembrane System

Surrounding the nucleus within the interior of the eukaryotic cell is a tightly packed mass of membranes. They fill the cell, dividing it into compartments, channeling the transport of molecules through the interior of the cell and providing the surfaces on which enzymes act. The system of internal compartments created by these membranes in eukaryotic cells constitutes the most fundamental distinction between the cells of eukaryotes and prokaryotes.

Endoplasmic Reticulum: The Transportation System

The extensive system of internal membranes is called the **endoplasmic reticulum,** often abbreviated **ER.** The ER creates a series of channels and interconnections, and it also isolates some spaces as membrane-enclosed sacs called **vesicles.** The surface of the ER is the place where the cell makes proteins intended for export (such as enzymes secreted from the cell surface). The surface of those regions of the ER devoted to the synthesis of transported proteins is heavily studded with ribosomes and appears pebbly, like the surface of sandpaper, when seen through an electron microscope. For this reason, these regions are called **rough ER.** Regions in which ER-bound ribosomes are relatively scarce are correspondingly called **smooth ER.** The surface of the smooth ER is embedded with enzymes that aid in the manufacture of carbohydrates and lipids.

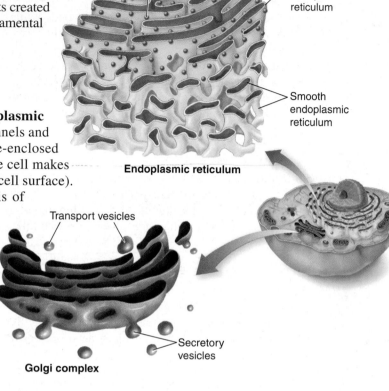

Endoplasmic reticulum

Transport vesicles

Secretory vesicles

Golgi complex

The Golgi Complex: The Delivery System

As new molecules are made on the surface of the ER, they are passed from the ER to flattened stacks of membranes called **Golgi bodies.** The number of Golgi bodies a cell contains ranges from 1 or a few in protists, to 20 or more in animal cells, and several hundred in certain plant cells. Golgi bodies function in the collection, packaging, and distribution of molecules manufactured in the cell. Scattered throughout the cytoplasm, Golgi bodies are collectively referred to as the **Golgi complex.**

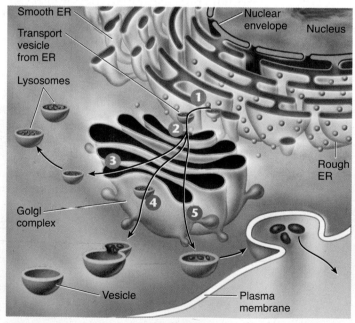

How the endomembrane system works

The rough ER, smooth ER, and Golgi work together as a transport system in the cell. Proteins and lipids that are manufactured on the ER membranes are transported throughout the channels of the ER and are packaged into transport vesicles that bud off from the ER ❶. The vesicles fuse with the membrane of the Golgi bodies, dumping their contents into the Golgi ❷. Within the Golgi bodies the molecules may take one of many paths, indicated by the branching arrows in the figure. Many of these molecules become tagged with carbohydrates. The molecules collect at the ends of the membranous folds of the Golgi bodies. Vesicles that pinch off from the ends carry the molecules to the different compartments of the cell ❸ and ❹, or to the inner surface of the plasma membrane, where molecules to be secreted are released to the outside ❺.

Lysosomes: Recycling Centers

Other organelles called **lysosomes** arise from the Golgi complex (the light orange vesicle budding at ❸) and contain a concentrated mix of the powerful enzymes manufactured in the rough ER that break down macromolecules. Lysosomes are also the recycling centers of the cell, digesting worn-out cell components to make way for newly formed ones while recycling the proteins and other materials of the old parts. Large organelles called mitochondria are replaced in some human tissues every 10 days, with lysosomes digesting the old ones as the new ones are produced. In addition to breaking down organelles and other structures within cells, lysosomes also eliminate particles (including other cells) that the cell has engulfed.

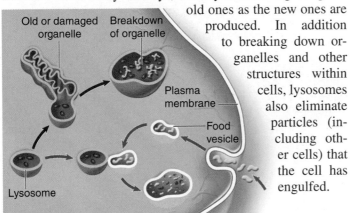

Peroxisomes: Chemical Specialty Shops

The interior of the eukaryotic cell contains a variety of membrane-bounded spherical organelles derived from the ER that carry out particular chemical functions. Almost all eukaryotic cells, for example, contain peroxisomes. Peroxisomes are spherical organelles that may contain a large crystal structure composed of protein (a peroxisome containing a crystal has been colored purple in the micrograph below). Peroxisomes contain two sets of enzymes. One set found in plant seeds converts fats to carbohydrates, and the other set found in all eukaryotes detoxifies various potentially harmful molecules—strong oxidants— that form in cells. They do this by using molecular oxygen to remove hydrogen atoms from specific molecules. These chemical reactions would be very destructive to the cell if not confined to the peroxisomes.

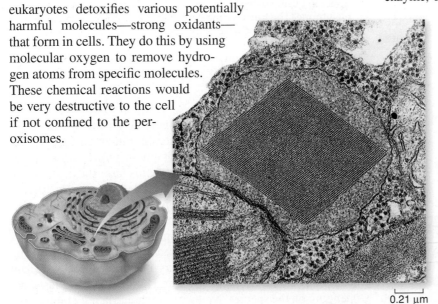

0.21 μm

The Endomembrane System and Your Health

Several serious human diseases result when the endomembrane system does not work as it should. One of the most serious of these involves a breakdown in the ability of the Golgi complex to properly address proteins destined for delivery to the cell's lysosomes. In a normal cell, the address label is a modified sugar called mannose-6-phosphate, which the Golgi complex attaches to lysosome-bound proteins with a special enzyme. In individuals lacking this enzyme, the proteins that should have gone to lysosomes lack the mannose-6-phosphate address tag, and instead are delivered to the plasma membrane and secreted from the cell. Because almost all the recycling enzymes normally found in lysosomes are missing, the lysosomes swell with undegraded materials. Viewed under a microscope, the lysosomes appear to contain numerous inclusions where the undegraded material has clumped together, giving this condition its name, *inclusion-cell disease*. The swollen lysosomes ultimately cause serious damage to the cells of a developing human embryo, leading to facial and skeletal abnormalities and mental retardation.

About 40 different human diseases are caused by failure of the endomembrane system to deliver particular enzymes to the lysosomes. Called *lysosome storage disorders,* these disorders follow the pattern you have seen for inclusion-cell disease. For lack of the necessary recycling enzyme, a particular cell material accumulates undegraded in the lysosomes, eventually leading to cell damage and death. Most of these disorders are fatal in early childhood.

In the lysosome storage disorder called *Pompe's disease,* lysosomes lack an enzyme necessary to digest glycogen. Glycogen, which the body uses as a ready source of energy, is stored in muscle and liver cells. When energy is required, lysosomes digest the glycogen, chopping off sugar units that the cell can use to produce energy (we explore how this happens in chapter 7). Pompe's disease lysosomes, because they lack the necessary enzyme, do not degrade glycogen. Instead, harmful amounts of glycogen accumulate in the muscle and liver cells.

In the lysosome storage disease called *Tay-Sachs disease,* lysosomes are missing the enzyme necessary to degrade specific glycolipids called gangliosides that are abundant in the plasma membranes of brain cells. In Tay-Sachs individuals, the glycolipids accumulate in brain cells, which swell and eventually burst, releasing oxidative enzymes that kill the brain cells. The affected individual begins to exhibit rapid mental deterioration at about six to eight months of age. Within a year after birth, affected children are blind. Paralysis and death follow within five years.

> **Key Learning Outcome 4.6** An extensive system of interior membranes organizes the interior of the cell into functional compartments that manufacture and deliver proteins and carry out a variety of specialized chemical processes.

4.7 Organelles That Contain DNA

Eukaryotic cells contain several kinds of complex, cell-like organelles that contain their own DNA and appear to have been derived from ancient bacteria assimilated by ancestral eukaryotes in the distant past. The two principal kinds are mitochondria (which occur in the cells of all but a very few eukaryotes) and chloroplasts (which do not occur in animal or fungi cells—they occur only in algae and plants).

Mitochondria: Powerhouses of the Cell

Eukaryotic organisms extract energy from organic molecules ("food") in a complex series of chemical reactions called **oxidative metabolism,** which takes place only in their mitochondria. **Mitochondria** (singular, **mitochondrion**) are sausage-shaped organelles about the size of a bacterial cell. Mitochondria are bounded by two membranes. The outer membrane, shown partially cut away in **figure 4.9,** is smooth and apparently derives from the plasma membrane of the host cell that first took up the bacterium long ago. The inner membrane, apparently the plasma membrane of the bacterium that gave rise to the mitochondrion, is bent into numerous folds called **cristae** (singular, **crista**) that resemble the folded plasma membranes in various groups of bacteria. The cutaway view of the figure shows how the cristae partition the mitochondrion into two compartments, an inner **matrix** and an outer compartment, called the *intermembrane space.* As you will learn in chapter 7, this architecture is critical to successfully carrying out oxidative metabolism.

During the 1.5 billion years in which mitochondria have existed in eukaryotic cells, most of their genes have been transferred to the chromosomes of the host cells. But mitochondria still have some of their original genes, contained in a circular, closed, naked molecule of DNA (called *mitochondrial DNA,* or *mtDNA*) that closely resembles the circular DNA molecule of a bacterium. On this mtDNA are several genes that produce some of the proteins essential for oxidative metabolism. In both mitochondria and bacteria, the circular DNA molecule is replicated during the process of division. When a mitochondrion divides, it copies its DNA located in the matrix and splits into two by simple fission, dividing much as bacteria do.

Chloroplasts: Energy-Capturing Centers

All photosynthesis in plants and algae takes place within another bacteria-like organelle, the **chloroplast (figure 4.10).** There is strong evidence that chloroplasts, like mitochondria, were derived from bacteria by symbiosis. A chloroplast is bounded, like a mitochondrion, by two membranes, the inner derived from the original bacterium and the outer resembling the host cell's ER. Chloroplasts are larger than mitochondria, and have a more complex organization. Inside the chloroplast, another series of membranes are fused to form stacks

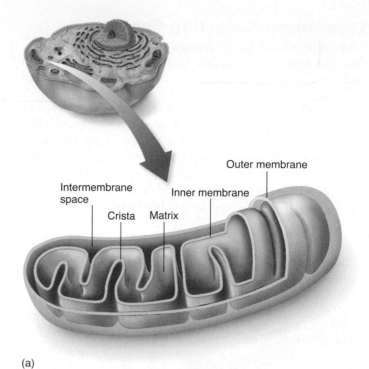

(a)

(b)

Figure 4.9 Mitochondria.

The mitochondria of a cell are sausage-shaped organelles within which oxidative metabolism takes place, and energy is extracted from food using oxygen. (*a*) A mitochondrion has a double membrane. The inner membrane is shaped into folds called cristae. The space within the cristae is called the matrix. The cristae greatly increase the surface area for oxidative metabolism. (*b*) Micrograph of two mitochondria, one in cross section, the other cut lengthwise.

of closed vesicles called **thylakoids,** the green disklike structures visible in the interior of the chloroplast in **figure 4.10.** The light-dependent reactions of photosynthesis take place within the thylakoids. The thylakoids are stacked on top of one another to form a column called a **granum** (plural, **grana**). The interior of a chloroplast is bathed with a semiliquid substance called the **stroma.**

Like mitochondria, chloroplasts have a circular DNA molecule. On this DNA are located many of the genes coding for the proteins necessary to carry out photosynthesis. Plant cells can contain from one to several hundred chloroplasts, depending on the species. Neither mitochondria nor chloroplasts can be grown in a cell-free culture; they are totally dependent on the cells within which they occur.

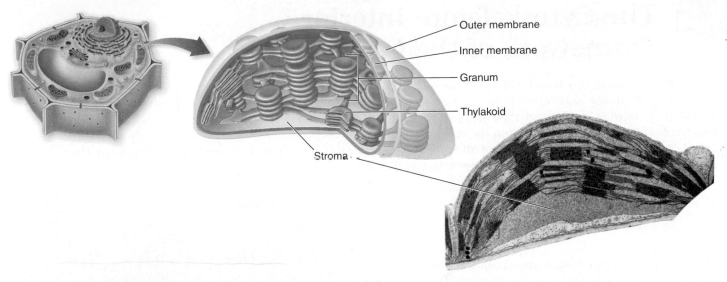

Figure 4.10 A chloroplast.

Bacteria-like organelles called chloroplasts are the sites of photosynthesis in photosynthetic eukaryotes. Like mitochondria, they have a complex system of internal membranes on which chemical reactions take place. The internal membranes of a chloroplast are fused to form stacks of closed vesicles called thylakoids. Photosynthesis occurs within these thylakoids. Thylakoids are stacked one on top of the other in columns called grana. The interior of the chloroplast is bathed in a semiliquid substance called the stroma.

Endosymbiosis

Symbiosis is a close relationship between organisms of different species that live together. The theory of **endosymbiosis** proposes that some of today's eukaryotic organelles evolved by a symbiosis in which one cell of a prokaryotic species was engulfed by and lived inside the cell of another species of prokaryote that was a precursor to eukaryotes. **Figure 4.11** shows how this is thought to have occurred. Many cells take up food or other substances through endocytosis, a process whereby the plasma membrane of a cell wraps around the substance, enclosing it within a vesicle inside the cell. Oftentimes, the contents in the vesicle are broken down with digestive enzymes. According to the endosymbiont theory this did not occur; instead, the engulfed prokaryotes provided their hosts with certain advantages associated with their special metabolic abilities. Two key eukaryotic organelles just described are believed to be the descendants of these endosymbiotic prokaryotes: mitochondria, which are thought to have originated as bacteria capable of carrying out oxidative metabolism; and chloroplasts, which apparently arose from photosynthetic bacteria.

The endosymbiont theory is supported by a wealth of evidence. Both mitochondria and chloroplasts are surrounded by two membranes; the inner membrane probably evolved from the plasma membrane of the engulfed bacterium, while the outer membrane is probably derived from the plasma membrane or endoplasmic reticulum of the host cell. Mitochondria are about the same size as most bacteria, and the cristae formed by their inner membranes resemble the folded membranes in various groups of bacteria. Mitochondrial ribosomes are also similar to bacterial ribosomes in size and structure. Both mitochondria and chloroplasts contain circular molecules of DNA similar to those in bacteria. Finally, mitochondria divide by simple fission, splitting in two just as bacterial cells do, and they apparently replicate and partition their DNA in much the same way as bacteria do.

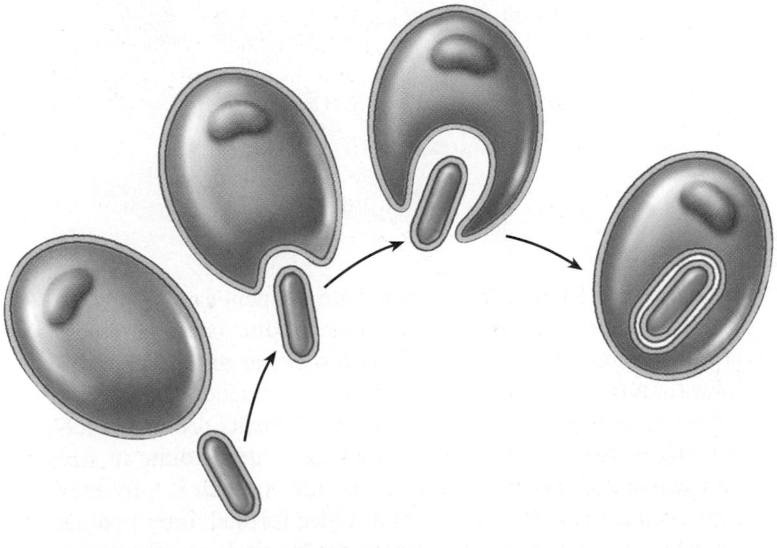

Figure 4.11 Endosymbiosis.

This figure shows how a double membrane may have been created during the symbiotic origin of mitochondria or chloroplasts.

> **Key Learning Outcome 4.7** Eukaryotic cells contain complex organelles that have their own DNA and are thought to have arisen by endosymbiosis from ancient bacteria.

4.8 The Cytoskeleton: Interior Framework of the Cell

If you were to shrink down and enter into the interior of a eukaryotic cell, your view would be similar to what you see in the illustration shown here: A dense network of protein fibers called the **cytoskeleton** provides a framework that supports the shape of the cell. The cytoskeleton also anchors organelles like the mitochondria to fixed locations within the cell interior. The protein fibers of the cytoskeleton are a dynamic system, constantly being formed and disassembled. There are three different kinds of protein fibers that make up the cytoskeleton, shown as enlargements below and in **table 4.2**.

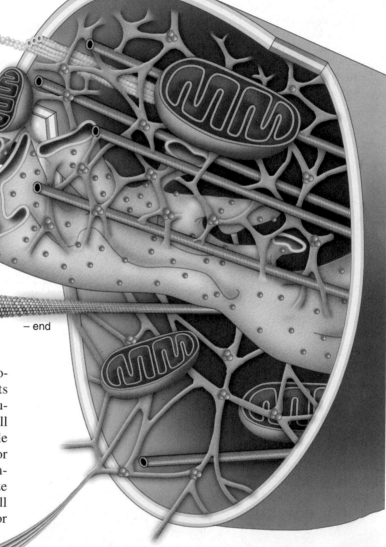

Actin filament

7 nm

Actin subunit

Microfilaments (Actin Filaments) Actin filaments are long fibers about 7 nanometers in diameter. Each filament is composed of two protein chains loosely twined together like two strands of pearls. Each "pearl," or subunit, on the chains is the globular protein **actin.** Actin filaments are found throughout the cell but are most highly concentrated just inside the plasma membrane. Actin filaments are responsible for cellular movements such as contraction, crawling, "pinching" during division, and formation of cellular extensions.

Microtubule

25 nm

Tubulin subunit

+ end

− end

Microtubules Microtubules are hollow tubes about 25 nanometers in diameter composed of *tubulin* protein subunits arranged side by side to form a tube. In many cells, microtubules form from nucleation centers near the center of the cell and radiate toward the periphery. The ends of the microtubule are designated as "+" (away from the nucleation center) or "−" (toward the nucleation center). Microtubules are comparatively stiff cytoskeletal elements that serve to organize metabolism and intracellular transport in the nondividing cell and to stabilize cell structure. They are also responsible for the movement of chromosomes in mitosis.

Intermediate filament

10 nm

Fibrous protein

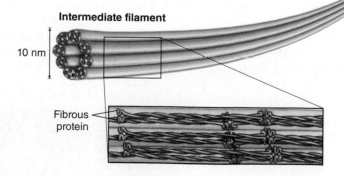

Intermediate Filaments Intermediate filaments are composed of overlapping staggered tetramers of protein. These tetramers are then bundled into cables. This molecular arrangement allows for a ropelike structure that imparts tremendous mechanical strength to the cell. Intermediate filaments are characteristically 8 to 10 nanometers in diameter, intermediate in size between actin filaments and microtubules (which is why they are called *intermediate* filaments). Once formed, intermediate filaments are stable and usually do not break down. They provide structural reinforcement to the cell and organelles.

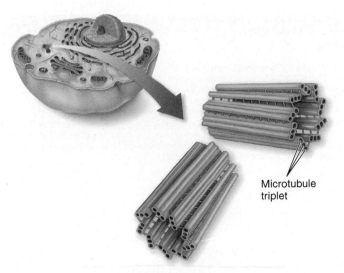

Figure 4.12   Centrioles.
Centrioles anchor and assemble microtubules. Centrioles usually occur in pairs and are composed of nine triplets of microtubules.

The cytoskeleton plays a major role in determining the shape of animal cells, which lack rigid cell walls. Because filaments can form and dissolve readily, the shape of an animal cell can change rapidly. If you examine the surface of an animal cell with a microscope, you will often find it alive with motion, projections shooting out from the surface and then retracting, only to shoot out elsewhere moments later.

The cytoskeleton is not only responsible for the cell's shape, but it also provides a scaffold both for ribosomes to carry out protein synthesis and for enzymes to be localized within defined areas of the cytoplasm. By anchoring particular enzymes near one another, the cytoskeleton participates with organelles in organizing the cell's activities.

Centrioles

Complex structures called **centrioles** assemble microtubules from tubulin subunits in the cells of animals and most protists. Centrioles occur in pairs within the cytoplasm, usually located at right angles to one another as you can see in **figure 4.12**. They are usually found near the nuclear envelope and are among the most structurally complex microtubular assemblies of the cell. In cells that contain flagella or cilia, each flagellum or cilium is anchored by a form of centriole called a *basal body*. Most animal and protist cells have both centrioles and basal bodies; higher plants and fungi lack them, instead organizing microtubules without such structures. Although they lack a membrane, centrioles resemble spirochete bacteria in many other respects. Some biologists believe that centrioles, like mitochondria and chloroplasts, originated as symbiotic bacteria.

Vacuoles: Storage Compartments

Within the interiors of plant and many protist cells, the cytoskeleton positions not only organelles, but also storage compartments that are membrane-bounded, called **vacuoles.** The center of the plant cell shown above in **figure 4.13**, as

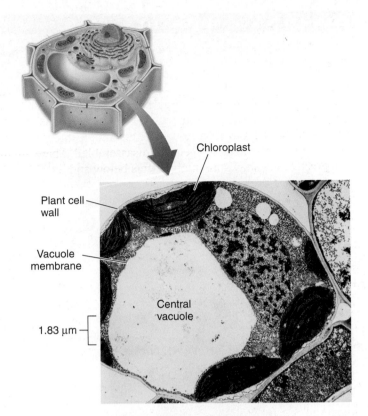

Figure 4.13   A plant central vacuole.
A plant's central vacuole stores dissolved substances and can increase in size to increase the surface area of a plant cell.

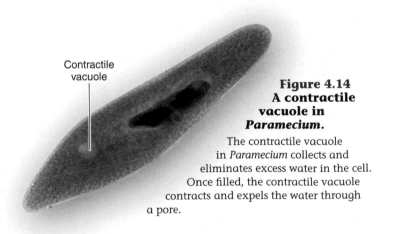

**Figure 4.14
A contractile vacuole in *Paramecium*.**

The contractile vacuole in *Paramecium* collects and eliminates excess water in the cell. Once filled, the contractile vacuole contracts and expels the water through a pore.

in all plant cells, contains a large, apparently empty space, called the *central vacuole*. This vacuole is not really empty; it contains large amounts of water and other materials, such as sugars, ions, and pigments. The central vacuole functions as a storage center for these important substances.

Certain protists, like *Paramecium* in **figure 4.14**, contain a **contractile vacuole** near the cell surface that accumulates excess water. This vacuole is bounded by actin filaments and has a small pore that opens to the outside of the cell. By rhythmic ATP-powered contractions, it pumps accumulated water out through the pore.

TABLE 4.2   EUKARYOTIC CELL STRUCTURES AND THEIR FUNCTIONS

Structure	Description

Plasma Membrane

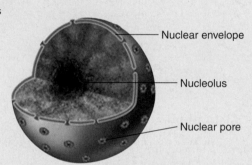

The plasma membrane is a phospholipid bilayer embedded with proteins that encloses a cell and separates its contents from its surroundings. The bilayer results from the tail-to-tail packing of the phospholipid molecules that make up the membrane. The proteins embedded in the lipid bilayer are in large part responsible for a cell's ability to interact with its environment. Transport proteins provide channels through which molecules and ions enter and leave the cell across the plasma membrane. Receptor proteins induce changes within the cell when they come in contact with specific molecules in the environment, such as hormones, or with molecules on the surface of neighboring cells.

Nucleus

Every cell contains DNA, the hereditary material. The DNA of eukaryotes is isolated within the nucleus, a spherical organelle surrounded by a double membrane structure called the nuclear envelope. This envelope is studded with pores that control traffic into and out of the nucleus. The DNA contains the genes that code for the proteins synthesized by the cell. Stabilized by proteins, it forms chromatin, the major component of the nucleus. When the cell prepares to divide, the chromatin of the nucleus condenses into threadlike chromosomes.

Endoplasmic Reticulum

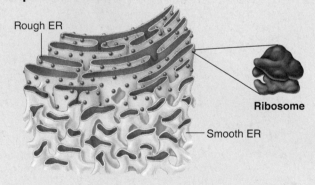

The hallmark of the eukaryotic cell is compartmentalization, achieved by an extensive endomembrane system that weaves through the cell interior. The membrane network is called the endoplasmic reticulum, or ER. The ER begins at the nuclear envelope and extends out into the cytoplasm, its sheets of membrane weaving through the cell interior. Rough ER contains numerous ribosomes that give it a bumpy appearance. These ribosomes manufacture proteins destined for the ER or for other parts of the cell. ER without these attached ribosomes is called smooth ER, which often functions to detoxify harmful substances or to aid in the synthesis of lipids. Sugar side chains are added to molecules as they pass through the ER. Delivery of molecules to other parts of the cell is via vesicles that pinch off from the borders of the rough ER.

Golgi Complex

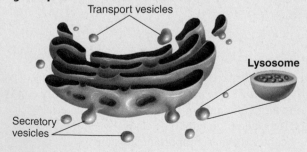

At various locations within the cytoplasm flattened stacks of membranes occur. Animal cells may contain 20, plant cells several hundred. Collectively, they are referred to as the Golgi complex. Molecules manufactured in the ER pass to the Golgi complex within vesicles. The Golgi sorts and packages these molecules and also synthesizes carbohydrates. The Golgi adds sugar side chains to molecules as they pass through the stacks of membranes. The Golgi then directs the molecules to lysosomes, secretory vesicles, or the plasma membrane.

Structure	Description

Mitochondrion

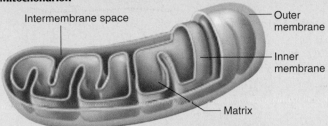

Intermembrane space

Outer membrane

Inner membrane

Matrix

Mitochondria are bacteria-like organelles that are responsible for extracting most of the energy a cell derives from the food molecules it consumes. Two membranes encase each mitochondrion, separated by an intermembrane space. The key energy-harvesting chemical reactions occur within the interior matrix. The energy is used to pump protons from the matrix into the intermembrane space; their return across this membrane drives the synthesis of ATP, the energy currency of the cell.

Chloroplast

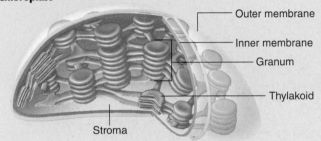

Outer membrane

Inner membrane

Granum

Thylakoid

Stroma

The green color of plants and algae results from cell organelles called chloroplasts rich in the photosynthetic green pigment chlorophyll. Photosynthesis is the sunlight-powered process that converts CO_2 in the air to the organic molecules of which all living things are composed. Chloroplasts, like mitochondria, are composed of two membranes separated by an intermembrane space. In a chloroplast, the inner membrane pinches into a series of sacs called thylakoids, which pile up in columns called grana. The chlorophyll-facilitated light reactions of photosynthesis take place within the thylakoids. These are suspended in a semiliquid substance called the stroma.

Cytoskeleton

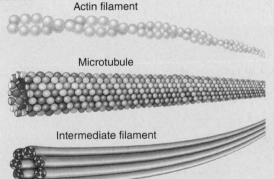

Actin filament

Microtubule

Intermediate filament

The cytoplasm of all eukaryotic cells is crisscrossed by a network of protein fibers called the cytoskeleton that supports the shape of the cell and anchors organelles to fixed locations. The cytoskeleton is a dynamic structure, composed of three kinds of fibers. Long actin filaments are responsible for cellular movements such as contraction, crawling, and the "pinching" that occurs as cells divide. Hollow microtubule tubes, constantly forming and disassembling, facilitate cellular movements and are responsible for moving materials within the cell. Special motor proteins move organelles around the cell on microtubular "tracks." Durable intermediate filaments provide the cell with structural stability.

Centrioles

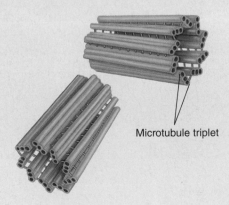

Microtubule triplet

Centrioles are barrel-shaped organelles found in the cells of animals and most protists. They occur in pairs, usually located at right angles to each other near the nucleus. Centrioles help assemble the animal cell's microtubules, playing a key role in producing the microtubules that move chromosomes during cell division. Centrioles are also involved in the formation of cilia and flagella, which are composed of sets of microtubules. Cells of plants and fungi lack centrioles, and cell biologists are still in the process of characterizing their microtubule-organizing centers.

Cell Movement

Essentially, all cell motion is tied to the movement of actin filaments, microtubules, or both. Intermediate filaments act as intracellular tendons, preventing excessive stretching of cells, and actin filaments play a major role in determining the shape of cells. Because actin filaments can form and dissolve so readily, they enable some cells to change shape quickly. If you look at the surfaces of such cells under a microscope, you will find them moving and changing shape.

Some Cells Crawl It is the arrangement of actin filaments within the cell cytoplasm that allows cells to "crawl," literally! Crawling is a significant cellular phenomenon, essential to inflammation, clotting, wound healing, and the spread of cancer. White blood cells in particular exhibit this ability. Produced in the bone marrow, these cells are released into the circulatory system and then eventually crawl out of capillaries and into the tissues to destroy potential pathogens. The crawling mechanism is an exquisite example of cellular coordination.

Cytoskeletal fibers play a role in other types of cell movement. For example, during animal cell reproduction (see chapters 8 and 9), chromosomes move to opposite sides of a dividing cell because they are attached to shortening microtubules. The cell then pinches in two when a belt of actin filaments contracts like a purse string. Muscle cells also use actin filaments to contract their cytoskeletons. The fluttering of an eyelash, the flight of an eagle, and the awkward crawling of a baby all depend on these cytoskeletal movements within muscle cells.

Swimming with Flagella and Cilia Some eukaryotic cells contain **flagella** (singular, **flagellum**), fine, long, threadlike organelles protruding from the cell surface. The cutaway view in **figure 4.15** shows how a flagellum arises from a microtubular structure called a **basal body,** with groups of microtubules arranged in rows of three, as you can see in the cross-sectional view. Some of these microtubules extend up into the flagellum, which consists of a circle of nine microtubule pairs surrounding two central ones (again seen in the cross-sectional view). This **9 + 2 arrangement** is a fundamental feature of eukaryotes and apparently evolved early in their history. Even in cells that lack

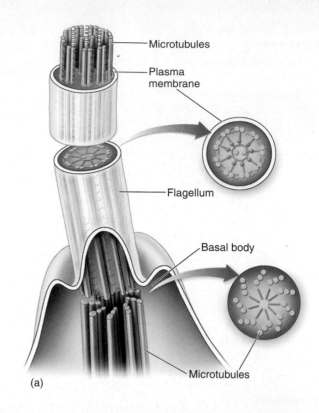

(a)

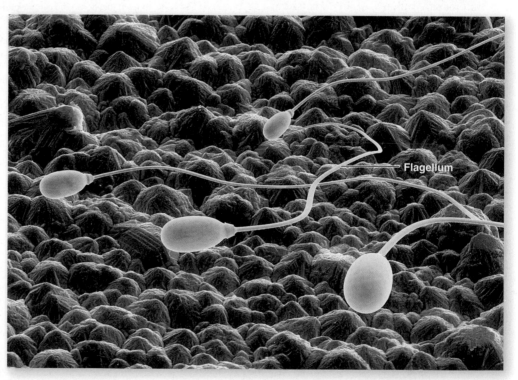

(b) Human sperm cells

Figure 4.15 Flagella.

(*a*) A eukaryotic flagellum springs directly from a basal body and is composed of a ring of nine pairs of microtubules with two microtubules in its core. (*b*) Human sperm cells have one flagellum.

flagella, derived structures with the same 9 + 2 arrangement often occur, like in the sensory hairs of the human ear. In humans, we find a single long flagellum on each sperm cell that propels the cell in a swimming motion. If flagella are numerous and organized in dense rows, they are called **cilia.** Cilia do not differ from flagella in their structure, but cilia are usually short. The *Paramecium* in **figure 4.16** is covered with cilia, giving it a furry appearance. In humans, dense mats of cilia project from cells that line our breathing tube, the trachea, to move mucus and dust particles out of the respiratory tract into the throat (where we can expel these unneeded contaminants by spitting or swallowing). Eukaryotic flagella serve a similar function as flagella found in prokaryotes, discussed in section 4.3, but they are very different structurally.

Moving Materials Within the Cell

All eukaryotic cells must move materials from one place to another in the cytoplasm. Most cells use the endomembrane system as an intracellular highway. As you saw on page 82, the Golgi complex packages materials into vesicles that come from the channels of the endoplasmic reticulum to the far reaches of the cell. However, this highway is only effective over short distances. When a cell has to transport materials through long extensions like the axon of a nerve cell, the endomembrane system highways are too slow. For these situations, eukaryotic cells have developed high-speed locomotives that run along microtubular tracks. Lysosomes move along such microtubular tracks to reach a food vacuole, and mitochondria travel down them to the far-away tips of long axons.

For long-distance intracellular transport, four components are required, illustrated in **figure 4.17**: a vesicle or organelle (the light tan structure) that is to be transported; a motor molecule, in this case dynein, that provides the energy-driven motion; connector molecules that connect the vesicle to the motor molecule (in the figure, they are the dynactin complex and other associated proteins); and microtubules (the green tube) on which the vesicle will ride like a train on a rail, pulled by a locomotive. As nature's tiniest motors, these motor proteins literally pull the transport vesicles along the microtubular tracks.

How does such a tiny motor work? A motor protein uses ATP to power its movement. Scientists have proposed that they use a type of stepping action. Also, the *direction* of movement is different for different motor proteins. The **dynein** motor protein shown in **figure 4.17** moves inward toward the cell's center, dragging the vesicle with it as it travels along toward the "−" end of a microtubule. Another motor protein, **kinesin,** directs movement in the opposite direction, toward the "+" end of a microtubule, which is toward the periphery of the cell. The destination of a particular transport vesicle and its contents is thus determined by the nature of the linking protein embedded within the

Figure 4.16   Cilia.
The surface of this *Paramecium* is covered with a dense forest of cilia.

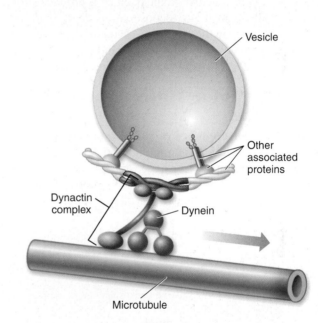

Figure 4.17   Molecular motors.
Vesicles that are transported within cells are attached with connector molecules, such as the dynactin complex shown here, to motor molecules, like dynein, which move along microtubules.

vesicle's membrane. Like possessing a ticket to one of two destinations, if a vesicle links to kinesin, it moves outward; if it links to dynein, it moves inward.

Key Learning Outcome 4.8   The cytoskeleton is a latticework of protein fibers that determines a cell's shape and anchors organelles to particular locations within the cytoplasm. Cells can move by changing their shape and move substances around the cell using molecular motors.

4.9 Outside the Plasma Membrane

Cell Walls Offer Protection and Support

Plants, fungi, and many protists cells share a characteristic with bacteria that is not shared with animal cells—that is, they have **cell walls,** which protect and support their cells. Eukaryotic cell walls are chemically and structurally different from bacterial cell walls. In plants, cell walls are composed of fibers of the polysaccharide cellulose, while in fungi they are composed of chitin. The **primary walls** of plant cells are laid down when the cell is still growing. These are the thinner, outer walls of the cells shown below in **figure 4.18**. Between the walls of adjacent cells is a sticky substance called the **middle lamella,** which glues the cells together. Some plant cells produce strong **secondary walls,** which are deposited inside the primary walls. As you can see in the photo, the secondary cell walls are very thick compared to the primary walls and therefore are not deposited until the cell has finished increasing in size.

An Extracellular Matrix Surrounds Animal Cells

As we discussed, many types of eukaryotic cells possess a cell wall exterior to the plasma membrane. The wall acts to protect the cell, maintain its shape, and prevent excessive water uptake. Animal cells are the great exception, lacking the cell walls that encase the cells of plants, fungi, and most protists. Animal cells secrete an elaborate mixture of **glycoproteins** (proteins with short chains of sugars attached to them) into the space around them, forming the **extracellular matrix (ECM),** which performs a function different than cell walls.

The fibrous protein collagen, the same protein in cartilage and ligaments, is abundant in the ECM. **Figure 4.19** shows how these fibers of collagen and another fibrous protein, elastin, are embedded within a complex web of other glycoproteins called proteoglycans, which form a protective layer over the cell surface.

The ECM is attached to the plasma membrane by a third kind of glycoprotein, **fibronectin.** As you can see in the figure, fibronectin molecules bind not only to ECM glycoproteins but also to proteins called **integrins,** which are an integral part of the plasma membrane. Integrins extend into the cytoplasm, where they are attached to the microfilaments of the cytoskeleton. Linking ECM and cytoskeleton, integrins allow the ECM to influence cell behavior in important ways, altering gene expression and cell migration patterns by a combination of mechanical and chemical signaling pathways. In this way, the ECM can help coordinate the behavior of all the cells in a particular tissue.

> **Key Learning Outcome 4.9** Plant and protist cells encase themselves within a strong cell wall. In animal cells, which lack a cell wall, the cytoskeleton is linked by integrin proteins to a web of glycoproteins called the extracellular matrix.

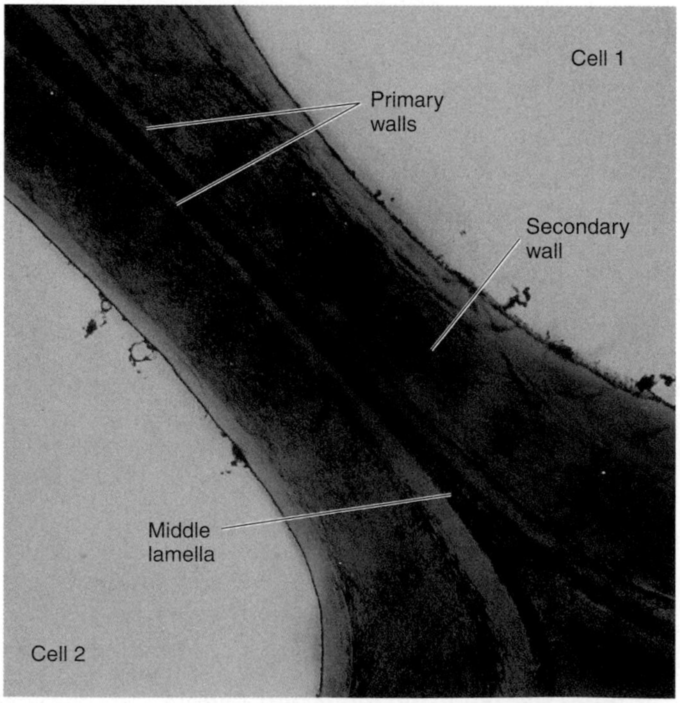

Figure 4.18   Cell walls in plants.

Plant cell walls are thick, strong, and rigid. Primary cell walls are laid down when the cell is young. Thicker secondary cell walls may be added later when the cell is fully grown. The middle lamella lies between the walls of adjacent cells and glues the cells together.

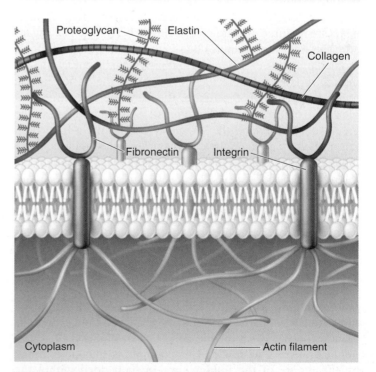

Figure 4.19   The extracellular matrix.

Animal cells are surrounded by an extracellular matrix (ECM) composed of various glycoproteins. The ECM carries out a variety of functions that influence cell behavior, including cell migration, gene expression, and the coordination of signaling between cells.

4.10 Diffusion

For cells to survive, food particles, water, and other materials must pass into the cell, and waste materials must be eliminated. All of this moving back and forth across the cell's plasma membrane occurs in one of three ways: (1) water and other substances diffuse through the membrane; (2) proteins in the membrane act as doors that admit certain molecules only; or (3) food particles and sometimes liquids are engulfed by the membrane folding around them. First we will examine diffusion.

Diffusion

Most molecules are in constant motion. How a molecule moves—just where it goes—is totally random, like shaking marbles in a cup. So, if two kinds of molecules are added together, they soon mix. The random motion of molecules always tends to produce uniform mixtures. To see how this works, visualize the simple experiment shown in the Key Biological Process illustration below, in which a lump of sugar is dropped into a beaker of water. The lump slowly breaks apart into individual sugar molecules, which move about randomly until eventually the sugar molecules become evenly distributed throughout the water in the beaker (**panel 4** below) as individual molecules take long random journeys. This process of random molecular mixing is called **diffusion.**

Selective Permeability

Diffusion becomes important to the life of a cell because of the chemical nature of biological membranes. As **figure 4.20** illustrates, the nonpolar nature of the lipid bilayer determines which substances can pass and which ones cannot. Oxygen gas and carbon dioxide are not repelled by the bilayer and can freely cross, and so can small nonpolar fats and lipids. However, sugars like glucose cannot, and neither can proteins. In fact, no polar molecule can freely travel across a biological membrane because of the barrier to diffusion imposed by the lipid bilayer. This is true of charged ions like Na^+, Cl^-, and H^+ and is also true of water molecules, which are very polar. It was once thought that water somehow "leaked" across the plasma membrane, slipping through gaps that open up when the hydrocarbon tails of the bilayer flex and bend, but biologists now discount this, as discussed on page 76. Later in this chapter we will further explore the movement of water molecules across plasma membranes.

Because the plasma membrane admits some substances and not others, it is said to be **selectively permeable.** The selective permeability of a cell's plasma membrane is arguably its most important property. What a cell is and how it functions are largely determined by which molecules it admits and which it refuses. As you proceed through this chapter, you will encounter a variety of ways in which particular cell types exercise control over their admission policy.

KEY BIOLOGICAL PROCESS: Diffusion

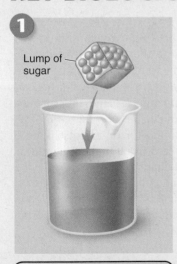

1

Lump of sugar

A lump of sugar is dropped into a beaker of water.

2

Sugar molecule

Sugar molecules begin to break off from the lump.

3

More and more sugar molecules move away and randomly bounce around.

4

Eventually, all of the sugar molecules become evenly distributed throughout the water.

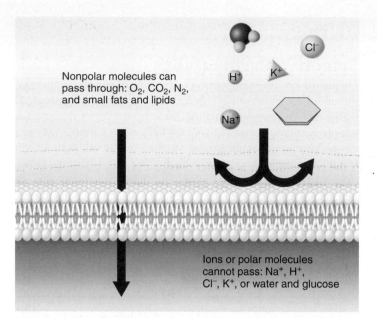

Nonpolar molecules can pass through: O_2, CO_2, N_2, and small fats and lipids

Cl^-

H^+ K^+

Na^+

Ions or polar molecules cannot pass: Na^+, H^+, Cl^-, K^+, or water and glucose

Figure 4.20 Membranes are selectively permeable.
Nonpolar molecules, such as those on the *left*, can pass through the membrane, but polar molecules and ions (on the *right*) cannot.

Concentration Gradients

The number of molecules present per unit volume on one side of a membrane is its *concentration* there. As a direct consequence of a cell membrane's selective permeability (**figure 4.20**), polar and charged molecules and ions often have a different concentration on one side of a membrane than on the other. This difference in concentration levels is called a **concentration gradient.**

When a substance moves from regions where its concentration is high to regions where its concentration is lower, it is said to move *down* its concentration gradient. How does a molecule "know" in what direction to move? It doesn't—molecules don't "know" anything. A molecule is equally likely to move in any direction and is constantly changing course in random ways. There are simply more molecules able to move from where they are common than from where they are scarce. This mixing process is simply the result of diffusion. In fact, diffusion is formally defined as *the net movement of molecules down a concentration gradient toward regions of lower concentration* (that is, where there are relatively fewer of them) *as a result of random motion.*

Importantly, each substance tends to diffuse in the direction established by its own concentration gradient, not the gradients established by other substances present in the same fluid. Thus the rate at which oxygen diffuses across a plant cell's plasma membrane into the cell is affected by that cell's oxygen concentration relative to the air, but not by its carbon dioxide concentration.

Molecules in Motion

If you think about it, a concentration gradient is a form of energy. Just as a boulder perched on a hilltop stores the energy it took to lift it there—energy released if the boulder rolls downhill, so a concentration gradient across a membrane can drive the movement of molecules across the membrane in the direction of lower concentration.

How fast molecules diffuse across a membrane (what physiologists call the rate of diffusion) is determined by two characteristics of the cell, and also by the physical characteristics of the environment in which the cell finds itself.

The Steepness of the Concentration Gradient Diffusion is most rapid when gradients are steepest. You can see why this must be so: Many more molecules randomly move away from a region of high concentration than into it from a region of low concentration. As this process continues, the number of molecules in the region of high concentration continually decreases, and the rate of diffusion slows down as fewer molecules leave and more arrive. Eventually, the same number is arriving as leaving. At this point, diffusion slows to a halt. Molecules are still moving, but the relative concentrations of the two regions no longer change—what a physiologist calls a state of dynamic equilibrium.

The Area of the Membrane Available for Diffusion Some gases like oxygen and a variety of small lipid-soluble molecules diffuse readily across the lipid bilayer of biological membranes, and the rate of their diffusion into or out of a cell is most rapid when the proportion of membrane surface occupied by the lipid bilayer is greatest. Membranes with larger portions of embedded proteins will exhibit lower rates of diffusion of gases and lipids moving through the bilayer, as the area available for their diffusion is less. Similarly, the rate of diffusion of charged ions like Na^+ and polar molecules like sugars and amino acids is greatest in membranes with the greatest number of protein channels with passages through which these molecules can pass. Because these channels are often quite specific, allowing only a particular ion or molecule to pass, the rate of diffusion of a substance is only affected by the number of channels actually available to it, rather than by the overall proportion of the membrane surface taken up by protein.

Physical Characteristics of the Cell Environment Temperature has a strong influence on the rate of diffusion, for the simple reason that higher temperatures cause molecules to move faster. In general, the rate of diffusion is greater for cells of organisms living at higher temperatures. High pressure also encourages faster diffusion, as molecules collide more often. This effect becomes quite important for organisms living in the deep ocean, where pressures are much higher than on the earth's surface. A third physical characteristic that influences the rate of diffusion across some cells in an important way is the electrical field in which a cell finds itself. An electric gradient across a nerve cell membrane can have an important influence on the rate at which ions diffuse in and out.

> **Key Learning Outcome 4.10** Random movements of molecules cause them to mix uniformly in solution, a process called diffusion. Molecules tend to diffuse down their concentration gradients, faster when the gradient is steeper.

4.11 Facilitated Diffusion

The selective permeability of biological membranes is perhaps their most important property. Ions and polar molecules can only cross the lipid core of membranes by passing through protein channels that bridge the bilayer.

Open Channels

The simplest of these channels are so-called *open channels,* shaped like tubes and functioning as open doors. As long as a molecule fits the channel, it is free to pass through in either direction, like a marble through a donut. Diffusion tends to equalize the concentration of such molecules on both sides of the membrane, with the molecules moving toward the side where they are scarcest. Many of the cell's water and ion channels are open channels, simple open pores that span the membrane. The open channels are selective, as only ions and molecules that precisely fit the pore can diffuse through it, in either direction. Often the pore has a "gate," a door that must be opened before an ion can pass through. Open ion channels with gates that swing open or shut in response to electrical charge play an essential role in signaling by the nervous system.

Carrier Proteins

There are limits to the specificity of open channels, as many different polar molecules are roughly the same size, shape, and charge. To increase the selectivity of membrane transport, cells employ a more complex channel that requires the diffusing molecule to bind to the surface of a "carrier" protein. Such hand-in-glove binding can be very specific. Once having bound their cargo, a carrier protein then physically carries the diffusing molecule across the membrane.

Each carrier protein binds with only a certain molecule, such as a particular sugar, amino acid, or ion, physically binding them on one side of the membrane and releasing them on the other. The direction of the molecule's net movement depends only on its concentration gradient across the membrane. If the concentration is greater outside the cell, the molecule is more likely to bind to the carrier on the extracellular side of the membrane, as in **panel 1** of the Key Biological Process illustration below, and be released on the cytoplasmic side, as in **panel 3**. The net movement always occurs from high concentration to low, just as it does in simple diffusion, but the process is facilitated by the carriers. For this reason, this mechanism of transport is given a special name, **facilitated diffusion.**

A characteristic feature of transport by carrier proteins is that its rate can be saturated. If the concentration of a substance is progressively increased, the rate of transport of the substance increases up to a certain point and then levels off. There are a limited number of carrier proteins in the membrane, and when the concentration of the transported substance is raised high enough, all the carriers will be in use. The transport system is then said to be "saturated."

> **Key Learning Outcome 4.11** Facilitated diffusion is the selective transport of substances across a membrane using a protein channel or carrier in the direction of lower concentration.

KEY BIOLOGICAL PROCESS: Facilitated Diffusion

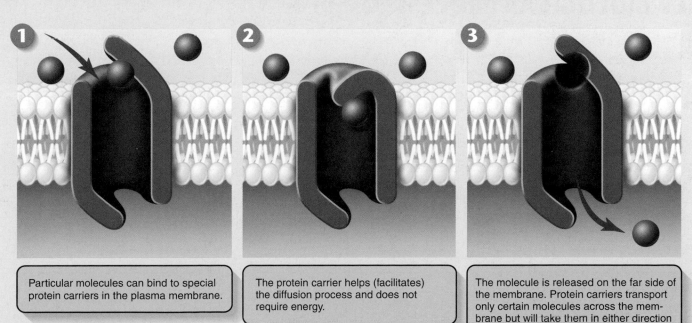

1 Particular molecules can bind to special protein carriers in the plasma membrane.

2 The protein carrier helps (facilitates) the diffusion process and does not require energy.

3 The molecule is released on the far side of the membrane. Protein carriers transport only certain molecules across the membrane but will take them in either direction down their concentration gradients.

4.12 Osmosis

Diffusion allows molecules like oxygen, carbon dioxide, and nonpolar lipids to cross the plasma membrane. The movement of water molecules is not blocked because there are many small channels, called aquaporins, that allow water to pass freely through the membrane (see page 76).

Because water is so important, biologically, the diffusion of water molecules from an area of high concentration to an area of lower concentration is given a specific name, called **osmosis.** However, the number of water molecules that are free to diffuse across the membrane is dependent upon the concentration of other substances in solution. To understand how water moves into and out of a cell, let's focus on the water molecules already present inside a cell. What are they doing? Many of them are interacting with the sugars, proteins, and other polar molecules inside. Remember, water is very polar itself and readily interacts with other polar molecules. These "social" water molecules are not randomly moving about as they were outside; instead, they remain clustered around the polar molecules they are interacting with. As a result, while water molecules keep coming into the cell by random motion, they don't randomly come out again. Because more water molecules come in than go out, there is a net movement of water into the cell. The experiment presented in the Key Biological Process illustration below shows what happens. The right side of the beaker represents the inside of a cell and the left side is a watery environment. When the polar molecule urea is present in the cell, as in **panel 2** below, water molecules cluster around each urea molecule and are no longer able to pass through the membrane to the "outside." In effect, the polar solute has reduced the number of free water molecules.

Because the "outside" of the cell (on the left) has more unbound water molecules, water moves by diffusion into the cell (from the left to the right).

The concentration of all particles dissolved in a solution (called **solutes**) is called the osmotic concentration of the solution. If two solutions have unequal osmotic concentrations, the solution with the higher solute concentration, like the right side of the beaker below, is said to be **hypertonic** (Greek *hyper,* more than), and the solution with the lower concentration, like the left side of the beaker, is **hypotonic** (Greek *hypo,* less than). If the osmotic concentrations of the two solutions are equal, the solutions are **isotonic** (Greek *iso,* the same).

In cells, the plasma membrane separates two aqueous solutions, one inside the cell (the cytoplasm) and one outside (the extracellular fluid). The direction of the net diffusion of water across this membrane is determined by the osmotic concentrations of the solutions on either side. For example, if the cytoplasm of a cell was hypotonic to the extracellular fluid, water would diffuse out of the cell, toward the solution with the higher concentration of solutes (and, therefore, the lower concentration of unbound water molecules). This loss of water from the cytoplasm would cause the cell to shrink until the osmotic concentrations of the cytoplasm and the extracellular fluid become equal.

Osmotic Pressure

What would happen if the cell's cytoplasm were hypertonic to the extracellular fluid? In this situation, water would diffuse into the cell from the extracellular fluid, causing the cell to swell. The pressure of the cytoplasm pushing out against the cell membrane, called **hydrostatic pressure,** would increase. On the other hand, the **osmotic pressure,** defined as

KEY BIOLOGICAL PROCESS: Osmosis

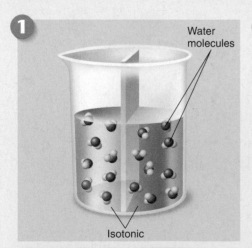

1

Water molecules

Isotonic

Diffusion causes water molecules to distribute themselves equally on both sides of a semipermeable membrane.

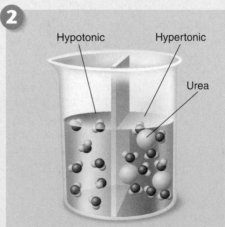

2

Hypotonic   Hypertonic

Urea

Addition of solute molecules that cannot cross the membrane reduces the number of free water molecules on that side, as they bind to the solute.

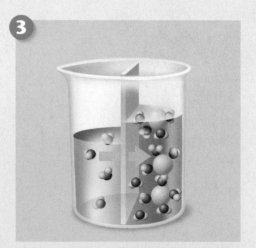

3

Diffusion then causes free water molecules to move from the side where their concentration is higher to the solute side, where their concentration is lower.

the pressure that must be applied to stop the osmotic movement of water across a membrane, would also be at work. If the membrane were strong enough, the cell would reach an equilibrium, at which the osmotic pressure, which tends to drive water into the cell, is exactly counterbalanced by the hydrostatic pressure, which tends to drive water back out of the cell. However, a plasma membrane by itself cannot withstand large internal pressures, and an isolated cell under such conditions would burst like an overinflated balloon. Accordingly, it is important for animal cells to maintain isotonic conditions.

Figure 4.21 illustrates how solutes create osmotic pressure. Look first at the red blood cells at the top of the figure. On the left, in a hypertonic solution like ocean water, there is a net movement of water molecules out of the red blood cell toward the higher concentration of solutes outside, causing the cell to shrivel. In an isotonic solution (in the middle), the concentration of solutes on either side of the red blood cell membrane is the same. Osmosis still occurs, but water diffuses into and out of the cell at the same rate, and the cell doesn't change size. In a hypotonic solution on the right, the concentration of solutes is higher within the cell than outside, so the net movement of water is into the cell. This is the situation utilized by Peter Agre in the experiment described on page 76 in which he demonstrated that aquaporin was a functioning water channel. A red blood cell is an enclosed structure, so as water entered a cell placed in a hypotonic solution (in his case, pure water), pressure is applied to the cell membrane causing the cell to swell, becoming spherical. This swelling can continue until the cell membrane can stretch no more and ruptures.

Now look at the plant cells at the bottom of figure 4.21. In these cells, unlike animal cells, the hydrostatic pressure generated by osmosis is counterbalanced by osmotic pressure, the force required to stop the flow of water into the cell. Plant cells have strong cell walls that can apply adequate osmotic pressure to keep the cell from rupturing.

Maintaining Osmotic Balance

Organisms have developed many solutions to the osmotic dilemma posed by being hypertonic to their environment.

Extrusion Some single-celled eukaryotes like the protist *Paramecium* use organelles called contractile vacuoles to remove water. Each vacuole collects water from various parts of the cytoplasm and transports it to the central part of the vacuole, near the cell surface. The vacuole possesses a small pore that opens to the outside of the cell. By contracting rhythmically, the vacuole pumps water out through the pore that is continuously seeping into the cell by osmosis.

Isosmotic Solutions Some organisms that live in the ocean adjust their internal concentration of solutes to match that of the surrounding seawater. Isotonic with respect to their environment, there is no net flow of water into or out of these cells. Many terrestrial animals solve the problem in a similar way, by circulating a fluid through their bodies that bathes cells in an isotonic solution. The blood in your body, for example, contains a high concentration of the protein albumin, which elevates the solute concentration of the blood to match your

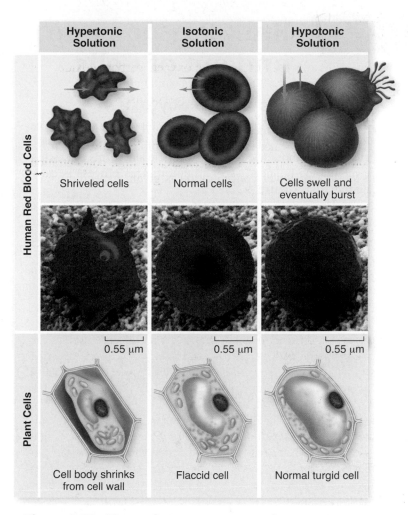

Figure 4.21 How solutes create osmotic pressure.
A cell is an enclosed structure, and so as water enters the cell from a hypotonic solution, pressure is applied to the plasma membrane until the cell ruptures. This hydrostatic pressure is counterbalanced by osmotic pressure, the force required to stop the flow of water into the cell. Plant cells have strong cell walls that can apply adequate osmotic pressure to keep the cell from rupturing.

cells. Sharks maintain a high concentration of urea in their blood and body fluids, keeping their cells isosmotic with respect to the sea water in which they swim.

Turgor Most plant cells are hypertonic to their immediate environment, containing a high concentration of solutes in their central vacuoles. The resulting internal hydrostatic pressure, known as **turgor pressure**, presses the plasma membrane firmly against the interior of the cell wall as you saw in **figure 4.7**, making the cell rigid. Most green plants depend on turgor pressure to maintain their shape, and wilt when they lack sufficient water.

> **Key Learning Outcome 4.12** Water molecules associated with polar solutes are not free to diffuse, so there is a net movement of water across a membrane toward the side with less "free" water. Osmosis is the diffusion of water, but not solutes, across a membrane. Cells must maintain an osmotic balance to function properly.

Bulk Passage into and out of Cells

Endocytosis and Exocytosis

The cells of many eukaryotes take in food and liquids by extending their plasma membranes outward toward food particles. The membrane engulfs the particle and forms a vesicle—a membrane-bounded sac—around it. This process is called **endocytosis** (figure 4.22).

The reverse of endocytosis is **exocytosis,** the discharge of material from vesicles at the cell surface. The vesicle in **figure 4.23** contains a substance to be discharged, or released, from the cell. The purple particles remain suspended in the vesicle as it fuses with the plasma membrane. The membrane that forms the vesicle is made of phospholipids, and as it comes in contact with the plasma membrane, the phospholipids of both membranes interact, forming a pore through which the contents leave the vesicle to the outside. In plant cells, exocytosis is an important means of exporting the materials needed to construct the cell wall through the plasma membrane. Among protists, the discharge of a contractile vacuole is a form of exocytosis. In animal cells, exocytosis provides a mechanism for secreting many hormones, neurotransmitters, digestive enzymes, and other substances.

Phagocytosis and Pinocytosis

If the material the cell takes in is particulate (made up of discrete particles), such as an organism, like the red bacterium in **figure 4.22a**, or some other fragment of organic matter, the process is called **phagocytosis** (Greek *phagein,* to eat, and *cytos,* cell). If the material the cell takes in is liquid or substances dissolved in a liquid like the small particles in **figure 4.22b**, it

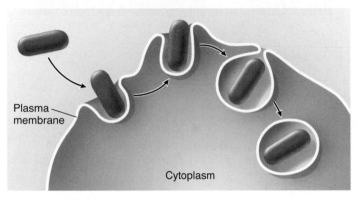

(a) Phagocytosis

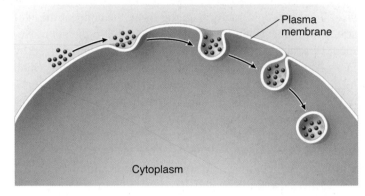

(b) Pinocytosis

Figure 4.22   Endocytosis.
Endocytosis is the process of engulfing material by folding the plasma membrane around it, forming a vesicle. (*a*) When the material is an organism or some other relatively large fragment of organic matter, the process is called phagocytosis. (*b*) When the material is a liquid, the process is called pinocytosis.

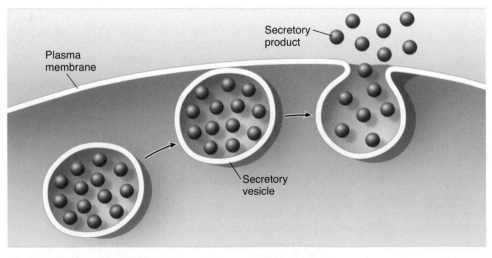

(a)

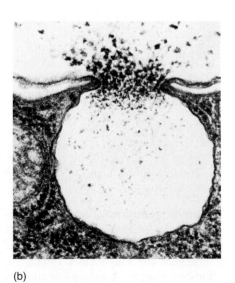

(b)

Figure 4.23   Exocytosis.
Exocytosis is the discharge of material from vesicles at the cell surface. (*a*) Proteins and other molecules are secreted from cells in small pockets called secretory vesicles, whose membranes fuse with the plasma membrane, thereby allowing the secretory vesicles to release their contents to the cell surface. (*b*) In the photomicrograph, you can see exocytosis taking place explosively.

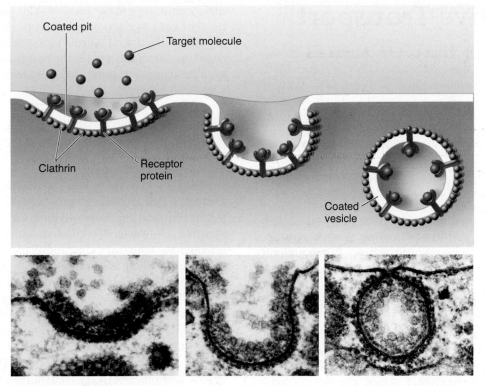

Figure 4.24 Receptor-mediated endocytosis.
Cells that undergo receptor-mediated endocytosis have pits coated with the protein clathrin that initiate endocytosis when target molecules bind to receptor proteins in the plasma membrane. In the photomicrographs, a coated pit appears in the plasma membrane of a developing egg cell, covered with a layer of proteins (80,000×). When an appropriate collection of molecules gathers in the coated pit, the pit deepens, and eventually seals off to form a vesicle.

is called **pinocytosis** (Greek *pinein,* to drink). Pinocytosis is common among animal cells. Mammalian egg cells, for example, "nurse" from surrounding cells; the nearby cells secrete nutrients that the maturing egg cell takes up by pinocytosis. Virtually all eukaryotic cells constantly carry out these kinds of endocytosis, trapping particles and extracellular fluid in vesicles and ingesting them. Endocytosis rates vary from one cell type to another. They can be surprisingly high: Some types of white blood cells ingest 25% of their cell volume each hour!

Receptor-Mediated Endocytosis

Specific molecules are often transported into eukaryotic cells through **receptor-mediated endocytosis,** illustrated in **figure 4.24.** Molecules to be transported into the cell, the red balls in the figure, first bind to specific receptors in the plasma membrane. The transport process is specific to only molecules that have a shape that fits snugly into the receptor. The plasma membrane of a particular kind of cell contains a characteristic battery of receptor types, each for a different kind of molecule.

The portion of the receptor molecule inside the membrane is trapped in an indented pit coated with the protein clathrin, visible in the photos as well as in the drawing above. The pits act like molecular mousetraps, closing over to form an internal vesicle when the right molecule enters the pit. The trigger that releases the trap is the binding of the properly fit-

ted target molecule to a receptor embedded in the membrane of the pit. When binding occurs, the cell reacts by initiating endocytosis. The process is highly specific and very fast.

One type of molecule that is taken up by receptor-mediated endocytosis is called low-density lipoprotein (LDL). The LDL molecules bring cholesterol into the cell where it can be incorporated into membranes. Cholesterol plays a key role in determining the stiffness of the cell's membrane. In the human genetic disease called hypercholesterolemia, the receptors lack tails and so are never caught in the clathrin-coated pits and, thus, are never taken up by the cells. The cholesterol stays in the bloodstream of affected individuals, coating their arteries and leading to heart attacks.

It is important to understand that receptor-mediated endocytosis in itself does not bring substances directly into the cytoplasm of a cell. The material taken in is still separated from the cytoplasm by the membrane of the vesicle. Other processes break down or release the contents.

> **Key Learning Outcome 4.13** The plasma membrane can engulf materials by endocytosis, folding the membrane around the material to encase it within a vesicle. Exocytosis is essentially this process in reverse, expelling substances using vesicles. Receptor-mediated endocytosis brings in only selected substances.

4.14 Active Transport

Other channels through the plasma membrane are closed doors. These channels open only when energy is provided. They are designed to enable the cell to maintain high or low concentrations of certain molecules, much more or less than exists outside the cell. Like motor-driven turnstiles, the channels operate to move a certain substance *up* its concentration gradient. The operation of these one-way, energy-requiring channels results in **active transport**, the movement of molecules across a membrane to a region of higher concentration by the expenditure of energy.

The Sodium-Potassium Pump The most important active transport channel is the **sodium-potassium (Na^+-K^+) pump,** which expends metabolic energy to actively pump sodium ions (Na^+) in one direction, out of cells, and potassium ions (K^+) in one direction, into cells. More than one-third of all the energy expended by your body's cells is spent driving Na^+-K^+ pump channels. This energy is derived from *adenosine triphosphate (ATP)*, a molecule we will learn more about in chapter 5. The transportation of two different ions in opposite directions happens because energy causes a change in the shape of the membrane protein carrier. The Key Biological Process illustration below walks you through one cycle of the pump. Each channel can move over 300 sodium ions per second when working full tilt. As a result of all this pumping, there are far fewer sodium ions in the cell. This concentration gradient, paid for by the expenditure of considerable metabolic energy in the form of ATP molecules, is exploited by your cells in many ways. Two of the most important are (1) the conduction of signals along nerve cells (discussed in detail in chapter 28) and (2) the pulling of valuable molecules such as sugars and amino acids into the cell *against* their concentration gradient!

We will focus for a moment on this second process. The plasma membranes of many cells are studded with facilitated diffusion channels, which offer a path for sodium ions that have been pumped out by the Na^+-K^+ pump to diffuse back in. There is a catch, however; these channels require that the sodium ions have a partner in order to pass through—like a dancing party where only couples are admitted through the door—which is why these are called **coupled channels.** Coupled channels won't let sodium ions across unless another molecule tags along, crossing hand in hand with the sodium ion. In some cases the partner molecule is a sugar (see the last entry in **table 4.3**), in others an amino acid or other molecule. Because the concentration gradient for sodium is so large, many sodium ions are trying to get back in, and this diffusion pressure drags in the partner molecules as well, even if they are already in high concentration within the cell. In this way, sugars and other actively transported molecules enter the cell—via special coupled channels.

> **Key Learning Outcome 4.14** Active transport is energy-driven transport across a membrane toward a region of higher concentration.

KEY BIOLOGICAL PROCESS: Sodium-Potassium Pump

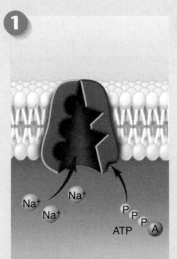

1 The sodium-potassium pump binds three sodium ions and a molecule of ATP.

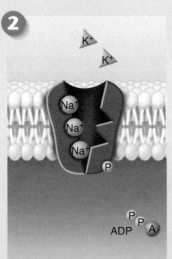

2 The splitting of ATP provides energy to change the shape of the channel. The sodium ions are driven through the channel.

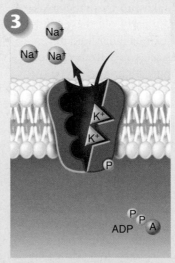

3 The sodium ions are released to the outside of the membrane, and the new shape of the channel allows two potassium ions to bind.

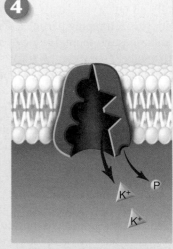

4 Release of the phosphate allows the channel to revert to its original form, releasing the potassium ions on the inside of the membrane.

TABLE 4.3 MECHANISMS FOR TRANSPORT ACROSS CELL MEMBRANES

Process	Passage Through Membrane	How it Works	Example
Passive Processes			
Diffusion			
Direct		Random molecular motion produces net migration of molecules toward region of lower concentration.	Movement of oxygen into cells
Protein channel		Polar molecules pass through a protein channel.	Movement of ions in or out of cell
Facilitated Diffusion			
Protein carrier		Molecule binds to carrier protein in membrane and is transported across; net movement is toward region of lower concentration.	Movement of glucose into cells
Osmosis			
Aquaporins		Diffusion of water across selectively permeable membrane.	Movement of water into cells placed in a hypotonic solution
Active Processes			
Endocytosis			
Membrane vesicle			
Phagocytosis		Particle is engulfed by membrane, which folds around it and forms a vesicle.	Ingestion of bacteria by white blood cells
Pinocytosis		Fluid droplets are engulfed by membrane, which forms vesicles around them.	"Nursing" of human egg cells
Receptor-mediated endocytosis		Endocytosis is triggered by the binding of a target molecule to a specific receptor.	Cholesterol uptake
Exocytosis			
Membrane vesicle		Vesicles fuse with plasma membrane and eject contents.	Secretion of mucus
Active Transport			
Protein carrier			
Na+-K+ pump		Carrier expends energy to transport a substance across a membrane against its concentration gradient.	Movement of Na+ and K+ against their concentration gradients
Coupled transport		Molecules are transported across a membrane against their concentration gradients by the cotransport of another substance down its concentration gradient.	Coupled uptake of glucose into cells against its concentration gradient

Why Does a Cell's Disposal of Damaged Proteins Consume Energy?

Much of modern biology is devoted to learning how cells build things—how the information encoded in DNA is used by cells to manufacture the proteins that make us what we are. The Nobel Prize in Chemistry was awarded in 2004 to researchers for their discovery of how the opposite, less glamorous process works: how cells break down and recycle proteins that are damaged or have outlived their usefulness.

It turns out that a cell's recycling of proteins is much more than just "taking out the trash." Particular proteins are removed, often quite quickly, and cells use such targeted removals to control a lot of their activities, timing when a cell carries out particular functions, when it divides, and even when it dies. Of the 25,000 genes in your DNA, about 1,000 take part in this protein recycling system.

Our understanding of how this system works begins with a puzzle first noted in the 1950s. Most enzymes that break down proteins, including those that digest food, do not need energy to work. But a cell's recycling of its own proteins does consume energy. Researchers had no idea why energy was needed.

The answer to this puzzle came from an unexpected direction. In 1975 scientists discovered a small protein in calves' brains consisting of just 76 amino acids. Soon they realized that exactly the same protein is found in all eukaryotes, from yeasts to humans. They called this ubiquitous ("found everywhere") protein *ubiquitin*. In the early 1980s researchers worked out that ubiquitin was a label that the cell attaches to proteins to mark them for destruction, a sort of molecular "kiss of death." The process of attaching ubiquitin takes energy, solving the puzzle of why protein recycling requires energy. The tagged proteins are taken to a barrel-shaped chamber in the cell's cytoplasm called a *proteasome,* which slices the proteins into bits that are then recycled by the cell into new protein.

The graph above displays the sort of protein recycling experiment that revealed ubiquitin's key role. The experiment monitors levels of a particular protein involved in cell division (the "target" protein) within human cells growing in culture in a laboratory flask. Two cultures are monitored in side-by-side experiments: In the culture indicated by red dots, cells contain functional copies of the ubiquitin gene (*ubi*+); in the culture indicated by blue dots the ubiquitin gene has been deleted from the DNA (*ubi*-). After 20 minutes, energy in the form of ATP is made available to the growing cells, which until then had been energy-starved.

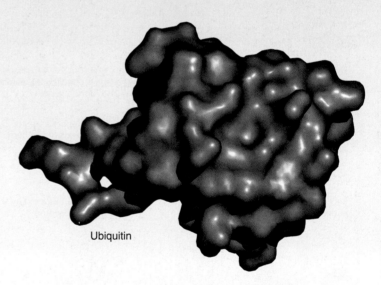

Effect of Ubiquitin on Protein Breakdown

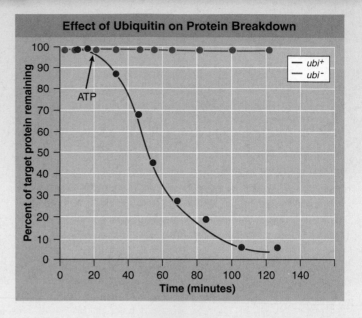
Ubiquitin

1. **Applying Concepts**
   a. Variable. In the graph, what is the dependent variable?
   b. Concentration. After 100 minutes, which of the two cultures represents the higher concentration of target protein?
2. **Interpreting Data** Does the addition of ATP affect the level of target protein in either culture? Which one?
3. **Making Inferences** How does this culture differ from the other? Why might ATP stimulate removal of target protein from this culture, but not the other?
4. **Drawing Conclusions** Using the information in the graph, suggest why the functioning of ubiquitin requires ATP energy for the effective removal of the target protein.

The World of Cells

4.1 Cells

- Cells are the smallest living structure. They consist of cytoplasm enclosed in a plasma membrane. Organisms can be composed of a single cell or multiple cells.

- Materials pass into and out of cells across the plasma membrane. A smaller cell has a larger surface-to-volume ratio, which increases the area through which materials may pass (**figure 4.2**). Because cells are so small (**figure 4.3**), a microscope is needed to view and study them (**table 4.1**).

4.2 The Plasma Membrane

- The plasma membrane that encloses all cells consists of a double layer of lipids, called the lipid bilayer, in which proteins are embedded. The structure of the plasma membrane is called the fluid mosaic model.

- The lipid bilayer (**integrated art, pages 74-75**) is made up of special lipid molecules called phospholipids, which have a polar (water-soluble) end and a nonpolar (water-insoluble) end. The bilayer forms because the nonpolar ends move away from the watery surroundings, forming the two layers. There are many different types of proteins associated with the plasma membrane.

Kinds of Cells

4.3 Prokaryotic Cells

- Prokaryotic cells are simple unicellular organisms that lack nuclei or other internal organelles and are usually encased in a rigid cell wall (**figure 4.4**). They vary in shape and may contain external structures (**figure 4.5**).

4.4 Eukaryotic Cells

- Eukaryotic cells are larger and more structurally complex than prokaryotic cells. They contain nuclei, organelles, and internal membrane systems (**figures 4.6** and **4.7**).

Tour of a Eukaryotic Cell

4.5 The Nucleus: The Cell's Control Center

- The nucleus is the command and control center of the cell. It contains the cell's DNA, which encodes the hereditary information that runs the cell (**figure 4.8**). The nucleolus is a darker area inside the nucleus where ribosomal RNA is made.

4.6 The Endomembrane System

- The endomembrane system (**integrated art, pages 82-83**) is a collection of interior membranes that organize and divide the cell's interior into functional areas. The endoplasmic reticulum is a transport system that modifies and moves proteins and other molecules produced in the ER to the Golgi complex. There are two types of ER in the cell, rough and smooth. The Golgi complex is a delivery system that carries molecules to the surface of the cell where they are released to the outside.

4.7 Organelles That Contain DNA

- The mitochondrion (**figure 4.9**) is called the powerhouse of the cell because it is the site of oxidative metabolism, an energy-

extracting process. Chloroplasts are the site of photosynthesis and are present in plants and algal cells (**figure 4.10**). Mitochondria and chloroplasts are cell-like organelles that appear to be ancient bacteria that formed endosymbiotic relationships with early eukaryotic cells (**figure 4.11**).

4.8 The Cytoskeleton: Interior Framework of the Cell

- The interior of the cell contains a network of protein fibers that make up the cytoskeleton (**integrated art, page 86**). The cytoskeleton supports the shape of the cell and anchors organelles in place.

- Centrioles assemble microtubules (**figure 4.12**). Vacuoles are storage compartments (**figures 4.13** and **4.14**).

- Cilia and flagella propel the cell through its environment (**figures 4.15** and **4.16**). Motor proteins move materials throughout the cell (**figure 4.17**).

4.9 Outside the Plasma Membrane

- The cells of plants, fungi, and many protists have cell walls that serve a similar function as prokaryotic cell walls, but are composed of different molecules (**figure 4.18**). Animal cells lack cell walls but contain an outer layer of glycoproteins, called the extracellular matrix (**figure 4.19**).

Transport Across Plasma Membranes

4.10 Diffusion

- Materials pass into and out of the cell passively through diffusion (**Key Biological Process, page 93**). Diffusion is the movement of molecules from an area of high concentration to an area of low concentration, down their concentration gradients.

4.11 Facilitated Diffusion

- In facilitated diffusion (**Key Biological Process, page 95**) a substance travels down its concentration gradient, but must bind to a protein carrier in order to pass across the membrane.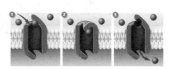

4.12 Osmosis

- Osmosis (**Key Biological Process, page 96**) is the movement of water into and out of the cell, driven by differing concentrations of solute. Water molecules move to areas of higher solute concentrations (**figure 4.21**).

4.13 Bulk Passage into and out of Cells

- Larger structures or larger quantities of material move into and out of the cell through endocytosis and exocytosis, respectively (**figures 4.22** and **4.23**). Receptor-mediated endocytosis is a selective transport process, bringing in only those substances that are able to bind to specific receptors (**figure 4.24**).

4.14 Active Transport

- Active transport involves the input of energy to transport a substance against (or up) its concentration gradient. Examples include the sodium-potassium pump (**Key Biological Process, page 100**) and coupled channels, where one substance travels down its concentration gradient but is cotransported with another substance against its concentration gradient.

1. Cell theory includes the principle that
   a. cells are the smallest living things; nothing smaller than a cell is considered alive.
   b. all cells are surrounded by cell walls that protect them.
   c. all organisms are made up of many cells arranged in specialized, functional groups.
   d. all cells contain membrane-bounded structures called organelles.

2. The plasma membrane is
   a. a carbohydrate layer that surrounds groups of cells to protect them.
   b. a double lipid layer with proteins inserted in it, which surrounds every cell individually.
   c. a thin sheet of structural proteins that encloses cytoplasm.
   d. composed of proteins that form a protective barrier.

3. Organisms that have cells with a relatively uniform cytoplasm and no organelles are called _____, and organisms whose cells have organelles and a nucleus are called _____.
   a. cellulose, nuclear
   b. eukaryotes, prokaryotes
   c. flagellated, streptococcal
   d. prokaryotes, eukaryotes

4. Within the nucleus of a cell you can find
   a. a nucleolus.
   b. a cytoskeleton.
   c. mitochondria.
   d. All of these.

5. The endomembrane system within a cell includes the
   a. cytoskeleton and the ribosomes.
   b. prokaryotes and the eukaryotes.
   c. endoplasmic reticulum and the Golgi complex.
   d. mitochondria and the chloroplasts.

6. It was once thought that only the nucleus of each cell contained DNA. We now know that DNA is also found in the
   a. cytoskeleton and the ribosomes.
   b. prokaryotes and the eukaryotes.
   c. endoplasmic reticulum and the Golgi bodies.
   d. mitochondria and the chloroplasts.

7. Which of the following statements is true?
   a. All cells have a cell wall for protection and structure.
   b. Eukaryotic cells in plants and fungi, and all prokaryotes, have a cell wall.
   c. There is a second membrane composed of structural carbohydrates surrounding all cells.
   d. Prokaryotes and all cells of eukaryotic animals have a cell wall.

8. If you put a drop of food coloring into a glass of water, the drop of color will
   a. fall to the bottom of the glass and sit there unless you stir the water; this is because of hydrogen bonding.
   b. float on the top of the water, like oil, unless you stir the water; this is because of surface tension.
   c. instantly disperse throughout the water; this is because of osmosis.
   d. slowly disperse throughout the water; this is because of diffusion.

9. When large molecules, such as food particles, need to get into a cell, they cannot easily pass through the plasma membrane, and so they move across the membrane through the processes of
   a. diffusion and osmosis.
   b. endocytosis and phagocytosis.
   c. exocytosis and pinocytosis.
   d. facilitated diffusion and active transport.

10. Active transport of specific molecules involves
    a. facilitated diffusion.
    b. endocytosis and phagocytosis.
    c. energy and specialized pumps or channels.
    d. permeability and a concentration gradient.

1. **Figure 4.3** The first microscope was used in about 1590. Electron microscopes came into common use about 70 years ago. Just over 100 years ago most physicians did not wash up between patients, even when someone had just died, or was very sick. Explain why it took so long to convince doctors to wash their hands.

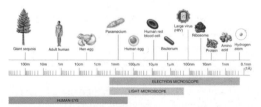

2. In the lungs, there are steep concentration gradients for oxygen and carbon dioxide molecules such that large numbers of these molecules move across the plasma membrane of the cells that line the lungs. These molecules pass through the plasma membranes by simple diffusion. This process is fast and efficient. Would this process be just as efficient if the oxygen and carbon dioxide molecules passed through the membranes by facilitated diffusion? Why or why not?

1. You are using a computer program to design a new single-celled organism. Discuss why a flat, platelike cell will be more efficient in transporting materials than a spherical, ball-like cell of the same volume.

2. Antibiotics are medicines that target bacterial infections in vertebrates. How can the antibiotic penicillin kill all the bacterial cells and not harm vertebrate cells? Hint: What part of the bacterial cell might the antibiotic be targeting?

3. Compare the cellular organelles and other structures to the parts of a city—for example, the nucleus is city hall and the DNA is all the city's laws and instructions.

4. A ribosome contains two subunits. The ribosome subunits are assembled within the nucleus, but ribosomes act in the cytoplasm. How do you imagine the subunits get out of the nucleus and into the cytoplasm?

5

Energy and Life

All of life is driven by energy. It took energy for these mice to climb up the wheat stalks. Their mousely activities while perched on the stalks—looking for danger, generating body heat, wiggling their whiskers—take energy. The energy comes from the wheat kernels and other foods these mice eat. By breaking the chemical bonds of carbohydrates and other molecules in the wheat kernels, and transferring the energy of these bonds to those of a "molecular currency" called ATP, the mice are able to capture the chemical energy in their food and put it to work. The cells of the mice perform this feat with the aid of enzymes, which are macromolecules with highly specialized shapes. Each enzyme's shape has a surface cavity called an active site, into which some specific chemical in the cell fits precisely, like a foot fits into a shoe of the proper size. When the chemical nudges in, the enzyme responds by bending, stressing particular covalent bonds in the chemical and so triggering a specific chemical reaction. The chemistry of life is enzyme chemistry.

5.1 The Flow of Energy in Living Things

We are about to begin our discussion of energy and cellular chemistry. Although these subjects may seem difficult at first, remember that all life is driven by energy. The concepts and processes discussed in the next three chapters are key to life. We are chemical machines, powered by chemical energy, and for the same reason that a successful race car driver must learn how the engine of a car works, we must look at cell chemistry. Indeed, if we are to understand ourselves, we must "look under the hood" at the chemical machinery of our cells and see how it operates.

As described in chapter 2, **energy** is defined as the ability to do work. It can be considered to exist in two states: kinetic energy and potential energy. **Kinetic energy** is the energy of motion. Objects that are not in the process of moving but have the capacity to do so are said to possess **potential energy,** or stored energy. The difference in the two states of energy is being experienced by the young man in **figure 5.1.** A boulder perched on a hilltop (**figure 5.1a**) has potential energy; after the man pushes the boulder and it begins to roll downhill (**figure 5.1b**), some of the boulder's potential energy is converted into kinetic energy. All of the work carried out by living organisms also involves the transformation of potential energy to kinetic energy.

Energy exists in many forms: mechanical energy, heat, sound, electric current, light, or radioactive radiation. Because it can exist in so many forms, there are many ways to measure energy. The most convenient is in terms of heat, because all other forms of energy can be converted to heat. Thus the study of energy is called **thermodynamics,** meaning "heat changes."

Energy flows into the biological world from the sun, which shines a constant beam of light on the earth. It is estimated that the sun provides the earth with more than 13×10^{23} calories per year, or 40 million billion calories per second! Plants, algae, and certain kinds of bacteria capture a fraction of this energy through photosynthesis. In photosynthesis, energy garnered from sunlight is used to combine small molecules (water and carbon dioxide) into more complex molecules (sugars). These complex sugar molecules have potential energy due to the arrangement of their atoms. This potential energy, in the form of chemical energy, does the work in cells. Recall from chapter 2 that an atom consists of a central nucleus surrounded by one or more orbiting electrons, and a covalent bond forms when two atomic nuclei share electrons. Breaking such a bond requires energy to pull the nuclei apart. Indeed, the strength of a covalent bond is measured by the amount of energy required to break it. For example, it takes 98.8 kcal to break 1 mole (6.023×10^{23}) of carbon–hydrogen (C–H) bonds.

All the chemical activities within cells can be viewed as a series of chemical reactions between molecules. A **chemical reaction** is the making or breaking of chemical bonds—gluing atoms together to form new molecules or tearing molecules apart and sometimes sticking the pieces onto other molecules.

> **Key Learning Outcome 5.1** Energy is the capacity to do work, either actively (kinetic energy) or stored for later use (potential energy). Chemical reactions occur when the covalent bonds linking atoms together are formed or broken.

(a) Potential energy

(b) Kinetic energy

Figure 5.1 Potential and kinetic energy.

Objects that have the capacity to move but are not moving have potential energy, while objects that are in motion have kinetic energy. (*a*) The energy required to move the ball up the hill is stored as potential energy. (*b*) This stored energy is released as kinetic energy as the ball rolls down the hill.

5.2 The Laws of Thermodynamics

Running, thinking, singing, reading these words—all activities of living organisms involve changes in energy. A set of universal laws we call the laws of thermodynamics govern these and all other energy changes in the universe.

The First Law of Thermodynamics

The first of these universal laws, the **first law of thermodynamics,** concerns the amount of energy in the universe. It states that energy can change from one state to another (from potential to kinetic, for example) but it can never be destroyed, nor can new energy be made. The total amount of energy in the universe remains constant.

A lion eating a giraffe is in the process of acquiring energy. Rather than creating new energy or capturing the energy in sunlight, the lion is merely transferring some of the potential energy stored in the giraffe's tissues to its own body (just as the giraffe obtained the potential energy stored in the plants it ate while it was alive). Within any living organism, this chemical potential energy can be shifted to other molecules and stored in chemical bonds, or it can be converted into kinetic energy, or into other forms of energy such as light or electrical energy. During each conversion, some of the energy dissipates into the environment as **heat energy,** a measure of the random motions of molecules (and, hence, a measure of one form of kinetic energy). Energy continuously flows through the biological world in one direction, with new energy from the sun constantly entering the system to replace the energy dissipated as heat.

Heat can be harnessed to do work only when there is a heat gradient—that is, a temperature difference between two areas. This is how a steam engine functions. In old steam locomotives like you see in **figure 5.2**, heat was used to move the wheels. First, a boiler (not shown) heats up water to create steam. The steam is then pumped into the cylinder of the steam engine, where it moves the piston to the right. The moving of this piston then does the work of the steam engine by moving a lever that turns the wheel. Cells are too small to maintain significant internal temperature differences, so heat

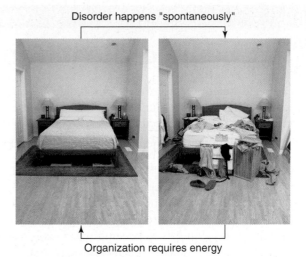

Disorder happens "spontaneously"

Organization requires energy

Figure 5.3 Entropy in action.
As time elapses, a teenager's room becomes more disorganized. It takes energy to clean it up.

energy is incapable of doing the work of cells. Thus, although the total amount of energy in the universe remains constant, the energy available to do useful work in a cell decreases, as progressively more of it dissipates as heat.

The Second Law of Thermodynamics

The **second law of thermodynamics** concerns this transformation of potential energy into heat, or random molecular motion. It states that the disorder in a closed system like the universe is continuously increasing. Put simply, disorder is more likely than order. For example, it is much more likely that a column of bricks will tumble over than that a pile of bricks will arrange themselves spontaneously to form a column. In general, energy transformations proceed spontaneously to convert matter from a more ordered, less stable form, to a less ordered, more stable form. Without an input of energy from the teenager (or a parent), the ordered room in **figure 5.3** falls into disorder.

Entropy

Entropy is a measure of the degree of disorder of a system, so the second law of thermodynamics can also be stated simply as "entropy increases." When the universe formed 10 to 20 billion years ago, it held all the potential energy it will ever have. It has become progressively more disordered ever since, with every energy exchange increasing the entropy of the universe.

> **Key Learning Outcome 5.2** The first law of thermodynamics states that energy cannot be created or destroyed; it can only undergo conversion from one form to another. The second law states that disorder (entropy) in the universe tends to increase. Life converts energy from the sun to other forms of energy that drive life processes; the energy is never lost, but as it is used, more and more of it is converted to heat.

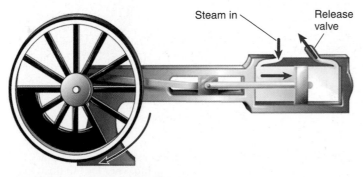

Steam in

Release valve

Figure 5.2 A steam engine.
In a steam engine, heat is used to produce steam. The expanding steam pushes against a piston that causes the wheel to turn.

5.3 Chemical Reactions

In a chemical reaction, the original molecules before the chemical reaction occurs are called **reactants,** or sometimes **substrates,** whereas the molecules that result after the reaction has taken place are called the **products** of the reaction. Not all chemical reactions are equally likely to occur. Just as a boulder is more likely to roll downhill than uphill, so a reaction is more likely to occur if it releases energy than if it needs to have energy supplied. Consider how the chemical reaction proceeds in **figure 5.4 ❶**. Like when rolling a boulder uphill, energy needs to be supplied. This is because the product of the reaction contains more energy than the reactant. This type of chemical reaction, called **endergonic,** does not occur spontaneously. By contrast, an **exergonic** reaction, shown in **❷**, tends to occur spontaneously because the product has less energy than the reactant, like a boulder that has rolled downhill.

Activation Energy

If all chemical reactions that release energy tend to occur spontaneously, it is fair to ask, "Why haven't all exergonic reactions occurred already?" Clearly they have not. If you ignite gasoline, it burns with a release of energy. So why doesn't all the gasoline in all the automobiles in the world just burn up right now? It doesn't because the burning of gasoline, and almost all other chemical reactions, requires an input of energy to get it started—a kick in the pants such as a match or spark plug. Even in exergonic reactions where the product contains or stores less energy than the reactants, it is first necessary to break existing chemical bonds in the reactants, and this takes energy. The extra energy required to destabilize existing chemical bonds and so initiate a chemical reaction is called **activation energy,** indicated by brackets in **figure 5.4 ❷** and **❸**. You must first nudge a boulder out of the hole it sits in before it can roll downhill. Activation energy is simply a chemical nudge.

Catalysis

One way to make an exergonic reaction more likely to happen is to lower the necessary activation energy. Like digging away the ground below your boulder, lowering activation energy reduces the nudge needed to get things started. The process of lowering the activation energy of a reaction is called **catalysis.** Catalysis cannot make an endergonic reaction occur spontaneously—you cannot avoid the need to supply energy—but it can make a reaction, endergonic or exergonic, proceed much faster. Compare the activation energy levels (the red arched arrows) in **❷** and **❸** below: The catalyzed reaction has a lower barrier to overcome.

> **Key Learning Outcome 5.3** Endergonic reactions require an input of energy. Exergonic reactions release energy. Activation energy that initiates chemical reactions can be lowered by catalysis.

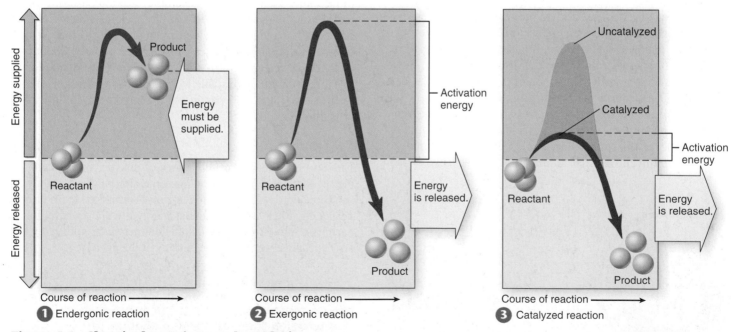

❶ Endergonic reaction

❷ Exergonic reaction

❸ Catalyzed reaction

Figure 5.4 Chemical reactions and catalysis.
❶ The products of endergonic reactions contain more energy than the reactants. **❷** The products of exergonic reactions contain less energy than the reactants, but exergonic reactions do not necessarily proceed rapidly because it takes energy to get them going. The "hill" in this energy diagram represents energy that must be supplied to destabilize existing chemical bonds. **❸** Catalyzed reactions occur faster because the amount of activation energy required to initiate the reaction—the height of the energy hill that must be overcome—is lowered, and the reaction proceeds to its end faster.

5.4 How Enzymes Work

Enzymes, which can be made of proteins or nucleic acids, are the catalysts used by cells to touch off particular chemical reactions. By controlling which enzymes are present, and when they are active, cells are able to control what happens within themselves, just as a conductor controls the music an orchestra produces by dictating which instruments play when.

An enzyme works by binding to a specific molecule and stressing the bonds of that molecule in such a way as to make a particular reaction more likely. The key to this activity is the shape of the enzyme. An enzyme is specific for a particular reactant, or substrate, because the enzyme surface provides a mold that very closely fits the shape of the desired reactant. For example, the blue-colored lysozyme enzyme in **figure 5.5** is contoured to fit a specific sugar molecule (the yellow reactant). Other molecules that fit less perfectly simply don't adhere to the enzyme's surface. The site on the enzyme surface where the reactant fits is called the **active site** (**panel 1** below). The site on the reactant that binds to an enzyme is called the **binding site.** Enzymes are not rigid. The binding of the reactant induces the enzyme to change its shape slightly. In **figure 5.5b** and in **panel 2** of the Key Biological Process illustration below, the edges of the enzyme now hug the reactant(s), leading to an "induced fit" between the enzyme and its reactant, like a hand wrapping around a baseball.

An enzyme lowers the activation energy of a particular reaction. In the case of lysozyme, an enzyme found in human

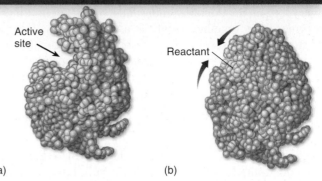

(a)   (b)

Figure 5.5   Enzyme shape determines its activity.
(*a*) A groove runs through the lysozyme enzyme (*blue* in this diagram) that fits the shape of the reactant (in this case, a chain of sugars). (*b*) When such a chain of sugars, indicated in *yellow*, slides into the groove, it induces the protein to change its shape slightly and embrace the substrate more intimately. This induced fit causes a chemical bond between two sugar molecules within the chain to break.

tears, the enzyme has an antibacterial function, encouraging the breaking of a particular chemical bond in molecules that make up the cell wall of bacteria (**figure 5.5**). The enzyme weakens the bond by drawing away some of its electrons. Alternatively, an enzyme may encourage the formation of a link between two reactants, like the blue and red colored molecules in **panel 2** below by holding them near each other. Regardless of the type of reaction, the enzyme is not affected by the chemical reaction and is available to be used again.

KEY BIOLOGICAL PROCESS: How Enzymes Work

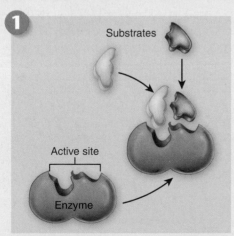

1

Substrates

Active site

Enzyme

Enzymes have a complex three-dimensional surface to which particular reactants (called substrates of that enzyme) fit, like a hand in a glove.

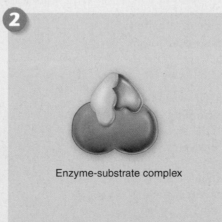

2

Enzyme-substrate complex

An enzyme and its substrate(s) bind tightly together, forming an enzyme-substrate complex. The binding brings key atoms near each other and stresses key covalent bonds.

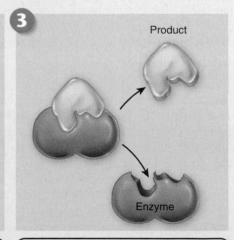

3

Product

Enzyme

As a result, a chemical reaction occurs within the active site, forming the product. The product then diffuses away, freeing the enzyme to work again.

Biochemical Pathways

Every organism contains thousands of different kinds of enzymes that together catalyze a bewildering variety of reactions. Often several of these reactions occur in a fixed sequence called a **biochemical pathway,** the product of one reaction becoming the substrate for the next. You can see in the biochemical pathway shown in **figure 5.6** how the initial substrate is altered by enzyme 1 so that it now fits into the active site of another enzyme, becoming the substrate for enzyme 2, and so on until the final product is produced. Because these reactions occur in sequence, the enzymes involved are often positioned near each other in the cell. For example, the enzymes involved in this biochemical pathway are all embedded in a membrane near each other. Many biochemical pathways occur in membranes, although enzyme assemblies also occur in organelles and within the cytoplasm. The close proximity of the enzymes allows the reactions of the biochemical pathway to proceed faster. Biochemical pathways are the organizational units of metabolism. We will discuss them more in chapters 6 and 7.

Factors Affecting Enzyme Activity

Temperature and pH can have a major influence on the action of enzymes. Enzyme activity is affected by any change in condition that alters the enzyme's three-dimensional shape.

Temperature When the temperature increases, the bonds that determine enzyme shape are too weak to hold the enzyme's peptide chains in the proper position, and the enzyme denatures. As a result, enzymes function best within an optimum temperature range, which is relatively narrow for most human enzymes. In the human body, enzymes work best at temperatures near the normal body temperature of 37°C, as shown by the brown curve in **figure 5.7a.** Also notice that the rates of enzyme reactions tend to drop quickly at higher temperatures, when the enzyme begins to unfold. This is why an extremely high fever in humans can be fatal. However, the shapes of the enzymes found in hotsprings bacteria (the red curve) are more stable, allowing the enzymes to function at much higher temperatures. This allows the bacteria to live in water that is near 70°C.

pH In addition, most enzymes also function within an optimal pH range, because the shape-determining polar interactions of enzymes are quite sensitive to hydrogen ion (H^+) concentration. Most human enzymes, such as the protein-degrading enzyme trypsin (the dark blue curve in **figure 5.7b**) work best within the range of pH 6 to 8. Blood has a pH of 7.4. However, some enzymes, such as the digestive enzyme pepsin (the light blue curve) are able to function in very acidic environments such as the stomach, but can't function at higher pHs.

> **Key Learning Outcome 5.4** Enzymes catalyze chemical reactions within cells and can be organized into biochemical pathways. Enzymes are sensitive to temperature and pH because both of these variables influence enzyme shape.

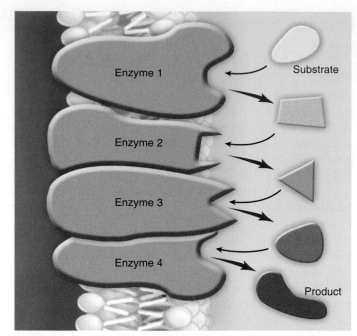

Figure 5.6 A biochemical pathway.
The original substrate is acted on by enzyme 1, changing the substrate to a new form recognized by enzyme 2. Each enzyme in the pathway acts on the product of the previous stage.

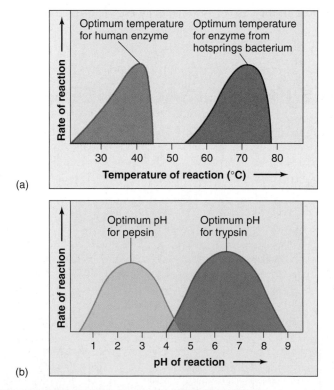

Figure 5.7 Enzymes are sensitive to their environment.
The activity of an enzyme is influenced by both (a) temperature and (b) pH. Most human enzymes work best at temperatures of about 40°C and within a pH range of 6 to 8.

KEY BIOLOGICAL PROCESS: Allosteric Enzyme Regulation

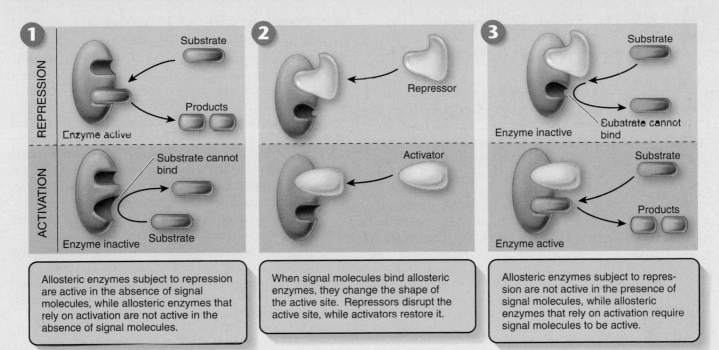

1

REPRESSION

Substrate

Products

Enzyme active

ACTIVATION

Substrate cannot bind

Enzyme inactive   Substrate

Allosteric enzymes subject to repression are active in the absence of signal molecules, while allosteric enzymes that rely on activation are not active in the absence of signal molecules.

2

Repressor

Activator

When signal molecules bind allosteric enzymes, they change the shape of the active site. Repressors disrupt the active site, while activators restore it.

3

Substrate

Substrate cannot bind

Enzyme inactive

Substrate

Products

Enzyme active

Allosteric enzymes subject to repression are not active in the presence of signal molecules, while allosteric enzymes that rely on activation require signal molecules to be active.

5.5 How Cells Regulate Enzymes

Because an enzyme must have a precise shape to work correctly, it is possible for the cell to control when an enzyme is active by altering its shape. Many enzymes have shapes that can be altered by the binding of "signal" molecules to their surfaces. Such enzymes are called *allosteric* (Latin, other shape). Enzymes can be inhibited or activated by the binding of signal molecules. For example, the upper tan panels in the Key Biological Process illustration above show an enzyme that is inhibited. The binding of a signal molecule, called a **repressor (panel 2)**, alters the shape of the enzyme's active site such that it cannot bind the substrate. In other cases, the enzyme may not be able to bind the reactants *unless* the signal molecule is bound to the enzyme. The lower set of panels shows a signal molecule serving as an **activator.** The red substrate cannot bind to the enzyme's active site unless the activator (the yellow molecule) is in place, altering the shape of the active site. The site where the signal molecule binds to the enzyme surface is called the **allosteric site.**

Enzymes are often regulated by a mechanism called **feedback inhibition,** where the product of the reaction acts as the repressor. Feedback inhibition can occur in two ways: *competitive inhibitors* and *noncompetitive inhibitors*. The blue molecule in **figure 5.8a** functions as a competitive inhibitor, blocking the active site so that the substrate cannot bind. The yellow molecule in **figure 5.8b** functions as a noncompetitive inhibitor. It binds to an allosteric site, changing the shape of the enzyme such that it is unable to bind to the substrate.

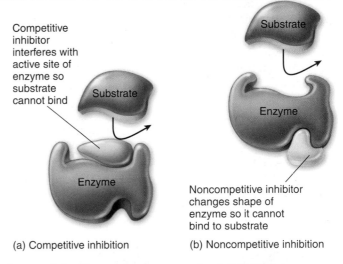

Competitive inhibitor interferes with active site of enzyme so substrate cannot bind

Substrate

Enzyme

Substrate

Enzyme

Noncompetitive inhibitor changes shape of enzyme so it cannot bind to substrate

(a) Competitive inhibition

(b) Noncompetitive inhibition

Figure 5.8   How enzymes can be inhibited.

(*a*) In competitive inhibition, the inhibitor interferes with the active site of the enzyme. (*b*) In noncompetitive inhibition, the inhibitor binds to the enzyme at a place away from the active site, effecting a conformational change in the enzyme so that it can no longer bind to its substrate. In feedback inhibition, the inhibitor molecule is the product of the reaction.

Many drugs and antibiotics work by inhibiting enzymes. Statin drugs like Lipitor lower cholesterol by inhibiting a key enzyme cells used to make cholesterol. The antibiotic penicillin inhibits an enzyme bacteria used in making cell walls. Because humans lack this enzyme, we are not harmed by the drug.

Key Learning Outcome 5.5 An enzyme's activity can be affected by signal molecules that bind to it, changing its shape.

5.6 ATP: The Energy Currency of the Cell

Cells use energy to do all those things that require work, but how does the cell use energy from the sun or the potential energy stored in molecules to power its activities? The sun's radiant energy and the energy stored in molecules are energy sources, but like money that is invested in stocks and bonds or real estate, these energy sources cannot be used directly to run a cell. To be useful, the energy from the sun or food molecules must first be converted to a source of energy that a cell can use, like someone converting stocks and bonds to ready cash. The "cash" molecule in the body is **adenosine triphosphate (ATP)**. ATP is the energy currency of the cell.

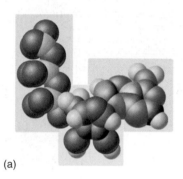

(a)

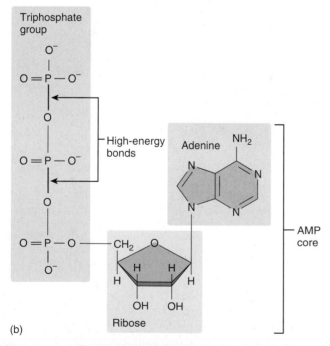

(b)

Figure 5.9 The parts of an ATP molecule.

The model (*a*) and structural diagram (*b*) both show that ATP consists of three phosphate groups attached to a ribose (five-carbon sugar) molecule. The ribose molecule is also attached to an adenine molecule (also one of the nitrogenous bases of DNA and RNA). When the endmost phosphate group is split off from the ATP molecule, considerable energy is released.

Structure of the ATP Molecule

Each ATP molecule is composed of the three parts shown in **figure 5.9**: (1) a sugar (colored blue) serves as the backbone to which the other two parts are attached, (2) adenine (colored peach) is one of the four nitrogenous bases in DNA and RNA, and (3) a chain of three phosphates (colored yellow) contain high-energy bonds.

As you can see in the figure, the phosphates carry negative electrical charges, and so it takes considerable chemical energy to hold the line of three phosphates next to one another at the end of ATP. Like a coiled spring, the phosphates are poised to push apart. It is for this reason that the chemical bonds linking the phosphates are such chemically reactive bonds.

When the endmost phosphate is broken off an ATP molecule, a sizable packet of energy is released. The reaction converts ATP to adenosine diphosphate, ADP. The second phosphate group can also be removed, yielding additional energy and leaving adenosine monophosphate (AMP). Most energy exchanges in cells involve cleavage of only the outermost bond, converting ATP into ADP and P_i, inorganic phosphate:

$$ATP \longleftrightarrow ADP + P_i + energy$$

Exergonic reactions require activation energy, and endergonic reactions require the input of even more energy, and so these reactions in the cell are usually coupled with the breaking of the phosphate bond in ATP, called *coupled reactions*. Because almost all chemical reactions in cells require less energy than is released by this reaction, ATP is able to power many of the cell's activities, producing heat as a by-product. **Table 5.1** introduces you to some of the key cellular activities powered by the breakdown of ATP. ATP is continually recycled from ADP and P_i through the ATP-ADP cycle.

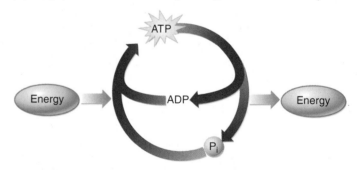

Cells use two different but complementary processes to convert energy from the sun and potential energy found in food molecules into ATP. Some cells convert energy from the sun into molecules of ATP through the process of **photosynthesis**, the subject of chapter 6. This ATP is then used to manufacture sugar molecules, converting the energy from ATP into potential energy stored in the bonds that hold the atoms together. All cells convert the potential energy found in food molecules into ATP through **cellular respiration**, the subject of chapter 7.

> **Key Learning Outcome 5.6** Cells use the energy in ATP molecules to drive chemical reactions.

TABLE 5.1   HOW CELLS USE ATP ENERGY TO POWER CELLULAR WORK

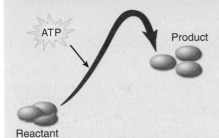

Biosynthesis

Cells use the energy released from the exergonic hydrolysis of ATP to drive endergonic reactions like those of protein synthesis, an approach called energy coupling.

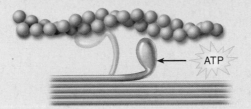

Contraction

In muscle cells, filaments of protein repeatedly slide past each other to achieve contraction of the cell. An input of ATP is required for the filaments to reset and slide again.

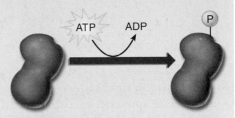

Chemical Activation

Proteins can become activated when a high-energy phosphate from ATP attaches to the protein, activating it. Other types of molecules can also become phosphorylated by transfer of a phosphate from ATP.

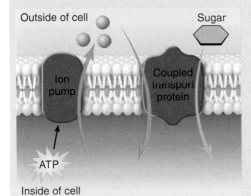

Importing Metabolites

Metabolite molecules such as amino acids and sugars can be transported into cells against their concentration gradients by coupling the intake of the metabolite to the inward movement of an ion moving down its concentration gradient, this ion gradient being established using ATP.

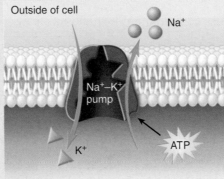

Active Transport: Na$^+$–K$^+$ Pump

Most animal cells maintain a low internal concentration of Na$^+$ relative to their surroundings, and a high internal concentration of K$^+$. This is achieved using a protein called the sodium-potassium pump, which actively pumps Na$^+$ out of the cell and K$^+$ in, using energy from ATP.

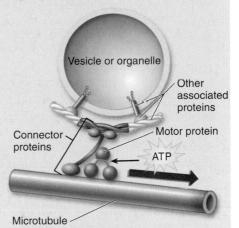

Cytoplasmic Transport

Within a cell's cytoplasm, vesicles or organelles can be dragged along microtubular tracks using molecular motor proteins, which are attached to the vesicle or organelle with connector proteins. The motor proteins use ATP to power their movement.

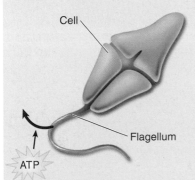

Flagellar Movements

Microtubules within flagella slide past each other to produce flagellar movements. ATP powers the sliding of the microtubules.

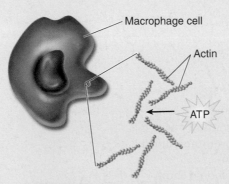

Cell Crawling

Actin filaments in a cell's cytoskeleton continually assemble and disassemble to achieve changes in cell shape and to allow cells to crawl over substrates or engulf materials. The dynamic character of actin is controlled by ATP molecules bound to actin filaments.

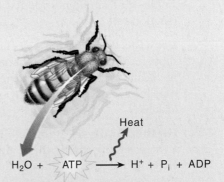

Heat Production

The hydrolysis of the ATP molecule releases heat. Reactions that hydrolyze ATP often take place in mitochondria or in contracting muscle cells and may be coupled to other reactions. The heat generated by these reactions can be used to maintain an organism's temperature.

Do Enzymes Physically Attach to Their Substrates?

When scientists first began to examine the chemical activities of organisms, no one knew that biochemical reactions were catalyzed by enzymes. The first enzyme was discovered in 1833 by French chemist Anselme Payen. He was studying how beer is made from barley: First barley is pressed and gently heated so its starches break down into simple 2-sugar units; then yeasts convert these units into ethanol. Payen found that the initial breakdown requires a chemical factor that is not alive, and that does not seem to be used up during the process—a catalyst. He called this first enzyme *diastase* (we call it *amylase* today).

Did this catalyst operate at a distance, increasing reaction rate all around it, much as raising the temperature of nearby molecules might do? Or did it operate in physical contact, actually attaching to the molecules whose reaction it catalyzed (its "substrate")?

The answer was discovered in 1903 by French chemist Victor Henri. He saw that the hypothesis that an enzyme physically binds to its substrate makes a clear and testable prediction: In a solution of substrate and enzyme, there must be a maximum reaction rate, faster than which the reaction cannot proceed. When all the enzyme molecules are working full tilt, the reaction simply cannot go any faster, no matter how much more substrate you add to the solution. To test this prediction, Henri carried out the experiment whose results you see in the graph, measuring the reaction rate (*V*) of diastase at different substrate concentrations (*S*).

1. **Making Inferences** As *S* increases, does *V* increase? If so, in what manner—steadily, or by smaller and smaller amounts? Is there a maximum reaction rate?

2. **Drawing Conclusions** Does this result provide support for the hypothesis that an enzyme binds physically to its substrate? Explain. If the hypothesis were incorrect, what would you expect the graph to look like?

3. **Further Analysis** If the smaller amounts by which *V* increases are strictly the result of fewer unoccupied enzymes being available at higher values of *S*, then the curve in Henri's experiment should show a pure exponential decline in *V*—mathematically, meaning a reciprocal plot (1/*V* versus 1/*S*) should be a straight line. If some other factor is also at work that reacts differently to substrate concentration, then the reciprocal plot would curve upward or downward. Fill in the reciprocal values in the table to the right, and then plot the values on the lower graph (1/*S* on the *x* axis and 1/*V* on the *y* axis). Is a reciprocal plot of Henri's data a straight line?

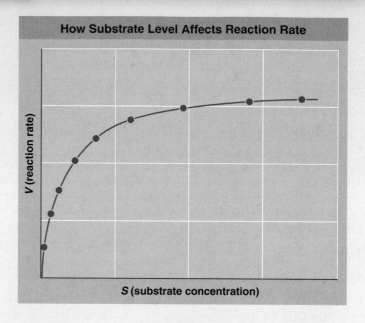

How Substrate Level Affects Reaction Rate

V (reaction rate) vs *S* (substrate concentration)

Trial	S	1/S	V	1/V
1	5	0.200	7.7	0.130
2	10		15.4	
3	25		23.1	
4	50		30.8	
5	75		38.5	
6	125		40.7	
7	200		46.2	
8	275		47.7	
9	350		48.5	

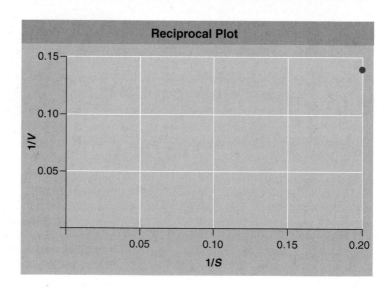

Reciprocal Plot

1/*V* vs 1/*S*

Cells and Energy

5.1 The Flow of Energy in Living Things

- Energy is the ability to do work. Energy exists in two states: kinetic energy and potential energy.

- Kinetic energy is the energy of motion. Potential energy is stored energy, which exists in objects that aren't in motion but have the capacity to move, like a ball poised at the top of a hill shown here from **figure 5.1.** All of the work carried out by living things involves the transformation of potential energy into kinetic energy.

- Energy flows from the sun to the earth, where it is trapped by photosynthetic organisms and stored in carbohydrates as potential energy. This energy is transferred during chemical reactions.

5.2 The Laws of Thermodynamics

- The laws of thermodynamics describe changes in energy in our universe. The first law of thermodynamics explains that energy can never be created or destroyed, only changed from one state to another. The total amount of energy in the universe remains constant.

- Energy exists in different forms in the universe, such as light, electrical, or heat energy. This energy, such as heat energy, can be harnessed to do work (**figure 5.2**).

- The second law of thermodynamics explains that the conversion of potential energy into random molecular motion is constantly increasing. This conversion of energy progresses from an ordered but less stable form to a disordered but stable form. Entropy, which is a measure of disorder in a system, is constantly increasing such that disorder is more likely than order. Energy must be used to maintain order (**figure 5.3**).

Cell Chemistry

5.3 Chemical Reactions

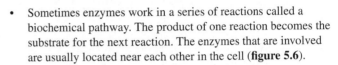

- Chemical reactions involve the breaking or formation of covalent bonds. The starting molecules are called the reactants, and the molecules produced by the reaction are called the products. Chemical reactions in which the products contain more potential energy than the reactants are called endergonic reactions. Chemical reactions that release energy are called exergonic reactions, as shown here from **figure 5.4,** and are more likely to occur.

- All chemical reactions require an input of energy. The energy required to start a reaction is called activation energy. A chemical reaction proceeds faster when its activation energy is lowered, and this occurs through a process called catalysis.

Enzymes

5.4 How Enzymes Work

- Enzymes are macromolecules that lower the activation energy of chemical reactions in the cell. Enzymes are catalysts.

- An enzyme, like the lysozyme shown here from **figure 5.5**, binds the reactant, or substrate. The substrate binds to the enzyme's active site. The enzyme forms around the reactant and acts to stress covalent bonds or bring atoms into closer proximity. The actions of the enzyme increase the likelihood that chemical bonds will break or form. An enzyme lowers the activation energy of the reaction. The enzyme is not affected by the reaction and can be used over and over again (**Key Biological Process, page 109**).

- Sometimes enzymes work in a series of reactions called a biochemical pathway. The product of one reaction becomes the substrate for the next reaction. The enzymes that are involved are usually located near each other in the cell (**figure 5.6**).

- In the cell, chemical reactions are regulated by controlling which enzymes are active. Other factors, such as temperature and pH, can also affect enzyme function, and so most enzymes have an optimal temperature and pH range (**figure 5.7**).

- Higher temperatures can disrupt the bonds that hold the enzyme in its proper shape, decreasing its ability to catalyze a chemical reaction. The bonds that hold the enzyme's shape are also affected by hydrogen ion concentrations, and so increasing or decreasing the pH can disrupt the enzyme's function.

5.5 How Cells Regulate Enzymes

- An enzyme can be inhibited or activated in the cell as a means of regulation by temporarily altering the enzyme's shape (**Key Biological Process, page 111**). An enzyme can be inhibited when a molecule, called a repressor, binds to the enzyme, altering the shape of the active site so that it cannot bind the substrate. Some enzymes need to be activated, or turned on, in order to bind to their substrate. A molecule called an activator binds to the enzyme, changing the shape of the active site so that it is able to bind the substrate. Enzymes that are controlled in this way are allosteric enzymes.

- A repressor molecule can bind to the active site of the enzyme, blocking it. This is called a competitive inhibition. In noncompetitive inhibition, the repressor binds to a different site on the enzyme, altering the shape of the active site so it cannot bind its substrate (**figure 5.8**). Enzymes are often regulated by feedback inhibition, a process where the product of the reaction functions as a repressor, shutting down its own synthesis.

How Cells Use Energy

5.6 ATP: The Energy Currency of the Cell

- Cells require energy to do work in the form of ATP (**table 5.1**). ATP contains a sugar, an adenine, and a chain of three phosphates, as shown here from **figure 5.9**. The three phosphates are held together with high-energy bonds. When the endmost phosphate bond breaks, considerable energy is released. A cell uses this energy to drive reactions in the cell by coupling the breakdown of ATP with other chemical reactions in the cell.

1. The ability to do work is the definition for
   a. thermodynamics.
   b. radiation.
   c. energy.
   d. entropy.

2. The first law of thermodynamics
   a. says that energy recycles constantly, as organisms use and reuse it.
   b. says that entropy, or disorder, continually increases in a closed system.
   c. is a formula for measuring entropy.
   d. says that energy can change forms, but cannot be made or destroyed.

3. The second law of thermodynamics
   a. says that energy recycles constantly, as organisms use and reuse it.
   b. says that entropy, or disorder, continually increases in a closed system.
   c. is a formula for measuring entropy.
   d. says that energy can change forms, but cannot be made or destroyed.

4. Chemical reactions that occur spontaneously are called
   a. exergonic and release energy.
   b. exergonic and their products contain more energy.
   c. endergonic and release energy.
   d. endergonic and their products contain more energy.

5. The catalysts that help an organism carry out needed chemical reactions are called
   a. hormones.
   b. enzymes.
   c. reactants.
   d. substrates.

6. Factors that affect the activity of an enzyme molecule include
   a. peptides and energy.
   b. thermodynamics.
   c. temperature and pH.
   d. entropy.

7. In order for an enzyme to work properly
   a. it must have a particular shape.
   b. the temperature must be within certain limits.
   c. the pH must be within certain limits.
   d. All of the above.

8. In competitive inhibition
   a. an enzyme molecule has to compete with other enzyme molecules for the necessary substrate.
   b. an enzyme molecule has to compete with other enzyme molecules for the necessary energy.
   c. an inhibitor molecule competes with the substrate for the active site on the enzyme.
   d. two different products compete for the same binding site on the enzyme.

9. Which of the following is not a component of ATP?
   a. active site
   b. ribose
   c. adenine
   d. phosphate groups

10. Endergonic reactions can occur in the cell because they are coupled with
    a. the breaking of phosphate bonds in ATP.
    b. uncatalyzed reactions.
    c. activators.
    d. all of the above.

Apply Your Understanding

1. **Figure 5.7** Examine the graphs shown here. Describe what happens to a human enzyme when the body's temperature is raised to 50°C. Looking at part (b), what happens to trypsin's ability to function as the surrounding concentration of H$^+$ ions increases? How do you think the enzyme pepsin would respond to a change in pH from 4 to 3?

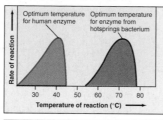

(a)

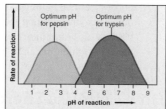

(b)

2. **Table 5.1** ATP forms primarily from the breakdown of glucose. If your blood glucose level were to drop precipitously, what sorts of problems would you expect this to cause in your body?

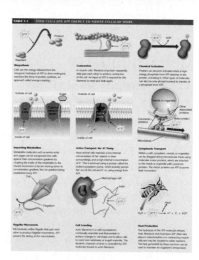

Synthesize What You Have Learned

1. Photosynthetic organisms, such as plants, algae, and bacteria, capture the sun's energy and transform it into sugar molecules that other organisms like us can harvest for energy. Explain where the sun's energy is stored in these sugar molecules, and how it is harvested by animals that eat the sugar.

2. A restriction endonuclease is an enzyme that cuts DNA at a specific, unique sequence, like GAATTC. How does a particular restriction enzyme "know" when it has found its target sequence?

3. When a baseball thrown by a pitcher encounters the swinging bat of a slugger, what happens to the ball's kinetic energy? What happens to the bat's kinetic energy?

6

Photosynthesis: Acquiring Energy from the Sun

I n this forest glade, you can literally see the pulse of life flowing through the organisms of an ecosystem. Sunlight beams down, a stream of energy in the form of packets of light called photons. Everywhere the light falls, there are plants—trees and shrubs and flowers and grasses, all with green leaves intercepting the energy as it rains down. In the cells of each leaf are organelles called chloroplasts that contain light-gathering pigments in their membranes. These pigments, notably the pigment chlorophyll, which makes leaves green, absorb photons of light and use the energy to strip electrons from water molecules. The chloroplasts use these electrons to reduce CO_2—that is, to add hydrogens (a hydrogen atom, you will recall, is just a proton with an associated electron)—and so make organic molecules. This process of capturing the sun's energy to build molecules is called photosynthesis—literally, using "light to build." In this chapter, we will delve into photosynthesis, tracing how light energy is captured, converted to chemical energy, and put to work assembling organic molecules. In the roots and other tissues of the plants, the opposite process is taking place. Organic molecules are being broken down in the process of cellular respiration to provide energy to power growth and cellular activities. These reactions, which take place largely in another kind of organelle called a mitochondrion, are the subject of the following chapter. Together, chloroplasts and mitochondria carry out a flow of energy driven by the power of sunlight.

6.1 An Overview of Photosynthesis

Life is powered by sunshine. All of the energy used by almost all living cells comes ultimately from the sun, captured by plants, algae, and some bacteria through the process of **photosynthesis.** Every oxygen atom in the air we breathe was once part of a water molecule, liberated by photosynthesis as you will discover in this chapter. Life as we know it is only possible because our earth is awash in energy streaming inward from the sun. Each day, the radiant energy that reaches the earth is equal to that of about 1 million Hiroshima-sized atomic bombs. About 1% of it is captured by photosynthesis and provides the energy that drives almost all life on earth. Use the arrows on this page and the next three pages to follow the path of energy from the sun through photosynthesis.

Trees Many kinds of organisms carry out photosynthesis, not only the many kinds of plants that make our world green, but also bacteria and algae. Photosynthesis is somewhat different in bacteria, but we will focus our attention on photosynthesis in plants, starting with this maple tree crowned with green leaves. Later we will look at the grass growing beneath the maple tree—it turns out that grasses and other related plants sometimes take a different approach to photosynthesis depending on the conditions.

Leaves To learn how this maple tree captures energy from sunlight, follow the light. It comes beaming in from the sun, down through earth's atmosphere, bathing the top of the tree in light. What part of the maple tree is actually being struck by this light? The green leaves are. Each branch at the top of the tree ends in a spread of these leaves, each leaf flat and thin like the page of a book. Within these green leaves is where photosynthesis occurs. No photosynthesis occurs within this tree's stem, covered with bark, and none in the roots, covered with soil—no light reaches these parts of the plant. The tree has a very efficient internal plumbing system that transports the products of photosynthesis to the stem, roots, and other parts of the plant so that they too may benefit from the capture of the sun's energy.

The Leaf Surface Now follow the light as it passes into a leaf. The beam of light first encounters a waxy protective layer called the cuticle. The cuticle acts a bit like a layer of clear fingernail polish, providing a thin, watertight and surprisingly strong layer of protection. Light passes right through this transparent wax, and then proceeds to pass right on through a layer of cells immediately beneath the cuticle called the epidermis. Only one cell thick, this epidermis acts as the "skin" of the leaf, providing more protection from damage and, very importantly, controlling how gases and water enter and leave the leaf. Very little of the light has been absorbed by the leaf at this point—neither the cuticle nor the epidermis absorb much.

Cross-section of leaf

Cuticle

Epidermis

Mesophyll

Vascular bundle

Bundle sheath

Stoma

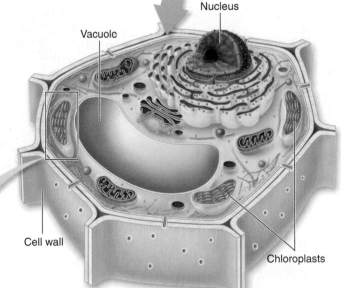

Nucleus

Vacuole

Cell wall

Chloroplasts

Mesophyll cell

Mesophyll Cells Passing through the epidermis, the light immediately encounters layer after layer of mesophyll cells. These cells fill the interior of the leaf. Unlike the cells of the epidermis, mesophyll cells contain numerous *chloroplasts,* which as you recall from chapter 4 are organelles found in all plants and algae. They are visible as green specks in the mesophyll cells in the cross section of the leaf above. It is here, within the mesophyll cells penetrated by the light beam, that photosynthesis occurs.

Thylakoid

Inner membrane

Outer membrane

Granum

Stroma

Chloroplast

Chloroplasts Light penetrates into mesophyll cells. The cell walls of the mesophyll cells don't absorb it, nor does the plasma membrane or nucleus or mitochondria. Why not? Because these elements of the mesophyll cell contain few if any molecules that absorb visible light. If chloroplasts were not also present in these cells, most of this light would pass right through, just as it passed through the epidermis. But chloroplasts are present, lots of them. One chloroplast is highlighted by a box in the mesophyll cell above. Light passes into the cell to the chloroplast, and when it reaches the chloroplast, it passes through the outer and inner membranes to reach the thylakoid structures within the chloroplast, clearly seen as the green disks in the cutaway chloroplast shown here.

Inside the Chloroplast

All the important events of photosynthesis happen inside the chloroplast. The journey of light into the chloroplasts ends when the light beam encounters a series of internal membranes within the chloroplast organized into flattened sacs called *thylakoids*. Often, numerous thylakoids are stacked on top of one another in columns called *grana*. In the drawing below, the grana look not unlike piles of dishes. While each thylakoid is a separate compartment that functions more-or-less independently, the membranes of the individual thylakoids are all connected, part of a single continuous membrane system. Occupying much of the interior of the chloroplast, this thylakoid membrane system is submerged within a semiliquid substance called *stroma*, which fills the interior of the chloroplast in much the same way that cytoplasm fills the interior of a cell. Suspended within the stroma are many enzymes and other proteins, including the enzymes that act later in photosynthesis to assemble organic molecules from carbon dioxide (CO_2) in reactions that do not require light and which are discussed later.

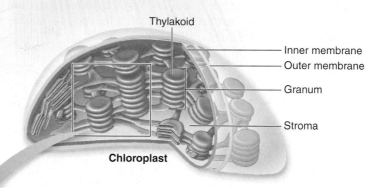

Chloroplast

Thylakoid
Inner membrane
Outer membrane
Granum
Stroma

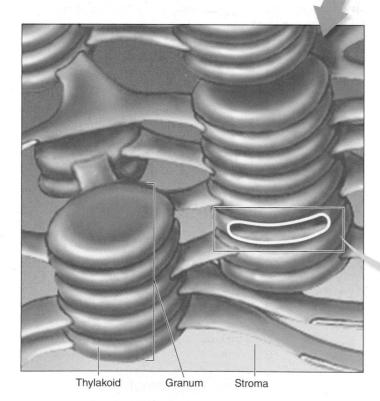

Thylakoid   Granum   Stroma

Penetrating the Thylakoid Surface The first key event of photosynthesis occurs when a beam of sunlight strikes the surface membrane of a thylakoid. Embedded within this membrane, like icebergs on an ocean, are clusters of light-absorbing pigments. A pigment molecule is a molecule that absorbs light energy. The primary pigment molecule in most photosystems is **chlorophyll,** an organic molecule that absorbs red and blue light, but does not absorb green wavelengths. The green light is instead reflected, giving the thylakoid and the chloroplast that contains it an intense green color. Plants are green because they are rich in green chloroplasts. Except for some alternative pigments also present in thylakoids, which we will discuss later, no other parts of the plant absorb visible light with such intensity.

Striking the Photosystem Within each pigment cluster, the chlorophyll molecules are arranged in a network called a *photosystem*. The light-absorbing chlorophyll molecules of a photosystem act together as an antenna to capture photons (units of light energy). A lattice of structural proteins, indicated by the purple element inserted into the thylakoid membrane in the diagram on the facing page, anchors each of the chlorophyll molecules of a photosystem into a precise position, such that every chlorophyll molecule is touching several others. Wherever a photon of light strikes the photosystem, some chlorophyll molecule will be in position to receive it.

Energy Absorption When a photon of sunlight strikes any chlorophyll molecule in the photosystem, the chlorophyll molecule it hits absorbs that photon's energy. The energy becomes part of the chlorophyll molecule, boosting some of its electrons to higher energy levels. Possessing these more energetic electrons, the chlorophyll molecule is said to now be "excited." With this key event, the biological world has captured energy from the sun.

Excitation of the Photosystem The excitation that the absorption of light creates is then passed from the chlorophyll molecule that was hit to another, and then to another, like a hot potato being passed down a line of people. This shuttling of excitation is not a chemical reaction, in which an electron physically passes between atoms. Rather, it is energy that passes from one chlorophyll molecule to its neighbor. A crude analogy to this form of energy transfer is the initial "break" in a game of pool. If the cue ball squarely hits the point of the triangular array of 15 billiard balls, the two balls at the far corners of the triangle fly off, and none of the central balls move at all. The kinetic energy is transferred through the central balls to the most distant ones. In much the same way, the photon's excitation energy moves through the photosystem from one chlorophyll to the next.

Energy Capture As the energy shuttles from one chlorophyll molecule to another within the photosystem network, it eventually arrives at a key chlorophyll molecule, the only one that is touching a membrane-bound protein. Like shaking a marble in a box with a walnut-sized hole in it, the excitation energy will find its way to this special chlorophyll just as sure as the marble will eventually find its way to and through the hole in the box. The special chlorophyll then transfers an excited (high-energy) electron to the acceptor molecule it is touching.

The Light-Dependent Reactions Like a baton being passed from one runner to another in a relay race, the electron is then passed from that acceptor protein to a series of other proteins in the membrane that put the energy of the electron to work making ATP and NADPH. In a way you will explore later in this chapter, the energy is used to power the movement of protons across the thylakoid membrane to make ATP and another key molecule, NADPH. So far, photosynthesis has consisted of two stages, indicated by numbers in the diagram to the lower left: ❶ capturing energy from sunlight—accomplished by the photosystem; and ❷ using the energy to make ATP and NADPH. These first two stages of photosynthesis take place only in the presence of light, and together are traditionally called the **light-dependent reactions.** ATP and NADPH are important energy-rich chemicals, and after this, the rest of photosynthesis becomes a chemical process.

The Light-Independent Reactions The ATP and NADPH molecules generated by the light-dependent reactions described above are then used to power a series of chemical reactions in the stroma of the chloroplast, each catalyzed by an enzyme present there. Acting together like the many stages of a manufacturing assembly line, these reactions accomplish the synthesis of carbohydrates from CO_2 in the air ❸. This third stage of photosynthesis, the formation of organic molecules like glucose from atmospheric CO_2, is called the **Calvin cycle,** but is also referred to as the **light-independent reactions** because it doesn't require light directly. We will examine the Calvin cycle in detail later in this chapter.

This completes our brief overview of photosynthesis. In the rest of the chapter we will revisit each stage and consider its elements in more detail. For now, the overall process may be summarized by the following simple equation:

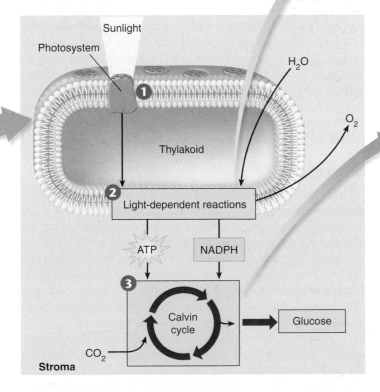

$$6\ CO_2 + 12\ H_2O + light\ energy \longrightarrow C_6H_{12}O_6 + 6\ H_2O + 6\ O_2$$

carbon dioxide   water   energy   glucose   water   oxygen

Key Learning Outcome 6.1 Photosynthesis uses energy from sunlight to power the synthesis of organic molecules from CO_2 in the air. In plants, photosynthesis takes place in specialized compartments within chloroplasts.

6.2 How Plants Capture Energy from Sunlight

Where is the energy in light? What is there about sunlight that a plant can use to create chemical bonds? The revolution in physics in the twentieth century taught us that light actually consists of tiny packets of energy called **photons,** which have properties both of particles and of waves. When light shines on your hand, your skin is being bombarded by a stream of these photons smashing onto its surface.

Sunlight contains photons of many energy levels, only some of which we "see." We call the full range of these photons the **electromagnetic spectrum.** As you can see in **figure 6.1,** some of the photons in sunlight have shorter wavelengths (toward the left side of the spectrum) and carry a great deal of energy—for example, gamma rays and ultraviolet (UV) light. Others such as radio waves carry very little energy and have longer wavelengths (hundreds to thousands of meters long). Our eyes perceive photons carrying intermediate amounts of energy as **visible light,** because the retinal pigment molecules in our eyes, which are different from chlorophyll, absorb only those photons of intermediate wavelength. Plants are even more picky, absorbing mainly blue and red light and reflecting back what is

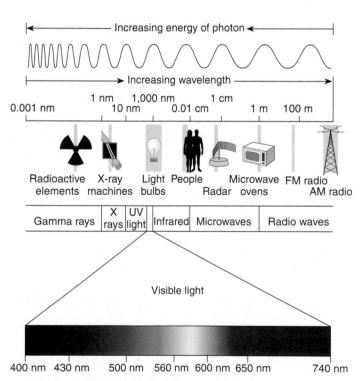

Figure 6.1 Photons of different energy: the electromagnetic spectrum.

Light is composed of packets of energy called photons. Some of the photons in light carry more energy than others. Light, a form of electromagnetic energy, is conveniently thought of as a wave. The shorter the wavelength of light, the greater the energy of its photons. Visible light represents only a small part of the electromagnetic spectrum, that with wavelengths between about 400 and 740 nanometers.

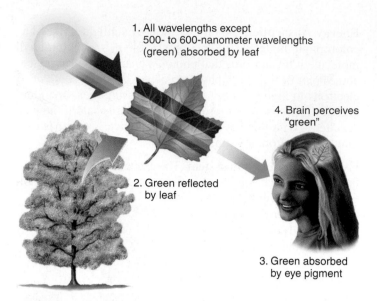

Figure 6.2 Why are plants green?

A leaf containing chlorophyll absorbs a broad range of photons—all the colors in the spectrum except for the photons around 500 to 600 nanometers. The leaf reflects these colors. These reflected wavelengths are absorbed by the visual pigments in our eyes, and our brains perceive the reflected wavelengths as "green."

left of the visible light. To understand why plants are green, look at the green tree in **figure 6.2.** The full spectrum of visible light shines on the leaves of this tree, and only the green wavelengths of light are not absorbed. They are reflected off the leaf, which is why our eyes perceive leaves as green.

How can a leaf or a human eye choose which photons to absorb? The answer to this important question has to do with the nature of atoms. Remember that electrons spin in particular orbits around the atomic nucleus, at different energy levels. Atoms absorb light by boosting electrons to higher energy levels, using the energy in the photon to power the move. Boosting the electron requires just the right amount of energy, no more and no less, just as when climbing a ladder you must raise your foot just so far to climb a rung. A particular kind of atom absorbs only certain photons of light, those with the appropriate amount of energy.

Pigments

As mentioned earlier, molecules that absorb light energy are called **pigments.** When we speak of visible light, we refer to those wavelengths that the pigment within human eyes, called *retinal,* can absorb—roughly from 380 nanometers (violet) to 750 nanometers (red). Other animals use different pigments for vision and thus "see" a different portion of the electromagnetic spectrum. For example, the pigment in insect eyes absorbs at shorter wavelengths than retinal. That is why bees can see ultraviolet light, which we cannot see, but are blind to red light, which we can see.

As noted, the main pigment in plants that absorbs light is chlorophyll. Its two forms, chlorophyll *a* and chlorophyll *b,* are similar in structure, but slight differences in their chemical "side groups" produce slight differences in their absorption spectra. An absorption spectrum is a graph indicating how effectively a pigment absorbs different wavelengths of visible light. For example, chlorophyll molecules will absorb photons

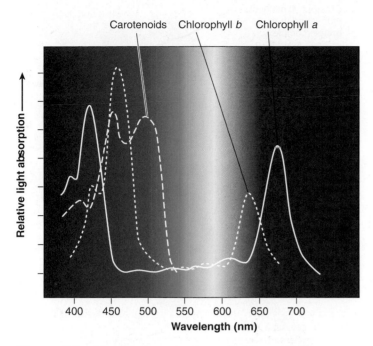

Figure 6.3 Absorption spectra of chlorophylls and carotenoids.

The peaks represent wavelengths of sunlight strongly absorbed by the two common forms of photosynthetic pigment, chlorophyll *a* and chlorophyll *b*, and by accessory carotenoid pigments. Chlorophylls absorb predominantly violet-blue and red light, in two narrow bands of the spectrum, while they reflect the green light in the middle of the spectrum. Carotenoids absorb mostly blue and green light and reflect orange and yellow light.

at the ends of the visible spectrum, the peaks you see in **figure 6.3**. While chlorophyll absorbs fewer kinds of photons than our visual pigment retinal, it is much more efficient at capturing them. Chlorophyll molecules capture photons with a metal ion (magnesium) that lies at the center of a complex carbon ring. Photons excite electrons of the magnesium ion, which are then channeled away by the carbon atoms.

While chlorophyll is the primary pigment involved in photosynthesis, plants also contain other pigments called *accessory pigments* that absorb light of wavelengths not captured by chlorophyll. **Carotenoids** are a group of accessory pigments that capture violet to blue-green light. As you can see in **figure 6.3**, these wavelengths of light are not efficiently absorbed by chlorophyll.

Accessory pigments give color to flowers, fruits, and vegetables but are also present in leaves, their presence usually masked by chlorophyll. During the warm months, when plants are actively producing food through photosynthesis, their cells are filled with chlorophyll-containing chloroplasts that cause the leaves to appear green, like the oak leaves on the left side of **figure 6.4**. In the fall, the days become shorter and cooler and for many species, leaves stop their food-making processes. Their chlorophyll molecules break down and are not replaced. When this happens, the colors reflected by accessory pigments become visible. The leaves turn colors of yellow, orange, and red, like the oak leaves on the right.

> **Key Learning Outcome 6.2** Plants use pigments like chlorophyll to capture photons of blue and red light, reflecting photons of green wavelengths.

Oak leaf in summer

Oak leaf in autumn

Figure 6.4 Fall colors are produced by pigments such as carotenoids.

During the spring and summer, chlorophyll masks the presence in leaves of other pigments called carotenoids. Cool temperatures in the fall cause leaves of deciduous trees to cease manufacturing chlorophyll. With chlorophyll no longer present to reflect green light, the orange and yellow light reflected by carotenoids gives bright colors to the autumn leaves.

6.3 Organizing Pigments into Photosystems

The light-dependent reactions of photosynthesis occur on membranes. In most photosynthetic bacteria, the proteins involved in the light-dependent reactions are embedded within the plasma membrane. In algae, intracellular membranes contain the proteins that drive the light-dependent reactions. In plants, photosynthesis occurs in specialized organelles called chloroplasts. The chlorophyll molecules and proteins involved in the light-dependent reactions are embedded in the thylakoid membranes inside the chloroplasts. A portion of a thylakoid membrane is enlarged in **figure 6.5.** The chlorophyll molecules can be seen as the green spheres embedded along with accessory pigment molecules within a matrix of proteins (the purple area) within the thylakoid membrane. This complex of protein and pigment makes up the **photosystem.**

The light-dependent reactions take place in five stages, illustrated in **figure 6.6.** Each stage will be discussed in detail later in this chapter:

1. **Capturing light.** In stage ❶, a photon of light of the appropriate wavelength is captured by a pigment molecule, and the excitation energy is passed from one chlorophyll molecule to another.
2. **Exciting an electron.** In stage ❷, the excitation energy is funneled to a key chlorophyll *a* molecule called the **reaction center.** The excitation energy causes the transfer of an excited electron from the reaction center to another molecule that is an electron acceptor. The reaction center replaces this "lost" electron with an electron from the breakdown of a water molecule. Oxygen is produced as a by-product of this reaction.
3. **Electron transport.** In stage ❸, the excited electron is then shuttled along a series of electron-carrier molecules embedded in the membrane. This is called the **electron transport system (ETS).** As the electron passes along the electron transport system, the energy from the electron is "siphoned" out in small amounts. This energy is used to pump hydrogen ions (protons) across the membrane, indicated by the blue arrow, eventually building up a high concentration of protons inside the thylakoid.
4. **Making ATP.** In stage ❹, the high concentration of protons can be used as an energy source to make ATP. Protons are only able to move back across the membrane via special channels, the protons flooding through them like water through a dam. The kinetic energy that is released by the movement of protons is transferred to potential energy in the building of ATP molecules from ADP. This process, called **chemiosmosis,** makes the ATP that will be used in the Calvin cycle to make carbohydrates.

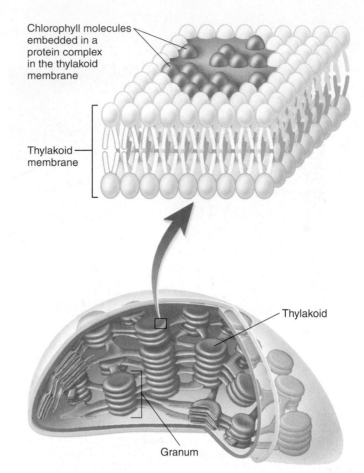

Figure 6.5 Chlorophyll embedded in a membrane.
Chlorophyll molecules are embedded in a network of proteins that hold the pigment molecules in place. The proteins are embedded within the membranes of thylakoids.

5. **Making NADPH.** The electron leaves the electron transport system and enters another photosystem where it is "reenergized" by the absorption of another photon of light. In ❺, this energized electron enters another electron transport system, where it is again shuttled along a series of electron-carrier molecules. The result of this electron transport system is not the synthesis of ATP, but rather the formation of NADPH. The electron is transferred to a molecule, NADP⁺, and a hydrogen ion that forms NADPH. This molecule is important in the synthesis of carbohydrates in the Calvin cycle.

Architecture of a Photosystem

In all but the most primitive bacteria, light is captured by photosystems. Like a magnifying glass focusing light on a precise point, a photosystem channels the excitation energy gathered by any one of its pigment molecules to a specific chlorophyll *a* molecule, the reaction center chlorophyll. For example, in **figure 6.7,** a chlorophyll molecule on the outer edge of the photosystem is excited by the photon, and this energy passes from one chlorophyll molecule to another, indicated by the yellow zig-zag arrow, until it reaches the reaction center molecule. This molecule then passes the energy, in the form of an excited electron, out of the photosystem to drive the synthesis of ATP and organic molecules.

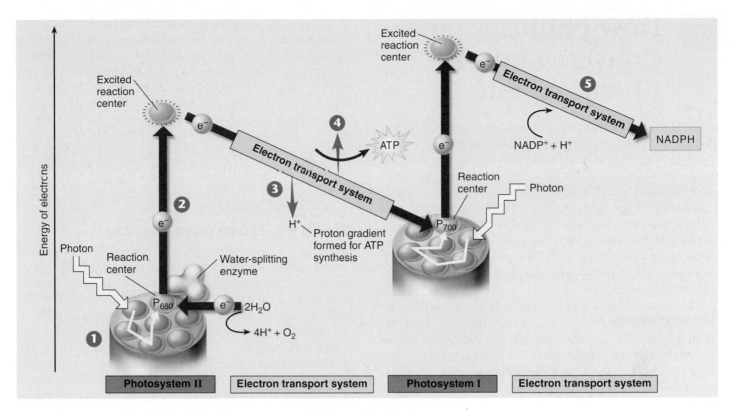

Figure 6.6 **Plants use two photosystems.**

In stage ❶, a photon excites pigment molecules in photosystem II. In stage ❷, a high-energy electron from photosystem II is transferred to the electron transport system. In stage ❸, the excited electron is used to pump a proton across the membrane. In stage ❹, the concentration gradient of protons is used to produce a molecule of ATP. In stage ❺, the ejected electron then passes to photosystem I, which uses it, with a photon of light energy, to drive the formation of NADPH.

Using Two Photosystems

Plants and algae use two photosystems, photosystems I and II, indicated by the two purple cylinders in **figure 6.6.** Photosystem II captures the energy that is used to produce the ATP needed to build sugar molecules. The light energy that it captures is used to transfer the energy of a photon of light ❶ to an excited electron ❷; the energy of this electron is then used by the electron transport system ❸ to produce ATP ❹.

Photosystem I powers the production of the hydrogen atoms needed to build sugars and other organic molecules from CO_2 (which has no hydrogen atoms). Photosystem I is used to energize an electron that, carried by a hydrogen ion (a proton), forms NADPH from NADP+ ❺. NADPH shuttles hydrogens to the Calvin cycle where sugars are made.

The photosystems are not numbered in the order in which they are used. Photosystem II actually acts first in the series, and photosystem I acts second. The confusion arises because the photosystems were named in the order in which they were discovered, and photosystem I was discovered before photosystem II.

Key Learning Outcome 6.3 Photon energy is captured by pigments that employ it to excite electrons that are channeled away to do the chemical work of producing ATP and NADPH.

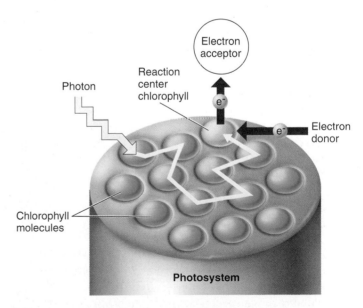

Figure 6.7 **How a photosystem works.**

When light of the proper wavelength strikes any pigment molecule within a photosystem, the light is absorbed and its excitation energy is then transferred from one molecule to another within the cluster of pigment molecules until it encounters the reaction center, which exports the energy as high-energy electrons to an acceptor molecule.

How Photosystems Convert Light to Chemical Energy

Plants use the two photosystems discussed in series, first one and then the other, to produce both ATP and NADPH. This two-stage process is called **noncyclic photophosphorylation,** because the path of the electrons is not a circle—the electrons ejected from the photosystems do not return to them, but rather end up in NADPH. The photosystems are replenished instead with electrons obtained by splitting water. As described earlier, photosystem II acts first. High-energy electrons generated by photosystem II are used to synthesize ATP and then passed to photosystem I to drive the production of NADPH.

Photosystem II

Within photosystem II (represented by the first purple structure you see on the left in **figure 6.8**), the reaction center consists of more than 10 transmembrane protein subunits. The *antenna complex,* which is the portion of the photosystem that contains all the pigment molecules, consists of some 250 molecules of chlorophyll *a* and accessory pigments bound to several protein chains. The antenna complex captures energy from a photon and funnels it to a reaction center chlorophyll. You can also see the antenna complex in the photosystem illustrated in **figure 6.7**. The reaction center gives up an excited electron to a primary electron acceptor in the electron transport system. The path of the excited electron is indicated with the red arrow. After the reaction center gives up an electron to the electron transport system, there is an empty electron orbital that needs to be filled. This electron is replaced with an electron from a water molecule. In photosystem II the oxygen atoms of two water molecules bind to a cluster of manganese atoms embedded within an enzyme and bound to the reaction center (notice the light gray water-splitting enzyme at the bottom left of photosystem II). This enzyme splits water, removing electrons one at a time to fill the holes left in the reaction center by the departure of light-energized electrons. As soon as four electrons have been removed from the two water molecules, O_2 is released.

Electron Transport System

The primary electron acceptor for the light-energized electrons leaving photosystem II passes the excited electron to a series of electron-carrier molecules called the *electron transport system.* These proteins are embedded within the thylakoid membrane; one of them is a "proton pump" protein, a type of active transport channel. The energy of the electron is used by this protein to pump a proton from the stroma into the thylakoid space (indicated by the blue arrow through the electron transport system). A nearby protein in the membrane then carries the now energy-depleted electron on to photosystem I.

Making ATP: Chemiosmosis

Before progressing onto photosystem I, let's see what happens with the protons that were pumped into the thylakoid by the electron transport system. Each thylakoid is a closed compartment into which protons are pumped. The thylakoid membrane is impermeable to protons, so protons build up inside the thylakoid space, creating a very large concentra-

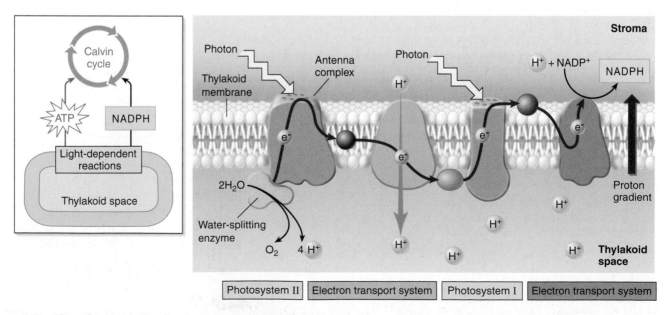

Figure 6.8 **The photosynthetic electron transport system.**

tion gradient. As you may recall from chapter 4, molecules in solution diffuse from areas of higher concentration to areas of lower concentration. Here, protons diffuse back out of the thylakoid space, down their concentration gradient, passing through special protein channels called ATP synthases. *ATP synthase* is an enzyme that can use the concentration gradient of protons to drive the synthesis of ATP from ADP. ATP synthase channels protrude like knobs on the external surface of the thylakoid membrane (**figure 6.9**). As protons pass out of the thylakoid through the ATP synthase channels, ADP is phosphorylated to ATP and released into the stroma (the fluid matrix inside the chloroplast). Because the chemical formation of ATP is driven by a diffusion process similar to osmosis, this type of ATP formation is called *chemiosmosis.*

Photosystem I

Now, with ATP formed, let's return our attention to the right half of **figure 6.8**, with photosystem I accepting an electron from the electron transport system. The reaction center of photosystem I is a membrane complex consisting of at least 13 protein subunits. Energy is fed to it by an antenna complex consisting of 130 chlorophyll *a* and accessory pigment molecules. The electron arriving from the first electron transport system has by no means lost all of its light-excited energy; almost half remains. Thus, the absorption of another photon of light energy by photosystem I boosts the electron leaving its reaction center to a very high energy level.

Making NADPH

Like photosystem II, photosystem I passes electrons to an electron transport system. When two of these electrons reach the end of this electron transport system, they are then donated to a molecule of NADP+ to form NADPH (one electron is transferred with a proton as a hydrogen atom). This reaction, which takes place on the stromal side of the thylakoid (as shown in **figure 6.8**), involves an NADP+, two electrons, and a proton. Because the reaction occurs on the stromal side of the membrane and involves the uptake of a proton in forming NADPH, it contributes further to the proton concentration gradient established during photosynthetic electron transport.

Products of the Light-Dependent Reactions

The light-dependent reactions can be seen more as a stepping stone, rather than an end point of photosynthesis. All of the products of the light-dependent reactions are either waste products, such as oxygen, or are ultimately used elsewhere in the cell. The ATP and NADPH produced in the light-dependent reactions end up being passed on to the Calvin cycle in the stroma of the chloroplast. The stroma contains the enzymes that catalyze the light-independent reactions, in which ATP is used to power chemical reactions that build carbohydrates. NADPH is used as the source of "reducing power," providing the hydrogens and electrons used in building carbohydrates. The next section discusses the Calvin cycle of photosynthesis.

> **Key Learning Outcome 6.4** The light-dependent reactions of photosynthesis produce the ATP and NADPH needed to build organic molecules, and release O_2 as a by-product of stripping hydrogen atoms and their associated electrons from water molecules.

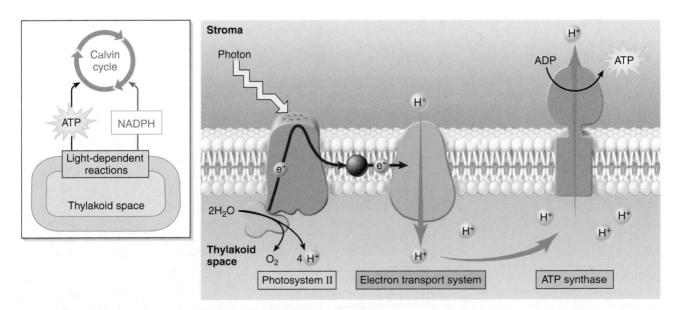

Figure 6.9 Chemiosmosis in a chloroplast.
The energy of the electron absorbed by photosystem II powers the pumping of protons into the thylakoid space. These protons then pass back out through ATP synthase channels, their movement powering the production of ATP.

6.5 Building New Molecules

The Calvin Cycle

Stated very simply, photosynthesis is a way of making organic molecules from carbon dioxide (CO_2). To build organic molecules, cells use raw materials provided by the light-dependent reactions:

1. **Energy.** ATP (provided by the ETS of photosystem II) drives the endergonic reactions.
2. **Reducing power.** NADPH (provided by the ETS of photosystem I) provides a source of hydrogens and the energetic electrons needed to bind them to carbon atoms. A molecule that accepts an electron is said to be reduced, as will be discussed in detail in chapter 7.

The actual assembly of new molecules employs a complex battery of enzymes in what is called the **Calvin cycle,** or **C_3 photosynthesis** (C_3 because the first molecule produced in the process is a three-carbon molecule). The Calvin cycle takes place in the stroma of the chloroplasts. The NADPH and the ATP that were generated by the light-dependent reactions are used in the Calvin cycle to build carbohydrate molecules. In the Key Biological Process illustration below, the number of carbon atoms at each stage is indicated by the number of balls. It takes six turns of the cycle to make one six-carbon molecule of glucose. The process takes place in three stages, highlighted in the three panels of the Key Biological Process illustration below. These three stages are also indicated by different-colored pie-shaped pieces in the more detailed look at the Calvin cycle provided in **figure 6.10.** Both figures indicate that three turns of the cycle are needed to produce one molecule of glyceraldehyde 3-phosphate. In any *one* turn of the cycle, a carbon atom from a carbon dioxide molecule

KEY BIOLOGICAL PROCESS: The Calvin Cycle

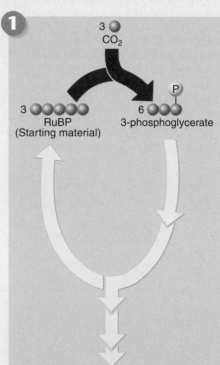

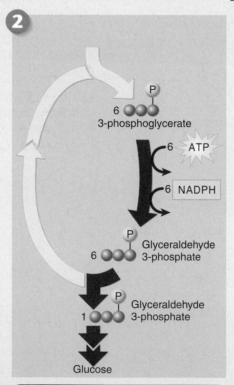

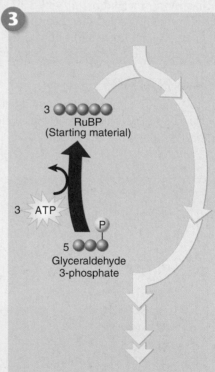

The Calvin cycle begins when a carbon atom from a CO_2 molecule is added to a five-carbon molecule (the starting material). The resulting six-carbon molecule is unstable and immediately splits into three-carbon molecules. (Three "turns" of the cycle are indicated here with three molecules of CO_2 entering the cycle.)

Then, through a series of reactions, energy from ATP and hydrogens from NADPH (the products of the light-dependent reactions) are added to the three-carbon molecules. The now-reduced three-carbon molecules either combine to make glucose or are used to make other molecules.

Most of the reduced three-carbon molecules are used to regenerate the five-carbon starting material, thus completing the cycle.

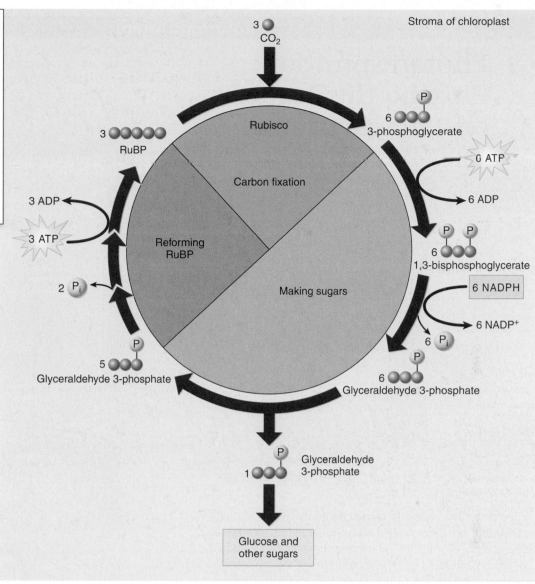

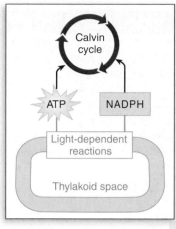

Figure 6.10
Reactions of the Calvin cycle.

For every three molecules of CO_2 that enter the cycle, one molecule of the three-carbon compound glyceraldehyde 3-phosphate (G3P) is produced. Notice that the process requires energy stored in ATP and NADPH, which are generated by the light-dependent reactions. This process occurs in the stroma of the chloroplast. The large 16-subunit enzyme that catalyzes the reaction, RuBP carboxylase, or **rubisco,** is the most abundant protein in chloroplasts and is thought to be the most abundant protein on earth.

is first added to a five-carbon sugar, producing two three-carbon sugars. This process, highlighted by the dark blue arrow in **panel 1** of the Key Biological Process illustration and the blue pie-shaped area in **figure 6.10**, is called *carbon fixation* because it attaches a carbon atom that was in a gas to an organic molecule.

Then, in a long series of reactions, the carbons are shuffled about. Eventually some of the resulting molecules are channeled off to make sugars (shown by the dark blue arrows in **panel 2** of the Key Biological Process illustration and at the bottom of the cycle within the purple colored area in **figure 6.10**). Other molecules are used to re-form the original five-carbon sugar (the dark blue arrow in **panel 3** of the Key Biological Process illustration and the light-red-colored area in **figure 6.10**), which is then available to restart the cycle. The cycle has to "turn" six times in order to form a new glucose molecule, because each turn of the cycle adds only one carbon atom from CO_2, and glucose is a six-carbon sugar.

Recycling ADP and NADP+

The products of the light-dependent reactions, ATP and NADPH, feed into the light-independent reactions of the Calvin cycle to make sugar molecules. To keep photosynthesis moving along, the cells must continually supply the light-dependent reactions with more ADP and NADP+. This is accomplished by recycling these products from the Calvin cycle. After the phosphate bonds are broken in ATP, ADP is available for chemiosmosis. After the hydrogens and electrons are stripped from NADPH, NADP+ is available to cycle back to the electron transport system of photosystem I.

> **Key Learning Outcome 6.5** In a series of reactions that do not directly require light, cells use ATP and NADPH provided by photosystems II and I to assemble new organic molecules.

6.6 Photorespiration: Putting the Brakes on Photosynthesis

Many plants have trouble carrying out C_3 photosynthesis when the weather is hot. A cross section of a leaf here shows how it responds to hot, arid weather:

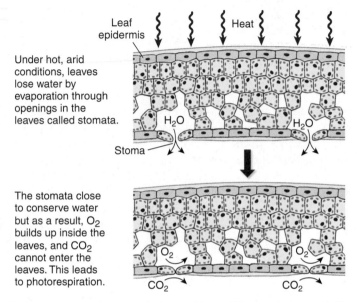

As temperatures increase in hot, arid conditions, plants partially close their leaf openings, called **stomata** (singular, **stoma**), to conserve water. As a result, you can see above that CO_2 and O_2 are not able to enter and exit the leaves through these openings. The concentration of CO_2 in the leaves falls, while the concentration of O_2 in the leaves rises. Under these conditions rubisco, the enzyme that carries out the first step of the Calvin cycle, engages in **photorespiration,** where the enzyme incorporates O_2, not CO_2, into the cycle and when this occurs, CO_2 is ultimately released as a by-product. Photorespiration thus short-circuits the successful performance of the Calvin cycle.

C_4 Photosynthesis

Some plants are able to adapt to climates with higher temperatures by performing **C_4 photosynthesis.** In this process, plants such as sugarcane, corn, and many grasses are able to fix carbon using different types of cells and chemical reactions within their leaves, thereby avoiding a reduction in photosynthesis due to higher temperatures.

A cross section of a leaf from a C_4 plant is shown in **figure 6.11**. Examining it, you can see how these plants solve the problem of photorespiration. In the enlargement, you see two cell types: The green cell is a mesophyll cell and the tan cell is a bundle-sheath cell. In the mesophyll cell, CO_2 combines with a three-carbon molecule instead of RuBP as it did in **figure 6.10**, producing a four-carbon molecule, oxaloacetate (hence the name, C_4 photosynthesis), rather than the three-carbon molecule

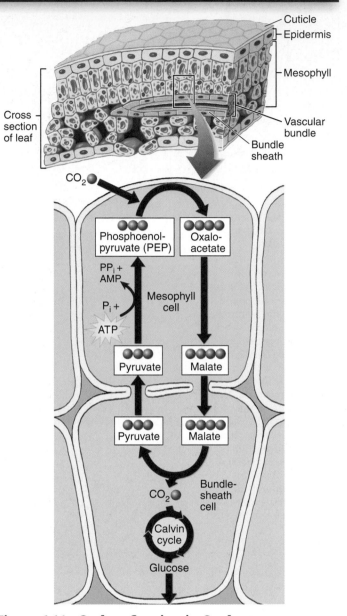

Figure 6.11   Carbon fixation in C_4 plants.
This process is called the C_4 pathway because the first molecule formed in the pathway is a four-carbon sugar, oxaloacetate. This molecule is converted into malate that is transported into bundle-sheath cells. Once there, malate undergoes a chemical reaction producing carbon dioxide. The carbon dioxide is trapped in the bundle-sheath cell, where it enters the Calvin cycle.

phosphoglycerate you saw in **figure 6.10**. C_4 plants carry out this process in the mesophyll cells of their leaves, using a different enzyme. The oxaloacetate is then converted to malate, which is transferred to the bundle-sheath cells of the leaf. In the tan bundle-sheath cell, malate is broken down to regenerate CO_2, which enters the Calvin cycle you are familiar with from **figure 6.10**, and sugars are synthesized. Why go to all this trouble? Because the bundle-sheath cells are impermeable to CO_2 and so the concentration of CO_2 increases within them, so much that the rate of photorespiration is substantially lowered.

Today's *Biology*

Cold-Tolerant C₄ Photosynthesis

Corn (*Zea mays*), one of humanity's most important agricultural crops, is highly productive when grown at warm temperatures. However, its commercial use in northern areas is severely limited by its much poorer performance at low temperatures. Much of corn's high productivity results from its use of the C_4 photosynthetic pathway, which has the highest efficiency of photosynthesis known. However, much of this efficiency is lost below 20° C. At 5° C, 80% of photosynthesis is lost.

In C_4 species like corn, sugarcane, sorghum, and switchgrass, sensitivity to low temperatures appears to depend on the sensitivity of key C_4 photosynthetic enzymes, particularly the Calvin cycle enzyme catalyzing the final stage illustrated in **figure 6.11**. This enzyme has the imposing name pyruvate orthophosphate dikinase and is abbreviated PPDK. PPDK, which appears to be the rate-limiting step in corn C_4 photosynthesis, is very sensitive to low temperature, with little activity remaining when temperatures fall below 10° C.

One relative of corn recently has been shown to be strikingly different. Chinese silver grass (*Miscanthus giganteus*) is a perennial grass that uses the same C_4 pathway as corn. However, in marked contrast to corn, it produces efficiently at temperatures as low as 5° C. With its greater tolerance of low temperatures, this species thrives at chilling temperatures, with individual stalks growing as high as 13 feet! Similar temperatures severely limit C_4 photosynthesis in its relative.

What is the cause of *Miscanthus*'s tolerance of cold? At low temperatures, when amounts of PPDK fall in corn, PPDK activity actually rises in *Miscanthus*. Researchers are currently examining the *Miscanthus* PPDK gene to better understand the cold-tolerance it confers. If these early results are confirmed, genetic engineers can explore the possibility of replacing the corn PPDK gene with the *Miscanthus* version, in the hope of greatly extending the northern range of corn, a key agricultural crop.

A second strategy to decrease photorespiration is used by many succulent (water-storing) plants such as cacti and pineapples. This mode of initial carbon fixation is called **crassulacean acid metabolism (CAM)** after the plant family Crassulaceae in which it was first discovered. In these plants, the stomata open during the night when it's cooler, and close during the day. CAM plants initially fix CO_2 into organic compounds at night, using the C_4 pathway. These organic compounds accumulate at night and are subsequently broken down during the following day, releasing CO_2. These high levels of CO_2 drive the Calvin cycle and decrease photorespiration. To understand how photosynthesis differs in CAM plants and C_4 plants, examine **figure 6.12**. In C_4 plants (on the left), the C_4 pathway occurs in mesophyll cells, while the Calvin cycle occurs in bundle-sheath cells. In CAM plants (on the right), the C_4 pathway and the Calvin cycle occur in the same cell, a mesophyll cell, but they occur at different times of the day, the C_4 cycle at night and the Calvin cycle during the day.

> **Key Learning Outcome 6.6** Photorespiration occurs due to a buildup of oxygen within photosynthetic cells. C_4 plants get around photorespiration by synthesizing sugars in bundle-sheath cells, and CAM plants delay the light-independent reactions until night, when stomata are open.

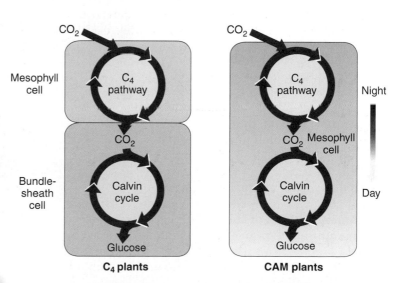

Figure 6.12 Comparing carbon fixation in C_4 and CAM plants.

Both C_4 and CAM plants utilize the C_4 and C_3 pathways. In C_4 plants, the pathways are separated spatially; the C_4 pathway takes place in the mesophyll cells and the C_3 pathway (the Calvin cycle) in the bundle-sheath cells. In CAM plants, the two pathways occur in mesophyll cells but are separated temporally; the C_4 pathway is utilized at night and the C_3 pathway during the day.

Does Iron Limit the Growth of Ocean Phytoplankton?

Phytoplankton are microscopic organisms that live in the oceans, carrying out much of the earth's photosynthesis. The photo below is of *Chaetoceros,* a phytoplankton. Decades ago, scientists noticed "dead zones" in the ocean where little photosynthesis occurred. Looking more closely, they found that phytoplankton collected from these waters are not able to efficiently fix CO_2 into carbohydrates. In an attempt to understand why not, the scientists hypothesized that lack of iron (needed by the ETS) was the problem, and predicted that fertilizing these ocean waters with iron could trigger an explosively rapid growth of phytoplankton.

To test this idea, they carried out a field experiment, seeding large areas of phytoplankton-poor ocean waters with iron crystals to see if this triggered phytoplankton growth. Other similarly phytoplankton-poor areas of ocean were not seeded with iron and served as controls.

In one such experiment, the results of which are presented in the graph to the right, a 72-km² grid of phytoplankton-deficient ocean water was seeded with iron crystals and a tracer substance in three successive treatments, indicated with arrows on the *x* axis of the graph (on days 0, 3, and 7). The multiple seedings were carried out to reduce the effect of the iron crystals dissipating over time. A smaller control grid, 24 km², was seeded with just the tracer substance.

To assess the numbers of phytoplankton organisms carrying out photosynthesis in the ocean water, investigators did not actually count organisms. Instead, they estimated the amount of chlorophyll *a* in water samples as an easier-to-measure index. An **index** is a parameter that accurately reflects the quantity of another less-easily-measured parameter. In this instance, the level of chlorophyll *a*, easily measured by monitoring the wavelengths of light absorbed by a liquid sample, is a suitable index of phytoplankton, as this pigment is found nowhere else in the ocean other than within phytoplankton.

Chlorophyll *a* measurements were made periodically on both test and control grids for 14 days. The results are plotted on the graph. Red points indicate chlorophyll *a* concentrations in iron-seeded waters; blue points indicate chlorophyll *a* levels in the control grid waters that were not seeded.

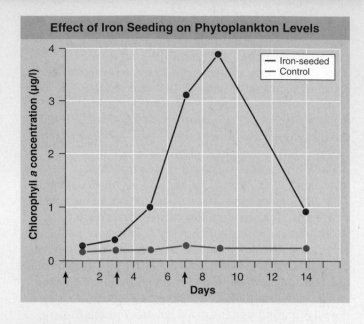

Effect of Iron Seeding on Phytoplankton Levels

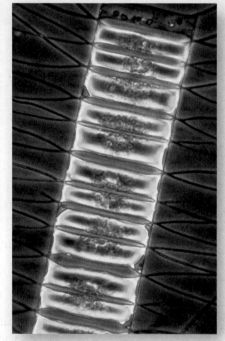

1. **Applying Concepts**
   a. **Variable.** In the graph above, which is the dependent variable?
   b. **Index.** What does the increase in levels of chlorophyll *a* say about numbers of phytoplankton?
   c. **Control.** What substance is lacking in the waters sampled in the blue-dot plots ?

2. **Interpreting Data**
   a. What happened to the levels of chlorophyll *a* in the test areas of the ocean (red dots)?
   b. What happened to the levels of chlorophyll *a* in the control areas (blue dots)?
   c. Comparing the red line to the blue line, about how many times more numerous are phytoplankton in iron-seeded waters on the three days of seeding?

3. **Making Inferences**
   a. What general statement can be made regarding the effect of seeding phytoplankton-poor regions of the ocean with iron?
   b. Why did chlorophyll *a* levels drop by day 14?

4. **Drawing Conclusions** Do these results support the claim that lack of iron is limiting the growth of phytoplankton, and thus of photosynthesis, in certain areas of the oceans?

5. **Further Analysis** Based on this experiment, what would be a potential drawback of using this method of seeding with iron to increase levels of ocean photosynthesis?

Photosynthesis

6.1 An Overview of Photosynthesis

- Photosynthesis is a biochemical process whereby energy from the sun is captured and used to build carbohydrates from CO_2 gas and water (**integrated art, pages 118-119**).

- Photosynthesis consists of a series of chemical reactions that occurs in two stages: The light-dependent reactions that produce ATP and NADPH occur on the thylakoid membranes of chloroplasts in plants, while the light-independent reactions (the Calvin cycle) that synthesize carbohydrates occur in the stroma (**integrated art, pages 120-121**).

6.2 How Plants Capture Energy from Sunlight

- Pigments are molecules that capture light energy. Energy present in visible light is captured by the pigment chlorophyll and other accessory pigments present in chloroplasts (**figure 6.1**).

- Plants appear green because of their chlorophyll pigments (**figure 6.2**). Chlorophyll absorbs wavelengths in the far ends of the visual spectrum (the blue and red wavelengths) and reflect the green wavelengths, which is why leaves appear green.

- Accessory pigments, such as carotenoids, capture energy from different areas of the spectrum than chlorophyll and give different colors to flowers, fruits, and other parts of plants that are not green (**figure 6.4**).

6.3 Organizing Pigments into Photosystems

- The light-dependent reactions occur on the thylakoid membranes of chloroplasts in plants. The chlorophyll molecules and other pigments involved in photosynthesis are embedded in a complex of proteins within the membrane called a photosystem (**figure 6.5**).

- Light energy is captured by the photosystem, where it excites an electron that is passed to an electron transport system. There, it is used to generate ATP and NADPH, both of which power the Calvin cycle. Plants utilize two photosystems that occur in series (**figure 6.6**). Photosystem II leads to the formation of ATP, and photosystem I leads to the formation of NADPH.

- The energy from a photon of light is absorbed by a chlorophyll molecule and is transferred between chlorophyll molecules in the photosystem, as shown here from **figure 6.7**. Once the energy is passed to the reaction center, it excites an electron, which is transferred to the electron transport system.

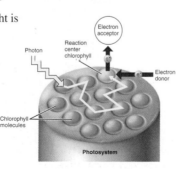

6.4 How Photosystems Convert Light to Chemical Energy

- The excited electron that leaves the reaction center of photosystem II and enters the electron transport system is replenished with an electron from the breakdown of a water molecule.

- The excited electron is passed from one protein to another in the electron transport system, where energy from the electron is used to operate a proton pump that pumps hydrogen ions across the membrane against a concentration gradient (**figure 6.8**).

- The hydrogen ion concentration gradient is used as a source of energy to generate molecules of ATP. This energy is used to drive H^+ back across the membrane through a specialized channel protein called ATP synthase, which catalyzes the formation of ATP, as shown here from **figure 6.9**.

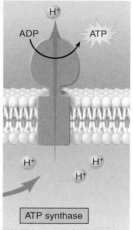
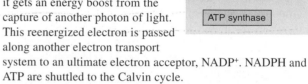

- After the electron passes along the first electron transport system, it is then transferred to a second photosystem, photosystem I, where it gets an energy boost from the capture of another photon of light. This reenergized electron is passed along another electron transport system to an ultimate electron acceptor, $NADP^+$. NADPH and ATP are shuttled to the Calvin cycle.

6.5 Building New Molecules

- The Calvin cycle is carried out by a series of enzymes that use the energy from ATP and electrons and hydrogen ions from NADPH to build molecules of carbohydrates by reducing CO_2 (**Key Biological Process, page 128** and **figure 6.10**).

- During the Calvin cycle, ADP and $NADP^+$ are recycled as by-products and feed back into the light-dependent reactions.

Photorespiration

6.6 Photorespiration: Putting the Brakes on Photosynthesis

- In hot, dry weather, plants will close the stomata in their leaves to conserve water. As a result, the levels of O_2 increase in the leaves, and CO_2 levels drop, as shown here from **page 130**. Under these conditions, the Calvin cycle, also called C_3 photosynthesis, is disrupted. When there is a higher internal concentration of oxygen, O_2 rather than CO_2 enters the Calvin cycle in a process called photorespiration. In this case, the first enzyme in the Calvin cycle, rubisco, binds oxygen instead of carbon dioxide.

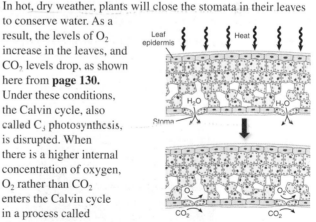

- C_4 plants reduce the effects of photorespiration by modifying the carbon-fixation step, splitting it into two steps that take place in different cells. The C_4 pathway produces malate in mesophyll cells (**figure 6.11**).

- In CAM plants, carbon dioxide is processed into intermediate organic molecules through the C_4 pathway during the night when stomata are open (**figure 6.12**).

1. The energy that is used by almost all living things on our planet comes from the sun. It is captured by plants, algae, and some bacteria through the process of
   a. thylakoid.
   c. photosynthesis.
   b. chloroplasts.
   d. the Calvin cycle.
2. Plants capture sunlight
   a. through photorespiration.
   b. with molecules called pigments that absorb photons and use their energy.
   c. with the light-independent reactions.
   d. with the electron transport system.
3. Visible light occupies what part of the electromagnetic spectrum?
   a. the entire spectrum
   b. the upper half of the spectrum (with longer wavelengths)
   c. a small portion in the middle of the spectrum
   d. the lower half of the spectrum (with shorter wavelengths)
4. The colors of light that are absorbed by chlorophyll are
   a. red and blue.
   b. green and yellow.
   c. infrared and ultraviolet.
   d. All colors are equally effective.
5. Once a plant has initially captured the energy of a photon,
   a. a series of reactions occurs in thylakoid membranes of the cell.
   b. the energy is transferred through several steps into a molecule of ATP.
   c. a water molecule is broken down, releasing oxygen.
   d. All of the above.

6. Plants use two photosystems to capture energy used to produce ATP and NADPH. The electrons used in these photosystems
   a. recycle through the system constantly, with energy added from the photons.
   b. recycle through the system several times and then are lost due to entropy.
   c. only go through the system once; they are obtained by splitting a water molecule.
   d. only go through the system once; they are obtained from the photon.
7. During photosynthesis, ATP molecules are generated by
   a. the Calvin cycle.
   b. chemiosmosis.
   c. the splitting of a water molecule.
   d. photons of light being absorbed by chlorophyll molecules.
8. NADPH is recycled during photosynthesis. It is produced during the _____ and used in the_____.
   a. electron transport system of photosystem I, Calvin cycle
   b. process of chemiosmosis, Calvin cycle
   c. electron transport system of photosystem II, electron transport system of photosystem I
   d. light-independent reactions, light-dependent reactions
9. The overall purpose of the Calvin cycle is to
   a. generate molecules of ATP.
   b. generate NADPH.
   c. build sugar molecules.
   d. produce oxygen.
10. Many plants cannot carry out the typical C_3 photosynthesis in hot weather, so some plants
    a. use the ATP cycle.
    b. use C_4 photosynthesis or CAM.
    c. shut down photosynthesis completely.
    d. All of these are true for different plants.

Apply Your Understanding

1. **Figure 6.3** Why do most leaves have more than one type of pigment?

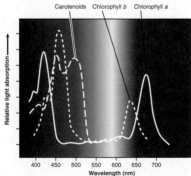

2. **Figure 6.9** Could a plant cell produce ATP through chemiosmosis if the thylakoid membrane was "leaky" with regard to protons? Explain.

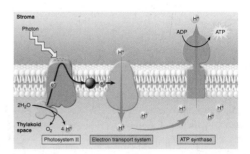

Synthesize What You Have Learned

1. To reduce six molecules of carbon dioxide to one molecule of glucose via photosynthesis, how many molecules of NADPH and ATP are required?
2. In theory, a plant kept in total darkness could still manufacture glucose, if it were supplied with which molecules?

3. If you were going to design a plant that would survive in the deserts of Arizona and New Mexico, how would you balance its need for CO_2 with its need to avoid water loss in the hot summer temperatures?

7

How Cells Harvest Energy from Food

Animals such as this chipmunk depend on the energy stored in the chemical bonds of the food they eat to power their life processes. Their lives are driven by energy. All the activities this chipmunk carries out—climbing trees, chewing on acorns, seeing and smelling and hearing its surroundings, thinking the thoughts that chipmunks think—use energy. But unlike the oak tree that produces the nuts on which this chipmunk is dining, no part of the chipmunk is green. It cannot carry out photosynthesis like an oak tree, and so cannot harvest energy from the sun as the tree does. Instead, it must get its energy secondhand, by consuming organic molecules manufactured by plants. The chemical energy that the oak tree invested in making its molecules is harvested by the chipmunk in a process called cellular respiration. The same processes are used by all animals to harvest energy from molecules—and by plants too. There is no sunlight under the soil where the oak tree's roots penetrate, and like the cells of the chipmunk, these plant root cells obtain the energy to fuel their lives from cellular respiration. In this chapter, we examine cellular respiration up close. As you will see, cellular respiration and photosynthesis have much in common.

7.1 Where Is the Energy in Food?

In both plants and animals, and in fact in almost all organisms, the energy for living is obtained by breaking down the organic molecules originally produced in plants. The ATP energy and reducing power invested in building the organic molecules are retrieved by stripping away the energetic electrons and using them to make ATP. When electrons are stripped away from chemical bonds, the food molecules are being oxidized (remember, oxidation is the loss of electrons). The oxidation of foodstuffs to obtain energy is called **cellular respiration.** Do not confuse the term cellular respiration with the breathing of oxygen gas that your lungs carry out, which is called simply respiration.

The cells of plants fuel their activities with sugars and other molecules that they produce through photosynthesis and breakdown in cellular respiration. Nonphotosynthetic organisms eat plants, extracting energy from plant tissue in cellular respiration. Other animals, like the lion gnawing with such relish on a giraffe leg in **figure 7.1,** eat these animals.

Eukaryotes produce the majority of their ATP by harvesting electrons from chemical bonds of the food molecule glucose. The electrons are transferred along an electron transport chain (similar to the electron transport system in photosynthesis), and eventually donated to oxygen gas. Chemically, there is little difference between this oxidation of carbohydrates in a cell and the burning of wood in a fireplace. In both instances, the reactants are carbohydrates and oxygen, and the products are carbon dioxide, water, and energy:

$$C_6H_{12}O_6 + 6\ O_2 \longrightarrow 6\ CO_2 + 6\ H_2O + energy$$
$$\text{(heat or ATP)}$$

In many of the reactions of photosynthesis and cellular respiration, electrons pass from one atom or molecule to another. When an atom or molecule loses an electron, it is said to be *oxidized*, and the process by which this occurs is called **oxidation.** The name reflects the fact that in biological systems, oxygen, which attracts electrons strongly, is the most common electron acceptor. Conversely, when an atom or molecule gains an electron, it is said to be *reduced,* and the process is called **reduction.** Oxidation and reduction always take place together, because every electron that is lost by an atom through oxidation is gained by some other atom through reduction. Therefore, chemical reactions of this sort are called **oxidation-reduction (redox) reactions.** In redox reactions, energy follows the electron, as shown in **figure 7.2.**

Cellular respiration is carried out in two stages, illustrated in **figure 7.3.** The first stage uses coupled reactions to make ATP. This stage, *glycolysis,* takes place in the cell's cytoplasm. Importantly, it is anaerobic (that is, it does not require oxygen). This ancient energy-extracting process is thought to have evolved over 2 billion years ago, when there was no oxygen in the earth's atmosphere.

Figure 7.1   Lion at lunch.
Energy that this lion extracts from its meal of giraffe will be used to power its roar, fuel its running, and build a bigger lion.

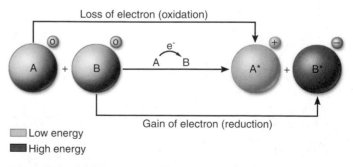

Low energy
High energy

Figure 7.2   Redox reactions.
Oxidation is the loss of an electron; reduction is the gain of an electron. Here the charges of molecules A and B are shown in small circles to the upper right of each molecule. Molecule A loses energy as it loses an electron, while molecule B gains energy as it gains an electron.

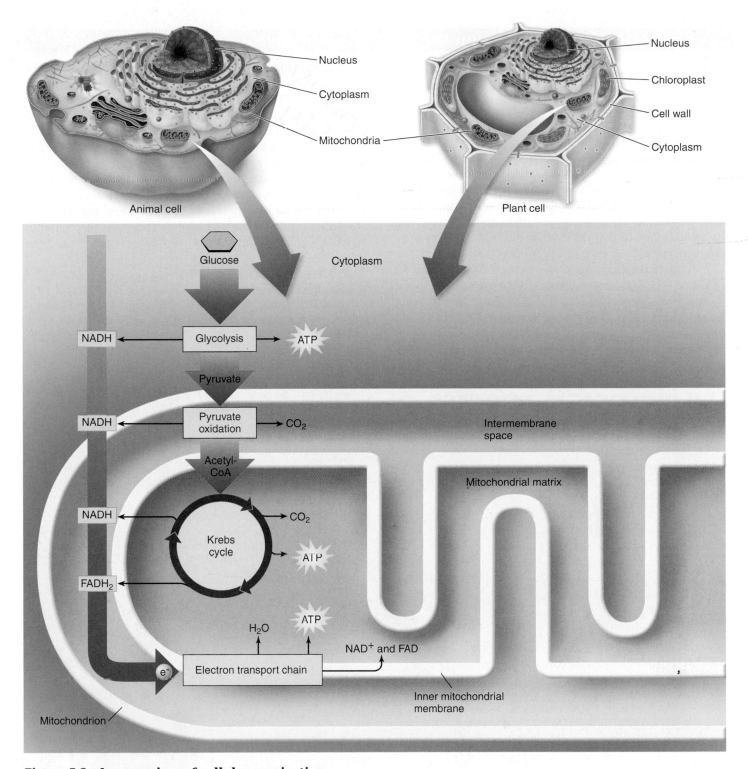

Figure 7.3   An overview of cellular respiration.

The second stage is aerobic (requires oxygen) and takes place within the mitochondrion. The focal point of this stage is the *Krebs cycle,* a cycle of chemical reactions that harvests electrons from C—H chemical bonds and passes the energy-rich electrons to carrier molecules, NADH and FADH$_2$. These molecules deliver the electrons to the electron transport chain, which uses their energy to power the production of ATP. This harvesting of electrons, a form of *oxidation,* is far more powerful than glycolysis at recovering energy from food mole-cules, and is how the bulk of the energy used by eukaryotic cells is extracted from food molecules.

Key Learning Outcome 7.1   Cellular respiration is the dismantling of food molecules to obtain energy. In aerobic respiration, the cell harvests energy from glucose molecules in two stages, glycolysis and oxidation.

7.2 Using Coupled Reactions to Make ATP

The first stage in cellular respiration, called **glycolysis,** is a series of sequential biochemical reactions, a *biochemical pathway.* In 10 enzyme-catalyzed reactions, the six-carbon sugar glucose is cleaved into two three-carbon molecules called pyruvate. The Key Biological Process illustration below presents a conceptual overview of the process, while **figure 7.4** provides a more detailed look at the series of 10 biochemical reactions. Where is the energy extracted? In each of two "coupled" reactions (steps 7 and 10 in **figure 7.4**), the breaking of a chemical bond in an exergonic reaction releases enough energy to drive the formation of an ATP molecule from ADP (an endergonic reaction). This transfer of a high-energy phosphate group from a substrate to ADP is called **substrate-level phosphorylation.** In the process, electrons and hydrogen atoms are extracted and donated to a carrier molecule called NAD$^+$. The NAD$^+$ carries the electrons as NADH to join the other electrons extracted during oxidative respiration, discussed in the following section. Only a small number of ATP molecules are made in glycolysis itself, two for each molecule of glucose, but in the absence of oxygen this is the only way organisms can get energy from food.

Glycolysis is thought to have been one of the earliest of all biochemical processes to evolve. Every living creature is capable of carrying out glycolysis.

Key Learning Outcome 7.2 In the first stage of cellular respiration, called glycolysis, cells shuffle chemical bonds in glucose so that two coupled reactions can occur, producing ATP by substrate-level phosphorylation.

KEY BIOLOGICAL PROCESS: Overview of Glycolysis

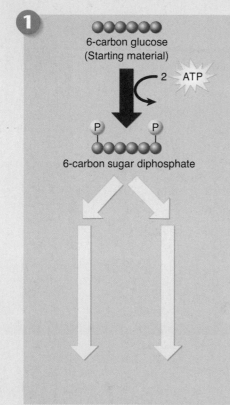

1

6-carbon glucose
(Starting material)

2 ATP

6-carbon sugar diphosphate

Priming reactions. Glycolysis begins with the addition of energy. Two high-energy phosphates from two molecules of ATP are added to the six-carbon molecule glucose, producing a six-carbon molecule with two phosphates.

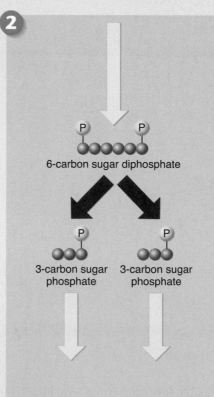

2

6-carbon sugar diphosphate

3-carbon sugar phosphate   3-carbon sugar phosphate

Cleavage reactions. Then, the phosphorylated six-carbon molecule is split in two, forming two three-carbon sugar phosphates.

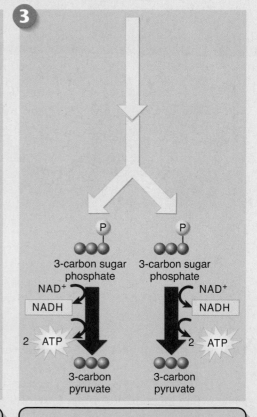

3

3-carbon sugar phosphate   3-carbon sugar phosphate

NAD$^+$ NAD$^+$
NADH   NADH

2 ATP   2 ATP

3-carbon pyruvate   3-carbon pyruvate

Energy-harvesting reactions. Finally, in a series of reactions, each of the two three-carbon sugar phosphates is converted to pyruvate. In the process, an energy-rich hydrogen is harvested as NADH, and two ATP molecules are formed for each pyruvate.

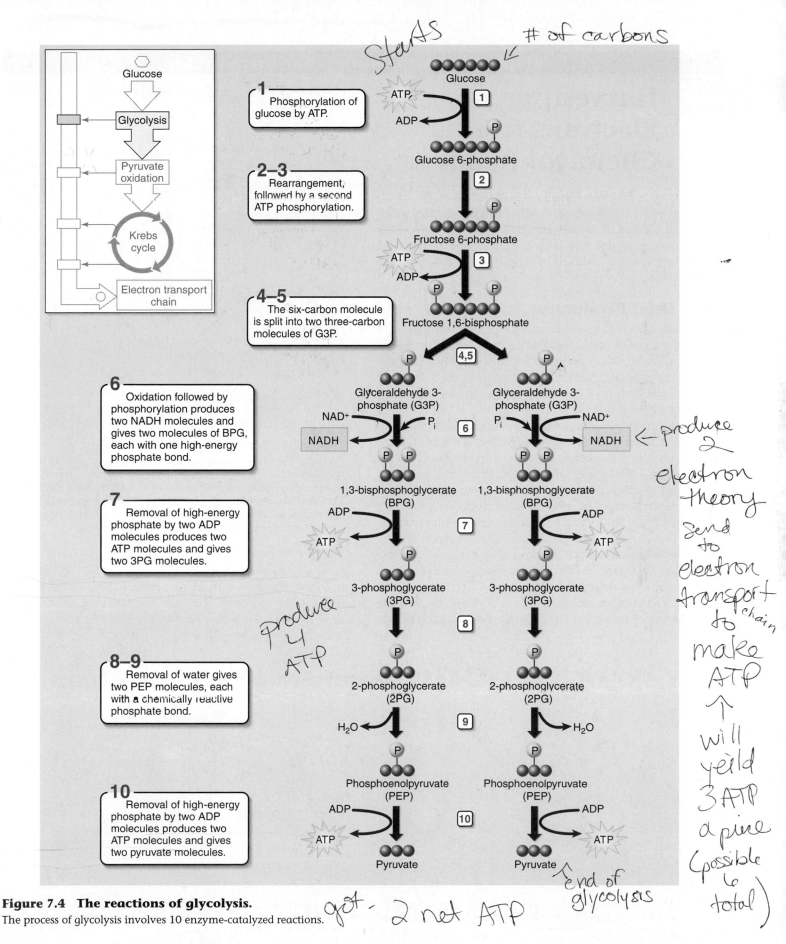

Figure 7.4 The reactions of glycolysis.

The process of glycolysis involves 10 enzyme-catalyzed reactions.

Handwritten annotations:
- starts
- # of carbons
- ← produce 2 electron theory send to electron transport chain to make ATP ↑ will yeild 3 ATP a piece (possible 6 total)
- Produce 4 ATP
- got - 2 net ATP
- End of glycolysis
- FADH2 worth 2 ATP a piece

7.3 Harvesting Electrons from Chemical Bonds

The first step of oxidative respiration in the mitochondrion is the oxidation of the three-carbon molecule called pyruvate, which is the end product of glycolysis. The cell harvests pyruvate's considerable energy in two steps: first, by oxidizing pyruvate to form acetyl-CoA, and then by oxidizing acetyl-CoA in the Krebs cycle.

Step One: Producing Acetyl-CoA

Pyruvate is oxidized in a single reaction that cleaves off one of pyruvate's three carbons. This carbon then departs as part of a CO_2 molecule, shown in **figure 7.5** coming off the pathway with the green arrow. Pyruvate dehydrogenase, the complex of enzymes that removes CO_2 from pyruvate, is one of the largest enzymes known. It contains 60 subunits! In the course of the reaction, a hydrogen and electrons are removed from pyruvate and donated to NAD^+ to form NADH. The Key Biological Process illustration below shows how an enzyme catalyzes this reaction, bringing the substrate (pyruvate) into proximity with NAD^+. Cells use NAD^+ to carry hydrogen atoms and energetic electrons from one molecule to another. NAD^+ oxidizes energy-rich molecules by acquiring their hydrogens (this proceeds $1 \longrightarrow 2 \longrightarrow 3$ in the figure) and then reduces other molecules by giving the hydrogens to them (this proceeds $3 \longrightarrow 2 \longrightarrow 1$). Now focus again on **figure 7.5**. The two-carbon fragment (called an acetyl group) that remains after removing CO_2 from pyruvate is joined to a cofactor called coenzyme A (CoA) by pyruvate dehydrogenase, forming a compound known as **acetyl-CoA.**

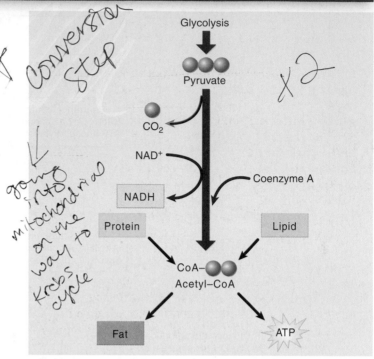

Figure 7.5 Producing acetyl-CoA.
Pyruvate, the three-carbon product of glycolysis, is oxidized to the two-carbon molecule acetyl-CoA, in the process losing one carbon atom as CO_2 and an electron (donated to NAD^+ to form NADH). Almost all the molecules you use as foodstuffs are converted to acetyl-CoA; the acetyl-CoA is then channeled into fat synthesis or into ATP production, depending on your body's needs.

If the cell has plentiful supplies of ATP, acetyl-CoA is funneled into fat synthesis, with its energetic electrons preserved for later needs. If the cell needs ATP now, the fragment is directed instead into ATP production through the Krebs cycle.

KEY BIOLOGICAL PROCESS: Transferring Hydrogen Atoms

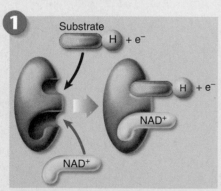
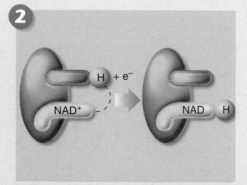
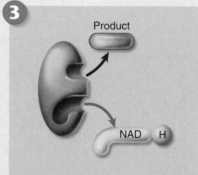

1. Enzymes that harvest hydrogen atoms have a binding site for NAD^+ located near the substrate binding site.

2. In an oxidation-reduction reaction, the hydrogen atom and an electron are transferred to NAD^+, forming NADH.

3. NADH then diffuses away and is available to donate the hydrogen to other molecules.

A Closer Look

Metabolic Efficiency and the Length of Food Chains

In the earth's ecosystems, the organisms that carry out photosynthesis are often consumed as food by other organisms. We call these "organism-eaters" *heterotrophs*. Humans are heterotrophs, as no human photosynthesizes.

It is thought that the first heterotrophs were ancient bacteria living in a world where photosynthesis had not yet introduced much oxygen into the oceans or atmosphere. The only mechanism they possessed to harvest chemical energy from their food was glycolysis. Neither oxygen-generating photosynthesis nor the oxidative stage of cellular respiration had evolved yet. It has been estimated that a heterotroph limited to glycolysis, as these ancient bacteria were, captures only 3.5% of the energy in the food it consumes. Hence, if such a heterotroph preserves 3.5% of the energy in the photosynthesizers it consumes, then any other heterotrophs that consume the first heterotroph will capture through glycolysis 3.5% of the energy in it, or 0.12% of the energy available in the original photosynthetic organisms. A very large base of photosynthesizers would thus be needed to support a small number of heterotrophs.

When organisms became able to extract energy from organic molecules by oxidative cellular respiration, which we discuss on the next page, this constraint became far less severe, because the efficiency of oxidative respiration is estimated to be about 32%. This increased efficiency results in the transmission of much more energy from one trophic level to another than does glycolysis. (A *trophic level* is a step in the movement of energy through an ecosystem.) The efficiency of oxidative cellular respiration has made possible the evolution of food chains, in which photosynthesizers are consumed by heterotrophs, which are consumed by other heterotrophs, and so on. You will read more about food chains in chapter 36.

Even with this very efficient oxidative metabolism, approximately two-thirds of the available energy is lost at each trophic level, and that puts a limit on how long a food chain can be. Most food chains, like the East African grassland ecosystem illustrated here, involve only three or rarely four trophic levels. Too much energy is lost at each transfer to allow chains to be much longer than that. For example, it would be impossible for a large human population to subsist by eating lions captured from the grasslands of East Africa; the amount of grass available there would not support enough zebras and other herbivores to maintain the number of lions needed to feed the human population. Thus, the ecological complexity of our world is fixed in a fundamental way by the chemistry of oxidative cellular respiration.

Photosynthesizers. The grass under this yellow fever tree grows actively during the hot, rainy season, capturing the energy of the sun and storing it in molecules of glucose, which are then converted into starch and stored in the grass.

Herbivores. These zebras consume the grass and transfer some of its stored energy into their own bodies.

Carnivores. The lion feeds on zebras and other animals, capturing part of their stored energy and storing it in its own body.

Scavengers. This hyena and the vultures occupy the same stage in the food chain as the lion. They also consume the body of the dead zebra, after it has been abandoned by the lion.

Refuse utilizers. These butterflies, mostly *Precis octavia*, are feeding on the material left in the hyena's dung after the food the hyena consumed had passed through its digestive tract.

A food chain in the savannas, or open grasslands, of East Africa.
At each of these levels in the food chain, only about a third or less of the energy present is used by the recipient.

Step Two: The Krebs Cycle

The next stage in oxidative respiration is called the **Krebs cycle,** named after the man who discovered it. The Krebs cycle (not to be confused with the Calvin cycle in photosynthesis) takes place within the mitochondrion. While a complex process, its nine reactions can be broken down into three stages, as indicated by the overview presented in the Key Biological Process illustration below:

Stage 1. Acetyl-CoA joins the cycle, binding to a four-carbon molecule and producing a six-carbon molecule.

Stage 2. Two carbons are removed as CO_2, their electrons donated to NAD^+, and a four-carbon molecule is left. A molecule of ATP is also produced.

Stage 3. More electrons are extracted, forming NADH and $FADH_2$; the four-carbon starting material is regenerated.

To examine the Krebs cycle in more detail, follow along the series of individual reactions illustrated in **figure 7.6**. The cycle starts when the two-carbon acetyl-CoA fragment produced from pyruvate is stuck onto a four-carbon sugar called oxaloacetate. Then, in rapid-fire order, a series of eight additional reactions occur (steps 2 through 9). When it is all over, two carbon atoms have been expelled as CO_2, one ATP molecule has been made in a coupled reaction, eight more energetic electrons have been harvested and taken away as NADH or on other carriers, such as $FADH_2$, which serves the same function as NADH, and we are left with the same four-carbon sugar we started with. The process of reactions is a cycle—that is, a circle of reactions. In each turn of the cycle, a new acetyl group replaces the two CO_2 molecules lost, and more electrons are extracted. Note that a single glucose molecule produces *two* turns of the cycle, one for each of the two pyruvate molecules generated by glycolysis.

In the process of cellular respiration, glucose is entirely consumed. The six-carbon glucose molecule is first cleaved into a pair of three-carbon pyruvate molecules during glycolysis. One of the carbons of each pyruvate is then lost as CO_2 in the conversion of pyruvate to acetyl-CoA, and the other two carbons are lost as CO_2 during the oxidations of the Krebs cycle. All that is left to mark the passing of the glucose molecule into six CO_2 molecules is its energy, preserved in four ATP molecules and electrons carried by 10 NADH and two $FADH_2$ carriers.

> **Key Learning Outcome 7.3** The end product of glycolysis, pyruvate, is oxidized to the two-carbon acetyl-CoA, yielding a pair of electrons plus CO_2. Acetyl-CoA then enters the Krebs cycle, yielding ATP, many energized electrons, and two CO_2 molecules.

KEY BIOLOGICAL PROCESS: Overview of the Krebs Cycle

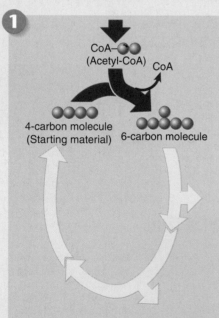

1

CoA—○○ (Acetyl-CoA) CoA
4-carbon molecule (Starting material)   6-carbon molecule

The Krebs cycle begins when a two-carbon fragment is transferred from acetyl-CoA to a four-carbon molecule (the starting material).

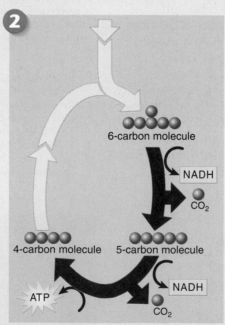

2

6-carbon molecule
NADH
CO_2
4-carbon molecule   5-carbon molecule
ATP
NADH
CO_2

Then, the resulting six-carbon molecule is oxidized (a hydrogen removed to form NADH) and decarboxylated (a carbon removed to form CO_2). Next, the five-carbon molecule is oxidized and decarboxylated again, and a coupled reaction generates ATP.

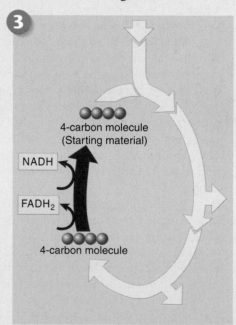

3

4-carbon molecule (Starting material)
NADH
$FADH_2$
4-carbon molecule

Finally, the resulting four-carbon molecule is further oxidized (hydrogens removed to form $FADH_2$ and NADH). This regenerates the four-carbon starting material, completing the cycle.

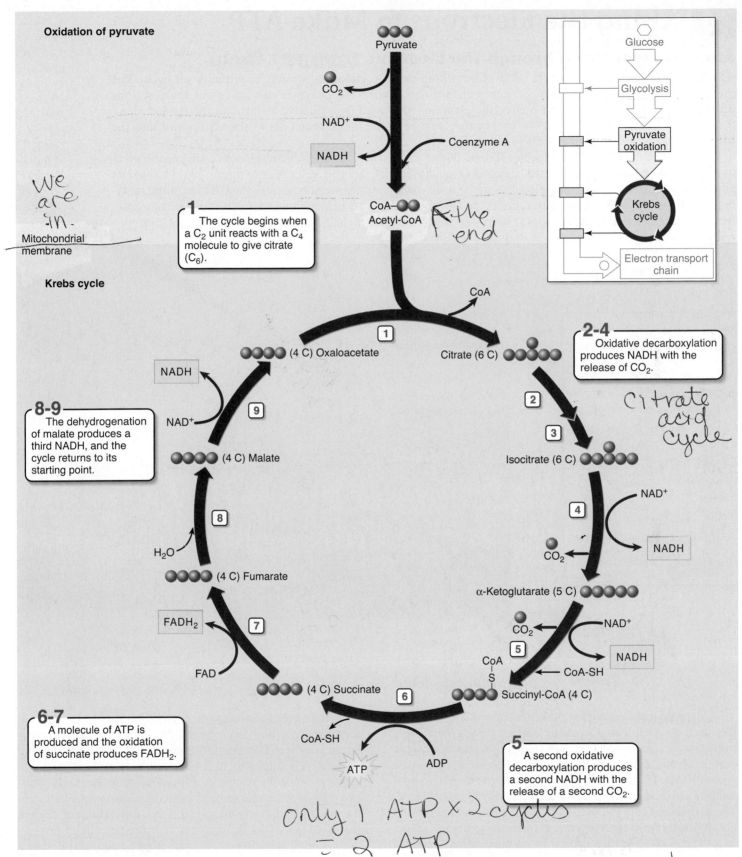

Oxidation of pyruvate

Pyruvate

CO_2

NAD^+

NADH

Coenzyme A

we are in Mitochondrial membrane

Krebs cycle

CoA—Acetyl-CoA

F the end

Glucose

Glycolysis

Pyruvate oxidation

Krebs cycle

Electron transport chain

1 The cycle begins when a C_2 unit reacts with a C_4 molecule to give citrate (C_6).

CoA

1

(4 C) Oxaloacetate

NADH

9

NAD^+

8-9 The dehydrogenation of malate produces a third NADH, and the cycle returns to its starting point.

(4 C) Malate

8

H_2O

(4 C) Fumarate

FADH₂

7

FAD

6-7 A molecule of ATP is produced and the oxidation of succinate produces FADH₂.

(4 C) Succinate

CoA-SH

ATP

ADP

6

Citrate (6 C)

Citrate acid cycle

2

3

2-4 Oxidative decarboxylation produces NADH with the release of CO_2.

Isocitrate (6 C)

4

NAD^+

NADH

CO_2

α-Ketoglutarate (5 C)

CO_2

NAD^+

5

NADH

CoA—S

CoA-SH

Succinyl-CoA (4 C)

5 A second oxidative decarboxylation produces a second NADH with the release of a second CO_2.

only 1 ATP × 2 cycles
= 2 ATP

Figure 7.6 The Krebs cycle.
This series of nine enzyme-catalyzed reactions takes place within the mitochondrion.

NADH = 3 × 2 = 6
FADH₂ = 1 × 2 = 2

Using the Electrons to Make ATP

Moving Electrons Through the Electron Transport Chain

In eukaryotes, aerobic respiration takes place within the mitochondria present in virtually all cells. The internal compartment, or **matrix,** of a mitochondrion contains the enzymes that carry out the reactions of the Krebs cycle. As described earlier, the electrons harvested by oxidative respiration are passed along the electron transport chain, and the energy they release transports protons out of the matrix and into the **intermembrane space.**

The NADH and FADH$_2$ molecules formed during the first stages of aerobic respiration each contain electrons and hydrogens that were gained when NAD$^+$ and FAD were reduced (refer back to **figure 7.3**). The NADH and FADH$_2$ molecules carry their electrons to the inner mitochondrial membrane (an enlarged area of the membrane is shown below), where they transfer the electrons to a series of membrane-associated molecules collectively called the **electron transport chain.** The electron transport chain works much as does the electron transport system you encountered in studying photosynthesis.

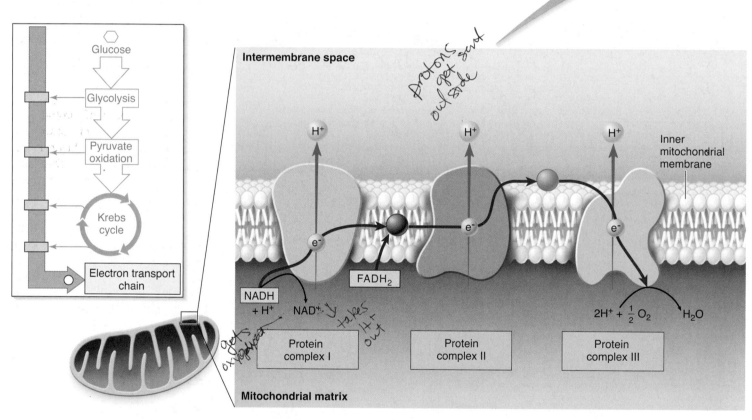

A protein complex (the pink structure above) receives the electrons and, using a mobile carrier, passes these electrons to a second protein complex (the purple structure). This protein complex, along with others in the chain, operates as a proton pump, using the energy of the electron to drive a proton out across the membrane into the intermembrane space. The arrows indicate the transport of the protons into the top half of the figure, which represents the intermembrane space.

The electron is then carried by another carrier to a third protein complex (the light blue structure). This complex uses electrons such as this one to link oxygen atoms with hydrogen ions to form molecules of water.

It is the availability of a plentiful supply of electron acceptor molecules like oxygen that makes oxidative respiration possible. The electron transport chain used in aerobic respiration is similar to, and may well have evolved from, the electron transport system employed in photosynthesis. Photosynthesis is thought to have preceded cellular respiration in the evolution of biochemical pathways, generating the oxygen that is necessary as the electron acceptor in cellular respiration. Natural selection didn't start from scratch and design a new biochemical pathway for cellular respiration; instead, it built on the photosynthetic pathway that already existed, and uses many of the same reactions.

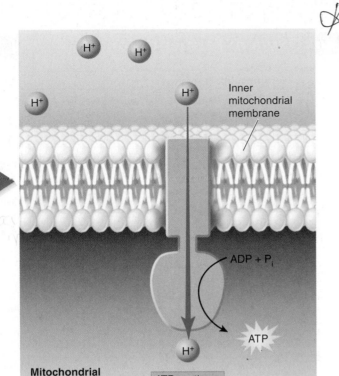

Mitochondrial matrix

ATP synthase

Producing ATP: Chemiosmosis

As the proton concentration in the intermembrane space rises above that in the matrix, the concentration gradient induces the protons to reenter the matrix by diffusion through a special proton channel called **ATP synthase.** ATP synthase channels are embedded in the inner mitochondrial membrane, as shown in the figure to the left. As the protons pass through, these channels synthesize ATP from ADP and P$_i$ within the matrix. The ATP is then transported by facilitated diffusion out of the mitochondrion and into the cell's cytoplasm. This ATP synthesizing process is the same chemiosmosis process that you encountered in studying photosynthesis in chapter 6.

Although we have discussed electron transport and chemiosmosis as separate processes, in a cell they are integrated, as shown below, left. The electron transport chain uses two electrons harvested in glycolysis, two harvested in pyruvate oxidation, and eight harvested in aerobic respiration (red arrows) to pump a large number of protons out across the inner mitochondrial membrane (shown in the upper right). Their subsequent reentry back into the mitochondrial matrix drives the synthesis of 34 ATP molecules by chemiosmosis (shown in the lower right). Two additional ATPs were harvested by a coupled reaction in glycolysis, and two more in the Krebs cycle. As two ATPs must be expended to transport NADH into the mitochondria by active transport, the grand total of ATPs harvested is thus 36 molecules.

Pyruvate from cytoplasm

Inner mitochondrial membrane

Intermembrane space

Electron transport chain

NADH

1 Electrons are harvested and carried to the transport chain.

2 Electrons provide energy to pump protons across the membrane.

Acetyl-CoA

NADH

Krebs cycle

FADH$_2$

H$_2$O

$\frac{1}{2}$ O$_2$ + 2 H$^+$

O$_2$

3 Oxygen joins with protons and electrons to form water.

CO$_2$

2 ATP

4 Protons diffuse back in down their concentration gradient, driving the synthesis of ATP.

34 ATP

ATP synthase

Mitochondrial matrix

Key Learning Outcome 7.4 The electrons harvested by oxidizing food molecules are used to power proton pumps that chemiosmotically drive the production of ATP.

7.5 Cells Can Metabolize Food Without Oxygen

Fermentation

In the absence of oxygen, aerobic metabolism cannot occur, and cells must rely exclusively on glycolysis to produce ATP. Under these conditions, the hydrogen atoms generated by glycolysis are donated to organic molecules in a process called **fermentation,** a process that recycles NAD^+, the electron acceptor required for glycolysis to proceed.

Bacteria carry out more than a dozen kinds of fermentations, all using some form of organic molecule to accept the hydrogen atom from NADH and thus recycle NAD^+:

$$\text{Organic molecule} \longleftrightarrow \text{Reduced organic molecule}$$
$$+ \text{NADH} \qquad\qquad + NAD^+$$

Often the reduced organic compound is an organic acid—such as acetic acid, butyric acid, propionic acid, or lactic acid—or an alcohol.

Ethanol Fermentation Eukaryotic cells are capable of only a few types of fermentation. In one type, which occurs in single-celled fungi called yeast, the molecule that accepts hydrogen from NADH is pyruvate, the end product of glycolysis itself. Yeast enzymes remove a CO_2 group from pyruvate through decarboxylation, producing a two-carbon molecule called acetaldehyde. The CO_2 released causes bread made with yeast to rise, while bread made without yeast (unleavened bread) does not. The acetaldehyde accepts a hydrogen atom from NADH, producing NAD^+ and ethanol (**figure 7.7**). This particular type of fermentation is of great interest to humans because it is the source of the ethanol in wine and beer. Ethanol is a by-product of fermentation that is actually toxic to yeast; as it approaches a concentration of about 12%, it begins to kill the yeast. That explains why naturally fermented wine contains only about 12% ethanol.

Lactic Acid Fermentation Most animal cells regenerate NAD^+ without decarboxylation. Muscle cells, for example, use an enzyme called lactate dehydrogenase to transfer a hydrogen atom from NADH back to the pyruvate that is produced by glycolysis. This reaction converts pyruvate into lactic acid and regenerates NAD^+ from NADH (**figure 7.7**). It therefore closes the metabolic circle, allowing glycolysis to continue as long as glucose is available. Circulating blood removes excess lactate (the ionized form of lactic acid) from muscles. It was once thought that during strenuous exercise, when the removal of lactic acid cannot keep pace with its

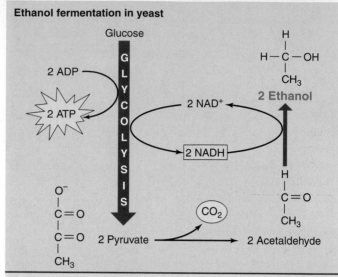

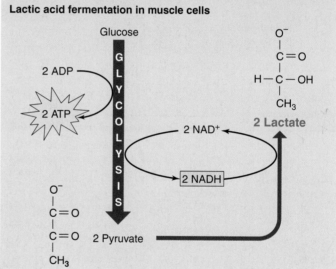

Figure 7.7 Fermentation.
Yeasts carry out the conversion of pyruvate to ethanol. Muscle cells convert pyruvate into lactate, which is less toxic than ethanol. In both cases, NAD^+ is regenerated to allow glycolysis to continue.

production, the accumulation of lactic acid induces muscle fatigue. However, as you will learn in chapter 22, muscle fatigue actually has a quite different cause, involving the leakage of calcium ions within muscles.

> **Key Learning Outcome 7.5** In fermentation, which occurs in the absence of oxygen, the electrons that result from the glycolytic breakdown of glucose are donated to an organic molecule, regenerating NAD^+ from NADH.

Beer and Wine— Products of Fermentation

Alcoholic fermentation, the anaerobic conversion of sugars into alcohol discussed on the facing page, predates human history. Like many other natural processes, humans seem to have first learned to control this process by stumbling across its benefit—the pleasures of beer and wine. Artifacts from the third millennium B.C. contain wine residue, and we know from the pollen record that domesticated grapes first became abundant at that time. The production of beer started even earlier, back in the fifth millennium B.C., making beer possibly one of the oldest beverages produced by humans.

Beer fermentors.

The process of making beer is called brewing, and involves the fermentation of cereal grains. Grains are rich in starches, which you will recall from chapter 3 are long chains of glucose molecules linked together. First, the cereal grains are crushed and soaked in warm water, creating an extract called *mash*. Enzymes called amylases are added to the mash to break down the starch within the grains into free sugars. The mash is filtered, yielding a darker, sugary liquid called wort. The wort is boiled in large tanks, like the ones shown in the photo here, called fermentors. This helps break down the starch and kills any bacteria or other microorganisms that might be present before they begin to feast on all the sugar. Hops, another plant product, is added during the boiling stage. Hops gives the solution a bitter taste that complements the sweet flavor of the sugar in the wort.

Now we are ready to make beer. The temperature of the wort is brought down and yeast is added. This begins the fermentation process. It can be done in either of two ways, yielding the two basic kinds of beer.

Lager. The yeast *Saccharomyces uvarum* is a bottom-fermenting yeast that settles on the bottom of the vat during fermentation. This yeast produces a lighter, pale beer called lager. Bottom fermentation is the most widespread method of brewing. Bottom fermentation occurs at low temperatures (5–8°C). After fermentation is complete, the beer is cooled further to 0°C, allowing for the beer to mature before the yeast is filtered out.

Ale. The yeast *S. cerevisiae*, often called "brewer's yeast," is a top-fermenting yeast. It rises to the top during fermentation and produces a darker, more aromatic beer called ale (also called porter or stout). In top fermentation, less yeast is used. The temperature is higher, around 15–25°C, and so fermentation occurs more quickly but less sugar is converted into alcohol, giving the beer a sweeter taste.

Most yeasts used in beer brewing are alcohol-tolerant only up to about 5% alcohol—alcohol levels above that kill the yeast. That is why commercial beer is typically 5% alcohol.

Carbon dioxide is also a product of fermentation, and gives beer the bubbles that form the head of the beer. However, because only a little CO_2 forms naturally, the last step in most brewing involves artificial carbonation, with CO_2 gas being injected into the beer.

Wine. Wine fermentation is similar to beer fermentation in that yeasts are used to ferment sugars into alcohol. Wine fermentation, however, uses grapes. Rather than the starches found in cereal grains, grapes are fruits rich in the sugar sucrose, a disaccharide combination of glucose and fructose. Grapes are crushed and placed into barrels. There is no need for boiling to break down the starch, and grapes have their own amylase enzymes to cleave sucrose into free glucose and fructose sugars. To start the fermentation, yeast is added directly to the crush. *Saccharomyces cerevisiae* and *S. bayanus* are common wine yeasts. They differ from the yeast used in beer fermentation in that they have higher alcohol tolerances—up to 12% alcohol (some wine yeasts have even higher alcohol tolerances).

A common misconception about wine is that the color of the wine results from the color of the grape juice used, red wine using juice from red grapes and white wines using juice from white grapes. This is not true. The juice of all grapes is similar in color, usually light. The color of red wine is produced by leaving the skins of red or black grapes in the crush during the fermentation process. White wines can be made from any color of grapes, and are white simply because the skins that contribute red coloring are removed before fermentation.

White wines are typically fermented at low temperatures (8–19°C). Red wines are fermented at higher temperatures of 25–32°C with yeasts that have a higher heat tolerance. During wine fermentation, the carbon dioxide is vented out, leaving no carbonation in the wine. Champagnes have some natural carbonation that results from a two-step fermentation process: In the first fermentation step, the carbon dioxide is allowed to escape; in a second fermentation step, the container is sealed, trapping the carbon dioxide. But, like beer brewing, the natural carbonation is often supplemented with artificial carbonation.

7.6 Glucose Is Not the Only Food Molecule

We have considered in detail the fate of a molecule of glucose, a simple sugar, in cellular respiration. But how much of what you eat is sugar? As a more realistic example of the food you eat, consider the fate of a fast-food hamburger. The hamburger you eat is composed of carbohydrates, fats, protein, and many other molecules. This diverse collection of complex molecules is broken down by the process of digestion in your stomach and intestines into simpler molecules. Carbohydrates are broken down into simple sugars, fats into fatty acids, and proteins into amino acids. These breakdown reactions produce little or no energy themselves, but prepare the way for cellular respiration—that is, glycolysis and oxidative metabolism. Nucleic acids are also present in the food you eat and are broken down during digestion, but these macromolecules store little energy that the body actually uses.

We have seen what happens to the glucose. What happens to the amino acids and fatty acids? These subunits undergo chemical modifications that convert them into products that feed into cellular respiration.

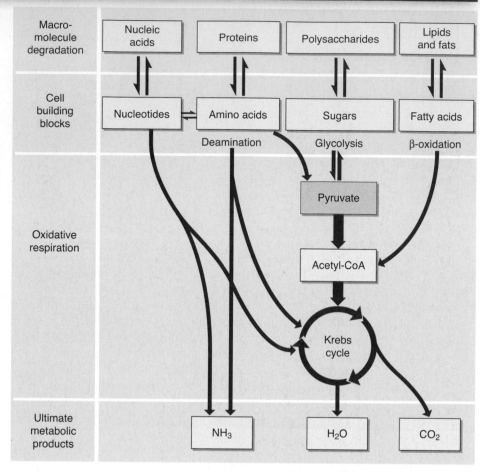

Figure 7.8 How cells obtain energy from foods.

Most organisms extract energy from organic molecules by oxidizing them. The first stage of this process, breaking down macromolecules into their subunits, yields little energy. The second stage, cellular respiration, extracts energy, primarily in the form of high-energy electrons. The subunit of many carbohydrates, glucose, readily enters glycolysis and passes through the biochemical pathways of oxidative respiration. However, the subunits of other macromolecules must be converted into products that can enter the biochemical pathways found in oxidative respiration.

Cellular Respiration of Protein

Proteins (the second category in **figure 7.8**) are first broken down into their individual amino acids. A series of *deamination* reactions removes the nitrogen side groups (called amino groups) and converts the rest of the amino acid into a molecule that takes part in the Krebs cycle. For example, alanine is converted into pyruvate, glutamate into α-ketoglutarate, and aspartate into oxaloacetate. The reactions of the Krebs cycle then extract the high-energy electrons from these molecules and put them to work making ATP.

Cellular Respiration of Fat

Lipids and fats (the fourth category in **figure 7.8**) are first broken down into fatty acids. A fatty acid typically has a long tail of sixteen or more —CH_2 links, and the many hydrogen atoms in these long tails provide a rich harvest of energy. Enzymes in the matrix of the mitochondrion first remove one two-carbon acetyl group from the end of a fatty acid tail, and then another, and then another, in effect chewing down the length of the tail in two-carbon bites. Eventually the entire fatty acid tail is converted into acetyl groups. Each acetyl group then combines with coenzyme A to form acetyl-CoA, which feeds into the Krebs cycle. This process is known as *β-oxidation*.

Thus, in addition to the carbohydrates, the proteins and fats in the hamburger also become important sources of energy.

Key Learning Outcome 7.6 Cells garner energy from proteins and fats, which are broken down into products that feed into cellular respiration.

Biology and *Staying Healthy*

Fad Diets and Impossible Dreams

Most Americans put on weight in middle age, slowly adding 30 or more pounds. They did not ask for that weight, do not want it, and are constantly looking for a way to get rid of it. It is not a lonely search—it seems like everyone past the flush of youth is trying to lose weight. Many have been seduced by fad diets, investing hope only to harvest frustration. The much discussed Atkins diet is the fad diet most have tried—*Dr. Atkins' Diet Revolution* is one of the 10 best-selling books in history, and was (and is) prominently displayed in bookstores. The reason this diet doesn't deliver on its promise of pain-free weight loss is well understood by science, but not by the general public. Only hope and hype make it a perpetual best seller.

The secret of the Atkins diet, stated simply, is to avoid carbohydrates. Atkins's basic proposition is that your body, if it does not detect blood glucose (from metabolizing carbohydrates), will think it is starving and start to burn body fat, even if there is lots of fat already circulating in your bloodstream. You may eat all the fat and protein you want, all the steak and eggs and butter and cheese, and you will still burn fat and lose weight—just don't eat any carbohydrates, any bread or pasta or potatoes or fruit or candy. Despite the title of Atkins's book, this diet is hardly revolutionary. A basic low-carbohydrate diet was first promoted over a century ago in the 1860s by William Banting, an English casket maker, in his best-selling book *Letter on Corpulence*. Books promoting low-carbohydrate diets have continued to be best sellers ever since.

Those who try the Atkins diet often lose 10 pounds in two to three weeks. In three months it is all back, and then some. So what happened? Where did the pounds go, and why did they come back? The temporary weight loss turns out to have a simple explanation. Carbohydrates act as water sponges in your body, and so forcing your body to become depleted of carbohydrates causes your body to lose water. The 10 pounds lost on this diet was not fat weight but water weight, quickly regained with the first starchy foods eaten.

The Atkins' diet is the sort of diet the American Heart Association tells us to avoid (all those saturated fats and cholesterol), and it is difficult to stay on. If you do hang in there, you will lose weight, simply because you eat less. Other popular diets these days, *The Zone* diet of Dr. Barry Sears and *The South Beach Diet* of Dr. Arthur Agatston, are also low-carbohydrate diets, although not as extreme as the Atkins diet. Like the Atkins diet, they work not for the bizarre reasons claimed by their promoters, but simply because they are low-calorie diets.

There are two basic laws that no diet can successfully violate:
1. All calories are equal.
2. (calories in) – (calories out) = fat.

The fundamental fallacy of the Atkins diet, the Zone diet, the South Beach diet, and indeed of all fad diets, is the idea that somehow carbohydrate calories are different from fat and protein calories. This is scientific foolishness. Every calorie you eat contributes equally to your eventual weight, whether it comes from carbohydrate, fat, or protein.

To the extent these diets work at all, they do so because they obey the second law. By reducing calories in, they reduce fat. If that were all there was to it, we should all go out and buy a diet book. Unfortunately, losing weight isn't that simple, as anyone who has seriously tried already knows. The problem is that your body will not cooperate.

If you try to lose weight by exercising and eating less, your body will attempt to compensate by metabolizing more efficiently. It has a fixed weight, what obesity researchers call a "set point," a weight to which it will keep trying to return. A few years ago, a group of researchers at Rockefeller University in New York, in a landmark study, found that if you lose weight, your metabolism slows down and becomes more efficient, burning fewer calories to do the same work—your body will do everything it can to gain the weight back! Similarly, if you gain weight, your metabolism speeds up. In this way your body uses its own natural weight control system to keep your weight at its set point. No wonder it's so hard to lose weight!

Clearly our bodies don't keep us at one weight all our adult lives. It turns out your body adjusts its fat thermostat—its set point—depending on your age, food intake and amount of physical activity. Adjustments are slow, however, and it seems to be a great deal easier to move the body's set point up than to move it down. Apparently higher levels of fat reduce the body's sensitivity to the leptin hormone that governs how efficiently we burn fat. That is why you can gain weight, despite your set point resisting the gain—your body still issues leptin alarm calls to speed metabolism, but your brain doesn't respond with as much sensitivity as it used to. Thus the fatter you get, the less effective your weight control system becomes.

This doesn't mean that we should give up and learn to love our fat. Rather, now that we are beginning to understand the biology of weight gain, we must accept the hard fact that we cannot beat the requirements of the two diet laws. The real trick is not to give up. Eat less and exercise more, and keep at it. In one year, or two, or three, your body will readjust its set point to reflect the new reality you have imposed by constant struggle. There simply isn't any easy way to lose weight.

How Do Swimming Fish Avoid Low Blood pH?

Animals that live in oxygen-poor environments, like worms living in the oxygen-free mud at the bottom of lakes, are not able to obtain the energy required for muscle movement from the Krebs cycle. Their cells lack the oxygen needed to accept the electrons stripped from food molecules. Instead, these animals rely on glycolysis to obtain ATP, donating the electron to pyruvate, forming lactic acid. While much less efficient than the Krebs cycle, glycolysis does not require oxygen. Even when oxygen is plentiful, the muscles of an active animal may use up oxygen more quickly than it can be supplied by the bloodstream and so be forced to temporarily rely on glycolysis to generate the ATP for continued contraction.

This presents a particular problem for fish. Fish blood is much lower in carbon dioxide than yours is, and as a consequence, the amount of sodium bicarbonate acting as a buffer in fish blood is also quite low. Now imagine you are a trout, and need to suddenly swim very fast to catch a mayfly for dinner. The vigorous swimming will cause your muscles to release large amounts of lactic acid into your poorly-buffered blood; this could severely disturb the blood's acid-base balance and so impede contraction of your swimming muscles before the prey is captured.

The graph to the right presents the results of an experiment designed to explore how a trout solves this dilemma. In the experiment, the trout was made to swim vigorously for 15 minutes in a laboratory tank, and then allowed a day's recovery. The lactic acid concentration in its blood was monitored periodically during swimming and recovery phases.

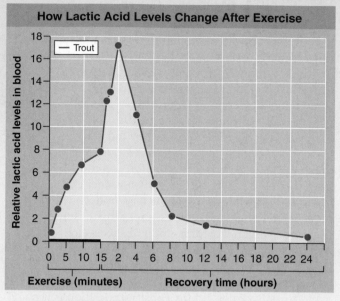

How Lactic Acid Levels Change After Exercise

1. **Applying Concepts**
   a. Variable. What is the dependent variable?

   b. Recording Data. Lactic acid levels are presented for both swimming and recovery periods. In what time units are the swimming data presented? The recovery data?

2. **Interpreting Data**
   a. What is the effect of exercise on the level of lactic acid in the trout's blood?
   b. Does the level of lactic acid change after exercise stops? How?

3. **Making Inferences** About how much of the total lactic acid created by vigorous swimming is released after this exercise stops? [Hint: Notice the x axis scale changes from minutes to hours, so replot all points to minutes and compare areas under curve.]

4. **Drawing Conclusions** Is this result consistent with the hypothesis that fish maintain blood pH levels by delaying the release of lactic acid from muscles? Why might this be beneficial to the fish?

An Overview of Cellular Respiration
7.1 Where Is the Energy in Food?

- Nonphotosynthetic organisms acquire energy from the breakdown of food, either by eating plants that store the food or by eating animals that have eaten plants (**figure 7.1**). Energy stored in carbohydrate molecules is extracted through the process of cellular respiration and is stored in the cell as ATP.

- Cellular respiration is carried out in two stages: glycolysis occurring in the cytoplasm and oxidation occurring in the mitochondria (**figure 7.3**).

Respiration Without Oxygen: Glycolysis
7.2 Using Coupled Reactions to Make ATP

- Glycolysis is an energy-extracting process. It is a series of 10 chemical reactions in which glucose is broken down into two three-carbon pyruvate molecules. The energy is extracted from glucose by two exergonic reactions that are coupled with an endergonic reaction that leads to the formation of ATP. This is called substrate-level phosphorylation.

- The initial steps in glycolysis, called the priming reactions, require the input of energy from ATP. After priming, the glucose molecule splits into two smaller molecules in the cleavage reactions (**Key Biological Process, page 138**). The energy harvesting reactions that follow produce ATP, but very little net ATP is produced (**figure 7.4**).

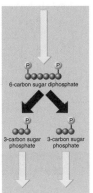

6-carbon sugar diphosphate

3-carbon sugar phosphate   3-carbon sugar phosphate

- Electrons extracted from glucose are donated to a carrier molecule, NAD$^+$, which becomes NADH. NADH carries electrons and hydrogen atoms to be used in a later stage of oxidative respiration.

Respiration With Oxygen: The Krebs Cycle
7.3 Harvesting Electrons from Chemical Bonds

- The two molecules of pyruvate formed in glycolysis are passed into the mitochondrion where they are converted into two molecules of acetyl-coenzyme A (**figure 7.5**). What the cell does with acetyl-CoA depends on the needs of the cell. If the cell has enough ATP, acetyl-CoA is used in synthesizing fat molecules. If the cell needs energy, acetyl-CoA is directed to the Krebs cycle.

- The formation of NADH is an enzyme catalyzed reaction (**Key Biological Process, page 140**). The enzyme brings the substrate and NAD$^+$ into close proximity. Through a redox reaction, a hydrogen atom and an electron are transferred to NAD$^+$, reducing it to NADH. NADH then carries the electrons and hydrogen to a later step in oxidative respiration.

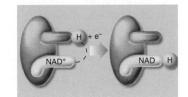

H$^+$ + e$^-$

NAD$^+$ NAD   H

- Acetyl-CoA enters a series of chemical reactions called the Krebs cycle, where one molecule of ATP is produced in a coupled reaction. Also, energy is harvested in the form of electrons that are transferred to molecules of NAD$^+$ and FAD to produce NADH and FADH$_2$, respectively (**Key Biological Process, page 142** and **figure 7.6**).

- The Krebs cycle makes two turns for every molecule of glucose that is oxidized.

7.4 Using the Electrons to Make ATP

- The molecules of NADH and FADH$_2$ that were produced during glycolysis and the Krebs cycle carry electrons to the inner mitochondrial membrane. Here they give up electrons to the electron transport chain. The electrons, along with their energy, are passed along the electron transport chain. The energy from the electrons drives proton pumps that pump H$^+$ across the inner membrane from the matrix to the intermembrane space, creating an H$^+$ concentration gradient (**integrated art, page 144**).

- When the electrons reach the end of the electron transport chain, they bind with oxygen and hydrogen to form water molecules.

- ATP is produced in the mitochondrion through chemiosmosis. The H$^+$ concentration gradient in the intermembrane space drives H$^+$ back across the membrane through ATP synthase channels (**integrated art, top of page 145**). The energy from the movement of H$^+$ through the channel is transferred to the chemical bonds in ATP. Thus, the energy stored in the glucose molecule is harvested through glycolysis and the Krebs cycle with the formation of NADH and FADH$_2$ and ultimately the formation of ATP. The energy carried by NADH and FADH$_2$ is transferred to the electron transport chain and is stored in ATP (**integrated art, bottom of page 145**).

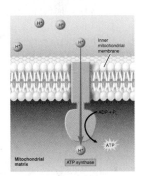

Harvesting Electrons Without Oxygen: Fermentation
7.5 Cells Can Metabolize Food Without Oxygen

- In the absence of oxygen, other molecules can be used as electron acceptors. When the electron acceptor is an organic molecule, the process is called fermentation (**figure 7.7**). Depending on what type of organic molecule accepts the electrons, either ethanol or lactic acid, in the form of lactate, is formed.

Other Sources of Energy
7.6 Glucose Is Not the Only Food Molecule

- Food sources other than glucose are also used in oxidative respiration. Macromolecules, such as proteins, lipids, and nucleic acids, are broken down into intermediate products that enter cellular respiration in different reaction steps, as shown here from **figure 7.8**.

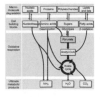

Test Your Understanding

1. In animals, the energy for life is obtained by cellular respiration. This involves
   a. breaking down the organic molecules that were consumed.
   b. capturing photons from plants.
   c. obtaining ATP from plants.
   d. breaking down CO_2 that was produced by plants.
2. During glycolysis, ATP forms by
   a. the breakdown of pyruvate.
   b. chemiosmosis.
   c. substrate-level phosphorylation.
   d. NAD^+.
3. Which of the following processes can occur in the absence of oxygen?
   a. the Krebs cycle
   b. glycolysis
   c. chemiosmosis
   d. All of the above.
4. Every living creature on this planet is capable of carrying out the rather inefficient biochemical process of glycolysis, which
   a. makes glucose, using the energy from ATP.
   b. makes ATP by splitting glucose and capturing the energy.
   c. phosphorylates ATP to make ADP.
   d. makes glucose, using oxygen and carbon dioxide and water.
5. The electrons generated from the Krebs cycle are transferred to _____, which then carries them to _____.
   a. NAD^+, oxygen
   b. NAD^+, the electron transport chain
   c. NADH, oxygen
   d. NADH, the electron transport chain

6. After glycolysis, the pyruvate molecules go to the
   a. nucleus of the cell and provide energy.
   b. membranes of the cell and are broken down in the presence of CO_2 to make more ATP.
   c. mitochondria of the cell and are broken down in the presence of O_2 to make more ATP.
   d. Golgi bodies and are packaged and stored until needed.
7. The vast majority of the ATP molecules produced within a cell are produced
   a. during pyruvate oxidation.
   b. during glycolysis.
   c. during the Krebs cycle.
   d. during the electron transport chain.
8. NAD^+ is recycled during
   a. glycolysis.
   b. fermentation.
   c. the Krebs cycle.
   d. the formation of acetyl-CoA.
9. The final electron acceptor in lactic acid fermentation is
   a. pyruvate.   c. lactic acid.
   b. NAD^+.   d. O_2.
10. Cells can extract energy from foodstuffs other than glucose because
    a. proteins, fatty acids, and nucleic acids get converted to glucose and then enter oxidative respiration.
    b. each type of macromolecule has its own oxidative respiration pathway.
    c. each type of macromolecule is broken down into its subunits, which enter the oxidative respiration pathway.
    d. they can all enter the glycolytic pathway.

Apply Your Understanding

1. Consider the structure of a mitochondrion, shown here in a cutaway view. If you were able to poke a hole in this mitochondrion, do you think it still would be able to perform oxidative respiration? Explain.

2. **Figure 7.8** Your friend Yevgeny wants to go on a low-carbohydrate diet so that he can lose some of the "baby fat" he's still carrying. He asks your advice; what do you tell him?

Synthesize What You Have Learned

1. The electron carrier cytochrome c is one of many different cytochromes, but unlike the others, the sequence of *cytochrome c* is nearly identical in all species. Why do you suppose this is so? Among humans, no genetic disorder affecting cytochrome c has ever been reported. Why do you suppose this is so?
2. How much less ATP would be generated in the cells of a person who consumed a diet of pyruvate instead of glucose (use one molecule of each for your calculation)?

3. Which of the following food molecules would generate the most ATP molecules, assuming that glycolysis, the Krebs cycle, and the electron transport chain were all functioning and that the foods were consumed in equal amounts: carbohydrates, proteins, or fats? Explain your answer.
4. Soft drinks are artificially carbonated, which is what causes them to fizz. Beer and sparkling wines are naturally carbonated. How does this natural carbonation occur?

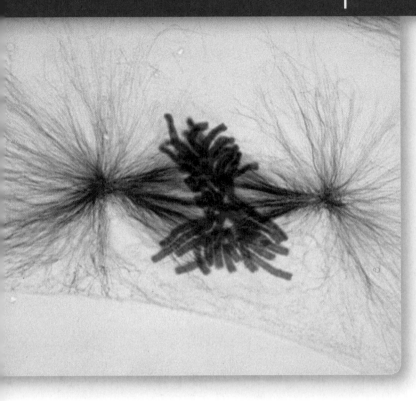

8

Mitosis

The cell you see above is a dividing cell of the Oregon newt *Taricha granulosa,* a kind of salamander. The micrograph (that is, a photo taken through a microscope) captures the cell in the metaphase stage of mitosis, when all the blue-stained chromosomes are lined up on the metaphase plate. Soon the red-stained spindle fibers will draw duplicates of the homologous chromosomes to opposite poles of the cell. When cell division is complete, two daughter cells will result, each containing the same amount of DNA as the parent cell. Different types of cells divide at different rates. Some human cells divide frequently, particularly those subjected to a lot of wear and tear. The epithelial cells of your skin divide so often that your skin replaces itself every two weeks. The lining of your stomach is replaced every few days! Nerve cells, on the other hand, can live for 100 years without dividing. Cells use a battery of genes to regulate when and how frequently they divide. If some of these genes become disabled, a cell may begin to divide ceaselessly, a condition we call cancer. Exposure to DNA-damaging chemicals such as those in cigarette smoke greatly increases the chance of this sort of event occurring in the tissues exposed to the smoke, which is why smokers will more likely get lung cancer than colon cancer.

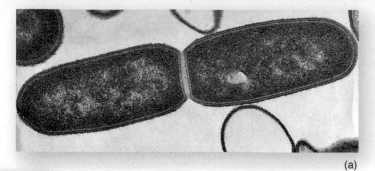

(a)

8.1 Prokaryotes Have a Simple Cell Cycle

All species reproduce, passing their hereditary information on to their offspring. In this chapter, we begin our consideration of heredity with a look at how cells reproduce. Cell division in prokaryotes takes place in two stages, which together make up a simple cell cycle. First the DNA is copied, and then the cell splits in half by a process called **binary fission.** The cell in **figure 8.1***a* is undergoing binary fission.

In prokaryotes, the hereditary information—that is, the genes that specify the prokaryote—is encoded in a single circle of DNA, called a prokaryotic chromosome. Before the cell itself divides, the DNA circle makes a copy of itself, a process called *replication.* Starting at one point, the origin of replication (the point where the two strands of DNA are connected at the top of **figure 8.1***b*), the double helix of DNA begins to unzip, exposing the two strands. The enlargement on the right of **figure 8.1***b* shows how the DNA replicates. The purple strand is from the original DNA and the red strand is the newly formed DNA. The new double helix is formed from each naked strand by placing on each exposed nucleotide its complementary nucleotide (that is, A with T, G with C, as discussed in chapter 3). DNA replication is discussed in more detail in chapter 11. When the unzipping has gone all the way around the circle, the cell possesses two copies of its hereditary information.

When the DNA has been copied, the cell grows, resulting in elongation. The newly replicated DNA molecules are partitioned toward each end of the cell. This partitioning process involves DNA sequences near the origin of replication, and results in these sequences being attached to the membrane. When the cell reaches an appropriate size, the prokaryotic cell begins to split into two equal halves. New plasma membrane and cell wall are added at a point between where the two DNA copies are partitioned, indicated by the green divider in **figure 8.1***b*. As the growing plasma membrane pushes inward, the cell is constricted in two, eventually forming two *daughter cells.* Each contains one prokaryotic chromosome and is a complete living cell in its own right.

Key Learning Outcome 8.1 Prokaryotes divide by binary fission after the DNA has replicated.

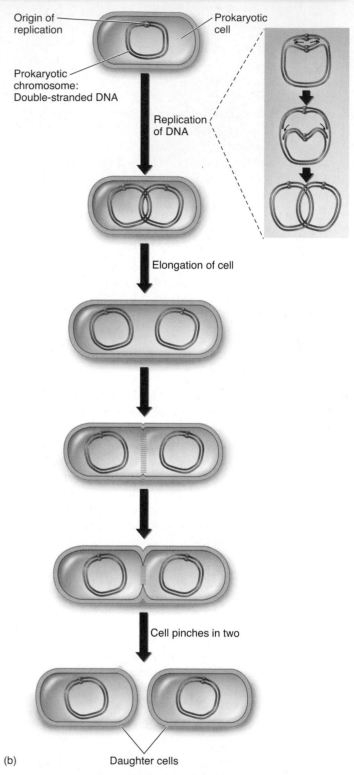

Origin of replication

Prokaryotic cell

Prokaryotic chromosome: Double-stranded DNA

Replication of DNA

Elongation of cell

Cell pinches in two

(b)   Daughter cells

Figure 8.1 Cell division in prokaryotes.

(*a*) Prokaryotes divide by a process of binary fission. Here, a cell has divided in two and is about to be pinched apart by the growing plasma membrane. (*b*) Before the cell splits, the circular DNA molecule of a prokaryote initiates replication at a single site, called the origin of replication, moving out in both directions. When the two moving replication points meet on the far side of the molecule, its replication is complete. The cell then undergoes binary fission, where the cell divides into two daughter cells.

Eukaryotes Have a Complex Cell Cycle

The evolution of the eukaryotes introduced several additional factors into the process of cell division. Eukaryotic cells are much larger than prokaryotic cells, and they contain much more DNA. Eukaryotic DNA is contained in a number of linear chromosomes, whose organization is much more complex than that of the single, circular DNA molecules in prokaryotes. A eukaryotic **chromosome** is a single, long DNA molecule wound tightly around proteins, called *histones,* into a compact shape.

Cell division in eukaryotes is more complex than in prokaryotes, both because eukaryotes contain far more DNA and because it is packaged differently. The cells of eukaryotic organisms either undergo mitosis or meiosis to divide up the DNA. **Mitosis** is the mechanism of cell division that occurs in an organism's nonreproductive cells, or *somatic cells.* An alternate process, called **meiosis,** divides the DNA in cells that participate in sexual reproduction, or *germ cells.* Meiosis results in the production of gametes, such as sperm and eggs, and is discussed in chapter 9.

The events that prepare the eukaryotic cell for division and the division process itself constitute a **complex cell cycle.** The Key Biological Process illustration below walks you through the phases of the cell cycle:

Interphase. This is the first phase of the cell cycle, **step 1** in the figure below, and is usually considered a resting phase, but the cell is far from resting. Interphase is itself made up of three phases:

G_1 phase. This "first gap" phase is the cell's primary growth phase. For most organisms, this phase occupies the major portion of the cell's life span.

S phase. In this "synthesis" phase, the DNA replicates, producing two copies of each chromosome.

G_2 phase. Cell division preparation continues in the "second gap" phase with the replication of mitochondria, chromosome condensation, and the synthesis of microtubules.

M phase. In mitosis, a microtubular apparatus binds to the chromosomes and moves them apart, shown in **steps 2–5.**

C phase. In cytokinesis, the cytoplasm divides, creating two daughter cells, shown in **step 6.**

Human cells growing in culture typically have a 22-hour cell cycle. Most cell types take about 80 minutes in this 22 hours to complete cell division: prophase—23 minutes, metaphase—29 minutes, anaphase—10 minutes, telophase—14 minutes, and cytokinesis—4 minutes. The proportion of the cell cycle spent in any one phase of mitosis varies considerably in different tissues.

> **Key Learning Outcome 8.2** Eukaryotic cells divide by separating duplicate copies of their chromosomes into daughter cells.

KEY BIOLOGICAL PROCESS: The Cell Cycle

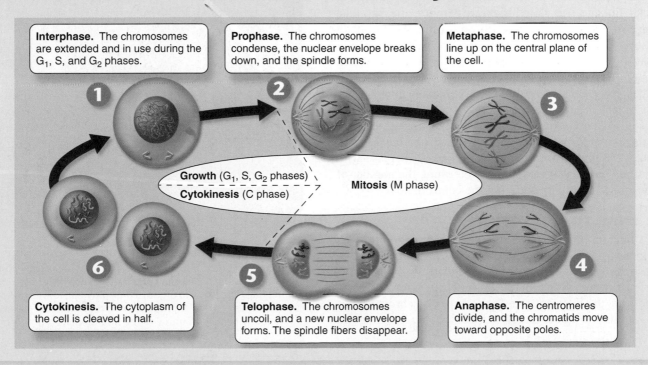

Interphase. The chromosomes are extended and in use during the G_1, S, and G_2 phases.

Prophase. The chromosomes condense, the nuclear envelope breaks down, and the spindle forms.

Metaphase. The chromosomes line up on the central plane of the cell.

Growth (G_1, S, G_2 phases)

Cytokinesis (C phase)

Mitosis (M phase)

Cytokinesis. The cytoplasm of the cell is cleaved in half.

Telophase. The chromosomes uncoil, and a new nuclear envelope forms. The spindle fibers disappear.

Anaphase. The centromeres divide, and the chromatids move toward opposite poles.

8.3 Chromosomes

Chromosomes were first observed by the German embryologist Walther Fleming in 1879 while he was examining the rapidly dividing cells of salamander larvae. When Fleming looked at the cells through what would now be a rather primitive light microscope, he saw minute threads within their nuclei that appeared to be dividing lengthwise. Fleming called their division *mitosis*, based on the Greek word *mitos*, meaning "thread."

Chromosome Number

Since their initial discovery, chromosomes have been found in the cells of all eukaryotes examined. Their number may vary enormously from one species to another. A few kinds of organisms—such as the Australian ant *Myrmecia* spp.; the plant *Haplopappus gracilis,* a relative of the sunflower that grows in North American deserts; and the fungus *Penicillium*—have only 1 pair of chromosomes, while some ferns have more than 500 pairs. Most eukaryotes have between 10 and 50 chromosomes in their body cells.

Homologous Chromosomes

Chromosomes exist in somatic cells as pairs, called **homologous chromosomes,** or **homologues.** Homologues carry information about the same traits at the same locations on each chromosome but the information can vary between homologues, which will be discussed in chapter 10. Cells that have two of each type of chromosome are called **diploid cells.** One chromosome of each pair is inherited from the mother (colored green in **figure 8.2**) and the other from the father (colored purple). Before cell division, each homologous chromosome replicates, resulting in two identical copies, called **sister chromatids.** You see in **figure 8.2** that the sister chromatids remain joined together after replication at a special linkage site called the **centromere,** the knoblike structure in the middle of each chromosome. Human body cells have a total of 46 chromosomes, which are actually 23 pairs of homologous chromosomes. In their duplicated state, before mitosis, there are still only 23 pairs of chromosomes, but each chromosome has duplicated and consists of two sister chromatids, for a total of 92 chromatids. The duplicated sister chromatids can make it confusing to count the number of chromosomes in an organism, but keep in mind that the number of centromeres doesn't increase with replication, and so you can always determine the number of chromosomes simply by counting the centromeres.

The Human Karyotype

The 46 human chromosomes can be paired as homologues by comparing size, shape, location of centromeres, and so on. This arrangement of chromosomes is called a *karyotype.* An example of a human karyotype is shown in **figure 8.3.** You can see how the different sizes and shapes of chromosomes allow scientists to pair together the ones that are homologous.

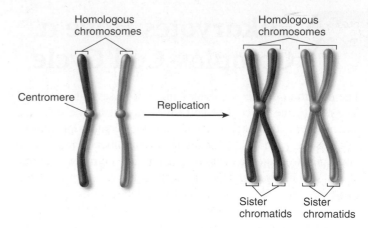

Figure 8.2   The difference between homologous chromosomes and sister chromatids.

Homologous chromosomes are a pair of the same chromosome—say, chromosome number 16. Sister chromatids are the two replicas of a single chromosome held together by the centromere after DNA replication. A duplicated chromosome looks somewhat like an X.

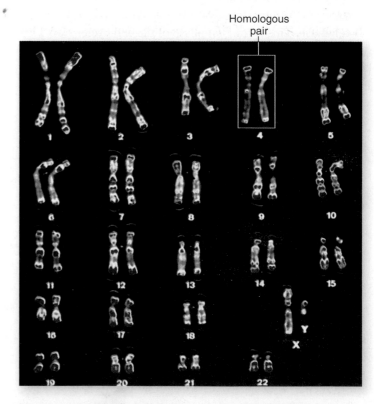

Figure 8.3   The 46 chromosomes of a human.

In this presentation, photographs of the individual chromosomes of a human male have been cut out and paired with their homologues, creating an organized display called a *karyotype.* The chromosomes are in a duplicated state, and the sister chromatids can actually be seen in many of the homologous pairs.

For example, chromosome 1 is much larger than chromosome 14, and its centromere is more centrally located on the chromosome. Each chromosome contains thousands of genes that play important roles in determining how a person's body develops and functions. For this reason, possession of all the chromosomes is essential to survival. Humans missing even one chromosome, a condition called monosomy, do not usually survive embryonic development. Nor does the human embryo develop properly with an extra copy of any one chromosome, a condition called trisomy. For all but a few of the smallest chromosomes, trisomy is fatal; even in those cases, serious problems result. We will revisit this issue of differences in chromosome number in chapter 10.

Chromosome Structure

Chromosomes are composed of **chromatin,** a complex of DNA and protein; most are about 40% DNA and 60% protein. A significant amount of RNA is also associated with chromosomes because chromosomes are the sites of RNA synthesis. The DNA of a chromosome is one very long, double-stranded fiber that extends unbroken through the entire length of the chromosome. A typical human chromosome contains about 140 million (1.4×10^8) nucleotides in its DNA. Furthermore, if the strand of DNA from a single chromosome were laid out in a straight line, it would be about 5 centimeters (2 inches) long. The amount of information in one human chromosome would fill about 2,000 printed books of 1,000 pages each! Fitting such a strand into a nucleus is like cramming a string the length of a football field into a baseball—and that's only 1 of 46 chromosomes! In the cell, however, the DNA is coiled, allowing it to fit into a much smaller space than would otherwise be possible.

Chromosome Coiling

The DNA of eukaryotes is divided into several chromosomes, although the chromosomes you see in **figure 8.3** hardly look like long, double-stranded molecules of DNA. These chromosomes, duplicated as sister chromatids, are formed by winding and twisting the long DNA strands into a much more compact structure. Winding up DNA presents an interesting challenge. Because the phosphate groups of DNA molecules have negative charges, it is impossible to just tightly wind up DNA because all the negative charges would simply repel one another. As you can see in **figure 8.4**, the DNA helix wraps around proteins with positive charges called **histones.** The positive charges of the histones counteract the negative charges of the DNA, so that the complex has no net charge. Every 200 nucleotides, the DNA duplex is coiled around a core of eight histone proteins, forming a complex known as a **nucleosome.** The nucleosomes, which resemble beads on a string in **figure 8.4**, are further coiled into a solenoid. This solenoid is then organized into looped domains. The final organization of the chromosome is not known, but it appears to involve further radial looping into rosettes around a preexisting scaffolding of protein. This complex of DNA and histone proteins, coiled tightly, forms a compact chromosome.

Key Learning Outcome 8.3 All eukaryotic cells store their hereditary information in chromosomes, but different kinds of organisms use very different numbers of chromosomes to store this information. Coiling of the DNA into chromosomes allows it to fit in the nucleus.

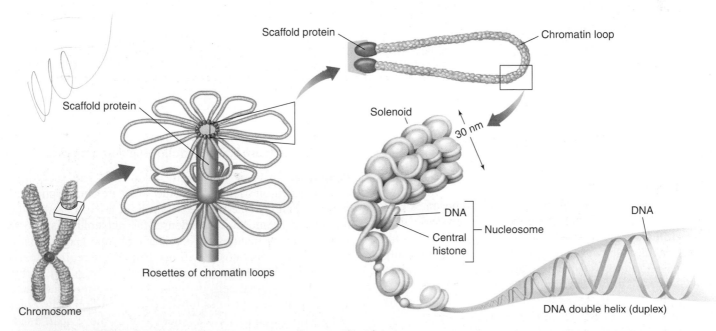

Figure 8.4 Levels of eukaryotic chromosomal organization.
Compact, rod-shaped chromosomes are in fact highly wound-up molecules of DNA. The arrangement illustrated here is one of many possibilities.

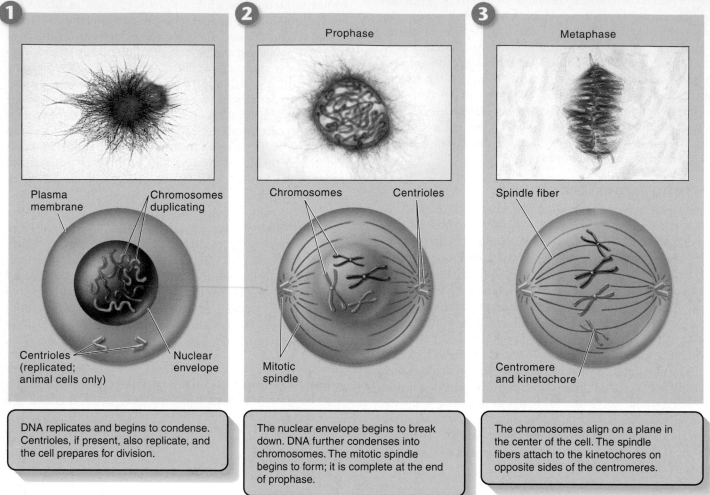

1

Plasma membrane

Chromosomes duplicating

Centrioles (replicated; animal cells only)

Nuclear envelope

DNA replicates and begins to condense. Centrioles, if present, also replicate, and the cell prepares for division.

2

Prophase

Chromosomes

Centrioles

Mitotic spindle

The nuclear envelope begins to break down. DNA further condenses into chromosomes. The mitotic spindle begins to form; it is complete at the end of prophase.

3

Metaphase

Spindle fiber

Centromere and kinetochore

The chromosomes align on a plane in the center of the cell. The spindle fibers attach to the kinetochores on opposite sides of the centromeres.

Figure 8.5 How cell division works.

Cell division in eukaryotes begins in interphase, carries through the four stages of mitosis, and ends with cytokinesis. Several features of the spindle illustrated in the drawings above appear in dividing animal cells but not in plant cells, and cannot be seen in the photographs, which are of the African blood lily *Haemanthus katharinae*. (In these exceptional photographs, the chromosomes are stained *blue* and microtubules stained *red*.)

8.4 Cell Division

Interphase

When cell division begins in interphase, chromosomes first replicate, and then begin to wind up tightly, a process called **condensation.** Sister chromatids are held together by a complex of proteins called *cohesin*. Chromosomes are not usually visible during interphase, but to clarify what is happening, they are shown in **panel 1** of **figure 8.5** as if they were.

Mitosis

Interphase is not a phase of mitosis, but it sets the stage for cell division. It is followed by nuclear division, called *mitosis*. Although the process of mitosis is continuous, with the stages flowing smoothly one into another, for ease of study, mitosis is traditionally subdivided into four stages: prophase,

metaphase, anaphase, and telophase. We will be referring to the panels in **figure 8.5** in the following descriptions.

Prophase: Mitosis Begins In **prophase,** the individual condensed chromosomes, the blue structures in the photo of **panel 2,** first become visible with a light microscope. As the replicated chromosomes condense, the nucleolus disappears and the cell dismantles the nuclear envelope and begins to assemble the apparatus it will use to pull the replicated sister chromatids to opposite ends ("poles") of the cell. In the center of an animal cell, the centrioles have replicated, and the two pairs of centrioles move apart toward opposite poles of the cell, forming between them as they move apart a network of protein cables called the **spindle.** In **panel 2,** the centrioles are positioned at the poles; the red structures in the drawing and photo are the protein cables that make up the spindle. Each cable is called a *spindle fiber* and is made of microtubules, which are long, hollow tubes of protein. Plant

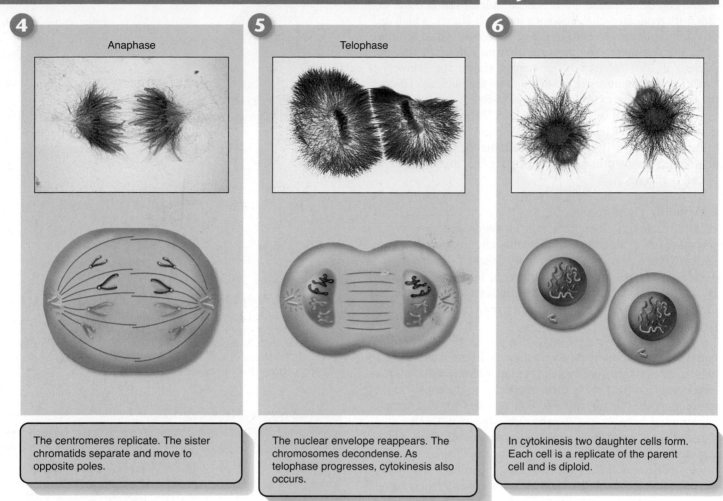

4 Anaphase

The centromeres replicate. The sister chromatids separate and move to opposite poles.

5 Telophase

The nuclear envelope reappears. The chromosomes decondense. As telophase progresses, cytokinesis also occurs.

6

In cytokinesis two daughter cells form. Each cell is a replicate of the parent cell and is diploid.

cells lack centrioles and instead brace the ends of the spindle toward the poles.

As condensation of the chromosomes continues, a second group of microtubules extends out from the poles toward the centromeres of the chromosomes. Each set of microtubules continues to grow longer until it makes contact with a disk of protein, called a *kinetochore,* associated with each side of the centromere. When the process is complete, one sister chromatid of each pair is attached by microtubules to one pole and the other sister chromatid to the other pole.

Metaphase: Alignment of the Chromosomes The second phase of mitosis, **metaphase,** begins when the chromosomes, each consisting of a pair of sister chromatids, align in the center of the cell along an imaginary plane that divides the cell in half, referred to as the equatorial plane. **Panel 3** shows the chromosomes beginning to align along the equatorial plane. Microtubules attached to the kinetochores of the centromeres are fully extended back toward the opposite poles of the cell.

Anaphase: Separation of the Chromatids In **anaphase,** enzymes cleave the cohesin link holding sister chromatids together, the kinetochores split, and the sister chromatids are freed from each other. Cell division is now simply a matter of reeling in the microtubules, dragging to the poles the sister chromatids, now referred to as daughter chromosomes. In **panel 4** you see the daughter chromosomes being pulled by their centromeres, the arms of the chromosomes dangling behind. The ends of the microtubules are dismantled, one bit after another, making the tubes shorter and shorter and so drawing the chromosome attached to the far end closer and closer to the opposite poles of the cell. When they finally arrive, each pole has one complete set of chromosomes.

Telophase: Re-formation of the Nuclei The only tasks that remain in **telophase** are the dismantling of the stage and the removal of the props. The mitotic spindle is disassembled, and a nuclear envelope forms around each set of chromosomes while they begin to uncoil, as shown in **panel 5**, and the nucleolus reappears.

Cytokinesis

At the end of telophase, mitosis is complete. The cell has divided its replicated chromosomes into two nuclei, which are positioned at opposite ends of the cell. Mitosis is also referred to as **karyokinesis.** You may recall from chapter 4 that the nucleus is also referred to as *karyon* (Latin for "kernel"); therefore, karyokinesis is the division of the nucleus. Toward the end of mitosis, **cytokinesis,** the division of the cytoplasm, occurs, and the cell is cleaved into roughly equal halves. Cytoplasmic organelles have already been replicated and resorted to the areas that will separate and become the daughter cells. Cytokinesis, shown in **panel 6** of **figure 8.5,** signals the end of cell division.

In animal cells, which lack cell walls, cytokinesis is achieved by pinching the cell in two with a contracting belt of actin filaments. As contraction proceeds, a *cleavage furrow* becomes evident around the cell's circumference, where the cytoplasm is being progressively pinched inward by the decreasing diameter of the actin belt. In **figure 8.6a** you see an animal cell pinching in half during cytokinesis. Imagine the cleavage furrow deepening further, until the cell is literally pinched in two.

Plant cells have rigid walls that are far too strong to be deformed by actin filament contraction. A different approach to cytokinesis has therefore evolved in plants. Plant cells assemble membrane components in their interior, at right angles to the mitotic spindle. In **figure 8.6b,** you can see how membrane is deposited between the daughter cells by vesicles that fuse together. This expanding partition, called a *cell plate,* grows outward until it reaches the interior surface of the plasma membrane and fuses with it, at which point it has effectively divided the cell in two. Cellulose is then laid down over the new membranes, forming the cell walls of the two new cells.

Cell Death

Despite the ability to divide, no cell lives forever. The ravages of living slowly tear away at a cell's machinery. To some degree, damaged parts can be replaced, but no replacement process is perfect. And sometimes the environment intervenes. If food supplies are cut off, for example, animal cells cannot obtain the energy necessary to maintain their lysosome membranes. The cells die, digested from within by their own enzymes.

During fetal development, many cells are programmed to die. In human embryos, hands and feet appear first as "paddles," but the skin cells between bones die on schedule to form the separated toes and fingers. **Figure 8.7** shows a developing human hand looking like a paddle. The cells in the tissue between the bones will later die, leaving behind a set of fingers. In ducks, this cell death is not part of the developmental program, which is why ducks have webbed feet and you don't.

Human cells appear to be programmed to undergo only so many cell divisions and then die, following a plan written into the genes. In tissue culture, cell lines divide about 50 times, and then the entire population of cells dies off. Even if some of the cells are frozen for years, when they are thawed they simply resume where they left off and die on schedule.

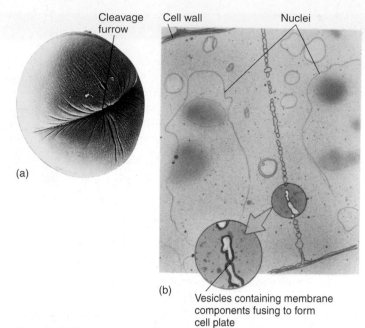

Figure 8.6 Cytokinesis.
The division of cytoplasm that occurs after mitosis is called cytokinesis and cleaves the cell into roughly equal halves. (*a*) In an animal cell, such as this sea urchin egg, a cleavage furrow forms around the dividing cell. (*b*) In this dividing plant cell, a cell plate is forming between the two newly forming daughter cells.

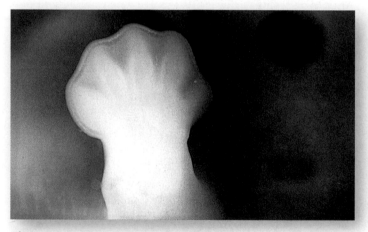

Figure 8.7 Programmed cell death.
In the human embryo, programmed cell death results in the formation of fingers and toes from paddlelike hands and feet.

Only cancer cells appear to thwart these instructions, dividing endlessly. All other cells in your body contain a hidden clock that keeps time by counting cell divisions, and when the alarm goes off the cells die.

> **Key Learning Outcome 8.4** The eukaryotic cell cycle starts in interphase with the condensation of replicated chromosomes; in mitosis, these chromosomes are drawn by microtubules to opposite ends of the cell; in cytokinesis, the cell is split into two daughter cells.

8.5 | Controlling the Cell Cycle

The events of the cell cycle are coordinated in much the same way in all eukaryotes. The control system that human cells use first evolved among the protists over a billion years ago; today, it operates in essentially the same way in fungi as it does in humans.

Check Points

The goal of controlling any cyclic process is to adjust the duration of the cycle to allow sufficient time for all events to occur. In principle, a variety of methods can achieve this goal. For example, an internal clock can be employed to allow adequate time for each phase of the cycle to be completed. This is how many organisms control their daily activity cycles. The disadvantage of using such a clock to control the cell cycle is that it is not very flexible. One way to achieve a more flexible and sensitive regulation of a cycle is simply to let the completion of each phase of the cycle trigger the beginning of the next phase, as a runner passing a baton starts the next leg in a relay race. Until recently, biologists thought this type of mechanism controlled the cell division cycle. However, we now know that eukaryotic cells employ a separate, centralized controller to regulate the process: At critical points in the cell cycle, further progress depends upon a central set of "go/no-go" switches that are regulated by feedback from the cell.

This mechanism is the same one engineers use to control many processes. For example, the furnace that heats a home in the winter typically goes through a daily heating cycle. When the daily cycle reaches the morning "turn on" checkpoint, sensors report whether the house temperature is below the set point (for example, 70°F). If it is, the thermostat triggers the furnace, which warms the house. If the house is already at least that warm, the thermostat does not start the furnace. Similarly, the cell cycle has key checkpoints where feedback signals from the cell about its size and the condition of its chromosomes can either trigger subsequent phases of the cycle or delay them to allow more time for the current phase to be completed.

Three principal checkpoints control the cell cycle in eukaryotes (**figure 8.8**):

1. **Cell growth is assessed at the G_1 checkpoint.** Located near the end of G_1 and just before entry into S phase, the G_1 checkpoint makes the key decision of whether the cell should divide, delay division, or enter a resting stage. In yeasts, where researchers first studied this checkpoint, it is called START. If conditions are favorable for division, the cell begins to copy its DNA, initiating S phase. The G_1 checkpoint is where the more complex eukaryotes typically arrest the cell cycle if environmental conditions make cell division impossible or if the cell passes into an extended resting period called G_0 (**figure 8.9**).

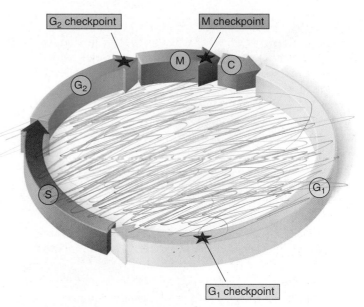

Figure 8.8 Control of the cell cycle.
Cells use a centralized control system to check whether proper conditions have been achieved before passing three key checkpoints in the cell cycle.

Figure 8.9 The G_1 checkpoint.
Feedback from the cell determines whether the cell cycle will proceed to the S phase, pause, or withdraw into G_0 for an extended rest period.

2. **DNA replication is assessed at the G_2 checkpoint.** The second checkpoint, the G_2 checkpoint, triggers the start of M phase. If this checkpoint is passed, the cell initiates the many molecular processes that signal the beginning of mitosis.
3. **Mitosis is assessed at the M checkpoint.** The third checkpoint, the M checkpoint, occurs at metaphase and triggers the exit from mitosis and cytokinesis and the beginning of G_1.

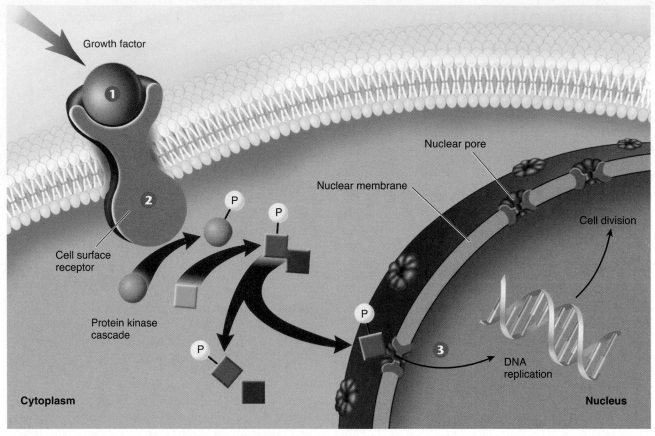

Figure 8.10 The cell proliferation-signaling pathway.
Binding of a growth factor sets in motion a cascading intracellular signaling pathway, which activates proteins in the nucleus that trigger cell division.

Growth Factors Trigger Cell Division

Cell division is initiated by small proteins called **growth factors.** Growth factors work by binding to the plasma membrane and triggering intracellular signaling systems. Fibroblasts are cells that form connective tissue in the body and possess numerous receptors on their plasma membranes for one of the first growth factors to be identified: platelet-derived growth factor (PDGF). When PDGF binds to a membrane receptor, it initiates an amplifying chain of internal cell signals that stimulates cell division.

PDGF was discovered when investigators found that fibroblasts would grow and divide in tissue culture only if the growth medium contained blood serum (the liquid that remains after blood clots); blood plasma (blood from which the cells have been removed without clotting) would not work. The researchers hypothesized that platelets in the blood clots were releasing into the serum one or more factors required for fibroblast growth. Eventually, they isolated such a factor and named it PDGF.

Growth factors such as PDGF override cellular controls that otherwise inhibit cell division. When a tissue is injured, a blood clot forms and the release of PDGF triggers neighboring cells to divide, helping to heal the wound. Only a tiny amount of PDGF (approximately 10^{-10} M) is required to stimulate cell division.

Characteristics of Growth Factors. Over 50 different proteins that function as growth factors have been isolated, and more undoubtedly exist. A specific cell surface receptor "recognizes" each growth factor, its shape fitting that growth factor precisely. **Figure 8.10** shows what happens when the growth factor ❶ binds with its receptor ❷. The receptor is activated and reacts by triggering a series of events within the cell, indicated by the arrows, that end with the replication of DNA and cell division ❸. The cellular selectivity of a particular growth factor depends upon which target cells bear its unique receptor. Some growth factors, like PDGF and epidermal growth factor (EGF), affect a broad range of cell types, while others affect only specific types. For example, nerve growth factor (NGF) promotes the growth of certain classes of neurons, and erythropoietin triggers cell division in red blood cell precursors. Most animal cells need a combination of several different growth factors to overcome the various controls that inhibit cell division.

The G_0 Phase. If cells are deprived of appropriate growth factors, they stop at the G_1 checkpoint of the cell cycle. With their growth and division arrested, they remain in the G_0 phase, as mentioned earlier. This nongrowing state is distinct from the interphase stages of the cell cycle, G_1, S, and G_2.

It is the ability to enter G_0 that accounts for the incredible diversity seen in the length of the cell cycle among different tissues. Epithelial cells lining the gut divide more than twice a day, constantly renewing the lining of the digestive tract. By contrast, liver cells divide only once every year or two, spending most of their time in G_0 phase. Mature neurons and muscle cells usually never leave G_0.

Aging and the Cell Cycle

All humans die. However, while each of us knows we shall someday die, few of us can escape wishing we could delay the process. Some succeed. The oldest documented living person, Jeanne Calment of France, reached the age of 122 years in 1997. The tantalizing possibility of long life that she represents is one reason why there is such interest in the aging process; if we knew enough about it perhaps we could slow it. A wide variety of theories have been advanced to explain why we age. In recent years, scientists have come a long way toward unraveling the puzzle.

The first clue was the discovery that cells appear to die on schedule, as if following a script. In a famous experiment carried out in 1961, geneticist Leonard Hayflick demonstrated that fibroblast cells growing in tissue culture will divide only a certain number of times. As you can see in **figure 8.11**, after about 50 population doublings, cell division stops, the cell cycle blocked just before DNA replication. If a cell sample is frozen after the cell has undergone 20 doublings, when thawed the cell resumes growth for 30 more doublings, then stops.

An explanation of the "Hayflick limit" was suggested in 1978 when Elizabeth Blackburn of the University of California, San Francisco, first glimpsed an extra length of DNA at the ends of chromosomes. These telomeric regions, about 5,000 nucleotides long, are each composed of several thousand repeats of the sequence TTAGGG. Blackburn found the telomeric region to be substantially shorter in body tissue chromosomes than in those of germ-line cells, the egg and sperm. She speculated that in body cells a portion of the telomere cap was lost by a chromosome during each cycle of DNA replication.

Blackburn was right. The cell machinery that replicates the DNA of each chromosome sits on the last 100 units of DNA at the chromosome's tip, and so cannot copy that bit. So each time the cell divides, its chromosomes get a little shorter. Eventually, after some 50 replication cycles, the protective telomeric cap is used up, and the cell line then enters senescence, no longer able to proliferate.

How do sperm and egg cells avoid this trap, dividing continuously for decades? Blackburn and collaborator Jack Szostak proposed that cells must possess a special enzyme that lengthens telomeres. In 1984 Blackburn's graduate student Carol Greider found the enzyme, dubbed "telomerase." Using it, egg and sperm maintain their chromosomes at a constant length of 5,000 nucleotide units. In body cells, by contrast, the telomerase gene is silent. For their discoveries, Blackburn, Greider, and Szostak received the 2009 Nobel Prize in Physiology or Medicine.

Later research has provided direct evidence for a causal relation between telomeric shortening and cell senescence. Using genetic engineering, teams of researchers from California and Texas in 1998 transferred into human body cell cul-

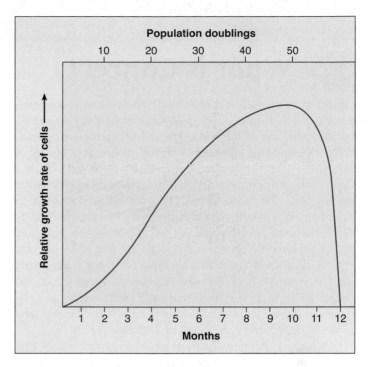

Figure 8.11 The Hayflick limit.
Normal human fibroblast (connective tissue) cells stop growing in culture after about 40 doublings, and within 10 more doublings all the cells are dead (*blue line*). When genetic engineers induce fibroblast cells to express the enzyme telomerase, proliferation continues long after 40 doublings.

tures a DNA fragment that unleashes each cell's telomerase gene. The result was unequivocal. New telomeric caps were added to the chromosomes of the cells, and the cells with the artificially elongated telomeres did not senesce at the Hayflick limit, continuing to divide in a healthy and vigorous manner for more than 20 additional generations.

This research shows clearly that loss of telomere DNA eventually restrains the ability of human cells to proliferate. And yet every human cell possesses a copy of the telomerase gene that, if expressed, would rebuild the telomere. Why do our cells accept aging if they need not? The answer, it seems, is to avoid cancer. By limiting the number of divisions allotted to human cell lines, the body ensures that no cell can continue to divide indefinitely. Suppression of the telomerase gene is, in a very real sense, cancer suppression. When scientists examine cancer cells, they commonly find their telomerase genes have been activated and are maintaining telomeres at full length. Thus telomere shortening is a tumor-suppressing mechanism, one of your body's key safeguards against cancer.

Key Learning Outcome 8.5 The complex cell cycle of eukaryotes is controlled with feedback at three checkpoints by protein signals called growth factors that initiate cell division. Telomeres play a key role in limiting cell proliferation.

8.6 What Is Cancer?

Cancer is a growth disorder of cells. It starts when an apparently normal cell begins to grow in an uncontrolled way, spreading out to other parts of the body. The result is a cluster of cells, called a **tumor,** that constantly expands in size. The cluster of pink lung cells in the photo in **figure 8.12** have begun to form a tumor. Benign tumors are completely enclosed by normal tissue and are said to be encapsulated. These tumors do not spread to other parts of the body and are therefore noninvasive. Malignant tumors are invasive and not encapsulated. Because they are not enclosed by normal tissue, cells are able to break away from the tumor and spread to other areas of the body. The tumor you see in **figure 8.12**, called a *carcinoma,* grows larger and eventually begins to shed cells that enter the bloodstream. Cells that leave a tumor and spread throughout the body, forming new tumors at distant sites, are called **metastases.**

Cancer is perhaps the most devastating and deadly disease. Most of us have had family or friends affected by the disease. In 2010, about 1.5 million American men and women were diagnosed with cancer. One in every two Americans born will be diagnosed with some form of cancer during their lifetime; nearly one in four are projected to die from cancer.

In the U.S., the three deadliest human cancers are lung cancer, cancer of the colon and rectum, and breast cancer. Lung cancer, responsible for the most cancer deaths, is largely preventable; most cases result from smoking cigarettes. Colorectal cancers appear to be fostered by the high-meat diets so favored in the United States. The cause of breast cancer is still a mystery.

Not surprisingly, researchers are expending a great deal of effort to learn the cause of cancer. Scientists have made considerable progress in the last 30 years using molecular biological techniques, and the rough outlines of understanding are now emerging. We now know that cancer is a gene disorder of somatic tissue, in which damaged genes fail to properly control cell growth and division. The cell division cycle is regulated by a sophisticated group of proteins called growth factors. Cancer results from the damage of these genes encod-

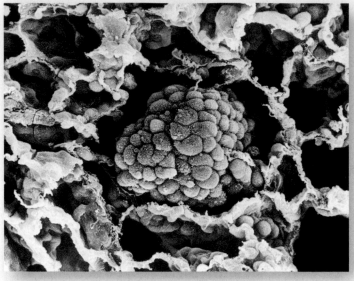

Figure 8.12   Lung cancer cells (300×).
These cells are from a tumor located in the alveolus (air sac) of a human lung.

ing these proteins. Damage to DNA, such as damage to these genes, is called **mutation.**

There are two general classes of growth factor genes that are usually involved in cancer: proto-oncogenes and tumor-suppressor genes. Genes known as **proto-oncogenes** encode proteins that stimulate cell division. Mutations to these genes can cause cells to divide excessively. Mutated proto-oncogenes become cancer-causing genes called **oncogenes.**

The second class of cancer-causing genes is called **tumor-suppressor genes.** Cell division is normally turned off in healthy cells by proteins encoded by tumor-suppressor genes. Mutations to these genes essentially "release the brakes," allowing the cell containing the mutated gene to divide uncontrolled.

> **Key Learning Outcome 8.6**   Cancer is unrestrained cell growth and division caused by damage to genes regulating the cell division cycle.

8.7 Cancer and Control of the Cell Cycle

Cancer can be caused by chemicals like those in cigarette smoke, by environmental factors such as UV rays that damage DNA, or in some instances by viruses that circumvent the cell's normal growth and division controls. Whatever the immediate cause, however, all cancers are characterized by unrestrained cell growth and division. The cell cycle never stops in a cancerous line of cells.

Cancer results from damaged genes failing to control cell division. Researchers have identified several of these

genes. One particular gene seems to be a key regulator of the cell cycle. Officially dubbed *p53* (researchers italicize the gene symbol to differentiate it from the protein), this gene plays a key role in the G_1 checkpoint of cell division. **Figure 8.13** illustrates how the product of this gene, the p53 protein, monitors the integrity of DNA, checking that it has been successfully replicated and is undamaged. If the p53 protein detects damaged DNA, as it does in the upper panel, it halts cell division and stimulates the activity of special enzymes to repair the damage. Once the DNA has been repaired, p53 allows cell division to continue, indicated by the upper path of arrows. In cases where the DNA cannot be repaired, p53 then directs the cell to kill itself, activating an apoptosis (cell suicide) program, indicated by the lower path of arrows.

By halting division in damaged cells, *p53* prevents the formation of tumors (even though its activities are not limited to cancer prevention). Scientists have found that *p53* is itself damaged beyond use in most human cancers they have examined. It is precisely because *p53* is nonfunctional that these cancer cells are able to repeatedly undergo cell division without being halted at the G₁ checkpoint. The lower panel of **figure 8.13** shows what happens when p53 doesn't function properly. The abnormal p53 does not stop cell division and the damaged strand is replicated, which results in damaged cells. As more and more damage occurs to these cells, they become cancerous. To test this, scientists administered healthy p53 protein to rapidly dividing cancer cells in a petri dish: The cells soon ceased dividing and died. Scientists have further reported that cigarette smoke causes mutations in the *p53* gene, reinforcing the strong link between smoking and cancer that you will encounter in chapter 24, section 24.6.

In about 50% of cancers, the *p53* cancer defense malfunctions because the *p53* gene itself has been damaged by chemicals or radiation, so that the protein the gene encodes no longer functions properly. In the other 50%, however, the defects lie in other genes. In many instances, the DNA is damaged at a site suppressing the production of a small molecule called MDM2 that is a potent natural inhibitor of the p53 protein. When the *MDM2* gene becomes overactive as a result of mutation, its protein product suppresses p53's activity.

A very promising approach to cancer prevention involves this second kind of p53 malfunction. Researchers studying the MDM2 protein found a relatively deep and well-defined pocket on its surface that proved to be the site that makes contact with the p53 protein. Perhaps, they hypothesized, a small molecule could be found that would fit into the site, and in doing so prevent p53 from binding. Such a molecule might prevent 50% of cancers! Searching for the key to fit this lock, they identified a family of synthetic chemicals they called "nutlins." When tumor cells that had functional *p53* genes were treated with nutlin, levels of p53 protein in the treated cells went up, and the tumor cells were killed, while normal cells treated in the same way were not killed. Nutlin is one of a host of new cancer therapies under active investigation.

> **Key Learning Outcome 8.7** Mutations disabling key elements of the G₁ checkpoint are associated with many cancers.

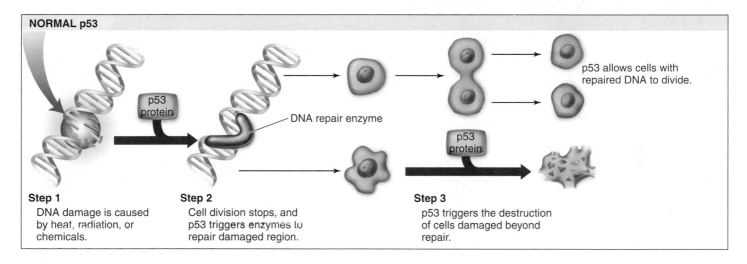

NORMAL p53

Step 1
DNA damage is caused by heat, radiation, or chemicals.

Step 2
Cell division stops, and p53 triggers enzymes to repair damaged region.

Step 3
p53 triggers the destruction of cells damaged beyond repair.

p53 allows cells with repaired DNA to divide.

DNA repair enzyme

p53 protein

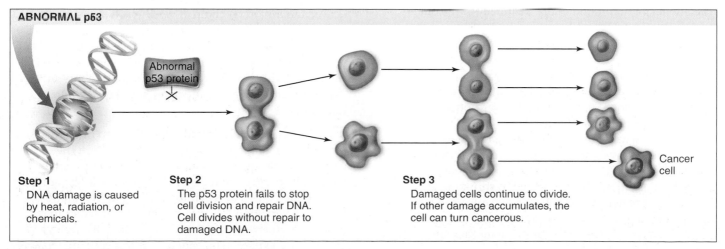

ABNORMAL p53

Step 1
DNA damage is caused by heat, radiation, or chemicals.

Step 2
The p53 protein fails to stop cell division and repair DNA. Cell divides without repair to damaged DNA.

Step 3
Damaged cells continue to divide. If other damage accumulates, the cell can turn cancerous.

Abnormal p53 protein

Cancer cell

Figure 8.13 Cell division and p53 protein.

Normal p53 protein monitors DNA, destroying cells with irreparable damage to their DNA. Abnormal p53 protein fails to stop cell division and repair DNA. As damaged cells proliferate, cancer develops.

Biology and *Staying Healthy*

Curing Cancer

Half of all Americans will face cancer at some point in their lives. Potential cancer therapies are being developed on many fronts. Some act to prevent the start of cancer within cells. Others act outside cancer cells, preventing tumors from growing and spreading. The figure on the right indicates targeted areas for the development of cancer treatments. The following discussion will examine each of these areas.

Preventing the Start of Cancer

Many promising cancer therapies act within potential cancer cells, focusing on different stages of the cell's "Shall I divide?" decision-making process.

[1] Receiving the Signal to Divide The first step in the decision process is receiving a "divide" signal, usually a small protein called a growth factor released from a neighboring cell. The growth factor, the red ball at #1 in the figure, is received by a protein receptor on the cell surface. Like banging on a door, its arrival signals that it's time to divide. Mutations that increase the number of receptors on the cell surface amplify the division signal and so lead to cancer. Over 20% of breast cancer tumors prove to overproduce a protein called HER2 associated with the receptor for epidermal growth factor (EGF).

Therapies directed at this stage of the decision process utilize the human immune system to attack cancer cells. Special protein molecules called *monoclonal antibodies,* created by genetic engineering, are the therapeutic agents. These monoclonal antibodies are designed to seek out and stick to HER2. Like waving a red flag, the presence of the monoclonal antibody calls down attack by the immune system on the HER2 cell. Because breast cancer cells overproduce HER2, they are killed preferentially. The biotechnology research company Genentech's recently approved monoclonal antibody, called herceptin, has given promising results in clinical tests.

Up to 70% of colon, prostate, lung, and head/neck cancers have excess copies of a related receptor, epidermal growth factor 1 (HER1). The monoclonal antibody C225, directed against HER1, has succeeded in shrinking 22% of advanced, previously incurable colon cancers in early clinical trials. Apparently blocking HER1 interferes with the ability of tumor cells to recover from chemotherapy or radiation.

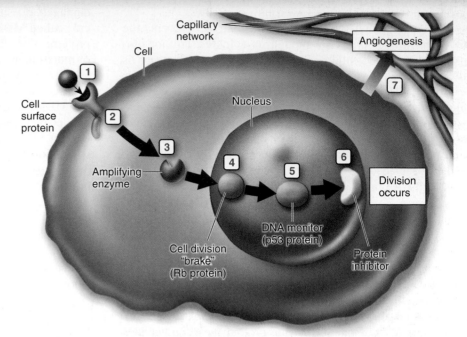

Seven different stages in the cancer process.

(1) On the cell surface, a growth factor's signal to divide is increased. (2) Just inside the cell, a protein relay switch that passes on the divide signal gets stuck in the "ON" position. (3) In the cytoplasm, enzymes that amplify the signal are amplified even more. In the nucleus, (4) a "brake" preventing DNA replication is inoperable, (5) proteins that check for damage in the DNA are inactivated, and (6) other proteins that inhibit the elongation of chromosome tips are destroyed. (7) The new tumor promotes angiogenesis, the formation of new blood vessels that promote growth.

[2] Passing the Signal via a Relay Switch The second step in the decision process is the passage of the signal into the cell's interior, the cytoplasm. This is carried out in normal cells by a protein called Ras that acts as a relay switch, #2 in the figure. When growth factor binds to a receptor like EGF, the adjacent Ras protein acts like it has been "goosed," contorting into a new shape. This new shape is chemically active, and initiates a chain of reactions that passes the "divide" signal inward toward the nucleus. Mutated forms of the Ras protein behave like a relay switch stuck in the "ON" position, continually instructing the cell to divide when it should not. Thirty percent of all cancers have a mutant form of Ras. So far, no effective therapies have been developed targeting this step.

[3] Amplifying the Signal The third step in the decision process is the amplification of the signal within the cytoplasm. Just as a TV signal needs to be amplified in order to be received at a distance, so a "divide" signal must be amplified if it is to reach the nucleus at the interior of the cell, a very long journey at a molecular scale. To get a signal all the way into the nucleus, the cell employs a sort of pony express. The "ponies" in this case are enzymes called *tyrosine kinases,* #3 in the figure. These enzymes add phosphate groups to proteins, but only at a particular amino acid, tyrosine. No other enzymes in the cell do this, so the tyrosine kinases form an elite core of signal carriers not confused by the myriad of other molecular activities going on around them.

Cells use an ingenious trick to amplify the signal as it moves toward the nucleus. Ras, when "ON," activates the initial protein kinase. This protein kinase activates other protein kinases that in their turn activate still others. The trick is that once a protein kinase enzyme is activated, it goes to work like a demon, activating hoards of others every second! And each and every one it activates behaves the same way too, activating still more, in a cascade of ever-widening effect. At each stage of the relay, the signal is amplified a thousandfold.

Mutations stimulating any of the protein kinases can dangerously increase the already amplified signal and lead to cancer. Some 15 of the cell's 32 internal tyrosine kinases have been implicated in cancer. Five percent of all cancers, for example, have a mutant hyperactive form of the protein kinase Src. The trouble begins when a mutation causes one of the tyrosine kinases to become locked into the "ON" position, sort of like a stuck doorbell that keeps ringing and ringing.

To cure the cancer, you have to find a way to shut the bell off. Each of the signal carriers presents a different problem, as you must quiet it without knocking out all the other signal pathways the cell needs. The cancer therapy drug Gleevec, a monoclonal antibody, has just the right shape to fit into a groove on the surface of the tyrosine kinase called "abl." Mutations locking abl "ON" are responsible for chronic myelogenous leukemia, a lethal form of white blood cell cancer. Gleevec totally disables abl. In clinical trials, blood counts revert to normal in more than 90% of cases.

4 **Releasing the Brake** The fourth step in the decision process is the removal of the "brake" the cell uses to restrain cell division. In healthy cells this brake, a tumor-suppressor protein called Rb, blocks the activity of a protein called E2F, #4 in the figure. When free, E2F enables the cell to copy its DNA. Normal cell division is triggered to begin when Rb is inhibited, unleashing E2F. Mutations that destroy Rb release E2F from its control completely, leading to ceaseless cell division. Forty percent of all cancers have a defective form of Rb.

Therapies directed at this stage of the decision process are only now being attempted. They focus on drugs able to inhibit E2F, which should halt the growth of tumors arising from inactive Rb. Experiments in mice in which the E2F genes have been destroyed provide a model system to study such drugs, which are being actively investigated.

5 **Checking That Everything Is Ready** The fifth step in the decision process is the mechanism used by the cell to ensure that its DNA is undamaged and ready to

divide. This job is carried out in healthy cells by the tumor-suppressor protein p53, which inspects the integrity of the DNA, #5 in the figure. When it detects damaged or foreign DNA, p53 stops cell division and activates the cell's DNA repair systems. If the damage doesn't get repaired in a reasonable time, p53 pulls the plug, triggering events that kill the cell. In this way, mutations such as those that cause cancer are either repaired or the cells containing them eliminated. If p53 is itself destroyed by mutation, future damage accumulates unrepaired. Among this damage are mutations that lead to cancer. Fifty percent of all cancers have a disabled p53. Fully 70% to 80% of lung cancers have a mutant inactive p53—the chemical benzo[a]pyrene in cigarette smoke is a potent mutagen of p53.

6 **Stepping on the Gas** Cell division starts with replication of the DNA. In healthy cells, another tumor suppressor "keeps the gas tank nearly empty" for the DNA replication process by inhibiting production of an enzyme called telomerase. Without this enzyme, a cell's chromosomes lose material from their tips, called telomeres. Every time a chromosome is copied, more tip material is lost. After some 30 divisions, so much is lost that copying is no longer possible. Cells in the tissues of an adult human have typically undergone 25 or more divisions. Cancer can't get very far with only the five remaining cell divisions, so inhibiting telomerase is a very effective natural brake on the cancer process, #6 in the figure. It is thought that almost all cancers involve a mutation that destroys the telomerase inhibitor, releasing this brake and making cancer possible. It should be possible to block cancer by reapplying this inhibition. Cancer therapies that inhibit telomerase are just beginning clinical trials.

Preventing the Spread of Cancer

7 **Stopping Tumor Growth** Once a cell begins cancerous growth, it forms an expanding tumor. As the tumor grows ever-larger, it requires an increasing supply of food and nutrients, obtained from the body's blood supply. To facilitate this necessary grocery shopping, tumors leak out substances into the surrounding tissues that encourage the formation of small blood vessels, a process called angiogenesis, #7 in the figure. Chemicals that inhibit this process are called angiogenesis inhibitors. Two such natural angiogenesis inhibitors, angiostatin and endostatin, caused tumors to regress to microscopic size in mice, but initial human trials were disappointing.

Laboratory drugs are more promising. A monoclonal antibody drug called Avastin, targeted against a blood vessel growth-promoting substance called vascular endothelial growth factor (VEGF), destroys the ability of VEGF to carry out its blood-vessel-forming job. Given to hundreds of advanced colon cancer patients in 2003 as part of a large clinical trial, Avastin improved colon cancer patients's chance of survival by 50% over chemotherapy.

Why Do Human Cells Age?

Human cells appear to have built-in life spans. As you learned in this chapter, cell biologist Leonard Hayflick reported in 1961 the startling result that skin cells growing in tissue culture, such as those growing in culture flasks in the photo below, will divide only a certain number of times. After about 50 population doublings, cell division stops (a **doubling** is a round of cell division producing two daughter cells for each dividing cell; for example, going from a population of 30 cells to 60 cells). If a cell sample is taken after 20 doublings and frozen, when thawed it resumes growth for 30 more doublings, and then stops. An explanation of the "Hayflick limit" was suggested in 1978 when researchers first glimpsed an extra length of DNA at the end of chromosomes. Dubbed telomeres, these lengths proved to be composed of the simple DNA sequence TTAGGG, repeated nearly a thousand times. Importantly, telomeres were found to be substantially shorter in the cells of older body tissues. This led

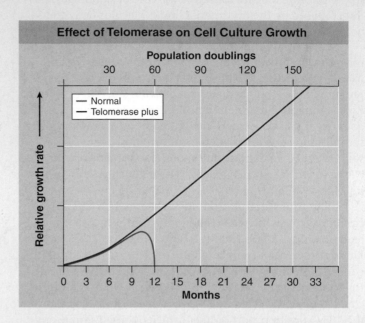

Effect of Telomerase on Cell Culture Growth

— Normal
— Telomerase plus

TTAGGG TTAGGG TTAGGG TTAGGG TTAGGG--------

to the hypothesis that a run of some 16 TTAGGGs was where the DNA replicating enzyme, called *polymerase,* first sat down on the DNA (16 TTAGGGs being the size of the enzyme's "footprint"), and because of being its docking spot, the polymerase was unable to copy that bit. Thus a 100-base portion of the telomere was lost by a chromosome during each doubling as DNA replicated. Eventually, after some 50 doubling cycles, each with a round of DNA replication, the telomere would be used up and there would be no place for the DNA replication enzyme to sit. The cell line would then enter senescence, no longer able to proliferate.

This hypothesis was tested in 1998. Using genetic engineering, researchers transferred into newly established human cell cultures a gene that leads to expression of an enzyme called *telomerase* that all cells possess but no body cell uses. This enzyme adds TTAGGG sequences back to the end of telomeres, in effect rebuilding the lost portions of the telomere. Laboratory cultures of cell lines with (telomerase plus) and without (normal) this gene were then monitored for many generations. The graph above displays the results.

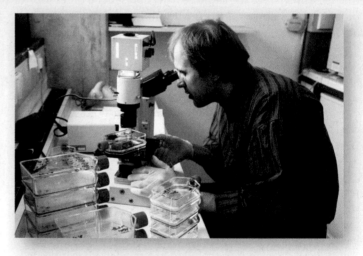

1. **Applying Concepts**
   a. Variable. In the graph, what is the dependent variable?
   b. Comparing Continuous Processes. How do normal skin cells (blue line) differ in their growth history from telomerase plus cells with the telomerase gene (red line)?
2. **Interpreting Data**
   a. After how many doublings do the normal cells cease to divide? Is this consistent with the telomerase hypothesis?
   b. After how many doublings do the telomerase plus cells cease to divide in this experiment?
3. **Making Inferences** After nine population doublings, would the rate of cell division be different between the two cultures? After 15? Why?
4. **Drawing Conclusions** How does the addition of the telomerase gene affect the senescence of skin cells growing in culture? Does this result confirm the telomerase hypothesis this experiment had set out to test?
5. **Further Analysis**
   a. Cancer cells are thought to possess mutations disabling the cell's ability to keep the telomerase gene shut off. How would you test this hypothesis?
   b. Sperm-producing cells continue to divide throughout a male's adult life. How might this be possible? How would you test this idea?

Cell Division

8.1 Prokaryotes Have a Simple Cell Cycle

- Prokaryotic cells divide in a two-step process: DNA replication and binary fission. The DNA is present as a single loop called a chromosome. The DNA begins replication at a site called the origin of replication. The DNA double strand unzips, and new strands form along the original strands, producing two circular chromosomes that separate to the ends of the cell. New plasma membrane and cell wall is added down  the middle of the cell, as shown here from **figure 8.1**, splitting the cell in two. This cell division, called binary fission, produces two daughter cells that are genetically identical to the parent cell.

8.2 Eukaryotes Have a Complex Cell Cycle

- Cell division in eukaryotes is more complex than in prokaryotes. The eukaryotic cell cycle occurs in several phases: interphase, M phase, and C phase. Interphase is considered a resting phase, but during interphase the cell is growing and preparing for cell division. Interphase is further broken down into its own phases. The G_1 phase is the growing phase and takes up the major portion of the cell's life cycle. The S phase is the synthesis phase and is when the DNA is replicated. The G_2 phase involves the final preparations for cell division.

- During the M and C phases, the chromosomes are distributed to opposite sides of the cell, which then divides its cytoplasm into two separate daughter cells (**Key Biological Process, page 155**).

8.3 Chromosomes

- Eukaryotic DNA is organized into chromosomes, and two chromosomes that carry copies of the same genes, like the two shown here from **figure 8.2**, are called homologous chromosomes. Before cells divide, the DNA replicates forming two identical copies of each chromosome, called sister chromatids. Sister chromatids are held together at the centromere region. Human somatic cells have 46 chromosomes (**figure 8.3**).

- The DNA in a chromosome is one long double-stranded fiber called chromatin. After the DNA is replicated, it begins to coil up in a process called condensation. The DNA wraps around histone proteins with each DNA-histone complex forming a nucleosome. The string of nucleosome then folds and loops on itself forming a compact chromosome (**figure 8.4**).

8.4 Cell Division

- Interphase (**figure 8.5, panel 1**) begins the cell cycle, followed by mitosis, which consists of four phases: prophase, metaphase, anaphase, and telophase.

- Prophase (**figure 8.5, panel 2**) signals the beginning of mitosis. The DNA that was replicated during interphase condenses into chromosomes. The sister chromatids stay attached at the centromeres. The nucleolus and nuclear envelope disappear. Centrioles migrate to opposite sides of the cell and begin forming the spindle. Microtubules that form the spindle extend from the poles and attach to the kinetochores, which are proteins located at the centromere areas of the chromosomes.

- Metaphase (**figure 8.5, panel 3**) involves the alignment of sister chromatids along the equatorial plane. Microtubules connect the kinetochores of sister chromatids to each of the poles.

- During anaphase (**figure 8.5, panel 4**), the kinetochores split, and enzymes break down the cohesin, freeing the sister chromatids. The microtubules shorten, pulling the sister chromatids apart and toward opposite poles.

- Telophase (**figure 8.5, panel 5**) signals the completion of nuclear division. The microtubule spindle is dismantled, the chromosomes begin to uncoil; the nuclear envelope and nucleolus re-form.

- Following mitosis, the cell separates into two daughter cells in a process called cytokinesis (**figure 8.6**). Cytokinesis in animal cells involves a pinching in of the cell around its equatorial plane until the cell eventually splits into two cells. Cytokinesis in plant cells involves the assembly of cell membranes and cell walls between the two poles, eventually forming two separate cells.

- Many cells are programmed to die, either as part of development (**figure 8.7**) or after a set number of cell divisions (usually about 50 divisions). Only cancer cells appear to divide endlessly.

8.5 Controlling the Cell Cycle

- The cell cycle is controlled at three checkpoints. The G_1 and G_2 checkpoints occur during interphase, and the third occurs during mitosis (**figure 8.8**). At the G_1 checkpoint the cell either initiates division or enters a period of rest called G_0 (**figure 8.9**). If cell division is initiated at the G_1 checkpoint, DNA replication begins and is checked at the G_2 checkpoint. Mitosis is initiated if the DNA has been accurately replicated.

- Cell division is initiated by proteins called growth factors (**figure 8.10**). After about 50 cell doublings, regions at the ends of chromosomes called telomeres become too short to allow cell division; this acts to suppress cancer (**figure 8.11**).

Cancer and the Cell Cycle

8.6 What Is Cancer?

- Cancer is a growth disorder of cells, where there is a loss of control over cell division. Cells begin to divide in an uncontrolled way, forming a mass of cells called a tumor (**figure 8.12**). A tumor in which cells break away from the mass and spread to other tissues is called metastasis. Cancer results when genes that encode proteins that control the cell cycle, such as proto-oncogenes and tumor-suppressor genes, are damaged.

8.7 Cancer and Control of the Cell Cycle

- The *p53* gene plays a key role in the G_1 checkpoint, checking the condition of the DNA. If the DNA is damaged, the p53 protein will stop cell division so the damaged DNA can be repaired. If the DNA cannot be repaired, the p53 protein will trigger the destruction of the cell. When the *p53* gene is damaged by mutations, the DNA is not checked, and cells with damaged DNA can divide. Further mutations to the DNA can accumulate in the cells and result in cancerous cells (**figure 8.13**).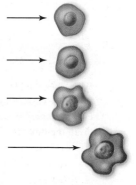

1. Prokaryotes reproduce new cells by
   a. copying DNA then undergoing binary fission.
   b. splitting in half.
   c. undergoing mitosis.
   d. copying DNA then undergoing the M phase.
2. The eukaryotic cell cycle is different from prokaryotic cell division in all the following ways *except*
   a. the amount of DNA present in the cells.
   b. how the DNA is packaged.
   c. in the production of genetically identical daughter cells.
   d. the involvement of microtubules.
3. In eukaryotes, the genetic material is found in chromosomes
   a. and the more complex the organism, the more pairs of chromosomes it has.
   b. and many organisms have only one chromosome.
   c. and most eukaryotes have between 10 and 50 pairs of chromosomes.
   d. and most eukaryotes have between 2 and 10 pairs of chromosomes.
4. Homologous chromosomes
   a. are also referred to as sister chromatids.
   b. are genetically identical.
   c. carry information about the same traits located in the same places on the chromosomes.
   d. are connected to each other at their centromeres.
5. Eukaryotic chromosomes are composed of
   a. DNA
   b. proteins.
   c. histones.
   d. All of the above.

6. In mitosis, when the duplicated chromosomes line up in the center of the cell, that stage is called
   a. prophase.
   b. metaphase.
   c. anaphase.
   d. telophase.
7. The division of the cytoplasm in the eukaryotic cell cycle is called
   a. interphase.
   b. mitosis.
   c. cytokinesis.
   d. binary fission.
8. The cell cycle is controlled by
   a. a series of checkpoints.
   b. telomerase.
   c. the G_0 phase.
   d. All of the above.
9. When cell division becomes unregulated, and a cluster of cells begins to grow without regard for the normal controls, that is called
   a. a mutation.
   b. cancer.
   c. metastases.
   d. oncogenes.
10. The normal function of the *p53* gene in the cell is
    a. as a tumor-suppressor gene.
    b. to monitor the DNA for damage.
    c. to trigger the destruction of cells not capable of DNA repair.
    d. All of the above.

1. **Figure 8.3** This karyotype shows a complete set of human chromosomes of an individual. At what stage of the cell cycle are such photos taken? Explain.

2. **Figure 8.4** During interphase the DNA is not visible through a microscope. Why isn't the DNA visible during interphase, and why would you expect this to be the case?

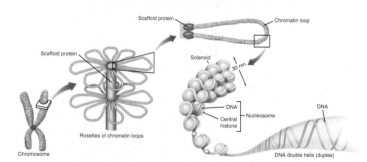

1. Why does the DNA in a cell need to change periodically from a long, double-helix chromatin molecule into a tightly wound-up chromosome? What does it do in one configuration that it cannot do in the other?

2. Despite all we know about cancer today, some types of cancers are still increasing in frequency. Lung cancer in women is one of those. What reason(s) might there be for this increasing problem? Can you suggest a solution?

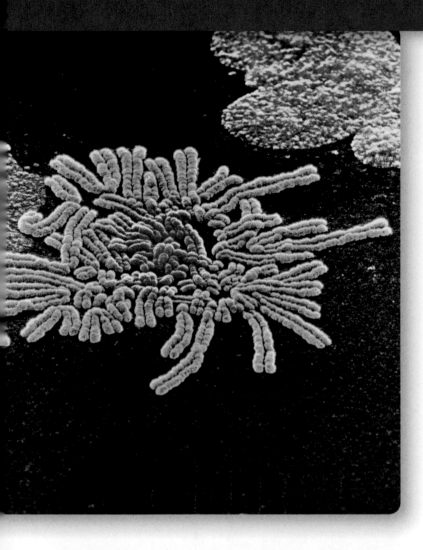

9
Meiosis

Humans, like most animals and plants, reproduce sexually. That is how you came into being: Your father contributed a sperm cell that united with an egg cell from your mother to form a cell called a zygote, containing both sets of chromosomes. Dividing repeatedly by mitosis, this zygote cell eventually gave rise to your adult body, made up of an astonishing number of cells—some 10 to 100 trillion. The sperm and egg that joined to form your initial cell were the products of a special form of cell division called meiosis, the subject of this chapter. Far more intricate than mitosis, the details of meiosis are not as well understood. The basic process, however, is clear. A cell dividing by meiosis goes through two nuclear divisions, replicating the DNA before the first division but not between the two divisions. In the photo above, chromosomes are lining up and getting ready to be pulled to opposite ends of the cell by microtubules too tiny to be visible to our eyes. When the two meiotic divisions are all over, there will be four cells, each with only half as much DNA as the initial cell. Confused? So were biologists when they first discovered meiosis. Hopefully, this chapter will make things clearer. It is important that you understand meiosis clearly, because meiosis and sexual reproduction play key roles in generating the tremendous genetic diversity that is the raw material of evolution.

9.1 Discovery of Meiosis

Only a few years after Walther Fleming's discovery of chromosomes in 1879, Belgian cytologist Pierre-Joseph van Beneden was surprised to find different numbers of chromosomes in different types of cells in the roundworm *Ascaris.* Specifically, he observed that the **gametes** (eggs and sperm) each contained two chromosomes, whereas the somatic (non-reproductive) cells of embryos and mature individuals each contained four.

Fertilization

From his observations, van Beneden proposed in 1887 that an egg and a sperm, each containing half the complement of chromosomes found in other cells, fuse to produce a single cell called a **zygote.** The zygote, like all of the somatic cells ultimately derived from it, contains two copies of each chromosome. The fusion of gametes to form a new cell is called **fertilization,** or **syngamy.**

Meiosis

It was clear even to early investigators that gamete formation must involve some mechanism that reduces the number of chromosomes to half the number found in other cells. If it did not, the chromosome number would double with each fertilization, and after only a few generations, the number of chromosomes in each cell would become impossibly large. For example, in just 10 generations, the 46 chromosomes present in human cells would increase to over 47,000 (46×2^{10}) chromosomes.

The number of chromosomes does not explode in this way because of a special reduction division that occurs during gamete formation, producing cells with half the normal number of chromosomes. The subsequent fusion of two of these cells ensures a consistent chromosome number from one generation to the next. This reduction division process, known as **meiosis,** is the subject of this chapter.

The Sexual Life Cycle

Meiosis and fertilization together constitute a cycle of reproduction. Two sets of chromosomes are present in the somatic cells of adult individuals, making them **diploid** cells (Greek, *di,* two), but only one set is present in the gametes, which are thus **haploid** (Greek, *haploos,* one). **Figure 9.1** shows how two haploid cells, a sperm cell containing three chromosomes contributed by the father and an egg cell containing three chromosomes contributed by the mother, fuse to form a diploid zygote with six chromosomes. Reproduction that involves this alternation of meiosis and fertilization is called **sexual reproduction.** Some organisms however, reproduce by mitotic division and don't involve the fusion of gametes. Reproduction in these organisms is referred to as **asexual reproduction.** Binary fission of prokaryotes shown in chapter 8 is an example of asexual reproduction. Some organisms are able to reproduce both asexu-

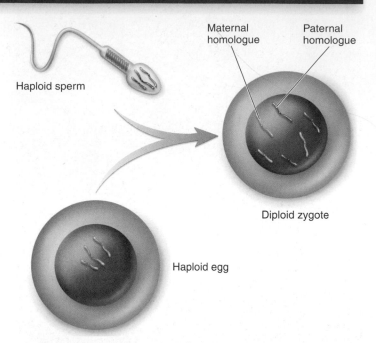

Figure 9.1 Diploid cells carry chromosomes from two parents.

A diploid cell contains two versions of each chromosome, a maternal homologue contributed by the mother's haploid egg, and a paternal homologue contributed by the father's haploid sperm.

Figure 9.2   Sexual and asexual reproduction.

Reproduction in an organism is not always either sexual or asexual. The strawberry plant reproduces both asexually (runners) and sexually (flowers).

ally and sexually. The strawberry plant pictured in **figure 9.2** reproduces sexually by fertilization that occurs in its flowers. Strawberries also reproduce asexually by sending out runners, stems that grow along the ground and produce new roots and shoots that give rise to genetically identical plants.

> **Key Learning Outcome 9.1**   Meiosis is a process of cell division in which the number of chromosomes in certain cells is halved during gamete formation.

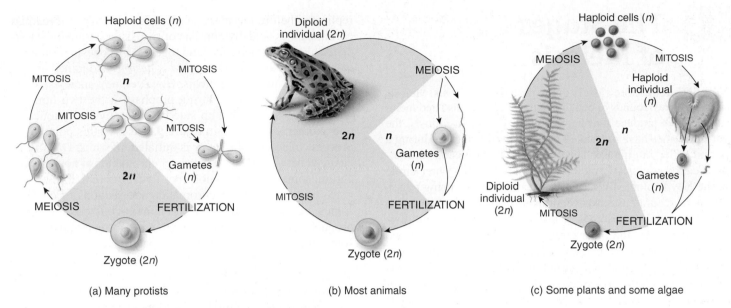

| (a) Many protists | (b) Most animals | (c) Some plants and some algae |

Figure 9.3 Three types of sexual life cycles.
In sexual reproduction, haploid cells or organisms alternate with diploid cells or organisms.

9.2 The Sexual Life Cycle

Somatic Tissues

The life cycles of all sexually reproducing organisms follow the same basic pattern of alternation between diploid chromosome numbers (the blue areas of the life cycles illustrated in **figure 9.3**) and haploid ones (the yellow areas). In most animals, fertilization results in the formation of a diploid zygote, shown in **figure 9.3b**, that begins to divide by mitosis. This single diploid cell eventually gives rise to all of the cells in the adult frog shown in the figure. These cells are called **somatic** cells, from the Latin word for "body." Each is genetically identical to the zygote.

In unicellular eukaryotic organisms like the protists shown in **figure 9.3a**, individual haploid cells function as gametes, fusing with other gamete cells. In plants like the fern shown in **figure 9.3c**, the haploid cells that meiosis produces divide by mitosis, forming a multicellular haploid phase, the heart-shaped structure in the figure. Some cells of this haploid phase eventually differentiate into eggs or sperm, which fuse to form a diploid zygote.

Germ-Line Tissues

In animals, the cells that will eventually undergo meiosis to produce gametes are set aside from somatic cells early in the course of development. These cells are often referred to as **germ-line** cells. Both the somatic cells and the gamete-producing germ-line cells are diploid, as indicated by blue arrows in **figure 9.4**. Somatic cells undergo mitosis to form genetically identical, diploid daughter cells. The germ-line cells undergo meiosis, indicated by the yellow arrows, producing haploid gametes.

Key Learning Outcome 9.2 In the sexual life cycle, there is an alternation of diploid and haploid phases.

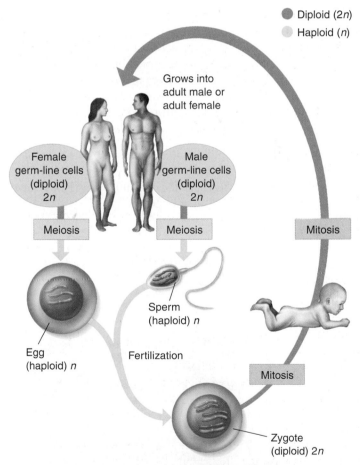

Figure 9.4 The sexual life cycle in animals.
In animals, the completion of meiosis is followed soon by fertilization. Thus, the vast majority of the life cycle is spent in the diploid stage. In this text, *n* stands for haploid and 2*n* stands for diploid. Germ-line cells are set aside early in development and undergo meiosis to form haploid gametes (eggs or sperm). The rest of the body cells are called somatic cells.

9.3 The Stages of Meiosis

Now, let's look more closely at the process of meiosis. Meiosis consists of two rounds of cell division, called meiosis I and meiosis II, which produce four haploid cells. Just as in mitosis, the chromosomes have replicated before meiosis begins, during a period called interphase. The first of the two divisions of meiosis, called **meiosis I** (meiosis I is shown in the outer circle of the Key Biological Process illustration on the facing page), serves to separate the two versions of each chromosome (the homologous chromosomes or homologues); the second division, **meiosis II** (the inner circle), serves to separate the two replicas of each version, called *sister chromatids*. Thus when meiosis is complete, what started out as one diploid cell ends up as four haploid cells. Because there was one replication of DNA but *two* cell divisions, the process reduces the number of chromosomes by half.

Meiosis I

Meiosis I is traditionally divided into four stages:

1. **Prophase I.** The two versions of each chromosome (the two homologues) pair up and exchange segments.
2. **Metaphase I.** The chromosomes align on a central plane.
3. **Anaphase I.** One homologue with its two sister chromatids still attached moves to a pole of the cell, and the other homologue moves to the opposite pole.
4. **Telophase I.** Individual chromosomes gather together at each of the two poles.

In **prophase I,** individual chromosomes first become visible, as viewed with a light microscope, as their DNA coils more and more tightly. Because the chromosomes (DNA) have replicated before the onset of meiosis, each of the threadlike chromosomes actually consists of two sister chromatids associated along their lengths (held together by cohesin proteins in a process called *sister chromatid cohesion*) and joined at their centromeres, just as in mitosis. However, now meiosis begins to differ from mitosis. During prophase I, the two homologous chromosomes line up side by side, physically touching one another, as you see in **figure 9.5**. It is at this point that a process called **crossing over** is initiated, in which DNA is exchanged between the two nonsister chromatids of homologous chromosomes. The chromosomes actually break in the same place on both nonsister chromatids and sections of chromosomes are swapped between the homologous chromosomes, producing a hybrid chromosome that is part maternal chromosome (the green sections) and part paternal chromosome (the purple sections). Two elements hold the homologous chromosomes together: (1) cohesion between sister chromatids; and (2) crossovers between nonsister chromatids (homologues). Late in prophase, the nuclear envelope disperses.

In **metaphase I,** the spindle apparatus forms, but because homologues are held close together by crossovers, spindle fibers can attach to only the outward-facing kinetochore of each centromere. For each pair of homologues, the orientation on the metaphase plate is random; which homologue is oriented toward which pole is a matter of chance. Like shuffling a deck of cards, many combinations are possible—in fact, 2 raised to a power equal to the number of chromosome pairs. For example, in a hypothetical cell that has three chromosome pairs, there are eight possible orientations (2^3). Each orientation results in gametes with different combinations of parental chromosomes. This process is called **independent assortment.** The chromosomes in **figure 9.6** line up along the metaphase plate, but whether the maternal chromosome (the green chromosomes) is on the right or left of the plate is completely random.

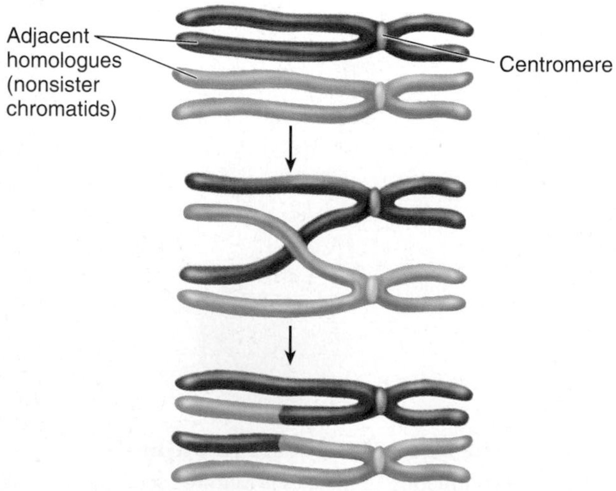

Figure 9.5 Crossing over.

In crossing over, the two homologues of each chromosome exchange portions. During the crossing over process, nonsister chromatids that are next to each other exchange chromosome arms or segments.

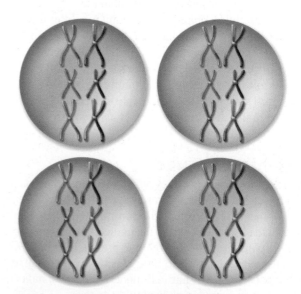

Figure 9.6 Independent assortment.
Independent assortment occurs because the orientation of chromosomes on the metaphase plate is random. Shown here are four possible orientations of chromosomes in a hypothetical cell. Each of the many possible orientations results in gametes with different combinations of parental chromosomes.

KEY BIOLOGICAL PROCESS: Meiosis

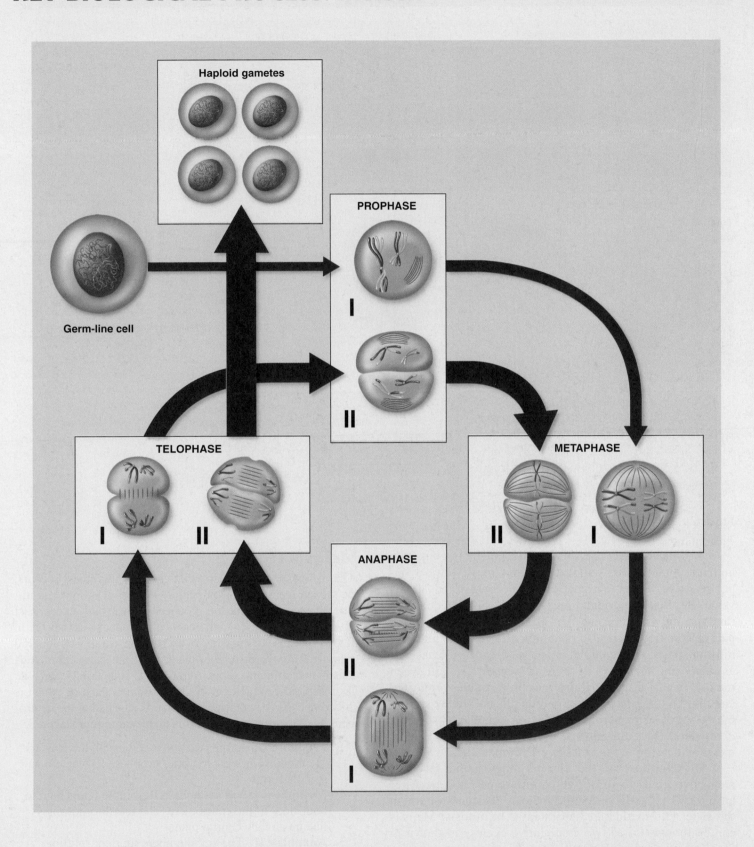

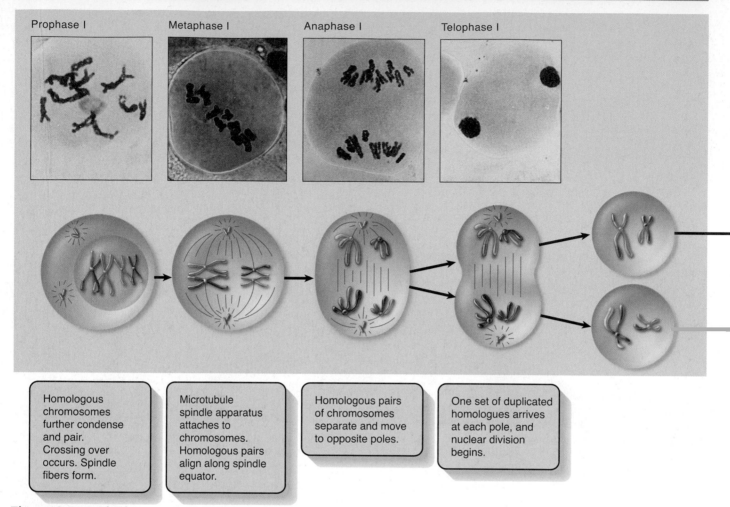

Prophase I Metaphase I Anaphase I Telophase I

Homologous chromosomes further condense and pair. Crossing over occurs. Spindle fibers form.

Microtubule spindle apparatus attaches to chromosomes. Homologous pairs align along spindle equator.

Homologous pairs of chromosomes separate and move to opposite poles.

One set of duplicated homologues arrives at each pole, and nuclear division begins.

Figure 9.7 Meiosis.

In **anaphase I,** the spindle attachment is complete, and homologues are pulled apart and move toward opposite poles. Sister chromatids are not separated at this stage. Because the orientation along the spindle equator is random, the chromosome that a pole receives from each pair of homologues is also random with respect to all chromosome pairs. At the end of anaphase I, each pole has half as many chromosomes as were present in the cell when meiosis began. Remember that the chromosomes replicated and thus contained two sister chromatids before the start of meiosis, but sister chromatids are not counted as separate chromosomes. As in mitosis, count the number of centromeres to determine the number of chromosomes.

In **telophase I,** the chromosomes gather at their respective poles to form two chromosome clusters. After an interval of variable length, meiosis II occurs in which the sister chromatids are separated as in mitosis. Meiosis can be thought of as two consecutive cycles, as shown in the Key Biological Process illustration on the previous page. The outer cycle contains the phases of meiosis I and the inner cycle contains the phases of meiosis II, discussed next.

Meiosis II

After a brief interphase, in which no DNA synthesis occurs, the second meiotic division begins. Meiosis II is simply a mitotic division involving the products of meiosis I, except that the sister chromatids are not genetically identical, as they are in mitosis, because of crossing over. You can see this by looking at figure 9.7, where some of the arms of the sister chromatids contain two different colors. At the end of anaphase I, each pole has a haploid complement of chromosomes, each of which is still composed of two sister chromatids attached at the centromere. Like meiosis I, meiosis II is divided into four stages:

1. **Prophase II.** At the two poles of the cell, the clusters of chromosomes enter a brief prophase II, where a new spindle forms.
2. **Metaphase II.** In metaphase II, spindle fibers bind to both sides of the centromeres and the chromosomes line up along a central plane.
3. **Anaphase II.** The spindle fibers shorten, splitting the centromeres and moving the sister chromatids to opposite poles.
4. **Telophase II.** Finally, the nuclear envelope re-forms around the four sets of daughter chromosomes.

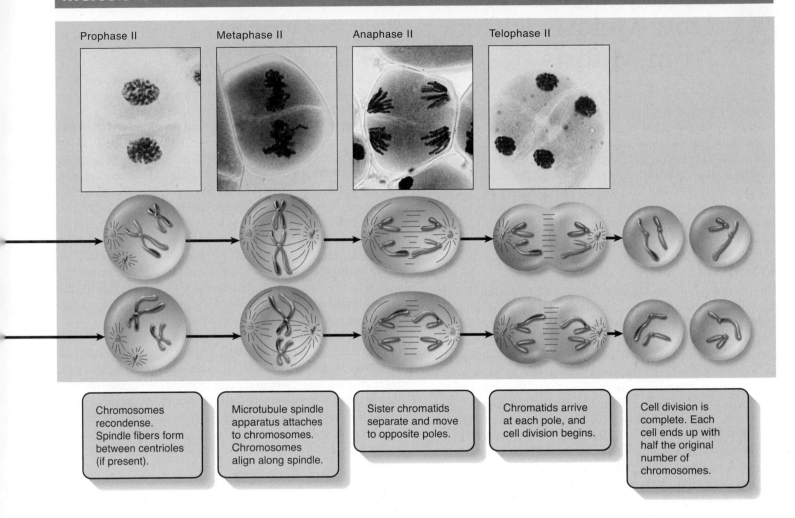

Prophase II — Chromosomes recondense. Spindle fibers form between centrioles (if present).

Metaphase II — Microtubule spindle apparatus attaches to chromosomes. Chromosomes align along spindle.

Anaphase II — Sister chromatids separate and move to opposite poles.

Telophase II — Chromatids arrive at each pole, and cell division begins.

Cell division is complete. Each cell ends up with half the original number of chromosomes.

The main outcome of the four stages of meiosis II—prophase II, metaphase II, anaphase II, and telophase II—is to separate the sister chromatids. The final result of this division is four cells containing haploid sets of chromosomes. No two are alike because of the crossing over in prophase I. The nuclei are then reorganized, and nuclear envelopes form around each haploid set of chromosomes. The cells that contain these haploid nuclei may develop directly into gametes, as they do in most animals. Alternatively, they may themselves divide mitotically, as they do in plants, fungi, and many protists, eventually producing greater numbers of gametes or, as in the case of some plants and insects, adult haploid individuals.

The Important Role of Crossing Over

If you think about it, the key to meiosis is that the sister chromatids of each chromosome are not separated from each other in the first division. Why not? What prevents microtubules from attaching to them and pulling them to opposite poles of the cell, just as eventually happens later in the second meiotic division? The answer is the crossing over that occurred early in the first division. By exchanging segments, the two homologues are tied together by strands of DNA. It is because microtubules can gain access to only one side of each homologue that they cannot pull the two sister chromatids apart! Imagine two people dancing closely—you can tie a rope to the back of each person's belt, but you cannot tie a second rope to their belt buckles because the two dancers are facing each other and are very close. In just the same way, microtubules cannot attach to the inner sides of the homologues because crossing over holds the homologous chromosomes together like dancing partners.

Key Learning Outcome 9.3 During meiosis I, homologous chromosomes move to opposite poles of the cell. At the end of meiosis II, each of the four haploid cells contains one copy of every chromosome in the set, rather than two. Because of crossing over, no two cells are the same.

9.4 How Meiosis Differs from Mitosis

While the general features of meiosis that you have just reviewed are similar in all eukaryotes, the detailed mechanism of meiosis varies somewhat in different organisms. This is particularly true of chromosomal separation mechanisms, which differ substantially in protists and fungi from the process in plants and animals that we describe here. Despite such differences in detail, however, two consistent features are seen in the meiotic processes of every eukaryote: synapsis and reduction division. Indeed, these two unique features are the key differences that distinguish meiosis from mitosis, which you studied in chapter 8.

Synapsis

The first of these two features happens early during the first nuclear division. Following chromosome replication, homologous chromosomes or homologues *pair all along their lengths,* with sister chromatids being held together, as mentioned earlier, by cohesin proteins. While homologues are thus physically joined, *genetic exchange occurs at one or more points between them.* The process of forming these complexes of homologous chromosomes is called **synapsis,** and the exchange process between paired homologues is called crossing over. **Figure 9.8a** shows how the homologous chromosomes are held together close enough that they are able to physically exchange segments of their DNA. Chromosomes are then drawn together along the equatorial plane of the dividing cell; subsequently, homologues are pulled apart by microtubules toward opposite poles of the cell. When this process is complete, the cluster of chromosomes at each pole contains one of the two homologues of each chromosome. Each pole is haploid, containing half the number of chromosomes present in the original diploid cell. Sister chromatids do not separate from each other in the first nuclear division, so each homologue is still composed of two chromatids joined at the centromere, and still considered one chromosome.

Reduction Division

The second unique feature of meiosis is that *the chromosome homologues do not replicate between the two nuclear divisions,* so that chromosome assortment in the second division separates sister chromatids of each chromosome into different daughter cells.

In most respects, the second meiotic division is identical to a normal mitotic division. However, because of the crossing over that occurred during the first division, the sister chromatids in meiosis II are not identical to each other. Also, there are only half the number of chromosomes in each cell at the beginning of meiosis II because only one of the homologues is present. **Figure 9.8b** shows how reduction division occurs. The diploid cell contains four chromosomes (two homologous pairs). After meiosis I, the cells contain just two chromosomes (remember to count the number of *centromeres,* because sister chromatids are not considered separate chromosomes). During meiosis II, the sister chromatids separate, but each gamete still only contains two chromosomes, half as many of the germ-line cell.

Because mitosis and meiosis use similar terminology, it is easy to confuse the two processes. **Figure 9.9** compares the two processes side-by-side. Both processes start with a diploid cell, but during meiosis, you can see that crossing over occurs and that during the first division in meiosis, the homologous pairs line up along the metaphase plate, while in mitosis, centromeres line up along the metaphase plate. These two differences result in haploid cells in meiosis and diploid cells in mitosis.

> **Key Learning Outcome 9.4** In meiosis, homologous chromosomes become intimately associated and do not replicate between the two nuclear divisions.

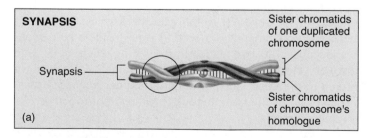

SYNAPSIS

Synapsis

Sister chromatids of one duplicated chromosome

Sister chromatids of chromosome's homologue

(a)

Figure 9.8 Unique features of meiosis.

(a) Synapsis draws homologous chromosomes together, all along their lengths, creating a situation (indicated by the circle) where two homologues can physically exchange portions of arms, a process called crossing over. (b) Reduction division, omitting a chromosome duplication before meiosis II, produces haploid gametes, thus ensuring that the chromosome number remains the same as that of the parents, following fertilization.

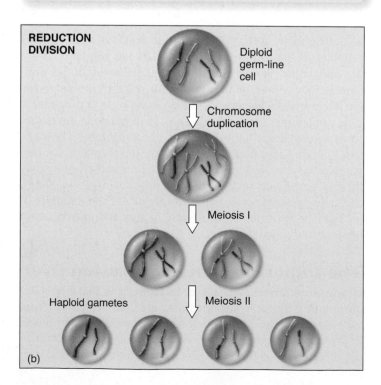

REDUCTION DIVISION

Diploid germ-line cell

Chromosome duplication

Meiosis I

Haploid gametes

Meiosis II

(b)

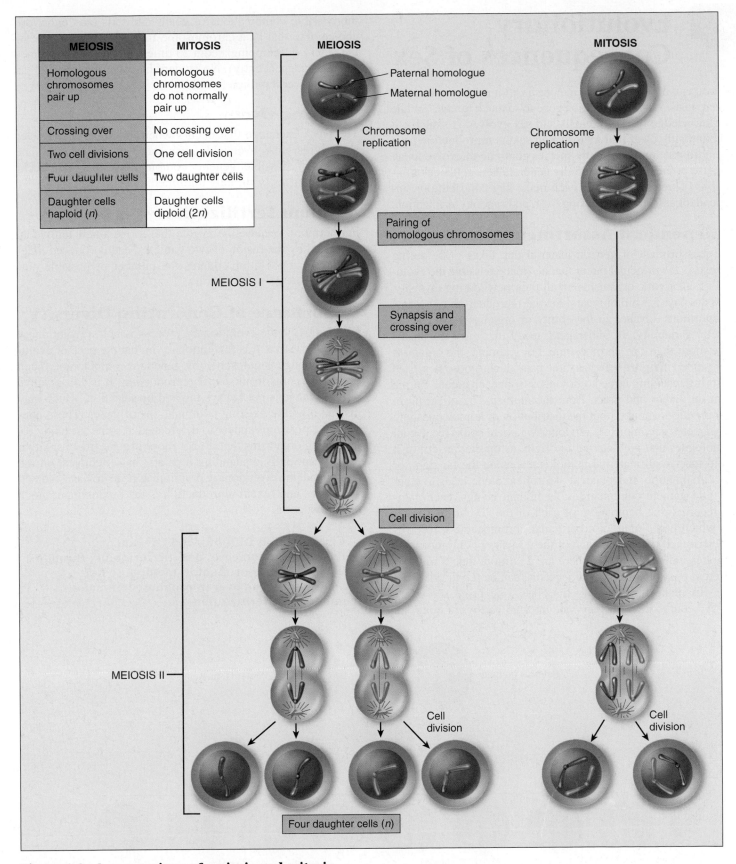

MEIOSIS	MITOSIS
Homologous chromosomes pair up	Homologous chromosomes do not normally pair up
Crossing over	No crossing over
Two cell divisions	One cell division
Four daughter cells	Two daughter cells
Daughter cells haploid (n)	Daughter cells diploid (2n)

MEIOSIS

Paternal homologue
Maternal homologue

Chromosome replication

Pairing of homologous chromosomes

MEIOSIS I

Synapsis and crossing over

Cell division

MEIOSIS II

Cell division

Four daughter cells (n)

MITOSIS

Chromosome replication

Cell division

Figure 9.9 A comparison of meiosis and mitosis.
Meiosis differs from mitosis in several key ways, highlighted by the orange boxes. Meiosis involves two nuclear divisions with no DNA replication between them. It thus produces four daughter cells, each with half the original number of chromosomes. Also, crossing over occurs in prophase I of meiosis. Mitosis involves a single nuclear division after DNA replication. Thus, it produces two daughter cells, each containing the original number of chromosomes, which are genetically identical to those in the parent cell.

9.5 Evolutionary Consequences of Sex

As you can now appreciate, meiosis is a lot more complicated than mitosis. Why has evolution gone to so much trouble? While our knowledge of how meiosis and sex evolved is sketchy, it is abundantly clear that meiosis and sexual reproduction have an enormous impact on how species continue to evolve today because of their ability to rapidly generate new genetic combinations. Three mechanisms each make key contributions: independent assortment, crossing over, and random fertilization.

Independent Assortment

The reassortment of genetic material that takes place during meiosis is the principal factor that has made possible the evolution of eukaryotic organisms, in all their bewildering diversity, over the past 1.5 billion years. Sexual reproduction represents an enormous advance in the ability of organisms to generate genetic variability. To understand, recall that most organisms have more than one chromosome. For example, the organism represented in figure 9.10 has three pairs of chromosomes, each offspring receiving three homologues from each parent, purple from the father and green from the mother. The offspring in turn produces gametes, but the distribution of homologues into the gametes is completely random. A gamete could receive all homologues that are paternal in origin, as on the far left; or it could receive all maternal homologues, as on the far right, or any combination. Independent assortment alone leads to eight possible gamete combinations. In human beings, each gamete receives one homologue of each of the 23 chromosomes, but which homologue of a particular chromosome it receives is determined randomly. Each of the 23 pairs of chromosomes migrates independently, so there are 2^{23} (more than 8 million) different possible kinds of gametes that can be produced.

To make this point to his class, one professor offers an "A" course grade to any student who can write down all the possible combinations of heads and tails (an "either/or" choice, like that of a chromosome migrating to one pole or the other) with flipping a coin 23 times (like 23 chromosomes moving independently). No student has ever won an "A," as there are over 8 million possibilities.

Crossing Over

The DNA exchange that occurs when the arms of nonsister chromatids cross over adds even more recombination. The number of possible genetic combinations that can occur among gametes is virtually unlimited.

Random Fertilization

Furthermore, because the zygote that forms a new individual is created by the fusion of two gametes, each produced independently, fertilization squares the number of possible outcomes ($2^{23} \times 2^{23} = 70$ trillion).

Importance of Generating Diversity

Paradoxically, the evolutionary process is both revolutionary and conservative. It is revolutionary in that the pace of evolutionary change is quickened by genetic recombination, much of which results from sexual reproduction. It is conservative in that change is not always favored by selection, which may instead preserve existing combinations of genes. These conservative pressures appear to be greatest in some asexually reproducing organisms that do not move around freely and that live in especially demanding habitats. In vertebrates, on the other hand, the evolutionary premium appears to have been on versatility, and sexual reproduction is the predominant mode of reproduction.

> **Key Learning Outcome 9.5** Sexual reproduction increases genetic variability through independent assortment in metaphase I of meiosis, crossing over in prophase I of meiosis, and random fertilization.

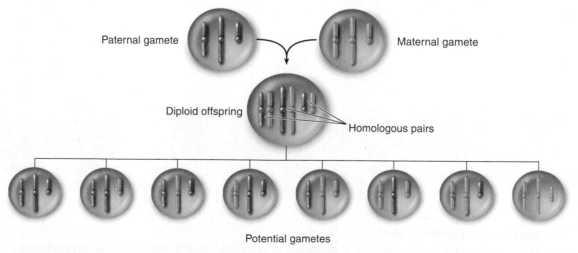

Figure 9.10 Independent assortment increases genetic variability.
Independent assortment contributes new gene combinations to the next generation because the orientation of chromosomes on the metaphase plate is random. In the cell shown here with three chromosome pairs, there are eight different gametes that can result, each with different combinations of parental chromosomes.

Why Sex?

Not all reproduction is sexual. In **asexual reproduction**, an individual inherits all of its chromosomes from a single parent and is, therefore, genetically identical to its parent. Prokaryotic cells reproduce asexually, undergoing binary fission to produce two daughter cells containing the same genetic information.

Most protists reproduce asexually except under conditions of stress; then they switch to sexual reproduction. Among plants and fungi, asexual reproduction is common.

In animals, asexual reproduction often involves the budding off of a localized mass of cells, which grows by mitosis to form a new individual.

Even when meiosis and the production of gametes occur, there may still be reproduction without sex. The development of an adult from an unfertilized egg, called **parthenogenesis,** is a common form of reproduction in arthropods. Among bees, for example, fertilized eggs develop into diploid females, but unfertilized eggs develop into haploid males. Parthenogenesis even occurs among the vertebrates. Some lizards, fishes, and amphibians are capable of reproducing in this way; their unfertilized eggs undergo a mitotic nuclear division without cell cleavage to produce a diploid cell, which then develops into an adult. In some plants, such as hawkweeds, dandelions, and blackberries, a process similar to parthenogenesis called *apomixis* can occur.

If reproduction can occur without sex, why does sex occur at all? This question has generated considerable discussion, particularly among evolutionary biologists. Sex is of great evolutionary advantage for populations or species, which benefit from the variability generated in meiosis by random orientation of chromosomes and by crossing over. However, evolution occurs because of changes at the level of *individual* survival and reproduction, rather than at the population level, and no obvious advantage accrues to the progeny of an individual that engages in sexual reproduction. In fact, recombination is a destructive as well as a constructive process in evolution. The segregation of chromosomes during meiosis tends to disrupt advantageous combinations of genes more often than it creates new, better adapted combinations; as a result, some of the diverse progeny produced by sexual reproduction will not be as well adapted as their parents were. In fact, the more complex the adaptation of an individual organism, the less likely that recombination will improve it, and the more likely that recombination will disrupt

it. It is, therefore, a puzzle to know what a well-adapted individual gains from participating in sexual reproduction, as *all* of its progeny could maintain its successful gene combinations if that individual simply reproduced asexually.

The DNA Repair Hypothesis. Several geneticists have suggested that sex occurs because only a diploid cell can effectively repair certain kinds of chromosome damage, particularly double-strand breaks in DNA. Both radiation and chemical events within cells can induce such breaks. As organisms became larger and longer-lived, it must have become increasingly important for them to be able to repair such damage. Synapsis, which in early stages of meiosis precisely aligns pairs of homologous chromosomes, may well have evolved originally as a mechanism for repairing double-strand damage to DNA. The undamaged homologous chromosome could be used as a template to repair the damaged chromosome. A transient diploid phase would have provided an opportunity for such repair. In yeast, mutations that inactivate the repair system for double-strand breaks of the chromosomes also prevent crossing over, suggesting a common mechanism for both synapsis and repair processes.

Muller's Ratchet. The geneticist Herman Muller pointed out in 1965 that asexual populations incorporate a kind of mutational ratchet mechanism—once harmful mutations arise, asexual populations have no way of eliminating them, and they accumulate over time, like turning a ratchet. Sexual populations, on the other hand, can employ recombination to generate individuals carrying fewer mutations, which selection can then favor. Sex may just be a way to keep the mutational load down.

The Red Queen Hypothesis. One evolutionary advantage of sex may be that it allows populations to "store" forms of a trait that are currently bad but have promise for reuse at some time in the future. Because populations are constrained by a changing physical and biological environment, selection is constantly acting against such traits. But in sexual species, selection can never get rid of those variants sheltered by more dominant forms of the trait.

The evolution of most sexual species, most of the time, thus manages to keep pace with ever-changing physical and biological constraints. This "treadmill evolution" is sometimes called the "Red Queen hypothesis," after the Queen of Hearts in Lewis Carroll's *Through the Looking Glass,* who tells Alice, "Now, here, you see, it takes all the running you can do, to keep in the same place."

Are New Microtubules Made When the Spindle Forms?

During interphase, before the beginning of meiosis, relatively few long microtubules extend from the centrosome (a zone around the centrioles of animal cells where microtubules are organized) to the cell periphery. Like most microtubules, these are refreshed at a low rate with resynthesis. Late in prophase, however, a dramatic change is seen—the centrosome divides into two, and a large increase is seen in the number of microtubules radiating from each of the two daughter centrosomes. The two clusters of new microtubules are easily seen as the green fibers connecting to the two sets of purple chromosomes in the micrograph of early prophase below (a **micrograph** is a photo taken through a microscope). This burst of microtubule assembly marks the beginning of the formation of the spindle characteristic of metaphase. When it first became known to cell biologists, they asked whether these were existing microtubules being repositioned in the spindle, or newly synthesized microtubules only produced just before metaphase begins.

The graph to the upper right displays the results of an experiment designed to answer this question. Mammalian cells in culture (cells **in culture** are growing in the laboratory on artificial medium) were injected with microtubule subunits (tubulin) to which a fluorescent dye had been attached (a **fluorescent dye** is one that glows when exposed to ultraviolet or short-wavelength visual light). After the fluorescent subunits had become incorporated into the cells's microtubules, all the fluorescence in a small region of a cell was bleached by an intense laser beam, destroying the microtubules there. Any subsequent rebuilding of microtubules in the bleached region would have to employ the fluorescent subunits present in the cell, causing recovery of fluorescence in the bleached region. The graph reports this recovery as a function of time, for interphase and metaphase cells. The dotted line represents the time for 50% recovery of fluorescence ($t_{1/2}$) (that is, $t_{1/2}$ is the time required for half of the microtubules in the region to be resynthesized).

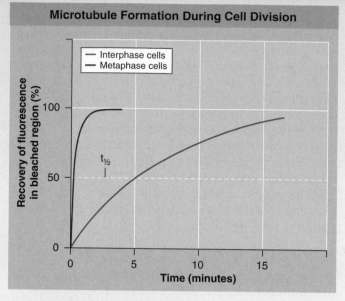

Microtubule Formation During Cell Division

1. **Applying Concepts**
   a. Variable. In the graph, what is the dependent variable?
   b. $t_{1/2}$. Are new microtubules synthesized during interphase? What is the $t_{1/2}$ of this replacement synthesis? Are new microtubules synthesized during metaphase? What is the $t_{1/2}$ of this replacement synthesis?

2. **Interpreting Data** Is there a difference in the rate at which microtubules are synthesized during interphase and metaphase? How big is the difference? What might account for it?

3. **Making Inferences**
   a. What general statement can be made regarding the relative rates of microtubule production before and during meiosis?
   b. Is there any difference in the final amount of microtubule synthesis that would occur if this experiment were to be continued for an additional 15 minutes?

4. **Drawing Conclusions** When are the microtubules of the spindle assembled?

5. **Further Analysis** The spindle breaks down after cell division is completed. Design an experiment to test whether the tubulin subunits of the spindle microtubules are recycled into other cell components, or destroyed, after meiosis.

10 µm

Meiosis

9.1 Discovery of Meiosis

- In sexually reproducing organisms, a gamete from the male fuses with a gamete from the female in a process called fertilization, or syngamy. The number of chromosomes in gametes must be halved to maintain the correct number of chromosomes in offspring (**figure 9.1**). Organisms accomplish this through a cell division process called meiosis.

- A cell that contains two copies of each chromosome is called a diploid cell. Cells, such as gametes, that contain only one copy of each chromosome are haploid cells.

- Sexual reproduction involves meiosis, but some organisms also undergo asexual reproduction, which is reproducing by mitosis or binary fission (**figure 9.2**).

9.2 The Sexual Life Cycle

- Sexual life cycles alternate between diploid and haploid stages, with variation in the amount of time devoted to each stage. Three types of sexual life cycles exist: In many protists the majority of the life cycle is devoted to the haploid stage; in most animals the majority of the life cycle is devoted to the diploid stage; and in plants and some algae the life cycle is split more equally between a haploid stage and a diploid stage (**figure 9.3**). Germ-line cells are diploid but produce haploid gametes (**figure 9.4**).

9.3 The Stages of Meiosis

- Meiosis (**Key Biological Process, page 175**) involves two nuclear divisions, meiosis I and meiosis II, each containing a prophase, metaphase, anaphase, and telophase. Like mitosis, the DNA replicates itself during interphase, before meiosis begins. Because there are two nuclear divisions but only one round of DNA replication, the four daughter cells contain half the number of chromosomes as the parent cell.

- Prophase I is distinguished by the exchange of genetic material between homologous chromosomes, a process called crossing over. In this process, homologous chromosomes align with each other along their  lengths, and sections of homologues are physically exchanged, as shown here from **figure 9.5**. This recombines the genetic information contained in the chromosomes.

- During metaphase I, microtubules in the spindle apparatus attach to homologous chromosomes, and chromosome pairs align along the metaphase plate. The alignment of the chromosomes is random, leading to the independent assortment of chromosomes into the gametes (**figure 9.6**).

- The homologous chromosomes separate during anaphase I, being pulled apart by the spindle apparatus toward their respective poles. This differs from mitosis and later in meiosis II, where sister chromatids separate in anaphase.

- In telophase I, the chromosomes cluster at the poles. This leads to meiosis II.

- Meiosis II mirrors mitosis in that it involves the separation of sister chromatids through the phases of prophase II, metaphase II, anaphase II, and telophase II. Meiosis II differs from mitosis in that there is no DNA replication before meiosis II. Homologous pairs are separated during meiosis I, such that each daughter cell, shown forming here in telophase II from **figure 9.7**, has only one-half the number of chromosomes. Also, the chromosomes in the daughter cells at the end of meiosis II are not genetically identical because of crossing over.

Comparing Meiosis and Mitosis

9.4 How Meiosis Differs from Mitosis

- Two processes that distinguish meiosis from mitosis are crossing over through synapsis and reduction division.

- When homologous chromosomes come together during prophase I, they associate with each other along their lengths, a process called synapsis (**figure 9.8a**). Synapsis does not occur in mitosis. During synapsis, sections of homologous chromosomes are physically exchanged in crossing over. Crossing over results in daughter cells that are not genetically identical to the parent cell or to each other. In contrast, mitosis results in daughter cells that are genetically identical to the parent cell and to each other.

- In meiosis, the daughter cells contain half the number of chromosomes as the parent cell due to reduction division. As shown here from **figure 9.8b**, reduction division occurs because meiosis contains two nuclear divisions but only one round of DNA replication during interphase.

- The primary reasons for the differences in meiosis and mitosis stem from the synapsis of homologous chromosomes in prophase I. Because of synapsis, the arms of homologous chromosomes are close enough to undergo crossing over. Synapsis also  blocks the inner kinetochores from attaching to the spindle. As a result, sister chromatids do not separate during meiosis I, resulting in reduction division (**figure 9.9**).

9.5 Evolutionary Consequences of Sex

- Sexual reproduction results in the introduction of genetic variation in future generations through independent assortment, crossing over, and random fertilization.

- Independent assortment results in the distribution of chromosomes into gametes, which creates many different combinations (**figure 9.10**).

- Crossing over provides even more genetic variability in gametes, such that the genetic combinations are virtually unlimited.

- The fusion of two gametes results in new genetic combinations that were created randomly, further increasing genetic diversity.

1. An egg and a sperm unite to form a new organism. To prevent the new organism from having twice as many chromosomes as its parents,
   a. half of the chromosomes in the new organism quickly disassemble, leaving the correct number.
   b. half of the chromosomes from the egg and half from the sperm are ejected from the new cell.
   c. the large egg contains all the chromosomes, the tiny sperm only contributes some DNA.
   d. the egg and sperm cells only have half the number of chromosomes found in the parents due to meiosis.

2. The diploid number of chromosomes in humans is 46. The haploid number is
   a. 138.   c. 46.
   b. 92.   d. 23.

3. In organisms that have sexual life cycles, there is a time when there are
   a. 1*n* gametes (haploid), followed by 2*n* zygotes (diploid).
   b. 2*n* gametes (haploid), followed by 1*n* zygotes (diploid).
   c. 2*n* gametes (diploid), followed by 1*n* zygotes (haploid).
   d. 1*n* gametes (diploid), followed by 2*n* zygotes (haploid).

4. Which of the following occurs in meiosis I?
   a. All chromosomes duplicate.
   b. Homologous chromosomes randomly orient themselves on the metaphase plate, called independent assortment.
   c. The duplicated sister chromatids separate.
   d. The original cell divides into four diploid cells.

5. Which of the following occurs in meiosis II?
   a. All chromosomes duplicate.
   b. Homologous chromosomes randomly separate, called independent assortment.
   c. The duplicated sister chromatids separate.
   d. Genetically identical daughter cells are produced.

6. During which stage of meiosis is crossing over initiated?
   a. prophase I   c. metaphase II
   b. anaphase I   d. interphase

7. Synapsis is the process whereby
   a. homologous pairs of chromosomes separate and migrate toward a pole.
   b. homologous chromosomes exchange chromosomal material.
   c. homologous chromosomes become closely associated along their lengths.
   d. the daughter cells contain half the number of chromosomes as the parent cell.

8. Crossing over is the process whereby
   a. homologous chromosomes cross over to opposite sides of the cell.
   b. homologous chromosomes exchange chromosomal material.
   c. homologous chromosomes become closely associated along their lengths.
   d. kinetochore fibers attach to both sides of a centromere.

9. Mitosis results in ____, while meiosis results in ____.
   a. cells that are genetically identical to the parent cell/haploid cells
   b. haploid cells/diploid cells
   c. four daughter cells/two daughter cells
   d. cells with half the number of chromosomes as the parent cell/cells that vary in chromosome number

10. A major consequence of sex and meiosis is that each species
    a. remains pretty much the same because the chromosomes are carefully duplicated and passed to the next generation.
    b. has a lot of genetic reassortment due to processes in meiosis II.
    c. has a lot of genetic reassortment due to processes in meiosis I.
    d. has a lot of genetic reassortment due to processes in telophase II.

Apply Your Understanding

1. **Figure 9.5** How is it that in meiosis you can end up with four "daughter cells" that are all genetically different from one another?

Adjacent homologues (nonsister chromatids)   Centromere

2. **Figure 9.8a** Referring to the homologous chromosomes shown here during prophase I, and knowing that they stay in synapsis during metaphase I, explain why it is that the sister chromatids don't separate as they do in mitosis.

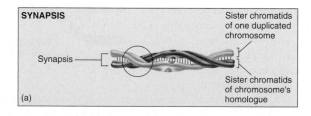

SYNAPSIS

Synapsis

Sister chromatids of one duplicated chromosome

Sister chromatids of chromosome's homologue

(a)

Synthesize What You Have Learned

1. It would seem that you only need one set of instructions for your body to do all the jobs it needs to carry out. So why aren't organisms simply haploid all their lives?

2. Are the gamete cells of your body haploid or diploid? Why not the alternative?

3. An organism has 56 chromosomes in its diploid stage. Indicate how many chromosomes are present in the following, and explain your reasoning:
   a. somatic cells   c. metaphase II
   b. metaphase I   d. gametes

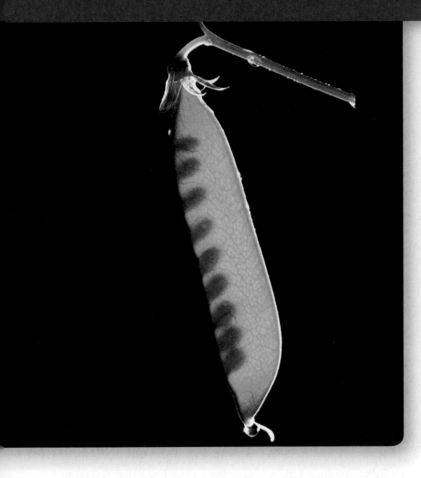

10
Foundations of Genetics

I n this pea pod, you can see the shadowy outlines of seeds that will form part of the next generation of this pea plant. While the seeds appear similar to one another, the plants they produce may differ in significant ways. This is because the gametes that produced the seeds contribute chromosomes from both parents, in effect "shuffling the deck of cards" so that a progeny plant will have some characteristics from one parent and some from the other. About 150 years ago, Gregor Mendel first described this process, before anyone knew what genes or chromosomes were. We now understand the process of heredity in considerable detail, and can begin to devise ways of treating some of the disorders that arise in people when particular genes are damaged in germ-line tissue. In this chapter you will watch as Mendel experiments with pea plants like the one above. Unlike researchers before him, Mendel carefully counted the number of each kind of pea plant his experiments produced and, looking at his results, saw a beautiful simplicity. The theory he proposed to explain it has become one of the key principles of biology.

10.1 Mendel and the Garden Pea

When you were born, many things about you resembled your mother or father. This tendency for traits to be passed from parent to offspring is called **heredity.** *Traits* are alternative forms of a character, or heritable feature. How does heredity happen? Before DNA and chromosomes were discovered, this puzzle was one of the greatest mysteries of science. The key to understanding the puzzle of heredity was found in the garden of an Austrian monastery over a century ago by a monk named Gregor Mendel (**figure 10.1**). Mendel used the scientific process described in chapter 1 as a powerful way of analyzing the problem. Crossing pea plants with one another, Mendel made observations that allowed him to form a simple but powerful hypothesis that accurately predicted patterns of heredity—that is, how many offspring would be like one parent and how many like the other. When Mendel's rules, introduced in chapter 1 as the theory of heredity, became widely known, investigators all over the world set out to discover the physical mechanism responsible for them. They learned that hereditary traits are instructions carefully laid out in the DNA a child receives from each parent. Mendel's solution to the puzzle of heredity was the first step on this journey of understanding and one of the greatest intellectual accomplishments in the history of science.

Figure 10.1   Gregor Mendel.
The key to understanding the puzzle of heredity was solved by Mendel, who cultivated pea plants in the garden of his monastery in Brünn, Austria.

Early Ideas About Heredity

Mendel was not the first person to try to understand heredity by crossing pea plants. Over 200 years earlier, British farmers had performed similar crosses and obtained results similar to Mendel's. They observed that in crosses between two types—tall and short plants, say—one type would disappear in one generation, only to reappear in the next. In the 1790s, for example, the British farmer T. A. Knight crossed a variety of the garden pea that had purple flowers with one that had white flowers. All the offspring of the cross had purple flowers. If two of these offspring were crossed, however, some of *their* offspring were purple and some were white. Knight noted that the purple had a "stronger tendency" to appear than white, but he did not count the numbers of each kind of offspring.

Mendel's Experiments

Gregor Mendel was born in 1822 to peasant parents and was educated in a monastery. He became a monk and was sent to the University of Vienna to study science and mathematics. Although he aspired to become a scientist and teacher, he failed his university exams for a teaching certificate and returned to the monastery, where he spent the rest of his life, eventually becoming abbot. Upon his return, Mendel joined an informal neighborhood science club, a group of farmers and others interested in science. Under the patronage of a local nobleman, each member set out to undertake scientific investigations, which were then discussed at meetings and published in the club's own journal. Mendel undertook to repeat the classic series of crosses with pea plants done by Knight and others, but this time he intended to count the numbers of each kind of offspring in the hope that the numbers would give some hint of what was going on. Quantitative approaches to science—measuring and counting—were just becoming fashionable in Europe.

Mendel's Experimental System: The Garden Pea

Mendel chose to study the garden pea because several of its characteristics made it easy to work with:

1. Many varieties were available. Mendel selected seven pairs of lines that differed in easily distinguished traits (including the white versus purple flowers that Knight had studied 60 years earlier).

2. Mendel knew from the work of Knight and others that he could expect the infrequent version of a character to disappear in one generation and reappear in the next. He knew, in other words, that he would have something to count.

3. Pea plants are small, easy to grow, produce large numbers of offspring, and mature quickly.

4. The reproductive organs of peas are enclosed within their flowers. **Figure 10.2** shows a cutaway view of the flower so that you can see the anther that holds the pollen and the carpel that holds the egg. Left alone, the flowers do not open. They simply fertilize themselves with their own pollen (male gametes). To carry out a cross, Mendel had only to pry the petals apart, reach in with a scissors, and snip off the male

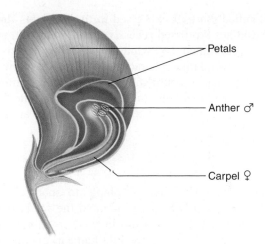

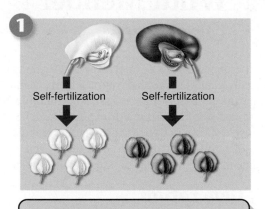

Figure 10.2   The garden pea.

Because it is easy to cultivate and because there are many distinctive varieties, the garden pea, *Pisum sativum,* was a popular choice as an experimental subject in investigations of heredity for as long as a century before Mendel's studies.

organs (anthers); he could then dust the female organs (the tip of the carpel) with pollen from another plant to make the cross.

Mendel's Experimental Design

Mendel's experimental design was the same as Knight's, only Mendel counted his plants. The crosses were carried out in three steps that are presented in the three panels in figure 10.3:

1. Mendel began by letting each variety self-fertilize for several generations. This ensured that each variety was **true-breeding,** meaning that it contained no other varieties of the trait, and so would produce only offspring of the same variety when it self-pollinated. The white flower variety, for example, produced only white flowers and no purple ones in each generation. Mendel called these lines the **P generation** (P for parental).

2. Mendel then conducted his experiment: He crossed two pea varieties exhibiting alternative traits, such as white versus purple flowers. The offspring that resulted he called the **F_1 generation** (F_1 for "first filial" generation, from the Latin word for "son" or "daughter").

3. Finally, Mendel allowed the plants produced in the crosses of step 2 to self-fertilize, and he counted the numbers of each kind of offspring that resulted in this **F_2** ("second filial") **generation.** As reported by Knight, the white flower trait reappeared in the F_2 generation, although not as frequently as the purple flower trait.

Key Learning Outcome 10.1 Mendel studied heredity by crossing true-breeding garden peas that differed in easily scored alternative traits and then allowing the offspring to self-fertilize.

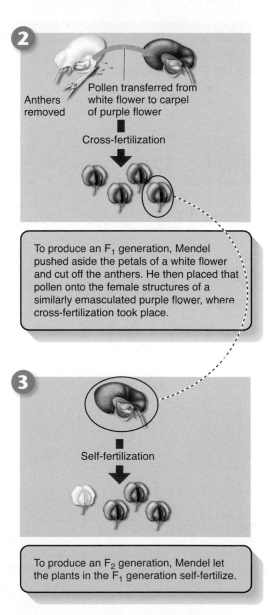

Figure 10.3   How Mendel conducted his experiments.

10.2 What Mendel Observed

Mendel experimented with a variety of traits in the garden pea and repeatedly made similar observations. In all, Mendel examined seven pairs of contrasting traits as shown in **table 10.1**. For each pair of contrasting traits that Mendel crossed he observed the same result, shown in **figure 10.3**, where a trait disappeared in the F$_1$ generation only to reappear in the F$_2$ generation. We will examine in detail Mendel's crosses with flower color.

The F$_1$ Generation

In the case of flower color, when Mendel crossed purple and white flowers, all the F$_1$ generation plants he observed were purple; he did not see the contrasting trait, white flowers.

Mendel called the trait expressed in the F$_1$ plants **dominant** and the trait not expressed **recessive.** In this case, purple flower color was dominant and white flower color recessive. Mendel studied several other characters in addition to flower color, and for every pair of contrasting traits Mendel examined, one proved to be dominant and the other recessive. The dominant and recessive traits for each character he studied are indicated in **table 10.1**.

The F$_2$ Generation

After allowing individual F$_1$ plants to mature and self-fertilize, Mendel collected and planted the seeds from each plant to see what the offspring in the F$_2$ generation would look like. Mendel found (as Knight had earlier) that some F$_2$ plants exhibited white flowers, the recessive trait. The recessive trait had disappeared in the F$_1$ generation, only to reap-

TABLE 10.1	SEVEN CHARACTERS MENDEL STUDIED IN HIS EXPERIMENTS					
	Character				F$_2$ Generation	
	Dominant Form	×	Recessive Form		Dominant: Recessive	Ratio
	Purple flowers	×	White flowers		705:224	3.15:1 (3/4:1/4)
	Yellow seeds	×	Green seeds		6,022:2,001	3.01:1 (3/4:1/4)
	Round seeds	×	Wrinkled seeds		5,474:1,850	2.96:1 (3/4:1/4)
	Green pods	×	Yellow pods		428:152	2.82:1 (3/4:1/4)
	Inflated pods	×	Constricted pods		882:299	2.95:1 (3/4:1/4)
	Axial flowers	×	Terminal flowers		651:207	3.14:1 (3/4:1/4)
	Tall plants	×	Dwarf plants		787:277	2.84:1 (3/4:1/4)

Figure 10.4 Round versus wrinkled seeds.
One of the differences among varieties of pea plants that Mendel studied was the shape of the seed. In some varieties the seeds were round, whereas in others they were wrinkled.

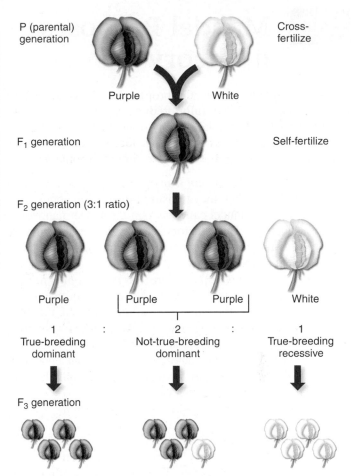

Figure 10.5 The F$_2$ generation is a disguised 1:2:1 ratio.
By allowing the F$_2$ generation to self-fertilize, Mendel found from the offspring (F$_3$) that the ratio of F$_2$ plants was one true-breeding dominant, two not-true-breeding dominant, and one true-breeding recessive.

pear in the F$_2$ generation. It must somehow have been present in the F$_1$ individuals but unexpressed!

At this stage Mendel instituted his radical change in experimental design. He *counted* the number of each type among the F$_2$ offspring. He believed the proportions of the F$_2$ types would provide some clue about the mechanism of heredity. In the cross between the purple-flowered F$_1$ plants, he counted a total of 929 F$_2$ individuals (see **table 10.1**). Of these, 705 (75.9%) had purple flowers and 224 (24.1%) had white flowers. Approximately one-fourth of the F$_2$ individuals exhibited the recessive form of the trait. Mendel carried out similar experiments with other traits, such as round versus wrinkled seeds (**figure 10.4**) and obtained the same result: Three-fourths of the F$_2$ individuals exhibited the dominant form of the character, and one-fourth displayed the recessive form. In other words, the dominant:recessive ratio among the F$_2$ plants was always close to 3:1.

A Disguised 1:2:1 Ratio

Mendel let the F$_2$ plants self-fertilize for another generation and found that the one-fourth that were recessive were true-breeding—future generations showed nothing but the recessive trait. Thus, the white F$_2$ individuals described previously showed only white flowers in the F$_3$ generation (as shown on the right in **figure 10.5**). Among the three-fourths of the plants that had shown the dominant trait in the F$_2$ generation, only one-third of the individuals were true-breeding in the F$_3$ generation (as shown on the left). The others showed both traits in the F$_3$ generation (as shown in the center)—and when Mendel counted their numbers, he found the ratio of dominant to recessive to again be 3:1! From these results Mendel concluded

that the 3:1 ratio he had observed in the F$_2$ generation was in fact a disguised 1:2:1 ratio:

1   **2**   **1**

true-breeding : not-true-breeding : true-breeding
dominant   **dominant**   **recessive**

Key Learning Outcome 10.2 When Mendel crossed two contrasting traits and counted the offspring in the subsequent generations, he observed that all of the offspring in the first generation exhibited one (dominant) trait, and none exhibited the other (recessive) trait. In the following generation, 25% were true-breeding for the dominant trait, 50% were not-true-breeding and appeared dominant, and 25% were true-breeding for the recessive trait.

10.3 Mendel Proposes a Theory

To explain his results, Mendel proposed a simple set of hypotheses that would faithfully predict the results he had observed. Now called Mendel's theory of heredity, it has become one of the most famous theories in the history of science. Mendel's theory is composed of five simple hypotheses:

Hypothesis 1: *Parents do not transmit traits directly to their offspring.* Rather, they transmit information about the traits, what Mendel called *merkmal* (the German word for "factor"). These factors act later, in the offspring, to produce the trait. In modern terminology, we call Mendel's factors **genes.**

Hypothesis 2: *Each parent contains two copies of the factor governing each trait.* The two copies may or may not be the same. If the two copies of the factor are the same (both encoding purple or both white flowers, for example), the individual is said to be **homozygous.** If the two copies of the factor are different (one encoding purple, the other white, for example), the individual is said to be **heterozygous.**

Hypothesis 3: *Alternative forms of a factor lead to alternative traits.* Alternative forms of a factor are called **alleles.** Mendel used lowercase letters to represent recessive alleles and uppercase letters to represent dominant ones. Thus, in the case of purple flowers, the dominant purple flower allele is represented as *P* and the recessive white flower allele is represented as *p*. In modern terms, we call the appearance of an individual, such as possessing white flowers, its **phenotype.** Appearance is determined by which alleles of the flower-color gene the plant receives from its parents, and we call those particular alleles the individual's **genotype.** Thus a pea plant might have the phenotype "white flower" and the genotype *pp*.

Hypothesis 4: *The two alleles that an individual possesses do not affect each other,* any more than two letters in a mailbox alter each other's contents. Each allele is passed on unchanged when the individual matures and produces its own gametes (egg and sperm). At the time, Mendel did not know that his factors were carried from parent to offspring on chromosomes. **Figure 10.6** shows a modern view of how genes are carried on chromosomes, with homologous chromosome carrying the same genes but not necessarily the same alleles. The location of a gene on a chromosome is called its *locus* (plural, *loci*).

Hypothesis 5: *The presence of an allele does not ensure that a trait will be expressed in the individual that carries it.* In heterozygous individuals, only the dominant allele achieves expression; the recessive allele is present but unexpressed.

These five hypotheses, taken together, constitute Mendel's model of the hereditary process. Many traits in humans exhibit dominant or recessive inheritance similar to the traits Mendel studied in peas (**table 10.2**).

Analyzing Mendel's Results

To analyze Mendel's results, it is important to remember that each trait is determined by the inheritance of alleles from the parents, one allele from the mother and the other from the father. These alleles, present on chromosomes, are distributed to gametes during meiosis. Each gamete receives one copy of each chromosome, and therefore one of the alleles.

Consider again Mendel's cross of purple-flowered with white-flowered plants. Like Mendel, we will assign the symbol *P* to the dominant allele, associated with the production of purple flowers, and the symbol *p* to the recessive allele, associated with the production of white flowers. As described

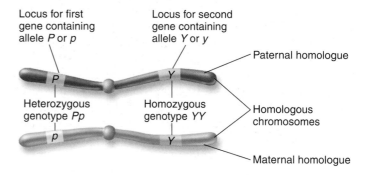

Figure 10.6 Alternative alleles of genes are located on homologous chromosomes.

TABLE 10.2	SOME DOMINANT AND RECESSIVE TRAITS IN HUMANS		
Recessive Traits	**Phenotypes**	**Dominant Traits**	**Phenotypes**
Common baldness	M-shaped hairline receding with age	Mid-digital hair	Presence of hair on middle segment of fingers
Albinism	Lack of melanin pigmentation	Brachydactyly	Short fingers
Alkaptonuria	Inability to metabolize homogentisic acid	Phenylthiocarbamide (PTC) sensitivity	Ability to taste PTC as bitter
Red-green color blindness	Inability to distinguish red and green wavelengths of light	Camptodactyly	Inability to straighten the little finger
		Polydactyly	Extra fingers and toes

earlier, by convention, genetic traits are usually assigned a letter symbol referring to their more common forms, in this case "*P*" for purple flower color. The dominant allele is written in uppercase, as *P;* the recessive allele (white flower color) is assigned the same symbol in lowercase, *p.*

In this system, the genotype of an individual true-breeding for the recessive white-flowered trait would be designated *pp.* In such an individual, both copies of the allele specify the white-flowered phenotype. Similarly, the genotype of a true-breeding purple-flowered individual would be designated *PP,* and a heterozygote would be designated *Pp* (dominant allele first). Using these conventions, and denoting a cross between two strains with ×, we can symbolize Mendel's original cross as *pp × PP.*

Punnett Squares

The possible results from a cross between a true-breeding, white-flowered plant (*pp*) and a true-breeding, purple-flowered plant (*PP*) can be visualized with a **Punnett square.** In a Punnett square, the possible gametes of one individual are listed along the horizontal side of the square, while the possible gametes of the other individual are listed along the vertical side. The genotypes of potential offspring are represented by the cells within the square. **Figure 10.7** walks you through the set-up of a Punnett square crossing two individual plants that are heterozygous for flower color (*Pp × Pp*). The genotypes of the parents are placed along the top and side and the genotypes of potential offspring appear in the cells.

The frequency that these genotypes occur in the offspring is usually expressed by a **probability.** For example, in a cross between a homozygous white-flowered plant (*pp*) and a homozygous purple-flowered plant (*PP*), *Pp* is the only possible genotype for all individuals in the F_1 generation as shown by the Punnett square on the left of **figure 10.8.** Because *P* is dominant to *p,* all individuals in the F_1 generation have purple flowers. When individuals from the F_1 generation

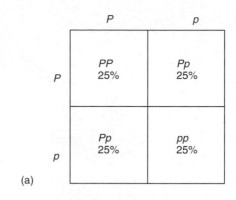

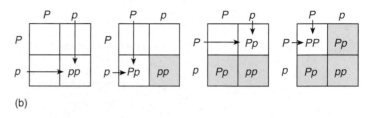

(b)

Figure 10.7 A Punnett square analysis.

(*a*) Each square represents 1/4 or 25% of the offspring from the cross. The squares in (*b*) show how the square is used to predict the genotypes of all potential offspring.

are crossed, as shown by the Punnett square on the right, the probability of obtaining a homozygous dominant (*PP*) individual in the F_2 is 25% because one-fourth of the possible genotypes are *PP.* Similarly, the probability of an individual in the F_2 generation being homozygous recessive (*pp*) is 25%. Because the heterozygous genotype has two possible ways of occurring (*Pp* and *pP*), it occurs in half of the cells within the square; the probability of obtaining a heterozygous (*Pp*) individual in the F_2 is 50% (25% + 25%).

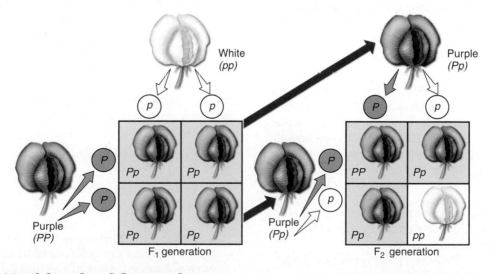

Figure 10.8 How Mendel analyzed flower color.

The only possible offspring of the first cross are *Pp* heterozygotes, purple in color. These individuals are known as the F_1 generation. When two heterozygous F_1 individuals cross, three kinds of offspring are possible: *PP* homozygotes (purple flowers); *Pp* heterozygotes (also purple flowers), which may form two ways; and *pp* homozygotes (white flowers). Among these individuals, known as the F_2 generation, the ratio of dominant phenotype to recessive phenotype is 3:1.

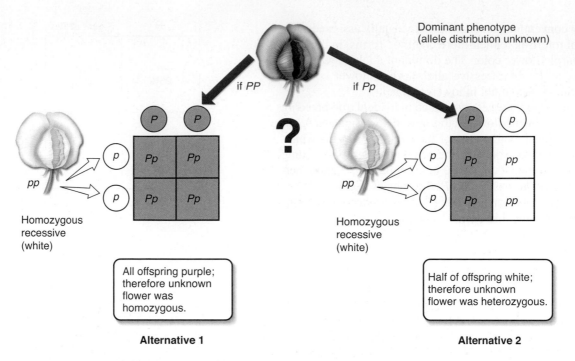

Figure 10.9 How Mendel used the testcross to detect heterozygotes.

To determine whether an individual exhibiting a dominant phenotype, such as purple flowers, was homozygous (*PP*) or heterozygous (*Pp*), Mendel devised the testcross. He crossed the individual with a known homozygous recessive (*pp*)—in this case, a plant with white flowers.

The Testcross

How did Mendel know which of the purple-flowered individuals in the F₂ generation (or the P generation) were homozygous (*PP*) and which were heterozygous (*Pp*)? It is not possible to tell simply by looking at them. For this reason, Mendel devised a simple and powerful procedure called the **testcross** to determine an individual's actual genetic composition. Consider a purple-flowered plant. It is impossible to tell whether such a plant is homozygous or heterozygous simply by looking at its phenotype. To learn its genotype, you must cross it with some other plant. What kind of cross would provide the answer? If you cross it with a homozygous dominant individual, all of the progeny will show the dominant phenotype whether the test plant is homozygous or heterozygous. It is also difficult (but not impossible) to distinguish between the two possible test plant genotypes by crossing with a heterozygous individual. However, if you cross the test plant with a homozygous recessive individual, the two possible test plant genotypes will give totally different results. To see how this works, step through a testcross of a purple-flowered plant with a white-flowered plant. **Figure 10.9** shows you the two possible alternatives:

Alternative 1 (on left): Unknown plant is homozygous (*PP*). *PP* × *pp:* All offspring have purple flowers (*Pp*) as shown by the four purple squares.

Alternative 2 (on right): Unknown plant is heterozygous (*Pp*). *Pp* × *pp:* One-half of offspring have white flowers (*pp*) and one-half have purple flowers (*Pp*) as shown by the two white and two purple squares.

To perform his testcross, Mendel crossed heterozygous individuals exhibiting the dominant trait back to the parent homozygous for the recessive trait. He predicted that the dominant and recessive traits would appear in a 1:1 ratio, and that is what he observed, as you can see illustrated in alternative 2 above.

For each pair of alleles he investigated, Mendel observed phenotypic F₂ ratios of 3:1 (see **table 10.1**) and testcross ratios very close to 1:1, just as his model predicted.

Testcrosses can also be used to determine the genotype of an individual when two genes are involved. Mendel carried out many two-gene crosses, some of which we will soon discuss. He often used testcrosses to verify the genotypes of particular dominant-appearing F₂ individuals. Thus an F₂ individual showing both dominant traits (*A_ B_*) might have any of the following genotypes: *AABB*, *AaBB*, *AABb*, or *AaBb*. By crossing dominant-appearing F₂ individuals with homozygous recessive individuals (that is, *A_ B_* × *aabb*), Mendel was able to determine if either or both of the traits bred true among the progeny and so determine the genotype of the F₂ parent.

AABB	trait A breeds true	trait B breeds true
AaBB		trait B breeds true
AABb	trait A breeds true	
AaBb		

Key Learning Outcome 10.3 The genes that an individual has are referred to as its genotype; the outward appearance of the individual is referred to as its phenotype. The phenotype is determined by the alleles inherited from the parents. Analyses using Punnett squares determine all possible genotypes of a particular cross. A test cross determines the genotype of a dominant trait.

10.4 Mendel's Laws

Mendel's First Law: Segregation

Mendel's model brilliantly predicts the results of his crosses, accounting in a neat and satisfying way for the ratios he observed. Similar patterns of heredity have since been observed in countless other organisms. Traits exhibiting this pattern of heredity are called *Mendelian traits.* Because of its overwhelming importance, Mendel's theory is often referred to as Mendel's first law, or the **law of segregation.** In modern terms, Mendel's first law states that *the two alleles of a trait separate from each other during the formation of gametes, so that half of the gametes will carry one copy and half will carry the other copy.*

Mendel's Second Law: Independent Assortment

Mendel went on to ask if the inheritance of one factor, such as flower color, influences the inheritance of other factors, such as plant height. To investigate this question, he first established a series of true-breeding lines of peas that differed from one another with respect to two of the seven pairs of characteristics he had studied. He then crossed contrasting pairs of true-breeding lines. **Figure 10.10** shows an experiment in which the P generation consists of homozygous individuals with round, yellow seeds (*RRYY* in the figure) that are crossed with individuals that are homozygous for wrinkled, green seeds (*rryy*). This cross produces offspring that have round, yellow seeds and are heterozygous for both of these traits (*RrYy*). Such F_1 individuals are said to be **dihybrid.** The chromosomes are then allocated to the gametes during meiosis such that there are four types of gametes for these two traits.

Mendel then allowed the dihybrid individuals to self-fertilize. If the segregation of alleles affecting seed shape and alleles affecting seed color were independent, the probability that a particular pair of seed-shape alleles would occur together with a particular pair of seed-color alleles would simply be a product of the two individual probabilities that each pair would occur separately. For example, the probability of an individual with wrinkled, green seeds appearing in the F_2 generation would be equal to the probability of an individual with wrinkled seeds (1 in 4) multiplied by the probability of an individual with green seeds (1 in 4), or 1 in 16.

In his dihybrid crosses, Mendel found that the frequency of phenotypes in the F_2 offspring closely matched the 9:3:3:1 ratio predicted by the Punnett square analysis shown in **figure 10.10.** He concluded that for the pairs of traits he studied, the inheritance of one trait does not influence the inheritance of the other trait, a result often referred to as Mendel's second law, or the **law of independent assortment.** We now know that this result is only valid for genes not located near one another on the same chromosome. Thus in modern terms, Mendel's second law is often stated as follows: *Genes located on different chromosomes are inherited independently of one another.*

Mendel's paper describing his results was published in the journal of his local scientific society in 1866. Unfortunately,

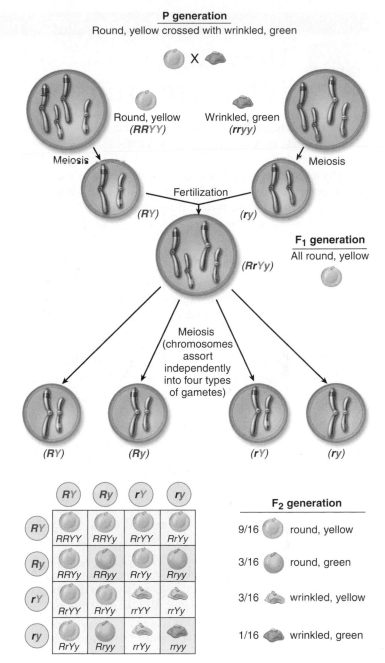

Figure 10.10 Analysis of a dihybrid cross.
This dihybrid cross shows round (*R*) versus wrinkled (*r*) seeds and yellow (*Y*) versus green (*y*) seeds. The ratio of the four possible phenotypes in the F_2 generation is predicted to be 9:3:3:1.

his paper failed to arouse much interest, and his work was forgotten. Sixteen years after his death, in 1900, several investigators independently rediscovered Mendel's pioneering paper while searching the literature in preparation for publishing their own findings, which were similar to those Mendel had quietly presented more than three decades earlier.

> **Key Learning Outcome 10.4** Mendel's theories of segregation and independent assortment are so well supported by experimental results that they are considered "laws."

10.5 How Genes Influence Traits

It is useful, before considering Mendelian genetics further, to gain a brief overview of how genes work. With this in mind, we will sketch, in broad strokes, a picture of how a Mendelian trait is influenced by a particular gene, how a gene can be altered by mutation, and the potential long-term evolutionary consequences of such an alteration. We will use the protein hemoglobin as our example—you can follow along on **figure 10.11** starting at the bottom.

From DNA to Protein

Each body cell of an individual contains the same set of DNA molecules, called the genome of that individual. As you learned in chapter 3, DNA molecules are composed of two strands twisted about each other, each the mirror image of the other. Each strand is a long chain of nucleotide subunits that are linked together. There are four kinds of nucleotides (A, T, C, and G), and like an alphabet with four letters, the order of nucleotides determines the message encoded in the DNA of a gene.

The human genome contains 20,000 to 25,000 genes. The DNA of the human genome is parcelled out into 23 pairs of chromosomes, each chromosome containing from 1,000 to 2,000 different genes. The bands on the chromosome in **figure 10.11** indicate areas that are rich in genes. You can see in the figure that the hemoglobin gene is located on chromosome 11.

At the next level in the figure, individual genes are "read" from the chromosomal DNA by enzymes that create an RNA transcript of the nucleotide sequence (except U is substituted for T). This RNA transcript of the hemoglobin (*Hb*) gene leaves the cell nucleus and acts as a work order for protein production in other parts of the cell. But, in eukaryotic cells, the RNA transcript has more information than is needed, so it is first "edited" to remove unnecessary bits before it leaves the nucleus. For example, the initial RNA gene transcript encoding the beta-subunit of the protein hemoglobin is 1,660 nucleotides long; after "editing," the resulting "messenger" RNA is 1,000 nucleotides long—you can see in the figure that the Hb mRNA is shorter than the RNA transcript of *Hb* gene.

After an RNA transcript is edited, it leaves the nucleus as messenger RNA (mRNA) and is delivered to ribosomes in the cytoplasm. Each ribosome is a tiny protein-assembly plant, and uses the sequence of the messenger RNA to determine the amino acid sequence of a particular polypeptide. In the case of beta-hemoglobin, the messenger RNA encodes a polypeptide strand of 146 amino acids.

How Proteins Determine the Phenotype

As we saw in chapter 3, polypeptide chains of amino acids, which in the figure resemble beads on a string, spontaneously fold in water into complex three-dimensional shapes. The beta-hemoglobin polypeptide folds into a compact mass that associates with three others to form an active hemoglobin protein molecule that is present in red blood cells. In the figure, each hemoglobin molecule binds oxygen (a process described fully in chapter 24) in the oxygen-rich environment of the lungs, and releases oxygen in the oxygen-poor environment of active tissues.

The oxygen-binding efficiency of the hemoglobin proteins in a person's bloodstream has a great deal to do with how well the body functions, particularly under conditions of strenuous physical activity, when delivery of oxygen to the body's muscles is the chief factor limiting the activity.

As a general rule, genes influence the phenotype by specifying the kind of proteins present in the body, which determines in large measure how that body functions.

How Mutation Alters Phenotype

A change in the identity of a single nucleotide within a gene, called a mutation, can have a profound effect if the change alters the identity of the amino acid encoded there. When a mutation of this sort occurs, the new version of the protein may fold differently, altering or destroying its function. For example, how well the hemoglobin protein performs its oxygen-binding duties depends a great deal on the precise shape that the protein assumes when it folds. A change in the identity of a single amino acid can have a drastic impact on that final shape. In particular, a change in the sixth amino acid of beta-hemoglobin from glutamic acid to valine causes the hemoglobin molecules to aggregate into stiff rods that deform blood cells into a sickle shape that can no longer carry oxygen efficiently. The resulting sickle-cell disease can be fatal.

Natural Selection for Alternative Phenotypes Leads to Evolution

Because random mutations occur in all genes occasionally, populations usually contain several versions of a gene, usually all but one of them rare. Sometimes the environment changes in such a way that one of the rare versions functions better under the new conditions. When that happens, natural selection will favor the rare allele, which will then become more common. The sickle-cell version of the beta-hemoglobin gene, rare throughout most of the world, is common in Central Africa because heterozygous individuals obtain enough functional hemoglobin from their one normal allele to get along, but are resistant to malaria, a deadly disease common there, due to their other sickle-cell allele.

Key Learning Outcome 10.5 Genes determine phenotypes by specifying the amino acid sequences, and thus the functional shapes, of the proteins that carry out cell activities. Mutations, by altering protein sequence, can change a protein's function and thus alter the phenotype in evolutionarily significant ways.

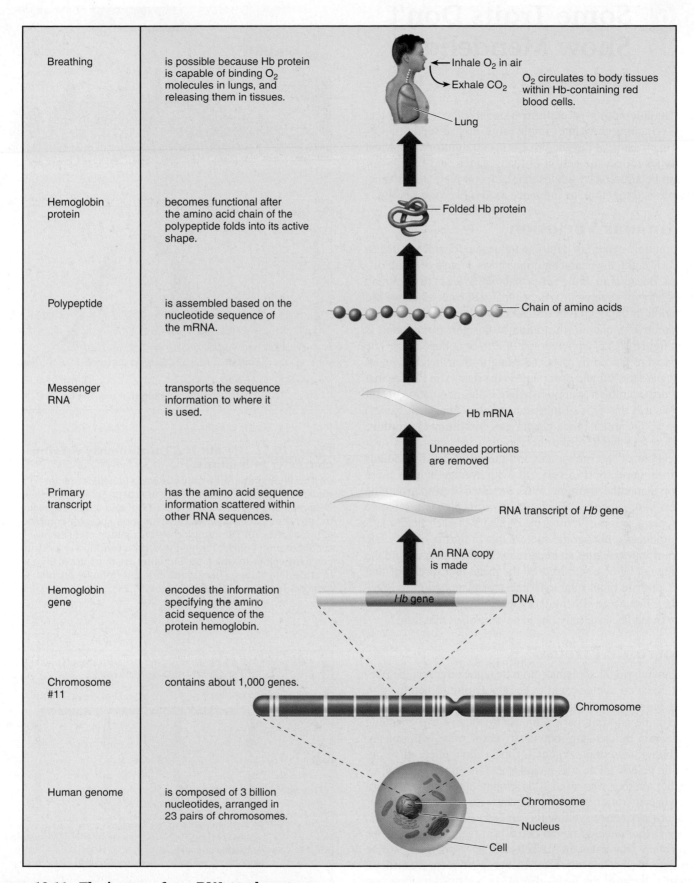

Figure 10.11 The journey from DNA to phenotype.

What an organism is like is determined in large measure by its genes. Here you see how one gene of the 20,000 to 25,000 in the human genome plays a key role in allowing oxygen to be carried throughout your body. The many steps on the journey from gene to trait are the subject of chapters 11 and 12.

10.6 Some Traits Don't Show Mendelian Inheritance

Scientists attempting to confirm Mendel's theory often had trouble obtaining the same simple ratios he had reported. Often the expression of the genotype is not straightforward. Most phenotypes reflect the action of many genes, and the phenotype can be affected by alleles that lack complete dominance, are expressed together, or influence each other's expression.

Continuous Variation

When multiple genes act jointly to influence a character such as height or weight, the character often shows a range of small differences. Because all of the genes that play a role in determining these phenotypes segregate independently of each other, we see a gradation in the degree of difference when many individuals are examined. A classic illustration of this sort of variation is seen in **figure 10.12**, a photograph of a 1914 college class. The students were placed in rows according to their heights, under 5 feet toward the left and over 6 feet to the right. You can see that there is considerable variation in height in this population of students. We call this type of inheritance **polygenic** (many genes) and we call this gradation in phenotypes **continuous variation.**

How can we describe the variation in a character such as the height of the individuals in **figure 10.12a**? Individuals range from quite short to very tall, with average heights more common than either extreme. What we often do is to group the variation into categories. Each height, in inches, is a separate phenotypic category. Plotting the numbers in each height category produces a histogram, such as that in **figure 10.12b**. The histogram approximates an idealized bell-shaped curve, and the variation can be characterized by the mean and spread of that curve. Compare this to the inheritance of plant height in Mendel's peas; they were either tall or dwarf, no intermediate height plants existed because only one gene controlled that trait.

Pleiotropic Effects

Often, an individual allele has more than one effect on the phenotype. Such an allele is said to be **pleiotropic.** When the pioneering French geneticist Lucien Cuenot studied yellow fur in mice, a dominant trait, he was unable to obtain a true-breeding yellow strain by crossing individual yellow mice with one another. Individuals homozygous for the yellow allele died, because the yellow allele was pleiotropic: One effect was yellow color, but another was a lethal developmental defect. A pleiotropic gene alteration may be dominant with respect to one phenotypic consequence (yellow fur) and recessive with respect to another (lethal developmental defect). In pleiotropy, one gene affects many characters, in marked contrast to polygeny, where many genes affect one character. Pleiotropic effects are difficult to predict, because the genes that affect a character often perform other functions we may know nothing about.

Pleiotropic effects are characteristic of many inherited disorders, such as cystic fibrosis and sickle-cell disease,

(a)

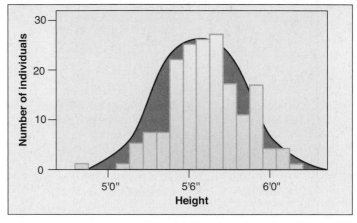

(b)

Figure 10.12   Height is a continuously varying character in humans.

(a) This photograph shows the variation in height among students of the 1914 class of the Connecticut Agricultural College. Because many genes contribute to height and tend to segregate independently of each other, there are many possible combinations of those genes. (b) The cumulative contribution of different combinations of alleles for height forms a continuous spectrum of possible heights, in which the extremes are much rarer than the intermediate values. This is quite different from the 3:1 ratio seen in Mendel's F_2 peas.

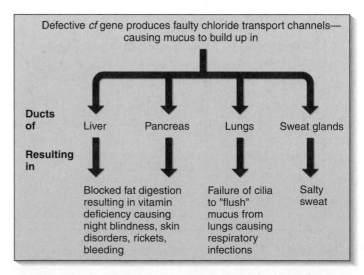

Figure 10.13   Pleiotropic effects of the cystic fibrosis gene, *cf*.

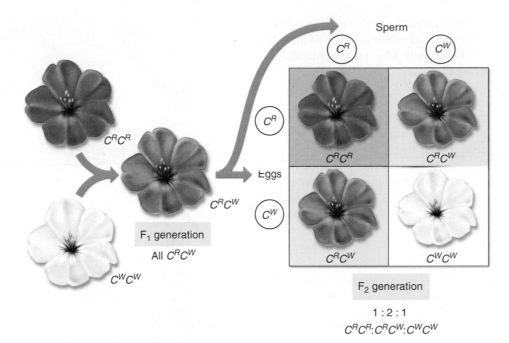

Sperm

C^R C^W

C^R

Eggs

C^W

C^RC^R C^RC^W

C^RC^W C^WC^W

C^RC^R

C^RC^W

F₁ generation

All C^RC^W

C^WC^W

F₂ generation

1 : 2 : 1

$C^RC^R:C^RC^W:C^WC^W$

Figure 10.14 Incomplete dominance.
In a cross between a red-flowered Japanese four o'clock, genotype C^RC^R, and a white-flowered one (C^WC^W), neither allele is dominant. The heterozygous progeny have pink flowers and the genotype C^RC^W. If two of these heterozygotes are crossed, the phenotypes of their progeny occur in a ratio of 1:2:1 (red:pink:white).

discussed later in this chapter. In these disorders, multiple symptoms can be traced back to a single gene defect. As shown in **figure 10.13,** cystic fibrosis patients exhibit overly sticky mucus, salty sweat, liver and pancreas failure, and a battery of other symptoms. All are pleiotropic effects of a single defect, a mutation in a gene that encodes a chloride ion transmembrane channel. In sickle-cell disease, a defect in the oxygen-carrying hemoglobin molecule causes anemia, heart failure, increased susceptibility to pneumonia, kidney failure, enlargement of the spleen, and many other symptoms. It is usually difficult to deduce the nature of the primary defect from the range of its pleiotropic effects.

Incomplete Dominance

Not all alternative alleles are fully dominant or fully recessive in heterozygotes. Some pairs of alleles exhibit **incomplete dominance** and produce a heterozygous phenotype that is intermediate between those of the parents. For example, the cross of red- and white-flowered Japanese four o'clocks described in **figure 10.14** produced red-, pink-, and white-flowered F₂ plants in a 1:2:1 ratio—heterozygotes are intermediate in color. This is different than in Mendel's pea plants that didn't exhibit incomplete dominance; the heterozygotes expressed the dominant phenotype.

Environmental Effects

The degree to which many alleles are expressed depends on the environment. Some alleles are heat-sensitive, for example. Traits influenced by such alleles are more sensitive to temperature or light than are the products of other alleles. The arctic fox in **figure 10.15,** for example, makes fur pigment only when the weather is warm. Can you see why

(a)

(b)

Figure 10.15 Environmental effects on an allele.
(a) An arctic fox in winter has a coat that is almost white, so it is difficult to see the fox against a snowy background. (b) In summer, the same fox's fur darkens to a reddish brown, so that it resembles the color of the surrounding tundra.

this trait would be an advantage for the fox? Imagine a fox that didn't possess this trait and was white all year round. It would be very visible to predators in the summer, standing out against its darker surroundings. Similarly, the *ch* allele in Himalayan rabbits and Siamese cats encodes a heat-sensitive version of tyrosinase, one of the enzymes mediating the production of melanin, a dark pigment. The ch version of the enzyme is inactivated at temperatures above about 33°C. At the surface of the main body and head, the temperature is above 33°C and the tyrosinase enzyme is inactive, while it is more active at body extremities such as the tips of the ears and tail, where the temperature is below 33°C. The dark melanin pigment this enzyme produces causes the ears, snout, feet, and tail to be black.

Epistasis

In some situations, two or more genes interact with each other, such that one gene contributes to or masks the expression of the other gene. This becomes apparent when analyzing dihybrid crosses involving these traits. Recall that when individuals heterozygous for two different genes mate (a dihybrid cross), offspring may display the dominant phenotype for both genes, either one of the genes, or for neither gene. Sometimes, however, an investigator cannot find four phenotype classes because two or more of the genotypes express the same phenotypes.

As was stated earlier, few phenotypes are the result of the action of one gene. Most traits reflect the action of many genes, some that act sequentially or jointly. **Epistasis** is an interaction between the products of two genes in which one of the genes modifies the phenotypic expression produced by the other. For example, some commercial varieties of corn, *Zea mays,* exhibit a purple pigment called *anthocyanin* in their seed coats, while others do not. In 1918, geneticist R. A. Emerson crossed two true-breeding corn varieties, neither exhibiting anthocyanin pigment. Surprisingly, all of the F_1 plants produced purple seeds.

When two of these pigment-producing F_1 plants were crossed to produce an F_2 generation, 56% were pigment producers and 44% were not. What was happening? Emerson correctly deduced that two genes were involved in producing pigment, and that the second cross had thus been a dihybrid cross like those performed by Mendel. Mendel had predicted 16 equally possible ways gametes could combine with each other, resulting in genotypes with a phenotypic ratio of 9:3:3:1 (9 + 3 + 3 + 1 = 16). How many of these were in each of the two types Emerson obtained? He multiplied the fraction that were pigment producers (0.56) by 16 to obtain 9 and multiplied the fraction that were not (0.44) by 16 to obtain 7. Thus, Emerson had a **modified ratio** of 9:7 instead of the usual 9:3:3:1 ratio. **Figure 10.16** shows the results of the dihybrid cross made by Emerson. Go back and compare these results with Mendel's dihybrid cross in **figure 10.10** and you can see that the F_2 genotypes in Emerson's results are consistent with what Mendel found; so why are the phenotypic ratios different?

Why Was Emerson's Ratio Modified? It turns out that in corn plants either one of the two genes that contribute

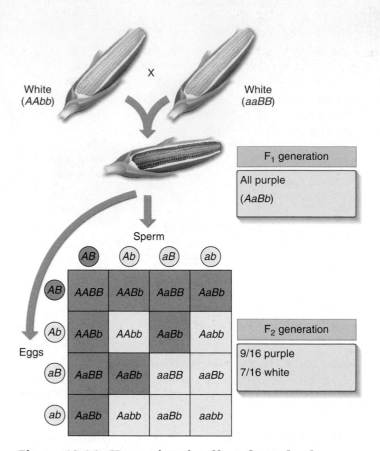

Figure 10.16 How epistasis affects kernel color.
The purple pigment found in some varieties of corn is the result of two genes. Unless a dominant allele is present at each of the two loci, no pigment is expressed.

to kernel color can block the expression of the other. One of the genes (*B*) produces an enzyme that permits colored pigment to be produced only if a dominant allele (*BB* or *Bb*) is present. The other gene (*A*) produces an enzyme that in its dominant form (*AA* or *Aa*) allows the pigment to be deposited on the seed coat color. Thus, an individual with two recessive alleles for gene *A* (no pigment deposition) will have white seed coats even though it is able to manufacture the pigment because it possesses dominant alleles for gene *B* (purple pigment production). Similarly, an individual with dominant alleles for gene *A* (pigment can be deposited) will also have white seed coats if it has only recessive alleles for gene *B* (pigment production) and cannot manufacture the pigment.

To produce and deposit pigment, a plant must possess at least one functional copy of each enzyme gene (*A_B_*). Of the 16 genotypes predicted by random assortment, 9 contain at least one dominant allele of both genes; they produce purple progeny and are colored darker in the Punnett square in **figure 10.16**. The remaining 7 genotypes lack dominant alleles at either or both loci (3 + 3 + 1 = 7) and so are phenotypically the same (nonpigmented—the light-color boxes in the Punnett square), giving the phenotypic ratio of 9:7 that Emerson observed.

Today's Biology

Does Environment Affect I.Q.?

Nowhere has the influence of environment on the expression of genetic traits led to more controversy than in studies of I.Q. scores. I.Q. is a controversial measure of general intelligence based on a written test that many feel to be biased toward white middle-class America. However well or poorly I.Q. scores measure intelligence, a person's I.Q. score has been believed for some time to be determined largely by his or her genes.

How did science come to that conclusion? Scientists measure the degree to which genes influence a multigene trait by using an off-putting statistical measure called the *variance*. Variance is defined as the square of the standard deviation (a measure of the degree-of-scatter of a group of numbers around their mean value), and has the very desirable property of being additive—that is, the total variance is equal to the sum of the variances of the factors influencing it.

What factors contribute to the total variance of I.Q. scores? There are three. The first factor is variation at the gene level, some gene combinations leading to higher I.Q. scores than others. The second factor is variation at the environmental level, some environments leading to higher I.Q. scores than others. The third factor is what a statistician calls the *covariance,* the degree to which environment affects genes.

The degree to which genes influence a trait like I.Q., the *heritability* of I.Q., is given the symbol H and is defined simply as the fraction of the total variance that is genetic.

So how heritable is I.Q.? Geneticists estimate the heritability of I.Q. by measuring the environmental and genetic contributions to the total variance of I.Q. scores. The environmental contributions to variance in I.Q. can be measured by comparing the I.Q. scores of identical twins reared together with those reared apart (any differences should reflect environmental influences). The genetic contributions can be measured by comparing identical twins reared together (which are 100% genetically identical) with fraternal twins reared together (which are 50% genetically identical). Any differences should reflect genes, as twins share identical prenatal conditions in the womb and are raised in virtually identical environmental circumstances, so when traits are more commonly shared between identical twins than fraternal twins, the difference is likely genetic.

When these sorts of "twin studies" have been done in the past, researchers have uniformly reported that I.Q. is highly heritable, with values of H typically reported as being around 0.7 (a very high value). While it didn't seem significant at the time, almost all the twins available for study over the years have come from middle-class or wealthy families.

The study of I.Q. has proven controversial, because I.Q. scores are often different when social and racial groups are compared. What is one to make of the observation that I.Q. scores of poor children measure lower as a group than do scores of children of middle-class and wealthy families? This difference has led to the controversial suggestion by some that the poor are genetically inferior.

What should we make of such a harsh conclusion? To make a judgement, we need to focus for a moment on the fact that these measures of the heritability of I.Q. have all made a critical assumption, one to which population geneticists, who specialize in these sorts of things, object strongly. The assumption is that environment does not affect gene expression, so that covariance makes no contribution to the total variance in I.Q. scores—that is, that the covariance contribution to H is zero.

Studies have allowed a direct assessment of this assumption. Importantly, it proves to be flat wrong.

In November of 2003, researchers reported an analysis of twin data from a study carried out in the late 1960s. The National Collaborative Prenatal Project, funded by the National Institutes of Health, enrolled nearly 50,000 pregnant women, most of them black and quite poor, in several major U.S. cities. Researchers collected abundant data, and gave the children I.Q. tests seven years later. Although not designed to study twins, this study was so big that many twins were born, 623 births. Seven years later, 320 of these pairs were located and given I.Q. tests. This thus constitutes a huge "twin study," the first ever conducted of I.Q. among the poor.

When the data were analyzed, the results were unlike any ever reported. The heritability of I.Q. was different in different environments! Most notably, the influence of genes on I.Q. was far less in conditions of poverty, where environmental limitations seem to block the expression of genetic potential. Specifically, for families of high socioeconomic status, H = 0.72, much as reported in previous studies, but for families raised in poverty, H = 0.10, a very low value, indicating that genes were making little contribution to observed I.Q. scores. The lower a child's socioeconomic status, the less impact genes had on I.Q.

These data say, with crystal clarity, that the genetic contributions to I.Q. don't mean much in an impoverished environment.

How does poverty in early childhood affect the brain? Neuroscientists reported in 2008 that many children growing up in very poor families experience poor nutrition and unhealthy levels of stress hormones, both of which impair their neural development. This affects language development and memory for the rest of their lives.

Clearly, improvements in the growing and learning environments of poor children can be expected to have a major impact on their I.Q. scores. Additionally, these data argue that the controversial differences reported in mean I.Q. scores between racial groups may well reflect no more than poverty, and are no more inevitable.

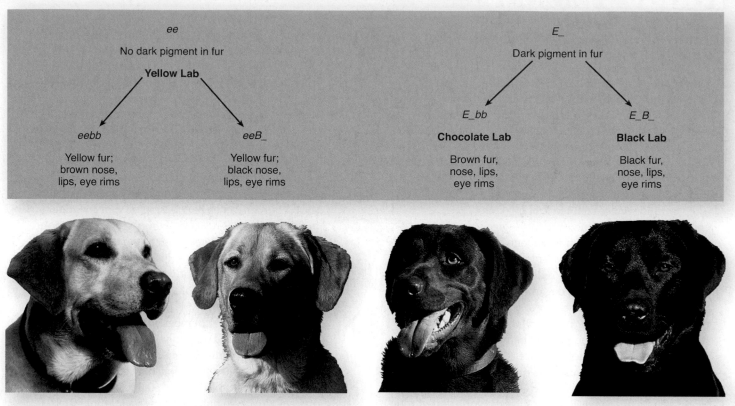

Figure 10.17 The effect of epistatic interactions on coat color in dogs.
The coat color seen in Labrador retrievers is an example of the interaction of two genes, each with two alleles. The *E* gene determines if the pigment will be deposited in the fur, and the *B* gene determines how dark the pigment will be.

Other Examples of Epistasis In many animals, coat color is the result of epistatic interactions among genes. Coat color in Labrador retrievers, a breed of dog, is due primarily to the interaction of two genes. The *E* gene determines if dark pigment will be deposited in the fur or not. If a dog has the genotype *ee* (like the two dogs on the left in **figure 10.17**), no pigment will be deposited in the fur, and it will be yellow. If a dog has the genotype *EE* or *Ee* (*E_*), pigment will be deposited in the fur (like the two dogs on the right).

A second gene, the *B* gene, determines how dark the pigment will be. Dogs with the genotype *E_bb* will have brown fur and are called chocolate labs. Dogs with the genotype *E_B_* are black labs with black fur. But, even in yellow dogs, the *B* gene does have some effect. Yellow dogs with the genotype *eebb* (on the far left) will have brown pigment on their nose, lips, and eye rims, while yellow dogs with the genotype *eeB_* (the second from the left) will have black pigment in these areas. The genes for coat color in this breed have been found, and a genetic test is available to determine the coat color in a litter of puppies.

Codominance

A gene may have more than two alleles in a population, and in fact most genes possess several different alleles. Often in heterozygotes there isn't a dominant allele; instead, the effects of both alleles are expressed. In these cases, the alleles are said to be **codominant.**

Codominance is seen in the color patterning of some animals. For example, the "roan" pattern is a coloring pattern exhibited in some varieties of horses and cattle. A roan animal expresses both white and colored hairs on at least part of its body. This intermingling of the different colored hairs creates either an overall lighter color, or patches of lighter and darker colors. The roan pattern results from a heterozygous genotype, such as produced by mating of a homozygous white and homozygous colored. Could the intermediate color be the result of incomplete dominance? No. The heterozygote that receives a white allele and a colored allele does not have individual hairs that are a mix of the two colors; rather, both alleles are being expressed, with the result that the animal has some hairs that are white and some that are colored. The gray horse in **figure 10.18** is exhibiting the roan pattern. It looks like it has gray hairs, but if you were able to examine its coat closely, you would see both white hairs and black hairs, giving it an overall gray color.

A human gene that exhibits more than one dominant allele is the gene that determines ABO blood type. This gene encodes an enzyme that adds sugar molecules to lipids on the surface of red blood cells. These sugars act as recognition markers for cells in the immune system and are called cell surface antigens. The gene that encodes the enzyme, designated *I*, has three common alleles: I^B, whose product adds the sugar galactose; I^A, whose product adds galactosamine; and *i*, which codes for a protein that does not add a sugar.

Different combinations of the three *I* gene alleles occur in different individuals because each person may

Figure 10.18 Codominance in color patterning.

This roan horse is heterozygous for coat color. The offspring of a cross between a white homozygote and a black homozygote, it expresses both phenotypes. Some of the hairs on its body are white and some are black.

be homozygous for any allele or heterozygous for any two. An individual heterozygous for the I^A and I^B alleles produces both forms of the enzyme and adds both galactose and galactosamine to the surfaces of red blood cells. Because both alleles are expressed simultaneously in heterozygotes, the I^A and I^B alleles are codominant. Both I^A and I^B are dominant over the i allele because both I^A or I^B alleles lead to sugar addition and the i allele does not. The different combinations of the three alleles produce four different phenotypes:

1. Type A individuals add only galactosamine. They are either $I^A I^A$ homozygotes or $I^A i$ heterozygotes (the three darkest boxes in **figure 10.19**).

2. Type B individuals add only galactose. They are either $I^B I^B$ homozygotes or $I^B i$ heterozygotes (the three lightest-colored boxes).

3. Type AB individuals add both sugars and are $I^A I^B$ heterozygotes (the two intermediate-colored boxes).

4. Type O individuals add neither sugar and are ii homozygotes (the one white box in **figure 10.19**).

These four different cell surface phenotypes are called the **ABO blood groups.** A person's immune system can distinguish between these four phenotypes. If a type A individual receives a transfusion of type B blood, the recipient's immune system recognizes that the type B blood cells possess a "foreign" antigen (galactose) and attacks the donated blood cells, causing the cells to clump or agglutinate. This also happens if the donated blood is type AB. However, if the donated blood is type O, it contains no galactose or galactosamine antigens on the surfaces of its blood cells, and so elicits no immune response to these antigens. For this reason, the type O individual is often referred to as a "universal donor." In general, any individual's immune system will tolerate a transfusion of type O blood. Because neither galactose nor galactosamine is foreign to type AB individuals (whose red blood cells have both sugars), those individuals may receive any type of blood.

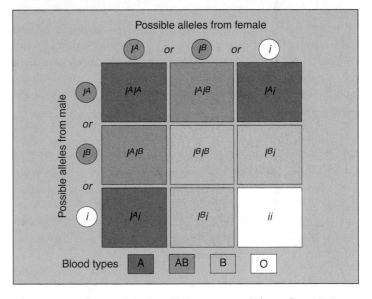

Figure 10.19 Multiple alleles controlling the ABO blood groups.

Three common alleles control the ABO blood groups. The different combinations of the three alleles result in four different blood type phenotypes: type A (either $I^A I^A$ homozygotes or $I^A i$ heterozygotes), type B (either $I^B I^B$ homozygotes or $I^B i$ heterozygotes), type AB ($I^A I^B$ heterozygotes), and type O (ii homozygotes).

Key Learning Outcome 10.6 A variety of factors can disguise the Mendelian segregation of alleles. Among them are continuous variation, which results when many genes contribute to a trait; pleiotropic effects, where one allele affects many phenotypes; incomplete dominance, which produces heterozygotes unlike either parent; environmental influences on the expression of phenotypes; and the interaction of more than one allele, as seen in epistasis and codominance.

10.7 Chromosomes Are the Vehicles of Mendelian Inheritance

The Chromosomal Theory of Inheritance

In the early twentieth century it was by no means obvious that chromosomes were the vehicles of hereditary information. A central role for chromosomes in heredity was first suggested in 1900 by the German geneticist Karl Correns, in one of the papers announcing the rediscovery of Mendel's work. Soon observations that similar chromosomes paired with one another during meiosis led to the *chromosomal theory of inheritance,* first formulated by American Walter Sutton in 1902.

Several pieces of evidence supported Sutton's theory. One was that reproduction involves the initial union of only two cells, egg and sperm. If Mendel's model was correct, then these two gametes must make equal hereditary contributions. Sperm, however, contain little cytoplasm, suggesting that the hereditary material must reside within the nuclei of the gametes. Furthermore, while diploid individuals have two copies of each pair of homologous chromosomes, gametes have only one. This observation was consistent with Mendel's model, in which diploid individuals have two copies of each heritable gene and gametes have one. Finally, chromosomes segregate during meiosis, and each pair of homologues orients on the metaphase plate independently of every other pair. Segregation and independent assortment were two characteristics of the genes in Mendel's model.

Problems with the Chromosomal Theory

Investigators soon pointed out one problem with this theory, however. If Mendelian traits are determined by genes located on the chromosomes, and if the independent assortment of Mendelian traits reflects the independent assortment of chromosomes in meiosis, why does the number of traits that assort independently in a given kind of organism often greatly exceed the number of chromosome pairs the organism possesses? This seemed a fatal objection, and it led many early researchers to have serious reservations about Sutton's theory.

Morgan's White-Eyed Fly

The essential correctness of the chromosomal theory of heredity was demonstrated by a single small fly. In 1910 Thomas Hunt Morgan, studying the fruit fly *Drosophila melanogaster,* detected a mutant male fly that differed strikingly from normal fruit flies: Its eyes were white instead of red (**figure 10.20**).

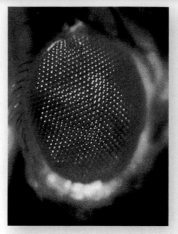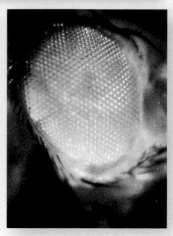

Figure 10.20   Red-eyed (wild type) and white-eyed (mutant) *Drosophila*.
The white-eye defect is hereditary, the result of a mutation in a gene located on the X chromosome. By studying this mutation, Morgan first demonstrated that genes are on chromosomes.

Morgan immediately set out to determine if this new trait would be inherited in a Mendelian fashion. He first crossed the mutant male with a normal female to see if red or white eyes were dominant. All of the F_1 progeny had red eyes, so Morgan concluded that red eye color was dominant over white. Following the experimental procedure that Mendel had established long ago, Morgan then crossed the red-eyed flies from the F_1 generation with each other. Of the 4,252 F_2 progeny Morgan examined, 782 (18%) had white eyes. Although the ratio of red eyes to white eyes in the F_2 progeny was greater than 3:1, the results of the cross nevertheless provided clear evidence that eye color segregates. However, there was something about the outcome that was strange and totally unpredicted by Mendel's theory—*all of the white-eyed F_2 flies were males!*

How could this result be explained? Perhaps it was impossible for a white-eyed female fly to exist; such individuals might not be viable for some unknown reason. To test this idea, Morgan testcrossed the female F_1 progeny with the original white-eyed male. He obtained white-eyed and red-eyed males and females in a 1:1:1:1 ratio, just as Mendelian theory predicted. Hence, a female could have white eyes. Why, then, were there no white-eyed females among the progeny of the original cross?

Sex Linkage Confirms the Chromosomal Theory

The solution to this puzzle involved sex. In *Drosophila,* the sex of an individual is determined by the number of copies of a particular chromosome, the X chromosome, that an individual possesses. A fly with two X chromosomes is a female, and a fly with only one X chromosome is a male. In males, the single X chromosome pairs in meiosis with a large, dissimilar partner called the Y chromosome. The female thus produces only X gametes, while the male produces both X and Y gametes. When fertilization involves an X sperm, the result is an XX zygote, which develops into a female; when fertilization involves a Y sperm, the result is an XY zygote, which develops into a male.

The solution to Morgan's puzzle is that the gene causing the white-eye trait in *Drosophila* resides only on the X chromosome—it is absent from the Y chromosome. (We now know that the Y chromosome in flies carries almost no functional genes.) A trait determined by a gene on the sex chromosome is said to be **sex-linked.** Knowing the white-eye trait is recessive to the red-eye trait, we can now see that Morgan's result was a natural consequence of the Mendelian assortment of chromosomes. **Figure 10.21** steps you through

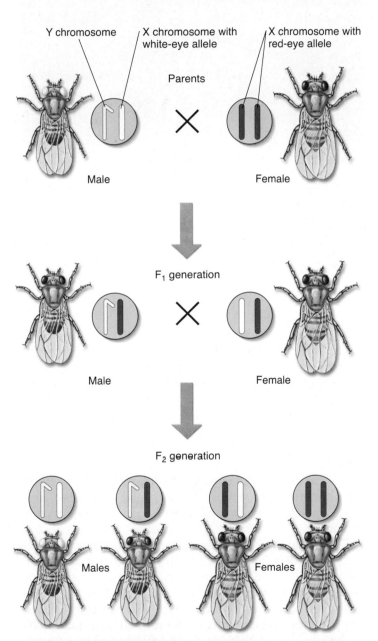

Figure 10.21 Morgan's experiment demonstrating the chromosomal basis of sex linkage.

The white-eyed mutant male fly was crossed with a normal female. The F₁ generation flies all exhibited red eyes, as expected for flies heterozygous for a recessive white-eye allele. In the F₂ generation, all of the white-eyed flies were male.

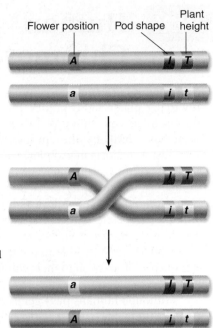

Figure 10.22 Linkage.

Genes that are located farther apart on a chromosome, like the genes for flower position (*A*) and pod shape (*I*) in Mendel's peas, will assort independently because crossing over results in recombination of these alleles. Pod shape (*I*) and plant height (*T*), however, are positioned very near each other, such that crossing over usually would not occur. These genes are said to be linked and do not undergo independent assortment.

Morgan's experiment, showing both the eye color alleles and the sex chromosomes. In this experiment, the F₁ generation all had red eyes, while the F₂ generation contained flies with white eyes—but they were all males. This at-first-surprising result happens because the segregation of the white-eye trait has a one-to-one correspondence with the segregation of the X chromosome. In other words, the white-eye gene is on the X chromosome. In humans, traits such as color-blindness (see page 208 and chapter 29) and hemophilia (a blood-clotting disease discussed later in this chapter) are sex-linked.

Morgan's experiment presented the first clear evidence that the genes determining Mendelian traits reside on chromosomes, just as Sutton had proposed. Now we can see that the reason Mendelian traits assort independently is because chromosomes assort independently. When Mendel observed the segregation of alternative traits in pea plants, he was observing a reflection of the meiotic segregation of the chromosomes, which contained the characters he was observing.

If genes are located on chromosomes, you might expect that two genes on the same chromosome would segregate together. However, if the two genes are located far from each other on the chromosome, like genes *A* and *I* in **figure 10.22**, the likelihood of crossing over occurring between them is very high, leading to independent segregation. Conversely, the closer two genes are to each other on a chromosome, like genes *I* and *T*, the less likely it is that a cross-over event will occur between them. Genes that are located quite close to each other almost always segregate together, meaning that they are inherited together. The tendency of close-together genes to segregate together is called **linkage.**

Key Learning Outcome 10.7 Mendelian traits assort independently because they are determined by genes located on chromosomes that assort independently in meiosis.

10.8 Human Chromosomes

The principles of genetics first proposed by Mendel apply not only to pea plants and fruit flies, but also to you. How closely you resemble your father or mother was largely established before your birth by the chromosomes you received from them, just as meiosis in pea plants determined the segregation of Mendel's traits. But many of the alleles found in human populations demand more serious concern than the color of a pea. Some of the most devastating human disorders result from alleles specifying defective forms of proteins that have important functions in our bodies. By studying human heredity, scientists are more able to predict which disorders parents might expect to pass on to their children, and with what probability.

Although humans pass genes on to the next generation in much the same way that other organisms do, we naturally have a special curiosity about ourselves. Because we know that some illnesses are hereditary and others are not, we cannot escape concern when a member of our family becomes ill. If a family member has had a stroke, we tend to worry about our own future health because we know that the propensity to suffer strokes can be hereditary. Few parents have babies without worrying about the possibility of birth defects. Genes are also clearly involved in such conditions as diabetes, depression, and alcoholism. The way in which genes interact with the environment to produce individuals with differing personalities is the subject of continuing intensive study. Because of the importance of genes in determining the course of our lives, we are all human geneticists, interested in what the laws of genetics reveal about ourselves and our families.

Human Chromosomes

Although chromosomes were discovered more than a century ago, the exact number of chromosomes that humans possess (46) was not established until 1956, when new techniques for accurately determining the number and form of human and other mammalian chromosomes were developed.

Biologists examine human chromosomes by collecting a blood sample, adding chemicals that induce the white blood cells in the sample to divide (red blood cells have lost their nuclei and cannot divide), and then adding other chemicals that arrest cell division at metaphase. Metaphase is the stage of mitosis when the chromosomes are most condensed and thus most easily distinguished from one another. The cells are then flattened, spreading out their contents, and the individual chromosomes are separated for examination. The chromosomes are stained and photographed, and a chromosomal "portrait" called a karyotype is prepared with the photographs of the individual chromosomes. A human karyotype is presented in **figure 10.23**. By convention, the chromosomes in a karyotype are presented with

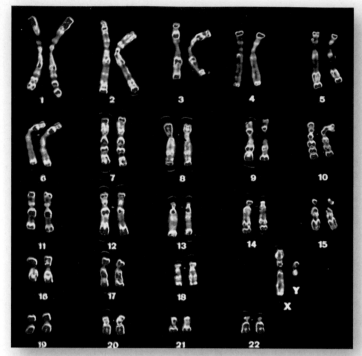

Figure 10.23 A human karyotype.
Photographs of each of the 46 chromosomes have been arranged in descending order of size. The banding patterns revealed by staining allow the investigator to identify homologues and pair them together.

homologues together, and chromosomes arranged in order of descending size.

Of the 23 pairs of human chromosomes you see in **figure 10.23**, 22 consist of members that are similar in size and morphology in both males and females. These chromosomes are called **autosomes.** In many plants and animals, including peas, fruit flies, and humans, the two members of the remaining pair—the so-called **sex chromosomes**—are unlike each other in males and similar in females. In humans, females are designated XX and males are designated XY. The Y chromosome is much smaller than the X chromosome and carries only a tenth the number of genes. Among the genes present on the Y chromosome are those that determine "maleness," and, therefore, humans who inherit the Y chromosome develop into males.

The karyotypes of individuals are often examined to detect genetic abnormalities arising from extra or lost chromosomes. For example, the human birth defect Down syndrome (discussed on the facing page) is associated with the presence of an extra copy of chromosome 21, which can be recognized easily in karyotypes, as there are 47 chromosomes rather than 46, and the extra chromosome can be identified by its banding pattern as a third copy of chromosome 21. Karyotypes of fetal cells taken before birth can reveal genetic abnormalities of this sort.

Nondisjunction

Some of the most significant human hereditary disorders arise as a result of problems with how human chromosomes sort during meiosis.

Sometimes during meiosis, sister chromatids or homologous chromosomes that paired up during metaphase remain stuck together instead of separating. The failure of chromosomes to separate correctly during either meiosis I or II is called **nondisjunction.** Nondisjunction leads to **aneuploidy,** an abnormal number of chromosomes. The nondisjunction you see in **figure 10.24** occurs because the homologous pair of larger chromosomes failed to separate in anaphase I. The gametes that result from this division have unequal numbers of chromosomes. Under normal meiosis, all gametes would be expected to have two chromosomes, but as you can see, two of these gametes have three chromosomes, while two others have just one.

Almost all humans of the same sex have the same karyotype simply because other arrangements don't work well. Humans who have lost even one copy of an autosome (called **monosomics**) do not survive development. In all but a few cases, humans who have gained an extra autosome (called **trisomics**) also do not survive. However, five of the smallest chromosomes—those numbered 13, 15, 18, 21, and 22—can be present in humans as three copies and still allow the individual to survive for a time. The presence of an extra chromosome 13, 15, or 18 causes severe developmental defects, and infants with such a genetic makeup die within a few months. In contrast, individuals who have an extra copy of chromosome 21 or, more rarely, chromosome 22, usually survive to adulthood. In such individuals, the maturation of the skeletal system is delayed, so they generally are short and have poor muscle tone. Their mental development is also affected.

Down Syndrome   The developmental defect produced by the trisomy 21 seen in **figure 10.25** was first described in 1866 by J. Langdon Down; for this reason, it is called **Down syndrome.** About 1 in every 750 children exhibits Down syndrome, and the frequency is similar in all racial groups. It is much more common in children of older mothers. The graph in **figure 10.26** (see next page) shows the increasing incidence in older mothers. In mothers under 30 years old, the incidence is only about 0.6 per 1,000 (or 1 in 1,500 births), while in mothers 30 to 35 years old, the incidence doubles to about 1.3 per 1,000 births (or 1 in 750 births). In mothers over 45, the risk is as high as 63 per 1,000 births (or 1 in 16 births). The reason that older mothers are more prone to Down syndrome babies is that all the eggs that a woman will ever produce are present in her ovaries by the time she is born, and as she gets older they may accumulate damage that can result in nondisjunction.

Figure 10.25   Down syndrome.

(a) In this karyotype of a male individual with Down syndrome, the trisomy at position 21 can be clearly seen.
(b) A person with Down syndrome.

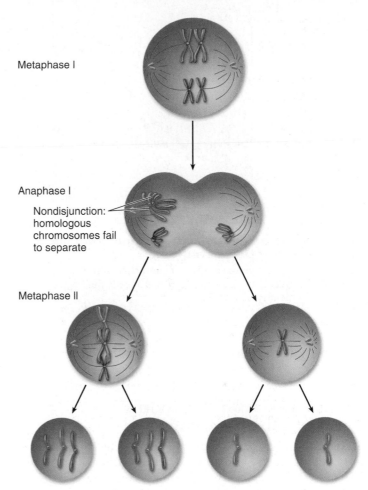

Metaphase I

Anaphase I

Nondisjunction: homologous chromosomes fail to separate

Metaphase II

Results in four gametes: two are $n+1$ and two are $n-1$

Figure 10.24   Nondisjunction in anaphase I.

In nondisjunction that occurs during meiosis I, one pair of homologous chromosomes fails to separate in anaphase I, and the gametes that result have one too many or one too few chromosomes. Nondisjunction can also occur in meiosis II, when sister chromatids fail to separate during anaphase II.

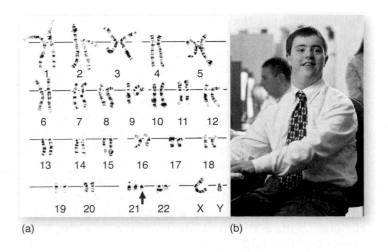

(a)

(b)

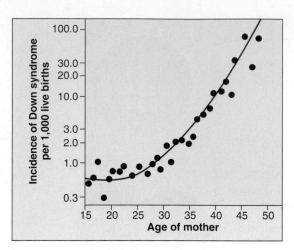

Figure 10.26 Correlation between maternal age and the incidence of Down syndrome.
As women age, the chances they will bear a child with Down syndrome increase. After a woman reaches age 35, the frequency of Down syndrome increases rapidly.

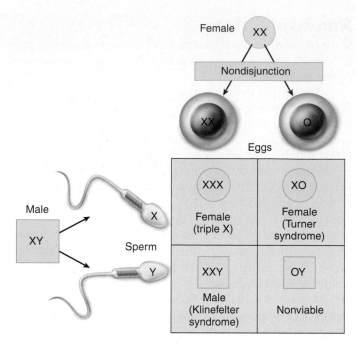

Figure 10.27 Nondisjunction of the X chromosome.
Nondisjunction of the X chromosome can produce sex chromosome aneuploidy—that is, abnormalities in the number of sex chromosomes.

Nondisjunction Involving the Sex Chromosomes

As noted, 22 of the 23 pairs of human chromosomes are perfectly matched in both males and females and are called autosomes. The remaining pair are the sex chromosomes, X and Y. In humans, as in *Drosophila* (but by no means in all diploid species), females are XX and males XY; any individual with at least one Y chromosome is male. The Y chromosome is highly condensed and bears few functional genes in most organisms. Some of the active genes the Y chromosome does possess are responsible for the features associated with "maleness." Individuals who gain or lose a sex chromosome do not generally experience the severe developmental abnormalities caused by changes in autosome numbers. Such individuals may reach maturity, but with somewhat abnormal features.

Nondisjunction of the X Chromosome When X chromosomes fail to separate during meiosis, some of the gametes that are produced possess both X chromosomes and so are XX gametes; the other gametes that result from such an event have no sex chromosome and are designated "O."

Figure 10.27 shows what happens if gametes from X chromosome nondisjunction combine with sperm. If an XX gamete combines with an X gamete, the resulting XXX zygote (in the upper left of the Punnett square) develops into a female who is taller than average but other symptoms can vary greatly. Some are normal in most respects, others may have lower reading and verbal skills, and still others are mentally retarded. If an XX gamete combines with a Y gamete (in the lower left), the XXY zygote develops into a sterile male who has many female body characteristics and, in some cases, diminished mental

capacity. This condition, called *Klinefelter syndrome,* occurs in about 1 in 500 male births.

If an O gamete fuses with a Y gamete (in the lower right), the OY zygote is nonviable and fails to develop further because humans cannot survive when they lack the genes on the X chromosome. If an O gamete fuses with an X gamete (in the upper right), the XO zygote develops into a sterile female of short stature, with a webbed neck and immature sex organs that do not undergo changes during puberty. The mental abilities of XO individuals are normal in verbal learning but lower in nonverbal/math-based problem solving. This condition, called *Turner syndrome,* occurs roughly once in every 5,000 female births.

Nondisjunction of the Y Chromosome The Y chromosome can also fail to separate in meiosis, leading to the formation of YY gametes. When these gametes combine with X gametes, the XYY zygotes develop into fertile males of normal appearance. The frequency of the XYY genotype is about 1 per 1,000 newborn males.

> **Key Learning Outcome 10.8** The particular array of chromosomes that an individual possesses is called the karyotype. The human karyotype usually contains 23 pairs of chromosomes. Autosome loss is always lethal, and an extra autosome is with few exceptions lethal too.

10.9 Studying Pedigrees

To study human heredity, scientists look at the results of crosses that have already been made. They study family trees, or **pedigrees,** to identify which relatives exhibit a trait. Then they can often determine whether the gene producing the trait is sex-linked (that is, located on the X chromosome) or autosomal, and whether the expression of the trait is dominant or recessive. Frequently the pedigree will also help an investigator infer which individuals in a family are homozygous and which are heterozygous for the allele specifying the trait.

Analyzing a Pedigree for Albinism

Albino individuals lack all pigmentation; their hair and skin are completely white. In the United States about 1 in 38,000 Caucasians and 1 in 22,000 African-Americans are albino. In the pedigree of albinism among a family of Hopi Indians presented in **figure 10.28,** each symbol represents one individual in the family history, with the circles representing females and the squares, males. In this pedigree, individuals that exhibit a trait being studied—in this case, albinism—are indicated by solid symbols; heterozygote "carriers" exhibiting normal phenotypes are indicated by half-filled symbols. Marriages are represented by horizontal lines connecting a circle and a square, from which a cluster of vertical lines descend indicating the children, arranged from left to right in order of their birth.

To analyze this pedigree of albinism, a geneticist traditionally asks three questions:

1. *Is albinism sex-linked or autosomal?* If the trait is sex-linked, it is usually seen only in males; if it is autosomal, it appears in both sexes fairly equally. In the pedigree below, the proportion of affected males (4 of 12, or 33%) is reasonably similar to the proportion of affected females (8 of 19, or 42%). (When counting numbers of affected individuals in a pedigree, exclude the parents in generation I, as well as any "outsiders" who marry into the family.) From this result, it is reasonable to conclude the trait is autosomal.

2. *Is albinism dominant or recessive?* If the trait is dominant, every albino child will have an albino parent. If recessive, however, an albino child's parents can appear normal, since both parents may be heterozygous "carriers." In the pedigree below, parents of most of the albino children do not exhibit the trait, which indicates that albinism is recessive. Four children in one family *do* have albino parents. The allele is very common among the Hopi Indians, from which this pedigree was derived, and thus homozygous individuals such as these albino parents are present in the Hopis in sufficient numbers that they sometimes marry. In this family, *both* parents are albino and *all* four children are albino, which is consistent with the finding that the trait albinism is recessive, with both parents homozygous for the allele.

3. *Is the albinism trait determined by a single gene, or by several?* If the trait is determined by a single gene, then a ratio of 3:1 (normal to albino) offspring should be born to heterozygous parents (indicated by half-filled symbols), reflecting Mendelian segregation in a cross. Thus about 25% of these children should be albino. But if the trait is determined by several genes, albinism would only be present in a few percent. In this pedigree, 8 of 24 children born to heterozygotes exhibit albinism, or 33%, strongly suggesting that only one gene is segregating in these crosses.

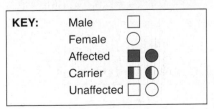

KEY:		
Male		□
Female		○
Affected	■	●
Carrier	◧	◐
Unaffected	□	○

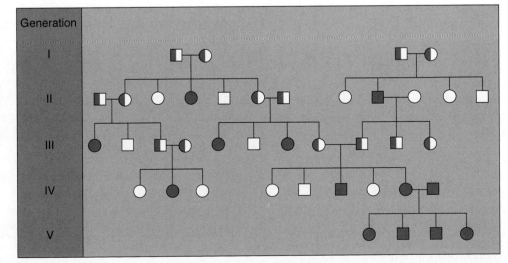

Figure 10.28 A pedigree of albinism.
In the photo, one of three girls from a Hopi Indian family (the left-most family in generation IV of the pedigree) is albino. The pedigree shows the inheritance of the gene causing albinism in this family, with the solid blue symbols indicating persons who are albino.

Analyzing a Pedigree for Color Blindness

The albinism pedigree analysis you have just examined indicates that albinism is an autosomal recessive trait controlled by a single gene. The inheritance of other human traits is studied in a similar way, although sometimes with different results. As an example, let us analyze a different trait. Red-green color blindness is an infrequent, although not rare, inherited trait in humans, affecting 5% to 9% of males. Color blindness is a group of eye disorders in which a person is not able to distinguish certain colors or shades of colors. It doesn't mean that they see only in black and white, but rather that they see colors but some different colors look the same to them. Special types of cells in the retina of the eye detect different colors of light and different shades. Recall from the discussion of the electromagnetic spectrum in chapter 6 that visible light contains different wavelengths of photons that appear as the spectrum of visible light shown here:

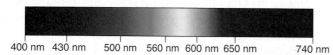

Our eyes contain three types of color receptors: one absorbs red light, one green light, and a third absorbs blue light. People with red-green color blindness have deficiencies in their ability to detect red and green light as being different, and so these colors appear the same to them. Test samples called Ishihara plates are used to determine if a person is color blind. The test plates contain different colored dots arranged to reveal a shape, often a number. People with normal vision are able to see the number while a person who is color blind for those colors is not able to see it. An Ishihara test for red-green color blindness is shown in **figure 10.29**.

Like albinism, a pedigree can be used to reveal the pattern of inheritance of color blindness. In the pedigree shown below, a red-green color blind man has five children with a woman who is heterozygous for the allele. Again, the solid-color symbols indicate an affected individual, in this case red-green color blind. Half-filled symbols indicate a heterozygous individual who carries the trait but does not express it.

To analyze this pedigree, you ask the same three questions as before:

1. *Is red-green color blindness sex-linked or autosomal?* Of the five affected individuals, all are male. The trait is clearly sex-linked.

2. *Is red-green color blindness dominant or recessive?* If the trait is dominant, then every color-blind child should have a color-blind parent. In this pedigree, however, that is not true in any family after that of the original male. The trait is clearly recessive.

3. *Is the red-green color blindness trait determined by a single gene?* If it is, then children born to heterozygous parents should be color-blind in about 25% of cases, reflecting a 3:1 Mendelian segregation of the trait. In this pedigree, 4 of 14, or 28%, of the children of heterozygous parents are color blind, indicating that a single gene is segregating (do not count the five children of the generation I parents because the father in this case is homozygous for the trait).

The results of this pedigree indicate that color blindness is caused by a single sex-linked, recessive gene. This doesn't mean that females can't be color blind, but in order for a female to be color blind, both X chromosomes would have to carry the color blind gene, and this only occurs in 0.5% of females.

Key Learning Outcome 10.9 The study of family trees can often reveal if an inherited trait is caused by a single gene, if that gene is located on the X chromosome, and if its alleles are recessive.

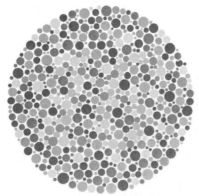

Source: This image has been reproduced from Ishihara's *Tests for Color Deficiency* published by KANEHARA TRADING INC., located in Tokyo, Japan. But tests for color deficiency cannot be conducted with this material. For accurate testing, the original plates should be used.

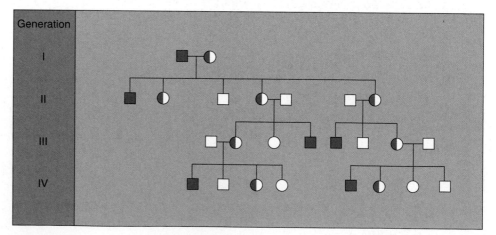

Figure 10.29 Pedigree of color blindness.

Individuals who are red-green color blind cannot see the number, as all the dots appear the same color. The pedigree traces red-green color blindness through four generations of a family.

10.10 | The Role of Mutation

Hemophilia: A Sex-Linked Trait

Blood in a cut clots as a result of the polymerization of protein fibers circulating in the blood. A dozen proteins are involved in this process, and all must function properly for a blood clot to form. A mutation causing any of these proteins to lose their activity leads to a form of **hemophilia,** a hereditary condition in which the blood clots slowly or not at all.

Hemophilias are recessive disorders, expressed only when an individual does not possess any copy of the normal allele and so cannot produce one of the proteins necessary for clotting. Most of the genes that encode the blood-clotting proteins are on autosomes, but two (designated VIII and IX) are on the X chromosome. These two genes are sex-linked (see section 10.7): Any male who inherits a mutant allele will develop hemophilia because his other sex chromosome is a Y chromosome that lacks any alleles of those genes.

The most famous instance of hemophilia, often called the Royal hemophilia, is a sex-linked form that arose in the royal family of England. This hemophilia was caused by a mutation in gene IX that occurred in one of the parents of Queen Victoria of England (1819–1901). The pedigree in **figure 10.30** shows that in the six generations since Queen Victoria, 10 of her male descendants have had hemophilia (the solid

squares). The present British royal family has escaped the disorder because Queen Victoria's son, King Edward VII, did not inherit the defective allele, and all the subsequent rulers of England are his descendants. Three of Victoria's nine

children did receive the defective allele, however, and they carried it by marriage into many of the other royal families of Europe. In this photo, Queen Victoria of England is surrounded by some of her descendants in 1894. Standing behind Victoria and wearing feathered boas are two of Victoria's granddaughters, Alice's daughter's: Princess Irene of Prussia (right), and Alexandra (left), who would soon become Czarina of Russia. Both Irene and Alexandra were also carriers of hemophilia.

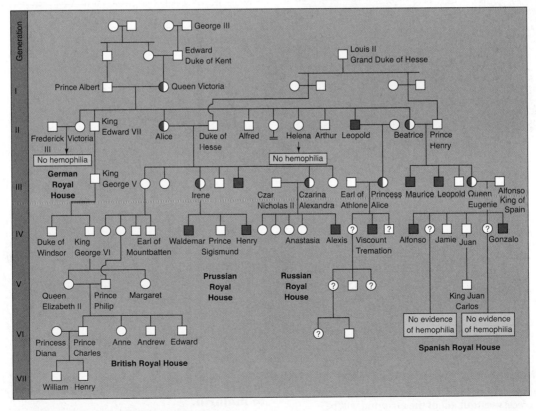

Figure 10.30 The Royal hemophilia pedigree.
Queen Victoria's daughter Alice introduced hemophilia into the Russian and Prussian royal houses, and her daughter Beatrice introduced it into the Spanish royal house. Victoria's son Leopold, himself a victim, also transmitted the disorder in a third line of descent. Half-shaded symbols represent carriers with one normal allele and one defective allele; fully shaded symbols represent affected individuals. Squares represent males; circles represent females.

Sickle-Cell Disease: A Recessive Trait

Sickle-cell disease is a recessive hereditary disorder. Its inheritance is shown in the pedigree in **figure 10.31**, where affected individuals are homozygous, carrying two copies of the mutated gene. Affected individuals have defective molecules of hemoglobin, the protein within red blood cells that carries oxygen. Consequently, these individuals are unable to properly transport oxygen to their tissues. The defective hemoglobin molecules stick to one another, forming stiff, rodlike structures and resulting in the formation of sickle-shaped red blood cells (**figure 10.31**). As a result of their stiffness and irregular shape, these cells have difficulty moving through the smallest blood vessels; they tend to accumulate in those vessels and form clots. People who have large proportions of sickle-shaped red blood cells tend to have intermittent illness and a shortened life span.

The hemoglobin in the defective red blood cells differs from that in normal red blood cells in only one of hemoglobin's 574 amino acid subunits. In the defective hemoglobin, the amino acid valine replaces a glutamic acid at a single position in the protein. Interestingly, the position of the change is far from the active site of hemoglobin where the iron-bearing heme group binds oxygen. Instead, the change occurs on the outer edge of the protein. Why then is the result so catastrophic? The sickle-cell mutation puts a very nonpolar amino acid on the surface of the hemoglobin protein, creating a "sticky patch" that sticks to other such patches—nonpolar amino acids tend to associate with one another in polar environments like water. As one hemoglobin adheres to another, chains of hemoglobin molecules form.

Individuals heterozygous for the sickle-cell allele are generally indistinguishable from normal persons. However, some of their red blood cells show the sickling characteristic when they are exposed to low levels of oxygen. The allele responsible for sickle-cell disease is particularly common among people of African descent because the sickle-cell allele is more common in Africa. About 9% of African Americans are heterozygous for this allele, and about 0.2% are homozygous and therefore have the disorder. In some groups of people in Africa, up to 45% of all individuals are heterozygous for this allele, and fully 6% are homozygous and express

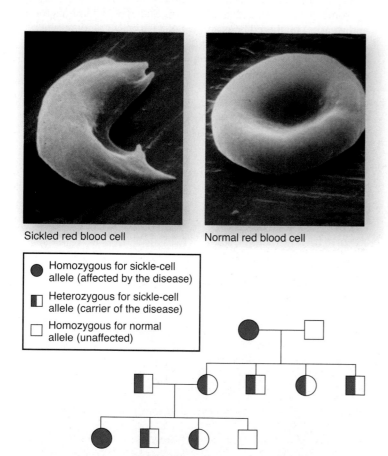

Sickled red blood cell   Normal red blood cell

● Homozygous for sickle-cell
   allele (affected by the disease)

◧ Heterozygous for sickle-cell
   allele (carrier of the disease)

□ Homozygous for normal
   allele (unaffected)

Figure 10.31   Inheritance of sickle-cell disease.

Sickle-cell disease is a recessive autosomal disorder. If one parent is homozygous for the recessive trait, all of the offspring will be carriers (heterozygotes), like the F$_1$ generation of Mendel's testcross. A normal red blood cell is shaped like a flattened sphere. In individuals homozygous for the sickle-cell trait, many of the red blood cells have sickle shapes.

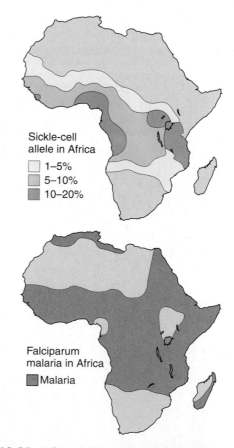

Sickle-cell
allele in Africa

□ 1–5%
▨ 5–10%
▨ 10–20%

Falciparum
malaria in Africa

▨ Malaria

Figure 10.32   The sickle-cell allele confers resistance to malaria.

The distribution of sickle-cell disease closely matches the occurrence of malaria in central Africa. This is not a coincidence. The sickle-cell allele, when heterozygous, confers resistance to malaria, a very serious disease.

the disorder. What factors determine the high frequency of sickle-cell disease in Africa? It turns out that heterozygosity for the sickle-cell allele increases resistance to malaria, a common and serious disease in Central Africa. Comparing the two maps shown in **figure 10.32**, you can see that the area of the sickle-cell trait matches well with the incidence of malaria. The interactions of sickle-cell disease and malaria are discussed further in chapter 14.

Tay-Sachs Disease: A Recessive Trait

Tay-Sachs disease is an incurable hereditary disorder in which the brain deteriorates. Affected children appear normal at birth and usually do not develop symptoms until about the eighth month, when signs of mental deterioration appear. The children are blind within a year after birth, and they rarely live past five years of age.

The Tay-Sachs allele produces the disease by encoding a nonfunctional form of the enzyme hexosaminidase A. This enzyme breaks down *gangliosides,* a class of lipids occurring within the lysosomes of brain cells. As a result, the lysosomes fill with gangliosides, swell, and eventually burst, releasing oxidative enzymes that kill the cells. There is no known cure for this disorder.

Tay-Sachs disease is rare in most human populations, occurring in only 1 in 300,000 births in the United States. However, the disease has a high incidence among Jews of Eastern and Central Europe (Ashkenazi) and among American Jews, 90% of whom trace their ancestry to Eastern and Central Europe. In these populations, it is estimated that 1 in 28 individuals is a heterozygous carrier of the disease, and approximately 1 in 3,500 infants has the disease. Because the disease is caused by a recessive allele, most of the people who carry the defective allele do not themselves develop symptoms of the disease because, as shown by the middle bar in **figure 10.33**, their one normal gene produces enough enzyme activity (50%) to keep the body functioning normally.

Huntington's Disease: A Dominant Trait

Not all hereditary disorders are recessive. **Huntington's disease** is a hereditary condition caused by a dominant allele that causes the progressive deterioration of brain cells. Perhaps 1 in 24,000 individuals develops the disorder. Because the allele is dominant, every individual who carries the allele expresses the disorder. Nevertheless, the disorder persists in human populations because its symptoms usually do not develop until the affected individuals are more than 30 years old, and by that time most of those individuals have already had children. Consequently, as illustrated by the pedigree in **figure 10.34**, the allele is often transmitted before the lethal condition develops.

> **Key Learning Outcome 10.10** Many human hereditary disorders reflect the presence of rare (and sometimes not so rare) mutations within human populations.

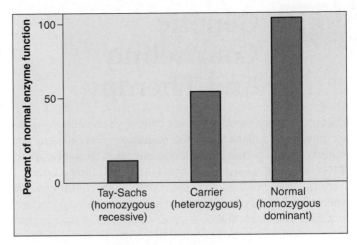

Figure 10.33 Tay-Sachs disease.

Homozygous individuals *(left bar)* typically have less than 10% of the normal level of hexosaminidase A *(right bar),* while heterozygous individuals *(middle bar)* have about 50% of the normal level—enough to prevent deterioration of the central nervous system.

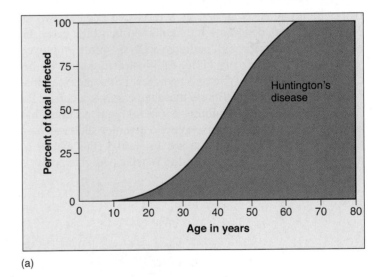

(a)

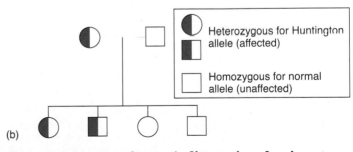

(b)

Figure 10.34 Huntington's disease is a dominant genetic disorder.

(a) Because of the late age of onset of Huntington's disease, the allele causing it persists despite being both dominant and fatal. (b) The pedigree illustrates how a dominant lethal allele can be passed from one generation to the next. Although the mother was affected, we can tell that she was heterozygous because if she were homozygous dominant, all of her children would have been affected. However, by the time she found out that she had the disease, she had probably already given birth to her children. In this way the trait passes on to the next generation even though it is fatal.

10.11 Genetic Counseling and Therapy

Although most genetic disorders cannot yet be cured, we are learning a great deal about them, and progress toward successful therapy is being made in many cases. However, in the absence of a cure, some parents may feel their only recourse is to try to avoid producing children with these conditions. The process of identifying parents at risk of producing children with genetic defects and of assessing the genetic state of early embryos is called **genetic counseling.** Genetic counseling can help prospective parents determine their risk of having a child with a genetic disorder and advise them on medical treatments or options if a genetic disorder is determined to exist in an unborn child.

High-Risk Pregnancies

If a genetic defect is caused by a recessive allele, how can potential parents determine the likelihood that they carry the allele? One way is through pedigree analysis, often employed as an aid in genetic counseling. As illustrated earlier in this chapter, by analyzing a person's pedigree, it is sometimes possible to estimate the likelihood that the person is a carrier for certain disorders. For example, if one of your relatives has been afflicted with a recessive genetic disorder such as cystic fibrosis, it is possible that you are a heterozygous carrier of the recessive allele for that disorder. When a pedigree analysis indicates that both parents of an expected child have a significant probability of being heterozygous carriers of a recessive allele responsible for a serious genetic disorder, the pregnancy is said to be a high-risk pregnancy. In such cases, there is a significant probability that the child will exhibit the clinical disorder.

Another class of high-risk pregnancies is that in which the mothers are more than 35 years old. As we have seen, the frequency of birth of infants with Down syndrome increases dramatically in the pregnancies of older women (see **figure 10.26**).

Genetic Screening

When a pregnancy is determined to be high risk, many women elect to undergo **amniocentesis,** a procedure that permits the prenatal diagnosis of many genetic disorders. **Figure 10.35** shows how an amniocentesis is performed. In the fourth month of pregnancy, a sterile hypodermic needle is inserted into the expanded uterus of the mother, and a small sample of the amniotic fluid bathing the fetus is removed. Within the fluid are free-floating cells derived from the fetus; once removed, these cells can be grown in cultures in the laboratory. During amniocentesis, the position of the needle and that of the fetus are usually observed by means of **ultrasound.** The ultrasound image in **figure 10.36** clearly reveals the fetus's position in the uterus. You can see its head and a hand extending up, maybe sucking its thumb. The sound waves used in ultrasound generate a live image that permits the person withdrawing the amniotic fluid to do so without damaging the fetus. In addition, ultrasound can be used to examine the fetus for signs of major abnormalities.

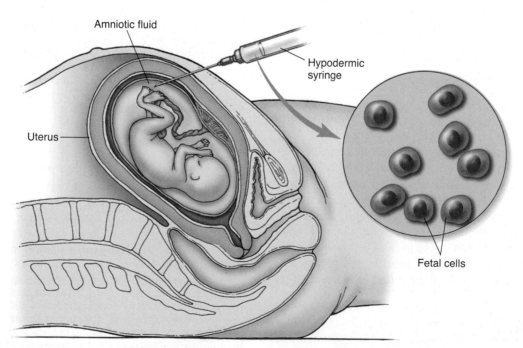

Figure 10.35   Amniocentesis.

A needle is inserted into the amniotic cavity, and a sample of amniotic fluid, containing some free cells derived from the fetus, is withdrawn into a syringe. The fetal cells are then grown in culture and their karyotype and many of their metabolic functions are examined.

In recent years, physicians have increasingly turned to another invasive procedure for genetic screening called **chorionic villus sampling.** In this procedure, the physician removes cells from the chorion, a membranous part of the placenta that nourishes the fetus. This procedure can be used earlier in pregnancy (by the eighth week) and yields results much more rapidly than does amniocentesis, but can increase the risk of miscarriage.

Genetic counselors look at three things in the cultures of cells obtained from amniocentesis or chorionic villus sampling:

1. **Chromosomal karyotype.** Analysis of the karyotype can reveal aneuploidy (extra or missing chromosomes) and gross chromosomal alterations.

2. **Enzyme activity.** In many cases, it is possible to test directly for the proper functioning of enzymes involved in genetic disorders. The lack of normal enzymatic activity signals the presence of the disorder. Thus, the lack of the enzyme responsible for breaking down phenylalanine signals PKU (phenylketonuria), the absence of the enzyme responsible for the breakdown of gangliosides indicates Tay-Sachs disease, and so forth.

3. **Genetic markers.** Genetic counselors can look for an association with known genetic markers. For sickle-cell anemia, Huntington's disease, and one form of muscular dystrophy (a genetic disorder characterized by weakened muscles), investigators have found other mutations on the same chromosomes that, by chance, occur at about the same place as the mutations that cause those disorders. By testing for the presence of these other mutations, a genetic counselor can identify individuals with a high probability of possessing the disorder-causing mutations. Finding such mutations in the first place is a little like searching for a needle in a haystack, but persistent efforts have proved successful in these three disorders. The associated mutations are detectable because they alter the length of the DNA segments that DNA-cleaving enzymes produce when they cut strands of DNA at particular places, an approach described in more detail in chapter 13.

DNA Screening

The mutations that cause hereditary defects are frequently caused by alteration of a single DNA nucleotide within a key gene. Such spot differences between the version of a gene you have and the one another person has are called "single nucleotide polymorphisms," or SNPs. With the completion of the Human Genome Project (described in detail in chapter 13), researchers have begun assembling a huge database of hundreds of thousands of SNPs. Each of us differs from the standard "type sequence" in several thousand gene-altering SNPs. Screening SNPs and comparing them to known SNP databases should soon allow genetic counselors to screen each patient for copies of genes leading to hereditary disorders such as cystic fibrosis and muscular dystrophy.

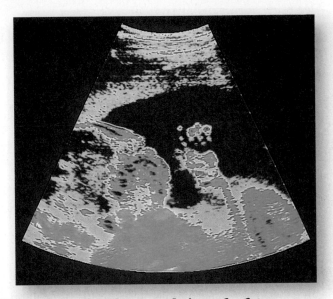

Figure 10.36   An ultrasound view of a fetus.
During the fourth month of pregnancy, when amniocentesis is normally performed, the fetus usually moves about actively. The head of the fetus (visualized in *green*) is to the *left*.

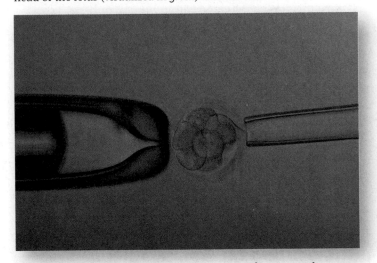

Figure 10.37   Preimplantation genetic screening.
The photograph shows a human embryo at the eight-cell stage, just before one of the eight cells is to be extracted for genetic testing by researchers.

Parents conceiving by in vitro fertilization have available a well-established screening procedure known as **preimplantation genetic screening.** In this test, the egg is fertilized outside the mother, in glassware, and allowed to divide three times, until it contains eight cells. One of the eight cells is then removed from each of several such 8-cell embryos (**figure 10.37**) and tested for any of 150 genetic defects. The remaining 7-cell embryos are each able to develop into normal fetuses, giving the parents the choice of identifying and implanting an embryo that is disease free.

> **Key Learning Outcome 10.11**   It has recently become possible to detect genetic defects early in pregnancy, allowing for appropriate planning by the prospective parents.

Why Woolly Hair Runs in Families

The woman in the photo on the right does not cut her hair. Her hair breaks off naturally as it grows, keeping it from getting long. Other members of her family have the same sort of hair, suggesting it is a hereditary trait. Because of its curly, fuzzy texture, this trait has been given the name "woolly hair."

While the woolly-hair trait is rare, it flares up in certain families. The extensive pedigree below (drawn curved so as to fit in the large families produced by the second and subsequent generations) records the incidence of woolly hair in five generations (the Roman numerals on the left) of a Norwegian family. As is the convention, affected individuals are indicated by solid symbols, with circles females and squares males. The pedigree below will provide you with the information you need to discover how this trait is inherited within human families.

1. **Applying Concepts** In the diagram below, how many individuals are documented? Are all of them related?

2. **Interpreting Data**
   a. Does the woolly-hair trait appear in both sexes equally?
   b. Does every woolly-hair child have a woolly-hair parent?
   c. What percentage of the offspring born to a woolly-haired parent are also woolly haired?

3. **Making Inferences**
   a. Is woolly hair sex-linked or autosomal?
   b. Is woolly hair dominant or recessive?
   c. Is the woolly-hair trait determined by a single gene, or by several?

4. **Drawing Conclusions**
   a. How many copies of the woolly-hair allele are necessary to produce a detectable change in a person's hair?
   b. Are there any woolly-hair homozygous individuals in the pedigree? Explain.

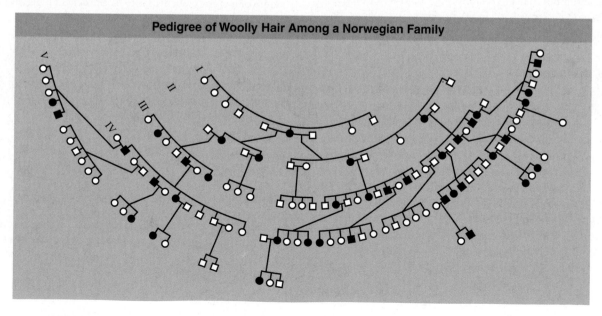

Pedigree of Woolly Hair Among a Norwegian Family

Mendel

10.1 Mendel and the Garden Pea

- Mendel, shown here from **figure 10.1**, studied inheritance using the garden pea (**figure 10.2**) and the scientific method.

- Mendel used plants that were true-breeding for a particular characteristic; these plants were the P generation. He then crossed two P generation plants that expressed alternate traits (different forms of a characteristic). Their offspring were called the F₁ generation. He then allowed the F₁ plants to self-fertilize, giving rise to the F₂ generation (**figure 10.3**).

10.2 What Mendel Observed

- In Mendel's experiments, the F₁ generation plants all expressed the same alternative form, called the dominant trait. In the F₂ generation, 3/4 of the offspring expressed the dominant trait and 1/4 expressed the other trait, called the recessive trait. Mendel found this 3:1 ratio in the F₂ generation in all the seven traits he studied (**table 10.1**). Mendel then found that this 3:1 ratio was actually a 1:2:1 ratio—1 true-breeding dominant: 2 not-true-breeding dominant: 1 true-breeding recessive (**figure 10.5**).

10.3 Mendel Proposes a Theory

- Mendel's theory of heredity explains that characteristics are passed from parent to offspring as alleles (alternate forms of a characteristic), one allele inherited from each parent. If both of the alleles are the same, the individual is homozygous for the trait. If the individual has one dominant and one recessive allele, it is heterozygous for the trait. An individual's alleles are referred to as its genotype, and the expression of those alleles is its phenotype.

- A Punnett square can be used to predict the probabilities of inheriting certain genotypes and phenotypes in the offspring of a cross (**figures 10.7** and **10.8**).

- A testcross is the mating of an individual of unknown genotype with an individual that is homozygous recessive. It is done to determine if the unknown genotype is homozygous or heterozygous for the dominant trait (**figure 10.9**).

10.4 Mendel's Laws

- Mendel's law of segregation states that alleles are distributed into gametes so that half of the gametes will carry one copy of a trait and the remaining gametes carry the other copy of the trait. Mendel's law of independent assortment states that the inheritance of one trait does not influence the inheritance of other traits. Genes located on different chromosomes are inherited independent of each other (**figure 10.10**).

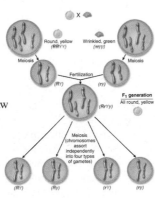

From Genotype to Phenotype

10.5 How Genes Influence Traits

- Genes coded in DNA determine phenotype because DNA encodes the amino acid sequences of proteins, and proteins are the outward expression of genes (**figure 10.11**). Alternative forms of a gene, called alleles, result from mutations.

10.6 Some Traits Don't Show Mendelian Inheritance

- Not all traits follow the inheritance patterns outlined by Mendel. Continuous variation results when more than one gene contributes in a cumulative way to a phenotype, resulting in a continuous array of phenotypes. This pattern of inheritance is called polygenic (**figure 10.12**). Pleiotropic effects result when one gene influences more than one trait (**figure 10.13**). Incomplete dominance results when alternative alleles are not fully dominant or fully recessive such that heterozygous individuals express a phenotype that is intermediate between the dominant and recessive phenotypes (**figure 10.14**). The expression of some genes is influenced by environmental factors (**figure 10.15**), such as the changing of fur color triggered by heat-sensitive alleles. Epistasis occurs when two or more genes interact, having an additive or masking effect (**figure 10.16**) or resulting in several different phenotypes (**figure 10.17**). Codominance occurs when there isn't a dominant allele—two alleles are expressed resulting in phenotypic expression of both alleles (**figures 10.18** and **10.19**).

Chromosomes and Heredity

10.7 Chromosomes Are the Vehicles of Mendelian Inheritance

- Genes assort independently because they are located on chromosomes that assort independently during meiosis. Morgan demonstrated this using an X-linked gene in fruit flies (**figure 10.21**). However, the farther apart two genes are on a chromosome, the more likely they are to segregate independently because of crossing over (**figure 10.22**).

10.8 Human Chromosomes

- Humans have 23 pairs of homologous chromosomes: 22 pairs of autosomes and one pair of sex chromosomes. Nondisjunction occurs when sister chromatids or homologous pairs fail to separate during meiosis (**figure 10.24**), resulting in gametes with too many or too few chromosomes. Nondisjunction of autosomes is usually fatal, Down Syndrome being an exception (**figures 10.25** and **10.26**), but the effects of nondisjunction of sex chromosomes are less severe (**figure 10.27**).

Human Hereditary Disorders

10.9 Studying Pedigrees

- By looking at pedigrees, scientists can determine various aspects about the genetics of a trait (**figures 10.28** and **10.29**).

10.10 The Role of Mutation

- Mutations can lead to genetic disorders such as hemophilia (**figure 10.30**), sickle-cell disease (**figures 10.31–10.32**), Tay-Sachs disease (**figure 10.33**), and Huntington's disease (**figure 10.34**).

10.11 Genetic Counseling and Therapy

- Some genetic disorders can be detected during pregnancy using methods such as amniocentesis (**figure 10.35**) and chorionic villus sampling.

Test Your Understanding

1. Gregor Mendel studied the garden pea plants because
   a. pea plants are small, easy to grow, grow quickly, and produce lots of flowers and seeds.
   b. he knew about studies with the garden pea that had been done for hundreds of years, and wanted to continue them, using math—counting and recording differences.
   c. he knew that there were many varieties available with distinctive characteristics.
   d. All of the above.

2. Mendel examined seven characteristics, such as flower color. He crossed plants with two different forms of a character (purple flowers and white flowers). In every case the first generation of offspring (F_1) were
   a. all purple flowers.
   b. half purple flowers and half white flowers.
   c. 3/4 purple and 1/4 white flowers.
   d. all white flowers.

3. Following question 2, when Mendel allowed the F_1 generation to self-fertilize, the offspring in the F_2 generation were
   a. all purple flowers.
   b. half purple flowers and half white flowers.
   c. 3/4 purple and 1/4 white flowers.
   d. all white flowers.

4. Mendel then studied his results, and proposed a set of hypotheses stating that parents transmit
   a. traits directly to their offspring and they are expressed.
   b. some factor, or information, about traits to their offspring and it may or may not be expressed.
   c. some factor, or information, about traits to their offspring and it will always be expressed.
   d. some factor, or information, about traits to their offspring and both traits are expressed in every generation, perhaps in a "blended" form with information from the other parent.

5. A cross between two individuals results in a ratio of 9:3:3:1 for four possible phenotypes. This is an example of a
   a. dihybrid cross.   c. testcross.
   b. monohybrid cross.   d. None of these is correct.

6. Human height shows a continuous variation from the very short to the very tall. Height is most likely controlled by
   a. epistatic genes.
   b. environmental factors.
   c. sex-linked genes.
   d. multiple genes.

7. In the human ABO blood grouping, the four basic blood types are type A, type B, type AB, and type O. The blood proteins A and B are
   a. simple dominant and recessive traits.
   b. incomplete dominant traits.
   c. codominant traits.
   d. sex-linked traits.

8. What finding finally determined that genes were carried on chromosomes?
   a. heat sensitivity of certain enzymes that determined coat color
   b. sex-linked eye color in fruit flies
   c. the finding of complete dominance
   d. establishing pedigrees

9. Nondisjunction
   a. occurs when homologous chromosomes or sister chromatids fail to separate during meiosis.
   b. may lead to Down syndrome.
   c. results in aneuploidy.
   d. All of the above.

10. Which of the following analyses can detect aneuploidy?
    a. enzyme activity
    b. chromosomal karyotyping
    c. pedigrees
    d. genetic markers

Apply Your Understanding: Additional Genetics Problems

1. Silky feathers in chickens is a single-gene recessive trait whose effect is to produce shiny plumage. If you had a normal-feathered bird, what would be the easiest cross to perform to determine if a bird is homozygous or heterozygous for the silky allele?

2. Among Hereford cattle there is a dominant allele called *polled;* the individuals that have this allele lack horns. Suppose you acquire a herd consisting entirely of polled cattle, and you carefully determine that no cow in the herd has horns. Some of the calves born that year, however, grow horns. You remove them from the herd and make certain that no horned adult has gotten into your pasture. Despite your efforts, more horned calves are born the next year. What is the reason for the appearance of the horned calves? If your goal is to maintain a herd consisting entirely of polled cattle, what should you do?

3. Brachydactyly is a rare human trait that causes a shortening of the length of the fingers by a third. A review of medical records reveals that the progeny of marriages between a brachydactylous person and a normal person are approximately half brachydactylous. What proportion of offspring in matings between two brachydactylous individuals would be expected to be brachydactylous?

4. Your instructor presents you with a *Drosophila* (fruit fly) with red eyes, as well as a stock of white-eyed flies and another stock of flies homozygous for the red-eye allele. You know that the presence of white eyes in *Drosophila* is caused by homozygosity for a recessive allele. How would you determine whether the single red-eyed fly was heterozygous for the white-eye allele?

5. Hemophilia is a recessive sex-linked human blood disease that leads to failure of blood to clot normally. One form of hemophilia has been traced to the royal family of England, from which it spread throughout the royal families of Europe. For the purposes of this problem, assume that it originated as a mutation either in Prince Albert or in his wife, Queen Victoria.

   a. Prince Albert did not have hemophilia. If the disease is a sex-linked recessive abnormality, how could it have originated in Prince Albert, a male, who would have been expected to exhibit sex-linked recessive traits?

   b. Alexis, the son of Czar Nicholas II of Russia and Empress Alexandra (a granddaughter of Victoria), had hemophilia, but their daughter Anastasia did not. Anastasia died, a victim of the Russian revolution, before she had any children. Can we assume that Anastasia would have been a carrier of the disease? Would your answer be different if the disease had been present in Nicholas II or in Alexandra?

6. A normally pigmented man marries an albino woman. They have three children, one of whom is an albino. What is the genotype of the father?

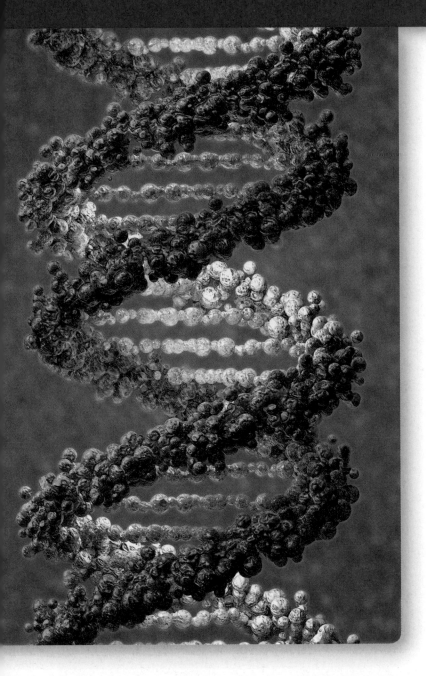

11

DNA: The Genetic Material

Learning Objectives

Genes Are Made of DNA

11.1 The Discovery of Transformation
1. Identify the two macromolecules found in chromosomes.
2. Describe Griffith's experiment demonstrating transformation.
3. Explain why Griffith's mixture of harmless dead bacteria with capsules and harmless live bacteria lacking capsules is deadly.

11.2 Experiments Identifying DNA as the Genetic Material
1. List the five ways that Avery's transforming principle resembles DNA.
2. Describe the Hershey-Chase experiment, and explain how it demonstrated that DNA is the hereditary material.

11.3 Discovering the Structure of DNA
1. Name the four DNA nucleotides, and identify which are purines.
2. Distinguish purine from pyrimidine.
3. State Chargaff's rule.
4. Describe Watson and Crick's proposed structure for the DNA molecule.
5. Explain why each nucleotide in Watson and Crick's DNA structure can form a base pair with only one of the four potential nucleotides.

DNA Replication

11.4 How the DNA Molecule Copies Itself
1. Diagram three alternative mechanisms of DNA replication.
2. Describe the Meselson-Stahl experiment, and explain how it confirms one of these mechanisms.
3. Name and describe the four stages of DNA replication.
4. Distinguish between the leading and lagging strands; define Okazaki fragments.
5. Explain the function of RNA primers and of ligases.

Altering the Genetic Message

11.5 Mutation
1. Contrast the inheritance of somatic and germ-line mutations.
2. List four molecular events that can alter the sequence of DNA.
3. Distinguish between transposition and chromosomal rearrangement.

Today's Biology: DNA Fingerprinting
Today's Biology: Tracing the DNA of Irish Kings
Biology and Staying Healthy: Protecting Your Genes

Inquiry & Analysis: Are Mutations Random or Directed by the Environment?

The realization that patterns of heredity can be explained by the segregation of chromosomes in meiosis raised a question that occupied biologists for over 50 years: What is the exact nature of the connection between hereditary traits and chromosomes? In this chapter you will examine some of the chain of experiments that have led to our current understanding of the molecular mechanisms of heredity. The experiments determining that DNA is the genetic material are among the most elegant in science. Just as in a good detective story, each conclusion has led to new questions. The intellectual path taken has not always been a straight one, the best questions not always obvious. But however erratic and lurching the course of the experimental journey, our picture of heredity has become progressively clearer, the image more sharply defined. We now understand in considerable detail how the DNA molecule copies itself, and how changes to it lead to hereditary gene mutations.

11.1 The Discovery of Transformation

The Griffith Experiment

As we learned in chapters 8, 9, and 10, chromosomes contain genes, which, in turn, contain hereditary information. However, Mendel's work left a key question unanswered: What *is* a gene? When biologists began to examine chromosomes in their search for genes, they soon learned that chromosomes are made of two kinds of macromolecules, both of which you encountered in chapter 3: **proteins** (long chains of *amino acid* subunits linked together in a string) and **DNA** (deoxyribonucleic acid—long chains of *nucleotide* subunits linked together in a string). It was possible to imagine that either of the two was the stuff that genes are made of—information might be stored in a sequence of different amino acids, or in a sequence of different nucleotides. But which one is the stuff of genes, protein or DNA? This question was answered clearly in a variety of different experiments, all of which shared the same basic design: If you separate the DNA in an individual's chromosomes from the protein, which of the two materials is able to change another individual's genes?

In 1928, British microbiologist Frederick Griffith made a series of unexpected observations while experimenting with pathogenic (disease-causing) bacteria. **Figure 11.1** takes you stepwise through his discoveries. When he infected mice with a virulent strain of *Streptococcus pneumoniae* bacteria (then known as *Pneumococcus*), the mice died of blood poisoning, as you can see in **panel 1**. However, when he infected similar mice with a mutant strain of *S. pneumoniae* that lacked the virulent strain's polysaccharide capsule, the mice showed no ill effects, as you can see in **panel 2**. The capsule was apparently necessary for infection. The normal pathogenic form of this bacterium is referred to as the *S form* because it forms smooth colonies in a culture dish. The mutant form, which lacks an enzyme needed to manufacture the polysaccharide capsule, is called the *R form* because it forms rough colonies.

To determine whether the polysaccharide capsule itself had a toxic effect, Griffith injected dead bacteria of the virulent S strain into mice and as **panel 3** shows, the mice remained perfectly healthy. Finally, as shown in **panel 4**, he injected mice with a mixture containing dead S bacteria of the virulent strain and live, capsuleless R bacteria, each of which by itself did not harm the mice. Unexpectedly, the mice developed disease symptoms and many of them died. The blood of the dead mice was found to contain high levels of live, virulent *Streptococcus* type S bacteria, which had surface proteins characteristic of the live (previously R) strain. Somehow, the information specifying the polysaccharide capsule had passed from the dead, virulent S bacteria to the live, capsuleless R bacteria in the mixture, permanently transforming the capsuleless R bacteria into the virulent S variety.

> **Key Learning Outcome 11.1** Hereditary information can pass from dead cells to living ones and transform them.

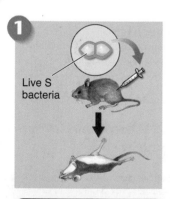

Live S bacteria

S bacteria have a polysaccharide capsule and are pathogenic. When they are injected into mice, the mice die.

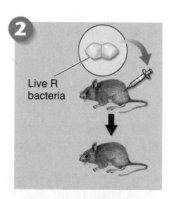

Live R bacteria

R bacteria do not have the capsule and do not kill mice.

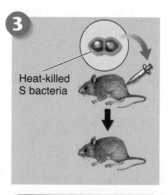

Heat-killed S bacteria

Heat-killed bacteria are dead but still have the capsule. They do not kill mice.

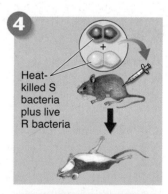

Heat-killed S bacteria plus live R bacteria

A mixture of live R bacteria and heat-killed S bacteria does cause mice to die.

Figure 11.1 How Griffith discovered transformation.

Transformation, the movement of a gene from one organism to another, provided some of the key evidence that DNA is the genetic material. Griffith found that extracts of dead pathogenic strains of the bacterium *Streptococcus pneumoniae* can "transform" live harmless strains into live pathogenic strains.

11.2 Experiments Identifying DNA as the Genetic Material

The Avery Experiments

The agent responsible for transforming *Streptococcus* went undiscovered until 1944. In a classic series of experiments, Oswald Avery and his coworkers Colin MacLeod and Maclyn McCarty characterized what they referred to as the "transforming principle." Avery and his colleagues prepared the same mixture of dead S *Streptococcus* and live R *Streptococcus* that Griffith had used, but first they removed as much of the protein as they could from their preparation of dead S *Streptococcus*, eventually achieving 99.98% purity. Despite the removal of nearly all protein from the dead S *Streptococcus*, the transforming activity was not reduced. Moreover, the properties of the transforming principle resembled those of DNA in several ways:

Same chemistry as DNA. When the purified principle was analyzed chemically, the array of elements agreed closely with DNA.

Same behavior as DNA. In an ultracentrifuge, the transforming principle migrated like DNA; in electrophoresis and other chemical and physical procedures, it also acted like DNA.

Not affected by lipid and protein extraction. Extracting the lipid and protein from the purified transforming principle did not reduce its activity.

Not destroyed by protein- or RNA-digesting enzymes. Protein-digesting enzymes did not affect the principle's activity, nor did RNA-digesting enzymes.

Destroyed by DNA-digesting enzymes. The DNA-digesting enzyme destroyed all transforming activity.

The evidence was overwhelming. They concluded that "a nucleic acid of the deoxyribose type is the fundamental unit of the transforming principle of *Pneumococcus* Type III"—in essence, that DNA is the hereditary material.

The Hershey-Chase Experiment

Avery's result was not widely appreciated at first because most biologists still preferred to think that genes were made of proteins. In 1952, however, a simple experiment carried out by Alfred Hershey and Martha Chase was impossible to ignore. The team studied the genes of viruses that infect bacteria. These viruses attach themselves to the surface of bacterial cells and inject their genes into the interior; once inside, the genes take over the genetic machinery of the cell and order the manufacturing of hundreds of new viruses. When mature, the progeny viruses burst out to infect other cells. These bacteria-infecting viruses have a very simple structure: a core of DNA surrounded by a coat of protein.

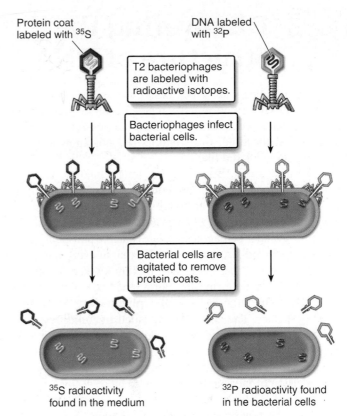

Protein coat labeled with ^{35}S

DNA labeled with ^{32}P

T2 bacteriophages are labeled with radioactive isotopes.

Bacteriophages infect bacterial cells.

Bacterial cells are agitated to remove protein coats.

^{35}S radioactivity found in the medium

^{32}P radioactivity found in the bacterial cells

Figure 11.2 The Hershey-Chase experiment.
The experiment that convinced most biologists that DNA is the genetic material was carried out soon after World War II, when radioactive isotopes were first becoming commonly available to researchers. Hershey and Chase used different radioactive labels to "tag" and track protein and DNA. They found that when bacterial viruses inserted their genes into bacteria to guide the production of new viruses, ^{35}S radioactivity did not enter infected bacterial cells and ^{32}P radioactivity did. Clearly the virus DNA, not the virus protein, was responsible for directing the production of new viruses.

In this experiment, shown in **figure 11.2**, Hershey and Chase used radioactive isotopes to "label" the DNA and protein of the viruses. Radioactively tagged molecules are indicated in red in the figure. In the preparation on the right, the viruses were grown so that their DNA contained radioactive phosphorus (^{32}P); in another preparation, on the left side of the figure, the viruses were grown so that their protein coats contained radioactive sulfur (^{35}S). After the labeled viruses were allowed to infect bacteria, Hershey and Chase shook the suspensions forcefully to dislodge attacking viruses from the surface of bacteria, used a rapidly spinning centrifuge to isolate the bacteria, and then asked a very simple question: What did the viruses inject into the bacterial cells, protein or DNA? They found that the bacterial cells infected by viruses containing the ^{32}P label had labeled tracer in their interiors; cells infected by viruses containing the ^{35}S labeled tracer did not. The conclusion was clear: The genes that viruses use to specify new viruses are made of DNA and not protein.

Key Learning Outcome 11.2 Several key experiments demonstrated conclusively that DNA, not protein, is the hereditary material.

11.3 Discovering the Structure of DNA

As it became clear that DNA was the molecule that stored the hereditary information, researchers began to question how this nucleic acid could carry out the complex function of inheritance. At the time, investigators did not know what the DNA molecule looked like.

We now know that DNA is a long, chainlike molecule made up of subunits called **nucleotides.** As you can see in **figure 11.3**, each nucleotide has three parts: a central sugar called *deoxyribose,* a phosphate (PO_4) group, and an organic base. The sugar (the lavender pentagon structure) and the phosphate group (the yellow-circled structure) are the same in every nucleotide of DNA. However, there are four different kinds of bases: two large ones with double-ring structures, and two small ones with single rings. The large bases, called **purines,** are **A** (adenine) and **G** (guanine). The small bases, called **pyrimidines,** are **C** (cytosine) and **T** (thymine). A key observation, made by Erwin Chargaff, was that DNA molecules always had equal amounts of purines and pyrimidines. In fact, with slight variations due to imprecision of measurement, the amount of A always equals the amount of T, and the amount of G always equals the amount of C. This observation (A = T, G = C), known as **Chargaff's rule,** suggested that DNA had a regular structure.

The significance of Chargaff's rule became clear in 1953 when the British chemist Rosalind Franklin carried out an X-ray diffraction experiment. In these experiments, DNA molecules are bombarded with X-ray beams, and when individual rays encounter atoms, their paths are bent or diffracted like a thrown ball bounces off or around an object. Each atomic encounter creates a pattern on photographic film, shown in **figure 11.4a**, that looks like the ripples created by tossing a rock into a smooth lake. Franklin's results suggested that the DNA molecule had the shape of a coiled spring or a corkscrew, a form called a **helix,** with the image in the photo from the viewpoint of looking down the center of the molecule.

Franklin's work was shared with two researchers at Cambridge University, Francis Crick and James Watson, before it was published. Using Tinkertoy-like models of the bases, Watson and Crick deduced the true structure of DNA (**figure 11.4b**): The DNA molecule is a **double helix,** a winding staircase of two strands whose bases face one another (**figure 11.4c**). Chargaff's rule is a direct reflection of this structure—every bulky purine on one strand is paired with a slender pyrimidine on the other strand. Specifically, A (the blue bases) pairs with T (the orange bases), and G (the purple bases) pairs with C (the pink bases). Because hydrogen bonds, shown as dotted lines, can form between the **base pairs,** the molecule keeps a constant thickness.

Key Learning Outcome 11.3 The DNA molecule consists of two strands of nucleotides held together by hydrogen bonds between bases. The two strands wind into a double helix.

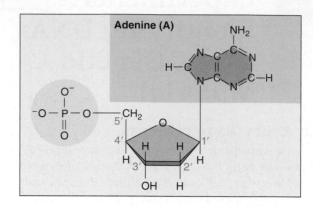

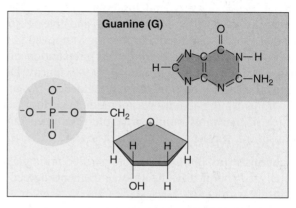

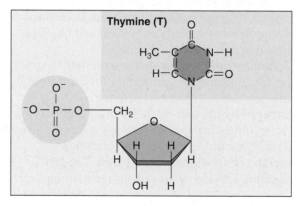

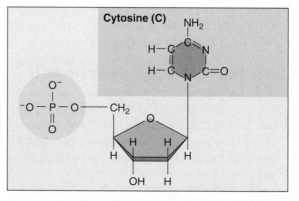

Figure 11.3 The four nucleotide subunits that make up DNA.

The nucleotide subunits of DNA are composed of three parts: a central five-carbon sugar called deoxyribose, a phosphate group, and an organic, nitrogen-containing base.

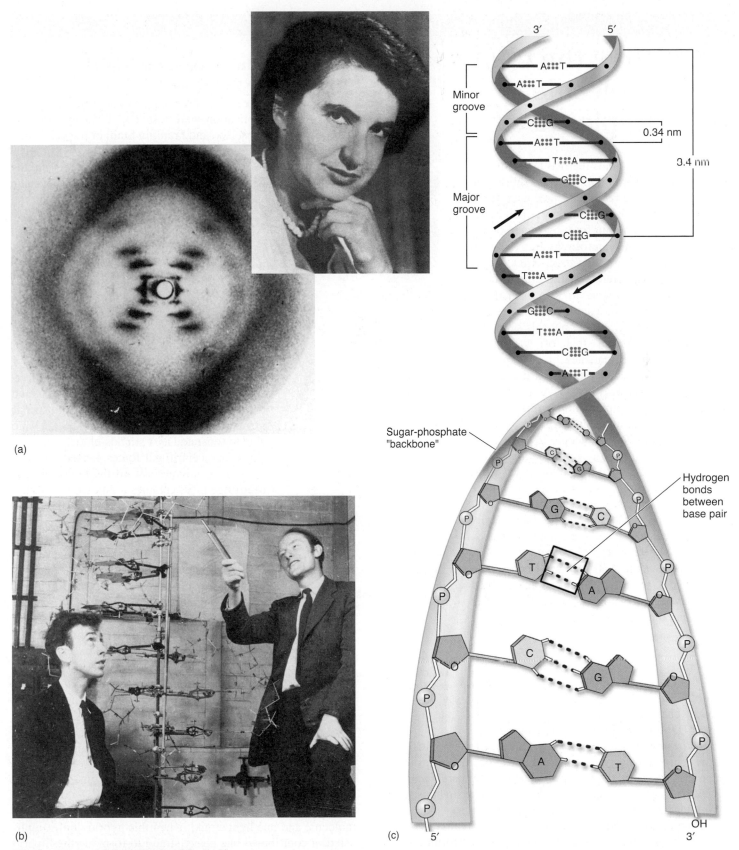

(a)

(b)

(c)

Figure 11.4   The DNA double helix.

(a) This X-ray diffraction photograph was made in 1953 by Rosalind Franklin (inset) in the laboratory of Maurice Wilkins. It suggested to Watson and Crick that the DNA molecule was a helix, like a winding staircase. (b) In 1953 Watson and Crick deduced the structure of DNA. James Watson (seated and peering up at their homemade model of the DNA molecule) was a young American postdoctoral student, and Francis Crick (pointing) was an English scientist. (c) The dimensions of the double helix were suggested by the X-ray diffraction studies. In a DNA duplex molecule, only two base pairs are possible: adenine (A) with thymine (T) and guanine (G) with cytosine (C). A G–C base pair has three hydrogen bonds; an A–T base pair has only two.

11.4 How the DNA Molecule Copies Itself

The attraction that holds the two DNA strands together is the formation of weak hydrogen bonds between the bases that face each other from the two strands. That is why A pairs with T and not C; A can only form hydrogen bonds with T. Similarly, G can form hydrogen bonds with C but not T. In the Watson-Crick model of DNA, the two strands of the double helix are said to be *complementary* to each other. One chain of the helix can have any sequence of bases, of A, T, G, and C, but this sequence completely determines that of its partner in the helix. If the sequence of one chain is ATTGCAT, the sequence of its partner in the double helix must be TAACGTA. Each chain in the helix is a complementary mirror image of the other. This **complementarity** makes it possible for the DNA molecule to copy itself during cell division in a very direct manner. But, there are three possible alternatives as to how the DNA could serve as a template for the assembly of new DNA molecules.

First, the two strands of the double helix could separate and serve as templates for the assembly of two new strands by base pairing A with T and G with C. This is what happens in **figure 11.5a**, with the original strand colored blue and the newly formed strands red. After replicating, the original strands rejoin, preserving the original strand of DNA and forming an entirely new strand. This is called *conservative replication*.

In the second alternative, the double helix need only "unzip" and assemble a new complementary chain along each single strand. This form of DNA replication is called *semiconservative replication*, because while the sequence of the original duplex is conserved after one round of replication, the duplex itself is not. Instead, each strand of the duplex becomes part of another duplex. You can see in **figure 11.5b** that the blue strand is from the original helix and the red strand is newly formed.

In the third alternative, called *dispersive replication*, the original DNA would serve as a template for the formation of new DNA strands but the new and old DNA would be dispersed among the two daughter strands. As shown in **figure 11.5c**, each daughter strand is made up of sections of original (blue) strands and new (red) strands.

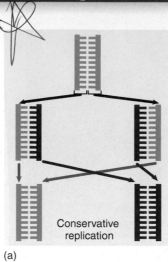

(a) Conservative replication

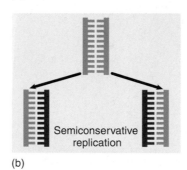

(b) Semiconservative replication

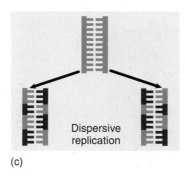

(c) Dispersive replication

Figure 11.5 Alternative mechanisms of DNA replication.

The Meselson-Stahl Experiment

The three alternative hypotheses of DNA replication were tested in 1958 by Matthew Meselson and Franklin Stahl of the California Institute of Technology. These two scientists grew bacteria in a medium containing the heavy isotope of nitrogen, ^{15}N, which became incorporated into the bases of the bacterial DNA (the upper petri dish in **figure 11.6**). After several generations, samples were taken from this culture and grown in a medium containing the normal lighter isotope ^{14}N, which became incorporated into the newly replicating DNA. Bacterial samples were taken from the ^{14}N media at 20 minute intervals (❷ through ❹). DNA was extracted from all three samples and a fourth sample, ❶, that served as a control.

By dissolving the DNA they had collected in a heavy salt called cesium chloride, and then spinning the solution at very high speeds in an ultracentrifuge, Meselson and Stahl were able to separate DNA strands of different densities. The centrifugal forces caused the cesium ions to migrate toward the bottom of the centrifuge tube, creating a gradient of cesium concentration, and thus a gradation of density. Each DNA strand floats or sinks in the gradient until it reaches the position where its density exactly matches the density of the cesium there. Because ^{15}N strands are denser than ^{14}N strands, they migrate farther down the tube to a denser region of cesium.

The DNA collected immediately after the transfer was all dense, as shown in test tube ❷. However, after the bacteria completed their first round of DNA replication in the ^{14}N medium, the density of their DNA had decreased to a value intermediate between ^{14}N-DNA and ^{15}N-DNA, as shown in test tube ❸. After the second round of replication, two density classes of DNA were observed, one intermediate and one equal to that of ^{14}N-DNA, as shown in test tube ❹.

Meselson and Stahl interpreted their results as follows: After the first round of replication, each daughter DNA duplex was a hybrid possessing one of the heavy strands of the parent molecule and one light strand; when this hybrid duplex replicated, it contributed one heavy strand to form another hybrid duplex and one light strand to form a light duplex. Thus, this experiment clearly ruled out conservative and dispersive DNA replication, and confirmed the prediction of the Watson-Crick model that DNA replicates in a semiconservative manner.

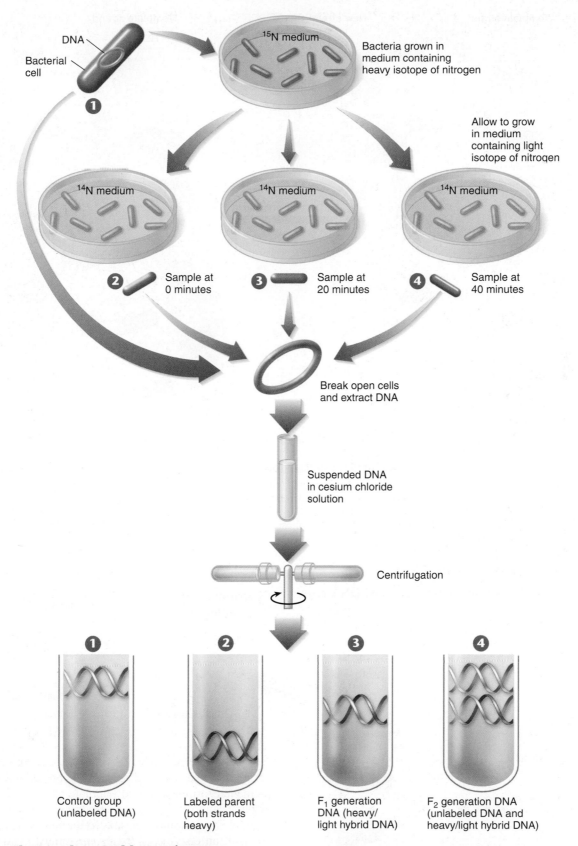

Figure 11.6 The Meselson-Stahl experiment.

Bacterial cells were grown for several generations in a medium containing a heavy isotope of nitrogen (^{15}N) and then were transferred to a new medium containing the normal lighter isotope (^{14}N). (The bacteria shown here are not drawn to scale, as tens of thousands of bacterial cells grow on even a tiny portion of a plate in culture.) At various times thereafter, samples of the bacteria were collected, and their DNA was dissolved in a solution of cesium chloride, which was spun rapidly in a centrifuge. The labeled and unlabeled DNA settled in different areas of the tube because they differed in weight. The DNA with two heavy strands settled down toward the bottom of the tube. The DNA with two light strands settled higher up in the tube. The DNA with one heavy and one light strand settled in between the other two.

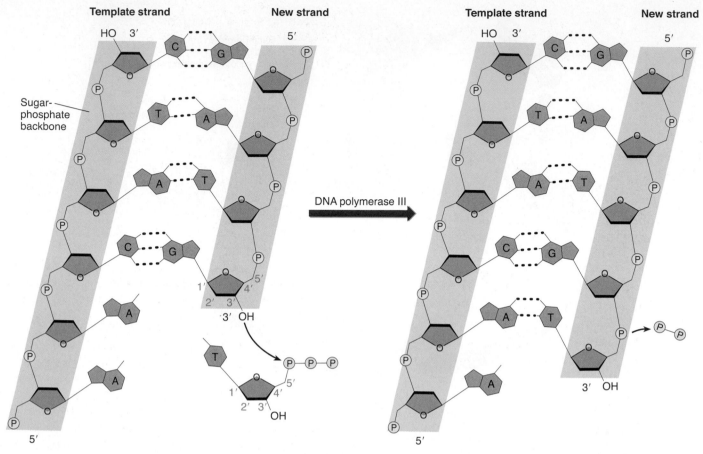

Figure 11.7   How nucleotides are added in DNA replication.

In a nucleotide, the phosphate group is attached to the 5' carbon atom of the sugar, and an OH group is attached to the 3' carbon atom. So, on a DNA strand, there will be a 5' phosphate at one end of the chain and a 3' OH at the other end. In the DNA double helix, the two strands of nucleotides pair up in opposite orientations, with one strand running 5' to 3' and the other running 3' to 5'. When nucleotides are added to a growing strand of DNA by the enzyme DNA polymerase III, the first phosphate group of the incoming nucleotide attaches to the OH group on the end nucleotide of the existing strand.

How DNA Copies Itself

The copying of DNA before cell division is called **DNA replication.** This process is overseen by six proteins in prokaryotes (in eukaryotes, some of the enzymes are different). These proteins coordinate the unwinding of the DNA duplex and assembly of new complementary DNA strands by the addition of nucleotides to existing strands (**figure 11.7**). Here is how the process works.

❶ Unwinding

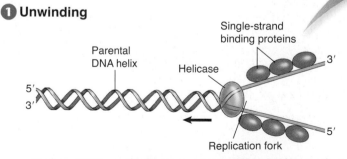

Before DNA replication can begin, an enzyme called *helicase* separates and unwinds the strands of the parental DNA. Single-strand binding proteins stabilize the single-stranded regions of DNA before they are replicated. Helicase moves up the DNA helix, unwinding as it goes.

❷ Priming the Leading Strand

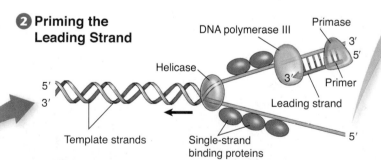

After the parental DNA duplex is unwound, an enzyme complex called *DNA polymerase III* can add nucleotides complementary to the exposed template strand of DNA. But this enzyme cannot begin a new strand; it can only add nucleotides to an existing strand. Thus, before a new strand of DNA can be built, an enzyme called *primase* must synthesize a short section of joined RNA nucleotides, called a *primer,* complementary to the single-stranded template. DNA polymerase III can then add onto the primer and assemble a complementary new strand of DNA on each old strand. One of the new strands of DNA is called the *leading strand;* it is the one that is extended in a 5' to 3' direction toward the replication fork.

conservative ✓

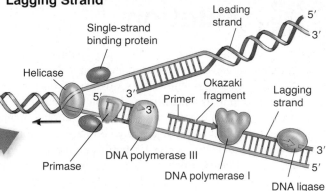

④ Priming and Building the Lagging Strand

Single-strand binding protein

Helicase

Leading strand

5′
3′

5′
3′

Primase

Primer

Okazaki fragment

Lagging strand

DNA polymerase III

DNA polymerase I

DNA ligase

3′
5′

③ Building the Leading Strand

DNA polymerase I

DNA polymerase III

Helicase

5′
3′

Leading strand

Single-strand binding proteins

5′

DNA polymerase III builds the leading strand by adding nucleotides to the 3′ end of the strand: the 5′ phosphate end of a nucleotide to be added attaches to the 3′ sugar end of the existing strand. DNA polymerase III moves toward the replication fork and builds the leading strand as one continuous strand. Before the leading strand can be completed, another polymerase enzyme called *DNA polymerase I* removes the RNA primer and fills in the gap with DNA nucleotides. The newly synthesized hybrid DNA can then rewind into a helix.

Because DNA polymerase III can only assemble new strands in the 5′ to 3′ direction, the other strand, called the *lagging strand,* is assembled in short 5′ to 3′ segments, moving away from the replication fork. Each lagging strand segment begins with an RNA primer, and then DNA polymerase III builds away from the replication fork until it encounters the previously synthesized section. These short stretches of newly synthesized DNA on the lagging strand are called *Okazaki fragments.* As the helix opens further, a new RNA primer is added, and DNA polymerase III must release the template it has completed and begin with the "new" template. The Okazaki fragments are then linked when DNA polymerase I removes the RNA primers and an enzyme called *DNA ligase* joins the ends of the newly synthesized segments of DNA. The entire lagging strand can only be replicated in this discontinuous fashion.

Eukaryotic chromosomes each contain a single, very long molecule of DNA (**figure 11.8**), one far too long to copy all the way from one end to the other with a single replication fork. Each eukaryotic chromosome is instead copied in sections of about 100,000 nucleotides, each with its own replication origin and fork.

The enormous amount of DNA that resides within the cells of your body represents a long series of DNA replications, starting with the DNA of a single cell—the fertilized egg. Living cells have evolved many mechanisms to avoid errors during DNA replication and to preserve the DNA from damage. These mechanisms of **DNA repair** proofread the strands of each daughter cell against one another for accuracy and correct any mistakes. But the proofreading is not perfect. If it were, no mistakes such as mutations would occur, no variation in gene sequence would result, and evolution would come to a halt. Mutation will be discussed in more detail in the next section and in chapter 14.

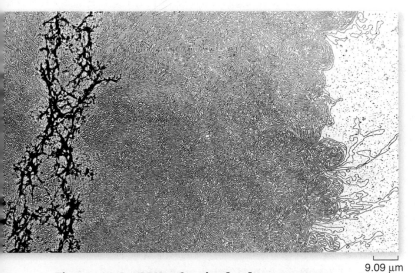

9.09 µm

Figure 11.8   DNA of a single chromosome.
This chromosome has been relieved of most of its packaging proteins, leaving the DNA in its extended form. The residual protein scaffolding appears as the dark material on the left side of the micrograph.

Key Learning Outcome 11.4   The basis for the great accuracy of DNA replication is complementarity. DNA's two strands are complementary mirror images of each other, so either one can be used as a template to reconstruct the other.

11.5 Mutation

There are two general ways in which the genetic message is altered: mutation and recombination. A change in the content of the genetic message—the base sequence of one or more genes—is referred to as a **mutation.** As you learned in the previous section, DNA copies itself by forming complementary strands along single strands of DNA when they are separated. The template strand directs the formation of the new strand. However, this replication process is not foolproof. Sometimes errors are made and these are called mutations. Some mutations alter the identity of a particular nucleotide, while others remove or add nucleotides to a gene. A change in the position of a portion of the genetic message is referred to as **recombination.** Some recombination events move a gene to a different chromosome; others alter the location of only part of a gene. The cells of eukaryotes contain an enormous amount of DNA, and the mechanisms that protect and proofread the DNA are not perfect. If they were, no variation would be generated.

Mistakes Happen

In fact, cells do make mistakes during replication, as shown in **figure 11.9**. And mutations can also occur because of DNA alteration by chemicals, like those in cigarette smoke, or by radiation, like the ultraviolet light from the sun or tanning beds. However, mutations are rare. In humans, sequencing the genomes of an entire family has revealed that only about 60 out of the 3 billion nucleotides of the genome are altered by mutation each generation. If changes were common, the genetic instructions encoded in DNA would soon degrade into meaningless gibberish. Limited as it might seem, the steady trickle of change that does occur is the very stuff of evolution. Every difference in the genetic messages that specify different organisms arose as the result of genetic change.

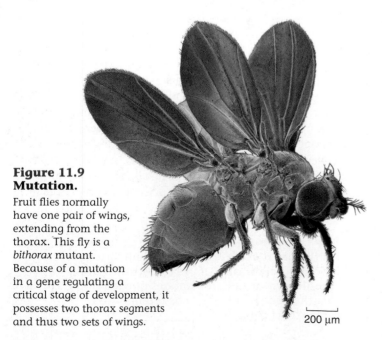

Figure 11.9 Mutation.

Fruit flies normally have one pair of wings, extending from the thorax. This fly is a *bithorax* mutant. Because of a mutation in a gene regulating a critical stage of development, it possesses two thorax segments and thus two sets of wings.

200 μm

Kinds of Mutation

The message that DNA carries in its genes is the "instructions" of how to make proteins. The sequence of nucleotides in a strand of DNA translates into the sequence of amino acids that makes up a protein. This process was introduced in section 10.5 on page 194 and will be described in more detail in chapter 12. If the core message in the DNA is altered through mutation, as shown by the substitution of T (in red) for G in **figure 11.10**, then the protein product can also be altered, sometimes to the point where it can no longer function properly. Because mutations can occur randomly in a cell's DNA, most mutations are detrimental, just as making a random change in a computer program usually worsens performance. The consequences of a detrimental mutation may be minor or catastrophic, depending on the function of the altered gene.

Mutations in Germ-Line Tissues The effect of a mutation depends critically on the identity of the cell in which the mutation occurs. During the embryonic development of all multicellular organisms, there comes a point when cells destined to form gametes (germ-line cells) are segregated from those that will form the other cells of the body (somatic cells). Only when a mutation occurs within a germ-line cell is it passed to subsequent generations as part of the hereditary endowment of the gametes derived from that cell. Mutations in germ-line

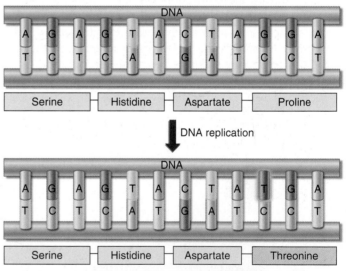

(a) Base substitution (red) in DNA: changes G to T in the DNA strand and, as a result, proline to threonine in the protein.

Normal protein

Mutated protein

(b) The mutated protein with the amino acid substitute folds differently than the normal protein and its function will most likely be affected.

Figure 11.10 Base substitution mutation.

(*a*) Some changes in a DNA sequence can result in a change in a single amino acid. (*b*) This results in a mutated protein that may not function the same as the normal protein.

Today's *Biology*

DNA Fingerprinting

Only identical twins have exactly the same DNA sequence. All other people differ from one another at many sites. In 1985 British geneticist Alec Jeffreys took advantage of this to develop a new forensic (that is, crime scene investigation) tool, DNA fingerprinting. The procedure involves cleaving an individual's DNA into small bits, which are spread apart on a gel to yield a pattern of bands, a "DNA fingerprint" characteristic of that person. The photo on the right shows the DNA fingerprints a prosecuting attorney presented in a rape trial in 1987. They consisted of autoradiographs, parallel bars on X-ray film. Each bar represents the position of a DNA fragment produced by techniques that will be described in more detail in chapter 13. The dark lanes with many bars represent standardized controls. Two different ways of producing the DNA fragments are shown, each highlighting particular sequences. A sample had been taken from the victim within hours of her attack; from it semen was collected and the semen DNA analyzed for its patterns.

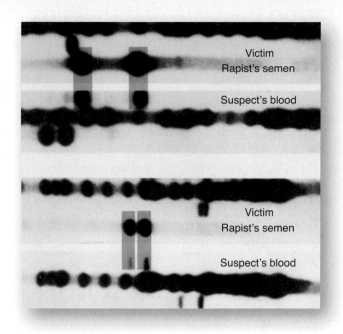

Compare the DNA fingerprint patterns of the semen to that of the suspect. You can see that the suspect's two patterns match that of the rapist, and these patterns are quite different from those of the victim. Clearly the semen collected from the rape victim and the blood sample from the suspect came from the same person. The suspect was Tommie Lee Andrews, and on November 6, 1987, the jury returned a verdict of guilty.

Since the Andrews verdict, DNA fingerprinting has been used as evidence in many court cases. DNA can be obtained at a crime scene from several different sources, such as small amounts of blood, hair, or semen. As the man who analyzed Andrews's DNA says: "It's like leaving your name, address, and social security number at the scene of the crime. It's that precise."

While some ways of detecting DNA differences highlight profiles shared by many people, others are quite rare. Using several, identity can be clearly established or ruled out. The DNA profiles of O. J. Simpson and blood samples from the murder scene of his former wife from his highly publicized and controversial murder trial in 1995 are presented on the right.

DNA fingerprinting is certainly not restricted to prosecution. It can also be used to establish innocence. More than 120 convicted people have been freed in the last 14 years using DNA evidence presented by The Innocence Project lawyers, for example.

Of course, the procedures involved in creating the DNA fingerprints and in analyzing them must be carried out properly—sloppy procedures could lead to a wrongful conviction. After widely publicized instances of questionable lab procedures, national standards have been developed.

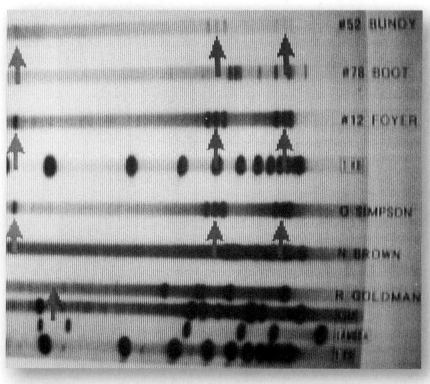

Today's Biology

Tracing the DNA of Irish Kings

Every time a mutation occurs in germ-line DNA—the DNA producing egg or sperm—there is the possibility that it will be passed on to future generations. However, there are a lot of "ifs": if the DNA change is not corrected by the cell's error-detecting machinery; if that particular egg or sperm is used to make a child; if that child survives and has children. Still, despite all the "ifs," we humans have over the centuries accumulated lots of mutations in our DNA. "Our DNA is a history book," geneticists say.

With the molecular tools that modern genetics provides, scientists are beginning to read that book, to trace the course of our species's history by tracking the changes that have occurred in our DNA. The National Geographic Society, for example, has been conducting a Geno-graphic Project comparing over 100,000 DNA samples from people all over the world, from Arctic Inuit Eskimos and Kenya's Masai to Australian aborigines and North American Pueblo Indians. Their hope is to create a picture of ancestral migratory routes, the historical paths people have taken as they populated the globe.

To gain the clearest possible picture of the past, gene researchers focus on DNA of the Y chromosome and the mitochondria. The Y chromosome of males does not recombine with other chromosomes. This has the effect of keeping mutations together once they occur. Similarly, the mitochondrial DNA of females is passed down from mother to child without recombination (sperm contribute no mitochondria to the fertilized egg).

To compare large numbers of DNA samples, it is not practical to sequence the entire Y chromosome or mitochondrial DNA of each individual. Instead, investigators monitor several dozen highly variable DNA locations, short bits of the chromosome within which lots of mutations have occurred. Because DNA mutations are rare events and there has been no recombination to shuffle the changes, when the same particular combination of mutations (what gene researchers call a haplotype) occurs in two people, they almost certainly are related, having inherited their Y chromosome or mitochondrial DNA from a common ancestor. Looking at many individuals in this way, investigators can build up a

picture of who is related to whom—a portrait of the past, inscribed on our genes.

To see how this works, consider a study carried out in 2006 by Daniel Bradley and colleagues at Trinity College in Dublin, Ireland. They set out to apply DNA studies to Irish history, which has always been a bit of a muddle. Writing did not become common in Ireland until 600 A.D., and little is known for sure of earlier events in Irish history. This has not, of course, prevented the Irish from preserving a rich story of those times.

Much as the British tell of a mythical King Arthur who few historians believe was a real person, so the Irish recount the tale of an Irish high king of the fifth century A.D. from whom an alarming number of Irishmen claim descent. Niall Noigiallach—Niall of the Nine Hostages—was so named because early in his reign he consolidated his power by taking hostages from the royal families of each of the five provinces that then constituted Ireland, as well as from Scotland, the Saxons, the Britons, and the Franks.

He founded a dynasty, the Ui Neill ("the descendants of Niall"), which ruled the northwest of Ireland from about 600 to 900 A.D. When the Irish took surnames around 1000 A.D., some chose names associated with the Ui Neill dynasties, names like Gallagher, Boyle, Doherty, O'Conner, Reilly, Flynn, Devlin, Donnelly, McLoughlin, Molloy, O'Rourke, and of course O'Neill (the prefix "O" is often added).

Did Niall of the Nine Hostages really exist? The DNA evidence gathered by Bradley argues yes. Some 20% of men in northwest Ireland have a distinctive genetic signature on their Y chromosomes, a haplotype carried down for over a thousand years. As you can see on the map to the left, this signature haplotype predominates in the northwest, the seat of Ui Neill power.

Indeed, wherever in the world you look (the Irish were particularly adept at migrating—over 400,000 residents of New York City claim Irish ancestry), this haplotype is much more common among Irishmen with the Ui Neill surnames than among Irishmen as a whole.

Niall is said to have had 14 sons, a large number even for those days, which might go a long way towards explaining that some 2 million men worldwide now carry his distinctive Y chromosome. Like Genghis Khan, ancestor of 16 million men in Asia, Niall of the Nine Hostages seems to have left quite a genetic footprint.

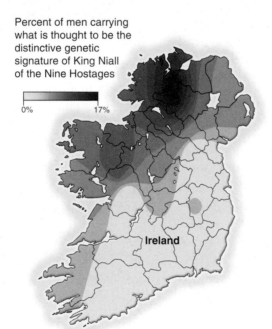

Percent of men carrying what is thought to be the distinctive genetic signature of King Niall of the Nine Hostages

0%   17%

Ireland

tissue are of enormous biological importance because they provide the raw material from which natural selection produces evolutionary change.

Mutations in Somatic Tissues Change can occur only if there are new, different allele combinations available to replace the old. Mutation produces new alleles, and recombination puts the alleles together in different combinations. In animals, it is the occurrence of these two processes in germline tissue that is important to evolution because mutations in somatic cells (somatic mutations) are not passed from one generation to the next. However, a somatic mutation may have drastic effects on the individual organism in which it occurs, because it is passed on to all of the cells that are descended from the original mutant cell. Thus, if a mutant lung cell divides, all cells derived from it will carry the mutation. Somatic mutations of lung cells are, as we shall see, the principal cause of lung cancer in humans.

Altering the Sequence of DNA One category of mutational changes affects the message itself, producing alterations in the sequence of DNA nucleotides (**table 11.1**). If alterations involve only one or a few base pairs in the coding sequence, they are called **point mutations.** Sometimes the identity of a base changes (*base substitution*), while other times one or a few bases are added (*insertion*) or lost (*deletion*). If an insertion or deletion throws the reading of the gene message out of register, a **frame-shift mutation** results. **Figure 11.10** shows a base substitution mutation that results in the change of an amino acid, from proline to threonine. This could be a minor change or catastrophic. However, suppose that this had been the deletion of a nucleotide, that the cytosine base nucleotide had been skipped during replication. This would shift the register of the DNA message (imagine removing the w from this sentence, yielding *"This oulds hiftt her egistero fth eDN Amessag"*) and you can see the problem. Many point mutations result from damage to the DNA caused by **mutagens,** usually radiation or chemicals. The latter class of mutations is of particular importance because modern industrial societies often release many chemical mutagens into the environment.

Changes in Gene Position Another category of mutations affects the way the genetic message is organized. In both prokaryotes and eukaryotes, individual genes may move from one place in the genome to another by **transposition** (see also page 259). When a particular gene moves to a different location, its expression or the expression of neighboring genes may be altered. In addition, large segments of chromosomes in eukaryotes may change their relative locations or undergo duplication. Such **chromosomal rearrangements** often have drastic effects on the expression of the genetic message.

The Importance of Genetic Change

All evolution begins with alterations in the genetic message that create new alleles or alter the organization of genes on chromosome. Some changes in germ-line tissue produce alterations that enable an organism to leave more offspring, and those changes tend to be preserved as the genetic endowment of future generations. Other changes reduce the ability of an organism to leave offspring. Those changes tend to be lost, as the organisms that

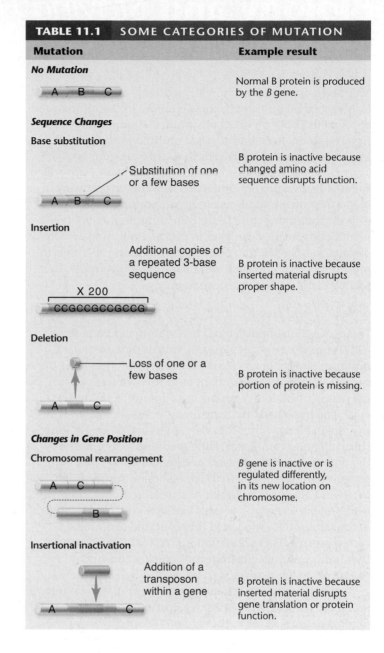

TABLE 11.1	SOME CATEGORIES OF MUTATION
Mutation	**Example result**
No Mutation	Normal B protein is produced by the *B* gene.
Sequence Changes	
Base substitution — Substitution of one or a few bases	B protein is inactive because changed amino acid sequence disrupts function.
Insertion — Additional copies of a repeated 3-base sequence, X 200, CCGCCGCCGCCG	B protein is inactive because inserted material disrupts proper shape.
Deletion — Loss of one or a few bases	B protein is inactive because portion of protein is missing.
Changes in Gene Position	
Chromosomal rearrangement	*B* gene is inactive or is regulated differently, in its new location on chromosome.
Insertional inactivation — Addition of a transposon within a gene	B protein is inactive because inserted material disrupts gene translation or protein function.

carry them contribute fewer members to future generations. Evolution can be viewed as the selection of particular combinations of alleles from a pool of alternatives. The rate of evolution is ultimately limited by the rate at which these alternatives are generated. Genetic change through mutation and recombination provides the raw material for evolution.

Genetic changes in somatic cells do not pass on to offspring, and so they have no direct evolutionary consequence. However, changes in the genes of somatic cells can have an important immediate impact if the gene affects development or is involved with regulation of cell proliferation.

Key Learning Outcome 11.5 Rare changes in genes, called mutations, can have significant effects on the individual when they occur in somatic tissue, but they are inherited only if they occur in germ-line tissue. Inherited changes provide the raw material for evolution.

Biology and *Staying Healthy*

Protecting Your Genes

This text's discussion of changes in genes—mutations—has largely focused on heredity, how changes in the information encoded in DNA can affect offspring. It is important, however, to realize that inherited mutations occur only in germ-line tissue, in the cells that generate your eggs or sperm. Mutations in the other cells of your body, in so-called somatic tissues, are not inherited. This does not, however, mean that such mutations are not important. In fact, somatic mutations can have a disastrous impact upon your health because they can lead to cancer. Protecting the DNA of your body's cells from damaging mutation is perhaps the most important thing you can do to prolong your life. Here we will examine two potential threats.

Smoking and Lung Cancer

The association of particular chemicals in cigarette smoke with lung cancer, particularly chemicals that are potent mutagens (see chapters 8 and 24), led researchers early on to suspect that lung cancer might be caused, at least in part, by the action of chemicals on the cells lining the lung.

The hypothesis that chemicals in tobacco cause cancer was first advanced over 200 years ago in 1761 by Dr. John Hill, an English physician. Hill noted unusual tumors of the nose in heavy snuff users and suggested tobacco had produced these cancers. In 1775, a London surgeon, Sir Percivall Pott, made a similar observation, noting that men who had been chimney sweeps exhibited frequent cancer of the scrotum. He suggested that soot and tars might be responsible. These observations led to the hypothesis that lung cancer results from the action of tars and other chemicals in tobacco smoke.

It was over a century before this hypothesis was directly tested. In 1915, Japanese doctor Katsusaburo Yamagiwa applied extracts of tar to the skin of 137 rabbits every two or three days for three months. Then he waited to see what would happen. After a year, cancers appeared at the site of application in seven of the rabbits. Yamagiwa had induced cancer with the tar, the first direct demonstration of chemical carcinogenesis. In the decades that followed, this approach demonstrated that many chemicals can cause cancer.

But do these lab studies apply to people? Do tars in cigarette smoke in fact induce lung cancer in humans? In 1949, the American physician Ernst Winder and the British epidemiologist Richard Doll independently reported that lung cancer showed a strong link to the smoking of cigarettes, which introduces tars into the lungs. Winder interviewed 684 lung cancer patients and 600 normal controls, asking whether each had ever smoked. Cancer rates were 40 times higher in heavy smokers than in nonsmokers. From these studies, it seemed likely as long as 50 years ago

that tars and other chemicals in cigarette smoke induce cancer in the lungs of persistent smokers. While this suggestion was resisted by the tobacco industry, the evidence that has accumulated since these pioneering studies makes a clear case, and there is no longer any real doubt. Chemicals in cigarette smoke cause cancer.

As you will learn in chapter 24 (page 525), tars and other chemicals in cigarette smoke cause lung cancer by mutating DNA, disabling genes that in normal lung cells restrain cell division. Lacking these restraints, the altered lung cells begin to divide ceaselessly, and lung cancer results. Just under 160,000 Americans died of lung cancer last year, and almost all of them were cigarette smokers.

If cigarette smoking is so dangerous, why do so many Americans smoke? Fully 23% of American men smoke, and 18% of women. Are they not aware of the danger? Of course they are. But they are not able to quit. Tobacco smoke, you see, also contains another chemical, nicotine, which is highly addictive. The nature of the addiction is discussed in detail in chapter 28 (page 592). Basically, what happens is that a smoker's brain makes physiological compensations to overcome the effects of nicotine, and once these adjustments are made the brain does not function normally without nicotine. The body's physiological response to nicotine is profound and unavoidable; there is no way to prevent addiction to nicotine with willpower.

Many people attempting to quit smoking use patches containing nicotine to help them, the idea being that providing nicotine removes the craving for cigarettes. This is true, it does—as long as you keep using the patch. Actually, using such patches simply substitutes one (admittedly less dangerous) nicotine source for another. If you are going to quit smoking, there is no way to avoid the necessity of eliminating the drug to which you are addicted, nicotine. There is no easy way out. The only way to quit is to quit.

Clearly, if you do not smoke, you should not start. Asked what three things were most important to improve Americans' health, a prominent physician replied: "Don't smoke. Don't smoke. Don't smoke."

Tanning and Skin Cancer

Almost all cells in the human body undergo cell division, replacing themselves as they wear out. Some adult cells do this quite frequently, others rarely if ever. Skin cells divide quite frequently. Exposed to a lot of wear-and-tear, they divide about every 27 days to replace dead or damaged cells. The skin sloughs off dead cells from the surface and replaces these with new cells from beneath. The average person will lose about 105 pounds of skin by the time he or she turns 70.

While skin can be damaged in many ways, the damage that seems to have the most long-term affect is caused by the sun. The skin contains cells called melanocytes that produce a pigment called melanin when exposed to UV light. Melanin produces a yellow to brown color in the skin. The type of melanin and the amount produced is genetically determined. People with darker skin types have more melanocytes and produce a melanin that is dark brown in color. Protected by UV-absorbing melanin, they almost never burn. Fair-skinned people have fewer melanocytes and produce melanin that is more yellow in color. Unprotected by melanin, these people sunburn easily and rarely tan. When cells on the body's surface are badly damaged by the sun—what we call a sunburn, the cells slough off. Recall the peeling that you experienced if you have ever had a bad sunburn.

Up until the early 20th century, a tan was a condition that people went to great lengths to avoid. A tanned body was a sign of the working class, people who had to work in the sun. The wealthy elite avoided the sun with pale skin being in fashion. All of this changed in the 1920s, when tans became a status symbol, with the wealthy able to travel to warm, sunny destinations, even in the middle of winter. That tan, bronzed glow that people would sit in the sun for hours to achieve was thought to be both healthy and attractive.

During the 1970s, doctors started to see an uptick in the number of cases of melanoma, a deadly form of skin cancer. New cases were increasing about 6% each year. Researchers proposed that UV rays from the sun were the underlying cause of this epidemic of skin cancer and warned people to avoid the sun when possible and protect themselves with sunscreen.

Malignant melanoma is the most deadly of skin cancers, although treatable if caught early. Melanoma is cancer of melanocyte cells. Melanoma lesions usually appear as shades of tan, brown, and black and often begin in or near a mole, and so changes in a mole are a symptom of melanoma. Melanoma is most prevalent in fair-skinned people, but unlike the other forms of skin cancer, it can also affect people with darker complexions.

The public has been slow to respond to warnings about avoiding sun exposure, perhaps because the cosmetic benefits of tanning are immediate while the health hazards are much delayed. The desire to achieve that tanned, bronzed body is as strong as ever.

A good tan requires regular exposure to the sun to maintain it, so indoor tanning salons have become popular. Tanning booths emit concentrated UV rays from two sides, allowing a person to tan in less time and in all weather conditions (sun, rain, snow). The indoor tanning business has grown in the United States to a $2 billion-a-year industry with an estimated 28 million Americans tanning annually.

People thought that building up a tan through the use of tanning booths would protect a person's skin from burning and would reduce the time exposed to the UV radiation, both leading to a reduced risk of skin cancer. However, recent research does not support these assumptions. A 2003 study of 106,000 Scandinavian women showed that exposure to UV rays in a tanning booth as little as once a month can increase your risk of melanoma by 55%, especially when the exposure is during early adulthood. Those women who were in their 20s and used sun lamps to tan were at the highest risk, about 150% higher than those who didn't use a tanning bed. As with other studies, fair-skinned women were at the greatest risk. In fact, tanning booths, even for those people who tan more easily, heighten the risk for skin cancer because people use the tanning booths year-round, increasing their cumulative exposure.

It is difficult to avoid the conclusion that to protect your genes you should avoid tanning booths. Like smoking cigarettes, excessive tanning is gambling with your life.

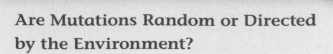
Are Mutations Random or Directed by the Environment?

Once biologists appreciated that Mendelian traits were in fact alternative versions of DNA sequences that resulted from mutations, a very important question arose and needed to be answered—are mutations random events that might happen anywhere on the DNA in a chromosome, or are they directed to some degree by the environment? Do the mutagens in cigarettes, for example, damage DNA at random locations, or do they preferentially seek out and alter specific sites such as those regulating the cell cycle?

This key question was addressed and answered in an elegant, deceptively simple experiment carried out in 1943 by two of the pioneers of molecular genetics, Salvadore Luria and Max Delbruck. They chose to examine a particular mutation that occurs in laboratory strains of the bacterium *E. coli*. These bacterial cells are susceptible to T1 viruses, tiny chemical parasites that infect, multiply within, and kill the bacteria. If 10^5 bacterial cells are exposed to 10^{10} T1 viruses, and the mixture spread on a culture dish, not one cell grows—every single *E. coli* cell is infected and killed. However, if you repeat the experiment using 10^9 bacterial cells, lots of cells survive! When tested, these surviving cells prove to be mutants, resistant to T1 infection. The question is, did the T1 virus cause the mutations, or were they present all along, too rare to be present in a sample of only 10^5 cells but common enough to be present in 10^9 cells?

To answer this question, Luria and Delbruck devised a simple experiment they called a "fluctuation test," illustrated here. Five cell generations are shown for each of four independent bacterial cultures, all tested for resistance in the fifth generation. If the T1 virus causes the mutations (top row), then each culture will have more or less the same number of resistant cells, with only a little fluctuation (that is, variation among the four). If, on the other hand, mutations are spontaneous and so equally likely to occur in any generation, then bacterial cultures in which the T1-resistance mutation occurs in earlier generations will possess far more resistant cells by the fifth generation than cultures in which the mutation occurs in later generations, resulting in wide fluctuation among the four cultures. The table presents the data they obtained for 20 individual cultures.

Number of Bacteria Resistant to T1 Virus			
Culture number	Resistant colonies found	Culture number	Resistant colonies found
1	1	11	107
2	0	12	0
3	3	13	0
4	0	14	0
5	0	15	1
6	5	16	0
7	0	17	0
8	5	18	64
9	0	19	0
10	6	20	35

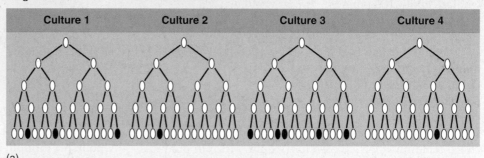
(a)

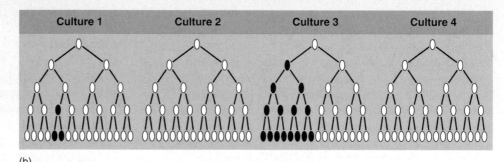
(b)

1. **Applying Concepts** Is there a dependent variable in this experiment? Explain.
2. **Interpreting Data** What is the mean number of T1-resistant colonies found in the 20 individual cultures?
3. **Making inferences**
   a. Comparing the 20 individual cultures, do the cultures exhibit similar numbers of T1-resistant bacterial cells?
   b. Which of the two alternative outcomes illustrated above, (a) or (b), is more similar to the outcome obtained by Luria and Delbruck in this experiment?
4. **Drawing Conclusions** Are these data consistent with the hypothesis that the mutation for T1 resistance among *E. coli* bacteria is caused by exposure to T1 virus? Explain.

Genes Are Made of DNA

11.1 The Discovery of Transformation

- Using *Streptococcus pneumoniae* bacteria, Griffith showed that information that controls physical characteristics can be passed from one bacterium to another, even from a dead bacterium.

- By injecting mice with different strains of *S. pneumoniae,* Griffith determined that some strains were pathogenic, resulting in the mice's deaths. Bacteria of the pathogenic strains contained polysaccharide capsules (S strain), like the bacteria shown here from **figure 11.1**, while those without the coats (R strain) were nonlethal. When he mixed dead pathogenic bacteria (S), which usually would not cause death, and live nonpathogenic bacteria (R) and injected them into mice, the mice died. The dead mice contained living S strains.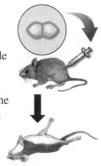

- Something passed from the dead lethal bacteria to the live nonlethal bacteria, causing them to turn deadly. Griffith did not determine whether this "transforming principle" was protein or DNA.

11.2 Experiments Identifying DNA as the Genetic Material

- Avery and colleagues showed that protein was not the source of this transformation. They replicated Griffith's experiment but removed all protein from the preparation. The virulent strain with its protein coat removed was still able to transform nonvirulent bacteria. This experimental result supported the hypothesis that DNA, not protein, was the transforming principle.

- Using bacterial viruses, Hershey and Chase showed that genes were carried on DNA and not proteins. They used two different radioactively tagged preparations, DNA in one and protein in the other (**figure 11.2**). Each preparation was used to infect bacteria. When they screened the two bacterial cultures, they discovered that the infected bacteria contained radioactively tagged DNA.

11.3 Discovering the Structure of DNA

- The structure of DNA was not known. The basic chemical components of DNA were determined to be nucleotides, like the one shown here from **figure 11.3**. Each nucleotide has a similar structure: a deoxyribose sugar attached to a phosphate group and one of four organic bases.

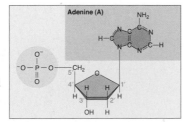

- Erwin Chargaff observed that two sets of bases are always present in equal amounts in a molecule of DNA (A nucleotides equal T nucleotides, and C nucleotides equal G nucleotides). This observation, called Chargaff's rule, gave some insight into the structure of DNA—that there was some regularity to the structure.

- Using X-ray diffraction, Rosalind Franklin was able to form a "picture" of DNA. The image suggested that the DNA molecule was coiled, a form called a helix.

- Using Chargaff's and Franklin's research, Watson and Crick determined that DNA is a double helix, two strands that are connected by base pairing between the nucleotide bases. An A nucleotide on one strand pairs with T on the other, and similarly G pairs with C (**figure 11.4**).

DNA Replication

11.4 How the DNA Molecule Copies Itself

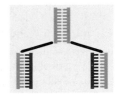

- The complementarity of DNA (A pairs with T and C with G) implies a method of replication where a single strand of DNA can serve as the template for production of another strand, but there are several ways in which this could occur (**figure 11.5**).

- Meselson and Stahl showed that DNA replicates semiconservatively, using each of the original strands as templates to form new strands. In semiconservative replication, each new strand of DNA consists of a template strand from the parent DNA and a newly synthesized strand that is complementary to the template strand (**figure 11.6**).

- In replication, the DNA molecule first unwinds by the actions of an enzyme called helicase. Each DNA strand is then copied by the actions of an enzyme called DNA polymerase. The two original strands serve as templates to the new DNA strands. DNA polymerase adds nucleotides to the new DNA strands that are complementary to the original single strands. DNA polymerase can only add on to an existing strand, and so the new strand begins after a section of nucleic acids called a primer is added. A different enzyme builds the primer. Nucleotides are added to the growing strand in a 5′ to 3′ direction (**figure 11.7**).

- The point where the DNA separates is called the replication fork. Because nucleotides can only be added onto the 3′ end of the growing DNA strand, DNA copies in a continuous manner on one of the strands, called the leading strand, and in a discontinuous manner on the other strand, called the lagging strand (**integrated art, page 224**). On the lagging strand, primers are inserted at the replication fork, and nucleotides are added in sections (**integrated art, page 225**). Before the new DNA strands rewind, the primers are removed and the DNA segments are linked together with another enzyme called DNA ligase.

- Errors can occur during the replication of DNA. The cell has many mechanisms to correct damage to the DNA or mistakes made during replication. This proofreading process compares one strand against its complementary strand and corrects errors, but this system is not foolproof.

Altering the Genetic Message

11.5 Mutation

- A mutation is a change in the nucleotide sequence of the genetic message. Mutations that change one or only a few nucleotides are called point mutations (**figure 11.10**). Some mutations occur through the movement of sections of DNA from one place to another, a process called transposition (**table 11.1**).

Test Your Understanding

1. In his experiments, Frederick Griffith found that
   a. hereditary information within a cell cannot be changed.
   b. hereditary information can be added to cells from other cells.
   c. mice infected with live R strains die.
   d. mice infected with heat-killed S strains die.
2. The experiment performed by Alfred Hershey and Martha Chase showed that the molecule viruses use to specify new viruses is
   a. a protein.
   b. a carbohydrate.
   c. ATP.
   d. DNA.
3. Erwin Chargaff, Rosalind Franklin, Francis Crick, and James Watson all worked on pieces of information relating to the
   a. structure of DNA.
   b. function of DNA.
   c. inheritance of DNA.
   d. mutations of DNA.
4. The four DNA nucleotides are all different in terms of
   a. their sizes.
   b. the number of hydrogen bonds they can form with their base pair.
   c. the type of nitrogen base.
   d. the type of sugar.
5. Which of the following lists the purine nucleotide bases?
   a. adenine and cytosine
   b. guanine and thymine
   c. cytosine and thymine
   d. adenine and guanine
6. If one strand of a DNA molecule has the base sequence ATTGCAT, its complementary strand will have the sequence
   a. ATTGCAT.   c. GCCATGC.
   b. TAACGTA.   d. CGGTACG.
7. Regarding the duplication of DNA, we now know that each double helix
   a. rejoins after replicating.
   b. splits down the middle into two single strands, and each one then acts as a template to build its complement.
   c. fragments into small chunks that duplicate and reassemble.
   d. All of these are true for different types of DNA.
8. DNA polymerase can only add nucleotides to an existing chain, so _____ is required.
   a. a primer
   b. helicase
   c. a lagging strand
   d. a leading strand
9. Genetic messages can be altered in two ways:
   a. through semiconservative replication or conservative replication.
   b. through the chromosome or through the protein.
   c. by mutation or by recombination.
   d. by activation or by repression.
10. Mutations can occur in
    a. germ-line tissues and are passed on to future generations.
    b. somatic tissues and are passed on to future generations.
    c. germ-line tissues but not in somatic tissues.
    d. somatic tissues but not in germ-line tissues.

Apply Your Understanding

1. **Figure 11.4c** What are some of the possible problems that could occur if the C nucleotide indicated with the red arrow is accidentally replaced with an A nucleotide?

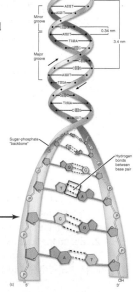

2. **Table 11.1** What types of mutations in the table result in a shift in the reading frame of the DNA? Explain.

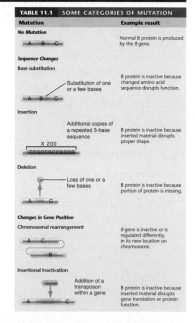

Synthesize What You Have Learned

1. The discovery that DNA is the genetic material was an experimental journey rather than a flash of insight. Highlighting individual experiments, use this journey to defend the statement attributed to Sir Isaac Newton in 1676 (though some say that Bernard of Chartres said it first, way back in about 1130!) that scientists build new ideas in science by "standing on the shoulders of giants."

2. Certain strains of bacteria are resistant to the antibiotic tetracycline, while other strains are sensitive to it. Design an experiment you would carry out to determine whether or not tetracycline resistance is an inherited trait specified by the DNA of the resistant strain.

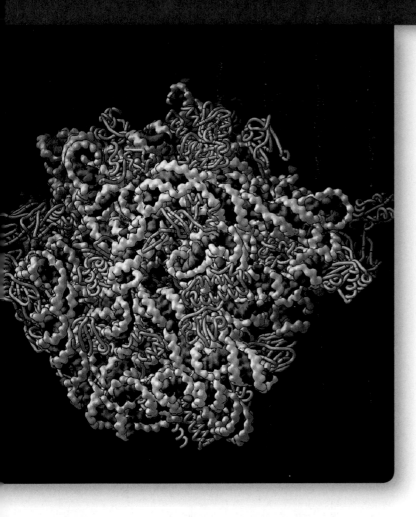

12

How Genes Work

Learning Objectives

From Gene to Protein

12.1 The Central Dogma
1. State the so-called "Central Dogma."
2. List the four kinds of RNA used in the synthesis of proteins.
3. Name the two stages of gene expression.
4. Describe the location and function of a promoter site.

12.2 Transcription
1. State the chemical difference between DNA and RNA.
2. Indicate in what chemical direction an mRNA chain is assembled during transcription.

12.3 Translation
1. Distinguish between codon and anticodon.
2. State how many codons do not code for an amino acid.
3. Describe the composition of a ribosome.
4. Contrast the roles of tRNA and activating enzymes in reading the genetic code.
5. List the order in which the A, P, and E sites of a ribosome are occupied by an amino acid.

12.4 Gene Expression
1. Compare the architecture of prokaryotic and eukaryotic genes.
2. Explain alternative splicing.
3. Compare the cellular locations of transcription and translation in prokaryotes and eukaryotes.
4. Describe the six stages of eukaryotic protein synthesis.

Regulating Gene Expression in Prokaryotes

12.5 How Prokaryotes Control Transcription
1. Define operon.
2. Diagram how the *lac* operon works.
3. Contrast the regulatory action of repressors and activators.
4. Describe how the CAP protein acts as an activator.

Regulating Gene Expression in Eukaryotes

12.6 Transcriptional Control in Eukaryotes
1. Describe the regulatory roles of histones and of DNA methylation in eukaryotes.

12.7 Controlling Transcription from a Distance
1. Distinguish between basal transcription factors and specific transcription factors.
2. Use a diagram to explain how enhancers regulate gene expression at a distance.

12.8 RNA-Level Control
1. Define RNA interference.
2. Explain how siRNA can regulate gene expression.
Biology and Staying Healthy: Silencing Genes to Treat Disease

12.9 Complex Regulation of Gene Expression
1. Name and describe six points where eukaryotic gene expression is controlled.

Inquiry & Analysis: Building Proteins in a Test Tube

R ibosomes, like the one you see here, are very complex cellular machines that assemble the polypeptide segments of proteins, using information that has been copied from genes onto RNA molecules. The ribosomes read the gene information copied onto these messenger RNA transcripts, and use it to determine the amino acid sequence of the new polypeptide that the ribosome is assembling. Each ribosome is made of over 50 different proteins (shown here in gold), as well as three chains of RNA composed of some 3,000 nucleotides (shown here in gray). It has been traditionally assumed that the proteins in a ribosome act as enzymes to catalyze the amino acid assembly process, with the RNA acting as a scaffold to position the proteins. In the year 2000 we learned the reverse to be true. Powerful X-ray diffraction studies revealed the complete detailed structure of a ribosome at atomic resolution. Unexpectedly, the many proteins of a ribosome are scattered over its surface like decorations on a Christmas tree. The role of these proteins seems to be to stabilize the many bends and twists of the RNA chains, the proteins acting like spot-welds between the RNA strands they touch. Importantly, there are no proteins on the inside of the ribosome where the chemistry of protein synthesis takes place—just twists of RNA. Thus, it is the ribosome's RNA, not its proteins, that catalyzes the joining together of amino acids! Clearly, our knowledge of how genes work is still increasing, often adjusting what seem to be fundamental concepts.

12.1 The Central Dogma

The discovery that genes are made of DNA, discussed in chapter 11, left unanswered the question of how the information in DNA is used. How does a string of nucleotides in a spiral molecule determine if you have red hair? We now know that the information in DNA is arrayed in little blocks, like entries in a dictionary, and each block is a gene that specifies the sequence of amino acids for a polypeptide. These polypeptides form the proteins that determine what a particular cell will be like.

All organisms, from the simplest bacteria to ourselves, use the same basic mechanism of reading and expressing genes, so fundamental to life as we know it that it is often referred to as the "Central Dogma": Information passes from the genes (DNA) to an RNA copy of the gene, and the RNA copy directs the sequential assembly of a chain of amino acids. Said briefly, **DNA ⟶ RNA ⟶ protein.**

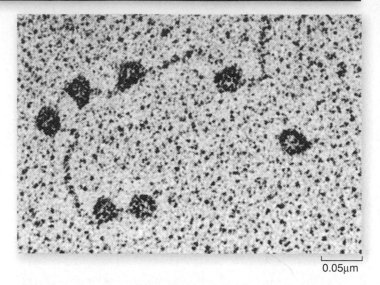

Figure 12.1 RNA polymerase.
In this electron micrograph, the dark circles are RNA polymerase molecules synthesizing RNA from a DNA template.

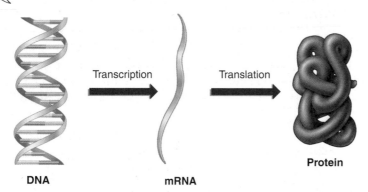

Transcription   Translation

DNA   **mRNA**   **Protein**

A cell uses four kinds of RNA in the synthesis of proteins: messenger RNA (mRNA), silencing RNA (siRNA), ribosomal RNA (rRNA), and transfer RNA (tRNA). These types of RNA are described in more detail later in this chapter.

The use of information in DNA to direct the production of particular proteins is called **gene expression.** Gene expression occurs in two stages: In the first stage, *transcription,* mRNA molecules are synthesized from genes within the DNA; in the second stage, *translation,* the mRNA is used to direct the production of polypeptides, the components of proteins.

Transcription: An Overview

The first step of the Central Dogma is the transfer of information from DNA to RNA, which occurs when an mRNA copy of the gene is produced. Because the DNA sequence in the gene is transcribed into an RNA sequence, this stage is called transcription. Transcription is initiated when the enzyme *RNA polymerase* binds to a special nucleotide sequence called a *promoter* located at the beginning of a gene. Starting there, the RNA polymerase moves along the strand into the gene (**figure 12.1**). As it encounters each DNA nucleotide, it adds the corresponding complementary RNA nucleotide to a grow-

ing mRNA strand. Thus, guanine (G), cytosine (C), thymine (T), and adenine (A) in the DNA would signal the addition of C, G, A, and uracil (U), respectively, to the mRNA.

When the RNA polymerase arrives at a transcriptional "stop" signal at the opposite end of the gene, it disengages from the DNA and releases the newly assembled RNA chain. This chain is a complementary transcript of the gene from which it was copied.

Translation: An Overview

The second step of the Central Dogma is the transfer of information from RNA to protein, which occurs when the information contained in the mRNA transcript is used to direct the sequence of amino acids during the synthesis of polypeptides by ribosomes. This process is called translation because the nucleotide sequence of the mRNA transcript is translated into an amino acid sequence in the polypeptide. Translation begins when an rRNA molecule within the ribosome recognizes and binds to a "start" sequence on the mRNA. The ribosome then moves along the mRNA molecule, three nucleotides at a time. Each group of three nucleotides is a codeword that specifies which amino acid will be added to the growing polypeptide chain, and is recognized by a specific tRNA molecule. The ribosome continues in this fashion until it encounters a translational "stop" signal; then it disengages from the mRNA and releases the completed polypeptide.

Key Learning Outcome 12.1 The information encoded in genes is expressed in two phases: transcription, which produces an mRNA molecule whose sequence is complementary to the DNA sequence of the gene; and translation, which assembles a polypeptide.

Transcription

Just as an architect protects building plans from loss or damage by keeping them safe in a central place and issuing only blueprint copies to on-site workers, so your cells protect their DNA instructions by keeping them safe within a central DNA storage area, the nucleus. The DNA never leaves the nucleus. Instead, the process of **transcription** creates "blueprint" copies of particular genes that are sent out into the cell to direct the assembly of proteins (**figure 12.2**).

These working copies of genes are made of ribonucleic acid (RNA) rather than DNA. Recall that RNA is the same as DNA except that the sugars in RNA have an extra oxygen atom and the pyrimidine base thymine (T) is replaced by a similar pyrimidine base called uracil, U (see **figure 3.9**).

The Transcription Process

The RNA copy of a gene used in the cell to produce a polypeptide is called **messenger RNA (mRNA)**—it is the messenger that conveys the information from the nucleus to the cytoplasm. The copying process that makes the mRNA is called transcription—just as monks in monasteries used to make copies of manuscripts by faithfully transcribing each letter, so enzymes within the nuclei of your cells make mRNA copies of your genes by faithfully complementing each nucleotide.

In your cells, the transcriber is a large and very sophisticated protein called **RNA polymerase.** It binds to one strand of a DNA double helix at the promoter site and then moves along the DNA strand like a train engine on a track. Although DNA is double-stranded, the two strands have complementary rather than identical sequences, so RNA polymerase is only able to bind one of the two DNA strands (the one with the promoter-site sequence it recognizes). As RNA polymerase goes along the DNA strand it is copying, it pairs each nucleotide with its complementary RNA version (G with C, A with U), building an mRNA chain in the 5′ to 3′ direction as it moves along the strand (**figure 12.3**).

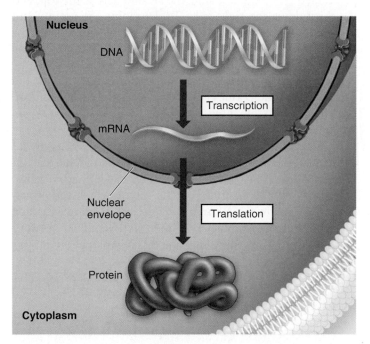

Figure 12.2   Overview of gene expression in a eukaryotic cell.

> **Key Learning Outcome 12.2**   Transcription is the production of an mRNA copy of a gene by the enzyme RNA polymerase.

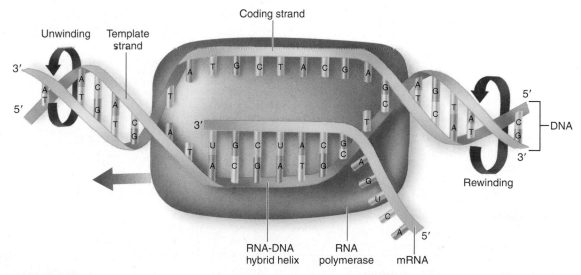

Figure 12.3   Transcription.
One of the strands of DNA functions as a template on which nucleotide building blocks are assembled into mRNA by RNA polymerase as it moves along the DNA strand.

12.3 | Translation

The Genetic Code

The essence of Mendelian genetics is that information determining hereditary traits, traits passed from parent to child, is encoded information. The information is written within the chromosomes in blocks called **genes.** Genes affect Mendelian traits by directing the production of particular proteins. The essence of gene expression, of using your genes, is reading the information encoded within DNA and using that information to produce proteins.

To correctly read a gene, a cell must translate the information encoded in DNA into the language of proteins—that is, it must convert the order of the gene's nucleotides into the order of amino acids in a polypeptide, a process called **translation.** The rules that govern this translation are called the **genetic code.**

The mRNA is transcribed from a gene in a linear sequence, one nucleotide following another, beginning at the promoter. There, RNA polymerase binds to the DNA and begins its assembly of the mRNA. Transcription ends when the RNA polymerase reaches a certain nucleotide sequence that signals it to stop.

However, the mRNA is not translated in this same way. The mRNA is "read" by a ribosome in three-nucleotide units. Each three-nucleotide sequence of the mRNA is called a **codon.** Each codon, with the exception of three, codes for a particular amino acid. Biologists worked out which codon corresponds to which amino acid by trial-and-error experiments carried out in test tubes. In these experiments, investigators used artificial mRNAs to direct the synthesis of polypeptides in the tube, and then looked to see the sequence of amino acids in the newly formed polypeptides. An mRNA that was a string of UUUUUU . . . , for example, produced a polypeptide that was a string of phenylalanine (Phe) amino acids, telling investigators that the codon UUU corresponded to the amino acid Phe. The entire genetic code dictionary is presented in **figure 12.4.** The first letters of the codon are positioned down the left side, the second across the top, and the third down the right side. To determine the amino acid encoded by a codon, say AGC, go to "A" on the left, follow the row over to the "G" column, and go down to the C on the right. As you discover, AGC encodes the amino acid serine. Because at each position of a three-letter codon, any of the four different nucleotides (U, C, A, G) may be used, there are 64 different possible three-letter codons (4 × 4 × 4 = 64) in the genetic code.

The genetic code is universal, the same in practically all organisms. GUC codes for valine in bacteria, in fruit flies, in eagles, and in your own cells. The only exception biologists have ever found to this rule is in the way in which cell organelles that contain DNA (mitochondria and chloroplasts) and a few microscopic protists read the "stop" codons. In every other instance, the same genetic code is employed by all living things.

The Genetic Code

First Letter	Second Letter								Third Letter
	U		**C**		**A**		**G**		
U	UUU UUC	Phenylalanine	UCU UCC UCA UCG	Serine	UAU UAC	Tyrosine	UGU UGC	Cysteine	U C
	UUA UUG	Leucine			UAA	Stop	UGA	Stop	A
					UAG	Stop	UGG	Tryptophan	G
C	CUU CUC CUA CUG	Leucine	CCU CCC CCA CCG	Proline	CAU CAC	Histidine	CGU CGC CGA CGG	Arginine	U C A G
					CAA CAG	Glutamine			
A	AUU AUC AUA	Isoleucine	ACU ACC ACA ACG	Threonine	AAU AAC	Asparagine	AGU AGC	Serine	U C A
	AUG	Methionine; Start			AAA AAG	Lysine	AGA AGG	Arginine	G
G	GUU GUC GUA GUG	Valine	GCU GCC GCA GCG	Alanine	GAU GAC	Aspartate	GGU GGC GGA GGG	Glycine	U C A G
					GAA GAG	Glutamate			

Figure 12.4 The genetic code (RNA codons).

A codon consists of three nucleotides read in sequence. For example, ACU codes for threonine. The first letter, A, is in the First Letter column; the second letter, C, is in the Second Letter row; and the third letter, U, is in the Third Letter column. Most amino acids are specified by more than one codon. For example, threonine is specified by four codons, which differ only in the third nucleotide (ACU, ACC, ACA, and ACG).

Translating the RNA Message into Proteins

The final result of the transcription process is the production of an mRNA copy of a gene. Like a photocopy, the mRNA can be used without damage or wear and tear on the original. After transcription of a gene is finished, the mRNA passes out of the nucleus (in eukaryotes) and into the cytoplasm through pores in the nuclear membrane. There, translation of the genetic message occurs. In translation, organelles called **ribosomes** use the mRNA produced by transcription to direct the synthesis of a polypeptide following the genetic code.

The Protein-Making Factory Ribosomes are the polypeptide-making factories of the cell. Each is very complex, containing over 50 different proteins and several segments of **ribosomal RNA (rRNA).** Ribosomes use mRNA, the "blueprint" copies of nuclear genes, to direct the assembly of polypeptides, which are then combined into proteins.

Ribosomes are composed of two pieces, or subunits, one nested into the other like a fist in the palm of your hand. The "fist" is the smaller of the two subunits, the pink structure in **figure 12.5**. Its rRNA has a short nucleotide sequence exposed on the surface of the subunit. This exposed sequence is identical to a sequence called the leader region that occurs at the beginning of all genes. Because of this, an mRNA molecule binds to the exposed rRNA of the small subunit like a fly sticking to flypaper.

The Key Role of tRNA Directly adjacent to the exposed rRNA sequence are three small pockets or dents, called the A, P, and E sites, in the surface of the ribosome (shown in **figure 12.5** and discussed shortly). These sites have just the right shape to bind yet a third kind of RNA molecule, **transfer RNA (tRNA).** It is tRNA molecules that bring amino acids to the ribosome used in making proteins. The tRNA molecules are chains about 80 nucleotides long. The string of nucleotides folds back on itself, forming a three-looped structure shown in **figure 12.6a.** The looped structure further folds into a compact shape shown in **figure 12.6b**, with a three-nucleotide sequence at one end (the pink loop) and an amino acid attachment site on the other end (the 3′ end).

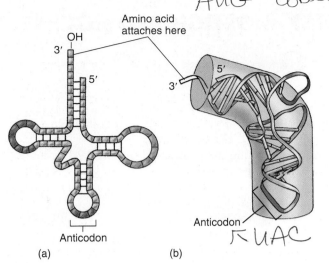

(a) (b)

AUG – codon

←UAC

Figure 12.6 The structure of tRNA.
tRNA, like mRNA, is a long strand of nucleotides. However, unlike mRNA, hydrogen bonding occurs between its nucleotides, causing the strand to form hairpin loops, as seen in (a). The loops then fold up on each other to create the compact, three-dimensional shape seen in (b). Amino acids attach to the free, single-stranded —OH end of a tRNA molecule. A three-nucleotide sequence called the anticodon in the lower loop of tRNA interacts with a complementary codon on the mRNA.

The three-nucleotide sequence, called the **anticodon,** is very important: It is the complementary sequence to 1 of the 64 codons of the genetic code! Special enzymes, called *activating enzymes,* match amino acids in the cytoplasm with their proper tRNAs. The anticodon determines which amino acid will attach to a particular tRNA.

Because the first dent in the ribosome, called the *A site* (the attachment site where amino-acid-bearing tRNAs will bind) is directly adjacent to where the mRNA binds to the rRNA, three nucleotides of the mRNA are positioned directly facing the anticodon of the tRNA. Like the address on a letter, the anticodon ensures that an amino acid is delivered to its correct "address" on the mRNA where the ribosome is assembling the polypeptide.

Making the Polypeptide Once an mRNA molecule has bound to the small ribosomal subunit, the other larger ribosomal subunit binds as well, forming a complete ribosome. The ribosome then begins the process of translation, illustrated in the panels of the Key Biological Process illustration on the next page. **Panel 1** shows how the mRNA begins to thread through the ribosome like a string passing through the hole in a doughnut. The mRNA passes through in short spurts, three nucleotides at a time, and at each burst of movement a new three-nucleotide codon on the mRNA is positioned opposite the A site in the ribosome, where a tRNA molecule first binds, as shown in **panel 2.**

As each new tRNA brings in an amino acid to each new codon presented at the A site, the old tRNA paired with the previous codon is passed over to the *P site* where peptide bonds form between the incoming amino acid and the growing polypeptide chain. The tRNA in the P site eventually shifts

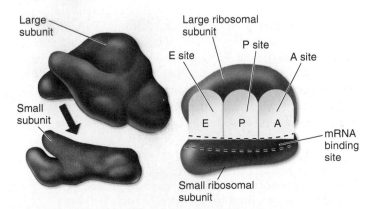

Figure 12.5 A ribosome is composed of two subunits.
The smaller subunit fits into a depression on the surface of the larger one. The A, P, and E sites on the ribosome play key roles in protein synthesis.

KEY BIOLOGICAL PROCESS: Translation

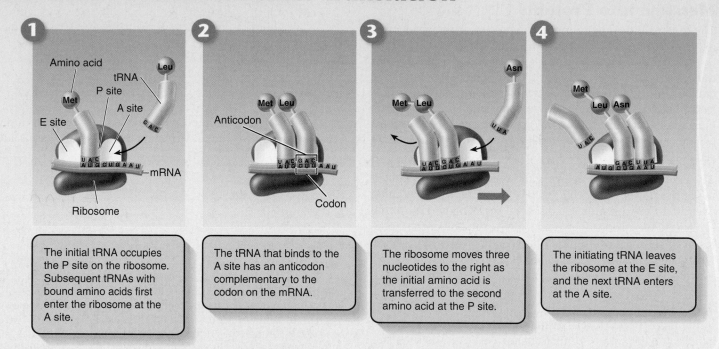

1 The initial tRNA occupies the P site on the ribosome. Subsequent tRNAs with bound amino acids first enter the ribosome at the A site.

2 The tRNA that binds to the A site has an anticodon complementary to the codon on the mRNA.

3 The ribosome moves three nucleotides to the right as the initial amino acid is transferred to the second amino acid at the P site.

4 The initiating tRNA leaves the ribosome at the E site, and the next tRNA enters at the A site.

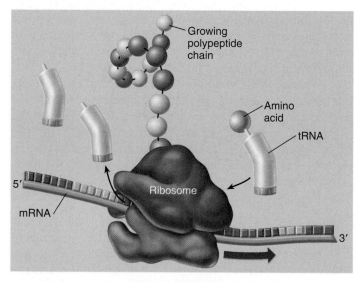

Figure 12.7 Ribosomes guide the translation process.

tRNA binds to an amino acid as determined by the anticodon sequence. Ribosomes bind the loaded tRNAs to their complementary sequences on the strand of mRNA. tRNA adds its amino acid to the growing polypeptide chain, which is released as the completed protein.

to the *E site* (the exit site), as shown in **panel 3**, and the amino acid it carried is attached to the end of a growing amino acid chain. The tRNA is then released (**panel 4**). So as the ribosome proceeds down the mRNA, one tRNA after another is selected to match the sequence of mRNA codons. In **figure 12.7** you can see the ribosome traveling along the length of the mRNA, the tRNAs bringing the amino acids into the ribosome and the growing polypeptide chain extending out from the ribosome. Translation continues until a "stop" codon is encountered, which signals the end of the polypeptide. The ribosome complex falls apart, and the newly made polypeptide is released into the cell.

As explained earlier, the overall flow of genetic information, the so-called "Central Dogma," is from DNA to mRNA to protein. For example, the polypeptide that is being formed in the Key Biological Process illustration above began with the DNA nucleotide sequence TACGACTTA, which is first transcribed into the mRNA sequence AUGCUGAAU. This sequence is then translated by the tRNAs into a polypeptide composed of the amino acids methionine—leucine—asparagine.

> **Key Learning Outcome 12.3**   The genetic code dictates how a particular nucleotide sequence specifies a particular amino acid sequence. A gene is transcribed into mRNA, which is then translated into a polypeptide. The sequence of mRNA codons dictates the corresponding sequence of amino acids in a growing polypeptide chain.

12.4 | Gene Expression

The Central Dogma, discussed in section 12.1, is the same in all organisms. **Figure 12.8** overviews the components needed for the key processes of DNA replication, transcription, and translation and the products that are formed in each. In general, the components are the same, the processes are the same, and the products are the same whether in prokaryotes or eukaryotes. However, there are some differences in gene expression between the two types of cells.

Architecture of the Gene

In prokaryotes, a gene is an uninterrupted stretch of DNA nucleotides whose transcript is read three nucleotides at a time to make a chain of amino acids. In eukaryotes, by contrast, genes are fragmented. In these more complex genes, the DNA nucleotide sequences encoding the amino acid sequence of a polypeptide are called **exons,** and the exons are interrupted frequently by extraneous nucleotides, "extra stuff" called **introns.** You can see them in the segment of DNA illustrated in **figure 12.9**; the exons are

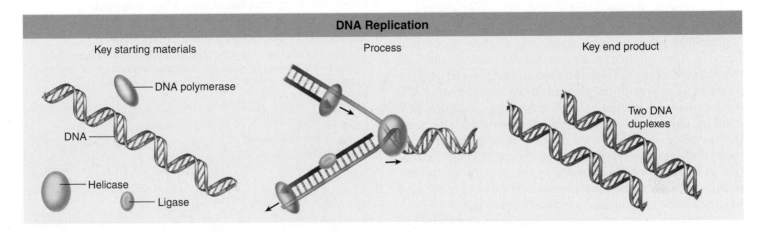

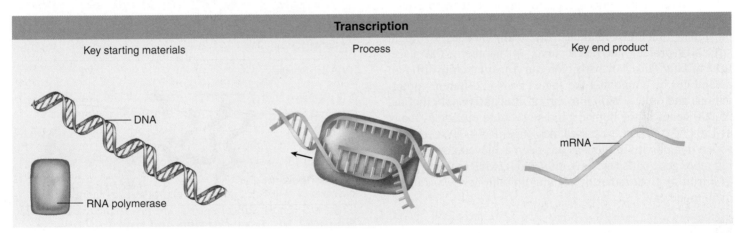

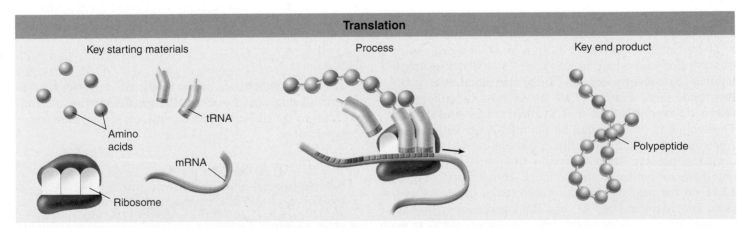

Figure 12.8   The processes of DNA replication, transcription, and translation.
These processes are generally the same in prokaryotes and eukaryotes.

the blue areas and the introns are the orange areas. Imagine looking at an interstate highway from a satellite. Scattered randomly along the thread of concrete would be cars, some moving in clusters, others individually; most of the road would be bare. That is what a eukaryotic gene is like: scattered exons embedded within much longer sequences of introns. In humans, only 1% to 1.5% of the genome is devoted to the exons that encode polypeptides, while 24% is devoted to the noncoding introns.

When a eukaryotic cell transcribes a gene, it first produces a **primary RNA transcript** of the entire gene, shown in **figure 12.9** with the exons in green and the introns in orange. Enzymes add modifications called a *5′ cap* and a *3′ poly-A tail,* which protect the RNA transcript from degradation. The primary transcript is then processed. Enzyme-RNA complexes excise out the introns and join together the exons to form the shorter, mature mRNA transcript that is actually translated into an amino acid chain. Notice that the mature mRNA transcript in **figure 12.9** contains only exons (green segments), no introns. Because introns are excised from the RNA transcript before it is translated into a polypeptide, they do not affect the structure of the polypeptide encoded by the gene in which they occur, despite the fact that introns represent over 90% of the nucleotide sequence of a typical human gene.

Why this crazy organization? It appears that many human genes can be spliced together in more than one way. In many instances, exons are not just random fragments, but rather functional modules. One exon encodes a straight stretch of protein, another a curve, yet another a flat place. Like mixing Tinkertoy parts, you can construct quite different assemblies by employing the same exons in different combinations and orders. With this sort of **alternative splicing,** the 25,000 genes of the human genome seem to encode as many as 120,000 different expressed messenger RNAs. It seems that added complexity in humans has been achieved not by gaining more gene parts (we have only about twice as many genes as a fruit fly), but rather by coming up with new ways to put them together.

Protein Synthesis

Protein synthesis in eukaryotes is more complex than in prokaryotes. Prokaryotic cells lack a nucleus and so there is no barrier between where mRNA is synthesized during transcription and where proteins are formed during translation. Consequently, a gene can be translated as it is being transcribed. **Figure 12.10** shows how the ribosomes attach to the mRNA as it is synthesized in prokaryotes. These clusters of ribosomes on the mRNA are called *polyribosomes.* In eukaryotic cells, a nuclear membrane separates the process of transcription from translation, making protein synthesis much more complicated. **Figure 12.11** on the next page walks you through the entire process. Transcription (step ❶) and RNA processing (step ❷) occur within the nucleus. In step ❸, the mRNA travels to the cytoplasm where it binds to the ribosome. In step ❹, tRNAs

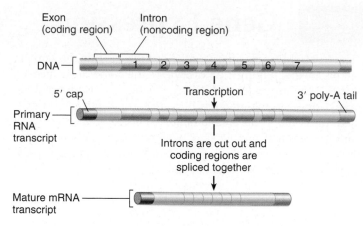

Figure 12.9   Processing eukaryotic RNA.
The gene shown here codes for a protein called ovalbumin. The ovalbumin gene and its primary transcript contain seven segments not present in the mRNA used by the ribosomes to direct the synthesis of the protein.

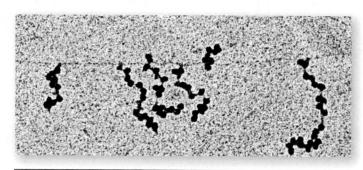

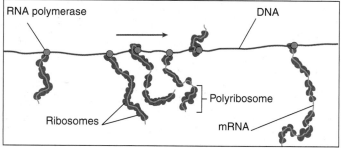

Figure 12.10   Transcription and translation in prokaryotes.
Ribosomes attach to an mRNA as it is formed, producing polyribosomes that translate the gene soon after it is transcribed.

bind to their appropriate amino acids, which correspond to their anticodons. In steps ❺ and ❻, the tRNAs bring the amino acids to the ribosome and the mRNA is translated into a polypeptide.

> **Key Learning Outcome 12.4   The general process of gene expression is similar in prokaryotes and eukaryotes, but differences exist in the architecture of the gene and the location of transcription and translation in the cell.**

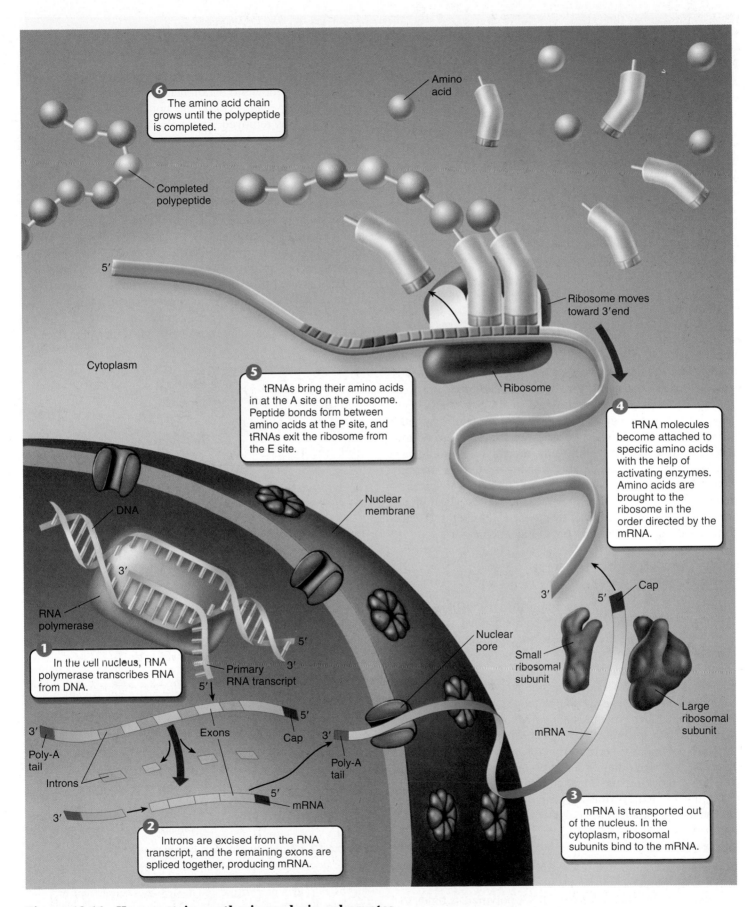

Figure 12.11 How protein synthesis works in eukaryotes.

12.5 How Prokaryotes Control Transcription

Being able to translate a gene into a polypeptide is only part of gene expression. Every cell must also be able to regulate when particular genes are used. Imagine if every instrument in a symphony played at full volume all the time, all the horns blowing full blast and each drum beating as fast and loudly as it could! No symphony plays that way because music is more than noise—it is the controlled expression of sound. In the same way, growth and development are due to the controlled expression of genes, each brought into play at the proper moment to achieve precise and delicate effects.

Control of gene expression is accomplished very differently in prokaryotes than in the cells of complex multicellular organisms. Prokaryotic cells have been shaped by evolution to grow and divide as rapidly as possible, enabling them to exploit transient resources. Proteins in prokaryotes turn over rapidly. This allows them to respond quickly to changes in their external environment by changing patterns of gene expression. In prokaryotes, the primary function of gene control is to adjust the cell's activities to its immediate environment. Changes in gene expression alter which enzymes are present in the cell in response to the quantity and type of available nutrients and the amount of oxygen present. Almost all of these changes are fully reversible, allowing the cell to adjust its enzyme levels up or down as the environment changes.

How Prokaryotes Turn Genes Off and On

Prokaryotes control the expression of their genes largely by saying *when* individual genes are to be transcribed. At the beginning of each gene are special regulatory sites that act as points of control. Specific regulatory proteins within the cell bind to these sites, turning transcription of the gene off or on.

For a gene to be transcribed, the RNA polymerase has to bind to a **promoter,** a specific sequence of nucleotides on the DNA that signals the beginning of a gene. In prokaryotes, gene expression is controlled by either blocking or allowing the RNA polymerase access to the promoter. Genes can be turned off by the binding of a **repressor,** a protein that binds to the DNA blocking the promoter. Genes can be turned on by the binding of an **activator,** a protein that makes the promoter more accessible to the RNA polymerase.

Repressors

Many genes are "negatively" controlled: They are turned off except when needed. In these genes, the regulatory site is located between the place where the RNA polymerase binds to the DNA (the promoter site) and the beginning edge of the gene. When a regulatory protein called a repressor is bound

to its regulatory site, called the *operator,* its presence blocks the movement of the polymerase toward the gene. Imagine if you went to sit down to eat dinner and someone was already sitting in your chair—you could not begin your meal until this person was removed from your chair. In the same way, the polymerase cannot begin transcribing the gene until the repressor protein is removed.

To turn on a gene whose transcription is blocked by a repressor, all that is required is to remove the repressor. Cells do this by binding special "signal" molecules to the repressor protein; the binding causes the repressor protein to contort into a shape that doesn't fit DNA, and it falls off, removing the barrier to transcription. A specific example demonstrating how repressor proteins work is the set of genes called the *lac* operon in the bacterium *Escherichia coli.* An **operon** is a segment of DNA containing a cluster of genes that are transcribed as a unit. The *lac* operon, shown in **figure 12.12,** consists of both polypeptide-encoding genes (labeled genes 1, 2,

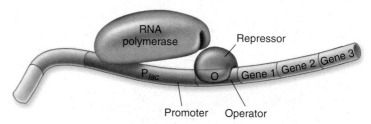

(a) *lac* operon is "repressed"

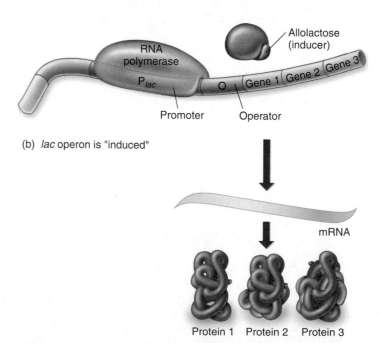

(b) *lac* operon is "induced"

Figure 12.12   How the lac operon works.
(*a*) The *lac* operon is shut down ("repressed") when the repressor protein is bound to the operator site. Because promoter and operator sites overlap, RNA polymerase and the repressor cannot bind at the same time. (*b*) The *lac* operon is transcribed ("induced") when allolactose binds to the repressor protein changing its shape so that it can no longer sit on the operator site and block polymerase binding.

and 3, which code for enzymes involved in breaking down the sugar lactose) and associated regulatory elements—the operator (the purple segment) and promoter (the orange segment). Transcription is turned off when a repressor molecule binds to the operator such that RNA polymerase cannot bind to the promoter. When *E. coli* encounters the sugar lactose, a metabolite of lactose called allolactose binds to the repressor protein and induces a twist in its shape that causes it to fall from the DNA. As you can see in **figure 12.12b**, RNA polymerase is no longer blocked, so it starts to transcribe the genes needed to break down the lactose to get energy.

Activators

Because RNA polymerase binds to a specific promoter site on one strand of the DNA double helix, it is necessary that the DNA double helix unzip in the vicinity of this site for the polymerase protein to be able to sit down properly. In many genes, this unzipping cannot take place without the assistance of a regulatory protein called an activator that binds to the DNA in this region and helps it unwind. Just as in the case of the repressor protein described previously, cells can turn genes on and off by binding "signal" molecules to the activator protein. These molecules either prevent the activator from binding to the DNA or enable it to do so. In the *lac* operon, a protein called *catabolite activator protein* (CAP) acts as an activator. CAP has to bind a signal molecule, cAMP, before it can associate with the DNA. Once the CAP/cAMP complex forms, as shown in **figure 12.13**, it binds to the DNA and makes the promoter more accessible to RNA polymerase.

Why bother with activators? Imagine if you had to eat every time you encountered food! Activator proteins enable a cell to cope with this sort of problem. Activators and repressors work together to control transcription. To understand how, let's consider the *lac* operon again, now shown in **figure 12.14**. When a bacterium encounters the sugar lactose, it may already have lots of energy in the form of glucose, as shown in panel ❶, and so does not need to break down more lactose.

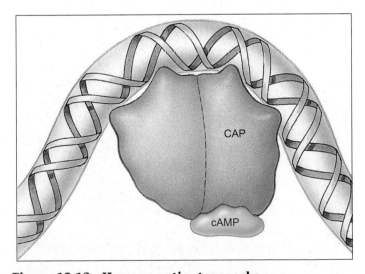

Figure 12.13 How an activator works.
Binding of the catabolite activator protein (CAP)/cAMP complex to DNA causes the DNA to bend around it. This increases the activity of RNA polymerase.

CAP can only bind and activate gene transcription when glucose levels are low. Because RNA polymerase requires the activator to function, the *lac* operon is not expressed. Also if glucose is present and lactose is absent, not only is the activator CAP unable to bind, but also a repressor blocks the promoter, as shown in panel ❷ below and in **figure 12.12a**. In the absence of both glucose and lactose, cAMP, the "low glucose" signal molecule (the green pie-shaped piece in panels ❸ and ❹) binds to CAP, and CAP is able to bind to the DNA. However, the repressor is still blocking transcription, as shown in panel ❸. Only in the absence of glucose and in the presence of lactose, the repressor is removed, the activator (CAP) is bound, and transcription proceeds, as shown in panel ❹.

> **Key Learning Outcome 12.5** Cells control the expression of genes by determining when they are transcribed. Some regulatory proteins block the binding of RNA polymerase, and others facilitate it.

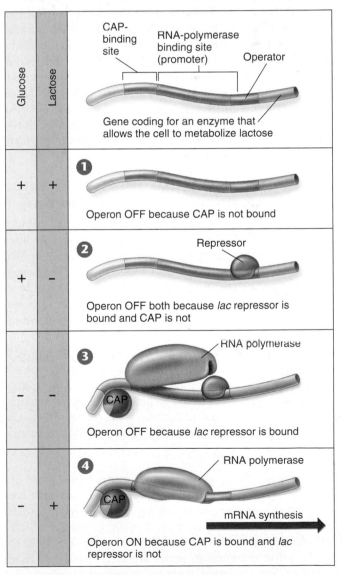

Figure 12.14 Activators and repressors at the *lac* operon.

12.6 Transcriptional Control in Eukaryotes

The Goals of Gene Expression Are Different in Eukaryotes

In multicellular organisms with relatively constant internal environments, the primary function of gene control in a cell is not to respond to that cell's immediate environment, like a prokaryote does, but rather to participate in regulating the body as a whole. Some of these changes in gene expression compensate for changes in the physiological condition of the body. Others mediate the decisions that ultimately produce the body, ensuring that the right genes are expressed in the right cells at the right time during development. The growth and development of multicellular organisms entail a long series of biochemical reactions. To produce the necessary enzymes, genes are transcribed in a carefully prescribed order, each for a specified period of time, following a fixed genetic program. The one-time expression of the genes that guide such a program is fundamentally different from the reversible metabolic adjustments prokaryotic cells make to the environment, like the turning on and off of the *lac* operon. In all multicellular organisms, changes in gene expression within particular cells serve the needs of the whole organism, rather than the survival of individual cells.

The Structure of Chromosomes Can Affect Eukaryotic Gene Expression

The first hurdle faced by RNA polymerase in transcribing a eukaryotic gene is gaining access to it. The DNA of eukaryotes is packaged into chromosomes. The packaging of DNA into nucleosomes and then into higher-order chromosome structures (refer back to **figure 8.4**) is directly related to the control of gene expression. Chromosome structure at its lowest level is the organization of DNA and histone proteins into nucleosomes, as shown in **figure 12.15**. These nucleosomes may block binding of RNA polymerase and other proteins called transcription factors at the promoter. The higher-order organization of chromosomes is not completely understood. It involves modifying histones to produce a greater condensation of the chromosomal material, called *chromatin,* making promoters even less accessible for protein-DNA interactions.

DNA Methylation

Chemical methylation of the DNA was once thought to play a major role in regulating gene expression in vertebrate cells. Methylation is the process of adding a methyl group ($—CH_3$) to cytosine nucleotides, creating 5-methylcytosine, which is still able to participate in base pairs. Scientists observed that many inactive mammalian genes are methylated, and it was tempting to conclude that methylation caused the inactivation. However, methylation is now viewed as having a less direct role in transcriptional regulation. It appears instead to block

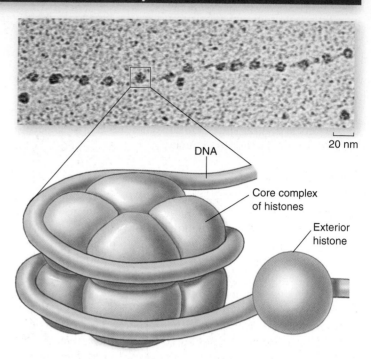

Figure 12.15 DNA coils around histones.
Within chromosomes, DNA is packaged into nucleosomes. In the electron micrograph (*top*), the DNA is partially unwound, and individual nucleosomes can be seen. In a nucleosome, the DNA double helix coils around a core complex of eight histones; one additional histone binds to the outside of the nucleosome.

accidental transcription of "turned-off" genes. DNA methylation thus ensures that once a gene is turned off, it stays off.

Chromatin Structure and Transcriptional Activators

As in prokaryotes, not all gene regulation in eukaryotes involves repression of transcription. In at least some instances, activation of a gene is needed before it can be transcribed. Most activating factors seem to act directly on transcription, helping the RNA polymerase to bind to the promoter. Other "coactivators" act by modifying the structure of chromatin to make DNA accessible. Recently, some coactivators have been shown to interact with histones. In these cases, it appears that transcription is increased by adding methyl groups to the histones. The methylation of histones disrupts the higher-order chromatin structure, making the DNA more accessible. This control also appears to work in the opposite way, with corepressors having been shown to remove methyl groups from the histones, making the DNA coil more tightly around the histones and restricting access to the DNA.

> **Key Learning Outcome 12.6** Transcriptional control in eukaryotes can be effected by the tight packaging of DNA into nucleosomes. Activators can loosen DNA's association with histones, making the DNA more accessible to RNA polymerase.

Controlling Transcription from a Distance

In eukaryotes, transcription is considerably more complex, and the amount of DNA involved in regulating eukaryotic genes is much greater.

Eukaryotic Transcription Factors

Eukaryotic transcription requires not only the RNA polymerase molecule, but also a variety of other proteins, called *transcription factors,* that interact with the polymerase.

Basal transcription factors are necessary for the assembly of a transcription apparatus and recruitment of RNA polymerase to a promoter. While these factors are required for transcription to occur, they do not increase the rate of transcription above a low rate, the so-called basal rate. These factors, colored green in **figure 12.16**, all come together to form the *initiation complex*. This is clearly much more complex than a bacterial RNA polymerase, which is a single enzyme protein.

The initiation complex, once assembled, will not achieve transcription at a high level without the participation of other gene-specific factors. The number and diversity of these **specific transcription factors,** colored tan in **figure 12.16**, is overwhelming. Multicellular organisms control which genes are expressed by regulating which specific transcription factors are available at a particular time and place.

Enhancers

While prokaryotic gene control regions, such as the operator, are positioned immediately upstream of the coding region, this is not true in eukaryotes. It turns out that far away sites called **enhancers** can have a major impact on the rate of transcription. Enhancers are nucleotide sequences where specific transcription factors, acting as activators, bind the DNA. The ability of enhancers to act over large

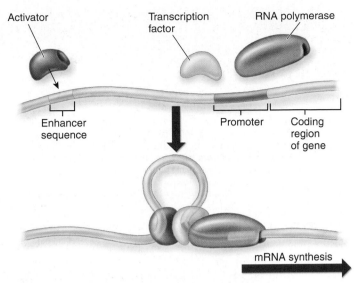

Figure 12.17   How enhancers work.
The activator binding site, or enhancer, is often located far from the gene. The binding of an activator protein brings the enhancer in contact with the gene.

distances is accomplished by DNA bending to form a loop. In **figure 12.17**, an activator binds to the yellow-colored enhancer far from the promoter. The DNA loops, bringing the activator in contact with the RNA polymerase/ initiation complex so that transcription can begin.

Complicating the process even more, several different additional transcription factors may modulate the action of a particular specific transcription factor. These **coactivators** and **mediators** act by first binding the transcription factor, and then binding to the transcription apparatus.

Tying It All Together

How can we make sense of this extremely complicated situation? Virtually all genes that are transcribed by RNA polymerase in eukaryotes need the same group of basal factors to assemble an initiation complex, but its ultimate level of transcription depends in each instance on the other specific factors involved that make up a *transcription complex*. This kind of combined gene regulation leads to great flexibility in the control of gene expression. It provides the cell the ability to produce finely graded responses to the many environmental and developmental signals that it may receive. The eukaryotic cell achieves a higher level of control because of the interaction of a large number of protein regulatory elements (**figure 12.18**). This control is more sophisticated but not fundamentally different from the integration achieved by the prokaryotic *lac* operon using two regulatory proteins.

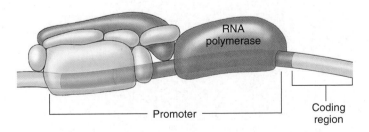

Figure 12.16   Formation of a eukaryotic initiation complex.
The basal transcription factors (*green*) bind to the promoter region of the DNA and form an initiation complex. A number of specific transcription factors (*tan*) bind to the basal transcription factor complex (the initiation complex) and together recruit the RNA polymerase molecule to the promoter.

Key Learning Outcome 12.7   Transcription factors and enhancers give eukaryotic cells great flexibility in controlling gene expression.

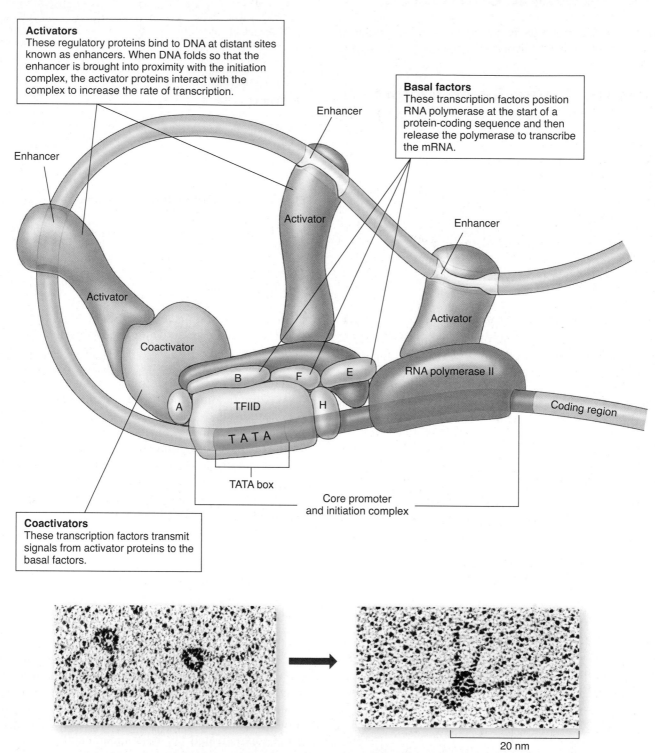

Activators
These regulatory proteins bind to DNA at distant sites known as enhancers. When DNA folds so that the enhancer is brought into proximity with the initiation complex, the activator proteins interact with the complex to increase the rate of transcription.

Basal factors
These transcription factors position RNA polymerase at the start of a protein-coding sequence and then release the polymerase to transcribe the mRNA.

Enhancer

Enhancer

Activator

Enhancer

Activator

Activator

Coactivator

Activator

B

F

E

RNA polymerase II

A

TFIID

H

Coding region

T A T A

TATA box

Core promoter
and initiation complex

Coactivators
These transcription factors transmit signals from activator proteins to the basal factors.

20 nm

Figure 12.18 Interactions of various factors within the transcription complex.
All specific transcription factors bind to enhancer sequences that may be distant from the promoter. These proteins can then interact with the initiation complex by DNA looping to bring the factors into proximity with the initiation complex. As detailed in the text, some transcription factors, activators, can directly interact with the RNA polymerase II or the initiation complex while others require additional coactivators. The electron micrographs are of the bacterial activator NtrC. When it binds to an enhancer, you can see how this causes the DNA to loop over to a distant site where RNA polymerase is bound, activating transcription. While such enhancers are rare in prokaryotes, they are common in eukaryotes.

12.8 RNA-Level Control

Thus far we have discussed gene regulation entirely in terms of proteins that regulate the start of transcription by blocking or activating the "reading" of a particular gene by RNA polymerase. Within the last decade, however, it has become increasingly clear that RNA molecules can regulate the expression of genes, acting after transcription as a second level of control.

Discovery of RNA Interference

As will be discussed in chapter 13, the bulk of the eukaryotic genome is not translated into proteins. This finding was puzzling at first, but biologists now suspect that RNA transcripts of these regions might play an important role in gene regulation. The finding that almost all the differences between human and chimpanzee DNA occur in such regions only adds to the suspicion.

All this began to make sense in 1998, when a simple experiment was carried out, for which Americans Andrew Fire and Craig Mello later won the Nobel Prize in Physiology or Medicine in 2006. These investigators injected double-stranded RNA molecules into the nematode worm *Caenorhabditis elegans*. This resulted in the silencing of the gene whose sequence was complementary to the double-stranded RNA, and of no other gene. The investigators called this very specific effect **gene silencing,** or **RNA interference.** What is going on here? As you will learn in chapter 16, RNA viruses replicate themselves through double-stranded intermediates—at a critical stage, a virus enzyme called *reverse transcriptase* travels along the virus RNA and assembles a complementary strand. Because the life cycle of many viruses involves a double-stranded RNA stage, RNA interference may have evolved as a cellular defense mechanism against these viruses; the evolution of this adaptation would have predated the evolutionary divergence of plants and animals. Indeed, double-stranded viral RNAs can be targeted for destruction by RNA interference machinery. Without intending to do so, the nematode researchers had stumbled across this defense.

How RNA Interference Works

Investigating interference, researchers noted that in the process of silencing a gene, plants produced short RNA molecules (ranging in length from 21 to 28 nucleotides) that matched the gene being silenced. Earlier researchers were focusing on far larger messenger RNA (mRNA), transfer RNA (tRNA), and ribosomal RNA (rRNA) and had not noticed these far smaller bits, tossing them out during experiments. These small RNAs appeared to regulate the activity of specific genes.

Soon researchers found evidence of similar small RNAs in a wide range of other organisms. In the plant *Arabidopsis thaliana,* small RNAs seemed to be involved in the regulation of genes critical to early development, while in yeasts they were identified as the agents that silence genes in tightly packed regions of the genome. In the ciliated protozoan *Tetrahymena thermophila,* the loss of major blocks of DNA during development seems guided by small RNA molecules.

The first clue of how small fragments of RNA can act to regulate gene expression emerged when researchers noted that stretches of double-stranded RNA injected into *C. elegans* can

**Figure 12.19
How RNA interference works.**
Double-stranded RNA is cut by dicer. The resulting siRNA associates with proteins forming a complex called RISC. siRNA becomes single stranded and binds to targeted mRNAs with the same or similar sequences, which blocks translation of the gene.

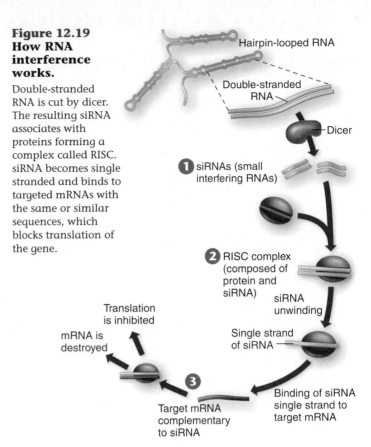

Hairpin-looped RNA

Double-stranded RNA

Dicer

❶ siRNAs (small interfering RNAs)

❷ RISC complex (composed of protein and siRNA)

siRNA unwinding

Single strand of siRNA

Binding of siRNA single strand to target mRNA

Target mRNA complementary to siRNA

❸

Translation is inhibited

mRNA is destroyed

dissociate. Each single strand can then form a double-stranded RNA by folding back in a hairpin loop, like the three sections of RNA shown toward the top of **figure 12.19**. This occurs because the two ends of the strand have a complementary nucleotide sequence. When the RNA loops, the complementary bases form base pairings that hold the strands together much as they do in the strands of a DNA duplex.

Exactly how does such a double-stranded RNA inhibit the expression of the gene from which the double-stranded RNA has been generated? In the first stage of RNA interference, an enzyme called *dicer* recognizes long, double-stranded RNA molecules and cuts them into short, small RNA segments called *siRNAs (small interfering RNAs)* ❶. In the next step, the siRNAs can assemble into a ribonucleoprotein complex called *RISC (RNA Interference Silencing Complex)* ❷. RISC then unwinds the siRNA duplex, which leaves one single strand of RNA that is able to bind to mRNAs complementary to it ❸ and thus silence the genes that produced those mRNA molecules.

Once the siRNA has bound to mRNA, the silencing is achieved in one of two ways: Either the mRNA is inhibited by blocking its translation into protein, or the mRNA is destroyed. The choice between inhibition and destruction is thought to be governed by how closely the sequence of the siRNA matches the mRNA sequence, with destruction being the outcome for best-matched targets.

> **Key Learning Outcome 12.8** Small interfering RNAs, called siRNAs, are formed from double-stranded sections of RNA molecules. These siRNAs bind to mRNA molecules in the cell and block their translation.

Silencing Genes to Treat Disease

The recent discovery that eukaryotes control their genes by selectively "silencing" particular gene transcripts has electrified biologists, as it opens exciting possibilities for treating disease and infection. Many diseases are caused by the expression of one or more genes. AIDS, for example, requires the expression of several genes of the HIV virus. Many chronic human diseases result from excessively active genes. What if doctors could somehow shut these genes off?

The idea is simple. If you can isolate a gene involved in the disorder and determine its sequence, then in principle you could synthesize an RNA molecule with the sequence of the opposite or "anti-sense" strand. This RNA would thus have a sequence complementary to the messenger RNA produced by that gene. Introduced into cells, this synthesized RNA might be able to bind to the messenger RNA, creating a double-stranded RNA that could not be read by ribosomes. If an anti-sense therapy could be made to work and be delivered practically and inexpensively, the AIDS epidemic could be halted in its tracks. Indeed, any viral infection could be combatted in this way. Influenza is perhaps the greatest killer of all infectious diseases. A workable anti-sense therapy could provide a means of stamping out a bird flu epidemic before the virus spreads.

By far the most exciting promise of anti-sense gene silencing therapy is the possibility of practical cancer therapy. Discussed in chapter 8, cancer kills more Americans than any other disease. We now know in considerable detail how cancer comes about. It results from damage to genes that regulate the cell cycle. The great promise of RNA gene silencing therapy comes from those cancer-causing gene mutations that increase the effectiveness of one or more "divide" signals. If these mutant genes could be silenced, the cancer could be shut down.

The possibility of using complementary RNA to silence troublesome genes has gotten a huge boost in the last few years from the discovery of a unique virus defense system in eukaryotes. In order to protect themselves from RNA virus infection, cells have a complex system for detecting, attacking, and destroying viral RNA. The system takes advantage of a subtle vulnerability of the infecting virus: At some point, in order to multiply within the infected cell, the virus must express its genes—it must make complementary copies of them that can serve as messenger RNAs to direct production of virus proteins. At that point, while the viral RNA molecule is double-stranded, the virus is vulnerable to attack: At no place in the cell is double-stranded RNA usually found, so by targeting double-stranded RNAs for immediate destruction, a cell can defeat virus infections.

Silencing genes with complementary RNA, dubbed "RNA interference," offers the exciting hope that successful treatment of many diseases may be literally at our doorstep.

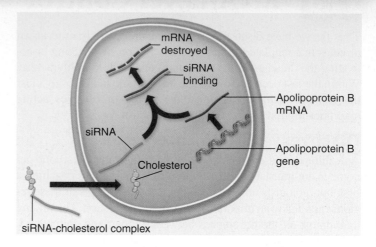

siRNA-cholesterol complex

First, however, scientists must figure out how to make RNA interference therapies work. They are facing some formidable technical problems, not the least of which is to find a way to deliver the interfering RNA to, and into, the target cells. The problem is that RNA is rapidly broken down in the bloodstream, and most of the body's cells don't readily absorb it, even if it does reach them. Some researchers are attempting to package the RNA into viruses, although as you will learn in chapter 13, gene therapies that have attempted this approach can trigger an immune response and could even cause cancer. Gene therapy researchers have been seeking safer virus gene-delivery vehicles; what they learn will surely be put to good effect.

One interesting alternative approach is to modify the RNA to protect it and make it more easily taken up by cells. This work focuses on the mRNA that encodes apolipoprotein B, a molecule involved in the metabolism of cholesterol. High levels of apolipoprotein are found in people with high levels of cholesterol, associated with increased risk of coronary heart disease. Interfering RNAs that target apolipoprotein B mRNA result in destruction of the mRNA, and lower levels of cholesterol. To effectively deliver it to the body's tissues, researchers simply attached a molecule of cholesterol to each interfering RNA molecule, as shown above. Levels of apolipoprotein B were reduced 50% to 70%, and blood cholesterol levels plummeted downwards, to the same levels seen in cells from which the apolipoprotein B gene had been deleted. It is not clear if this approach will work for many other RNAs, but it looks promising.

A second major problem confronting those seeking to develop successful therapies based on RNA silencing of troublesome genes is one of specificity. It is very important that only the target gene be silenced. Before carrying out clinical trials involving large numbers of people, it is imperative that we be sure the interfering RNA will not shut down vital human genes as well as the targeted virus or cancer genes. Some studies suggest this will not be a problem, while in others a range of "off-target" genes seem to be affected. This possibility will have to be carefully evaluated for each new therapy being developed.

12.9 Complex Regulation of Gene Expression

As you have seen, the eukaryotic gene is structurally more complex than the prokaryotic gene, and the regulation of gene expression is also more complex. Eukaryotic gene expression is controlled at many stages, reviewed in figure 12.20. Chromatin structure can affect gene expression by determining if a gene will be accessible to RNA polymerase. The availability of many factors influences the rate at which a particular gene is transcribed. Once transcribed, a gene's expression can be altered by alternative splicing, or it can be silenced by RNA interference. Although gene regulation often occurs earlier in the process of gene expression, some control mechanisms act later. The availability of translational proteins can affect protein synthesis, and a protein can also be chemically modified after it is produced.

> **Key Learning Outcome 12.9** A eukaryotic gene is controlled at many points during gene expression.

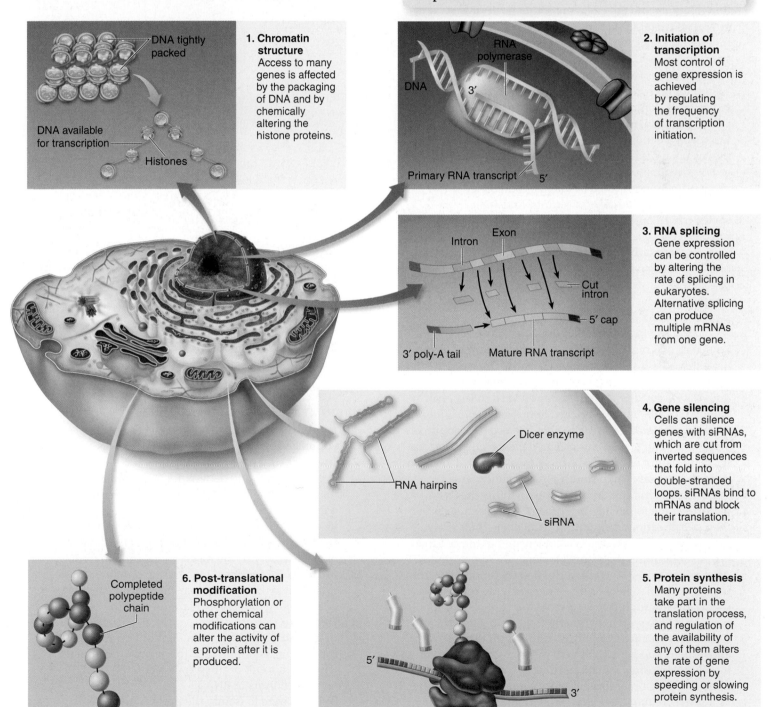

1. Chromatin structure
Access to many genes is affected by the packaging of DNA and by chemically altering the histone proteins.

DNA tightly packed
DNA available for transcription
Histones

2. Initiation of transcription
Most control of gene expression is achieved by regulating the frequency of transcription initiation.

RNA polymerase
DNA
3′
Primary RNA transcript
5′

3. RNA splicing
Gene expression can be controlled by altering the rate of splicing in eukaryotes. Alternative splicing can produce multiple mRNAs from one gene.

Intron   Exon
Cut intron
5′ cap
3′ poly-A tail   Mature RNA transcript

4. Gene silencing
Cells can silence genes with siRNAs, which are cut from inverted sequences that fold into double-stranded loops. siRNAs bind to mRNAs and block their translation.

Dicer enzyme
RNA hairpins
siRNA

6. Post-translational modification
Phosphorylation or other chemical modifications can alter the activity of a protein after it is produced.

Completed polypeptide chain

5. Protein synthesis
Many proteins take part in the translation process, and regulation of the availability of any of them alters the rate of gene expression by speeding or slowing protein synthesis.

5′
3′

Figure 12.20 The control of gene expression in eukaryotes.

Building Proteins in a Test Tube

The complex mechanisms used by cells to build proteins were not discovered all at once. Our understanding came slowly, accumulating through a long series of experiments, each telling us a little bit more. To gain some sense of the incremental nature of this experimental journey, and to appreciate the excitement that each step gave, it is useful to step into the shoes of an investigator back when little was known and the way forward was not clear.

The shoes we will step into are those of Paul Zamecnik, an early pioneer in protein synthesis research. Working with colleagues at Massachusetts General Hospital in the early 1950s, Zamecnik first asked the most direct of questions: Where in the cell are proteins synthesized? To find out, they injected radioactive amino acids into rats. After a few hours, the labeled amino acids could be found as part of newly made proteins in the livers of the rats. And, if the livers were removed and checked only minutes after injection, radioactive-labeled proteins were found only associated with small particles in the cytoplasm. Composed of protein and RNA, these particles, later named ribosomes, had been discovered years earlier by electron microscope studies of cell components. This experiment identified them as the sites of protein synthesis in the cell.

After several years of trial-and-error tinkering, Zamecnik and his colleagues had worked out a "cell-free" protein-synthesis system that would lead to the synthesis of proteins in a test tube. It included ribosomes, mRNA, and ATP to provide energy. It also included a collection of required soluble "factors" isolated from homogenized rat cells that somehow worked with the ribosome to get the job done. When Zamecnik's team characterized these required factors, they found most of them to be proteins, as expected, but also present in the mix was a small RNA, very unexpected.

To see what this small RNA was doing, they performed the following experiment. In a test tube, they added various amounts of ^{14}C-leucine (that is, the radioactively labeled amino acid leucine) to the cell-free system containing the soluble factors, ribosomes, and ATP. After waiting a bit, they then isolated the small RNA from the mixture and checked it for radioactivity. You can see the results in graph (a).

In a follow-up experiment, they mixed the radioactive leucine-small RNA complex that this experiment had generated with cell extracts containing intact endoplasmic reticulum (that is, a cell system of ribosomes on membranes quite capable of making protein). Looking to see where the radioactive label now went, they then isolated the newly made protein as well as the small RNA [see graph (b)].

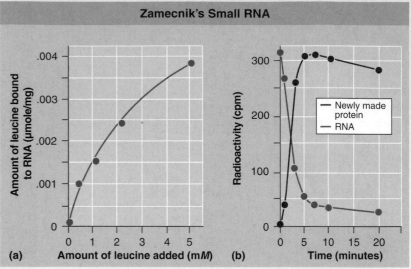

Zamecnik's Small RNA

(a) Amount of leucine added (mM)

(b) Time (minutes)

EXPERIMENT A, shown in graph (a)

1. **Applying Concepts** What is the dependent variable?
2. **Interpreting Data** Does the amount of leucine added to the test tube have an effect on the amount of leucine found bound to the small RNA?
3. **Making Inferences** Is the amount of leucine bound to small RNA proportional to the amount of leucine added to the mixture?
4. **Drawing Conclusions** Can you reasonably conclude from this result that the amino acid leucine is binding to the small RNA?

EXPERIMENT B, shown in graph (b)

1. **Applying Concepts** What is the dependent variable?
2. **Interpreting Data**
   a. Monitoring radioactivity for 20 minutes after the addition of the radioactive leucine-small RNA complex to the cell extract, what happens to the level of radioactivity in the small RNA (blue)?
   b. Over the same period, what happens to the level of radioactivity in the newly made protein (red)?
3. **Making Inferences** Is the same amount of radioactivity being lost from the small RNA that is being gained by the newly made protein?
4. **Drawing Conclusions**
   a. Is it reasonable to conclude that the small RNA is donating its amino acid to the growing protein? [Hint: As a result of this experiment, the small RNA was called transfer RNA.]
   b. If you were to isolate the protein from this experiment made after 20 minutes, which amino acids would be radioactively labeled? Explain.

From Gene to Protein

12.1 The Central Dogma

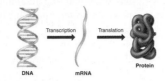

- DNA is the storage site of genetic information in the cell. The process of gene expression, DNA to RNA to protein, is called the "Central Dogma."

- Gene expression occurs in two stages using different types of RNA: transcription, where an mRNA copy is made from the DNA, and translation, where the information on the mRNA is translated into a protein using rRNA and tRNA.

12.2 Transcription

- In transcription, DNA serves as a template for mRNA synthesis by a protein called RNA polymerase. RNA polymerase binds to one strand of the DNA at a site called the promoter. RNA polymerase adds complementary nucleotides onto the growing mRNA in a way that is similar to the actions of DNA polymerase (**figure 12.3**).

12.3 Translation

- The message within the mRNA is coded in the order of the nucleotides, which are read in three-nucleotide units called codons. Each codon corresponds to a particular amino acid. The rules that govern the translation of codons on the mRNA into amino acids is called the genetic code (**figure 12.4**).

- The genetic code contains 64 codons but codes for only 20 amino acids; therefore, there is duplication. In many instances, two or more different codons encode the same amino acid.

- In translation, the mRNA carries the message to the cytoplasm. rRNA combines with proteins to form a structure called a ribosome (**figure 12.5**), the platform on which proteins are assembled. tRNAs, like one shown here from **figure 12.6**, carry amino acids to the ribosome to build the polypeptide chain.

- A ribosome contains a small rRNA subunit and a large rRNA subunit. Translation begins when the mRNA binds to the small ribosomal subunit, which triggers the binding of the large subunit to the small subunit, forming a complete ribosome.

- A ribosome moves along mRNA, while the tRNA molecules that contain anticodon sequences complementary to the codons bring amino acids to the ribosome. The amino acids add to the growing polypeptide chain (**Key Biological Process, page 240**).

12.4 Gene Expression

- Prokaryotic genes are contained within a stretch of DNA that is transcribed into mRNA and then is translated in its entirety. Eukaryotic genes are fragmented, containing coding regions called exons, and noncoding regions called introns.

- The entire eukaryotic gene is transcribed into RNA, but the introns are spliced out before translation. This editing process is shown here from **figure 12.9**.

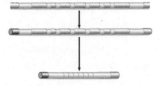

- Exons can be spliced together in different ways, a process called alternative splicing, producing different protein products from the same sections of DNA.

- In prokaryotes, transcription and translation occur simultaneously in the cytoplasm (**figure 12.10**). In eukaryotes, the RNA transcript is first produced and then processed (introns spliced out) in the nucleus. The mRNA then travels to the cytoplasm to be translated into a polypeptide (**figure 12.11**).

Regulating Gene Expression in Prokaryotes

12.5 How Prokaryotes Control Transcription

- In prokaryotic cells, genes are turned off when a repressor protein blocks the promoter, as shown here from **figure 12.12**, binding to a site called an operator.

- Some genes can be turned on only when a protein called an activator binds to the DNA and opens up the double helix so that RNA polymerase can bind to the promoter.

- The *lac* operon contains a cluster of genes that are involved in the breakdown of the sugar lactose. When the proteins produced by the *lac* operon genes are needed, an inducer molecule will bind to the repressor protein so it can't attach to the DNA, thereby freeing the promoter so that RNA polymerase can bind (**figure 12.12**).

- The *lac* operon is also controlled by an activator. The activator alters the shape of DNA, which allows the RNA polymerase to bind to the DNA (**figure 12.13**). It is only when the activator binds to the DNA and the repressor is removed that the RNA polymerase can bind to the promoter (**figure 12.14**).

Regulating Gene Expression in Eukaryotes

12.6 Transcriptional Control in Eukaryotes

- The coiling of DNA around histones (**figure 12.15**) restricts RNA polymerase's access to the DNA. Controlling gene expression may involve chemical modification of histones that make the DNA more accessible.

12.7 Controlling Transcription from a Distance

- In eukaryotic cells, transcription requires the binding of transcription factors before RNA polymerase can bind to the promoter (**figure 12.16**). Eukaryotic genes are controlled from distant locations called enhancers (**figures 12.17** and **12.18**).

12.8 RNA-Level Control

- RNA interference blocks translation. Small sections of RNA, called siRNA, bind to mRNA in the cytoplasm, which blocks the expression of the gene (**figure 12.19**).

12.9 Complex Regulation of Gene Expression

- Eukaryotic gene expression is controlled at many levels (**figure 12.20**).

Test Your Understanding

1. Which of the following is not a type of RNA?
   a. nRNA (nuclear RNA)
   b. mRNA (messenger RNA)
   c. rRNA (ribosomal RNA)
   d. tRNA (transfer RNA)

2. Each amino acid in a polypeptide is specified by
   a. an enhancer.
   b. a promoter.
   c. an rRNA molecule.
   d. a codon.

3. The three-nucleotide codon system can be arranged into _____ combinations.
   a. 16
   b. 20
   c. 64
   d. 128

4. The process of obtaining a copy of the information in a gene as a strand of messenger RNA is called
   a. polymerase.
   b. expression.
   c. transcription.
   d. translation.

5. The site where RNA polymerase attaches to the DNA molecule to start the formation of an RNA molecule is called a(n)
   a. promoter.
   b. exon.
   c. intron.
   d. enhancer.

6. The process of taking the information on a strand of messenger RNA and building an amino acid chain, which will become all or part of a protein molecule, is called
   a. polymerase.   c. transcription.
   b. expression.   d. translation.

7. If an mRNA codon reads UAC, its complementary anticodon will be
   a. TUC.
   b. ATG.
   c. AUG.
   d. CAG.

8. Which of the following accurately describes gene expression in prokaryotic cells?
   a. All genes are on all the time in all cells, making the needed amino acid sequences.
   b. Some genes are always off unless a promoter turns them on.
   c. Some genes are always on unless a promoter turns them off.
   d. Some genes remain off as long as a repressor is bound.

9. Which of the following statements is correct about eukaryotic gene expression?
   a. mRNAs must have introns spliced out.
   b. mRNAs contain the transcript of only one gene.
   c. Enhancers act from a distance.
   d. All of the above.

10. Which of the following is *not* a mechanism of controlling gene expression in eukaryotic cells?
    a. blocking translation with siRNA
    b. activating an enhancer
    c. translating a gene as it is being transcribed
    d. alternative splicing of the primary RNA transcript

Apply Your Understanding

1. **Page 236** Assume that this figure is showing gene expression in a eukaryotic cell. What step is missing in the process?

2. **Figure 12.12** Can genes 1, 2, and 3 be transcribed? Describe what would happen if an inducer molecule was added to the complex.

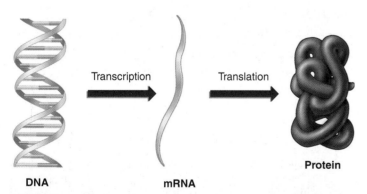

DNA — Transcription → mRNA — Translation → Protein

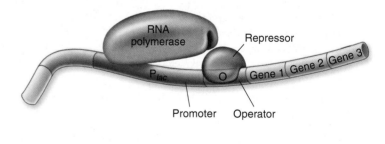

RNA polymerase, Repressor, P*lac*, O, Gene 1, Gene 2, Gene 3, Promoter, Operator

Synthesize What You Have Learned

1. On the television program *The X Files*, Agent Scully discovers an extraterrestrial life form that has a DNA genome like ours, but with a four-letter genetic code instead of the triplet genetic code that we earthly inhabitants possess. How many different amino acids would this extraterrestrial code be able to specify (assuming there were that many kinds of amino acids available on the extraterrestrial planet)? Why do you think the terrestrial code allows 64 combinations, when only 22 amino acids are common here on earth? Why do you think only 20 amino acids are common in proteins here on earth?

2. What would happen if all the genes in a cell were always active?

3. The nucleotide sequence of a hypothetical gene is:
   TACATACTTAGTTACGTCGCCCGGAAATAT
   a. What will be the sequence on the mRNA when it is transcribed?
   b. What will be the amino acid sequence of the protein when it's translated?
   c. What would happen to the amino acid chain if the highlighted nucleotide underwent a mutation and was changed to an A nucleotide?

13

The New Biology

This sheep, Dolly, was the first animal to be cloned from a single adult cell. The lamb you see beside her is her offspring, normal in every respect. From Dolly we learn that genes are not lost during development. If a single adult cell can be induced to switch the proper combination of genes on and off, that one cell can develop into a normal adult individual. Embryonic stem cells are like this—poised to become any cell of the body as the embryo develops. It may be possible to replace damaged tissues with healthy tissue grown from a patient's own embryonic stem cells, so long as the disorder is not an inherited one. The approach has been used successfully in laboratory mice to cure a variety of disorders. However, its use in humans would require employing stem cells derived from the patient, and because this may involve destroying a human embryo the approach is controversial. Another approach, when the damaged tissue does result from a defective gene, is to repair rather than replace, using a virus to transfer a healthy gene into those tissues that lack it. In this chapter you will explore genomic screening, the application of gene technology to medicine and agriculture, reproductive cloning, stem cell tissue replacement, and gene therapy, all areas in which a revolution is reshaping biology.

13.1 Genomics

Recent years have seen an explosion of interest in comparing the entire DNA content of different organisms, a new field of biology called **genomics.** While initial successes focused on organisms with relatively small numbers of genes, researchers have recently completed the sequencing of several large eukaryotic genomes, including our own.

The full complement of genetic information of an organism—all of its genes and other DNA—is called its **genome.** To study a genome, the DNA is first sequenced, a process that allows each nucleotide of a DNA strand to be read in order. The first genome to be sequenced was a very simple one: a small bacterial virus called φ-X174 (φ is the Greek letter phi). Frederick Sanger, inventor of the first practical way to sequence DNA, obtained the sequence of this 5,375-nucleotide genome in 1977. This was followed by the sequencing of dozens of prokaryotic genomes. The advent of automated DNA sequencing machines in recent years has made the DNA sequencing of much larger eukaryotic genomes practical, including our own (**table 13.1**).

Sequencing DNA

In sequencing DNA, the DNA of unknown sequence is first cut into fragments. Each DNA fragment is then copied (amplified), so there are thousands of copies of the fragment. The DNA fragments are then mixed with copies of DNA polymerase, copies of a primer (recall from chapter 11 that DNA polymerase can only add nucleotides onto an existing strand of nucleotides), a supply of the four nucleotide bases, and a supply of four different chain-terminating chemical tags. The chemical tags act as one of the four nucleotide bases in DNA synthesis, undergoing complementary base pairing. First, heat is applied to denature the double-stranded DNA fragments.

The solution is then allowed to cool, allowing the primer (the lighter blue box in **figure 13.1❶**) to bind to a single strand of the DNA, and synthesis of the complementary strand proceeds. Whenever a chemical tag is added instead of a nucleotide base, the synthesis stops, as shown in the figure. For example, the terminating red "T" was added after three normal nucleotides and synthesis stopped. Because of the relatively low concentration of the chemical tags compared with the nucleotides, a tag that binds to G on the DNA fragment, for example, will not necessarily be added to the first G site. Thus, the mixture will contain a series of double-stranded DNA fragments of different lengths, corresponding to the different distances the polymerase traveled from the primer before a chain-terminating tag was incorporated (six are shown in ❶).

The series of fragments are then separated according to size by gel electrophoresis. The fragments become arrayed like the rungs of a ladder, each rung being one base longer than the one below it. Compare the lengths of the fragments in ❶ and their positions on the gel in ❷. The shortest fragment has only one nucleotide (G) added to the primer, so it is the lowest rung on the gel. In automated DNA sequencing, fluorescently colored chemical tags are used to label the fragments, one color for each type of nucleotide. Computers read the colors on the gel to determine the DNA sequence and display this sequence as a series of colored peaks (❸ and ❹). The development of automated sequencers in the mid-1990s has allowed the sequencing of large eukaryotic genomes. A research institute with several hundred such instruments can sequence 100 million base pairs every day, with only 15 minutes of human attention!

> **Key Learning Outcome 13.1** Powerful automated DNA sequencing technology is now revealing the DNA sequences of entire genomes.

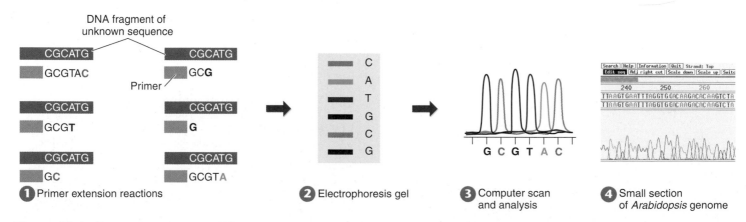

Figure 13.1 How to sequence DNA.

❶ DNA is sequenced by adding complementary bases to a single-stranded fragment. DNA synthesis stops when a chemical tag is inserted instead of a nucleotide, resulting in different sizes of DNA fragments. ❷ The DNA fragments of varying lengths are separated by gel electrophoresis, the smaller fragments migrating farther down the gel. (The boldface letters indicate the chemical tags added in step ❶ that stopped the replication process.) ❸ Computers scan the gel, from smallest to largest fragments, and display the DNA sequence as a series of colored peaks. ❹ Data from an automated DNA-sequencing run show the nucleotide sequence for a small section of the *Arabidopsis* (plant) genome.

TABLE 13.1	SOME EUKARYOTIC GENOMES			
Organism		**Estimated Genome Size (Mbp)**	**Number of Genes (×1,000)**	**Nature of Genome**
Vertebrates				
	Homo sapiens (human)	3,200	20–25	The first large genome to be sequenced; the number of transcribable genes is far less than expected; much of the genome is occupied by repeated DNA sequences.
	Pan troglodytes (chimpanzee)	2,800	20–25	There are few base substitutions between chimp and human genomes, less than 2%, but many small sequences of DNA have been lost as the two species diverged, often with significant effects.
	Mus musculus (mouse)	2,500	25	Roughly 80% of mouse genes have a functional equivalent in the human genome; importantly, large portions of the noncoding DNA of mouse and human have been conserved; overall, rodent genomes (mouse and rat) appear to be evolving more than twice as fast as primate genomes (humans and chimpanzees).
	Gallus gallus (chicken)	1,000	20–23	One-third the size of the human genome; genetic variation among domestic chickens seems much higher than in humans.
	Fugu rubripes (pufferfish)	365	35	The *Fugu* genome is only one-ninth the size of the human genome, yet it contains 10,000 more genes.
Invertebrates				
	Caenorhabditis elegans (nematode)	97	21	The fact that every cell of *C. elegans* has been identified makes its genome a particularly powerful tool in developmental biology.
	Drosophila melanogaster (fruit fly)	137	13	*Drosophila* telomere regions lack the simple repeated segments that are characteristic of most eukaryotic telomeres. About one-third of the genome consists of gene-poor centric heterochromatin.
	Anopheles gambiae (mosquito)	278	15	The extent of similarity between *Anopheles* and *Drosophila* is approximately equal to that between human and pufferfish.
	Nematostella vectensis (sea anemone)	450	18	The genome of this cnidarian is much more like vertebrate genomes than nematode or insect genomes that appear to have become streamlined by evolution.
Plants				
	Oryza sativa (rice)	430	33–50	The rice genome contains only 13% as much DNA as the human genome, but roughly twice as many genes; like the human genome, it is rich in repetitive DNA.
	Populus trichocarpa (cottonwood tree)	500	45	This fast-growing tree is widely used by the timber and paper industries. Its genome, fifty times smaller than the pine genome, is one-third heterochromatin.
Fungi				
	Saccharomyces cerevisiae (brewer's yeast)	13	6	*S. cerevisiae* was the first eukaryotic cell to have its genome fully sequenced.
Protists				
	Plasmodium falciparum (malaria parasite)	23	5	The *Plasmodium* genome has an unusually high proportion of adenine and thymine. Scarcely 5,000 genes contain the bare essentials of the eukaryotic cell.

13.2 The Human Genome

On June 26, 2000, geneticists announced that the entire human genome had been sequenced. This effort presented no small challenge, as the human genome is huge—more than 3 billion base pairs, which is the largest genome sequenced to date. To get an idea of the magnitude of the task, consider that if all 3.2 billion base pairs were written down on the pages of this book, the book would be 500,000 pages long, and it would take you about 60 years, working eight hours a day, every day, at five bases a second, to read it all.

Reading the human genome for the first time, geneticists encountered four big surprises.

1. The Number of Genes Is Quite Low

The human genome sequence contains only 20,000 to 25,000 protein-encoding genes, only 1% of the genome. As you can see in **figure 13.2**, this is scarcely more genes than in a nematode worm (21,000 genes), not quite double the number in a fruit fly (13,000 genes). Researchers had confidently anticipated at least four times as many genes because over 100,000 unique messenger RNA (mRNA) molecules can be found in human cells—surely, they argued, it would take as many genes to make them.

How can human cells contain more mRNAs than genes? Recall from chapter 12 that in a typical human gene, the sequence of DNA nucleotides that specifies a protein is broken into many bits called exons, scattered among much longer segments of nontranslated DNA called introns. Imagine this paragraph was a human gene; all the occurrences of the letter "e" could be considered exons, while the rest would be non-coding introns, which make up 24% of the human genome.

When a cell uses a human gene to make a protein, it first manufactures mRNA copies of the gene, then splices the exons together, getting rid of the intron sequences in the process. Now here's the turn of events researchers had not anticipated: The exon portions of human gene transcripts are often spliced together in different ways, called *alternative splicing*. As we discussed in chapter 12, each exon is actually a module; one exon may code for one part of a protein, another for a different part of a protein. When the exon transcripts are mixed in different ways, very different protein shapes can be built.

With alternative mRNA splicing, it is easy to see how 25,000 genes can encode four times as many proteins. The added complexity of human proteins occurs because the gene parts are put together in new ways. Great music is made from simple tunes in much the same way.

2. Some Chromosomes Have Few Genes

In addition to the fragmenting of genes by the scattering of exons throughout the genome, there is another interesting "organizational" aspect of the genome. Genes are not distributed evenly over the genome. The small chromosome number 19 is packed densely with genes, transcription factors, and other functional elements. The much larger chromosome numbers 4 and 8, by contrast, have few genes, scattered like isolated hamlets in a desert. On most chromosomes, vast stretches of seemingly barren DNA fill the chromosomes between clusters rich in genes.

3. Genes Exist in Many Copy Numbers

Four different classes of protein-encoding genes are found in the human genome, differing largely in gene copy number.

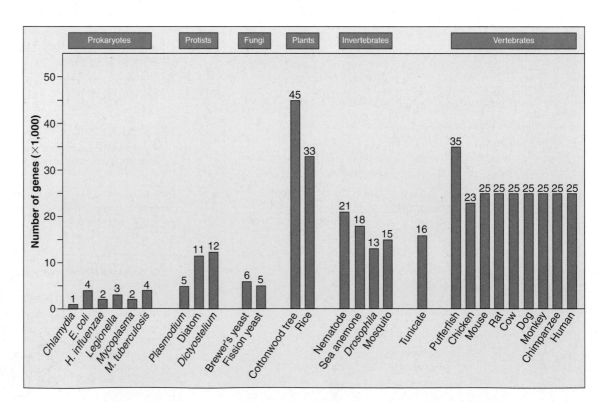

Figure 13.2
Comparing genome size.

All mammals have the same size genome, 20,000 to 25,000 protein-encoding nuclear genes. The unexpectedly larger sizes of the plant and pufferfish genomes are thought to reflect whole-genome duplications rather than increased complexity.

Single-copy genes. Many eukaryotic genes exist as single copies at a particular location on a chromosome. Mutations in these genes produce recessive Mendelian inheritance of those traits. Silent copies inactivated by mutation, called *pseudogenes,* are as common as protein-encoding genes.

Segmental duplications. Human chromosomes contain many segmental duplications, whole blocks of genes that have been copied over from one chromosome to another. Blocks of similar genes in the same order are found throughout the genome. Chromosome 19 seems to have been the biggest borrower, with blocks of genes shared with 16 other chromosomes.

Multigene families. Many genes exist as parts of multigene families, groups of related but distinctly different genes that often occur together in a cluster. Multigene families contain from three to several dozen genes. Although they differ from each other, the genes of a multigene family are clearly related in their sequences, making it likely that they arose from a single ancestral sequence.

Tandem clusters. These groups of repeated genes consist of DNA sequences that are repeated many thousands of times, one copy following another in tandem array. By transcribing all of the copies in these tandem clusters simultaneously, a cell can rapidly obtain large amounts of the product they encode. For example, the genes encoding rRNA are present in clusters of several hundred copies.

4. Most Genome DNA Is Noncoding

The fourth notable characteristic of the human genome is the startling amount of noncoding DNA it possesses. Only 1% to 1.5% of the human genome is coding DNA, devoted to genes encoding proteins. Each of your cells has about 6 feet of DNA stuffed into it, but of that, less than 1 inch is devoted to genes! Nearly 99% of the DNA in your cells seems to have little or nothing to do with the instructions that make you who you are (figure 13.3).

There are four major types of noncoding human DNA:

Noncoding DNA within genes. As discussed earlier, a human gene is made up of numerous fragments of protein-encoding information (exons) embedded within a much larger matrix of noncoding DNA (introns). Introns make up 24% of the human genome—exons only 1%!

Structural DNA. Some regions of the chromosomes remain highly condensed, tightly coiled, and untranscribed throughout the cell cycle. These portions—about 20% of the DNA—tend to be localized around the centromere, or located near the telomeres, or ends, of the chromosome.

Repeated sequences. Scattered about chromosomes are simple sequence repeats of two or three nucleotides like CA or CGG, repeated like a broken record thousands and thousands of times. These make up about 3% of the human genome. An additional 7%

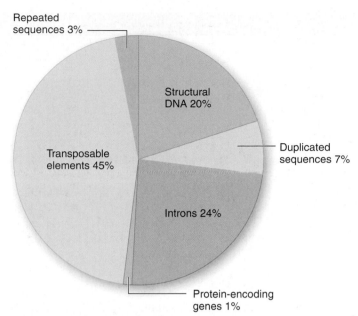

Figure 13.3 The human genome.
Very little of the human genome is devoted to protein-encoding genes, indicated by the light blue section in this pie chart.

is devoted to other sorts of duplicated sequences. Repetitive sequences with excess C and G tend to be found in the neighborhood of genes, while A- and T-rich repeats dominate the nongene deserts. The light bands on chromosome karyotypes now have an explanation—they are regions rich in GC and genes. Dark bands signal neighborhoods rich in A and T, which are thin on genes. Chromosome 8, for example, contains many nongene areas that are indicated by dark bands, while chromosome 19 is dense with genes and so it has few dark bands.

Transposable elements. Fully 45% of the human genome consists of mobile parasitic bits of DNA called *transposable elements.* Discovered by Barbara McClintock in 1950 (she won the Nobel Prize in Physiology or Medicine for her discovery in 1983), transposable elements are bits of DNA that are able to jump from one location on a chromosome to another— tiny molecular versions of Mexican jumping beans. Because they leave a copy of themselves behind when they jump, their numbers in the genome increase as generations pass. Nested within the human genome are over half a million copies of an ancient transposable element called *Alu,* composing fully 10% of the entire human genome. Often jumping right into genes, *Alu* transpositions cause many harmful mutations.

Key Learning Outcome 13.2 The entire 3.2-billion-base pair human genome has been sequenced. Only about 1% to 1.5% of the human genome is devoted to protein-encoding genes. Much of the rest is composed of transposable elements.

Curing disease. One of two young girls who were the first humans "cured" of a hereditary disorder by transferring into their bodies healthy versions of a defective gene. The transfer was successfully carried out in 1990, and twenty years later the girls remain healthy.

Increasing yields. The genetically engineered salmon on the *right* have shortened production cycles and are heavier than the nontransgenic salmon of the same age on the *left*.

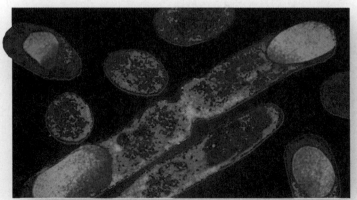

Pest-proofing plants. The genetically engineered cotton plants on the *right* have a gene that inhibits feeding by weevils; the cotton plants on the *left* lack this gene, and produce far fewer cotton bolls.

Figure 13.4 Examples of genetic engineering.

13.3 A Scientific Revolution

In recent years, **genetic engineering**—the ability to manipulate genes and move them from one organism to another—has led to great advances in medicine and agriculture (**figure 13.4**). Most of the insulin used to treat diabetes is now obtained from bacteria that contain a human insulin gene. In late 1990, the first transfers of genes from one human to another were carried out to correct the effects of defective genes in a rare genetic immune disorder called *Severe Combined Immunodeficiency* (SCID)—often called the "Bubble Boy Disorder" based on a young boy who lived out his life in an enclosed, germ-free environment. In addition, cultivated plants and animals can be genetically engineered to resist pests, grow bigger, or grow faster.

Restriction Enzymes

The first stage in any genetic engineering experiment is to chop up the "source" DNA to get a copy of the gene you wish to transfer. This first stage is the key to successful transfer of the gene, and learning how to do it is what has led to the genetic revolution. The trick is in how the DNA molecules are cut. The cutting must be done in such a way that the resulting DNA fragments have "sticky ends" that can later be joined with another molecule of DNA.

This special form of molecular surgery is carried out by **restriction enzymes,** also called *restriction endonucleases,* which are special enzymes that bind to specific short sequences (typically four to six nucleotides long) on the DNA. These sequences are very unusual in that they are symmetrical—the two strands of the DNA duplex have the same nucleotide sequence, running in opposite directions! The sequence in **figure 13.5**, for example, is GAATTC. Try writing down the sequence of the opposite strand: it is CTTAAG—the same sequence, written backward. This sequence is recognized by

Producing insulin. The common bacteria *Escherichia coli* (*E. coli*) can be genetically engineered to contain the gene that codes for the protein insulin. The bacteria are turned into insulin-producing factories and can produce large quantities of insulin for diabetic patients. In the image above, insulin-producing sites inside genetically-altered *E. coli* cells are orange.

the restriction enzyme *Eco*RI. Other restriction enzymes recognize other sequences.

What makes the DNA fragments "sticky" is that most restriction enzymes do not make their incision in the center of the sequence; rather, the cut is made to one side. In the sequence in **figure 13.5 ❶**, the cut is made on both strands between the G and A nucleotides, G/AATTC. This produces a break, with short, single strands of DNA dangling from each end. Because the two single-stranded ends are complementary in sequence, they could pair up and heal the break, with the aid of a sealing enzyme—*or* they could pair with *any other DNA fragment cut by the same enzyme* because all would have the same single-stranded sticky ends. **Figure 13.5 ❷** shows how DNA from another source (the orange DNA) also cut with *Eco*RI has the same sticky ends as the original source DNA. Any gene in any organism cut by the enzyme that attacks GAATTC sequences will have the same sticky ends, and can be joined to any other with the aid of a sealing enzyme called *DNA ligase* ❸.

Formation of cDNA

As already mentioned, eukaryotic genes are encoded in segments called exons separated from one another by numerous nontranslated sequences called introns. The entire gene is transcribed by RNA polymerase, producing what is called the primary RNA transcript (**figure 13.6**). Before a eukaryotic gene can be translated into a protein, the introns must be cut out of this primary transcript. The fragments that remain are then spliced together to form the mRNA, which is eventually translated in the cytoplasm. When transferring eukaryotic genes into bacteria (discussed in section 13.4), it is necessary to transfer DNA that has had the intron information removed because bacteria lack the enzymes to carry out this processing. Bacterial genes do not contain introns. To produce eukaryotic DNA without introns, genetic engineers first isolate from the cytoplasm the processed mRNA corresponding to a particular gene. The cytoplasmic mRNA has *only* exons, properly spliced together. An enzyme called *reverse transcriptase* is then used to make a complementary DNA strand of the mRNA. A double-stranded DNA molecule is then produced. Such a version of a gene is called *complementary DNA*, or **cDNA.**

cDNA technology is being used in other ways, such as determining the patterns of gene expression in different cells. Every cell in an organism contains the same DNA, but genes are selectively turned on and off in any given cell. Researchers can see what genes are actively expressed using cDNAs.

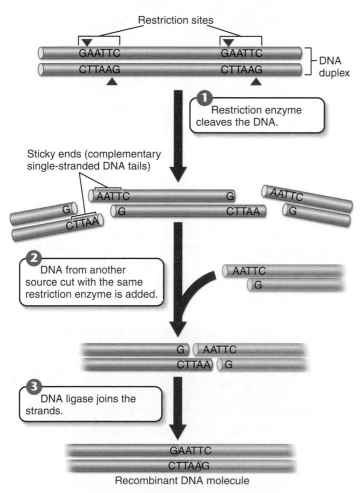

Figure 13.5 How restriction enzymes produce DNA fragments with sticky ends.

The restriction enzyme *Eco*RI always cleaves the sequence GAATTC between G and A. Because the same sequence occurs on both strands, both are cut. However, the two sequences run in opposite directions on the two strands. As a result, single-stranded tails are produced that are complementary to each other, or "sticky."

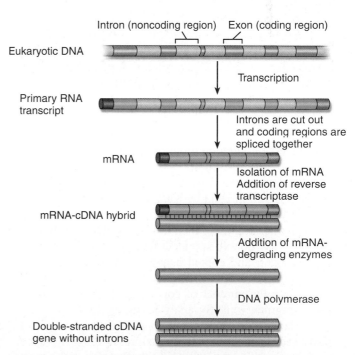

Figure 13.6 cDNA: Producing an intron-free version of a eukaryotic gene for genetic engineering.

In eukaryotic cells, a primary RNA transcript is processed into the mRNA. The mRNA is isolated and converted into cDNA.

DNA Fingerprinting and Forensic Science

DNA fingerprinting is a process used to compare samples of DNA. Just as fingerprinting revolutionized forensic evidence in the early 1900s, so DNA fingerprinting is revolutionizing it today. A hair, a minute speck of blood, a drop of semen, all can serve as sources of DNA to convict or clear a suspect.

The process of DNA fingerprinting uses probes to fish out particular sequences to be compared from the thousands of other sequences in the human genome. Because different people have different DNA sequences throughout their genome, they will have different restriction enzyme cutting sites, and so produce different-sized fragments that move to different locations on a gel. Radioactive probes bind to particular locations that occur randomly many times in the genome, "lighting up" the fragments that contain that sequence. If several different probes are used, the chance of any two individuals having the same gel pattern is less than one in a billion. DNA fragments that bind the probes are then visible on autoradiographic film as dark bands. The autoradiograph gel patterns are in essence DNA "fingerprints" that can be used in criminal investigations and other identification applications.

In the first time DNA evidence was used in a court of law, DNA probes were used to characterize DNA isolated from a rape victim's blood, from semen left by the rapist, and from the suspect's blood:

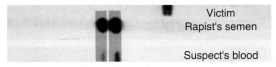

You can see that the suspect's pattern (indicated in pink) matches that of the rapist, and is not at all like that of the victim. Other probes produced similar results. On November 6, 1987, the jury returned a verdict of guilty, the first time a person in the United States was convicted of a crime based on DNA evidence. Since this verdict, DNA fingerprinting has been admitted as evidence in thousands of court cases.

PCR Amplification

Tiny DNA samples, as little as that found in a single human hair, can be magnified to millions of copies by a process called **PCR** (for **polymerase chain reaction**). The use of DNA in forensic analysis often depends critically on this PCR technology, developed in 1983 by Kary Mullis. In PCR, a double-stranded DNA fragment is heated so it becomes single stranded; each of the strands is then copied by DNA polymerase to produce two double-stranded fragments. The fragments are heated again and copied again to produce four double-stranded fragments. This cycle is repeated many times, each time doubling the number of copies, until enough copies of the DNA fragment exist for analysis (**figure 13.7**).

The amount of DNA found in a single hair will do. Biologists used to think that DNA was present only in the cells at the root of a hair, not in the hair shaft made of the protein keratin. But we now know follicle cells become incorporated into the growing shaft and their DNA is sealed in by the keratin, protecting it from being degraded by bacteria and fungi.

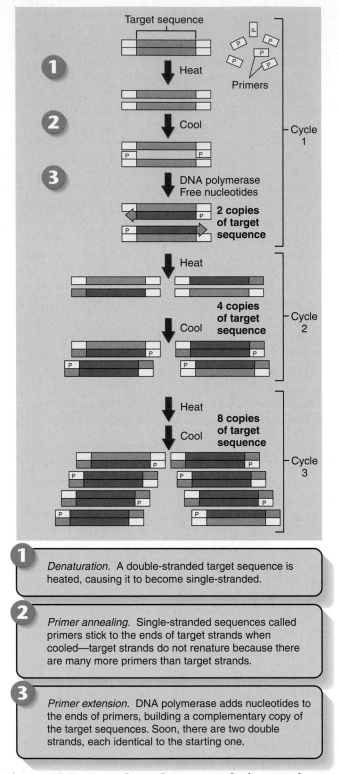

Figure 13.7   How the polymerase chain reaction works.

1 *Denaturation.* A double-stranded target sequence is heated, causing it to become single-stranded.

2 *Primer annealing.* Single-stranded sequences called primers stick to the ends of target strands when cooled—target strands do not renature because there are many more primers than target strands.

3 *Primer extension.* DNA polymerase adds nucleotides to the ends of primers, building a complementary copy of the target sequences. Soon, there are two double strands, each identical to the starting one.

Key Learning Outcome 13.3 Restriction enzymes, the key tools that make genetic engineering possible, bind to specific short sequences of DNA and cut the DNA. This produces fragments with "sticky ends," which can be rejoined in different combinations.

Today's Biology

DNA and the Innocence Project

Every person's DNA is uniquely their own, a sequence of nucleotides found in no other person. Like a molecular social security number a billion digits long, the nucleotide sequence of an individual's genes can provide proof-positive identification of a rapist or murderer from DNA left at the scene of a crime—proof more reliable than fingerprints, more reliable than an eye witness, even more reliable than a confession by the suspect.

Never was this demonstrated more clearly than on May 16, 2006 in a Rochester, New York courtroom. There a judge freed convicted murderer Douglas Warney after 10 years in prison.

The murder occurred on New Year's Day in 1996. The bloody body of one William Beason, a prominent community activist, was found in his bed. Police called in all the usual suspects, in this case everyone known to be an acquaintance of the victim. Warney, an unemployed 34-year-old who had dropped out of school in the eighth grade, committed robberies, and worked as a male hustler, learned that detectives wanted to speak to him about the killing, and went to the police station for questioning. Within hours he was charged with murder.

Warney's interrogation was not recorded, but it resulted in a signed confession based on the words that the detective sergeant said Warney uttered, a confession that contained accurate details about the murder scene that had not been made public. Warney said that the victim was wearing a nightgown and had been cooking chicken in a pot, and that the murderer had used a 12-inch serrated knife. The case against him in court rested almost entirely on these vivid details. Even though Warney recanted his confession at his trial, the accuracy of his confession was damning. The details that Warney provided to the sergeant could have come only from someone who was present at the crime scene.

There were problems with his confession, brought out at trial. Three elements of the signed statement's account of that night were clearly not true. It said Warney had driven to the victim's house in his brother's car, but his brother did not own a car; it said Warney disposed of his bloody clothes after the stabbing in the garbage can behind the house, but the can, buried in snow from the day of the crime, did not contain bloody clothes; it named a relative of Warney as an accomplice, but that relative was in a secure rehabilitation center on that day.

The most difficult bit of evidence to match to the written confession was blood found at the scene that was not that of the victim. Drops of a second person's blood were found on the floor and on a towel. The difficult bit was that the blood was a different blood type than Warney's.

At trial, prosecutors pointed out that the blood could have come from the accomplice mentioned in the confession, and pounded away on the point that the details in the confession could only have come from first-hand knowledge of the crime.

After a short trial, Douglas Warney was convicted of the murder of William Beason and sentenced to 25 years.

When appeals failed, Mr. Warney in 2004 sought help from the Innocence Project, a nonprofit legal clinic that helps identify wrongly convicted individuals and secure their freedom. Set up at the Benjamin N. Cardozo School of Law in New York by lawyers Barry Scheck and Peter Neufeld, the Innocence Project specializes in using DNA technology to establish innocence.

The Innocence Project staff petitioned the court for additional DNA testing using new sensitive DNA probes, arguing that this might disclose evidence that would have resulted in a different verdict. The judge refused, ruling that the possibility that the blood found at the scene of the crime might match that of a criminal already in the state databank was "too speculative and improbable" to warrant the new tests.

Someone in the prosecutor's office must have been persuaded, however. Without notifying Warney's legal team, Project Innocence, or the court, this good Samaritan arranged for new DNA tests on the blood drops found at the crime scene. When compared to the New York State criminal DNA database, they hit a strong match. The blood was that of Eldred Johnson, in prison for slitting his landlady's throat in Utica two weeks before Beason was killed.

When confronted, the prisoner Johnson readily admitted to the stabbing of Beason. He said he was the sole killer, and had never met Douglas Warney.

This leaves the interesting question of how Warney's signed confession came to include such accurate information about the crime scene. It now appears he may have been fed critical details about the crime scene by the homicide detective leading the investigation, who has since died.

DNA testing has become a pillar of the American criminal justice system. It has provided key evidence that has established the guilt of thousands of suspects beyond any reasonable doubt—and, as you see here, also provided the evidence that our criminal justice system sometimes convicts and sentences innocent people. Over more than a decade, the Innocence Project and other similar efforts have cleared hundreds of convicted people, strong proof that wrongful convictions are not isolated or rare events. DNA testing opens a window of hope for the wrongly convicted.

13.4 Genetic Engineering and Medicine

Much of the excitement about genetic engineering has focused on its potential to improve medicine—to aid in curing and preventing illness. Major advances have been made in the production of proteins used to treat illness, and in the creation of new vaccines to combat infections.

Making "Magic Bullets"

Many illnesses occur because of gene defects that prevent our bodies from making critical proteins. Juvenile diabetes is such an illness. The body is unable to control levels of sugar in the blood because a critical protein, **insulin,** cannot be made. This failure can be overcome if the body can be supplied with the protein it lacks. The donated protein is in a very real sense a "magic bullet" to combat the body's inability to regulate itself.

Until recently, the principal problem with using regulatory proteins as drugs was in manufacturing the protein. Proteins that regulate the body's functions are typically present in the body in very low amounts, and this makes them difficult and expensive to obtain in quantity. With genetic engineering techniques, the problem of obtaining large amounts of rare proteins has been largely overcome. The cDNA of genes encoding medically important proteins are now introduced into bacteria (**table 13.2**). Because the host bacteria can be grown cheaply, large amounts of the desired protein can be easily isolated. In 1982, the U.S. Food and Drug Administration approved the use of human insulin produced from genetically engineered bacteria, the first commercial product of genetic engineering.

The use of genetic engineering techniques in bacteria has provided ample sources of therapeutic proteins, but the application extends beyond bacteria. Today hundreds of pharmaceutical companies around the world are busy producing other medically important proteins, expanding the use of these genetic engineering techniques. A gene added to the DNA of the mouse on the right in **figure 13.8** produces human growth hormone, allowing the mouse to grow larger than its twin.

The advantage of using genetic engineering is clearly seen with **factor VIII,** a protein that promotes blood clotting. A deficiency in factor VIII leads to hemophilia, an inherited disorder (discussed in chapter 10) that is characterized by prolonged bleeding. For a long time, hemophiliacs received blood factor VIII that had been isolated from donated blood. Unfortunately, some of the donated blood had been infected with viruses such as HIV and hepatitis B, which were then unknowingly transmitted to those people who received blood transfusions. Today the use of genetically engineered factor VIII produced in the laboratory eliminates the risks associated with blood products obtained from other individuals.

TABLE 13.2	GENETICALLY ENGINEERED DRUGS
Product	**Effects and Uses**
Anticoagulants	Involved in dissolving blood clots; used to treat heart attack patients
Colony-stimulating factors	Stimulate white blood cell production; used to treat infections and immune system deficiencies
Erythropoietin	Stimulates red blood cell production; used to treat anemia in individuals with kidney disorders
Factor VIII	Promotes blood clotting; used to treat hemophilia
Growth factors	Stimulate differentiation and growth of various cell types; used to aid wound healing
Human growth hormone	Used to treat dwarfism
Insulin	Involved in controlling blood sugar levels; used in treating diabetes
Interferons	Disrupt the reproduction of viruses; used to treat some cancers
Interleukins	Activate and stimulate white blood cells; used to treat wounds, HIV infections, cancer, immune deficiencies

Figure 13.8 Genetically engineered human growth hormone.

These two mice are genetically identical, but the large one has one extra gene: the gene encoding human growth hormone. The gene was added to the mouse's genome by genetic engineers and is now a stable part of the mouse's genetic make-up. In humans, growth hormone is used to treat various forms of dwarfism.

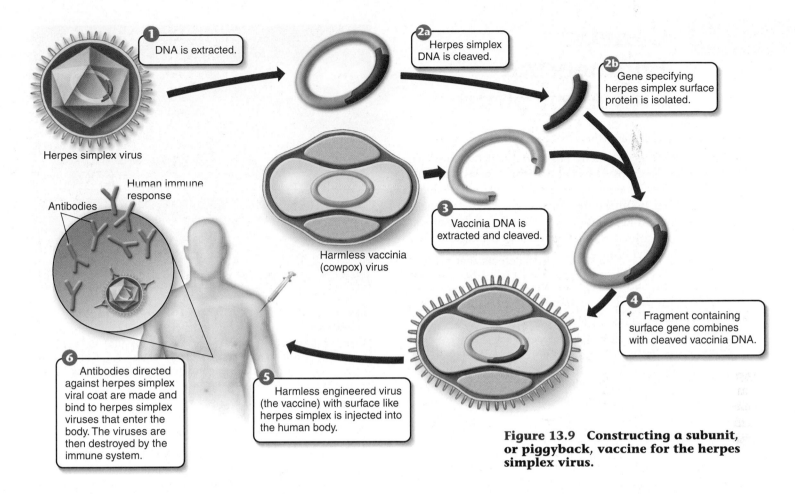

Figure 13.9 **Constructing a subunit, or piggyback, vaccine for the herpes simplex virus.**

Image labels:
1. DNA is extracted.
2a. Herpes simplex DNA is cleaved.
2b. Gene specifying herpes simplex surface protein is isolated.
3. Vaccinia DNA is extracted and cleaved.
4. Fragment containing surface gene combines with cleaved vaccinia DNA.
5. Harmless engineered virus (the vaccine) with surface like herpes simplex is injected into the human body.
6. Antibodies directed against herpes simplex viral coat are made and bind to herpes simplex viruses that enter the body. The viruses are then destroyed by the immune system.

Herpes simplex virus
Human immune response
Antibodies
Harmless vaccinia (cowpox) virus

Piggyback Vaccines

Another area of potential significance involves the use of genetic engineering to produce subunit vaccines against viruses such as those that cause herpes and hepatitis. Genes encoding part of the protein-polysaccharide coat of the herpes simplex virus or hepatitis B virus are spliced into a fragment of the vaccinia (cowpox) virus genome. The vaccinia virus, which is essentially harmless to humans and was used by British physician Edward Jenner more than 200 years ago in his pioneering vaccinations against smallpox, is now used as a vector to carry a viral coat gene into cultured mammalian cells. As shown in figure 13.9, the steps in constructing a subunit vaccine for herpes simplex begin with ❶ extracting the herpes simplex viral DNA and ❷ isolating a gene that codes for a protein on the surface of the virus. The cowpox viral DNA is extracted and cleaved ❸, and the herpes gene is then combined with the cowpox DNA ❹. The recombinant DNA is inserted into a cowpox virus. Many copies of the recombinant virus, which have the outside coat of a herpes virus, are produced. When this recombinant virus is injected into a human ❺, the immune system produces antibodies directed against the coat of the recombinant virus ❻. The person therefore develops an immunity to the virus. Vaccines produced in this way, also known as **piggyback vaccines,** are harmless because the vaccinia virus is benign, and only a small fragment of the DNA from the disease-causing virus is introduced via the recombinant virus.

In 1995, the first clinical trial began of a new kind of vaccine, called a DNA vaccine. DNA containing a viral gene is injected and taken up by cells of the body, where the gene is expressed. The infected cells trigger a cellular immune response, in which blood cells known as killer T cells attack the infected cells. The first DNA vaccines spliced an influenza virus gene encoding an internal nucleoprotein into a plasmid, which was then injected into mice. The mice developed strong cellular immune responses against influenza. The approach offers great promise.

In 2010, the first effective **cancer vaccines** were announced. A cancer vaccine is therapeutic rather than preventive, stimulating the immune system to attack a tumor in the same way invading microbes are attacked. The first cancer vaccine approved for clinical use employs proteins from prostate cancer cells to induce the immune system to attack prostate cancer tumors. Another cancer vaccine, very effective in mice but not yet approved for humans, uses a protein called alpha-lactalbumin to trigger an attack on breast cancer cells. The protein is not found on normal breast cancer cells except when women are breast-feeding. As 96% of the lifetime risk of a woman for breast cancer is after child-bearing years, post-menopausal use of this vaccine may prove to be a powerful therapy against early undetected breast cancers.

Key Learning Outcome 13.4 Genetic engineering has facilitated the production of medically important proteins and led to novel vaccines.

13.5 Genetic Engineering and Agriculture

Pest Resistance

An important effort of genetic engineers in agriculture has involved making crops resistant to insect pests without spraying with pesticides, a great saving to the environment. Consider cotton. Its fibers are a major source of raw material for clothing throughout the world, yet the plant itself can hardly survive in a field because many insects attack it. Over 40% of the chemical insecticides used today are employed to kill insects that eat cotton plants. The world's environment would greatly benefit if these thousands of tons of insecticide were not needed. Biologists are now in the process of producing cotton plants that are resistant to attack by insects.

One successful approach uses a kind of soil bacterium, *Bacillus thuringiensis* (Bt), that produces a protein that is toxic when eaten by crop pests, such as larvae (caterpillars) of butterflies. When the gene producing the Bt protein is inserted into the chromosomes of tomatoes, the plants begin to manufacture Bt protein. While not harmful to humans, it makes the tomatoes highly toxic to hornworms (one of the most serious pests of commercial tomato crops).

Many important plant pests also attack roots. To combat these pests, genetic engineers are introducing the *Bt* gene into different kinds of bacteria, ones that colonize the roots of crop plants. Any insects eating such roots consume the bacteria and so are lethally attacked by the Bt protein.

Figure 13.11 Genetically engineered herbicide resistance.

All four of these petunia plants were exposed to equal doses of an herbicide. The two on *top* were genetically engineered to be resistant to glyphosate, the active ingredient in the herbicide, whereas the two dead ones on the *bottom* were not.

Herbicide Resistance

A major advance has been the creation of crop plants that are resistant to the herbicide *glyphosate,* a powerful biodegradable herbicide that kills most actively growing plants. Glyphosate is used in orchards and agricultural fields to control weeds. Growing plants need to make a lot of protein, and glyphosate stops them from making protein by destroying an enzyme necessary for the manufacture of so-called aromatic amino acids (that is, amino acids that contain a ring structure, like phenylalanine—see **figure 3.5**). Humans are unaffected by glyphosate because we don't make aromatic amino acids—we obtain them from plants we eat! To make crop plants resistant to this powerful plant killer, genetic engineers screened thousands of organisms until they found a species of bacteria that could make aromatic amino acids in the presence of glyphosate. They then isolated the gene encoding the resistant enzyme and successfully introduced the gene into plants. They inserted the gene into the plants using DNA particle guns, also called gene guns. You can see in **figure 13.10** how a DNA particle gun works. Small tungsten or gold pellets are coated with DNA (red in the figure) that contains the gene of interest and placed in the DNA particle gun. The DNA gun literally shoots the gene into plant cells in culture where the gene can be incorporated into the plant genome and then expressed. Plants that have been genetically engineered in this way are shown in **figure 13.11**. The two plants on top were genetically engineered to be resistant to glyphosate, the herbicide that killed the two plants at the bottom of the photo.

The creation of glyphosate-tolerant crops is of major benefit to the environment. Glyphosate is quickly broken down in the environment, which makes its use a great improvement over long-lasting chemical herbicides. Also, not having to plow to remove weeds reduces the loss of fertile topsoil to erosion.

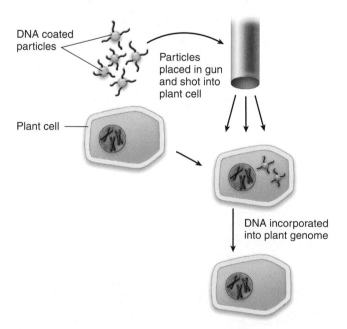

DNA coated particles

Particles placed in gun and shot into plant cell

Plant cell

DNA incorporated into plant genome

Figure 13.10 Shooting genes into cells.

A DNA particle gun, also called a gene gun, fires tungsten or gold particles coated with DNA into plant cells. The DNA-coated particles pass through the cell wall and into the cell, where the DNA is incorporated into the plant cell's DNA. The gene encoded by the DNA is expressed.

More Nutritious Crops

The cultivation of genetically modified (GM) crops of corn, cotton, soybeans, and other plants (**table 13.3**) has become commonplace in the United States. In 2004, 85% of soybeans in the United States were planted with seeds genetically modified to be herbicide resistant. The result has been that less tillage was needed and, as a consequence, soil erosion was greatly lessened. Pest-resistant GM corn in 2004 comprised over 50% of all corn planted in the United States, and pest-resistant GM cotton comprised 81% of all cotton. In both cases, the change greatly lessens the amount of chemical pesticide used on the crops. These benefits of soil preservation and chemical pesticide reduction, while significant, have been largely bestowed upon farmers, making their cultivation of crops cheaper and more efficient.

Like the first act of a play, these developments have served mainly to set the stage for the real action, which is only now beginning to happen. The real promise of plant genetic engineering is to produce genetically modified plants with desirable traits that directly benefit the consumer.

One recent advance, nutritionally improved "golden" rice, gives us a hint of what is to come. In developing countries, large numbers of people live on simple diets that are poor sources of vitamins and minerals (what botanists called "micronutrients"). Worldwide, the two major micronutrient deficiencies are iron, which affects 1.4 billion women, 24% of the world population, and vitamin A, affecting 40 million children, 7% of the world population. The deficiencies are especially severe in developing countries where the major staple food is rice. In recent research, Swiss bioengineer Ingo Potrykus and his team at the Institute of Plant Sciences, Zurich, have gone a long way toward solving this problem. Supported by the Rockefeller Foundation and with results to be made free to developing countries, the work is a model of what plant genetic engineering can achieve.

To solve the problem of dietary iron deficiency among rice eaters, Potrykus first asked why rice is such a poor source of dietary iron. The problem, and the answer, proved to have three parts:

1. *Too little iron.* The proteins of rice endosperm have unusually low amounts of iron. To solve this problem, a ferritin gene (abbreviated as **Fe** in **figure 13.12**) was transferred into rice from beans. Ferritin is a protein with an extraordinarily high iron content, and so greatly increased the iron content of the rice.
2. *Inhibition of iron absorption by the intestine.* Rice contains an unusually high concentration of a chemical called phytate, which inhibits iron absorption in the intestine—it stops your body from taking up the iron in the rice. To solve this problem, a gene encoding an enzyme called phytase (abbreviated as **Pt**) that destroys phytate was transferred into rice from a fungus.
3. *Too little sulfur for efficient iron absorption.* The human body requires sulfur for the uptake of iron, and rice has very little of it. To solve this problem, a gene encoding a sulfur-rich protein (abbreviated as **S**) was transferred into rice from wild rice.

To solve the problem of vitamin A deficiency, the same approach was taken. First, the problem was identified. It turns out rice only goes partway toward making vitamin A; there are no enzymes in rice to catalyze the last four steps. To solve the problem, genes encoding these four enzymes (abbreviated **A₁ A₂ A₃ A₄**) were added to rice from a flower, the daffodil.

The development of transgenic rice is only the first step in the battle to combat dietary deficiencies. The added nutritional value only makes up for half a person's requirements, and many years will be required to breed the genes into lines adapted to local conditions, but it is a promising start, representative of the very real promise of genetic engineering.

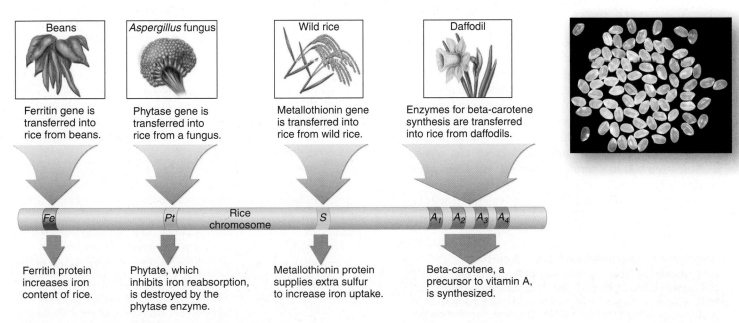

Beans	*Aspergillus* fungus	Wild rice	Daffodil
Ferritin gene is transferred into rice from beans.	Phytase gene is transferred into rice from a fungus.	Metallothionin gene is transferred into rice from wild rice.	Enzymes for beta-carotene synthesis are transferred into rice from daffodils.

Rice chromosome: Fe — Pt — S — A₁ A₂ A₃ A₄

| Ferritin protein increases iron content of rice. | Phytate, which inhibits iron reabsorption, is destroyed by the phytase enzyme. | Metallothionin protein supplies extra sulfur to increase iron uptake. | Beta-carotene, a precursor to vitamin A, is synthesized. |

Figure 13.12 Transgenic "golden" rice.

How Do We Measure the Potential Risks of Genetically Modified Crops?

Is Eating Genetically Modified Food Dangerous? Many consumers worry that when bioengineers introduce novel genes into genetically modified (GM) crops, there may be dangerous consequences for the food we eat. The introduction of glyphosate-resistance into soybeans is an example. Could introduced proteins like the enzyme making the GM soybeans glyphosate-tolerant cause a fatal immune reaction in some people? Because the potential danger of allergic reactions is quite real, every time a protein-encoding gene is introduced into a GM crop it is necessary to carry out extensive tests of the introduced protein's allergen potential. No GM crop currently being produced in the United States contains a protein that acts as an allergen to humans. On this score, then, the risk of genetic engineering to the food supply seems to be slight.

Are GM Crops Harmful to the Environment? Those concerned about the widespread use of GM crops raise three legitimate concerns:

1. *Harm to Other Organisms.* Might pollen from Bt corn harm non-pest insects that chance to eat it? Studies suggest little possibility of harm.
2. *Resistance.* All insecticides and herbicides used in agriculture share the problem that pests eventually evolve resistance to them, in much the same way that bacterial populations evolve resistance to antibiotics. To prevent this, farmers are required to plant at least 20% non-Bt crops alongside Bt crops to provide refuges where insect populations are not under selection pressure and in this way to slow the development of resistance. As a result, despite the widespread use of Bt crops like corn, soybeans, and cotton since 1996, there are as of yet only a few cases of insects developing resistance to Bt plants in the field. Unfortunately, the same restrictions have not been required for farmers using the herbicide glyphosate, leading to a different result: By the year 2010, glyphosate-resistant weeds have been reported by upset farmers in 22 states.
3. *Gene Flow.* How about the possibility that introduced genes will pass from GM crops to their wild relatives? For the major GM crops, there is usually no potential relative around to receive the modified gene from the GM crop. There are no wild relatives of soybeans in Europe, for example. Thus there can be no gene escape from GM soybeans in Europe, any more than genes can flow from you to your pet dog or cat. However—and this is a big however—for secondary crops only now being genetically modified, studies suggest it will be difficult to prevent GM crops from interbreeding with surrounding relatives to create new hybrids.

> **Key Learning Outcome 13.5** Genetic engineering affords great opportunities for progress in food production. On balance, the risks appear slight, and the potential benefits substantial.

TABLE 13.3	GENETICALLY MODIFIED CROPS
Rice 	Genes have been added to commercial rice from daffodils for vitamin A, and from beans, fungi, and wild rice to supply dietary iron; transgenic strains that are cold-tolerant are under development.
Wheat 	New strains of wheat, resistant to the herbicide glyphosate, greatly reduce the need for tilling and so reduce loss of topsoil.
Soybean 	A major animal feed crop, soybeans tolerant of the herbicide glyphosate were used in 90% of U.S. soybean acreage in 2010. Varieties are being developed that contain the *Bt* gene, to protect the crop from insect pests without chemical pesticides. The nutritional value of soybean crops is being improved by genetic engineers in several ways, including transgenic varieties with high tryptophan (soybeans are poor in this essential amino acid), reduced trans-fatty acids, and enhanced omega-3 (beneficial) fatty acids, common in fish oil but low in plants.
Corn 	Corn varieties resistant to insect pests (Bt corn) are widely planted (40% of U.S. acreage); varieties also tolerant of the herbicide glyphosate have been recently developed. Varieties that are drought resistant are being developed, as well as nutritionally improved lines with high lysine, vitamin A, and high levels of the unsaturated fat oleic acid, which reduces harmful cholesterol and so prevents clogged arteries.
Cotton 	Cotton crops are attacked by cotton bollworm, budworm, and other lepidopteran insects; more than 40% of all chemical pesticide tonnage worldwide is applied to cotton. A form of the *Bt* gene toxic to all lepidopterans but harmless to other insects has transformed cotton to a crop that requires few chemical pesticides. 81% of U.S. acreage is Bt cotton.
Peanut	The lesser cornstalk borer causes serious damage to peanut crops. An insect-resistant variety is under development by gene engineers to control this pest.
Potato 	Verticillium wilt (a fungal disease) infects the water-conducting tissues of potatoes, reducing crop yields 40%. An antifungal gene from alfalfa reduces infections sixfold.
Canola 	Canola, a major vegetable oil and animal feed crop, is typically grown in narrow rows with little cultivation, requiring extensive application of chemical herbicides to keep down weeds. New glyphosate-tolerant varieties require far less chemical treatment. 80% of U.S. canola acreage planted is gene-modified canola.

Today's *Biology*

A DNA Timeline

In 2000, Craig Venter of Celera, President Clinton, and Francis Collins of the Human Genome Project announce the human genome.

2006 Japanese cell biologist Shinya Yamanaka uses only four transcription factors to reprogram adult skin cells into embryonic stem cells, opening the possibility of ethical therapeutic cloning.

2000 Two teams, led by Craig Venter and Francis Collins, complete draft sequences of the human genome.

1998 Andrew Fire and Craig Mello discover RNA interference, leading to a Nobel Prize only eight years later.

1996 Ian Wilmut uses the nucleus of an adult cell to successfully clone a sheep, "Dolly."

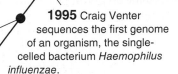

1995 Craig Venter sequences the first genome of an organism, the single-celled bacterium *Haemophilus influenzae*.

1992 Lawyers Barry Scheck and Peter Neufeld start the Innocence Project, whose efforts have cleared more than 120 wrongly convicted people through the use of DNA technology.

1985 British geneticist Alec Jeffreys invents DNA fingerprinting, the use of DNA in forensic analysis to match people to biological tissue found at crime scenes.

1973 Herbert Boyer and Stanley Cohen invent genetic engineering, successfully inserting an amphibian RNA gene into a different organism.

1983 Kary Mullis develops the polymerase chain reaction (PCR), allowing amplification and analysis of minute traces of DNA, such as that found in a single human hair.

1964 Marshall Nirenberg and Har Khorana break the genetic code, learning which three-letter code words of DNA correspond to each amino acid in proteins.

1956 Vernon Ingram shows that sickle cell disease is due to a DNA mutation leading to a single amino acid change in the protein hemoglobin.

1952 Alfred Hershey and Martha Chase demonstrate that viruses inject DNA into bacteria to reproduce, not protein; this experiment convinces most biologists that DNA is the genetic material.

1953 James Watson and Francis Crick propose that the DNA molecule is a double helix, each strand's nucleotide sequence complementary to the other.

1950 Graduate student Ray Gosling, working in the lab of British biochemist Maurice Wilkins, obtains the first clear X-ray diffraction patterns of DNA; over the next two years, Rosalind Franklin and he produce ever-clearer pictures.

1928 British microbiologist Frederick Griffith discovers transformation of living bacteria by material from dead ones.

1944 American biochemist Oswald Avery purifies Griffith's transforming principle, and demonstrates conclusively that it is DNA, although this conclusion was not appreciated at first.

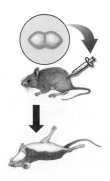

1869 German chemist Friedrich Miescher discovers DNA, called "nucleic acid" because it was isolated from sperm nuclei and is slightly acidic.

13.6 Reproductive Cloning

One of the most active and exciting areas of biology involves recently developed approaches to manipulating animal cells. In this section, you will encounter three areas where landmark progress is being made in cell technology: reproductive cloning of farm animals, stem cell research, and gene therapy. Advances in cell technology hold the promise of revolutionizing our lives.

The idea of cloning animals was first suggested in 1938 by German embryologist Hans Spemann (called the "father of modern embryology"), who proposed what he called a "fantastical experiment": remove the nucleus from an egg cell (creating an enucleated egg) and put in its place a nucleus from another cell. When attempted many years later (**figure 13.13**), this experiment actually succeeded in frogs, sheep, monkeys, and many other animals. However, only donor nuclei extracted from early embryos seemed to work. After repeated failures using nuclei from adult cells, many researchers became convinced that the nuclei of animal cells become irreversibly committed to a developmental pathway after the first few cell divisions of the developing embryo.

Wilmut's Lamb

Then, in the 1990s, a key insight was made in Scotland by geneticist Keith Campbell, a specialist in studying the cell cycle of agricultural animals. Recall from chapter 8 that the division cycle of eukaryotic cells progresses in several stages. Campbell reasoned, "Maybe the egg and the donated nucleus need to be at the same stage in the cell cycle." This proved to be a

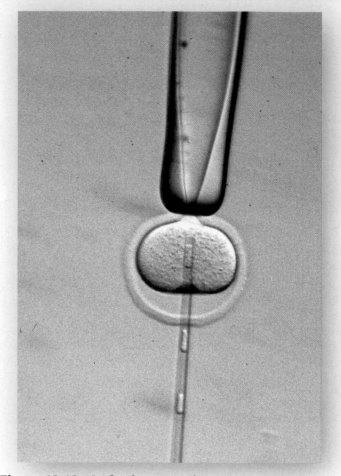

Figure 13.13   A cloning experiment.
In this photo, a nucleus is being injected from a micropipette *(bottom)* into an enucleated egg cell held in place by a pipette.

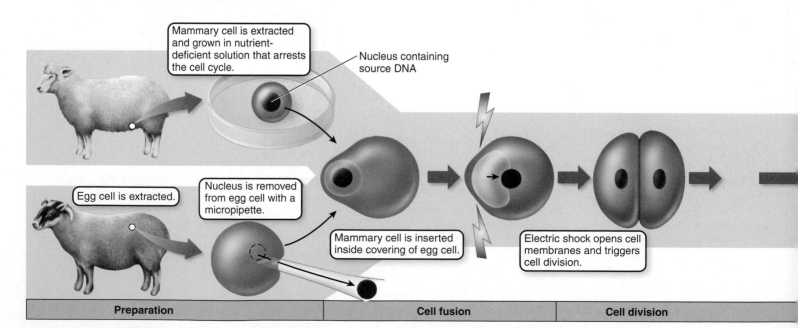

Mammary cell is extracted and grown in nutrient-deficient solution that arrests the cell cycle.

Nucleus containing source DNA

Egg cell is extracted.

Nucleus is removed from egg cell with a micropipette.

Mammary cell is inserted inside covering of egg cell.

Electric shock opens cell membranes and triggers cell division.

| Preparation | Cell fusion | Cell division |

Figure 13.14   Wilmut's animal cloning experiment.

key insight. In 1994 researchers succeeded in cloning farm animals from advanced embryos by first starving the cells, so that they paused at the beginning of the cell cycle. Two starved cells are thus synchronized at the same point in the cell cycle.

Campbell's colleague Ian Wilmut then attempted the key breakthrough, the experiment that had been eluding researchers: He set out to transfer the nucleus from an adult differentiated cell into an enucleated egg, and to allow the resulting embryo to grow and develop in a surrogate mother, hopefully producing a healthy animal (**figure 13.14**). Approximately five months later, on July 5, 1996, the mother gave birth to a lamb. This lamb, "Dolly," was the first successful clone generated from an adult animal cell. Dolly grew into a healthy adult, and as you can see in the photo at the beginning of this chapter, she went on to have healthy offspring normal in every respect.

Progress with Reproductive Cloning

Since Dolly's birth in 1996, scientists have successfully cloned a wide variety of farm animals with desired characteristics, including cows, pigs, goats, horses, and donkeys, as well as pets like cats and dogs. Snuppy, the puppy in **figure 13.15**, was the first dog to be cloned. For most farm animals, cloning procedures have become increasingly efficient since Dolly was cloned. However, the development of clones into adults tends to go unexpectedly haywire. Almost none survive to live a normal life span. Even Dolly died prematurely in 2003, having lived only half a normal sheep life span.

The Importance of Gene Reprogramming

What is going wrong? It turns out that as mammalian eggs and sperm mature, their DNA is conditioned by the parent female or male, a process called reprogramming. Chemical changes are made to the DNA that alter when particular genes are expressed without changing the nucleotide sequences. In the years since Dolly, scientists have learned a lot about gene reprogramming, also called **epigenetics.** Epigenetics works by blocking the cell's ability to read certain genes. A gene is locked in the off

Figure 13.15   Cloning the family pet.
This puppy named "Snuppy" is the first dog cloned. Beside him to the left, is the adult male dog that provided the skin cell from which Snuppy was cloned. The dog in the photo on the right was Snuppy's surrogate mother.

position by adding a –CH_3 (methyl) group to some of its cytosine nucleotides. After a gene has been altered like this, the polymerase protein that is supposed to "read" the gene can no longer recognize it. The gene has been shut off.

We are only beginning to learn how to reprogram human DNA, so any attempt to clone a human is simply throwing stones in the dark, hoping to hit a target we cannot see. For this and many other reasons, human reproductive cloning is regarded as highly unethical.

> **Key Learning Outcome 13.6**   Although recent experiments have demonstrated the possibility of cloning animals from adult tissue, the cloning of farm animals often fails for lack of proper epigenetic reprogramming.

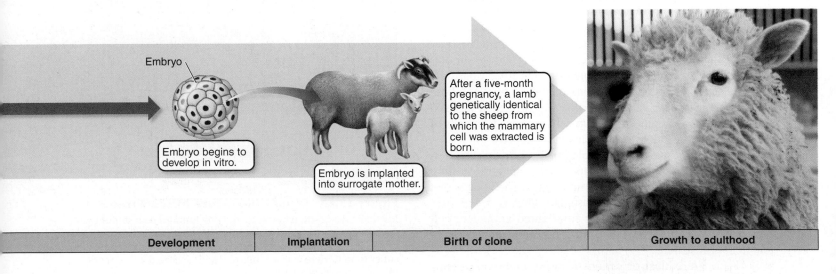

Embryo

Embryo begins to develop in vitro.

Embryo is implanted into surrogate mother.

After a five-month pregnancy, a lamb genetically identical to the sheep from which the mammary cell was extracted is born.

| Development | Implantation | Birth of clone | Growth to adulthood |

13.7 Stem Cell Therapy

You can see a mass of human embryonic stem cells in **figure 13.16**. Many are **totipotent**—able to form any body tissue, and even an entire adult animal. What is an embryonic stem cell, and why is it totipotent? To answer this question, we need to consider for a moment where an embryo comes from. At the dawn of a human life, a sperm fertilizes an egg to create a single cell destined to become a child. As development commences, that cell begins to divide, producing after four divisions a small mass of 16 **embryonic stem cells.** Each of these embryonic stem cells has all of the genes needed to produce a normal individual.

As development proceeds, some of these embryonic stem cells become committed to forming specific types of tissues, such as nerve tissues, and, after this step is taken, cannot ever produce any other kind of cell. In the case of nerve tissue, they are then called *nerve stem cells*. Others become specialized to produce blood cells, others to produce muscle tissue, and still others to form the other tissues of the body. Each major tissue is formed from its own kind of tissue-specific **adult stem cell.** Because an adult stem cell forms only that one kind of tissue, it is not totipotent.

Using Stem Cells to Repair Damaged Tissues

Embryonic stem cells offer the exciting possibility of restoring damaged tissues. To understand how, follow along in **figure 13.18**. A few days after fertilization, an embryonic stage called the *blastocyst* forms ❶. Embryonic stem cells are harvested from its inner cell mass or from cells of the embryo at a later stage ❷. These embryonic stem cells can be grown in tissue culture as seen in **figure 13.16**, and in principle be induced to form any type of tissue in the body ❸. The resulting healthy tissue can then be injected into the patient where it will grow and replace damaged tissue ❹. Alternatively, where possible, adult stem cells can be isolated and when injected back into the body, can form certain types of tissue cells.

Both adult and embryonic stem cell transfer experiments have been carried out successfully in mice. Adult blood stem cells have been used to cure leukemia. Heart muscle cells grown from mouse embryonic stem cells have successfully replaced the damaged heart tissue of a living mouse. In other experiments, damaged spinal neurons have been partially repaired. DOPA-producing neurons of mouse brains, whose loss is responsible for Parkinson's disease, have been successfully replaced with embryonic stem cells, as have islet cells of the pancreas, whose loss leads to juvenile diabetes.

Because the course of development is broadly similar in all mammals, these experiments in mice suggest exciting possibilities for stem cell therapy in humans. The hope is that individuals with Parkinson's disease, like Michael J. Fox (**figure 13.17**), might be partially or fully cured with stem cell therapy. As you might imagine, work proceeds intensively in this field of research.

There are ethical objections to using embryonic stems cells but new experimental results hint at ways around this

Figure 13.16   Human embryonic stem cells (×20).
This mass is a colony of undifferentiated human embryonic stem cells growing in tissue culture and surrounded by fibroblasts (elongated cells) that serve as a "feeder layer."

Figure 13.17   Promoting a cure for Parkinson's.
Michael J. Fox, with whom you may be familiar as a star of the *Back to the Future* film series and the TV show *Family Ties*, is a victim of Parkinson's disease, and a prominent spokesman for those who suffer from it. Here you see him testifying before the U.S. Senate (along with fellow advocate Mary Tyler Moore) on the need for vigorous efforts to support research seeking a cure.

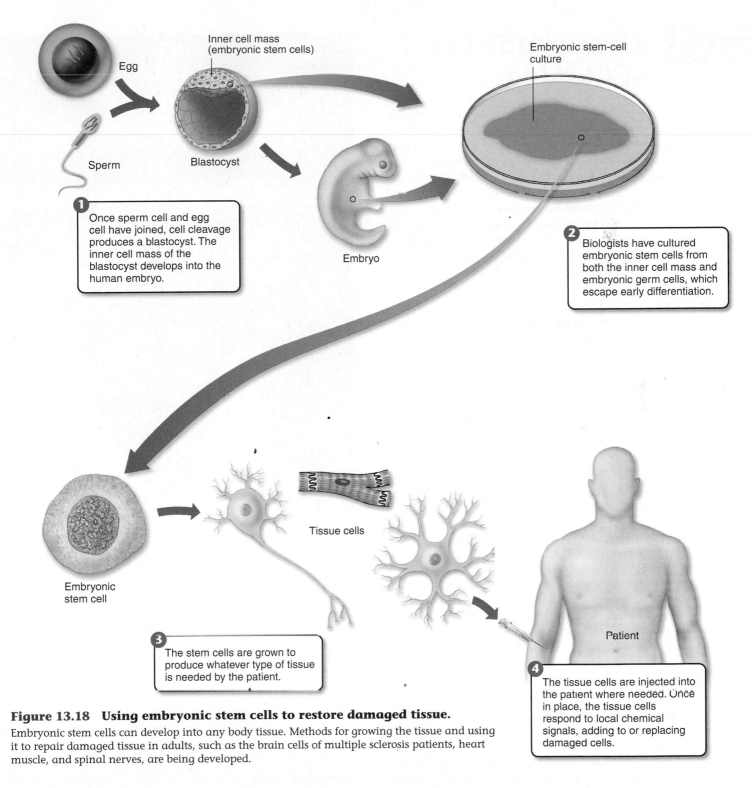

Figure 13.18 Using embryonic stem cells to restore damaged tissue.

Embryonic stem cells can develop into any body tissue. Methods for growing the tissue and using it to repair damaged tissue in adults, such as the brain cells of multiple sclerosis patients, heart muscle, and spinal nerves, are being developed.

Labels in figure:

Egg

Sperm

Inner cell mass (embryonic stem cells)

Blastocyst

Embryo

Embryonic stem-cell culture

1 Once sperm cell and egg cell have joined, cell cleavage produces a blastocyst. The inner cell mass of the blastocyst develops into the human embryo.

2 Biologists have cultured embryonic stem cells from both the inner cell mass and embryonic germ cells, which escape early differentiation.

Embryonic stem cell

Tissue cells

3 The stem cells are grown to produce whatever type of tissue is needed by the patient.

Patient

4 The tissue cells are injected into the patient where needed. Once in place, the tissue cells respond to local chemical signals, adding to or replacing damaged cells.

ethical maze. In 2007, researchers in two independent laboratories reported that they had engineered embryonic stemlike cells from normal adult human skin cells. The cells they created were pluripotent—they could differentiate into many different cell types. Whether pluripotency extends to totipotency is still being investigated. How were these cells transformed? The essential clue came six years earlier, when fusing adult cells with embryonic stem cells transformed the adult cells into pluripotent cells, as if factors had been transmitted to the adult cells that conferred pluripotency. Then, in a crucial advance in 2006, Japanese cell biologist Shinya Yamanaka introduced into adult human skin cells not the entire contents of an embryonic stem cell, but just the genes for four transcription factors. Once inside, these four factors induced a series of events that converted the adult cell to pluripotency. In effect, he had found a way to reprogram the adult cells to be embryonic stem cells. From proof-of-principle in a laboratory culture dish to actual medical application is still a leap, but the possibility is exciting.

Key Learning Outcome 13.7   Human adult and embryonic stem cells offer the possibility of replacing damaged or lost human tissues.

Therapeutic Use of Cloning

While exciting, the therapeutic uses of stem cells to cure leukemia, type I diabetes, Parkinson's disease, damaged heart muscle, and injured nerve tissue were all achieved in experiments carried out using strains of mice without functioning immune systems. Why is this important? Because had these mice possessed fully functional immune systems, they almost certainly would have rejected the implanted stem cells as foreign. Humans with normal immune systems might well refuse to accept transplanted stem cells simply because they are from another individual. For such stem cell therapy to work in humans, this problem needs to be addressed and solved.

Cloning to Achieve Immune Acceptance

Early in 2001, a research team at the Rockefeller University reported a way around this potentially serious problem. Their solution? They first isolated skin cells from a mouse, then using the same procedure that created Dolly, they created a 120-cell embryo from them. The embryo was then destroyed, its embryonic stem cells harvested and cultured (**figure 13.19**) for transfer to replace injured tissue. This procedure is called **therapeutic cloning.** Therapeutic cloning and the procedure that was used to create Dolly, called **reproductive cloning,** are contrasted in **figure 13.20**. You can see that steps ❶ through ❺ are essentially the same for both procedures, but the two methods proceed differently after that. In reproductive cloning, the blastocyst from step ❺ is implanted in a surrogate mother in step ❻ₐ, developing into a baby that is genetically identical to the nucleus donor, step ❼ₐ. In therapeutic cloning, by contrast, stem cells from the blastocyst of step ❺ are removed and grown in culture, step ❻. These stem cells are developed into particular tissue types, such as pancreatic islet cells in step ❼, and can then be injected or transplanted into a patient that needs them, such as a diabetic patient, where the new islet cells can begin producing insulin.

Therapeutic cloning, or, more technically, *somatic cell nuclear transfer,* successfully addresses the key problem that must be solved before embryonic stem cells can be used to repair damaged human tissues, which is immune acceptance. Because stem cells are cloned from the body's own tissues in therapeutic cloning, they pass the immune system's "self" identity check, and the body readily accepts them.

Gene Reprogramming to Achieve Immune Acceptance

In therapeutic cloning, the cloned embryo is destroyed to obtain embryonic stem cells. What is the moral standing of a six-day human embryo? Considering it a living individual, many people regard therapeutic cloning to be ethically unacceptable. Recent research discussed on the previous page suggests

Figure 13.19 Embryonic stem cells growing in cell culture.

Embryonic stem cells derived from early human embryos will grow indefinitely in tissue culture. When transplanted, they can sometimes be induced to form new cells of the adult tissue into which they have been placed. This suggests exciting therapeutic uses.

an alternative approach that avoids this problem: reprogramming adult cells into embryonic stem cells by introducing just a few genes into the adult cells. The genes are so-called transcription factors, turning on key genes that act to reverse the "shut off" epigenetic changes that have occurred during development of the adult cells. Human applications, if even possible, are probably far into the future, but the possibility of reprogramming adult cells is exciting.

> **Key Learning Outcome 13.8** Therapeutic cloning involves initiating blastocyst development from a patient's tissue using nuclear transplant procedures, then using the blastocyst's embryonic stem cells to replace the patient's damaged or lost tissue. Gene reprogramming of adult tissue cells may allow a less controversial approach.

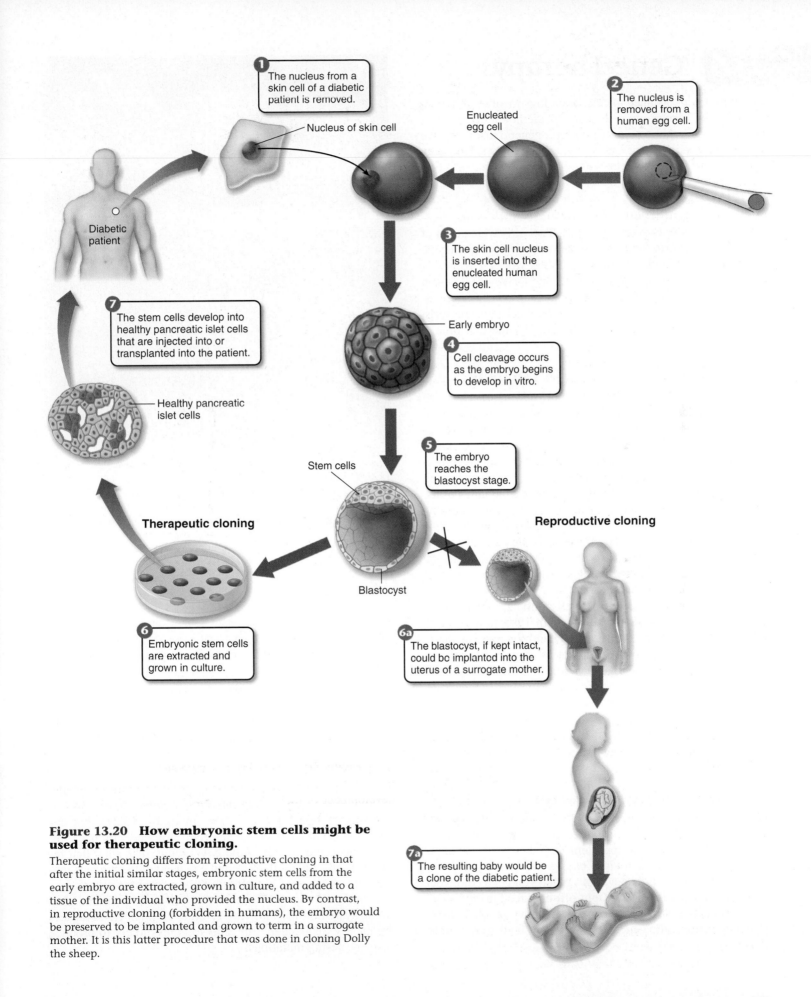

1 The nucleus from a skin cell of a diabetic patient is removed.

Nucleus of skin cell

Enucleated egg cell

2 The nucleus is removed from a human egg cell.

Diabetic patient

3 The skin cell nucleus is inserted into the enucleated human egg cell.

7 The stem cells develop into healthy pancreatic islet cells that are injected into or transplanted into the patient.

Early embryo

4 Cell cleavage occurs as the embryo begins to develop in vitro.

Healthy pancreatic islet cells

Stem cells

5 The embryo reaches the blastocyst stage.

Therapeutic cloning

Reproductive cloning

Blastocyst

6 Embryonic stem cells are extracted and grown in culture.

6a The blastocyst, if kept intact, could be implanted into the uterus of a surrogate mother.

Figure 13.20 How embryonic stem cells might be used for therapeutic cloning.

Therapeutic cloning differs from reproductive cloning in that after the initial similar stages, embryonic stem cells from the early embryo are extracted, grown in culture, and added to a tissue of the individual who provided the nucleus. By contrast, in reproductive cloning (forbidden in humans), the embryo would be preserved to be implanted and grown to term in a surrogate mother. It is this latter procedure that was done in cloning Dolly the sheep.

7a The resulting baby would be a clone of the diabetic patient.

13.9 Gene Therapy

The third major advance in cell technology involves introducing "healthy" genes into cells that lack them. For decades scientists have sought to cure often-fatal genetic disorders like cystic fibrosis, muscular dystrophy, and multiple sclerosis by replacing the defective gene with a functional one.

Early Success

A successful **gene transfer therapy** procedure was first demonstrated in 1990 (see section 13.3). Two girls were cured of a rare blood disorder due to a defective gene for the enzyme adenosine deaminase. Scientists isolated working copies of this gene and introduced them into bone marrow cells taken from the girls. The gene-modified bone marrow cells were allowed to proliferate, then were injected back into the girls. The girls recovered and stayed healthy. For the first time, a genetic disorder was cured by gene therapy.

The Rush to Cure Cystic Fibrosis

Researchers quickly set out to apply the new approach to one of the big killers, cystic fibrosis. The defective gene, labelled *cf*, had been isolated in 1989. Five years later, in 1994, researchers successfully transferred a healthy *cf* gene into a mouse that had a defective one—they in effect had cured cystic fibrosis in a mouse. They achieved this remarkable result by adding the *cf* gene to a virus that infected the lungs of the mouse, carrying the gene with it "piggyback" into the lung cells. The virus chosen as the "vector" was adenovirus (the red viruses in **figure 13.21**), a virus that causes colds and is very infective of lung cells. To avoid any complications, the lab mice used in the experiment had their immune systems disabled.

Very encouraged by these well-publicized preliminary trials with mice, several labs set out in 1995 to attempt to cure cystic fibrosis by transferring healthy copies of the *cf* gene into human patients. Confident of success, researchers added the human *cf* gene to adenovirus then administered the gene-bearing virus into the lungs of cystic fibrosis patients. For eight weeks the gene therapy did seem successful, but then disaster struck. The gene-modified cells in the patients' lungs came under attack by the patients' own immune systems. The "healthy" *cf* genes were lost and with them any chance of a cure.

Problems with the Vector

Other attempts at gene therapy met with similar results, eight weeks of hope followed by failure. In retrospect, although it was not obvious then, the problem with these early attempts seems predictable. Adenovirus causes colds. Do you know anyone who has never had a cold? When you get a cold, your body produces antibodies to fight off the infection, and so all of us have antibodies directed against adenovirus. We were introducing therapeutic genes in a vector our bodies are primed to destroy.

Figure 13.21 Adenovirus and AAV vectors (×200,000).

Adenovirus, the *red* virus particles above, has been used to carry healthy genes in clinical trials of gene therapy. Its use as a vector is problematic, however. AAV, the much smaller *bluish-green* virus particles seen in association with adenovirus here, lacks the many problems of adenovirus and is a much more promising gene transfer vector.

A second serious problem is that when the adenovirus infects a cell, it inserts its DNA into the human chromosome. Unfortunately, it does so at a random location. This means that the insertion events could cause mutations—if the viral DNA inserts into the middle of a gene, it could inactivate that gene. Because the spot where the adenovirus inserts is random, some of the mutations that result can be expected to cause cancer, certainly an unacceptable consequence.

A More Promising Vector

Researchers are now investigating a much more promising vector, a tiny virus called *adeno-associated virus* (AAV—the smaller bluish-green viruses in **figure 13.21**) that has only two genes. To create a vector for gene transfer, researchers remove both of the AAV genes. The shell that remains is still quite infective and can carry human genes into patients. AAV does not elicit a strong immune response—cells infected with AAV are not eliminated by a patient's immune system. Importantly, AAV enters human DNA far less frequently than adenovirus, and so is less likely to produce cancer-causing mutations.

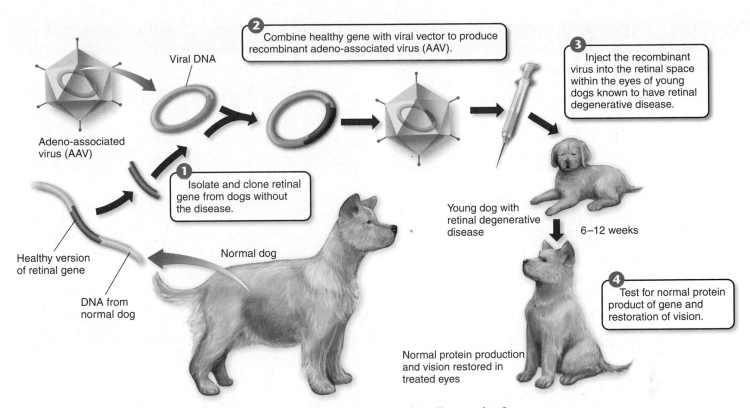

Figure 13.22 Using gene therapy to cure a retinal degenerative disease in dogs.
Researchers were able to use genes from healthy dogs to restore vision in dogs blinded by an inherited retinal degenerative disease. This disease also occurs in human infants and is caused by a defective gene that leads to early vision loss, degeneration of the retinas, and blindness. In the gene therapy experiments, genes from dogs without the disease were inserted into 3-month-old dogs that were known to carry the defective gene and that had been blind since birth. Six weeks after the treatment, the dogs' eyes were producing the normal form of the gene's protein product, and by three months, tests showed that the dogs' vision was restored.

Success with New Vectors

In 1999, AAV successfully cured anemia in rhesus monkeys. In monkeys, humans, and other mammals, red blood cell production is stimulated by a protein called *erythropoietin* (EPO). People with a type of anemia caused by low red blood cell counts, like dialysis patients, get regular injections of EPO. Using AAV to carry a souped-up EPO gene into the monkeys, scientists were able to greatly elevate their red blood cell counts, curing the monkeys of anemia—and they stayed cured.

A similar experiment using AAV cured dogs of a hereditary disorder leading to retinal degeneration and blindness. These dogs had a defective gene that produced a mutant form of a protein associated with the retina of the eye and were blind. Recombinant viral DNA was made using a healthy version of the gene, shown in steps ❶ and ❷ in **figure 13.22**. Injection of AAV bearing the needed gene into the fluid-filled compartment behind the retina, step ❸, restored sight in the dogs, step ❹. This procedure was recently tried on human patients with some success.

In 2003, gene therapy clinical trials attempting to cure severe combined immune deficiency (SCID) were halted when 5 of the 20 patients in the trial developed leukemia. Apparently the vector had contained a small segment of DNA homologous to a leukemia-causing human gene. When the vector inserted there, the leukemia-causing genes were activated.

Researchers stripped out the leukemia-causing segment of the vector, and launched yet another series of new gene therapy clinical trials. In 2009, a team used an improved vector to successfully treat 12 patients suffering from Leber's congenital blindness. All patients had some improvement in eyesight. In another study reported in 2009, two patients were treated for a rare, fatal brain disease called adrenoleukodystrophy (ALD), the disease featured in the film "Lorenzo's Oil." The treatment stopped the progression of the disease in its tracks. Three years later, the patients remain stable and can attend school.

In 2010, researchers began a new SCID trial with the improved vector, encouraged by the fact that all the patients in the 2003 trial who did not develop leukemia were completely cured of SCID. Trials are also underway for a wide variety of other disorders.

> **Key Learning Outcome 13.9** In principle, it should be possible to cure hereditary disorders like cystic fibrosis by transferring a healthy gene into the cells of affected tissues. Early attempts using adenovirus vectors were not often successful. New virus vectors avoid the problems of earlier vectors and offer promise of gene transfer therapy cures.

Can Modified Genes Escape from GM Crops?

On page 268, the question of whether gene flow of GM crops posed a problem to the environment was discussed. A field experiment conducted in 2004 by the Environmental Protection Agency assessed the possibility that introduced genes could pass from genetically modified golf course grass to other plants. Investigators introduced a gene conferring herbicide resistance (the EPSP synthetase gene for resistance to glyphosate) into golf course bentgrass, *Agrostis stolonifera,* and then looked to see if the gene passed from the GM grass to other plants of the same species, and also if it passed to other related species.

The map below displays the setup of this elaborate field study. A total of 178 *A. stolonifera* plants were placed outside the golf course, many of them downwind. An additional 69 bentgrass plants were found to be already growing downwind, most of them the related species *A. gigantea.* Seeds were collected from each of these plants, and the DNA of resulting seedlings tested for the presence of the gene introduced into the GM golf course grass. In the graph, the upper red histogram (a **histogram** is a "bar graph" that sorts data into a series of discontinuous categories, the value of each bar representing the number of individuals in a category, or, as in this case, the average value of entries in that category) presents the relative frequency with which the gene was found in *A. stolonifera* plants located at various distances from the golf course. The lower blue histogram does the same for *A. gigantea* plants.

1. **Applying Concepts**
   a. Reading a Histogram. Does the gene conferring resistance to herbicide pass to other plants of this species, *A. stolonifera*? To individuals of the related species *A. gigantea*?
   b. What is the maximal distance over which the herbicide resistance gene is transferred to other

plants of this species? Of the related species? What are these distances, expressed in miles?

2. **Interpreting Data**
   a. What general statement can be made about the effect of distance on the likelihood that the herbicide resistance gene will pass to another plant?
   b. Are there any significant differences in the gene flow to individuals of *A. stolonifera* and to individuals of the related species *A. gigantea*?

3. **Making Inferences** What mechanism do you propose to account for this gene flow?

4. **Drawing Conclusions** Is it fair to conclude that genetically modified traits can pass from crops to other plants? What qualifications would you place on your conclusion?

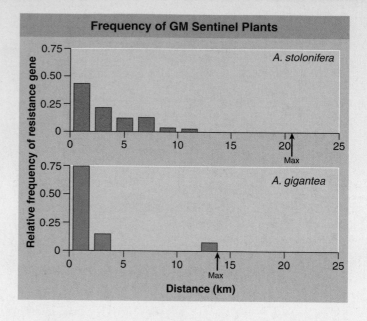

Sequencing Entire Genomes

13.1 Genomics

- The genetic information of an organism, its genes and other DNA, is called its genome. The sequencing and study of genomes is an area of biology called genomics (**table 13.1**).

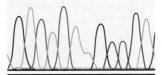

- The sequencing of entire genomes, a once long and tedious process, has been made faster and easier with automated systems (**figure 13.1**).

13.2 The Human Genome

- The human genome contains about 20,000 to 25,000 genes (**figure 13.2**), far less than what was expected based on the number of unique mRNA molecules present in our cells.

- Genes are organized in different ways in the genome, with nearly 99% of the human genome containing noncoding segments of DNA (**figure 13.3**).

Genetic Engineering

13.3 A Scientific Revolution

- Genetic engineering is the process of moving genes from one organism to another. It is having a major impact on medicine and agriculture (**figure 13.4**).

- Restriction enzymes are a special kind of enzyme that binds to short sequences of DNA and cuts them at specific locations. When two different molecules of DNA are cut with the same restriction enzyme, sticky ends form, allowing the segments of different DNA to be joined (**figure 13.5**).

- Before transferring a eukaryotic gene into a bacterial cell, the intron regions must be removed. This is accomplished with the use of cDNA, a complementary copy of the gene using the processed mRNA to make double-stranded DNA that doesn't contain the introns (**figure 13.6**).

- DNA fingerprinting is a process using probes to compare two samples of DNA. The probes bind to the DNA samples, creating restriction patterns that can be compared.

- The polymerase chain reaction (PCR) is a procedure used to amplify small amounts of DNA (**figure 13.7**).

13.4 Genetic Engineering and Medicine

- Genetic engineering is used in the production of medically important proteins used to treat illnesses (**table 13.2**).

- Vaccines are developed using genetic engineering. A gene that encodes a viral protein of a pathogenic virus is inserted into the DNA of a harmless virus that serves as a vector, as shown here from **figure 13.9**. The vector carrying the recombinant DNA is injected into a human. The vector infects the body, replicates, and the recombinant DNA is translated producing the viral proteins. The body elicits an immune response against the proteins, which protects the person from an infection by the pathogenic virus in the future.

13.5 Genetic Engineering and Agriculture

- Genetic engineering has been used in crop plants to make them more cost effective to grow (**figure 13.11**) or more nutritious (**figure 13.12**). However, GM plants are a source of controversy because of potential dangers that may result from the genetic manipulation of crop plants.

The Revolution in Cell Technology

13.6 Reproductive Cloning

- Ian Wilmut succeeded in cloning a sheep by synchronizing the donated nucleus and the egg cell to the same stage of the cell cycle (**figure 13.14**).

- Other animals have been successfully cloned, but problems and complications arise, often causing premature death. The problems with cloning appear to be caused by the lack of modifications that need to be made to the DNA, which turns certain genes on or off, a process called epigenetic reprogramming.

13.7 Stem Cell Therapy

- Embryonic stem cells are totipotent cells, which are cells that are able to divide and develop into any type of cell in the body or develop into an entire individual. These cells are present in the early embryo. Because of the totipotent nature of embryonic stem cells, they could be used to replace tissues lost or damaged due to accident or disease (**figure 13.18**).

13.8 Therapeutic Use of Cloning

- The use of embryonic stem cells to replace damaged tissue has one major drawback: tissue rejection. The embryonic stem cells are treated as foreign cells by the patient's body and are rejected. Therapeutic cloning could alleviate this problem.

- Therapeutic cloning is the process whereby a cell from an individual who has lost tissue function is cloned, producing an embryo that is genetically identical to the person. Embryonic stem cells are then harvested from the cloned embryo and injected into the same individual. The embryonic stem cells regrow the lost or damaged tissue without eliciting an immune response (**figure 13.20**). However, this procedure, like others using embryonic stem cells, is controversial. Adult cells epigenetically reprogrammed to behave like embryonic stem cells offers great promise of a more acceptable treatment.

13.9 Gene Therapy

- Using gene therapy, a patient with a genetic disorder is cured by replacing a defective gene with a "healthy" gene. In theory, this should work—but early attempts to cure cystic fibrosis failed because of immunological reactions to the adenovirus vector used to carry the healthy genes into the patient.

- The recent focus of gene therapy has been to identify a vector that avoids the problems encountered with the adenovirus vector. Promising results in experiments using a parvovirus called adeno-associated virus (AAV) has scientists hopeful that new vectors will eliminate the problems seen with adenovirus (**figure 13.22**).

1. The total amount of DNA in an organism, including all of its genes and other DNA, is its
   a. heredity.
   b. genetics.
   c. genome.
   d. genomics.

2. A possible reason why humans have such a small number of genes as opposed to what was anticipated by scientists is that
   a. humans don't need more than 25,000 genes to function.
   b. the exons used to make a specific mRNA can be rearranged to form different proteins.
   c. the sample size used to sequence the human genome was not big enough, so the number of genes estimated could be low.
   d. the number of genes will increase as scientists find out what all of the noncoding DNA actually does.

3. A protein that can cut DNA at specific DNA base sequences is called a
   a. DNase.
   b. DNA ligase.
   c. restriction enzyme.
   d. DNA polymerase.

4. Complementary DNA or cDNA is produced by
   a. inserting a gene into a bacterial cell.
   b. exposing the mRNA of the desired eukaryotic gene to reverse transcriptase.
   c. exposing the source DNA to restriction enzymes.
   d. exposing the source DNA to a probe.

5. Which of the following statements is correct?
   a. DNA fingerprinting is not admissible in court.
   b. DNA fingerprinting can prove with 100% certainty that two samples of DNA are from the same person.
   c. DNA fingerprinting becomes more and more reliable as more probes are used.
   d. No two people will ever have the same restriction pattern.

6. Using drugs produced by genetically engineered bacteria allows
   a. the drug to be produced in far larger amounts than in the past.
   b. humans to permanently correct the effect of a missing gene from their own systems.
   c. humans to cure cystic fibrosis.
   d. All of the above.

7. Some of the advantages to using genetically modified organisms in agriculture include
   a. increased yield.
   b. maintaining current nutritive value.
   c. mass-producing proteins.
   d. curing genetic diseases.

8. Which of the following is *not* a concern about the use of genetically modified crops?
   a. possible danger to humans after consumption
   b. insecticide resistance developing in pest species
   c. gene flow into natural relatives of GM crops
   d. harm to the crop itself from mutations

9. One of the main biological problems with replacing damaged tissue through the use of embryonic stem cells is
   a. immunological rejection of the tissue by the patient.
   b. that stem cells may not target appropriate tissue.
   c. the time needed to grow sufficient amounts of tissue.
   d. that genetic mutation of chosen stem cells may cause future problems.

10. In gene therapy, healthy genes are placed into animal cells that have defective genes by using
   a. a DNA particle gun.
   b. micropipettes (needles).
   c. viruses.
   d. Cells are not modified genetically. Instead, healthy tissue is grown and transplanted into the patient.

1. **Figure 13.1** Can you sequence the unknown section of DNA with the DNA fragments obtained from the following DNA analysis?

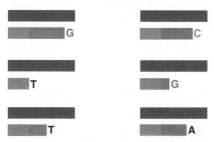

2. **Figure 13.14** Your friend Thomas wants to know why scientists don't just take an egg cell, with its own nucleus intact, and shock it to begin cell division. How do you answer him?

1. The goal behind therapeutic cloning is to replace tissue that is damaged due to an accident or nongenetic disease. Do you think therapeutic cloning as shown in figure 13.20 would work to replace tissue damage caused by genetic disorders?

2. If a person has a genetic disease such as cystic fibrosis, the hope is that we will be able to use gene therapy to cure him or her. When that happens, will the patient no longer be able to pass the *cf* gene on to his or her children?

3. Much of the technology for producing GM foods is owned by multinational corporations, which seek to maintain intellectual ownership of their creations. As one example, Monsanto Corporation requires farmers to sign contracts for glyphosate-tolerant soybeans that prevent the farmers from saving seed for replanting the next year. The company has aggressively brought suit against violators. On the one hand, companies need to be able to profit from their products, and the development costs of GM foods are enormous. Without potential profit, future GM crops will not be developed. On the other hand, in many highly populated regions of the world, people who face famine when their crops fail simply cannot afford to pay the price of seeds every year. How would you want to see this challenging issue handled?

14

Evolution and Natural Selection

These four finches live on the Galápagos Islands, a cluster of volcanic islands far out to sea off the coast of South America. All descendants of a single ancestral migrant, blown to the islands from the mainland long ago, the Galápagos finches gave Darwin valuable clues about how natural selection shapes the evolution of species. The two upper finches are ground finches, their different beaks adapting them to eat different-sized seeds. The finch on the left consumes smaller, slender seeds. The stouter beak of the finch on the right enables it to crack open larger, drier seeds. On the lower left is a woodpecker finch, a kind of tree finch that carries around a cactus spine, which it uses to probe for insects in deep crevices. On the lower right is a warbler finch that like its namesake, eats crawling insects. Each of these species utilizes food resources differently. Their different ways of interacting with the community within which they live generate the selective pressures that shape the evolution of groups like Darwin's finches.

14.1 Darwin's Voyage on HMS Beagle

The great diversity of life on earth—ranging from bacteria to elephants and roses—is the result of a long process of **evolution,** the change that occurs in organisms's characteristics through time. In 1859, the English naturalist Charles Darwin (1809–82; **figure 14.1**) first suggested an explanation for why evolution occurs, a process he called *natural selection.* Biologists soon became convinced Darwin was right and now consider evolution one of the central concepts of the science of biology. In this chapter, we examine Darwin and evolution in detail, as the concepts we encounter will provide a solid foundation for your exploration of the living world.

The theory of evolution proposes that a population can change over time, sometimes forming a new species. A **species** is a population or group of populations that possess similar characteristics and can interbreed and produce fertile offspring. This famous theory provides a good example of how a scientist develops a hypothesis—in this case, a hypothesis of how evolution occurs—and how, after much testing, the hypothesis is eventually accepted as a theory.

Charles Robert Darwin was an English naturalist who, after 30 years of study and observation, wrote one of the most famous and influential books of all time. This book, *On the Origin of Species by Means of Natural Selection, or The Preservation of Favoured Races in the Struggle for Life,* created a sensation when it was published, and the ideas Darwin expressed in it have played a central role in the development of human thought ever since.

In Darwin's time, most people believed that the various kinds of organisms and their individual structures resulted from direct actions of the Creator. Species were thought to be specially created and unchangeable over the course of time. In contrast to these views, a number of earlier philosophers had presented the view that living things must have changed during the history of life on earth. Darwin proposed a concept he called natural selection as a coherent, logical explanation for this process. Darwin's book, as its title indicates, presented a conclusion that differed sharply from conventional wisdom. Although his theory did not challenge the existence of a Divine Creator, Darwin argued that this Creator did not simply create things and then leave them forever unchanged. Instead, Darwin's God expressed Himself through the operation of natural laws that produced change over time—evolution.

The story of Darwin and his theory begins in 1831, when he was 22 years old. The small British naval vessel HMS *Beagle* that you see in **figure 14.2** was about to set sail on a five-year navigational mapping expedition around the coasts of South America. The red arrows in **figure 14.3** indicate the route taken by HMS *Beagle.* The young (26-year-old) captain of HMS *Beagle,* unable by British naval tradition to

Figure 14.1 The theory of evolution by natural selection was proposed by Charles Darwin.
This rediscovered photograph appears to be the last ever taken of the great biologist. It was taken in 1881, the year before Darwin died.

have social contact with his crew, and anticipating a voyage that would last many years, wanted a gentleman companion, someone to talk to. Indeed, the *Beagle*'s previous skipper had broken down and shot himself to death after three solitary years away from home.

On the recommendation of one of his professors at Cambridge University, Darwin, the son of a wealthy doctor and very much a gentleman, was selected to serve as the captain's companion, primarily to share his table at mealtime during every shipboard dinner of the long voyage. Darwin paid his own expenses, and even brought along a manservant.

Darwin took on the role of ship's naturalist (the official naturalist, a man named Robert McKormick, left the ship before the first year was out). During this long voyage, Darwin had the chance to study a wide variety of plants and animals on continents and islands and in distant seas. He was able to explore the biological richness of the tropical forests, examine the extraordinary fossils of huge extinct mammals in Patagonia at the southern tip of South America, and observe the remarkable series of related but distinct forms of life on the **Galápagos Islands.** Such an opportunity clearly played an important role in the development of his thoughts about the nature of life on earth.

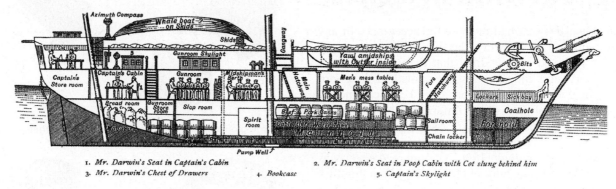

1. *Mr. Darwin's Seat in Captain's Cabin*   2. *Mr. Darwin's Seat in Poop Cabin with Cot slung behind him*
3. *Mr. Darwin's Chest of Drawers*   4. *Bookcase*   5. *Captain's Skylight*

Figure 14.2   Cross section of HMS *Beagle*.

HMS *Beagle*, a 10-gun brig of 242 tons, only 90 feet in length, had a crew of 74 people! After he first saw the ship, Darwin wrote to his college professor Henslow: "The absolute want of room is an evil that nothing can surmount."

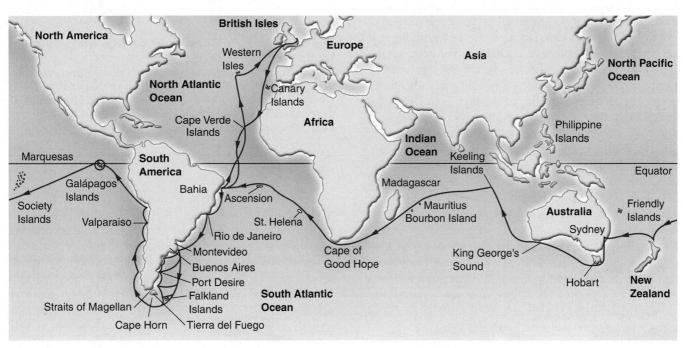

Figure 14.3   The five-year voyage of HMS *Beagle*.

Although the ship sailed around the world, most of its time was spent exploring the coasts and coastal islands of South America, such as the Galápagos Islands. Darwin's studies of the animals of these islands played a key role in the eventual development of his theory of evolution by means of natural selection.

When Darwin returned from the voyage at the age of 27, he began a long period of study and contemplation. During the next 10 years, he published important books on several different subjects, including the formation of oceanic islands from coral reefs and the geology of South America. He also devoted eight years of study to barnacles, a group of small marine animals with shells that inhabit rocks and pilings, eventually writing a four-volume work on their classification and natural history. In 1842, Darwin and his family moved out of London to a country home at Down, in the county of Kent. In these pleasant surroundings, Darwin lived, studied, and wrote for the next 40 years.

> **Key Learning Outcome 14.1   Darwin was the first to propose natural selection as the mechanism of evolution that produced the diversity of life on earth.**

14.2 Darwin's Evidence

One of the obstacles that had blocked the acceptance of any theory of evolution in Darwin's day was the incorrect notion, widely believed at that time, that the earth was only a few thousand years old. The discovery of thick layers of rocks, evidences of extensive and prolonged erosion, and the increasing numbers of diverse and unfamiliar fossils discovered during Darwin's time made this assertion seem less and less likely. The great geologist Charles Lyell (1797–1875), whose *Principles of Geology* (1830) Darwin read eagerly as he sailed on HMS *Beagle,* outlined for the first time the story of an ancient world in which plant and animal species were constantly becoming extinct while others were emerging. It was this world that Darwin sought to explain.

What Darwin Saw

When HMS *Beagle* set sail, Darwin was fully convinced that species were immutable, meaning that they were not subject to being changed. Indeed, it was not until two or three years after his return that he began to seriously consider the possibility that they could change. Nevertheless, during his five years on the ship, Darwin observed a number of phenomena that were of central importance to him in reaching his ultimate conclusion. For example, in the rich fossil beds of southern South America, he observed fossils of the extinct armadillo shown on the right in **figure 14.4**. They were surprisingly similar in form to the armadillos that still lived in the same area, shown on the left. Why would similar living and fossil organisms be in the same area unless the earlier form had given rise to the other? Later, Darwin's observations would be strengthened by the discovery of other examples of fossils that show intermediate characteristics, pointing to successive change.

Repeatedly, Darwin saw that the characteristics of similar species varied somewhat from place to place. These geographical patterns suggested to him that organismal lineages change gradually as individuals move into new habitats. On the Galápagos Islands, 900 kilometers (540 miles) off the

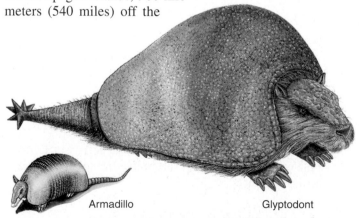

Armadillo   Glyptodont

Figure 14.4 Fossil evidence of evolution.
The now-extinct glyptodont was a large 2,000-kilogram South American armadillo (about the size of a small car), much larger than the modern armadillo, which weighs an average of about 4.5 kilograms and is about the size of a house cat. The similarity of fossils such as the glyptodonts to living organisms found in the same regions suggested to Darwin that evolution had taken place.

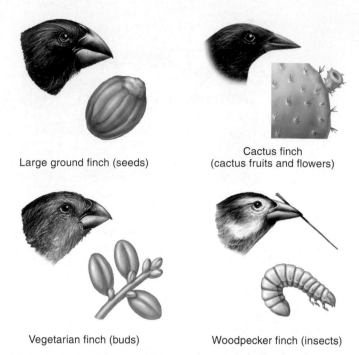

Large ground finch (seeds)

Cactus finch (cactus fruits and flowers)

Vegetarian finch (buds)

Woodpecker finch (insects)

Figure 14.5 Four Galápagos finches and what they eat.
Darwin observed 14 different species of finches on the Galápagos Islands, differing mainly in their beaks and feeding habits. These four finches eat very different food items, and Darwin surmised that the very different shapes of their beaks represented evolutionary adaptations improving their ability to do so.

coast of Ecuador, Darwin encountered a variety of different finches on the islands. The 14 species, although related, differed slightly in appearance. Darwin felt it most reasonable to assume all these birds had descended from a common ancestor blown by winds from the South American mainland several million years ago. Eating different foods, on different islands, the species had changed in different ways, most notably in the size of their beaks. The larger beak of the ground finch in the upper left of **figure 14.5** is better suited to crack open the large seeds it eats. As the generations descended from the common ancestor, these ground finches changed and adapted, what Darwin referred to as "descent with modification"—evolution.

In a more general sense, Darwin was struck by the fact that the plants and animals on these relatively young volcanic islands resembled those on the nearby coast of South America. If each one of these plants and animals had been created independently and simply placed on the Galápagos Islands, why didn't they resemble the plants and animals of islands with similar climates, such as those off the coast of Africa, for example? Why did they resemble those of the adjacent South American coast instead?

Key Learning Outcome 14.2 The fossils and patterns of life that Darwin observed on the voyage of HMS *Beagle* eventually convinced him that evolution had taken place.

The Theory of Natural Selection

It is one thing to observe the results of evolution but quite another to understand how it happens. Darwin's great achievement lies in his formulation of the hypothesis that evolution occurs because of natural selection.

Darwin and Malthus

Of key importance to the development of Darwin's insight was his study of Thomas Malthus's *Essay on the Principle of Population* (1798). In his book, Malthus pointed out that populations of plants and animals (including human beings) tend to increase geometrically, while the ability of humans to increase their food supply increases only arithmetically. A geometric progression is one in which the elements increase by a constant factor; the blue line in **figure 14.6** shows the progression 2, 6, 18, 54, . . . and each number is three times the preceding one. An arithmetic progression, in contrast, is one in which the elements increase by a constant difference; the red line shows the progression 2, 4, 6, 8, . . . and each number is two greater than the preceding one.

Because populations increase geometrically, virtually any kind of animal or plant, if it could reproduce unchecked, would cover the entire surface of the world within a surprisingly short time. Instead, population sizes of species remain fairly constant year after year, because death limits population numbers. Malthus's conclusion provided the key ingredient that was necessary for Darwin to develop the hypothesis that evolution occurs by natural selection.

Natural Selection

Sparked by Malthus's ideas, Darwin saw that although every organism has the potential to produce more offspring than can survive, only a limited number actually do survive and produce further offspring. Many examples appear in nature. Sea turtles, for instance, will return to the beaches where they hatched to lay their eggs. Each female will lay about 100 eggs. The beach could be covered with thousands of hatchlings, like in **figure 14.7**, trying to make it to water's edge. Less than 10% will actually reach adulthood and return to this beach to reproduce. Darwin combined his observation with what he had seen on the voyage of HMS *Beagle*, as well as with his own experiences in breeding domestic animals, and made an important association: Those individuals that possess physical, behavioral, or other attributes that help them live in their environment are more likely to survive than those that do not have these characteristics. By surviving, they gain the opportunity to pass on their favorable characteristics to their offspring. As the frequency of these characteristics increases in the population, the nature of the population as a whole will gradually change. Darwin called this process **natural selection.** The driving force he identified has often been referred to as *survival of the fittest*. However, this is not to say the biggest or the strongest always survive. These characteristics may be favorable in one environment but less favorable in another.

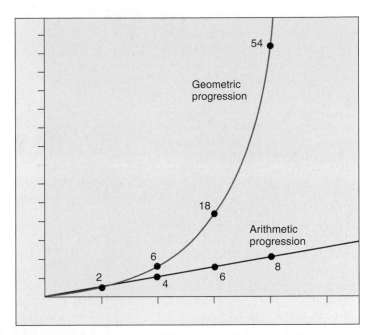

Figure 14.6   Geometric and arithmetic progressions.
An arithmetic progression increases by a constant difference (for example, units of 1 or 2 or 3), while a geometric progression increases by a constant factor (for example, by 2 or by 3 or by 4). Malthus contended that the human growth curve was geometric, but the human food production curve was only arithmetic. Can you see the problems this difference would cause?

Figure 14.7   Sea turtle hatchlings.
These newly-hatched sea turtles make their way to the ocean from their nests on the beach. Thousands of eggs may be laid on a beach during a spawning, but less than 10% will survive to adulthood. Natural predators, human egg poachers, and environmental challenges prevent the majority of offspring from surviving. As Darwin observed, sea turtles produce more offspring than will actually survive to reproduce.

The organisms that are "best suited" to their particular environment survive more often, and therefore produce more offspring than others in the population, and in this sense are the "fittest."

Darwin was thoroughly familiar with variation in domesticated animals and began *On the Origin of Species* with a detailed discussion of pigeon breeding. He knew that breeders selected certain varieties of pigeons and other animals, such as dogs, to produce certain characteristics, a process Darwin called **artificial selection.** Once this had been done, the animals would breed true for the characteristics that had been selected. Darwin had also observed that the differences purposely developed between domesticated races or breeds were often greater than those that separated wild species. Domestic pigeon breeds, for example, show much greater variety than all of the hundreds of wild species of pigeons found throughout the world. Such relationships suggested to Darwin that evolutionary change could occur in nature too. Surely if pigeon breeders could foster such variation by "artificial selection," nature through environmental pressures could do the same, playing the breeder's role in selecting the next generation—a process Darwin called *natural selection.*

Darwin's theory provides a simple and direct explanation of biological diversity, or why animals are different in different places—because habitats differ in their requirements and opportunities, the organisms with characteristics favored locally by natural selection will tend to vary in different places. As we will discuss later in this chapter in section 14.9, there are five evolutionary forces that can affect biological diversity, although natural selection is the only evolutionary force that produces *adaptive* changes.

Darwin Drafts His Argument

Darwin drafted the overall argument for evolution by natural selection in a preliminary manuscript in 1842. After showing the manuscript to a few of his closest scientific friends, however, Darwin put it in a drawer and for 16 years turned to other research. No one knows for sure why Darwin did not publish his initial manuscript—it is very thorough and outlines his ideas in detail. Some historians have suggested that Darwin was wary of igniting public, and even private, criticism of his evolutionary ideas—there could have been little doubt in his mind that his theory of evolution by natural selection would spark controversy. Others have proposed that Darwin was

Figure 14.8 Darwin greets his monkey ancestor.

In his time, Darwin was often portrayed unsympathetically, as in this drawing from an 1874 publication.

simply refining his theory, although there is little evidence he altered his initial manuscript in all that time.

Wallace Has the Same Idea

The stimulus that finally brought Darwin's theory into print was an essay he received in 1858. A young English naturalist named Alfred Russel Wallace (1823–1913) sent the essay to Darwin from Malaysia; it concisely set forth the theory of evolution by means of natural selection, a theory Wallace had developed independently of Darwin. Like Darwin, Wallace had been greatly influenced by Malthus's 1798 book. Colleagues of Wallace, knowing of Darwin's work, encouraged him to communicate with Darwin. After receiving Wallace's essay, Darwin arranged for a joint presentation of their ideas at a seminar in London. Darwin then completed his own book, expanding the 1842 manuscript that he had written so long ago, and submitted it for publication.

Publication of Darwin's Theory

Darwin's book appeared in November 1859 and caused an immediate sensation. Although people had long accepted that humans closely resembled apes in many characteristics, the possibility that there might be a direct evolutionary relationship was unacceptable to many. Darwin did not actually discuss this idea in his book, but it followed directly from the principles he outlined. In a subsequent book, *The Descent of Man,* Darwin presented the argument directly, building a powerful case that humans and living apes have common ancestors. Many people were deeply disturbed with the suggestion that human beings were descended from the same ancestor as apes, and Darwin's book on evolution caused him to become a victim of the satirists of his day—the cartoon in **figure 14.8** is a vivid example. Darwin's arguments for the theory of evolution by natural selection were so compelling, however, that his views were almost completely accepted within the intellectual community of Great Britain after the 1860s.

Key Learning Outcome 14.3 The fact that populations do not really expand geometrically implies that nature acts to limit population numbers. The traits of organisms that survive to produce more offspring will be more common in future generations—a process Darwin called natural selection.

14.4 | The Beaks of Darwin's Finches

Darwin's Galápagos finches played a key role in his argument for evolution by natural selection. He collected 31 specimens of finches from three islands when he visited the Galápagos Islands in 1835. Darwin, not an expert on birds, had trouble identifying the specimens. He believed by examining their beaks that his collection contained wrens, "gross-beaks," and blackbirds.

The Importance of the Beak

Upon Darwin's return to England, ornithologist John Gould examined the finches. Gould recognized that Darwin's collection was in fact a closely related group of distinct species, all similar to one another except for their beaks. In all, 14 species are now recognized, 13 from the Galápagos and one from far-distant Cocos Island. The ground finches with the larger beaks in **figure 14.9** feed on seeds that they crush in their beaks, whereas those with narrower beaks eat insects, including the warbler finch (named for its resemblance to a mainland bird). Other species include fruit and bud eaters, and species that feed on cactus fruits and the insects they attract; some populations of the sharp-beaked ground finch even include "vampires" that creep up on seabirds and use their sharp beaks to drink their blood. Perhaps most remarkable are the tool users, like the woodpecker finch you see in the upper left

Figure 14.10   A gene shapes the beaks of Darwin's finches.

A cell signalling molecule called "bone morphogenic protein 4" (BMP4) has been shown by DNA researchers to tailor the shape of the beak in Darwin's finches.

of the figure, that picks up a twig, cactus spine, or leaf stalk, trims it into shape with its beak, and then pokes it into dead branches to pry out grubs.

The differences in the beaks of Darwin's finches are due to differences in the genes of the birds. When biologists compare the DNA of large ground finches (with stout beaks for cracking large seeds) to the DNA of small ground finches (with more slender beaks), the only growth factor gene that is different in the DNA of the two species is *BMP4* (**figure 14.10** and **figure 1.15**). The difference is in how the gene is used. The large ground finches, with larger beaks, make more BMP4 protein than do the small ground finches.

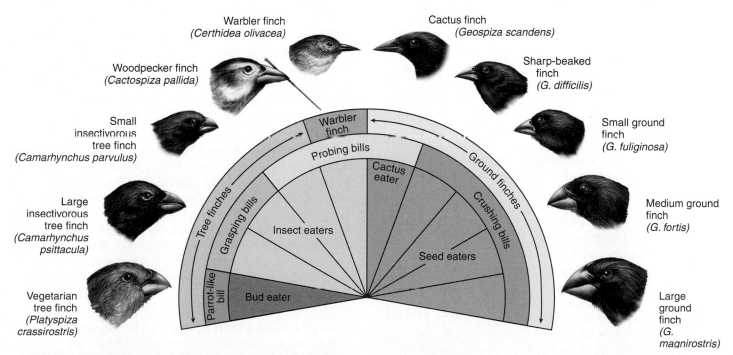

Figure 14.9   A diversity of finches on a single island.

Ten species of Darwin's finches from Isla Santa Cruz, one of the Galápagos Islands. The 10 species show differences in beaks and feeding habits. These differences presumably arose when the finches arrived and encountered habitats lacking small birds. Scientists concluded that all of these birds derived from a single common ancestor.

The correspondence between the beaks of the 14 finch species and their food source immediately suggested to Darwin that evolution had shaped them:

"Seeing this gradation and diversity of structure in one small, intimately related group of birds, one might really fancy that from an original paucity of birds in this archipelago, one species has been taken and modified for different ends."

Checking to See if Darwin Was Right

If Darwin's suggestion that the beak of an ancestral finch had been "modified for different ends" is correct, then it ought to be possible to see the different species of finches acting out their evolutionary roles, each using its beak to acquire its particular food specialty. The four species that crush seeds within their beaks, for example, should feed on different seeds, with those with stouter beaks specializing on harder-to-crush seeds.

Many biologists visited the Galápagos after Darwin, but it was 100 years before any tried this key test of his hypothesis. When the great naturalist David Lack finally set out to do this in 1938, observing the birds closely for a full five months, his observations seemed to contradict Darwin's proposal! Lack often observed many different species of finch feeding together on the same seeds. His data indicated that the stout-beaked species and the slender-beaked species were feeding on the very same array of seeds.

We now know that it was Lack's misfortune to study the birds during a wet year, when food was plentiful. The size of the finch's beak is of little importance in such flush times; slender and stout beaks work equally well to gather the abundant tender small seeds. Later work revealed a very different picture during dry years, when few seeds are available.

A Closer Look

Starting in 1973, Peter and Rosemary Grant of Princeton University and generations of their students have studied the medium ground finch, *Geospiza fortis*, on a tiny island in the center of the Galápagos called Daphne Major. These finches feed preferentially on small, tender seeds, abundantly available in wet years. The birds resort to larger, drier seeds that are harder to crush when small seeds are hard to find. Such lean times come during periods of dry weather, when plants produce few seeds, large or small.

By carefully measuring the beak shape of many birds every year, the Grants were able to assemble for the first time a detailed portrait of evolution in action. The Grants found that beak depth changed from one year to the next in a predictable fashion. During droughts, plants produced few seeds, and all available small seeds quickly were eaten, leaving large seeds as the major remaining source of food. As a result, birds with large beaks survived better, because they were better able to break open these large seeds. Consequently, the average beak depth of birds in the population increased the next year because this next generation included offspring of the large-beaked birds that survived. The off-

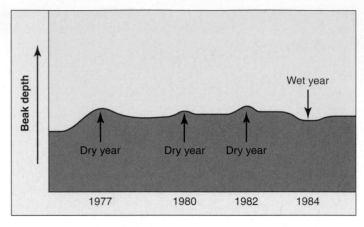

Figure 14.11 Evidence that natural selection alters beak size in *Geospiza fortis*.

In dry years, when only large, tough seeds were available, the mean beak size increased. In wet years, when many small seeds were available, smaller beaks became more common.

spring of the surviving "dry year" birds had larger beaks, an evolutionary response which led to the peaks you see in the graph in **figure 14.11**. The reason there are peaks and not plateaus is that the average beak size decreased again when wet seasons returned because the larger beak size was no longer more favorable when seeds were plentiful and so smaller-beaked birds survived to reproduce.

Could these changes in beak dimension reflect the action of natural selection? An alternative possibility might be that the changes in beak depth do not reflect changes in gene frequencies but rather are simply a response to diet, with poorly fed birds having stouter beaks. To rule out this possibility, the Grants measured the relation of parent beak size to offspring beak size, examining many broods over several years. The depth of the beak was passed down faithfully from one generation to the next, suggesting the differences in beak size indeed reflected gene differences.

Support for Darwin

If the year-to-year changes in beak depth can be predicted by the pattern of dry years, then Darwin was right after all—natural selection influences beak size based on available food supply. In the study discussed here, birds with stout beaks have an advantage during dry periods, for they can break the large, dry seeds that are the only food available. When small seeds become plentiful once again with the return of wet weather, a smaller beak proves a more efficient tool for harvesting smaller seeds.

> **Key Learning Outcome 14.4** In Darwin's finches, natural selection adjusts the shape of the beak in response to the nature of the food supply, adjustments that are occurring even today.

How Natural Selection Produces Diversity

Darwin believed that each Galápagos finch species had adapted to the particular foods and other conditions on the particular island it inhabited. Because the islands presented different opportunities, a cluster of species resulted. Presumably, the ancestor of Darwin's finches reached these newly formed islands before other land birds, so that when it arrived, all of the niches where birds occur on the mainland were unoccupied. A *niche* is what a biologist calls the way a species makes a living—the biological (that is, other organisms) and physical (climate, food, shelter, etc.) conditions with which an organism interacts as it attempts to survive and reproduce. As the new arrivals to the Galápagos moved into vacant niches and adopted new lifestyles, they were subjected to diverse sets of selective pressures. Under these circumstances, the ancestral finches rapidly split into a series of populations, some of which evolved into separate species.

The phenomenon by which a cluster of species change, as they occupy a series of different habitats within a region, is called *adaptive radiation*. **Figure 14.12** shows how the 14 species of Darwin's finches on the Galápagos Islands and Cocos Island are thought to have evolved. The ancestral population, indicated by the base of the brackets, migrated to the islands about 2 million years ago and underwent adaptive radiation giving rise to the 14 different species.

The descendants of the original finches that reached the Galápagos Islands now occupy many different kinds of habitats on the islands. The 14 species that inhabit the Galápagos Islands and Cocos Island occupy four types of niches:

1. **Ground finches.** There are six species of *Geospiza* ground finches. Most of the ground finches feed on seeds. The size of their beaks is related to the size of the seeds they eat. Some of the ground finches feed primarily on cactus flowers and fruits and have longer, larger, more pointed beaks.
2. **Tree finches.** There are five species of insect-eating tree finches. Four species have beaks that are suitable for feeding on insects. The woodpecker finch has a chisel-like beak. This unique bird carries around a twig or a cactus spine, which it uses to probe for insects in deep crevices.
3. **Vegetarian finch.** The very heavy beak of this bud-eating bird is used to wrench buds from branches.
4. **Warbler finches.** These unusual birds play the same ecological role in the Galápagos woods that warblers play on the mainland, searching continually over the leaves and branches for insects. They have a slender, warblerlike beak.

> **Key Learning Outcome 14.5** Darwin's finches, all derived from one similar mainland species, have radiated widely on the Galápagos Islands, filling unoccupied niches in a variety of ways.

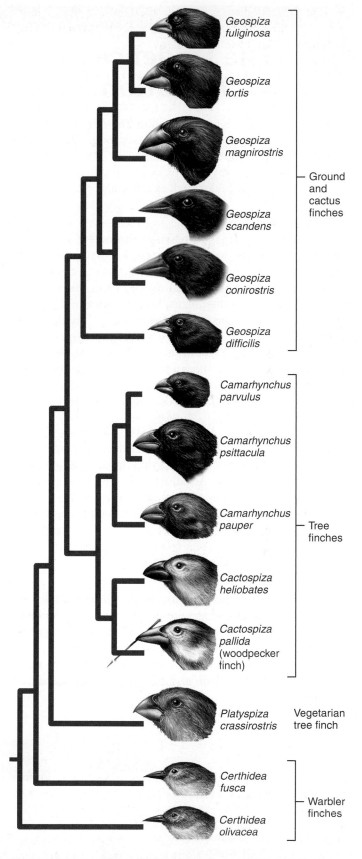

Figure 14.12 An evolutionary tree of Darwin's finches.

This family tree was constructed by comparing DNA of the 14 species. Their position at the base of the finch tree suggests that warbler finches were among the first adaptive types to evolve in the Galápagos.

The Evidence for Evolution

The evidence that Darwin presented in *The Origin of Species* to support his theory of evolution was strong. We will now examine other lines of evidence supporting Darwin's theory, including information revealed by examining fossils, anatomical features, and molecules such as DNA and proteins.

The Fossil Record

The most direct evidence of macroevolution is found in the fossil record. **Fossils** are the preserved remains, tracks, or traces of once-living organisms. Fossils are created when organisms become buried in sediment. The calcium in bone or other hard tissue mineralizes, and the surrounding sediment eventually hardens to form rock. Most fossils are, in effect, skeletons. In the rare cases when fossils form in very fine sediment, feathers may also be preserved. When remains are frozen or become suspended in amber (fossilized plant resin), however, the entire body may be preserved. The fossils contained in layers of sedimentary rock reveal a history of life on earth.

By dating the rock in which a fossil occurs, like the one in **figure 14.13**, we can get an accurate idea of how old the fossil is. Rocks are dated by measuring the amount of certain radioisotopes in the rock. A radioisotope will break down, or decay, into other isotopes or elements. This occurs at a constant rate and so the amount of a radioisotope present in the rock is an indication of the rock's age.

Using Fossils to Test the Theory of Evolution

If the theory of evolution is correct, then the fossils we see preserved in rock should represent a history of evolutionary change. The theory makes the clear prediction that a parade of successive changes should be seen, as first one change occurs and then another. If the theory of evolution is not correct, on the other hand, then such orderly change is not expected.

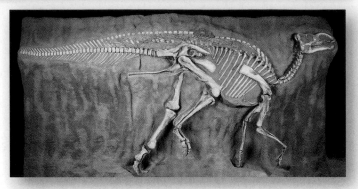

Figure 14.13   Dinosaur fossil of *Parasaurolophus*.

To test this prediction, we follow a logical procedure:

1. *Assemble a collection of fossils of a particular group of organisms.* You might, for example, gather together a collection of fossil titanotheres, a hoofed mammal that lived between about 50 million and 35 million years ago.
2. *Date each of the fossils.* In dating the fossils, it is important to make no reference to what the fossil is like. Imagine it as being concealed in a black box of rock, with only the box being dated.
3. *Order the fossils by their age.* Without looking in the "black boxes," place them in a series, beginning with the oldest and proceeding to the youngest.
4. *Now examine the fossils.* Do the differences between the fossils appear jumbled, or is there evidence of successive change as evolution predicts? You can judge for yourself in **figure 14.14**. During the 15 million years spanned by this collection of titanothere fossils, the small, bony protuberance located above the nose 50 million years ago evolved in a series of continuous changes into relatively large blunt horns.

It is important not to miss the key point of the result you see illustrated in **figure 14.14**: Evolution is an observation, not a conclusion. Because the dating of the samples is independent of what the samples are like, *successive change through time is a data statement*. While the statement that evolution

Figure 14.14   Testing the theory of evolution with fossil titanotheres.

Here you see illustrated changes in a group of hoofed mammals known as titanotheres, which lived between about 50 million and 35 million years ago. During this time, the small, bony protuberance located above the nose 50 million years ago evolved into relatively large, blunt horns.

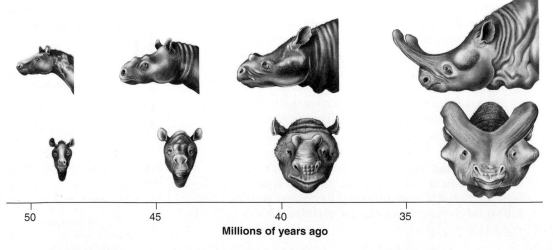

50   45   40   35

Millions of years ago

Today's Biology

Darwin and Moby Dick

Moby Dick, the white whale hunted by Captain Ahab in Melville's novel, was a sperm whale. One of the ocean's great predators, a large sperm whale is a voracious meat-eater that may span over 60 feet and weigh 50 tons. A sperm whale is not a fish, though. Unlike the great white shark in *Jaws*, a whale has hairs (not many), and a female whale has milk-producing mammary glands with which it feeds its young. A sperm whale is a mammal, just as you are! This raises an interesting question. If Darwin is right about the fossil record reflecting life's evolutionary past, then fossils tell us mammals evolved from reptiles on land at about the time of the dinosaurs. How did they end up back in the water?

The evolutionary history of whales has long fascinated biologists, but only in recent years have fossils been discovered that reveal the answer to this intriguing question. A series of discoveries now allows biologists to trace the evolutionary history of the most colossal animals ever to live on earth back to their beginnings at the dawn of the Age of Mammals. Whales, it turns out, are the descendants of four-legged land mammals that reinvaded the sea some 50 million years ago, much as seals and walruses are doing today. It's pretty startling to realize that Moby Dick's evolutionary ancestor lived on the steppes of Asia and looked like a modest-sized pig a few feet long and weighing perhaps 50 pounds, with four toes on each foot.

From what land mammal did whales arise? Researchers had long speculated that it might be a hoofed meat-eater with three toes known as a *mesonychid*, related to rhinoceroses. Subtle clues suggested this—the arrangement of ridges on the molar teeth, the positioning of the ear bones in the skull. But findings announced in 2001 reveal these subtle clues to have been misleading. Ankle bones from two newly described 50 million-year-old whale species discovered by Philip Gingerich of the University of Michigan are those of an artiodactyl, a four-toed mammal related to hippos, cattle, and pigs. Even more recently, Japanese researchers studying DNA have discovered unique genetic markers shared today only by whales and hippos.

Biologists now conclude that whales, like hippos, are descended from a group of early four-hoofed mammals called *anthracotheres*, modest-sized grazing animals with a piggish appearance abundant in Europe and Asia 50 million years ago.

In Pakistan in 1994, biologists discovered its descendant, the oldest known whale. The fossil was 49 million years old, had four legs, each with four-toed feet and a little hoof at the tip of each toe. Dubbed *Ambulocetus* (walking whale), it was sharp-toothed and about the size of a large sea lion. Analysis of the minerals in its teeth reveal it drank fresh water, so like a seal it was not yet completely a marine animal. Its nostrils were on the end of the snout, like a dog's.

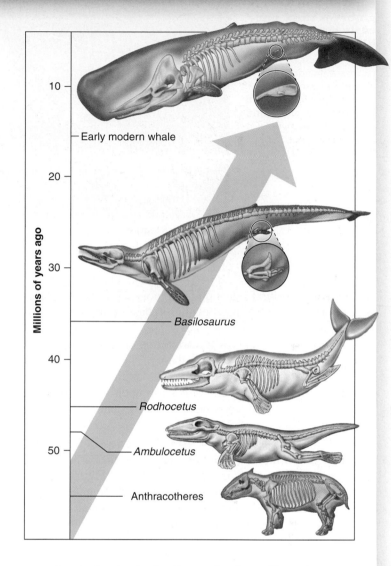

Appearing in the fossil record a few million years later is *Rodhocetus,* also seal-like but with smaller hind limbs and the teeth of an ocean water drinker. Its nostrils are shifted higher on the skull, halfway towards the top of the head.

Almost 10 million years later, about 37 million years ago, we see the first representatives of *Basilosaurus,* a giant 60-foot-long serpentlike whale with shrunken hind legs still complete down to jointed knees and toes.

The earliest modern whales appear in the fossil record 15 million years ago. The nostrils are now in the top of the head, a "blowhole" that allows it to break the surface, inhale, and resubmerge without having to stop or tilt the head up. The hind legs are gone, with vestigial tiny bones remaining that are unattached to the pelvis. Still, today's whales retain all the genes used to code for legs—occasionally a whale is born having sprouted a leg or two.

So it seems to have taken 35 million years to evolve a whale from the piglike ancestor of a hippopotamus—intermediate steps preserved in the fossil record for us to see. Darwin, who always believed that gaps in the vertebrate fossil record would eventually be filled in, would have been delighted.

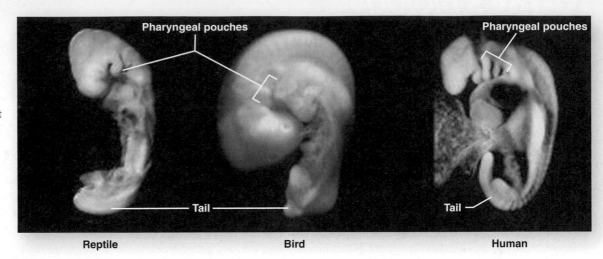

**Figure 14.15
Embryos show our early evolutionary history.**

These embryos, representing various vertebrate animals, show the primitive features that all vertebrates share early in their development, such as pharyngeal pouches and a tail.

Pharyngeal pouches

Pharyngeal pouches

Tail

Tail

Reptile

Bird

Human

is the result of natural selection is a theory advanced by Darwin, the statement that macroevolution has occurred is a factual observation.

Many other examples illustrate this clear confirmation of the key prediction of Darwin's theory. The evolution of today's large, single-hoof horse with complex molar teeth from a much smaller four-toed ancestor with much simpler molar teeth is a familiar and clearly documented instance.

The Anatomical Record

Much of the evolutionary history of vertebrates can be seen in the way in which their embryos develop. **Figure 14.15** shows three different embryos early in development, and as you can see, all vertebrate embryos have pharyngeal pouches (that develop into gill slits in fish); also every vertebrate embryo has a long bony tail, even if the tail is not present in the fully developed animal. These relict developmental forms strongly suggest that all vertebrates share a basic set of developmental instructions.

As vertebrates have evolved, the same bones are sometimes still there but put to different uses, their presence betraying their evolutionary past. For example, the forelimbs of vertebrates are all **homologous structures;** that is, although the structure and function of the bones have diverged, they are derived from the same body part present in a common ancestor. You can see in **figure 14.16** how the bones of the forelimb have been modified for different functions. The yellow- and purple-colored bones, which correspond in humans to the bones of the forearm and wrist and fingers, respectively, are modified to make up the wings in the bat, the full leg of the horse, and the paddle in the fin of the porpoise.

Not all similar features are homologous. Sometimes features found in different lineages come to resemble each other as a result of parallel evolutionary adaptations to similar environments. This form of evolutionary change is referred to as *convergent evolution,* and these similar-looking features are called **analogous structures.** For example, the wings of birds, pterosaurs, and bats are analogous structures, modified through natural selection to serve the same function and

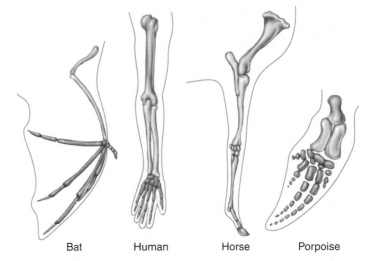

Bat   Human   Horse   Porpoise

Figure 14.16   Homology among vertebrate limbs.

Homologies among the forelimbs of four mammals show the ways in which the proportions of the bones have changed in relation to the particular way of life of each organism. Although considerable differences can be seen in form and function, the same basic bones are present in each forelimb.

therefore look the same (**figure 14.17**). Similarly, the marsupial mammals of Australia evolved in isolation from placental mammals, but similar selective pressures have generated very similar kinds of animals.

Sometimes structures are put to no use at all! In living whales, which evolved from hoofed mammals, the bones of the pelvis that formerly anchored the two hind limbs are all that remain of the rear legs, unattached to any other bones and serving no apparent purpose (the reduced pelvic bone can be seen in the figure in the "Today's Biology" reading on the previous page). Another example of what are called *vestigial organs* is the human appendix. The great apes, our closest relatives, have an appendix much larger than ours attached to the gut tube, which holds bacteria used in digesting the cellulose cell walls of the plants eaten by these primates. The human appendix is a vestigial version of this structure that now serves no function in digestion (although it may have acquired an alternate function in the lymphatic system).

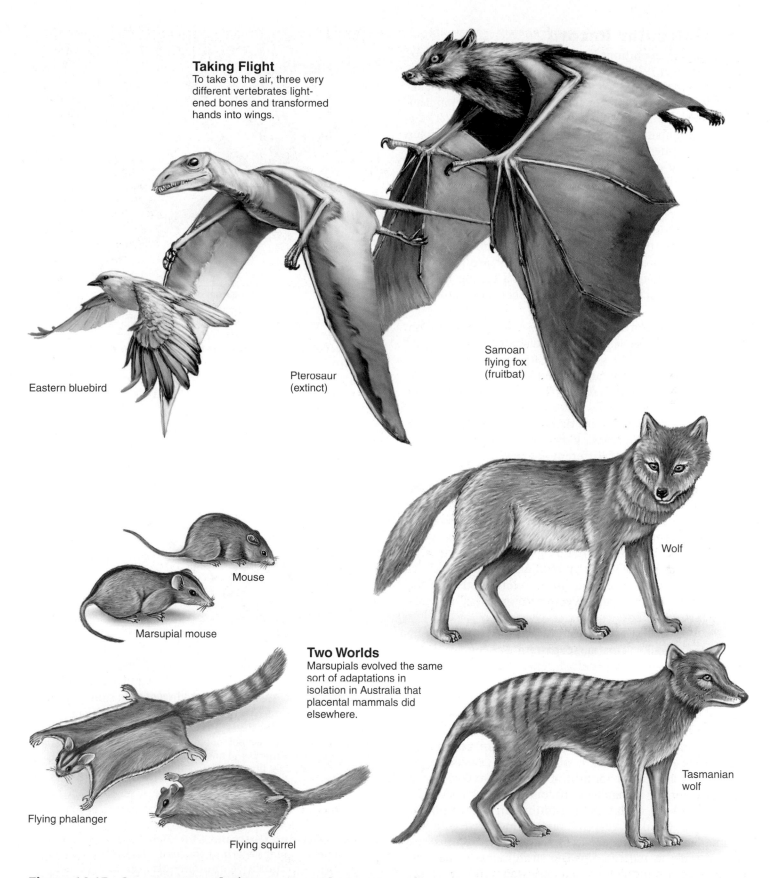

Taking Flight
To take to the air, three very different vertebrates lightened bones and transformed hands into wings.

Eastern bluebird

Pterosaur (extinct)

Samoan flying fox (fruitbat)

Mouse

Marsupial mouse

Wolf

Two Worlds
Marsupials evolved the same sort of adaptations in isolation in Australia that placental mammals did elsewhere.

Flying phalanger

Flying squirrel

Tasmanian wolf

Figure 14.17 Convergent evolution: many paths to one goal.
Over the course of evolution, form often follows function. Members of very different animal groups often adapt in similar fashions when challenged by similar opportunities. These are but a few of many examples of such convergent evolution. The flying vertebrates represent mammals (bat), reptiles (pterosaur), and birds (bluebird). The three pairs of terrestrial vertebrates each contrast a North American placental mammal with an Australian marsupial one.

Source: "Taking Flight" image from The New York Times, *December 15, 1988.* The New York Times. *Reprinted with permission.*

The Molecular Record

Traces of our evolutionary past are also evident at the molecular level. We possess the same set of color vision genes as our ancestors, only more complex, and we employ pattern formation genes during early development that all animals share. Indeed, if you think about it, the fact that organisms have evolved from a series of simpler ancestors implies that a record of evolutionary change is present in the cells of each of us, in our DNA. According to evolutionary theory, new alleles arise from older ones by mutation and come to predominance through favorable selection. A series of evolutionary changes thus implies a continual accumulation of genetic changes in the DNA. From this you can see that evolutionary theory makes a clear prediction: Organisms that are more distantly related should have accumulated a greater number of evolutionary differences than two species that are more closely related.

This prediction is now subject to direct test. Recent DNA research allows us to directly compare the genomes of different organisms. The result is clear: For a broad array of vertebrates, the more distantly related two organisms are, the greater their genomic difference. This research is described later in this chapter on pages 298 and 299.

This same pattern of divergence can be clearly seen at the protein level. Comparing the hemoglobin amino acid sequence of different species with the human sequence in **figure 14.18**, you can see that species more closely related to humans have fewer differences in the amino acid structure of their hemoglobin. Macaques, primates closely related to humans, have fewer differences from humans (only 8 different amino acids) than do more distantly related mammals like dogs (which have 32 different amino acids). Nonmammalian terrestrial vertebrates differ even more, and marine vertebrates are the most different of all. Again, the prediction of evolutionary theory is strongly confirmed.

Molecular Clocks This same pattern is seen when the DNA sequence of an individual gene is compared over a much broader array of organisms. One well-studied case is the mammalian *cytochrome c* gene (cytochrome *c* is a protein that plays a key role in oxidative metabolism). **Figure 14.19** compares the time when two species diverged on the *x* axis to the number of differences in their *cytochrome c* gene on the *y* axis. To practice using this data set, go back about 75 million years ago to find a common ancestor for humans and rodents—in that time there have been about 60 base substitutions in cytochrome *c*. This graph reveals a very important finding: Evolutionary changes appear to accumulate in cytochrome *c* at a constant rate, as indicated by the straightness of the blue line connecting the points. This constancy is sometimes referred to as a molecular clock. Most proteins for which data are available appear to accumulate changes over time in this fashion, although different proteins can evolve at very different rates.

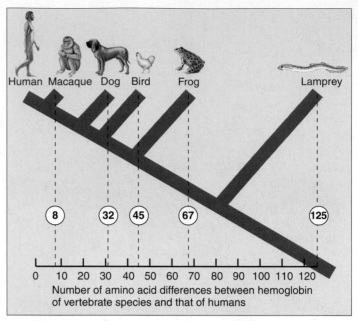

Figure 14.18   Molecules reflect evolutionary divergence.

The greater the evolutionary distance from humans (as revealed by the *blue* evolutionary tree based on the fossil record), the greater the number of amino acid differences in the vertebrate hemoglobin polypeptide.

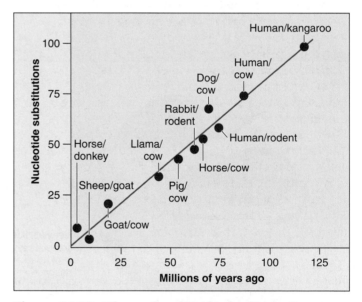

Figure 14.19   The molecular clock of cytochrome c.

When the time since each pair of organisms presumably diverged is plotted against the number of nucleotide differences in cytochrome *c*, the result is a straight line, suggesting that the *cytochrome c* gene is evolving at a constant rate.

Key Learning Outcome 14.6   The fossil record provides a clear record of successive evolutionary change. Comparative anatomy also offers evidence that evolution has occurred. Finally, the genetic record exhibits successive evolution, the DNA of organisms accumulating increasing numbers of changes over time.

14.7 Evolution's Critics

Of all the major ideas of biology, evolution is perhaps the best known to the general public because many people mistakenly believe that evolution represents a challenge to their religious beliefs. A person can have a spiritual faith in God and still be an excellent scientist—and evolutionist. Because Darwin's theory of evolution is the subject of often-bitter public controversy, we will examine the objections of evolution's critics in detail to see why there is such a disconnect between science and public opinion.

History of the Controversy

An Old Conflict Immediately after publication of *The Origin of Species,* English clergymen attacked Darwin's book as heretical; Gladstone, England's prime minister and a famous statesman, condemned it. The book was defended by Thomas Huxley and other scientists, who gradually won over the scientific establishment, and by the turn of the century evolution was generally accepted by the world's scientific community.

The Fundamentalist Movement By the 1920s the teaching of evolution had become frequent enough in American public schools to alarm conservative critics of evolution who saw Darwinism as a threat to their Christian beliefs. Between 1921 and 1929, fundamentalists introduced bills outlawing the teaching of evolution in 37 state legislatures. Four passed: Tennessee, Mississippi, Arkansas, and Texas.

Civil rights groups used the case of high school teacher John Scopes to challenge the Tennessee law within months of its being passed in 1925. The trial attracted national attention—you might have seen it portrayed in the film *Inherit the Wind.* Scopes, who had indeed violated the new law, lost.

After the 1920s, there were few other attempts to pass state laws preventing the teaching of evolution. Only one bill was introduced between 1930 and 1963. Why? Because Darwin's fundamentalist critics had succeeded quietly in winning their way. Textbooks published throughout the 1930s ignored evolution, new editions of texts removing the words *evolution* and *Darwin* from their indices. In the 1920–29 period, for example, the average number of words about the evolution of humans in 93 secondary school texts was 1,339; in the 1930–39 period it had dropped to 439. As late as 1950–59 it was 614. To quote biologist Ernst Mayr, "The word EVOLUTION simply disappeared from American schoolbooks."

These antievolution laws remained on the books for many years. Then in 1965 teacher Susan Epperson was convicted of teaching evolution under the 1928 Arkansas law. In 1968 the United States Supreme Court found the Arkansas antievolution law to be unconstitutional; the 1920s laws were soon repealed.

Russian advances in space in the early 1960s created a public outcry for better American science education. New biology textbooks reintroduced evolution, and gave it renewed emphasis. The average number of words per text devoted to the evolution of man, for example, rose from 614 in the period 1950–59 to 8,977 in the period 1960–69. By the 1970s, evolution again formed the core of most biology schoolbooks.

The Scientific Creationism Movement Again alarmed by the prevalence of evolution in public school biology classes, Darwin's critics took a new approach. It began with a proposal by the Institute for Creationism Research in 1964, which said, "Creationism is just as much a science as is evolution, and evolution is just as much a religion as is creation." This proposal has become known as *creationism science.* It was soon followed by the introduction of legislation in state legislatures mandating that "all theories of origins be accorded equal time." Creationism was represented as being as much a scientific theory as evolution, to which students had a right to be exposed.

In 1981 the state legislatures of Arkansas and Louisiana passed "equal-time" bills into law. The Louisiana equal-time law requiring "balanced treatment of creation-science and evolution-science in public schools" was struck down by the Supreme Court in 1987, which judged that creation science is not, in fact, science but rather a religious view that has no place in public science classrooms.

Local Action In the following decades, Darwin's critics began to substitute the school board for the legislature. Unlike most European countries, which set their school curricula through a central education ministry, U.S. education is highly decentralized, with elected education boards setting science standards at the state and local levels. These standards determine the content of statewide assessment tests, and have a major impact on what is taught in classrooms.

Critics of Darwin have run successfully for seats on local and state education boards across the United States, and from these positions have begun to alter standards to lessen the impact of evolution in the classroom. Great publicity followed the removal of evolution from the Kansas state standards in 1999 and again in 2005, but in many other states the same effect has been achieved more quietly. Only 22 states today mandate the teaching of natural selection, for example. Four states fail to mention evolution at all.

Intelligent Design In recent years, critics of Darwin have begun new attempts to combat the teaching of evolution in the classroom, arguing before state and local school boards that life is too complex for natural selection and so must reflect intelligent design. They go on to argue that this "theory of intelligent design" (ID) should be presented in the science classroom as an alternative to the theory of evolution.

Scientists object strongly to dubbing ID a scientific theory. The essence of science is seeking explanations in what can be observed, tested, replicated by others, and possibly falsified. Explanations that cannot be tested and potentially rejected simply aren't science. If someone invokes a nonnatural cause—a supernatural force—in their research, and you decide to test it, can you think of any way to do so such that it could be falsified? Supernatural causation is not science.

The source of sharp public controversy, intelligent design has been overwhelmingly rejected by the scientific community, which does not regard intelligent design to be science at all, but rather thinly disguised creationism, a religious view that has no place in the science classroom.

Arguments Advanced by Darwin's Critics

Critics of evolution have raised a variety of objections to Darwin's theory of evolution by natural selection:

1. **Evolution is not solidly demonstrated.** *"Evolution is just a theory,"* critics point out, as if theory meant lack of knowledge, some kind of guess. Scientists, however, use the word theory in a very different sense than the general public does (see section 1.8). Theories are the solid ground of science, supported with much experimental evidence and that of which we are most certain. Few of us doubt the theory of gravity because it is "just a theory."

2. **The intelligent design argument.** *"The organs of living creatures are too complex for a random process to have produced."* This classic "argument from design" was first proposed nearly 200 years ago by William Paley in his book *Natural Theology*—the existence of a clock is evidence of the existence of a clockmaker, Paley argues. Similarly, Darwin's critics argue that organs like the mammalian ear are too complex to be due to blind evolution. There must have been a designer. Biologists do not agree. The intermediates in the evolution of the mammalian ear are well documented in the fossil record. These intermediate forms were each favored by natural selection because they each had value—being able to amplify sound a little is better than not being able to amplify it at all. Complex structures like the mammalian ear evolved as a progression of slight improvements. Nor is the solution always optimal. The vertebrate eye, for example, is poorly designed.

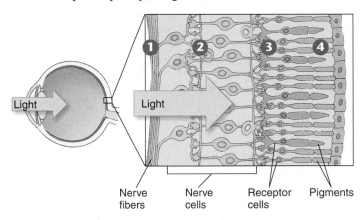

Nerve fibers   Nerve cells   Receptor cells   Pigments

The visual pigments in a vertebrate eye that are stimulated by light are embedded in the retinal tissue, facing backward to the direction of the light. The light has to pass through nerve fibers ❶, nerve cells ❷, and receptor cells ❸, before reaching the pigments ❹. No intelligent designer would design an eye backwards!

3. **There are no fossil intermediates.** *"No one ever saw a fin on the way to becoming a leg,"* critics claim, pointing to the many gaps in the fossil record in Darwin's day. Since then, however, most fossil intermediates in vertebrate evolution have indeed been found. A clear line of fossils now traces the transition between fish and amphibians, between reptiles and mammals, and between apes and humans. The animal that produced the fossil shown below is an extinct lobe-finned fish (genus *Tiktaalik*) that lived approximately 375 million years ago. Coined by its discoverer as a "fishopod," it clearly has some characteristics that are fishlike, similar to fish that lived about 380 million years ago, and others that are more like early tetrapods, which lived about 365 million years ago. *Tiktaalik* appears to be a transitional animal, between fish and amphibians.

4. **Evolution violates the second law of thermodynamics.** *"A jumble of soda cans doesn't by itself jump neatly into a stack—things become more disorganized due to random events, not more organized."* Biologists point out that this argument ignores what the second law really says: Disorder increases in a closed system, which the earth most certainly is not. Energy enters the biosphere from the sun, fueling life and all the processes that organize it.

5. **Natural selection does not imply evolution.** *"No scientist has come up with an experiment where fish evolve into frogs and leap away from predators."* Is microevolution (evolution within a species) the mechanism that has produced macroevolution (evolution among species)? Most biologists that have studied the problem think so. The differences between breeds produced by artificial selection—such as chihuahuas, dachshunds, and greyhounds—are more distinctive than differences between wild canine species. Laboratory selection experiments with insects easily create forms that cannot interbreed and thus would in nature be considered different species. Thus, production of radically different forms has indeed been observed, repeatedly.

6. **Life could not have evolved in water.** *"Because the peptide bond does not form spontaneously in water, amino acids could never have spontaneously linked together to form proteins; nor is there any chemical reason why biological proteins contain only the L-isomer and not the D-isomer."* Both of these contentions are valid, but do not require rejecting evolution. Rather, they suggest that the early evolution of life took place on a surface rather than in solution. Amino acids link up spontaneously on the surface of clays, for example, which can have a shape that selects the L-isomer.

The Irreducible Complexity Fallacy

The century-and-a-half-old "intelligent design" argument of William Paley has been recently articulated in a new molecular guise by Lehigh University biochemistry professor Michael Behe. In his 1996 book *Darwin's Black Box: The Biochemical Challenge to Evolution,* Behe argues that the intricate molecular machinery of our cells is so elaborate, our body processes so interconnected, that they cannot be explained by evolution from simpler stages in the way that Darwinists explain the evolution of the mammalian ear. The molecular machinery of the cell is "irreducibly complex." Behe defines an irreducibly complex system as "a single system composed of several well-matched, interacting parts that contribute to the basic function, wherein the removal of any one of the parts causes the system to effectively cease functioning." Each part plays a vital role. Remove just one, Behe emphasizes, and cell molecular machinery cannot function.

As an example of such an irreducibly complex system, Behe describes the series of more than a dozen blood clotting proteins that act in our body to cause blood to clot around a wound. Take out any step in the complex cascade of reactions that leads to coagulation of blood, says Behe, and your body's blood would leak out from a cut like water from a ruptured pipe. Remove a single enzyme from the complementary system that confines the clotting process to the immediate vicinity of the wound, and all your lifeblood would harden. Either condition would be fatal. The need for *all* the parts of such complex systems to work leads directly to Behe's criticism of Darwin's theory of evolution by natural selection. Behe writes that "irreducibly complex systems cannot evolve in a Darwinian fashion." If dozens of different proteins all must work correctly to clot blood, how could natural selection act to fashion any one of the individual proteins? No one protein does anything on its own, just as a portion of a watch doesn't tell time. Behe argues that, like Paley's watch, the blood clotting system must have been designed all at once, as a single functioning machine.

What's wrong with Behe's argument, as evolutionary scientists have been quick to point out, is that each part of a complex molecular machine does not evolve by itself, despite Behe's claim that it must. The several parts evolve together, in concert, precisely because evolution acts on the system, not its parts. That's the fundamental fallacy in Behe's argument. Natural selection can act on a complex system because at every stage of its evolution, the system functions. Parts that improve function are added, and, because of later changes, eventually become essential, in the same way that the second rung of a ladder becomes essential once you have added a third.

The mammalian blood clotting system, for example, has evolved in stages from much simpler systems. By comparing the amino acid sequences of the many proteins, biochemist Russell Doolittle has estimated how long it has been since each protein evolved. You can see what he has learned in **figure 14.20**. The core of the vertebrate clotting system, called the "common pathway" (highlighted in blue), evolved at the dawn of the vertebrates approximately 600 million

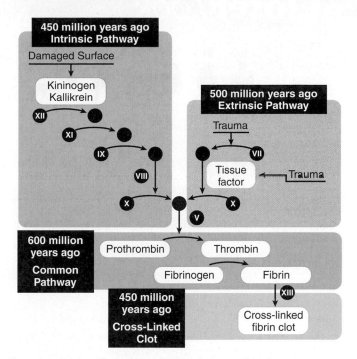

Figure 14.20 How blood clotting evolved.
The blood clotting system evolved in steps, with new proteins adding on to the preceding step.

years ago, and is found today in lampreys, the most primitive fish. As vertebrates evolved, proteins were added to the clotting system, improving its efficiency. The so-called *extrinsic pathway* (highlighted in pink), triggered by substances released from damaged tissues, was added 500 million years ago. Each step in the pathway amplifies what goes before, so adding the extrinsic pathway greatly increases the amplification and thus the sensitivity of the system. Fifty million years later, a third component was added, the so-called *intrinsic pathway* (highlighted in tan). It is triggered by contact with the jagged surfaces produced by injury. Again, amplification and sensitivity were increased to ultimately end up with blood clots formed by the cross linking of fibrin (highlighted in green). At each stage as the clotting system evolved to become more complex, its overall performance came to depend on the added elements. Mammalian clotting, which utilizes all three pathways, no longer functions if any one of them is disabled. Blood clotting has become "irreducibly complex"—as the result of Darwinian evolution. Behe's claim that complex cellular and molecular processes can't be explained by Darwinism is wrong. Indeed, examination of the human genome reveals that the cluster of blood clotting genes arose through duplication of genes, with increasing amounts of change. The evolution of the blood clotting system is an observation, not a surmise. Its irreducible complexity is a fallacy.

Key Learning Outcome 14.7 Darwin's theory of evolution, while accepted overwhelmingly by scientists, has its objectors. Their criticisms are without scientific merit.

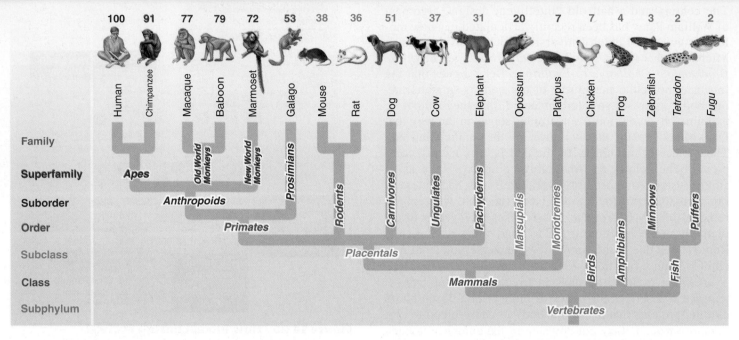

Putting Intelligent Design to the Test

In the spring of 2006 the South Carolina Board of Education rejected a state panel's proposal to change high school standards by calling on students to critically analyze evolution. The Board stated it felt the proposal was a ploy to promote the avoidance of teaching evolution. Similar proposals to add a requirement that students critically analyze evolution had been rejected earlier in the year by the Utah and Ohio Boards of Education, and are currently under consideration in several other states.

What are we to make of this? Surely no scientist can object to critically analyzing any theory. That is what science is all about, seeking explanations for what can be observed, tested, replicated, and possibly falsified. Indeed, biologists claim that Darwin's theory of evolution has been subjected to as much critical analysis as any theory in the history of science.

So why the objection to this change in high school standards? Because many scientists and teachers, apparently including the South Carolina Board of Education, feel the change is simply intended to promote the teaching of a nonscientific alternative to evolution in classrooms.

This distinction between an assertion that can be tested and one that cannot goes to the very nature of science. Actually, nothing makes this difference more clear cut than the critical analysis so sought after by South Carolina's critics of evolution. So let's do it. Let's put Darwin to the test.

As explained earlier in the chapter, if Darwin's assertion is correct, that organisms evolved from ancestral species, then we should be able to track evolutionary changes in our DNA. The variation that we see between species reflects adaptations to environmental challenges, adaptations that result from changes in DNA. Therefore, a series of evolutionary changes should be reflected in an accumulation of genetic changes in the DNA. This hypothesis, that evolutionary changes reflect accumulated changes in DNA, leads to the following prediction: Two species that are more distantly related (for example, humans and mice) should have accumulated a greater number of evolutionary differences than two species that are more closely related (say, humans and chimpanzees).

So have they? Let's compare vertebrate species to see. The "family tree" above shows how biologists believe 18 different vertebrate species are related. Apes and monkeys, because they are in the same order (primates), are considered more closely related to each other than either are to members of another order, such as mice and rats (rodents).

The wealth of genomes (a genome is all the DNA that an organism possesses) that have been sequenced since completion of the human genome project allows us to directly compare the DNA of these 18 vertebrates. To reduce the size of the task, investigators at the National Human Genome Research Institute working at the University of California, Santa Cruz, focused on 44 so-called ENCODE regions scattered around the vertebrate genomes. These regions, corresponding to 30 Mb (megabase, or million bases) or roughly 1% of the total human genome, were selected to be representative of the genome as a whole, containing protein-encoding genes as well as noncoding DNA.

For each vertebrate species, the investigators determined the similarity of its DNA to that of humans—that is, the percent of the nucleotides in that organism's 44 ENCODE regions that match those of the human genome.

You can see the result in each instance presented as a number above the picture of each organism on the vertebrate family tree. As Darwin's theory predicts, the closer the relatives, the less the genomic difference we see. The chimpanzee genome is more like the human genome (91%) than the monkey genomes are (72 to 79%). Furthermore, these five genomes, all in the primate order, are more like each other than any are to those of another order, such as rodents (mouse and rat).

In general, as you proceed through the taxonomic categories of the vertebrate family tree from very distant relatives on the right (some in the same class as humans) to very close ones on the left (in the same family), you can see clearly that genomic similarity increases as taxonomic distance decreases—just as Darwin's theory predicts. The prediction of evolutionary theory is solidly confirmed.

The analysis does not have to stop here. The evolutionary history of the vertebrates is quite well known from fossils, and because many of these fossils have been independently dated using tools such as radioisotope dating, it is possible to recast the analysis in terms of concrete intervals of time, and assess directly whether or not vertebrate genomes accumulate more differences over longer periods of time as Darwin's theory predicts.

For each of the 18 vertebrates being analyzed, the graph above plots genomic similarity—how alike the DNA sequence of the vertebrate's ENCODE regions are to those of the human genome—against divergence time (that is, how many millions of years have elapsed since that vertebrate and humans shared a common ancestor in the fossil record). Thus the last common ancestor shared by chickens and humans was an early reptile called a *dicynodont* that lived some 250 million years ago; since then the genomes of the two species have changed so much that only 7% of their ENCODE sequences are still the same.

The result seen in the graph is striking and very clear: Over their more than 300 million year history, vertebrates have accumulated more and more genetic change in their DNA. "Descent with modification" was Darwin's

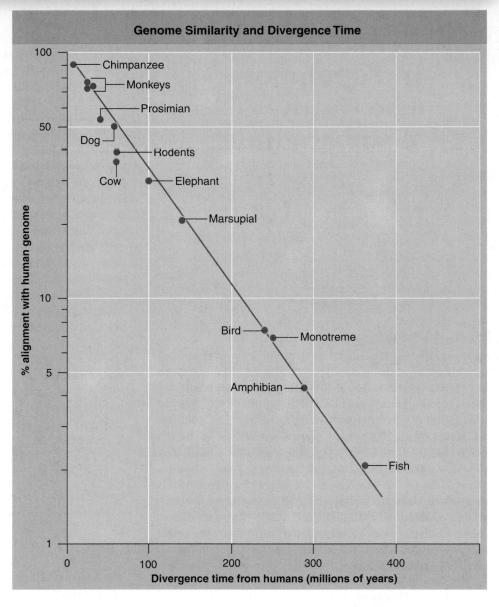

definition of evolution, and that is exactly what we see in the graph. The evolution of the vertebrate genome is not a theory, but an observation.

The wealth of data made available by the human genome project has allowed a powerful test of Darwin's prediction. The conclusion to which the test leads us is that evolution is an observed fact, clearly revealed in the DNA of vertebrates.

This is the sort of critical analysis that science requires, and that the theory of evolution has again passed. Anyone suggesting that a nonscientific alternative to evolution, such as Intelligent Design, offers an alternative scientific explanation to evolution is welcome to subject it to the same sort of critical analysis you have seen employed here. Can you think of a way to do so? It is precisely because the assertion of intelligent design cannot be critically analyzed—it does not make any testable prediction—that it is not science and has no place in science classrooms.

14.8 Genetic Change in Populations: The Hardy-Weinberg Rule

Population genetics is the study of the properties of genes in populations. Genetic variation within natural populations could not be explained by Darwin and his contemporaries. The way in which meiosis produces genetic segregation among the progeny of a hybrid had not yet been discovered. And, although Mendel performed his experiments during this same time period, his work was largely unknown. Selection, scientists then thought, should always favor an optimal form, and so tend to eliminate variation.

Hardy-Weinberg Equilibrium

Indeed, variation within populations puzzled many scientists; **alleles** (alternative forms of a gene) that were dominant were believed to drive recessive alleles out of populations, with selection favoring an optimal form. The solution to the puzzle of why genetic variation persists was developed in 1908 by G. H. Hardy and W. Weinberg. Hardy and Weinberg studied **allele frequencies** (the proportion of alleles of a particular type in a population) in a population's *gene pool,* which is the sum of all of the genes in a population, including all alleles in all individuals. Hardy and Weinberg pointed out that in a large population in which there is random mating, and in the absence of forces that change allele frequencies, the original genotype proportions remain constant from generation to generation. Dominant alleles do not, in fact, replace recessive ones. Because their proportions do not change, the genotypes are said to be in **Hardy-Weinberg equilibrium.**

The Hardy-Weinberg rule is viewed as a baseline to which the frequencies of alleles in a population can be compared. If the allele frequencies are not changing (they are in Hardy-Weinberg equilibrium), the population is not evolving. If, however, allele frequencies are sampled at one point in time and they differ greatly from what would be expected under Hardy-Weinberg equilibrium, then the population is undergoing evolutionary change.

Hardy and Weinberg came to their conclusion by analyzing the frequencies of alleles in successive generations. The **frequency** of something is defined as the proportion of individuals with a certain characteristic, compared to the entire population. Thus, in the population of 1,000 cats shown in **figure 14.21**, there are 840 black cats and 160 white cats. To determine the frequency of black cats, divide 840 by 1,000 (840/1,000), which is 0.84. The frequency of white cats is 160/1,000 = 0.16.

Knowing the frequency of the phenotype, one can calculate the frequency of the genotypes and alleles in the population. By convention, the frequency of the more common of two alleles (in this case B for the black allele) is designated by the letter p and that of the less common allele (b for the white allele) by the letter q. Because there are only two alleles, the sum of p and q must always equal 1 ($p + q = 1$).

In algebraic terms, the Hardy-Weinberg equilibrium is written as an equation. For a gene with two alternative alleles B (frequency p) and b (frequency q), the equation looks like this:

$$p^2 + 2pq + q^2 = 1$$

p^2	$2pq$	q^2
Individuals homozygous for allele B	Individuals heterozygous for alleles B and b	Individuals homozygous for allele b

You will notice that not only does the sum of the alleles add up to 1 but so does the sum of the frequencies of genotypes.

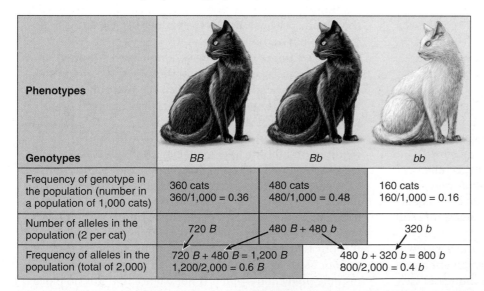

Phenotypes			
Genotypes	*BB*	*Bb*	*bb*
Frequency of genotype in the population (number in a population of 1,000 cats)	360 cats 360/1,000 = 0.36	480 cats 480/1,000 = 0.48	160 cats 160/1,000 = 0.16
Number of alleles in the population (2 per cat)	720 *B*	480 *B* + 480 *b*	320 *b*
Frequency of alleles in the population (total of 2,000)	720 *B* + 480 *B* = 1,200 *B* 1,200/2,000 = 0.6 *B*	480 *b* + 320 *b* = 800 *b* 800/2,000 = 0.4 *b*	

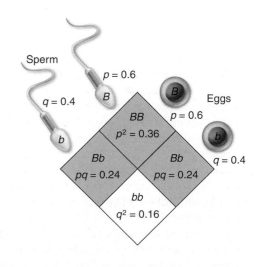

Figure 14.21 Calculating allele frequencies at Hardy-Weinberg equilibrium.

The example here is a population of 1,000 cats, in which 160 are white and 840 are black. White cats are *bb,* and black cats are *BB* or *Bb.*

Knowing the frequencies of the alleles in a population doesn't reveal whether the population is evolving or not. We need to look at future generations to determine this. Using the allele frequencies calculated for our population of cats, we can predict what the genotypic and phenotypic frequencies will be in future generations. The Punnett square shown on the right in figure 14.21 is constructed with allele frequencies of 0.6 for the *B* allele and 0.4 for the *b* allele, taken from the bottom row of the chart. It might help you to consider these frequencies as percentages, with a 0.6 representing 60% of the population and 0.4 representing 40% of the population. According to the Hardy-Weinberg rule, 60% of the sperm in the population will carry the *B* allele (indicated as $p = 0.6$ in the Punnett square) and 40% of the sperm will carry the *b* allele ($q = 0.4$). When these are crossed with eggs carrying the same allele frequencies (60% or $p = 0.6$ *B* allele and 40% or $q = 0.4$ *b* allele), the predicted genotypic frequencies can be simply calculated. The genotypic ratio for *BB,* in the upper quadrant, equals the frequency of *B* (0.6) multiplied by the frequency of *B* (0.6) or (0.6 x 0.6 = 0.36). So, if the population is not evolving, the genotypic ratio for *BB* would stay the same, and 0.36 or 36% of the cats in future generations would be homozygous dominant (*BB*) for coat color. Likewise, 0.48 or 48% of the cats would be heterozygous *Bb* (0.24 + 0.24 = 0.48), and 0.16 or 16% of the cats would be homozygous recessive *bb*.

Hardy-Weinberg Assumptions

The Hardy-Weinberg rule is based on certain assumptions. The equation on page 300 is true only if the following five assumptions are met:

1. The size of the population is very large or effectively infinite.
2. Individuals mate with one another at random.
3. There is no mutation.
4. There is no input of new copies of any allele from any extraneous source (such as migration from a nearby population) or losses of copies of alleles through emigration (individuals leaving the population).
5. All alleles are replaced equally from generation to generation (natural selection is not occurring).

Hardy-Weinberg: A Null Hypothesis

Many populations, and most human populations, are large and randomly mating with respect to most traits (a few traits affecting appearance undergo strong sexual selection in humans). Thus, many populations are similar to the ideal population envisioned by Hardy and Weinberg. For some genes, however, the observed proportion of heterozygotes does not match the value calculated from the allele frequencies. When this occurs, it indicates that something is acting on the population to alter one or more of the genotypic frequencies, whether it is selection, nonrandom mating, migration, or some other factor. Viewed in this light, Hardy-Weinberg can be viewed as a *null hypothesis*. A null hypothesis is a prediction that is made stating there will be no differences in the parameters being measured. If over several generations, the genotypic frequencies in the population do not match those predicted by the Hardy-Weinberg equation, the null hypothesis would be rejected and the assumption made that some force is acting on the population to change the frequencies of alleles. The factors that can affect the frequencies of alleles in a population are discussed in detail later in this chapter.

Case-Study: Cystic Fibrosis in Humans

How valid are the predictions made by the Hardy-Weinberg equation? For many genes, they prove to be very accurate. As an example, consider the recessive allele responsible for the serious human disease cystic fibrosis. This allele (q) is present in Caucasians in North America at a frequency of 0.022. What proportion of Caucasian North Americans, therefore, is expected to express this trait? The frequency of double-recessive individuals (q^2) is expected to be:

$$q^2 = 0.022 \times 0.022 = 0.00048$$

which equals 0.48 in every 1,000 individuals or about 1 in every 2,000 individuals, very close to real estimates.

What proportion is expected to be heterozygous carriers? If the frequency of the recessive allele q is 0.022, then the frequency of the dominant allele p must be $p = 1 - q$ or:

$$p = 1 - 0.022 = 0.978$$

The frequency of heterozygous individuals ($2pq$) is thus expected to be:

$$2 \times 0.978 \times 0.022 = 0.043$$

It is estimated that 12 million individuals in the United States are carriers of the cystic fibrosis allele. In a population of 292 million people, that is a frequency of 0.041, very close to projections using the Hardy-Weinberg equation. However, if the frequency of the cystic fibrosis allele in the United States were to change, this would suggest that the population is no longer following the assumptions of the Hardy-Weinberg rule. For example, if prospective parents who were carriers of the allele chose not to have children, the frequency of the allele would decrease in future generations. Mating would no longer be random, because those carrying the allele would not mate. Consider another scenario. If gene therapies were developed that were able to cure the symptoms of cystic fibrosis, patients would survive longer and would have more of an opportunity to reproduce. This would increase the frequency of the allele in future generations. An increase could also result from an influx of the allele into the population by migration, if the allele were more frequent among individuals migrating into the country.

> **Key Learning Outcome 14.8** In a large, randomly-mating population that fulfills the other Hardy-Weinberg assumptions, allele frequencies can be expected to be in Hardy-Weinberg equilibrium. If they are not, then the population is undergoing evolutionary change.

Agents of Evolution

Many factors can alter allele frequencies. But only five alter the proportions of homozygotes and heterozygotes enough to produce significant deviations from the proportions predicted by the Hardy-Weinberg rule:

1. Mutation
2. Nonrandom mating
3. Genetic drift
4. Migration
5. Selection

Mutation

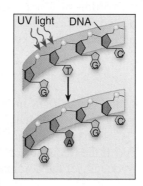

A **mutation** is a change in a nucleotide sequence in DNA. For example, a T nucleotide could undergo a mutation and be replaced with an A nucleotide. Mutation from one allele to another obviously can change the proportions of particular alleles in a population. But mutation rates are generally too low to significantly alter Hardy-Weinberg proportions of common alleles. Many genes mutate 1 to 10 times per 100,000 cell divisions. Some of these mutations are harmful, while others are neutral or, even rarer, beneficial. Also, the mutations must affect the DNA of the germ cells (egg and sperm), or the mutation will not be passed on to offspring. The mutation rate is so slow that few populations are around long enough to accumulate significant numbers of mutations. However, no matter how rare, mutation is the ultimate source of genetic variation in a population.

Nonrandom Mating

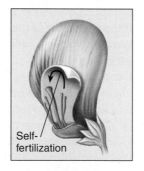

Individuals with certain genotypes sometimes mate with one another either more or less commonly than would be expected on a random basis, a phenomenon known as **nonrandom mating.** One type of nonrandom mating is **sexual selection,** choosing a mate often based on certain physical characteristics. Another type of nonrandom mating is inbreeding, or mating with relatives, such as in the self-fertilization of a flower. Inbreeding increases the proportions of individuals that are homozygous because no individuals mate with any genotype but their own. As a result, inbred populations contain more homozygous individuals than predicted by the Hardy-Weinberg rule. For this reason, populations of self-fertilizing plants consist primarily of homozygous individuals, whereas outcrossing plants, which interbreed with individuals different from themselves, have a higher proportion of heterozygous individuals. Nonrandom mating alters genotype frequencies but not allele frequencies. The allele frequencies remain the same—the alleles are just distributed differently among the offspring.

Genetic Drift

In small populations, the frequencies of particular alleles may be changed drastically by chance alone. In an extreme case, individual alleles of a given gene may all be represented in few individuals, and some of the alleles may be accidentally lost if those individuals fail to reproduce or die. This loss of individuals and their alleles is due to random events rather than the fitness of the individuals carrying those alleles. This is not to say that alleles are always lost with genetic drift, but allele frequencies appear to change randomly, as if the frequencies were drifting; thus, random changes in allele frequencies is known as **genetic drift.** A series of small populations that are isolated from one another may come to differ strongly as a result of genetic drift.

When one or a few individuals migrate and become the founders of a new, isolated population at some distance from their place of origin, the alleles that they carry are of special significance in the new population. Even if these alleles are rare in the source population, they will become a significant fraction of the new population's genetic endowment. This is called the **founder effect.** As a result of the founder effect, rare alleles and combinations often become more common in new, isolated populations. The founder effect is particularly important in the evolution of organisms that occur on oceanic islands, such as the Galápagos Islands that Darwin visited. Most of the kinds of organisms that occur in such areas were probably derived from one or a few initial founders. In a similar way, isolated human populations are often dominated by the genetic features that were characteristic of their founders, particularly if only a few individuals were involved initially (**figure 14.22**).

Even if organisms do not move from place to place, occasionally their populations may be drastically reduced in size. This may result from flooding, drought,

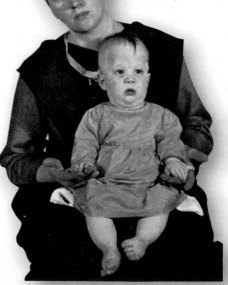

**Figure 14.22
The founder effect.**

This Amish woman is holding her child, who has Ellis-van Creveld syndrome. The characteristic symptoms are short limbs, dwarfed stature, and extra fingers. This disorder was introduced in the Amish community by one of its founders in the 18th century and persists to this day because of reproductive isolation.

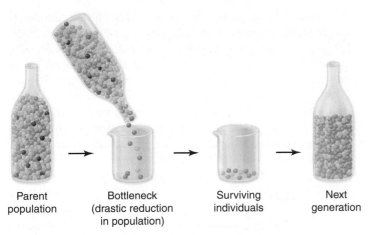

Figure 14.23 Genetic drift: a bottleneck effect.
The parent population contains roughly equal numbers of green and yellow individuals and a small number of red individuals. By chance, the few remaining individuals that contribute to the next generation are mostly green. The bottleneck occurs because so few individuals form the next generation, as might happen after an epidemic or a catastrophic storm.

Parent population

Bottleneck (drastic reduction in population)

Surviving individuals

Next generation

earthquakes, and other natural forces or from progressive changes in the environment. The surviving individuals constitute a random genetic sample of the original population. Such a restriction in genetic variability has been termed the **bottleneck effect** (figure 14.23). The very low levels of genetic variability seen in African cheetahs today is thought to reflect a near-extinction event in the past.

Migration

Migration is defined in genetic terms as the movement of individuals between populations. It can be a powerful force, upsetting the genetic stability of natural populations. Migration includes movement of individuals into a population, called *immigration,* or the movement of individuals out of a population, called *emigration.* If the characteristics of the newly arrived individuals differ from those already there, and if the newly arrived individuals adapt to survive in the new area and mate successfully, then the genetic composition of the receiving population may be altered.

Sometimes migration is not obvious. Subtle movements include the drifting of gametes of plants, or of the immature stages of marine organisms, from one place to another. For example, a bee can carry pollen from a flower in one population to a flower in another population. By doing this, the bee may be introducing new alleles into a population. However it occurs, migration can alter the genetic characteristics of populations and cause a population to be out of Hardy-Weinberg equilibrium. Thus, migration can cause evolutionary change. The magnitude of effects of migration is based on two factors: (1) the proportion

of migrants in the population, and (2) the difference in allele frequencies between the migrants and the original population. The actual evolutionary impact of migration is difficult to assess, and depends heavily on the selective forces prevailing at the different places where the populations occur.

Selection

As Darwin pointed out, some individuals leave behind more progeny than others, and the likelihood they will do so is affected by their inherited characteristics. The result of this process is called **selection** and was familiar even in Darwin's day to breeders of horses and farm animals. In so-called **artificial selection,** the breeder selects for the desired characteristics. For example, mating larger animals with each other produces offspring that are larger. In **natural selection,** Darwin suggested the environment plays this role, with conditions in nature determining which kinds of individuals in a population are the most fit (meaning individuals that are best suited to their environment; see section 14.3) and so affecting the proportions of genes among individuals of future populations. The environment imposes the conditions that determine the results of selection and, thus, the direction of evolution (**figure 14.24**).

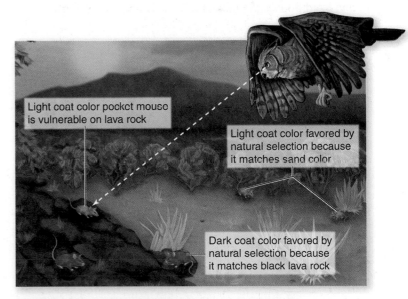

Light coat color pocket mouse is vulnerable on lava rock

Light coat color favored by natural selection because it matches sand color

Dark coat color favored by natural selection because it matches black lava rock

Figure 14.24 Selection for coat color in mice.
In the American southwest, ancient lava flows have produced black rock formations that contrast starkly with the surrounding light-colored desert sand. Populations of many species of animals occurring on these rocks are dark in color, whereas sand-dwelling populations are much lighter. For example, in pocket mice, selection favors coat color that matches their surroundings. The close match between coat color and background color camouflages the mice and provides protection from avian predators. These mice are very visible when placed in the opposite habitats.

Forms of Selection

Selection operates in natural populations of a species as skill does in a football game. In any individual game, it can be difficult to predict the winner because chance can play an important role in the outcome. But over a long season, the teams with the most skillful players usually win the most games. In nature, those individuals best suited to their environments tend to win the evolutionary game by leaving the most offspring, although chance can play a major role in the life of any one individual. While you cannot predict the fate of any one individual, or any one coin toss, it is possible to predict which kind of individual will tend to become more common in populations of a species, as it is possible to predict the proportion of heads after many coin tosses.

In nature, many traits, perhaps most, are affected by more than one gene. The interactions between genes are typically complex, as you saw in chapter 10. For example, alleles of many different genes play a role in determining human height (see **figure 10.12**). In such cases, selection operates on all the genes, influencing most strongly those that make the greatest contribution to the phenotype. How selection changes the population depends on which genotypes are favored. Three types of natural selection have been identified: stabilizing selection, disruptive selection, and directional selection.

Stabilizing Selection

When selection acts to eliminate both extremes from an array of phenotypes—for example, eliminating larger and smaller body sizes—the result is an increase in the frequency of the already common intermediate phenotype (such as a midsized body). This is called **stabilizing selection:**

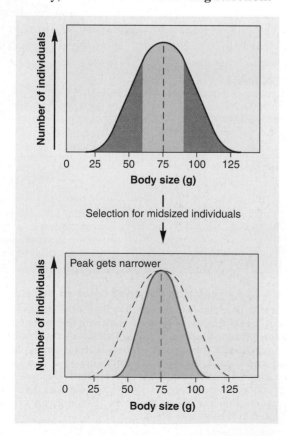

In a classic study carried out after an "uncommonly severe storm of snow, rain, and sleet" on February 1, 1898, 136 starving English sparrows were collected and brought to the laboratory of H. C. Bumpus at Brown University in Providence, Rhode Island. Of these, 64 died and 72 survived. Bumpus took standard measurements on all the birds. He found that among males, the surviving birds tended to be bigger, as one might expect from the action of directional selection (discussed later). However, among females, the birds that survived were those that were more average in size. Among the female birds that perished were many more individuals that had extreme measurements, either very large or very small.

In Bumpus' quaint phrasing, "The process of selective elimination is most severe with extremely varying individuals no matter in what directions the variation may occur. It is quite as dangerous to be conspicuously above a certain standard of organic excellence as it is to be conspicuously below the standard. It is the *type* that nature favors."

In the Bumpus study, selection had acted most strongly against these "extreme-sized" female birds. Stabilizing selection does not change which phenotype is the most common of the population—the average-sized birds were already the most common phenotype—but rather makes it even more common by eliminating extremes. In effect, selection is operating to prevent change away from the middle range of values.

Many examples similar to Bumpus's female sparrows are known. For example, in humans, infants with intermediate weight at birth have the highest survival rate:

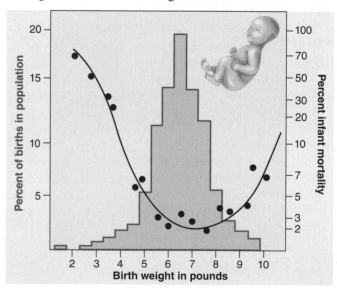

More specifically, the death rate among human babies is lowest at an intermediate birth weight between 7 and 8 pounds indicated by the red line in the graph above, which consists of data compiled from U.S. birth records over many years. The intermediate weights are also the most common in the population, indicated by the blue area. Larger and smaller babies both occur less frequently and have a greater tendency to die at or near birth. In a similar way, chickens eggs of intermediate weight have the highest hatching success.

Disruptive Selection

In some situations, selection acts to eliminate the intermediate type, resulting in the two more extreme phenotypes becoming more common in the population. This type of selection is called **disruptive selection:**

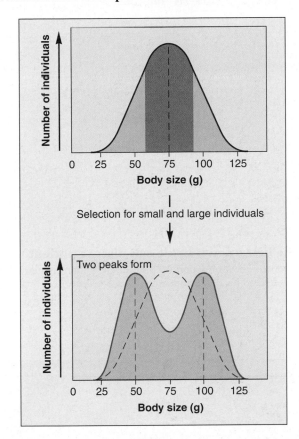

A clear example is the different beak sizes of the African black-bellied seedcracker finch *Pyrenestes ostrinus*. Populations of these birds contain individuals with large and small beaks, but very few individuals with intermediate-sized beaks. As their name implies, these birds feed on seeds, and the available seeds fall into two size categories: large and small. Only large-beaked birds, like the one on the left in the figure below, can open the tough shells of large seeds, whereas birds with the smallest beaks, like the one on the right, are more adept at handling small seeds. Birds with intermediate-sized beaks are at a disadvantage with both seed types: unable to open large seeds and too clumsy to efficiently process small seeds. Consequently, selection acts to eliminate the intermediate phenotypes, in effect partitioning the population into two phenotypically distinct groups.

Directional Selection

In other situations, selection acts to eliminate one extreme from an array of phenotypes, resulting in the other extreme phenotype becoming more common in the population. This form of selection is called **directional selection:**

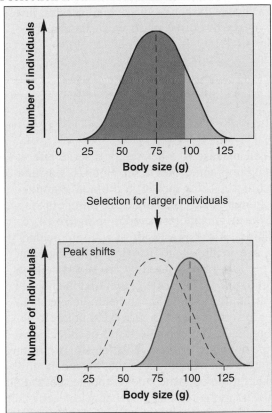

For example, in the experiment below, flies (*Drosophila*) that moved toward light were eliminated from the population, and only flies that moved away from light were used as parents for the next generation. After 20 generations of selected matings, flies that flew toward light were far less frequent in the population.

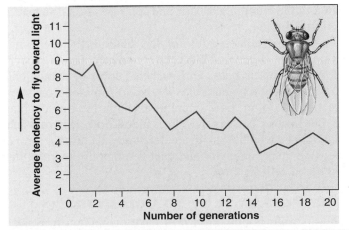

Key Learning Outcome 14.9 Five evolutionary forces have the potential to significantly alter allele and genotype frequencies in populations: mutation, nonrandom mating, genetic drift, migration, and selection. Selection can favor intermediate values, or one or both extremes.

In the time since Darwin suggested the pivotal role of natural selection in evolution, many examples have been found in which natural selection is clearly acting to change the genetic makeup of species, just as Darwin predicted. Here we will examine three examples.

14.10 Sickle-Cell Anemia

Sickle-cell disease is a hereditary disease affecting hemoglobin molecules in the blood. It was first detected in 1904 in Chicago in a blood examination of an individual complaining of tiredness. You can see the original doctor's report in **figure 14.25**. The disorder arises as a result of a single nucleotide change in the gene encoding β-hemoglobin, one of the key proteins used by red blood cells to transport oxygen. The sickle-cell mutation changes the sixth amino acid in the β-hemoglobin chain (position B6) from glutamic acid (very polar) to valine (nonpolar). The unhappy result of this change is that the nonpolar *valine* at position B6, protruding from a corner of the hemoglobin molecule, fits nicely into a nonpolar pocket on the opposite side of another hemoglobin molecule; the nonpolar regions associate with each other. As the two-molecule unit that forms still has both a B6 valine and an opposite nonpolar pocket, other hemoglobins clump on, and long chains form as in **figure 14.26a**. The result is the deformed "sickle-shaped" red blood cell you see in **figure 14.26b**. In normal everyday hemoglobin, by contrast, the polar amino acid *glutamic acid* occurs at position B6. This polar amino acid is not attracted to the nonpolar pocket, so no hemoglobin clumping occurs, and cells are normal shaped as in **figure 14.26c**.

Persons homozygous for the sickle-cell genetic mutation in the β-hemoglobin gene frequently have a reduced life span. This is because the sickled form of hemoglobin does not carry oxygen atoms well, and red blood cells that are sickled do not flow smoothly through the tiny capillaries but instead jam up and block blood flow. Heterozygous individuals, who have both a defective and a normal form of the gene, make enough functional hemoglobin to keep their red blood cells healthy.

The Puzzle: Why So Common?

The disorder is now known to have originated in Central Africa, where the frequency of the sickle-cell allele is about 0.12. One in 100 people is homozygous for the defective allele and develops the fatal disorder. Sickle-cell disease affects roughly two African Americans out of every thousand but is almost unknown among other racial groups.

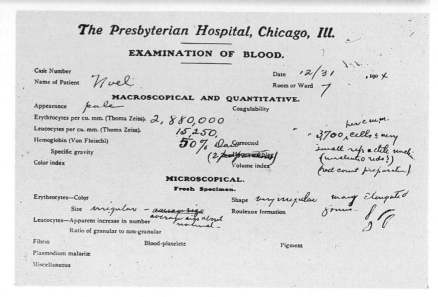

Figure 14.25 The first known sickle-cell disease patient.
Dr. Ernest Irons's blood examination report on his patient Walter Clement Noel, December 31, 1904, described his oddly shaped red blood cells.

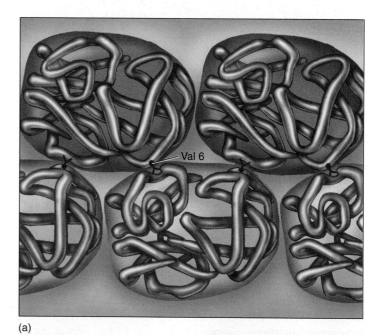

(a)

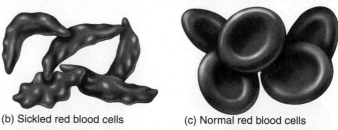

(b) Sickled red blood cells

(c) Normal red blood cells

Figure 14.26 Why the sickle-cell mutation causes hemoglobin to clump.

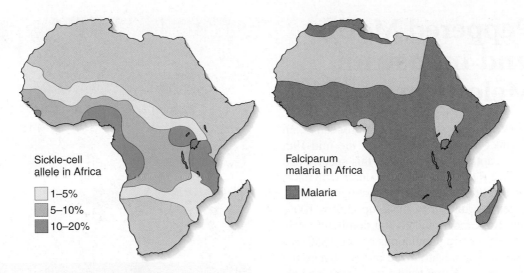

Figure 14.27 How stabilizing selection maintains sickle-cell disease.
The diagrams show the frequency of the sickle-cell allele (*left*) and the distribution of falciparum malaria (*right*). Falciparum malaria is one of the most devastating forms of the often fatal disease. As you can see, its distribution in Africa is closely correlated with that of the allele of the sickle-cell characteristic.

Sickle-cell
allele in Africa

- 1–5%
- 5–10%
- 10–20%

Falciparum
malaria in Africa

- Malaria

If Darwin is right, and natural selection drives evolution, then why has natural selection not acted against the defective allele in Africa and eliminated it from the human population there? Why is this potentially fatal allele instead very common there?

The Answer: Stabilizing Selection

The defective allele has not been eliminated from Central Africa because people who are heterozygous for the sickle-cell allele are much less susceptible to malaria, one of the leading causes of death in Central Africa. Examine the maps in **figure 14.27**, and you will see the relationship between sickle-cell disease and malaria clearly. The map on the left shows the frequency of the sickle-cell allele, the darker green areas indicating a 10% to 20% frequency of the allele. The map on the right indicates the distribution of malaria in dark orange. Clearly, the areas that are colored in darker green on the left map overlap many of the dark orange areas in the map on the right. Even though the population pays a high price—the many individuals in each generation who are homozygous for the sickle-cell allele die—the deaths are far fewer than would occur due to malaria if the heterozygous individuals were not malaria resistant. One in 5 individuals (20%) are heterozygous and survive malaria, while only 1 in 100 (1%) are homozygous and die of sickle-cell disease. Similar inheritance patterns of the sickle-cell allele are found in other countries frequently exposed to malaria, such as areas around the Mediterranean, India, and Indonesia. Natural selection has favored the sickle-cell allele in Central Africa and other areas hit by malaria because the payoff in survival of heterozygotes more than makes up for the price in death of homozygotes. This phenomenon is an example of **heterozygote advantage.**

Stabilizing selection (also called *balancing selection*) is thus acting on the sickle-cell allele: (1) Selection tends to eliminate the sickle-cell allele because of its lethal effects on homozygous individuals, and (2) selection tends to favor the sickle-cell allele because it protects heterozygotes from malaria. Like a manager balancing a store's inventory, natural selection increases the frequency of an allele in a species as long as there is something to be gained by it, until the cost balances the benefit.

Stabilizing selection occurs because malarial resistance counterbalances lethal sickle-cell disease. Malaria is a tropical disease that has essentially been eradicated in the United States since the early 1950s, and stabilizing selection has not favored the sickle-cell allele here. Africans brought to America several centuries ago have not gained any evolutionary advantage in all that time from being heterozygous for the sickle-cell allele. There is no benefit to being resistant to malaria if there is no danger of getting malaria anyway. As a result, the selection against the sickle-cell allele in America is not counterbalanced by any advantage, and the allele has become far less common among African Americans than among native Africans in Central Africa.

Stabilizing selection is thought to have influenced many other human genes in a similar fashion. The recessive *cf* allele causing cystic fibrosis is unusually common in northwestern Europeans. People that are heterozygous for the *cf* allele are protected from the dehydration caused by cholera, and the *cf* allele may provide protection against typhoid fever too. Apparently, the bacterium causing typhoid fever uses the healthy version of the CFTR protein (see page 80) to enter the cells it infects, but it cannot use the cystic fibrosis version of the protein. As with sickle-cell disease, heterozygotes are protected.

Key Learning Outcome 14.10 The prevalence of sickle-cell disease in African populations is thought to reflect the action of natural selection. Natural selection favors individuals carrying one copy of the sickle-cell allele, because they are resistant to malaria, common in Africa.

14.11 | Peppered Moths and Industrial Melanism

The peppered moth, *Biston betularia,* is a European moth that rests on tree trunks during the day. Until the mid-19th century, almost every captured individual of this species had light-colored wings. From that time on, individuals with dark-colored wings increased in frequency in the moth populations near industrialized centers, until they made up almost 100% of these populations. Dark individuals had a dominant allele that was present but very rare in populations before 1850. Biologists soon noticed that in industrialized regions where the dark moths were common, the tree trunks were darkened almost black by the soot of pollution. Dark moths were much less conspicuous resting on them than light moths were. In addition, air pollution that was spreading in the industrialized regions had killed many of the light-colored lichens on tree trunks, making the trunks darker.

Selection for Melanism

Can Darwin's theory explain the increase in the frequency of the dark allele? Why did dark moths gain a survival advantage around 1850? An amateur moth collector named J. W. Tutt proposed in 1896 what became the most commonly accepted hypothesis explaining the decline of the light-colored moths. He suggested that light forms were more visible to predators on sooty trees that have lost their lichens. Consequently, birds ate the peppered moths resting on the trunks of trees during the day. The dark forms, in contrast, were at an advantage because they were camouflaged (**figure 14.28**). Although Tutt initially had no evidence, British ecologist Bernard Kettlewell tested the hypothesis in the 1950s by rearing populations of peppered moths with equal numbers of dark and light individuals. Kettlewell then released these populations into two sets of woods: one, near heavily polluted Birmingham, the other, in unpolluted Dorset. Kettlewell set up traps in the woods to see how many of both kinds of moths survived. To evaluate his results, he had marked the released moths with a dot of paint on the underside of their wings, where birds could not see it.

In the polluted area near Birmingham, Kettlewell trapped 19% of the light moths, but 40% of the dark ones. This indicated that dark moths had a far better chance of surviving in these polluted woods where the tree trunks were dark. In the relatively unpolluted Dorset woods, Kettlewell recovered 12.5% of the light moths but only 6% of the dark ones. This indicated that where the tree trunks were still light-colored, light moths had a much better chance of survival. Kettlewell later solidified his argument by placing dead moths on trees and filming birds looking for food. Sometimes the birds actually passed right over a moth that was the same color as its background.

Industrial Melanism

Industrial melanism is a term used to describe the evolutionary process in which darker individuals come to predominate

Figure 14.28   Tutt's hypothesis explaining industrial melanism.

Color variants of the peppered moth (*Biston betularia*). Tutt proposed that the dark moth is more visible to predators on unpolluted trees (*top*), while the light moth is more visible to predators on bark blackened by industrial pollution (*bottom*).

over lighter individuals since the industrial revolution as a result of natural selection. The process is widely believed to have taken place because the dark organisms are better concealed from their predators in habitats that have been darkened by soot and other forms of industrial pollution, as suggested by Kettlewell.

Dozens of other species of moths have changed in the same way as the peppered moth in industrialized areas throughout Eurasia and North America, with dark forms becoming more common from the mid-19th century onward as industrialization spread.

Selection Against Melanism

As of the second half of the 20th century, with the widespread implementation of pollution controls, these trends are reversing, not only for the peppered moth in many areas in England but also for many other species of moths throughout the northern continents. These examples provide some of the best-documented instances of changes in allelic frequencies of natural populations as a result of natural selection due to specific factors in the environment.

In England, the air pollution promoting industrial melanism began to reverse following enactment of Clean Air legislation in 1956. Beginning in 1959, the *Biston* population at Caldy Common outside Liverpool has been sampled each year. The frequency of the melanic (dark) form dropped from a high of 94% in 1960 to a low of 19% in 1995 (**figure 14.29**). Similar reversals have been documented at numerous other locations throughout England. The drop correlates well with a drop in air pollution, particularly with tree-darkening sulfur dioxide and suspended particulates.

Interestingly, the same reversal of industrial melanism appears to have occurred in America during the same time that it was happening in England. Industrial melanism in the American subspecies of the peppered moth was not as widespread as in England, but it has been well documented at a rural field station near Detroit. Of 576 peppered moths collected there from 1959 to 1961, 515 were melanic, a frequency of 89%. The American Clean Air Act, passed in 1963, led to significant reductions in air pollution. Resampled in 1994, the Detroit field station peppered moth population had only 15% melanic moths! The moths in Liverpool and Detroit, both part of the same natural experiment, exhibit strong evidence of natural selection.

Reconsidering the Target of Natural Selection

Tutt's hypothesis, widely accepted in the light of Kettlewell's studies, is currently being reevaluated. The problem is that the recent selection against melanism does not appear to correlate with changes in tree lichens. At Caldy Common, the light form of the peppered moth began its increase in frequency long before lichens began to reappear on the trees. At the Detroit field station, the lichens never changed significantly as the dark moths first became dominant and then declined over the last 30 years. In fact, investigators have not been able to find peppered moths on Detroit trees at all, whether covered with lichens or not. Wherever the moths rest during the day, it does not appear to be on tree bark. Some evidence suggests they rest on leaves on the treetops, but no one is sure.

The action of selection may depend on other differences between light and dark forms of the peppered moth as well as their wing coloration. Researchers report, for example, a clear difference in their ability to survive as caterpillars under a variety of conditions. Perhaps natural selection is also targeting the caterpillars rather than the adults. While we can't yet say exactly what the targets of selection are, researchers are actively investigating this, one of the best-documented instances of natural selection in action.

Natural Selection for Melanism in Mice

Melanism is not restricted to insects. Cats and many other mammals have melanic forms that are subject to natural selection in much the same way as moths. The coat color of desert pocket mice that live on differently colored rock habitats provides a clear-cut example of natural selection acting on melanism. In Arizona and New Mexico, these small, wild pocket mice live in isolated black volcanic lava beds and the pale

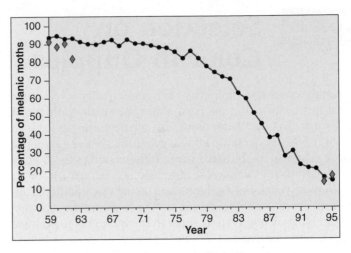

Figure 14.29 Selection against melanism.
The circles indicate the frequency of melanic *Biston betularia* moths at Caldy Common in England, sampled continuously from 1959 to 1995. Red diamonds indicate frequencies of melanic *B. betularia* in Michigan from 1959 to 1962 and from 1994 to 1995.

soils between them. Melanin synthesis during hair development of pocket mice is regulated by the receptor gene *MC1R*. Mutations that disable *MC1R* lead to melanism. Such mutations are dominant alleles, so whenever they are present in a population, dark pocket mice are seen. When wild populations of pocket mice were surveyed by biologists from the University of Arizona, there was a striking correlation between coat color and the color of the rock on which the population of pocket mice lived.

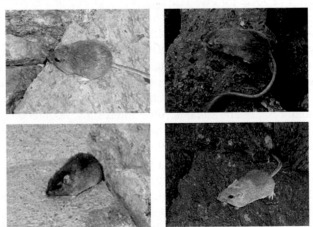

As you can see in the two upper photographs, the close match between coat color and background color gives the mice cryptic protection from avian predators, particularly owls. These mice are very visible when placed in the opposite habitats (lower photos).

> **Key Learning Outcome 14.11** Natural selection has favored the dark form of the peppered moth in areas subject to severe air pollution, perhaps because on darkened trees they are less easily seen by moth-eating birds. Selection has in turn favored the light form as pollution has abated.

Selection on Color in Guppies

To study evolution, biologists have traditionally investigated what has happened in the past, sometimes many millions of years ago. To learn about dinosaurs, a paleontologist looks at dinosaur fossils. To study human evolution, an anthropologist looks at human fossils and, increasingly, examines the "family tree" of mutations that have accumulated in human DNA over millions of years. For the biologists taking this traditional approach, evolutionary biology is similar to astronomy and history, relying on observation rather than experiment to examine ideas about past events.

Nonetheless, evolutionary biology is not entirely an observational science. Darwin was right about many things, but one area in which he was mistaken concerns the pace at which evolution occurs. Darwin thought that evolution occurred at a very slow, almost imperceptible, pace. However, in recent years many case studies have demonstrated that, in some circumstances, evolutionary change can occur rapidly. Consequently, it is possible to establish experimental studies to test evolutionary hypotheses. Although laboratory studies on fruit flies and other organisms have been common for more than 50 years, it has only been in recent years that scientists have started conducting experimental studies of evolution in nature. One excellent example of how observations of the natural world can be combined with rigorous experiments in the lab and in the field concerns research on the guppy, *Poecilia reticulata.*

Guppies Live in Different Environments

The guppy is a popular aquarium fish because of its bright coloration and prolific reproduction. In nature, guppies are found in small streams in northeastern South America and the nearby island of Trinidad. In Trinidad, guppies are found in many mountain streams. One interesting feature of several streams is that they have waterfalls. Amazingly, guppies and some other fish are capable of colonizing portions of the stream above the waterfall. The killifish, *Rivulus hartii,* is a particularly good colonizer; apparently on rainy nights, it will wriggle out of the stream and move through the damp leaf litter. Guppies are not so proficient, but they are good at swimming upstream. During flood seasons, rivers sometimes overflow their banks, creating secondary channels that move through the forest. During these occasions, guppies may be able to move upstream and invade pools above waterfalls. By contrast, not all species are capable of such dispersal and thus are only found in these streams below the first waterfall. One species whose distribution is restricted by waterfalls is the pike cichlid, *Crenicichla alta,* a voracious predator that feeds on other fish, including guppies.

Because of these barriers to dispersal, guppies can be found in two very different environments. The guppies you see living in pools just below the waterfalls in **figure 14.30** are faced with predation by the pike cichlid. This substantial risk keeps rates of survival relatively low. By contrast, in similar pools just above the waterfall, the only predator present is the killifish, which only rarely preys on guppies. Guppy populations above and below waterfalls exhibit many differences. In the high-predation pools, male guppies exhibit the drab coloration you see in the guppies below the waterfall in **figure 14.30**. Moreover, they tend to reproduce at a younger age and attain relatively smaller adult sizes. By contrast, male fish above the waterfall in the figure display gaudy colors that they use to court females. Adults mature later and grow to larger sizes.

These differences suggest the function of natural selection. In the low-predation environment, males display gaudy colors and spots that help in mating. Moreover, larger males are most successful at holding territories and mating with females, and larger females lay more eggs. Thus, in the absence of predators, larger and more colorful fish may have produced more offspring, leading to the evolution of those traits. In pools below the waterfall, however, natural selection would favor different traits. Colorful males are likely to attract the attention of the pike cichlid, and high predation

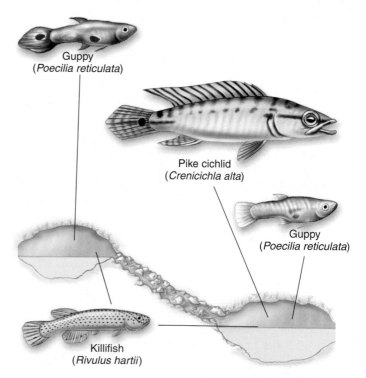

Figure 14.30 The evolution of protective coloration in guppies.

In pools below waterfalls where predation is high, male guppies (*Poecilia reticulata*) are drab colored. In the absence of the highly predatory pike cichlid (*Crenicichla alta*), male guppies in pools above waterfalls are much more colorful and attractive to females. The killifish (*Rivulus hartii*) is also a predator but only rarely eats guppies. The evolution of these differences in guppies can be experimentally tested.

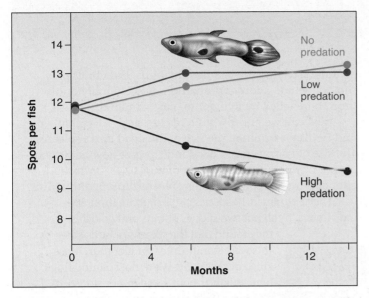

Figure 14.31 Evolutionary change in spot number.
Guppies raised in low-predation or no predation environments in laboratory greenhouses had a greater number of spots, whereas selection in more dangerous environments, like the pools with the highly predatory pike cichlid, led to less conspicuous fish. The same results are seen in field experiments conducted in pools above and below waterfalls (photo).

rates mean that most fish live short lives; thus, individuals that are more drab and shunt energy into early reproduction, rather than into growth to a larger size, are likely to be favored by natural selection.

The Experiments

Although the differences between guppies living above and below the waterfalls suggest that they represent evolutionary responses to differences in the strength of predation, alternative explanations are possible. Perhaps, for example, only very large fish are capable of swimming upstream past the waterfall to colonize pools. If this were the case, then a founder effect would occur in which the new population was established solely by individuals with genes for large size.

Laboratory Experiment The only way to rule out such alternative possibilities is to conduct a controlled experiment. John Endler, now of the University of California, Santa Barbara, conducted the first experiments in large pools in laboratory greenhouses. At the start of the experiment, a group of 2,000 guppies was divided equally among 10 large pools. Six months later, pike cichlids were added to four of the pools and killifish to another four, with the remaining two pools left to serve as "no predator" controls. Fourteen months later (which corresponds to 10 guppy generations), the scientists compared the populations. You can see their results in **figure 14.31**. The guppies in the killifish pool (the blue line) and control pools (the green line) were notably large, brightly colored fish with about 13 colorful spots per individual. In contrast, the guppies in the pike cichlid pools (the red line) were smaller and drab in coloration, with a reduced number of spots (about 9 per fish). These results clearly suggest that predation can lead to rapid

evolutionary change, but do these laboratory experiments reflect what occurs in nature?

Field Experiment To find out, Endler and colleagues—including David Reznick, now at the University of California, Riverside—located two streams that had guppies in pools below a waterfall, but not above it (see photograph in **figure 14.31**). As in other Trinidadian streams, the pike cichlid was present in the lower pools, but only the killifish was found above the waterfalls. The scientists then transplanted guppies to the upper pools and returned at several-year intervals to monitor the populations. Despite originating from populations in which predation levels were high, the transplanted populations rapidly evolved the traits characteristic of low-predation guppies: They matured late, attained greater size, and had brighter colors. Control populations in the lower pools, by contrast, continued to be drab and matured early and at smaller sizes. Laboratory studies confirmed that the differences between the populations were the result of genetic differences. These results demonstrate that substantial evolutionary change can occur in less than 12 years. More generally, these studies indicate how scientists can formulate hypotheses about how evolution occurs and then test these hypotheses in natural conditions. The results give strong support to the theory of evolution by natural selection.

> **Key Learning Outcome 14.12** Experiments can be conducted in nature to test hypotheses about how evolution occurs. Such studies reveal that natural selection can lead to rapid evolutionary change.

Are Bird-Killing Cats Nature's Way of Making Better Birds?

Death is not pretty, early in the morning on the doorstep. A small dead bird was left at our front door one morning, lying by the newspaper as if it might at any moment fly away. I knew it would not. Like other birds before it, it was a gift to our household by Feisty, a cat who lives with us. Feisty is a killer of birds, and every so often he leaves one for us, like rent.

We have four cats, and the other three, true housecats, would not know what to do with a bird. Feisty is different, a long-haired gray Persian with the soul of a hunter. While the other three cats sleep safely in the house with us, Feisty spends most nights outside, prowling.

Feisty's nocturnal donations are not well received by my family. More than once it has been suggested, as we donate the bird to the trashman, that perhaps Feisty would be happier living in the country.

As a biologist I try to take a more scientific view. I tell my girls that getting rid of Feisty is unwarranted because hunting cats like Feisty actually help birds, in a Darwinian sort of way. Like an evolutionary quality control check, I explain, predators ensure that only those individuals of a population that are better-suited to their environment contribute to the next generation, by the simple expedient of removing the lesser-suited. By taking the birds who are least able to escape predation—the sick and the old—Feisty culls the local bird population, leaving it on average a little better off.

That's what I tell my girls. It all makes sense, from a biological point of view, and it is a story they have heard before, in movies like *Never Cry Wolf*, and *The Lion King*. So Feisty is given a reprieve, and survives to hunt another night.

What I haven't told my girls is how little evidence actually backs up this pretty defense of Feisty's behavior. My explanation may be couched in scientific language, but without proof this "predator-as-purifier" tale is no more than a hypothesis. It might be true, and then again it might not. By such thin string has Feisty's future with our family hung.

Recently the string became a strong cable. Two French biologists put the hypothesis I had been using to defend Feisty to the test. To my great relief, it was supported.

Drs. Anders Møller and Johannes Erritzoe of the Université Pierre et Marie Curie in Paris devised a simple way to test the hypothesis. They compared the health of birds killed by domestic cats like Feisty with that of birds killed in accidents such as flying into glass windows or moving cars. Glass windows do not select for the weak or infirm—a healthy bird flies into a glass window and breaks its neck just as easily as a sickly bird. If cats are actually selecting the less-healthy birds, then their prey should include a larger proportion of sickly individuals than those felled by flying into glass windows.

How can we know what birds are sickly? Drs. Møller and Erritzoe examined the size of the dead bird's spleens. The size of its spleen is a good indicator of how healthy a bird is. Birds experiencing a lot of infections, or harboring a lot of parasites, have smaller spleens than healthy birds.

They examined 18 species of birds, more than 500 individuals. In all but two species (robins and goldcrests) they found that the spleens of birds killed by cats were significantly smaller than those killed accidentally. We're not splitting hairs here, talking about some minor statistical difference. Spleens were on average a third smaller in cat-killed birds. In five bird species (blackcaps, house sparrows, lesser whitethroats, skylarks, and spotted flycatchers), the spleens of birds pounced on by cats were less than half the size of those killed by flying at speed into glass windows or moving cars.

As a control to be sure that additional factors were not operating, the Paris biologists checked for other differences between birds killed by cats and birds killed accidentally. Weight, sex, and wing length, all of which you could imagine might be important, were not significant. Cat-killed birds had, on average, the same weight, proportion of females, and wing length as accident-killed birds.

One other factor did make a difference: age. About 50% of the birds killed accidentally were young, while fully 70% of the birds killed by cats were. Apparently it's not quite so easy to catch an experienced old codger as it is a callow youth.

So Feisty was just doing Darwin's duty, I pleaded, informing my girls that the birds he catches would soon have died anyway. But a dead bird on a doorstep argues louder than any science, and they remained unconvinced.

They are my daughters, and thus not ones to give in without a fight. Scouring the Internet, they assembled this counter-argument: Predatory house cats not unlike Feisty, as well as feral cats (domesticated cats that have been abandoned to the wild), are causing major problems for native bird populations of England, New Zealand, and Australia, as well as here in the United States. Although house cats like Feisty have the predatory instincts of their ancestors, they seem to lack the restraint that their wild relatives have. Most wild cats hunt only when hungry, but pet and feral cats seem to "love the kill," not killing for food but for sport.

So Darwin and I lost this argument. It seems I must restrict Feisty's hunting expeditions after all. While a little pruning may benefit a bird population, wholesale slaughter only devastates it. I will always see a lion whenever I look at Feisty on the prowl, but it will be a lion restricted to indoor hunting.

14.13 The Biological Species Concept

A key aspect of Darwin's theory of evolution is his proposal that adaptation (microevolution) leads ultimately to large-scale changes leading to species formation and higher taxonomic groups (macroevolution). The way natural selection leads to the formation of new species has been thoroughly documented by biologists, who have observed the stages of the species-forming process, or **speciation,** in many different plants, animals, and microorganisms. Speciation usually involves successive change: First, local populations become increasingly specialized; then, if they become different enough, natural selection may act to keep them that way.

Before we can discuss how one species gives rise to another, we need to understand exactly what a species is. The evolutionary biologist Ernst Mayr coined the **biological species concept,** which defines species as "groups of actually or potentially interbreeding natural populations which are reproductively isolated from other such groups."

In other words, the biological species concept says that a species is composed of populations whose members mate with each other and produce fertile offspring—or would do so if they came into contact. Conversely, populations whose members do not mate with each other or who cannot produce fertile offspring are said to be **reproductively isolated** and, thus, members of different species.

What causes reproductive isolation? If organisms cannot interbreed or cannot produce fertile offspring, they clearly belong to different species. However, some populations that are considered to be separate species can interbreed and produce fertile offspring, but they ordinarily do not do so under natural conditions. They are still considered to be reproductively isolated in that genes from one species generally will not be able to enter the gene pool of the other species. **Table 14.1** summarizes the steps at which barriers to successful reproduction may occur. Such barriers are termed **reproductive isolating mechanisms** because they prevent genetic exchange between species. We will first discuss *prezygotic isolating mechanisms,* those that prevent the formation of zygotes. Then we will examine *postzygotic isolating mechanisms,* those that prevent the proper functioning of zygotes after they have formed.

Even though the definition of what constitutes a species is of fundamental importance to evolutionary biology, this issue has still not been completely settled and is currently the subject of considerable research and debate. For example, the biological species concept has had a number of problems. Plants of different species cross fertilize and produce fertile hybrids at much higher frequencies than first thought. Hybridization is common enough to cast doubt about whether reproductive isolation is the only force maintaining the integrity of plant species.

TABLE 14.1 ISOLATING MECHANISMS

Mechanism		Description
Prezygotic Isolating Mechanisms		
Geographic isolation		Species occur in different areas, which are often separated by a physical barrier such as a river or mountain range.
Ecological isolation		Species occur in the same area, but they occupy different habitats. Survival of hybrids is low because they are not adapted to either environment of their parents.
Temporal isolation		Species reproduce in different seasons or at different times of the day.
Behavioral isolation		Species differ in their mating rituals.
Mechanical isolation		Structural differences between species prevent mating.
Prevention of gamete fusion		Gametes of one species function poorly with the gametes of another species or within the reproductive tract of another species.
Postzygotic Isolating Mechanisms		
Hybrid inviability or infertility		Hybrid embryos do not develop properly, hybrid adults do not survive in nature, or hybrid adults are sterile or have reduced fertility.

Key Learning Outcome 14.13 A species is generally defined as a group of similar organisms that does not exchange genes extensively with other groups in nature.

14.14 Isolating Mechanisms

Prezygotic Isolating Mechanisms

Geographical Isolation This mechanism is perhaps the easiest to understand. Species that exist in different areas are not able to interbreed. The two populations of flowers in the first panel of **table 14.1** are separated by a mountain range and so would not be capable of interbreeding.

Ecological Isolation Even if two species occur in the same area, they may utilize different portions of the environment and thus not hybridize because they do not encounter each other, like the lizards in the second panel of **table 14.1**. One lives on the ground and the other in the trees. Another example in nature is the ranges of lions and tigers in India. Their ranges overlapped until about 150 years ago. Even when they did overlap, however, there were no records of natural hybrids. Lions stayed mainly in the open grassland and hunted in groups called prides; tigers tended to be solitary creatures of the forest. Because of their ecological and behavioral differences, lions and tigers rarely came into direct contact with each other, even though their ranges overlapped thousands of square kilometers. **Figure 14.32** shows that hybrids are possible; the tigon shown in **figure 14.32c** is a hybrid of a lion and tiger. These matings do not occur in the wild but can happen in artificial environments such as zoos.

Temporal Isolation *Lactuca graminifolia* and *L. canadensis*, two species of wild lettuce, grow together along roadsides throughout the southeastern United States. Hybrids between these two species are easily made experimentally and are completely fertile. But such hybrids are rare in nature because *L. graminifolia* flowers in early spring and *L. canadensis* flowers in summer. This is called temporal isolation and is shown in the third panel in **table 14.1**. When the blooming periods of these two species overlap, as they do occasionally, the two species do form hybrids, which may become locally abundant.

Behavioral Isolation In chapter 37, we will consider the often elaborate courtship and mating rituals of some groups of animals, which tend to keep these species distinct in nature even if they inhabit the same places. This behavioral isolation is discussed in the fourth panel of **table 14.1**. For example, mallard and pintail ducks are perhaps the two most common freshwater ducks in North America. In captivity, they produce completely fertile offspring, but in nature they nest side-by-side and rarely hybridize.

Mechanical Isolation Structural differences that prevent mating between related species of animals and plants is called mechanical isolation and is shown in panel five of **table 14.1**.

(a)

(b)

(c)

Figure 14.32 Lions and tigers are ecologically isolated.

The ranges of lions and tigers used to overlap in India. However, lions and tigers do not hybridize in the wild because they utilize different portions of the habitat. (*a*) Tigers are solitary animals that live in the forest, whereas (*b*) lions live in open grassland. (*c*) Hybrids, such as this tigon, have been successfully produced in captivity, but hybridization does not occur in the wild.

Figure 14.33 Postzygotic isolation in leopard frogs.

Numbers indicate the following species in the geographic ranges shown: (1) *Rana pipiens*; (2) *Rana blairi*; (3) *Rana sphenocephala*; (4) *Rana berlandieri*. These four species resemble one another closely in their external features. Their status as separate species was first suspected when hybrids between them were found to produce defective embryos in the laboratory. Subsequent research revealed that the mating calls of the four species differ substantially, indicating that the species have both pre- and postzygotic isolating mechanisms.

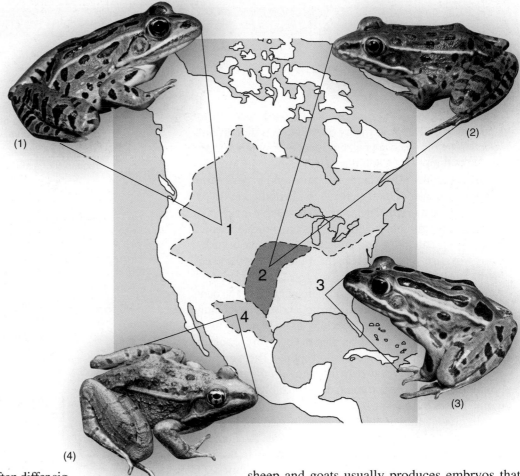

Flowers of related species of plants often differ significantly in their proportions and structures. Some of these differences limit the transfer of pollen from one plant species to another. For example, bees may pick up the pollen of one species on a certain place on their bodies; if this area does not come into contact with the receptive structures of the flowers of another plant species, the pollen is not transferred.

Prevention of Gamete Fusion In animals that shed their gametes directly into water, eggs and sperm derived from different species may not attract one another. Many land animals may not hybridize successfully because the sperm of one species may function so poorly within the reproductive tract of another that fertilization never takes place. In plants, the growth of pollen tubes may be impeded in hybrids between different species. In both plants and animals, the operation of such isolating mechanisms prevents the union of gametes even following successful mating. The sixth panel in **table 14.1** discusses this isolating mechanism.

Postzygotic Isolating Mechanisms

All of the factors we have discussed up to this point tend to prevent hybridization. If hybrid matings do occur, and zygotes are produced, many factors may still prevent those zygotes from developing into normally functioning, fertile individuals. Development in any species is a complex process. In hybrids, the genetic complements of two species may be so different that they cannot function together normally in embryonic development. For example, hybridization between

sheep and goats usually produces embryos that die in the earliest developmental stages.

Figure 14.33 shows four species of leopard frogs (genus *Rana*) and their ranges throughout North America. It was assumed for a long time that they constituted a single species. However, careful examination revealed that although the frogs appear similar, successful mating between them is rare because of problems that occur as the fertilized eggs develop. Many of the hybrid combinations cannot be produced even in the laboratory. Examples of this kind, in which similar species have been recognized only as a result of hybridization experiments, are common in plants.

Even if hybrids survive the embryo stage, however, they may not develop normally. If the hybrids are weaker than their parents, they will almost certainly be eliminated in nature. Even if they are vigorous and strong, as in the case of the mule, a hybrid between a female horse and a male donkey, they may still be sterile and thus incapable of contributing to succeeding generations. Sterility may result in hybrids because the development of sex organs may be abnormal, because the chromosomes derived from the respective parents may not pair properly, or from a variety of other causes.

> **Key Learning Outcome 14.14** Prezygotic isolating mechanisms lead to reproductive isolation by preventing the formation of hybrid zygotes. Postzygotic mechanisms lead to the failure of hybrid zygotes to develop normally, or they prevent hybrids from becoming established in nature.

Does Natural Selection Act on Enzyme Polymorphism?

The essence of Darwin's theory of evolution is that, in nature, selection favors some gene alternatives over others. Many studies of natural selection have focused on genes encoding enzymes because populations in nature tend to possess many alternative alleles of their enzymes (a phenomenon called *enzyme polymorphism*). Often investigators have looked to see if weather influences which alleles are more common in natural populations. A particularly nice example of such a study was carried out on a fish, the mummichog (*Fundulus heteroclitus*), which ranges along the East Coast of North America. Researchers studied allele frequencies of the gene encoding the enzyme lactate dehydrogenase, which catalyzes the conversion of pyruvate to lactate. As you learned in chapter 7, this reaction is a key step in energy metabolism, particularly when oxygen is in short supply. There are two common alleles of lactate dehydrogenase in these fish populations, with allele *a* being a better catalyst at lower temperatures than allele *b*.

In an experiment, investigators sampled the frequency of allele *a* in 41 fish populations located over 14 degrees of latitude, from Jacksonville, Florida (31° North), to Bar Harbor, Maine (44° North). Annual mean water temperatures change 1° C per degree change in latitude. The survey is designed to test a prediction of the hypothesis that natural selection acts on this enzyme polymorphism. If it does, then you would expect that allele *a*, producing a better "low-temperature" enzyme, would be more common in the colder waters of the more northern latitudes. The graph on the right presents the results of this survey. The points on the graph are derived from pie chart data such as shown for 20 populations in the map (a **pie chart diagram** assigns a slice of the pie to each variable; the size of the slice is proportional to the contribution made by that variable to the total). The blue line on the graph is the line that best fits the data (a **"best-fit" line**, also called a **regression line**, is determined statistically by a process called *regression analysis*).

1. **Applying Concepts**
   a. Variable. In the graph, what is the dependent variable?
   b. Reading pie charts. In the fish population located at 35° N latitude, what is the frequency of the *a* allele? Locate this point on the graph.
   c. Analyzing a continuous variable. Compare the frequency of allele *a* among fish captured in waters at 44° N latitude with the frequency among fish captured at 31° N latitude. Is there a pattern? Describe it.
2. **Interpreting Data** At what latitude do fish populations exhibit the greatest variability in allele *a* frequency?
3. **Making Inferences**
   a. Are fish populations in cold waters at 44° N latitude

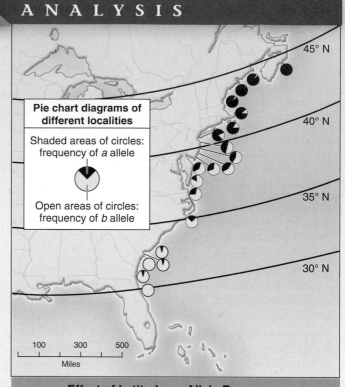

Pie chart diagrams of different localities

Shaded areas of circles: frequency of *a* allele

Open areas of circles: frequency of *b* allele

45° N
40° N
35° N
30° N

100 300 500
Miles

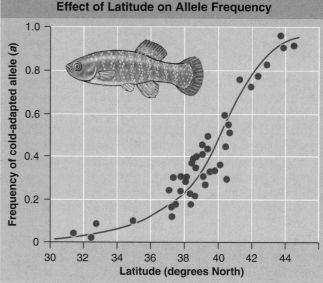

Effect of Latitude on Allele Frequency

Frequency of cold-adapted allele (*a*)

Latitude (degrees North)

more or less likely to contain heterozygous individuals than fish populations in warm waters at 31° N latitude? Why this difference, or lack of it?
   b. Where along this latitudinal gradient in the frequency of allele *a* would you expect to find the highest frequency of heterozygous individuals? Why?
4. **Drawing Conclusions** Are the differences in population frequencies of allele *a* consistent with the hypothesis that natural selection is acting on the alleles encoding this enzyme? Explain.
5. **Further Analysis** If you were to release fish captured at 32° N into populations located at 44° N, so that the local population now had equal frequencies of the two alleles, what would you expect to happen in future generations? How might you test this prediction?

Evolution

14.1 Darwin's Voyage on HMS *Beagle*

- The theory of evolution through natural selection, a theory proposed by Darwin, is overwhelmingly accepted by scientists and is considered to be the backbone of the science of biology.

14.2 Darwin's Evidence

- Darwin observed fossils in South America of extinct species that resembled living species (**figure 14.4**). On the Galápagos Islands, Darwin observed finches that differed slightly in appearance between islands but resembled finches found on the South American mainland (**figure 14.5**).

14.3 The Theory of Natural Selection

- Key to Darwin's hypothesis was the observation by Malthus that the food supply limits population growth. A population grows only as large as that which can live off of the available food that limits geometric growth of a population (**figure 14.6**).

- Using Malthus's observations and his own, Darwin proposed that individuals that are better suited to their environments survive to produce offspring, gaining the opportunity to pass their characteristics on to future generations, what Darwin called natural selection.

Darwin's Finches: Evolution in Action

14.4 The Beaks of Darwin's Finches

- By observing the different sizes and shapes of beaks in the closely related finches of the Galápagos Islands (**figure 14.9**), and correlating the beaks with the types of food consumed, Darwin concluded that the birds's beaks were modified from an ancestral species based on the food available, each suited to its food supply. Scientists have identified a gene, *BMP4*, that is expressed differently in birds with different shaped beaks.

14.5 How Natural Selection Produces Diversity

- The 14 species of finches found on the islands off the coast of South America descended from a mainland species that adapted to different niches, a process called adaptive radiation (**figure 14.12**).

The Theory of Evolution

14.6 The Evidence for Evolution

- The evidence for evolution includes the fossil record. The titanothere and its ancestors are known only from the fossil record (**figure 14.14**). The fossil record reveals organisms that are intermediate in form.

- The evidence for evolution also includes the anatomical record, which reveals similarities in structures between species (**figure 14.15**). Homologous structures are similar in structure but differ in their functions (**figure 14.16**). Analogous structures are similar in function but differ in their underlying structure.

- The molecular record traces changes in the genomes and proteins of species over time (**figures 14.18** and **14.19**).

14.7 Evolution's Critics

- Darwin's theory of evolution through natural selection has always had its critics. Their criticisms of evolution, however, are without scientific merit (**integrated art, page 296** and **figure 14.20**).

How Populations Evolve

14.8 Genetic Change in Populations: The Hardy-Weinberg Rule

- If a population follows the five assumptions of Hardy-Weinberg, the frequencies of alleles within the population will not change (**figure 14.21**). However, if a population is small, has selective mating, experiences mutations or migration, or is under the influence of natural selection, the allele frequencies will be different from those predicted by the Hardy-Weinberg Rule.

14.9 Agents of Evolution

- Five factors act on populations to change their allele and genotype frequencies (**integrated art, pages 302–303**). Mutations are changes in DNA. Nonrandom mating occurs when individuals seek out mates based on certain traits. Genetic drift is the random loss of alleles in a population due to chance occurrences, not due to fitness (**figures 14.22** and **14.23**). Migrations are the movements of individuals or alleles into or out of a population. Selection occurs when individuals with certain traits leave more offspring because their traits allow them to better respond to the challenges of their environment (**figure 14.24**).

- Selection can act on the genes in that population in several different ways (**integrated art, pages 304–305**). Stabilizing selection tends to reduce extreme phenotypes. Disruptive selection tends to reduce intermediate phenotypes. Directional selection tends to reduce one extreme phenotype from the population.

Adaptation Within Populations

14.10 Sickle-Cell Anemia

- Sickle-cell disease is an example of heterozygote advantage, where individuals who are heterozygous for a trait tend to survive better in areas with malaria (**figures 14.26** and **14.27**).

14.11 Peppered Moths and Industrial Melanism

- Natural selection favors dark-colored (melanic) organisms in areas of heavy pollution or in other cases of background matching (**figures 14.28** and **14.29**).

14.12 Selection on Color in Guppies

- Experiments have shown evolutionary change in guppy populations due to natural selection (**figures 14.30** and **14.31**).

How Species Form

14.13 The Biological Species Concept

- The biological species concept states that a species is a group of organisms that mate with each other and produce fertile offspring, or would do so if in contact with each other. If they cannot mate, or mate but cannot produce fertile offspring, they are said to be reproductively isolated (**table 14.1**).

14.14 Isolating Mechanisms

- There are two types of isolating mechanisms: prezygotic and postzygotic. Prezygotic isolating mechanisms prevent the formation of a hybrid zygote. Postzygotic isolating mechanisms prevent normal development of a hybrid zygote or result in sterile offspring (**figure 14.33**).

1. Darwin was greatly influenced by Thomas Malthus, who pointed out that
   a. food supplies increase geometrically.
   b. populations increase arithmetically.
   c. populations are capable of geometric increase, yet remain at constant levels.
   d. the food supply usually increases faster than the population that depends on it.

2. Darwin proposed that individuals with traits that help them live in their immediate environment are more likely to survive and reproduce than individuals without those traits. He called this
   a. natural selection.   c. the theory of evolution.
   b. arithmetic progression.   d. geometric progression.

3. A great deal of research has been done on Darwin's finches over the last 70 years. The research
   a. seems to often contradict Darwin's original ideas.
   b. seems to agree with Darwin's original ideas.
   c. does not show any clear patterns that support or refute Darwin's original ideas.
   d. suggests a different explanation for the evolution of finches.

4. One of the major sources of evidence for evolution is in the comparative anatomy of organisms. Features that look different but have similar structural origin are called
   a. homologous structures.   c. vestigial structures.
   b. analogous structures.   d. equivalent structures.

5. A large group of organisms lives in a large, stable ecosystem. There is no competition for resources. Individuals show no mate preferences. All organisms appear to be identical except for a few individuals in the most recent generation of offspring that exhibit a different fur coat color and pattern. The ecosystem and population are geographically isolated from other populations of the same organism. Which Hardy-Weinberg assumption seems to have been violated?
   a. large population size
   b. random mating within the population
   c. no mutation within the population
   d. no input of new alleles from outside or loss of alleles

6. A population of 1,000 individuals has 200 individuals who show a homozygous recessive phenotype and 800 individuals who express the dominant phenotype. What is the frequency of homozygous recessive individuals in this population?
   a. 0.20   c. 0.45
   b. 0.30   d. 0.55

7. A chance event occurs that causes a population to lose some individuals (they died)—hence, a loss of alleles in the population results from
   a. mutation.   c. selection.
   b. migration.   d. genetic drift.

8. Selection that causes one extreme phenotype to be more frequent in a population is an example of
   a. disruptive selection.   c. directional selection.
   b. stabilizing selection.   d. equivalent selection.

9. A key element of Ernst Mayr's biological species concept is
   a. homologous isolation.   c. convergent isolation.
   b. divergent isolation.   d. reproductive isolation.

10. Which of the following is *not* a prezygotic isolating mechanism?
    a. behavioral isolation
    b. ecological isolation
    c. hybrid infertility
    d. None of the above.

1. **Pages 304-305** Because of prolonged drought, the trees on an island are producing nuts that are much smaller with thicker and harder shells. What would you predict would happen to birds that depend on the nuts for food? What type of selection will result?

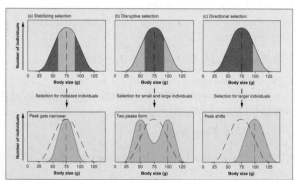

2. **Table 14.1** A very heavy rainstorm floods a mountain river, changing its course and digging a deep canyon through the soft soils of the meadow in the valley downstream. How might mice populations on the two sides of the valley be affected?

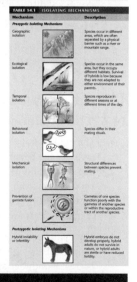

1. Can natural selection occur among genetically identical clones? Explain your reasoning.

2. The evolutionary pathways of some groups of related organisms seem to have moved from larger to smaller organisms over time, such as the glyptodont to the armadillo, the mammoth to the elephant. Yet other groups of related organisms have exhibited a trend of increased sizes of species over time, such as the tiny eohippus to the horse. Explain how this can occur.

3. In a courtroom in 2005, biologist Ken Miller criticized the claims of intelligent design. After noting that 99.9% of the organisms that have ever lived on earth are now extinct, he said that "an intelligent designer who designed things, 99.9% of which didn't last, certainly wouldn't be very intelligent." Evaluate Miller's criticism.

15

How We Name Living Things

I n 1799, the skin of a most unusual animal was sent to England by Captain John Hunter, governor of the British penal colony in New South Wales (Australia). Covered in soft fur, it was less than 2 feet long. As it had mammary glands with which to suckle its young, it was clearly a mammal, but in other ways it seemed much like a reptile. Males have internal testes, and females have a shared urinary and reproductive tract opening called a cloaca, lay eggs as reptiles do, and like reptilian eggs, the yolk of the fertilized egg does not divide. It thus seemed a confusing mixture of mammalian and reptilian traits. Adding to this impression was its appearance: It has a tail not unlike that of a beaver, a bill not unlike that of a duck, and webbed feet! It was as if a child had mixed together body parts at random—a most unusual animal. Individuals like the one pictured here are abundant in freshwater streams of eastern Australia today. What does one call such a beast? In its original 1799 description, it was named *Platypus anatinus* (flatfooted, ducklike animal), which was later changed to *Ornithorhynchus anatinus* (ducklike animal with a bird's snout)— informally, the duckbill platypus. How biologists assign names to the organisms they discover is the subject of this chapter. You will be surprised at how much information is crammed into the two words of a scientific name.

15.1 Invention of the Linnaean System

It is estimated that our world is populated by some 10 to 100 million different kinds of organisms. To talk about them and study them, it is necessary to give them names, just as it is necessary that people have names. Of course, no one can remember the name of every kind of organism, so biologists use a kind of multilevel grouping of individuals called **classification.**

Organisms were first classified more than 2,000 years ago by the Greek philosopher Aristotle, who categorized living things as either plants or animals. He classified animals as either land, water, or air dwellers, and he divided plants into three kinds based on stem differences. This simple classification system was expanded by the Greeks and Romans, who grouped animals and plants into basic units such as cats, horses, and oaks. Eventually, these units began to be called **genera** (singular, **genus**), the Latin word for "group." Starting in the Middle Ages, these names began to be systematically written down, using Latin, the language used by scholars at that time. Thus, cats were assigned to the genus *Felis,* horses to *Equus,* and oaks to *Quercus*—names that the Romans had applied to these groups.

The classification system of the Middle Ages, called the *polynomial system,* was used virtually unchanged for hundreds of years, until it was replaced about 250 years ago by the **binomial system** introduced by Linnaeus.

The Polynomial System

Until the mid-1700s, biologists usually added a series of descriptive terms to the name of the genus when they wanted to refer to a particular kind of organism, which they called a **species.** These phrases, starting with the name of the genus, came to be known as **polynomials** (*poly,* many, and *nomial,* name), strings of Latin words and phrases consisting of up to 12 or more words. For example, the common wild briar rose was called *Rosa sylvestris inodora seu canina* by some and *Rosa sylvestris alba cum rubore, folio glabro* by others. This would be like the mayor of New York referring to a particular citizen as "Brooklyn resident: Democrat, male, Caucasian, middle income, Protestant, elderly, likely voter, short, bald, heavyset, wears glasses, works in the Bronx selling shoes." As you can imagine, these polynomial names were cumbersome. Even more worrisome, the names were altered at will by later authors, so that a given organism really did not have a single name that was its alone, as was the case with the briar rose.

The Binomial System

A much simpler system of naming animals, plants, and other organisms stems from the work of the Swedish biologist Carolus Linnaeus (1707–1778). Linnaeus devoted his life to a challenge that had defeated many biologists before him—cataloging all the different kinds of organisms. Linnaeus, a

(a) *Quercus phellos*
(Willow oak)

(b) *Quercus rubra*
(Red oak)

Figure 15.1 How Linnaeus named two species of oaks.

(*a*) Willow oak, *Quercus phellos.* (*b*) Red oak, *Quercus rubra.* Although they are clearly oaks (members of the genus *Quercus*), these two species differ sharply in the shapes and sizes of their leaves and in many other features, including their overall geographical distributions.

botanist studying the plants of Sweden and from around the world, developed a plant classification system that grouped plants based on their reproductive structures. This system resulted in some seemingly unnatural groupings and therefore was never universally accepted. However, in the 1750s he produced several major works that, like his earlier books, employed the polynomial system. But as a kind of shorthand, Linnaeus also included in these books a two-part name for each species (others had also occasionally done this, but Linnaeus used these shorthand names consistently). These two-part names, or **binomials** (*bi* is the Latin prefix for "two"), have become our standard way of designating species. For example, he designated the willow oak (shown in **figure 15.1a** with its smaller, unlobed leaves) *Quercus phellos* and the red oak (with the larger, deeply lobed leaves in **figure 15.1b**) *Quercus rubra,* even though he also included the polynomial name for these species. We also use binomial names for ourselves, our so-called given and family names. So, this naming system is like the mayor of New York calling the Brooklyn resident Sylvester Kingston.

Linnaeus took the naming of organisms a step further, grouping similar organisms into higher-level categories based on similar characteristics (discussed later). Although not intended to show evolutionary connections between different organisms, this hierarchical system acknowledged that there were broad similarities shared by groups of species that distinguished them from other groups.

> **Key Learning Outcome 15.1   Two-part (binomial) Latin names, first used by Linnaeus, are now universally employed by biologists to name organisms.**

15.2 Species Names

A group of organisms at a particular level in a classification system is called a **taxon** (plural, **taxa**), and the branch of biology that identifies and names such groups of organisms is called **taxonomy.** Taxonomists are in a real sense detectives, biologists who must use clues of appearance and behavior to identify and assign names to organisms.

By formal agreement among taxonomists throughout the world, no two organisms can have the same name. So that no one country is favored, a language spoken by no country—Latin—is used for the names. Because the scientific name of an organism is the same anywhere in the world, this system provides a standard and precise way of communicating, whether the language of a particular biologist is Chinese, Arabic, Spanish, or English. This is a great improvement over the use of common names, which often vary from one place to the next. As you can see in **figure 15.2,** in America corn refers to the plant in the upper photo on the left, but in Europe it refers to the plant Americans call wheat, the lower photo on the left.

A bear is a large placental omnivore in the United States (the upper middle photo), but in Australia it is a koala, a vegetarian marsupial (the lower middle photo). A robin in North America (the upper right photo) is a very different bird in Europe (the lower right photo).

By convention, the first word of the binomial name is the genus to which the organism belongs. This word is always capitalized. The second word, called the *specific epithet*, refers to the particular species and is not capitalized. The two words together are called the **scientific name,** or species name, and are written in italics. The system of naming animals, plants, and other organisms established by Linnaeus has served the science of biology well for nearly 250 years.

> **Key Learning Outcome 15.2** By convention, the first part of a binomial species name identifies the genus to which the species belongs, and the second part distinguishes that particular species from other species in the genus.

(a)   (b)   (c)

Figure 15.2 Common names make poor labels.

The common names corn (*a*), bear (*b*), and robin (*c*) bring clear images to our minds (photos on *top*), but the images would be very different to someone living in Europe or Australia (photos on *bottom*). There, the same common names are used to label very different species.

15.3 Higher Categories

Like the mayor of New York, a biologist needs more than two categories to classify all the world's living things. Taxonomists group the genera with similar properties into a cluster called a **family.** For example, the eastern gray squirrel at the bottom of **figure 15.3** is placed in a family with other squirrel-like animals including prairie dogs, marmots, and chipmunks. Similarly, families that share major characteristics are placed into the same **order** (for example, squirrels placed in with other rodents). Orders with common properties are placed into the same **class** (squirrels in the class Mammalia), and classes with similar characteristics into the same **phylum** (plural, phyla) such as the Chordata. Botanists (that is, those who study plants) also call plant phyla "divisions." Finally, the phyla are assigned to one of several gigantic groups, the **kingdoms.** Biologists currently recognize six kingdoms: two kinds of prokaryotes (Archaea and Bacteria), a largely unicellular group of eukaryotes (Protista), and three multicellular groups (Fungi, Plantae, and Animalia). To remember the seven categories in their proper order, it may prove useful to memorize a phrase such as "**k**indly **p**ay **c**ash **o**r **f**urnish **g**ood **s**ecurity" or "**K**ing **P**hilip **c**ame **o**ver **f**or **g**reen **s**paghetti" (kingdom–phylum–class–order–family–genus–species).

In addition, an eighth level of classification, called *domains,* is sometimes used. Domains are the broadest and most inclusive taxa, and biologists recognize three of them, Bacteria, Archaea, and Eukarya—which are discussed later in this chapter.

Each of the categories in this **Linnaean system of classification** is loaded with information. For example, consider a honeybee:

Level 1: Its species name, *Apis mellifera,* identifies the particular species of honey bee.
Level 2: Its genus name, *Apis,* tells you it is a honey bee.
Level 3: Its family, Apidae, are all bees, some solitary, others living in hives as *A. mellifera* does.
Level 4: Its order, Hymenoptera, tells you that it is likely able to sting and may live in colonies.
Level 5: Its class, Insecta, says that *A. mellifera* has three major body segments, with wings and three pairs of legs attached to the middle segment.
Level 6: Its phylum, Arthropoda, tells us that it has a hard cuticle of chitin and jointed appendages.
Level 7: Its kingdom, Animalia, says that it is a multicellular heterotroph whose cells lack cell walls.
Level 8: An addition to the Linnaean system, its domain, Eukarya, says that its cells contain membrane-bounded organelles.

> **Key Learning Outcome 15.3** A hierarchical system is used to classify organisms in which higher categories convey more general information about the group.

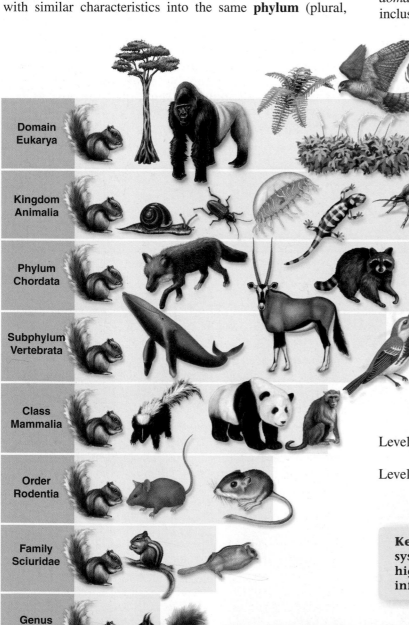

Domain Eukarya

Kingdom Animalia

Phylum Chordata

Subphylum Vertebrata

Class Mammalia

Order Rodentia

Family Sciuridae

Genus *Sciurus*

Species *Sciurus carolinensis*

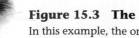

Sciurus carolinensis

Figure 15.3 The hierarchical system used to classify an organism.

In this example, the organism is first recognized as a eukaryote (domain: Eukarya). Second, within this domain, it is an animal (kingdom: Animalia). Among the different phyla of animals, it is a vertebrate (phylum: Chordata, subphylum: Vertebrata). The organism's fur characterizes it as a mammal (class: Mammalia). Within this class, it is distinguished by its gnawing teeth (order: Rodentia). Next, because it has four front toes and five back toes, it is a squirrel (family: Sciuridae). Within this family, it is a tree squirrel (genus: *Sciurus*), with gray fur and white-tipped hairs on the tail (species: *Sciurus carolinensis*, the eastern gray squirrel).

15.4 What Is a Species?

The basic biological unit in the Linnaean system of classification is the species. John Ray (1627–1705), an English clergyman and scientist, was one of the first to propose a general definition of species. In about 1700, he suggested a simple way to recognize a species: All the individuals that belong to it can breed with one another and produce fertile offspring. By Ray's definition, the offspring of a single mating were all considered to belong to the same species, even if they contained different-looking individuals, as long as these individuals could interbreed. All domestic cats are one species (they can all interbreed), while carp are not the same species as goldfish (they cannot interbreed). The donkey you see in **figure 15.4** is not the same species as the horse, because when they interbreed, the offspring—mules—are sterile.

Horse

Donkey

Mule

Figure 15.4 Ray's definition of a species.
According to Ray, donkeys and horses are not the same species. Even though they produce very hardy offspring (mules) when they mate, the mules are sterile, meaning that they cannot produce offspring.

The Biological Species Concept

With Ray's observation, the species began to be regarded as an important biological unit that could be cataloged and understood, the task that Linnaeus set himself a generation later. Where information was available, Linnaeus used Ray's species concept, and it is still widely used today. When the evolutionary ideas of Darwin were joined to the genetic ideas of Mendel in the 1920s to form the field of population genetics, it became desirable to define the category species more precisely. The definition that emerged, the so-called *biological species concept,* discussed in chapter 14, defined *species* as groups that are reproductively isolated. In other words, hybrids (offspring of different species that mate) occur rarely in nature, whereas individuals that belong to the same species are able to interbreed freely.

As we learned in chapter 14, the biological species concept works fairly well for animals, where strong barriers to hybridization between species exist, but very poorly for members of the other kingdoms. The problem is that the biological species concept assumes that organisms regularly outcross—that is, interbreed with individuals other than themselves, individuals with a different genetic makeup from their own but within their same species. Animals regularly outcross, and so the concept works well for animals. However, outcrossing is less common in the other five kingdoms. In prokaryotes and many protists, fungi, and some plants, asexual reproduction, reproduction without sex, predominates. These species clearly cannot be characterized in the same way as outcrossing animals and plants—they do not interbreed with one another, much less with the individuals of other species.

Complicating matters further, the reproductive barriers that are the key element in the biological species concept,
although common among animal species, are not typical of other kinds of organisms. In fact, there are essentially no barriers to hybridization between the species in many groups of trees, such as oaks, and other plants, such as orchids. Even among animals, fish species are able to form fertile hybrids with one another, though they may not do so in nature.

In practice, biologists today recognize species in different groups in much the way they always did, as groups that differ from one another in their visible features. Within animals, the biological species concept is still widely employed, while among plants and other kingdoms it is not. Molecular data are causing a reevaluation of traditional classification systems, and, taking into account morphology, life cycles, metabolism, and other characteristics, they are changing the way scientists classify plants, protists, fungi, prokaryotes, and even animals.

How Many Kinds of Species Are There?

Since the time of Linnaeus, about 1.5 million species have been named. But the actual number of species in the world is undoubtedly much greater, judging from the very large numbers that are still being discovered. Some scientists estimate that at least 10 million species exist on earth, and at least two-thirds of these occur in the tropics.

Key Learning Outcome 15.4 Among animals, species are generally defined as reproductively isolated groups; among the other kingdoms such a definition is less useful, as their species typically have weaker barriers to hybridization.

15.5 How to Build a Family Tree

After naming and classifying some 1.5 million organisms, what have biologists learned? One very important advantage of being able to classify particular species of plants, animals, and other organisms is that we can identify species that are useful to humans as sources of food and medicine. For example, if you cannot tell the fungus *Penicillium* from *Aspergillus,* you have little chance of producing the antibiotic penicillin. In a thousand ways, just having names for organisms is of immense importance in our modern world.

Taxonomy also enables us to glimpse the evolutionary history of life on earth. The more similar two taxa are, the more closely related they are likely to be, for the same reason that you are more like your brothers and sisters than like strangers selected from a crowd. By looking at the differences and similarities between organisms, biologists can attempt to reconstruct the tree of life, inferring which organisms evolved from which other ones, in what order, and when. The evolutionary history of an organism and its relationship to other species is called **phylogeny.** The reconstruction and study of evolutionary trees, or **phylogenetic trees,** including the naming and classifying of organisms, is an area of study called **systematics.**

Cladistics

A simple and objective way to construct a phylogenetic tree is to focus on key characters that some organisms share because they have inherited them from a common ancestor. A **clade** is a group of organisms related by descent, and this approach to constructing a phylogeny is called **cladistics.** Cladistics infers phylogeny (that is, builds family trees) according to similarities derived from a common ancestor, so-called **derived characters.** Derived characters are defined as characters that are present in a group of organisms that arose from a common ancestor that lacked the character. The key to the approach is being able to identify morphological, physiological, or behavioral traits that differ among the organisms being studied and can be attributed to a common ancestor. By examining the distribution of these traits among the organisms, it is possible to construct a **cladogram,** a branching diagram that represents the phylogeny. A cladogram of the vertebrates is shown in figure 15.5.

Cladograms are not true family trees, derived directly from data that document ancestors and descendants like the fossil record does. Instead, cladograms convey comparative information about *relative* relationships. Organisms that are closer together on a cladogram simply share a more recent common ancestor than those that are farther apart. Because the analysis is comparative, it is necessary to have something to anchor the comparison to, some solid ground against which the comparisons can be made. To achieve this, each cladogram must contain an **outgroup,** a rather different organism (but not *too* different) to serve as a baseline for comparisons among the other organisms being evaluated, called the **ingroup.** For example, in **figure 15.5** the lamprey is the outgroup to the clade of animals that have jaws. Comparisons are then made up the cladogram, beginning with lampreys and sharks, based on the emergence of derived characters. For example, the shark differs from the lamprey in that it has jaws, the derived character missing in the lamprey. The derived characters are in the colored boxes along the main line of the cladogram. Salamanders differ from sharks in that they have lungs, and so on up the cladogram.

Cladistics is a relatively new approach in biology and has become popular among students of evolution. This is because it does a very good job of portraying the *order* in which a series of evolutionary events have occurred. The great strength of a cladogram is that it can be completely objective. A computer fed the data will generate exactly the same cladogram time and again. In fact, most cladistic analyses involve many characters, and computers are required to make the comparisons. Although objective, the phylogenetic trees are not absolute. Phylogenetic trees are hypotheses, proposed explanations of how organisms may have evolved.

Sometimes cladograms are adjusted to "weight" characters, or take into account the variation in the "strength" (importance) of a character—the size or location of a fin, the

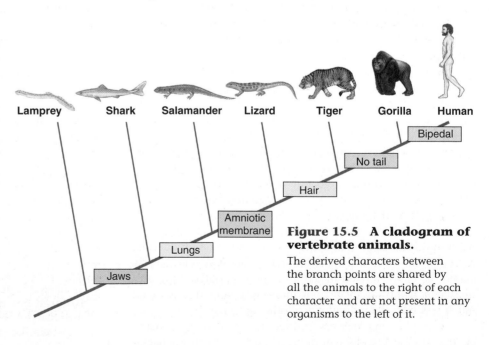

Figure 15.5 A cladogram of vertebrate animals.

The derived characters between the branch points are shared by all the animals to the right of each character and are not present in any organisms to the left of it.

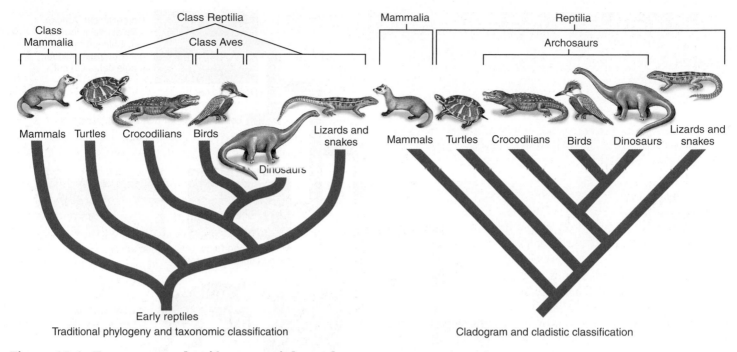

Figure 15.6 Two ways to classify terrestrial vertebrates.
Traditional taxonomic analyses place birds in their own class (Aves) because birds have evolved several unique adaptations that separate them from the reptiles. Cladistic analyses, however, place crocodiles, dinosaurs, and birds together (as archosaurs) because they share many derived characters, indicating a recent shared ancestry. In practice, most biologists adopt the traditional approach and consider birds as members of the class Aves rather than Reptilia.

effectiveness of a lung. For example, let's say that the following are five unique events that occurred on September 11, 2001: (1) My cat was declawed, (2) I had a wisdom tooth pulled, (3) I sold my first car, (4) terrorists attacked the United States using commercial airplanes, and (5) I passed physics. Without weighting the events, each one is assigned equal importance. In a nonweighted cladistic sense, they are equal (all happened only once, and on that day), but in a practical, real-world sense, they certainly are not. One event, the terrorist attack, had a far greater impact and importance than the others. Because evolutionary success depends so critically on just such high-impact events, these weighted cladograms attempt to assign extra weight to the evolutionary significance of key characters.

Weighted cladograms are controversial. The problem with them is the systematist usually cannot always know how important each character is. The history of systematics has many examples of overemphasis or reliance on characters that later turned out to be less informative than had been thought. This is why many systematists now choose to weight all characters equally in cladograms.

Traditional Taxonomy

Weighting characters lies at the core of **traditional taxonomy.** In this approach, phylogenies are constructed based on a vast amount of information about the morphology and biology of the organism gathered over a long period of time. Traditional

taxonomists use both ancestral and derived characters to construct their trees, whereas cladists use only derived characters. The large amount of information used by traditional taxonomists permits a knowledgeable weighting of characters according to their biological significance. In traditional taxonomy, the full observational power and judgment of the biologist is brought to bear—and also any biases he or she may have. For example, in classifying the terrestrial vertebrates, traditional taxonomists, shown by the phylogeny on the left in **figure 15.6**, place birds in their own class (Aves), giving great weight to the characters that made powered flight possible, such as feathers. However, a cladogram of vertebrate evolution, as shown on the right in **figure 15.6**, lumps birds in among the reptiles with crocodiles and dinosaurs. This accurately reflects their ancestry but ignores the immense evolutionary impact of a derived character such as feathers.

Overall, phylogenetic trees based on traditional taxonomy are information-rich, while cladograms often do a better job of deciphering evolutionary histories. Traditional taxonomy is the better approach when a great deal of information is available to guide character weighting. For example, the cat family tree in **figure 15.7** on the next page reflects a lot of knowledge about the different groups of felines. However, cladistics is the preferred approach when little information is available about how the character affects the life of the organism.

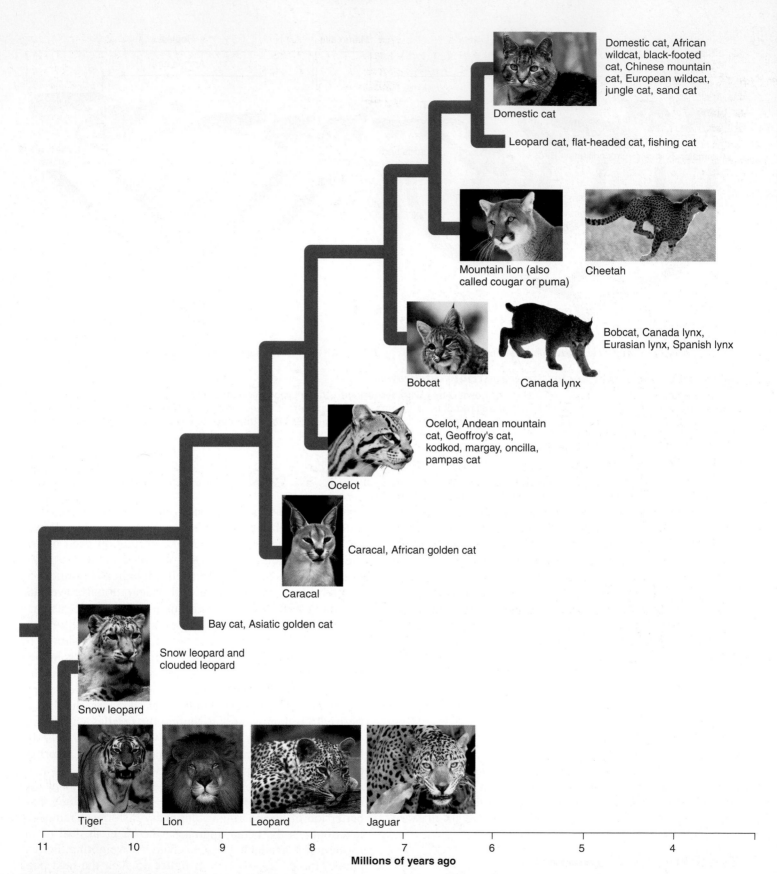

Domestic cat, African wildcat, black-footed cat, Chinese mountain cat, European wildcat, jungle cat, sand cat

Domestic cat

Leopard cat, flat-headed cat, fishing cat

Mountain lion (also called cougar or puma)

Cheetah

Bobcat, Canada lynx, Eurasian lynx, Spanish lynx

Bobcat

Canada lynx

Ocelot, Andean mountain cat, Geoffroy's cat, kodkod, margay, oncilla, pampas cat

Ocelot

Caracal, African golden cat

Caracal

Bay cat, Asiatic golden cat

Snow leopard and clouded leopard

Snow leopard

Tiger

Lion

Leopard

Jaguar

11 10 9 8 7 6 5 4
Millions of years ago

Figure 15.7 The cat family tree.

Recent studies of DNA similarities reported in 2006 have allowed biologists to construct this feline family tree of the eight major cat lineages and their individual species. Among the oldest of all cats are the four big panthers: tiger, lion, leopard, and jaguar. The other big cats, the cheetah and mountain lion, are members of a much younger lineage and are not close relatives of the big four. Domestic cats evolved most recently.

How Do You Read a Family Tree?

Evolutionary trees, more formally called phylogenies, have become an essential tool of modern biology, used to track the spread of mad cow disease, to trace an individual's ancestry, and even to predict which horses might win the Kentucky Derby. Most importantly, evolutionary trees provide the main framework within which evidence for evolution is evaluated.

Given their central role in biology, it is important that you learn how to "read" a tree properly. Said simply, a phylogeny or evolutionary tree is a depiction of lines of descent (**figure 15.8**). Its function is to communicate the evolutionary relationships among its elements. In a typical tree, individual genes, species, or other elements occupy the branch tips. Underneath, a network of branches connects to the base.

The essential point in reading such a tree is to understand that the nodes (branching points) correspond to actual organisms that lived in the past. The tree does *not* illustrate the degree of similarity among the branch tips, but rather shows actual historical relationships. Although closely related organisms tend to be similar to one another, this is not the case if the rate of evolution is not uniform. As you saw in **figure 15.6** previously, crocodiles are more closely related to birds than they are to lizards, even though anyone can see that crocodiles look a lot more like lizards than birds.

Once a tree is seen as a story, an historical account, it is easy to avoid confusion about relatives. The rule is simple: The more recently two species share a common ancestor, the more closely related they are. There is nothing new about this. This is how you refer to your relatives. You are more closely related to your first cousin than to your second cousin because your last common ancestor with your first cousin lived two generations ago (grandparents), while your last common ancestor with your second cousin lived three generations ago (great grandparents).

Now look how this works in an evolutionary tree depicting ancestry. Consider the tree diagram shown below. Some people erroneously conclude that a frog is more closely related to a shark than to a human. A frog is actually more closely related to a human than to a shark because the last common ancestor of a frog and a human (labeled *x* in the figure) is a descendant of the last common ancestor of a frog and a shark (labeled *y* in the figure), and thus lived more recently. It is that simple. Most problems with reading evolutionary trees come when one reads a tree along the tips. In the tree pictured below, this approach yields an orderly sequence from sharks to frogs to humans. This sequential way of reading a phylogeny is in-

Figure 15.8 Tree as icon.
Darwin developed the metaphor of the "tree of life," living species tracing back through time to common ancestors in the same way that separate twigs on a tree trace back to the same branches.

correct because it suggests a linear progression from primitive to advanced species, which in no way is justified by the tree. If so, the frog would be the ancestor of a living human.

The correct way to read a tree is as a set of hierarchically nested groups, each of them a clade such as you encountered in **figure 15.5**. In the tree shown here, there are three meaningful clades: human-tiger, human-tiger-lizard, and human-tiger-lizard-frog.

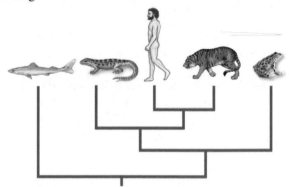

The difference between reading branch tips and reading clades becomes apparent if the branches are rotated so that the order of the tips is changed, as on the tree above. Although the order of branch tips is different, the branching patterns of descent—and the clade composition—is identical to the arrangement on the left. Evolutionary trees should be read by focusing on clade structure, which helps to emphasize that evolution is not a linear narrative.

> **Key Learning Outcome 15.5** An evolutionary tree depicts lines of descent, and is best read by focusing on clades. A cladogram is based on the order in which groups evolved, while a traditional taxonomic tree weights characters according to assumed importance.

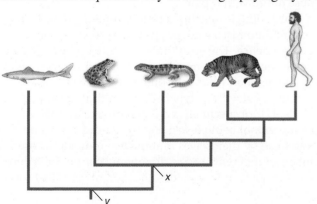

Today's *Biology*

DNA "Bar Codes"

The great diversity of life on earth is one of the glories of our planet. Wherever we look around us, we are surrounded by a profusion of life. A typical backyard contains hundreds of species of animals and plants, the same size slice of tropical rainforest contains orders of magnitude more. In North America alone, there are 709 identified species of birds, ranging in size from eagles with wingspans as long as your car to hummingbirds smaller than your thumb.

This profusion of species creates a problem that you might not at first anticipate. With so many species, how are you to know which is which? Animals and plants don't come with easy-to-read labels telling you to which species an individual belongs. Consider for example the two marsh wrens on the facing page. The darker individual on the right is from New Jersey, the lighter individual to its left from California. The difference in color between these two individuals tempts you to leap to an unwarranted conclusion, that the New Jersey specimen is an Eastern Marsh Wren—while the California specimen is a Western Marsh Wren. The conclusion is unwarranted because body color varies widely in both groups. There are lighter colored Eastern Marsh Wrens, and darker Western Marsh Wrens, leaving you in a quandary: Confronted with an individual marsh wren, how are you to determine which kind it is?

One solution to this dilemma is to take the specimen to a professional bird taxonomist, who will evaluate beak shape, plumage, and many other characters to correctly identify the bird. This is the approach Charles Darwin took in his study of the Galápagos. The finches, mockingbirds, and other birds he collected on the islands were studied and identified years later at the British Museum in London. However, there are not all that many taxonomists, and an awful lot of organisms whose identity we need to know.

Enter Dr. Paul Hebert of the University of Guelph in Ontario, Canada, with a deceptively simple suggestion, which is that organisms do in fact come with easy-to-read labels. Certain genes vary little among individuals of a species, but different species have different versions—why not let these genes serve as ID tags? DNA sequencing machines read off the order of nucleotides in batches of about 650 bases at a time, so he proposed examining the first 648 nucleotides of a gene called cytochrome *c* oxidase subunit 1 (*CO1*). Why this particular gene? For four reasons: First, because this gene is located on the mitochondrial DNA rather than on the nuclear chromosome, it is inherited solely from the mother and so escapes the shuffling of genetic material between generations that meiosis creates. Second, mitochondrial DNA is more stable than nuclear DNA, and can be obtained from museum specimens up to 20 years old. Third, in most animal species this gene has no inserted or deleted DNA, allowing all *CO1* sequences to be lined up side-by-side for direct comparison.

Fourth, and most importantly, *CO1* differences between individuals within a species are surprisingly small—just 2% of individuals differ at all along the 648-nucleotide stretch. This within-species uniformity is unusual—in a typical gene, many differences would be found between members of a species, so many as to overlap with individuals of closely related species. Not so for *CO1*. Perhaps because cytochrome oxidase plays such a critical role in oxidative metabolism, any changes within a species are rare, and when they do occur, spread rapidly through all members of the species. However, after two species have diverged, rare changes that occur in one species do not spread to the other, and so the two species accumulate defining differences.

How well does Hebert's suggestion work in practice? As a practical test, he set out to compare the first 648 nucleotides of the *CO1* gene in mitochondrial DNA obtained from museum specimens of birds. He characterized 341 of the 709 known species of North American birds, and in every instance found a unique sequence characteristic of the species. While much larger samples will have to be examined to prove *CO1* provides a distinctive signature, especially for closely related species, these initial results are very promising. You can see on the facing page the sequences Hebert found for the two marsh wrens we were discussing. Eastern and Western Marsh Wrens differ in 21 places, indicated by the lines between the two colored "bar codes." Just as Hebert proposed, the *CO1* sequences are easy-to-read labels that clearly identify an individual as being one kind of marsh wren, and not the other.

Dr. Hebert calls his approach "DNA bar coding" in analogy with the bar codes on supermarket items. The great potential of bar coding is that it solves the problem with which we introduced this essay, promising to allow anyone to correctly identify an unknown specimen in a direct and unambiguous fashion.

Of course, not every organism is a bird. How well does *CO1* work as a bar code for other kinds of organisms? So far, it seems to work well for animals, but not as well for plants. Plant biologists have already begun to examine instead two genes found on chloroplast DNA, which seems better suited to distinguishing between plant species. While the central idea of bar coding is the use of a standard gene as a reference, the use of different genes for different major groups presents no major problem—few taxonomists would confuse a bird with a plant.

Bar coding is meeting a certain amount of resistance among taxonomists, who fear its widespread use will lead to sloppy science. The error that gives them pause is easily demonstrated with the marsh wren bar codes shown on the facing page: The 21 bar code differences serve to associate each individual with a "type" specimen that has a particular bar code. If that specimen is a different species, then so is the individual being tested. But the 21 differences do *not* establish by themselves that the two individuals are members of different species. Taxonomy cannot be reduced to a single gene, although identification of described species apparently can.

Western Marsh Wren (California)

Eastern Marsh Wren (New Jersey)

KEY:
Adenine ——
Thymine ——
Cytosine ——
Guanine ——

DNA from Western Marsh Wren

DNA DIFFERENCES BETWEEN THE TWO BIRDS

DNA from Eastern Marsh Wren

While DNA bar codes based on the *CO1* gene appear to provide a useful ID tag for birds and potentially many other animal species, it is important to be very clear about the nature of the groups the bar codes identify. Any reproductively isolated line of descent can, and indeed would be expected to, develop a unique *CO1* bar code. That does not mean the group is a separate species. For example, the aboriginal peoples of Australia have been reproductively isolated for many centuries, but if they were to have developed a unique *CO1* bar code, that would not make them a separate species. It would imply no more and no less than genetic isolation.

This important distinction can be clearly seen in a recent study of Costa Rican butterflies. The skipper butterfly *Astraptes fulgerator* was first described in 1775 and ranges from Texas to Argentina. Over the course of 25 years of study, University of Pennsylvania ecologist Daniel Janzen has raised some 2,500 caterpillars of the *Astraptes* skipper that, as you can see in the photographs, come in 10 versions, each of which feeds on a different plant but all of which give

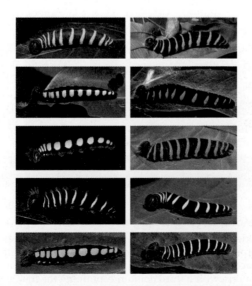

rise to adults that look identical to one another and are all considered members of the same species.

Hearing of the bar code technology being developed by Hebert, Janzen removed one leg from each of his numerous preserved adult skipper butterflies and sent the samples to Dr. Hebert for analysis. Using the same *CO1* bar code analysis that had proven so powerful in identifying bird species, Hebert found that Janzen's skipper collection fell neatly into 10 separate bar code clusters. All members of a cluster had nearly the same bar code, different from the other nine clusters. Importantly, the clusters matched the caterpillar groups! Each type of caterpillar had its own *CO1* signature. The researchers concluded that in Costa Rica there is not one but 10 species of skipper, each with a strikingly different caterpillar. Janzen speculates that the groups diverged perhaps 4 million years ago, based on the amount of *CO1* difference seen, each group specializing on a different caterpillar food plant. Is this a valid way to identify species? Some taxonomists are excited by the approach, others more cautious.

15.6 The Kingdoms of Life

Classification systems have gone through their own evolution of sorts, as illustrated in **figure 15.9**. The earliest classification systems recognized only two kingdoms of living things: animals, shown in blue in **figure 15.9a**, and plants, shown in green. But as biologists discovered microorganisms (the yellow-colored boxes in **figure 15.9b**) and learned more about other organisms like the protists (in teal) and the fungi (in light brown), they added kingdoms in recognition of fundamental differences. Most biologists now use a six-kingdom system (indicated by the six different-colored boxes in **figure 15.9c**) first proposed by Carl Woese of the University of Illinois.

In this system, four kingdoms consist of eukaryotic organisms. The two most familiar kingdoms, **Animalia** and **Plantae,** contain only organisms that are multicellular during most of their life cycle. These groups of animals and plants are no doubt familiar to you. The kingdom **Fungi** contains multicellular forms, such as mushrooms and molds, and single-celled yeasts, which are thought to have multicellular ancestors. Fundamental differences divide these three kingdoms. Plants are mainly stationary, but some have motile sperm; fungi have no motile cells; animals are mainly motile. Animals ingest their food, plants manufacture it, and fungi digest it by means of secreted extracellular enzymes. Each of these kingdoms probably evolved from a different single-celled ancestor.

The large number of unicellular eukaryotes are arbitrarily grouped into a single kingdom called **Protista** (see chapter 17). They include the algae and many kinds of microscopic aquatic organisms. This kingdom is an artificial group in that many of these organisms are only distantly related, and the classification of the protists is in flux.

The remaining two kingdoms, **Archaea** and **Bacteria,** consist of prokaryotic organisms, which are vastly different from all other living things (see chapter 16). The prokaryotes with which you are most familiar, those that cause disease or are used in industry, are members of the kingdom Bacteria. Archaea are a diverse group including the methanogens and extreme thermophiles, and they differ greatly from bacteria in many ways. The characteristics of these six kingdoms are presented in **table 15.1.**

Domains

As biologists have learned more about the archaea, it has become increasingly clear that this ancient group is very different from all other organisms. When the full genomic DNA sequences of an archaean and a bacterium were first compared in 1996, the differences proved striking. Archaea are as different from bacteria as bacteria are from eukaryotes. Recognizing this, biologists have in recent years adopted a taxonomic level higher than kingdom that recognizes three **domains** (**figure 15.9d**). Bacteria (yellow-colored box) are in one domain, archaea (red-colored box) are in a second, and eukaryotes (the four purple boxes representing the four eukaryotic kingdoms) in the third. While the domain Eukarya contains four kingdoms of organisms, the domains Bacteria and Archaea contain only one kingdom in each. Because of this, the kingdom level of classification for Bacteria and Archaea is now often omitted, biologists using just their domain and phyla names.

> **Key Learning Outcome 15.6** Living organisms are grouped into three categories called domains. One of the domains, Eukarya, is divided into four kingdoms: Protista, Fungi, Plantae, and Animalia.

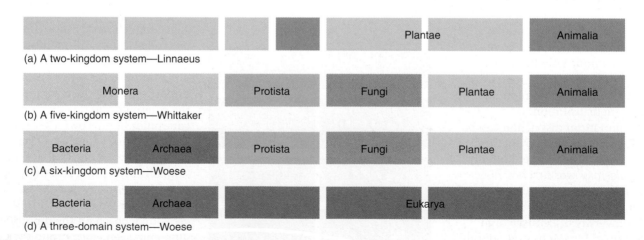

(a) A two-kingdom system—Linnaeus

(b) A five-kingdom system—Whittaker

(c) A six-kingdom system—Woese

(d) A three-domain system—Woese

Figure 15.9 Different approaches to classifying living organisms.

(a) Linnaeus popularized a two-kingdom approach, in which the fungi and the photosynthetic protists were classified as plants and the nonphotosynthetic protists as animals; when prokaryotes were described, they too were considered plants. (b) Whittaker in 1969 proposed a five-kingdom system that soon became widely accepted. (c) Woese has championed splitting the prokaryotes into two kingdoms for a total of six kingdoms or even assigning them separate domains, with a third domain containing the four eukaryotic kingdoms (d).

Domain Bacteria

The domain Bacteria contains one kingdom of the same name, Bacteria. The bacteria are the most abundant organisms on earth. There are more living bacteria in your mouth than there are mammals living on earth. Although too tiny to see with the unaided eye, bacteria play critical roles throughout the biosphere. They extract from the air all the nitrogen used by organisms, and they play key roles in cycling carbon and sulfur.

There are many different kinds of bacteria, and the evolutionary links between them are not well understood. Although there is considerable disagreement among taxonomists about the details of bacterial classification, most recognize 12 to 15 major groups of bacteria. Comparisons of the nucleotide sequences of rRNA molecules are beginning to reveal how these groups are related to each other and to the other two domains. The archaea and eukaryotes are more closely related to each other than to bacteria and are on a separate evolutionary branch of the tree (as seen in **figure 15.10**), even though archaea and bacteria are both prokaryotes.

Key Learning Outcome 15.7 Bacteria are as different from archaea as they are from eukaryotes.

TABLE 15.1	CHARACTERISTICS OF THE SIX KINGDOMS					
Domain	**Bacteria**	**Archaea**	**Eukarya**			
Kingdom	**Bacteria**	**Archaea**	**Protista**	**Plantae**	**Fungi**	**Animalia**
Cell type	Prokaryotic	Prokaryotic	Eukaryotic	Eukaryotic	Eukaryotic	Eukaryotic
Nuclear envelope	Absent	Absent	Present	Present	Present	Present
Mitochondria	Absent	Absent	Present or absent	Present	Present or absent	Present
Chloroplasts	None (photosynthetic membranes in some types)	None (bacteriorhodopsin in one species)	Present in some forms	Present	Absent	Absent
Cell wall	Present in most; peptidoglycan	Present in most; polysaccharide, glycoprotein, or protein	Present in some forms; various types	Cellulose and other polysaccharides	Chitin and other noncellulose polysaccharides	Absent
Means of genetic recombination, if present	Conjugation, transduction, transformation	Conjugation, transduction, transformation	Fertilization and meiosis	Fertilization and meiosis	Fertilization and meiosis	Fertilization and meiosis
Mode of nutrition	Autotrophic (chemosynthetic, photosynthetic) or heterotrophic	Autotrophic (photosynthesis in one species) or heterotrophic	Photosynthetic or heterotrophic or combination of both	Photosynthetic, chlorophylls *a* and *b*	Absorption	Digestion
Motility	Bacterial flagella, gliding, or nonmotile	Unique flagella in some	9 + 2 cilia and flagella; amoeboid, contractile fibrils	None in most forms, 9 + 2 cilia and flagella in gametes of some forms	Nonmotile	9 + 2 cilia and flagella, contractile fibrils
Multicellularity	Absent	Absent	Absent in most forms	Present in all forms	Present in most forms	Present in all forms

15.8 Domain Archaea

The domain Archaea contains one kingdom by the same name, the Archaea. The term *archaea* (Greek, *archaio*, ancient) refers to the ancient origin of this group of prokaryotes, which most likely diverged very early from the bacteria. Notice in **figure 15.10** that the Archaea, in red, branched off from a line of prokaryotic ancestors that led to the evolution of eukaryotes. Today, archaea inhabit some of the most extreme environments on earth. Though a diverse group, all archaea share certain key characteristics. Their cell walls lack the peptidoglycan characteristic of the cell walls of bacteria. They possess very unusual lipids and characteristic ribosomal RNA (rRNA) sequences. Also, some of their genes possess introns, unlike those of bacteria.

Archaea are grouped into three general categories: methanogens, extremophiles, and nonextreme archaea.

Methanogens (such as *Methanococcus*) obtain their energy by using hydrogen gas (H_2) to reduce carbon dioxide (CO_2) to methane gas (CH_4). They are strict anaerobes, poisoned by even traces of oxygen. They live in swamps, marshes, and the intestines of mammals. Methanogens release about 2 billion tons of methane gas into the atmosphere each year.

Extremophiles are able to grow under conditions that seem extreme to us.

Thermophiles ("heat lovers") live in very hot places, typically from 60° to 80°C. Many thermophiles have metabolisms based on sulfur. Thus, the *Sulfolobus* inhabiting the hot sulfur springs of Yellowstone National Park at 70° to 75°C obtain their energy by oxidizing elemental sulfur to sulfuric acid. The recently described *Pyrolobus fumarii* holds the current record for heat stability, temperature optimum (106°C), and temperature maximum (113°C). These extreme temperatures are characteristic of the deep-sea hydrothermal vents where this organism was discovered. *P. fumarii* is so heat-tolerant that it is not killed by a one-hour treatment in an autoclave (121°C)!

Halophiles ("salt lovers") live in very salty places like the Great Salt Lake in Utah, Mono Lake in California, and the Dead Sea in Israel. Whereas the salinity of seawater is around 3%, these prokaryotes thrive in, and indeed require, water with a salinity of 15% to 20%.

pH-tolerant archaea grow in highly acidic (pH = 0.7) and very basic (pH = 11) environments.

Pressure-tolerant archaea have been isolated from ocean depths that require at least 300 atmospheres of pressure to survive, and tolerate up to 800 atmospheres!

Nonextreme archaea grow in the same environments bacteria do. As the genomes of archaea have become better known, microbiologists have been able to identify **signature sequences** of DNA present in all archaea and in no other organisms. When samples from soil or seawater are tested for genes matching these signature sequences, many of the prokaryotes living there prove to be archaea. Clearly, archaea are not restricted to extreme habitats, as microbiologists used to think.

> **Key Learning Outcome 15.8**   Archaea are unique prokaryotes that inhabit diverse environments, some of them extreme.

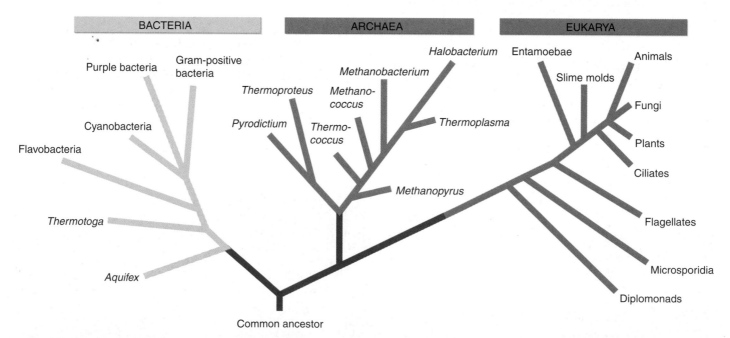

Figure 15.10   A tree of life.
This phylogeny, prepared from rRNA analyses, shows the evolutionary relationships among the three domains. The base of the tree was determined by examining genes that are duplicated in all three domains, the duplication presumably having occurred in the common ancestor. When one of the duplicates is used to construct the tree, the other can be used to root it. This approach clearly indicates that the root of the tree is within the bacterial domain. Archaea and eukaryotes diverged later and are more closely related to each other than either is to bacteria.

Domain Eukarya

For at least 1 billion years, prokaryotes ruled the earth. No other organisms existed to eat them or compete with them, and their tiny cells formed the world's oldest fossils. The third great domain of life, the eukaryotes, appear in the fossil record much later, only about 1.5 billion years ago. Metabolically, eukaryotes are more uniform than prokaryotes. Each of the two domains of prokaryotic organisms has far more metabolic diversity than all eukaryotic organisms taken together.

Three Largely Multicellular Kingdoms

Fungi, plants, and animals are well-defined evolutionary groups, each of them clearly stemming from a different single-celled eukaryotic ancestor. They are largely multicellular, each a distinct evolutionary line from an ancestor that would be classified in the kingdom Protista.

The amount of diversity among the protists, however, is much greater than that within or between the three largely multicellular kingdoms derived from the protists. Because of the size and ecological dominance of plants, animals, and fungi, and because they are predominantly multicellular, we recognize them as kingdoms distinct from Protista.

A Fourth Very Diverse Kingdom

When multicellularity evolved, the diverse kinds of single-celled organisms that existed at that time did not simply become extinct. A wide variety of unicellular eukaryotes and their relatives exists today, grouped together in the kingdom Protista solely because they are not fungi, plants, or animals. Protists are a fascinating group containing many organisms of intense interest and great biological significance.

Symbiosis and the Origin of Eukaryotes

The hallmark of eukaryotes is complex cellular organization, highlighted by an extensive endomembrane system that subdivides the eukaryotic cell into functional compartments called organelles (see chapter 4). Not all of these organelles, however, are derived from the endomembrane system. Mitochondria and chloroplasts are both believed to have entered early eukaryotic cells by a process called endosymbiosis (endo, inside) where an organism such as a bacterium is taken into the cell and remains functional inside the cell.

With few exceptions, all modern eukaryotic cells possess energy-producing organelles, the mitochondria. Mitochondria are about the size of bacteria and contain DNA. Comparison of the nucleotide sequence of this DNA with that of a variety of organisms indicates clearly that mitochondria are the descendants of purple bacteria that were incorporated into eukaryotic cells early in the history of the group. Some protist phyla have in addition acquired chloroplasts during the course of their evolution and thus are photosynthetic. These chloroplasts are derived from cyanobacteria that became symbiotic in several groups of protists early in their history. **Figure 15.11a** shows how this could have happened, with the green cyanobacterium being engulfed by an early protist. Some of these photosynthetic protists gave rise to land plants. Endosymbiosis is not strictly an ancient process but still happens today. Some photosynthetic protists are endosymbionts of some eukaryotic organisms, such as certain species of sponges, jellyfish, corals (the green structures inside the coral in **figure 15.11b** are endosymbiotic protists), octopuses, and others. We discussed the theory of the endosymbiotic origin of mitochondria and chloroplasts in chapter 4, and we will revisit it in chapter 17.

> **Key Learning Outcome 15.9** Eukaryotic cells acquired mitochondria and chloroplasts by endosymbiosis. The organisms in the domain Eukarya are divided into four kingdoms: fungi, plants, animals, and protists.

Figure 15.11 Endosymbiosis.

(a) This figure shows how an organelle could have arisen in early eukaryotic cells through a process called endosymbiosis. An organism, such as a bacterium, is taken into the cell through a process similar to endocytosis but remains functional inside the host cell. (b) Many corals contain endosymbionts, algae called zooxanthellae that carry out photosynthesis and provide the coral with nutrients. In this photograph, the zooxanthellae are the greenish-brown spheres packed into the tentacles of a coral animal.

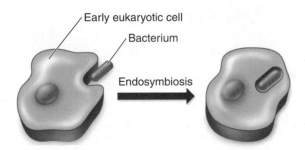

(a)

(b)

What Causes New Forms to Arise?

Biologists once presumed that new forms—genera, families, and orders—arose most often during times of massive geological disturbance, stimulated by the resulting environmental changes. But no such relationship exists. An alternative hypothesis was proposed by evolutionist George Simpson in 1953. He proposed that diversification followed new evolutionary innovations, "inventions" that permitted an organism to occupy a new "adaptive zone." After a burst of new orders that define the major groups, subsequent specialization would lead to new genera.

The early bony fishes, typified by the sturgeon (see *lower right*), had feeble jaws and long, sharklike tails. They dominated the Devonian (the Age of Fishes), to be succeeded in the Triassic (the period when dinosaurs appeared) by fishes like the gar pike, with a shorter, more powerful jaw that improved feeding and a shortened, more maneuverable tail that improved locomotion. They were in turn succeeded by teleost fishes like the perch, with an even better tail for fast, maneuverable swimming, and a complex mouth with a mobile upper jaw that slides forward as the mouth opens.

This history allows a clear test of Simpson's hypothesis. Was the appearance of these three orders followed by a burst of evolution as Simpson predicts, the new innovations in feeding and locomotion opening wide the door of opportunity? If so, many new genera should be seen in the fossil record soon after the appearance of each new order. If not, the pattern of when new genera appear should not track the appearance of new orders.

The graph shows the evolutionary history of the class Osteichthyes, the bony fishes, since they first appeared in the Silurian some 420 million years ago.

1. **Applying Concepts** What is the dependent variable?
2. **Interpreting Data** Three great innovations in jaw and tail occur during the history of the bony fishes, producing the superorders represented by sturgeons, then gars, and then teleost fishes. In what period did each innovation occur?
3. **Making Inferences** Do bursts of new genera appear at these same three times, or later?
4. **Drawing Conclusions** Do the data presented in the graph support Simpson's hypothesis? Explain.
5. **Further Analysis** If you were to plot on the graph the rate at which new families of fishes appeared, what general pattern would you expect to see, relative to new orders, if Simpson is right? Explain.

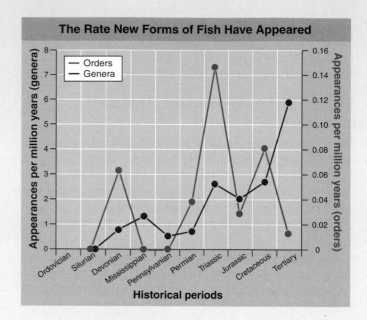

The Rate New Forms of Fish Have Appeared

Perch

Gar pike

Sturgeon

The Classification of Organisms

15.1 Invention of the Linnaean System

- Scientists use a system of grouping similar organisms together, called classification. Latin is used because it was the language used by earlier philosophers and scientists.

- The polynomial system of classification named an organism by using a list of adjectives that described the organism. The binomial system, using a two-part name, was originally developed as a "shorthand" reference to the polynomial name. Linnaeus used this two-part naming system consistently and its use became widespread (**figure 15.1**).

15.2 Species Names

- Taxonomy is the area of biology involved in identifying, naming, and grouping organisms. Scientific names consist of two parts—the genus and species. The genus is capitalized but the species is not. The two parts of the name are italicized. Scientific names are standardized, universal names that are less confusing than common names (**figure 15.2**).

15.3 Higher Categories

- In addition to the genus and species names, an organism is also assigned to higher levels of classification. The higher categories convey more general information about the organisms in a particular group. The most general category, domain, is the largest grouping, followed by ever-increasingly specific information that is used to group organisms into a kingdom, phylum, class, order, family, genus, and species (**figure 15.3**).

15.4 What Is a Species?

- The biological species concept states that a species is a group of organisms that is reproductively isolated, meaning that the individuals mate and produce fertile offspring with each other but not with those of other species.

- This concept works well to define animal species because animals regularly outcross (mate with other individuals—**figure 15.4**). However, the concept does not apply to other organisms (fungi, protists, plants, and prokaryotes) that regularly reproduce without mating through asexual reproduction. The classification of these organisms relies more on physical, behavioral, and molecular characteristics.

Inferring Phylogeny

15.5 How to Build a Family Tree

- In addition to organizing a great number of organisms, the study of taxonomy also gives us a glimpse of the evolutionary history of life on earth. Organisms with similar characteristics are more likely to be related to each other. The evolutionary history of an organism and its relationship to other species is called phylogeny.

- Phylogenetic trees can be created using key characteristics that are shared by some organisms, presumably having been inherited from a common ancestor. A group of organisms that are related by descent is called a clade, and a phylogenetic tree organized in this manner is called a cladogram (**figure 15.5**).

- Cladograms can sometimes be misleading when the characteristics are weighted, placing more importance on a characteristic that seems to have a more significant impact on evolution. The problem with this system is that some characteristics may turn out to be less important than first thought. For this reason, cladograms work best when all characteristics are weighted equally.

- Traditional taxonomy focuses more on the significance or evolutionary impact of a characteristic and not just on the commonality of the characteristic. For example, in traditional taxonomy, birds are placed in their own class, as shown here from **figure 15.6**, even though, evolutionarily, birds fall within the reptile group. Traditional taxonomy is used when more information is available to weight characteristics that seem more significant (**figure 15.7**), while cladistics places more emphasis on the order or timing in which unique, or derived, characteristics appear.

Kingdoms and Domains

15.6 The Kingdoms of Life

- The designation of kingdoms, the second-highest category used in classification, has changed over the years as more and more information about organisms has been uncovered. Currently six kingdoms have been identified: Bacteria, Archaea, Protista, Fungi, Plantae, and Animalia (**figure 15.9** and **table 15.1**).

- The domain level of classification was added in the mid-1990s, recognizing three fundamentally different types of cells: Eukarya (eukaryotic cells), Archaea (prokaryotic archaea), and Bacteria (prokaryotic bacteria) (**figure 15.10**).

15.7 Domain Bacteria

- The domain Bacteria contains prokaryotic organisms in the kingdom Bacteria. These single-celled organisms are the most abundant organisms on earth, and play key roles in ecology.

15.8 Domain Archaea

- The domain Archaea contains prokaryotic organisms in the kingdom Archaea. Although they are prokaryotes, archaea are as different from bacteria as they are from eukaryotes. These single-celled organisms are found in diverse environments but most interestingly, in very extreme environments.

15.9 Domain Eukarya

- The domain Eukarya contains very diverse organisms from four kingdoms, which are similar in that they are all eukaryotes. Fungi, plants, and animals are multicellular organisms, and the protists are primarily single celled but very diverse. Eukaryotes contain cellular organelles that were most likely acquired through endosymbiosis (**figures 15.11**).

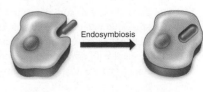

1. The wolf, domestic dog, and red fox are all in the same family, Canidae. The scientific name for the wolf is *Canis lupus,* the domestic dog is *Canis familiaris,* and the red fox is *Vulpes vulpes.* This means that
   a. the red fox is in the same family, but different genus than dogs and wolves.
   b. the dog is in the same family, but different genus than red foxes and wolves.
   c. the wolf is in the same family, but different genus than dogs and red foxes.
   d. all three organisms are in different genera.

2. The evolutionary relationship of organisms, and their relationships to other species, are its
   a. taxonomy.
   b. phylogeny.
   c. ontogeny.
   d. systematics.

3. Organisms are classified based on
   a. physical, behavioral, and molecular characteristics.
   b where the organism lives.
   c. what the organism eats.
   d. the size of the organism.

4. Organisms that are closer together on a cladogram
   a. are in the same family.
   b. comprise an outgroup.
   c. share a more recent common ancestor than those organisms that are farther apart.
   d. share fewer derived characters than organisms that are farther apart.

5. The six kingdoms of organisms can be organized into three domains based on
   a. where the organism lives.
   b. what the organism eats.
   c. cell structure.
   d. cell structure and DNA sequence.

6. All of the extremophiles belong to the domain of
   a. Bacteria.
   b. Archaea.
   c. Prokarya.
   d. Eukarya.

7. Bacteria are similar to Archaea in that they
   a. arose through endosymbiosis.
   b. are multicellular.
   c. live in extreme environments.
   d. are prokaryotes.

8. It is theorized that the ancestral organisms that gave rise to the plants, animals, and fungi originated in the kingdom
   a. Bacteria.
   b. Archaea.
   c. Protista.
   d. All of the above, each giving rise to one of the three kingdoms listed.

9. One difference between the kingdom Protista and the other three kingdoms in the domain Eukarya is that the other kingdoms are mostly
   a. prokaryotic.
   b. multicellular.
   c. eukaryotic.
   d. unicellular.

10. It is thought that the mitochondria and chloroplasts of eukaryotic cells came from
    a. the development of the internal membrane system.
    b. protists.
    c. mutation.
    d. endosymbiosis of bacteria.

1. **Figure 15.5** Which of the organisms shown have amniotic membranes that surround the fetus?

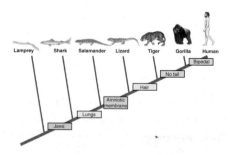

2. **Figure 15.10** Compared to the degree of difference between bacteria and archaea, how similar are animals and plants?

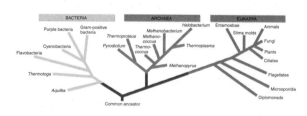

1. Classification systems summarize much information about organisms, although often a key trait allows us to immediately distinguish between two different ones. Most of us are very familiar with dogs and with cats, which are common pets. What key traits can you think of that would always distinguish a dog from a cat?

2. Birds and mammals have four-chambered hearts, while most living reptiles have three-chambered hearts. Such a fundamental difference suggests birds and mammals should be placed in the same clade, and yet many biologists instead lump birds in with the reptile clade. Evaluate their decision to do so.

16

Prokaryotes: The First Single-Celled Creatures

These two children wait in May 1995 outside the hospital in the Congo (formerly Zaire) town of Kikwit, where their parents and others infected with Ebola virus are being isolated. Seventy-eight percent of the infected died. Although viruses are not organisms—just fragments of DNA or RNA encased in protein—they can have a deadly impact on living organisms. Even the simplest living creatures, prokaryotes, are subject to viral infection; multiplying within infected cells, the viruses eventually burst forth, killing the cell. A person infected with Ebola virus suffers a similar fate, the virus invading connective tissue and fatally rupturing blood vessels. It used to be popular to talk of viruses as somehow transitional between living and nonliving, but biologists no longer hold this view. Rather, viruses are viewed as renegade segments of genomes, bits of DNA or RNA that have broken away from chromosomes but, using a host cell's machinery, are still able to produce copies of themselves. In this chapter, we will examine the simplest cellular organisms, the prokaryotes, and the viruses that infect them. We begin with a discussion on the origins of life and an examination of bacteria and archaea. We finish by taking a closer look at the viruses that infect animals and plants. Many of them have a major impact on human health; for example, influenza has been responsible for millions of human deaths.

16.1 Origin of Life

All living organisms are constructed of the same four kinds of macromolecules, discussed in chapter 3, the bricks and mortar of cells. Where the first macromolecules came from and how they came to be assembled together into cells are among the least understood questions in biology—questions that address the very origin of life itself.

No one knows for sure where the first organisms (thought to be like today's bacteria) came from. It is not possible to go back in time and watch how life originated, nor are there any witnesses. Nevertheless, it is difficult to avoid being curious about the origin of life—about what, or who, is responsible for the appearance of the first living organisms on earth. There are, in principle, at least three possibilities:

1. **Extraterrestrial origin.** Life may not have originated on earth at all but instead may have been carried to it, perhaps as an extraterrestrial infection of spores originating on a planet of a distant star. How life came to exist on that planet is a question we cannot hope to answer soon.

2. **Special creation.** Life-forms may have been put on earth by supernatural or divine forces. This viewpoint, called *creationism* or *intelligent design,* is common to most Western religions. However, almost all scientists reject creationism and intelligent design because to accept its supernatural explanation requires abandoning the scientific approach.

3. **Evolution.** Life may have evolved from inanimate matter, with associations among molecules becoming more and more complex. In this view, the force leading to life was selection; changes in molecules that increased their stability caused the molecules to persist longer.

In this text, we focus on the third possibility and attempt to understand whether the forces of evolution could have led to the origin of life and, if so, how the process might have occurred. This is not to say that the third possibility, evolution, is definitely the correct one. Any one of the three possibilities might be true. Nor does the third possibility preclude religion: A divine agency might have acted via evolution. Rather, we are limiting the scope of our inquiry to scientific matters. Of the three possibilities, only the third permits testable hypotheses to be constructed and so provides the only scientific explanation—that is, one that could potentially be disproved by experiment.

Forming Life's Building Blocks

How can we learn about the origin of the first cells? One way is to try to reconstruct what the earth was like when life originated 2.5 billion years ago. We know from rocks that there was little or no oxygen in the earth's atmosphere then and more of the hydrogen-rich gases hydrogen sulfide (SH_2), ammonia (NH_3), and methane (CH_4). Electrons in these gases would have been frequently pushed to higher energy levels by photons crashing into them from the sun or by electrical energy in lightning (**figure 16.1**). Today, high-energy electrons are quickly soaked up by the oxygen in earth's atmosphere (air is 21% oxygen, all of it contributed by photosynthesis) because oxygen atoms have a great "thirst" for such electrons. But in the absence of oxygen, high-energy electrons would have been free to help form biological molecules.

When the scientists Stanley Miller and Harold Urey reconstructed the oxygen-free atmosphere of the early earth in their laboratory and subjected it to the lightning and UV radiation it would have experienced then, they found that many of the building blocks of organisms, such as amino acids and nucleotides, formed spontaneously. They concluded that life may have evolved in a "primordial soup" of biological molecules formed in the ancient earth's oceans.

Figure 16.1 Lightning provides energy to form molecules.

Before life evolved, the simple molecules in the earth's atmosphere combined to form more complex molecules. The energy that drove these chemical reactions is thought to have come from UV radiation, lightning, and other forms of geothermal energy.

Figure 16.2 A chemical process involving bubbles may have preceded the origin of life.
In 1986, geophysicist Louis Lerman proposed that the chemical processes leading to the evolution of life took place within bubbles on the ocean's surface.

Recently, concerns have been raised regarding the "primordial soup" hypothesis as the origin of life on earth. If the earth's atmosphere had no oxygen soon after it was formed, as Miller and Urey assumed (and most evidence supports this assumption), then there would have been no protective layer of ozone to shield the earth's surface from the sun's damaging UV radiation. Without an ozone layer, scientists think UV radiation would have destroyed any ammonia and methane present in the atmosphere. When these gases are missing, the **Miller-Urey experiment** does not produce key biological molecules such as amino acids. If the necessary ammonia and methane were not in the atmosphere, where were they?

In the last two decades, support has grown among scientists for what has been called the **bubble model.** This model, proposed by geophysicist Louis Lerman in 1986, suggests that the problems with the primordial soup hypothesis disappear if the model is "stirred up" a bit. The bubble model, shown in **figure 16.2**, proposes that the key chemical processes generating the building blocks of life took place not in a primordial soup but rather within bubbles on the ocean's surface. Bubbles produced by erupting volcanoes under the sea ❶ contain various gases. Because water molecules are polar, water bubbles

tend to attract other polar molecules, in effect concentrating them within the bubbles ❷. Chemical reactions would proceed much faster in bubbles, where polar reactants would be concentrated. The bubble model solves a key problem with the primordial soup hypothesis. Inside the bubbles, the methane and ammonia required to produce amino acids would have been protected from destruction by UV radiation, with the surface of the bubble reflecting the UV rays. The bubbles pop when they reach the surface ❸ and release their chemical contents into the atmosphere ❹. Eventually the molecules reenter the oceans packaged in raindrops ❺.

If you have ever watched the ocean surge upon the shore, you may have noticed the foamy froth created by the agitated water. The edges of the primitive oceans were more than likely very frothy places bombarded by ultraviolet and other ionizing radiation, and exposed to an atmosphere that may have contained methane and other simple organic molecules.

Key Learning Outcome 16.1 Life appeared on earth 2.5 billion years ago. It may have arisen spontaneously, although the nature of the process is not clearly understood.

How Cells Arose

It is one thing to make amino acids spontaneously and quite another to link them together into proteins. Recall from **figure 3.6** that making a peptide bond involves producing a molecule of water as one of the products of the reaction. Because this chemical reaction is freely reversible, it should not occur spontaneously in water (an excess of water would push it in the opposite direction). Scientists now suspect that the first macromolecules to form were not proteins but RNA molecules. When "primed" with high-energy phosphate groups (available in many minerals), RNA nucleotides spontaneously form polynucleotide chains that might, folded up, have been capable of catalyzing the formation of the first proteins.

The First Cells

We don't know how the first cells formed, but most scientists suspect they aggregated spontaneously. When complex carbon-containing macromolecules are present in water, they tend to gather together, sometimes forming aggregations big enough to see without a microscope. Try vigorously shaking a bottle of oil-and-vinegar salad dressing—tiny bubbles called **microspheres** form spontaneously, suspended in the vinegar. Similar microspheres might have represented the first step in the evolution of cellular organization. A bubble, such as those produced by soap solutions, is a hollow, spherical structure. Certain molecules, particularly those with hydrophobic regions, will spontaneously form bubbles in water. The structure of the bubble shields the hydrophobic regions of the molecules from contact with water. Such microspheres have many cell-like properties—their outer boundary resembles the membranes of a cell in that it has two layers (see pages 74 and 75), and the microspheres can increase in size and divide. Bubble models, such as the Lerman model discussed on the previous page, propose that over millions of years, those microspheres better able to incorporate molecules and energy would have tended to persist longer than others. Although it is true that lipid microspheres will form readily in water, there appears to be no hereditary mechanism to transfer these improvements from parent microsphere to offspring.

As we learned earlier, scientists suspect that the first macromolecules to form were RNA molecules, and with the recent discovery that RNA molecules can behave as enzymes, catalyzing their own assembly, this provides a possible early mechanism of inheritance. Perhaps the first cellular components were RNA molecules, and initial steps on the evolutionary journey led to increasingly complex and stable RNA molecules. Later, the stability of RNA might have improved even more when surrounded by a microsphere. Eventually DNA may have taken the place of RNA as the storage molecule for genetic information because the double-stranded DNA would have been more stable than single-stranded RNA.

When we speak of it having taken millions of years for a cell to develop, it is hard to believe there would be enough time for an organism as complicated as a human to develop. But in the scheme of things, human beings are recent additions. If we look at the development of living organisms as a 24-hour clock

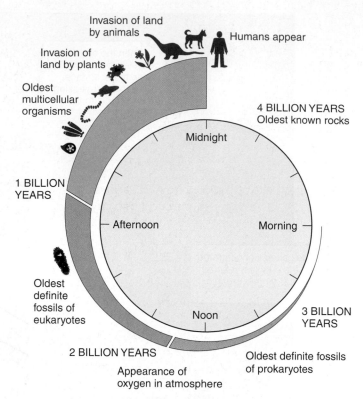

Figure 16.3 A clock of biological time.
A billion seconds ago, most students using this text had not yet been born. A billion minutes ago, Jesus was alive and walking in Galilee. A billion hours ago, the first modern humans were beginning to appear. A billion days ago, the ancestors of humans were beginning to use tools. A billion months ago, the last dinosaurs had not yet been hatched. A billion years ago, no creature had ever walked on the surface of the earth.

of biological time as shown in **figure 16.3**, with the formation of the earth 4.5 billion years ago being midnight, humans do not appear until the day is almost all over, only minutes before its end.

As you can see, the scientific vision of life's origin is at best a hazy outline. Although scientists have not disproven the hypothesis that life originated naturally and spontaneously, little is known about what actually happened. Many different scenarios seem possible, and some have solid support from experiments. Deep-sea hydrothermal vents are an interesting possibility; the prokaryotes populating these vents are among the most primitive of living organisms. Other researchers have proposed that life originated deep in the earth's crust.

Because we know so little about how DNA, RNA, and hereditary mechanisms first developed, science is currently unable to resolve disputes concerning the origin of life. How life might have originated naturally and spontaneously remains a subject of intense interest, research, and discussion among scientists.

Key Learning Outcome 16.2 Little is known about how the first cells originated. Current hypotheses involve chemical evolution within bubbles, but this is an area of intense interest in research.

Today's *Biology*

Has Life Evolved Elsewhere?

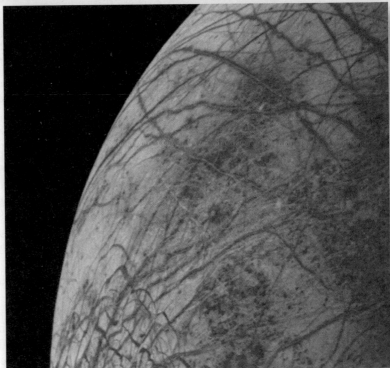

We should not overlook the possibility that life processes might have evolved in different ways on other planets. A functional genetic system, capable of accumulating and replicating changes and thus of adaptation and evolution, could theoretically evolve from molecules other than carbon, hydrogen, nitrogen, and oxygen in a different environment. Silicon, like carbon, needs four electrons to fill its outer energy level, and ammonia is even more polar than water. Perhaps under radically different temperatures and pressures, these elements might form molecules as diverse and flexible as those carbon has formed on earth.

The universe has 10^{20} (100,000,000,000,000,000,000) stars similar to our sun. We don't know how many of these stars have planets, but it seems increasingly likely that many do. Since 1996, astronomers have been detecting planets orbiting distant stars. At least 10% of stars are thought to have planetary systems. If only 1 in 10,000 of these planets is the right size and at the right distance from its star to duplicate the conditions in which life originated on earth, the "life experiment" will have been repeated 10^{15} times (that is, a million billion times). It does not seem likely that we are alone.

A dull gray chunk of rock collected in 1984 in Antarctica ignited an uproar about ancient life on Mars with the report that the rock contains evidence of possible life. Analysis of gases trapped within small pockets of the rock indicate it is a meteorite from Mars. It is, in fact, the oldest rock known to science—fully 4.5 billion years old. Evidence collected by the 2004 NASA Mars mission (the photo below of the Martian surface was taken by the rover, Spirit) suggests that the surface, now cold and arid, was much warmer when the Antarctic meteorite formed 4.5 billion years ago, that water flowed over its surface, and that it had a carbon dioxide atmosphere—conditions not too different from those that spawned life on earth.

When examined with powerful electron microscopes, carbonate patches within the meteorite exhibit what look like microfossils, some 20 to 100 nanometers in length. One hundred times smaller than any known bacteria, it is not clear they actually are fossils, but the resemblance to bacteria is striking.

Viewed as a whole, the evidence of bacterial life associated with the Mars meteorite is not compelling. Clearly, more painstaking research remains to be done before the discovery can claim a scientific consensus. However, while there is no conclusive evidence of bacterial life associated with this meteorite, it seems very possible that life has evolved on other worlds in addition to our own.

There are planets other than ancient Mars with conditions not unlike those on earth. Europa, a large moon of Jupiter, is a promising candidate (photo above). Europa is covered with ice, and photos taken in close orbit in the winter of 1998 reveal seas of liquid water beneath a thin skin of ice. Additional satellite photos taken in 1999 suggest that a few miles under the ice lies a liquid ocean of water larger than earth's, warmed by the push and pull of the gravitational attraction of Jupiter's many large satellite moons. The conditions on Europa now are far less hostile to life than the conditions that existed in the oceans of the primitive earth. In coming decades, satellite missions are scheduled to explore this ocean for life.

16.3 The Simplest Organisms

Judging from fossils in ancient rocks, prokaryotes have been plentiful on earth for over 2.5 billion years. From the diverse array of early living forms, a few became the ancestors of the great majority of organisms alive today. Several ancient forms including cyanobacteria have survived; others gave rise to other prokaryotes and to the second great group of prokaryotes, the Archaea; still others probably became extinct millions or even billions of years ago. The fossil record indicates that eukaryotic cells, being much larger than prokaryotes and exhibiting elaborate shapes in some cases, did not appear until about 1.5 billion years ago. Therefore, for at least 1 billion years prokaryotes were the only organisms that existed.

Today, prokaryotes are the simplest and most abundant form of life on earth. In a spoonful of farmland soil, 2.5 billion bacteria may be present. In 1 hectare (about 2.5 acres) of wheat land in England, the weight of bacteria in the soil is approximately equal to that of 100 sheep!

It is not surprising, then, that prokaryotes occupy a very important place in the web of life on earth. They play a key role in cycling minerals within the earth's ecosystems. In fact, photosynthetic bacteria were, in large measure, responsible for the introduction of oxygen into the earth's atmosphere. Bacteria are responsible for some of the most deadly animal and plant diseases, including many human diseases. Bacteria and archaea are our constant companions, present in everything we eat and on everything we touch.

The Structure of a Prokaryote

The essential character of prokaryotes can be conveyed in a simple sentence: **Prokaryotes** are small, simply organized, single cells that lack an organized nucleus. Therefore, bacteria and archaea are prokaryotes; their single circle of DNA is not confined by a nuclear membrane in a nucleus, as in the cells of eukaryotes. Too tiny to see with the naked eye, the cells of prokaryotes are simple in form. **Figure 16.4** illustrates the various shapes of prokaryotes. Many exist as single cells, either rod-shaped (bacilli), spherical (cocci), or spirally coiled (spirilla), some with large flagella. Other kinds of prokaryotes aggregate into filaments, and a few even form stalked structures.

The prokaryotic cell's plasma membrane is encased within a cell wall. The cell wall of bacteria is made of peptidoglycan, a network of polysaccharide molecules linked together by peptide interbridges. Many species of bacteria have a cell wall composed of layers of peptidoglycan, represented by the purple rodlike structures in the diagram shown here.

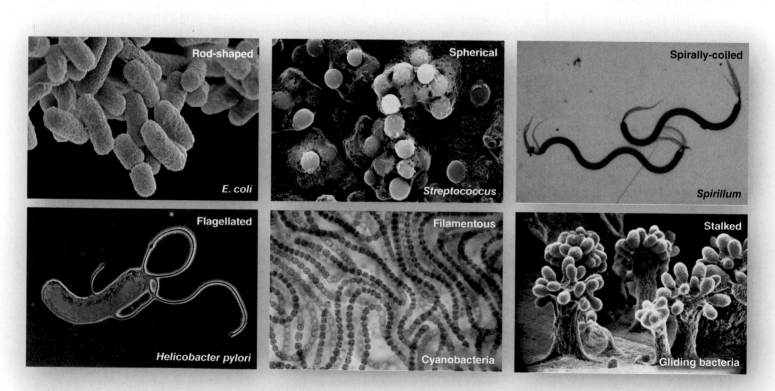

Figure 16.4   Prokaryotes come in many shapes.

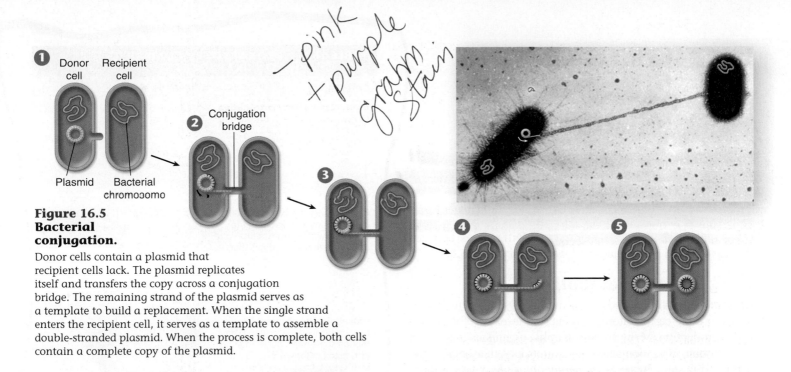

pink + purple grahm stain

Figure 16.5 Bacterial conjugation.

Donor cells contain a plasmid that recipient cells lack. The plasmid replicates itself and transfers the copy across a conjugation bridge. The remaining strand of the plasmid serves as a template to build a replacement. When the single strand enters the recipient cell, it serves as a template to assemble a double-stranded plasmid. When the process is complete, both cells contain a complete copy of the plasmid.

Other species have an outer membrane composed of large molecules of lipopolysaccharide (the red lipids in the diagram to the right) with chains of sugars attached to them covering a thinner peptidoglycan cell wall. Bacteria are commonly classified by the presence or absence of this membrane as **gram-positive,** with no outer membrane (as on the facing page), or **gram-negative,** possessing an outer membrane (as shown here). The name refers to the Danish microbiologist Hans Gram, who developed a staining process that stains the cell types differently. A purple dye is retained in the thicker peptidoglycan layer in the cell walls of gram-positive bacteria and they stain purple. In bacteria with an outer membrane, the peptidoglycan layer is thinner and does not retain the purple dye, which is washed away easily. A counterstain with a red dye is retained and so the cells stain red, not purple. The outer membranes of gram-negative bacteria make them resistant to antibiotics that attack the bacterial cell wall. That is why penicillin, which targets the protein cross-links of the bacterial cell wall, is effective only against gram-positive bacteria. Outside the cell wall and membrane, many bacteria have a gelatinous layer called a **capsule.**

Many kinds of bacteria possess threadlike **flagella,** long strands of protein that may extend out several times the length of the cell body. Bacteria swim by twisting these flagella in a corkscrew motion. Some bacteria also possess multiple shorter flagella called **pili** (singular, **pilus**), which act as docking cables, helping the cell to attach to surfaces or other cells. When exposed to harsh conditions (dryness or high temperature), some bacteria form thick-walled **endospores** around their DNA and a small bit of cytoplasm. These endospores are highly resistant to environmental stress and may germinate to form new active bacteria even after centuries.

How Prokaryotes Reproduce

Prokaryotes reproduce using a process called **binary fission,** in which an individual cell simply increases in size and divides in two. Following replication of the prokaryotic DNA, the plasma membrane and cell wall grow inward and eventually divide the cell by forming a new wall from the outside.

Some bacteria can exchange genetic information by passing plasmids from one cell to another in a process called **conjugation.** A plasmid is a small, circular fragment of DNA that replicates outside the main bacterial chromosome. In bacterial conjugation, seen in **figure 16.5,** the pilus of the donor cell extends out and contacts a recipient cell ❶, forming a passageway called a *conjugation bridge* between the two cells. The pilus draws the two cells close together. The plasmid in the donor cell begins to replicate its DNA ❷, passing the replicated copy out across the bridge and into the recipient cell ❸ where a complementary strand is synthesized ❹. The recipient cell now contains some of the genetic material found in the donor cell ❺. Genes that produce antibiotic resistance in bacteria are often transferred from one bacterial cell to another through conjugation. In addition to conjugation, bacteria can also obtain genetic information by taking up DNA from the environment (transformation; see **figure 11.1**) or from bacterial viruses (discussed later in this chapter; see **figure 16.11**).

> **Key Learning Outcome 16.3** Prokaryotes are the smallest and simplest organisms, a single cell with no internal compartments or organelles. They divide by binary fission.

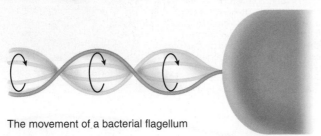

The movement of a bacterial flagellum

Comparing Prokaryotes to Eukaryotes

Prokaryotes differ from eukaryotes in many respects: The cytoplasm of prokaryotes has very little internal organization, prokaryotes are unicellular and much smaller than eukaryotes, the prokaryotic chromosome is a single circle of DNA, cell division and flagella are simple, and prokaryotes are far more metabolically diverse than eukaryotes. Differences between prokaryotes and eukaryotes are illustrated in **table 16.1**.

Prokaryotic Metabolism

Prokaryotes have evolved many more ways than eukaryotes to acquire the carbon atoms and energy necessary for growth and reproduction. Many are **autotrophs,** organisms that obtain their carbon from inorganic CO_2. Autotrophs that obtain their energy from sunlight are called *photoautotrophs,* whereas those that harvest energy from inorganic chemicals are called *chemoautotrophs.* Other prokaryotes are **heterotrophs,** organisms that obtain at least some of their carbon from organic molecules like glucose. Heterotrophs that obtain their energy from sunlight are called *photoheterotrophs,* whereas those that harvest energy from organic molecules are called *chemoheterotrophs.*

Photoautotrophs Many prokaryotes carry out photosynthesis, using the energy of sunlight to build organic molecules from carbon dioxide. The cyanobacteria use chlorophyll *a* as the key light-capturing pigment and use H_2O as an electron donor, leaving oxygen gas as a by-product. Other prokaryotes use bacteriochlorophyll as their pigment and H_2S as an electron donor, leaving elemental sulfur as the by-product.

Chemoautotrophs Some prokaryotes obtain their energy by oxidizing inorganic substances. Nitrifiers, for example, oxidize ammonia or nitrite to form the nitrate that is taken up by plants. Other prokaryotes oxidize sulfur, hydrogen gas, and other inorganic molecules. On the dark ocean floor at depths of 2,500 meters, entire ecosystems subsist on prokaryotes that oxidize hydrogen sulfide as it escapes from volcanic vents.

Photoheterotrophs The so-called purple nonsulfur bacteria use light as their source of energy but obtain carbon from organic molecules such as carbohydrates or alcohols that have been produced by other organisms.

Chemoheterotrophs Most prokaryotes obtain both carbon atoms and energy from organic molecules. These include decomposers and most pathogens (disease-causing bacteria).

> **Key Learning Outcome 16.4** Prokaryotes differ from eukaryotes in having no nucleus or other interior compartments, in being far more metabolically diverse, and in many other fundamental respects.

TABLE 16.1 PROKARYOTES COMPARED TO EUKARYOTES

Feature	Example
Internal compartmentalization. Unlike eukaryotic cells, prokaryotic cells contain no internal compartments, no internal membrane system, and no cell nucleus.	 Prokaryotic cell
Cell size. Most prokaryotic cells are only about 1 micrometer in diameter, whereas most eukaryotic cells are well over 10 times that size.	 Prokaryotic cell Eukaryotic cell
Unicellularity. All prokaryotes are fundamentally single celled. Even though some may adhere together in a matrix or form filaments, their cytoplasms are not directly interconnected, and their activities are not integrated and coordinated, as is the case in multicellular eukaryotes.	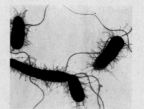 Unicellular bacteria
Chromosomes. Prokaryotes do not possess chromosomes in which proteins are complexed with the DNA, as eukaryotes do. Instead, their DNA exists as a single circle in the cytoplasm.	 Prokaryotic chromosome Eukaryotic chromosomes
Cell division. Cell division in prokaryotes takes place by binary fission (see chapter 8). The cells simply pinch in two. In eukaryotes, microtubules pull chromosomes to opposite poles during the cell division process, called mitosis.	 Binary fission in prokaryotes Mitosis in eukaryotes
Flagella. Prokaryotic flagella are simple, composed of a single fiber of protein that is spun like a propeller. Eukaryotic flagella are more complex structures, with a 9 + 2 arrangement of microtubules, that whip back and forth rather than rotate.	 Simple bacterial flagellum
Metabolic diversity. Prokaryotes possess many metabolic abilities that eukaryotes do not: Prokaryotes perform several different kinds of anaerobic and aerobic photosynthesis; prokaryotes can obtain their energy from oxidizing inorganic compounds (so-called chemoautotrophs); and prokaryotes can fix atmospheric nitrogen.	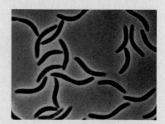 Chemoautotrophs

Importance of Prokaryotes

Prokaryotes and the Environment

Prokaryotes were largely responsible for creating the properties of the atmosphere and the soil for over 2 billion years. They are metabolically much more diverse than eukaryotes, which is why they are able to exist in such a wide range of habitats. The many autotrophic bacteria—either photoautotrophic or chemoautotrophic—make major contributions to the world carbon balance in terrestrial, freshwater, and marine habitats. Other heterotrophic bacteria play a key role in world ecology by breaking down organic compounds. The carbon, nitrogen, phosphorus, sulfur, and other atoms that make up living organisms come from the environment, and when organisms die and decay, they all return to the environment. The prokaryotes, and other organisms such as fungi, that carry out the breakdown portion of this cycle are called *decomposers*. Another key role of prokaryotes in the world ecosystem relates to the fact that only a few genera of bacteria—and no other organisms—have the ability to fix atmospheric nitrogen and thus make it available for use by other organisms.

Bacteria and Gene Engineering

Applying genetic engineering methods to produce improved strains of bacteria for commercial use holds enormous promise for the future. Bacteria are under intense investigation, for example, as nonpolluting insect control agents. *Bacillus thuringiensis* produces a protein that is toxic when ingested by certain insects, and improved, highly specific strains of B. *thuringiensis* have greatly increased its usefulness as a biological control agent. Genetically modified bacteria have also been extraordinarily useful in producing insulin and other therapeutic proteins. Gene modified bacteria are also playing a part in removing environmental pollutants. Oil-degrading bacteria were used to clean up the *Exxon Valdez* oil spill off the coast of Alaska. The rocks on the left in figure 16.6 below show the oil contamination, while the rocks on the right show an area that was cleaned up using bacteria.

Bacteria, Disease, and Bioterrorism

Some bacteria cause major diseases in plants and animals, including humans. Among important human bacterial diseases that can be lethal are anthrax, cholera, plague, pneumonia, tuberculosis (TB), and typhus. Many pathogenic (disease-causing) bacteria like cholera are dispersed in food and water. Some like typhus and plague, spread among rodents and humans by fleas. Others, like TB, are spread through the air in water droplets (from a cough or sneeze), infecting those who inhale the droplets. Among these inhalation pathogens is anthrax, a disease associated with livestock that rarely kills humans. Most human infections are cutaneous, infecting a cut in the skin, but if a significant number of anthrax endospores are inhaled, the pulmonary (lung) infection is often fatal. Biological warfare programs in the United States and the former Soviet Union focused on anthrax as a near-ideal biological weapon, although it has never been used in war. Bioterrorists struck at the United States with anthrax endospores in 2001.

> **Key Learning Outcome 16.5** Prokaryotes make many important contributions to the world ecosystem, including occupying key roles in cycling carbon and nitrogen.

Figure 16.6 Using bacteria to clean up oil spills.
Bacteria can be used to remove environmental pollutants, such as the hydrocarbons released into the Gulf of Mexico in the 2010 oil spill. In areas contaminated by the *Exxon Valdez* oil spill (*rocks on the left*), oil-degrading bacteria produced dramatic results (*rocks on the right*).

16.6 Prokaryotic Lifestyles

Archaea

Many of the archaea that survive today are **methanogens,** prokaryotes that use hydrogen (H_2) gas to reduce carbon dioxide (CO_2) to methane (CH_4). Methanogens are strict anaerobes, poisoned by oxygen gas. They live in swamps and marshes, where other microbes have consumed all the oxygen. The methane that they produce bubbles up as "marsh gas." Methanogens also live in the guts of cows and other herbivores that live on a diet of cellulose, converting the CO_2 produced by its digestion to methane gas. The best understood archaea are extremophiles, which live in unusually harsh environments, such as the very salty Dead Sea and the Great Salt Lake (over 10 times saltier than seawater). **Thermoacidophiles** favor hot, acidic springs such as the sulfur springs of Yellowstone National Park (figure 16.7), where the water is nearly 80°C, with an acidic pH of 2 or 3.

Bacteria

Almost all prokaryotes that have been described by scientists are members of the kingdom Bacteria. Many are heterotrophs that power their lives by consuming organic molecules, whereas others are photosynthetic, gaining their energy from the sun. **Cyanobacteria** are among the most prominent of the photosynthetic bacteria. We have already mentioned the critical role that the members of this ancient phylum played in the history of the earth by generating the oxygen in our atmosphere. Cyanobacteria are filamentous bacteria, like the *Anabaena* pictured in **figure 16.8.** Nitrogen fixation occurs in almost all cyanobacteria, within specialized cells called **heterocysts** (the enlarged cells that appear along the filament of *Anabaena* and in many of the other filamentous members of this phylum). In **nitrogen fixation,** atmospheric nitrogen is converted to a form that can be used by living organisms.

There are numerous phyla of nonphotosynthetic bacteria. Some are chemoautotrophs, but most are heterotrophs. Some of these heterotrophs are decomposers, breaking down organic material. Bacteria and fungi play the leading role in breaking down organic molecules formed by biological processes, thereby making the nutrients in these molecules available once more for recycling. Decomposition is just as indispensable to the continuation of life on earth as is photosynthesis.

Although bacteria are unicellular organisms, they sometimes form associations, as in *Anabaena;* also layers of bacterial cells, called **biofilms,** can form on the surface of a substrate. By forming biofilms, the bacteria create a microenvironment that facilitates their growth. Biofilms are found in nature but also impact humans because they can form on teeth and on medical equipment such as catheters and contact lenses. Biofilms can protect the bacteria from disinfectants.

Figure 16.7 Thermoacidophiles live in hot springs.
These archaea growing in Sulfide Spring, Yellowstone National Park, Wyoming, are able to tolerate high acid levels and very hot temperatures.

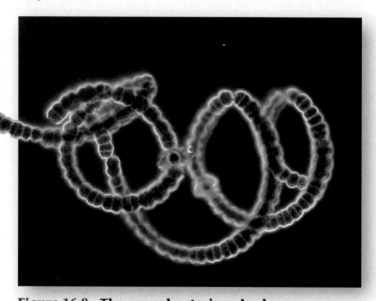

Figure 16.8 The cyanobacterium *Anabaena*.
Individual cells adhere in filaments. The larger cells (areas on the filament that seem to be bulging) are heterocysts, specialized cells in which nitrogen fixation occurs. These organisms exhibit one of the closest approaches to multicellularity among the bacteria.

Bacteria cause many diseases in humans (**table 16.2**), including cholera, diphtheria, and leprosy. Among the most serious of bacterial diseases is **tuberculosis (TB),** a disease of the respiratory tract caused by the bacterium *Mycobacterium tuberculosis*. TB is a leading cause of death throughout the world. Spread through the air, tuberculosis is quite infectious. TB was a major health risk in the United States until the discovery of effective drugs to suppress it in the 1950s. The appearance of drug-resistant strains in the 1990s has raised serious concern within the medical community, and the search is on for new types of anti-TB drugs.

Key Learning Outcome 16.6 Most commonly encountered prokaryotes are bacteria; some cause significant diseases in humans.

TABLE 16.2 — IMPORTANT HUMAN BACTERIAL DISEASES

Disease	Pathogen	Vector/ Reservoir	Symptoms and Mode of Transmission
Anthrax	Bacillus anthracis	Farm animals	Bacterial infection that can be transmitted through inhaled endospores, by contact, or ingestion. Rare except in sporadic outbreaks. Pulmonary (inhaled) anthrax is often fatal, while cutaneous anthrax (infection through cuts) is readily treated with antibiotics. Anthrax endospores have been used as a biological weapon.
Botulism	Clostridium botulinum	Improperly prepared food	Contracted through ingestion of contaminated food; endospores can sometimes persist in cans and bottles if the containers have not been heated at a high enough temperature to kill the spores. Produces acutely toxic poison; can be fatal.
Chlamydia	Chlamydia trachomatis	Humans (STD)	Urogenital infections with possible spread to eyes and respiratory tract. Occurs worldwide; increasingly common over past 20 years.
Cholera	Vibrio cholerae	Humans (feces), plankton	Causes severe diarrhea that can lead to death by dehydration; 50% peak mortality if the disease goes untreated. A major killer in times of crowding and poor sanitation; over 100,000 died in Rwanda in 1994 during a cholera outbreak.
Dental caries	Streptococcus	Humans	A dense collection of this bacteria on the surface of teeth leads to secretion of acids that destroy minerals in tooth enamel—sugar will not cause caries but bacteria feeding on it will.
Diphtheria	Corynebacterium diphtheriae	Humans	Acute inflammation and lesions of mucous membranes. Spread through contact with infected individual. Vaccine available.
Gonorrhea	Neisseria gonorrhoeae	Humans only	STD, on the increase worldwide. Usually not fatal.
Hansen's disease (leprosy)	Mycobacterium leprae	Humans, feral armadillos	Chronic infection of the skin; worldwide incidence about 10 to 12 million, especially in Southeast Asia. Spread through contact with infected individuals.
Lyme disease	Borrelia burgdorferi	Ticks, deer, small rodents	Spread through bite of infected tick. Lesion followed by malaise, fever, fatigue, pain, stiff neck, and headache.
Peptic ulcers	Helicobacter pylori	Humans	Originally thought to be caused by stress or diet, most peptic ulcers now appear to be caused by this bacterium; good news for ulcer sufferers as it can be treated with antibiotics.
Plague	Yersinia pestis	Fleas of wild rodents: rats and squirrels	Killed one-fourth of the population of Europe in the 14th century; endemic in wild rodent populations of the western United States in the 1990s.
Pneumonia	Streptococcus, Mycoplasma, Chlamydia, Klebsiella	Humans	Acute infection of the lungs, often fatal if not treated.
Tuberculosis	Mycobacterium tuberculosis	Humans	An acute bacterial infection of the lungs, lymph, and meninges. Its incidence is on the rise, complicated by the emergence of new strains of the bacteria that are resistant to antibiotics.
Typhoid fever	Salmonella typhi	Humans	A systemic bacterial disease of worldwide incidence. Less than 500 cases a year are reported in the United States. The disease is spread through contaminated water or foods (such as improperly washed fruits and vegetables). Vaccines are available for travelers.
Typhus	Rickettsia	Lice, rat fleas, humans	Historically, a major killer in times of crowding and poor sanitation; transmitted from human to human through the bite of infected lice and fleas. Typhus has a peak untreated mortality rate of 70%.

16.7 The Structure of Viruses

The border between the living and the nonliving is very clear to a biologist. Living organisms are cellular and able to grow and reproduce independently, guided by information encoded within DNA. The simplest creatures living on earth today that satisfy these criteria are prokaryotes. Viruses, on the other hand, do not satisfy the criteria for "living" because they possess only a portion of the properties of organisms. **Viruses** are literally "parasitic" chemicals, segments of DNA (or sometimes RNA) wrapped in a protein coat. They cannot reproduce on their own, and for this reason they are not considered alive by biologists. They can, however, reproduce within cells, often with disastrous results to the host organism.

Viruses are very small. The smallest viruses are only about 17 nanometers in diameter. Viruses are so small that they are smaller than many of the molecules in a cell. Most viruses can be detected only by using the higher resolution of an electron microscope.

The true nature of viruses was discovered in 1935, when the biologist Wendell Stanley prepared an extract of a plant virus called *tobacco mosaic virus (TMV)* and attempted to purify it. To his great surprise, the purified TMV preparation precipitated (that is, separated from solution) in the form of crystals. This was surprising because precipitation is something that only chemicals do—the TMV virus was acting like a chemical rather than an organism. Stanley concluded that TMV is best regarded as just that— a chemical matter rather than a living organism.

Each particle of TMV virus is in fact a mixture of two chemicals: RNA and protein. The TMV virus, pictured in **figure 16.9b**, has the structure of a Twinkie, a tube made of an RNA core (the green springlike structure) surrounded by a coat of protein (the purple structures that encircle the RNA). Later workers were able to separate the RNA from the protein and purify and store each chemical. Then, when they reassembled the two components, the reconstructed TMV particles were fully able to infect healthy tobacco plants and so clearly *were* the virus itself, not merely chemicals derived from it.

Viruses occur in all organisms, from bacteria to humans, and in every case the basic structure of the virus is the same, a core of nucleic acid surrounded by protein. There is considerable difference, however, in the details. In **figure 16.9** you can compare the structure of bacterial, plant, and animal viruses—they are clearly quite different from one another and there is even a wide variety of shapes and structures within each group of viruses. Bacterial viruses, called *bacteriophages,* can have elaborate structures, like the bacteriophage in **figure 16.9a** that looks more like a lunar module than a virus. Many plant viruses like TMV have a core of RNA, and some animal viruses like HIV (**figure 16.9c**) do too. Several different segments of DNA or RNA may be present in animal virus particles, along with many different kinds of protein. Like TMV, most viruses form a protein sheath, or **capsid,** around their nucleic acid core. In addition, many viruses (like HIV) form a membranelike **envelope,** rich in proteins, lipids, and glycoprotein molecules, around the capsid.

> **Key Learning Outcome 16.7** Viruses are genomes of DNA or RNA, encased in a protein shell, that can infect cells and replicate within them. They are chemical assemblies, not cells, and are not alive.

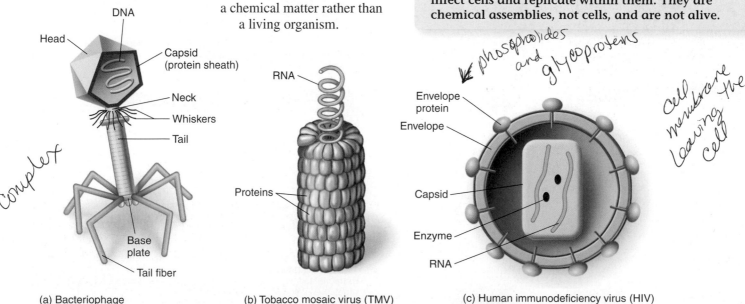

(a) Bacteriophage   (b) Tobacco mosaic virus (TMV)   (c) Human immunodeficiency virus (HIV)

Figure 16.9 The structure of bacterial, plant, and animal viruses.
(*a*) Bacterial viruses, called bacteriophages, often have a complex structure. (*b*) TMV infects plants and consists of 2,130 identical protein molecules (*purple*) that form a cylindrical coat around the single strand of RNA (*green*). The RNA backbone determines the shape of the virus and is protected by the identical protein molecules packed tightly around it. (*c*) In the human immunodeficiency virus (HIV), the RNA core is held within a capsid that is encased by a protein envelope.

How Bacteriophages Enter Prokaryotic Cells

Bacteriophages are viruses that infect bacteria. They are diverse both structurally and functionally, and are united solely by their occurrence in bacterial hosts. Bacteriophages with double-stranded DNA have played a key role in molecular biology. Many of these bacteriophages are large and complex, with relatively large amounts of DNA and proteins. Some of them have been named as members of a "T" series (T1, T2, and so forth); others have been given different kinds of names. To illustrate the diversity of these viruses, T3 and T7 bacteriophages are icosahedral and have short tails. In contrast, the so-called T-even bacteriophages (T2, T4, and T6) are more complex, as shown by the T4 bacteriophage in **figure 16.9a**. T-even phages have an icosahedral head that holds the DNA (shown in the cutaway view), a capsid that consists primarily of three proteins, a connecting neck with a collar and "whiskers," a long tail, and a complex base plate. Many of these structures are also visible in the micrograph of a T4 bacteriophage in **figure 16.10a**.

The Lytic Cycle

During the process of bacterial infection by a T4 bacteriophage, at least one of the tail fibers contacts the lipopolysaccarides of the host bacterial cell wall. The other tail fibers set the bacteriophage perpendicular to the surface of the bacterium and bring the base plate into contact with the cell surface, as seen on the left side in **figure 16.10b**. After the bacteriophage is in place, the tail contracts, and the tail tube passes through an opening that appears in the base plate, piercing the bacterial cell wall (as shown on the right side of **figure 16.10b**). The contents of the head, mostly DNA, are then injected into the host cytoplasm.

The T-series bacteriophages and other phages such as lambda (λ) are all virulent viruses, multiplying within infected cells and eventually lysing (rupturing) them. When a virus kills the infected host cell in which it is replicating, the reproductive cycle is referred to as a *lytic cycle* (see **figure 16.11**). The viral DNA that is injected in the cell is transcribed and translated by the host cell into viral components that are assembled in the host cell. Eventually, the host cell ruptures and the new lambda phages are released, ready to infect more cells.

The Lysogenic Cycle

Many bacteriophages do not immediately kill the cells they infect, instead integrating their nucleic acid into the genome of the infected host cell (see the lower cycle shown in **figure 16.11**). While residing there, it is called a **prophage.** Among the bacteriophages that do this is the lambda phage of

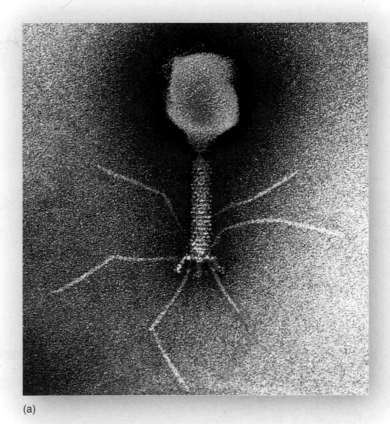

(a)

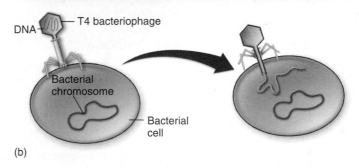

(b)

Figure 16.10 A T4 bacteriophage.
(a) Electron micrograph of T4 and (b) diagram of a T4 bacteriophage infecting a bacterial cell.

Escherichia coli, which is also a lytic phage. We know as much about this bacteriophage as we do about virtually any other biological particle; the complete sequence of its 48,502 bases has been determined. At least 23 proteins are associated with the development and maturation of lambda phage, and many other enzymes are involved in the integration of these viruses into the host genome.

The integration of a virus into a cellular genome is called **lysogeny.** At a later time, the prophage may exit the genome and initiate virus replication. This sort of reproductive cycle, involving a period of genome integration, is called a *lysogenic cycle.* Viruses that become stably integrated within the genome of their host cells are called lysogenic viruses or temperate viruses.

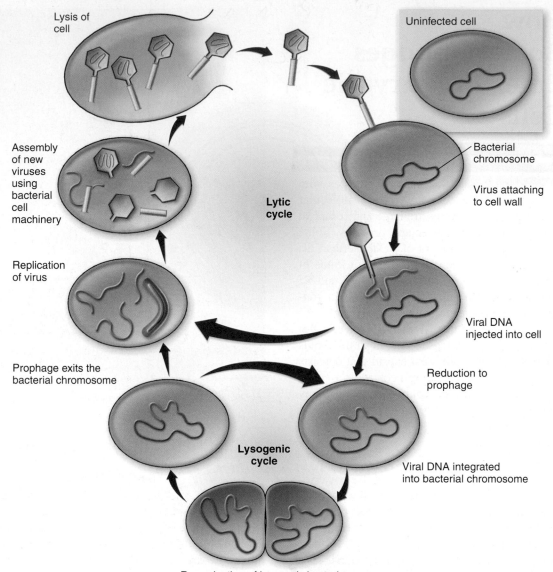

Figure 16.11 Lytic and lysogenic cycles of a bacteriophage.

In the lytic cycle, the bacteriophage exists as viral DNA, free in the bacterial host cell's cytoplasm. The viral DNA directs the production of new viral particles by the host cell until the virus kills the cell by lysis. In the lysogenic cycle, the bacteriophage DNA is integrated into the large, circular DNA molecule of the host bacterium and reproduces. It may continue to replicate and produce lysogenic bacteria or enter the lytic cycle and kill the cell. Bacteriophages are much smaller, relative to their hosts, than illustrated in this diagram.

Gene Conversion and Cholera

The expression of viral genes integrated into bacterial chromosomes is called **gene conversion.** An important example of the potential serious effects of viral genes becoming integrated into a bacterial chromosome is provided by the bacterium responsible for the often-fatal human disease cholera. The bacteria *Vibrio cholerae* usually exist in a harmless form, but a second disease-causing, virulent form also occurs. In this latter form, the bacterium causes the deadly disease cholera. Research now shows that a bacteriophage that infects *V. cholerae* introduces into the host bacterial cell a gene that codes for the cholera toxin. This gene becomes incorporated into the bacterial DNA, where it is translated along with the other host genes, thereby transforming the benign bacterium to a disease-causing agent.

Lysogenic conversion is also responsible for the presence of toxin genes in (and much of the virulence of) other pathogens like *Corynebacterium diphtheriae*, which causes diphtheria, *Streptococcus pyogenes*, which causes scarlet fever, and *Clostridium botulinum*, which causes botulism.

Key Learning Outcome 16.8 **Bacteriophages are a diverse group of viruses that attack bacteria. Some kill their host in a lytic cycle; others integrate into the host's genome, initiating a lysogenic cycle. Bacteriophages transform *Vibrio cholerae* and other bacteria into disease-causing agents.**

Biology and *Staying Healthy*

Bird and Swine Flu

The influenza virus has been one of the most lethal viruses in human history. Flu viruses are animal RNA viruses containing 11 genes. An individual flu virus resembles a sphere studded with spikes composed of two kinds of protein. Different strains of flu virus, called subtypes, differ in their protein spikes. One of these proteins, hemagglutinin (H), aids the virus in gaining access to the cell interior. The other, neuraminidase (N), helps the daughter virus break free of the host cell once virus replication has been completed. Flu viruses are currently classified into 13 distinct H subtypes and 9 distinct N subtypes, each of which requires a different vaccine to protect against infection. Thus, the virus that caused the Hong Kong flu epidemic of 1968 has type 3 H molecules and type 2 N molecules, and is called H3N2.

Recombination within humans
A person infected with a flu virus can become infected with another type of flu virus by direct contact with birds. The two viruses can undergo genetic recombination to produce a third type of virus, which can spread from human to human.

Recombination within pigs
Pigs can contract flu viruses from both birds and humans. The flu viruses can undergo genetic recombination in the pig, to produce a new kind of flu virus, which can spread from pigs to humans.

How New Flu Strains Arise

Worldwide epidemics of the flu in the last century have been caused by shifts in flu virus H-N combinations. The "killer flu" of 1918, H1N1, thought to have passed directly from birds to humans, killed between 40 and 100 million people worldwide. The Asian flu of 1957, H2N2, killed over 100,000 Americans, and the Hong Kong flu of 1968, H3N2, killed 70,000 Americans.

It is no accident that new strains of flu usually originate in the Far East. The most common hosts of influenza virus are ducks, chickens, and pigs, which in Asia often live in close proximity to each other and to humans. Pigs are subject to infection by both bird and human strains of the virus, and individual animals are often simultaneously infected with multiple strains. This creates conditions favoring genetic recombination between strains, as illustrated above, sometimes putting together novel combinations of H and N spikes unrecognizable by human immune defenses specific for the old configuration. The Hong Kong flu, for example, arose from recombination between H3N8 from ducks and H2N2 from humans. The new strain of influenza, in this case H3N2, then passed back to humans, creating an epidemic because the human population had never experienced that H-N combination before.

Conditions for a Pandemic

Not every new strain of influenza creates a worldwide flu epidemic. Three conditions are necessary: (1) The new strain must contain a novel combination of H and N spikes, so that the human population has no significant immunity from infection; (2) the new strain must be able to replicate in humans and cause death—many bird influenza viruses are harmless to people because they cannot multiply in human cells; (3) the new strain must be efficiently transmitted between humans. The H1N1 killer flu of 1918 spread in water droplets exhaled from infected individuals and subsequently inhaled into the lower respiratory tract of nearby people.

The new strain need not be deadly to every infected person in order to produce a pandemic—the H1N1 flu of 1918 had an overall mortality rate of only 2%, and yet killed 40 to 100 million people. Why did so many die? Because so much of the world's population was infected.

Bird Flu

A potentially deadly new strain of flu virus emerged in Hong Kong in 1997, H5N1. Like the 1918 pandemic strain, H5N1 passes to humans directly from infected birds, usually chickens or ducks, and for this reason has been dubbed "bird flu." Bird flu satisfies the first two conditions of a pandemic: H5N1 is a novel combination of H and N spikes for which humans have little immunity, and the resulting strain is particularly deadly, with a mortality of 59% (much higher than the 2% mortality of the 1918 H1N1 strain). Fortunately, the third condition for a pandemic is not yet met: The H5N1 strain of flu virus does not spread easily from person to person, and the number of human infections remains small.

Swine Flu

A second potentially pandemic form of flu virus, H1N1, emerged in Mexico in 2009, passing to humans from infected pigs. It seems to have arisen by multiple genetic recombination events between humans, birds, and pigs. Like the 1918 H1N1 virus, this flu (dubbed "swine flu") passes easily from person to person, and within a year it spread around the world. Also like the 1918 virus, the initial wave of swine flu infection triggered only mild symptoms in most people. The 1918 H1N1 virus only became deadly in a second wave of infection, after the virus had better adapted to living in the human body. Like bird flu, public health officials continue to watch swine flu carefully, for fear that a subsequent wave of infection may also become more lethal.

16.9 How Animal Viruses Enter Cells

As we just discussed, bacterial viruses punch a hole in the bacterial cell wall and inject their DNA inside. Plant viruses like TMV enter plant cells through tiny rips in the cell wall at points of injury. Animal viruses typically enter their host cells by membrane fusion, or sometimes by endocytosis, a process described in chapter 4, in which the cell's plasma membrane dimples inward, surrounding and engulfing the virus particle.

A diverse array of viruses occur among animals. A good way to gain a general idea of how they enter cells is to look at one animal virus in detail. Here we will look at the virus responsible for a relatively new and fatal disease, acquired immunodeficiency syndrome (AIDS). AIDS was first reported in the United States in 1981. It was not long before the infectious agent, human immunodeficiency virus (HIV), was identified by laboratories. HIV is shown budding off of a cell in figure 16.12. The cell is the purple and yellow structure at the bottom, and the HIVs are the circular structures suspended above the surface of the cell. HIV's genes are closely related to those of a chimpanzee virus, suggesting that HIV first entered humans in Africa from chimpanzees.

One of the cruelest aspects of AIDS is that clinical symptoms typically do not begin to develop until long after infection by the HIV virus, generally 8 to 10 years after the initial exposure to HIV. During this long interval, carriers of HIV have no clinical symptoms but are typically fully infectious, making the spread of HIV very difficult to control.

Attachment

When HIV is introduced into the human bloodstream, the virus particle circulates throughout the entire body but will only infect certain cells, ones called *macrophages* (Latin, big eaters). Macrophages are the garbage collectors of the body, taking up and recycling fragments of ruptured cells and other bits of organic debris. It is not surprising that HIV specializes in infecting this one kind of cell—many other animal viruses are similarly narrow in their requirements. For example, poliovirus has an affinity for motor nerve cells, and hepatitis virus infects primarily liver cells.

How does a virus such as HIV recognize a specific kind of target cell such as a macrophage? Every kind of cell in the human body has a specific array of cell surface marker proteins, molecules that serve to identify the cells. HIV viruses are able to recognize the macrophage cell surface markers. Studding the surface of each HIV virus are spikes that bang into any cell the virus encounters. Look back to **figure 16.9c**, a drawing of HIV that shows these spikes (the lollipop-looking structures embedded in the envelope). Each spike is composed of a protein called *gp120*. Only when gp120 happens onto a cell surface marker that matches its shape does the HIV virus adhere to an animal cell and infect it. It turns out that gp120 precisely fits a cell surface marker called **CD4,** and that CD4 occurs on the surfaces of macrophages. **Panel 1** in **figure 16.13** shows the gp120 protein of HIV docking onto the CD4 surface marker on the macrophage.

Entry into Macrophages

Certain cells of the immune system, called T lymphocytes, or *T cells,* also possess CD4 markers. Why are they not infected right away, as macrophages are? This is the key question underlying the mystery of the long AIDS latent period. When T lymphocytes become infected and killed, AIDS commences. So what holds off T cell infection so long?

Researchers have learned that after docking onto the CD4 receptor of a macrophage, the HIV virus requires a second receptor protein, called **CCR5,** to pull itself across the plasma membrane. After gp120 binds to CD4, its shape becomes twisted (a chemist would say it goes through a conformational change) into a new form that fits the CCR5 coreceptor molecule. Investigators speculate that after the conformational change, the coreceptor CCR5 passes the gp120-CD4 complex through the plasma membrane by triggering membrane fusion. Macrophages have the CCR5 coreceptor, as shown in **panel 1,** but T lymphocytes do not.

Replication

Panel 1 also shows that once inside the macrophage cell, the HIV virus particle sheds its protective coat. This leaves the virus nucleic acid (RNA in this case) floating in the cell's cytoplasm, along with a viral enzyme that was also within the virus shell. This enzyme, called **reverse transcriptase,** binds to the tip of the virus RNA and slides down it, synthesizing DNA that matches the information contained in the virus RNA, shown in **panel 2.** Importantly, the HIV reverse transcriptase enzyme doesn't do its job very accurately. It often makes mistakes in reading the HIV RNA, and so creates many new mutations. The mistake-ridden double-stranded DNA that it produces may integrate itself into the host cell's DNA, as in **panel 2;** it can then direct the host cell's machinery to produce many copies of the virus, shown in **panel 3.**

In all of this process, no lasting damage is done to the host cell. HIV does not rupture and kill the macrophage cells it infects. Instead, the new viruses are released from the

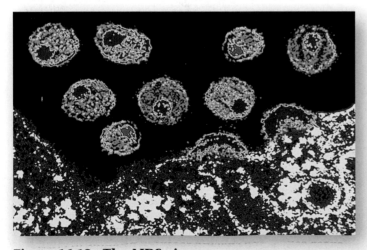

Figure 16.12 The AIDS virus.
HIV particles exit a cell to spread and infect neighboring cells.

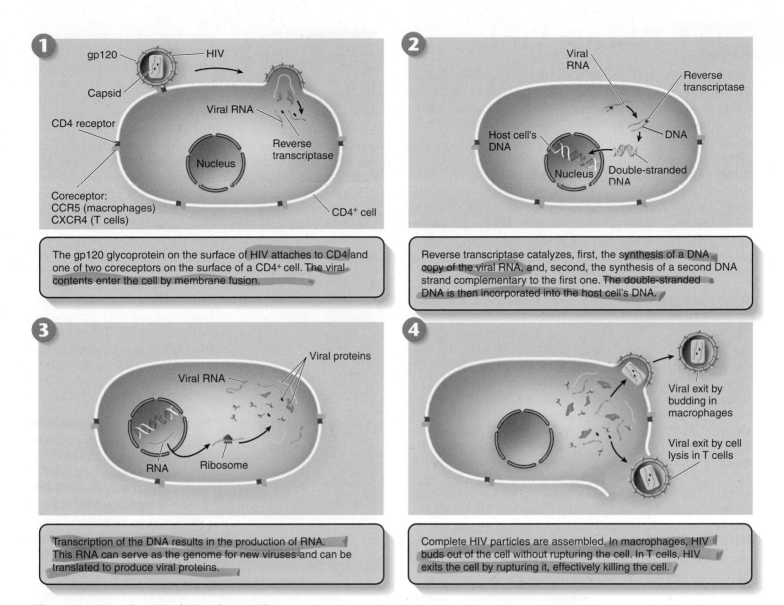

1 The gp120 glycoprotein on the surface of HIV attaches to CD4 and one of two coreceptors on the surface of a CD4+ cell. The viral contents enter the cell by membrane fusion.

2 Reverse transcriptase catalyzes, first, the synthesis of a DNA copy of the viral RNA, and, second, the synthesis of a second DNA strand complementary to the first one. The double-stranded DNA is then incorporated into the host cell's DNA.

3 Transcription of the DNA results in the production of RNA. This RNA can serve as the genome for new viruses and can be translated to produce viral proteins.

4 Complete HIV particles are assembled. In macrophages, HIV buds out of the cell without rupturing the cell. In T cells, HIV exits the cell by rupturing it, effectively killing the cell.

Figure 16.13 The HIV infection cycle.

cell by budding (shown in the upper right of **panel 4**), a process much like exocytosis, in which the new viruses fold out opposite to the way that HIV initially gained entry into the cell at the start of the infection.

This, then, is the basis of the long latency period characteristic of AIDS. The HIV virus cycles through macrophages over a period of years, multiplying powerfully but doing little apparent damage to the body.

Starting AIDS: Entry into T Cells

During this long latent period, HIV is constantly replicating and mutating as it cycles through successive generations of macrophages. Eventually, by chance, HIV alters the gene for gp120 in a way that causes the gp120 protein to change its coreceptor allegiance. This new form of gp120 protein prefers to bind instead to a different coreceptor, **CXCR4,** a receptor that occurs on the surface of T cells that have the CD4 cell surface marker. Soon the body's T cells become infected with HIV.

This has deadly consequences, as new viruses exit T cells by bursting through the plasma membrane. This rupturing destroys the T cell's physical integrity and kills it (as shown at the lower right in **panel 4**). Therefore, HIV can either bud off the cell as in macrophages or it can cause cell lysis in T cells. In the case of T cells, as the released viruses infect nearby CD4+ T cells, they in turn are ruptured, in a widening circle of cell death. It is this destruction of the body's T cells, which fight other infections in the body, that blocks the body's immune response and leads directly to the onset of AIDS. Cancers and opportunistic infections are free to invade the defenseless body.

Key Learning Outcome 16.9 Animal viruses enter cells using specific receptor proteins to cross the plasma membrane.

Disease Viruses

Humans have known and feared diseases caused by viruses for thousands of years. Among the diseases that viruses cause (**table 16.3**) are AIDS, influenza, yellow fever, polio, chicken pox, measles, herpes, infectious hepatitis, and smallpox, as well as many other diseases not as well known.

The Origin of Viral Diseases

Sometimes viruses that originate in one organism pass to another, causing a disease in the new host. New pathogens arising in this way, called *emerging viruses,* represent a greater threat today than in the past, as air travel and world trade in animals allow infected individuals and animals to move about the world quickly, spreading an infection.

Influenza Perhaps the most lethal virus in human history has been the influenza virus. Between 40 and 100 million worldwide died of flu within 19 months in 1918 and 1919—an astonishing number. The natural reservoir of influenza virus is in ducks, chickens, and pigs in central Asia. Major flu pandemics (that is, worldwide epidemics) have arisen in Asian ducks through recombination within multiple infected individuals, putting together novel combinations of virus surface proteins unrecognizable by human immune defenses. The Asian flu of 1957 killed over 100,000 Americans. The Hong Kong flu of 1968 infected 50 million people in the United States alone, of which 70,000 died. The swine flu epidemic of 2009, discussed on page 351, killed thousands of American children.

AIDS (HIV, human immunodeficieny virus) The virus that causes AIDS first entered humans from chimpanzees somewhere in Central Africa, probably between 1910 and 1950. The chimpanzee virus, called simian immunodeficiency virus, or SIV, mutates rapidly, at a rate of 1% a year, and after it entered humans it continued to do so, soon becoming what we now know as HIV and spreading to human populations worldwide, mostly through sexual contact with an infected person. AIDS was first reported as a disease in 1981. In the following 30 years, 25 million people have died of AIDS, and 33 million more are currently infected. Where did chimpanzees acquire SIV? SIV viruses are rampant in African monkeys, and chimpanzees eat monkeys. Study of the nucleotide sequences of monkey SIVs reveal that one end of the chimp virus RNA closely resembles the SIV found in red-capped mangabey monkeys, while the other end resembles the virus from the greater spot-nosed monkey. It thus seems certain that chimpanzees acquired SIV from monkeys they ate. Humans ate infected monkeys giving us HIV.

Ebola virus Among the most lethal of emerging viruses are a collection of filamentous viruses arising in Central Africa that attack human connective tissue. With lethality rates in excess of 50%, these so-called filoviruses cause some of the most lethal infectious diseases known. One, Ebola virus, has exhibited lethality rates in excess of 90% in isolated outbreaks in Central Africa. Luckily, victims die too fast to spread the disease very far. Researchers have reported evidence implicating fruit bats as Ebola hosts. These large bats are eaten for food everywhere in Central Africa that outbreaks have occurred.

Hantavirus A sudden outbreak of a highly fatal hemorrhagic infection in the southwestern United States in 1993 was soon attributed to a strain of hantavirus, an RNA virus associated with rodents. This strain was eventually traced to deer mice. The deer mouse hantavirus is transmitted to humans through fecal contamination in areas of human habitation. Control of deer mouse populations has limited the disease.

SARS A recently emerged strain of coronavirus was responsible for a worldwide outbreak in 2003 of *severe acute respiratory syndrome* (SARS), a respiratory infection with pneumonia-like symptoms that in over 8% of cases is fatal. When the 29,751-nucleotide RNA genome of the SARS virus was sequenced, it proved to be a completely new form of coronavirus, not closely related to any of the three previously described forms. Virologists in 2005 identified the Chinese horseshoe bat as the natural host of the SARS virus. Because these bats are healthy carriers not sickened by the virus, and occur commonly throughout Asia, it will be difficult to prevent future outbreaks.

West Nile Virus A mosquito-borne virus, West Nile virus first infected people in North America in 1999. Carried by infected crows and other birds, the virus proceeded to spread across the country, with 4,156 cases at its peak in 2002, 284 of whom died. By 2005 the wave of infection had greatly lessened. The virus is thought to be transmitted to humans by mosquitoes that have previously bitten infected birds. Earlier spread of the virus through Europe also abated after several years.

> **Key Learning Outcome 16.10** Viruses are responsible for some of the most lethal diseases of humans. Some of the most serious examples are viruses that have transferred to humans from some other host.

TABLE 16.3	IMPORTANT HUMAN VIRAL DISEASES		
Disease	**Pathogen**	**Vector/ Reservoir**	**Symptoms and Mode of Transmission**
AIDS	HIV	Humans	Destroys immune defenses, resulting in death by infection or cancer. About 33 million are infected with HIV worldwide.
Chicken pox	Human herpes- virus 3 (HHV-3 or varicella-zoster)	Humans	Spread through contact with infected individuals. No cure. Rarely fatal. Vaccine approved in United States in early 1995.
Ebola hemorrhagic fever	Filoviruses (such as Ebola virus)	Unknown	Acute hemorrhagic fever; virus attacks connective tissue, leading to massive hemorrhaging and death. Peak mortality is 50% to 90% if the disease goes untreated. Outbreaks confined to local regions of Central Africa.
Hepatitis B (viral)	Hepatitis B virus (HBV)	Humans	Highly infectious through contact with infected body fluids. Approximately 1% of U.S. population infected. Vaccine available, no cure. Can be fatal.
Herpes	Herpes simplex virus (HSV or HHV-1/2)	Humans	Fever blisters; spread primarily through contact with infected saliva. Very prevalent worldwide. No cure. Exhibits latency—the disease can be dormant for several years.
Influenza	Influenza viruses	Humans, ducks, pigs	Historically a major killer (between 40 and 100 million died in 1918–19); wild Asian ducks, chickens, and pigs are major reservoirs. The ducks are not affected by the flu virus, which shuffles its antigen genes while multiplying within them, leading to new flu strains.
Measles	Paramyxoviruses	Humans	Extremely contagious through contact with infected individuals. Vaccine available. Usually contracted in childhood, when it is not serious; more dangerous to adults.
Polio	Poliovirus	Humans	Acute viral infection of the central nervous system that can lead to paralysis and is often fatal. Prior to the development of Salk's vaccine in 1954, 60,000 people a year contracted the disease in the United States alone.
Rabies	Rhabdovirus	Wild and domestic Canidae (dogs, foxes, wolves, coyotes, etc.)	An acute viral encephalomyelitis transmitted by the bite of an infected animal. Fatal if untreated.
SARS	Coronavirus	Small mammals	Acute respiratory infection; can be fatal, but as with any emerging disease, this can change rapidly.
Smallpox	Variola virus	Formerly humans, now thought to exist only in government labs	Historically a major killer, the last recorded case of smallpox was in 1977. A worldwide vaccination campaign wiped out the disease completely. There is current debate as to whether the virus has been removed from government labs of the former Soviet Union and could now be made available to terrorists.
Yellow fever	Flavivirus	Humans, mosquitoes	Spread from individual to individual by mosquito bites; a notable cause of death during the construction of the Panama Canal. If untreated, this disease has a peak mortality rate of 60%.

Does HIV Infect All White Blood Cells?

Humans are protected from microbial infections by their immune system, a collection of cells that circulate in the blood. Loosely called "white blood cells," this collection actually contains a variety of different cell types. Some of them possess CD4 cell surface identification markers (think of them as ID tags). Cells that trigger antibody production when they detect virus-infected cells and macrophage cells that initially attack invading bacteria both carry CD4 ID tags. Other cells possess CD8 ID tags, such as killer cells, which are immune cells that bore holes into virus-infected cells. In an AIDS patient, neither CD4 nor CD8 cells actively defend against HIV infection. Are either or both of these cell types killed by the HIV virus?

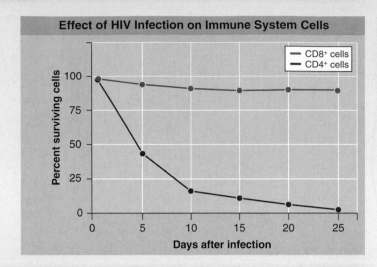

Effect of HIV Infection on Immune System Cells

To investigate this issue, researchers mixed together CD4-tagged cells (called CD4$^+$ cells) and CD8-tagged cells (called CD8$^+$ cells), and then added HIV to the mixture. HIV, colored red in the electron micrograph shown here, was then able to infect either kind of cell. The white blood cell culture was monitored at 5-day intervals for 25 days, taking a sample at each interval and scoring it for how many CD4$^+$ cells and how many CD8$^+$ cells it contained. The surviving percentage of each cell type in each sample is presented in the graph on the right above.

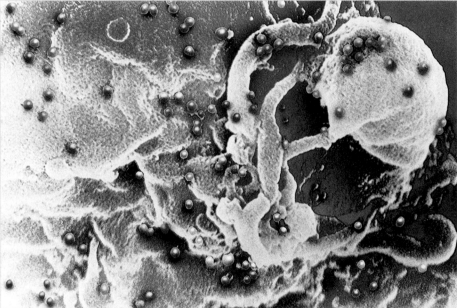

1. **Applying Concepts**
   a. Variable. In the graph, what is the dependent variable?
   b. Percentage. If the percentage of surviving cells decreases, what does this say about the absolute number of cells? Can the absolute number of cells increase if the surviving percent decreases? Explain.

2. **Interpreting Data**
   a. Does the percentage of surviving cells change over the course of three weeks for CD4$^+$ cells? For CD8$^+$ cells?
   b. Over the course of the three weeks, is there any obvious difference in the percentage survival of the two cell types? Describe it. How would you quantify this difference? [Hint: Plot the *ratio* of surviving CD4$^+$ to surviving CD8$^+$ cells versus days after infection.]

3. **Making Inferences** What would you say is responsible for the difference in percentage of surviving cells between the two cell types? How might you test this inference?

4. **Drawing Conclusions**
   a. Is either type of white blood cell totally eliminated by HIV infection over the course of this experiment?
   b. Is either type of cell virtually eliminated? If so, which one?
   c. Is either type of cell not strongly affected by HIV infection? If so, which one? Can you think of a reason why the percent surviving cells of this cell type changes at all? How might you test this hypothesis?

5. **Further Analysis** Neither CD4$^+$ nor CD8$^+$ cells actively defend AIDS patients. If one of them is not eliminated by HIV, why do you suppose it ceases to defend HIV-infected AIDS patients? Can you think of a way to investigate this possibility?

Origin of the First Cells

16.1 Origin of Life

- Life on earth may have originated from an extraterrestrial source, may have been put on this earth by a divine being, or may have evolved from inanimate matter. Only the third explanation is scientifically testable at this point.

- In experiments, reconstructing the conditions of early earth led to the hypothesis that life evolved spontaneously in a "primordial soup" of biological molecules. This hypothesis is in question, but the "bubble model" suggests that biological molecules were captured in bubbles, as shown here from **figure 16.2**, where they underwent chemical reactions, leading to the origin of life.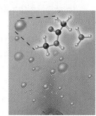

16.2 How Cells Arose

- We don't know how the first cells formed, but current hypotheses suggest that they may have formed spontaneously, from bubble-enclosed molecules such as RNA.

- Bubbles, called microspheres, form spontaneously, and scientists proposed that if organic molecules, such as RNA that has enzymatic capabilities, were trapped inside a microsphere, the structure would have the ability to carry hereditary information and to make copies of itself.

Prokaryotes

16.3 The Simplest Organisms

- Bacteria are the most ancient form of life, existing more than 2.5 billion years ago. These early prokaryotes gave rise to the present-day bacteria and archaea.

- Prokaryotes have very simple internal structures, lacking nuclei and other membrane-bounded compartments.

- The plasma membrane of prokaryotes is encased in a cell wall. The cell wall of bacteria is made of peptidoglycan, and the cell wall of archaea lacks peptidoglycan, being made of protein and/or polysaccharides. Bacteria are divided into two groups, gram-positive and gram-negative, based on the construction of their cell walls (**integrated art, pages 342-343**).

- Bacteria may have flagella or pili and may form endospores. They reproduce by splitting in two, called binary fission, and may exchange genetic information through conjugation.

- Conjugation occurs when two bacterial cells are brought into contact, and the pilus of one cell, the donor, contacts another cell, called a recipient. A conjugation bridge forms between the two cells. The plasmid in the donor cell is copied, and a single strand of the DNA is transferred to the recipient cell across the conjugation bridge. Once inside the recipient cell, a complementary strand is synthesized, completing the plasmid. Now, the recipient cell contains the same genetic information found on the plasmid of the donor cell (**figure 16.5**).

16.4 Comparing Prokaryotes to Eukaryotes

- Prokaryotes, like the one shown here from **table 16.1**, are different  from eukaryotes in many ways, including that they lack interior compartments, including nuclei, and are more metabolically diverse.

16.5 Importance of Prokaryotes

- Bacteria were instrumental in creating the properties of the atmosphere and soil found on earth. They are major participants in the carbon and nitrogen cycles. They were key to the development of genetic engineering; however, they are responsible for many diseases.

16.6 Prokaryotic Lifestyles

- Although archaea live in many different environments, the best understood are the extremophiles, which live in unusually harsh environments.

- Bacteria are the most abundant organisms on the planet. They are a very diverse group: Some are photosynthetic, some are able to fix nitrogen, and some are decomposers. Some bacteria are also pathogenic, causing human diseases (**table 16.2**).

Viruses

16.7 The Structure of Viruses

- Viruses are not living organisms, but rather are parasitic chemicals that enter and replicate inside cells. They contain a nucleic acid core surrounded by a protein coat called a capsid, and some have an outer membranelike envelope. Viruses infect bacteria, plants, and animals, and while all viruses have the same general structure, they vary greatly in shape and size (**figure 16.9**).

16.8 How Bacteriophages Enter Prokaryotic Cells

- Viruses that infect bacteria are called bacteriophages. They do not enter the host cell but instead inject their nucleic acid into the host as shown here from **figure 16.10**. The virus then enters either a lytic cycle or a lysogenic cycle. In the lytic cycle, the viral DNA directs the production of multiple copies of the virus in the host cell, which eventually ruptures, releasing the viruses to infect other cells. In the lysogenic cycle, the viral DNA becomes incorporated into the host DNA. Viral DNA is replicated with the host DNA and is passed on to offspring. At some point, the viral DNA enters the lytic cycle (**figure 16.11**).

16.9 How Animal Viruses Enter Cells

- Animal viruses enter host cells through endocytosis or membrane fusion. The HIV virus attaches to surface receptors on the host where it is engulfed. Once inside the cell, HIV produces DNA from its viral RNA using the enzyme reverse transcriptase. The viral DNA enters the host DNA and directs the formation of new viruses that bud off the host cell (**figure 16.13**). At some point, HIV is altered so that it binds to receptors on $CD4^+$ T cells. The $CD4^+$ infection ends in lysis of the cells, rapidly killing T cells needed to fight other infections.

16.10 Disease Viruses

- Viruses, like the influenza virus shown here from **table 16.3**, cause many diseases, often spreading from animals to humans. Influenza, a deadly virus that killed millions in 1918, poses a continuing threat.

1. Which is *not* an assumption used in the Miller and Urey experiment?
   a. The primordial atmosphere of the earth contained methane.
   b. The primordial atmosphere of the earth contained the same amount of oxygen as today.
   c. Lightning provided energy for chemical reactions.
   d. Small inorganic chemicals used to make larger organic molecules were present in the atmosphere and the water.

2. While it is still unknown how the first cells formed, scientists suspect that the first active biological macromolecule was
   a. protein.   c. RNA.
   b. DNA.   d. carbohydrates.

3. Bacteria
   a. are prokaryotic.
   b. have been on the earth for at least 2.5 billion years.
   c. are the most abundant life-form on earth.
   d. All of the above.

4. Some species of prokaryotes are able to obtain carbon from CO_2 and energy by oxidizing inorganic chemicals. These species are also called
   a. photoautotrophs.   c. photoheterotrophs.
   b. chemoautotrophs.   d. chemoheterotrophs.

5. Which of the following can be attributed to bacteria?
   a. decomposition of dead organic matter
   b. increasing oxygen levels in the atmosphere
   c. insect resistance in plants
   d. All of the above.

6. Cyanobacteria are thought to have been very prominent in earth's history by
   a. making nucleic acids.
   b. making proteins.
   c. producing the carbon dioxide that is in the atmosphere.
   d. producing the oxygen that is in the atmosphere.

7. Viruses are
   a. protein coats that contain DNA or RNA.
   b. simple eukaryotic cells.
   c. simple prokaryotic cells.
   d. alive.

8. A viral reproductive cycle in which the virus enters the cell, uses the cell structures to make more viruses, then breaks open the cell to release the new viruses is called the
   a. lysogenic cycle.
   b. lambda cycle.
   c. lytic cycle.
   d. prophage cycle.

9. Animal viruses enter animal cells by
   a. exocytosis.
   b. matching a marker on the surface of the virus to a complementary marker on the surface of a cell.
   c. contacting the host cell with protein tail fibers.
   d. contacting any place on the cell's membrane with the virus's protein coat.

10. HIV is a virus that contains RNA. To insert its genetic material into the cell's genome, HIV must
    a. use the cell's ribosomes and enzymes to read the RNA backward to produce DNA.
    b. do nothing; the virus RNA will insert itself into the cell DNA without changing.
    c. use an enzyme called reverse transcriptase to convert the virus RNA to DNA, which is then integrated into the cell's chromosome.
    d. No answer is correct.

1. **Figure 16.2** Life doesn't violate the Second Law of Thermodynamics because earth is not a closed system. Energy is constantly being added. What provides the energy to form the molecules shown in this figure?

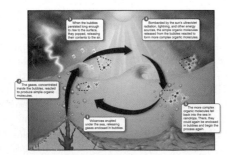

2. **Figure 16.9** Viruses have dramatically different forms and shapes. What are the consistent features of all the viruses shown?

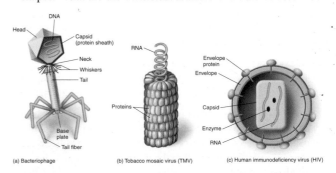

(a) Bacteriophage   (b) Tobacco mosaic virus (TMV)   (c) Human immunodeficiency virus (HIV)

1. *Pelomyxa palustris* is an unusual single-celled eukaryote (it is assigned its own phylum!) with no mitochondria. *Pelomyxa* is able to live in the presence of oxygen, as its cytoplasm contains symbiotic bacteria able to carry out oxidative metabolism. How does *Pelomyxa*'s ability to oxidize food molecules argue for or against the endosymbiont theory of the origin of eukaryotes?

17

Protists: Advent of the Eukaryotes

Y ou are a eukaryote, an organism composed of cells that contain a nucleus. All the organisms you see around you are eukaryotes, too, as prokaryote organisms are too small for you to see without a microscope to magnify them. Biologists sort the eukaryotes of the living world into four great groups, called kingdoms: animals, plants, fungi, and everything else. This chapter concerns the fourth catch-all group, the protists (kingdom Protista). The beautiful flowerlike creature you see here is a protist, the green algae *Acetabularia*. It is photosynthetic and grows as long slender stalks as long as your thumb. In the last century some biologists considered it to be a very simple sort of plant. Today, however, most biologists consider *Acetabularia* to be a protist, restricting the plant kingdom to multicellular terrestrial photosynthetic organisms (and a few marine and aquatic species like water lilies clearly derived from terrestrial ancestors). *Acetabularia* is marine, not terrestrial, and it is unicellular, with a single nucleus found in the base of its stalk. In this chapter, we will explore how protists are thought to have evolved, and the sorts of creatures found among this most diverse of biological kingdoms. Multicellularity evolved many times within the protists, producing the ancestors of the animal, plant, and fungi kingdoms, as well as several kinds of multicellular algae, some as large as trees.

17.1 | Origin of Eukaryotic Cells

The First Eukaryotic Cells

All fossils more than 1.7 billion years old are small, simple cells, similar to the bacteria of today. In rocks about 1.7 billion years old, we begin to see the first microfossils, which are noticeably larger than bacteria and have internal membranes and thicker walls. A new kind of organism had appeared, called a **eukaryote** (Greek *eu*, "true," and *karyon*, "nut"). One of the main features of a eukaryotic cell is the presence of an internal structure called a nucleus (see section 4.5). As discussed in chapter 15, animals, plants, fungi, and protists are all eukaryotes. In this chapter, we will explore the protists, from which all other eukaryotes evolved. But first we will examine some of the unifying characteristics of eukaryotes, and how they might have originated.

To begin, how might a nucleus have arisen? Many bacteria have infoldings of their outer membranes extending into the interior that serve as passageways between the surface and the cell's interior. The network of internal membranes in eukaryotes, called the endoplasmic reticulum (ER), is thought to have evolved from such infoldings, as is the nuclear envelope (**figure 17.1**). The prokaryotic cell shown on the far left has infoldings of the plasma membrane, and the DNA resides in the center of the cell. In ancestral eukaryotic cells, these internal membrane extensions evolved to project farther into the cell, continuing their function as passageways between the interior and exterior of the cell. Eventually, these membranes came to form an enclosure surrounding the DNA, shown on the right, which became the nuclear envelope.

What was the first eukaryote like? We cannot be sure, but a good model is *Pelomyxa palustris,* a single-celled, non-photosynthetic organism that some scientists feel represents an early stage in the evolution of eukaryotic cells. The cells of *Pelomyxa* are much larger than bacterial cells and contain a complex system of internal membranes. Although they resemble some of the largest early fossil eukaryotes, these cells are unlike those of any other eukaryote: *Pelomyxa* lacks mitochondria and only rarely undergoes mitosis. However, biologists know very little of the origin of *Pelomyxa*. It may have lost mitochondria rather than never having had them at all. This primitive eukaryote is so distinctive that it is assigned a phylum all its own, Caryoblastea.

Because of similarities in their DNA, it is widely assumed that the first eukaryotic cells were nonphotosynthetic descendants of archaea.

Endosymbiosis

In addition to an internal system of membranes and a nucleus, eukaryotic cells contain several other distinctive organelles. These organelles were discussed in chapter 4. Two of these organelles, mitochondria and chloroplasts, are especially unique because they resemble bacterial cells and even contain their own DNA. As discussed in section 4.7 and section 15.9, mitochondria and chloroplasts are thought to have arisen by endosymbiosis, where one organism comes to live inside another. The **endosymbiotic theory,** now widely accepted, suggests that at a critical stage in the evolution of eukaryotic cells, energy-producing aerobic bacteria came to reside symbiotically (that is, cooperatively) within larger early eukaryotic cells, eventually evolving into the cell organelles we now know as mitochondria. Similarly, photosynthetic bacteria came to live within some of these early eukaryotic cells, leading to the evolution of chloroplasts (**figure 17.2**), the photosynthetic organelles of plants and algae. Now, let's examine the evidence supporting the endosymbiotic theory a little more closely.

Mitochondria Mitochondria, the energy-generating organelles in eukaryotic cells, are sausage-shaped organelles about 1 to 3 micrometers long, about the same size as most bacteria. Mitochondria are bounded by *two* membranes. The outer membrane is smooth and was apparently derived from the

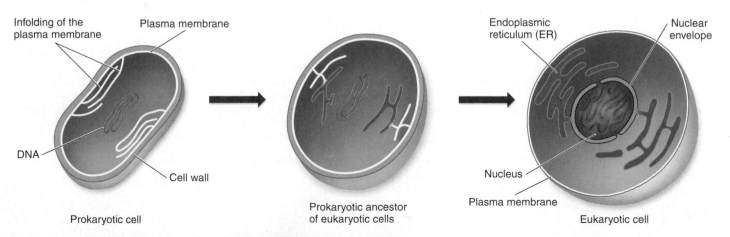

Infolding of the plasma membrane

Plasma membrane

Endoplasmic reticulum (ER)

Nuclear envelope

DNA

Cell wall

Nucleus

Plasma membrane

Prokaryotic cell

Prokaryotic ancestor of eukaryotic cells

Eukaryotic cell

Figure 17.1 Origin of the nucleus and endoplasmic reticulum.
Many bacteria today have infoldings of the plasma membrane. The eukaryotic internal membrane system called the endoplasmic reticulum (ER) and the nuclear envelope may have evolved from such infoldings of the plasma membrane of prokaryotic cells that gave rise to eukaryotic cells.

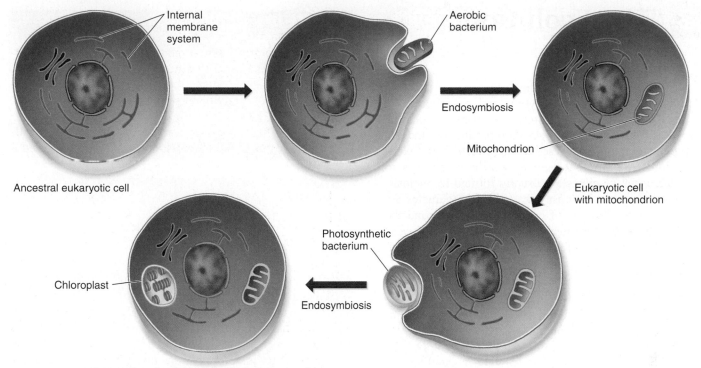

Figure 17.2 **The theory of endosymbiosis.**

Scientists propose that ancestral eukaryotic cells engulfed aerobic bacteria, which then became mitochondria in the eukaryotic cell. Chloroplasts may also have originated in this way, with eukaryotic cells engulfing photosynthetic bacteria that became chloroplasts.

host cell as it wrapped around the bacterium. The inner membrane is folded into numerous layers, embedded within which are the proteins of oxidative metabolism.

During the billion-and-a-half years in which mitochondria have existed as endosymbionts within eukaryotic cells, most of their genes have been transferred to the chromosomes of the host cells—but not all. Each mitochondrion still has its own genome, a circular, closed molecule of DNA similar to that found in bacteria, on which is located genes encoding some of the essential proteins of oxidative metabolism. These genes are transcribed within the mitochondrion, using mitochondrial ribosomes that are smaller than those of eukaryotic cells, very much like bacterial ribosomes in size and structure. Mitochondria divide by simple fission, just as bacteria do, and can divide on their own without the cell nucleus dividing. Mitochondria also replicate and sort their DNA much as bacteria do. However, the cell's nuclear genes direct the process, and mitochondria cannot be grown outside of the eukaryotic cell, in cell-free culture.

Chloroplasts Many eukaryotic cells contain other endosymbiotic bacteria in addition to mitochondria. Plants and algae contain chloroplasts, bacteria-like organelles that were apparently derived from symbiotic photosynthetic bacteria. Chloroplasts have a complex system of inner membranes and a circle of DNA. While all mitochondria are thought to have arisen from a single symbiotic event, it is difficult to be sure with chloroplasts. Three biochemically distinct classes of chloroplasts exist, but all appear to have their origin in the cyanobacteria.

Red algae and green algae seem to have acquired cyanobacteria directly as endosymbionts, and may be sister groups. Other algae have chloroplasts of secondary origin, having taken up one of these algae in their past. The chloroplasts of euglenoids are thought to be green algal in origin, while those of brown algae and diatoms are likely of red algal origin. The chloroplasts of dinoflagellates seem to be of complex origins, which might include diatoms.

Mitosis

As mentioned earlier, the primitive eukaryote *Pelomyxa* does not exhibit mitosis, the eukaryotic process of cell division. How did mitosis evolve? The mechanism of mitosis, now so common among eukaryotes, did not evolve all at once. Traces of very different, and possibly intermediate, mechanisms survive today in some of the eukaryotes. In fungi and some groups of protists, for example, the nuclear membrane does not dissolve, and mitosis is confined to the nucleus. When mitosis is complete in these organisms, the nucleus divides into two daughter nuclei, and only then does the rest of the cell divide. This separate nuclear division phase of mitosis does not occur in most protists, or in plants or animals. We do not know if it represents an intermediate step on the evolutionary journey to the form of mitosis that is characteristic of most eukaryotes today, or if it is simply a different way of solving the same problem. There are no fossils in which we can see the interiors of dividing cells well enough to be able to trace the history of mitosis.

Key Learning Outcome 17.1 The theory of endosymbiosis proposes that mitochondria originated as symbiotic aerobic bacteria and chloroplasts originated from a second endosymbiotic event with photosynthetic bacteria.

The Evolution of Sex

In the previous section, we mentioned some of the structural differences between prokaryotes and eukaryotes. But one of the most profoundly important characteristics of eukaryotes is the capacity for sexual reproduction. Indeed, many types of protists undergo sexual reproduction. In **sexual reproduction,** two different parents contribute gametes to form the offspring. Gametes are usually formed by meiosis, discussed in chapter 9. In most eukaryotes, the gametes are haploid (have a single copy of each chromosome), and the offspring produced by their fusion are diploid (have two copies of each chromosome). In this section, we examine sexual reproduction among the eukaryotes and how it evolved.

Life Without Sex

To fully understand sexual reproduction, we must first examine asexual reproduction among the eukaryotes. Consider, for example, a sponge. A sponge can reproduce by simply fragmenting its body, a process called *budding.* Each small portion grows and gives rise to a new sponge. This is an example of **asexual reproduction,** reproduction without forming gametes. In asexual reproduction, the offspring are genetically identical to the parent, barring mutation. The majority of protists reproduce asexually most of the time. Some protists such as the green algae exhibit a true sexual cycle, but only transiently. Asexual reproduction in a protist called *Paramecium* is shown in **figure 17.3a.** The single cell duplicates its DNA, grows larger, and then splits in two. The fusion of two haploid cells to create a diploid zygote, the essential act of sexual reproduction, occurs only under stress. *Paramecium* is again shown in **figure 17.3b** but now undergoing sexual reproduction. In this case, the cell is not splitting in half; rather two cells are coming into close contact. In a process called *conjugation,* they exchange genetic information in their haploid nuclei.

The development of an adult from an unfertilized egg is a form of asexual reproduction called **parthenogenesis.** Parthenogenesis is a common form of reproduction among insects. In bees, for example, fertilized eggs develop into females, while unfertilized eggs become males. Some lizards, fishes, and amphibians reproduce by parthenogenesis; an unfertilized egg undergoes mitosis without cytokinesis to produce a diploid cell, which then undergoes development as if it had been produced by sexual union of two gametes.

Many plants and marine fishes undergo a form of sexual reproduction that does not involve partners. In **self-fertilization,** one individual provides both male and female gametes. Mendel's peas, discussed in chapter 10, produced their F_2 generations by "selfing." Why isn't this asexual reproduction (after all, there is only one parent)? This is considered to be sexual rather than asexual reproduction because the offspring are not genetically identical to the parent. During the production of the gametes by meiosis, considerable genetic reassortment occurs—that is why Mendel's F_2 plants were not all the same!

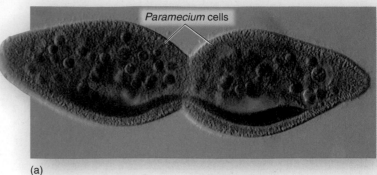

(a)

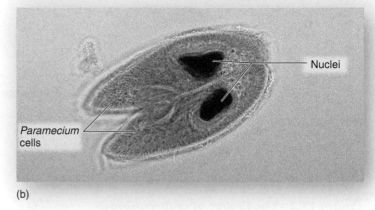

(b)

Figure 17.3 Reproduction among paramecia.
(a) When *Paramecium* reproduces asexually, a mature individual divides, and two genetically identical individuals result. (b) In sexual reproduction, two mature cells fuse in a process called conjugation (×100) and exchange haploid nuclei.

Why Sex?

If reproduction without sex is so common among eukaryotes today, it is a fair question to ask why sex occurs at all. Evolution is the result of changes that occur at the level of *individual* survival and reproduction, and it is not immediately obvious what advantage is gained by the progeny of an individual that engages in sexual reproduction. Indeed, the segregation of chromosomes that occurs in meiosis tends to disrupt advantageous combinations of genes more often than it assembles new, better-adapted ones. Because all the progeny could maintain a parent's successful gene combinations if the parent employed asexual reproduction, the widespread use of sexual reproduction among eukaryotes raises a puzzle: Where is the benefit from sex that promoted the evolution of sexual reproduction?

How Sex Evolved

In attempting to answer this question, biologists have looked more carefully at where sex first evolved—among the protists. Why do many protists form a diploid cell in response to stress? Biologists think this occurs because only in a diploid cell can certain kinds of chromosome damage be repaired effectively, particularly double-strand breaks in DNA. Such breaks are induced, for example, by desiccation—drying out. The early stages of meiosis, in which the two copies of each chromosome line up and pair with each other, seems to

Key: ☐ Haploid ▨ Diploid

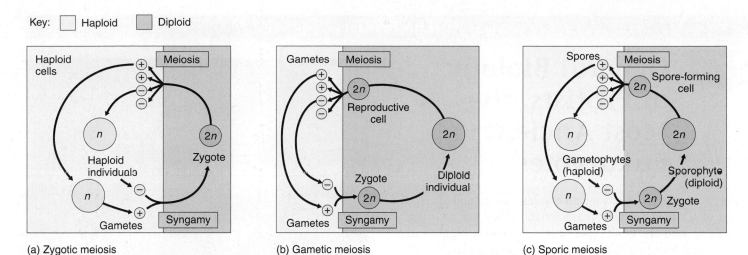

(a) Zygotic meiosis   (b) Gametic meiosis   (c) Sporic meiosis

Figure 17.4   Three types of eukaryotic life cycles.
(*a*) Zygotic meiosis, a life cycle found in most protists. (*b*) Gametic meiosis, a life cycle typical of animals. (*c*) Sporic meiosis, a life cycle found in plants.

have evolved originally as a mechanism for repairing double-strand damage to DNA by using the undamaged version of the chromosome as a template to guide the fixing of the damaged one. In yeasts, mutations that inactivate the system that repairs double-strand breaks of the chromosomes also prevent crossing over. Thus, it seems likely that sexual reproduction and the close association between pairs of chromosomes that occurs during meiosis first evolved as mechanisms to repair chromosomal damage by using the second copy of the chromosome as a template.

Why Sex Is Important

One of the most important evolutionary innovations of eukaryotes was the invention of sex. Sexual reproduction provides a powerful means of shuffling genes, quickly generating different combinations of genes among individuals. Genetic diversity is the raw material for evolution. In many cases the pace of evolution appears to be geared to the level of genetic variation available for selection to act upon—the greater the genetic diversity, the more rapid the evolutionary pace. Programs for selecting larger domestic cattle and sheep, for example, proceed rapidly at first but then slow as all of the existing genetic combinations are exhausted; further progress must then await the generation of new gene combinations. The genetic recombination produced by sexual reproduction has had an enormous evolutionary impact because of its ability to rapidly generate extensive genetic diversity.

Sexual Life Cycles

Many protists are haploid all their lives, but with few exceptions, animals and plants are diploid at some stage of their lives. That is, the body cells of most animals and plants have two sets of chromosomes, one from the male and one from the female parent. The production of haploid gametes by meiosis, followed by the union of two gametes in sexual reproduction, is called the **sexual life cycle.**

Eukaryotes are characterized by three major types of sexual life cycles (**figure 17.4**):

1. In the simplest of these, found in many algae, the zygote formed by the fusion of gametes is the only diploid cell. This sort of life cycle, which you can see in **figure 17.4***a*, is said to represent **zygotic meiosis,** because in algae the zygote undergoes meiosis. Haploid cells occupy the major portion of the life cycle, as indicated by the larger yellow box; the diploid zygote undergoes meiosis immediately after it is formed.

2. In most animals, the gametes are the only haploid cells. They exhibit **gametic meiosis,** because in animals meiosis produces the gametes. Here the diploid cells occupy the major portion of the life cycle, as indicated by the larger blue box in **figure 17.4***b*.

3. Plants exhibit **sporic meiosis,** because in plants the spore-forming cells undergo meiosis. In plants there is a regular **alternation of generations** between a haploid phase (the yellow boxed area in **figure 17.4***c*) and a diploid phase (the blue boxed area in **figure 17.4***c*). The diploid phase produces spores that give rise to the haploid phase, and the haploid phase produces gametes that fuse to give rise to the diploid phase.

The genesis of sex, then, involved meiosis and fertilization with the participation of two parents. We have previously said that bacteria lack true sexual reproduction, although in some groups, two bacteria do pair up in conjugation and exchange parts of their genome. The evolution of true sexual reproduction among the protists has no doubt contributed importantly to their tremendous diversification and adaptation to an extraordinary range of ways of life, as we shall see in section 17.3.

> **Key Learning Outcome 17.2**   Sex evolved among eukaryotes as a mechanism to repair chromosomal damage, but its importance is as a means of generating diversity.

17.3 General Biology of Protists, the Most Ancient Eukaryotes

Protists are the most ancient eukaryotes and are united on the basis of a single negative characteristic: They are not fungi, plants, or animals. In all other respects, they are highly variable with no uniting features. Many are unicellular, like the *Vorticella* you see in **figure 17.5** with its contractible stalk, but there are numerous colonial and multicellular groups. Most are microscopic, but some are as large as trees. We will start our discussion of the protists with an overview of some of their important features.

The Cell Surface

Protists possess varied types of cell surfaces. All protists have plasma membranes. But some protists, like algae and molds, are additionally encased within strong cell walls. Still others, like diatoms and radiolarians, secrete glassy shells of silica.

Locomotor Organelles

Movement in protists is also accomplished by diverse mechanisms. Protists move by cilia, flagella, pseudopods, or gliding mechanisms. Many protists wave one or more flagella to propel themselves through the water, whereas others use banks of short, flagella-like structures called cilia to create water currents for their feeding or propulsion. *Pseudopodia* are the chief means of locomotion among amoebas, whose pseudopods are large, blunt extensions of the cell body called *lobopodia*. Other related protists extend thin, branching protrusions called *filopodia*. Still other protists extend long, thin pseudopodia called *axopodia* supported by axial rods of microtubules. Axopodia can be extended or retracted. Because the tips can adhere to adjacent surfaces, the cell can move by a rolling motion, shortening the axopodia in front and extending those in the rear.

Cyst Formation

Many protists with delicate surfaces are successful in quite harsh habitats. How do they manage to survive so well? They survive inhospitable conditions by forming **cysts.** A cyst is a dormant form of a cell with a resistant outer covering in which cell metabolism is more or less completely shut down. Amoebic parasites in vertebrates, for example, form cysts that are quite resistant to gastric acidity (although they will not tolerate desiccation or high temperature).

Nutrition

Protists employ every form of nutritional acquisition except chemoautotrophy, which has so far been observed only in prokaryotes. Some protists are photosynthetic autotrophs and

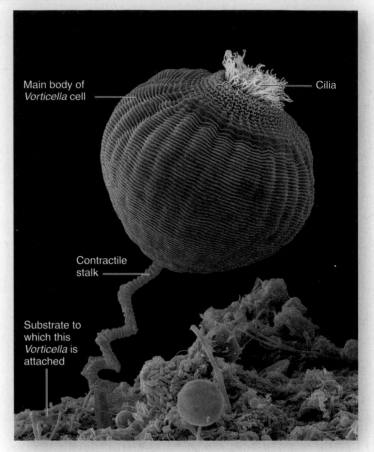

Figure 17.5   A unicellular protist.
The protist kingdom is a catch-all kingdom for many different groups of unicellular organisms, such as this *Vorticella* (phylum Ciliophora), which is heterotrophic, feeds on bacteria, and has a contractible stalk.

are called **phototrophs.** Others are heterotrophs that obtain energy from organic molecules synthesized by other organisms. Among heterotrophic protists, those that ingest visible particles of food are called **phagotrophs,** or **holozoic feeders.** Those ingesting food in soluble form are called **osmotrophs,** or **saprozoic feeders.**

Phagotrophs ingest food particles into intracellular vesicles called **food vacuoles,** or **phagosomes.** Lysosomes fuse with the food vacuoles, introducing enzymes that digest the food particles within. As the digested molecules are absorbed across the vacuolar membrane, the food vacuole becomes progressively smaller.

Reproduction

Protists typically reproduce asexually, most reproducing sexually only in times of stress. Asexual reproduction involves mitosis, but the process is often somewhat different from the mitosis that occurs in multicellular animals. The nuclear membrane, for example, often persists throughout mitosis, with the microtubular spindle forming within it. In some groups, asexual reproduction involves spore formation, in others fission. The most common type of fission is **binary,** in which a cell simply splits into nearly equal halves. When the progeny cell is considerably smaller than its parent, and then grows to adult size, the fission is called **budding.** In multiple fission, or **schizogony,** common among some protists, fission is preceded by several nuclear divisions, so that fission produces several individuals almost simultaneously.

Sexual reproduction also takes place in many forms among the protists. In ciliates, **gametic meiosis** occurs just before gamete formation, as it does in most animals. In the sporozoans, **zygotic meiosis** occurs directly *after* fertilization, and all the individuals that are produced are haploid until the next zygote is formed. In algae, there is **sporic meiosis,** producing an alternation of generations similar to that seen in plants, with significant portions of the life cycle spent as haploid as well as diploid.

Multicellularity

A single cell has limits. It can only be so big without encountering serious surface-to-volume problems. Said simply, as a cell becomes larger, there is too little surface area for so much volume. The evolution of multicellular individuals composed of many cells solved this problem. **Multicellularity** is a condition in which an organism is composed of many cells, permanently associated with one another, that integrate their activities. The key advantage of multicellularity is that it allows specialization—distinct types of cells, tissues, and organs can be differentiated within an individual's body, each with a different function. With such functional "division of labor" within its body, a multicellular organism can possess cells devoted specifically to protecting the body, others to moving it about, still others to seeking mates and prey, and yet others to carry on a host of other activities. This allows the organism to function on a scale and with a complexity that would have been impossible for its unicellular ancestors. In just this way, a small city of 50,000 inhabitants is vastly more complex and capable than a crowd of 50,000 people in a football stadium— each city dweller is specialized in a particular activity that is interrelated to everyone else's, rather than just being another body in a crowd.

Colonies A **colonial organism** is a collection of cells that are permanently associated but in which little or no integration of cell activities occurs. Many protists form colonial assemblies, consisting of many cells with little differentiation or integration. In some protists, the distinction between colonial and multicellular is blurred. For example, in the green algae *Volvox* shown in **figure 17.6,** individual motile cells aggregate into a hollow ball of cells that moves by a coordinated beating of the flagella of the individual cells—like scores of rowers all pulling their oars in concert. A few cells near the rear of the moving colony are reproductive cells, but most are relatively undifferentiated.

Aggregates An **aggregation** is a more transient collection of cells that come together for a period of time and then separate. Cellular slime molds, for example, are unicellular organisms that spend most of their lives moving about and feeding as single-celled amoebas. They are common in damp soil and on rotting logs, where they move around and ingest bacteria and other small organisms. When the individual amoebas exhaust the supply of bacteria in a given area and are near starvation, all of the individual organisms in that immediate area aggregate into a large moving mass of cells called a slug. By moving to a different location, the aggregation increases the chance that food will be found.

Multicellular Individuals True multicellularity, in which the activities of the individual cells are coordinated and the cells themselves are in contact, occurs only in eukaryotes and is one of their major characteristics. Three groups of protists have independently attained true but simple multicellularity—the brown algae (phylum Phaeophyta), green algae (phylum Chlorophyta), and red algae (phylum Rhodophyta). In **multicellular organisms,** individuals are composed of many cells that interact with one another and coordinate their activities.

Simple multicellularity does not imply small size or limited adaptability. Some marine algae grow to be enormous. An individual kelp, one of the brown algae, may grow to tens of meters in length—some taller than a redwood! Red algae grow at great depths in the sea, far below where kelp or other algae are found. But not all algae are multicellular. Green algae, for example, include many kinds of multicellular organisms but an even larger number of unicellular ones.

> **Key Learning Outcome 17.3** Protists exhibit a wide range of forms, locomotion, nutrition, and reproduction. Their cells form clusters with varying degrees of specialization, from transient aggregations to more persistent colonies to permanently multicellular organisms.

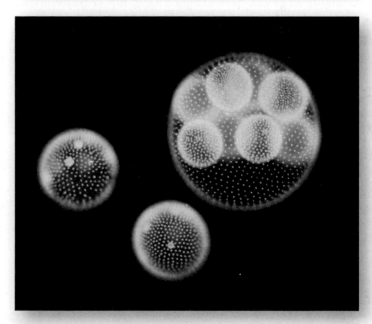

Figure 17.6 A colonial protist.
Individual, motile, unicellular green algae are united in the protist *Volvox* as a hollow colony of cells that moves by the beating of the flagella of its individual cells. Some species of *Volvox* have cytoplasmic connections between the cells that help coordinate colony activities. The *Volvox* colony is a highly complex form that has many of the properties of multicellular life.

17.4 Classifying the Protists

Protists are the most diverse of the four kingdoms in the domain Eukarya. The 200,000 different forms in the kingdom Protista include many unicellular, colonial, and multicellular groups. Protists were the first cells to contain a nucleus, as described in section 17.1—indeed, an organized internal membrane system that creates organelle compartments is the key feature that distinguished protists and other eukaryotes from archaea and bacteria. The evolution of early protists, like the fossil algae seen in **figure 17.7**, was one of the most important steps in life's evolutionary journey.

Probably the most important statement we can make about classifying the Kingdom Protista is that it is an artificial group; as a matter of convenience, single-celled eukaryotic organisms have typically been grouped together into this kingdom. This lumps many very different and only distantly related forms together. A taxonomist would say that the kingdom Protista is not monophyletic—that it contains many groups that do not share a common ancestor.

Traditionally, biologists have grouped protists artificially into functionally related categories, much as was done in the 19th century. Protists were typically grouped into photosynthesizers (algae), heterotrophs (protozoa), and absorbers (funguslike protists).

New applications of a wide variety of molecular methods are providing important new insights into the evolutionary relationships among the different groups of protists. For the first time, we can begin to see how many are related. Molecular taxonomists have assigned 12 of the 17 major protist phyla to 7 monophyletic groups, or "clades." All members within each clade share the same common ancestor. The seven

Figure 17.7 Early eukaryotic fossil.
Fossil algae that lived in Siberia 1 billion years ago.

groups can be tentatively arrayed on the protist phylogenetic tree as pictured in **figure 17.8**, although much remains uncertain. The lineages connecting the seven groups may change as we learn more.

Five major protist phyla cannot yet be placed on this tree with any confidence, but they are described in **table 17.1**, along with the other protist phyla. These five phyla include some of the more familiar protists such as amoebas. As researchers carry out more detailed DNA-level comparisons, our understanding of these five groups will increase.

It seems likely that over the next few years, the traditional "matter of convenience" kingdom Protista will be replaced by a more illuminating arrangement that expresses the evolutionary relationships among the members of this very diverse kingdom.

Key Learning Outcome 17.4   12 of the 17 protist phyla can be assigned positions on the protist phylogenetic tree; the relationship of five others is still being worked out.

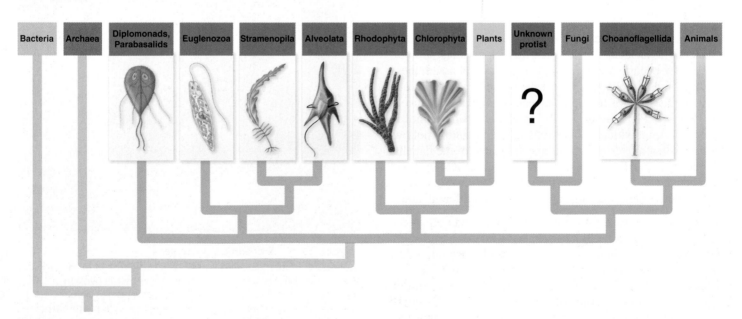

Figure 17.8 The major protist clades.

TABLE 17.1		KINDS OF PROTISTS		
Group	**Phylum**	**Typical Examples**		**Key Characteristics**
DIPLOMONADS				
Diplomonads	Diplomonadida	*Giardia intestinalis*		Move by flagella; have two nuclei
PARABASALIDS				
Parabasalids	Parabasalida	*Trichomonas vaginalis*		Undulating membrane; some are pathogens, others digest cellulose in gut of termites
EUGLENOZOA				
Euglenoids	Euglenozoa	*Euglena*		Unicellular; some photosynthetic; others heterotrophic; contain chlorophylls *a* and *b* or none
Trypanosomes	Euglenozoa	Trypanosomes		Heterotrophic; unicellular
STRAMENOPILA				
Brown algae	Phaeophyta	Kelp		Multicellular; contain chlorophylls *a* and *c*
Diatoms	Chrysophyta	*Diatoma*		Unicellular; manufacture the carbohydrate chrysolaminarin; unique double shells of silica; contain chlorophylls *a* and *c*
Water molds	Oomycota	*Phytophthora infestans*		Terrestrial and freshwater
ALVEOLATA				
Ciliates	Ciliophora	*Paramecium*		Heterotrophic unicellular protists with cells of fixed shape possessing two nuclei and many cilia; many contain highly complex and specialized organelles
Dinoflagellates	Pyrrhophyta	Red tides		Unicellular; two flagella; contain chlorophylls *a* and *b*
Sporozoans	Apicomplexa	*Plasmodium*		Nonmotile; unicellular; the apical end of the spores contains a complex mass of organelles
RHODOPHYTA				
Red algae	Rhodophyta	Coralline algae		Most multicellular; contain chlorophyll *a* and a red pigment
CHLOROPHYTA				
Green algae	Chlorophyta	*Chlamydomonas, Ulva*		Unicellular or multicellular; contain chlorophylls *a* and *b*; ancestor of plants
CHOANOFLAGELLIDA				
Choanoflagellates	Choanozoa	Choanoflagellates		Flagellated feeding funnel; ancestor of animals
PHYLOGENY NOT YET DETERMINED				
Amoebas	Rhizopoda	*Amoeba*		Move by pseudopodia
Forams	Foraminifera	Forams		Rigid shells; move by protoplasmic streaming
Radiolarians	Actinopoda	Radiolarians		Glassy skeletons; needlelike pseudopods
Cellular slime molds	Acrasiomycota	*Dictyostelium*		Colonial aggregations of individual cells; most closely related to amoebas
Plasmodial slime molds	Myxomycota	*Fuligo*		Stream along as a multinucleate mass of cytoplasm

The Base of the Protist Tree

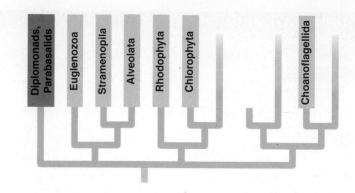

Among the 17 phyla of living protists, two unicellular groups seem more closely linked to early eukaryotes, at the very base of the protist phylogenetic tree. Both groups possess flagella, but lack mitochondria. Because mitochondrial genes are found in their nuclei, it seems that these two groups lost their mitochondria, rather than never having had them.

Diplomonads Have Two Nuclei

Diplomonads propel themselves through the water with flagella, and are unusual in that each single-celled individual has two nuclei. The diplomonad *Giardia intestinalis,* a common parasite, passes from one human to another via feces-contaminated water, causing diarrhea.

Parabasalids Swim with Undulating Membranes

Parabasalids propel themselves through the water with undulating membranes as well as flagella. The *Trichomonas vaginalis* seen in this photograph is a parasite

Trichomonas vaginalis

.83 μm

that causes vaginitis, a sexually transmitted disease in humans. Other species of parabasalids play key roles in forest ecosystems. They live in the guts of termites and digest cellulose, something the termites themselves cannot do; this symbiosis aids forests in recycling the carbon tied up in fallen trees.

Key Learning Outcome 17.5
Molecular evidence suggests that diplomonads and parabasalids are the closest living protists to now-extinct early eukaryotes.

A Diverse Kingdom

As we have mentioned, molecular comparisons allow us to place 12 of the 17 protist phyla with some confidence on the protist phylogenetic tree (we will get to the other five later). The largest branch of the protist phylogenetic tree contains seven phyla clustered within three monophyletic groups.

Euglenozoa Are Free-living Protists with Anterior Flagella

Euglenozoa are freshwater protists with the majority having two flagella. There are two major groups in the phylum Euglenozoa, the euglenoids and the trypanosomes.

Euglenoids like *Euglena* are euglenozoans with two flagella. As shown in **figure 17.9,** the flagella are attached at the base of a flask-shaped opening called the reservoir located at the anterior end of the cell. One of the flagella is long and has a row of very fine, short, hairlike projections along one side. A second, shorter flagellum is located within the reservoir but does not emerge from it. Contractile vacuoles collect excess water from all parts of the organism and empty it into the reservoir, which apparently helps regulate the osmotic pressure within the organism. The stigma is light-sensitive and helps these photosynthetic organisms move toward light. Reproduction is by mitotic cell division; no sexual reproduction is known in this group.

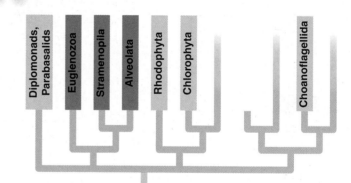

Figure 17.9 *Euglena.*
In euglenoids such as *Euglena,* starch forms around pyrenoids; other food reserves are stored in paramylon granules.

Flagellum
Contractile vacuole
Second flagellum
Pellicle
Reservoir
Paramylon granule
Stigma Basal body Nucleus Pyrenoid Chloroplast

Euglenoids clearly illustrate the folly of attempting to classify protists as tiny animals or plants. About one-third of the 1,000 known species have chloroplasts and are photosynthetic; the others lack chloroplasts, ingest their food, and are heterotrophic. In the dark, many photosynthetic euglenoids reduce the size of their chloroplasts (they may appear to disappear!) and become heterotrophs until they return to the light.

Trypanosomes are euglenozoans with a unique single mitochondrion that contains two circles of DNA. Trypanosomes are important pathogens of human beings, responsible for many serious diseases. Perhaps the most devastating is African sleeping sickness, in which infected individuals experience extreme lethargy and fatigue. The trypanosomes spend some of their life cycle in the blood and saliva of a carrier, such as the tsetse fly shown in **figure 17.10a**. When the fly feeds on a human, the trypanosomes are spread to the human host, where they circulate in its blood. You can see a trypanosome in **figure 17.10b**, the wormlike shape among the circular red blood cells.

Leishmaniasis is a trypanosome infection transmitted by sand flies that causes severe skin sores. If the protists reach internal organs, death can result. About 1.5 million new cases are reported each year.

Stramenopila Are Protists with Fine Hairs

Another major group on the protist phylogenetic tree contains three phyla that possess fine hairs on their flagella. These flagellated cells may only appear at certain times in the life cycle, or may have been lost completely, as in the diatoms.

Brown algae, members of the phylum Phaeophyta, contain the longest, fastest-growing, and most photosynthetically productive living organisms, the giant kelp. Kelp form underwater forests with individuals over 100 meters long. The 1,500 species of brown algae are all multicellular and almost all marine. They are the most conspicuous seaweeds in the ocean. The larger brown algae, like the one in **figure 17.11a**, have flattened blades, stalks, and anchoring bases and often contain complex internal plumbing like that of plants. The life cycle has alternating generations, the large individuals being the sporophyte (diploid) generation.

Diatoms, members of the phylum Chrysophyta, are photosynthetic unicellular protists with a unique double shell of silica. Like tiny oysters, the shells resemble small boxes with lids, one fitting inside the other.

Diatoms are abundant in both oceans and lakes. There are over 11,500 species, of two sorts: some with radial symmetry that look like tiny wheels and others with bilateral (two-sided) symmetry (**figure 17.11b**). The shells of fossil diatoms form thick deposits that are mined commercially as "diatomaceous earth," used as an abrasive or to make paint sparkle. Diatoms move in a complex manner that is still being inves-

Figure 17.10 Trypanosomes cause sleeping sickness.

(a) A tsetse fly is sucking blood from a human arm in Tanzania, East Africa. The fly's saliva can transmit the trypanosomes that cause sleeping sickness to the human it bites. (b) In this photograph, the undulating, changeable shape of *Trypanosoma* is visible among human red blood cells.

(a)

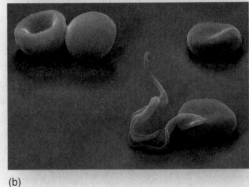

(b)

(b)

(a)

Figure 17.11 Two of the more common stramenopila.

(a) Massive "groves" of giant kelp, a kind of brown algae, contain some of the largest organisms on earth. (b) These diatoms have unique silica, two-part shells.

tigated. Their movement involves protoplasmic streaming along grooves in their shell and the waving of tiny vibrating fibrils within the grooves.

Water molds, phylum Oomycota, are the downy mildews that are often seen in moist environments. There are 580 named species, all of which either parasitize living organisms or feed on dead organic matter. Many oomycetes are important plant pathogens, including *Phytophthora infestans,* which causes late blight in potatoes. This mold was responsible for the Irish potato famine of 1845–47, during which about 400,000 Irish people starved.

Alveolata Are Protists with Submembrane Vesicles

Another main "branch" on the protist phylogenetic tree contains three phyla, all of which have a layer of flattened vesicles called alveoli beneath their plasma membrane. The alveoli are thought to function in the transport of materials out of the cell, similar to Golgi bodies.

Ciliates, members of the phylum Ciliophora, are very complex and unusual unicellular heterotrophs with large numbers of cilia (tiny beating hairs) covering the outside of the body, and two nuclei per cell (the micronucleus and macronucleus). Ciliates are so different from other eukaryotes (they even use the genetic code differently!) that many taxonomists argue they should be placed in a separate kingdom of their own. About 8,000 species have been named.

Ciliates have a pellicle, a protein scaffold inside the plasma membrane that can change shape, which makes the body wall tough but flexible. The body interior is extremely complex, inspiring some biologists to consider ciliates multicellular organisms without cell boundaries rather than unicellular. The *Paramecium* in **figure 17.12** is a typical ciliate. It has a complex digestive process, with a gullet ("mouth") that serves as an intake channel for bacteria and food particles. Once ingested, they are then enclosed in membrane bubbles called food vacuoles and digested by enzymes. Reproduction is usually by fission, with the body splitting in half, but ciliates also undergo a form of sexual reproduction called *conjugation* (see page 362, **figure 17.3***b*), in which haploid nuclei that have arisen by meiosis are exchanged.

Dinoflagellates, members of the phylum Pyrrhophyta, are photosynthetic unicellular protists, most with two flagella of unequal length. There are about 1,000 species. Some occur in freshwater, but most are marine. Bioluminescent dinoflagellates produce the twinkling light sometimes seen in marine waters at night. Most dinoflagellates have a stiff coat of cellulose, often encrusted with silica, giving them unusual shapes. The four genera of dinoflagellates in **figure 17.13** show how their flagella are unique, unlike those of any other phylum.

One beats in a groove circling the body like a belt, the other in a groove perpendicular to it. Their beating rotates the body like a top.

A few dinoflagellates produce powerful toxins such as the poisonous "red tides," which are population explosions of such dinoflagellates. You can see in **figure 17.14** why they are called red tides, coloring the water a reddish color. The toxins can affect humans when they eat seafood taken from red tide contaminated waters. Dinoflagellates reproduce by splitting in half.

Sporozoans are spore-forming unicellular parasites of animals, all members of the phylum Apicomplexa. They are named after a unique arrangement of microtubules and other cell organelles at one end of the cell called an apical complex. This complex is used to facilitate invading a host cell.

Sporozoans are responsible for many diseases in humans and domestic animals. Sporozoans infect animals with small spores that are transmitted from host to host.

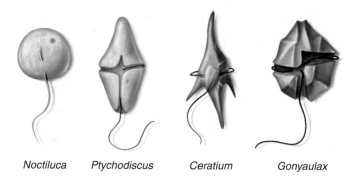

Noctiluca *Ptychodiscus* *Ceratium* *Gonyaulax*

Figure 17.13 Dinoflagellates.

Noctiluca, which lacks the heavy cellulosic armor characteristic of most dinoflagellates, is one of the bioluminescent organisms that causes the waves to sparkle in warm seas at certain times of the year. In the other three genera, the shorter encircling flagellum may be seen in its groove, and the longer one projects away.

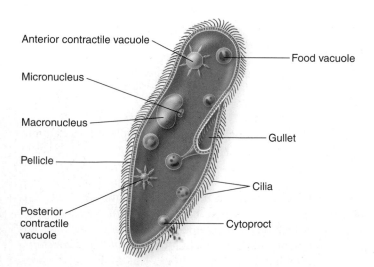

Figure 17.12 A ciliate.

The main features of the familiar ciliate *Paramecium* are shown.

Figure 17.14 Red tide.

Red tides are caused by population explosions of dinoflagellates. The pigments in the dinoflagellates, or in some cases other organisms, color the water.

Sporozoans have complex life cycles that involve both asexual and sexual phases, as illustrated in **figure 17.15**. Sporozoans of the genus *Plasmodium* cause malaria and are spread among humans by mosquitoes of the genus *Anopheles*. When a mosquito inserts its proboscis into a human, it injects about a thousand sporozoites into the blood ❶. They travel to the liver within a few minutes. If even one sporozoite reaches the liver, it will multiply rapidly there and still cause malaria ❷. The sporozoites transform inside the liver and spread to the blood where they progress through several more stages ❸, some of which develop into gametocytes ❹. The gametocytes are ingested by a mosquito ❺, where fertilization takes place forming sporozoites ❻, and the cycle begins anew.

Malaria is one of the most serious diseases in the world. About 500 million people are affected by it at any one time, and approximately 2 million of them, mostly children, die each year.

> **Key Learning Outcome 17.6** Seven phyla clustered in the largest branch of the protist phylogenetic tree exhibit an amazing diversity of form and function.

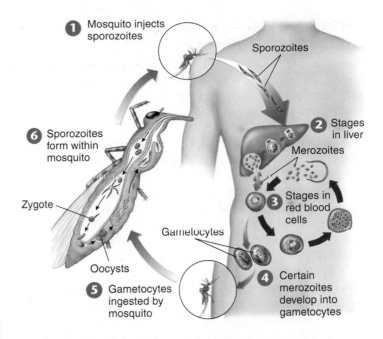

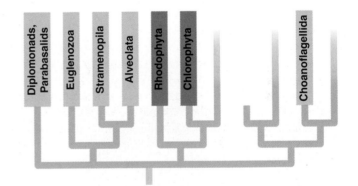

Figure 17.15 A sporozoan life cycle.
Plasmodium is the sporozoan that causes malaria. *Plasmodium* has a complex life cycle that alternates between mosquitoes and mammals.

17.7 The Road to Plants

Another key lineage on the protist phylogenetic tree, the red and green algae, marks the path that has led to the evolution of plants. Both red and green algae contain similar chloroplasts, and molecular analysis indicates a single endosymbiotic origin for both, confirming a common ancestry. The red algae appear to have arisen before the evolutionary lineage that led from green algae to plants.

Rhodophyta Are Ancient Photosynthesizers

Red algae, members of the phylum Rhodophyta, possess red pigments called phycobilins that give them their characteristic color, as shown in **figure 17.16**. Almost all of the 4,000 species of red algae are multicellular and live in the sea, where they grow more deeply than any other photosynthetic organism. Red algae have complex bodies made of interwoven filaments of cells. The laboratory media agar is made from the cell walls of red algae. The life cycle of red algae is complex, usually involving alternation of generations, like in plants. None of the red algae have flagella or centrioles, suggesting that red algae may be one of the more ancient groups of eukaryotes.

Figure 17.16 Red algae.
These red algae have their cellulose cell walls heavily impregnated with calcium carbonate, the same material of which oyster shells are made. Because they are hard and occur on coral reefs, they are called coralline algae.

Chlorophyta Are the Direct Ancestors of Plants

Green algae, members of the phylum Chlorophyta, are of special interest because the ancestor of terrestrial plants was a member of this group. Green algae chloroplasts are similar to plant chloroplasts, and like them, they contain chlorophylls *a* and *b*.

Green algae are an extremely varied group of more than 7,000 species, mostly mobile and aquatic like *Chlamydomonas,* shown in **figure 17.17a,** but a few (like *Chlorella*) are immobile in moist soil or on tree trunks. Although most green algae are microscopic and unicellular, some are intermediate, colonial, or truly multicellular. Some of the most elaborate colonies are seen in *Volvox,* a species which forms a hollow sphere made of 500 or more cells (**figure 17.17b**). The two flagella of each cell beat in time with all the others to rotate the colony, which has reproductive cells at one end. While *Volvox* borders on multicellularity, the green algae, *Ulva* (sea lettuce), shows true multicellularity. The sexual life cycle in *Ulva* alternates between a multicellular haploid structure called a *gametophyte,* and a multicellular diploid structure called a *sporophyte* (**figure 17.18**). The sporophyte is similar in appearance to the gametophyte.

> **Key Learning Outcome 17.7** The red and green algae share a common ancestor, but the green algae gave rise to terrestrial plants.

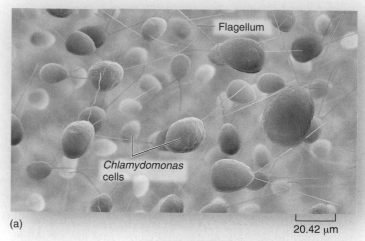

(a)

20.42 μm

Figure 17.17 Green algae.

(a) *Chlamydomonas* is a unicellular mobile green algae. (b) *Volvox* forms colonies, an intermediate stage on the way to multicellularity.

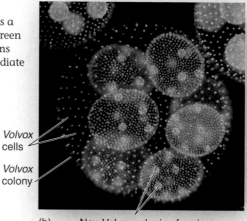

(b)   New *Volvox* colonies forming

Figure 17.18   A green algae life cycle: *Ulva.*

Individuals of this green algae exhibit a life cycle that is somewhat unique among algae in that they alternate between a haploid form called the gametophyte and a diploid form called the sporophyte, which are identical in appearance and consist of flattened sheets two cells thick. In the haploid (*n*) gametophyte, gametangia give rise to haploid gametes, which fuse to form a diploid (*2n*) zygote. The zygote germinates to form the diploid sporophyte. Sporangia within the sporophyte give rise to haploid spores by meiosis. The haploid spores develop into haploid gametophytes.

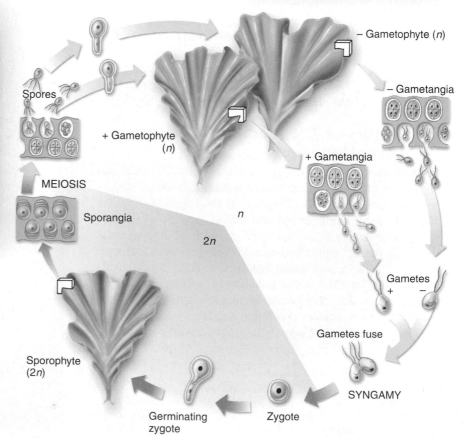

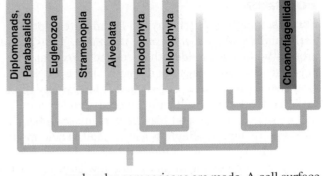

17.8 The Road to Animals

Just as plants (and undoubtedly fungi) had protist ancestors, so too did animals, in this case an unusual single-celled heterotroph called a choanoflagellate.

Choanoflagellida Are the Direct Ancestors of Animals

Choanoflagellates are unicellular, heterotrophic protists with a single, long flagellum. The flagellum is surrounded by a funnel-shaped contractible collar of closely placed filaments, not unlike the woven strands of a basket. The whipping motion of the flagellum draws water into the funnel, forcing it past the filaments, which strain bacteria from the water. The choanoflagellate protist feeds on these bacteria. Precisely this same type of filtering cell is found in sponges, the most primitive of animals (discussed in chapter 19).

While no choanoflagellate is multicellular, some are colonial, forming spherical assemblies (**figure 17.19**) that look very much like freshwater sponges. This close relationship between choanoflagellates and animals is also seen when molecular comparisons are made. A cell surface receptor used to initiate an intercellular signal pathway is the same in both choanoflagellates and sponges. Detailed genomic comparisons are not yet available, but can be expected to further confirm this relationship.

> **Key Learning Outcome 17.8**
> Choanoflagellates have a unique cell structure also found in sponges, and are believed to be the direct ancestors of animals.

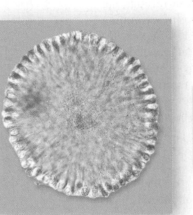

25 μm

Figure 17.19 Colonial choanoflagellates.

Colonial choanoflagellates resemble their close animal relatives, the sponges.

17.9 "Not Yet Located on the Protist Phylogenetic Tree"

Five major phyla of protists cannot yet be located on the protist phylogenetic tree. Amoebas, forams, radiolarians, and both cellular and plasmodial slime molds have no permanent locomotor apparatus, and instead use their cytoplasm to aid movement.

Amoebas

Amoebas, members of the phylum Rhizopoda, lack flagella and cell walls. There are several hundred species. They move from place to place by **pseudopodia** (Greek, *pseudo*, false, and *podium*, foot), flowing projections of cytoplasm that extend outward. In **figure 17.20a**, you can see an amoeba putting a pseudopod forward and then flowing into it. Amoebas are abundant in soil, and many are parasites of animals. Reproduction in amoebas occurs by simple fission (**figure 17.20b**). They lack meiosis and any form of sexual reproduction.

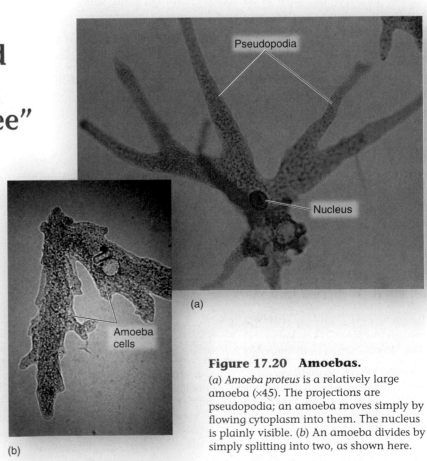

Figure 17.20 Amoebas.

(a) *Amoeba proteus* is a relatively large amoeba (×45). The projections are pseudopodia; an amoeba moves simply by flowing cytoplasm into them. The nucleus is plainly visible. (b) An amoeba divides by simply splitting into two, as shown here.

Forams

Forams, members of the phylum Foraminifera, possess rigid shells and move by *cytoplasmic streaming*. They are marine protists with pore-studded shells called *tests* that may be as big as several centimeters in diameter. There are several hundred species of forams. Their shells, built largely of calcium carbonate, are often brilliantly colored—vivid yellow, bright red, or salmon pink—and may have many chambers arrayed in a spiral shape resembling a tiny snail. Their multichambered body can be seen in **figure 17.21.**

Most forams live in sand, but a few are free-floating organisms, part of the ocean's plankton community. Long, thin cytoplasmic projections called *podia* radiate out through the pores in the shells of these protists and are used for swimming and capturing prey. The life cycle of forams is complex, involving alternation between haploid and diploid generations. Forams have deposited massive accumulations of their shells for more than 200 million years. Limestone is often rich in forams's remains—the White Cliffs of Dover, the famous landmark on the southern England seacoast shown in **figure 17.22,** is made almost entirely of foram shells.

Radiolarians

Radiolarians are unusual amoeboid protists that belong to another phylum, Actinopoda. While most amoeboid cells have an amorphous shape, radiolarians secrete a glassy exoskeleton made of silica that gives their body a distinctive shape. Either radially or bilaterally symmetrical, the shells of different species form elaborate shapes. The pseudopods of *Actinosphaerium*, seen in **figure 17.23,** extrude outward along spiky projections of the glassy exoskeleton like thorns radiating out from the cell body.

Figure 17.22   White cliffs of Dover.
The limestone that forms these cliffs is composed almost entirely of fossil shells of protists, including foraminifera.

Figure 17.21   A foram.
In this living foram, a representative of phylum Foraminifera, the podia—thin cytoplasmic projections—extend through pores in the calcareous test, or shell, of the organism.

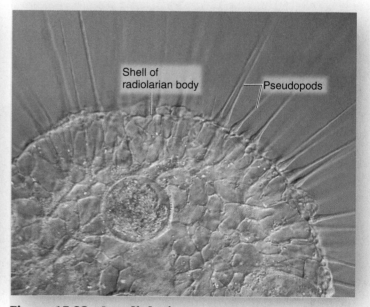

Figure 17.23   A radiolarian.
This amoeba-like radiolarian *Actinosphaerium* (×300), of the phylum Actinopoda, has striking needlelike pseudopods.

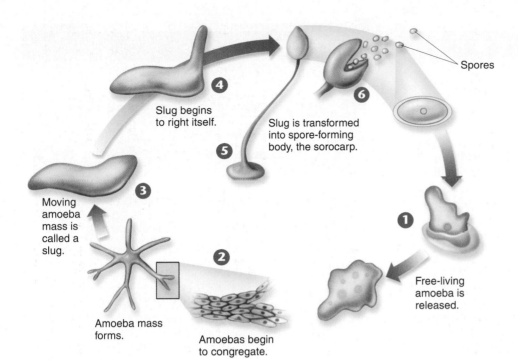

Figure 17.24 Slime mold.
The life cycle of the cellular slime mold *Dictyostelium discoideum* (phylum Acrasiomycota). ❶ Germinating spores form amoebas. ❷ The amoebas aggregate and move toward a fixed center. ❸ They form a multicellular slug 2 to 3 millimeters long that migrates toward light. ❹ The slug stops moving and begins to differentiate into a spore-forming body, called a sorocarp ❺. Within the heads of the sorocarps, the amoebas become encysted as spores ❻.

Spores

Slug begins to right itself.

Slug is transformed into spore-forming body, the sorocarp.

Moving amoeba mass is called a slug.

Amoeba mass forms.

Amoebas begin to congregate.

Free-living amoeba is released.

Cellular Slime Molds

Slime molds are heterotrophic protists that are sometimes confused with fungi, although they do not resemble them in any significant respect. For example, slime molds have cell walls made of cellulose, whereas fungal walls are made of chitin. Also, slime molds carry out normal mitosis, while fungal mitosis is unusual and will be discussed in chapter 18.

The two major kinds of slime molds that originated at different times are only distantly related. In the **cellular slime molds,** phylum Acrasiomycota, individual cells aggregate and differentiate into complex associations called *slugs.* Slugs are mobile and while not truly multicellular, they are far along that evolutionary path.

Cellular slime molds are more closely related to amoebas than to any other phylum. There are 70 named species, the best known of which is *Dictyostelium discoideum. Dictyostelium* is basically a unicellular scavenger with the interesting life cycle shown in **figure 17.24.** When deprived of food, thousands of individual *Dictyostelium* amoebas ❶ come together forming a slug (❷ through ❹) that moves to a new habitat. There, the colony differentiates into a base, a stalk, and a swollen tip ❺ that develops **spores.** Each of these spores, when released ❻, becomes a new amoeba, which begins to feed and so restarts the life cycle.

Plasmodial Slime Molds

Plasmodial slime molds, phylum Myxomycota, comprise a group of about 500 species that stream along as a **plasmodium,** a nonwalled multinucleate mass of cytoplasm. The yellow mass you see in **figure 17.25** is composed of a group of cells without cell walls separating the individual cells. They move together as a single unit. Plasmodia can flow around obstacles and even pass through a mesh cloth. Extending pseudo-

Figure 17.25 Plasmodial slime mold.
This multinucleate plasmodium (phylum Myxomycota) moves about in search of the bacteria and other organic particles that it ingests.

podia as they move, they engulf and digest bacteria and other organic material. If the plasmodium begins to dry or starve, it migrates away rapidly and then stops and often divides into many small mounds, each of which produces a spore-laden structure. Spores germinate when favorable conditions return.

Key Learning Outcome 17.9 Amoebas, forams, radiolarians, and molds are heterotrophs with restricted mobility, many of which form aggregates. Their position on the protist phylogenetic tree is not yet firmly established.

Defining a Treatment Window for Malaria

While malaria kills more people each year than any other infectious disease, the combination of mosquito control and effective treatment has virtually eliminated this disease from the United States. In 1941, more than 4,000 Americans died of malaria; in the year 2006, by contrast, fewer than five people died of malaria.

The key to controlling malaria has come from understanding its life cycle. The first critical advance came in 1897 in a remote field hospital in Secunderabad, India, when English physician Ronald Ross observed that hospital patients who did not have malaria were more likely to develop the disease in the open wards (those without screens or netting) than in wards with closed windows or screens. Observing closely, he saw that patients in the open wards were being bitten by mosquitoes of the genus *Anopheles*. Dissecting mosquitoes who had bitten malaria patients, he found the plasmodium parasite. Newly hatched mosquitoes who had not yet fed, when allowed to feed on malaria-free blood, did not acquire the parasite. Ross reached the conclusion that mosquitoes were spreading the disease from one person to another, passing along the parasite while feeding. In every country where it has been possible to eliminate the *Anopheles* mosquitoes, the incidence of the disease malaria has plummeted.

The second critical advance came with the development of drugs to treat malaria victims. The British had discovered in India in the mid-1800s that a bitter substance called quinine taken from the bark of cinchona trees was useful in suppressing attacks of malaria. The boys in the photograph are being treated with an intravenous solution of quinine. Quinine also reduces the fever during attacks, but does not cure the disease. Today physicians instead use the synthetic drugs chloroquine and primaquine, which are much more effective than quinine, with fewer side effects.

Unlike quinine, these two drugs can cure patients completely because they attack and destroy one of the phases of the plasmodium life cycle, the merozoites released into the bloodstream several days after

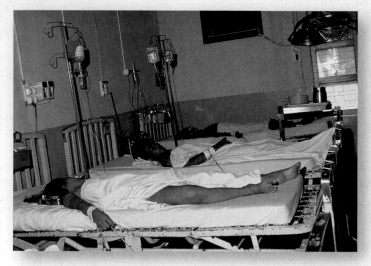

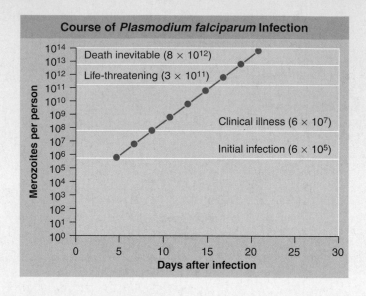

Course of *Plasmodium falciparum* Infection

Death inevitable (8×10^{12})
Life-threatening (3×10^{11})
Clinical illness (6×10^7)
Initial infection (6×10^5)

y axis: Merozoites per person
x axis: Days after infection

infection—but only if the drugs are administered soon enough after the bite that starts the infection.

In order to determine the time frame for successful treatment, doctors have carefully studied the time course of a malarial infection. The graph above presents what they have found. Numbers of merozoites are presented on the *y* axis on a log scale—each step reflects a 10-fold increase in numbers. The infection becomes life-threatening if 1% of red blood cells become infected, and death is almost inevitable if 20% of red blood cells are infected.

1. **Applying Concepts** What is the dependent variable?
2. **Making Inferences**
   a. How long after infection is it before the liver releases merozoites into the bloodstream (that is, initial infection by merozoites)? Before the disease becomes life-threatening? Before death is inevitable?
   b. How long does it take merozoites to multiply 10-fold?
3. **Drawing Conclusions** After the first appearance of clinical illness symptoms, how many days can the disease be treated before it becomes life-threatening? Before treatment has little or no chance of saving the patient's life?

The Evolution of Eukaryotes

17.1 Origin of Eukaryotic Cells

- Eukaryotic cells, with internal organelles, first appear as microfossils about 1.7 billion years ago.

- Many bacteria have infoldings of their plasma membranes that extend toward the interior of the cell. It is thought that this type of membrane infolding gave rise to the endoplasmic reticulum and formed the nuclear envelope (**figure 17.1**).

- Some eukaryotic organelles may have arisen from endosymbiotic events when energy-producing bacteria became incorporated into early eukaryotic cells, giving rise to mitochondria (**figure 17.2**). Similarly, photosynthetic bacteria may have become incorporated into some eukaryotic cells, giving rise to chloroplasts.

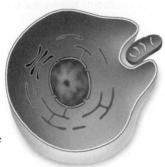

17.2 The Evolution of Sex

- A key characteristic of eukaryotes is the process of sexual reproduction, forming offspring from the fusion of two gametes. Many eukaryotes are capable of asexual reproduction through budding, parthenogenesis, and reproduction that involves only one parent (self-fertilization). However, biologists think that sexual reproduction first evolved in eukaryotes not as a means of reproduction but rather as a mechanism for repairing damaged chromosomes during meiosis when homologues align.

- Sexual reproduction permits genetic recombination that leads to genetic diversity in future generations.

- There are three major types of sexual life cycles: (1) zygotic meiosis, where the haploid phase makes up the majority of the life cycle, (2) gametic meiosis, where the diploid phase is the dominant form, and (3) sporic meiosis, where the cycle alternates equally between the haploid and diploid phases (**figure 17.4**).

The Protists

17.3 General Biology of Protists, the Most Ancient Eukaryotes

- Protists were the earliest eukaryotic cells, but they are a very diverse kingdom of organisms.

- Protists have diverse cell surfaces. They exhibit diverse modes of locomotion. Some protists form cysts to protect themselves in harsh environmental conditions. Protists have diverse modes of acquiring nutrients. Most protists reproduce asexually except for times of stress, when they engage in sexual reproduction. Some protists exist as single cells (**figure 17.5**), whereas others form colonies (**figure 17.6**) or aggregates of cells. True multicellularity is present in the algae.

17.4 Classifying the Protists

- The classification of the organisms in the kingdom Protista is in a state of flux. Traditionally, they were grouped into three functionally related categories: photosynthesizers (algae), heterotrophs (protozoa), and absorbers (funguslike protists). These artificial categories did not imply evolutionary relationships. Molecular methods of analyzing the various types of protists are providing insight into the evolutionary relationships among the protist groups, aiding in classification.

- Twelve of the 17 major protist phyla have been assigned to one of seven monophyletic groups. The tentative protist phylogenetic tree will likely change (**figure 17.8** and **table 17.1**).

- Five major protist phyla have not yet been placed on the phylogenetic tree, but as researchers learn more, the evolutionary relationships between these groups will become clearer.

17.5 The Base of the Protist Tree

- Two protists groups, Diplomonads and Parabasalids, are unicellular organisms that may be the closest living relatives to the ancestral eukaryote. Both groups have flagella but lack mitochondria. Molecular analysis reveals that mitochondrial genes are present in their nuclei, and so it seems they lost their mitochondria rather than never having had these organelles.

17.6 A Diverse Kingdom

- The largest branch of the protist phylogenetic tree contains the Euglenozoa, Stramenopila, and Alveolata. The Euglenozoa sub-branch and phylum contains two groups, the euglenoids and the trypanosomes. Both groups are freshwater protists that have two flagella (**figures 17.9** and **17.10**).

- The Stramenopila subbranch contains three protist phyla (Phaeophyta, Chrysophyta, and Oomycota) that have fine hairs on their flagella. Brown algae (phylum Phaeophyta) include fast-growing giant kelp (**figure 17.11a**). Diatoms (phylum Chrysophyta) are unicellular photosynthetic protists with a unique double shell made of silica (**figure 17.11b**). The water molds (phylum Oomycota) feed on dead organisms.

- The Alveolata subbranch contains the ciliates, the dinoflagellates, and the sporozoans. The ciliates (phylum Ciliophora) have large numbers of cilia covering their bodies (**figure 17.12**). The dinoflagellates (phylum Pyrrhophyta) are photosynthetic, unicellular protists with two flagella (**figure 17.13**). Some are bioluminescent, and a few produce powerful toxins, such as poisonous red tides (**figure 17.14**). The sporozoans, phylum Apicomplexa, are unicellular, nonmotile spore-forming protists (**figure 17.15**).

17.7 The Road to Plants

- Another branch on the protist phylogenetic tree contains the photosynthetic red algae (phylum Rhodophyta, **figure 17.16**) and green algae (phylum Chlorophyta, **figures 17.17** and **17.18**). Green algae are the direct ancestors of terrestrial plants.

17.8 The Road to Animals

- The branch on the protist phylogenetic tree that gave rise to animals contains a group called the choanoflagellates. Some choanoflagellates are colonial, forming spherical assemblies of cells (**figure 17.19**).

17.9 "Not Yet Located on the Protist Phylogenetic Tree"

- Five major phyla have not yet been placed on the protist phylogenetic tree: Amoebas, forams, radiolarians, and cellular and plasmodial slime molds (**figures 17.20–17.25**).

1. One piece of supporting evidence for the endosymbiotic theory for the origin of eukaryotic cells is that
   a. eukaryotic cells have internal membranes.
   b. mitochondria and chloroplasts have their own DNA.
   c. Golgi bodies and endoplasmic reticulum were present in ancestral cells.
   d. the nuclear membrane could only have come from another cell.

2. Many protists survive unfavorable environmental conditions by forming
   a. gametes.
   b. zygotes.
   c. cysts.
   d. aggregations.

3. Protists do *not* include
   a. algae.
   b. amoebas.
   c. multicellular organisms.
   d. mushrooms.

4. Some members of the phylum Euglenozoa
   a. do not have chloroplasts.
   b. conduct photosynthesis.
   c. are human pathogens.
   d. All of the above.

5. Kelp, which sometimes forms large, underwater forests, are actually protists called
   a. diatoms.
   b. brown algae.
   c. dinoflagellates.
   d. red algae.

6. Sporozoans have life cycles that
   a. have both a sexual and an asexual phase.
   b. have only an asexual phase.
   c. have only a sexual phase.
   d. only require fragmentation to produce new individuals.

7. The ancestor of plants was a member of what group of protists?
   a. Brown algae
   b. Alveolata
   c. Green algae
   d. Both b and c

8. Scientists believe that the ancestral animal cell comes from the protist group
   a. ciliates.
   b. choanoflagellates.
   c. dinoflagellates.
   d. euglenoids.

9. Amoebas, foraminifera, and radiolarians move using their
   a. cytoplasm.
   b. flagella.
   c. cilia.
   d. setae.

10. Protists that form aggregates, have cellulose cell walls, and are heterotrophic are probably
    a. ciliates.
    b. radiolarians.
    c. slime molds.
    d. euglenoids.

Apply Your Understanding

1. **Figure 17.18** What is the difference between the leaf-shaped structures in the yellow portion of the cycle below and the leaf-shaped structures in the blue portion of the cycle?

2. **Figure 17.24** Is the slime mold shown here a multicelled organism or numerous single-celled organisms working together? Discuss.

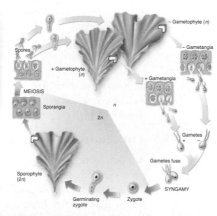

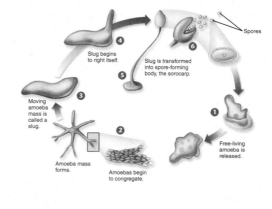

Synthesize What You Have Learned

1. In plants, the mitochondrial, chloroplast, and nuclear genomes all contain ribosomal RNA genes. Would you expect the ribosomal RNA genes of a plant's mitochondria to be more like those of its chloroplasts, or more closely related to its nuclear ribosomal RNA genes? How about to the ribosomal RNA genes in cyanobacteria? Explain.

2. The cellular slime molds, the plasmodial slime molds, and the water molds all exhibit a similar means of restricted mobility. Can we conclude that they are closely related? Discuss.

3. The flagellated protist *Giardia* does not have true mitochondria but does have tiny organelles called mitosomes. What features would you look for to test the hypothesis that mitosomes are derived from mitochondria?

18

Fungi Invade the Land

L ife evolved in the sea, and for more than a billion years was confined to the oceans of the earth. On land, there was only bare rock. This bleak terrestrial picture changed over half a billion years ago, when the first fungi are thought to have invaded the land. The difficulty of the challenge this first invasion posed cannot be overstated. Animals could not be the first to invade. Animals are heterotrophs—what would they eat? Fungi, also heterotrophs, faced the same problem. Algae are photosynthetic, and so food did not present such a challenge for their invasion of land. Sunlight would be able to provide all the energy they might need. But how would they obtain nutrients? Phosphorus, nitrogen, iron, and many other chemical elements critical to life cannot be obtained by algae from bare rock. The solution to this dilemma was a sort of "You scratch my back, I'll scratch yours" cooperation. Fungi called ascomycetes formed associations with photosynthetic algae, forming a two-member partnership called a lichen. The lichen you see in this photo is growing on bare rock. The algae within it harvest energy from sunlight, while the fungal cells derive minerals from the rock. In this chapter you will become better acquainted with fungi and then explore their partnerships with algae and plants.

18.1 Complex Multicellularity

The algae are structurally simple multicellular organisms that fill the evolutionary gap between unicellular protists and more complex multicellular organisms (fungi, plants, and animals). In **complex multicellular organisms,** individuals are composed of many highly specialized kinds of cells that coordinate their activities. There are three kingdoms that exhibit complex multicellularity:

1. **Plants.** Multicellular green algae were almost certainly the direct ancestors of the plants (see chapter 17) and were themselves considered plants in the 19th century. However, green algae are basically aquatic and much simpler in structure than plants and are considered protists in the six-kingdom system used widely today.
2. **Animals.** Animals arose from a unicellular protist ancestor. The simplest (and seemingly most primitive) animals today, the sponges, seem clearly to have evolved from a kind of flagellate.
3. **Fungi.** Fungi also arose from a unicellular protist ancestor, one different from the ancestor of animals. Certain protists, including slime molds and water molds, have been considered fungi ("molds"), although they are usually classified as protists and are not thought to resemble ancestors of fungi. The true protist ancestor of fungi is as yet unknown. This is one of the great unsolved problems of taxonomy.

Perhaps the most important characteristic of complex multicellular organisms is *cell specialization.* If you think about it, having a variety of different sorts of cells within the same individual implies something very important about the genes of the individual: *Different cells are using different genes!* The process whereby a single cell (in humans, a fertilized egg) becomes a multicellular individual with many different kinds of cells is called **development.** The cell specialization that is the hallmark of complex multicellular life is the direct result of cells developing in different ways by activating different genes.

A second key characteristic of complex multicellular organisms is *intercellular coordination,* the adjustment of a cell's activity in response to what other cells are doing. The cells of all complex multicellular organisms communicate with one another with chemical signals called hormones. In some organisms like sponges, there is relatively little coordination between the cells; in other organisms like humans, almost every cell is under complex coordination.

> **Key Learning Outcome 18.1** Fungi, plants, and animals are complexly multicellular, with specialized cell types and coordination between cells.

18.2 A Fungus Is Not a Plant

The fungi are a distinct kingdom of organisms, comprising about 74,000 named species. **Mycologists,** scientists who study fungi, believe there may be many more species in existence. Although fungi were at one time included in the plant kingdom, they lack chlorophyll and resemble plants only in their general appearance and lack of mobility. Significant differences between fungi and plants include the following:

Fungi are heterotrophs. Perhaps most obviously, a mushroom is not green because it does not contain chlorophyll. Virtually all plants are photosynthesizers, whereas no fungi carry out photosynthesis. Instead, fungi obtain their food by secreting digestive enzymes onto whatever they are attached to and then absorbing into their bodies the organic molecules that are released by the enzymes.

Fungi have filamentous bodies. A plant is built of groups of functionally different cells called tissues, with different parts typically made of several different tissues. Fungi by contrast are basically filamentous in their growth form (that is, their bodies consist entirely of cells organized into long, slender filaments called *hyphae*), even though these filaments may be packed together to form a mass, called a *mycelium* (figure 18.1).

Fungi have nonmotile sperm. Some plants have motile sperm with flagella. The majority of fungi do not.

Fungi have cell walls made of chitin. The cell walls of fungi contain chitin, the same tough material that a crab shell is made of. The cell walls of plants are made of cellulose, also a strong building material. Chitin, however, is far more resistant to microbial degradation than is cellulose.

Figure 18.1 Masses of hyphae form mycelia.

The body of a fungus is composed of strings of cells called hyphae that pack together to form a mycelium like the dense, interwoven mat you see here growing through leaves on a forest floor in Maryland. Most of the body of a fungus is occupied by its mycelium.

Fungi have nuclear mitosis. Mitosis in fungi is different than mitosis in plants and most other eukaryotes in one key respect: The nuclear envelope does not break down and re-form; instead, all of mitosis takes place *within* the nucleus. A spindle apparatus forms there, dragging chromosomes to opposite poles of the *nucleus* (not the cell, as in all other eukaryotes).

You could build a much longer list, but already the take-home lesson is clear: Fungi are not like plants at all! Their many unique features are strong evidence that fungi are not closely related to any other group of organisms.

The Body of a Fungus

Fungi exist mainly in the form of slender filaments, barely visible with the naked eye, called **hyphae** (singular, **hypha**). A hypha is basically a long string of cells. Different hyphae then associate with each other to form a much larger structure, like the shelf fungus you see growing on a tree in figure 18.2.

The main body of a fungus is not the mushroom, which is a temporary reproductive structure, but rather the extensive network of fine hyphae that penetrate the soil, wood, or flesh in which the fungus is growing. A mass of hyphae is called a **mycelium** (plural, **mycelia**) and may contain many meters of individual hyphae.

Fungal cells are able to exhibit a high degree of communication within such structures, because although most cells of fungal hyphae are separated by cross-walls called *septa* (singular, *septum*), these septa rarely form a complete barrier, and as a consequence, cytoplasm is able to flow from one cell to another throughout the hyphae. The photo in figure 18.3 shows a junction between two fungal cells and the septum that partially separates them. From one fungal cell to the next, cytoplasm flows, streaming freely down the hypha through openings in the septa. Keep in mind the differences in scale between the hyphae and mycelia; the hypha in **figure 18.3** is about 3.4 μm across whereas the mycelium in **figure 18.1** is visible with the naked eye.

Because of such cytoplasmic streaming, proteins synthesized throughout the hyphae can be carried to the hyphal tips. This novel body plan is perhaps the most important innovation of the fungal kingdom. As a result of it, fungi can respond quickly to environmental changes, growing very rapidly when food and water are plentiful and the temperature is optimal. This body organization creates a unique relationship between the fungus and its environment. All parts of the fungal body are metabolically active, secreting digestive enzymes and actively attempting to digest and absorb any organic material with which the fungus comes in contact.

Also due to cytoplasmic streaming, many nuclei may be connected by the shared cytoplasm of a fungal mycelium.

Figure 18.2   A shelf fungus, *Trametes versicolor*.

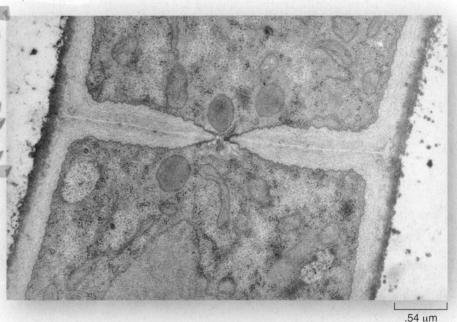

.54 μm

Figure 18.3   Septum and pore between cells in a hypha.
This photomicrograph clearly shows cytoplasmic streaming through a pore in the septum between two adjacent cells of a fungal hypha.

None of them (except for reproductive cells) are isolated in any one cell; all of them are linked cytoplasmically with every cell of the mycelium. Indeed, the entire concept of multicellularity takes on a new meaning among the fungi, the ultimate communal sharers among the multicellular organisms.

> **Key Learning Outcome 18.2   Fungi are not at all like plants. The fungal body is basically long strings of cells, often interconnected.**

18.3 Reproduction and Nutrition of Fungi

How Fungi Reproduce

Fungi reproduce both asexually and sexually. All fungal nuclei except for the zygote are haploid. Often in the sexual reproduction of fungi, individuals of different "mating types" must participate, much as two sexes are required for human reproduction. Sexual reproduction is initiated when two hyphae of genetically different mating types come in contact, and the hyphae fuse. What happens next? In animals and plants, when the two haploid gametes fuse, the two haploid nuclei immediately fuse to form the diploid nucleus of the zygote. As you might by now expect, fungi handle things differently. In most fungi, the two nuclei do not fuse immediately. Instead, they remain unmarried inhabitants of the same house, coexisting in a common cytoplasm for most of the life of the fungus! A fungal hypha that has two nuclei is called **dikaryotic.** If the nuclei are derived from two genetically different individuals, it is called a **heterokaryon** (Greek, *heteros,* other, and *karyon,* kernel or nucleus). A fungal hypha in which all the nuclei are genetically similar is said to be a **homokaryon** (Greek, *homo,* one).

When reproductive structures are formed in fungi, complete septa form between cells, the only exception to the free flow of cytoplasm between cells of the fungal body. There are three kinds of reproductive structures: (1) **gametangia** form haploid gametes, which fuse to give rise to a zygote that undergoes meiosis; (2) **sporangia** produce haploid spores that can be dispersed; and (3) **conidiophores** produce asexual spores called **conidia** that can be produced quickly and allow for the rapid colonization of a new food source.

Figure 18.5   The oyster mushroom.
This species, *Pleurotus ostreatus,* immobilizes nematodes, which the fungus uses as a source of food.

Spores are a common means of reproduction among the fungi. The puffball fungus in **figure 18.4** is releasing spores in a somewhat explosive manner. Spores are well suited to the needs of an organism anchored to one place. They are so small and light that they may remain suspended in the air for long periods of time and may be carried great distances. When a spore lands in a suitable place, it germinates and begins to divide, soon giving rise to a new fungal hypha.

How Fungi Obtain Nutrients

All fungi obtain their food by secreting digestive enzymes into their surroundings and then absorbing back into the fungus the organic molecules produced by this **external digestion.** Many fungi are able to break down the cellulose in wood, cleaving the linkages between glucose subunits and then absorbing the glucose molecules as food. That is why fungi are so often seen growing on trees.

Just as some plants like the Venus flytrap are active carnivores, so some fungi are active predators. For example, the edible oyster fungus *Pleurotus ostreatus,* shown growing on a tree in **figure 18.5**, attracts tiny roundworms known as nematodes that feed on it—and secretes a substance that anesthetizes the nematodes. When the worms become sluggish and inactive, the fungal hyphae envelop and penetrate their bodies and absorb their contents, a rich source of nitrogen (always in short supply in natural ecosystems).

> **Key Learning Outcome 18.3**   Fungi reproduce both asexually and sexually. They obtain their nutrients by secreting digestive enzymes into their surroundings and then absorbing the digested molecules back into the fungal body.

Figure 18.4   Many fungi produce spores.
Spores explode from the surface of a puffball fungus.

18.4 | Kinds of Fungi

Fungi are an ancient group of organisms at least 400 million years old. There are nearly 74,000 described species, in five groups (described in **table 18.1**), and many more awaiting discovery. Many fungi are harmful because they decay, rot, and spoil many different materials as they obtain food and because they cause serious diseases in animals and particularly in plants. Other fungi, however, are extremely useful. The manufacturing of both bread and beer depends on the biochemical activities of yeasts, single-celled fungi that produce abundant quantities of carbon dioxide and ethanol. Fungi are used on a major scale in industry to convert one complex organic molecule into another; many commercially important steroids are synthesized in this way.

The four major fungal phyla, distinguished from one another primarily by their mode of sexual reproduction, are the zygomycetes, the ascomycetes, the basidiomycetes, and the chytridiomycetes. A fifth group, the imperfect fungi, is an artificial grouping of fungi in which sexual reproduction has not been observed; these organisms are assigned to an appropriate group once their mode of sexual reproduction is identified. Molecular data are contributing to our understanding of the fungal phylogeny and as additional molecular evidence is acquired, a new fungal phylogeny is likely to appear. However, it already seems clear that fungi are more closely related to animals than to plants.

> **Key Learning Outcome 18.4** The fungal phyla are distinguished primarily by their modes of sexual reproduction. However, molecular sequence data will undoubtedly result in changes to the fungal phylogeny in the future.

TABLE 18.1 FUNGI

Phylum	Typical Examples	Key Characteristics	Approximate Number of Living Species
Zygomycota	*Rhizopus* (black bread mold)	Reproduce sexually and asexually; multinucleate hyphae lack septa except for reproductive structures; fusion of hyphae leads directly to formation of a zygote, in which meiosis occurs just before it germinates	1,050
Ascomycota	Yeasts, truffles, morels	Reproduce by sexual means; ascospores are formed inside a sac called an ascus; asexual reproduction is also common	32,000
Basidiomycota	Mushrooms, toadstools, rusts	Reproduce by sexual means; basidiospores are borne on club-shaped structures called basidia; the terminal hyphal cell that produces spores is called a basidium; asexual reproduction occurs occasionally	22,000
Chytridiomycota	*Allomyces*	Produce flagellated gametes (zoospores); predominately aquatic, some freshwater and some marine; oldest group of fungi	1,500
Imperfect fungi (not a phylum)	*Aspergillus*, *Penicillium*	Sexual reproduction has not been observed; most are thought to be ascomycetes that have lost the ability to reproduce sexually	17,000

18.5 Zygomycetes

The **zygomycetes,** members of the phylum Zygomycota, are unique among the fungi in that the fusion of hyphae does not produce a heterokaryon (a cell with two haploid nuclei). Instead, the two nuclei fuse and form a single diploid nucleus. Just as the fusion of sperm and egg produces a zygote in plants and animals, so this fusion produces a zygote. The name *zygomycetes* means "fungi that make zygotes."

Zygomycetes are the exception to the rule among fungi, and there are not many different kinds of them—only about 1,050 named species (about 1% of the named fungi). Included among them are some of the most frequent bread molds (the so-called black molds) and many microscopic fungi found on decaying organic material including strawberries and other fruits. Another important group of zygomycetes is called the *Glomales*. These soil-borne fungi form symbiotic relationships with roots of terrestrial plants and may have aided in the evolution of terrestrial plants, enhancing the uptake of minerals and water from the soil.

Reproduction among the zygomycetes is typically asexual. A cell at the tip of a hypha becomes walled off by a complete septum, forming an erect stalk tipped by a sporangium within which haploid spores are produced. These are the lollipop-shaped structures you see in the life cycle illustrated in **figure 18.6**. Their spores are shed into the wind and blown to new locations, where the spores germinate and attempt to start new mycelia. Sexual reproduction is unusual but may occur in times of stress. It is shown in the lower part of the figure, where hyphae from two different mating strains (+ is green and − is red) fuse and their nuclei also fuse, forming a diploid zygote. At the point where the two hyphae fuse, a sturdy and resistant structure called a **zygosporangium** forms. The zygosporangium is a very effective survival mechanism, a resting structure that allows the organism to remain dormant for long periods of time when conditions are not favorable. When conditions improve, the zygosporangium forms a stalked structure topped with a sporangium. Meiosis occurs within the sporangia and haploid spores are released, just as in the asexual portion of the life cycle.

> **Key Learning Outcome 18.5** Zygomycetes are unusual fungi that typically reproduce asexually; when hyphae do fuse, a zygote (one 2*n* nucleus), rather than a heterokaryon (two haploid nuclei), is produced.

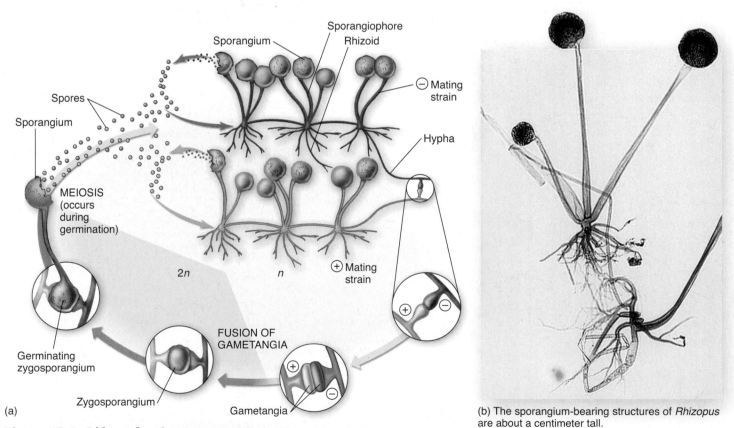

Figure 18.6 Life cycle of a zygomycete.

(a) In the life cycle of *Rhizopus,* a zygomycete that grows on moist bread and other similar substrates, the hyphae grow over the surface of the bread or other material on which the fungus feeds, producing erect, sporangium-bearing stalks in clumps, also shown in (b). If two hyphae grow together, their nuclei may fuse, producing a zygote. This zygote, which is the only diploid cell of the life cycle, acquires a thick, black coat (colored *purple* in the diagram above) and is then called a zygosporangium. Meiosis occurs during its germination, and normal, haploid hyphae grow from the haploid spores that result from this process.

(b) The sporangium-bearing structures of *Rhizopus* are about a centimeter tall.

18.6 Ascomycetes

Phylum Ascomycota, the **ascomycetes,** is the largest of the fungal phyla, with about 32,000 named species and many more being discovered each year. Among the ascomycetes are such familiar and economically important fungi as yeasts, morels, and truffles, as well as molds such as *Neurospora* (a historically important organism in genetic research) and many plant fungal pathogens, such as those that cause Dutch elm disease and chestnut blight.

Reproduction among the ascomycetes is usually asexual, just as it is among the zygomycetes. The hyphae of ascomycetes possess incomplete septa that separate the cells but have a large central pore in them, so that the flow of cytoplasm up and down the hypha is not impeded. Asexual reproduction occurs when the tips of hyphae become fully isolated from the rest of the mycelium by a complete septum, forming asexual spores called *conidia* (**figure 18.7a**, in the enlarged, circled area), each often containing several nuclei. When one of these conidia is released, air currents carry it to another place, where it may germinate to form a new mycelium.

It is important not to get confused by the number of nuclei in conidia. These multinucleate spores are *haploid,* not diploid, because there is only one version of the genome (one set of ascomycete chromosomes) present, whereas in a diploid cell there are two genetically different sets of chromosomes present. The actual number of nuclei is not what's important—it's the number of different genomes.

The ascomycetes are named for their sexual reproductive structure, the **ascus** (plural, **asci**), which differentiates within a larger structure called the *ascocarp.* The morel in **figure 18.7b** shows an ascocarp. The ascus is a microscopic cell that forms on the tips of the hyphae within the ascocarp, and it is where the zygote forms. The zygote is the only diploid nucleus of the ascomycete life cycle, indicated by the light blue section of the life cycle. The zygote undergoes meiosis, producing haploid spores called *ascospores.* When a mature ascus bursts, individual ascospores may be thrown as far as 30 centimeters. Considering how small the ascus is, this is truly an amazing distance. This would be like you hitting a baseball 1.25 kilometers, 10 times longer than Babe Ruth's longest home run!

> **Key Learning Outcome 18.6** Most fungi are ascomycetes, which form zygotes within reproductive structures called asci.

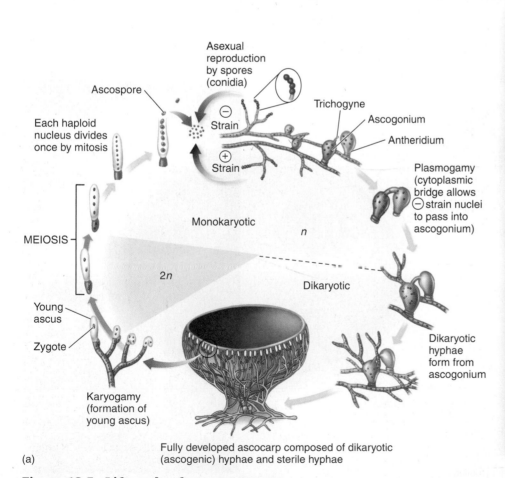

(b) This yellow morel is an ascomycete.

Figure 18.7 Life cycle of an ascomycete.

(*a*) Asexual reproduction takes place by means of conidia, spores cut off by septa at the ends of modified hyphae. Sexual reproduction occurs when the female gametangium, or *ascogonium*, fuses with the male gametangium, or *antheridium*, through a structure called the trichogyne. The ascocarp (*b*) develops from the dikaryotic hyphae and sterile hyphae. The nuclei fuse forming the diploid zygote inside the ascus. The zygote undergoes meiosis, leading to the formation of haploid ascospores.

18.7 Basidiomycetes

The phylum Basidiomycota contains the most familiar of the fungi among their 22,000 named species—the mushrooms, toadstools, puffballs, and shelf fungi. Many mushrooms are used as food, but others are deadly poisonous. Some species are cultivated as crops—the button mushroom *Agaricus bisporus* is grown in more than 70 countries, producing a crop in 1998 with a value of over $15 billion. Also among the **basidiomycetes** are bread yeasts and plant pathogens including rusts and smuts. Rust infections resemble rusting metal, whereas smut infections appear black and powdery due to their spores.

The life cycle of a basidiomycete (**figure 18.8a**) starts with the production of a hypha from a germinating spore. These hyphae lack septa at first, just as in zygomycetes. Eventually, however, septa are formed between each of the nuclei—but as in ascomycetes, there are holes in these cell separations, allowing cytoplasm to flow freely between cells. These hyphae grow, forming complex mycelia, and when hyphae of two different mating types (⊕ and ⊖) fuse, they form cells in which the nuclei remain separate—they do not fuse into one nucleus. Recall that if two distinct nuclei occur within each cell of the hypha, it is called dikaryotic, indicated by the *n + n* tan area of the cycle. The dikaryotic hypha that results goes on to form a dikaryotic mycelium. The mycelium forms a complex structure made of dikaryotic hyphae called the *basidiocarp,* or mushroom (**figure 18.8b**).

The two nuclei in each cell of a dikaryotic hypha can coexist together for a very long time without fusing. Unlike the other two fungal phyla, asexual reproduction is infrequent among the basidiomycetes, which typically reproduce sexually.

In sexual reproduction, zygotes (the only diploid cells of the life cycle) form when the two nuclei of dikaryotic cells fuse (on the right-hand side of the cycle). This occurs within a club-shaped reproductive structure called the **basidium** (plural, **basidia**). Meiosis occurs in each basidium, forming haploid spores called basidiospores. The basidia occur in a dense layer on the underside of the cap of the mushroom, where the surface is folded like an accordion. It has been estimated that a mushroom with an 8-centimeter cap can produce 40 million spores per hour!

> **Key Learning Outcome 18.7** Mushrooms are basidiomycetes, which form club-shaped reproductive structures called basidia.

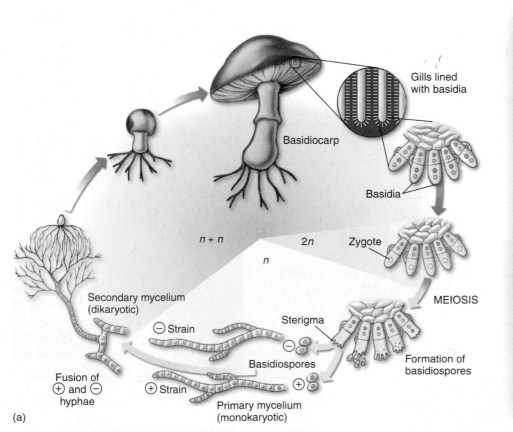

(b) An *Amanita* mushroom

Figure 18.8 Life cycle of a basidiomycete.

(a) Basidiomycetes usually reproduce sexually, with the fusion of nuclei in the basidia to produce a zygote. Meiosis follows syngamy and produces basidiospores that eventually form a basidiocarp (b).

18.8 Chytridiomycetes, Imperfect Fungi, and Yeasts

Chytridiomycetes

The phylum Chytridiomycota (chytrids) are mostly aquatic organisms, although some are found in soils and other terrestrial environments. Chytrids are the most primitive fungi, retaining flagellated gametes (called zoospores) from their protist ancestors. The other fungal groups are thought to have lost their flagellated stage at some point early in their evolutionary history as fungi. Several species of chytrids are plant pathogens that cause minor diseases. One species of chytrid, *Batrachochytrium dendrobatidis,* has been identified as a potential pathogen of frogs. The spores released by the fungi become embedded in the skin, where they seem to interfere with normal skin functions like respiration. A reddening of the skin on the legs and abdomen of the infected frog in **figure 18.9** is a symptom of a *B. dendrobatidis* infection. This fatal disease appears to be a contributing cause to the recent sharp declines in amphibian populations seen worldwide.

Imperfect Fungi

In addition to the four phyla of fungi that differ primarily in their mode of sexual reproduction, there are some 17,000 described species of fungi in whom sexual reproduction has not been observed. These cannot be formally assigned to one of the four sexually reproducing phyla and so are grouped for convenience as the so-called **imperfect fungi** (**figure 18.10**). The imperfect fungi are fungi that have lost the ability to reproduce sexually. Most of them appear to be ascomycetes, although some basidiomycetes are also included—these can be distinguished by features of the hyphae and asexual reproduction. Most of the fungi that cause skin diseases, including athlete's foot and ringworm, are caused by imperfect fungi. Fungal diseases are often difficult to treat pharmaceutically because of the ancestral relationship between animals and fungi. Medicines that kill fungal cells may also adversely affect animal cells.

Figure 18.10 Imperfect fungi.
Imperfect fungi are fungi in which sexual reproduction is unknown. *Verticillium alboatrum,* an important pathogen of alfalfa, has whorled conidia. The single-celled conidia of this member of the imperfect fungi are borne at the ends of the conidiophores.

Budding

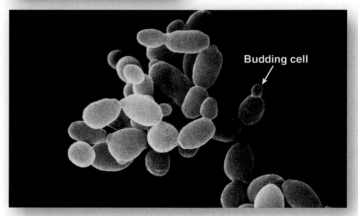

Budding cell

Figure 18.11 Budding in *Saccharomyces.*
These yeast cells tend to hang together in chains, a feature that calls to mind the derivation of single-celled yeast from multicellular ancestors.

Yeasts

Yeast is the generic (general) name given to any unicellular fungus. Although single-celled, yeasts appear almost certainly to have been derived from multicellular ancestors. There are about 250 named species of yeasts, including *Saccharomyces cerevisiae* (brewer's yeast), used for thousands of years in the production of bread, beer, and wine. Other yeasts are pathogens, including *Candida,* a common source of vaginal infection.

Just as in ascomycetes, most of yeast reproduction is asexual and takes place by cell fission or budding (the formation of a small cell from a portion of a larger one, as you see happening at the arrow in **figure 18.11**). Sexual reproduction among yeasts occurs when two yeast cells fuse. The new cell containing two nuclei functions as an ascus. After the two nuclei fuse, meiosis produces four ascospores, which develop directly into new yeast cells.

> **Key Learning Outcome 18.8** Chytridiomycetes are a group of fungi most closely related to ancestral fungi. Imperfect fungi may have lost the ability to reproduce sexually. Yeasts are unicellular fungi.

Figure 18.9 Chytrid infection.
The pathogenic chytrid, *Batrachochytrium dendrobatidis,* has infected this frog.

18.9 Ecological Roles of Fungi

Decomposers

Fungi, together with bacteria, are the principal decomposers in the biosphere. They break down organic materials and return the substances that had been locked in those molecules to circulation in the ecosystem. Fungi are virtually the only organisms capable of breaking down lignin, one of the major constituents of wood. By breaking down such substances, fungi make carbon, nitrogen, and phosphorus from the bodies of dead organisms available to other organisms.

In breaking down organic matter, some fungi attack living plants and animals as a source of organic molecules, whereas others attack dead ones. Fungi often act as disease-causing organisms for both animals and plants. The fungus *Armillaria,* shown in **figure 18.12**, is infecting a stand of conifers. The fungus originates in the center of an area indicated by the circles, and grows outward. Fungi are responsible for billions of dollars in agricultural losses every year.

Commercial Uses

The same aggressive metabolism that makes fungi ecologically important has been put to commercial use in many ways. The manufacture of both bread and beer depends on the biochemical activities of yeasts, single-celled fungi that produce abundant quantities of ethanol and carbon dioxide. Cheese and wine achieve their delicate flavors because of the metabolic processes of certain fungi. Vast industries depend on the biochemical manufacture of organic substances such as citric acid by fungi in culture. Many antibiotics, including penicillin, are derived from fungi.

Edible and Poisonous Fungi

Many types of ascomycete and basidiomycete fungi are edible (**figure 18.13***a, b*). They are commercially grown and can also be picked from the wild. The basidiomycete *Agaricus bisporus* grows in the wild but is also one of the most widely cultivated mushrooms in the world. Known as the "button mushroom" when it is small, it is also sold as the portobello mushroom when larger. Other examples of edible fungi include the yellow chanterelle (*Cantharellus cibarius*), morels (see **figure 18.7***b*), and shiitake (*Lentinula edodes*). A great deal of care must be taken when selecting mushrooms for consumption, as many species are poisonous due to toxins they contain. Poisonous mushrooms (**figure 18.13***c*) cause a range of symptoms, from slight allergic and digestive reactions, to hallucination, organ failure, and death.

Fungal Associations

Fungi are involved in a variety of intimate associations with algae and plants that play very important roles in the biological

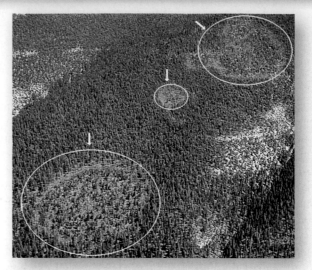

Figure 18.12   World's largest organism?
Armillaria, a pathogenic fungus shown here afflicting three discrete regions of coniferous forest in Montana, grows from a central focus as a single clone. The large patch at the bottom of the picture is almost 8 hectares.

(a)   (b)

(c)

Figure 18.13 Edible and poisonous mushrooms.

Edible mushrooms include (*a*) button mushrooms (*Agaricus bisporus*) and (*b*) yellow chanterelles (*Cantharellus cibarius*). Poisonous mushrooms include (*c*) *Amanita muscaria.*

world. These associations typically involve a sharing of abilities between a heterotroph (the fungus) and a photosynthesizer (the algae or plant). The fungus contributes the ability to absorb minerals and other nutrients very efficiently from the environment; the photosynthesizer contributes the ability to use sunlight to power the building of organic molecules. Alone, the fungus has no source of food, the photosynthesizer no source of nutrients. Together, each has access to both food and nutrients, a partnership in which both participants benefit.

Mycorrhizae Associations between fungi and plant roots are called **mycorrhizae** (Greek *myco,* fungus, and *rhizos,* roots). The roots of about 80% of all kinds of plants are involved in such associations. In fact, it has been estimated that fungi account for as much as 15% of the total weight of the world's plant roots! **Figure 18.14** shows how extensive this relationship can be. The roots on the left are roots from pine trees not associated with fungi. The roots in the middle and on the right exhibit mycorrhizae. You can see how the mycorrhizae greatly increase the surface area of the root. In a mycorrhiza, filaments of the fungus act as superefficient root hairs, projecting out from the epidermis, or outermost cell layer, of the terminal portions of the root. The fungal filaments aid in the direct transfer of phosphorus and other minerals from the soil into the roots of the plant, while the plant supplies organic carbon to the symbiotic fungus.

The earliest fossil plants often have mycorrhizal roots, which are thought to have played an important role in the invasion of land by plants. The soils of that time would have completely lacked organic matter, and mycorrhizal plants are particularly successful in such infertile soils. The most primitive vascular plants surviving today continue to depend strongly on mycorrhizae.

Lichens A **lichen** is an association between a fungus and a photosynthetic partner. Ascomycetes are the fungal partners in all but 20 of the 15,000 different species of lichens that have been characterized. Most of the visible body of a lichen consists of its fungus, but interwoven between layers of hyphae within the fungus are cyanobacteria, green algae, or sometimes both. Enough light penetrates the translucent layers of hyphae to make photosynthesis possible. Specialized fungal hyphae envelop and sometimes penetrate the photosynthetic cells, serving as highways to collect and transfer to the fungal body the sugars and other organic molecules manufactured by the photosynthetic cells. The fungus transmits special biochemical signals that direct the cyanobacteria or green algae to produce metabolic substances that they would not if growing independently of the fungus. Indeed, the fungus is not able to grow or survive without its photosynthetic partner. Many biologists characterize this particular symbiotic relationship as one of slavery rather than cooperation, a controlled parasitism of the photosynthetic organism by the fungal host.

The durable construction of the fungus, combined with the photosynthetic abilities of its partner, has enabled lichens to invade the harshest of habitats, from the tops of mountains to dry, bare rock faces in the desert. The orange substance growing on the rocks in **figure 18.15** is a lichen. In such harsh, exposed areas, lichens are often the first colonists, breaking down the rocks and setting the stage for the invasion of other organisms.

Lichens are extremely sensitive to pollutants in the atmosphere because they readily

Figure 18.14 Mycorrhizae on the roots of pines.
From left to right are pine roots not associated with a fungus, white mycorrhizae formed by *Rhizopogon,* and yellow-brown mycorrhizae formed by *Pisolithus.*

absorb substances dissolved in rain and dew. This is why lichens are generally absent in and around cities—they are acutely sensitive to sulfur dioxide produced by automobile traffic and industrial activity. Such pollutants destroy their chlorophyll molecules and thus decrease photosynthesis and upset the physiological balance between the fungus and the algae or cyanobacteria.

Key Learning Outcome 18.9 Fungi are key decomposers and play many other important ecological and commercial roles. Mycorrhizae are symbiotic associations between fungi and plant roots. Lichens are symbiotic associations between a fungus and a photosynthetic partner (a cyanobacterium or an alga).

Figure 18.15 Lichens growing on a rock.

Are Chytrids Killing The Frogs?

As you learned earlier in this chapter, chytrid fungi are thought to be playing a major role in a worldwide wave of amphibian extinctions, discussed in much more detail in chapter 38 (page 799). Our awareness of the possible role of chytrids began in Queensland (the northeastern portion of Australia) in 1993, when a mass die-off of frogs was reported. All different kinds of frogs seemed to be affected, and entire populations were wiped out. In the rainforests of northern Queensland, populations of the sharp-nosed torrent frog (*Taudactylus acutirostris*) were found to be so seriously affected as to be in danger of extinction. Captive colonies were set up at James Cook University and at the Melbourne and Taronga zoos in an attempt to preserve the species. Unfortunately, the preservation of the species failed. Every frog in the colonies died.

What was killing the frogs? The answer to this question came in 1998, when researchers examined the epithelium (skin) of sick frogs under the scanning electron microscope and saw what you can see in the photomicrographs to the right. Normally a relatively smooth surface, the epithelium of the dying frogs was roughened, with spherical bodies protruding from the surface.

These protrusions are zoosporangia, asexual reproductive structures of a chytrid fungus. One is shown up-close (*inset*). Each zoosporangium is roughly spherical, with one or more small projecting tubes. Millions of tiny zoospores develop in each zoosporangium. When the plug blocking the tip of a tube disappears, the spores are discharged onto the surface of adjacent skin cells, or into the water, where their flagella allow them to swim until they encounter another host. When one of the zoospores contacts the skin of another frog, it attaches and forms a new zoosporangium in the subsurface layer of the skin, renewing the infection cycle.

Study of the infecting chytrids revealed them to be members of the species *Batrachochytrium dendrobatidis*. This was unexpected. Chytrids are typically found in water and soil, and although there are several types known to infect plants and insects, no chytrid had ever been known to infect a vertebrate.

These initial scanning electron micrograph results seemed to make a pretty convincing case that chytrids had caused the mass die-off of frogs in Queensland. However, in order to provide more direct evidence, a series of experiments were carried out in which the ability of the chytrid fungus to kill frogs was directly assessed.

In one such experiment, typical of many, some frogs of the genus *Dendrobates* were exposed to chytrids and others were not. After three weeks, all frogs were examined for shed skin, a clinical sign of the frog-killing disease. The results are seen in the pie charts above.

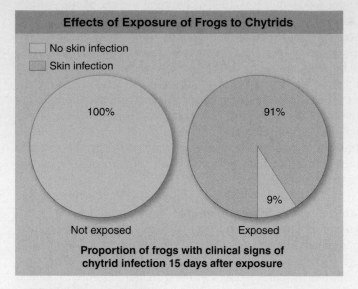

Effects of Exposure of Frogs to Chytrids

☐ No skin infection
▨ Skin infection

Not exposed: 100%

Exposed: 91% / 9%

Proportion of frogs with clinical signs of chytrid infection 15 days after exposure

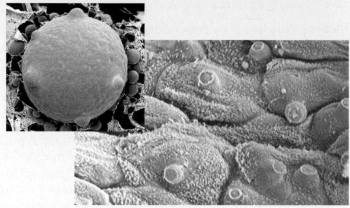

1. **Applying Concepts** In this study, is there a dependent variable? If so, what is it?
2. **Interpreting Data** What is the incidence of disease in nonexposed frogs? In exposed frogs?
3. **Making Inferences** Is there any association between exposure to the chytrid *B. dendrobatidis* and development of the skin infection that is a clinical sign of life-threatening illness in frogs?
4. **Drawing Conclusions** What is the impact of exposure to chytrids upon the likelihood of developing the frog-killing disease?
5. **Further Analysis**
   a. Many kinds of frogs and salamanders are dying all over the world. Does this experiment suggest a way to determine how general is the susceptibility of amphibians to chytrid infection?
   b. While a few frog die-offs have occurred in the past, none have been nearly this serious. Do you think *B. dendrobatidis* is a new species, or do you think environmental changes like global warming or increased UV radiation resulting from ozone depletion might be the cause? Discuss.

Fungi as Multicellular Organisms

18.1 Complex Multicellularity

- Fungi, plants, and animals are complex multicellular organisms, unlike the algae that are structurally simple multicellular organisms. Fungi contain highly specialized cell types and exhibit intercellular communication.

- Cell specialization requires that different cell types use different genes. During development, different genes are activated in different types of cells.

- Intercellular communication is accomplished when cells are able to communicate with each other by releasing chemical signals, such as hormones, that travel between cells.

18.2 A Fungus Is Not a Plant

- Fungi are sometimes compared to plants, but fungi are like plants only in their lack of mobility. They differ from plants in many ways. They are heterotrophs; unlike plants, they don't manufacture their own food through photosynthesis. Also, the body of a fungus is composed of long slender filaments called hyphae that pack together to form a mycelium (**figure 18.1**). Most fungi have nonmotile sperm, unlike some plants. Fungi have cell walls made of chitin, which is different from the cellulose cell walls found in plant cells. Fungi also undergo nuclear mitosis, where the nuclei divide but the cell doesn't. Most fungal cells are separated by an incomplete wall called a septum (**figure 18.3**), which allows cytoplasm to pass between cells for intercellular communication.

18.3 Reproduction and Nutrition of Fungi

- Fungi reproduce both sexually and asexually. Sexual reproduction takes place between two genetically different "mating types." In fungi, two hyphae of different mating types fuse. Their haploid nuclei may stay separate in the cell, as in the case of a heterokaryon. In specialized reproductive structures, these nuclei will fuse to form zygotes that immediately undergo meiosis.

- Three types of reproductive structures are found in fungi: gametangia, sporangia, and conidiophores.

- Fungi obtain nutrients through external digestion, in which enzymes are secreted onto food. The food is digested outside of the body, and the nutrients are absorbed by the fungal cells. Fungi can break down cellulose, which is why they are often found growing on trees, as demonstrated by this oyster mushroom from **figure 18.5**.

Fungal Diversity

18.4 Kinds of Fungi

- The fungi are an ancient group of organisms that impact life on earth. Many are harmful, causing rotting and decaying of food and diseases of plants and animals. Others are beneficial in industry, being used in the production of bread and beer and other commercial uses.

- There are four major phyla of fungi—Zygomycota, Ascomycota, Basidiomycota, and Chytridiomycota—classified according to their mode of sexual reproduction.

- A fifth group, the imperfect fungi, is a "catch-all" category of fungi in which sexual reproduction has not been observed (**table 18.1**).

18.5 Zygomycetes

- Zygomycetes usually reproduce asexually, releasing haploid spores from sporangia. However, they do reproduce sexually in times of stress. Hyphae of different mating strains fuse forming a diploid zygosporangium, which germinates and forms a sporangium, shown here at the tips from **figure 18.6**. Meiosis occurs inside the sporangium, and haploid spores that are genetically different from either mating strain are released. Bread molds are members of this phylum.

18.6 Ascomycetes

- Ascomycetes comprise the largest fungal phylum and include morels, truffles, and many plant fungal pathogens. They usually reproduce asexually through the release of haploid spores called conidia. The reproductive structure is the ascus, which forms on the tips of dikaryotic hyphae that result from the fusion of different mating strains. Haploid nuclei in the ascus fuse, forming a diploid zygote. The zygote undergoes meiosis, forming ascospores that are released from the ascus (**figure 18.7**).

18.7 Basidiomycetes

- Basidiomycetes are the recognizable mushrooms, toadstools, puffballs, and shelf fungi. They reproduce sexually through the fusion of hyphae of two different mating strains. The fused hyphae grow into a dikaryotic mycelium called a basidiocarp (**figure 18.8**). The reproductive structure, called a basidium, is contained in the cap of the mushroom. In the basidium, haploid nuclei fuse to form a diploid zygote that undergoes meiosis, producing haploid basidiospores that are released.

18.8 Chytridiomycetes, Imperfect Fungi, and Yeasts

- The Chytridiomycetes, the most primitive of the fungi, are mostly aquatic with flagellated gametes called zoospores. The chytrid species *Batrachochytrium dendrobatidis* is a pathogen of frogs, causing a fatal skin infection (**figure 18.9**).

- The imperfect fungi are a group of species in which sexual reproduction has not been observed (**figure 18.10**).

- Yeast is a general name given to unicellular fungi (**figure 18.11**). They usually reproduce asexually through budding.

The Ecology of Fungi

18.9 Ecological Roles of Fungi

- Fungi play key roles in the environment as decomposers and have commercial uses.

- Fungi are involved in intimate associations with the roots of some plants as mycorrhizae (**figures 18.14**), or with cyanobacteria or algae as lichens, shown here from **figure 18.15**.

Test Your Understanding

1. The most important characteristic of complex multicellular organisms is
   a. intercellular communication.
   b. cell development.
   c. cell specialization.
   d. cell reproduction.
2. Which of the following is *not* a characteristic of the fungi kingdom?
   a. heterotrophic
   b. cellulose cell walls
   c. nuclear mitosis
   d. nonmotile sperm
3. The main body of a fungus is the
   a. hyphae.
   b. septa.
   c. mushroom.
   d. mycelium.
4. Fungi reproduce
   a. both sexually and asexually.
   b. sexually only.
   c. asexually only.
   d. by fragmentation.
5. Morels and truffles belong to the fungus phylum
   a. Zygomycota.
   b. Ascomycota.
   c. Basidiomycota.
   d. Chytridiomycota.
6. Zygomycetes are different from other fungi because they do not produce
   a. mycelium.
   b. fruiting bodies.
   c. a heterokaryon.
   d. a sporangium.
7. Ascomycetes form reproductive spores in
   a. a special sac called the ascus.
   b. gills on the basidiocarp.
   c. sporangiophores.
   d. the mycelium.
8. Meiosis in basidiomycetes occurs in the
   a. hyphae.
   b. basidia.
   c. mycelium.
   d. basidiocarp.
9. Lichens are mutualistic associations between
   a. plants and fungi.
   b. algae and fungi.
   c. termites and fungi.
   d. coral and fungi.
10. Mycorrhizae help plants obtain
    a. water.
    b. oxygen.
    c. carbon dioxide.
    d. minerals.

Apply Your Understanding

1. **Figure 18.8a** Explain what part of the fungus life cycle is represented by the familiar mushroom that you see in stores.

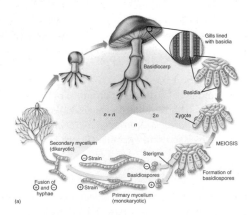

2. **Figure 18.14** Why is the surface area of the plant roots on the far left in the photo below much less than that of the plant roots in the middle and on the far right? How does this difference in surface area affect a plant's ability to absorb nutrients?

Synthesize What You Have Learned

1. Devise an experiment to test the hypothesis that fungi benefit more from a lichen relationship than do algae.
2. Protruding from the bark of a forest tree you detect an object that looks somewhat like a mushroom. Describe at least three criteria that you might use to determine whether the object is a piece of the tree or a piece of a fungus.
3. Your friend has athlete's foot and is unhappy with how long it is taking to get rid of it. Disgusted, he asks you why antibiotics like penicillin (derived from a fungus) and tetracycline are so much more effective in treating bacterial infections of humans than fungicides are in treating human fungal infections. What do you tell him?

19

Evolution of the Animal Phyla

nimals are the most diverse in appearance of all eukaryotes. *Polistes,* the common paper wasp you see here, is a member of the most diverse of all animal groups, the insects. It has been a major challenge for biologists to sort out the millions of kinds of animals. This wasp has a segmented external skeleton and jointed appendages, and so, based on these characteristics, is classified as an arthropod. But how are arthropods related to mollusks such as snails, and to segmented worms like earthworms? Until recently, biologists grouped all three kinds of animals together, as they all share a coelom body cavity, a character assumed to be so fundamental it could have evolved only once. Now, molecular analyses suggest this assumption may be wrong. Instead, mollusks and segmented worms are grouped together with other animals that grow the same way you do, by adding additional mass to an existing body, while arthropods are grouped with other molting animals. These animals increase in size by molting their external skeletons, an ability that seems to have evolved only once. Thus we learn that even in a long-established field like taxonomy, biology is constantly changing.

19.1 General Features of Animals

As best as biologists can determine, we and all other animals evolved from a kind of protist called a choanoflagellate. From these early animal ancestors, a great diversity of animals has since evolved. While the evolutionary relationships among the different types of animals are being debated (see section 19.2), all animals have several features in common (**table 19.1**): (1) Animals are heterotrophs and must ingest plants, algae, animals, or other organisms for nourishment. (2) All animals are multicellular, and unlike plants and protists, animal cells lack cell walls. (3) Animals are able to move from place to place. (4) Animals are very diverse in form and habitat. (5) Most animals reproduce sexually. (6) Animals have characteristic tissues and patterns of embryonic development.

> **Key Learning Outcome 19.1** Animals are complex, multicellular, heterotrophic organisms. Most animals also possess internal tissues.

TABLE 19.1	GENERAL FEATURES OF ANIMALS
Heterotrophs. Unlike autotrophic plants and algae, animals cannot construct organic molecules from inorganic chemicals. All animals are heterotrophs—that is, they obtain energy and organic molecules by ingesting other organisms. Some animals (herbivores) consume autotrophs, other animals (carnivores) consume heterotrophs; others, like the bear to the right, are omnivores that eat both autotrophs and heterotrophs, and still others (detritivores) consume decomposing organisms.	
Multicellular. All animals are multicellular, often with complex bodies like that of this brittlestar (*right*). The unicellular heterotrophic organisms called protozoa, which were at one time regarded as simple animals, are now considered members of the large and diverse kingdom Protista, discussed in chapter 17.	
No Cell Walls. Animal cells are distinct among those of multicellular organisms because they lack rigid cell walls and are usually quite flexible, like these cancer cells. The many cells of animal bodies are held together by extracellular lattices of structural proteins such as collagen. Other proteins form a collection of unique intercellular junctions between animal cells.	
Active Movement. The ability of animals to move more rapidly and in more complex ways than members of other kingdoms is perhaps their most striking characteristic, one that is directly related to the flexibility of their cells and the evolution of nerve and muscle tissues. A remarkable form of movement unique to animals is flying, an ability that is well developed among vertebrates and insects like this butterfly. The only terrestrial vertebrate group never to have evolved flight is amphibians.	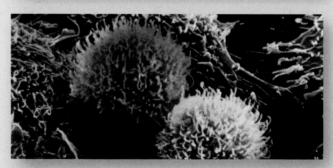

TABLE 19.1 *(continued)*

Diverse in Form. Almost all animals (99%) are **invertebrates**, which, like this millipede, lack a backbone. Of the estimated 10 million living animal species, only 42,500 have a backbone and are referred to as **vertebrates.** Animals are very diverse in form, ranging in size from organisms too small to see with the unaided eye to enormous whales and giant squids.

Diverse in Habitat. The animal kingdom includes about 35 phyla, most of which, like these jellyfish (phylum Cnidaria), occur in the sea. Far fewer phyla occur in freshwater and fewer still occur on land. Members of three successful marine phyla, Arthropoda (insects), Mollusca (snails), and Chordata (vertebrates), dominate animal life on land.

Sexual Reproduction. Most animals reproduce sexually, as these tortoises are doing. Animal eggs, which are nonmotile, are much larger than the small, usually flagellated sperm. In animals, cells formed in meiosis function directly as gametes. The haploid cells do not divide by mitosis first, as they do in plants and fungi, but rather fuse directly with each other to form the zygote. Consequently, with a few exceptions, there is no counterpart among animals to the alternation of haploid (gametophyte) and diploid (sporophyte) generations characteristic of plants.

Embryonic Development. Most animals have a similar pattern of embryonic development. The zygote first undergoes a series of mitotic divisions, called *cleavage,* and, like this dividing frog egg, becomes a solid ball of cells, the **morula,** then a hollow ball of cells, the **blastula.** In most animals, the blastula folds inward at one point to form a hollow sac with an opening at one end called the **blastopore.** An embryo at this stage is called a **gastrula.** The subsequent growth and movement of the cells of the gastrula differ widely from one phylum of animals to another.

Unique Tissues. The cells of all animals except sponges are organized into structural and functional units called **tissues,** collections of cells that have joined together and are specialized to perform a specific function. Animals are unique in having two tissues associated with movement: (1) muscle tissue, which powers animal movement, and (2) nervous tissue, which conducts signals among cells. Neuromuscular junctions, where nerves connect with muscle tissue, are shown here.

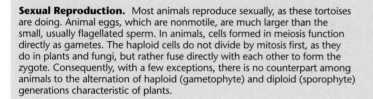

The Animal Family Tree

The features described in the previous section evolved over millions of years. The great diversity of animals today is the result of this long evolutionary journey. The multicellular animals, or metazoans, are traditionally divided into 35 distinct and very different phyla. How these phyla are related to each other has been the source of much discussion among biologists.

The Traditional Viewpoint

Taxonomists have traditionally attempted to create animal **phylogenies** (family trees; see section 15.5) by comparing anatomical features and aspects of embryological development. A broad consensus emerged over the last century about the main branches of the animal family tree.

The First Branch: Tissues The kingdom Animalia is traditionally divided by taxonomists into two main branches: (1) **Parazoa** ("beside animals")—animals that for the most part lack a definite symmetry and possess neither tissues nor organs, mostly composed of the sponges, phylum Porifera; and (2) **Eumetazoa** ("true animals")—animals that have a definite shape and symmetry and, in most cases, tissues organized into organs and organ systems. In **figure 19.1**, all animals shown to the right of Parazoa are eumetazoans.

All the branches in the animal family tree trace back to one ancestor at the base of the phylogeny. This shared ancestor was probably a choanoflagellate, a colonial flagellated protist that lived in the Precambrian era over 700 million years ago.

The Second Branch: Symmetry The eumetazoan branch of the animal family tree itself has two principal branches, differing in the nature of the embryonic layers that form during development and go on to differentiate into the tissues of the adult animal. Eumetazoans of the subgroup **Radiata** (having radial symmetry) have two layers, an outer *ectoderm* and an inner *endoderm,* and thus are called diploblastic. All other eumetazoans are the **Bilateria** (having bilateral symmetry) and are triploblastic, producing a third layer, the *mesoderm,* between the ectoderm and endoderm.

Further Branches Further branches of the animal family tree were assigned by taxonomists by comparing traits that seemed profoundly important to the evolutionary history of phyla, key features of the body plan shared by all animals belonging to that branch. Thus, the bilaterally symmetrical animals were split into groups with a body cavity and those without (acoelomates); animals with a body cavity were split into those with a true coelom (body cavity enclosed by mesoderm) and those without (the pseudocoelomates); animals with a coelom were split into those whose coelom derived from the digestive tube and those that did not, and so on.

Because of the either-or nature of the categories set up by traditional taxonomists, this approach has produced a family tree like the one in **figure 19.1**, with a lot of paired branches. The arbitrary nature of the divisions has always been obvious to biologists, but in spite of that, most biologists feel that it faithfully represents the general nature of the evolutionary history of metazoans.

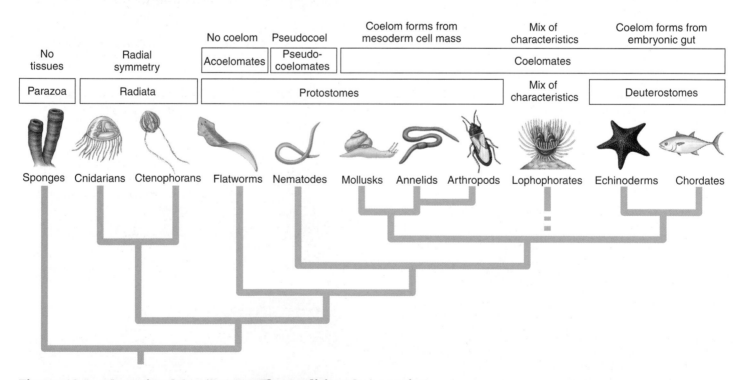

Figure 19.1 The animal family tree: The traditional viewpoint.
Biologists have traditionally divided the animals into 35 distinct phyla. The diagram above illustrates the relationships among some of the major animal phyla. The bilaterally symmetrical animals (those to the right of Radiata in the figure above) are sorted into three groups that differ with respect to their body cavity: acoelomates, pseudocoelomates, and coelomates.

A New Look at the Animal Family Tree

The traditional animal phylogeny, while accepted by a broad consensus of biologists for almost a century, is now being re-evaluated. Its simple either-or organization has always presented certain problems—puzzling minor groups do not fit well into the standard scheme. Results hint strongly that the key body-form characteristics that biologists have traditionally used to construct animal phylogenies—segmentation, coeloms, jointed appendages, and the like—are not the always preserved characters we had supposed. These features appear to have been gained and lost again during the course of the evolution of some unusual animals. If this pattern of change in what had been considered basic characters should prove general, our view of how the various animal phyla relate to one another is in need of major revision.

The last decade has seen a wealth of new molecular RNA and DNA sequence data on the various animal groups. The new field of **molecular systematics** uses unique sequences within certain genes to identify clusters of related groups. Using these sorts of molecular data, a variety of molecular phylogenies have been produced in the last decade. While differing from one another in many important respects, the new molecular phylogenies have the same deep branch structure as the traditional animal family tree (compare the lower branches in the "new" family tree in **figure 19.2** with the lower branches in **figure 19.1**). However, most agree on one revolutionary difference from the traditional phylogeny used in this text and presented in **figure 19.1**: The protostomes (which have a different pattern of development than the deuterostomes—a topic that will be discussed later in this chapter) are broken into two distinct clades. **Figure 19.2** is a consensus molecular phylogeny developed from DNA, ribosomal RNA, and protein studies. In it, the traditional protostome group is broken up into Lophotrochozoa and Ecdysozoa.

Lophotrochozoans are animals that grow by adding mass to an existing body. They are named for a distinctive feeding apparatus called a *lophophore* found in some phyla of the molecularly defined group. These animals—which usually live in water, have ciliary locomotion, and trochophore larvae —include flatworms, mollusks, and annelids.

Ecdysozoans have exoskeletons that must be shed for the animal to grow. This sort of molting process is called *ecdysis,* which is why these animals are called ecdysozoans. They include the roundworms (nematodes) and arthropods. How an animal grows wasn't a key characteristic when classifying animals in the traditional approach, but proves to be an important characteristic when comparing animals molecularly.

This new view of the metazoan Tree of Life is only a rough outline—at present, molecular phylogenetic analysis of the animal kingdom is in its infancy. Phylogenies developed from different molecules sometimes suggest quite different evolutionary relationships. For this reason, in this text you will explore animal diversity guided by the traditional animal family tree. However, the childhood of the new molecular approach is likely to be short. Over the next few years, a mountain of additional molecular data can be anticipated. As more data are brought to bear, the confusion can be expected to lessen and the relationships within groupings more confidently resolved.

> **Key Learning Outcome 19.2** Major groups are related in very different ways in molecular phylogenies than in the more traditional approach based on form and structure.

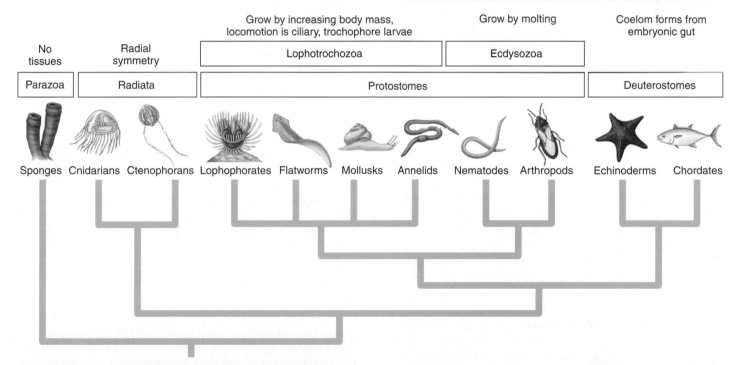

Figure 19.2 The animal family tree: A new look.

New phylogenies suggest that the protostomes might be better grouped according to whether they grow by adding mass to an existing body (Lophotrochozoa) or by molting (Ecdysozoa).

19.3 Six Key Transitions in Body Plan

The evolution of animals is marked by six key transitions: the evolution of tissues, bilateral symmetry, a body cavity, segmentation, molting, and deuterostome development. These six body transitions are indicated at the branchpoints of the animal evolutionary tree in **figure 19.3**.

1. Evolution of Tissues

The simplest animals, the Parazoa, lack both defined tissues and organs. Characterized by the sponges, these animals exist as aggregates of cells with minimal intercellular coordination. All other animals, the Eumetazoa, have distinct tissues with highly specialized cells.

2. Evolution of Bilateral Symmetry

Sponges also lack any definite symmetry, growing asymmetrically as irregular masses. Virtually all other animals have a definite shape and symmetry that can be defined along an imaginary axis drawn through the animal's body.

Radial Symmetry Symmetrical bodies first evolved in marine animals exhibiting **radial symmetry.** The parts of their bodies are arranged around a central axis in such a way that any plane passing through the central axis divides the organism into halves that are approximate mirror images.

Bilateral Symmetry The bodies of all other animals are marked by a fundamental **bilateral symmetry,** a body design in which the body has a right and a left half that are mirror images of each other. This unique form of organization allows parts of the body to evolve in different ways, permitting different organs to be located in different parts of the body. Also, bilaterally symmetrical animals move from place to place more efficiently than radially symmetrical ones, which, in general, lead a sessile or passively floating existence. Due to their increased mobility, bilaterally symmetrical animals are efficient in seeking food, locating mates, and avoiding predators.

3. Evolution of a Body Cavity

A third key transition in the evolution of the animal body plan was the evolution of the body cavity. The evolution of efficient organ systems within the animal body was not possible until a body cavity evolved for supporting organs, distributing materials, and fostering complex developmental interactions.

The presence of a body cavity allows the digestive tract to be larger and longer. This longer passage allows for storage of undigested food and longer exposure to enzymes for more complete digestion. Such an arrangement allows an animal to eat a great deal when it is safe to do so and then to hide during the digestive process, thus limiting the animal's exposure to predators.

An internal body cavity also provides space within which the gonads (ovaries and testes) can expand, allowing the accumulation of large numbers of eggs and sperm. Such storage capacity allows the diverse modifications of breeding strategy that characterize the more advanced phyla of animals. Furthermore, large numbers of gametes can be stored and released when the conditions are as favorable as possible for the survival of the young animals.

4. The Evolution of Segmentation

The fourth key transition in the animal body plan involved the subdivision of the body into **segments.** Just as it is efficient for workers to construct a tunnel from a series of identical prefabricated parts, so segmented animals are assembled from a succession of identical segments. Segmentation was assumed to have evolved only once among the invertebrates in the traditional taxonomy, as it seemed such a significant alteration of body plan.

5. The Evolution of Molting

Most coelomate animals grow by gradually adding mass to their body. However, this creates a serious problem for animals with a hard exoskeleton, which can hold only so much tissue. To grow further, the individual must shed its hard exoskeleton, a process called **molting** or, more formally, ecdysis.

Ecdysis occurs among both nematodes and arthropods. In the traditional taxonomy these are treated as two independent evolutionary events. The new phylogenies suggest ecdysis evolved only once. This would imply that arthropods and nematodes, both of which have hard exoskeletons and molt, are sister groups, and that segmentation rather than ecdysis must have evolved several times among the invertebrates, rather than once.

6. The Evolution of Deuterostome Development

Bilateral animals can be divided into two groups based on differences in the basic pattern of development. One group is called the **protostomes** (from the Greek words *protos,* first, and *stoma,* mouth) and includes the flatworms, nematodes, mollusks, annelids, and arthropods. Two outwardly dissimilar groups, the echinoderms and the chordates, together with a few other smaller related phyla, comprise the second group, the **deuterostomes** (Greek, *deuteros,* second, and *stoma,* mouth). Protostomes and deuterostomes differ in several aspects of embryo growth and will be discussed later in the chapter.

Deuterostomes evolved from protostomes more than 630 million years ago, and the consistency of deuterostome development, and its distinctiveness from that of the protostomes suggests that it evolved once, in a common ancestor to all of the phyla that exhibit it.

Characteristics of the major animal phyla are described in **table 19.2**.

> **Key Learning Outcome 19.3** Six key transitions in body design are responsible for most of the differences we see among the major animal phyla.

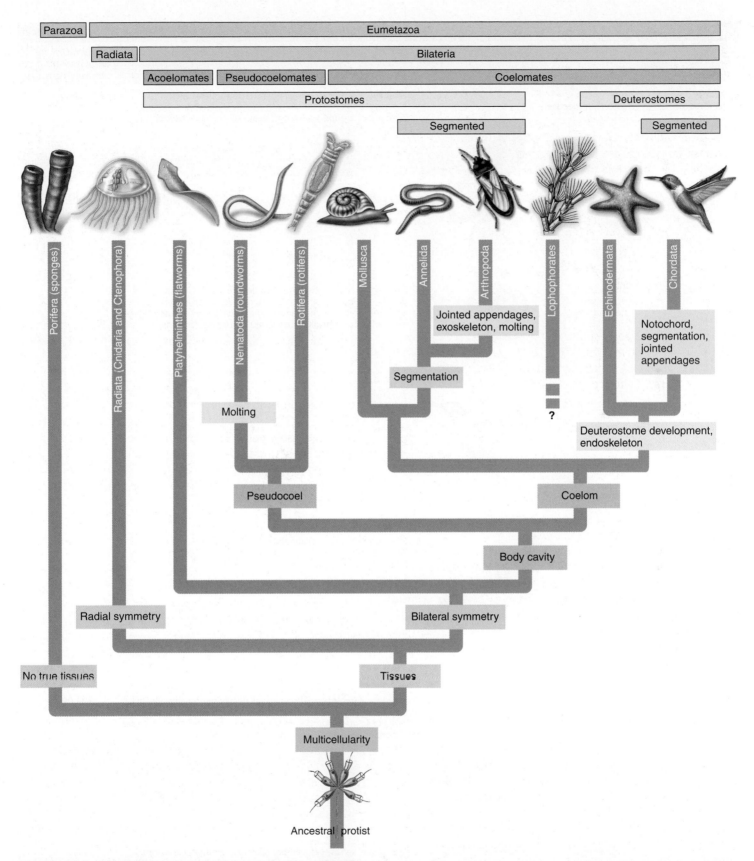

Figure 19.3 Evolutionary trends among the animals.

In this chapter, we examine a series of key evolutionary innovations in the animal body plan, shown here along the branches. Some of the major animal phyla are shown on this tree. Lophophorates exhibit a mix of protostome and deuterostome characteristics. The traditional tree shown here assumes segmentation arose only once among the invertebrates, while molting arose independently in nematodes and arthropods. The newly proposed molecular phylogenies assume molting arose only once, while segmentation arose independently in annelids, arthropods, and chordates.

TABLE 19.2   THE MAJOR ANIMAL PHYLA

Phylum	Typical Examples		Key Characteristics	Approximate Number of Named Species
Arthropoda (arthropods)	Insects, crabs, spiders, millipedes		Most successful of all animal phyla; chitinous exoskeleton covering segmented bodies with paired, jointed appendages; most insect groups have wings; nearly all are freshwater or terrestrial	1,000,000
Mollusca (mollusks)	Snails, clams, octopuses, nudibranchs		Soft-bodied coelomates whose bodies are divided into three parts: head-foot, visceral mass, and mantle; many have shells; almost all possess a unique rasping tongue called a radula; most are marine or freshwater but 35,000 species are terrestrial	110,000
Chordata (chordates)	Mammals, fish, reptiles, birds, amphibians		Segmented coelomates with a notochord; possess a dorsal nerve cord, pharyngeal pouches, and a tail at some stage of life; in vertebrates, the notochord is replaced during development by the spinal column; most are marine, many are freshwater, and 20,000 species are terrestrial	56,000
Platyhelminthes (flatworms)	*Planaria*, tapeworms, flukes		Solid, unsegmented, bilaterally symmetrical worms; no body cavity; digestive cavity, if present, has only one opening; marine, freshwater, or parasitic	20,000
Nematoda (roundworms)	*Ascaris*, pinworms, hookworms, *Filaria*		Pseudocoelomate, unsegmented, bilaterally symmetrical worms; tubular digestive tract passing from mouth to anus; tiny; without cilia; live in great numbers in soil and aquatic sediments; some are important animal parasites	20,000
Annelida (segmented worms)	Earthworms, marine worms, leeches		Coelomate, serially segmented, bilaterally symmetrical worms; complete digestive tract; most have bristles called setae on each segment that anchor them during crawling; marine, freshwater, and terrestrial	12,000

TABLE 19.2 *(continued)*

Phylum	Typical Examples		Key Characteristics	Approximate Number of Named Species
Cnidaria (cnidarians)	Jellyfish, hydra, corals, sea anemones		Soft, gelatinous, radially symmetrical bodies whose digestive cavity has a single opening; possess tentacles armed with stinging cells called cnidocytes that shoot sharp harpoons called nematocysts; almost entirely marine	10,000
Echinodermata (echinoderms)	Sea stars, sea urchins, sand dollars, sea cucumbers		Deuterostomes with radially symmetrical adult bodies; endoskeleton of calcium plates; pentamerous (five-part) body plan and unique water vascular system with tube feet; able to regenerate lost body parts; all are marine	6,000
Porifera (sponges)	Barrel sponges, boring sponges, basket sponges, vase sponges		Asymmetrical bodies without distinct tissues or organs; saclike body consists of two layers breached by many pores; internal cavity lined with food-filtering cells called choanocytes; most marine (150 species live in freshwater)	5,150
Lophophorates (Bryozoa, also called moss animals or Ectoprocta)	*Bowerbankia, Plumatella,* sea mats, sea moss		Microscopic, aquatic deuterostomes that form branching colonies, possess circular or U-shaped row of ciliated tentacles for feeding called a lophophore that usually protrudes through pores in a hard exoskeleton; Bryozoa are also called Ectoprocta because the anus, or proct, is external to the lophophore; marine or freshwater	4,000
Rotifera (wheel animals)	Rotifers		Small, aquatic pseudocoelomates with a crown of cilia around the mouth resembling a wheel; almost all live in freshwater	2,000

19.4 Sponges: Animals Without Tissues

Sponges, members of the phylum Porifera, are the simplest animals. Most sponges completely lack symmetry, and although some of their cells are highly specialized, they are not organized into tissues. The bodies of sponges consist of little more than masses of specialized cells embedded in a gel-like matrix, like chopped fruit in Jell-O. However, sponge cells do possess a key property of animal cells: cell recognition. For example, when a sponge is passed through a fine silk mesh, individual cells separate and then reaggregate on the other side to re-form the sponge. Clumps of cells disassociated from a sponge can give rise to entirely new sponges.

About 5,000 species exist, almost all in the sea. Some are tiny, and others are more than 2 meters in diameter (the diver in **figure 19.4a** could almost crawl inside the sponge shown). The body of an adult sponge is anchored in place on the seafloor and is shaped like a vase (as you can see in **figure 19.4b**). The outside of the sponge is covered with a skin of flattened cells called epithelial cells that protect the sponge.

The Phylum Facts illustration on the facing page takes you on a tour through a sponge. The body of the sponge is perforated by tiny holes. The name of the phylum, Porifera, refers to this system of pores. Unique flagellated cells called **choanocytes,** or collar cells, line the body cavity of the sponge (see the enlarged drawing of the choanocyte). The beating of the flagella of the many choanocytes draws water in through the pores (indicated by the black arrows) and through the cavity. One cubic centimeter of sponge tissue can propel more than 20 liters of water a day in and out of the sponge body! Why all this moving of water? The sponge is a "filter-feeder." The beating of each choanocyte's flagellum draws water through its collar, made of small, hairlike projections resembling a picket fence (you can see this in the enlarged view). Any food particles in the water, such as protists and tiny animals, are trapped in the fence and later ingested by the choanocyte or other cells of the sponge.

The choanocytes of sponges very closely resemble a kind of protist called choanoflagellates, which seem almost certain to have been the ancestors of sponges. Indeed, they may be the ancestors of *all* animals, although it is difficult to be certain that sponges are the direct ancestors of the other more complex phyla of animals.

> **Key Learning Outcome 19.4** Sponges have a multicellular body with specialized cells but lack definite symmetry and organized tissues.

(a)

(b)

Figure 19.4 Diversity in sponges.

These two marine sponges are barrel sponges. They are among the largest of sponges, with well-organized forms. Many are more than 2 meters in diameter (*a*), while others are smaller (*b*).

Phylum Porifera: Sponges

Key Evolutionary Innovation: MULTICELLULARITY

The body of a sponge (phylum Porifera) is **multicellular**—it contains many cells, of several distinctly different types, whose activities are coordinated with each other. The sponge body is not symmetrical and has no organized tissues.

Sponges · Cnidarians · Flatworms · Nematodes · Mollusks · Annelids · Arthropods · Echinoderms · Chordates

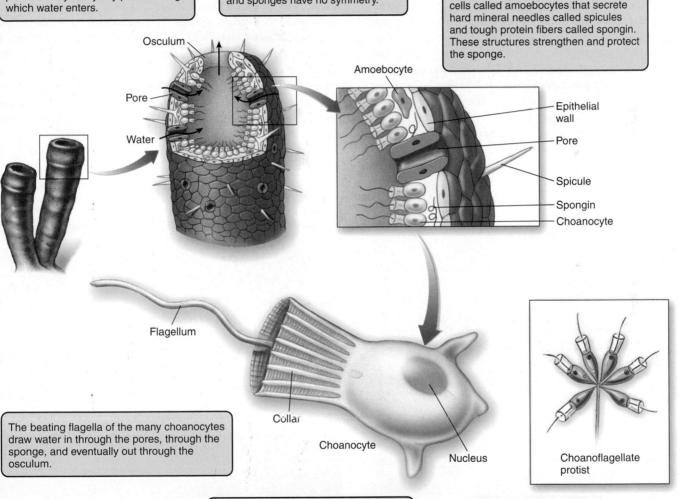

The body of a sponge is lined with cells called choanocytes and is perforated by many tiny pores through which water enters.

Sponges are multicellular, containing many different cell types. These cell types are not organized into tissues, and sponges have no symmetry.

Between the outer wall and the body cavity of the sponge are amoeboid cells called amoebocytes that secrete hard mineral needles called spicules and tough protein fibers called spongin. These structures strengthen and protect the sponge.

Osculum
Pore
Water
Amoebocyte
Epithelial wall
Pore
Spicule
Spongin
Choanocyte

Flagellum
Collar
Choanocyte
Nucleus
Choanoflagellate protist

The beating flagella of the many choanocytes draw water in through the pores, through the sponge, and eventually out through the osculum.

When a choanocyte beats its flagellum, water is drawn down through openings in its collar, where food particles become trapped. The particles are then devoured by endocytosis.

Each choanocyte very closely resembles a type of colonial protist called a choanoflagellate. It seems certain that these protists are the ancestors of the sponges, and probably of all animals.

19.5 Cnidarians: Tissues Lead to Greater Specialization

All animals other than sponges have both symmetry and tissues and thus are eumetazoans. The structure of eumetazoans is much more complex than that of sponges. All eumetazoans form distinct embryonic layers. The **radially symmetrical** (that is, with body parts arranged around a central axis) eumetazoans have two embryonic layers: an outer **ectoderm,** which gives rise to the epidermis (the outer layer of purple cells in the Phylum Facts illustration on the facing page), and an inner **endoderm** (the inner layer of yellow cells), which gives rise to the gastrodermis. A jellylike layer called the *mesoglea* (the red-colored area) forms between the epidermis and gastrodermis. These layers give rise to the basic body plan, differentiating into the many tissues of the body. No such tissues are present in sponges.

The most primitive eumetazoans to exhibit symmetry and tissues are two radially symmetrical phyla whose bodies are organized around an oral-aboral axis, like the petals of a daisy. The oral side of the animal contains the "mouth."

Radial symmetry offers advantages to animals that either remain attached or closely associated to the surface or to animals that are free-floating. These animals don't pass through their environment, but rather they interact with their environment on all sides. These two phyla are Cnidaria (pronounced ni-DAH-ree-ah), which includes hydra (**figure 19.5a**), jellyfish (**figure 19.5b**), corals (**figure 19.5c**), and sea anemones (**figure 19.5d**), and Ctenophora (pronounced tea-NO-fo-rah), a minor phylum that includes the comb jellies. These two phyla together are called the Radiata. The bodies of all other eumetazoans, the Bilateria, are marked by a fundamental bilateral symmetry (discussed in section 19.6). Even sea stars, which exhibit radial symmetry as adults, are bilaterally symmetrical when young.

A major evolutionary innovation among the radiates is the **extracellular digestion** of food. In sponges, food trapped by a choanocyte is taken directly by endocytosis into that cell, or into a circulating amoeboid cell, where the food is digested. In radiates, digestion begins *outside of cells,* in a gut cavity, called the *gastrovascular cavity.* After the food is broken down into smaller pieces, cells lining the gut cavity will complete digestion intracellularly. Extracellular digestion is the same heterotrophic strategy pursued by fungi, except that fungi digest food outside their bodies, while animals digest it within their bodies, in a cavity. This evolutionary advance has been retained by all of the more advanced groups of animals. For the first time it became possible to digest an animal larger than oneself.

(a) (b) (c) (d)

Figure 19.5 Representative cnidarians.

(*a*) Hydroids are a group of cnidarians that are mostly marine and colonial. However, *Hydra,* shown above, is a freshwater genus whose members exist as solitary polyps. (*b*) Jellyfish are translucent, marine cnidarians. Together, (*c*) corals and (*d*) sea anemones comprise the largest group of cnidarians.

Phylum Cnidaria: **Cnidarians**

Key Evolutionary Innovations: SYMMETRY and TISSUES

The cells of a cnidarian like *Hydra* (phylum Cnidaria) are organized into specialized **tissues**. The interior gut cavity is specialized for **extracellular digestion**—that is, digestion within a gut cavity rather than within individual cells. Unlike sponges, cnidarians are **radially symmetrical**, with parts arranged around a central axis like the petals of a daisy.

Sponges | Cnidarians | Flatworms | Nematodes | Mollusks | Annelids | Arthropods | Echinoderms | Chordates

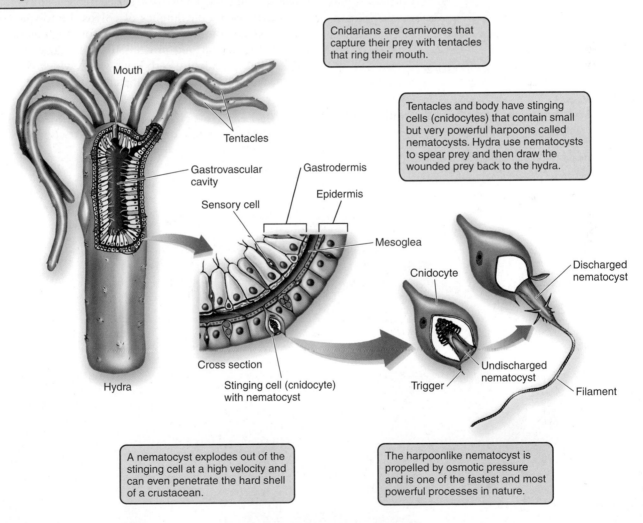

Hydra and other cnidarians are radially symmetrical, and the cells of cnidarians are organized into tissues.

A major innovation of cnidarians is extracellular digestion of food—that is, digestion within a gut cavity.

Cnidarians are carnivores that capture their prey with tentacles that ring their mouth.

Tentacles and body have stinging cells (cnidocytes) that contain small but very powerful harpoons called nematocysts. Hydra use nematocysts to spear prey and then draw the wounded prey back to the hydra.

Mouth

Tentacles

Gastrovascular cavity

Gastrodermis

Epidermis

Sensory cell

Mesoglea

Cnidocyte

Discharged nematocyst

Cross section

Undischarged nematocyst

Hydra

Stinging cell (cnidocyte) with nematocyst

Trigger

Filament

A nematocyst explodes out of the stinging cell at a high velocity and can even penetrate the hard shell of a crustacean.

The harpoonlike nematocyst is propelled by osmotic pressure and is one of the fastest and most powerful processes in nature.

Cnidarians

Cnidarians (phylum Cnidaria) are carnivores that capture their prey, such as fishes and shellfish, with tentacles that ring their mouths. The Phylum Facts illustration walks through the key characteristics of cnidarians, including the spaghetti-like tentacles that surround the mouth. These tentacles, and sometimes the body surface, bear stinging cells called **cnidocytes**, which are unique to this group and give the phylum its name. Within each cnidocyte is a small but powerful harpoon called a **nematocyst**, which cnidarians use to spear their prey and then draw the harpooned prey back to the tentacle containing the cnidocyte. An undischarged and a discharged nematocyst are shown enlarged in the Phylum Facts illustration. The cnidocyte builds up a very high internal osmotic pressure and uses it to push the nematocyst outward so explosively that the barb can penetrate the hard shell of a crab.

Cnidarians have two basic body forms: **medusae,** the floating form in **figure 19.6**, and **polyps,** the sessile form. Many cnidarians exist only as medusae, others only as polyps, and still others alternate between these two phases during the course of their life cycles. **Figure 19.7** shows the life cycle of a cnidarian that alternates between both forms. Medusae are free-floating, gelatinous, and often umbrella-shaped forms that produce gametes. Their mouths point downward, with a ring of tentacles hanging down around the edges (hence the radial symmetry). Medusae are commonly called "jellyfish" because of their gelatinous interior or "stinging nettles" because of their nematocysts. Polyps are cylindrical, pipe-shaped animals that usually attach to a rock. They also exhibit radial symmetry. The *Hydra,* sea anemones, and corals shown in **figure 19.5** are examples of polyps. In polyps, the mouth faces away from the rock and therefore is often directed upward. For shelter and protection, corals deposit an external "skeleton" of calcium carbonate within which they live. This is the structure usually identified as coral.

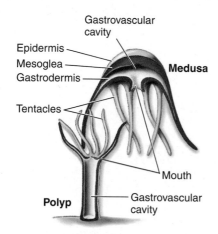

Figure 19.6 The two basic body forms of cnidarians.
The medusa (*top*) and the polyp (*bottom*) are the two phases that alternate in the life cycles of many cnidarians, but several species (corals and sea anemones, for example) exist only as polyps.

> **Key Learning Outcome 19.5** Cnidarians possess radial symmetry and specialized tissues and carry out extracellular digestion.

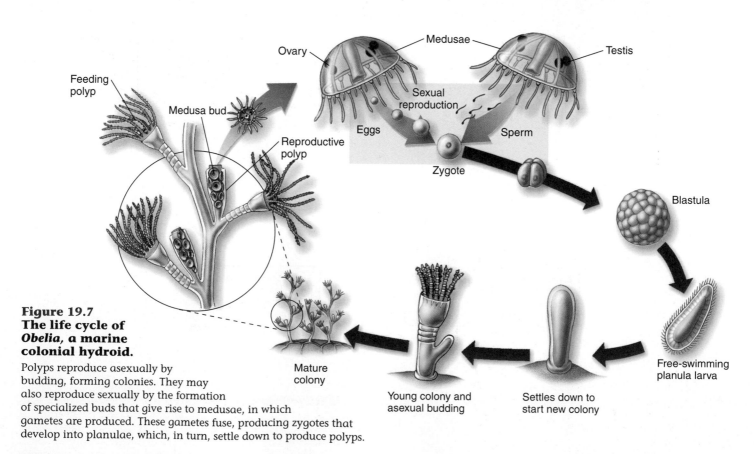

**Figure 19.7
The life cycle of
Obelia, a marine
colonial hydroid.**

Polyps reproduce asexually by budding, forming colonies. They may also reproduce sexually by the formation of specialized buds that give rise to medusae, in which gametes are produced. These gametes fuse, producing zygotes that develop into planulae, which, in turn, settle down to produce polyps.

19.6 | Solid Worms: Bilateral Symmetry

All eumetazoans other than cnidarians and ctenophores are **bilaterally symmetrical**—that is, they have a right half and a left half that are mirror images of each other. This is apparent when you compare the radially symmetrical sea anemone in **figure 19.8** with the bilaterally symmetrical squirrel. Any of the three planes that cut the sea anemone in half produce mirror images, but only one plane, the green sagittal plane, produces mirror images of the squirrel. In looking at a bilaterally symmetrical animal, you refer to the top half of the animal as **dorsal** and the bottom half as **ventral.** The front is called **anterior** and the back **posterior.** Bilateral symmetry was a major evolutionary advance among the animals because it allows different parts of the body to become specialized in different ways. For example, most bilaterally symmetrical animals have evolved a definite head end, a process called **cephalization.** Animals that have heads are often active and mobile, moving through their environment headfirst, with sensory organs concentrated in front so the animal can test for food, danger, and mates, as it enters new surroundings.

The bilaterally symmetrical eumetazoans produce three embryonic layers that develop into the tissues of the body: an outer ectoderm (colored blue in the drawing of a flatworm in **figure 19.9**), an inner endoderm (colored yellow), and a third layer, the **mesoderm** (colored red), between the ectoderm and endoderm. In general, the outer coverings of the body and the nervous system develop from the ectoderm, the digestive organs and intestines develop from the endoderm, and the skeleton and muscles develop from the mesoderm.

The simplest of all bilaterally symmetrical animals are the **solid worms.** By far the largest phylum of these, with about 20,000 species, is Platyhelminthes (pronounced plat-ee-hel-MIN-theeze), which includes the flatworms. Flatworms are the simplest animals in which organs occur. An organ is a collection of different tissues that function as a unit. The testes and uterus of flatworms are reproductive organs, for example. The dark spots on the head are eyespots that can detect light, although they cannot focus an image like your eyes can.

Solid worms lack any internal cavity other than the digestive tract. Flatworms are soft-bodied animals flattened from top to bottom, like a piece of tape or ribbon. If you were to cut a flatworm in half across its body, as in **figure 19.9**, you would see that the gut is completely surrounded by tissues and organs. This solid body construction is called **acoelomate,** meaning "without a body cavity."

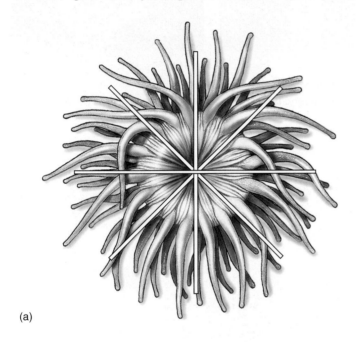

(a)

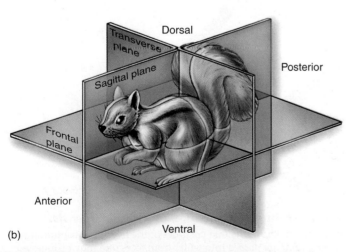

(b)

Figure 19.8 How radial and bilateral symmetry differ.

(a) Radial symmetry is the regular arrangement of parts around a central axis. (b) Bilateral symmetry is reflected in a body form that has a left and right half.

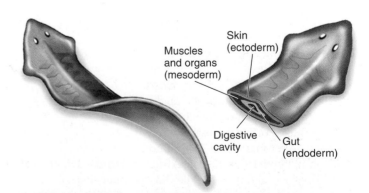

Figure 19.9 Body plan of a solid worm.

All bilaterally symmetrical eumetazoans produce three layers during embryonic development: an outer ectoderm, a middle mesoderm, and an inner endoderm. These layers differentiate to form the skin, muscles and organs, and gut, respectively, in the adult animal.

Flatworms

Although flatworms have a simple body design, they do have a definite head at the anterior end and they do possess organs. Flatworms range in size from a millimeter or less to many meters long, as in some tapeworms. Most species of flatworms are parasitic, occurring within the bodies of many other kinds of animals. Other flatworms are free-living, occurring in a wide variety of marine and freshwater habitats, as well as moist places on land (figure 19.10). Free-living flatworms are carnivores and scavengers; they eat various small animals and bits of organic debris. They move from place to place by means of ciliated epithelial cells concentrated on their ventral surfaces.

There are two classes of parasitic flatworms, which live within the bodies of other animals: flukes and tapeworms. Both groups of worms have epithelial layers resistant to the digestive enzymes and immune defenses produced by their hosts—an important feature in their parasitic way of life. Some parasitic flatworms require only one host, but many flukes require two or more hosts to complete their life cycles. The liver fluke shown in figure 19.11 requires two hosts besides humans (or some other mammal). The eggs are released from the mammal ❶ and ingested by a snail where the fluke develops into a tadpolelike larva ❷ that is released into the water. The larvae bore into the muscle of a fish where they form cysts. Mammals become infected when they eat raw, infected fish ❸. The parasitic lifestyle has resulted in the eventual loss of features not used or needed by the parasite. Parasitic flatworms lack certain features of the free-living flatworms, such as cilia in the adult stage, eyespots, and other sensory organs that lack adaptive significance for an organism that lives within the body of another animal, a loss sometimes dubbed "degenerative evolution." The tapeworm described in the Phylum Facts illustration is a classic example of degenerative evolution. As you read the characteristics in the blue boxes, notice that the tapeworm's body has been reduced to two functions, eating and reproducing.

Parasitic flatworms, like the human liver fluke *Clonorchis sinensis,* have had a significant impact on humans. Other very important flukes are the blood flukes of the genus *Schistosoma.* They afflict more than 200 million people throughout tropical Asia, Africa, Latin America, and the Middle East, about 1 in 20 of the world's population. Three species of *Schistosoma* cause the disease called schistosomiasis, or bilharzia. Over 20,000 people die each year from this disease.

(a)   (b)

Figure 19.10   Flatworms.
(*a*) A common flatworm, *Planaria.* (*b*) A marine free-living flatworm.

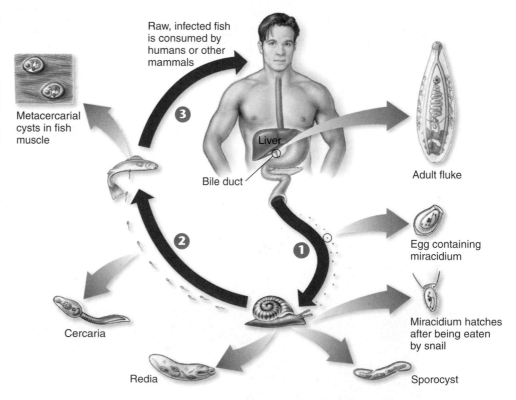

Raw, infected fish is consumed by humans or other mammals

Metacercarial cysts in fish muscle

Liver

Bile duct

Adult fluke

Egg containing miracidium

Miracidium hatches after being eaten by snail

Cercaria

Redia

Sporocyst

Figure 19.11   Life cycle of the human liver fluke, *Clonorchis sinensis.*

Adult flukes are 1–2 centimeters long and live in the bile passages of the liver. Eggs containing a complete, first-stage larva, or miracidium, are passed into water from feces and may be ingested by a snail ❶. Within the snail, the egg transforms into a sporocyst, which produces larvae called rediae. These grow into tadpolelike larvae called cercariae. Cercariae escape into the water ❷ where they bore into the muscles of certain fish (members of the goldfish and carp family) forming cysts. Mammals eating raw fish consume the cysts ❸. The flukes emerge from the cyst and travel to the bile duct where they mature and infest the liver, causing liver damage.

Phylum Platyhelminthes: Solid Worms

Key Evolutionary Innovation: BILATERAL SYMMETRY

Acoelomate solid worms such as the flatworms (phylum Platyhelminthes) were the first animals to be **bilaterally symmetrical** and to have a distinct **head**. The evolution of the mesoderm in solid worms allowed the formation of digestive and other organs.

Sponges | Cnidarians | Flatworms | Nematodes | Mollusks | Annelids | Arthropods | Echinoderms | Chordates

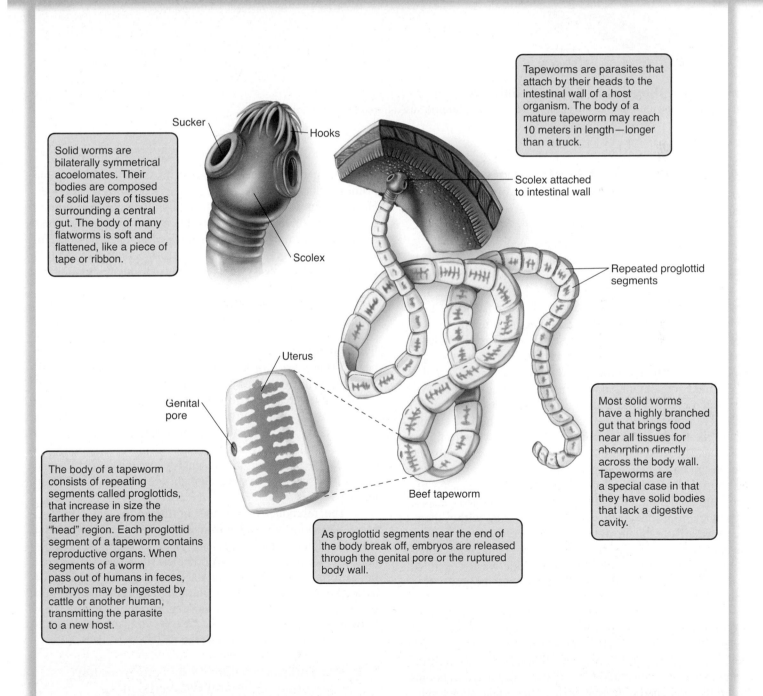

Sucker

Hooks

Solid worms are bilaterally symmetrical acoelomates. Their bodies are composed of solid layers of tissues surrounding a central gut. The body of many flatworms is soft and flattened, like a piece of tape or ribbon.

Scolex

Tapeworms are parasites that attach by their heads to the intestinal wall of a host organism. The body of a mature tapeworm may reach 10 meters in length—longer than a truck.

Scolex attached to intestinal wall

Repeated proglottid segments

Uterus

Genital pore

Beef tapeworm

Most solid worms have a highly branched gut that brings food near all tissues for absorption directly across the body wall. Tapeworms are a special case in that they have solid bodies that lack a digestive cavity.

The body of a tapeworm consists of repeating segments called proglottids, that increase in size the farther they are from the "head" region. Each proglottid segment of a tapeworm contains reproductive organs. When segments of a worm pass out of humans in feces, embryos may be ingested by cattle or another human, transmitting the parasite to a new host.

As proglottid segments near the end of the body break off, embryos are released through the genital pore or the ruptured body wall.

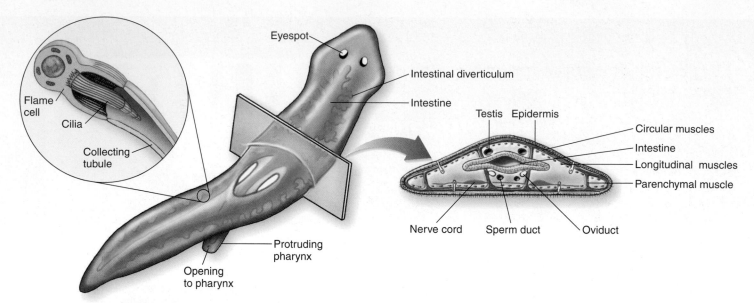

Figure 19.12 Diagram of flatworm anatomy.
The organism shown is *Dugesia,* the familiar freshwater "planaria" used in many biology laboratories.

Characteristics of Flatworms

Those flatworms that have a digestive cavity have an incomplete gut, one with only one opening. As a result, they cannot eat, digest, and eliminate undigested particles of food simultaneously. Thus, flatworms cannot feed continuously, as more advanced animals can. The gut is branched and extends throughout the body (the gut is the green structure in the *Planaria* in **figure 19.12**), functioning in both digestion and transport of food. Cells that line the gut engulf most of the food particles by phagocytosis and digest them; but, as in the cnidarians, some of these particles are partly digested extracellularly. Tapeworms, which are parasitic flatworms, lack digestive systems. They absorb their food directly through their body walls.

Unlike cnidarians, flatworms have an excretory system, which consists of a network of fine tubules (little tubes) that runs throughout the body. Cilia line the hollow centers of bulblike **flame cells** (shown in the enlarged view in **figure 19.12**), which are located on the side branches of the tubules. Cilia in the flame cells move water and excretory substances into the tubules and then out through exit pores located between the epidermal cells. Flame cells were named because of the flickering movements of the tuft of cilia within them. They primarily regulate the water balance of the organism. The excretory function of flame cells appears to be a secondary one. A large proportion of the metabolic wastes excreted by flatworms probably diffuses directly into the gut and is eliminated through the mouth.

Like sponges, cnidarians, and ctenophorans, flatworms lack a **circulatory system,** which is a network of vessels that carries fluids, oxygen, and food molecules to parts of the body. Consequently, all flatworm cells must be within diffusion distance of oxygen and food. Flatworms have thin bodies and highly branched digestive cavities that make such a relationship possible.

The nervous system of flatworms is very simple. Some primitive flatworms have only a loosely organized nerve net. However, most members of this phylum have longitudinal nerve cords (blue structures on the ventral side in the cross section above) that constitute a simple central nervous system. Between the longitudinal cords are cross connections, so that the flatworm nervous system resembles a ladder extending the length of the body.

Free-living flatworms use sensory pits or tentacles along the sides of their heads to detect food, chemicals, or movements of the fluid in which they are living. Free-living members of this phylum also have eyespots on their heads. These are inverted, pigmented cups containing light-sensitive cells connected to the nervous system. These eyespots enable the worms to distinguish light from dark. Flatworms are far more active than cnidarians or ctenophores. Such activity is characteristic of bilaterally symmetrical animals. In flatworms, this activity seems to be related to the greater concentration of sensory organs and, to some degree, the nervous system elements in the heads of these animals.

The reproductive systems of flatworms are complex. Most flatworms are **hermaphroditic,** with each individual containing both male and female sexual structures. In some parasitic flatworms, there is a complex succession of distinct larval forms. Some genera of flatworms are also capable of asexual regeneration; when a single individual is divided into two or more parts, each part can regenerate an entirely new flatworm.

> **Key Learning Outcome 19.6** Flatworms have internal organs, bilateral symmetry, and a distinct head. They do not have a body cavity.

19.7 Roundworms: The Evolution of a Body Cavity

A key transition in the evolution of the animal body plan was the evolution of the body cavity. All bilaterally symmetrical animals other than solid worms have a cavity within their body. The evolution of an internal body cavity was an important improvement in animal body design for three reasons:

1. **Circulation.** Fluids that move within the body cavity can serve as a circulatory system, permitting the rapid passage of materials from one part of the body to another and opening the way to larger bodies.
2. **Movement.** Fluid in the cavity makes the animal's body rigid, permitting resistance to muscle contraction and thus opening the way to muscle-driven body movement.
3. **Organ function.** In a fluid-filled enclosure, body organs can function without being deformed by surrounding muscles. For example, food can pass freely through a gut suspended within a cavity, at a rate not controlled by when the animal moves.

Kinds of Body Cavities

There are three basic kinds of body plans found in bilaterally symmetrical animals. Acoelomates, such as solid worms that we discussed in the previous section and that are shown at the top of **figure 19.13**, have no body cavity. **Pseudocoelomates,** shown in the middle of the figure, have a body cavity called the **pseudocoel** located between the mesoderm (red layer) and endoderm (yellow layer). A third way of organizing the body is one in which the fluid-filled body cavity develops not between endoderm and mesoderm but rather entirely within the mesoderm. Such a body cavity is called a **coelom** (the two arch-shaped cavities in the worm at the bottom of the figure), and animals that possess such a cavity are called **coelomates.** In coelomates, the gut is suspended, along with other organ systems of the animal, within the coelom; the coelom, in turn, is surrounded by a layer of epithelial cells entirely derived from the mesoderm.

The development of a body cavity poses a problem—circulation—solved in pseudocoelomates by churning the fluid within the body cavity. In coelomates, the gut is again surrounded by tissue that presents a barrier to diffusion, just as it was in solid worms. This problem is solved among coelomates by the development of a circulatory system. The circulating fluid, or blood, carries nutrients and oxygen to the tissues and removes wastes and carbon dioxide. Blood is usually pushed through the circulatory system by contraction of one or more muscular hearts. In an open circulatory system, the blood passes from vessels into sinuses, mixes with body fluid, and then

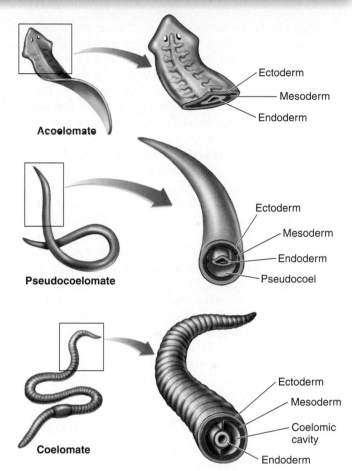

Figure 19.13 Three body plans for bilaterally symmetrical animals.
Acoelomates, such as flatworms, have no body cavity between the digestive tract (endoderm) and the outer body layer (ectoderm). Pseudocoelomates have a body cavity, the pseudocoel, between the endoderm and the mesoderm. Coelomates have a body cavity, the coelom, that develops entirely within the mesoderm, and so is lined on both sides by mesoderm tissue.

reenters the vessels later in another location. In a closed circulatory system, the blood remains separated from the body fluid in a network of vessels that can be separately controlled. Also, blood moves through a closed circulatory system faster and more efficiently than it does through an open system.

The evolutionary relationship among coelomates, pseudocoelomates, and acoelomates is not clear. Acoelomates, for example, could have given rise to coelomates, but scientists also cannot rule out the possibility that acoelomates were derived from coelomates. The two main phyla of pseudocoelomates do not appear to be closely related.

Roundworms: Pseudocoelomates

As we have noted, all bilaterally symmetrical animals except solid worms possess an internal body cavity. Among them, seven phyla are characterized by their possession of a pseudocoel. In all pseudocoelomates, the pseudocoel serves as a hydrostatic skeleton—one that gains its rigidity from being filled with fluid under pressure. The animal's muscles can work against this "skeleton," thus making the movements of pseudocoelomates far more efficient than those of the acoelomates.

Only one of the seven pseudocoelomate phyla includes a large number of species. This phylum, Nematoda, includes some 20,000 recognized species of **nematodes,** eelworms, and other roundworms. Scientists estimate that the actual number might approach 100 times that many. Members of this phylum are found everywhere. Nematodes are abundant and diverse in marine and freshwater habitats, and many members of this phylum are parasites of animals and plants, like the intestinal roundworm in **figure 19.14a.** Many nematodes are microscopic and live in soil. It has been estimated that a spadeful of fertile soil may contain, on the average, a million nematodes.

A second phylum consisting of animals with a pseudocoelomate body plan is Rotifera, the rotifers. **Rotifers** are common, small, basically aquatic animals that have a crown of cilia at their heads, which can just barely be seen in **figure 19.14b;** they range from 0.04 to 2 millimeters long. About 2,000 species exist throughout the world. Bilaterally symmetrical and covered with chitin, rotifers depend on their cilia for both locomotion and feeding, ingesting bacteria, protists, and small animals.

All pseudocoelomates lack a defined circulatory system; this role is performed by the fluids that move within the pseudocoel. Most pseudocoelomates have a complete, one-way digestive tract that acts like an assembly line. Food is broken down, absorbed, and then treated and stored.

Phylum Nematoda: The Roundworms

Nematodes are bilaterally symmetrical, cylindrical, unsegmented worms. Shown in longitudinal and cross sections in the Phylum Facts illustration on the facing page, they are covered by a flexible, thick cuticle, which is molted as they grow. Their muscles constitute a layer beneath the epidermis and extend along the length of the worm, rather than encircling its body. These longitudinal muscles, which can be seen in the cross section attaching to the outer layer of the body, pull both against the cuticle and the pseudocoel, which forms a hydrostatic skeleton. When nematodes move, their bodies whip about from side to side.

Near the mouth of a nematode, at its anterior end (toward the left side of the diagram), are usually 16 raised, hairlike sensory organs. The mouth is often equipped with piercing organs called *stylets.* Food passes through the mouth as a result of the sucking action of a muscular chamber called the **pharynx.** After passing through a short corridor into the pharynx, food continues through the other portions of the digestive tract, where it is broken down and then digested. Some of the water with which the food has been mixed is reabsorbed near the end of the digestive tract, and material that has not been digested is eliminated through the anus.

Nematodes completely lack flagella or cilia, even on sperm cells. Reproduction in nematodes is sexual, with sexes usually separate (a female with a uterus, ovary, and oviducts is shown in the Phylum Facts illustration). Their development is simple, and the adults consist of very few cells. For this reason, nematodes have become extremely important subjects for genetic and developmental studies. The 1-millimeter-long *Caenorhabditis elegans* matures in only three days, its body

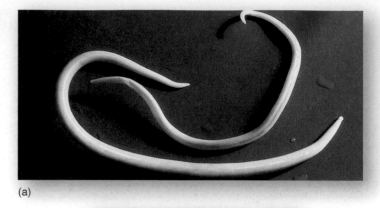

(a)

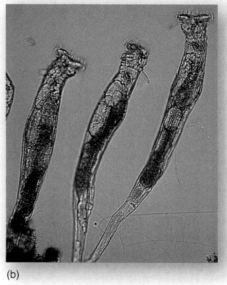

(b)

Figure 19.14   Pseudocoelomates.
(a) These nematodes (phylum Nematoda) are intestinal roundworms that infect humans and some other animals. Their fertilized eggs pass out with feces and can remain viable in soil for years.
(b) Rotifers (phylum Rotifera) are common aquatic animals that depend on their crown of cilia for feeding and locomotion.

is transparent, and it has only 959 cells. It is the only animal whose complete developmental cellular anatomy is known, and the first animal whose genome (97 million DNA bases encoding over 21,000 different genes) was fully sequenced.

Some nematodes are parasitic in humans, cats, dogs, and animals of economic importance, such as cows and sheep. Heartworm infections in dogs and cats are caused by a parasitic nematode that infests the heart of the animal. About 50 species of nematodes, including several that are rather common in the United States, regularly parasitize human beings. Trichinosis, a nematode-caused disease in temperate regions, is caused by worms of the genus *Trichinella.* These worms live in the small intestine of pigs, where fertilized female worms burrow into the intestinal wall. Once it has penetrated these tissues, each female produces about 1,500 live young. The young enter the lymph channels and travel to muscle tissue throughout the body, where they mature and form cysts. Infection in human beings or other animals arises from eating undercooked or raw pork in which the cysts of *Trichinella* are present. If the worms are abundant, a fatal infection can result, but such infections are rare.

Phylum Nematoda: Roundworms

Key Evolutionary Innovation: BODY CAVITY

The major innovation in body design in roundworms (phylum Nematoda) is a **body cavity** between the gut and the body wall. This cavity is the pseudocoel. It allows nutrients to circulate throughout the body and prevents organs from being deformed by muscle movements.

Sponges | Cnidarians | Flatworms | Nematodes | Mollusks | Annelids | Arthropods | Echinoderms | Chordates

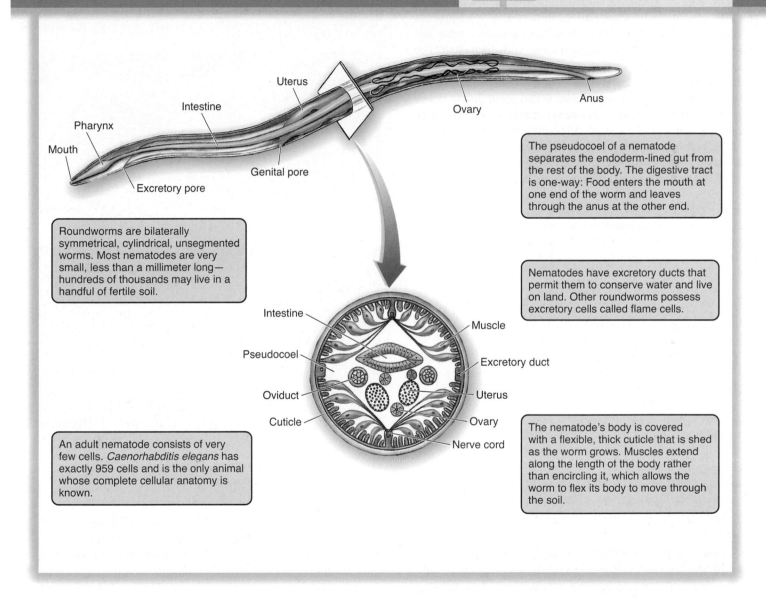

Uterus

Intestine

Pharynx

Mouth

Excretory pore

Genital pore

Ovary

Anus

The pseudocoel of a nematode separates the endoderm-lined gut from the rest of the body. The digestive tract is one-way: Food enters the mouth at one end of the worm and leaves through the anus at the other end.

Roundworms are bilaterally symmetrical, cylindrical, unsegmented worms. Most nematodes are very small, less than a millimeter long—hundreds of thousands may live in a handful of fertile soil.

Nematodes have excretory ducts that permit them to conserve water and live on land. Other roundworms possess excretory cells called flame cells.

Intestine

Pseudocoel

Oviduct

Cuticle

Muscle

Excretory duct

Uterus

Ovary

Nerve cord

An adult nematode consists of very few cells. *Caenorhabditis elegans* has exactly 959 cells and is the only animal whose complete cellular anatomy is known.

The nematode's body is covered with a flexible, thick cuticle that is shed as the worm grows. Muscles extend along the length of the body rather than encircling it, which allows the worm to flex its body to move through the soil.

A more prevalent human parasitic nematode is *Ascaris lumbricoides*. This intestinal worm infects approximately one in six people worldwide but is rare in areas with modern plumbing. These worms live in the intestines and spread their fertilized eggs in feces, which can remain viable in the soil for years. Adult females, which are up to 30 centimeters long, contain up to 30 million eggs, and can lay up to 20,000 of them each day.

Key Learning Outcome 19.7 Some body cavities develop between the endoderm and mesoderm (pseudocoelomates), others within the mesoderm (coelomates). Roundworms have a pseudocoel body cavity. Nematodes, a kind of roundworm, are very common in soil, and several are parasites.

Mollusks: Coelomates

Coelomates

Even though acoelomates and pseudocoelomates have proven very successful, the bulk of the animal kingdom consists of coelomates. Coelomates have a new body design that repositions the fluid. What is the functional difference between a pseudocoel and a coelom, and why has the latter kind of body cavity been so overwhelmingly successful? The answer has to do with the nature of animal embryonic development. In animals, development of specialized tissues involves a process called **primary induction,** in which the three primary tissues (endoderm, mesoderm, and ectoderm) interact with each other. The interaction requires physical contact. A major advantage of the coelomate body plan is that it allows contact between mesoderm and endoderm, so that primary induction can occur during development. For example, contact between mesoderm and endoderm permits localized portions of the digestive tract to develop into complex, highly specialized regions like the stomach. In pseudocoelomates, mesoderm and endoderm are separated by the body cavity, limiting developmental interactions between these tissues.

Mollusks

The only major phylum of coelomates without segmented bodies are the Mollusca. The **mollusks** are the largest animal phylum, except for the arthropods, with over 110,000 species. Mollusks are mostly marine, but occur almost everywhere.

Mollusks include three general groups with outwardly different body plans. However, the seeming differences hide a basically similar body design. The body of mollusks is composed of three distinct parts: a head-foot, a central section called the visceral mass that contains the body's organs, and a mantle. The foot of a mollusk is muscular and may be adapted for locomotion, attachment, food capture (in squids and octopuses), or various combinations of these functions. The **mantle** is a heavy fold of tissue wrapped around the visceral mass like a cape, with the gills positioned on its inner surface like the lining of a coat. The **gills** are filamentous projections of tissue, rich in blood vessels, that capture oxygen from the water circulating between the mantle and visceral mass and release carbon dioxide.

The three major groups of mollusks, all different variations upon this same basic design, are gastropods, bivalves, and cephalopods.

1. **Gastropods** (snails, like the one shown in **figure 19.15a**, and slugs) use the muscular foot to crawl, and their mantle often secretes a single, hard protective shell. All terrestrial mollusks are gastropods.
2. **Bivalves** (clams, oysters, and scallops) secrete a two-part shell with a hinge (**figure 19.15b**), as their name implies. They filter-feed by drawing water into their shell.
3. **Cephalopods** (octopuses, like the one shown in **figure 19.15c**, and squids) have modified the mantle cavity to create a jet propulsion system that can propel them rapidly through the water. In most groups, the shell is greatly reduced to an internal structure or is absent.

(a)

(b)

(c)

Figure 19.15 Three major groups of mollusks.
(a) A gastropod. (b) A bivalve. (c) A cephalopod.

The Phylum Facts illustration walks you through the characteristics of mollusks, including a unique feature of mollusks, the **radula,** which is a rasping, tonguelike organ. With rows of pointed, backward-curving teeth, the radula is used by some snails to scrape algae off rocks. The small holes often seen in oyster shells are produced by gastropods that have bored holes to kill the oyster and extract its body.

In most mollusks, as stated earlier, the outer surface of the mantle also secretes a protective shell, partially cut away in the facing Phylum Facts illustration. The shell consists of a horny outer layer, rich in protein, which protects the two underlying calcium-rich layers from erosion. The inner layer is pearly and is used as mother-of-pearl. Pearls themselves are formed when a foreign object, such as a grain of sand,

Phylum Mollusca: Mollusks

Key Evolutionary Innovation: COELOM

The body cavity of a mollusk like this snail (phylum Mollusca) is a **coelom**, completely enclosed within the mesoderm. This allows physical contact between the mesoderm and the endoderm, permitting interactions that lead to development of highly specialized organs such as a stomach.

Sponges | Cnidarians | Flatworms | Nematodes | Mollusks | Annelids | Arthropods | Echinoderms | Chordates

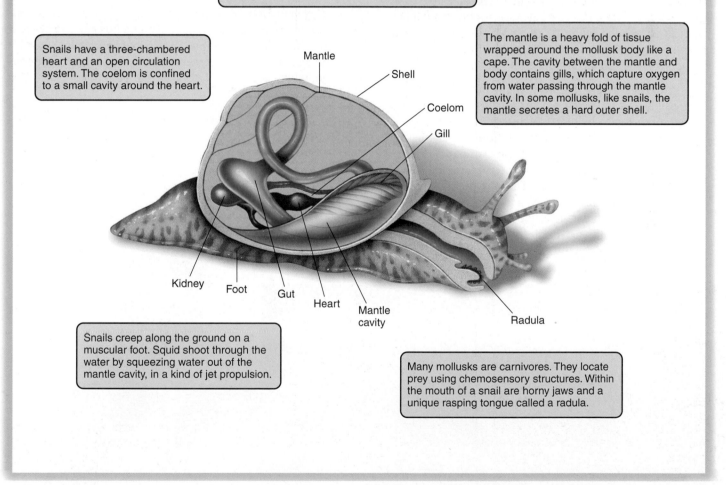

Mollusks were among the first animals to develop an efficient excretory system. Tubular structures called nephridia (a type of kidney) gather wastes from the coelom and discharge them into the mantle cavity.

Snails have a three-chambered heart and an open circulation system. The coelom is confined to a small cavity around the heart.

The mantle is a heavy fold of tissue wrapped around the mollusk body like a cape. The cavity between the mantle and body contains gills, which capture oxygen from water passing through the mantle cavity. In some mollusks, like snails, the mantle secretes a hard outer shell.

Mantle

Shell

Coelom

Gill

Kidney   Foot   Gut   Heart   Mantle cavity   Radula

Snails creep along the ground on a muscular foot. Squid shoot through the water by squeezing water out of the mantle cavity, in a kind of jet propulsion.

Many mollusks are carnivores. They locate prey using chemosensory structures. Within the mouth of a snail are horny jaws and a unique rasping tongue called a radula.

becomes lodged between the mantle and the inner shell layer of a bivalve, including clams and oysters. The mantle coats the foreign object with layer upon layer of shell material to reduce irritation. The shell serves primarily for protection, with some mollusks withdrawing into their shell when threatened.

Key Learning Outcome 19.8 Mollusks have a coelom body cavity but are not segmented. Although diverse, their basic body plans include a foot, the visceral mass, and a mantle.

19.9 Annelids: The Rise of Segmentation

One of the early key innovations in body plan to arise among the coelomates was **segmentation,** the building of a body from a series of similar segments. The first segmented animals to evolve were the **annelid worms,** phylum Annelida. These advanced coelomates are assembled as a chain of nearly identical segments, like the boxcars of a train. The great advantage of such segmentation is the evolutionary flexibility it offers—a small change in an existing segment can produce a new kind of segment with a different function. Thus, some segments are modified for reproduction, some for feeding, and others for eliminating wastes.

Two-thirds of all annelids live in the sea (about 8,000 species, including the bristle worm in **figure 19.16***b*); most of the rest—some 3,100 species—are earthworms (shown emerging from underground in **figure 19.16***a*). The basic body plan of an annelid is a tube within a tube: The digestive tract, the light pink tube in the Phylum Facts illustration on the facing page, is suspended within the coelom, which is itself a tube running from mouth to anus. There are three characteristics of annelid body organization:

1. **Repeated segments.** The body segments of an annelid are visible as a series of ringlike structures running the length of the body, looking like a stack of doughnuts. The segments are divided internally from one another by partitions, just as walls separate the rooms of a building. In each of the cylindrical segments, the excretory and locomotor organs are repeated. The body fluid within the coelom of each segment creates a hydrostatic (liquid-supported) skeleton that gives the segment rigidity, like an inflated balloon. Muscles within each segment pull against the fluid in the coelom. Because each segment is separate, each is able to expand or contract independently. This lets the worm's body move in ways that are quite complex. When an earthworm crawls on a flat surface, for example, it lengthens some parts of its body while shortening others (see **figure 22.8**).

2. **Specialized segments.** The anterior (front) segments of annelids contain the sensory organs of the worm. Elaborate eyes with lenses and retinas have evolved in some annelids. One anterior segment contains a well-developed cerebral ganglion, or brain (the yellow bulbed structures in the illustration).

3. **Connections.** Because partitions separate the segments, it is necessary to provide ways for materials and information to pass between segments. A circulatory system (the red vessels in the illustration) carries blood from one segment to another, while nerve cords (the yellow chainlike structure along the ventral wall in the figure) connect the nerve centers located in each segment with each other and the brain. The brain can then coordinate the worm's activities.

Segmentation underlies the body organization of all complex coelomate animals, not only annelids but also arthropods (crustaceans, spiders, and insects) and chordates (mostly vertebrates). Vertebrate segmentation is seen in the vertebral column, which is a stack of very similar vertebrae.

> **Key Learning Outcome 19.9** Annelids are segmented worms. Most species are marine, but some—about one-third of the species—are terrestrial.

Figure 19.16 Representative annelids.

(*a*) Earthworms are the terrestrial annelids. This night crawler, *Lumbricus terrestris,* is in its burrow. (*b*) This bristle worm is an aquatic annelid, a polychaete.

Phylum Annelida: Annelids

Key Evolutionary Innovation: SEGMENTATION

Marine polychaetes and earthworms (phylum Annelida) were the first organisms to evolve a body plan based on **repeated body segments**. Most segments are identical and are separated from other segments by partitions.

Sponges | Cnidarians | Flatworms | Nematodes | Mollusks | Annelids | Arthropods | Echinoderms | Chordates

Each segment contains a set of excretory organs (nephridia) and a nerve center.

Earthworms crawl by anchoring bristles called setae to the ground and pulling against them. Polychaete annelids have a flattened body and swim or crawl by flexing it.

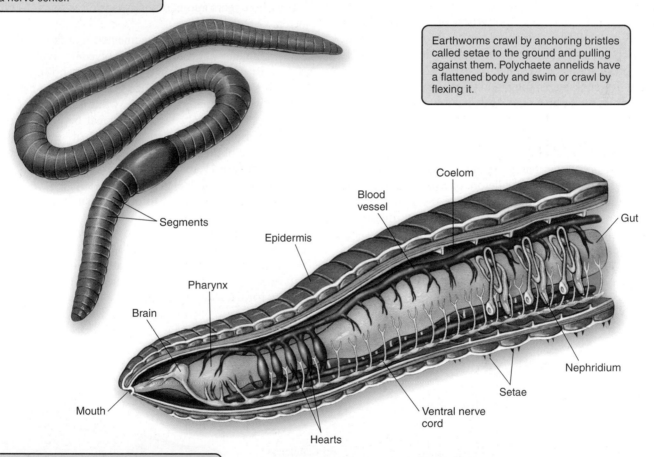

- Segments
- Coelom
- Blood vessel
- Gut
- Epidermis
- Pharynx
- Brain
- Nephridium
- Mouth
- Setae
- Ventral nerve cord
- Hearts

Segments are connected by the circulatory and nervous systems. A series of hearts at the anterior (front) end pump the blood. A well-developed brain located in an anterior segment coordinates the activities of all segments.

Each segment has a coelom. Muscles squeeze the fluid of the coelom, making each segment rigid, like an inflated balloon. Because each segment can contract independently, a worm can crawl by lengthening some segments while shortening others.

19.10 Arthropods: Advent of Jointed Appendages

A profound innovation marks the origin of the body plan characteristic of the most successful of all animal groups, the **arthropods,** phylum Arthropoda. This innovation was the development of jointed appendages.

Jointed Appendages

The name *arthropod* comes from two Greek words, *arthros,* jointed, and *podes,* feet. All arthropods (**figure 19.17**) have jointed appendages. Some are legs, and others may be modified for other uses. To gain some idea of the importance of jointed appendages, imagine yourself without them—no hips, knees, ankles, shoulders, elbows, wrists, or knuckles. Without jointed appendages, you could not walk or grasp an object. Arthropods use jointed appendages as legs and wings for moving, as antennae to sense their environment, and as mouthparts for sucking, ripping, and chewing prey. A scorpion, for example, seizes and tears apart its prey with mouthpart appendages modified as large pincers.

Rigid Exoskeleton

The arthropod body plan has a second great innovation: Arthropods have a rigid external skeleton, or **exoskeleton,** made of chitin. In any animal, a key function of the skeleton is to provide places for muscle attachment, and in arthropods the muscles attach to the interior surface of the hard chitin shell, which also protects the animal from predators and impedes water loss.

However, while chitin is hard and tough, it is also brittle and cannot support great weight. As a result, the exoskeleton must be much thicker to bear the pull of the muscles in large insects than in small ones, so there is a limit to how big an arthropod body can be. That is why you don't see beetles as big as birds or crabs the size of a cow—the exoskeleton would be so thick the animal couldn't move its great weight. Another limitation on size is the fact that in many arthropods, including insects, all parts of the body need to be near a respiratory passage to obtain oxygen. The reason for this is that the respiratory system (see section 24.1), not the circulatory system, carries oxygen to the tissues.

In fact, the great majority of arthropod species consist of small animals—mostly about a millimeter in length—but members of the phylum range in adult size from about 80 micrometers long (some parasitic mites) to 3.6 meters across (a gigantic crab found in the sea off Japan). Some lobsters are nearly a meter in length. The largest living insects are about 33 centimeters long, but the giant dragonflies that lived 300 million years ago had wingspans of as much as 60 centimeters (2 feet)!

Arthropod bodies are segmented like those of annelids, from which they almost certainly evolved. Individual segments often exist only during early development, however, and fuse into functional groups as adults. For example, a caterpillar (a larval stage) has many segments, while a butterfly (and other adult insects) has only three functional body regions—head, thorax, and abdomen—each composed of several fused segments. Some of the segmentation can still be seen in the grasshopper in **figure 19.18**, especially in the abdomen.

Arthropods have proven very successful due to the arthropod innovations of jointed appendages and exoskeletons. The Phylum Facts illustration on the facing page walks you through a succinct overview of arthropod characteristics. About two-thirds of all named species on earth are arthropods. Scientists estimate that a quintillion (a billion billion) insects are alive at any one time—200 million insects for each living human!

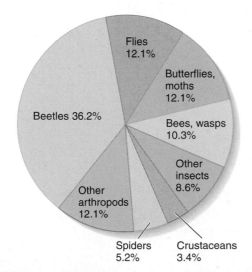

Figure 19.17   Arthropods are a successful group.
About two-thirds of all named species are arthropods. About 80% of all arthropods are insects, and about half of the named species of insects are beetles.

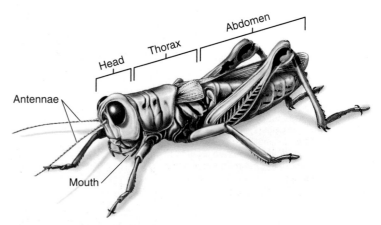

Figure 19.18   Segmentation in insects.
This grasshopper illustrates the body segmentation found in adult insects. The many segments found in most larval stages of insects become fused in the adult, giving rise to three adult body regions: the head, thorax, and abdomen. The appendages—legs, wings, mouthparts, antennae—are jointed.

Phylum Arthropoda: Arthropods

Key Evolutionary Innovations: JOINTED APPENDAGES and EXOSKELETON

Insects and other arthropods (phylum Arthropoda) have a coelom, segmented bodies, and **jointed appendages**. The three body regions of an insect (head, thorax, and abdomen) are each actually composed of a number of segments that fuse during development. All arthropods have a strong **exoskeleton** made of chitin. One class of arthropods, the insects, has evolved **wings**, which permit them to fly rapidly through the air.

Sponges · Cnidarians · Flatworms · Nematodes · Mollusks · Annelids · Arthropods · Echinoderms · Chordates

The jointed appendages of insects may be modified into antennae, mouthparts, legs, or wings. Attached to the central body region, the thorax, there are three pairs of legs and, most often, two pairs of wings (some insects like flies have retained only one wing pair). The wings are sheets of chitin.

Insects eliminate wastes by collecting circulatory fluid osmotically in Malpighian tubules that extend from the gut into the blood and then reabsorbing the fluid, but not the wastes.

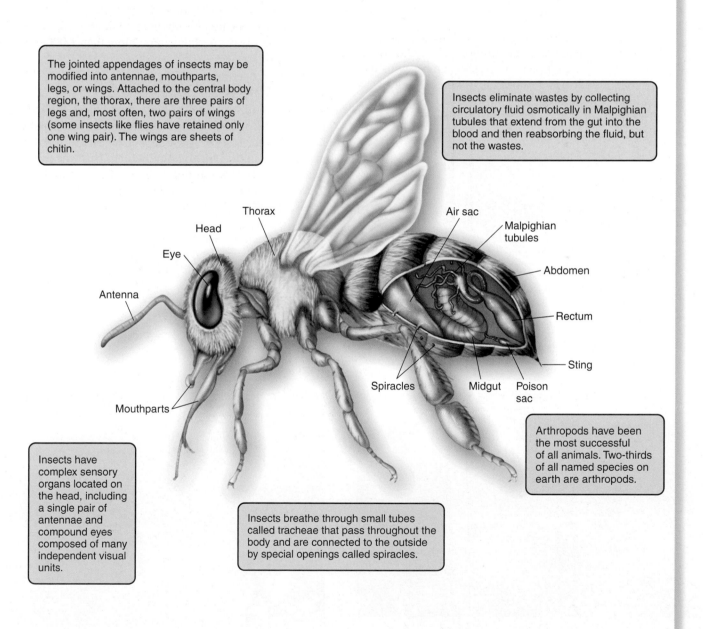

Thorax

Head

Eye

Antenna

Mouthparts

Air sac

Malpighian tubules

Abdomen

Rectum

Sting

Spiracles

Midgut

Poison sac

Insects have complex sensory organs located on the head, including a single pair of antennae and compound eyes composed of many independent visual units.

Insects breathe through small tubes called tracheae that pass throughout the body and are connected to the outside by special openings called spiracles.

Arthropods have been the most successful of all animals. Two-thirds of all named species on earth are arthropods.

Chelicerates

Arthropods such as spiders, mites, scorpions, and a few others lack jaws, or *mandibles,* and are called **chelicerates.** Their mouthparts, known as *chelicerae,* evolved from the appendages nearest the animal's anterior end, as in the jumping spider pictured below. The chelicerae are the foremost appendages located on the head.

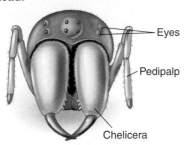

The chelicerate fossil record goes back as far as that of any multicellular animal, about 630 million years. One of the most ancient groups contains the horseshoe crabs, with five species surviving today. Horseshoe crabs swim on their backs by moving their abdominal plates and walk on their five pairs of legs. The body of a horseshoe crab is covered by a hard shell, which has a long tailpiece (**figure 19.19**); there are two compound eyes and two simple eyes on the shell. Horseshoe crabs feed at night, primarily on mollusks and annelids.

By far the largest of the three classes of chelicerates is the largely terrestrial class Arachnida, with some 57,000 named species, including the spiders (two poisonous spiders found in North America are pictured in **figure 19.20**), ticks, mites, scorpions, and daddy longlegs. Arachnids have a pair of chelicerae, a pair of pedipalps, and four pairs of walking legs. The chelicerae consist of a stout basal portion and a movable fang often connected to a poison gland. *Pedipalps,* the next pair of appendages, may resemble legs, but they have one less segment. In scorpions, the pedipalps are large and pinching. Most **arachnids** are carnivorous, although mites are largely herbivorous. Ticks are blood-feeding ectoparasites of vertebrates, and some ticks may carry diseases, such as Rocky Mountain spotted fever and Lyme disease.

Sea spiders are also chelicerates and are relatively common, especially in coastal waters. More than 1,000 species are in the class.

Mandibulates

The remaining arthropods have mandibles, formed by the modification of one of the pairs of anterior appendages, not necessarily the foremost set of appendages. The foremost set of appendages in the bullfrog ant pictured below is the antennae with the mandibles forming from the next set of appendages. These arthropods, called **mandibulates,** include the crustaceans, insects, centipedes, millipedes, and a few other groups.

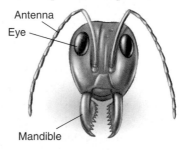

Figure 19.19 Horseshoe crabs.
These horseshoe crabs, *Limulus,* are emerging from the sea to mate.

(a)

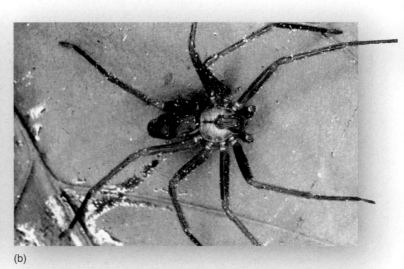
(b)

Figure 19.20 Arachnids.
(a) One of the poisonous spiders in the United States and Canada is the black widow spider, *Latrodectus mactans.* (b) Another of the poisonous spiders in this area is the brown recluse, *Loxosceles reclusa.* Both species are common throughout temperate and subtropical North America, but they rarely bite humans.

Crustaceans The **crustaceans** (subphylum Crustacea) are a large, diverse group of primarily aquatic organisms, including some 35,000 species of crabs, shrimps, lobsters, crayfish, water fleas, pillbugs, sowbugs, barnacles, and related groups (**figure 19.21**). Often incredibly abundant in marine and freshwater habitats, and playing a role of critical importance in virtually all aquatic ecosystems, crustaceans have been called "the insects of the water." Most crustaceans have two pairs of antennae (the first pair are shorter and often referred to as antennules, as labeled in **figure 19.22**), three pairs of chewing appendages (one pair being the mandibles), and various numbers of pairs of legs. The nauplius larva stage through which all crustaceans pass provides evidence that all members of this diverse group are descended from a common ancestor. The nauplius hatches with three pairs of appendages and metamorphoses through several stages before reaching maturity. In many groups, this nauplius stage is passed in the egg, and development of the hatchling to the adult form is direct.

Crustaceans differ from the insects in that the head and thorax are fused together, forming the cephalothorax, and they have legs on their abdomen as well as on their thorax. Many crustaceans have compound eyes. In addition, they have delicate tactile hairs that project from the cuticle all over the body. Larger crustaceans have feathery gills near the bases of their legs. In smaller members of this class, gas exchange takes place directly through the thinner areas of the cuticle or the entire body. Most crustaceans have separate sexes. Many different kinds of specialized copulation occur among the crustaceans, and the members of some orders carry their eggs with them, either singly or in egg pouches, until they hatch.

Crustaceans include marine, freshwater, and terrestrial forms. Crustaceans such as shrimp, lobsters, crabs, and crayfish are called decapods, meaning "ten-footed," like the lobster in **figure 19.22**. Pillbugs and sowbugs are terrestrial crustaceans but usually live in moist places. Barnacles are a group of crustaceans that are sessile as adults but have free-swimming

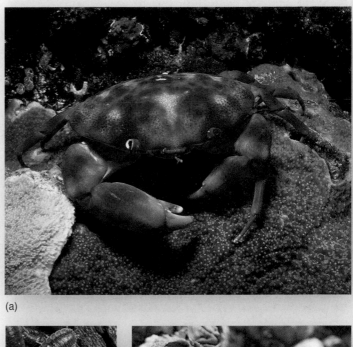
(a)

(b)   (c)

Figure 19.21   Crustaceans.
(*a*) Dark-fingered coral crab. (*b*) Sowbugs, *Porcellio scaber*.
(*c*) Barnacles are sessile animals that permanently attach themselves to a hard substrate.

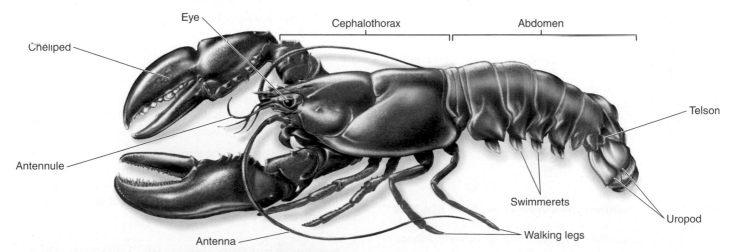

Figure 19.22   Body of a lobster, *Homarus americanus*.
Some of the specialized terms used to describe crustaceans are indicated. For example, the head and thorax are fused together into a cephalothorax. Appendages called swimmerets occur in lines along the sides of the abdomen and are used in reproduction and also for swimming. Flattened appendages known as uropods form a kind of compound "paddle" at the end of the abdomen. Lobsters may also have a telson, or tail spine.

(a)

(b)

Figure 19.23 Centipedes and millipedes.

Centipedes are active predators, whereas millipedes are sedentary herbivores. (a) Centipede, *Scolopendra.* (b) Millipede, *Sigmoria,* in North Carolina.

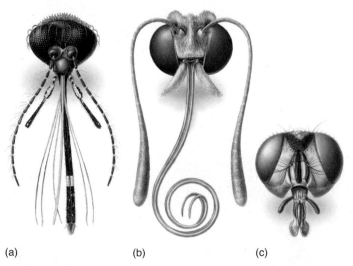

(a)   (b)   (c)

Figure 19.24 Modified mouthparts in three kinds of insects.

(a) Mosquito, *Culex;* (b) alfalfa butterfly, *Colias;* (c) housefly, *Musca domestica.*

larvae. The larvae attach their heads to rocks or other submerged objects and then stir food into their mouths with their feathery legs.

Millipedes and Centipedes Millipedes and centipedes have bodies that consist of a head region followed by numerous similar segments. Centipedes have one pair of legs on each body segment, and millipedes have two. This difference is apparent if you compare the leg arrangements in the centipede in **figure 19.23a** and the millipede in **figure 19.23b**. The centipedes are all carnivorous and feed mainly on insects. The appendages of the first trunk segment are modified into a pair of venomous fangs. In contrast, most millipedes are herbivores, feeding mostly on decaying vegetation. Millipedes live mainly in damp, protected places, such as under leaf litter, in rotting logs, under bark or stones, or in the soil. The first animal to have lived on land was a millipede; its 420-million-year-old fossil was reported in 2004.

Insects The **insects,** class Insecta, are by far the largest group of arthropods, whether measured in terms of numbers of species or numbers of individuals; as such, they are the most abundant group of eukaryotes on earth. Most insects are relatively small, ranging from 0.1 millimeters to about 30 centimeters in length. Insects have three body sections:

1. **Head.** The insect head is very elaborate, with a single pair of antennae and elaborate mouthparts that are well-suited to their diets. For example, the mouthparts in the mosquito in **figure 19.24a** are modified for piercing the skin; the long proboscis of the butterfly in **figure 19.24b** can uncoil to reach down into flowers; and the short mouthparts of the housefly in **figure 19.24c** are modified for sopping up liquids. Most insects have compound eyes that are composed of independent visual units.

2. **Thorax.** The thorax consists of three segments, each of which has a pair of legs. Most insects also have two pairs of wings attached to the thorax. In some insects, such as beetles, grasshoppers, and crickets, the outer pair of wings is adapted for protection rather than flight.

3. **Abdomen.** The abdomen consists of up to 12 segments. Digestion takes place primarily in the stomach, and excretion takes place through organs called *Malpighian tubules,* which constitute an efficient mechanism for water conservation and were a key adaptation facilitating invasion of the land by arthropods.

Although primarily a terrestrial group, insects live in every conceivable habitat on land and in freshwater, and a few have even invaded the sea. About 1 million species have been identified with many others awaiting detection and classification (**figure 19.25**).

Key Learning Outcome 19.10 Arthropods, the most successful animal phylum, have jointed appendages, a rigid exoskeleton, and, in the case of insects, wings.

Figure 19.25 Insect diversity.
(*a*) Some insects have a tough exoskeleton, like this stag beetle (order Coleoptera). (*b*) Human flea, *Pulex irritans* (order Siphonaptera). Fleas are flattened laterally, slipping easily through hair. (*c*) The honeybee, *Apis mellifera* (order Hymenoptera), is a widely domesticated and efficient pollinator of flowering plants. (*d*) This pot dragonfly (order Odonata) has a fragile exoskeleton. (*e*) A true bug, *Edessa rufomarginata* (order Hemiptera), in Panama. (*f*) Copulating grasshoppers (order Orthoptera). (*g*) Luna moth, *Actias luna,* in Virginia. Luna moths and their relatives are among the most spectacular insects (order Lepidoptera).

19.11 Protostomes and Deuterostomes

All the animals we have met so far have essentially the same kind of embryonic development. Cell divisions of the fertilized egg produce a hollow ball of cells, a blastula, which indents to form a two-layer-thick ball with a blastopore opening to the outside. In mollusks, annelids, and arthropods, the mouth (stoma) develops from or near the blastopore. An animal whose mouth develops in this way is called a **protostome** (figure 19.26, *top*). If such an animal has a distinct anus or anal pore, it develops later in another region of the embryo.

A second distinct pattern of embryological development occurs in the echinoderms and the chordates. In these animals, the anus forms from or near the blastopore, and the mouth forms subsequently on another part of the blastula. This group of phyla consists of animals that are called the **deuterostomes** (**figure 19.26**, *bottom*).

Deuterostomes represent a revolution in embryonic development. In addition to the fate of the blastopore, deuterostomes differ from protostomes in three other features:

1. The progressive division of cells during embryonic growth is called *cleavage.* The cleavage pattern relative to the embryo's polar axis determines how the cells array. In nearly all protostomes, each new cell buds off at an angle oblique to the polar axis. As a result, a new cell nestles into the space between the older ones in a closely packed array (see the 16-cell stage in the upper row of cells). This pattern is called **spiral cleavage** because a line drawn through a sequence of dividing cells spirals outward from the polar axis, (indicated by the curving blue arrow at the 32-cell stage).

In deuterostomes, the cells divide parallel to and at right angles to the polar axis. As a result, the pairs of cells from each division are positioned directly above and below one another (see the 16-cell stage in the lower row of cells); this process gives rise to a loosely packed array of cells. This pattern is called **radial cleavage** because a line drawn through a sequence of dividing cells describes a radius outward from the polar axis (indicated by the straight blue arrow at the 32-cell stage).

2. In protostomes, the developmental fate of each cell in the embryo is fixed when that cell first appears. Even at the four-celled stage, each cell is different, containing different chemical developmental signals and no one cell, if separated from the others, can develop into a complete animal. In deuterostomes, on the other hand, the first cleavage divisions of the fertilized embryo produce identical daughter cells, and any single cell, if separated, can develop into a complete organism.

3. In all coelomates, the coelom originates from mesoderm. In protostomes, this occurs simply and directly: The mesoderm cells simply move away from one another as the coelomic cavity expands within the mesoderm. However, in deuterostomes, the coelom is normally produced by an evagination of the **archenteron**—the main cavity within the gastrula, also called the primitive gut. This cavity, lined with endoderm, opens to the outside via the blastopore and eventually becomes the gut cavity. The evaginating cells give rise to the mesodermal cells, and the mesoderm expands to form the coelom.

> **Key Learning Outcome 19.11** In protostomes, the egg cleaves spirally, and the blastopore becomes the mouth. In deuterostomes, the egg cleaves radially, and the blastopore becomes the animal's anus.

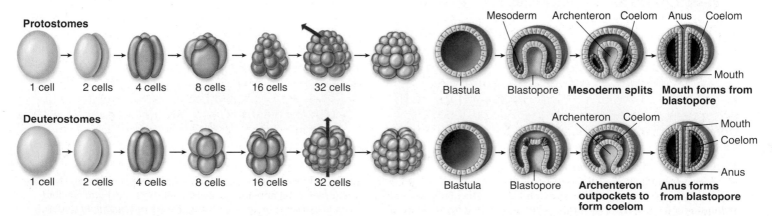

Figure 19.26 Embryonic development in protostomes and deuterostomes.
Cleavage of the egg produces a hollow ball of cells called the blastula. Invagination, or infolding, of the blastula produces the blastopore. In protostomes, embryonic cells cleave in a spiral pattern and become tightly packed. The blastopore becomes the animal's mouth, and the coelom originates from a mesodermal split. In deuterostomes, embryonic cells cleave radially and form a loosely packed array. The blastopore becomes the animal's anus, and the mouth develops at the other end. The coelom originates from an evagination, or outpouching, of the archenteron in deuterostomes.

A **Closer** Look

Diversity Is Only Skin Deep

Perhaps the most important lesson to emerge from the study of animal diversity is not the incredible variety of animal forms, from worms and spiders to sharks and antelopes, but rather their deep similarities. The body plans of all animals are assembled along a similar path, as if from the same basic blueprint. The same genes play critical roles throughout the animal kingdom, small changes in how they are activated leading to very different body forms.

The molecular mechanisms used to orchestrate development are thought to have evolved very early in the history of multicellular life. Animals utilize transcription factors, like those discussed on page 247, to turn on or off particular sets of genes as they develop, determining just what developmental processes occur, where, and when. In many cases the same gene is found to control the same developmental process in many, if not all, animals. For example, a gene in mice called *Pax6* encodes a transcription factor that initiates development of the eye. Mice without a functional copy of this gene do not make the transcription factor and are eyeless. When a gene was discovered in fruit flies that caused the flies to lack eyes, this gene was found to have essentially the same DNA sequence as the mouse gene—the same *Pax6* master regulator gene was responsible for triggering eye development in both insects and vertebrates. Indeed, when Swiss biologist Walter Gehring inserted the mouse version of *Pax6* into the fruitfly genome, a compound eye (the multifaceted kind that flies have) was formed on the leg of the fly! It seems that although insects and vertebrates diverged from a common ancestor more than 500 million years ago, they still control their development with genes so similar that the vertebrate gene seems to function quite normally in the insect genome.

Pax6 plays this same role of releasing eye development in many other animals. Even marine ribbon worms use it to initiate development of their eye spots. The *Pax6* genes of all these animals have similar gene sequences, suggesting that *Pax6* acquired its evolutionary role in eye development only a single time more than 500 million years ago, in the common ancestor of all animals that use *Pax6* today.

A more ancient master regulator gene called *Hox* determines basic body form. *Hox* genes appeared before the divergence of plants and animals; in plants they modulate shoot growth and leaf form, and in animals they establish the basic body plan.

All segmented animals appear to use organized clusters of *Hox* genes to control their development. After the sequential action of several "segmentation" genes, the body of these early embryos has a basically segmented body plan. This is true of the embryos of earthworms, fruit flies, mice—and human beings. The key to the further development of the animal body is to now give identity to each of the

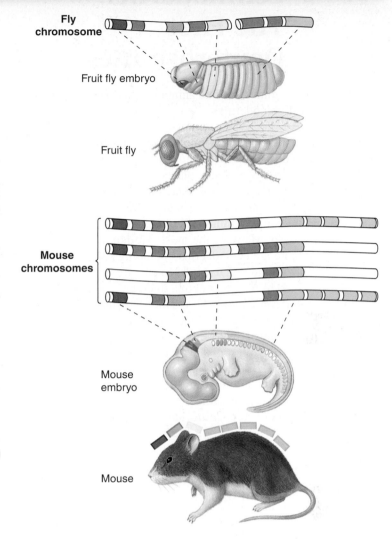

Fly chromosome

Fruit fly embryo

Fruit fly

Mouse chromosomes

Mouse embryo

Mouse

segments—to determine whether a particular segment will become back, or neck, or head, for example. In fruit flies and mice, similar clusters of *Hox* genes control this process. Flies have a single set of *Hox* genes, located on the same chromosome, while mice have four sets, each on a different chromosome (it appears the vertebrate genome underwent two entire duplications early in vertebrate evolution). In the illustration here, the genes are color-coded to match the parts of the body in which they are expressed.

How does a cluster of *Hox* genes work together to control segment development? Each *Hox* gene produces a protein with an identical 60-amino-acid segment that lets it bind to DNA as a transcription factor and, in so doing, activate the genes located where it binds. The differences between each *Hox* gene of a set determine where on DNA a Hox protein binds, and so which set of genes it activates.

Hox genes have also been found in clusters in radially symmetrical cnidarians such as hydra, suggesting that the ancestral *Hox* cluster preceded the divergence of radially and symmetrical animals in animal evolution.

Echinoderms: The First Deuterostomes

The first deuterostomes, marine animals called **echinoderms** in the phylum Echinodermata, appeared more than 650 million years ago. The term *echinoderm* means "spiny skin" and refers to an **endoskeleton** composed of hard, calcium-rich plates called ossicles that lie just beneath a delicate skin. When they are first formed, the plates are enclosed in living tissue, and so are truly an endoskeleton, although in adults they fuse, forming a hard shell. About 6,000 species of echinoderms are living today, almost all of them on the ocean bottom (**figure 19.27**). Many of the most familiar animals seen along the seashore are echinoderms, including sea stars (starfish), sea urchins, sand dollars, and sea cucumbers.

The body plan of echinoderms undergoes a fundamental shift during development: All echinoderms are bilaterally symmetrical as larvae but become radially symmetrical as adults. Many biologists believe that early echinoderms were sessile and evolved adult radial symmetry as an adaptation to the sessile existence. Bilateral symmetry is of adaptive value to an animal that travels through its environment, whereas radial symmetry is of value to an animal whose environment meets it on all sides. Adult echinoderms have a five-part body plan, easily seen in the five arms of a sea star. Its nervous system consists of a central ring of nerves from which five branches arise—while the animal is capable of complex response patterns, there is no centralization of function, no "brain." Some echinoderms like feather stars have 10 or 15 arms, but always multiples of five.

A key evolutionary innovation of echinoderms is the development of a hydraulic system to aid movement. Called a **water vascular system,** this fluid-filled system is composed of a central ring canal from which five radial canals extend out into the arms (see the Phylum Facts illustration on the facing page). From each radial canal, tiny vessels extend through short side branches into thousands of tiny, hollow tube feet. At the base of each tube foot is a fluid-filled muscular sac that acts as a valve (the yellow balls labeled "ampulla"). When a sac contracts, its fluid is prevented from reentering the radial canal and instead is forced into the tube foot, thus extending it. When extended, the tube foot attaches itself to the ocean bottom, often aided by suckers. The sea star can then pull against these tube feet and so haul itself over the seafloor.

Most echinoderms reproduce sexually, but they have the ability to regenerate lost parts, which can lead to asexual reproduction. In a few sea stars, asexual reproduction takes place by splitting, and the broken parts of the sea star can sometimes regenerate whole animals.

Key Learning Outcome 19.12 Echinoderms are deuterostomes with an endoskeleton of hard plates, often fused together. Adults are radially symmetrical.

Figure 19.27 Diversity in echinoderms.
(*a*) Warty sea cucumber, *Parastichopus parvimensis.* (*b*) Feather star (class Crinoidea) on the Great Barrier Reef in Australia. (*c*) Brittle star, *Ophiothrix* (class Ophiuroidea). (*d*) Sand dollar, *Echinarachnius parma.* (*e*) Giant red sea urchin, *Strongylocentrotus franciscanus.* (*f*) Sea star, *Oreaster occidentalis* (class Asteroidea), in the Gulf of California, Mexico.

Phylum Echinodermata: Echinoderms

Key Evolutionary Innovations: DEUTEROSTOME DEVELOPMENT and ENDOSKELETON

Echinoderms like sea stars (phylum Echinodermata) are coelomates with a **deuterostome** pattern of development. A delicate skin stretches over an **endoskeleton** made of calcium rich plates, often fused into a continuous, tough spiny layer.

Sponges | Cnidarians | Flatworms | Nematodes | Mollusks | Annelids | Arthropods | Echinoderms | Chordates

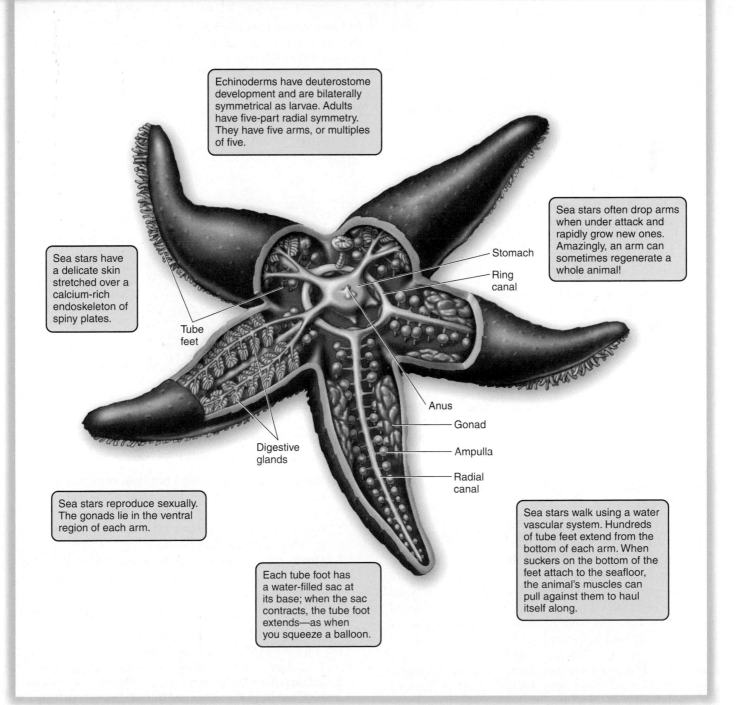

Echinoderms have deuterostome development and are bilaterally symmetrical as larvae. Adults have five-part radial symmetry. They have five arms, or multiples of five.

Sea stars often drop arms when under attack and rapidly grow new ones. Amazingly, an arm can sometimes regenerate a whole animal!

Sea stars have a delicate skin stretched over a calcium-rich endoskeleton of spiny plates.

Stomach

Ring canal

Tube feet

Anus

Gonad

Digestive glands

Ampulla

Radial canal

Sea stars reproduce sexually. The gonads lie in the ventral region of each arm.

Sea stars walk using a water vascular system. Hundreds of tube feet extend from the bottom of each arm. When suckers on the bottom of the feet attach to the seafloor, the animal's muscles can pull against them to haul itself along.

Each tube foot has a water-filled sac at its base; when the sac contracts, the tube foot extends—as when you squeeze a balloon.

Chordates: Improving the Skeleton

General Characteristics of Chordates

Chordates (phylum Chordata) are deuterostome coelomates whose nearest relations in the animal kingdom are the echinoderms, also deuterostomes. Chordates exhibit great improvements in the endoskeleton over what is seen in echinoderms. The endoskeleton of echinoderms is functionally similar to the exoskeleton of arthropods in that it is a hard shell that encases the body, with muscles attached to its inner surface. Chordates employ a very different kind of endoskeleton, one that is truly internal. Members of the phylum Chordata are characterized by a flexible rod called a **notochord** that develops along the back of the embryo. Muscles attached to this rod allowed early chordates to swing their bodies back and forth, swimming through the water. This key evolutionary innovation, attaching muscles to an internal element, started chordates along an evolutionary path that leads to the vertebrates and for the first time to truly large animals.

The approximately 56,000 species of chordates are distinguished by four principal features:

1. **Notochord.** A stiff, but flexible, rod that forms beneath the nerve cord in the early embryo (the yellow rod in the facing Phylum Facts illustration).
2. **Nerve cord.** A single, hollow, dorsal (along the back) nerve cord (the blue rod in the illustration), to which the nerves that reach the different parts of the body are attached.
3. **Pharyngeal pouches.** A series of pouches behind the mouth that develop into slits in some animals. The slits open into the pharynx, which is a muscular tube that connects the mouth to the digestive tract and windpipe (gill slits are labeled in the Phylum Facts illustration).
4. **Postanal tail.** Chordates have a postanal tail, a tail that extends beyond the anus. A postanal tail is present at least during their embryonic development if not in the adult. Nearly all other animals have a terminal anus.

All chordates have all four of these characteristics at some time in their lives. For example, the tunicates in **figure 19.28a** look more like sponges than chordates, but their larval stage, which resembles a tadpole, has all four features listed above. Human embryos have pharyngeal pouches, a nerve cord, a notochord, and a postanal tail as embryos. The nerve cord remains in the adult, differentiating into the brain and spinal cord. The pharyngeal pouches and postanal tail disappear during human development, and the notochord is replaced with the vertebral column. In their body plan, all chordates are segmented, and distinct blocks of muscles can be seen clearly in many forms (**figure 19.29**).

(a)   (b)

Figure 19.28   Nonvertebrate chordates.

(a) Beautiful blue and gold tunicates. (b) Two lancelets, *Branchiostoma lanceolatum*, partly buried in shell gravel, with their anterior ends protruding. The muscle segments are clearly visible in this photograph.

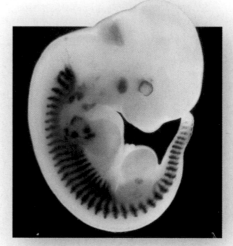

**Figure 19.29
A mouse embryo.**

At 11.5 days of development, the muscle is already divided into segments called somites (stained *dark* in this photo), reflecting the fundamentally segmented nature of all chordates.

Vertebrates

With the exception of tunicates (**figure 19.28a**) and lancelets (**figure 19.28b**), all chordates are **vertebrates.** Vertebrates differ from tunicates and lancelets in two important respects:

1. **Backbone.** The notochord becomes surrounded and then replaced during the course of the embryo's development by a bony vertebral column, a stack of bones called *vertebrae* that encloses the dorsal nerve cord like a sleeve and protects it.
2. **Head.** All vertebrates except the earliest fishes have a distinct and well-differentiated head, with a skull and brain.

All vertebrates have an internal skeleton made of bone or cartilage against which the muscles work. This endoskeleton makes possible the great size and extraordinary powers of movement that characterize the vertebrates.

Key Learning Outcome 19.13 Chordates have a notochord at some stage of their development. In adult vertebrates, the notochord is replaced by a backbone.

Phylum Chordata: Chordates

Key Evolutionary Innovation: NOTOCHORD

Vertebrates, tunicates, and lancelets are chordates (phylum Chordata), coelomate animals with a stiff, but flexible, rod, the **notochord**, that acts to anchor internal muscles, permitting rapid body movements. Chordates also possess **pharyngeal pouches** (relics of their aquatic ancestry) and a dorsal **hollow nerve cord**. In vertebrates, the notochord is replaced during embryonic development by the vertebral column.

Sponges · Cnidarians · Flatworms · Nematodes · Mollusks · Annelids · Arthropods · Echinoderms · Chordates

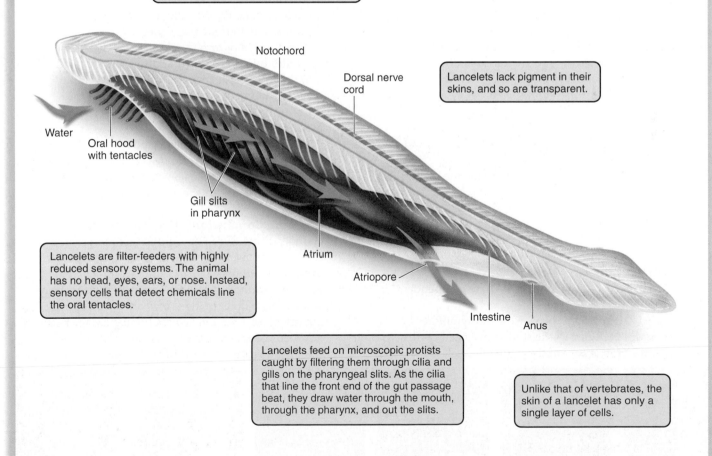

In a lancelet, the simplest chordate, the flexible notochord persists throughout life and aids swimming by giving muscles something to pull against. In the lancelet these muscles form a series of discrete blocks that can easily be seen.

Notochord

Dorsal nerve cord

Lancelets lack pigment in their skins, and so are transparent.

Water

Oral hood with tentacles

Gill slits in pharynx

Atrium

Atriopore

Intestine

Anus

Lancelets are filter-feeders with highly reduced sensory systems. The animal has no head, eyes, ears, or nose. Instead, sensory cells that detect chemicals line the oral tentacles.

Lancelets feed on microscopic protists caught by filtering them through cilia and gills on the pharyngeal slits. As the cilia that line the front end of the gut passage beat, they draw water through the mouth, through the pharynx, and out the slits.

Unlike that of vertebrates, the skin of a lancelet has only a single layer of cells.

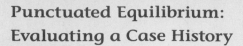
Punctuated Equilibrium:
Evaluating a Case History

Biologists have long argued over the rate at which evolution occurs. Some organisms appear to have evolved gradually (gradualism), while in others evolution seems to have occurred in spurts (punctuated equilibrium). There is evidence of both patterns in the fossil record. Perhaps the most famous claim of punctuated equilibrium has been made by researchers studying the fossil record of marine bryozoans. Bryozoans are microscopic aquatic animals that form branching colonies. You encountered them earlier in this chapter as lophophorates. The fossil record is particularly well documented for Caribbean bryozoan species of the genus *Metrarabdotos,* whose fossil record extends back more than 15 million years without interruption (a **fossil** is the mineralized stonelike remains of a long-dead organism; a **fossil record** is the total collection of fossils of that particular kind of organism known to science).

The graph to the upper right displays an analysis of the *Metrarabdotos* fossil record. Researchers first formulated a comprehensive character index based upon a broad array of bryozoan traits. (A **character index** is a number assigned to a specimen based on its morphology. Different characteristics are measured and assigned quantitative values, and the character index is determined by adding together the individual character values that apply to the specimen. The closer the character indices are for two specimens, the more closely related they are.) Then each fossil is measured for all of the traits. They then calculated the index number for that fossil and plotted it on the graph as a black dot. Each cluster of dots within an oval represents a distinct species.

1. **Applying Concepts**
   a. Variable. In the diagram, is there a dependent variable? If so, what is it?
   b. Analyzing Diagrams. How many different species are included in the study illustrated by this diagram? How many of these are extinct?

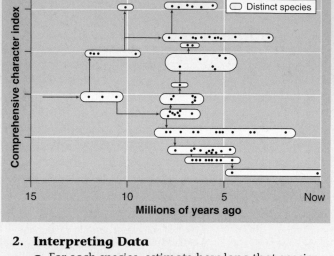

History of Evolutionary Change Among Bryozoa

(x axis: Millions of years ago — 15, 10, 5, Now)
(y axis: Comprehensive character index)
Legend: Distinct species

2. **Interpreting Data**
   a. For each species, estimate how long that species survives in the fossil record. For simplicity, a species found only once should be assigned a duration of 1 million years. What is the average evolutionary duration of a *Metrarabdotos* species?
   b. Create a histogram of your species-duration estimates (place the duration times on the *x* axis and the number of species on the *y* axis). What general statement can be made regarding the distribution of *Metrarabdotos* species durations?

3. **Making Inferences**
   a. How many of the species exhibit variation in the comprehensive character index?
   b. How does the magnitude of this variation *within* species compare with the variation seen *between* species?

4. **Drawing Conclusions** Does major evolutionary change, as measured by significant changes in this comprehensive character index, occur gradually or in occasional bursts?

5. **Further Analysis** Plot the number of *Metrarabdotos* species versus date (millions of years ago), in increments of 1 million years. Characterize the result. What do you suppose is responsible for this? How would you go about assessing this possibility?

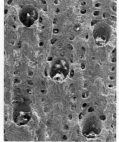

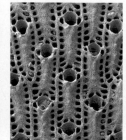
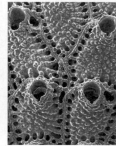

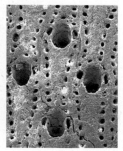

Six species of fossil *Metrarabdotos*

Introduction to the Animals

19.1 General Features of Animals

- Animals are complex multicellular heterotrophs. They are mobile and reproduce sexually. Animals cells do not have cell walls, and animal embryos have a similar pattern of development (**table 19.1**).

19.2 The Animal Family Tree

- Traditionally, animals have been classified based on morphological characteristics. Phylogeny was also determined using anatomical features and embryological development (**figure 19.1**). But, this traditional view of animal phylogeny is now being reevaluated. New molecular methods that compare RNA and DNA of animals are producing new phylogenies (**figure 19.2**). The classification of the protostomes has been reorganized based on how they grow.

19.3 Six Key Transitions in Body Plan

- The large diversity of animals can be traced to six key transitions in body plan—the evolution of tissues, bilateral symmetry, a body cavity, segmentation, molting, and a deuterostome pattern of development. The beetle shown here from **figure 19.3** exhibits five of these six transitions, lacking only deuterostome development.

The Simplest Animals

19.4 Sponges: Animals Without Tissues

- Sponges, in the subkingdom Parazoa, are aquatic, have specialized cells, but lack tissues. The vase-shaped adult is anchored to a substrate (**figure 19.4**). Sponges are filter-feeders. Specialized cells called choanocytes trap food particles that are filtered from the water (**Phylum Facts: Porifera, page 403**).

19.5 Cnidarians: Tissues Lead to Greater Specialization

- Cnidarians have radially symmetrical bodies and two embryonic cell layers, an ectoderm and an endoderm (**Phylum Facts: Cnidaria, page 405**). Cnidarians are carnivores, capturing their prey and digesting it extracellularly in the gastrovascular cavity. Many cnidarians exist only as polyps or medusae (**figure 19.6**), but others alternate forms during their life cycles (**figure 19.7**).

The Advent of Bilateral Symmetry

19.6 Solid Worms: Bilateral Symmetry

- All other animals exhibit bilateral symmetry (**figure 19.8**). In addition to the ectoderm and endoderm, the bilateral eumetazoans also have a mesoderm layer that forms between the other two.

- The simplest bilaterally symmetrical animals are the solid worms, including flatworms (**figure 19.10**), flukes (**figure 19.11**), and other parasitic worms. They have three embryonic tissue layers and a digestive cavity but are acoelomates, lacking a body cavity. In many flatworms, the gut is branched, and serves in both digestion and circulation (**Phylum Facts: Platyhelminthes, page 409**). Flatworms have an excretory system to get rid of waste using specialized cells called flame cells (**figure 19.12**).

The Advent of a Body Cavity

19.7 Roundworms: The Evolution of a Body Cavity

- The evolution of a body cavity improved circulation, movement, and organ function. The roundworms have a body cavity between the endoderm and mesoderm, which is not a true body cavity, so they are called pseudocoelomates (**figure 19.13** and **Phylum Facts: Nematoda, page 413**).

19.8 Mollusks: Coelomates

- Mollusks are coelomates: their body cavity forms within the mesoderm. There are three major groups: gastropods (snails and slugs), bivalves (clams and oysters), and cephalopods (octopuses and squids). All contain a head-foot, a visceral mass, and a mantle. Many also have a rasping tonguelike structure called a radula (**Phylum Facts: Mollusca, page 415**).

19.9 Annelids: The Rise of Segmentation

- Segmentation first evolved in annelid worms, which provided evolutionary flexibility, with different segments becoming specialized for different functions. The basic body plan of annelids is a tube-in-a-tube, the digestive tract and other organs suspended in the coelom (**Phylum Facts: Annelida, page 417**).

19.10 Arthropods: Advent of Jointed Appendages

- Arthropods are the most successful animal phylum. About two-thirds of all named species are arthropods. Arthropods are segmented, with segments fusing to form three body parts: head, thorax, and abdomen (**figure 19.18**). Jointed appendages first evolved in this group, which allowed for improved mobility, grasping, biting, and chewing. A rigid exoskeleton provides protection for the animal and serves as an anchor for muscles (**Phylum Facts: Arthropoda, page 419**). Arthropods include spiders, mites, scorpions, crustaceans, insects, centipedes, and millipedes (**figures 19.19–19.25**).

Redesigning the Embryo

19.11 Protostomes and Deuterostomes

- There are two different developmental patterns in coelomates. In protostomes, the blastopore develops into the mouth. Mollusks, annelids, and arthropods are protostomes. In deuterostomes, the echinoderms and chordates, the blastopore develops into the anus. Other aspects of development differ, including spiral cleavage in protostomes and radial cleavage in deuterostomes (**figure 19.26**).

19.12 Echinoderms: The First Deuterostomes

- Echinoderms have an endoskeleton made up of bony plates that lie under the skin. The adults are radially symmetrical, which appears to be an adaptation to their environment (**Phylum Facts: Echinodermata, page 427**).

19.13 Chordates: Improving the Skeleton

- The chordates have a truly internal endoskeleton and are distinguished by the presence of a notochord, a dorsal nerve cord, pharyngeal pouches, and a postanal tail (**Phylum Facts: Chordata, page 429**). The notochord is replaced with the backbone in vertebrates.

Test Your Understanding

1. The traditional organization of the animal family tree of classification is based mostly on
   a. DNA sequences.
   b. protein structure.
   c. conserved anatomical characteristics.
   d. ribosome structure.
2. Sponges possess unique, collared flagellated cells called
   a. cnidocytes.
   b. choanocytes.
   c. choanoflagellates.
   d. epithelial cells.
3. A characteristic that animals in the phylum Cnidaria share with fungi is
   a. chitinous structural support.
   b. sporulation.
   c. extracellular digestion.
   d. cnidocytes.
4. Which of the following characteristics is *not* seen in the phylum Platyhelminthes?
   a. cephalization
   b. the presence of a mesoderm
   c. specialization of digestive tract
   d. bilateral symmetry
5. One difference between the pseudocoel found in the phylum Nematoda (roundworms) and the coelom found in the phylum Annelida (segmented worms) is the pseudocoel develops between the mesoderm and the _____ in roundworms, and the coelom develops in the _____ in segmented worms.
   a. ectoderm; mesoderm
   b. endoderm; mesoderm
   c. ectoderm; endoderm
   d. endoderm; ectoderm
6. Unlike other coelomate animal phyla, mollusks lack
   a. segmentation.
   b. cephalization.
   c. three primary tissue layers.
   d. some type of body cavity.
7. Segmentation, first seen in annelids, allows evolutionary advantages by
   a. allowing more fluid movement for less energy expense.
   b. specialization of segments to carry out different functions.
   c. allowing coelom development.
   d. concentrating sensory and nervous tissues and organs in the direction of motion.
8. The main limiting factor of arthropod size is the
   a. inefficiency of open circulatory systems.
   b. weight of the muscles needed to move the organism.
   c. weight of the thick exoskeleton needed to support very large insects.
   d. weight of the entire organism, which if too great would crush the soft body of the arthropod during molting.
9. Echinoderms begin life as free-swimming larvae with bilateral symmetry and then become adults with radial symmetry. Some biologists explain this change in symmetry as an adaptation for
   a. an animal traveling through the environment, rather than one having a sessile lifestyle.
   b. an animal with a sessile lifestyle, rather than one that moves through the environment.
   c. a predator, rather than a filter-feeder.
   d. an animal living in a marine environment, rather than a freshwater environment.
10. Which of the following animals is *not* a chordate?
    a. dogs
    b. sea cucumbers
    c. tunicates
    d. lancelets

Apply Your Understanding

1. **Figure 19.17** A famous evolutionist, when asked the most important thing he has learned, said "God has an inordinate fondness for beetles." About two-thirds of all the named species on our planet are arthropods. From the arthropod pie chart shown here, estimate about what percentage of the named species on our planet are beetles.

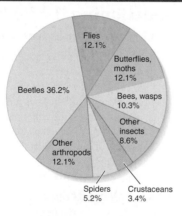

Flies 12.1%
Butterflies, moths 12.1%
Beetles 36.2%
Bees, wasps 10.3%
Other insects 8.6%
Other arthropods 12.1%
Spiders 5.2%
Crustaceans 3.4%

2. **Figures 19.18** and **19.22** What segment of the grasshopper, a typical insect, contains the legs? What segment of the lobster, a typical crustacean, contains the legs? Can you think of any arthropod for which this is not true?

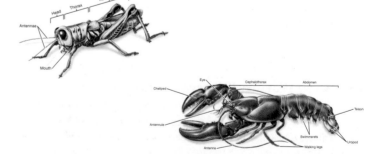

Synthesize What You Have Learned

1. You have discovered a new organism deep in the jungle. You are trying to decide if it is a very slow-moving animal or a plant that responds to light and touch. What characteristics can you investigate to help you decide?
2. Compare sponges with cnidarians. How do food particles and waste products enter and leave the two kinds of organisms? How might the similarity be explained?
3. While there are more described species of insects than all other animals put together, only a very few insects live in marine environments, and most of them on the ocean surface or shoreline. What factors do you think might be responsible for the insects's lack of evolutionary success within the oceans?
4. Do you think it is their internal skeletons that allow vertebrates to be so much larger than other animals? Explain.

20

History of the Vertebrates

A nimals with backbones are among the most visible organisms in the living world. With over 54,000 species, they come in an astonishing range of sizes, from pygmy moles the size of your thumb and hummingbirds that are even smaller, to elephants like those above and whales that are even larger. Vertebrates first evolved in the sea, and today over half of all vertebrate species are fishes. But the great success of vertebrates has come with their successful invasion of land about 350 million years ago. Vertebrates, along with arthropods, now dominate animal life on land. Much of this success is due to the increased complexity of their internal organ systems, and particularly to the internal skeleton that is characteristic of vertebrates, which allows them to grow to great size. Humans, like the elephants above, are mammals, vertebrates with hair that nurse their offspring with milk. The first mammals appeared at the same time as the dinosaurs, but for over 150 million years were mostly a minor group. When dinosaurs became extinct 65 million years ago, mammals survived and have since taken over many of the ecological roles once occupied by dinosaurs.

20.1 The Paleozoic Era

When scientists first began to study and date fossils, they had to find some way to organize the different time periods from which the fossils came. They divided the earth's past into large blocks of time called **eras** (the upper bars in **figure 20.1**). Eras are further subdivided into smaller blocks of time called **periods** (the darker blue bars below the era bars), and some

periods, in turn, are subdivided into **epochs,** which can be divided into **ages** (not shown in the figure). It might be helpful to refer back to this figure throughout the chapter when different eras and periods are discussed.

Virtually all of the major groups of animals that survive at the present time originated in the sea at the beginning of the **Paleozoic era** (the upper lavender bar), during or soon after the Cambrian period (545–490 M.Y.A.). Thus, the major diversification of animal life on earth occurred largely in

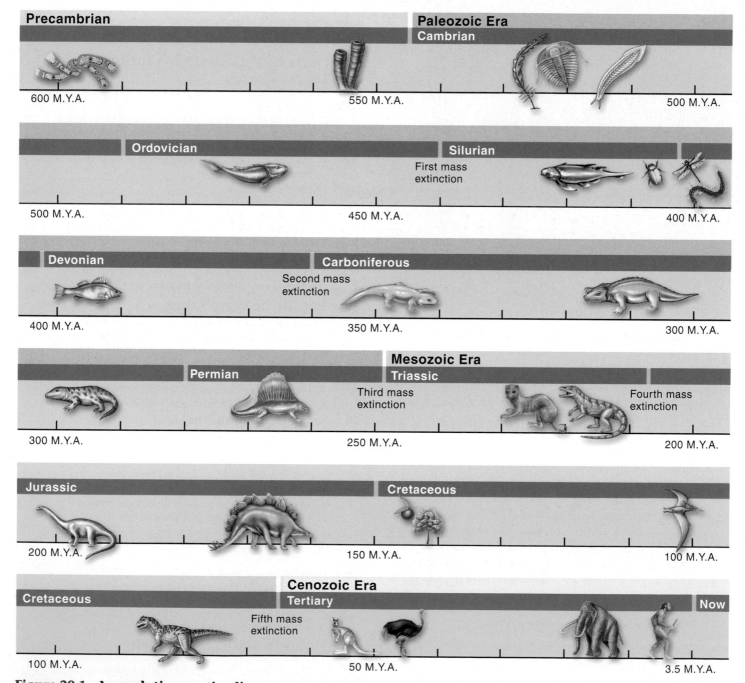

Figure 20.1 An evolutionary timeline.

Vertebrates evolved in the seas about 500 million years ago (M.Y.A.) and invaded land about 150 million years later. Dinosaurs and mammals evolved in the Triassic period about 220 million years ago. Dinosaurs dominated life on land for over 150 million years, until their sudden extinction 65 million years ago left mammals free to flourish.

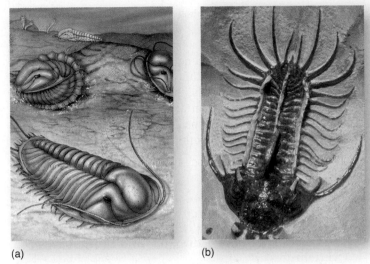

(a)

(b)

Figure 20.2 Life in the Cambrian.

(a) Trilobites are shown in this reconstruction of a community of marine organisms in the Cambrian period, 545 to 490 million years ago. (b) A fossil trilobite.

Figure 20.3 Ammonite fossil from the Jurassic.

Figure 20.4 An early reptile: the pelycosaur.

the sea, and fossils from the early Paleozoic era are found in the marine fossil record.

Many of the animal phyla that appeared in the Cambrian period, like the bizarre trilobites you see above in **figure 20.2**, have no surviving close relatives. Ammonites, shelled, cephalopod mollusks that originated in the Paleozoic era, were among the most abundant creatures on earth 100 million years ago (**figure 20.3**).

The first vertebrates evolved about 500 million years ago in the oceans—fishes without jaws. They didn't have paired fins either—many of them looked something like a flat hotdog with a hole at one end and a fin at the other. For over 100 million years, a parade of different kinds of fishes were the only vertebrates on earth. They became the dominant creatures in the sea, some as large as 10 meters, larger than most cars.

Invasion of the Land

Only a few of the animal phyla that evolved in the Cambrian seas have invaded the land successfully; most others have remained exclusively marine. The first organisms to colonize the land were fungi and plants, over 500 million years ago. It seems probable that plants first occupied the land in symbiotic association with fungi, as discussed in chapter 18.

The first invasion of the land by animals, and perhaps the most successful invasion of the land, was accomplished by the arthropods, a phylum of hard-shelled animals with jointed legs and a segmented body. This invasion occurred about 410 million years ago.

Vertebrates invaded the land during the Carboniferous period (360–280 M.Y.A.). The first vertebrates to live on land were the amphibians, represented today by frogs, toads, salamanders, and caecilians (legless amphibians). The earliest amphibians known are from the Devonian period. Then about 300 million years ago, the first reptiles appeared. Within 50 million years, the reptiles, better suited than amphibians to living out of water, replaced them as the dominant land animal on earth. The pelycosaur, the fan-backed animal pictured in **figure 20.4**, was an early reptile. By the Permian period, the major evolutionary lines on land had been established and were expanding.

Mass Extinctions

The history of life on earth has been marked by periodic episodes of extinction, where the loss of species outpaces the formation of new species. Particularly sharp declines in species diversity are called **mass extinctions.** Five mass extinctions have occurred, the first of them near the end of the Ordovician period about 438 million years ago. At that time, most of the existing families of trilobites (see **figure 20.2**), a very common type of marine arthropod, became extinct. Another mass extinction occurred about 360 million years ago at the end of the Devonian period.

The third and most drastic mass extinction in the history of life on earth happened during the last 10 million years of the Permian period, marking the end of the Paleozoic era. It is estimated that 96% of all species of marine animals that were living at that time became extinct! All of the trilobites disappeared forever. Brachiopods, marine animals resembling mollusks but with a different filter-feeding system, were extremely diverse and widespread during the Permian; only a few species survived.

Mass extinctions left vacant many ecological opportunities, and for this reason they were followed by rapid evolution among the relatively few plants, animals, and other organisms that survived the extinction. Little is known about the causes of major extinctions. In the case of the Permian mass extinction, some scientists argue that the extinction was brought on by a gradual accumulation of carbon dioxide in ocean waters, the result of large-scale volcanism due to the collision of the earth's landmasses during formation of the single large "super-continent" of Pangaea. Such an increase in carbon dioxide would have severely disrupted the ability of animals to carry out metabolism and form their shells.

The most famous and well-studied extinction, though not as drastic, occurred at the end of the Cretaceous period (65 million years ago), at which time the dinosaurs and a variety of other organisms went extinct. Recent findings have supported the hypothesis that this fifth mass extinction event was triggered when a large asteroid slammed into the earth, perhaps causing global forest fires and obscuring the sun for months by throwing particles into the air.

We are living during a new sixth mass extinction event. The number of species in the world is greater today than it has ever been. Unfortunately, that number is decreasing at an alarming rate due to human activity. Some estimate that as many as one-fourth of all species will become extinct in the near future, a rate of extinction not seen on earth since the Cretaceous mass extinction.

> **Key Learning Outcome 20.1** The diversification of the animal phyla occurred in the sea. The two animal phyla that have invaded the land most successfully are the arthropods and chordates (the vertebrates).

20.2 The Mesozoic Era

The **Mesozoic era** (248–65 M.Y.A.) has traditionally been divided into three periods: the Triassic, the Jurassic, and the Cretaceous (see **figure 20.1**). During the late Triassic period, all the continents were together in a single super-continent called Pangaea. There were few mountain ranges over this enormous stretch of land, the interior of which was arid and dry, with widespread deserts. During the Jurassic, the great super-continent of Pangaea began to break up. Long fingers of ocean began to separate the northern part, called Laurasia (the future continents of North America, Europe, and Asia) from the southern part, called Gondwana (the future continents of South America, Africa, Australia, and Antarctica). The two land masses became fully separated by the end of the Jurassic period. World sea levels began to rise, and much of Laurasia and Gondwana began to be inundated by seawater, forming shallow inland seas. The world climate became even warmer, and because so much of the land was nearer to the oceans, conditions became progressively less arid. During the Cretaceous, both Laurasia and Gondwana fragmented into the continents we now know. Sea levels continued to rise, so that by the mid-Cretaceous, sea levels had reached an all-time high, and the interior of North America was a vast inland sea. Much of the world's climate was tropical, hot and wet, like a greenhouse. Most importantly, the early Cretaceous saw the first appearance of flowering plants, the angiosperms.

The Mesozoic era was a time of intensive evolution of terrestrial plants and animals. With the success of the reptiles, vertebrates truly came to dominate the surface of the earth. Many kinds of reptiles evolved, from those smaller than a chicken to others even larger than a semitrailer truck. Just the leg of the sauropod pictured in **figure 20.5** is over 18 feet tall, indicating a massive animal. Some flew, and others swam.

Figure 20.5 Some dinosaurs were truly enormous.
Here the paleontologist Jim Jensen is standing by a reconstruction of the leg of a sauropod fossil he found. Sauropods were herbivores that had enormous barrel-shaped bodies with heavy columnlike legs and very long necks and tails. Some weighed 55 tons, stood 10 meters tall, and were over 30 meters long.

(a) Triassic period

(b) Jurassic period

(c) Cretaceous period

Figure 20.6 Dinosaurs.

Dinosaurs, the most successful of all terrestrial vertebrates, dominated life on land for 150 million years. These three images depict scenes during the three periods of the Mesozoic era. Dinosaurs changed greatly during their long evolutionary history from the Triassic period through the Jurassic period and Cretaceous period. These pictures give only a hint of the great diversity of forms seen in the fossil record.

From among reptile ancestors, dinosaurs, birds, and mammals evolved. Although dinosaurs and mammals appear at about the same time in the fossil record, 200 to 220 million years ago, the dinosaurs quickly filled the evolutionary niche for large animals. For over 150 million years, dinosaurs dominated the face of the earth (**figure 20.6**). Think of it—this time frame is over a *million centuries!* If you could look back to that distant time and have each century flash by your eye in a brief minute, it would take a thousand days to see it all. During this entire period, most mammals were no larger than a cat. Dinosaurs reached the height of their diversification and dominance of the land during the Jurassic and Cretaceous periods.

Because of the major extinction that ended the Paleozoic era (see **figure 20.1**), only 4% of species survived into the Mesozoic. However, as we have described, these survivors gave rise to new species that radiated to form new genera and families. Both on land and in the sea, the number of species of almost all groups of organisms has been climbing steadily for the past 250 million years and is now at an all-time high. This extended recovery from the great Permian extinction had one interruption. About 65 million years ago at the end of the Cretaceous period, dinosaurs disappeared, along with the flying reptiles called pterosaurs, shown as a fossil in **figure 20.7**, the great marine reptiles, and other animals, such as ammonites. This extinction marks the end of the Mesozoic era. Mammals quickly took their place, becoming in their turn abundant and diverse—as they are today.

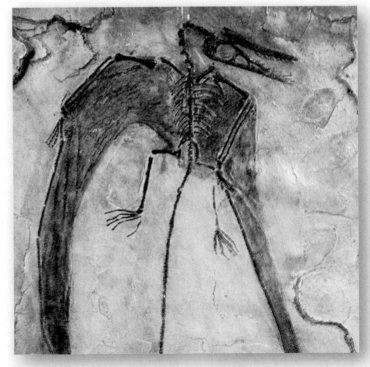

Figure 20.7 An extinct flying reptile.

Pterosaurs, such as the fossil pictured here, became extinct with the dinosaurs about 65 million years ago. Flight has evolved three separate times among vertebrates; however, birds and bats are the only representatives still in existence.

What Happened to the Dinosaurs?

Dinosaurs disappeared abruptly from the fossil record 65 million years ago, within a time frame of less than 2 million years. Their extinction marked the end of the Mesozoic era. **Figure 20.8** shows the animal groups that survived this mass extinction, which is indicated by the yellow bar; no dinosaurs are to the right side of the yellow bar. Why? Many explanations have been advanced, including volcanoes and infectious disease. The most widely accepted theory assigns the blame to an asteroid that hit the earth at that time. Physicist Luis W. Alvarez and his associates discovered that the usually rare element iridium was abundant in many parts of the world in a thin layer of sediment that marked the end of the Cretaceous period. Iridium is rare on earth but common in meteorites. Alvarez and his colleagues have proposed that if a large meteorite 10 kilometers in diameter struck the surface of the earth 65 million years ago, a dense cloud would have been thrown up. The cloud would have been rich in iridium, and as its particles settled, the iridium would have been incorporated into the layers of sedimentary rock being deposited at the time. By darkening the world, the cloud would have greatly slowed or temporarily halted photosynthesis and driven many kinds of organisms to extinction.

The Alvarez hypothesis has become widely accepted among biologists, although it remains controversial among some. These holdouts argue that it is not certain that dinosaurs became extinct suddenly, as they would have by a meteorite collision, and question whether other kinds of animals and plants show the types of patterns that would have been expected as the result of a meteorite collision. The issue was largely settled in the minds of most scientists when an impact crater was discovered in the sea just off the coast of the Yucatán peninsula. For hundreds of kilometers in all directions are signs of the meteor's impact, including quartz crystals with shock patterns that could only have been produced by an enormous impact (they are produced, for example, as a by-product of nuclear tests). Large amounts of soot were deposited in rock worldwide at the time of the extinction, indicating very widespread burning. At what date does radiodating place the Yucatán impact? Sixty-five million years ago.

> **Key Learning Outcome 20.2** The Mesozoic era was the Age of Dinosaurs. They became extinct abruptly 65 million years ago, probably as the result of a meteor impact.

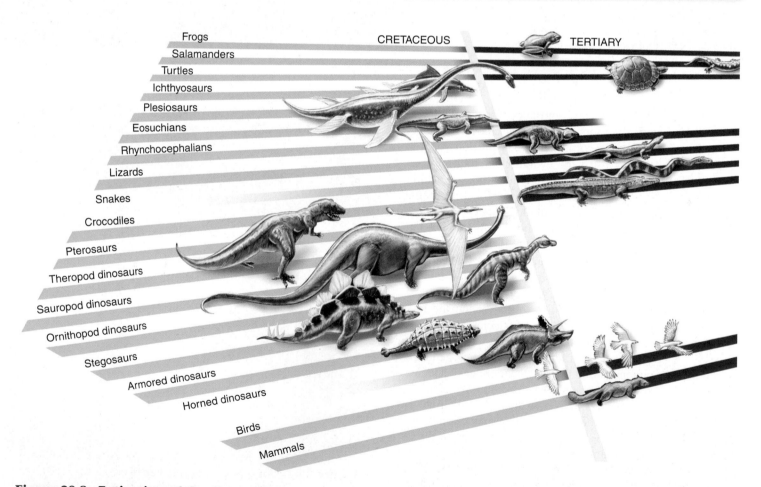

Figure 20.8 Extinction of the dinosaurs.

The dinosaurs became extinct at the end of the Mesozoic era some 65 million years ago (*yellow line*) in a major extinction event that also eliminated all the great marine reptiles (plesiosaurs and ichthyosaurs) as well as the largest of the primitive land mammals. The birds and smaller mammals survived and went on to occupy the aerial and terrestrial modes of living in the environment left vacant by the dinosaurs. Crocodiles, small lizards, and turtles also survived, but reptiles never again achieved the diversity of the Cretaceous period.

The Cenozoic Era

The relatively warm, moist climates of the early **Cenozoic era** (65 M.Y.A. to present) have gradually given way to today's colder and drier climate. The first half of the Cenozoic was very warm, with junglelike forests at the poles. With the extinction of the dinosaurs and many other organisms and a change in climate, new forms of life were able to invade new habitats. Mammals diversified from earlier, small nocturnal forms to many new forms. Most present-day orders of mammals appeared at this time, a period of great diversity.

About 40 million years ago, the climate began to cool, ice caps formed at the poles, and the world entered into an ice age. As glaciation in Antarctica and the Northern Hemisphere became fully established by about 13 million years ago, regional climates cooled dramatically. A series of glaciations followed, the most recent ending about 10,000 years ago. Many very large mammals evolved during the ice ages, including mastodons, mammoths, saber-toothed tigers, and enormous cave bears (**table 20.1**).

The Antarctic, Arctic, and Greenland ice masses that formed as a result of these glaciations have made the world's climate cooler near the poles, warmer near the equator, and drier in the middle latitudes than ever before. In general, forests covered most of the land area of continents, except for Antarctica, until about 15 million years ago, when the forests began to recede rapidly and modern plant communities appeared. During the past several million years, the formation of extensive deserts in northern Africa, the Middle East, and India made migration between Africa and Asia very difficult for organisms of tropical forests. Overall, the Cenozoic era has been characterized by sharp differences in habitat, even within small areas, and the regional evolution of distinct groups of plants and animals. Although there has been an overall decline in mammalian species, these factors have facilitated the rapid formation of many other new species.

Key Learning Outcome 20.3 We live in the Cenozoic era, the Age of Mammals. Many large mammals common in the ice ages are now extinct.

TABLE 20.1   SOME GROUPS OF EXTINCT MAMMALS

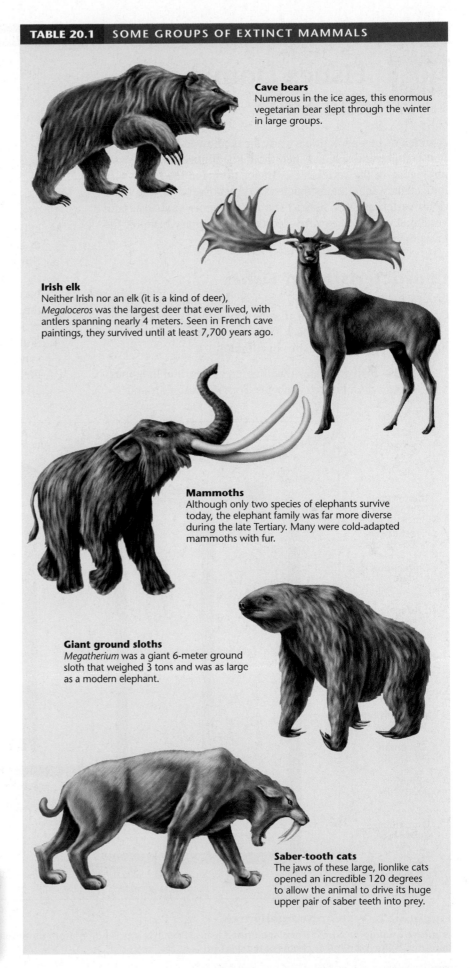

Cave bears
Numerous in the ice ages, this enormous vegetarian bear slept through the winter in large groups.

Irish elk
Neither Irish nor an elk (it is a kind of deer), *Megaloceros* was the largest deer that ever lived, with antlers spanning nearly 4 meters. Seen in French cave paintings, they survived until at least 7,700 years ago.

Mammoths
Although only two species of elephants survive today, the elephant family was far more diverse during the late Tertiary. Many were cold-adapted mammoths with fur.

Giant ground sloths
Megatherium was a giant 6-meter ground sloth that weighed 3 tons and was as large as a modern elephant.

Saber-tooth cats
The jaws of these large, lionlike cats opened an incredible 120 degrees to allow the animal to drive its huge upper pair of saber teeth into prey.

20.4 Fishes Dominate the Sea

A series of key evolutionary advances allowed vertebrates to first conquer the sea and then the land. **Figure 20.9** shows a phylogeny of the vertebrates. Branch points in the family tree indicate key adaptations that lead to great diversity. About half of all vertebrates are **fishes.** The most diverse and successful vertebrate group, they provided the evolutionary base for the invasion of land by amphibians.

Characteristics of Fishes

From whale sharks that are 12 meters long to tiny cichlids no larger than your fingernail, fishes vary considerably in size, shape, color, and appearance. However varied, all fishes have four important characteristics in common:

1. **Gills.** Fish are water-dwelling creatures, and they must extract dissolved oxygen gas from the water around them. They do this by directing a flow of water through their mouths and across their gills. Gills are composed of fine filaments of tissue rich in blood vessels. They are located at the back of the mouth. When water passes over the gills as the fish swallows water, oxygen gas diffuses from the water into the fish's blood.

2. **Vertebral column.** All fishes have an internal skeleton with a spine surrounding the dorsal nerve cord, although it may not necessarily be made of bone. The brain is fully encased within a protective box, called the skull or cranium, which is made of bone or cartilage.

3. **Single-loop blood circulation.** Blood is pumped from the heart to the gills. From the gills, the oxygenated blood passes to the rest of the body and then returns to the heart. The heart is a muscular tube-pump made up of four chambers that contract in sequence.

4. **Nutritional deficiencies.** Fishes are unable to synthesize the aromatic amino acids and must consume them in their diet. This inability has been inherited by all their vertebrate descendants.

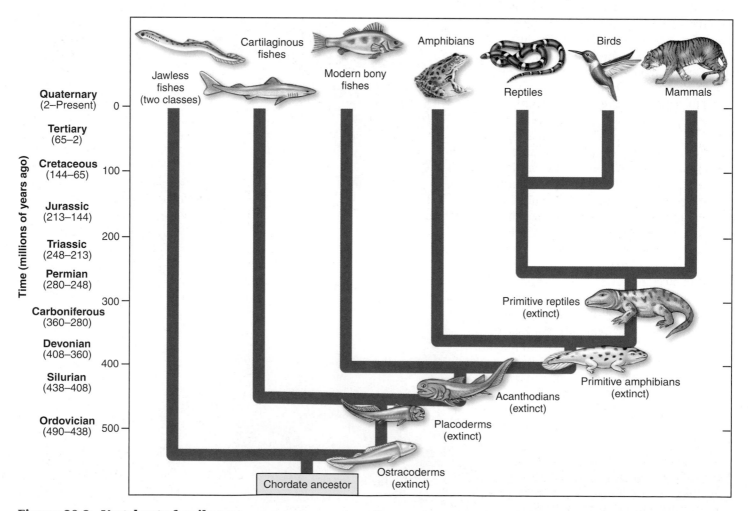

Figure 20.9 Vertebrate family tree.
Primitive amphibians arose from lobe-finned fishes. Primitive reptiles arose from amphibians and gave rise in turn to mammals and to dinosaurs, which are the ancestors of today's birds.

The First Fishes

The first backboned animals were jawless fishes that appeared in the sea about 500 million years ago. These fishes were members of a group called *ostraco-derms*, meaning "shell-skinned." Only their head-shields were made of bone; their elaborate internal skeletons were constructed of cartilage.

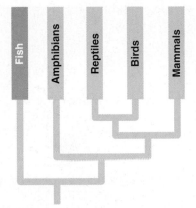

Wriggling through the water, jawless and toothless, these fishes sucked up small food particles from the ocean floor. Most less than a foot long, the earliest groups respired with gills but had no fins—just a primitive tail to push them through the water. These fishes were a great evolutionary success, dominating the world's oceans for about 100 million years. By the end of this period, some groups of ostracoderms had developed primitive fins to help them swim and massive shields of bone for protection. They were eventually replaced by new kinds of fishes that were hunters. One group of jawless fishes, the **agnathans,** survive today as hagfish and parasitic lampreys, shown in **figure 20.10**.

The Evolution of Jaws

The evolution of fishes has been dominated by adaptations to two challenges of surviving as a predator in water:

1. What is the best way to grab hold of potential prey?
2. What is the best way to pursue prey through water?

The fishes that replaced the jawless ones 360 million years ago were powerful predators with much better solutions to both evolutionary challenges. A fundamentally important evolutionary advance was achieved about 410 million years ago—the development of jaws. As illustrated in **figure 20.11**, jaws seem to have evolved from the frontmost of a series of arch supports (colored red and blue) made of cartilage that were used to reinforce the tissue between gill slits, holding the slits open. As you step through the figure from left to right, you can see how the gill arches evolved, reforming into the jaws.

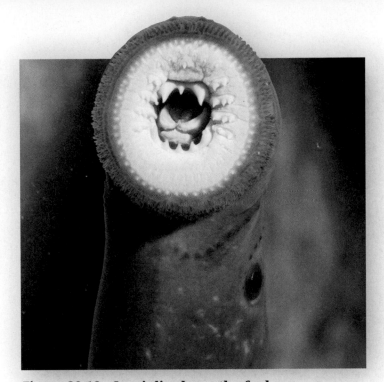

Figure 20.10 Specialized mouth of a lamprey.
Lampreys use their suckerlike mouths to attach themselves to the fish on which they prey. When they have done so, they bore a hole in the fish with their teeth and feed on its blood.

Extinct armored fishes called *placoderms* and spiny fishes called *acanthodians* both had jaws and paired fins. Spiny fishes were predators and far better swimmers than ostracoderms, with as many as seven paired fins to aid their swimming. The larger placoderm fishes had massive heads armored with heavy bony plates. Many placoderms grew to enormous sizes, some over 9 meters long!

Both spiny fishes and placoderms are extinct now, replaced in turn by fishes that evolved even better ways of moving through the water, the sharks and the bony fishes. The earliest sharks and bony fishes appear in the fossil record soon after spiny fishes and placoderms do. However, after sharing the seas for 150 million years, the long competition finally ended with the complete disappearance of the less maneuverable early jawed fishes. For the last 250 million years, all jawed fishes swimming in the world's oceans and rivers have been either sharks (and their relatives, the rays) or bony fishes.

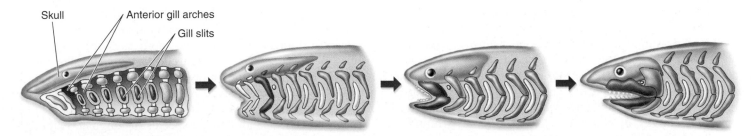

Figure 20.11 A key adaptation among fishes: evolution of the jaw.
Jaws evolved from the anterior gill arches of ancient, jawless fishes.

Sharks

The problem of improving speed and maneuverability in swimming was solved in sharks by the replacement of the heavy bony skeleton of the early fishes with a far lighter one made of strong, flexible cartilage. Members of this group, the class Chondrichthyes, consist of sharks, skates, and rays. Sharks are very powerful swimmers, with a back fin, a tail fin, and two sets of paired side fins for controlled thrusting through the water (**figure 20.12**). Skates and rays are flattened sharks that are bottom-dwellers; they evolved some 200 million years after the sharks first appeared. Today there are about 750 species of sharks, skates, and rays.

Some of the largest sharks filter their food from the water like jawless fishes, but most are predators, their mouths armed with rows of hard, sharp teeth. Sharks are well-adapted to their predatory life due to their sophisticated sensory systems. From a distance, sharks can detect prey using their highly developed sense of smell. Also, a sensory system called the *lateral line system* allows sharks to sense disturbances in the water. At close range, special electroreceptors located primarily on the shark's head allow a shark to detect electric fields that surround all animals. Reproduction among the Chondrichthyes is the most advanced of any fish. Shark eggs are fertilized internally. During mating, the male grasps the female with modified fins called claspers. Sperm pass from the male into the female through grooves in the claspers. About 40% of sharks, skates, and rays lay fertilized eggs. In some of these species, the eggs are laid in a hard capsule, sometimes called a "mermaid's purse." The eggs of other species develop within the female's body, and the pups are born alive. In still other species, embryos develop within the mother and are nourished by maternal secretions or by a placenta-like structure (mammalian placentas are discussed in section 20.8).

Bony Fishes

The problem of improving speed and maneuverability in swimming was solved in bony fishes (**figure 20.13**) in a very different way. Instead of gaining speed through lightness, as sharks did, bony fishes adopted a heavy internal skeleton made completely of bone. Such an internal skeleton is very strong, providing a base against which very strong muscles can pull. Bony fishes are still buoyant though because they possess a **swim bladder.** The swim bladder is a gas-filled sac that allows fish to regulate their buoyant density and so remain effortlessly suspended at any depth in the water. You can explore how a swim bladder works by examining the enlarged drawing in **figure 20.14.** The amount of air in the swim bladder is adjusted by extracting gases from the blood passing through blood vessels near the swim bladder, or by releasing gases back into the blood. Using a swim bladder, a bony fish can rise up and down in the water the same way a submarine does. The swim bladder solution to the challenge of swimming has proven to be a great success in bony fish. Sharks, by contrast, increase buoyancy with oil in their liver, but still must keep swimming (or moving through the water) or they will sink, because their bodies are denser than water.

Figure 20.12 Chondrichthyes.
The Galápagos shark is a member of the class Chondrichthyes, which are mainly predators or scavengers and spend most of their time in graceful motion. As they move, they create a flow of water past their gills, from which they extract oxygen.

Figure 20.13 Bony fishes.
Bony fishes are extremely diverse, containing more species than all other kinds of vertebrates combined. This Korean angelfish, *Pomacanthus semicircularis,* in Fiji, is one of the many striking fishes that live around coral reefs in tropical seas.

Bony fishes (class Osteichthyes) consist of the lobe-finned fishes (subclass Sarcopterygii) and the ray-finned fishes (subclass Actinopterygii), which include the vast majority of today's fishes. In ray-finned fishes, the fins contain only bony rays for support and no muscles; the fins are moved by muscles within the body. In lobe-finned fishes, fins consist of a fleshy muscular lobe that contains a core of bones that form joints with each other; bony rays are only at the tips of each lobed fin (see **figure 20.15 ❶**). Muscles within each lobe can move the fin rays independently of each other. Lobe-finned fishes evolved 390 million years ago, shortly after bony fishes appeared. Eight species survive today, two species of coelacanth and six species of lungfish. Although rare today, lobe-finned fishes played an important evolutionary role, as they gave rise to the first tetrapods (four-legged animals), the amphibians.

TABLE 20.2 MAJOR CLASSES OF FISHES

Class	Typical Examples		Key Characteristics	Approximate Number of Living Species
Acanthodii	Spiny fishes		Fishes with jaws; all now extinct; paired fins supported by sharp spines	Extinct
Placodermi	Armored fishes		Jawed fishes with heavily armored heads; often quite large	Extinct
Osteichthyes Actinopterygii (subclass)	Ray-finned fishes		Most diverse group of vertebrates; swim bladders and bony skeletons; paired fins supported by bony rays	30,000
Sarcopterygii (subclass)	Lobe-finned fishes		Largely extinct group of bony fishes; ancestral to amphibians; paired lobed fins	8
Chondrichthyes	Sharks, skates, rays		Streamlined hunters; cartilaginous skeletons; no swim bladders; internal fertilization	750
Myxini	Hagfishes		Jawless fishes with no paired appendages; scavengers; mostly blind, but a well-developed sense of smell	30
Cephalaspidomorphi	Lampreys		Largely extinct group of jawless fishes with no paired appendages; parasitic and nonparasitic types; all breed in freshwater	35

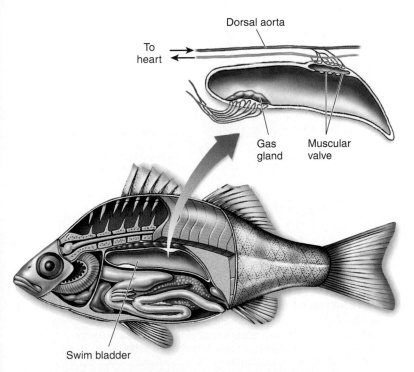

Figure 20.14 Diagram of a swim bladder.
The bony fishes use this structure, which evolved as a dorsal outpocketing of the pharynx, to control their buoyancy in water. The swim bladder can be filled with or drained of gas to allow the fish to control buoyancy. Gases are taken from the blood, and the gas gland secretes the gases into the swim bladder; gas is released from the bladder by a muscular valve.

The major groups of fishes, both living and extinct, are summarized in **table 20.2**. Bony fishes are the most successful of all fishes, indeed of all vertebrates. Of the approximately 30,800 living species of fishes in the world today, about 30,000 species are bony fishes with swim bladders. That's more species than all other kinds of vertebrates combined! In fact, if you could stand in one place and have one representative of every vertebrate animal species alive today pass by you, one after the other, half of them would be bony fishes.

The remarkable success of the bony fishes has resulted from a series of significant adaptations. In addition to the swim bladder, they have a highly developed **lateral line system,** a sensory system that enables them to detect changes in water pressure and thus the movement of predators and prey in the water. The lateral line system is discussed in more detail in chapter 29. Also, most bony fishes have a hard plate called the **operculum** that covers the gills on each side of the head. Flexing the operculum permits bony fishes to pump water over their gills. Using the operculum as very efficient bellows, bony fishes can pass water over the gills while stationary in the water. That is what a goldfish in a fish tank is doing when it seems to be gulping.

> **Key Learning Outcome 20.4** Fishes are characterized by gills, a simple, single-loop circulatory system, and a vertebral column. Sharks are fast swimmers, whereas the very successful bony fishes have unique characteristics such as swim bladders and lateral line systems.

20.5 Amphibians Invade the Land

Frogs, salamanders, and caecilians, the damp-skinned vertebrates, are direct descendants of fishes. They are the sole survivors of a very successful group, the **amphibians,** the first vertebrates to walk on land. Amphibians likely evolved from the lobe-finned fishes, fish with paired fins that consist of a long, fleshy, muscular lobe supported by a central core of bones that form fully articulated joints with one another. In 2006, the discovery of a new fossil fish (genus *Tiktaalik*) exhibited a further step in the transition between fish and amphibians (**figure 20.15 ❷**).

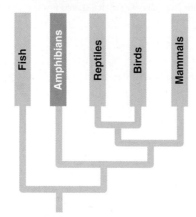

Chordate ancestor

Characteristics of Amphibians

Amphibians have five key characteristics that allowed them to successfully invade the land:

1. **Legs.** Frogs and salamanders have four legs and can move about on land quite well. The way in which legs are thought to have evolved from fins is illustrated in **figure 20.15**. Notice that the arrangement of the bones in the early amphibian limbs ❸ is similar to the arrangement found in the lobe-finned fish ❶ and in *Tiktaalik* ❷. Legs were one of the key adaptations to life on land.
2. **Lungs.** Most amphibians possess a pair of lungs, although the internal surfaces are poorly developed. Lungs were necessary because the delicate structure of fish gills requires the buoyancy of water to support them.
3. **Cutaneous respiration.** Frogs, salamanders, and caecilians all supplement the use of lungs by respiring directly across their skin, which is kept moist and provides an extensive surface area. This mode of respiration limits the body size of amphibians, because it is only efficient for a high surface-to-volume ratio.
4. **Pulmonary veins.** After blood is pumped through the lungs, two large veins called pulmonary veins return the aerated blood to the heart for repumping. This allows aerated blood to be pumped to tissues at a much higher pressure than when it leaves the lungs.
5. **Partially divided heart.** Greater amounts of oxygen are required by muscles for movement and support

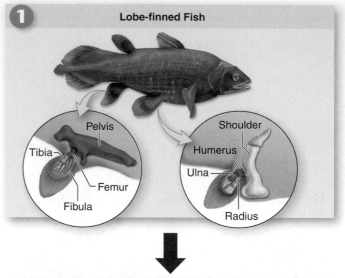

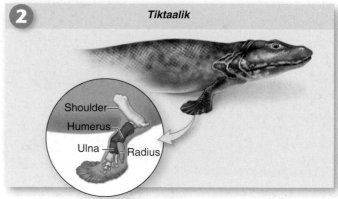

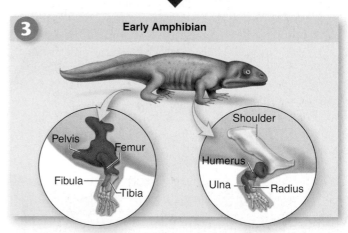

Figure 20.15   A key adaptation of amphibians: the evolution of legs.

In ray-finned fishes, the fins contain only bony rays. ❶ In lobe-finned fish, the fins have a central core of bones (inside a fleshy lobe) in addition to bony rays. Some lobe-finned fishes could move out onto land. ❷ In the incomplete fossil of *Tiktaalik* (which did not contain the hindlimbs), the shoulder, forearm, and wrist bones were like those of amphibians, but the end of the limb was like that of lobe-finned fishes. ❸ In primitive amphibians, the positions of the limb bones are shifted, and bony "toes" are present.

TABLE 20.3   ORDERS OF AMPHIBIANS

Order	Typical Examples		Key Characteristics	Approximate Number of Living Species
Anura	Frogs, toads		Compact tailless body; large head fused to the trunk; rear limbs specialized for jumping	4,200
Urodela (or Caudata)	Salamanders, newts		Slender body; long tail and limbs set out at right angles to the body	500
Apoda (or Gymnophiona)	Caecilians		Tropical group with a snakelike body; no limbs; little or no tail; internal fertilization	150

on land. The chambers of the amphibian heart are separated by a dividing wall that helps prevent aerated blood from the lungs from mixing with nonaerated blood being returned to the heart from the rest of the body. This separates the blood circulation into two separate paths, pulmonary and systemic. The separation is incomplete, however, and some mixing does occur.

History of Amphibians

Amphibians were the dominant land vertebrates for 100 million years (figure 20.16). They first became common in the Carboniferous period, when much of the land was covered by lowland tropical swamps. Amphibians reached their greatest diversity during the mid-Permian period, when 40 families existed. Sixty percent of them were fully terrestrial, with bony plates and armor covering their bodies. Many of these terrestrial amphibians grew to be very large—some as big as a pony! After the great Permian extinction, the terrestrial forms began to decline, and by the time dinosaurs evolved, only 15 families remained, all aquatic.

Approximately 4,850 species of amphibians exist today (figure 20.17), in 37 different families and 3 orders, all descended from three aquatic families that survived competition with reptiles by reinvading the water. The three living orders of the class Amphibia are: Anura, frogs, and toads; Urodela, salamanders, and newts; and Apoda, caecilians (table 20.3). Most of today's amphibians exhibit the ancestral reproductive cycle: Eggs laid in water hatch into aquatic larval forms with gills, which eventually undergo *metamorphosis* into adult forms with lungs. Many amphibians exhibit exceptions to this pattern, but all are tied to moist if not aquatic environments due to their thin skin. In moist habitats, particularly in the tropics, amphibians are often the most abundant and successful vertebrates to be found today.

> **Key Learning Outcome 20.5** Amphibians were the first vertebrates to successfully invade land. They developed legs, lungs, and the pulmonary vein, which allowed them to repump oxygenated blood and thus deliver oxygen far more efficiently to the body's muscles.

Figure 20.16 A terrestrial amphibian of the early Permian.

By the Permian period, many types of amphibians were fully terrestrial, and some, like this *Cacops*, had extensive body armor.

Figure 20.17 A representative of today's amphibians.

This red-eyed tree frog, *Agalychnis callidryas*, is a member of the group of amphibians that includes frogs and toads (order Anura).

20.6 Reptiles Conquer the Land

Characteristics of Reptiles

If we think of amphibians as the "first draft" of a manuscript about survival on land, then **reptiles** were the published book. All living reptiles share certain fundamental characteristics, features they retained from the time when they replaced amphibians as the dominant terrestrial vertebrates. Among the most important are:

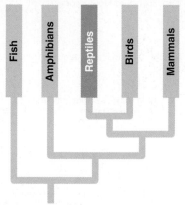

Chordate ancestor

1. **Amniotic egg.** Amphibians never succeeded in becoming fully terrestrial because amphibian eggs must be laid in water to avoid drying out. Most reptiles lay watertight eggs that offer various layers of protection from drying out. The reptilian **amniotic egg** (figure 20.18) contains a food source (the yolk) and a series of four membranes—the chorion (the outermost layer), the amnion (the membrane surrounding the embryo), the yolk sac (containing the yolk), and the allantois (the red structure).

2. **Dry skin.** Amphibians have a moist skin and must remain in moist places to avoid drying out. In reptiles, a layer of scales or armor covers their bodies, preventing water loss.

3. **Thoracic breathing.** Amphibians breathe by squeezing their throat to pump air into their lungs; this limits their breathing capacity to the volume of their mouth. Reptiles developed thoracic breathing, expanding and contracting the rib cage to suck air into the lungs and then force it out.

In addition, reptiles improved on the innovations first attempted by amphibians. Legs were arranged to more effectively support the body's weight, allowing reptile bodies to be

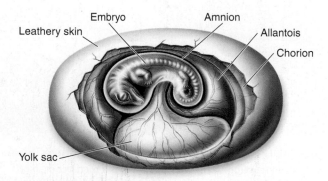

Embryo
Amnion
Leathery skin
Allantois
Chorion
Yolk sac

Figure 20.18 The watertight egg.
The watertight amniotic egg allows reptiles to live in a wide variety of terrestrial habitats.

Figure 20.19 An early archosaur.
Euparkeria, a thecodont, had rows of bony plates along the sides of the backbone, as seen in modern crocodiles and alligators.

bigger and to run. Also, the lungs and heart were altered to make them far more efficient. Today some 7,000 species in the class Reptilia (**table 20.4**) are found in practically every wet and dry habitat on earth. Modern reptiles include four groups: turtles and tortoises, crocodiles and alligators, snakes and lizards, and tuataras.

History of Reptiles

Reptiles first evolved about 300 million years ago when the world was entering a long, dry period. At about this time, three lineages of reptiles formed. In one lineage, "sail-back" **pelycosaurs** (see figure 20.4) rose to prominence. With long, sharp "steak knife" teeth, pelycosaurs were the first land vertebrates able to kill animals their own size. Pelycosaurs were dominant for 50 million years, at their height comprising 70% of all land vertebrates. The **therapsids** (see figure 20.22) replaced the pelycosaurs about 250 million years ago and had a more upright stance than the sprawling pelycosaurs. For 20 million years, therapsids were the dominant land vertebrate. Around 170 million years ago, most therapsids became extinct, but one group survived and eventually gave rise to the mammals.

A second lineage diverged and eventually gave rise to the ancestors of turtles around 200 million years ago. Turtles have remained largely unchanged since the Triassic, making them a very ancient reptile lineage.

As therapsids were becoming less common around 230 million years ago, a third lineage of reptiles gave rise to the ancestors of snakes, lizards, and tuataras, to marine reptiles, the **ichthyosaurs** and **plesiosaurs,** and to the **archosaurs.** Early archosaurs resembled crocodiles (figure 20.19), but later forms called **thecodonts** were the first reptiles to be bipedal—to stand on two feet. Early archosaurs rose to prominence during the Triassic and eventually gave rise to four significant groups: **dinosaurs,** a very diverse group, some of which grew to be larger than houses; **crocodiles,** which have changed little from that time until now; **pterosaurs,** which were flying reptiles; and birds, about which we will have more to say. Dinosaurs were the most successful of all land vertebrates, but around 65 million years ago, they became extinct, along with the marine reptiles and pterosaurs.

> **Key Learning Outcome 20.6** Reptiles have three characteristics that suit them well for life on land: a watertight (amniotic) egg, dry skin, and thoracic breathing.

TABLE 20.4 ORDERS OF REPTILES

Order	Typical Examples		Key Characteristics	Approximate Number of Living Species
Ornithischia	Stegosaur		Dinosaurs with two pelvic bones facing backward, like a bird's pelvis; herbivores, with turtlelike upper beak; legs positioned under body	Extinct
Saurischia	Tyrannosaur		Dinosaurs with one pelvic bone facing forward, the other back, like a lizard's pelvis; both plant and flesh eaters; legs positioned under body	Extinct
Pterosauria	Pterosaur		Flying reptiles; wings were made of skin stretched between fourth fingers and body; wingspans of early forms typically 60 centimeters, later forms nearly 12 meters, making them the largest flying animal ever; early forms had teeth and long tails, but some later forms showed features similar to those seen in birds, including hollow bones, keeled breast bone, reduced tail, lack of teeth, and insulatory filaments	Extinct
Plesiosauria	Plesiosaur		Barrel-shaped marine reptiles with sharp teeth and large, paddle-shaped fins; some had snakelike necks twice as long as their bodies	Extinct
Ichthyosauria	Ichthyosaur		Streamlined marine reptiles with many body similarities to sharks and modern fishes	Extinct
Squamata, suborder Sauria	Lizards		Lizards; limbs set at right angles to body; anus is in transverse (sideways) slit; most are terrestrial	3,800
Squamata, suborder Serpentes	Snakes		Snakes; no legs; move by slithering; scaly skin is shed periodically; most are terrestrial	3,000
Chelonia	Turtles, tortoises, sea turtles		Ancient armored reptiles with shell of bony plates to which vertebrae and ribs are fused; sharp, horny beak without teeth	250
Crocodilia	Crocodiles, alligators, gavials, caimans		Advanced reptiles with four-chambered heart and socketed teeth; anus is a longitudinal (lengthwise) slit; closest living relatives to birds	25
Rhynchocephalia	Tuataras		Sole survivors of a once successful group that largely disappeared before the dinosaurs; fused, wedgelike, socketless teeth; primitive third eye under skin of forehead	2

Dinosaurs

Dinosaurs are the most successful of all terrestrial vertebrates. Dinosaurs dominated life on land for 150 million years, an almost unimaginably long time—for comparison, humans have been on earth only 1 million years. During their long history, dinosaurs changed a great deal, because the world they lived in changed—the world's continents moved, radically altering the earth's climates. Thus, we cannot study dinosaurs as if they were a particular kind of animal, describing one "type" that is representative of the group. Rather, we have to look at dinosaurs more as a "story," a long parade of change and adaptation.

Origin of Dinosaurs

The first dinosaurs evolved from thecodonts, a crocodile-like group of now-extinct meat-eating reptiles. The oldest dinosaurs of which we have any clear evidence left their fossils in late Triassic period rock in Argentina, some 235 million years ago. These early dinosaurs, all meat-eaters, were the first vertebrates to exhibit the key dinosaurian improvement in body design—their legs are positioned directly under the body, allowing them to run swiftly after prey.

Golden Age of Dinosaurs

The Jurassic period (from 213 to 144 million years ago) is called the "Golden Age" of dinosaurs, because of the variety and abundance of dinosaurs that lived during this time. Among the largest land animals of all time, giant **sauropods** were the dominant herbivores (plant eaters). Some weighed 55 tons, stood over 10 meters tall (about 35 feet), and were over 30 meters long (about 100 feet), longer than a basketball court! They had enormous barrel-shaped bodies with heavy columnlike legs, and very long necks and tails.

By the late Jurassic, very sophisticated carnivorous **theropods** (flesh-eating dinosaurs) had evolved. Bipedal,

Stomach Design Sauropods were vegetarians that fed on cycads, palmlike plants with mushy interiors. They shredded the plants with thin spoon-shaped or pencil-like teeth, and ate the shreds without grinding. The grinding necessary to prepare the plant material for digestion took place within the stomach, where stones swallowed by the dinosaur mashed the plant material to a pulp. The cellulose plant material was then digested by microbes within stomachs that acted as huge digestion vats.

Neck Design Sauropods had extremely long necks, as much as half their length. These long necks did not have more vertebrae than the necks of other dinosaurs, but each was elongated to three times the length of a back vertebra. How did the neck muscles support this 35-ton weight? They didn't. Front and back legs acted like the towers of a suspension bridge. A groove ran along the top of the backbone, holding a massive tendon that ran like a cable between tail and neck. The cable allowed the weight of the long neck to be supported by a counterbalancing tail.

Tail Design Their tracks tell us that sauropods did not drag their tails along the ground, but rather held them stiffly out behind. This, and the fact that the neck bones of most sauropods butt up against each other squarely, suggests to many paleontologists that sauropods held their necks out straight, rather than "up" as often illustrated.

Leg Design The immense weight of the sauropod body was supported by four pillarlike legs that did not bend far when walking. Each leg ended with a broad, round "elephant-like" foot. Like all dinosaurs, sauropods stood on their toes—a wedge-shaped heel pad supported the great weight.

Heart Design If a sauropod stood with its head up, its brain would be about 25 feet higher than its heart. The weight of a column of blood this tall would produce enormous pressure in the heart. If it were a typical reptile heart, with an incomplete division of the ventricle, blood would push into the pulmonary circuit and blow out the thin-walled capillaries of the lungs. This argues that sauropods must have had four-chambered hearts, like modern birds and mammals do, with a complete separation of the ventricles.

Body design of a sauropod.

with powerful legs, short arms, and a big head, theropods were well-suited for rapid running and a quick, slashing attack. *Velociraptor*, with the largest brain for body size of any dinosaur, was a highly effective predator, killing its prey with a large, razor-sharp claw on its rear feet. *Tyrannosaurus* had one of the largest brains in the animal kingdom, was one of the largest land predators ever known, and had a massive skull with large, 8-inch teeth.

Triumph of the Chewers

With the rise of angiosperm plants at the beginning of the Cretaceous period (from 144 to 65 million years ago), sauropods were replaced by plant-eating dinosaurs with "chewing" teeth. The jaws of *Iguanodon, Triceratops,* and *Hadrosaurus* contain enormous batteries of grinding teeth that shred, pound, and grind even the toughest angiosperms. As long as 30 feet and weighing up to 5 tons, many of these chewing dinosaurs were larger than a modern battle tank.

Extinction of the Dinosaurs

65 million years ago all dinosaurs disappear from the fossil record. What caused the sudden extinction of the dinosaurs, after 150 million years of success? Most biologists now agree that the most likely cause was the impact of a gigantic meteor (it appears to have been 5–10 miles across!) off the coast of Yucatán. The impact created a huge crater 185 miles in diameter, throwing massive amounts of material into the atmosphere that would have blocked out all sunlight for a considerable period of time, creating a world-wide period of low temperature. Insulated with feathers or fur, warm-blooded (endothermic) birds and mammals survived, and cold-blooded (ectothermic) reptiles and amphibians did too—cold-blooded animals simply lower their activity levels. However, most if not all Cretaceous dinosaurs appear to have been warm-blooded, biologists now believe, but with no insulation, they had no way to retain body heat in the period of intense cold after the meteor impact.

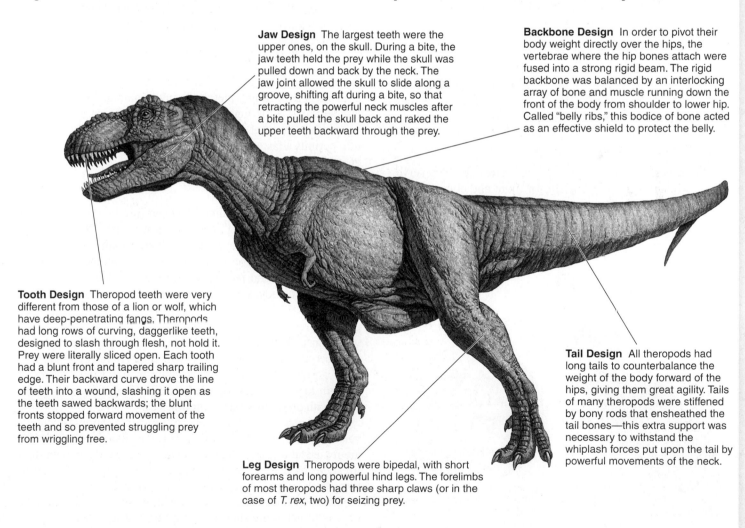

Jaw Design The largest teeth were the upper ones, on the skull. During a bite, the jaw teeth held the prey while the skull was pulled down and back by the neck. The jaw joint allowed the skull to slide along a groove, shifting aft during a bite, so that retracting the powerful neck muscles after a bite pulled the skull back and raked the upper teeth backward through the prey.

Backbone Design In order to pivot their body weight directly over the hips, the vertebrae where the hip bones attach were fused into a strong rigid beam. The rigid backbone was balanced by an interlocking array of bone and muscle running down the front of the body from shoulder to lower hip. Called "belly ribs," this bodice of bone acted as an effective shield to protect the belly.

Tooth Design Theropod teeth were very different from those of a lion or wolf, which have deep-penetrating fangs. Theropods had long rows of curving, daggerlike teeth, designed to slash through flesh, not hold it. Prey were literally sliced open. Each tooth had a blunt front and tapered sharp trailing edge. Their backward curve drove the line of teeth into a wound, slashing it open as the teeth sawed backwards; the blunt fronts stopped forward movement of the teeth and so prevented struggling prey from wriggling free.

Tail Design All theropods had long tails to counterbalance the weight of the body forward of the hips, giving them great agility. Tails of many theropods were stiffened by bony rods that ensheathed the tail bones—this extra support was necessary to withstand the whiplash forces put upon the tail by powerful movements of the neck.

Leg Design Theropods were bipedal, with short forearms and long powerful hind legs. The forelimbs of most theropods had three sharp claws (or in the case of *T. rex*, two) for seizing prey.

Body design of a theropod.

20.7 Birds Master the Air

Birds evolved from small bipedal dinosaurs about 150 million years ago, but they were not common until the flying reptiles called pterosaurs became extinct along with the dinosaurs. Unlike pterosaurs, birds are insulated with feathers. Birds are so structurally similar to dinosaurs in all other respects that there is little doubt that birds are their direct descendants. All but a few taxonomists,

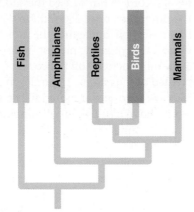

Chordate ancestor

however, still continue to place birds in their own class, Aves, rather than lumping them in with reptiles, because birds are such a distinct and important group.

Characteristics of Birds

Modern birds lack teeth and have only vestigial tails, but they still retain many reptilian characteristics. For instance, birds lay amniotic eggs, although the shells of bird eggs are hard rather than leathery. Also, reptilian scales are present on the feet and lower legs of birds. What makes birds unique? What distinguishes them from living reptiles?

1. **Feathers.** Derived from reptilian scales, feathers serve two functions: providing lift for flight and conserving heat. Feathers consist of a center shaft with barbs extending out (**figure 20.20**). The barbs are held together with secondary branches called *barbules* that hook over each other. This reinforces the structure of

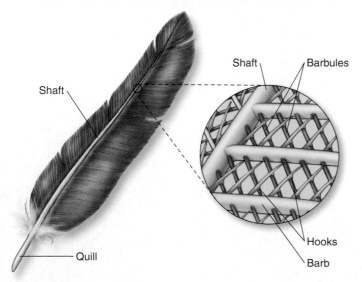

Shaft
Shaft
Barbules
Hooks
Barb
Quill

Figure 20.20 Feathers.
The barbs off the main shaft of a feather have secondary branches called barbules. The barbules of adjacent barbs are attached to one another by microscopic hooks.

Figure 20.21
Archaeopteryx.
Archaeopteryx lived about 150 million years ago and is the oldest fossil bird.

the feather without adding much weight to it. Like scales, feathers can be replaced. Among living animals, feathers are unique to birds. Several types of dinosaurs also possessed feathers, some with bands of color.

2. **Flight skeleton.** The bones of birds are thin and hollow. Many of the bones are fused, making the bird skeleton more rigid than a reptilian skeleton and forming a sturdy frame that anchors muscles during flight. The power for active flight comes from large breast muscles that can make up 30% of a bird's total body weight. They stretch down from the wing and attach to the breastbone, which is greatly enlarged and bears a prominent keel for muscle attachment. They also attach to the fused collarbones that form the so-called wishbone. No other living vertebrates have a fused collarbone or a keeled breastbone.

Birds, like mammals, are endothermic. They generate enough heat through metabolism to maintain a high body temperature. Birds maintain body temperatures significantly higher than those of most mammals. The high body temperature permits a faster metabolism, necessary to satisfy the large energy requirements of flight.

History of Birds

The oldest bird of which there is a clear fossil is ***Archaeopteryx*** (meaning "ancient wing" and shown in **figure 20.21**), which was about the size of a crow and shared many features with small, bipedal, carnivorous dinosaurs. For example, it had teeth and a long reptilian tail. And unlike the hollow bones of today's birds, its bones were solid. Fossils recently discovered in China show that several species of dinosaurs possessed feathers or structures similar to feathers. In these, however, the feathers were most likely used for insulation or display. By the early Cretaceous, only a few million years after *Archaeopteryx*, a diverse array of birds had evolved, with many of the features of modern birds. The diverse birds of the Cretaceous shared the skies with pterosaurs for 70 million years.

Today about 8,600 species of birds (class Aves) occupy a variety of habitats all over the world. The major orders of birds are reviewed in **table 20.5**. You can tell a great deal about their lifestyles by examining their beaks. For example, carnivorous birds such as hawks have a sharp beak for tearing apart meat, the beaks of ducks are flat for shoveling through mud, and the beaks of finches are short and thick for crushing seeds.

> **Key Learning Outcome 20.7** Birds are descendants of dinosaurs. Feathers and a strong, light skeleton make flight possible.

Are Birds Dinosaurs?

Archaeopteryx (pronounced "ahr-kee-OP-tuh-riks") is the first bird for which we have clear fossil evidence. About the size of a crow, the first fossil was found in a 150-million-year-old limestone quarry in Bavaria in 1861. It had the clawed fingers and long bony tail of a dinosaur, with the wishbone and feathered wings of a bird.

For more than a century people have argued about *Archaeopteryx*. Did *Archaeopteryx* evolve from a dinosaur, or from some other reptile? The preponderance of evidence favors a dinosaur ancestor. *Archaeopteryx* is remarkably like a theropod dinosaur called *Velociraptor*. You may remember velociraptors as the scary creatures that stalked the kids in the kitchen in the film *Jurassic Park*. Like *Velociraptor, Archaeopteryx* has an unusual swivel-jointed wrist; a long, very deep shoulder blade; a fused collar bone (familiar as the "wishbone" of Thanksgiving turkeys); and many other shared features.

The convincing proof of a dinosaur-bird direct link was the discovery in 1996 in China of dinosaurs with feathers. The first of these, called *Sinosauropteryx*, had no wings, but was covered with a light featherlike fuzz. A dinosaur with a feather coat, or just stubble?

In 2010, researchers succeeded in answering that question. Paleontologist Mike Benton and colleagues at the University of Bristol, England reported that the simple blunt feathers of *Sinosauropteryx* contain tiny baglike organelles called *melanosomes,* each stuffed with either the black pigment melanin (the same pigment that makes a crow black), or the reddish-brown pigment pheomelanin (which gives a chestnut color similar to that seen in many racehorses). The presence of melanosomes is a common occurrence in the animal kingdom. They are found within the exterior cells of all the major groups of vertebrates. The discovery of melanosomes by the Bristol group proves beyond dispute that the bristles studding *Sinosauropteryx* are indeed feathers and not collagen fiber fuzz.

In birds today, the shape and arrangement of melanosomes helps produce the color of bird feathers. Using scanning electron microscopes to examine the surfaces of the dinosaur feathers, the University of Bristol researchers found that the *Sinosauropteryx* melanosomes were not randomly located along the tail feathers, but rather occurred in broad bands—the dinosaur had alternating orange and white rings down its tail!

In the years since *Sinosauropteryx* was discovered, another very exciting fossil has come to light from the same Chinese fossil fields. Called *Caudipteryx* (that's "caw-DIP-ter-iks"), Greek for "tail feathers," the fossil dinosaur has large feathers on its tail and arms. Big feathers, not bristles, with much of the detailed architecture of modern bird feathers. Two of these feathered dinosaurs were discovered in 1998, and a third beautifully preserved specimen was reported a few years later.

While *Caudipteryx* has a handful of birdlike features, including feathers, it has many features in common with dinosaurs like *Velociraptor,* including short arms, serrated teeth, a velociraptor-like pelvis, and a bony bar behind the eye. Paleontologists who have studied the new fossils describe *Caudipteryx* as sitting on a branch of the dinosaur family tree between *Velociraptor* and *Archaeopteryx.*

Sinosauropteryx	*Velociraptor*	*Caudipteryx*	*Archaeopteryx*	Modern Birds
This theropod dinosaur had short arms and ran along the ground. Its body was covered with filaments that may have been used for insulation and that are the first evidence of feathers.	This larger, carnivorous theropod possessed a swiveling wrist bone, a type of joint that is also found in birds and is necessary for flight.	Recently discovered fossils of this theropod indicate that it is intermediate between dinosaurs and birds. This small, very fast runner was covered with primitive (symmetrical and therefore flightless) feathers.	This oldest known bird had asymmetrical feathers, with a narrower leading edge and streamlined trailing edge. It could probably fly short distances.	

Birds

Dinosaurs

It seems feathers are not a distinguishing trait of birds. They first evolved among the dinosaurs. Because the arms of *Caudipteryx* were too short to use as wings, feathers probably didn't evolve for flight. Instead, they could have served as a colorful display to attract mates (as they do in peacocks today), or as insulation (as they do in penguins today). Flight is something that certain kinds of dinosaurs achieved as they evolved longer arms. We call these dinosaurs birds.

Why then continue to assign birds to a separate class, Aves? If birds are dinosaurs, why not just toss them in with the reptiles, as dinosaurs are? Because classification must do more than just reflect the order in which organisms evolved. It also serves a very practical function. The feathers and flight of today's birds set them apart in such a fundamental way that most biologists, well aware of their dinosaur roots, still choose to assign them their own unique class, Aves. Others have abandoned Aves, and call birds another kind of reptile. Its a judgement call, and the jury is still out.

TABLE 20.5 MAJOR ORDERS OF BIRDS

Order	Typical Examples		Key Characteristics	Approximate Number of Living Species
Passeriformes	Crows, mockingbirds, robins, sparrows, starlings, warblers		*Songbirds* Well-developed vocal organs; perching feet; dependent young	5,276 (largest of all bird orders; contains over 60% of all species)
Apodiformes	Hummingbirds, swifts		*Fast fliers* Short legs; small bodies; rapid wing beat	428
Piciformes	Honeyguides, toucans, woodpeckers		*Woodpeckers or toucans* Grasping feet; chisel-like, sharp bills can break down wood	383
Psittaciformes	Cockatoos, parrots		*Parrots* Large, powerful bills for crushing seeds; well-developed vocal organs	340
Charadriiformes	Auks, gulls, plovers, sandpipers, terns		*Shorebirds* Long, stiltlike legs; slender, probing bills	331
Columbiformes	Doves, pigeons		*Pigeons* Perching feet; rounded, stout bodies	303
Falconiformes	Eagles, falcons, hawks, vultures		*Birds of prey* Carnivorous; keen vision; sharp, pointed beaks for tearing flesh; active during the day	288
Galliformes	Chickens, grouse, pheasants, quail		*Game birds* Often limited flying ability; rounded bodies	268
Gruiformes	Bitterns, coots, cranes, rails		*Marsh birds* Long, stiltlike legs; diverse body shapes; marsh-dwellers	209
Anseriformes	Ducks, geese, swans		*Waterfowl* Webbed toes; broad bill with filtering ridges	150
Strigiformes	Barn owls, screech owls		*Owls* Nocturnal birds of prey; strong beaks; powerful feet	146
Ciconiiformes	Herons, ibises, storks		*Waders* Long-legged; large bodies	114
Procellariiformes	Albatrosses, petrels		*Seabirds* Tube-shaped bills; capable of flying for long periods of time	104
Sphenisciformes	Emperor penguins, crested penguins		*Penguins* Marine; modified wings for swimming; flightless; found only in Southern Hemisphere; thick coats of insulating feathers	18
Dinornithiformes	Kiwis		*Kiwis* Flightless; small; primitive; confined to New Zealand	2
Struthioniformes	Ostriches		*Ostriches* Powerful running legs; flightless; only two toes; very large	1

20.8 Mammals Adapt to Colder Times

Characteristics of Mammals

Most large land-dwelling vertebrates are mammals. The **mammals** (class Mammalia) that first evolved about 220 million years ago side by side with the dinosaurs would look strange to you, not at all like modern-day lions and tigers and bears. They share three key characteristics with mammals living today:

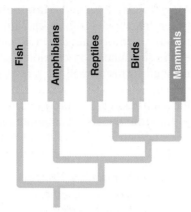

Chordate ancestor

1. **Mammary glands.** Female mammals have mammary glands, which produce milk to nurse the newborns. Even baby whales are nursed by their mother's milk. Milk is a very-high-calorie food (human milk has 750 kcal per liter), important because of the high energy needs of a rapidly growing newborn mammal.

2. **Hair.** Among living vertebrates, only mammals have hair (even whales and dolphins have a few sensitive bristles on their snout). A hair is a filament composed of dead cells filled with the protein keratin. The primary function of hair is insulation. The insulation provided by fur may have ensured the survival of mammals when the dinosaurs perished.

3. **Middle ear.** All mammals have three middle-ear bones, which evolved from bones in the reptile jaw. These bones play a key role in hearing by amplifying vibrations created by sound waves that beat upon the eardrum.

History of the Mammals

We have learned a lot about the evolutionary history of mammals from their fossils. The first mammals arose from therapsids, pictured above in **figure 20.22**, in the late Triassic about 220 million years ago, just as the first dinosaurs evolved from thecodont archosaurs. Tiny, shrewlike creatures that ate insects, most mammals were only a minor element in a land that quickly came to be dominated by dinosaurs. Fossils reveal that these early mammals had large eye sockets, evidence that they may have been active at night.

For 155 million years, all the time the dinosaurs flourished, mammals were a minor group. At the end of the Cretaceous period, 65 million years ago, when dinosaurs and many other land and marine animals became extinct, mammals rapidly diversified. Mammals reached their maximum diversity during the Tertiary period, about 15 million years ago.

Today, 4,500 species of mammals occupy all the large-body niches that dinosaurs once claimed. Present-day

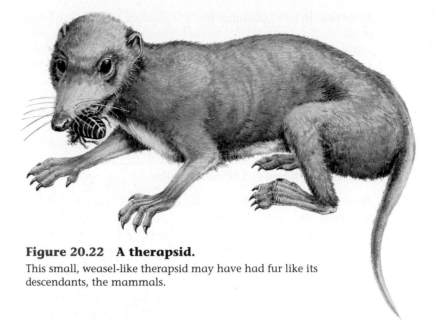

Figure 20.22 A therapsid.
This small, weasel-like therapsid may have had fur like its descendants, the mammals.

mammals range in size from 1.5-gram shrews to 100-ton whales. Almost half of all mammals are rodents—mice and their relatives. Almost one-quarter of all mammals are bats! Mammals have even invaded the seas, as plesiosaur and ichthyosaur reptiles did so successfully millions of years earlier—79 species of whales and dolphins live in today's oceans. The major orders of mammals are described in **table 20.6**.

Other Characteristics of Modern Mammals

Endothermy Mammals are endothermic, a crucial adaptation that allows them to be active at any time of the day or night and to colonize severe environments, from deserts to ice fields. Many characteristics, such as hair to provide insulation, play important roles in making endothermy possible. Also, the more efficient blood circulation provided by a four-chambered heart and the more efficient breathing provided by the *diaphragm* (a special sheet of muscles below the rib cage that aids breathing) make possible the higher metabolic rate upon which endothermy depends.

Teeth Reptiles have homodont dentition: All of an individual's teeth are the same. However, mammals have heterodont dentition, with different types of teeth that are highly specialized to match particular eating habits. It is usually possible to determine a mammal's diet simply by examining its teeth. A dog's long canine teeth are well suited for biting and holding prey, its molar teeth are sharp for ripping off chunks of flesh. In contrast, canine teeth are absent in horses; instead the horse clips off mouthfuls of plants with its flat, chisel-like incisors. Its molars are covered with ridges to effectively grind and break up tough plant tissues. Rodents, such as a squirrel, are gnawers and have long incisors for breaking open nuts and seeds. These incisors are ever-growing—that is, the ends may become sharp and wear down, but new incisor growth maintains the length.

Placenta In most mammal species, females carry their young in the uterus during development, nourishing them by a *placenta,* and give birth to live young. The placenta is a specialized organ within the womb of the mother that brings the bloodstream of the fetus into close contact with the bloodstream of the mother. The placenta evolved from membranes present in the amniotic egg. **Figure 20.23** shows a drawing of a fetus within the uterus. The placenta is to the right side, attached to the umbilical cord. Food, water, and oxygen can pass across from mother to child, and wastes can pass over to the mother's blood and be carried away.

Hooves and Horns Keratin, the protein of hair, is also the structural building material in claws, fingernails, and hooves. Hooves are specialized keratin pads on the toes of horses, cows, sheep, antelopes, and other running mammals. The pads are hard and horny, protecting the toe and cushioning it from impact.

The horns of cattle, sheep, and antelopes are composed of a core of bone surrounded by a sheath of compacted keratin. The bony core is attached to the skull, and the horn is not shed. Deer antlers are made not of keratin but of bone. Male deer grow and shed a set of antlers each year.

Today's Mammals

Monotremes: Egg-Laying Mammals The duck-billed platypus and two species of echidna, or spiny anteater (**figure 20.24a**), are the only living monotremes. The monotremes have many reptilian features, including laying shelled eggs, but they also have both of the defining mammalian features: hair and functioning mammary glands. Females lack well-developed nipples, so the newly hatched babies cannot suckle. Instead, the milk oozes onto the mother's fur, and the babies

lap it off with their tongues. The platypus, found only in Australia, lives much of its life in the water and is a good swimmer. It uses its bill much as a duck does, rooting in the mud for worms and other small animals.

Marsupials: Pouched Mammals The major difference between marsupials (**figure 20.24b**) and other mammals is their pattern of embryonic development. In marsupials, a fertilized egg is surrounded by chorionic and amniotic membranes, but no shell forms around the egg as it does in monotremes. The marsupial embryo is nourished by an abundant yolk within the shell-less egg. Shortly before birth, a short-lived placenta forms from the chorion membrane. After the embryo is born, tiny and hairless, it crawls into the marsupial pouch where it latches onto a nipple and continues its development.

Placental Mammals Mammals that produce a true placenta, which nourishes the embryo throughout its entire development, are called placental mammals (**figure 20.24c**). Most species of mammals living today, including humans, are in this group. Unlike marsupials, the young undergo a considerable period of development before they are born.

> **Key Learning Outcome 20.8** Mammals are endotherms that nurse their young with milk and exhibit a variety of different kinds of teeth. All mammals have at least some hair.

(a)

(b)

(c)

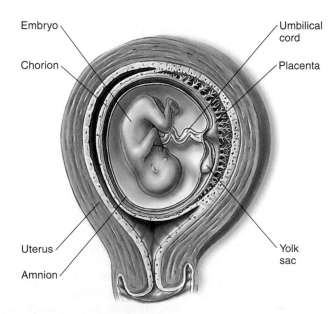

Figure 20.23 The placenta.

The placenta evolved from membranes in the amniotic egg. The umbilical cord evolved from the allantois. The chorion forms most of the placenta itself. The placenta serves as the provisional lungs, intestine, and kidneys of the embryo, without ever mixing maternal and fetal blood.

Labels on figure: Embryo, Chorion, Uterus, Amnion, Umbilical cord, Placenta, Yolk sac

Figure 20.24 Today's mammals.

(*a*) This echidna, *Tachyglossus aculeatus,* is a monotreme. (*b*) Marsupials include kangaroos, like this adult with young in its pouch. (*c*) This female African lion, *Panthera leo* (order Carnivora), is a placental mammal.

TABLE 20.6 MAJOR ORDERS OF MAMMALS

Order	Typical Examples	Key Characteristics	Approximate Number of Living Species
Rodentia	Beavers, mice, porcupines, rats	*Small plant eaters* Chisel-like incisor teeth	1,814
Chiroptera	Bats	*Flying mammals* Primarily fruit or insect eaters; elongated fingers; thin wing membrane; nocturnal; navigate by sonar	986
Insectivora	Moles, shrews	*Small, burrowing mammals* Insect eaters; most primitive placental mammals; spend most of their time underground	390
Marsupialia	Kangaroos, koalas	*Pouched mammals* Young develop in abdominal pouch	280
Carnivora	Bears, cats, raccoons, weasels, dogs	*Carnivorous predators* Teeth adapted for shearing flesh; no native families in Australia	240
Primates	Apes, humans, lemurs, monkeys	*Tree-dwellers* Large brain size; binocular vision; opposable thumb; end product of a line that branched off early from other mammals	233
Artiodactyla	Cattle, deer, giraffes, pigs	*Hoofed mammals* With two or four toes; mostly herbivores	211
Cetacea	Dolphins, porpoises, whales	*Fully marine mammals* Streamlined bodies; front limbs modified into flippers; no hind limbs; blowholes on top of head; no hair except on muzzle	79
Lagomorpha	Rabbits, hares, pikas	*Rodent-like jumpers* Four upper incisors (rather than the two seen in rodents); hindlegs often longer than forelegs, an adaptation for jumping	69
Suborder Pinnipedia	Sea lions, seals, walruses	*Marine carnivores* Feed mainly on fish; limbs modified for swimming	34
Edentata	Anteaters, armadillos, sloths	*Toothless insect eaters* Many are toothless, but some have degenerate, peglike teeth	30
Perissodactyla	Horses, rhinoceroses, zebras	*Hoofed mammals with one or three toes* Herbivorous teeth adapted for chewing	17
Proboscidea	Elephants	*Long-trunked herbivores* Two upper incisors elongated as tusks; largest living land animal	2

Are Extinction Rates Constant?

Since the time of the dinosaurs, the number of living species has risen steadily. Today there are over 700 families of marine animals, containing thousands of described species.

Interspersed, however, have been a number of major setbacks, termed mass extinctions, in which the number of species has greatly decreased. Five major mass extinctions have been identified, the most severe of which occurred at the end of the Permian Period, approximately 225 million years ago, at which time more than half of all families and as many as 96% of all species may have perished.

The most famous and well-studied extinction occurred 65 million years ago at the end of the Cretaceous Period. At that time, dinosaurs and a variety of other organisms went extinct, most likely due to the collision of a large meteor with earth. This mass extinction did have one positive effect, though: With the disappearance of dinosaurs, mammals that previously had been relatively small and inconspicuous, quickly experienced a vast evolutionary radiation, that ultimately produced a wide variety of organisms, including elephants, tigers, whales, and humans. Indeed, a general observation is that biological diversity tends to rebound quickly after mass extinctions, reaching comparable levels of species richness, even if the organisms making up that diversity are not the same.

Today, the number of species in the world is decreasing at an alarming rate due to human activities. We are living during a sixth mass extinction. Some estimate that species are becoming extinct at a rate not seen on earth since the Cretaceous mass extinction.

One thing that the Cretaceous mass extinction and the present-day mass extinction share is that species are becoming extinct for reasons that have nothing to do with what they themselves are like. Is this generally true of extinctions, or are mass extinctions a special case?

Evolutionist Lee Van Valen put forth the hypothesis in 1973 that extinction is indeed usually due to random events unrelated to a species's particular adaptations. If this is in fact the case, then the likelihood that a species will go extinct would be expected to be virtually constant, when viewed over long periods of time.

Van Valen's hypothesis has been tested by evolutionary biologists for a variety of groups. One of the most complete fossil records available for such a test is that of marine echinoids (sea urchins and sand dollars). The fossil echinoid you see in the photo, genus *Cidaris*, is from the Cretaceous some 75 million years ago. In the graph above you see an examination of the 200 million year fossil record of echinoids. Data are presented as the number of echinoid families that have survived for over a period of 200 million years. The red dashed line shows

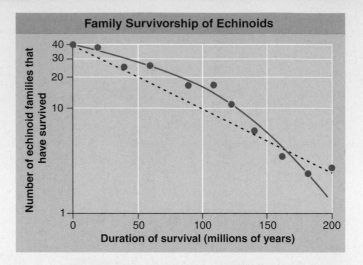

Family Survivorship of Echinoids

a theoretical constant extinction rate, as postulated by Van Valen. The blue line is a "best-fit" curve determined by statistical regression analysis.

1. **Applying Concepts** What is the dependent variable?
2. **Interpreting Data** Which curve best represents the number of echinoid families at the time of the fossil in the photo?
3. **Making Inferences** Over the 200-million-year fossil record of echinoids, which of the two lines best represents the data?
4. **Drawing Conclusions** Is Van Valen's hypothesis supported by this analysis?

Overview of Vertebrate Evolution

20.1 The Paleozoic Era

- The history of earth is broken into blocks of time in order to provide a standard to which scientists can refer. Large blocks of time are called eras. Eras are broken down into periods, which contain epochs. Epochs are broken down into ages.

- Animal life began in the sea primarily during the Paleozoic era (**figure 20.1**). Some animal phyla have no living relatives, like the trilobites shown here from **figure 20.2**, but all of the major animal groups that exist today have ancestors that trace back to this time.

- Mass extinctions, where the loss of species outpaces new species formation, have occurred throughout earth's history. The largest mass extinction occurred at the end of the Paleozoic era, where some 96% of marine species became extinct. The more well-known mass extinction, yet not as large, occurred at the end of the Cretaceous period, when the dinosaurs and a variety of other organisms became extinct.

20.2 The Mesozoic Era

- The Mesozoic era is divided into three periods: the Triassic, the Jurassic, and the Cretaceous. During this time, the single arid landmass called Pangaea broke apart into a northern part, called Laurasia, and a southern part, called Gondwana. These landmasses broke up further and formed the present-day continents. Much of the world's climate was tropical. Amphibians invaded the land and gave rise to the reptiles. Early reptiles gave rise to dinosaurs, birds, and mammals.

- The Mesozoic era ended with the mass extinction of the dinosaurs around 65 million years ago (**figure 20.8**), probably as a result of a meteor that struck the earth.

20.3 The Cenozoic Era

- During the current Cenozoic era, the earth has experienced changes in climate from a relatively warm, moist climate to a colder, drier climate. With the mass extinction of the dinosaurs and a colder climate, mammals were able to expand into empty niches, giving rise to many large animals (**table 20.1**).

The Parade of Vertebrates

20.4 Fishes Dominate the Sea

- Fishes are the ancestors of all vertebrates (**figure 20.9**). Although they are a very diverse group, all fishes have four characteristics in common: gills, a vertebral column, a single-loop circulatory system, and nutritional deficiencies.

- Early fishes lacked jaws, but the evolution of jaws (**figure 20.11**) was key to their survival as predators. Sharks, seen as the ultimate predator, have a flexible cartilaginous skeleton and are very fast swimmers. Bony fish, like the one shown here from **table 20.2**, have swim bladders to control buoyancy in the water (**figure 20.14**), and they are the most successful group of vertebrates.

20.5 Amphibians Invade the Land

- Amphibians, which include frogs and toads, salamanders and newts, and caecilians, were the first vertebrates to invade the land. They likely evolved from lobe-finned fishes that had fins that extended from fleshy muscular lobes. The lobes were supported by jointed bones. Fish fossils have been found that show some of the transitions from fin to leg (**figure 20.15**).

- Adaptations to a terrestrial environment included the development of legs, lungs, cutaneous respiration, pulmonary veins, and a partially divided heart. Most of the adaptations found in amphibians improve the extraction and delivery of oxygen from the air to the tissues.

20.6 Reptiles Conquer the Land

- The key adaptations found in reptiles that made them better suited than amphibians to a terrestrial environment were the evolution of a watertight amniotic egg (**figure 20.18**), watertight skin to keep the body from drying out, and thoracic breathing, which expanded the capacity of the lungs.

- The history of reptiles includes early animals that were prominent on land, like the pelycosaurs, followed by the therapsids, which gave rise to mammals. Other groups of early reptiles gave rise to modern reptiles, birds, and several extinct groups. Today, reptiles, like the snake shown here from **table 20.4**, are found in every wet and dry environment.

20.7 Birds Master the Air

- Birds evolved about 150 million years ago but didn't dominate the skies until the pterosaurs became extinct. Although they are closely related to reptiles, birds are placed into their own class, Aves.

- Two key adaptations are seen in birds: feathers and a skeleton modified for flight. A bird's skeleton is lightweight, with thin, hollow bones, but sturdy, allowing for the attachment of muscles. Birds are endothermic, which allowed them to adapt to the colder climates of the Cenozoic era.

- Birds evolved from dinosaurs and diversified during the Cretaceous period (**table 20.5**). *Archaeopteryx* is the oldest bird for which there is a clear fossil (**figure 20.21**).

20.8 Mammals Adapt to Colder Times

- Mammals (**table 20.6**) arose during the time of the dinosaurs, but did not reach their maximum diversity until the Tertiary period, after the dinosaurs became extinct.

- Mammals are distinguished by the presence of mammary glands for feeding their young, hair that insulates the body, and bones in the middle ear that amplify sound. Like birds, mammals are endothermic.

- The evolution of the placenta (**figure 20.23**) allowed mammals to nourish their young inside the mother's body. There are three main groups of mammals living today: monotremes, marsupials, and placental mammals (**figure 20.24**).

1. Paleontologists divide the earth's past into the following hierarchical organization, beginning with the largest block of time:
   a. age, era, period, epoch.
   b. era, period, epoch, age.
   c. epoch, age, era, period.
   d. period, epoch, age, era.
2. Of the animal phyla, the only two to successfully populate terrestrial habitats in large numbers of species and individuals are the
   a. arthropods and the segmented worms.
   b. sponges and the chordates.
   c. cnidarians and the arthropods.
   d. arthropods and the chordates.
3. In the Paleozoic era, the first vertebrate animal group to live successfully on land was the
   a. amphibians.
   b. reptiles.
   c. birds.
   d. mammals.
4. Mammals began to diversify, with larger forms evolving
   a. after the Cretaceous extinction.
   b. during the Triassic.
   c. at the same time that large-bodied dinosaurs evolved.
   d. All of the above.
5. All fish species, living and extinct, share all of the following characteristics *except*
   a. gills.
   b. jaws.
   c. internal skeleton with dorsal nerve cord.
   d. single-loop circulatory system.
6. Chondrichthyes (sharks) and Osteichthyes (bony fish) have evolved anatomical solutions to increase swimming speed and maneuverability. Which modification is *not* found in Osteichthyes?
   a. a lateral line system
   b. buoyancy control through swim bladders
   c. an internal skeleton made of cartilage
   d. an operculum
7. Adaptations in reptiles do *not* include
   a. an amniotic egg.
   b. a layer of scales on the skin.
   c. middle ear bones.
   d. modifications to the respiratory system.
8. Characteristics that evolved in birds to allow for flight include
   a. reptilian-like scales on the legs.
   b. a hard-shelled amniotic egg.
   c. internal fertilization.
   d. thin, hollow bones in the skeleton.
9. Both birds and mammals share the physiological characteristic of endothermy. How do these animals maintain a high body temperature?
   a. They live in warm environments.
   b. They have high metabolic rates.
   c. They fly, which produces heat.
   d. They eat a lot.
10. A characteristic unique to most species of mammals and no other vertebrates is
    a. endothermy.
    b. skin covering for insulation and protection from dehydration.
    c. hair.
    d. a notochord.

Apply Your Understanding

1. **Figure 20.18** What are the advantages for reptiles of having their eggs covered with a leathery outer shell? Given these advantages, why do you suppose birds evolved eggs with a hard outer shell?

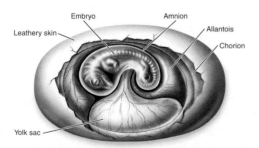

Embryo Amnion Allantois Chorion
Leathery skin
Yolk sac

2. **Figure 20.24** Echidnas eat insects, kangaroos eat plants or fungi, and lions are meat-eaters. What differences would you expect to find in the teeth of these three types of mammals?

(a) (b) (c)

Synthesize What You Have Learned

1. Among terrestrial vertebrates flight has evolved three times. Among what class of terrestrial vertebrates has flight never evolved? Can you think of a reason why not?
2. What characteristics of birds and mammals contribute to their endothermic lifestyles?
3. Amphibians are one of evolution's great success stories, evolving before the reptiles and still common worldwide today. Tragically, amphibians appear to be undergoing a precipitous decline in recent years. What features might make amphibians particularly vulnerable to man's modifications of their environment?

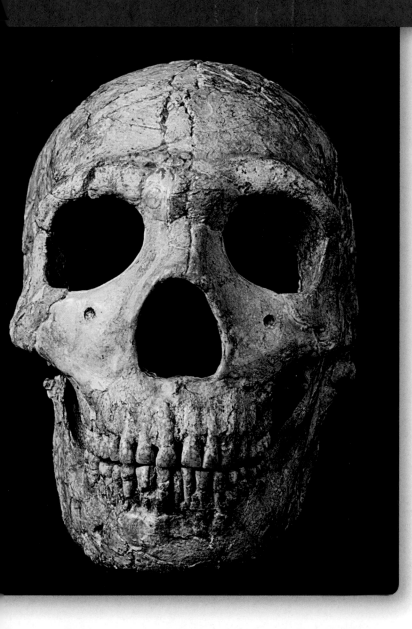

21

How Humans Evolved

The skull you see above is that of an adult male hominid found in the Amud cave in Israel in early summer of 1961. The cave is perched about 30 meters above a streambed through which water flows to the Sea of Galilee, and by a perennial spring. Modern estimates using electron spin resonance techniques date the hominid occupation of the cave at between 40,000 and 50,000 years ago. Perhaps the water attracted the hominids to the site. Judging from the degree of closure in the sutures (connections) between the cranial bones, the adult male died around age 25, apparently of a blow to the side of the head. He is clearly a Neanderthal, with a long narrow face, pronounced brow ridge above the eyes, and a perfectly round cranium, like a bowling ball. The skull's age is relatively late in the period when Neanderthals lived, but most remarkable is the size of his brain. Modern humans have brain sizes of about 1,500 cubic centimeters (cc). This individual's brain is 1,740 cc! While not as large as this, Neanderthal fossils typically have larger brains than modern humans, on the order of 1,650 cc. This raises a very interesting question: As hominids evolved, did their brains become progressively larger? And does this suggest that Neanderthals were smarter than us?

21.1 The Evolutionary Path to Humans

Humans are new arrivals on the biological scene. Fifty years ago, a visual image was proposed that makes this point in a powerful way. Imagine a motion picture of earth taken from space, beginning 757 million years ago, with one image being photographed each year. If you project this film at the normal speed of 24 images per second, it would take you a year to view it, with each day representing 2.1 million years. Starting on January first, there is no visible life on earth's surface for the first four months. In May the first plants cover the land. Dinosaurs dominate the world for 70 days, from late September to December first, when they disappear abruptly. By late December, the modern families of mammals appear, but not until midday on New Year's Eve are direct human ancestors seen. Between 9:30 and 10 p.m., *Homo sapiens* migrates out of Africa to Europe, Asia, and America. At 11:54 p.m., recorded human history begins.

The story of human evolution begins then, around 65 million years ago, with the explosive radiation of a group of small, arboreal (tree-dwelling) mammals called the *Archonta*. These primarily insectivorous mammals had large eyes and were most likely *nocturnal* (active at night). Their radiation gave rise to different types of mammals, including bats, tree shrews, and **primates,** the order of mammals that contains humans.

The Earliest Primates

Primates are mammals with two distinct features that allowed them to succeed in the arboreal, insect-eating environment:

1. **Grasping fingers and toes.** Primates have grasping hands and feet that let them grip limbs, hang from branches, seize food, and, in some primates, use tools. The first digit in many primates is opposable.

2. **Binocular vision.** The eyes of primates are shifted forward to the front of the face. This produces overlapping binocular vision that lets the brain judge distance precisely—important to an animal moving through the trees.

Other mammals have binocular vision, but only primates have both binocular vision and grasping hands, making them particularly well-adapted to their environment.

The Evolution of Prosimians and Anthropoids

About 40 million years ago, the earliest primates split into two groups: the prosimians and the anthropoids (**figure 21.1**). The **prosimians** looked something like a cross between a squirrel and a cat. Only a few prosimians survive today: the tarsiers, lemurs, and lorises. Most prosimians are nocturnal.

The **anthropoids,** or higher primates, include monkeys, apes, and humans. The early anthropoids, now extinct, are thought to have evolved in Africa. Their direct descendants are a very successful group of primates, the monkeys. About 30 million years ago, some anthropoids migrated to South America. Their descendants, the "New World" monkeys, are easy to identify: All are arboreal, they have flat spreading noses, and many of them grasp objects with long prehensile tails. By contrast, the "Old World" monkeys include ground-dwelling as well as arboreal species, have dog-like faces, and do not have prehensile tails.

> **Key Learning Outcome 21.1** The earliest primates arose from small, tree-dwelling insect eaters and gave rise to prosimians and then anthropoids.

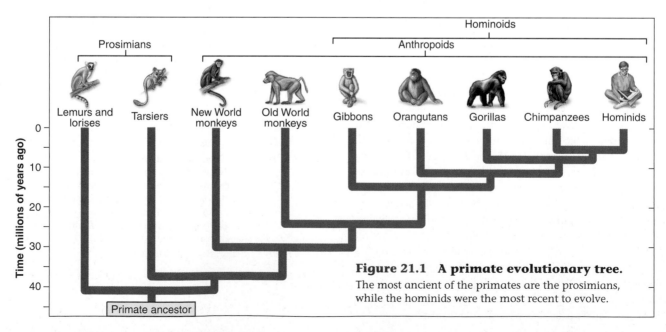

Figure 21.1 A primate evolutionary tree.
The most ancient of the primates are the prosimians, while the hominids were the most recent to evolve.

How the Apes Evolved

Hominoids evolved from anthropoid ancestors. **Hominoids** include the **apes** and the **hominids** (humans and their direct ancestors). The living apes consist of the gibbons (*Hylobates*), orangutans (*Pongo*), gorillas (genus *Gorilla*), and chimpanzees (*Pan*). Apes have larger brains than monkeys, and they lack tails. With the exception of the gibbon, which is small, all living apes are larger than any monkey. Apes exhibit the most adaptable behavior of any mammal except human beings. Once widespread in Africa and Asia, apes are rare today, living in relatively small areas. No apes ever occurred in North or South America.

Which Ape Is Our Closest Relative?

Studies of ape DNA have explained a great deal about how the living apes evolved. The Asian apes evolved first. The line of apes leading to gibbons diverged from other apes about 15 million years ago, whereas orangutans split off about 10 million years ago. Neither are closely related to humans.

The African apes evolved more recently, between 6 and 10 million years ago. These apes are the closest living relatives to humans. Chimpanzees are more closely related to humans than gorillas are; chimpanzees diverged from the ape line less than 6 million years ago. Because this split was so recent, the genes of humans and chimpanzees have not had time to evolve many differences—humans and chimpanzees share 98.6% of their nuclear DNA, a level of genetic similarity normally found between sibling species of the same genus! Gorilla DNA differs from human DNA by about 2.3%. This somewhat greater genetic difference reflects the greater time since the gorilla lineage evolved, around 8 million years ago.

Comparing Apes to Hominids

The common ancestor of apes and hominids is thought to have been an arboreal climber. Much of the subsequent evolution of the hominoids reflected different approaches to locomotion. Hominids became **bipedal,** walking upright, while the apes evolved knuckle-walking, supporting their weight on the back sides of their fingers (monkeys, by contrast, use the palms of their hands).

Humans depart from apes in several areas of anatomy related to bipedal locomotion. Because humans walk on two legs, their vertebral column (the bones highlighted in green in **figure 21.2**) is more curved than an ape's, and the human spinal cord exits from the bottom rather than the back of the skull (see where the green vertebral column joins to the yellow skull). The human pelvis (in blue) has become broader and more bowl-shaped, with the bones curving forward to center the weight of the body over the legs. The hip, knee, and foot (in which the human big toe no longer splays sideways) have all changed proportions.

Being bipedal, humans carry much of the body's weight on the lower limbs, which constitute 32% to 38% of the body's weight and are longer than the upper limbs; human upper limbs do not bear the body's weight and make up only 7% to 9% of human body weight. African apes walk on all fours, with the upper and lower limbs both bearing the body's weight; in gorillas, the longer upper limbs (in purple) account for 14% to 16% of body weight, the somewhat shorter lower limbs for about 18%.

> **Key Learning Outcome 21.2** Apes and hominids arose from anthropoid ancestors. Among living apes, chimpanzees are the most closely related to humans.

Chimpanzee

- Skull attaches posteriorly
- Spine slightly curved
- Arms longer than legs and also used for walking
- Long, narrow pelvis
- Femur angled out

Australopithecine

- Skull attaches inferiorly
- Spine S-shaped
- Arms shorter than legs and not used for walking
- Bowl-shaped pelvis
- Femur angled in

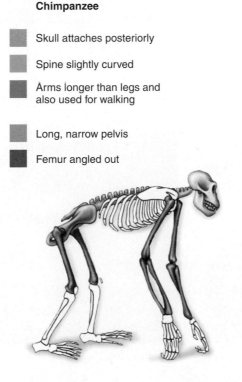
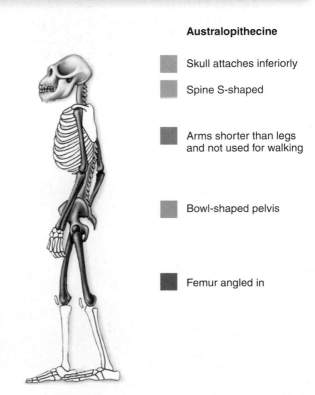

Figure 21.2
A comparison of ape and hominid skeletons.

Early humans, such as australopithecines, were able to walk upright because their arms were shorter, their spinal cord exited from the bottom of the skull, their pelvis was bowl-shaped and centered the body weight over the legs, and their femurs angled inward, directly below the body, to carry its weight.

21.3 Walking Upright

Five to 10 million years ago, the world's climate began to get cooler, and the great forests of Africa were largely replaced with savannas and open woodland. In response to these changes, a new kind of ape was evolving, one that was bipedal. These new apes are classified as hominids—that is, of the human line.

The major groups of hominids include three to seven species of the genus *Homo* (depending how you classify them), seven species of the older, smaller-brained genus *Australopithecus,* and several even older lineages. In every case where the fossils allow a determination to be made, the hominids are bipedal, walking upright. Bipedal locomotion, while not unique to humans (**figure 21.3**), seems to have set the genus *Homo* on a new evolutionary path.

The Origins of Bipedalism

A treasure trove of fossils unearthed in Africa demonstrate that bipedalism extended back 4 million years ago; knee joints, pelvis, and leg bones all exhibit the hallmarks of an upright stance. Substantial brain expansion, on the other hand, did not appear until roughly 2 million years ago. In hominid evolution, upright walking clearly preceded large brains.

Remarkable evidence that early hominids were bipedal is a set of some 69 hominid footprints found at Laetoli, East

Figure 21.4   The Laetoli footprints.

These *Australopithecus* footprints are 3.7 million years old. The impression in the ash reveals a strong heelstrike and a deep indentation made by the big toe, much as you might make in sand when pushing off to take a step. Importantly, the big toe is not splayed out to the side as in a monkey or ape—the footprints were clearly made by a hominid.

Africa (**figure 21.4**). Two individuals, one larger than the other, walked side by side for 27 meters, their footprints preserved in 3.7-million-year-old volcanic ash!

Key Learning Outcome 21.3   The evolution of bipedalism—walking upright—marks the beginning of hominid evolution.

Figure 21.3   Walking upright has evolved many times among vertebrates.

(*a*) *T. rex* (reptiles),
(*b*) king penguin (birds),
(*c*) ostrich (birds),
(*d*) kangaroo (mammals),
(*e*) *Australopithecus* (mammals).

(a)

(b)

(c)

(d)

(e)

21.4 The Hominid Family Tree

The Oldest Known Hominid

In recent years, anthropologists have found a remarkable series of early hominid fossils extending as far back as 6 to 7 million years. Often displaying a mixture of primitive and modern traits, these fossils have thrown the study of early hominids into turmoil. While the inclusion of these fossils among the hominids seems warranted, only a few specimens of these early genera have been discovered, and they do not provide enough information to determine with a degree of certainty their relationships to australopithecines and humans. The search for more early hominid fossils continues.

The First Australopithecine

In 1995, hominid fossils 4.2 million years old were found in the Rift Valley in Kenya. The fossils are fragmentary, but they include complete upper and lower jaws, a piece of the skull, arm bones, and a partial tibia (leg bone). The fossils were assigned to the species *Australopithecus anamensis; anam* is the Turkana word for "lake." While clearly australopithecine, the fossils are intermediate in many ways between apes and a 3-million-year-old, more complete, fossil called *Australopithecus afarensis*. Numerous fragmentary specimens of *Australopithecus anamensis* have since been found.

Most researchers agree that these slightly built *A. anamensis* individuals represent the true base of our family tree, the first members of the genus *Australopithecus*, and thus ancestor to *A. afarensis* and several other species of australopithecines (figure 21.5) whose fossils have been discovered.

Differing Views of the Hominid Family Tree

Investigators take two different philosophical approaches to characterizing the diverse group of African hominid fossils. One group focuses on common elements in different fossils, and tends to lump together fossils that share key characters. Differences between the fossils are attributed to diversity within the group. Other investigators focus more pointedly on the differences between hominid fossils. They are more inclined to assign fossils that exhibit differences to different species. For example, in the hominid evolutionary tree presented in **figure 21.5**, "lumpers" recognize three species of *Homo* (they group the red bars into one species, the dark orange bars into a second species, and the light orange bars into a third species). "Splitters," on the other hand, recognize no fewer than seven species (each of these bars indicating a separate species)! Until we have more fossils, it is not possible to decide which view is correct.

> **Key Learning Outcome 21.4** The earliest australopithecine yet described is *A. anamensis*, over 4 million years old. Some researchers assign all fossils of the genus *Homo* to three species, while others recognize at least seven species.

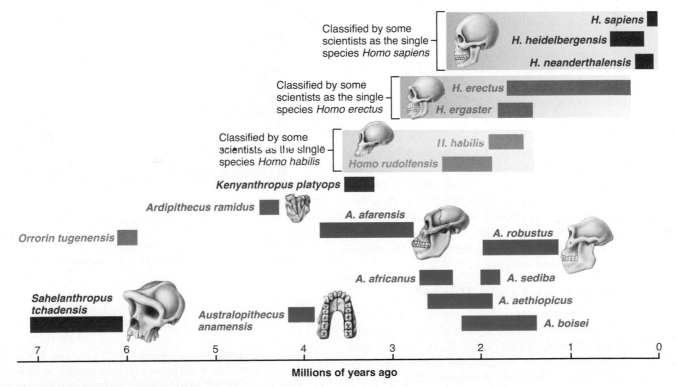

Figure 21.5 A hominid evolutionary tree.

In this tree, the most widely accepted, the horizontal bars show the dates of first and last appearances of proposed species. Seven species of *Australopithecus* and seven of *Homo* are included, as well as four other newly described early hominid genera.

21.5 | African Origin: Early *Homo*

The first humans evolved from australopithecine ancestors about 2 million years ago. The exact ancestor has not been clearly defined but is commonly thought to be *A. afarensis.* Only within the last 30 years have a significant number of fossils of early *Homo* been uncovered. An explosion of interest has fueled intensive field exploration in the last few years, and new finds are announced yearly; every year, our picture of the base of the human evolutionary tree grows clearer.

Homo habilis

In the early 1960s, stone tools were found scattered among hominid bones close to the site where *A. boisei* had been unearthed. Although the fossils were badly crushed, painstaking reconstruction of the many pieces suggested a skull with a brain volume of about 680 cubic centimeters (cc), larger than the australopithecine range of 400 to 550 cubic centimeters. Because of its association with tools (**figure 21.6**), this early human was called ***Homo habilis,*** meaning "handy man." Partial skeletons discovered in 1986 indicate that *H. habilis* was small in stature, with arms longer than legs and a skeleton so similar to *Australopithecus* that many researchers at first questioned whether this fossil was human.

Homo rudolfensis

In 1972, Richard Leakey, working east of Lake Rudolf in northern Kenya, discovered a virtually complete skull about the same age as *H. habilis.* The skull, 1.9 million years old, had a brain volume of 750 cubic centimeters and many of the characteristics of human skulls—it was clearly human and not australopithecine. Some anthropologists assign this skull to *H. habilis,* arguing it is a large male. Other anthropologists assign it to a separate species, *H. rudolfensis,* because of its substantial brain expansion.

Homo ergaster

Some of the early *Homo* fossils being discovered do not easily fit into either of these species. They tend to have even larger brains than *H. rudolfensis,* with skeletons less like an australopithecine and more like a modern human in both size and proportion. Interestingly, they also have small cheek teeth, as modern humans do. Some anthropologists have placed these specimens in a third species of early *Homo,* *H. ergaster* (*ergaster* is from the Greek word for "workman"), shown in **figure 21.7**.

How Diverse Was Early *Homo*?

Because so few fossils have been found of early *Homo,* there is lively debate about whether they should all be lumped into *H. habilis* or split into the two species *H. rudolfensis* and *H. habilis.* If the two species designations are accepted, as increasing numbers of researchers are doing, then it would appear that *Homo* underwent an adaptive radiation (as described in chapter 14) with *H. rudolfensis,* the most ancient species, followed by *H. habilis.* Because of its modern

Figure 21.6 *Homo habilis.*
An artist's rendition of what *Homo habilis* may have looked like. Larger brain size and the use of tools distinguish *Homo habilis* from australopithecines.

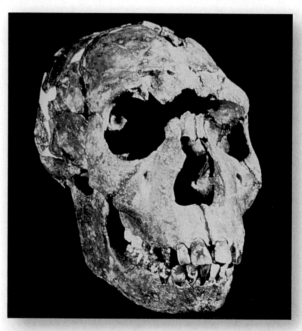

Figure 21.7 *Homo ergaster.*
This skull of a boy, who apparently died in early adolescence, is 1.6 million years old and has been assigned to the species *Homo ergaster.* Much larger than earlier hominids, he was about 1.5 meters in height and weighed approximately 47 kilograms.

skeleton, *H. ergaster* is lumped with *H. erectus* and is thought to be the most likely ancestor to later species of *Homo.*

> **Key Learning Outcome 21.5**   Early species of *Homo,* the oldest members of our genus, had a distinctly larger brain than australopithecines and most likely used tools. There may have been several different species.

Out of Africa:
Homo erectus

Our picture of what early *Homo* was like lacks detail because it is based only on a few specimens. Some scientists still dispute *H. habilis*'s qualifications as a true human because it had not moved far from its australopithecine roots. There is no such doubt about the species that replaced it, **Homo erectus.** Many specimens have been found, and *H. erectus* was without any doubt a true human.

Java Man

After the publication of Darwin's book *On the Origin of Species* in 1859, there was much public discussion about "the missing link," the fossil ancestor common to both humans and apes. Darwin was convinced our origin lay in Africa. Puzzling over the question, a Dutch doctor and anatomist named Eugene Dubois was swayed by Alfred Russel Wallace's conviction that our origins lay instead in southeast Asia. Dubois was particularly intrigued by the orangutans, the "old men of the forest" from Java and Borneo. Many of the orangutan's anatomical features seemed to fit the picture of a "missing link." So Dubois closed his medical practice to seek fossil evidence of the missing link in the home country of the orangutan, Java.

Dubois enrolled as an army surgeon in the Royal Dutch East Indies Army and set up practice, first in Sumatra, then in a small village on the Solo River in eastern Java. In 1891, while digging into a hill by the river (**figure 21.8**) that villagers claimed had "dragon bones," he unearthed a skull cap and, upstream, a thighbone. He was very excited by his find, informally called **Java man,** for three reasons:

1. The structure of the thighbone clearly indicated that the individual had long, straight legs and was an excellent walker.
2. The size of the skull cap suggested a very large brain, about 1,000 cubic centimeters.
3. Most surprising, the bones seemed as much as 500,000 years old, judged by other fossils Dubois unearthed with them.

The fossil hominid that Dubois had found was far older than any fossil hominid discovered up to that time, and few scientists were willing at the time to accept that it was an ancient species of human. In the years since Dubois died, about 40 individuals similar in character and age to his fossil have been found in Java, including in 1969 the nearly complete skull of an adult male.

Peking Man

Another generation passed before scientists were forced to admit that Dubois had been right all along. In the 1920s a skull was discovered in a cave on "Dragon Bone Hill" some 40 kilometers south of Peking (now Beijing), China, that closely resembled Java man. Continued excavation at the site eventually revealed 14 skulls, many excellently preserved, together with lower jaws and other bones. Crude tools were also found, and most important of all, the ashes of campfires.

Figure 21.8 Where *Homo erectus* was first discovered. Digging into a hill on the bank of the Solo River in 1891, Eugene Dubois discovered the first fossil evidence that man's origins go back more than a million years.

Casts of these fossils were distributed for study to laboratories around the world. The originals were loaded onto a truck and evacuated from Peking in December, 1941, at the beginning of Japan's invasion in World War II, only to disappear into the confusion of history. No one knows what happened to the truck or its priceless cargo.

A Very Successful Species

Java man and Peking man are now recognized as belonging to the same species, *H. erectus*. *H. erectus* was a lot larger than *H. habilis*—about 1.5 meters tall. It walked erect, as did *H. habilis,* but had a larger brain, about 1,000 cubic centimeters. The cranial capacity of *H. erectus* was about halfway between that of *Australopithecus* and *Homo sapiens*. Its skull had prominent brow ridges and, like modern humans, a rounded jaw. Most interesting of all, the shape of the skull interior suggests that *H. erectus* was able to talk.

Where did *H. erectus* come from? It should come as no surprise to you that it came out of Africa. In 1976 a complete *H. erectus* skull was discovered in East Africa. It was 1.5 million years old, a million years older than the Java and Peking finds. Far more successful than *H. habilis, H. erectus* quickly became widespread and abundant in Africa and within 1 million years had migrated into Asia and Europe. A social species, *H. erectus* lived in tribes of 20 to 50 people, often dwelling in caves. They successfully hunted large animals, butchered them using flint and bone tools, and cooked them over fires—the site in China contains the remains of horses, bears, elephants, deer, and rhinoceroses.

H. erectus survived for over a million years, longer than any other species of human. These very adaptable humans disappeared in Africa only about 500,000 years ago, as modern humans were emerging. Interestingly, they survived even longer in Asia.

> **Key Learning Outcome 21.6** *Homo erectus* evolved in Africa and migrated from there to Europe and Asia.

21.7 Our Own Species Also Evolved in Africa

The evolutionary journey to modern humans entered its final phase when modern humans first appeared in Africa about 600,000 years ago. Investigators who focus on human diversity consider there to have been three species of modern humans: *Homo heidelbergensis, H. neanderthalensis,* and *H. sapiens.*

The oldest modern human, *H. heidelbergensis,* is known from a 600,000-year-old fossil from Ethiopia. Although it coexisted with *H. erectus* in Africa, *H. heidelbergensis* has more advanced anatomical features, such as a bony keel running along the midline of the skull, a thick ridge over the eye sockets, and a large brain. Also, its forehead and nasal bones are very like those of *H. sapiens. H. heidelbergensis* seems to have spread to several areas in Africa, Europe, and western Asia.

As *H. erectus* was becoming rarer, about 130,000 years ago, a new species of human, *H. neanderthalensis,* appeared in Europe. *H. neanderthalensis* likely branched off from the ancestral line leading to modern humans 500,000 years ago. Compared to modern humans, Neanderthals were short, stocky, and powerfully built. Their skulls were massive, with protruding faces, heavy, bony ridges over the brows, and larger braincases.

Out of Africa—Again?

The oldest fossil known of ***Homo sapiens,*** our own species, is from Ethiopia and is about 130,000 years old. Outside of Africa and the Middle East, there are no clearly dated *H. sapiens* fossils older than roughly 40,000 years of age. The implication is that *H. sapiens* evolved in Africa and then migrated to Europe and Asia: a hypothesis called the **Recently-Out-of-Africa Model.** An opposing view, the **Multiregional Hypothesis,** argues that the human races independently evolved from *H. erectus* in different parts of the world.

Recently, scientists studying human DNA have helped clarify this controversy. Researchers sequencing mitochondrial DNA and a variety of nuclear genes on the Y and X chromosomes and on autosomes have consistently found that all *H. sapiens* shared a common ancestor 170,000 years ago. Scientists now generally accept the conclusions of this broad array of gene data: The Multiregional Hypothesis is wrong. Our family tree has a single stem.

The DNA data reveal a distinct branch on the *H. sapiens* family tree 52,000 years ago, separating Africans from nonAfricans. This is consistent with the hypothesis that *H. sapiens* originated in Africa, from there spreading to all parts of the world, retracing the path taken by *H. erectus* half a million years before. **Figure 21.9** traces the proposed paths taken by three different species of *Homo. H. erectus,* shown

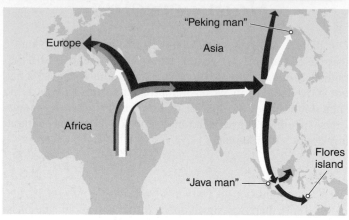

Figure 21.9 Out of Africa—many times.

Several lines of evidence indicate that *Homo* spread from Africa to Europe and Asia repeatedly. First, *H. erectus* (*white arrow*) spread as far as Java and China. Later, *H. heidelbergensis* (*orange arrow*) migrated out of Africa and into Europe and western Asia. And the pattern is repeated again still later by *H. sapiens* (*red arrow*) who migrated out of Africa and into Europe and Asia and eventually into Australia and North America. Lowering of sea levels due to Ice Age coolings made it possible to migrate between landmasses that are currently separated by water. Fossils of *H. neanderthalensis* are known only from Europe and Mediterranean regions, implying that this species may have evolved in Europe.

by the white arrows, evolved first and left Africa, spreading out across Europe and Asia. *H. heidelbergensis* evolved later and followed a similar path, indicated by the orange arrows. Still later, *H. sapiens* repeated the pattern, indicated by the red arrows, but traveled even farther.

A Fourth Species of Recent Human

Evidence has begun to accumulate suggesting that until as recently as 13,000 years ago another species of human existed, hidden away from us on a tiny island in Indonesia. Only 1 meter tall, *Homo floresiensis* has been a startling discovery, one that is still being verified, as described on the facing page.

A Fifth?

Even more recently, in 2010, evidence has emerged pointing to the possible existence of yet a fifth species of recent human, one that coexisted with Neanderthals and *H. sapiens* in Asia 40,000 years ago. The tantalizing evidence comes from DNA extracted from a human finger bone preserved in a cold mountain cave in southern Siberia. The cave is rich in stone tools, but reveals only scattered human bones. When researchers sequenced mitochondrial DNA, they found a human sequence, but one unlike either Neanderthals or *H. sapiens.* The data hint at an unknown type of human that left Africa in a previously unsuspected migration about 1 million years ago.

> **Key Learning Outcome 21.7** *Homo sapiens,* our species, seems to have evolved in Africa and then, like *H. erectus* and *H. heidelbergensis* before it, migrated to Europe and Asia.

Today's *Biology*

Meet Our Hobbit Cousin

For the last several years, the field of paleoanthropology, the study of human evolution, has been stood on its head by a tiny fossil from Indonesia.

This story begins in 1996, when Australian paleoanthropologist Michael Morwood reported that his team had excavated stone tools from a site on Flores island 300 miles south of Borneo in Indonesia. The site was almost a million years old, long before the period when scientists thought sophisticated humans arrived there. Even more to the point, Flores is isolated by a deep-water trench from the rest of Asia. This deep-water barrier is called the Wallace Line after Darwin's contemporary Alfred Russel Wallace who first noticed the marked differences between animals living on either side of it. On one side they resemble Asian species but on the other side they resemble Australian species. Might early humans have somehow managed to cross?

Morwood and Indonesian collaborators started looking for fossils on both sides of the Wallace Line. In 2001 they began to examine caves that had been explored years earlier by other archaeologists, and in 2003 they hit paydirt at Liang Bua ("cool cave"). Six meters under the floor of the cave, in 18,000-year-old sediment, Morwood's team found the nearly complete fossil skeleton of a hominid, an early human. The teeth are worn, and the skull bones are knitted together in an adult way, so the fossilized individual was an adult—but an adult hominid only 1 meter tall, with a brain of only 380 cc (cubic centimeters)! This is the stature and brain size of a chimpanzee. The fossil foot, essentially complete, is extremely large. Workers at the site nicknamed the newly discovered hominid "the hobbit" after the J.R.R. Tolkien characters in the book *The Lord of the Rings*.

A firestorm of controversy broke out among the scientific community. This *Homo floresiensis,* as Morwood dubbed the fossil, makes no sense when viewed against the detailed picture paleoanthropologists have

Homo floresiensis (far right) compared with a modern human female. *H. floresiensis* preyed upon a dwarf species of elephant, which also occurred on the island of Flores.

pieced together of human evolution. For one thing, it is too young. In a 3-million-year-old prehuman ancestor, the Flores fossil's tiny dimensions would not be surprising, but the fossil is only 18,000 years old. That means it was alive when *Homo sapiens* (modern human) had evolved and even crossed the Wallace Line to Australia.

For another thing, the Flores individual is too short. Early African human ancestors called australopithecines were this size, 20 kg and about a meter tall, but every adult human fossil ever found is much bigger and taller. The humans alive 18,000 years ago were not tiny but the same stature as humans today.

And yet, despite all these objections that the Flores fossil could not exist, there it is, hard data. What are we to make of it? Morwood's suggestion is that the Flores fossil is a member of a human species that evolved in isolation on Flores island from ancient humans that reached there long ago. The ancestor of *Homo sapiens,* called *Homo erectus,* shares certain characteristics with the Flores fossil and so is the most likely candidate. Morwood goes on to point out that large mammals colonizing remote small islands often tend to evolve into isolated dwarf species. Many examples are known, including pygmy hippos, ground sloths, deer, and even dinosaurs. Indeed, pygmy elephants no bigger than cattle lived on Flores island at that very time; fossils of hunted specimens are found in the Liang Bua cave! Why do dwarf species evolve? The theory is that on islands with few or no predators, there is no advantage to being big, and with limited food there is great advantage to being small. On Flores island, Morwood suggests, the human evolutionary story has played out differently than it did in Africa, evolution favoring the small.

"Wait a minute," cry researchers from the Field Museum in Chicago, who in 2006 challenged the classification of the Flores fossil as a new species. Its brain is simply too small for a dwarf species, they claim—judging from other instances in which pygmy species have developed in isolation, the Flores brain size of 380 cc would indicate a creature only 1 foot tall! They propose that instead it may be a modern human suffering from microencephaly, a genetic disorder that results in small brain size.

Morwood's team is not ready to concede the argument. In subsequent digs at the Liang Bua cave in 2004 they have found a total of eight more tiny individuals, all of them even smaller than the complete fossil unearthed the previous year. It stretches the imagination to believe that all suffer from microencephaly. Strengthening their claim that the Flores fossils are not modern humans was the discovery of two complete lower jaws, neither of which have chins. Chins are found on the lower jaws of all modern human fossils.

The Floresians of the Liang Bua region seem to have perished after an eruption of one of the island's many volcanoes about 13,000 years ago. But they may have survived much longer elsewhere on Flores, Dr. Morwood believes. The search for more fossils goes on.

The Only Surviving Hominid

The **Neanderthals** (classified by some as a separate human species, *H. neanderthalensis*) were named after the Neander Valley of Germany, where their fossils were first discovered in 1856. The closest evolutionary relatives of present-day humans, Neanderthals were common in large parts of Europe and Western Asia 70,000 years ago. The Neanderthals made diverse tools, including scrapers, spearheads, and hand axes. They lived in huts or caves.

In 2010, the bulk of the Neanderthal genome was determined using fragments of DNA extracted from fossil bones of three Neanderthal females who lived in Croatia more than 38,000 years ago. The genome sequence confirms that the Neanderthals have quite different DNA than *Homo sapiens,* supporting the view that Neanderthals were a separate species. Interestingly, comparing this composite Neanderthal genome with the complete genomes of five living humans from different parts of the world, both European and Asian *H. sapiens* today contain about 1% to 4% of sequences inherited from Neanderthals. This offers the tantalizing suggestion that early modern humans may have interbred with Neanderthals! Although the degree of gene sharing is slight, this suggestion that each of us carries a few Neanderthal genes is sure to provoke controversy.

Fossils of *H. neanderthalensis* abruptly disappear from the fossil record about 34,000 years ago and are replaced by fossils of *H. sapiens* called the **Cro-Magnons** (named after the valley in France where their fossils were first discovered). We can only speculate why this sudden replacement occurred, but it was complete all over Europe in a short period. There is some evidence that the Cro-Magnons came from Africa—fossils of essentially modern aspect, but as much as 100,000 years old, have been found there.

The Cro-Magnons used sophisticated stone tools, had a complex social organization, and are thought to have had full language capabilities. They lived by hunting. The world was cooler than it is now—the time of the last glaciation—and Europe was covered with grasslands inhabited by large herds of grazing animals.

Humans of modern appearance eventually spread across Siberia to North America, which they reached at least 13,000 years ago, after the ice had begun to retreat and a land bridge still connected Siberia and Alaska. By 10,000 years ago, about 5 million people inhabited the entire world (compared with over 6 *billion* today). A genomic survey of world populations carried out in 2002 (**figure 21.10**) provides clear evidence of the migration of our species out of Africa and across the globe. As discussed on the facing page, people cluster into five distinct groups that correspond to the populations living today on the five continents. Population-specific evolutionary changes seen in DNA comparisons include genes related to sucrose metabolism in East Asians, to skin pigmentation and lactose tolerance in Europeans, and in the metabolism of the sugar mannose in Africans. Clearly, humans are still evolving, locally.

Homo sapiens Are Unique

We humans are the current stage of a long evolutionary history. The evolution of our genus has been marked by a progressive increase in brain size. While not the only animal capable of conceptual thought, we have refined and extended this ability until it has become the hallmark of our species. We control our biological future in a way never before possible—an exciting potential and frightening responsibility.

> **Key Learning Outcome 21.8** Our species, *Homo sapiens,* is good at conceptual thought and tool use and is the only animal that uses symbolic language. *H. sapiens* continues to evolve.

Modern Genetic Clusters

- Africa
- Eurasia
- East Asia
- Oceania
- America

For example, the average genome of the Uygur people is:

52% East Asian
46% Eurasian
2% Other races

*arrows show migration routes of ancestral human populations

Figure 21.10 *Homo sapiens* **is still evolving.**

Researchers comparing the DNA genomes of 52 modern human populations have discovered five main genetic clusters, which largely correspond to major geographic regions.

Today's *Biology*

Race and Medicine

Few issues in biology have stirred more social controversy than race. Race has a deceptively simple definition, referring to groups of individuals related by ancestry that differ from other groups, but not enough to constitute separate species. The controversy arises because of the way people have used the concept of race to justify the abuse of humans. The African slave trade is but one obvious example. Perhaps in some measure responding to these sorts of injustice, scientists have largely abandoned the concept of race except as a social construct. In 1972 geneticists pointed out that if one looked at genes rather than faces, the differences between the genes of an African and a European would be hardly greater than the difference between those of any two Europeans. For 38 years, gene data has continually reinforced the validity of this observation, and the concept of genetic races has been largely abandoned.

Recent detailed comparisons of the genomes of people around the world reveal that particular alleles defining human features often occur in clusters on chromosomes. Because they are close together, the genes experience little recombination over the centuries. The descendants of a person who has a particular combination of alleles will almost always have that same combination. The set of alleles, technically called a "haplotype," reflects the common ancestry of these descendants from that ancestor.

Ignoring skin color and eye shape, and instead comparing the DNA sequences of hundreds of regions of the human genome, investigators have compared the haplotypes of a large sample of people from around the world. They found five large groups containing similar clusters of variation: Europe, East Asia, Africa, America, and Australasia. That these are more or less the major races of traditional anthropology is not the point. The point is that humanity evolved in these five regions in isolation from one another, that today each of these groups is composed of individuals with a shared ancestry, and that by analyzing the DNA of an individual we can deduce that ancestry.

Why bother making that rather arcane and socially controversial point? Because analysis of the human genome is revealing that many diseases are influenced by alleles that have arisen since the five major branches of the human family tree separated from one another, and are much more common in the human ancestral group within which the DNA mutation causing the disease first occurred. For example, a mutation causing hemochromatosis, a disorder of iron metabolism, is rare or absent among Indians or Chinese, but very common among northern Europeans (it occurs in 7.5% of Swedes), who also commonly possess

an allele leading to adult lactose intolerance (inability to digest lactose) not common in many other groups. Similarly, the hemoglobin S mutation causing sickle cell disease is common among Africans of Bantu ancestry, but seems to have arisen only there.

This is a pattern we see again and again as we compare genomes of people living in different parts of the world—and it has a very important consequence. Because of common ancestry, genetic diseases (disorders that are inherited) have a lot to do with geography. The risk that an African American man will be afflicted with hypertensive heart disease or prostate cancer is nearly three times greater than that for a European American man, while the European American is far more likely to develop cystic fibrosis or multiple sclerosis.

These differences in DNA variants carry over to genetic differences in how individuals respond to treatment. African Americans, for example, respond poorly to some of the main drugs used to treat heart conditions, such as beta-blockers and angiotensin enzyme inhibitors.

Scientists and doctors that recognize this unfortunate fact are not racists. They fully agree that while using the races of traditional anthropology to pin-point which therapeutic treatment to recommend is an improvement over treating all patients alike, it is still a very poor way of getting at these differences. It would be far better to simply ignore skin color and other single-gene differences and instead perform for each patient a broad "gene variation" analysis. Hopelessly difficult only a few years ago, this now seems an attractive avenue to improve medical treatment for all of us.

It is important to keep clearly in mind the goal of sorting out individual human ancestry, which is to identify common lines of descent that share common response to potential therapies. It is NOT to assign people to over-arching racial categories. This point is being made with great clarity in a course being taught at Pennsylvania State University, where about 90 students had their DNA sampled and compared with four of the five major human groups. Many of these students had thought of themselves as "100% white," but only a few were. One "white" student learned that 14% of his DNA came from Africa, and 6% from East Asia. Similarly, "black" students found that as much as half of their genetic material came from Europe, and a significant amount from Asia as well.

The point is, rigid ideas about the biological basis of identity are wrong. Humanity is much more complex than indicated by a few genes affecting skin color and eye shape. The more clearly we can understand that complexity, the better we can deal with the medical consequences, and the great potential that human diversity provides all of us.

Has Brain Size Increased as Hominids Evolved?

As noted in this chapter, brain size has become progressively larger as hominids evolved. Interestingly, Neanderthal fossils (*left* in the photo below) typically have larger brains than fossils of modern humans (*right* in photo below), about 1,650 cubic centimeters (cc) for *Homo neanderthalensis* versus about 1,500 cc for *H. sapiens*. Does this suggest that Neanderthals were smarter than us?

The graph to the right explores the evolution of hominid brain size by plotting the age of each major type of hominid versus its brain size (that is, the volume of the skull cranium's interior). For each type of hominid, there is some variation in cranial volume among the fossils that have been described, and a typical value is presented (the number in parentheses by each point). The value for *H. neanderthalensis*, for example, is plotted as a typical 1,650 cc, even though a skull found in the Amud cave of Israel (see page 459) is 90 cc larger. Some paleontologists consider *H. ergaster* to be a variant of *H. erectus*, and *H. heidelbergensis* and *H. neanderthalensis* to be variants of *H. sapiens*, but for the sake of this analysis, the "splitters" view is presented. While the question used to be controversial, most anthropologists now feel that *H. neanderthalensis* and *H. sapiens* are separate species, both descended from *H. heidelbergensis* (however it is named).

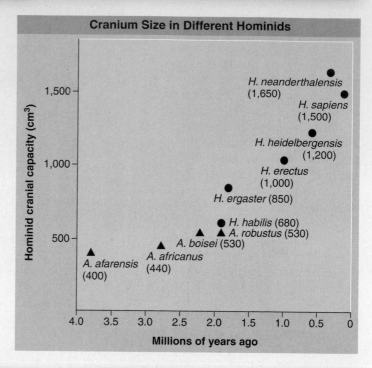

Cranium Size in Different Hominids

H. neanderthalensis (1,650)
H. sapiens (1,500)
H. heidelbergensis (1,200)
H. erectus (1,000)
H. ergaster (850)
H. habilis (680)
A. robustus (530)
A. boisei (530)
A. africanus (440)
A. afarensis (400)

Hominid cranial capacity (cm³) — 500, 1,000, 1,500

Millions of years ago — 4.0, 3.5, 3.0, 2.5, 2.0, 1.5, 1.0, 0.5, 0

Neanderthal

Homo sapiens

1. **Applying Concepts** In the graph, what is the dependent variable?
2. **Interpreting Data**
   a. Which human species of *Homo* has the biggest brain? The smallest?
   b. Which australopithecine has the biggest brain? The smallest?
   c. Does any australopithecine have a brain as large as a human?
3. **Making Inferences**
   a. Over 2 million years, does the brain size of australopithecines change? How much? What percent increase is this?
   b. Over 2 million years, does the brain size of humans change? How much? What percent increase is this?
4. **Drawing Conclusions**
   a. Does brain size appear to have evolved faster in the genus *Homo* than in the genus *Australopithecus*? How much faster?
   b. Given the clear and undisputed larger brain size of Neanderthals, and the conclusion you have drawn in **4a,** does this allow you to further conclude that Neanderthals were smarter than today's humans?
5. **Further Analysis** What key unverified assumption does this conclusion depend upon? If you do not accept this further conclusion, why then do you think brain size has evolved as rapidly as it has in the genus *Homo*?

The Evolution of Primates

21.1 The Evolutionary Path to Humans

- The history of primate evolution begins with a group of nocturnal, tree-dwelling, insect-eating mammals called the Archonta. The adaptive radiation of these animals beginning about 65 million years ago gave rise to many different mammals, including bats, tree shrews, and primates.

- Primates are mammals with two key features: grasping fingers and toes and binocular vision. The combination of these characteristics allowed primates to succeed as arboreal insectivores.

- About 40 million years ago, early primates split into two groups: the prosimians and the anthropoids. The anthropoids include monkeys, apes, and humans (**figure 21.1**).

- The anthropoids split further. In South America, they evolved into the New World monkeys, which are tree-dwelling monkeys with flat noses and prehensile tails. In Africa, anthropoids evolved into the Old World monkeys and hominoids.

21.2 How the Apes Evolved

- Hominoids include the apes (the gibbon, orangutan, gorilla, and chimpanzee) and the hominids, which are humans (*Homo sapiens*) and their direct but extinct ancestors. *H. sapiens*'s closest living relative is the chimpanzee.

- The divergence of apes and hominids occurred as they adapted different modes of locomotion. Hominids became bipedal, walking upright. Many of the changes in the skeletons of hominids compared to apes reflects this difference in locomotion (**figure 21.2**).

The First Hominids

21.3 Walking Upright

- Bipedalism is a key adaptation that contributed to hominid evolution (**figure 21.4**), but how or why this method of locomotion evolved is not clear.

- Bipedalism is not unique to humans, but walking upright and larger brain size set *Homo* on a new evolutionary path. Fossil evidence indicates that bipedalism preceded an increased brain size, but the question of *why* bipedalism evolved still hasn't been answered.

21.4 The Hominid Family Tree

- The hominid evolutionary tree is not complete and is interpreted differently (**figure 21.5**). Scientists nicknamed "lumpers" tend to group specimens together based on common characteristics, whereas scientists dubbed "splitters" focus on the differences in specimens and tend to recognize more individual species.

The First Humans

21.5 African Origin: Early *Homo*

- Early humans, genus *Homo,* evolved from ancestors of *Australopithecus* about 2 million years ago. *Homo habilis* found in Africa closely resembles *Australopithecus* except *H. habilis* has a larger brain (**figure 21.6**). *Homo rudolfensis,* also found in Africa, is clearly human, with a larger brain and more humanlike skull features. *Homo ergaster* (**figure 21.7**) was designated as a third species of early humans because it is more humanlike than are *H. habilis* and *H. rudolfensis. H. ergaster* is thought to be the most likely ancestor of later species of *Homo.*

21.6 Out of Africa: *Homo erectus*

- *Homo erectus* is without a doubt a true human species. Java man was the first fossil specimen of *H. erectus* found (**figure 21.8**). His leg bone indicated that he was bipedal, his skull indicated a brain about twice the size of *Australopithecus,* and the fossil was about 500,000 years old.

- Peking man found later in China resembled Java man, lending support to the evolutionary position of *H. erectus. H. erectus* fossils were also found in Africa, but the African fossils proved to be 1.5 million years old. This suggested that *H. erectus* originated in Africa and migrated out to Europe and Asia.

Modern Humans

21.7 Our Own Species Also Evolved in Africa

- Modern humans first appeared in Africa about 600,000 years ago. Some scientists suggest that a total of three species of modern human evolved: *Homo heidelbergensis, H. neanderthalensis,* and *H. sapiens. H. heidelbergensis* evolved in Africa but migrated to Europe and western Asia following the path that *H. erectus* took earlier.

- *H. neanderthalensis* appeared in Europe about the time *H. erectus* was becoming rarer. *H. neanderthalensis* is likely a branch off of the ancestral line of modern humans. *H. sapiens* appeared in Africa about 130,000 years ago. Like *H. erectus* and *H. heidelbergensis, H. sapiens* migrated out of Africa to Europe and Asia and eventually to Australia and North America (**figure 21.9**).

21.8 The Only Surviving Hominid

- The Neanderthals (*H. neanderthalensis*) were prevalent in Europe and Asia but abruptly disappeared about 34,000 years ago, being replaced by *H. sapiens* called Cro-Magnons. Cro-Magnons were sophisticated tool users, had social organization and language, and were hunters.

- The human species continues to evolve, with different adaptations appearing in different populations in response to local conditions (**figure 21.10**).

1. Although many mammals have binocular vision, the anatomical adaptation(s) that set primates apart from these mammals is/are
   a. prehensile tails.
   b. opposable digits on hands.
   c. mammary glands.
   d. hair-covered skin.
2. The earliest humans were thought to have evolved in
   a. Africa.
   b. Asia.
   c. Australia.
   d. Europe.
3. Anthropoids are primates that include all of the following except
   a. monkeys.
   b. apes.
   c. lemurs.
   d. humans.
4. Which of the following anatomical characteristics of hominids contributed to bipedalism?
   a. longer and heavier lower limbs
   b. curved vertebral column
   c. bowl-shaped pelvis
   d. All of the above.
5. What feature appeared first in early human ancestors?
   a. language
   b. increased brain size
   c. tool use
   d. bipedalism

6. A characteristic used to differentiate between *Australopithecus* and *Homo* is
   a. brain size.   c. walking upright.
   b. presence in Africa.   d. All of these are correct.
7. Some scientists believe the first hominid to be
   a. *Australopithecus boisei.*
   b. *Australopithecus anamensis.*
   c. *Australopithecus afarensis.*
   d. *Australopithecus robustus.*
8. Which of the following most closely resembles the australopithecines?
   a. *Homo sapiens*
   b. *Homo heidelbergensis*
   c. *Homo neanderthalensis*
   d. *Homo habilis*
9. The first hominid to migrate extensively to Europe and Asia was
   a. *Homo sapiens.*
   b. *Homo heidelbergensis.*
   c. *Homo neanderthalensis.*
   d. *Homo erectus.*
10. DNA and chromosomal studies seem to indicate that *Homo sapiens* originated in
    a. many different regions, wherever *Homo erectus* was found.
    b. Africa.
    c. Asia.
    d. Europe.

1. **Figure 21.2** Explain how the skeletal characteristics of an australopithecine improve the ability to stand and walk upright for long periods.

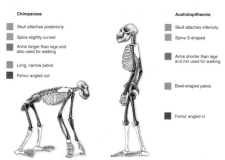

2. **Figure 21.9** Genomic evidence supports the view that modern humans migrated from Africa and spread throughout the globe, not once but several times, each time adapting to local environmental challenges. To what degree does the fossil evidence support this view? Explain.

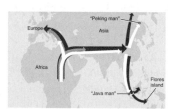

1. As early hominids evolved, drying and cooling of the climate caused expansion of African savannas. What evidence would you accept that this change favored upright walking? What are some of the advantages of bipedalism on the savanna?
2. Once common in Australia, a huge flightless bird called *Genyornis newtoni* is now extinct. Isotopic dating of its fossilized eggshells indicate that no *Genyornis* eggs are younger than 50,000 years old. Humans colonized Australia 50,000 years ago. What types of evidence would be needed to support the hypothesis that humans hunted this flightless bird to extinction?

3. Mitochondrial DNA (from mitochondria in the egg and so not subject to recombination) is passed, essentially unchanged, from a mother to all her children through the generations. Likewise, the genes on the Y chromosome (also not subject to recombination) are passed from father to son. Would you expect studies of mitochondrial DNA variation and of Y chromosome variation to both yield the same result when researchers attempt to assess the hypothesis that all human species evolved in Africa?

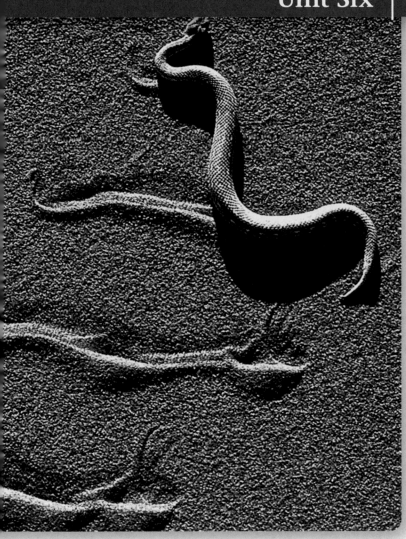

22

The Animal Body and How It Moves

When most people think of animals, they think of their pet dogs and cats, and of the animals that they've seen in a zoo, on a farm, in an aquarium, or out in the wild. Despite the differences among these animals, they all share the same basic body plan, with the same sorts of tissues and organs operating in much the same way. In this chapter, we will begin a detailed consideration of the biology of animals and of the fascinating structure and function of their bodies. After examining the major tissues of the vertebrate body, we will end the chapter by focusing in on how the animal body achieves a complex activity, locomotion, by coordinating the activities of nerves, muscles, and bones to propel itself through its environment. Animals are unrivaled among the inhabitants of the living world in their ability to move about from one place to another. Under water, on land, and in the air, animals swim, burrow, crawl, slither, slide, walk, jump, run, glide, soar, and fly. This sidewinder rattlesnake can move surprisingly rapidly over the desert sand by a coordinated series of muscle contractions, throwing its long body into a series of sinuous curves. On land, only vertebrates and arthropods have developed a means of rapid surface locomotion.

22.1 Innovations in Body Design

As described in chapter 19, six evolutionary innovations in the design of animals's bodies have led to the diversity seen in the kingdom Animalia.

1. Tissues

The first innovation in body design was the development of tissues (❶ in **table 22.1**). The simplest animals, the sponges, have some cell specialization but mostly lack tissues; a few seem to possess rudimentary tissues. All other animals possess tissues and highly specialized cells.

2. Radial Versus Bilateral Symmetry

All sponges lack any definite symmetry, instead growing asymmetrically as irregular masses. Symmetrical bodies, ❷ in the table, first evolved in cnidarians (jellyfish, sea anemones, and corals) and ctenophores (comb jellies). The bodies of these two types of animals exhibit **radial symmetry,** in which the parts of the body are arranged around a central axis.

The bodies of all other animals are characterized by **bilateral symmetry,** in which the body can be divided into left and right halves along only one plane. Bilateral symmetry allows parts of the body to evolve in different ways, permitting the segregation of organs in different parts of the body. In some higher animals like echinoderms (sea stars, also called starfish), the adults are radially symmetrical, but even in them, the larvae are bilaterally symmetrical.

3. Solid Body Versus Body Cavity

Another transition in the evolution of the animal body plan, shown in ❸ in the table, was the evolution of the body cavity. A body cavity allowed the evolution of efficient organ systems, more rapid circulation of fluids within the body, and complex movement of the animal body.

In the animal kingdom, we see three body arrangements: Some animals called **acoelomates** have no body cavity, whereas others have either of two different types of body cavities, distinguished primarily by where they develop within the three embryonic layers. A body cavity that forms between the endoderm and the mesoderm is called a pseudocoel, and the animals in which it occurs are called **pseudocoelomates.** A body cavity that forms entirely within the mesoderm is called a coelom, and animals in which it occurs are called **coelomates.**

4. Nonsegmented Versus Segmented Body

A further transition in animal body plan involved the subdivision of the body into **segments,** ❹ in **table 22.1**. Segmentation permits the following evolutionary advantages:

1. In annelids and other highly segmented animals, each segment may go on to develop a more or less complete set of adult organ systems. Damage to any one segment need not be fatal to the individual because the other segments duplicate that segment's functions.
2. Locomotion is far more effective when individual segments can move independently because the animal as a whole has more flexibility of movement. Segmentation can be seen in the bodies of all annelids, arthropods, and chordates, although it is not always obvious.

5. Incremental Growth Versus Molting

Most coelomate animals grow the way you do, by gradually adding mass to their body. Nematodes and arthropods grow very differently. Their bodies are encased within a hard exoskeleton, and this exterior shell must be shed periodically in a process called molting (❺ in **table 22.1**) to make way for a larger body.

6. Protostomes Versus Deuterostomes

The echinoderms (sea stars) and the chordates (vertebrates) have a series of key embryological features different from those shared by other animal phyla. Because it is extremely unlikely that these features evolved more than once, it is believed that these two seemingly quite different phyla share a common ancestor. They are members of a group called the **deuterostomes** (❻ in the table). All other coelomate animals are called **protostomes.** Deuterostomes differ fundamentally from protostomes in three aspects of embryonic growth.

1. **How cleavage forms a hollow ball of cells.** Deuterostomes differ from protostomes in one of the earliest steps of development, the divisions that determine the plane in which the cells divide. Most protostomes undergo spiral cleavage, whereas deuterostomes undergo radial cleavage.
2. **How the blastopore determines the body axis.** Deuterostomes differ from protostomes in the way in which the embryo grows. The blastopore of a protostome becomes the animal's mouth, and the anus develops at the other end. In a deuterostome, by contrast, the blastopore becomes the animal's anus, and the mouth develops at the other end.
3. **How the developmental fate of the embryo is fixed.** Most protostomes undergo determinate cleavage, which rigidly fixes the developmental fate of each cell very early. No one cell isolated at even the four-cell stage can go on to form a normal individual. In marked contrast, deuterostomes undergo indeterminate cleavage, with each cell retaining the capacity to develop into a complete individual.

Key Learning Outcome 22.1 Six key innovations in animal body design set the stage for the great diversity in the animal kingdom.

TABLE 22.1   INNOVATIONS IN BODY DESIGN

Innovation	Difference in Design	Organisms
❶ Specialized cells versus tissues No true tissues Tissues	In most sponges, specialized cell types exist, but they are not organized into tissues. In all other animals, cells are organized into tissues, and through intercellular coordination, these tissues are integrated with other cells and tissues of the organism.	No tissues: most sponges; tissues: all other animals
❷ Radial versus bilateral symmetry	In radial symmetry, the parts of the body are arranged around a central axis, usually a mouth. In bilateral symmetry, the body is divided into right and left halves that are mirror images of each other. Bilateral symmetry results in specialization in areas of the body such as a head region.	Radial symmetry: cnidarians and adult echinoderms; bilateral symmetry: all other eumetazoan animals
❸ Solid body versus body cavity Ectoderm Mesoderm Endoderm Ectoderm Mesoderm Endoderm Pseudocoel Ectoderm Mesoderm Coelom Endoderm	The evolution of a body cavity resulted in the expansion of organ systems in the body, supported within the body cavity. Animals that lack a body cavity are acoelomates; the body cavity of pseudocoelomates forms between germ layers; the body cavity of coelomates forms within the mesoderm.	Acoelomates: flatworms; pseudocoelomates: roundworms; coelomates: mollusks, annelids, arthropods, echinoderms, and chordates
❹ Nonsegmented versus segmented body Segments	In segmented animals, the body is subdivided into compartments, called segments. Segmentation resulted in redundant structures that serve as "backups" if some segments are damaged. Segmented animals also move around more efficiently.	Nonsegmented coelomate animals: mollusks and echinoderms; segmented animals: annelids, arthropods, and chordates
❺ Incremental growth versus molting Grows by molting Grows by incremental growth	Because their bodies are encased in rigid exoskeletons, arthropods and roundworms (nematodes) must shed their hard exterior exoskeletons periodically as they grow to accommodate the increase in body mass, a process of molting more formally called ecdysis. This contrasts with the gradual incremental body growth typical of humans and all other coelomate animals.	Molting: arthropods and nematodes; incremental growth: all other coelomates
❻ Protostome versus deuterostome development Spiral cleavage Radial cleavage Blastopore becomes mouth Blastopore becomes anus Protostomes Deuterostomes	In protostome development, the cells undergo spiral cleavage, the blastopore becomes the mouth, and the developmental fate of each embryonic cell is determined early. In deuterostome development, cells undergo radial cleavage, the blastopore becomes the anus, and the developmental fate of each embryonic cell is flexible (each cell retains the capacity to develop into a complete individual).	Deuterostomes: echinoderms and chordates; protostomes: all other bilaterally symmetrical animals

22.2 Organization of the Vertebrate Body

All vertebrates have the same general architecture: a long internal tube that extends from mouth to anus that is suspended within an internal body cavity called the *coelom.* The coelom of many terrestrial vertebrates is divided into two parts: the *thoracic cavity,* which contains the heart and lungs, and the *abdominal cavity,* which contains the stomach, intestines, and liver. The vertebrate body is supported by an internal scaffold, or skeleton, made up of jointed bones. A bony skull surrounds and protects the brain, while a column of bones, the vertebrae, surrounds the spinal cord.

Like all animals, the vertebrate body is composed of cells—over 10 to 100 trillion of them in your body. It's difficult to picture how large this number actually is. A line of 10 trillion cars would stretch from the earth to the sun and back 50 million times! Not all of these cells in your body are the same, of course. If they were, we would not be bodies but amorphous blobs. Vertebrate bodies contain over 100 different kinds of cells.

Tissues

Groups of cells of the same type are organized within the body into **tissues,** which are the structural and functional units of the vertebrate body. A tissue is a group of cells of the same type that performs a particular function in the body.

Tissues form as the vertebrate body develops. Early in development, the growing mass of cells that will become a mature animal differentiates into three fundamental layers of cells: endoderm, mesoderm, and ectoderm. These three kinds of embryonic cell layers, in turn, differentiate into the more than 100 different kinds of cells in the adult body.

It is possible to assemble many different kinds of tissue from 100 cell types, but biologists have traditionally grouped adult tissues into four general classes: *epithelial, connective, muscle,* and *nerve tissue.* The bird pictured in **figure 22.1** contains all four classes of tissues, and as you can see in the circled enlargements, each class of tissue contains different types of cells. Of these, connective tissues (indicated by the light green arrows) are particularly diverse.

Organs

Organs are body structures composed of several different tissues grouped together into a larger structural and functional unit, just as a factory is a group of people with different jobs who work together to make something. The heart is an organ. It contains cardiac muscle tissue wrapped in connective tissue and joined to many nerves. All of these tissues work together to pump blood through the body: The cardiac muscles contract, which squeezes the heart to push the blood; the connective tissues act as a bag to hold the heart in the proper shape and ensure that the different chambers of the heart squeeze in the proper order; and the nerves control the rate at which the heart beats. No single tissue can do the job of the heart, any more than one piston can do the job of an automobile engine.

You are probably familiar with many of the major organs of a vertebrate body. Lungs are organs that terrestrial vertebrates use to extract oxygen from the air. Fish use gills to accomplish the same task from water. The stomach is an organ that digests food, and the liver an organ that controls the level of sugar and other chemicals in the blood. Organs are the machines of the vertebrate body, each built from several different tissues and each doing a particular job. How many others can you name?

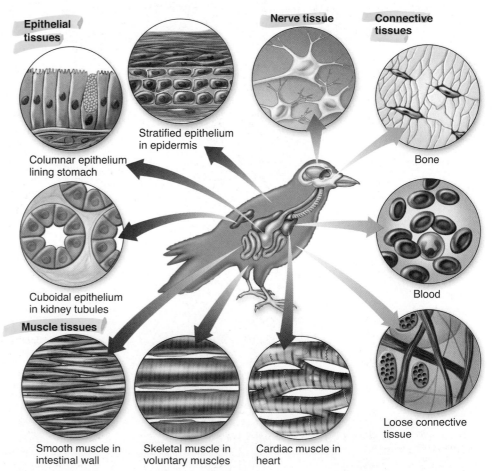

Epithelial tissues

Columnar epithelium lining stomach

Stratified epithelium in epidermis

Cuboidal epithelium in kidney tubules

Muscle tissues

Smooth muscle in intestinal wall

Skeletal muscle in voluntary muscles

Cardiac muscle in heart

Nerve tissue

Connective tissues

Bone

Blood

Loose connective tissue

Figure 22.1 Vertebrate tissue types.
The four basic classes of tissue are epithelial, nerve, connective, and muscle.

Organ Systems

An **organ system** is a group of organs that work together to carry out an important function. For example, the vertebrate digestive system is an organ system composed of individual organs that break up food (beaks or teeth), pass the food to the stomach (esophagus), break down the food (stomach and intestine), absorb the food (intestine), and expel the solid residue (rectum). If all of these organs do their job right, the body obtains energy and necessary building materials from food. The digestive system is a particularly complex organ system with many different organs consisting of many different types of cells, all working together to carry out a complex function. The circulatory system illustrated in **figure 22.2** involves fewer different types of organs, but the level of organization is the same—organ systems are made up of organs, that are made up of tissues, that are made up of cells.

The vertebrate body contains 11 principal organ systems:

1. **Skeletal.** Perhaps the most distinguishing feature of the vertebrate body is its bony internal skeleton.

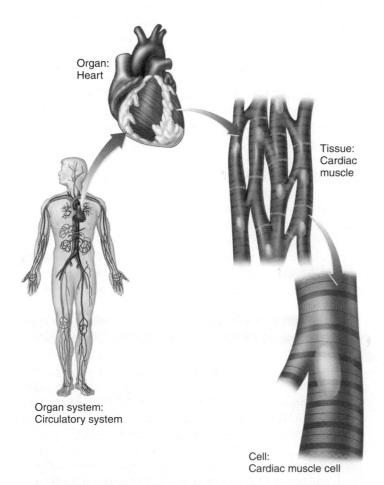

Organ:
Heart

Tissue:
Cardiac
muscle

Organ system:
Circulatory system

Cell:
Cardiac muscle cell

Figure 22.2 Levels of organization within the vertebrate body.

Similar cell types operate together and form tissues. Tissues functioning together form organs. Several organs working together to carry out a function for the body are called an organ system. The circulatory system is an example of an organ system.

The skeletal system protects the body and provides support for locomotion and movement. Its principal components are bones, skull, cartilage, and ligaments. Like arthropods, vertebrates have jointed appendages—the arms, hands, legs, and feet.

2. **Circulatory.** The circulatory system transports oxygen, nutrients, and chemical signals to the cells of the body and removes carbon dioxide, chemical wastes, and water. Its principal components are the heart, blood vessels, and blood.

3. **Endocrine.** The endocrine system coordinates and integrates the activities of the body through the release of hormones. Its principal components are the pituitary, adrenal, thyroid, and other ductless glands.

4. **Nervous.** The activities of the body are coordinated by the nervous system. Its principal components are the nerves, sense organs, brain, and spinal cord.

5. **Respiratory.** The respiratory system captures oxygen and exchanges gases and is composed of the lungs, trachea, and other air passageways.

6. **Immune and lymphatic.** The immune system removes foreign bodies from the bloodstream using special cells, such as lymphocytes, macrophages, and antibodies. The lymphatic system provides vessels that transport extracellular fluid and fats to the circulatory system but also provides sites (lymph nodes and thymus, tonsils, and spleen) for the storage of immune cells.

7. **Digestive.** The digestive system captures soluble nutrients from ingested food. Its principal components are the mouth, esophagus, stomach, intestines, liver, and pancreas.

8. **Urinary.** The urinary system removes metabolic wastes from the bloodstream. Its principal components are the kidneys, bladder, and associated ducts.

9. **Muscular.** The muscular system produces movement, both within the body and of its limbs. Its principal components are skeletal muscle, cardiac muscle, and smooth muscle.

10. **Reproductive.** The reproductive system carries out reproduction. Its principal components are the testes in males, ovaries in females, and associated reproductive structures.

11. **Integumentary.** The integumentary system covers and protects the body. Its principal components are the skin, hair, nails, and sweat glands.

Key Learning Outcome 22.2   Groups of cells of the same type are organized in the vertebrate body into tissues. Organs are body structures composed of several different tissues. An organ system is a group of organs that work together to carry out an important function.

22.3 Epithelium Is Protective Tissue

We begin our discussion of tissues on the outside. Epithelial cells are the guards and protectors of the body. They cover its surface and determine which substances enter it and which do not. The organization of the vertebrate body is fundamentally tubular, with one tube (the digestive tract) suspended inside another (the body cavity or coelom) like an inner tube inside a tire. The outside of the body is covered with cells (skin) that develop from embryonic *ectoderm* tissue; the body cavity is lined with cells that develop from embryonic *mesoderm* tissue; and the hollow inner core of the digestive tract (the gut) is lined with cells that develop from embryonic *endoderm* tissue. All three germ layers give rise to epithelial cells. Although different in embryonic origin, all epithelial cells are broadly similar in form and function and together are called the **epithelium.**

The body's epithelial layers function in three ways:

1. They *protect underlying tissues* from dehydration (water loss) and mechanical damage. Because epithelium encases all the body's surfaces, every substance that enters or leaves the body must cross an epithelial layer, even one as thick as the gila monster's in **figure 22.3**.
2. They *provide sensory surfaces.* Many of a vertebrate's sense organs are in fact modified epithelial cells.
3. They *secrete materials.* Most secretory glands are derived from pockets of epithelial cells that pinch together during embryonic development.

Types of Epithelial Cells and Epithelial Tissues

Epithelial cells are classified into three types according to their shapes: squamous, cuboidal, and columnar. Layers of epithelial tissue are usually only one or a few cells thick. Individual epithelial cells possess only a small amount of cytoplasm and have a relatively low metabolic rate. A characteristic of all epithelia is that sheets of cells are tightly bound together with very little space between them. This forms the barrier that is key to the functioning of the epithelium.

Epithelium possesses remarkable regenerative abilities. The cells of epithelial layers are constantly being replaced throughout the life of the organism. The cells lining the digestive tract, for example, are continuously replaced every few days. The epidermis, the epithelium that forms the skin, is renewed every two weeks. The liver, which is a football-sized gland formed of epithelial tissue, can readily regenerate substantial portions of itself removed during surgery.

There are two general kinds of epithelial tissue. First, the membranes that line the lungs and the major cavities of the body are a **simple epithelium** only a single cell layer thick. The first three entries in **table 22.2** (❶, ❷, ❸) are simple epithelium.

Figure 22.3 The epithelium prevents dehydration.
The tough, scaly skin of this gila monster provides a layer of protection against dehydration and injury. For all land-dwelling vertebrates, the relative impermeability of the surface epithelium (the epidermis) to water offers essential protection from dehydration and from airborne pathogens (disease-causing organisms).

When you read the functions of each, you can see why these layers are only one cell thick—these are surfaces across which many materials must pass, entering and leaving the body's compartments, and it is important that the "road" into and out of the body not be too long. Second, the skin, or epidermis, is a **stratified epithelium** composed of more complex epithelial cells several layers thick, ❹ in the table. Several layers are necessary to provide adequate cushioning and protection and to enable the skin to continuously replace its cells. The epithelium that lines parts of the respiratory tract ❺ also looks like stratified epithelium, but it is actually a single layer of **pseudostratified epithelium.** It looks like several layers because the nuclei are positioned in different places in the cells and so give the appearance of several layers of cells.

A type of simple epithelial tissue that has a secretory function is cuboidal epithelium, which is found in the **glands** of the body. Endocrine glands secrete hormones into the blood. Exocrine glands (those with ducts that open to the body's outside) secrete sweat, milk, saliva, and digestive enzymes out of the body. Exocrine glands also secrete digestive enzymes into the stomach. If you think about it, the stomach and digestive tract are *outside* the body, because they are the inner canal that passes right through the body. It is possible for a substance to pass all the way through this digestive tract, from mouth to anus, and never enter the body at all. A substance must cross an epithelial layer to truly enter the body.

Key Learning Outcome 22.3 Epithelial tissue is the protective tissue of the vertebrate body. In addition to providing protection and support, vertebrate epithelial tissues secrete key materials.

find tissue types + functions (handwritten)

TABLE 22.2 EPITHELIAL TISSUE

Tissue	Typical Location	Tissue Function
Simple Epithelium		
❶ Squamous Simple squamous epithelial cell Nucleus *round* (handwritten)	Lining of lungs, capillary walls, and blood vessels	Flat and thin cells; provides a thin layer across which diffusion can readily occur; the cells when viewed from the surface look like tiles on a floor
❷ Cuboidal Cuboidal epithelial cells Nucleus *round* (handwritten) Cytoplasm *cube* (handwritten)	Lining of some glands and kidney tubules; covering of ovaries	Cells rich in specific transport channels; functions in secretion and specific absorption
❸ Columnar Columnar epithelial cells Nucleus Goblet cell *elongated* (handwritten)	Surface lining of stomach, intestines, and parts of respiratory tract	Thicker cell layer; provides protection and functions in secretion and absorption
Stratified Epithelium		
❹ Squamous Stratified squamous cells Nuclei	Outer layer of skin; lining of mouth	Tough layer of cells; provides protection
Pseudostratified Epithelium		
❺ Columnar Cilia Pseudo–stratified columnar cell Goblet cell	Lining of parts of respiratory tract	Functions in secretion of mucus; dense with cilia (small, hairlike projections) that aid in movement of mucus; provides protection

22.4 Connective Tissue Supports the Body

The cells of **connective tissue** provide the vertebrate body with its structural building blocks and also with its most potent defenses. Derived from the mesoderm, these cells are sometimes densely packed together, and sometimes widely dispersed, just as the soldiers of an army are sometimes massed together in a formation and sometimes widely scattered as guerrillas. Connective tissue cells fall into three functional categories: (1) the cells of the immune system, which act to defend the body; (2) the cells of the skeletal system, which support the body; and (3) the blood and fat cells, which store and distribute substances throughout the body. While grouping these diverse types of cells together may seem odd, all connective tissues share a common structural feature: They all have abundant extracellular "matrix" material between widely spaced cells.

Immune Connective Tissue

The cells of the immune system, the many kinds of so-called "white blood cells" (❶ in **table 22.3**), roam the body within the bloodstream. They are mobile hunters of invading microorganisms and cancer cells. The two principal kinds of immune system cells are **macrophages,** which engulf and digest invading microorganisms (as shown in the photo in **table 22.3**), and **lymphocytes,** which make antibodies or attack virus-infected cells. Immune cells are carried through the body in a fluid matrix called *plasma.*

Skeletal Connective Tissue

Three kinds of connective tissue are the principal components of the skeletal system: fibrous connective tissue, cartilage, and bone. Although composed of similar cells, they differ in the nature of the matrix that is laid down between individual cells.

1. **Fibrous connective tissue.** The most common kind of connective tissue in the vertebrate body, fibrous connective tissue ❷, is composed of flat, irregularly branching cells called **fibroblasts** that secrete structurally strong proteins into the spaces between the cells. The different types of proteins they contain give these tissues different strengths. *Loose connective tissue,* pictured in **table 22.3,** provides the least amount of strength. *Dense connective tissue* is very strong, while *elastic connective tissue* provides strength and the ability to stretch and recoil. The most commonly secreted protein, collagen, is the most abundant protein in the human body—in fact, one-quarter of all the protein in your body is collagen! Fibroblasts are also active in wound healing; scar tissue, for example, possesses a collagen matrix.

2. **Cartilage.** In cartilage ❸, a collagen matrix between cartilage cells (technically called **chondrocytes**) forms in long parallel arrays along the lines of mechanical stress. What results is a firm and flexible tissue of great strength, just as strands of nylon molecules laid down in long, parallel arrays produce strong, flexible ropes. Cartilage makes up the entire skeletal system of the modern agnathans and cartilaginous fishes. In most adult vertebrates, however, cartilage is restricted to the articular (joint) surfaces of bones that form freely movable joints and to other specific locations.

3. **Bone.** Bone ❹ is similar to cartilage, except that the collagen fibers are coated with a calcium phosphate salt, making the tissue rigid. The structure of bone and the way it is formed are discussed shortly.

Storage and Transport Connective Tissue

The third general class of connective tissue is made up of cells that are specialized to accumulate and transport particular molecules. They include the fat-accumulating cells of **adipose tissue** ❺. The large, seemingly empty areas in the adipose tissue in **table 22.3** are actually fat-containing vacuoles within the adipose cells. Red blood cells ❻, called **erythrocytes,** also function in transport and storage. About 5 billion erythrocytes are present in every milliliter of your blood. Erythrocytes transport oxygen and carbon dioxide in blood. They are unusual in that during their maturation they lose most of their organelles, including the nucleus, mitochondria, and endoplasmic reticulum. Instead, occupying the interior of each erythrocyte are about 300 million molecules of hemoglobin, the protein that carries oxygen.

The fluid, or **plasma,** in which erythrocytes move is both the "banquet table" and the "refuse heap" of the vertebrate body. Practically every substance used by cells is dissolved in plasma, including inorganic salts like sodium and calcium, body wastes, and food molecules like sugars, lipids, and amino acids. Plasma also contains a wide variety of proteins, including antibodies and albumin, which gives the blood its viscosity.

A Closer Look at Bone

The vertebrate endoskeleton is strong because of the structural nature of bone. Bone consists of living bone cells embedded within an inert matrix composed of the structural protein collagen and several minerals, including a calcium phosphate salt called *hydroxyapatite.* Bone is produced by coating collagen fibers with hydroxyapatite; the result is a material that is strong without being brittle. To understand how coating collagen fibers with calcium salts makes such an ideal structural material, consider fiberglass. Fiberglass is composed of glass fibers embedded in epoxy glue. The individual fibers are rigid, giving great strength, but they are also brittle. The epoxy glue, on the other hand, is flexible but weak. The composite, fiberglass, is both strong and flexible because when stress causes an individual fiber to break, the crack runs into glue before it reaches another fiber. The glue distorts and reduces the concentration of the stress—in effect, the glue spreads the stress over many fibers.

The construction of bone is similar to that of fiberglass: Small, needle-shaped crystals of a calcium phosphate mineral,

TABLE 22.3 CONNECTIVE TISSUE

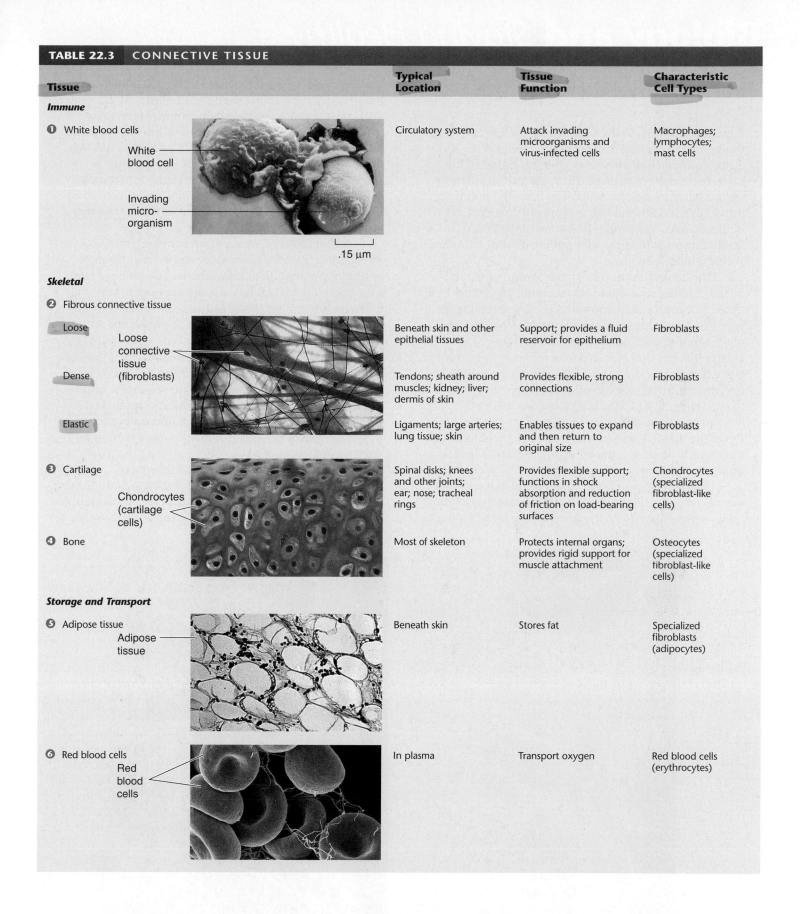

Tissue		Typical Location	Tissue Function	Characteristic Cell Types
Immune				
❶ White blood cells		Circulatory system	Attack invading microorganisms and virus-infected cells	Macrophages; lymphocytes; mast cells
White blood cell				
Invading micro-organism				
.15 μm				
Skeletal				
❷ Fibrous connective tissue				
Loose	Loose connective tissue (fibroblasts)	Beneath skin and other epithelial tissues	Support; provides a fluid reservoir for epithelium	Fibroblasts
Dense		Tendons; sheath around muscles; kidney; liver; dermis of skin	Provides flexible, strong connections	Fibroblasts
Elastic		Ligaments; large arteries; lung tissue; skin	Enables tissues to expand and then return to original size	Fibroblasts
❸ Cartilage	Chondrocytes (cartilage cells)	Spinal disks; knees and other joints; ear; nose; tracheal rings	Provides flexible support; functions in shock absorption and reduction of friction on load-bearing surfaces	Chondrocytes (specialized fibroblast-like cells)
❹ Bone		Most of skeleton	Protects internal organs; provides rigid support for muscle attachment	Osteocytes (specialized fibroblast-like cells)
Storage and Transport				
❺ Adipose tissue	Adipose tissue	Beneath skin	Stores fat	Specialized fibroblasts (adipocytes)
❻ Red blood cells	Red blood cells	In plasma	Transport oxygen	Red blood cells (erythrocytes)

hydroxyapatite, surround and impregnate collagen fibers within bone. No crack can penetrate far into bone because any stress that breaks a hard hydroxyapatite crystal passes into the collagenous matrix, which dissipates the stress. The hydroxyapatite mineral provides rigidity, whereas the collagen "glue" provides flexibility.

Most of us think of bones as solid and rocklike. But actually, bone is a dynamic tissue that is constantly being

Biology and *Staying Healthy*

Losing Bone: Osteoporosis

Running, jumping, swimming, flying, even squirming through the mud—all are things that animals do. A distinguishing characteristic of animals is this ability to move from place to place. Animals are the only true multicellular organisms capable of traveling around or through their environment. As you will discover later in this chapter, vertebrates like us accomplish this feat through the action of muscles pulling against an internal skeleton made of bone.

The bones that make up your skeleton are not lifeless, like the scaffolding of a building, but rather living tissue that continues to grow and change throughout your life. Just because you reach your full height does not mean that your bones stop growing. They stop elongating, but that is merely one aspect of bone growth. Every day of your life, your bones are adding new tissue and breaking down old tissue, reshaping your skeleton to meet the needs of your body. If you break a bone, your body heals the fracture by adding new bone tissue at the site of the break. However, when the process of forming new bone is impaired, the bone becomes weak and unable to heal itself.

A striking example of the impact of this failure of bone healing is exhibited by the little old lady that you might see walking down the street with a hunched back.

The bump on her back is caused by an imbalance in bone metabolism, rather than poor posture. As a person ages, the backbone, hips, and other bones tend to lose bone mass much faster than new bone mass is produced. Excessive bone loss is a condition called *osteoporosis* (also discussed on page 484). A section of normal bone is shown on the *left* below. A section of bone from a patient with osteoporosis is shown on the *right* below. You can see how much less bone tissue is present in the patient with osteoporosis.

Unfortunately for the person suffering from osteoporosis, even everyday activities such as bending or lifting can result in a broken bone. The little old lady with the humped back has already experienced multiple bone fractures in her spine. Although we often refer to it as the "backbone," the spine is not one bone but rather a column of many smaller bones stacked one on top of the other, like building blocks. The spine holds the upper body erect, and in a person with osteoporosis, just the stress of the body's weight on the weakened bones causes small fractures up and down the spine. Multiple spine fractures can result in a curvature of the back and a stooped posture. A loss of height of an inch or more is often an early sign of osteoporosis.

How does a person know if he or she has osteoporosis? Unfortunately, a fractured bone is usually the primary warning sign of osteoporosis. There are diagnostic tests that can check a person's bone density, but they often

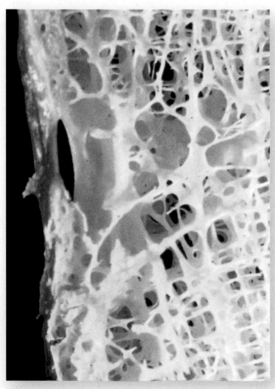

(a) Normal bone tissue

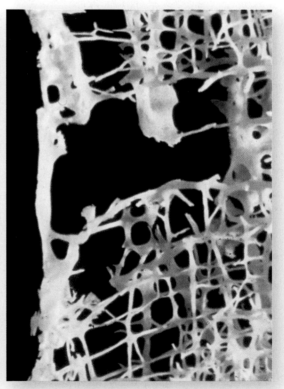

(b) Advanced osteoporosis

aren't done unless a person experiences a fracture. A bone density test should be performed when a person over age 45 breaks a bone or in a person less than age 45 who has had several fractures. The test used to diagnose osteoporosis, a bone density test, measures the density, or mass, of the bones. The hips, spine, and wrists are often used for evaluation because these are the bones most commonly broken due to osteoporosis.

A bone density test called DXA, for dual X-ray absorptiometry, is considered the gold standard test for diagnosing osteoporosis and monitoring bone loss over time. The test is a noninvasive low-dose X-ray. Actually, two X-rays of different energy levels (or wavelengths) are used to differentiate between bone and soft tissue. This provides a very accurate measurement of bone density. The bone density of the patient is compared to that of a healthy 30-year-old, which is when bone density is highest in both male and females, and a number known as a T-score is assigned. A T-score of 1 or above is normal. A negative T-score indicates that some bone loss has occurred. A number between –2.5 and –1 suggests an increased risk of fractures with a 10% to 30% loss in bone density, while a value below –2.5 indicates osteoporosis.

Osteoporosis is often called a "silent disease" because symptoms usually aren't apparent until a fracture occurs. However, one out of every two women over 50 years of age experiences clinical bone loss to some degree. The incidence of osteoporosis is considerably less in men, with about one out of every four men experiencing bone loss. Over 10 million people in the United States have osteoporosis, with twice that many at risk. Osteoporosis causes about 1.5 million fractures a year in men and women, with spine and hip fractures making up the majority of these.

Of the 10 million Americans affected by osteoporosis, over 80% are older women. Osteoporosis is most prevalent in postmenopausal women over the age of 50. As you will discover in chapter 30, the process of building up bone and breaking it down is controlled by various hormones, chemical signals released into the bloodstream that control body functions. As a woman's body experiences changes in various hormone levels after menopause, some body functions, including the cyclic building up and breaking down of bone tissue, are altered. Other factors also contribute to osteoporosis including family history, inadequate sources of calcium and vitamin D in the diet, smoking, ethnicity (Caucasian and Asian women have a higher risk), and lack of exercise.

How to Avoid Osteoporosis

Is osteoporosis unavoidable? Not at all. Doctors are learning more and more about osteoporosis and how to treat it, and importantly, how to prevent it. Although it is a disease found most frequently in the elderly, it can be prevented by actions taken when a person is younger.

How can you avoid osteoporosis later in life? The "sculpting" of bone occurs throughout your lifetime, but it is in your childhood and teenage years when new bone is added at a faster rate than it is lost by the breakdown of bone tissue. This net addition of bone tissue continues until a person reaches about 30 years of age, when the balance begins to tip toward the direction of bone loss. Osteoporosis develops when the rate of bone loss exceeds the replacement of bone to the point when the bone becomes weak and brittle.

As discussed on page 480, bone forms when calcium phosphate hardens around collagen fibers, and so an adequate supply of calcium is important to healthy bone development. Because of this, it is important that children and teenagers have enough calcium in their diets to fulfill the needs of developing bone, as well as other body activities that require calcium.

Therefore, the best way to avoid osteoporosis in your elder years is to get ample amounts of calcium before you're 30. Building up strong, healthy, dense bones before age 30 means that you have a larger pool of calcium in your bones to offset bone loss in your older years. As people age, their bodies are less efficient at absorbing calcium and so people over age 50 actually need to increase their calcium intake to 1,200 mg per day. Foods high in calcium include low-fat dairy products, dark green leafy vegetables, and orange juice. Many people take calcium supplements to ensure that they get enough calcium.

Physical activity is also key to preventing osteoporosis, but it is important that the activity involves weight-bearing exercise such as walking or jogging. New bone tissue is added to areas of the bone that are stressed to make those areas stronger. Weight-bearing exercises, both in young and old, help to strengthen areas of the bone that might be most susceptible to fractures.

While successful treatments are available, it is important to remember that the best treatment for osteoporosis is to avoid getting it by building up your bones when you're young.

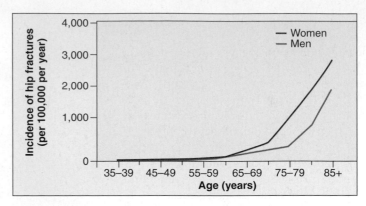

Figure 22.5   Osteoporosis.

More common in older women than in younger women or men, osteoporosis is a bone disorder in which bones progressively lose minerals. This graphs shows the incidence of osteoporotic hip fractures in men and women.

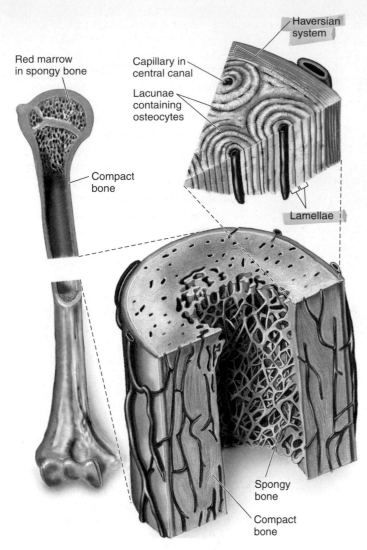

Figure 22.4   The structure of bone.

Some parts of bones are dense and compact, giving the bone strength. Other parts are spongy, with a more open lattice; red blood cells form in the bone marrow. Osteocytes, which are mature osteoblasts, lie in tight spaces called lacunae.

reconstructed. The cross section through a bone in **figure 22.4** shows that the outer layer of bone is very dense and compact and so is called **compact bone.** The interior is less compact, with a more open lattice structure, and is called **spongy bone.** Red blood cells form in the red marrow of spongy bone. New bone is formed in two stages: First, collagen is secreted by cells called **osteoblasts,** which lay down a matrix of fibers along lines of stress. Then calcium minerals impregnate the fibers. Bone is laid down in thin, concentric layers, like layers of paint on an old pipe. The layers form as a series of tubes around a narrow central channel called a **central canal,** also called a *Haversian canal,* which runs parallel to the length of the bone. The many central canals within a bone, all interconnected, contain blood vessels and nerves that provide a lifeline to its living, bone-forming cells.

When bone is first formed in the embryo, osteoblasts use a cartilage skeleton as a template for bone formation. During childhood, bones grow actively. The total bone mass in a healthy young adult, by contrast, does not change much from one year to the next. This does not mean change is not occurring. Large amounts of calcium and thousands of **osteocytes** (mature osteoblasts) are constantly being removed and replaced, but total bone mass does not change because deposit and removal take place at about the same rate.

Two cell types are responsible for this dynamic bone "remodeling": *Osteoblasts* deposit bone, and **osteoclasts** secrete enzymes that digest the organic matrix of bone, liberating calcium for reabsorption by the bloodstream (see **figure 30.9**). The dynamic remodeling of bone adjusts bone strength to workload, new bone being formed along lines of stress. When a bone is subjected to compression, mineral deposition by osteoblasts exceeds withdrawals by osteoclasts. That is why long-distance runners must slowly increase the distances they attempt, to allow their bones to strengthen along lines of stress; otherwise, stress fractures can cripple them.

As a person ages, the backbone and other bones tend to decline in mass. Excessive bone loss is a condition called **osteoporosis.** After the onset of osteoporosis, the replacement of calcium and other minerals lags behind withdrawal, causing the bone tissue to gradually erode. Eventually the bones become brittle and easily broken. **Figure 22.5** shows the effects of osteoporosis in both men and women. Women are more than twice as likely to develop osteoporosis compared to men.

Key Learning Outcome 22.4   Connective tissues support the vertebrate body and consist of cells embedded in an extracellular matrix. They include cells of the immune system, cells of the skeletal system, and cells found throughout the body like blood and fat cells. Bone is a type of connective tissue.

22.5 Muscle Tissue Lets the Body Move

Muscle cells are the motors of the vertebrate body. The distinguishing characteristic of muscle cells, the thing that makes them unique, is the abundance of contractible protein fibers within them. These fibers, called **myofilaments,** are made of the proteins actin and myosin. Vertebrate cells have a fine network of these myofilaments, but muscle cells have many more than other cells. Crammed in like the fibers of a rope, they take up practically the entire volume of the muscle cell. When actin and myosin slide past each other, the muscle contracts. Like slamming a spring-loaded door, the shortening of

all of these fibers together within a muscle cell can produce considerable force. The process of muscle contraction will be discussed later in this chapter.

The vertebrate body possesses three different kinds of muscle cells: smooth muscle, skeletal muscle, and cardiac muscle (**table 22.4**). In smooth muscle, the myofilaments are only loosely organized (seen under ❶ in the table). In skeletal and cardiac muscle, the myofilaments are bunched together into fibers called *myofibrils.* Each myofibril contains many thousands of myofilaments, all aligned to provide maximum force when they simultaneously slide past each other. Skeletal and cardiac muscle are often called striated muscles because, as you can see under ❷ and ❸ in the table, their cells appear to have transverse stripes when viewed in longitudinal section under the microscope.

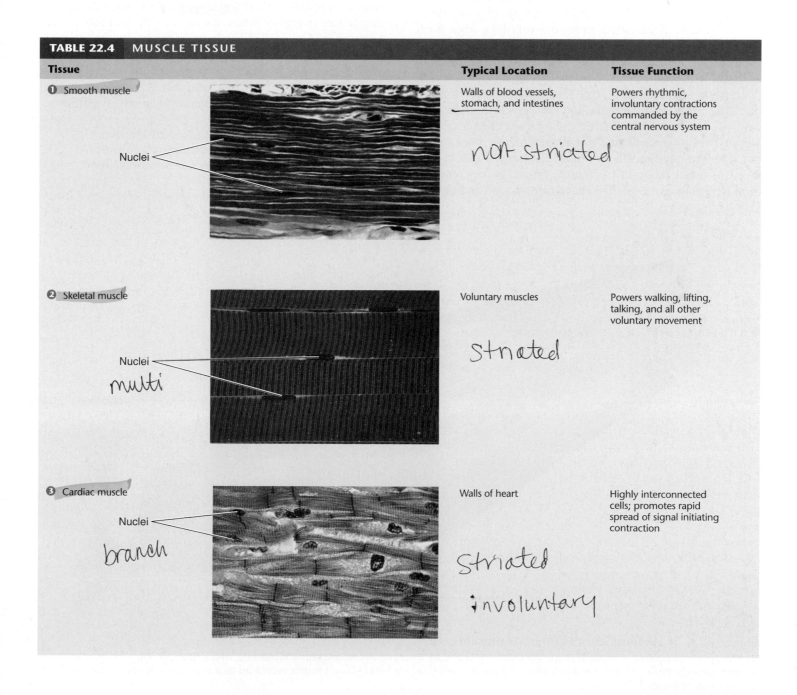

TABLE 22.4 MUSCLE TISSUE

Tissue		Typical Location	Tissue Function
❶ Smooth muscle	Nuclei	Walls of blood vessels, stomach, and intestines	Powers rhythmic, involuntary contractions commanded by the central nervous system
❷ Skeletal muscle	Nuclei	Voluntary muscles	Powers walking, lifting, talking, and all other voluntary movement
❸ Cardiac muscle	Nuclei	Walls of heart	Highly interconnected cells; promotes rapid spread of signal initiating contraction

Handwritten annotations: not striated; striated; multi; branch; striated; involuntary

Smooth Muscle

Smooth muscle cells are long and spindle-shaped, each containing a single nucleus. However, the individual myofilaments are not aligned into orderly assemblies as they are in skeletal and cardiac muscles. Smooth muscle tissue is organized into sheets of cells. In some tissues, smooth muscle cells contract only when they are stimulated by a nerve or hormone. Examples are the muscles that line the walls of many blood vessels and those that make up the iris of the vertebrate eye. In other smooth muscle tissue, such as that found in the wall of the gut, the individual cells contract spontaneously, leading to a slow, steady contraction of the tissue.

Skeletal Muscle

Skeletal muscles move the bones of the skeleton. Skeletal muscle cells are produced during development by the fusion of several cells at their ends to form a very long fiber. Each of these muscle cell fibers still contains all the original nuclei, pushed out to the periphery of the cytoplasm. Looking at **figure 22.6**, you can see how the nuclei are positioned to the outside of the muscle fiber. Each **muscle fiber** consists of many elongated **myofibrils,** and each myofibril is, in turn, composed of many myofilaments, the protein filaments actin and myosin. Myofibrils and myofilaments have been pulled out from the muscle cells in **figure 22.6** so you can see the levels of organization in the muscle fiber. The key property of muscle cells is the relative abundance of actin and myosin

within them, which enables a muscle cell to contract. These protein filaments are present as part of the cytoskeleton of all eukaryotic cells, but they are far more abundant and more highly organized in muscle cells.

Cardiac Muscle

The vertebrate heart is composed of striated **cardiac muscle** in which the fibers are arranged very differently from the fibers of skeletal muscle. Instead of very long, multinucleate cells running the length of the muscle, heart muscle is composed of chains of single cells, each with its own nucleus (refer back to the photo under ❸ in **table 22.4**). Chains of cells are organized into fibers that branch and interconnect, forming a latticework. This lattice structure is critical to the way heart muscle functions. Each heart muscle cell is coupled to its neighbors electrically by tiny holes called *gap junctions* that pierce the plasma membranes in regions where the cells touch each other. Heart contraction is initiated at one location by the opening of transmembrane channels that conduct ions across the membrane. This changes the electrical properties of the membrane. An electrical impulse then passes from cell to cell across the gap junctions, causing the heart to contract in an orderly pulsation.

> **Key Learning Outcome 22.5** Muscle tissue is the tool the vertebrate body uses to move its limbs, contract its organs, and pump blood through its circulatory system.

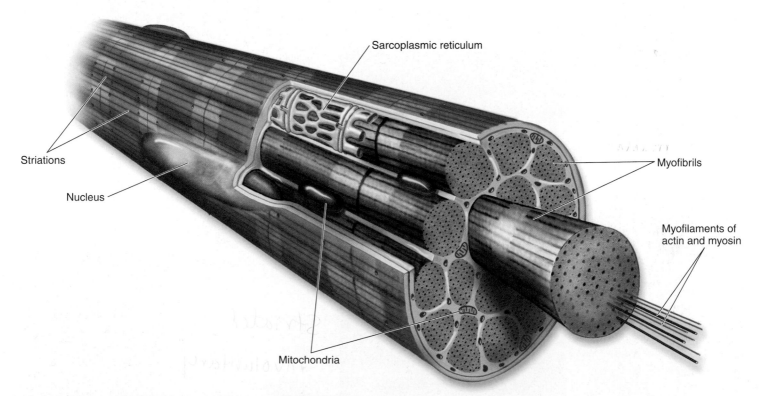

Figure 22.6 A skeletal muscle fiber, or muscle cell.

Each muscle is composed of bundles of muscle cells, or fibers. Each fiber is composed of many myofibrils, which are each, in turn, composed of myofilaments. Muscle cells have a modified endoplasmic reticulum called the sarcoplasmic reticulum that is involved in the regulation of calcium ions in muscles.

22.6 Nerve Tissue Conducts Signals Rapidly

Nerve cells carry information rapidly from one vertebrate organ to another. Nerve tissue, the fourth major class of vertebrate tissue, is composed of two kinds of cells: (1) **neurons,** which are specialized for the transmission of nerve impulses, and (2) supporting **glial cells,** which supply the neurons with nutrients, support, and insulation.

Neurons have a highly specialized cell architecture that enables them to conduct signals rapidly throughout the body. Their plasma membranes are rich in ion-selective channels that maintain a voltage difference between the interior and the exterior of the cell, the equivalent of a battery. When ion channels in a local area of the membrane open, ions flood in from the exterior, temporarily wiping out the charge difference. This process, called depolarization, tends to open nearby voltage-sensitive channels in the neuron membrane, resulting in a wave of electrical activity that travels down the entire length of the neuron as a nerve impulse.

Each neuron is composed of three parts, as illustrated in **figure 22.7**: (1) a **cell body,** which contains the nucleus; (2) threadlike extensions called **dendrites** extending from the cell body, which act as antennae, bringing nerve impulses to the cell body from other cells or sensory systems; and (3) a single, long extension called an **axon,** which carries nerve impulses away from the cell body. Axons often carry nerve impulses for considerable distances: The axons that extend from the skull to the pelvis in a giraffe are about 3 meters long!

The body contains neurons of various sizes and shapes. Some are tiny and have only a few projections, others are bushy and have more projections, and still others have extensions that are meters long. However, all fit into one of three general categories of neurons as shown in **table 22.5.** *Sensory neurons* ❶ generally carry electrical impulses from the body to the central nervous system (CNS), the brain and spinal cord. *Motor neurons* ❷ generally carry electrical impulses from the central nervous system to the muscles. *Association neurons* ❸ occur within the central nervous system and act as a "connector" between sensory and motor neurons. These will be discussed in more detail in chapter 28. Neurons are not normally in direct contact with one another. Instead, a tiny gap called a **synapse** separates them. Neurons communicate with other neurons by passing chemical signals called **neurotransmitters** across the gap.

Vertebrate nerves appear as fine white threads when viewed with the naked eye, but they are actually composed of bundles of axons. Like a telephone trunk cable, nerves include large numbers of independent communication channels—bundles composed of hundreds of axons, each connecting a nerve cell to a muscle fiber or other type of cell. It is important not to confuse a nerve with a neuron. A nerve is made up of the axons of many neurons, just as a cable is made of many wires.

Key Learning Outcome 22.6 Nerve tissue provides the vertebrate body with a means of communication and coordination.

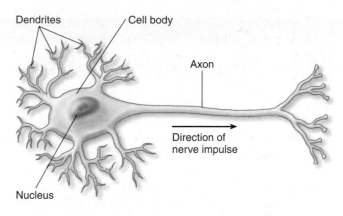

Figure 22.7 Neurons carry nerve impulses.
Neurons carry nerve impulses, which are electrical signals, from their initiation in dendrites, to the cell body, and down the length of the axon, where they may pass the signal to a neighboring cell.

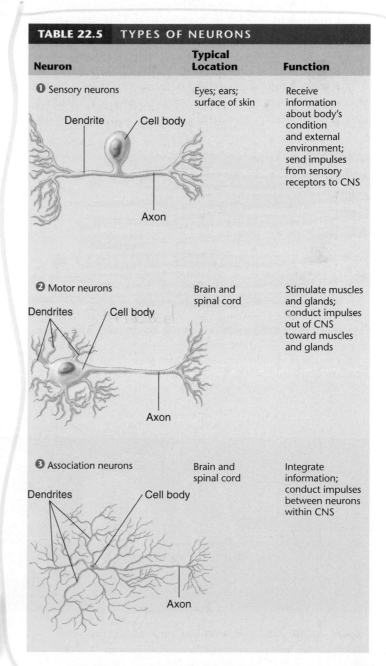

TABLE 22.5	TYPES OF NEURONS	
Neuron	**Typical Location**	**Function**
❶ Sensory neurons	Eyes; ears; surface of skin	Receive information about body's condition and external environment; send impulses from sensory receptors to CNS
❷ Motor neurons	Brain and spinal cord	Stimulate muscles and glands; conduct impulses out of CNS toward muscles and glands
❸ Association neurons	Brain and spinal cord	Integrate information; conduct impulses between neurons within CNS

22.7 | Types of Skeletons

With muscles alone, the animal body could not move—it would simply pulsate as its muscles contracted and relaxed in futile cycles. For a muscle to produce movement, it must direct its force against another object. Animals are able to move because the opposite ends of their muscles are attached to a rigid scaffold, or **skeleton,** so that the muscles have something to pull against. There are three types of skeletal systems in the animal kingdom: hydraulic skeletons, exoskeletons, and endoskeletons.

Hydraulic skeletons are found in soft-bodied invertebrates such as earthworms and jellyfish. In this case, a fluid-filled cavity is encircled by muscle fibers that raise the pressure of the fluid when they contract. The earthworm in **figure 22.8** moves forward by a wave of contractions of circular muscles that begins anteriorly and compresses the body, so that the fluid pressure pushes it forward. Contractions of longitudinal muscles then pull the rest of the body along.

Exoskeletons surround the body as a rigid hard case to which muscles attach internally. When a muscle contracts, it moves the section of exoskeleton to which it is attached.

Figure 22.9   Crustaceans have an exoskeleton.
The exoskeleton of this rock crab is bright orange.

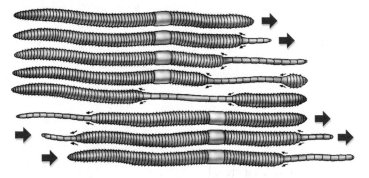

Figure 22.8   Earthworms have a hydraulic skeleton.
When an earthworm's circular muscles contract, the internal fluid presses on the longitudinal muscles, which then stretch to elongate segments of the earthworm. A wave of contractions down the body of the earthworm produces forward movement.

Arthropods, such as crustaceans (**figure 22.9**) and insects, have exoskeletons made of the polysaccharide *chitin.* An animal with an exoskeleton cannot get too large because its exoskeleton would have to become thicker and heavier to prevent collapse. If an insect were the size of an elephant, its exoskeleton would have to be so thick and heavy it would hardly be able to move.

Endoskeletons, found in vertebrates and echinoderms, are rigid internal skeletons to which muscles are attached. Vertebrates have a soft, flexible exterior that stretches to accommodate the movements of their skeleton. The endoskeleton of most vertebrates is composed of bone (**figure 22.10**). Unlike chitin, bone is a cellular, living tissue capable of growth, self-repair, and remodeling in response to physical stresses.

A Vertebrate Endoskeleton: The Human Skeleton

The human skeleton is made up of 206 individual bones. If you saw them as a pile of bones jumbled together, it would be hard to make any sense of them. To understand the skeleton, it is necessary to group the 206 bones according to

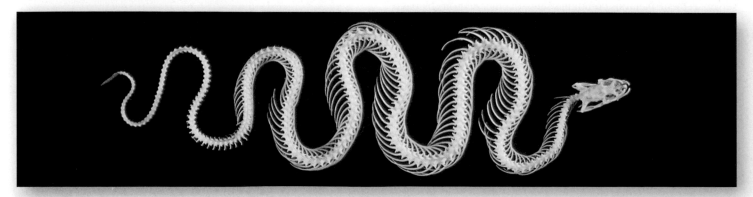

Figure 22.10   Snakes have an endoskeleton.
The endoskeleton of most vertebrates is made of bone. A snake's skeleton is specialized for quick lateral movement.

their function and position in the body. The 80 bones of the **axial skeleton,** the purple-colored bones in **figure 22.11,** support the main body axis, while the remaining 126 bones of the **appendicular skeleton,** the tan-colored bones, support the arms and legs. These two skeletons function more or less independently—that is, the muscles controlling the axial skeleton (postural muscles) are managed by the brain separately from those controlling the appendages (manipulatory muscles).

The Axial Skeleton

The axial skeleton is made up of the skull, backbone, and rib cage. Of the skull's 28 bones, only 8 form the cranium, which encases the brain; the rest are facial bones and middle ear bones.

The skull is attached to the upper end of the backbone, which is also called the *spine,* or **vertebral column.** The spine is made up of 26 vertebrae, stacked one on top of the other to provide a flexible column surrounding and protecting the spinal cord. Curving forward from the vertebrae are 12 pairs of ribs, attached at the front to the breastbone, or sternum, and forming a protective cage around the heart and lungs.

The Appendicular Skeleton

The 126 bones of the appendicular skeleton are attached to the axial skeleton at the shoulders and hips. The shoulder, or **pectoral girdle,** is composed of two large, flat shoulder blades, each connected to the top of the sternum by a slender, curved collarbone (clavicle). The arms are attached to the pectoral girdle; each arm and hand contains 30 bones. The clavicle is the most frequently broken bone of the body. Can you guess why? Because if you fall on an outstretched arm, a large component of the force is transmitted to the clavicle.

The **pelvic girdle** forms a bowl that provides strong connections for the legs, which must bear the weight of the body. Each leg and foot contains a total of 30 bones.

Joints, points where two bones come together confer flexibility to the rigid endoskeleton, allowing a range of motion determined by the type of joint. There are three main classes of joints based on mobility. *Immovable joints,* such as the sutures of the skull, are capable of little to no movement. *Slightly movable joints,* such as the joints between vertebrae in the spine, allow the bones some movement. And *freely movable joints* allow a range of motion and are seen in the joints of the limbs (for example, the shoulder, elbow, hip, knee), the jaw, and fingers and toes. Depending on the type of joint, bones are held together at the joint by cartilage, fibrous connective tissue, or a fibrous capsule filled with a lubricating fluid.

> **Key Learning Outcome 22.7** The animal skeletal system provides a framework against which the body's muscles can pull. Many soft-bodied invertebrates employ a hydraulic skeleton, whereas arthropods have a rigid, hard exoskeleton surrounding their body. Echinoderms and vertebrates have an internal endoskeleton to which muscles attach.

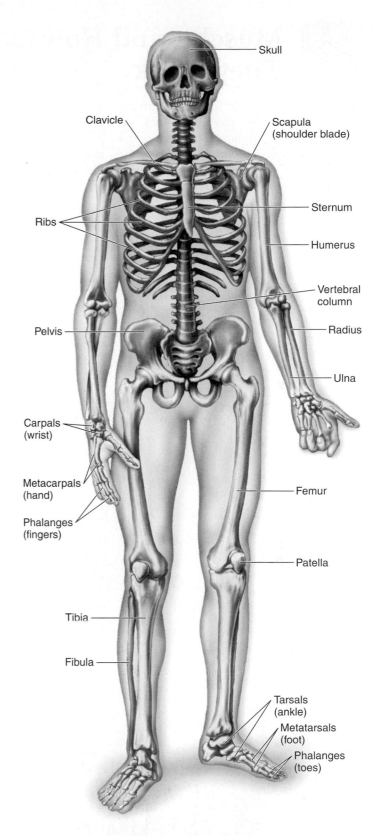

Figure 22.11 Axial and appendicular skeletons.
The axial skeleton is shown in *purple,* and the appendicular skeleton is shown in *tan.* Some of the joints are shown in *green.*

Muscles and How They Work

Three kinds of muscle together form the vertebrate muscular system. As we have discussed, the vertebrate body is able to move because *skeletal muscles* pull the bones with considerable force. The heart pumps because of the contraction of *cardiac muscle.* Food moves through the intestines because of the rhythmic contractions of *smooth muscle.*

Actions of Skeletal Muscle

Skeletal muscles move the bones of the skeleton. Some of the major human muscles are labeled on the right in **figure 22.12**. Muscles are attached to bones by straps of dense connective tissue called **tendons.** Bones pivot about flexible connections called *joints,* pulled back and forth by the muscles attached to them. Each muscle pulls on a specific bone. One end of the muscle, the *origin,* is attached by a tendon to a bone that remains stationary during a contraction. This provides an object against which the muscle can pull. The other end of the muscle, the *insertion,* is attached to a bone that moves if the muscle contracts. For example, origin and insertion for the sartorius muscle is labeled on the left in **figure 22.12**. This muscle helps bend the leg at the hip, bringing the knee to the chest. The origin of the muscle is at the hip and stays stationary. The insertion is at the knee, such that when the muscle contracts (gets shorter) the knee is pulled toward the chest.

Muscles can only pull, not push, because myofibrils contract rather than expand. For this reason, the muscles in the movable joints of vertebrate are attached in opposing pairs, called flexors and extensors that, when contracted, move the bones in different directions. As you can see in **figure 22.13**, when the **flexor** muscle at the back of your upper leg contracts, the lower leg is moved closer to the thigh. When the **extensor** muscle at the front of your upper leg contracts, the lower leg is moved in the opposite direction, away from the thigh.

All muscles contract, but there are two types of muscle contractions, isotonic and isometric contractions. In **isotonic contractions,** the muscle shortens, moving the bones as just described. In **isometric contractions,** a force is exerted by the muscle, but the muscle does not shorten. This occurs when you try to lift something very heavy. Eventually, if your muscles generate enough force and you are able to lift the object, the isometric contraction becomes isotonic.

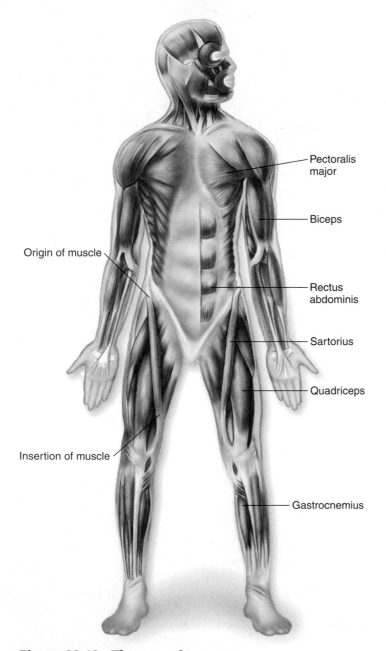

Pectoralis major

Biceps

Origin of muscle

Rectus abdominis

Sartorius

Quadriceps

Insertion of muscle

Gastrocnemius

Figure 22.12 The muscular system.
Some of the major muscles in the human body are labeled.

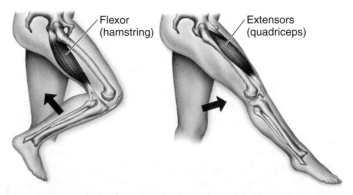

Flexor (hamstring)

Extensors (quadriceps)

Figure 22.13 Flexor and extensor muscles.
Limb movement is always the result of muscle contraction, never muscle extension. Muscles that retract limbs are called flexors; those that extend limbs are called extensors.

Author's Corner

The Author Works Out

No one seeing the ring of fat decorating my middle would take me for a runner. Only in my memory do I get up with the robins, lace up my running shoes, bounce out the front door, and run the streets around Washington University before going to work. Now my 5-K runs are 30-year-old memories. Any mention I make of my running in a race only evokes screams of laughter from my daughters, and an arch look from my wife. Memory is cruelest when it is accurate.

I remember clearly the day I stopped running. It was a cool fall morning in 1978, and I was part of a mob running a 5-K (that's 5 kilometers for the uninitiated) race, winding around the hills near the university. I started to get flashes of pain in my legs below the knees—like shin splints, but much worse. Imagine fire pouring on your bones. Did I stop running? No. Like a bonehead I kept going, "working through the pain," and finished the race. I have never run a race since.

I had pulled a muscle in my thigh, which caused part of the pain. But that wasn't all. The pain in my lower legs wasn't shin splints, and didn't go away. A trip to the doctor revealed multiple stress fractures in both legs. The X rays of my legs looked like tiny threads had been wrapped around the shaft of each bone, like the red stripe on a barber's pole. It was summer before I could walk without pain.

What went wrong? Isn't running supposed to be GOOD for you? Not if you run improperly. In my enthusiasm to be healthy, I ignored some simple rules and paid the price. The biology lesson I ignored had to do with how bones grow. The long bones of your legs are not made of stone, solid and permanent. They are dynamic structures, constantly being re-formed and strengthened in response to the stresses to which you subject them.

To understand how bone grows, we first need to recall a bit about what bone is like. Bone, as you have learned in this chapter, is made of fibers of a flexible protein called collagen stuck together to form cartilage. While an embryo, all your bones are made of cartilage. As your adult body develops, the collagen fibers become impregnated with tiny, needle-shaped crystals of calcium phosphate, turning the cartilage into bone. The crystals are brittle but rigid, giving bone great strength. Collagen is flexible but weak, but like the epoxy of fiberglass, it acts to spread any stress over many crystals, making bone resistant to fracture. As a result, bone is both strong and flexible.

When you subject a bone in your body to stress—say, by running—the bone grows so as to withstand the greater workload. How does the bone "know" just where to add more material? When stress deforms the collagen fibers of a leg bone, the interior of the collagen fibers becomes exposed, like opening your jacket and exposing your shirt. The fiber interior produces a minute electrical charge. Cells called fibroblasts are attracted to the electricity like bugs to night lights, and secrete more collagen there. As a result, new collagen fibers are laid down on a bone along the lines of stress. Slowly, over months, calcium phosphate crystals convert the new collagen to new bone. In your legs, the new bone forms along the long stress lines that curve down along the shank of the bone.

Now go back 30 years, and visualize me pounding happily down the concrete pavement each morning. I had only recently begun to run on the sidewalk, and for an hour or more at a stretch. Every stride I took those mornings was a blow to my shinbones, a stress to which my bones no doubt began to respond by forming collagen along the spiral lines of stress. Had I run on a softer surface, the daily stress would have been far less severe. Had I gradually increased my running, new bone would have had time to form properly in response to the added stress. I gave my leg bones a lot of stress, and no time to respond to it. I pushed them too hard, too fast, and they gave way.

Nor was my improper running limited to overstressed leg bones. Remember that pulled thigh muscle? In my excessive enthusiasm, I never warmed up before I ran. I was having too much fun to worry about such details. Wiser now, I am sure the pulled thigh muscle was a direct result of failing to properly stretch before running.

I was reminded of that pulled muscle recently, listening to a good friend of my wife's describe how she sets out early each morning for a long run without stretching or warming up. I can see her in my mind's eye, bundled up warmly on the cooler mornings, an enthusiastic gazelle pounding down the pavement in search of good health. Unless she uses more sense than I did, she may fail to find it.

Muscle Contraction

Recall from **figure 22.6** that myofibrils are composed of bundles of myofilaments. Far too fine to see with the naked eye, the individual myofilaments of vertebrate muscles are only 8 to 12 nanometers thick. Each is a long, threadlike filament of the proteins actin or myosin. An **actin filament** consists of two strings of actin molecules wrapped around one another, like two strands of pearls loosely wound together. A **myosin filament** is also composed of two strings of protein wound about each other, but a myosin filament is about twice as long as an actin filament, and the myosin strings have a very unusual shape. One end of a myosin filament consists of a very long rod, while the other end consists of a double-headed globular region, or "head." Overall, a myosin filament looks a bit like a two-headed snake. This odd structure is the key to how muscles work.

How Myofilaments Contract

The **sliding filament model** of muscle contraction, seen in the Key Biological Process illustration below, describes how actin and myosin cause muscles to contract. Focus on the knob-shaped myosin head in **panel 1**. When a muscle contraction begins, the heads of the myosin filaments move first. Like flexing your hand downward at the wrist, the heads bend backward and inward as in **panel 2**. This moves them closer to their rod-like backbones and several nanometers in the direction of the flex. In itself, this myosin head-flex accomplishes nothing—but the myosin head is attached to the actin filament! As a result, the actin filament is pulled along with the myosin head as it flexes, causing the actin filament to slide by the myosin filament in the direction of the flex (the dotted circles in **panel 2** indicate the movement of the actin filament). As one after another myosin head flexes, the myosin in effect "walks" step by step along the actin. Each step uses a molecule of ATP to recock the myosin head (in **panel 3**) before it attaches to the actin again (**panel 4**), ready for the next flex.

How does this sliding of actin past myosin lead to myofibril contraction and muscle cell movement? The actin filament is anchored at one end, at a position in striated muscle called the Z line, indicated by the lavender-colored bars toward the edges in the Key Biological Process illustration on the facing page. Two Z lines with the actin and myosin filaments in between make up a contractile unit called a **sarcomere.** Because it is tethered like this, the actin cannot simply move off. Instead, the actin pulls the anchor with it! As actin moves past myosin, it drags the Z line toward the myosin. The secret of muscle contraction is that each myosin is interposed between two pairs of actin filaments, which are anchored at both ends to Z lines, as shown in **panel 1**. One moving to the left and the other to the right, the two pairs of actin molecules drag the Z lines toward each other as they slide past the myosin core, shown progressively in **panel 2** and **panel 3**. As the Z lines are pulled closer together, the plasma membranes to which they are attached move toward one another, and the cell contracts.

KEY BIOLOGICAL PROCESS: Myofilament Contraction

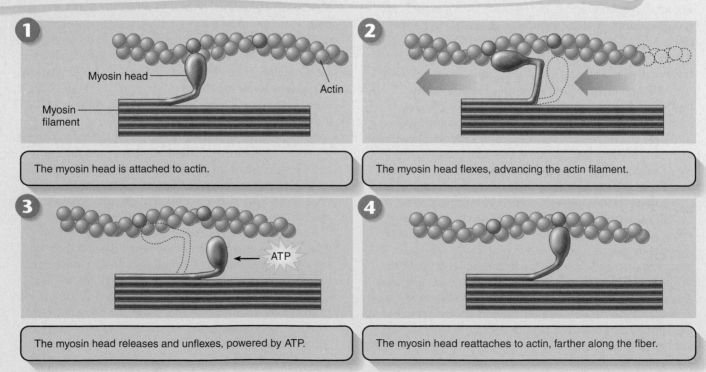

1. The myosin head is attached to actin.

2. The myosin head flexes, advancing the actin filament.

3. The myosin head releases and unflexes, powered by ATP.

4. The myosin head reattaches to actin, farther along the fiber.

KEY BIOLOGICAL PROCESS: The Sliding Filament Model

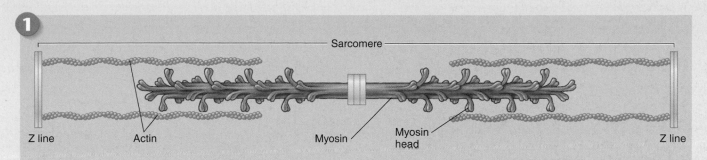

1

Sarcomere

Z line   Actin   Myosin   Myosin head   Z line

The heads on the two ends of the myosin filament are oriented in opposite directions.

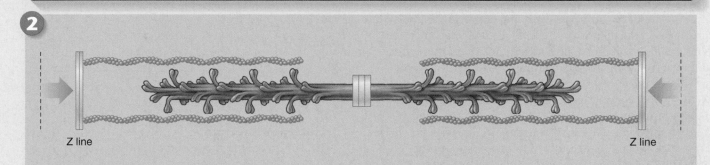

2

Z line   Z line

Thus, as the right-hand end of the myosin filament "walks" along the actin filaments, pulling them and their attached Z line leftward toward the center, the left-hand end of the same myosin filament "walks" along the actin filaments, pulling them and their attached Z line rightward toward the center.

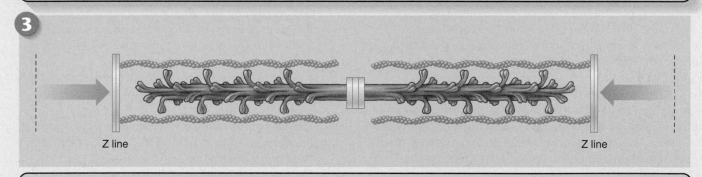

3

Z line   Z line

The result is that both Z lines move toward the center—and contraction occurs.

The Role of Calcium Ions

When a muscle is relaxed, its myosin heads are "cocked" and ready, but are unable to bind to actin. This is because the attachment sites for the myosin heads on the actin are physically blocked by another protein, known as **tropomyosin.** Myosin heads therefore cannot bind to actin in the relaxed muscle, and the filaments cannot slide.

For a muscle to contract, the tropomyosin must first be moved out of the way so that the myosin heads can bind to actin. This requires calcium ions (Ca^{++}). When the Ca^{++} concentration of the muscle cell cytoplasm increases, Ca^{++}, acting through another protein, causes the tropomyosin to move out of the way. When this repositioning has occurred, the myosin heads attach to actin and, using ATP energy, move along the actin in a stepwise fashion to shorten the myofibril.

Muscle fibers store Ca^{++} in a modified endoplasmic reticulum called the **sarcoplasmic reticulum.** When a muscle fiber is stimulated to contract, Ca^{++} is released from the sarcoplasmic reticulum and diffuses into the myofibrils, where it initiates contraction. When a muscle works too hard, the Ca^{++} channels become leaky, releasing small amounts of Ca^{++} that act to weaken muscle contractions and result in muscle fatigue.

Key Learning Outcome 22.8 Muscles are made of many tiny threadlike filaments of actin and myosin called myofilaments. Muscles work by using ATP to power the sliding of myosin along actin, causing the myofibrils to contract.

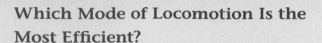

Which Mode of Locomotion Is the Most Efficient?

Running, flying, and swimming require more energy than sitting still, but how do they compare? The greatest differences between moving on land, in the air, and in water result from the differences in support and resistance to movement provided by water and air. The weight of swimming animals is fully supported by the surrounding water, and no effort goes into supporting the body, while running and flying animals must support the full weight of their bodies. On the other hand, water presents considerable resistance to movement, air much less, so that flying and running require less energy to push the medium out of the way.

A simple way to compare the costs of moving for different animals is to determine how much energy it takes to move. The energy cost to run, fly, or swim is in each case the energy required to move one unit of body mass over one unit of distance with that mode of locomotion. (Energy is measured in the **metric system** as a **kilocalorie** [**kcal**] or, technically, 4.184 kilojoules [note that the Calorie measured in food diets and written with a capital C is equivalent to 1 kcal]; body mass is measured in kilograms, where 1 kilogram [**kg**] is 2.2 pounds; distance is measured in kilometers, where 1 kilometer [**km**] is 0.62 miles). The graph to the right displays three such "cost-of-motion" studies. The blue squares represent running; the red circles, flying; and the green triangles, swimming. In each study, the line is drawn as the statistical "best-fit" for the points. Some animals like humans have data in two lines, as they both run (well) and swim (poorly). Ducks have data in all three lines, as they not only fly (very well), but also run and swim (poorly).

1. **Applying Concepts**
   a. Variables. In the graph, what is the dependent variable?
   b. Comparing Continuous Variables. Do the three modes of locomotion have the same or different costs?

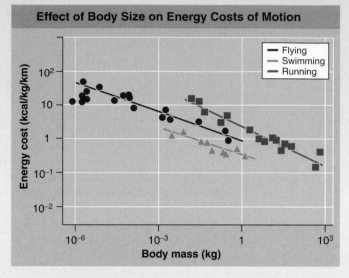

Effect of Body Size on Energy Costs of Motion

Legend: Flying, Swimming, Running

y-axis: Energy cost (kcal/kg/km) — 10^2, 10, 1, 10^{-1}, 10^{-2}
x-axis: Body mass (kg) — 10^{-6}, 10^{-3}, 1, 10^3

2. **Interpreting Data**
   a. For any given mode of locomotion, what is the impact of body size on cost of moving?
   b. Is the impact of body mass the same for all three modes of locomotion? If not, which mode's cost is least affected by body mass? Why do you think this is so?

3. **Making Inferences**
   a. Comparing the energy costs of running versus flying for animals of the same body size, which mode of locomotion is the most expensive? Why would you expect this to be so?
   b. Comparing the energy costs of swimming to flying, which uses the least energy? Why would you expect this to be so?

4. **Drawing Conclusions** In general, which mode of locomotion is the most efficient? The least efficient? Why do you think this is so?

5. **Further Analysis**
   a. How would you expect the slithering of a snake to compare to the three modes of locomotion examined here? Why?
   b. Do you think the costs of running by an athlete decrease with training? Why? How might you go about testing this?

The Animal Body Plan

22.1 Innovations in Body Design

- Several innovations have been key to the diversity seen in the animal kingdom: tissues, radial versus bilateral symmetry, a body cavity, segmentation, incremental growth versus molting, and protostome versus deuterostome development (**table 22.1**).

22.2 Organization of the Vertebrate Body

- All vertebrates have the same general architecture: a tube (the gut or digestive system) suspended in a cavity (the coelom) that in many terrestrial vertebrates is divided into a thoracic and an abdominal cavity.

- The vertebrate body is composed of cells that exhibit increasing levels of structural and functional complexity.

- Cells group together into tissues, which act as functional units (**figure 22.1**). Organs of the body are composed of several different tissues that act together to perform a higher level of function. Organs work together in an organ system to perform larger-scale body functions (**figure 22.2**). The vertebrate body contains 11 principal organ systems.

Tissues of the Vertebrate Body

22.3 Epithelium Is Protective Tissue

- Epithelial tissue is composed of different types of epithelial cells. It covers internal and external surfaces of the body providing protection.

- The structure of the epithelium determines its function (**table 22.2**). Some types of epithelium are a single layer of cells through which substances can pass. Some epithelium is stratified, providing protection. Pseudostratified epithelium appears to have several layers, but is actually just a single layer of cells with nonuniform placement of nuclei.

22.4 Connective Tissue Supports the Body

- The connective tissues of the body are very diverse in structure and function, but all are composed of cells embedded in an extracellular matrix. The matrix may be hard as in bone, flexible as in fibrous connective tissue, adipose tissue, and cartilage, or fluid as in blood, like the red blood cells shown here from **table 22.3** floating in the fluid matrix called plasma. Connective tissue functions in providing the body structural support, storage, and transportation of nutrients, gases, and immune cells. Bone is a living tissue. Bone cells called osteoblasts lay down a fibrous matrix. Calcium minerals then impregnate the fibers, causing the matrix to harden into compact bone (**figure 22.4**).

22.5 Muscle Tissue Lets the Body Move

- Muscle tissue is dynamic tissue; it contracts, causing the body to move. There are three types of muscle tissue: smooth, skeletal, and cardiac muscle (**table 22.4**). All three types of muscle contain actin and myosin myofilaments but differ in the organization of the myofilaments.

- Smooth muscle cells are long, spindle-shaped cells organized into sheets. In smooth muscle, contraction is initiated by a nerve or hormone. Smooth muscle is found in the walls of blood vessels and the digestive system.

- Skeletal muscle cells are fused into long fibers (**figure 22.6**). They contract as small units when stimulated by nerves. Skeletal muscle is attached to the skeleton, so when the muscle contracts, the skeleton moves.

- Cells of cardiac muscle found in the heart are interconnected so that they contract in an orderly pulsation.

22.6 Nerve Tissue Conducts Signals Rapidly

- Nerve tissue is composed of neurons (**table 22.5**) and supporting glial cells. Neurons (**figure 22.7**) carry electrical impulses from one area of the body to another, providing the body a means of communication and coordination.

The Skeletal and Muscular Systems

22.7 Types of Skeletons

- The skeletal system provides a framework on which muscles act to move the body. Soft-bodied invertebrates have hydraulic skeletons, where muscles act on a fluid-filled cavity (**figure 22.8**).

- Arthropods have exoskeletons, where muscles attach from within to the hard outer covering of the body (**figure 22.9**).

- Vertebrates and echinoderms have endoskeletons, where muscles attach to bones or cartilage inside the body (**figure 22.10**). The human skeleton has 206 individual bones that make up the axial and appendicular skeletons (**figure 22.11**).

22.8 Muscles and How They Work

- Skeletal muscles attach to the skeleton at two points. The end that attaches to the stationary bone is called the origin. The muscle passes over a joint and attaches to another bone at a point called the insertion (**figure 22.12**). As the muscle contracts, the insertion is brought closer to the origin and the joint flexes. Muscles act in opposing pairs to flex or extend a joint (**figure 22.13**).

- Strands of actin and myosin stack on top of each other, creating an orderly array. Myosin attaches to actin (**Key Biological Process, page 492**) and pulls the 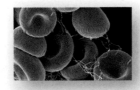 actin along its length, causing the myofilaments to slide past each other. Actin myofilaments are anchored at each end to a structure called the Z line. As the actin slides along the myosin, the anchor points are brought closer together, resulting in a shortening of the muscle (**Key Biological Process, page 493**). Energy from ATP causes the myofilaments to dissociate and reattach, triggering the sliding motion again. This is called the sliding filament model.

- This process is controlled by Ca^{++} in the muscle cells. Muscle contraction is inhibited by tropomyosin, a protein that blocks myosin binding sites on actin. When Ca^{++} is present, it causes the tropomyosin to reposition so that the myosin binding sites are exposed. With the binding sites exposed, myosin can bind actin, triggering a muscle contraction.

1. One of the innovations in animal body design, segmentation, allowed for
   a. development of efficient internal organ systems.
   b. more flexible movement as individual segments can move independently of each other.
   c. locating organs in different areas of the body.
   d. early determination of embryonic cells.

2. Which of the following is the correct organization sequence from smallest to largest in animals?
   a. cells, tissues, organs, organ systems, organism
   b. organism, organ systems, organs, tissues, cells
   c. tissues, organs, cells, organ systems, organism
   d. organs, tissues, cells, organism, organ systems

3. Which of the following is *not* a function of the epithelial tissue?
   a. secrete materials
   b. provide sensory surfaces
   c. move the body
   d. protect underlying tissue from damage and dehydration

4. An example of connective tissue is
   a. nerve cells in your fingers.
   b. skin cells.
   c. brain cells.
   d. red blood cells.

5. When a person has osteoporosis, the work of _____ falls behind the work of _____.
   a. osteoclasts; osteoblasts
   b. osteoclasts; collagen
   c. osteoblasts; osteoclasts
   d. osteoblasts; collagen

6. Nerve impulses pass from one nerve cell to another through the use of
   a. hormones.   c. pheromones.
   b. neurotransmitters.   d. calcium ions.

7. The type of muscle used to move the leg when walking is
   a. skeletal.   c. smooth.
   b. cardiac.   d. All of the these are correct.

8. The vertebral column is part of the
   a. appendicular skeleton.
   b. axial skeleton.
   c. hydrostatic skeleton.
   d. exoskeleton.

9. Movement of a limb in two directions requires a pair of muscles because
   a. a single muscle can only pull and not push.
   b. a single muscle can only push and not pull.
   c. moving a limb requires more force than one muscle can generate.
   d. None of the above.

10. The role of calcium in the process of muscle contraction is to
    a. gather ATP for the myosin to use.
    b. cause the myosin head to shift position, contracting the myofibril.
    c. cause the myosin head to detach from the actin, causing the muscle to relax.
    d. expose myosin attachment sites on actin.

Apply Your Understanding

1. **Page 482** Areas of bone are composed of an open lattice framework of minerals, including calcium. Why wouldn't it be more sensible for bones to be solid?

2. **Figure 22.11** The bones of your head and trunk form three cagelike structures: the skull, the rib cage, and the pelvic girdle. What is the functional importance of this arrangement?

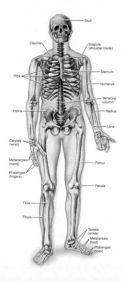

Synthesize What You Have Learned

1. The range of regulated body temperatures in mammals is about 36°C to 40°C, while in birds it is slightly higher, 38°C to 42°C, close to the limit compatible with life. Why do you imagine birds maintain higher body temperatures than mammals? Do you think eagles and hummingbirds maintain the same body temperature? Explain.

2. Imagine that you are designing a living organism, some type of vertebrate. Explain briefly how the major organ systems would interact with each other in your design.

3. When you are exercising strenuously, such as playing tennis, dancing to fast music, or doing aerobics, you begin to breathe rapidly and your heart rate increases. If you continue, you become "out of breath" and flushed. Why does your body respond in this fashion?

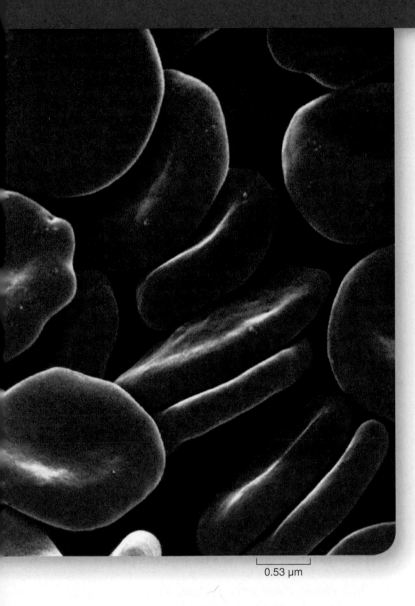

0.53 µm

23
Circulation

Blood has been called the river of life. Among all the vertebrate body's tissues, blood is the only liquid tissue, a fluid highway transporting gases, nutrients, hormones, antibodies, and wastes throughout the body. The material that makes blood a liquid is a protein-rich fluid called plasma that makes up approximately 55% of blood. The other 45% is made up mostly of red and white blood cells. Red blood cells like those seen above are the oxygen transporters of the vertebrate circulatory system. There are approximately 5 million of them in each microliter (1 µl) of blood! Each red blood cell is shaped like a rounded cushion, squashed in the center, and is crammed full of a protein called hemoglobin, an iron-containing molecule that gives blood its red color. Oxygen binds easily to the iron in hemoglobin, making red blood cells efficient oxygen carriers. A single red blood cell contains about 250 million hemoglobin molecules, and each red blood cell can carry about 1 billion molecules of oxygen at one time. The average life span of a red blood cell is only 120 days—about 2 million new ones are produced in the bone marrow every second to replace those that die or are worn out. In this chapter we will examine the ways in which vertebrates use cells like these, and the fluid surrounding them, to transport oxygen, food, and information throughout the body.

23.1 Open and Closed Circulatory Systems

Every cell in the vertebrate body must acquire the energy it needs for living from organic molecules outside the body. Like residents of a city whose food is imported from farms in the countryside, the cells of the body need trucks to carry the food, highways for the trucks to travel on, and a means to cook the food when it arrives. In the vertebrate body, the trucks are blood, the highways are blood vessels, and oxygen molecules are used to cook the food. Remember from chapter 7 that cells obtain energy by "burning" sugars like glucose, using up oxygen and generating carbon dioxide. In animals, the organ system that provides the trucks and highways is called the *circulatory system,* while the organ system that acquires the oxygen fuel and disposes of the carbon dioxide waste is called the *respiratory system.* We discuss the functions of the circulatory system in this chapter and of the respiratory system in chapter 24.

Not all eukaryotes have circulatory systems. Among the unicellular protists, oxygen and nutrients are obtained directly by simple diffusion from the aqueous external environment.

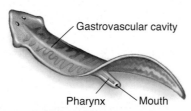

Planaria: **gastrovascular cavity**

Animals amplify the power of diffusion with a body cavity. Cnidarians, such as *Hydra,* and flatworms, such as *Planaria* (pictured above), have cells that are directly exposed to either the external environment or to a body cavity called the **gastrovascular cavity** that functions in both digestion and circulation, delivering nutrients and oxygen directly to the tissue cells by diffusion from the digestive cavity. The gastrovascular cavity of *Hydra* extends even into the tentacles (see page 532), and that of *Planaria,* colored green in the figure above, branches extensively to supply every cell with oxygen and the nourishment obtained by digestion.

Larger animals, however, have tissues that are several cell layers thick, so many cells are too far away from the body surface or digestive cavity to exchange materials directly with the environment. Instead, oxygen and nutrients are transported from the environment and digestive cavity to the body cells by an internal fluid within a circulatory system.

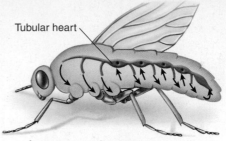

Insect: open circulation

There are two main types of circulatory systems: *open* and *closed.* In an **open circulatory system,** such as that found in arthropods and many mollusks, there is no distinction between the circulating fluid (blood) and the extracellular fluid of the body tissues. This fluid is thus called *hemolymph.* Insects, like the fly pictured here, have a muscular tube that serves as a heart to pump the hemolymph through a network of open-ended channels that empty into the cavities in the body (downward-pointing arrows). There, the hemolymph delivers nutrients to the cells of the body. It then reenters the circulatory system through pores in the heart (upward-pointing arrows). The pores close when the heart pumps to keep the hemolymph from flowing back out into the body cavity.

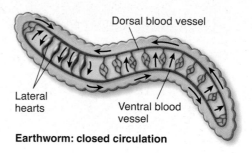

Earthworm: closed circulation

In a **closed circulatory system,** the circulating fluid, or *blood,* is always enclosed within blood vessels that transport blood away from and back to a pump, the *heart.* Blood pumped to and from the hearts always remains within this system of vessels. Annelids and all vertebrates have a closed circulatory system. In annelids, such as the earthworm seen here, a dorsal blood vessel contracts rhythmically to function as a pump. Blood is pumped through five small connecting vessels called lateral hearts, which also function as pumps, to a ventral blood vessel, which transports the blood posteriorly (lower arrows) until it eventually reenters the dorsal blood vessel (upper arrows). Smaller vessels branch between the ventral and dorsal blood vessels to supply the tissues of the earthworm with oxygen and nutrients and to carry away waste products.

In vertebrates, blood vessels form a tubular network that permits blood to flow from the heart to all the cells of the body and then back to the heart. *Arteries* carry blood away from the heart, whereas *veins* return blood to the heart. Blood passes from the arterial to the venous system in *capillaries,* which are the thinnest and most numerous of the blood vessels.

As blood plasma passes through capillaries, the pressure of the blood forces some of this fluid out of the capillary walls. Fluid derived this way is called **interstitial fluid.** Some of this fluid returns directly to capillaries, and some enters into **lymph vessels,** located in the connective tissues around the blood vessels. This fluid, now called *lymph,* is returned to the venous blood at specific sites. The lymphatic system is considered a part of the circulatory system and is discussed later in this chapter.

The Functions of Vertebrate Circulatory Systems

The functions of the circulatory system can be divided into three areas: transportation, regulation, and protection.

1. *Transportation.* Substances essential for cellular functions are transported by the circulatory system. These substances can be categorized as follows:

   Respiratory. Red blood cells, or erythrocytes, transport oxygen to the tissue cells. In the capillaries of lungs or gills, oxygen attaches to hemoglobin molecules within the erythrocytes and is transported to the cells for aerobic respiration. Carbon dioxide produced by cell respiration is carried by the blood to the lungs or gills for elimination.

   Nutritive. The digestive system is responsible for the breakdown of food so that nutrients can be absorbed through the intestinal wall and into the blood vessels of the circulatory system. The blood then carries these absorbed products of digestion through the liver and to the cells of the body.

   Excretory. Metabolic wastes, excessive water and ions, and other molecules in plasma (the fluid portion of blood) are filtered through the capillaries of the kidneys and excreted in urine.

   Endocrine. The blood carries hormones from the endocrine glands, where they are secreted, to the distant target organs they regulate.

2. *Regulation.* The cardiovascular system participates in temperature regulation in two ways:

   Temperature regulation. In warm-blooded vertebrates, or homeotherms, a constant body temperature is maintained, regardless of the surrounding temperature. This is accomplished in part by blood vessels located just under the epidermis. When the ambient temperature is cold, the superficial vessels constrict to divert the warm blood to deeper vessels. When the ambient temperature is warm, the superficial vessels dilate so that the warmth of the blood can be lost by radiation.

   Heat exchange. Some vertebrates also retain heat in a cold environment by using a **countercurrent heat exchange. Figure 23.1** shows how a countercurrent heat exchange system works in the flipper of a killer whale. In this process, a vessel carrying warm blood from deep within the body (colored red) passes next to a

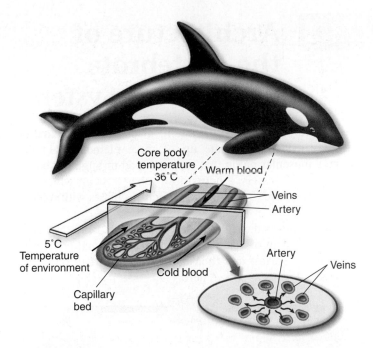

Figure 23.1 Countercurrent heat exchange.
Many marine mammals, such as this killer whale, limit heat loss in cold water by countercurrent flow that allows heat exchange between arteries and veins. The warm blood pumped from within the body in arteries warms the cold blood returning from the skin in veins, so that the core body temperature can remain constant in cold water. The cutaway portion in the figure shows how the veins surround the artery, maximizing the heat exchange between the artery and the veins.

vessel carrying cold blood from the surface of the body (colored blue). The warm blood going out heats the cold blood returning from the body surface (heat indicated by the red arrows in the cross section), so that this blood is no longer cold when it reaches the interior of the body, helping to maintain a stable core body temperature.

3. *Protection.* The circulatory system protects against injury and foreign microbes or toxins introduced into the body:

   Blood clotting. The clotting mechanism protects against blood loss when vessels are damaged. This clotting mechanism involves both proteins from the blood plasma and blood cell structures called platelets (discussed in section 23.4).

   Immune defense. The blood contains proteins and white blood cells, or leukocytes, that provide immunity against many disease-causing agents. Some white blood cells are phagocytic, some produce antibodies, and some act by other mechanisms to protect the body.

> **Key Learning Outcome 23.1** Circulatory systems may be open or closed. All vertebrates have a closed circulatory system, in which blood circulates away from the heart in arteries and back to the heart in veins. The circulatory system serves a variety of functions, including transportation, regulation, and protection.

Architecture of the Vertebrate Circulatory System

The vertebrate circulatory system, also called the **cardiovascular system,** is made up of three elements: (1) the **heart,** a muscular pump that pushes blood through the body; (2) the **blood vessels,** a network of tubes through which the blood moves; and (3) the **blood,** which circulates within these vessels.

Blood moves through the body in a cycle, from the heart, through a system of vessels: From the arteries and arterioles, into the capillaries, and then back to the heart through the venules and veins as shown here:

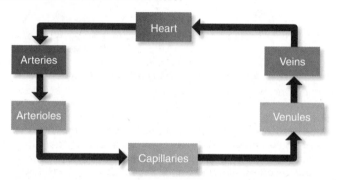

Blood leaves the heart through vessels known as **arteries.** From the arteries, the blood passes into a network of smaller arteries called **arterioles,** shown in **figure 23.2a ❶**. From these, it is eventually forced through a capillary bed **❷**, a fine

latticework of very narrow tubes called **capillaries** (from the Latin, *capillus,* "a hair"). While passing through the capillaries, the blood exchanges gases and metabolites (glucose, vitamins, hormones) with the cells of the body. After traversing the capillaries, the blood passes into a fourth kind of vessel, the **venules,** or small veins, **❸**. A network of venules empties into larger **veins** that collect the circulating blood and carry it back to the heart.

Capillary beds can be opened or closed, based on the physiological needs of the tissues. Smooth muscles (discussed later) can control blood flow to the arterioles, but, in addition, flow through the capillaries can be controlled by the relaxation or contraction of small circular muscles called *precapillary sphincters.* The closing of the precapillary sphincters by contraction is shown in **figure 23.2b** with the blood being diverted from the capillary bed.

The capillaries have a much smaller diameter than the other blood vessels of the body. Blood leaves the mammalian heart through a large artery, the aorta, a tube that in your body has a diameter of about 2 centimeters (about the same as your thumb). But when blood reaches the capillaries, it passes through vessels with an average diameter of only 8 micrometers, a reduction in radius of some 1,250 times!

This decrease in size of blood vessels has a very important consequence. Although each capillary is very narrow, there are so many of them that the capillaries have the greatest *total* cross-sectional area of any other type of vessel. Consequently, this allows more time for blood to exchange materials with the surrounding extracellular fluid. By the time the blood reaches the end of a capillary, it has released some of its oxygen and nutrients and picked up carbon dioxide and other waste products. Blood loses most of its pressure and velocity

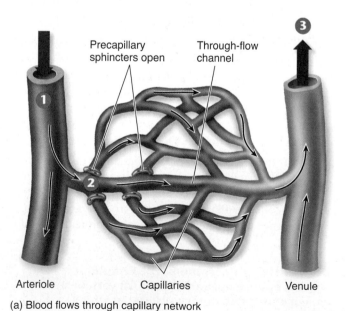

(a) Blood flows through capillary network

(b) Blood flow in capillary network is limited

Figure 23.2 The capillary network connects arteries with veins.
Through-flow channels connect arterioles directly to venules. Branching from these through-flow channels is a network of finer channels, the capillaries. Most of the exchange between the body tissues and the red blood cells occurs while they are in this capillary network. The flow of blood into the capillaries is controlled by bands of muscle called precapillary sphincters located at the entrance to each capillary. (a) When a sphincter is open, blood flows through that capillary. (b) When a sphincter contracts, it closes off the capillary.

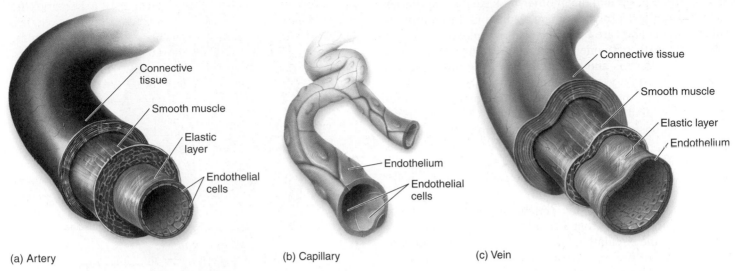

(a) Artery   (b) Capillary   (c) Vein

Figure 23.3   The structure of blood vessels.
(*a*) Arteries, which carry blood away from the heart, are expandable and are composed of layers of tissue. (*b*) Capillaries are simple tubes whose thin walls facilitate the exchange of materials between the blood and the cells of the body. (*c*) Veins, which transport blood back to the heart, do not need to be as sturdy as arteries. The walls of veins have thinner muscle layers than arteries, and they collapse when empty. Note, this drawing is not to scale; as stated in the text, arteries can be up to 2 centimeters in diameter, the capillaries are only about 8 micrometers in diameter, and the largest veins can be up to 3 centimeters in diameter.

in passing through the vast capillary networks, and so is under very low pressure when it enters the veins. The blood flow through the capillaries is like water flowing out of the sprinkler head of a watering can—the stream of blood spreads out to many small streams. These smaller streams don't flow with as much force, nor as quickly, as the larger stream that entered the capillary bed.

Arteries: Highways from the Heart

The arterial system, composed of arteries and arterioles, carries blood away from the heart. An artery is more than simply a pipe. Blood comes from the heart in pulses rather than in a smooth flow, slamming into the artery in great big slugs as the heart forcefully ejects its contents with each contraction. The artery has to be able to *expand* to withstand the pressure caused by each contraction of the heart. An artery, then, is designed as an expandable tube, with its walls made up of four layers of tissue. **Figure 23.3***a* shows these layers pulled out from the artery, like a telescope so they are more easily seen. The innermost thin layer is composed of endothelial cells. Surrounding them is a layer of elastic fibers and then a thick layer of smooth muscle, which in turn is encased within an envelope of protective connective tissue. Because this sheath and envelope are elastic, the artery is able to expand its volume considerably when the heart contracts, shoving a new volume of blood into the artery—just as a tubular balloon expands when you blow more air into it. The steady contraction of the smooth muscle layer strengthens the wall of the vessel against overexpansion.

Arterioles differ from arteries in two ways. They are smaller in diameter, and the muscle layer that surrounds an arteriole can be relaxed under the influence of hormones to enlarge the diameter. When the diameter increases, the blood flow also increases, an advantage during times of high body activity. Most arterioles are also in contact with nerve fibers. When stimulated by these nerves, the muscle lining of the arteriole con-

tracts, constricting the diameter of the vessel. Such contraction limits the flow of blood to the extremities during periods of low temperature or stress. You turn pale when you are scared or cold because the arterioles in your skin are constricting. You blush for just the opposite reason. When you overheat or are embarrassed, the nerve fibers connected to muscles surrounding the arterioles are inhibited, which relaxes the smooth muscle and causes the arterioles in the skin to expand, bringing heat to the surface for escape.

Capillaries: Where Exchange Takes Place

Capillaries are where oxygen and food molecules are transferred from the blood to the body's cells and where waste carbon dioxide is picked up. To facilitate this back-and-forth traffic, capillaries are narrow (**figure 23.4**) and have thin walls

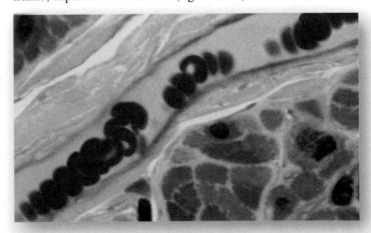

Figure 23.4   Red blood cells within a capillary.
The red blood cells in this capillary pass along in single file. Red blood cells can pass through capillaries even narrower than their own diameter, pushed along by the pressure of the pumping heart.

across which gases and metabolites pass easily. Capillaries have the simplest structure of any element in the cardiovascular system. They are built like a soft-drink straw, simple tubes with walls only one cell thick (see **figure 23.3b**). The average capillary is about 1 millimeter long and connects an arteriole with a venule. All capillaries are very narrow, with an internal diameter of about 8 micrometers, just bigger than the diameter of a red blood cell (5 to 7 micrometers). This design is critical to the function of capillaries. By bumping against the sides of the vessel as they pass through (like the cells in **figure 23.4**), the red blood cells are forced into close contact with the capillary walls, making exchange easier.

Almost all cells of the vertebrate body are no more than 100 micrometers from a capillary. At any one moment, about 5% of the circulating blood is in capillaries, a network that amounts to several thousand miles in overall length. If all the capillaries in your body were laid end to end, they would extend across the United States! Individual capillaries have high resistance to flow because of their small diameters. However, the total cross-sectional area of the extensive capillary network (that is, the sum of all the diameters of all the capillaries, expressed as area) is greater than that of the arteries leading to it. As a result, the blood pressure is actually far lower in the capillaries than in the arteries. This is important, because the walls of capillaries are not strong, and they would burst if exposed to the pressures that arteries routinely withstand.

Veins: Returning Blood to the Heart

Veins are vessels that return blood to the heart. Veins do not have to accommodate the pulsing pressures that arteries do because much of the force of the heartbeat is weakened by the high resistance and great cross-sectional area of the capillary network. For this reason, the walls of veins have much thinner layers of muscle and elastic fiber, as seen in **figure 23.3c**. An empty artery will stay open, like a pipe, but when a vein is empty, its walls collapse like an empty balloon. In **figure 23.5**

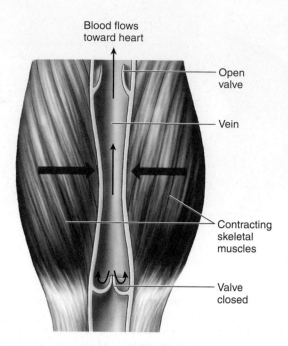

Figure 23.6   Flow of blood through veins.
Venous valves ensure that blood moves through the veins in only one direction back to the heart. This movement of blood is aided by the contraction of skeletal muscles surrounding the veins.

you can see a vein and an artery side-by-side. The vein on the left is partially collapsed, while the artery on the right still holds its shape.

Because the pressure of the blood flowing within veins is low, it becomes important to avoid any further resistance to flow, lest there not be enough pressure to get the blood back to the heart. Because a wide tube presents much less resistance to flow than a narrow one, the internal passageway of veins is often quite large, requiring only a small pressure difference to return blood to the heart. The diameters of the largest veins in the human body, the venae cavae, which lead into the heart, are fully 3 centimeters; this is wider than your thumb! Pressure alone cannot force the blood in the veins back to the heart but several features provide help. Most significantly, when skeletal muscles surrounding the veins contract, they move blood by squeezing the veins. Veins also have unidirectional valves (the small flaps within the vein in **figure 23.6**) that ensure the return of this blood by preventing it from flowing backward. These structural features keep the blood flowing in a cycle through the circulatory system.

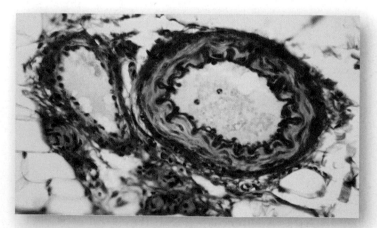

Figure 23.5   Veins and arteries.
The vein (*left*) has the same general structure as the artery (*right*), but an artery retains its shape when empty, while a vein collapses.

> **Key Learning Outcome 23.2** The vertebrate circulatory system is composed of arteries, which carry blood away from the heart; capillaries, a network of narrow tubes across whose thin walls the exchange of gases and food molecules takes place; and veins, which return blood from the capillaries to the heart.

23.3 The Lymphatic System: Recovering Lost Fluid

The cardiovascular system is very leaky. Fluids are forced out across the thin walls of the capillaries by the pumping pressure of the heart. Although this loss is unavoidable—the circulatory system could not do its job of gas and metabolite exchange without tiny vessels with thin walls—it is important that the loss be made up. In your body, about 4 liters of fluid leave your cardiovascular system in this way each day, more than half the body's total supply of about 5.6 liters of blood! To collect and recycle this fluid, the body uses a second circulatory system called the **lymphatic system.** Like the blood circulatory system, the lymphatic system is composed of a network of vessels; their distribution throughout the body is shown in **figure 23.7.**

Fluid is filtered from the capillaries near their arteriole ends where the blood pressure is higher, as shown by the red arrows in **figure 23.8** Most of the fluid returns to the blood capillaries by osmosis (the light purple arrows); however, excess fluid enters the open-ended lymphatic capillaries (indicated by the green

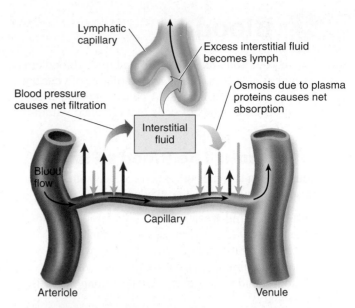

Figure 23.8   Lymphatic capillaries reclaim fluid from interstitial fluid.
Blood pressure forces fluid from capillaries into the interstitial fluid surrounding cells, where it reenters the blood or drains into lymphatic capillaries.

arrow). These capillaries gather up liquid from the spaces surrounding cells and carry it through a series of progressively larger vessels to two large lymphatic vessels, which drain into veins in the lower part of the neck through one-way valves. Once within the lymphatic system, this fluid is called **lymph.**

Fluid is driven through the lymphatic system when its vessels are squeezed by the movements of the body's muscles. The lymphatic vessels contain a series of one-way valves that permit movement only in the direction of the neck. In some cases, the lymphatic vessels also contract rhythmically. In many fishes, all amphibians and reptiles, bird embryos, and some adult birds, movement of lymph is propelled by **lymph hearts.**

The lymphatic system has three other important functions:

1. **It returns proteins to the circulation.** So much blood flows through the capillaries that there is significant leakage of blood proteins, even though the capillary walls are not very permeable to proteins (proteins are very large molecules). By recapturing the lost fluid, the lymphatic system also returns this lost protein.
2. **It transports fats absorbed from the intestine.** Lymph capillaries called *lacteals* penetrate the lining of the small intestine. These lacteals absorb fats from the digestive tract and eventually transport them to the circulatory system by way of the lymphatic system.
3. **It aids in the body's defense.** Along a lymph vessel are swellings called **lymph nodes,** seen in **figure 23.7,** filled with white blood cells specialized for defense. The lymphatic system carries bacteria to the lymph nodes and spleen for destruction.

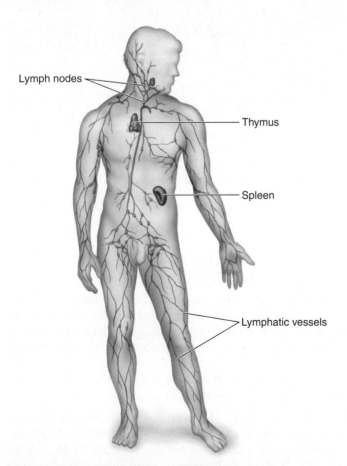

Figure 23.7   The human lymphatic system.
The lymphatic system consists of lymphatic vessels and capillaries, lymph nodes, and lymphatic organs, including the spleen and thymus gland.

Key Learning Outcome 23.3 The lymphatic system returns to the circulatory system fluids and proteins that leak through the capillaries.

Blood

About 5% of your body mass is composed of the blood circulating through the arteries, veins, and capillaries of your body. This blood is composed of a fluid called **plasma,** together with several different kinds of cells that circulate within that fluid.

Blood Plasma: The Blood's Fluid

Blood plasma is a complex solution of water with three very different sorts of substances dissolved within it:

1. **Metabolites and wastes.** If the circulatory system is the highway of the vertebrate body, the blood contains the traffic traveling on that highway. Dissolved within its plasma are glucose, vitamins, hormones, and wastes that circulate among the cells of the body.

2. **Salts and ions.** Like the seas in which life arose, plasma is a dilute salt solution. The chief plasma ions are sodium, chloride, and bicarbonate. In addition, trace amounts of other salts, such as calcium and magnesium, as well as metallic ions, including copper, potassium, and zinc, are present in plasma. The composition of the plasma is not unlike that of seawater.

3. **Proteins.** Blood plasma is 90% water. Passing by all the cells of the body, blood would soon lose most of its water to them by osmosis if it did not contain as high a concentration of proteins as the cells it passes. Some of the proteins blood plasma contains are antibodies that are active in the immune system. More than half the amount of protein that is necessary to balance the protein content of the cells of the body consists of a single protein, **serum albumin,** which circulates in the blood as an osmotic counterforce. Human blood contains 46 grams of serum albumin per liter—that's over half a pound of it in your body. Starvation and protein deficiency result in reduced levels of protein in the blood. This lack of plasma proteins produces swelling of the body because the body's cells, which now have a higher level of solutes than the blood, take up water from the albumin-deficient blood. A symptom of protein deficiency diseases such as kwashiorkor is edema, a swelling of tissues, although other factors can also result in edema.

   The liver produces most of the plasma proteins: albumin; the alpha and beta globulins, which serve as carriers of lipids and steroid hormones; and *fibrinogen,* which is required for blood clotting. When blood in a test tube clots, the fibrinogen is converted into insoluble threads of *fibrin* that become part of the clot. You can see a blood clot forming in **figure 23.9.** Red blood cells, the disk-shaped cells, are becoming trapped in and among the threads of fibrin. The fluid that's left, which lacks fibrinogen and so cannot clot, is called **serum.**

Figure 23.9 Threads of fibrin.
This scanning electron micrograph (×1,430) shows fibrin threads among red blood cells. Fibrin is formed from a soluble protein, fibrinogen, in the plasma to produce a blood clot when a blood vessel is damaged.

Blood Cells: Cells That Circulate Through the Body

Although blood is liquid, nearly half of its volume is actually occupied by cells. The three principal cellular components of blood are erythrocytes (red blood cells), leukocytes (white blood cells), and cell fragments called platelets. The fraction of the total volume of the blood that is occupied by red blood cells is referred to as the blood's **hematocrit.** In humans, the hematocrit is usually about 45%.

Erythrocytes Carry Hemoglobin Each microliter (1 µl) of blood contains about 5 million **erythrocytes.** Each human erythrocyte, pictured at the top of **figure 23.10,** is a flat disk with a central depression on both sides, something like a doughnut with a hole that doesn't go all the way through. Erythrocytes carry oxygen to the cells of the body. Almost the

Blood cell	Life span in blood	Function
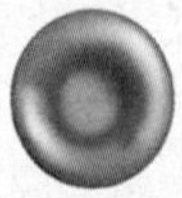 Erythrocyte	120 days	O_2 and CO_2 transport
 Neutrophil	7 hours	Immune defenses
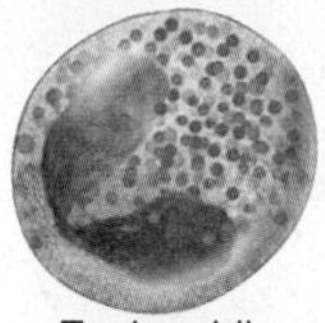 Eosinophil	Unknown	Defense against parasites
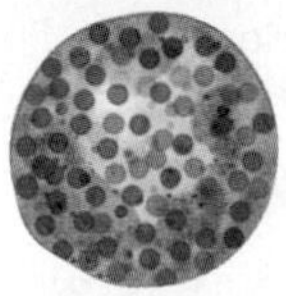 Basophil	Unknown	Inflammatory response
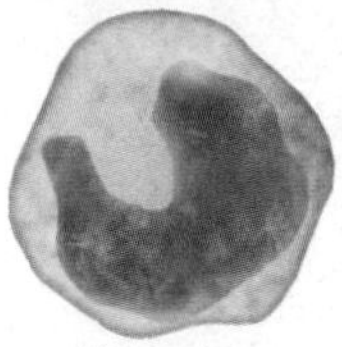 Monocyte	3 days	Immune surveillance (precursor of tissue macrophage)
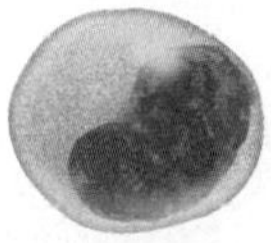 B lymphocyte	Unknown	Antibody production (precursor of plasma cells)
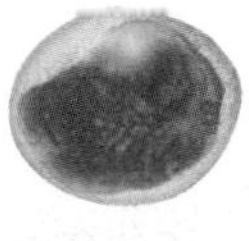 T lymphocyte	Unknown	Cellular immune response
 Platelets	7–8 days	Blood clotting

Figure 23.10 Types of blood cells.

Erythrocytes, leukocytes (neutrophils, eosinophils, basophils, monocytes, and lymphocytes), and platelets are the three principal cellular components of blood in vertebrates.

entire interior of an erythrocyte is packed with hemoglobin, a protein that binds oxygen in the lungs and delivers it to the cells of the body.

Mature mammalian erythrocytes function like box-cars rather than trucks. Like a vehicle without an engine, erythrocytes contain neither a nucleus nor the machinery to make proteins. Because they lack a nucleus, these cells are unable to repair themselves and therefore have a rather short life; any one human erythrocyte lives only about four months. New erythrocytes are constantly being synthesized and released into the blood by cells within the soft interior marrow of bones.

Leukocytes Defend the Body Less than 1% of the cells in mammalian blood are **leukocytes,** also called **white blood cells.** Leukocytes (the somewhat transparent cells shown in **figure 23.10** with large or odd-shaped nuclei) are larger than red blood cells. They contain no hemoglobin and are essentially colorless. There are several kinds of leukocytes, each with a different function. Neutrophils are the most numerous of the leukocytes, followed in order by lymphocytes, monocytes, eosinophils, and basophils. *Neutrophils* attack like kamikazes, responding to foreign cells by releasing chemicals that kill all the cells in the neighborhood—including themselves. Monocytes give rise to *macrophages,* which attack and kill foreign cells by ingesting them (the name means "large eater"). Lymphocytes include *B cells,* which produce antibodies, and *T cells,* which kill infected body cells.

All of these white blood cell types, and others, help defend the body against invading microorganisms and other foreign substances, as you will see in chapter 27. Unlike other blood cells, leukocytes are not confined to the bloodstream; they are mobile soldiers that also migrate out into the fluid surrounding cells.

Platelets Help Blood to Clot Certain large cells within the bone marrow, called **megakaryocytes,** regularly pinch off bits of their cytoplasm. These cell fragments, called **platelets** (shown at the bottom of **figure 23.10**), contain no nuclei. Entering the bloodstream, they play a key role in blood clotting. In a clot, a gluey mesh of fibrin protein fibers (shown in **figure 23.9**) sticks platelets together to form a mass that plugs the rupture in the blood vessel. The clot provides a tight, strong seal, much as the inner lining of a tubeless tire seals punctures. The fibrin that forms the clot is made in a series of reactions that start when circulating platelets first encounter the site of an injury. Responding to chemicals released by the damaged blood vessel, platelets release a protein factor into the blood that starts the clotting process.

Key Learning Outcome 23.4 Blood is a collection of cells that circulate within a protein-rich, salty fluid called plasma. Some of the cells circulating in the blood carry out gas transport; others are engaged in defending the body from infection.

23.5 Fish Circulation

In chapter 22, we described one of the greatest evolutionary achievements of animals—locomotion. As the body size and physiological abilities of animals increased, so did the need for more efficient mechanisms to both deliver nutrients and oxygen and remove wastes and carbon dioxide from the growing mass of tissues. Vertebrates have evolved a remarkable set of adaptations to meet this challenge.

The chordates ancestral to the vertebrates are thought to have had simple tubular hearts, similar to those now seen in lancelets (see chapter 19). The heart was little more than a specialized zone of the ventral artery, more heavily muscled than the rest of the arteries, which contracted in simple peristaltic waves.

The development of gills by fishes required a more efficient pump, and in fishes we see the evolution of a true chamber-pump heart. The fish heart, shown in **figure 23.11a**, is in essence, a tube with four chambers arrayed one after the other. The first two chambers—the **sinus venosus (SV)** and **atrium (A)**—are collection chambers, while the second two—the **ventricle (V)** and **conus arteriosus (CA)**—are pumping chambers. The SV and the CA chambers are greatly reduced in higher vertebrates.

As might be expected from the early chordate hearts from which the fish heart evolved, the sequence of the heartbeat in fishes is a peristaltic sequence, starting at the rear (the SV) and moving to the front (to the CA). The first of the four chambers to contract is the sinus venosus, followed by the atrium, the ventricle, and finally the conus arteriosus. Despite shifts in the relative positions of the chambers in the vertebrates that evolved later, this heartbeat sequence is maintained in all vertebrates. In fish, the electrical impulse that produces the contraction is initiated in the sinus venosus; in other vertebrates, the electrical impulse is initiated by their equivalent of the sinus venosus.

The fish heart is remarkably well suited to the gill respiratory apparatus and represents one of the major evolutionary innovations in the vertebrates. Perhaps its greatest advantage is that the blood it delivers to the tissues of the body is fully oxygenated because it is pumped through the gills first, as shown in the circulation cycle in **figure 23.11b**. Blood is pumped first through the gills, toward the right side of the cycle, where it becomes oxygenated; from the gills, it flows through a network of arteries and capillaries to the rest of the body; then it returns to the heart through the veins. This arrangement has one great limitation, however. Recall from the discussion in section 23.2 that blood loses pressure when it passes through capillaries, and so in fish, the blood loses much of the pressure developed by the contraction of the heart as it passes through the capillaries in the gills. Because of this, the circulation from the gills through the rest of the body is sluggish. This feature limits the rate of oxygen delivery to the rest of the body.

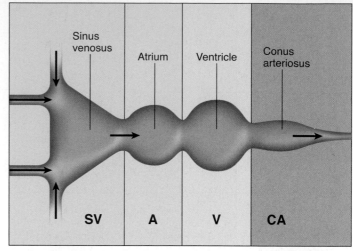

(a)

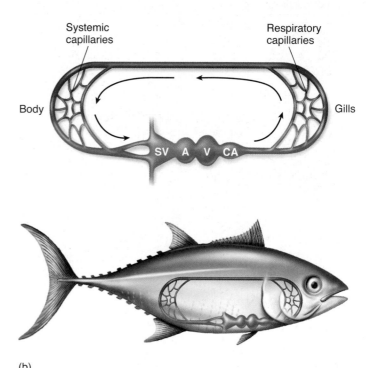

(b)

Figure 23.11   The heart and circulation of a fish.

(a) Diagram of a fish heart, showing the chambers in series with each other. (b) Diagram of the fish circulation, showing that blood is pumped to the gills, and then blood flows directly to the body. Blood rich in oxygen (oxygenated) is shown in *red*; blood low in oxygen (deoxygenated) is shown in *blue*.

> **Key Learning Outcome 23.5**   The fish heart is a modified tube, consisting of a series of four chambers. Blood first enters the heart at the sinus venosus, where the wavelike contraction of the heart begins.

23.6 Amphibian and Reptile Circulation

The advent of lungs involved a major change in the pattern of circulation. After blood is pumped by the heart to the lungs, it does not go directly to the tissues of the body but instead returns to the heart. This results in two circulations: one that goes between the heart and the lungs, called the **pulmonary circulation,** and one that goes between the heart and the rest of the body, called the **systemic circulation.**

If no changes had occurred in the structure of the heart, the oxygenated blood from the lungs would be mixed in the heart with the deoxygenated blood returning from the rest of the body. Consequently, the heart would pump a mixture of oxygenated and deoxygenated blood rather than fully oxygenated blood. The amphibian heart has several structural features that help reduce this mixing. First, the atrium is divided by a *septum,* or dividing wall, into two chambers: The right atrium (the blue-colored area in **figure 23.12a**) receives deoxygenated blood from the systemic circulation, and the left atrium (colored red) receives oxygenated blood from the lungs. The septum prevents the two stores of blood from mixing in the atria, but some mixing might be expected when the contents of each atrium enter the single, common ventricle (the purple-colored area). Surprisingly, however, little mixing actually occurs. Two factors contribute to the separation of the two blood supplies. First, the ventricle in some amphibians is lined with folds that help direct the flow of blood from the atria. Second, the conus arteriosus (the branched structure in the foreground) is partially separated by another septum that directs deoxygenated blood into the *pulmonary arteries* toward the lungs, and oxygenated blood into the *aorta,* the major artery of the systemic circulation to the body. To trace the blood flow through the heart, follow the deoxygenated blood, the blue arrows, from the body to the lungs and then the oxygenated blood, the red arrows, from the lungs out to the body.

Amphibians in water supplement the oxygenation of the blood by obtaining additional oxygen by diffusion through their skin. This process is called *cutaneous respiration.*

Among reptiles, additional modifications have reduced the mixing of blood in the heart still further. In addition to having two separate atria, reptiles have a septum that partially subdivides the ventricle, with red and blue halves of the ventricle in **figure 23.12b.** This results in an even greater separation of oxygenated and deoxygenated blood within the heart. The separation is complete in one order of reptiles, the crocodiles, which have two separate ventricles divided by a complete septum. Crocodiles therefore have a completely divided pulmonary and systemic circulation. Another change in the circulation of reptiles is that the conus arteriosus has become incorporated into the trunks of the large arteries leaving the heart.

> **Key Learning Outcome 23.6** Amphibians and reptiles have two circulations, pulmonary and systemic, that deliver blood to the lungs and to the rest of the body, respectively.

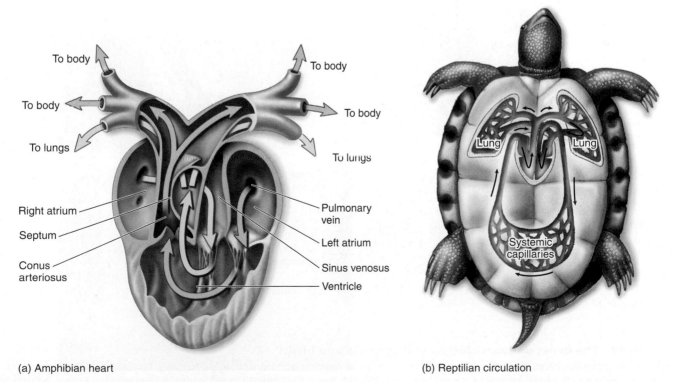

(a) Amphibian heart

(b) Reptilian circulation

Figure 23.12 The amphibian heart and reptilian circulation.
(*a*) The frog heart has two atria but only one ventricle, which pumps blood both to the lungs and to the body. Despite the potential for mixing, the oxygenated and deoxygenated bloods (*red* and *blue,* respectively) mix very little as they are pumped to the body and lungs. (*b*) In reptiles, not only are there two separate atria, but the ventricle is also partially divided.

23.7 Mammalian and Bird Circulation

Mammals, birds, and crocodiles have a four-chambered heart that is really two separate pumping systems operating together within a single unit. One of these pumps blood to the lungs, while the other pumps blood to the rest of the body. The left side has two connected chambers, and so does the right, but the two sides are not connected with one another. The increased efficiency of the double circulatory system in mammals and birds is thought to have been important in the evolution of endothermy (warm-bloodedness), because a more efficient circulation is necessary to support the high metabolic rate required.

Circulation Through the Heart

Let's follow the journey of blood through the mammalian heart in **figure 23.13a,** starting with the entry of oxygen-rich blood into the heart from the lungs. Oxygenated blood (the pink arrows) from the lungs enters the left side of the heart (which is on the right as you look at the figure), emptying directly into the **left atrium** through large vessels called the **pulmonary veins.** From the atrium, blood flows through an opening into the adjoining chamber, the **left ventricle.** Most of this flow, roughly 70%, occurs while the heart is relaxed. When the heart starts to contract, the atrium contracts first, pushing the remaining 30% of its blood into the ventricle.

After a slight delay, the ventricle contracts. The walls of the ventricle are far more muscular than those of the atrium (as seen in this cross section), and thus this contraction is much stronger. It forces most of the blood out of the ventricle in a single strong pulse. The blood is prevented from going back into the atrium by a large, one-way valve, the **bicuspid (mitral) valve,** or left atrioventricular valve, whose two flaps are pushed shut as the ventricle contracts.

Prevented from reentering the atrium, the blood within the contracting left ventricle takes the only other passage out (partially covered by the large blue vessel in the figure). It moves through a second opening that leads into a large blood vessel called the **aorta.** The aorta is separated from the left

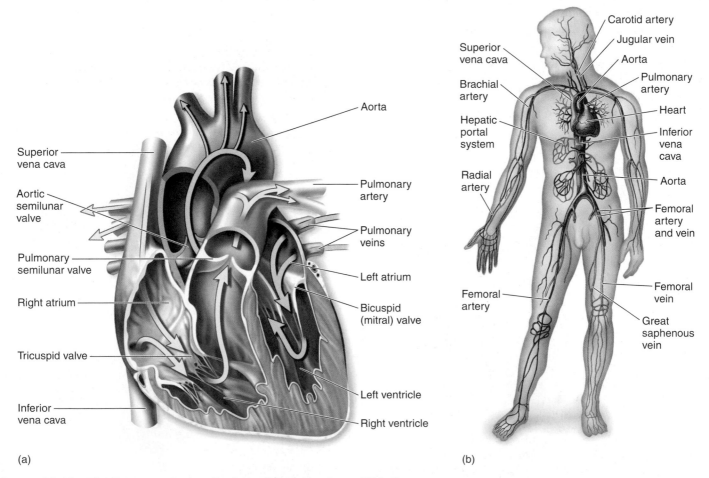

Figure 23.13   The heart and circulation of mammals and birds.

(*a*) Unlike the amphibian heart, this heart has a septum dividing the ventricle into left and right ventricles. Oxygenated blood from the lungs enters the left atrium of the heart by way of the pulmonary veins. This blood then enters the left ventricle, from which it passes into the aorta to circulate throughout the body and deliver oxygen to the tissues. When gas exchange has taken place at the tissues, veins return blood to the heart. After entering the right atrium by way of the superior and inferior venae cavae, deoxygenated blood passes into the right ventricle and then through the pulmonary valve to the lungs by way of the pulmonary artery. (*b*) Some of the major arteries and veins in the human circulatory system are shown.

ventricle by a one-way valve, the **aortic semilunar valve.** The aortic valve is oriented to permit the flow of the blood *out* of the ventricle. Once this outward flow has occurred, the aortic valve closes, preventing the reentry of blood from the aorta into the heart. The aorta and its many branches are systemic arteries and carry oxygen-rich blood to all parts of the body.

Eventually, this blood returns to the heart after delivering its cargo of oxygen to the cells of the body. In returning, it passes through a series of progressively larger veins, ending in two large veins that empty into the right atrium of the heart. The **superior vena cava** drains the upper body (extending from the heart to the top of the figure), and the **inferior vena cava** drains the lower body (extending from the heart downward).

The right side of the heart is similar in organization to the left side. Blood (the blue-colored arrows) passes from the **right atrium** into the **right ventricle** through a one-way valve, the **tricuspid valve** or right atrioventricular valve. It passes out of the contracting right ventricle through a second valve, the **pulmonary semilunar valve,** into the **pulmonary arteries,** which carry the deoxygenated blood to the lungs. The blood then returns from the lungs to the left side of the heart with a new cargo of oxygen, which is pumped to the rest of the body. **Figure 23.13***b* shows the major veins and arteries in the human body. Veins carry blood to the heart and arteries carry blood away from the heart.

Because the overall circulatory system is closed, the same volume of blood must move through the pulmonary circulation as through the much larger systemic circulation with each heartbeat. Therefore, the right and left ventricles must pump the same amount of blood each time they contract. If the output of one ventricle did not match that of the other, fluid would accumulate and pressure would increase in one of the circuits. The result would be increased filtration out of the capillaries and edema (as occurs in congestive heart failure, for example). Although the volume of blood pumped by the two ventricles is the same, the pressure they generate is not. The left ventricle, which pumps blood through the systemic pathway, is more muscular and generates more pressure than the right ventricle.

The evolution of multicellular organisms has depended critically on the ability to circulate materials throughout the body efficiently. Vertebrates carefully regulate the operation of their circulatory systems and are able to integrate their body activities. Indeed, the metabolic demands of different vertebrates have shaped the evolution of circulatory systems.

Monitoring the Heart's Performance

The simplest way to monitor heartbeat is to listen to the heart at work, using a stethoscope. The first sound you hear, a low-pitched *lub,* is the closing of the bicuspid and tricuspid valves at the start of ventricular contraction. A little later, you hear a higher-pitched *dub,* the closing of the pulmonary and aortic valves at the end of ventricular contraction. If the valves are not closing fully, or if they open incompletely, a turbulence is created within the heart. This turbulence can be heard as a **heart murmur,** a liquid sloshing sound.

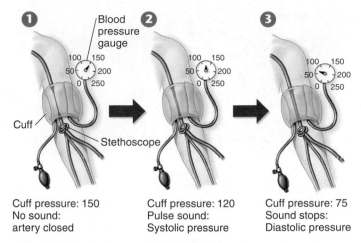

Figure 23.14 Measuring blood pressure.
The blood pressure cuff is tightened to stop the blood flow through the brachial artery ❶. As the cuff is loosened, the systolic pressure is recorded as the pressure at which a pulse is heard through a stethoscope ❷. The diastolic pressure is recorded as the pressure at which a sound is no longer heard ❸.

A second way to examine the events of the heartbeat is to monitor the blood pressure. This is done using a device called a sphygmomanometer, which measures the blood pressure in the brachial artery found on the inside part of the arm, at the elbow (**figure 23.14**). A cuff wrapped around the upper arm is tightened enough to stop the flow of blood to the lower part of the arm ❶. As the cuff is loosened, blood begins pulsating through the artery and can be detected using a stethoscope. Two measurements are recorded: The systolic pressure ❷ is recorded when a pulse is heard, and the diastolic pressure ❸ is recorded when the pressure in the cuff is so low that the sound stops.

To understand these measurements it is important to remember what is happening in the heart. During the first part of the heartbeat, the atria are filling. At this time the pressure in the arteries leading from the left side of the heart out to the tissues of the body decreases slightly. This low pressure is referred to as the **diastolic** pressure. During the contraction of the left ventricle, a pulse of blood is forced into the systemic arterial system, immediately raising the blood pressure within these vessels. The high blood pressure produced in this pushing period, which ends with the closing of the aortic valve, is referred to as the **systolic** pressure. Normal blood pressure values are 70 to 90 diastolic and 110 to 130 systolic. When the inner walls of the arteries accumulate fats, as they do in the condition known as *atherosclerosis,* the diameters of the passageways are narrowed. If this occurs, the systolic blood pressure is elevated.

How the Heart Contracts

Throughout the evolutionary history of the vertebrate heart, the sinus venosus has served as a pacemaker, the site where the impulses that produce the heartbeat originate. Although it constitutes a major chamber in the fish heart, it is reduced in size in amphibians and further reduced in reptiles. In mammals and birds, the sinus venosus is no longer a separate

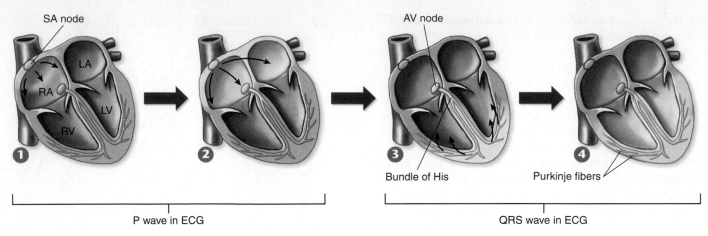

P wave in ECG   QRS wave in ECG

Figure 23.15 How the mammalian heart contracts.
❶ Contraction of the mammalian heart is initiated by an electrical signal (yellow highlighting) that begins at the SA node. The electrical activity of the heart can be recorded on an electrocardiogram (ECG or EKG), shown to the right. ❷ After passing over the right and left atria and causing their contraction (forming the P wave on the ECG), ❸ the signal reaches the AV node, from which it passes to the ventricles by the bundle of His. ❹ The signal is then conducted rapidly over the surface of the ventricles by a set of finer fibers called Purkinje fibers, causing the ventricles to contract (forming the QRS wave on the ECG). The T wave on the ECG corresponds to the relaxation of the ventricles. This ECG is showing a heart rate of 60 beats per minute. The time between peaks varies with changes in heart rates.

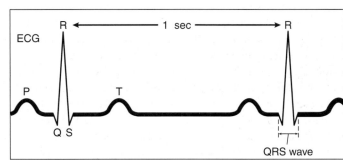

chamber, but some of its tissue remains in the wall of the right atrium, near the point where the systemic veins empty into the atrium. This tissue, which is called the **sinoatrial (SA) node** (indicated in panel ❶ of **figure 23.15**), is still the site where each heartbeat originates.

The contraction of the heart consists of a carefully orchestrated series of muscle contractions. First, the atria contract together, followed by the ventricles. Contraction is initiated by the sinoatrial (SA) node. Its membranes spontaneously depolarize (that is, admit ions that cause it to become more positively charged), creating electrical signals with a regular rhythm that determines the rhythm of the heart's beating. In this way, the SA node acts as a *pacemaker* for the heart. Each electrical signal initiated within this pacemaker region passes quickly from one heart muscle cell to another in a wave that envelops the left and the right atria almost simultaneously (indicated by the yellow coloring in **figure 23.15**, panels ❶ and ❷).

But the electrical signals do not immediately spread to the ventricles. There is a pause before the lower half of the heart starts to contract. The reason for the delay is that the atria of the heart are separated from the ventricles by connective tissue that does not propagate the electrical signal. The signal would not pass to the ventricles at all except for a slender connection of cardiac muscle cells known as the **atrioventricular (AV) node** (labeled in panel ❸), which connects across the gap to a strand of specialized muscle known as the atrioventricular bundle, or **bundle of His.** Bundle branches divide into fast-conducting **Purkinje fibers,** which initiate the almost simultaneous contraction of all the cells of the right and left ventricles about 0.1 seconds after the atria contract.

This delay permits the atria to finish emptying their contents into the corresponding ventricles before those ventricles start to contract. The contraction of the ventricles begins at the apex (the bottom of the heart) where depolarization of Purkinje fibers begins the electrical signal. The contraction then spreads up toward the atria (indicated by the yellow coloring in panels ❸ and ❹). This results in a "wringing out" of the ventricle, forcing the blood up and out of the heart.

Because the vertebrate body basically consists of water, it conducts electrical currents rather well. The electrical signal passing over the surface of the heart generates an electrical current that passes in a wave throughout the body. The magnitude of this electrical pulse is tiny, but it can be detected by sensors placed on the skin. The recording, called an **electrocardiogram (ECG or EKG),** shows how the cells of the heart respond electrically during the cardiac cycle (**figure 23.15**). A **cardiac cycle** consists of one sequence of contraction and relaxation of the heart. The first peak in the recording, P, is produced by the electrical activity of the atria. The second, larger peak, QRS, is produced by ventricular stimulation; during this time, the ventricles contract and eject blood into the arteries. The last peak, T, reflects ventricular relaxation.

> **Key Learning Outcome 23.7** The mammalian heart is a two-cycle pump. The left side pumps oxygenated blood to the body's tissues, while the right side pumps O_2-depleted blood to the lung. The performance of the heart can be monitored in many ways.

Circulation

23.1 Open and Closed Circulatory Systems

- Animal cells acquire oxygen from the environment and nutrients from the food they eat, but not all animals do this in the same way (**integrated art, page 498**). Cnidarians and flatworms have a gastrovascular cavity that circulates the products of gas exchange and digestion.

- Mollusks and arthropods have open circulatory systems. Hemolymph is pumped by tubular hearts through blood vessels that open into a body cavity. The hemolymph delivers oxygen and nutrients to the cells and then the hemolymph reenters the circulatory system through pores in the tubular heart.

- Annelids and all vertebrates have closed circulatory systems. The blood stays within closed vessels that extend throughout the body. The blood is propelled through the body by the pumping of a heart.

- In vertebrates, the circulatory system functions in the transportation of substances throughout the body, in the regulation of body temperature (**figure 23.1**), and in the protection of the body through blood clotting and immunity.

23.2 Architecture of the Vertebrate Circulatory System

- In vertebrates, blood circulates from the heart through arteries and arterioles to capillaries. Blood flows from the capillaries back to the heart through venules and larger veins.

- Capillaries are the sites of gas, nutrient, and waste exchange. A capillary bed can be closed, restricting the flow of blood to that area (**figure 23.2**).

- Blood vessels vary in size and structure from thick-walled arteries to capillaries that are a single cell layer (**figure 23.3**).

- Veins bring blood back to the heart. The blood moves through the veins with the help of contractions of skeletal muscles and one-way valves inside the veins that keep the blood from backing up in the vessel (**figure 23.6**).

23.3 The Lymphatic System: Recovering Lost Fluid

- Needed fluid is lost to the body tissues through leaky capillary walls. This interstitial fluid is recovered by the lymphatic system in lymphatic capillaries (**figures 23.7** and **23.8**). The fluid, now called lymph, drains into lymphatic vessels. Lymphatic vessels return the fluid and lost proteins back to the circulatory system. The lymphatic system has other functions related to digestion and immunity.

23.4 Blood

- Blood is a salty, protein-rich fluid that circulates through the body in the circulatory system. Blood contains fluid called plasma, red blood cells (like the one pictured here from **figure 23.10**) that are involved in gas exchange, and various types of white blood cells, called leukocytes (like the neutrophil cell shown here), that defend the body against infection. Platelets are fragments of cells involved in blood clotting.

Evolution of Vertebrate Circulatory Systems

23.5 Fish Circulation

- The fish heart consists of a series of chambers that contract in sequence. Blood enters the heart and is collected in the first two chambers, the sinus venosus and the atrium. Blood then passes into the third and fourth chambers, the ventricle and conus arteriosus, where it is pumped to the body. It first passes to the gills, where gas exchange occurs, and then oxygenated blood travels throughout the body (**figure 23.11**). Deoxygenated blood returns to the heart, where it is again pumped to the gills.

23.6 Amphibian and Reptile Circulation

- Amphibians and reptiles have lungs and so the blood pumps in two cycles (**figure 23.12**): between the heart and lungs (pulmonary circulation) and between the heart and the body (systemic circulation). In the first cycle, deoxygenated blood from the body enters the heart through the right atrium. From there, it passes to the ventricle and then to the lungs where gas exchange occurs. The oxygenated blood then cycles from the lungs into the left atrium and then passes to the ventricle, where it is pumped out to the body. The ventricle is partially separated in the reptilian heart, which reduces mixing of deoxygenated and oxygenated blood.

23.7 Mammalian and Bird Circulation

- Mammalian and bird hearts are two-cycle pumps—but, unlike amphibians and reptiles, the heart has four chambers (**figure 23.13**). This complete separation of the pulmonary and systemic circulatory systems improves the efficiency of the heart.

- Oxygenated blood from the lungs enters the left side of the heart through the pulmonary vein, emptying into the left atrium. From there the blood passes into the left ventricle where it is pumped out to the body. After delivering oxygen to the tissues of the body, the deoxygenated blood flows back to the heart and enters the right atrium. The blood then passes into the right ventricle. When the ventricle contracts, the blood is pumped to the lungs.

- A relatively easy way to monitor the activity of the heart is with a stethoscope. The lub-dub sounds heard are the closing of heart valves. The activity of the heart can also be monitored by measuring blood pressure. Measurements of the systolic and diastolic pressures indicate how hard the heart is having to work to pump blood through the body (**figure 23.14**).

- A pacemaker, called the sinoatrial (SA) node, controls heartbeat rate. An electrical impulse that begins in the SA node triggers muscles in the atrium to contract. The electrical impulse passes through the atria, as shown in this panel from **figure 23.15**, and stimulates the AV node. An electrical impulse then travels from the AV node down to the apex of the heart, where it initiates another wave of muscle contraction of the ventricles. The electrical impulses of the heart can be detected and recorded as an electrocardiogram (ECG).

1. Which of the following is *not* a function of the circulatory system?
   a. regulation of body temperature
   b. protection against injury, foreign toxins, and microbes
   c. transportation of materials in the body
   d. All of the above are functions of the circulatory system.

2. How do some vertebrates maintain body temperature in cold environments?
   a. by mixing cold blood with warm blood in the heart
   b. by pumping more warm blood to the extremities
   c. by passing warm blood near cold blood in the extremities to warm the blood
   d. All of the above are used by vertebrates in cold environments.

3. Exchange of waste material, oxygen, carbon dioxide, and metabolites such as salts and food molecules, takes place in the
   a. capillaries.
   b. venules.
   c. arterioles.
   d. arteries.

4. The lymphatic system is like the circulatory system in that they both
   a. have nodes that filter out pathogens.
   b. are made up of arteries.
   c. deliver blood to the heart.
   d. carry fluids.

5. The most numerous type of blood cell is the
   a. macrophage.
   b. leukocyte.
   c. platelet.
   d. erythrocyte.

6. What advancement in fish led to a more efficient circulatory system?
   a. A muscular pump is first seen in fishes.
   b. A closed circulatory system is first seen in fishes.
   c. A heart with separate chambers is first seen in fishes.
   d. A double-loop system is first seen in fishes.

7. Additional septa in the amphibian and reptile heart allow for
   a. higher blood pressure to move blood faster.
   b. better separation of oxygenated and deoxygenated blood.
   c. better body temperature regulation.
   d. better transport of food to needy tissues.

8. Which of the following statements is *false?*
   a. Only arteries carry oxygenated blood.
   b. Both arteries and veins have a layer of smooth muscle.
   c. Capillary beds lie between arteries and veins.
   d. Sphincters regulate the flow of blood through capillaries.

9. The four-chambered heart and double-loop vessel system are thought to be important in the evolution of
   a. locomotion.
   b. ectothermy.
   c. exothermy.
   d. endothermy.

10. High blood pressure can be detected using
    a. a stethoscope.
    b. an ECG.
    c. a sphygmomanometer.
    d. All of the above.

1. **Figure 23.6** It has been suggested that being a dedicated "couch potato" or "video game addict" can cause circulatory problems. Explain how this might happen.

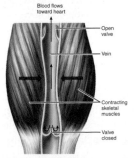

2. **Figure 23.11** By comparing the fish circulatory system to that of a mammal, explain how the fish's circulatory system could be more efficient than that of a mammal without the fish possessing lungs.

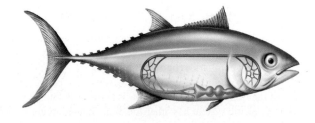

1. When major arteries become partially blocked by deposits of plaque, the heart's left ventricle must work harder and harder to pump enough blood, so that the heart eventually becomes weakened and begins to fail. This condition, known as congestive heart failure, often leads to fatal pulmonary edema (accumulation of fluid). Explain why.

2. Some babies, called "blue babies," are born with a small hole in the wall between their heart's left and right ventricle. Why do you suppose they are called "blue"? [Hint: How might this hole affect the oxygen content of the blood being pumped out into the body's general circulation by the baby's heart?]

24
Respiration

All animals obtain the energy that fuels their lives by consuming other organisms, harvesting energy-rich electrons from the organic molecules of these creatures, and then using these electrons to drive the synthesis of ATP and other molecules. Afterward, the spent electrons are donated to oxygen gas (O_2) to form water (H_2O), while the carbon atoms left over after the electrons were stripped from them combine with oxygen to form carbon dioxide (CO_2). Capturing energy by animals is in effect a process that utilizes oxygen and produces carbon dioxide. The uptake of oxygen and the release of carbon dioxide together are called respiration, neatly defining one of the principal evolutionary challenges facing all animals—how to obtain oxygen and dispose of carbon dioxide. The evolution of respiratory mechanisms among the vertebrates has favored changes that maximize the exchange of these two gases. The most efficient respiratory mechanism to evolve in water is the gill, used by bony fishes and sharks (see photo, *upper left*). The earliest land vertebrates, amphibians like the frog appearing at the *lower left,* utilized simple lungs, and also respired through their moist skin. Reptiles like the crocodile have expandable rib cages to draw air into the lungs. The lung of mammals like the squirrel seen at the *upper right* have a greatly increased interior surface area, making it a more powerful respiratory machine. Birds have an improved respiratory design of the lung with air sacs and crosscurrent flow.

24.1 Types of Respiratory Systems

As discussed in chapter 7, animals obtain the energy they need by oxidizing molecules rich in energy-laden carbon–hydrogen bonds. This oxidative metabolism requires a ready supply of oxygen. The uptake of oxygen and the simultaneous release of carbon dioxide together constitute a form of gas exchange called **respiration.**

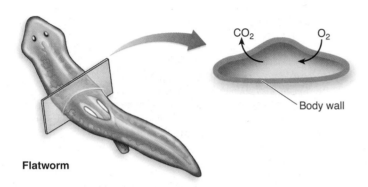

Flatworm

Most of the primitive phyla of organisms obtain oxygen by direct diffusion from their aquatic environments, which contains about 10 milliliters of dissolved oxygen per liter. Sponges, cnidarians, many flatworms and roundworms, and some annelid worms all obtain their oxygen by diffusion from surrounding water. Oxygen and carbon dioxide diffuse across the surface of the body, as shown in the flatworm above. Similarly, some members of the vertebrate class Amphibia conduct gas exchange by direct diffusion through their moist skin.

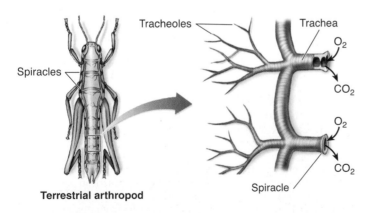

Terrestrial arthropod

Terrestrial arthropods do not have a major respiratory organ. Instead, a network of air ducts called **tracheae** (the purple tubes in the figure above), branching into smaller and smaller tubes, carries air to every part of the body. The tracheae open to the outside of the body through structures called **spiracles,** which can be closed and opened.

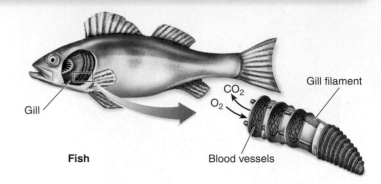

Fish

The more advanced marine invertebrates (mollusks, arthropods, and echinoderms) possess special respiratory organs called gills that increase the surface area available for diffusion of oxygen. A **gill** is basically a thin sheet of tissue that waves through the water. Gills can be simple, as in the papulae of echinoderms, or complex, as in the highly convoluted gills of fish. In fish, the gills are protected by a covering called an operculum (removed in the figure above). Because of this, their gills do not wave in the water, instead water is pumped over the gills and gas exchange occurs across the walls of capillaries contained in the gills.

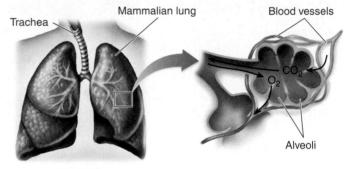

Mammal

Despite the high efficiency of gills as respiratory organs in aquatic environments, gills were replaced in most terrestrial vertebrates with a pair of respiratory organs called **lungs.** The fine filaments of gills lack structural strength and rely on water for support: A fish out of water soon suffocates because its gills collapse. The lungs of terrestrial vertebrates are basically large air sacs, with gas exchange occurring between air drawn into the lung and blood flowing through the capillaries in the wall of the lung sac. In mammals, the surface area of the lung—and thus its oxygen-gathering capabilities—is greatly increased by tiny sub-compartments called *alveoli* (shown in the enlargement above).

> **Key Learning Outcome 24.1** Aquatic animals extract oxygen dissolved in water, some by direct diffusion, others with gills. Terrestrial animals use tracheae or lungs.

24.2 Respiration in Aquatic Vertebrates

Have you ever seen the face of a swimming fish up close? A swimming fish continuously opens and closes its mouth, pushing water through the mouth cavity and out a slit at the rear of the mouth—and (this is the whole point) past the gills on its one-way journey.

This swallowing process, which seems so awkward, is at the heart of a great advance in gill design achieved by the fishes. What is important about the swallowing is that it causes the water to always move past the fish's gills *in one and always the same direction*. Moving the water past the gills in the same direction permits **countercurrent flow,** which is a supremely efficient way of extracting oxygen. Here is how it works:

Each gill is composed of two rows of gill filaments (two sections of gills are shown in the middle panel in **figure 24.1,** each with two rows of gill filaments). The gill filaments are made of thin membranous plates stacked one on top of the other and projecting out into the flow of water. As water flows past the filaments from front to back (indicated by the blue arrows in the enlargements), oxygen diffuses from the water into blood circulating within the gill filament. Within each filament the blood circulation is arranged so that the blood is carried in the direction opposite the movement of the water, from the back of the filament to the front. The advantage of the countercurrent flow system is that blood in the blood vessels of the gill filaments always encounters water that has a higher oxygen concentration, resulting in the diffusion of oxygen into the blood vessels. To understand this, compare the countercurrent exchange system in **figure 24.2a** to a concurrent exchange system in **figure 24.2b.** In the countercurrent system, when blood and water flow in opposite directions, the initial oxygen concentration difference at the bottom is not large (10% in the blood and 15% in water), but it is sufficient for oxygen diffusion. As the blood oxygen concentration increases as it travels upward, the blood continually encounters water with a higher oxygen concentration. Even at 85% oxygen concentration in the blood, it is still encountering oxygen concentrations in water of 100%. In the concurrent exchange system, oxygen diffusion is rapid at

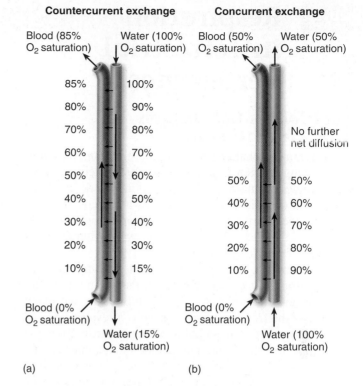

Figure 24.2 Countercurrent flow.

first, because of the large difference in oxygen concentrations between the blood and water (0% versus 100%), but quickly slows as the difference becomes less until it reaches equilibrium at 50% saturation. Thus, countercurrent flow ensures that an oxygen concentration gradient remains between blood and water throughout the flow, permitting oxygen diffusion along the entire length of the filament.

Because of the countercurrent flow, the blood in the fish's gills can build up oxygen concentrations as high as those of the water entering the gills. The gills of bony fishes are the most efficient respiratory machines that have ever evolved among organisms.

> **Key Learning Outcome 24.2** Fish gills achieve countercurrent flow, making them very efficient at extracting oxygen.

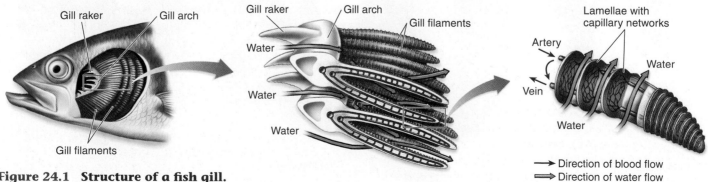

Figure 24.1 Structure of a fish gill.

Water passes from the gill arch over the filaments (from *left* to *right* in the diagram). Water always passes the lamellae in the same direction, which is opposite to the direction the blood circulates across the lamellae.

24.3 Respiration in Terrestrial Vertebrates

Amphibians Get Oxygen from Air with Lungs

One of the major challenges facing the first land vertebrates was obtaining oxygen from air. Fish gills, which are superb oxygen-gathering machines in water, don't work in air. The gill's system of delicate membranes has no means of support in air, and the membranes collapse on top of one another—that's why a fish dies when kept out of water, literally suffocating in air for lack of oxygen.

Unlike a fish, if you lift a frog out of water and place it on dry ground, it doesn't suffocate. Partly this is because the frog is able to respire through its moist skin, but mainly it is because the frog has lungs. A **lung** is a respiratory organ designed like a bag. The amphibian lung is hardly more than a sac with a convoluted internal membrane that opens up to a central cavity (the convoluted membrane is shown in **figure 24.3a**). The air moves into the sac through a tubular passage from the head and then back out again through the same passage. Lungs are not as efficient as gills because new air that is inhaled mixes with old air already in the lung. But air contains about 210 milliliters of oxygen per liter, over 20 times as much as seawater. So, because there is so much more oxygen *in* air, the lung doesn't have to be as efficient as the gill.

Reptiles and Mammals Increase the Lung Surface

Reptiles are far more active than amphibians, so they need more oxygen. But reptiles cannot rely on their skin for respiration the way amphibians can; their dry scaly skin is "watertight" to avoid water loss. Instead, the lungs of reptiles contain

a larger surface area. The internal membrane is also convoluted but the central cavity has many small air chambers, shown as partitions in **figure 24.3b**, which greatly increase the surface area of the lung available for diffusion of oxygen.

Because mammals maintain a constant body temperature by heating their bodies metabolically, they have even greater metabolic demands for oxygen than do reptiles. The problem of harvesting more oxygen is solved by increasing the diffusion surface area within the lung even more. The lungs of mammals possess on their inner surface many small chambers called **alveoli** that look like clusters of grapes in **figure 24.3c**. Each cluster is connected to the main air sac in the lung by a short passageway called a **bronchiole.** Air within the lung passes through the bronchioles to the alveoli, where all oxygen uptake and carbon dioxide disposal takes place. In more active mammals, the individual alveoli are smaller and more numerous, increasing the diffusion surface area even more. Humans have about 300 million alveoli in each of their lungs, for a total surface area devoted to diffusion of about 80 square meters (about 42 times the surface area of the body)!

Birds Perfect the Lung

There is a limit to how much efficiency can be improved by increasing the surface area of the lung, a limit that has already been reached by the more active mammals. This efficiency is not enough for the metabolic needs of birds. Flying creates a respiratory demand for oxygen that exceeds the capacity of the saclike lungs of even the most active mammal. Unlike bats, whose flight involves considerable gliding, most birds beat their wings rapidly as they fly, often for quite a long time. This intensive wing beating uses up a lot of energy quickly, because the wing muscles must contract very frequently. Flying birds thus must carry out very active oxidative respiration within their cells to replenish the ATP expended by their flight muscles, and this requires a great deal of oxygen.

A novel way to improve the efficiency of the lung, one that does not involve further increases in its surface area,

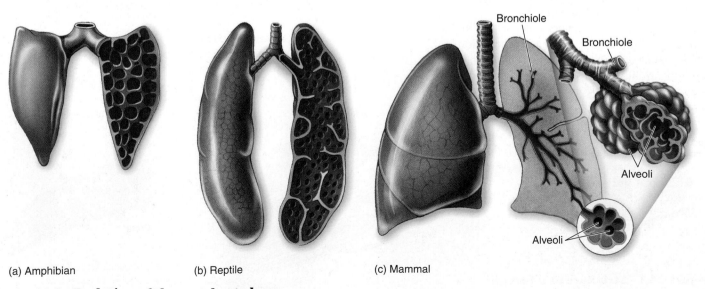

(a) Amphibian   (b) Reptile   (c) Mammal

Figure 24.3 Evolution of the vertebrate lung.

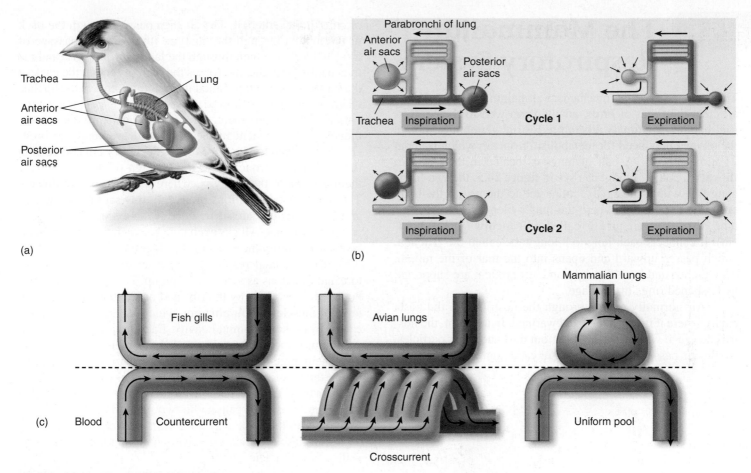

Figure 24.4 How a bird breathes.

(a) A bird's respiratory system is composed of the trachea, anterior air sacs, lungs, and posterior air sacs. (b) Breathing occurs in two cycles: In cycle 1, air is drawn from the trachea into the posterior air sacs and then is exhaled into a lung; in cycle 2, the air is drawn from the lung into the anterior air sacs and then is exhaled through the trachea. Passage of air through the lungs is always in the same direction, from posterior to anterior (*right* to *left* here). (c) Decreasing efficiency of respiratory systems, from fish *(left)*, to bird *(middle)*, to mammal *(right)*, the least efficient.

evolved in birds's lungs. This higher-efficiency lung copes with the demands of flight. Can you guess what it is? In effect, birds do what fishes do! An avian lung is connected to a series of air sacs outside of the lung. When a bird inhales, the air passes by the lung and directly to posterior air sacs you can see in **figure 24.4a**, which act as holding tanks. When the bird exhales, the air flows from these air sacs forward into the lung and then on through another set of anterior air sacs in front of the lungs and out of the body. **Figure 24.4b** shows the three components of the bird's respiratory system: the posterior air sacs, the air passageways through the lungs called the parabronchi, and the anterior air sacs. It takes two breathing cycles for air to pass through the bird's respiratory system. What is the advantage of this complicated pathway? It creates a unidirectional flow of air through the lungs.

Air flows through the lungs of birds in one direction only, from back to front. This one-way air flow results in two significant improvements: (1) There is no dead volume, as in the mammalian lung, so the air passing across the diffusion surface of the bird lung is always fully oxygenated; (2) just as in the gills of fishes, the flow of blood past the lung runs in a direction different from that of the unidirectional air flow. It is not opposite, as in fish; instead, the latticework of capillaries

is arranged across the air flow, at a 90-degree angle, called a **crosscurrent flow** (the middle panel of **figure 24.4c**). This is not as efficient as the 180-degree arrangement of fishes (on the left), but the blood leaving the lung can still contain more oxygen than exhaled air, which no mammalian lung (shown on the right) can do. That is why a sparrow has no trouble flying at an altitude of 6,000 meters on an Andean mountain peak, whereas a mouse of the same body mass and with a similar high metabolic rate will stand panting, unable even to walk. The sparrow is simply getting more oxygen than the mouse.

Just as fish gills are the most efficient aquatic respiratory machines, bird lungs are the most efficient atmospheric ones. Both achieve high efficiency by using different types of current flow systems.

> **Key Learning Outcome 24.3** Terrestrial vertebrates employ lungs to extract oxygen from air. Bird lungs are the most efficient atmospheric respiratory machines, achieving a form of crosscurrent flow.

24.4 The Mammalian Respiratory System

The oxygen-gathering mechanism of mammals, although less efficient than that of birds, adapts them well to their terrestrial habitat. Mammals, like all terrestrial vertebrates, obtain the oxygen they need for metabolism from air, which is about 21% oxygen gas. A pair of lungs is located in the chest, or **thoracic,** cavity. As you can see in **figure 24.5**, the two lungs hang free within the cavity, connected to the rest of the body only at one position, where the lung's blood vessels and air tube enter. This air tube is called a **bronchus** (plural, bronchi). It connects each lung to a long tube called the **trachea,** which passes upward and opens into the rear of the mouth. The trachea and both the right and left bronchi are supported by C-shaped rings of cartilage.

Air normally enters through the nostrils into the nasal cavity where it is moistened and warmed. In addition, the nostrils are lined with hairs that filter out dust and other particles. As the air passes through the nasal cavity, an extensive array of cilia further filters it. The air then passes through the back of the mouth, through the pharynx (the common passage of food and air), and then through the larynx (voice box) and the trachea. Because the air crosses the path of food at the back of the throat, a special flap called the epiglottis covers the trachea whenever food is swallowed, to keep it from "going down the wrong pipe." From there, air passes down through several branchings of bronchi in the lungs and eventually to bronchioles that lead to alveoli. Mucous secretory ciliated cells in the trachea and bronchi also trap foreign particles and carry them upward to the pharynx, where they can be swallowed. The lungs contain millions of alveoli, tiny sacs clustered like grapes. The alveoli are surrounded by an extremely extensive capillary network. All gas exchange between the air and blood takes place across the walls of the alveoli.

The mammal respiratory apparatus is simple in structure and functions as a one-cycle pump. The thoracic cavity is bounded on its sides by the ribs and on the bottom by a thick layer of muscle, the **diaphragm,** which separates the thoracic cavity from the abdominal cavity. Each lung is covered by a very thin, smooth membrane called the **pleural membrane.** This membrane also folds back on itself to line the interior of the thoracic cavity, into which the lungs hang. The space between these two layers of membrane is very small and filled with fluid. This fluid causes the two membranes to adhere to each other in the same way a thin film of water can hold two plates of glass together, effectively coupling the lungs to the walls of the thoracic cavity.

Similar to the way a layer of fluid causes the pleural membrane of the lungs to adhere to the inside of the thoracic cavity, fluid inside the alveoli can cause these air sacs to collapse. This doesn't occur because epithelial cells that line the

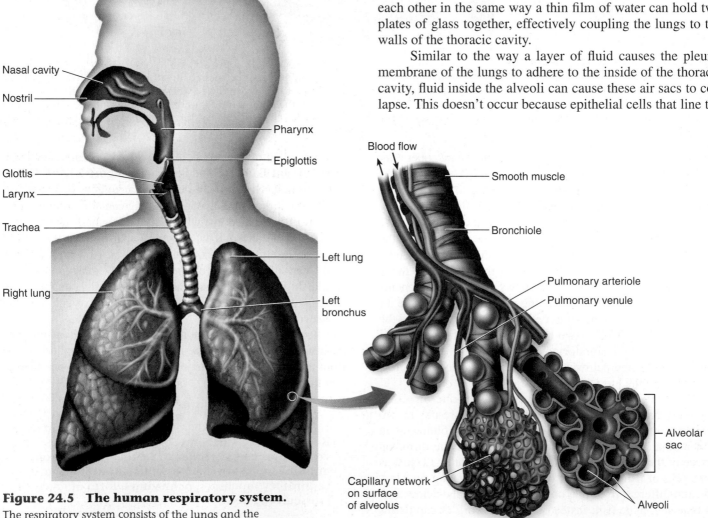

Figure 24.5 The human respiratory system.
The respiratory system consists of the lungs and the passages that lead to them.

KEY BIOLOGICAL PROCESS: Breathing

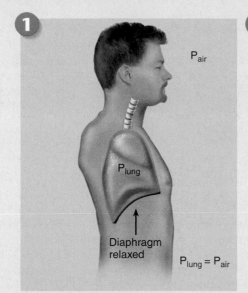

1

P_{air}

P_{lung}

Diaphragm
relaxed

$P_{lung} = P_{air}$

Before inhalation, the air pressure in the lungs (P_{lung}) is equal to the atmospheric pressure (P_{air}).

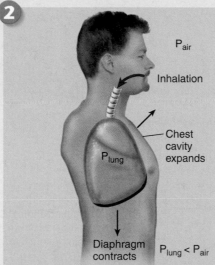

2

P_{air}

Inhalation

Chest
cavity
expands

P_{lung}

Diaphragm
contracts

$P_{lung} < P_{air}$

During inhalation, the diaphragm contracts, and the chest cavity expands downward and outward. This increases the volume of the chest cavity and lungs, which reduces the air pressure inside the lungs, and the air from outside the body flows into the lungs.

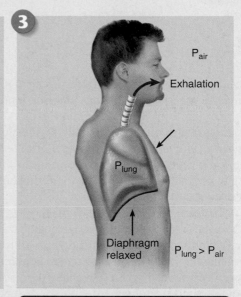

3

P_{air}

Exhalation

P_{lung}

Diaphragm
relaxed

$P_{lung} > P_{air}$

During exhalation, the diaphragm relaxes, decreasing the volume of the chest cavity. The pressure increases in the lungs, forcing air out of the lungs.

---

alveoli secrete a mixture of lipoprotein molecules called *surfactant*. Surfactant molecules form a thin layer on the inner lining of the alveoli and reduce the surface tension that would otherwise cause the alveoli to collapse. Premature babies sometimes suffer from respiratory distress syndrome because surfactant isn't produced in adequate amounts until the seventh month of gestation.

Air is drawn into the lungs by the creation of negative pressure—that is, pressure in the lungs is less than atmospheric pressure. The pressure in the lungs is reduced when the volume of the lungs is increased. This is similar to how a bellow pump or accordion works. In both cases, when the bellow is extended the volume inside increases, causing air to rush in. How does this occur in the lungs? The volume in the lungs increases when muscles that surround the thoracic cavity contract, causing the thoracic cavity to increase in size. Because the lungs adhere to the thoracic cavity, they also expand in size. This creates negative pressure in the lungs, and air rushes in.

The Mechanics of Breathing

The active pumping of air in and out of the lungs is called **breathing.** During *inhalation,* muscular contraction causes the walls of the chest cavity to expand so that the rib cage moves outward and upward. The diaphragm, the red-colored lower border of the lung in the Key Biological Process illustration above, is dome-shaped when relaxed, as in **panel 1**, but moves downward and flattens during this contraction, as in **panel 2**. In effect, we have enlarged the bellow in the accordion.

During *exhalation* (in **panel 3**) the ribs and diaphragm return to their original resting position. In doing so, they exert pressure on the lungs. This pressure is transmitted uniformly over the entire surface of the lung, forcing air from the inner cavity back out to the atmosphere. In a human, a typical breath at rest moves about 0.5 liters of air, called the **tidal volume.** The extra amount that can be forced into and out of the lung is called the **vital capacity** and is about 4.5 liters in men and 3.1 liters in women. The air remaining in the lung after such a maximal expiration is the **residual volume,** or dead volume, typically about 1.2 liters.

Because the diffusion surfaces of the lungs are not exposed to fully oxygenated air, but rather to a mixture of fresh and partly oxygenated air, the respiratory efficiency of mammalian lungs is far from maximal. As explained earlier, a bird, whose lungs do not retain a residual volume, is able to achieve far greater respiratory efficiency.

Key Learning Outcome 24.4 In mammals, the lungs are located within a thoracic cavity that is surrounded by muscles. By contracting and relaxing, these muscles expand or reduce the volume of the cavity, drawing air into the lungs or forcing it out.

24.5 How Respiration Works: Gas Exchange

When oxygen has diffused from the air into the moist cells lining the inner surface of the lung, its journey has just begun. Passing from these cells into the bloodstream, the oxygen is carried throughout the body by the circulatory system, described in chapter 23. It has been estimated that it would take a molecule of oxygen three years to diffuse from your lung to your toe if it moved only by diffusion, unassisted by a circulatory system.

Oxygen moves within the circulatory system carried piggyback on the protein **hemoglobin.** Hemoglobin molecules contain iron, which binds oxygen, as shown in **figure 24.6**. The oxygen binds in a reversible way, which is necessary so that the oxygen can unload when it reaches the tissues of the body. Hemoglobin is manufactured within red blood cells and gives them their color. Hemoglobin never leaves these cells, which circulate in the bloodstream like ships bearing cargo. Oxygen binds to hemoglobin within these cells as they pass through the capillaries surrounding the alveoli of the lungs. Carried away within red blood cells, the oxygen-bearing hemoglobin molecules later release their oxygen molecules to metabolizing cells at different locations in the body.

O₂ Transport

Hemoglobin molecules act like little oxygen (O_2) sponges, soaking up oxygen within red blood cells and causing more to diffuse in from the blood plasma. At the high O_2 levels that occur in the blood supply of the lung (**panel 1** in the Key Biological Process on the facing page), most hemoglobin molecules carry a full load of oxygen atoms. Later, in the tissues, the O_2 levels are much lower, so that hemoglobin gives up its bound oxygen (as in **panel 3**).

In tissue, the presence of carbon dioxide (CO_2) causes the hemoglobin molecule to assume a different shape, one that gives up its oxygen more easily. This speeds up the unloading of oxygen from hemoglobin even more. The effect of CO_2 on oxygen unloading is of real importance, because CO_2 is produced by the tissues at the site of cell metabolism. For this reason, the blood unloads oxygen more readily within those tissues undergoing metabolism and generating CO_2.

CO₂ Transport

At the same time the red blood cells are unloading oxygen (**panel 3**), they are also absorbing CO_2 from the tissue. About 8% of the CO_2 in blood is simply dissolved in plasma. Another 20% is bound to hemoglobin; however, it does not bind to the heme group but rather to another site on the hemoglobin molecule, and so it does not compete with oxygen binding. The remaining 72% of the CO_2 diffuses into the red blood cells. To maximize the amount of CO_2 that diffuses from the tissues into the plasma, it is important to keep the levels of CO_2 in the plasma low. The natural tendency is for the CO_2 to diffuse

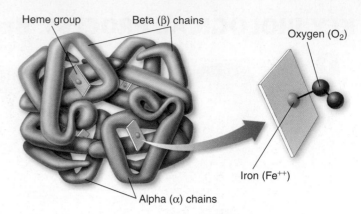

Figure 24.6 The hemoglobin molecule.
The hemoglobin molecule is actually composed of four protein chain subunits: two copies of the "alpha chain" and two copies of the "beta chain." Each chain is associated with a heme group, and each heme group has a central iron atom, which can bind to a molecule of oxygen.

out of the red blood cells back to the plasma where CO_2 levels are low. To keep this from happening, the enzyme *carbonic anhydrase* combines CO_2 molecules with water molecules in the red blood cells (**panel 4**) to form carbonic acid (H_2CO_3). This acid dissociates into **bicarbonate (HCO_3^-)** and hydrogen (H^+) ions. The H^+ binds to hemoglobin. A transporter protein in the membrane of the red blood cell moves the bicarbonate out of the red blood cell into the plasma. This transporter protein exchanges one chloride ion for a bicarbonate, a process called the *chloride shift*. This reaction keeps the levels of CO_2 in the blood plasma low, facilitating the diffusion of more CO_2 into it from the surrounding tissue. The facilitation is critical to CO_2 removal, because the difference in CO_2 concentration between blood and tissue is not large (only 5%). The formation of carbonic acid is also important in maintaining the acid-base balance of the blood, because bicarbonate serves as the major buffer of the blood plasma. You'll recall from chapter 2 that a buffer is a substance that can adjust the levels of hydrogen ions (H^+) in a solution, thereby controlling the pH of a solution.

The blood plasma carries bicarbonate ions back to the lungs. The lower CO_2 concentration in the air inside the lungs causes the carbonic anhydrase reaction to proceed in the reverse direction (**panel 6**), releasing gaseous CO_2, which diffuses outward from the blood into the alveoli. With the next exhalation, this CO_2 leaves the body. The one-fifth of the CO_2 bound to hemoglobin also leaves because hemoglobin has a greater affinity for O_2 than for CO_2 at low CO_2 concentrations. The diffusion of CO_2 outward from the red blood cells causes the hemoglobin within these cells to release its bound CO_2 and take up O_2 instead. The red blood cells, with a new load of O_2, then start their next respiratory journey.

NO Transport

Hemoglobin also has the ability to hold and release the gas nitric oxide (NO), which is produced in endothelial cells. Nitric oxide, although a noxious gas in the atmosphere, has an important physiological role in the body, acting on many kinds of cells to change their shape and function. For example, in blood vessels the presence of NO causes the blood vessels to

KEY BIOLOGICAL PROCESS: Gas Exchange During Respiration

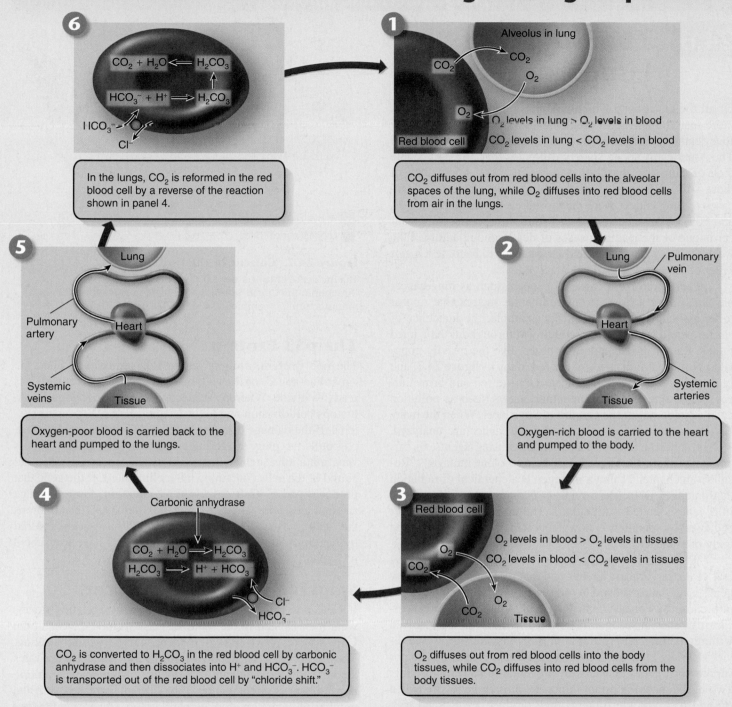

6 In the lungs, CO_2 is reformed in the red blood cell by a reverse of the reaction shown in panel 4.

$CO_2 + H_2O \longleftarrow H_2CO_3$

$HCO_3^- + H^+ \longrightarrow H_2CO_3$

HCO_3^- Cl^-

1 Alveolus in lung

CO_2 CO_2 O_2

O_2 O_2 levels in lung > O_2 levels in blood

Red blood cell CO_2 levels in lung < CO_2 levels in blood

CO_2 diffuses out from red blood cells into the alveolar spaces of the lung, while O_2 diffuses into red blood cells from air in the lungs.

5 Lung

Pulmonary artery Heart

Systemic veins Tissue

Oxygen-poor blood is carried back to the heart and pumped to the lungs.

2 Lung Pulmonary vein

Heart

Tissue Systemic arteries

Oxygen-rich blood is carried to the heart and pumped to the body.

4 Carbonic anhydrase

$CO_2 + H_2O \longrightarrow H_2CO_3$

$H_2CO_3 \longrightarrow H^+ + HCO_3$

Cl^-

HCO_3^-

CO_2 is converted to H_2CO_3 in the red blood cell by carbonic anhydrase and then dissociates into H^+ and HCO_3^-. HCO_3^- is transported out of the red blood cell by "chloride shift."

3 Red blood cell

O_2 levels in blood > O_2 levels in tissues

O_2 CO_2 levels in blood < CO_2 levels in tissues

CO_2

O_2

CO_2 Tissue

O_2 diffuses out from red blood cells into the body tissues, while CO_2 diffuses into red blood cells from the body tissues.

---

expand because it relaxes the surrounding muscle cells. Thus, blood flow and blood pressure are regulated by the amount of NO released into the bloodstream.

Hemoglobin carries NO in a special form called super nitric oxide. In this form, NO has acquired an extra electron and is able to bind to an amino acid, called cysteine, present in hemoglobin. In the lungs, hemoglobin that is dumping CO_2 and picking up O_2 also picks up NO as super nitric oxide. In tissues, hemoglobin that is releasing its O_2 and picking up CO_2 may also release super nitric oxide as NO into the blood, making blood vessels expand, thereby increasing blood flow to the tissue. Alternatively,

hemoglobin may trap any excesses of NO on its iron atoms left vacant by the release of oxygen, causing blood vessels to constrict, which reduces blood flow to the tissue. When the red blood cells return to the lungs, hemoglobin dumps its CO_2 and the NO bound to the iron atoms. It is then ready to pick up O_2 and super nitric oxide and continue the cycle.

Key Learning Outcome 24.5 Oxygen and NO move through the circulatory system carried by the protein hemoglobin within red blood cells. Most CO_2 is transported in the plasma as bicarbonate.

24.6 The Nature of Lung Cancer

Of all the diseases to which humans are susceptible, none is more feared than cancer (see chapter 8). Nearly one in every four deaths in the United States was caused by cancer in 2009. The American Cancer Society estimates that 562,340 people died of cancer in the United States in 2009. About 28% of these—157,910 people—died of **lung cancer.** About 140,000 cases of lung cancer were diagnosed each year in the 1980s, and 90% of these persons died within three years. Lung cancer is one of the leading causes of death among adults in the world today. What has caused lung cancer to become a major killer of Americans?

The search for a cause of cancers such as lung cancer has uncovered a host of environmental factors that appear to be associated with cancer. For example, the incidence of cancer per 1,000 people is not uniform throughout the United States. Rather, it is centered in cities, like the heavily populated Northeast indicated by the red areas in **figure 24.7**, and in the Mississippi Delta, indicated by the red and brown areas, suggesting that environmental factors such as pollution and pesticide runoff may contribute to cancer. When the many environmental factors associated with cancer are analyzed, a clear pattern emerges: Most cancer-causing agents, or carcinogens, share the property of being potent mutagens. Recall from chapter 11 that a mutagen is a chemical or radiation that damages DNA, destroying or changing genes (a change in a gene is called a mutation). The conclusion that cancer is caused by mutation is now supported by an overwhelming body of evidence.

What sort of genes are being mutated? In the last several years, researchers have found that mutation of only a few genes is all that is needed to transform normally dividing cells into cancerous ones. Identifying and isolating these cancer-causing genes, investigators have learned that all are involved with regulating cell proliferation (how fast cells grow and divide). A key element in this regulation are so-called tumor suppressors, genes that actively prevent tumors from forming. Two of the most important tumor-suppressor genes are called *Rb* and *p53,* and the proteins they produce are Rb and p53, respectively (recall from chapter 8 that genes are usually indicated in italics and proteins in regular type).

The Rb Protein

The **Rb protein** (named after retinoblastoma, the rare eye cancer in which it was first discovered) acts as a brake on cell division, attaching itself to the machinery the cell uses to replicate its DNA, and preventing it from doing so. When the cell wants to divide, a growth factor molecule ties up Rb so that it is not available to act as a brake on the division process. If the gene that produces Rb is disabled, there are no brakes to prevent the cell from replicating its DNA and dividing. The control switch is locked in the "ON" position.

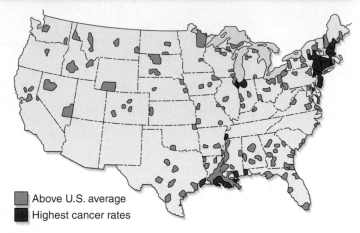

▢ Above U.S. average
◼ Highest cancer rates

Figure 24.7 Cancer in the United States.
The incidence of cancer per 1,000 people is not uniform throughout the United States. It is centered in cities where chemical manufacturing is common and in the Mississippi Delta.

The p53 Protein

The **p53 protein,** a tumor suppressor sometimes called the "guardian angel" of the cell, inspects the DNA to ensure it is ready to divide. When p53 detects damaged or foreign DNA, it stops cell division and activates the cell's DNA repair systems. If the damage doesn't get repaired in a reasonable time, p53 pulls the plug, triggering events that kill the cell. In this way, mutations such as those that cause cancer are either repaired or the cells containing them eliminated. If the gene that produces p53 is itself destroyed by mutation, future damage accumulates unrepaired. Among this damage are mutations that lead to cancer, mutations that would have been repaired by healthy p53. Fifty percent of all cancers have a disabled *p53* gene.

Smoking Causes Lung Cancer

If cancer is caused by damage to growth-regulating genes, what then has led to the rapid increase in lung cancer in the United States? Two lines of evidence are particularly telling. The first consists of detailed information about cancer rates among smokers. The annual incidence of lung cancer among nonsmokers is about 17 per 100,000 but increases with the number of cigarettes smoked per day to a staggering 300 per 100,000 for those smoking 30 cigarettes a day.

A second line of evidence consists of changes in the incidence of lung cancer that mirror changes in smoking habits. Look carefully at the data presented in **figure 24.8**. The upper graph is compiled from data on men and shows the incidence of smoking (blue line) and of lung cancer (red line) in the United States since 1900. In 1920, lung cancer was still a rare disease. About 30 years after the incidence of smoking began to increase among men, lung cancer also started to become more common. Now look at the lower graph, which presents data on women. Because of social mores, a significant number of women in the United States did not smoke until after World War II (see blue line), when many social conventions changed. As late as 1963, only 6,588 women had died of lung cancer. But as women's frequency of smoking has increased, so has their incidence of lung cancer (red line), again with a lag of about 30 years. Women today have achieved equality with men in the number of cigarettes they smoke, and their lung cancer death rates are now rapidly approaching those for men. In 2009, an estimated 70,000 American women died of lung cancer.

How does smoking cause cancer? Cigarette smoke contains many powerful mutagens, among them benzo[a]pyrene, and smoking introduces these mutagens to the lung tissues. Benzo[a]pyrene, for example, binds to three sites on the *p53* gene and causes mutations at these sites that inactivate the gene. In 1997, scientists studying this tumor-suppressor gene demonstrated a direct link between cigarettes and lung cancer. They found that the *p53* gene is inactivated in 70% of all lung cancers. When these inactivated *p53* genes are examined, they prove to have mutations at just the three sites where benzo[a]pyrene binds! Clearly, the chemical in cigarette smoke is responsible for the lung cancer.

In the face of these facts, why do so many people continue to smoke? Because the nicotine in cigarette smoke is an addictive drug. Researchers have identified the receptor on central nervous system neurons that it binds to, and they have demonstrated that smoking leads to a decrease in the brain's population of these receptors, leading to a craving for more cigarettes. This mechanism of nicotine addiction is very similar to that of cocaine addiction; once a person becomes addicted, it is difficult to quit. About half of those who try eventually succeed. Because your life is at stake, it is well worth the effort.

Clearly, an effective way to avoid lung cancer is not to smoke. Life insurance companies have computed that, on a statistical basis, smoking a single cigarette lowers your life expectancy 10.7 minutes (more than the time it takes to smoke the cigarette!). Every pack of 20 cigarettes bears an unwritten label: *The price of smoking this pack of cigarettes is 3 1/2 hours of your life.* Smoking a cigarette is very much like going into a totally dark room with a person who has a gun and standing still. The person with the gun cannot see you, does not know where you are, and shoots once in a random direction. A hit is unlikely, and most shots miss. As the person keeps shooting, however, the chance of eventually scoring a hit becomes more likely. Every time an individual smokes a cigarette, mutagens are being shot at his or her genes. Nor do statistics protect any one individual: Nothing says the first shot will not hit. Older people are not the only ones who die of lung cancer.

Importantly, nonsmokers represented fully 15% of newly diagnosed lung cancers in 2007, and two-thirds of the nonsmoking lung cancer patients were women. In that year, lung cancer among nonsmoking women killed three times as many women as did cervical cancer. Very little research is being done to explore why nonsmoking lung cancer is so much more common among women than men. More studies should be conducted.

Key Learning Outcome 24.6 Cancer results from the mutation of genes that, when healthy, enable the cell to regulate cell division. Many of these mutations are caused by cigarette smoke.

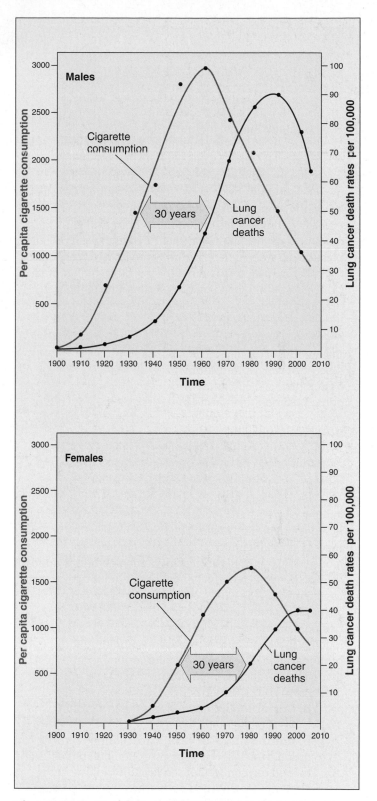

Figure 24.8 Incidence of lung cancer in men and women.

Lung cancer was a rare disease a century ago. As men in the United States increased smoking in the early 1900s, the incidence of lung cancer also increased. Women followed suit years later, and in 2008, more than 71,000 women died of lung cancer, a rate of 40 per 100,000. In that year, about 20% of women smoked and about 25% of men.

How Do Llamas Live So High Up?

Because of mixing, the air animals breathe is 21% oxygen everywhere, even way up into the sky 100 km above earth's surface. However, the amount of air (the number of molecules in a unit volume) decreases sharply with altitude, as shown in the upper graph. Air pressure at 5,000 meters is half that at sea level. This lack of air presents a serious problem to humans, as mountain climbers know. The amount of oxygen in the air (measured as oxygen partial pressure) is lower, so there is simply too little oxygen to fuel a climber's muscles. To combat this problem, high-altitude climbers typically spend months acclimating to high altitudes, a period in which their bodies greatly increase the amount of hemoglobin in their red blood cells and so increase the amount of oxygen the red blood cells can capture. Many mammals live their entire lives at high altitudes. The llama and the vicuna (pictured here) both live in the high Andes of South America, often above 5,000 meters. Do they stuff extra hemoglobin into their red blood cells too, or are they able to solve the problem of low oxygen in another way?

The graph on the lower right displays three "oxygen loading curves" that reveal the effectiveness with which hemoglobin binds oxygen. The more effective the binding, the less oxygen required before hemoglobin becomes fully loaded. In the graph, the percent hemoglobin saturation (that is, how much of the hemoglobin is bound to oxygen) is presented on the y axis, and the oxygen partial pressure (a measure of the amount of oxygen available to the hemoglobin molecules) is presented on the x axis. Oxygen-loading curves are presented for three mammalian species: humans living at sea level, and llamas and vicunas, each living in the Andes above 5,000 meters.

1. **Applying Concepts**
   a. Variable. In the graph on the lower right, what is the dependent variable(s)?
   b. Comparing Curves. Extrapolating on the lower graph, which species possesses hemoglobin able to load oxygen well at sea level partial pressures (160 mm Hg)? Which of the three species possesses hemoglobin better able to load oxygen on Mount Everest (use data from upper graph)?
2. **Interpreting Data**
   a. The partial pressure of oxygen in human muscle tissue at sea level is about 40 mm Hg. What is the percent hemoglobin bound to O_2 for each of the three species at this partial pressure? What percent of the human hemoglobin has released its oxygen? Of the llama? Of the vicuna?

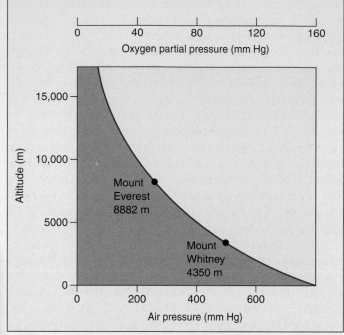

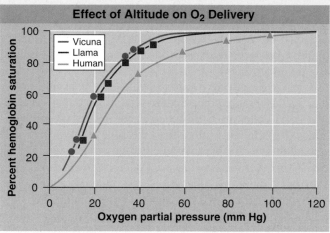

Effect of Altitude on O_2 Delivery

   b. Are there any significant differences in the hemoglobin saturation values for the two high-altitude species?
3. **Making Inferences** At an elevation of 5,000 meters, the partial pressure of oxygen is 80 mm Hg (half of what it is at sea level). At this elevation, how much of human hemoglobin has succeeded in binding oxygen? How much of llama hemoglobin? Of vicuna?
4. **Drawing Conclusions** What is the effect of shifting the oxygen loading curve to the left? What general statement can be made regarding the affinity of hemoglobin for oxygen in the three species?
5. **Further Analysis** What saturation values would you expect in llamas raised from birth in the National Zoo at Washington, D.C.? Why would you expect this? How might you test your prediction?

Respiration

24.1 Types of Respiratory Systems

- Animals employ a diverse array of respiratory systems (**integrated art, page 516**). Some aquatic animals extract oxygen and release carbon dioxide across the skin.

- The evolution of gills increased the surface area for gas exchange in higher aquatic animals. The complexity of gills varies greatly in these animal groups from simple papulae in echinoderms to the highly convoluted gills of fish.

- Terrestrial animals extract oxygen from the air with tracheae or lungs. The tracheae system in arthropods is a network of air ducts with openings to the outside, called spiracles. Other terrestrial vertebrates use lungs.

24.2 Respiration in Aquatic Vertebrates

- Fish gills, shown here from **figure 24.1**, are very efficient at extracting oxygen from water because of a countercurrent flow system. Water is pumped passed the gill filaments in a one-way direction that is opposite the flow of blood in the capillaries, called countercurrent flow. Because of this flow pattern, the water passing over the gills always has a higher concentration of oxygen compared to that in the underlying blood. This concentration gradient drives the diffusion of oxygen into the blood (**figure 24.2**).

24.3 Respiration in Terrestrial Vertebrates

- Lungs evolved in efficiency, from amphibian, to reptilian, to mammalian lungs, by increasing surface area (**figure 24.3**).

- Mammalian lungs contain many airways that branch out and end in small chambers called alveoli that look like clusters of grapes. The alveoli increase the surface area for the diffusion of gases.

- Bird lungs use a crosscurrent flow system that is more efficient at extracting oxygen than other terrestrial animals's lungs. Air flows through the bird lung in one direction, from a set of posterior air sacs through the lungs in a direction perpendicular to the flow of blood, called a crosscurrent flow. This allows the lung to extract more oxygen than the uniform pool lung found in mammals, but not as efficiently as the countercurrent system found in fish (**figure 24.4**). During a second inhalation, the air moves from the lungs into a set of anterior air sacs and from there it is exhaled out of the body.

24.4 The Mammalian Respiratory System

- Mammalian lungs are positioned within an internal cavity, called the thoracic cavity, as shown here from **figure 24.5**. An airway called a bronchus connects each lung to the trachea. The trachea is a tube that extends up to the rear of the mouth.

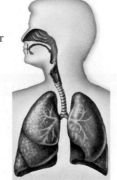

- The airways of the mammalian lung end in alveoli, which are surrounded by capillaries. Gas exchange occurs across the single cell layers of the alveoli and capillary walls.

- Membranes line the outside of the lungs and the inner layer of the thoracic cavity.

Fluid between these two membranes causes the lungs to stick to the walls of the thoracic cavity. Contraction of the muscles that line the thoracic cavity expands the space in the lungs, causing air to rush into the lungs. Relaxation of the muscles causes air to be exhaled (**Key Biological Process, page 521**).

- The tidal volume is the amount of air that moves in and out of the lungs at rest. The extra amount of air that can be forced into and out of the lungs is called the vital capacity. The air that remains in the lungs after the forced exhalation is called the residual volume, or dead volume.

24.5 How Respiration Works: Gas Exchange

- Hemoglobin, found in red blood cells and shown here from **figure 24.6**, carries oxygen from the lungs to the cells of the body. In the lungs where oxygen concentrations are high, oxygen diffuses into the red blood cells in the blood. Oxygen binds to iron atoms contained within the heme groups of hemoglobin. The oxygen is then carried to areas of the body that are low in oxygen. The oxygen diffuses down its concentration gradient entering cells in the surrounding tissues (**Key Biological Process, page 523**).

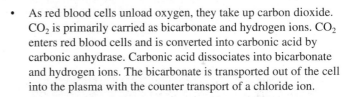

- As red blood cells unload oxygen, they take up carbon dioxide. CO_2 is primarily carried as bicarbonate and hydrogen ions. CO_2 enters red blood cells and is converted into carbonic acid by carbonic anhydrase. Carbonic acid dissociates into bicarbonate and hydrogen ions. The bicarbonate is transported out of the cell into the plasma with the counter transport of a chloride ion.

- In the lungs, the carbonic anhydrase reaction is reversed, converting bicarbonate and hydrogen ions back to carbon dioxide where it passes out of the blood and into the alveoli traveling down its concentration gradient. Once in the lungs, the CO_2 is exhaled.

- The blood also transports the gas nitric oxide (NO) as super nitric oxide. NO is produced in endothelial cells and acts in dilating blood vessels. It is carried by hemoglobin, but binds to a different area of the molecule, not to the oxygen binding site.

Lung Cancer and Smoking

24.6 The Nature of Lung Cancer

- Cancer results from mutations of the DNA, often caused by chemicals in the environment. Areas of the country that have a higher amount of environmental contaminants also have a higher incidence of cancer (**figure 24.7**).

- Many cancers have been linked to mutations of two key tumor suppressor genes, *Rb* and *p53*. The protein products of these two genes control cell division, keeping the cell from dividing when it is not supposed to. If mutations damage either of these genes, the cells are able to divide uncontrollably.

- Lung cancer results from mutations caused by chemicals in cigarette smoke. The incidence of cigarette smoking and lung cancer mirror each other with a separation of about 30 years, time for the accumulation of mutations that lead to cancer (**figure 24.8**).

Test Your Understanding

1. In the gills of fish, countercurrent flow allows
   a. the animal's blood to be continually exposed to water of lower oxygen concentration.
   b. the animal's blood to be continually exposed to water of equal oxygen concentration.
   c. the animal's blood to be continually exposed to water of higher oxygen concentration.
   d. for a larger surface area for gas exchange to occur.
2. The purpose of gill filaments and the lung alveoli is to
   a. decrease the surface area available for gas exchange.
   b. increase the surface area available for gas exchange.
   c. decrease the volume available for gas exchange.
   d. increase the volume available for gas exchange.
3. In general, the need for alveoli increases as the
   a. energy need for different vertebrate classes increases.
   b. energy need for different vertebrate classes decreases.
   c. habitat for different vertebrate classes changes.
   d. nutrition need for different vertebrate classes increases.
4. Which group of animals is the most efficient at extracting oxygen?
   a. reptiles   c. fish
   b. birds   d. mammals
5. Which of the following statements about the bird's respiratory system is *false*?
   a. Birds breathe using a crosscurrent flow of air and blood.
   b. Birds have air sacs in addition to a lung.
   c. It takes three cycles of breathing for air to pass through the bird's respiratory system.
   d. The bird's respiratory system has two locations where air is held during inhalation.

6. When you take a deep breath, your chest moves out because
   a. the incoming air expands the thoracic cavity.
   b. contracting your abdominal muscles forces the chest cavity outward.
   c. contracting the muscles around the thoracic cavity pulls your chest out.
   d. positive pressure builds up in the lungs, expanding the chest cavity and forcing the chest out.
7. Oxygen is transported by
   a. hemoglobin in red blood cells.
   b. dissolving it in the blood plasma.
   c. binding to proteins in the blood plasma.
   d. platelets in the blood plasma.
8. Most of the carbon dioxide is transported
   a. by hemoglobin in red blood cells.
   b. as bicarbonate in the blood plasma.
   c. by proteins in the blood plasma.
   d. as carbonic acid in red blood cells.
9. Which of the following is *not* carried by hemoglobin?
   a. bicarbonate
   b. carbon dioxide
   c. oxygen
   d. super nitric oxide
10. Which of the following can lead to cancer?
    a. smoking
    b. pollution
    c. mutations of *Rb* and *p53*
    d. All of the above.

Apply Your Understanding

1. **Key Biological Process: Gas Exchange During Respiration**
   Explain what would happen to carbon dioxide transport if a person was poisoned with a chemical that blocked the actions of carbonic anhydrase.

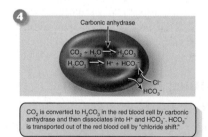

2. **Figure 24.8** What conclusions can you draw about cigarette smoking and health from these data? Explain.

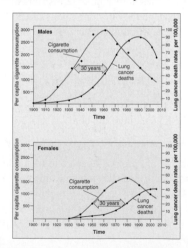

Synthesize What You Have Learned

1. Sometimes when people are eating, they take a bite that is too big, or is not completely chewed, and when they swallow, the food becomes stuck partway down the esophagus near the epiglottis (a flap that covers the trachea when a person is swallowing). When food is stuck in this location, a person usually can't breathe. In this case, people have been instructed to do the Heimlich maneuver, which is a method of pushing up rapidly on the diaphragm, compressing the lungs. Why might this help?
2. How is it that cigarette smoking can be linked to an increased incidence of many kinds of cancer, not just lung cancer?

25

The Path of Food Through the Animal Body

T here are no photosynthetic animals. Animals are heterotrophs, gaining the energy to power their lives by consuming and oxidizing organic molecules present in other organisms. All animals must continuously consume plant material or other animals in order to live. The grass in this prairie dog's mouth will be consumed and converted within its cells to body tissue, energy, and refuse. Most of the molecules in the grass are far too large to be conveniently absorbed by the prairie dog's cells, and so they are first broken down into smaller pieces: Carbohydrates are broken down into simple sugars, proteins into amino acids, fats into fatty acids. This process, called digestion, is the focus of this chapter. The prairie dog's digestive system is a long tube passing from mouth to anus, with specialized segments for digestion and for the subsequent absorption into its body of the resulting sugars, amino acids, and fatty acids. Whatever is left is excreted from the body as feces. Recently, researchers have found a sea slug that appears to take up chloroplasts from the algae it eats—might these sea slugs be capable of photosynthesis? No prairie dog, or any other vertebrate animal, has the benefit of this option. In this chapter you will follow the path of food as it moves through the vertebrate body. As you will see, it is a surprisingly interesting journey.

25.1 Food for Energy and Growth

The food animals eat provides both a source of energy and essential molecules such as certain amino acids and fats that the animal body is not able to manufacture for itself. An optimal diet contains more carbohydrates than fats and also a significant amount of protein, as recommended by the federal government's "pyramid of nutrition" in **figure 25.1**. The pyramid is intended as a general guideline of what a person should eat. A healthy diet should include more of the foods indicated by the larger sections—for example, the gold section indicates grains and cereals with an emphasis on whole grains. Fats (the yellow section) are recommended in much smaller amounts because they have a far greater number of energy-rich carbon–hydrogen bonds and thus a much higher energy content per gram than carbohydrates or proteins. Carbohydrates and proteins contain more carbon–oxygen bonds that are already oxidized. For this reason, fats are a very efficient way to store energy. When food is consumed, it is either metabolized by cells of the body, or it is converted into fat and stored in fat cells.

Carbohydrates are obtained primarily from cereals, grains, breads, fruits, and vegetables. On the average, carbohydrates contain 4.1 calories per gram; fats, by comparison, contain 9.3 calories per gram, over twice as much. Dietary fats are obtained from oils, margarine, and butter and are abundant in fried foods, meats, and processed snack foods, such as potato chips and crackers. Like carbohydrates, proteins have 4.1 calories per gram and can be obtained from many foods, including dairy products, poultry, fish, meat, and grains.

The body uses carbohydrates for energy and fats to construct cell membranes and other cell structures, to insulate nervous tissue, and to provide energy. Fat-soluble vitamins that are essential for proper health are also absorbed with fats. Proteins are used for energy and as building materials for cell structures, enzymes, hemoglobin, hormones, and muscle and bone tissue.

In wealthy countries, such as those of North America and Europe, being significantly overweight is common, the result of habitual overeating and high-fat diets, in which fats constitute over 35% of the total caloric intake. The international standard measure of appropriate body weight is the body mass index (BMI), estimated as your body weight in kilograms, divided by your height in meters squared. A BMI chart is presented in **figure 25.2**. To determine your BMI, find your height in the left hand column (in feet and inches) and trace it across to the column with your weight in pounds. A BMI value of 25 (dark blue boxes) and above is considered overweight and 30 or over is considered obese. In the United States, the National Institutes of Health estimated in 2004 that 66% of adults, 133.6 million Americans, were overweight, with a body mass index of 25 or more. Of those individuals, 63.6 million were considered obese with a body mass index of 30 or greater. Being overweight is highly correlated with coronary heart disease, diabetes, and many other disorders. However, starving yourself is also not the answer. A BMI of less than 18.5 is also unhealthy, often resulting from eating disorders including anorexia nervosa.

Even an animal that is completely at rest requires energy to support its metabolism. This minimum rate of energy consumption, called the *basal metabolic rate* (BMR), is relatively constant for a given individual. Exercise raises the metabolic rate above the basal levels, so the amount of energy the body requires per day is determined not only by the BMR but also by the level of physical activity. Therefore, energy requirements can be altered by the choice of diet (caloric intake) and the amount of energy expended in exercise.

One essential characteristic of food is its fiber content. Fiber is the part of plant food that cannot be digested by humans and is found in fruits, vegetables, and grains. Other animals, however, have evolved many different ways to process food that has a relatively high fiber content. Diets that are low in fiber, now common in the United States, result in a slower passage of food through the colon. This low dietary fiber content is thought to be associated with incidences of colon cancer in the United States, which are among the highest levels in the world.

Figure 25.1 The pyramid of nutrition.
The width of each section indicates how much you should consume of that food group. The gold section is grains, green is vegetables, red is fruits, yellow is fats, aqua is dairy, and purple is meats and beans. However, one size does not fit all; go to www.mypyramid.gov to customize the food pyramid that is right for you.

WEIGHT	100	105	110	115	120	125	130	135	140	145	150	155	160	165	170	175	180	185	190	195	200	205
HEIGHT																						
5'0"	20	21	21	22	23	24	25	26	27	28	29	30	31	32	33	34	35	36	37	38	39	40
5'1"	19	20	21	22	23	24	25	26	26	27	28	29	30	31	32	33	34	35	36	37	38	39
5'2"	18	19	20	21	22	23	24	25	26	27	27	28	29	30	31	32	33	34	35	36	37	37
5'3"	18	19	19	20	21	22	23	24	25	26	27	27	28	29	30	31	32	33	34	35	35	36
5'4"	17	18	19	20	21	21	22	23	24	25	26	27	27	28	29	30	31	32	33	33	34	35
5'5"	17	17	18	19	20	21	22	22	23	24	25	26	27	27	28	29	30	31	32	32	33	34
5'6"	16	17	18	19	19	20	21	22	23	23	24	25	26	27	27	28	29	30	31	31	32	33
5'7"	16	16	17	18	19	20	20	21	22	23	23	24	25	26	27	27	28	29	30	31	31	32
5'8"	15	16	17	17	18	19	20	21	21	22	23	23	24	25	26	27	27	28	29	30	30	31
5'9"	15	16	16	17	18	18	19	20	21	21	22	23	24	24	25	26	27	27	28	29	30	30
5'10"	14	15	16	17	17	18	19	19	20	21	22	22	23	24	24	25	26	27	27	28	29	29
5'11"	14	15	15	16	17	17	18	19	20	20	21	22	22	23	24	24	25	26	26	27	28	29
6'0"	14	14	15	16	16	17	18	18	19	20	20	21	22	22	23	24	24	25	26	26	27	28
6'1"	13	14	15	15	16	16	17	18	18	19	20	20	21	22	22	23	24	24	25	26	26	27
6'2"	13	13	14	15	15	16	17	17	18	19	19	20	21	21	22	23	24	24	25	26	26	
6'3"	12	13	14	14	15	16	16	17	17	18	19	19	20	21	21	22	22	23	24	24	25	26
6'4"	12	13	13	14	15	16	16	16	17	18	18	19	19	20	21	21	22	23	23	24	24	25

Figure 25.2 Are you overweight?

This chart presents the body mass index (BMI) values used by federal health authorities to determine who is overweight. Your body mass index is at the intersection of your height and weight.

Essential Substances for Growth

Over the course of their evolution, many animals have lost the ability to manufacture certain substances they need, substances that often play critical roles in their metabolism. Mosquitoes and many other blood-sucking insects, for example, cannot manufacture cholesterol, but they obtain it in their diet because human blood is rich in cholesterol. Many vertebrates are unable to manufacture one or more of the 20 amino acids used to make proteins. Humans are unable to synthesize eight amino acids: lysine, tryptophan, threonine, methionine, phenylalanine, leucine, isoleucine, and valine. These amino acids, called **essential amino acids,** must therefore be obtained from proteins in the food we eat. For this reason, it is important to eat so-called complete proteins—that is, ones containing all the essential amino acids. In addition, all vertebrates have also lost the ability to synthesize certain polyunsaturated fats that provide backbones for the many kinds of fats their bodies manufacture.

Trace Elements In addition to supplying energy, food that is consumed must also supply the body with essential minerals such as calcium and phosphorus, as well as a wide variety of **trace elements,** which are minerals required in very small amounts. Among the trace elements are iodine (a component of thyroid hormone), cobalt (a component of vitamin B_{12}),

zinc and molybdenum (components of enzymes), manganese, and selenium. All of these, with the possible exception of selenium, are also essential for plant growth; animals obtain them directly from plants that they eat or indirectly from animals that have eaten plants.

Vitamins Essential organic substances that are used in trace amounts are called **vitamins.** Humans require at least 13 different vitamins. Many vitamins are required cofactors for cellular enzymes. Humans, monkeys, and guinea pigs, for example, have lost the ability to synthesize ascorbic acid (vitamin C) and will develop the potentially fatal disease called scurvy—characterized by weakness, spongy gums, and bleeding of the skin and mucous membranes—if vitamin C is not supplied in their diets.

Key Learning Outcome 25.1 Food is an essential source of calories. It is important to maintain a proper balance of carbohydrate, protein, and fat. Individuals with a body mass index of 25 or more are considered overweight. Food also provides key amino acids that the body cannot manufacture for itself, as well as necessary trace elements and vitamins.

25.2 Types of Digestive Systems

Heterotrophs are divided into three groups on the basis of their food sources. Animals that eat plants exclusively are classified as **herbivores;** common examples include cows, horses, rabbits, and sparrows. Animals that are meat eaters, such as cats, eagles, trout, and frogs, are **carnivores. Omnivores** are animals that eat both plants and other animals. We humans are omnivores, as are pigs, bears, and crows.

Single-celled organisms (as well as sponges) digest their food intracellularly, breaking down food particles with digestive enzymes inside their cells. Other animals digest their food extracellularly, within a digestive cavity. In this case, the digestive enzymes are released into a cavity that is continuous with the animal's external environment. In flatworms (such as *Planaria*) and cnidarians, like the hydra in **figure 25.3**, the digestive cavity in the center of the body has only one opening at the top that serves as both mouth (the red arrow bringing food in) and anus (the blue arrow passing waste out). There can be no specialization within this type of digestive system, called a *gastrovascular cavity,* because every cell is exposed to all stages of food digestion.

Specialization occurs when the digestive tract, or alimentary canal, has a separate mouth and anus, so that transport of food is one way. Three examples are shown in **figure 25.4**. The most primitive digestive tract is seen in nematodes (phylum Nematoda), where it is simply a tubular *gut* lined by an epithelial membrane. Earthworms (phylum Annelida) have a digestive tract specialized in different regions for the ingestion, storage (crop), fragmentation (gizzard), digestion, and absorption of food (intestine). All higher animals, like the salamander, show similar specializations.

The ingested food may be stored in a specialized region of the digestive tract or may first be subjected to physical fragmentation through the chewing action of teeth (in the mouth of many vertebrates) or the grinding action of pebbles (in the gizzard of earthworms and birds). Chemical **digestion** then occurs primarily in the intestine, breaking down the larger food molecules of polysaccharides, fats, and proteins into smaller subunits. Carbohydrate digestion begins in the mouth of some animals, and protein digestion begins in the stomach in some animals. Chemical digestion involves hydrolysis reactions that liberate the subunits—primarily monosaccharides, amino acids, and fatty acids—from the food. These products of chemical digestion pass through the epithelial lining of the gut and ultimately into the blood, in a process known as absorption. Any molecules in the food that are not absorbed cannot be used by the animal. These wastes are excreted through the anus.

> **Key Learning Outcome 25.2** Most animals digest their food extracellularly. A digestive tract with a one-way transport of food allows specialization of regions for different functions.

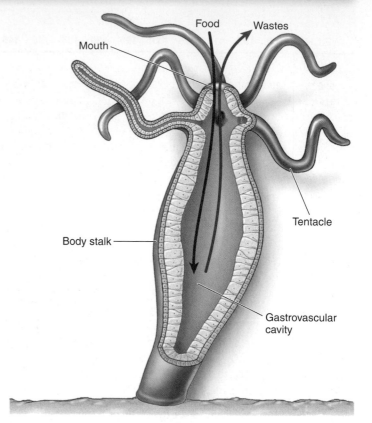

Figure 25.3 Two-way digestive tract.
Food particles enter and leave the gastrovascular cavity of *Hydra* through the same opening.

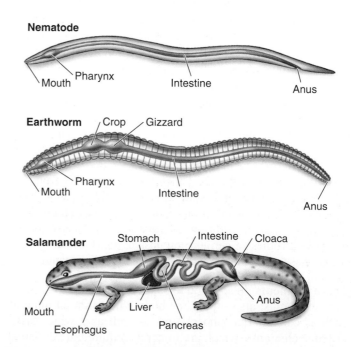

Figure 25.4 One-way digestive tracts.
One-way movement through the digestive tract allows different regions of the digestive system to become specialized for different functions.

25.3 Vertebrate Digestive Systems

In humans and other vertebrates, the digestive system consists of a tubular gastrointestinal tract and accessory digestive organs (**figure 25.5**). Working through the figure from the top down, the initial components of the gastrointestinal tract are the mouth and the pharynx, which is the common passage of the oral and nasal cavities. The pharynx leads to the esophagus, a muscular tube that delivers food to the stomach, where some preliminary digestion occurs. From the stomach, food passes to the first part of the small intestine, where a battery of digestive enzymes continues the digestive process. The products of digestion then pass across the wall of the small intestine into the bloodstream. The small intestine empties what remains into the large intestine, where water and minerals are absorbed. In most vertebrates other than mammals, the waste products emerge from the large intestine into a cavity called the cloaca (see the salamander in **figure 25.4**), which also receives the products of the urinary and reproductive systems. In mammals, the urogenital products are separated from the fecal material in the large intestine, also called the colon; the fecal material enters the rectum and is expelled through the anus.

In general, carnivores have shorter intestines for their size than do herbivores. A short intestine is adequate for a carnivore, but herbivores ingest a large amount of plant cellulose, which resists digestion. These animals have a long, convoluted small intestine. In addition, mammals called *ruminants* (such as cows) that consume grass and other vegetation have stomachs with multiple chambers, where bacteria aid in the digestion of cellulose. Other herbivores, including rabbits and horses, digest cellulose (with the aid of bacteria) in a blind pouch called the **cecum** located at the beginning of the large intestine (these will be discussed in more detail in section 25.7). Accessory digestive organs described later in this chapter include the liver, the gallbladder, and the pancreas.

The tubular gastrointestinal tract of a vertebrate, as you can see in **figure 25.6**, has a characteristic layered structure. Working from the inside (the lumen) outward, the innermost layer (light pink) is the mucosa, an epithelium that lines the lumen. The next major tissue layer, composed of connective tissue, is called the submucosa (darker pink). Just outside the submucosa is the muscularis, which consists of a double layer of smooth muscles. The muscles in the inner layer have a circular orientation, and those in the outer layer are arranged longitudinally. An outer connective tissue layer, the serosa, covers the external surface of the tract. Nerves, intertwined in regions called *plexuses,* are located in the submucosa and help regulate the gastrointestinal activities.

> **Key Learning Outcome 25.3** The vertebrate digestive system consists of a tubular gastrointestinal tract, which is modified in different animals, composed of a series of tissue layers.

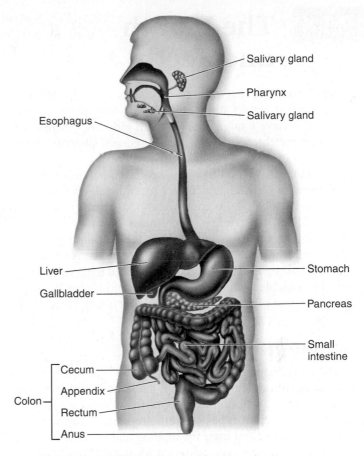

Figure 25.5 The human digestive system.
The tubular gastrointestinal tract and accessory digestive organs are shown. The colon extends from the cecum to the anus.

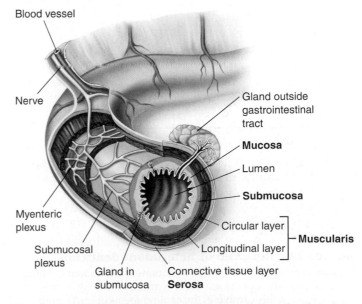

Figure 25.6 The layers of the gastrointestinal tract.
The mucosa contains a lining epithelium, the submucosa is composed of connective tissue (as is the outer serosa layer), and the muscularis consists of smooth muscles.

25.4 The Mouth and Teeth

Specializations of the digestive systems in different kinds of vertebrates reflect differences in the way these animals live. Fishes have a large pharynx with gill slits, whereas air-breathing vertebrates have a greatly reduced pharynx. Many vertebrates have teeth (discussed below), and chewing (mastication) breaks up food into small particles and mixes it with fluid secretions. Birds, which lack teeth, break up food in their two-chambered stomachs. The first chamber, the proventriculus (**figure 25.7**), produces digestive enzymes, which are passed along with the food into the second chamber, the gizzard. The gizzard contains small pebbles ingested by the bird, which are churned together with the food by muscular action. This churning grinds up the seeds and other hard plant material into smaller chunks that can be digested more easily in the intestine.

Vertebrate Teeth

Recall from chapter 20 that while reptiles and fish have homodont dentition (teeth that are all the same), most mammals have heterodont dentition, teeth of different specialized types. Up to four different types of teeth are seen: incisors, which are chisel-shaped and used for nipping and biting; "canines," which are sharp, pointed teeth used for tearing food; and premolars (bicuspids) and molars, which usually have flattened, ridged surfaces for grinding and crushing food. The front teeth in the upper and lower jaws of mammals are incisors. On each side of the incisors are the canines. Behind the canines are premolars and then molars.

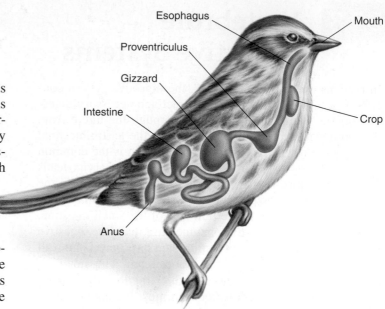

Figure 25.7 The digestive tract of birds.
In birds, food enters the mouth and is stored in the crop. Because birds lack teeth, they swallow gritty objects or pebbles, which lodge in the gizzard, to help pulverize food. Digestive enzymes produced in the proventriculus are churned up with the food and gritty objects in the gizzard before passing into the intestine.

This general pattern of heterodont dentition is modified in different mammals depending on their diet (**figure 25.8**). For example, in carnivorous mammals the canines are prominent, and the premolars and molars are more bladelike, with sharp edges adapted for cutting and shearing. Carnivores often tear off pieces of their prey but have little need to chew them, because digestive enzymes can act directly on animal cells. (Recall how a cat or dog gulps down its food.) By contrast, grass-eating herbivores, such as cows and horses, must pulverize the cellulose cell walls of plant tissue before digesting it. In these mammals, the incisors can be well-developed and are used to cut grass and other plants. The canines are reduced or absent, and the premolars and molars are large, flat teeth with complex ridges well suited for grinding.

Humans are omnivores, and human teeth are adapted for eating both plant and animal food. Viewed simply, humans are carnivores in the front of the mouth and herbivores in the back. Children have only 20 teeth, but these deciduous teeth are lost during childhood and are replaced by 32 adult teeth. The third molars are the wisdom teeth, which usually grow in during the late teens or early twenties, when a person is assumed to have gained a little "wisdom."

As you can see in **figure 25.9**, the tooth is a living organ, composed of connective tissue, nerves, and blood vessels, held in place by cementum, a bonelike substance that anchors the tooth in the jaw. The interior of the tooth contains connective tissue called pulp that extends into the root canals and contains nerves and blood vessels. A layer of calcified tissue called dentin surrounds the pulp cavity. The portion of the tooth that projects above the gums is called the crown and is covered with an extremely hard, nonliving substance

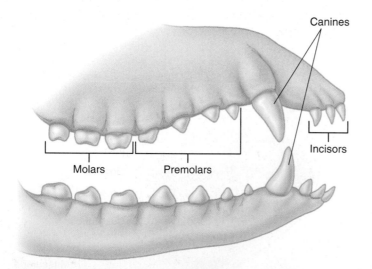

Figure 25.8 Diagram of heterodont dentition.
Different mammals have specific variations of heterodont dentition, depending on whether the mammal is an herbivore, carnivore, or omnivore. In this carnivore, the canines are prominent, and the premolars and molars are pointed—adaptations for tearing and ripping food. In herbivores, some of the incisors are large, the canines are reduced or absent, and the premolars and molars are flattened—adaptations for nipping and grinding vegetation.

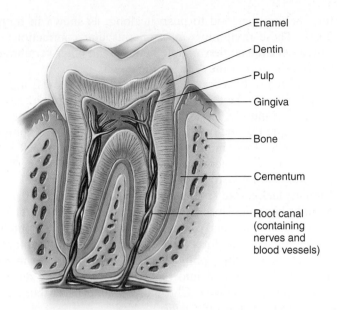

Figure 25.9 Human teeth.
Each vertebrate tooth is alive, with a central pulp containing nerves and blood vessels. The actual chewing surface is a hard enamel layered over the softer dentin, which forms the body of the tooth.

called enamel. Enamel protects the tooth against abrasion and acids that are produced by bacteria living in the mouth. Cavities form when bacterial acids break down the enamel, allowing bacteria to infect the inner tissues of the tooth.

Processing Food in the Mouth

Inside the mouth, the tongue mixes food with a mucous solution, **saliva.** In humans, three pairs of salivary glands secrete saliva into the mouth through ducts in the mouth's mucosal lining. Saliva moistens and lubricates the food so that it is easier to swallow and does not abrade the tissue it passes on its way through the esophagus. Saliva also contains the hydrolytic enzyme salivary **amylase,** which initiates the breakdown of the polysaccharide starch into the disaccharide maltose. This digestion is usually minimal in humans, however, because most people don't chew their food for very long.

The secretions of the salivary glands are controlled by the nervous system, which in humans maintains a constant flow of about half a milliliter of saliva per minute when the mouth is empty of food. This continuous secretion keeps the mouth moist. The presence of food in the mouth triggers an increased rate of saliva secretion, as taste-sensitive neurons in the mouth send impulses to the brain, which responds by stimulating the salivary glands. The most potent stimuli are acidic solutions; lemon juice, for example, can increase the rate of salivation eightfold. The sight, sound, or smell of food can stimulate salivation markedly in dogs, but in humans, these stimuli are much less effective than is thinking or talking about food.

Swallowing

When food is ready to be swallowed, the tongue moves it to the back of the mouth. In mammals, the process of swallowing begins when the soft palate elevates, pushing against the back wall of the pharynx (**figure 25.10**). Elevation of the soft palate seals off the nasal cavity and prevents food from entering it ❶. Pressure against the pharynx stimulates neurons within its walls, which send impulses to the swallowing center in the brain. In response, muscles are stimulated to contract and raise the *larynx* (voice box). This pushes the *glottis,* the opening from the larynx into the trachea (windpipe), against a flap of tissue called the *epiglottis* ❷. These actions keep food out of the respiratory tract, directing it instead into the esophagus ❸.

> **Key Learning Outcome 25.4** In many vertebrates, ingested food is fragmented through the tearing or grinding action of specialized teeth. In birds, this is accomplished through the grinding action of pebbles in the gizzard. Food mixed with saliva is swallowed and enters the esophagus.

Air
Hard palate
Tongue
Soft palate
Pharynx
Epiglottis
Glottis
Larynx
Trachea
Esophagus

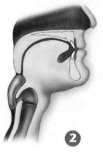
❶ ❷

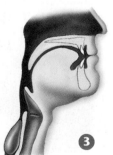
❸

Figure 25.10 **The human pharynx, palate, and larynx.**

The Esophagus and Stomach

Structure and Function of the Esophagus

Swallowed food enters a muscular tube called the **esophagus,** which connects the pharynx to the stomach. In adult humans, the esophagus is about 25 centimeters long; the upper third is enveloped in skeletal muscle, for voluntary control of swallowing, while the lower two-thirds is surrounded by involuntary smooth muscle. The swallowing center stimulates successive waves of contraction in these muscles that move food along the esophagus to the stomach. The muscles relax ahead of the food, allowing it to pass freely, and con-

tract behind the food to push it along, as shown in figure 25.11. These rhythmic waves of muscular contraction are called **peristalsis;** they enable humans and other vertebrates to swallow even if they are upside down.

In many vertebrates, the movement of food from the esophagus into the stomach is controlled by a ring of circular smooth muscle, the **sphincter,** that opens in response to the pressure exerted by the food. Contraction of this sphincter prevents food in the stomach from moving back into the esophagus. Rodents and horses have a true sphincter at this site, and thus stomach contents cannot move back out. Humans lack a true sphincter, and stomach contents can be brought back out during vomiting, when the sphincter between the stomach and esophagus is relaxed and the contents of the stomach are forcefully expelled through the mouth. The relaxing of this sphincter can also result in the movement of stomach acid into the esophagus, causing an irritation called *heartburn.* Chronic and severe heartburn is a condition known as *acid reflux.*

Structure and Function of the Stomach

The **stomach** is a saclike portion of the digestive tract. Its inner surface is highly convoluted, enabling it to fold up when empty and open out like an expanding balloon as it fills with food. Thus, while the human stomach has a volume of only about 50 milliliters when empty, it may expand to contain 2 to 4 liters of food when full.

The stomach contains an extra layer of smooth muscle for churning food and mixing it with *gastric juice,* an acidic secretion of the tubular gastric glands of the mucosa. The gastric glands lie at the bottom of deep depressions, the gastric pits shown in the enlargement in figure 25.12. These exocrine glands contain two kinds of secretory cells: *parietal cells,* which secrete hydrochloric acid (HCl); and *chief cells,* which secrete pepsinogen, a weak protease (protein-digesting enzyme) that requires a very low pH to be active. This low pH is provided by the HCl. Activated pepsinogen molecules then cleave each other at specific sites, producing a much more active protease, pepsin. This process of secreting a relatively inactive enzyme that is then converted into a more active enzyme outside the cell prevents the chief cells from digesting themselves. It should be noted that only proteins are partially digested in the stomach—there is no significant digestion of carbohydrates or fats.

Action of Acid

The human stomach produces about 2 liters of HCl and other gastric secretions every day, creating a very acidic solution inside the stomach. The concentration of HCl in this solution is about 10 millimolar, corresponding to a pH of 2. Thus, gastric juice is about 250,000 times more acidic than blood, whose normal pH is 7.4. The low pH in the stomach helps denature food proteins, making them easier to digest, and keeps pepsin maximally active. Active pepsin hydrolyzes food proteins into shorter chains of polypeptides that are not fully digested until the mixture enters the small intestine. The mixture of partially digested food and gastric juice is called **chyme.**

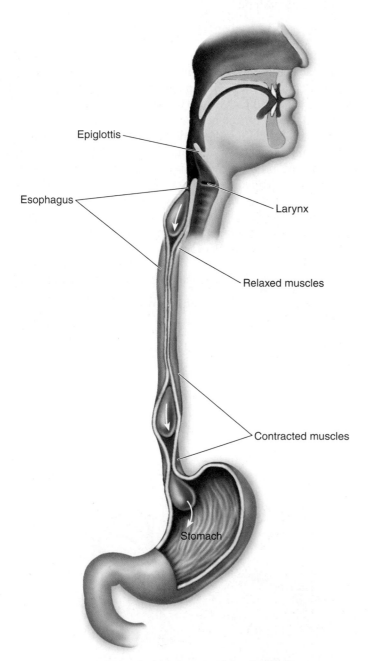

Epiglottis

Esophagus

Larynx

Relaxed muscles

Contracted muscles

Stomach

Figure 25.11 The esophagus and peristalsis.

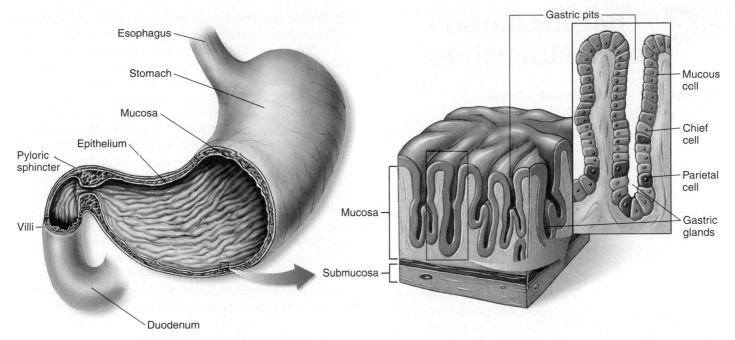

Figure 25.12 The stomach and gastric glands.

Food enters the stomach from the esophagus. The epithelial walls of the stomach are dotted with gastric pits, which contain glands that secrete hydrochloric acid (HCl) and the enzyme pepsinogen. The gastric glands consist of mucous cells, chief cells that secrete pepsinogen, and parietal cells that secrete HCl. Gastric pits are the openings of the gastric glands.

The acidic solution within the stomach also kills most of the bacteria that are ingested with the food. The few bacteria that survive the stomach and enter the intestine intact are able to grow and multiply there, particularly in the large intestine. In fact, most vertebrates harbor thriving colonies of bacteria within their intestines, and bacteria are a major component of feces. As we discuss later, bacteria that live within the digestive tract of cows and other ruminants play a key role in the ability of these mammals to digest cellulose.

Ulcers

It is important that the stomach not produce too much acid. If it did, the body could not neutralize the acid later in the small intestine, a step essential for the final stage of digestion. Production of acid is controlled by hormones. These hormones are produced by endocrine cells scattered within the walls of the stomach. The hormone gastrin regulates the synthesis of HCl by the parietal cells of the gastric pits, permitting HCl to be made only when the pH of the stomach is higher than about 1.5.

Overproduction of gastric acid can occasionally eat a hole through the wall of the stomach. Such gastric **ulcers** are rare, however, because epithelial cells in the mucosa of the stomach are protected somewhat by a layer of alkaline mucus, and because those cells are rapidly replaced by cell division if they become damaged (gastric epithelial cells are replaced every two to three days). Over 90% of gastrointestinal ulcers are duodenal ulcers, which are ulcers of the small intestine. These may be produced when excessive amounts of acidic chyme are delivered into the duodenum, so that the acid cannot be properly neutralized through the action of al-

kaline pancreatic juice (described later). Susceptibility to ulcers is increased when the mucosal barriers to self-digestion are weakened by an infection of the bacterium *Helicobacter pylori*. Modern antibiotic treatments can reduce symptoms and often cure the ulcer.

In addition to producing HCl, the parietal cells of the stomach also secrete intrinsic factor, a polypeptide needed for the intestinal absorption of vitamin B_{12}. Because this vitamin is required for the production of red blood cells, persons who lack sufficient intrinsic factor develop a type of anemia (low red blood cell count) called *pernicious anemia.*

Leaving the Stomach

Chyme leaves the stomach through the *pyloric sphincter,* shown at the base of the stomach in **figure 25.12**, to enter the small intestine. This is where all terminal digestion of carbohydrates, fats, and proteins occurs, and where the products of digestion—amino acids, glucose, and fatty acids—are absorbed into the blood. Only water from chyme and a few substances such as aspirin and alcohol are absorbed through the wall of the stomach.

> **Key Learning Outcome 25.5** Peristaltic waves of contraction propel food along the esophagus to the stomach. Gastric juice contains strong hydrochloric acid and the protein-digesting enzyme pepsin, which begins the digestion of proteins into shorter polypeptides. The acidic chyme is then transferred through the pyloric sphincter to the small intestine.

The Small and Large Intestines

Digestion and Absorption: The Small Intestine

The digestive tract exits from the stomach into the **small intestine (figure 25.13)**, where the breaking down of large molecules into small ones occurs. Only relatively small portions of food are introduced into the small intestine at one time, to allow time for acid to be neutralized and enzymes to act. The small intestine is the true digestive vat of the body. Within it, carbohydrates are broken down into simple sugars, proteins into amino acids, and fats into fatty acids. Once these small molecules have been produced, they pass across the epithelial wall of the small intestine into the bloodstream.

Some of the enzymes necessary for these digestive processes are secreted by the cells of the intestinal wall. Most, however, are made in a large gland called the *pancreas* (discussed in section 25.8), situated near the junction of the stomach and the small intestine. It is one of the body's major exocrine glands (secreting through ducts). The pancreas sends its secretions into the small intestine through a duct that empties into its initial segment, the **duodenum.** Your small intestine is approximately 6 meters long—unwound and stood on its end, it would be far taller than you are! Only the first 25 centimeters, about 4% of the total length, is the duodenum. It is within this initial segment, where the pancreatic enzymes enter the small intestine, that digestion occurs.

Much of the food energy the vertebrate body harvests is obtained from fats. The digestion of fats is carried out by a collection of molecules known as *bile salts* secreted into the duodenum from the *liver* (also discussed in section 25.8). Because fats are insoluble in water, they enter the intestine as drops within the watery chyme. The bile salts, which are partly lipid-soluble and partly water-soluble, work like detergents. They combine with fats to form microscopic droplets in a process called emulsification. These tiny droplets have greater surface areas upon which the enzyme that breaks down fats, called lipase, can work. This allows the digestion of fats to proceed more rapidly.

Two areas make up the rest of the small intestine (96% of its length), the **jejunum** and the **ileum.** Digestion continues into the jejunum, but the ileum is devoted to absorbing water and the products of digestion into the bloodstream. The lining of the small intestine is folded into ridges, as shown in **figure 25.13a.** The ridges are covered with fine fingerlike projections called **villi** (singular, **villus**), shown in the first enlarged view, but each too small to see with the naked eye. In turn, each of the cells covering a villus is covered on its outer surface by a field of cytoplasmic projections called **microvilli.** The enlargement of the villus shows epithelial cells lining the villus, and the further enlargement of these cells shows the microvilli on the surface side of the cells. Scanning and transmission electron micrographs in **figure 25.13b, c** give you different perspectives of the microvilli. Both villi and microvilli greatly increase the absorptive surface of the lining of the small intestine. The average surface area of the small intestine is about 300 square meters, more than the surface of many swimming pools!

The amount of material passing through the small intestine is startlingly large. Per day, an average human consumes about 800 grams of solid food, and 1,200 milliliters of water, for a total volume of about 2 liters. To this amount is added about 1.5 liters of fluid from the salivary glands, 2 liters from the gastric secretions of the stomach, 1.5 liters from the pancreas, 0.5 liters from the liver, and 1.5 liters of intestinal secretions. The total adds up to a remarkable 9 liters—more than 10% of the total volume of your body! However, although the flux is great, the *net* passage is small. Almost all these fluids and solids are reabsorbed during their passage through the small intestine—about 8.5 liters across the walls of the small intestine and 0.35 liters across the wall of the large intestine. Of the 800 grams of solids and 9 liters of liquids that enter the digestive tract each day, only about 50 grams of solids and 100 milliliters of liquids leave the body as feces. The fluid absorption efficiency of the digestive tract thus approaches 99%, very high indeed.

Concentration of Solids: The Large Intestine

The **large intestine,** or **colon,** is much shorter than the small intestine, approximately 1 meter long, but it is called the large intestine because of its larger diameter. The small intestine empties directly into the large intestine at a junction where the cecum and the appendix are located, which are two structures no longer actively used in humans. No digestion takes place within the large intestine, and only about 6% to 7% of fluid absorption occurs there. The large intestine is not convoluted, lying instead in three relatively straight segments, and its inner surface does not possess villi. As a consequence, the large intestine has only one-thirtieth the absorptive surface area of the small intestine. Although some water, sodium, and vitamin K are absorbed across its walls, the primary function of the large intestine is to act as a refuse dump. Within it, undigested material, including large amounts of plant fiber and cellulose, is compacted into the final excretory product, **feces,** and stored. Many bacteria live and actively divide within the large intestine. Bacterial fermentation produces gas within the colon at a rate of about 500 milliliters per day. This rate increases greatly after the consumption of beans or other vegetable matter because the passage of undigested plant material (fiber) into the large intestine provides substrates for fermentation.

The final segment of the digestive tract is a short extension of the large intestine called the **rectum.** Compact solids within the colon pass through the rectum as a result of the peristaltic contractions of the muscles encasing the large intestine, and then out of the body through the **anus.**

> **Key Learning Outcome 25.6** Most digestion occurs in the initial upper portion of the small intestine, called the duodenum. The rest of the small intestine is devoted to absorption of water and the products of digestion. The large intestine compacts residual solid wastes.

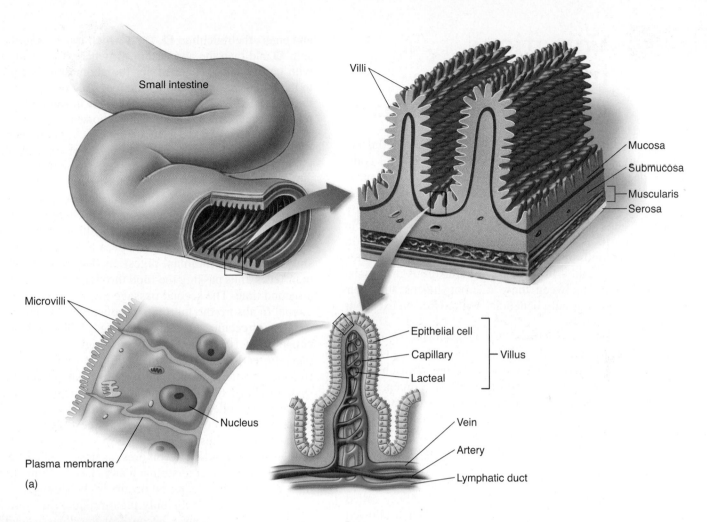

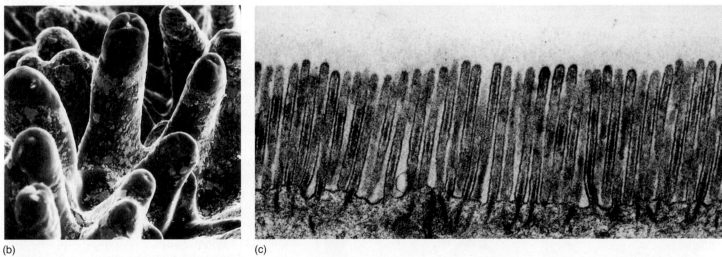

Figure 25.13 The small intestine.

(*a*) A cross section of the small intestine shows the structure of the villi and microvilli. (*b*) Microvilli, shown in a scanning electron micrograph, are very densely clustered, giving the small intestine an enormous surface area, which is very important for efficient absorption. (*c*) Intestinal microvilli as shown in a transmission electron micrograph.

25.7 Variations in Vertebrate Digestive Systems

Most animals lack the enzymes necessary to digest cellulose, the carbohydrate that functions as the chief structural component of plants. The digestive tracts of some animals, however, contain prokaryotes and protists that convert cellulose into substances the host can digest. Although digestion by gastrointestinal microorganisms plays a relatively small role in human nutrition, it is an essential element in the nutrition of many other kinds of animals, including insects like termites and cockroaches and a few groups of herbivorous mammals. The relationships between these microorganisms and their animal hosts are mutually beneficial and provide an excellent example of symbiosis.

Cows, deer, and other herbivores called *ruminants* have large, divided stomachs. By following the path food takes in **figure 25.14**, we can explore the areas of the stomach. Food enters the stomach by way of the rumen ❶. The rumen, which may hold up to 50 gallons, serves as a fermentation vat in which prokaryotes and protists convert cellulose and other molecules into a variety of simpler compounds. The location of the rumen at the front of the four chambers is important because it allows the animal to regurgitate and rechew the contents of the rumen (see how the arrow leaves the stomach after looping through the rumen and reenters), an activity called *rumination,* or "chewing the cud." The cud is then swallowed

and enters the reticulum ❷, from which it passes to the omasum ❸ and then the abomasum ❹, where it is finally mixed with gastric juice. Hence, only the abomasum is equivalent to the human stomach in its function. This process leads to a far more efficient digestion of cellulose in ruminants than in mammals that lack a rumen, such as horses.

In some animals such as rodents, horses, and lagomorphs (rabbits and hares), the digestion of cellulose by microorganisms takes place in the cecum, which is greatly enlarged in these animals (see the rabbit, a nonruminant herbivore, in **figure 25.15**). Because the cecum is located beyond the stomach, regurgitation of its contents is impossible. However, rodents and lagomorphs have evolved another way to digest cellulose that achieves a degree of efficiency similar to that of ruminant digestion. They do this by eating their feces, thus passing the food through their digestive tract a second time. The second passage makes it possible for the animal to absorb the nutrients produced by the microorganisms in its cecum. Animals that engage in this practice of **coprophagy** (from the Greek words *copros,* excrement, and *phagein,* eat) cannot remain healthy if they are prevented from eating their feces. The organization of the digestive system reflects the diet of the animal. Thus the large cecum of the rabbit reflects a diet of plants. In contrast, the insectivore and carnivore in **figure 25.15** digest primarily protein from animal bodies; therefore, they have a reduced or absent cecum. Ruminant herbivores, as described earlier, have a large four-chambered stomach and also a cecum, although most digestion of vegetation occurs in the stomach.

Cellulose is not the only plant product that vertebrates can use as a food source because of the digestive activities of intestinal microorganisms. Wax, a substance indigestible by most terrestrial animals, is digested by symbiotic bacteria living in the gut of honeyguides, African birds that eat the wax in bees' nests. In the marine food chain, wax is a major constituent of copepods (crustaceans in the plankton), and many marine fish and birds appear to be able to digest wax with the aid of symbiotic microorganisms.

Another example of the way intestinal microorganisms function in the metabolism of their animal hosts is provided by the synthesis of vitamin K. All mammals rely on intestinal bacteria to synthesize this vitamin, which is necessary for the clotting of blood. Birds, which lack these bacteria, must consume the required quantities of vitamin K in their food. In humans, prolonged treatment with antibiotics greatly reduces the populations of bacteria in the intestine; under such circumstances, it may be necessary to provide supplementary vitamin K.

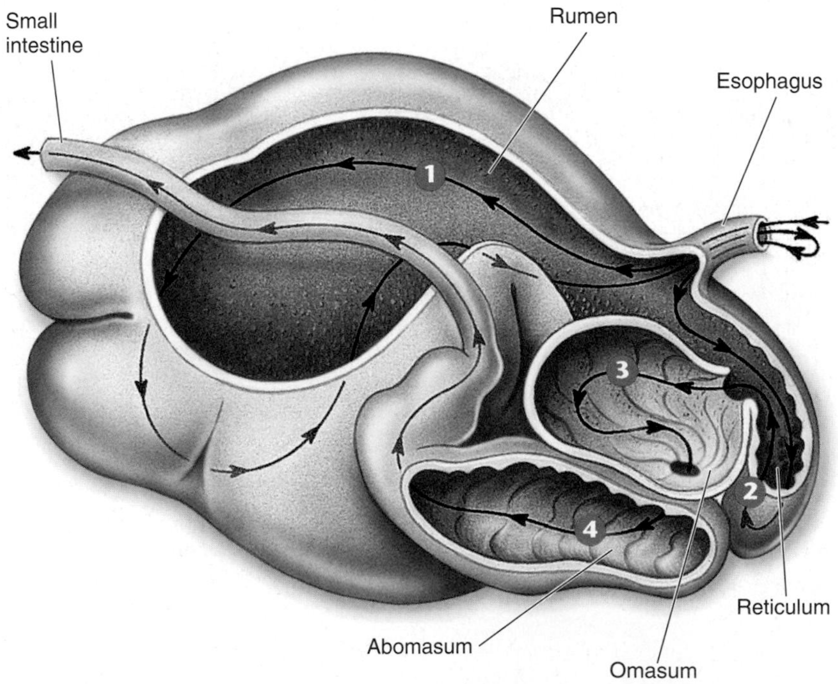

Figure 25.14 Four-chambered stomach of a ruminant.

The grass and other plants that a ruminant, such as a cow, eats enter the rumen, where they are partially digested. From there, the food may be regurgitated and rechewed. The food is then transferred through the last three chambers. Only the abomasum secretes gastric juice.

> **Key Learning Outcome 25.7** Much of the food value of plants is tied up in cellulose, and the digestive tract of many animals harbors colonies of cellulose-digestive microorganisms.

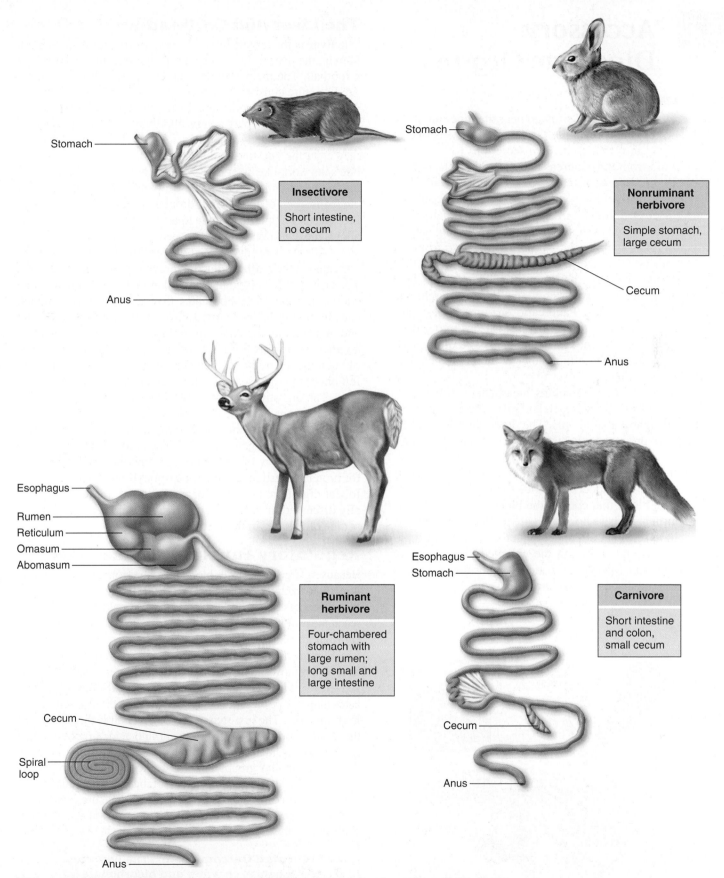

Figure 25.15 The digestive systems of different mammals reflect their diets.

Herbivores require long digestive tracts with specialized compartments for the breakdown of plant matter. Protein diets are more easily digested; thus, insectivorous and carnivorous mammals have shorter digestive tracts with few specialized pouches.

25.8 Accessory Digestive Organs

The Pancreas

The **pancreas,** a large gland situated near the junction of the stomach and the small intestine (see **figure 25.5**), is one of the accessory organs that contribute secretions to the digestive tract. Fluid from the pancreas is secreted into the duodenum through the *pancreatic duct* shown in **figure 25.16**. Note that the pancreatic duct joins with another duct, the common bile duct (discussed later), before entering the small intestine. This fluid contains a host of enzymes, including trypsin and chymotrypsin, which digest proteins. Inactive forms of these enzymes are released into the duodenum and are then activated by the enzymes of the intestine. Pancreatic fluid also contains pancreatic amylase, which digests starch; and lipase, which digests fats. Pancreatic enzymes digest proteins into smaller polypeptides, polysaccharides into shorter chains of sugars, and fat into free fatty acids and other products. The digestion of these molecules is then completed by the intestinal enzymes.

Pancreatic fluid also contains bicarbonate, which neutralizes the HCl from the stomach and gives the chyme in the duodenum a slightly alkaline pH. The digestive enzymes and bicarbonate are produced by clusters of secretory cells known as *acini.*

In addition to its exocrine role in digestion, the pancreas also functions as an endocrine gland, secreting several hormones into the blood that control the blood levels of glucose and other nutrients. These hormones are produced in the **islets of Langerhans,** clusters of endocrine cells scattered throughout the pancreas and shown in the enlarged view in **figure 25.16**. The two most important pancreatic hormones, insulin and glucagon, are discussed in chapters 26 and 30.

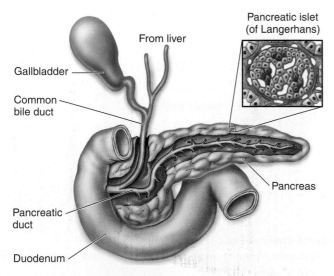

Figure 25.16 The pancreatic and bile ducts empty into the duodenum.

The Liver and Gallbladder

The **liver** is the largest internal organ of the body. In an adult human, the liver weighs about 1.5 kilograms and is the size of a football. The main exocrine secretion of the liver is **bile,** a fluid mixture consisting of *bile pigments* and *bile salts* that is delivered into the duodenum during the digestion of a meal.

The bile salts play a very important role in the digestion of fats. As explained earlier, fats are insoluble in water, and so they enter the intestine as drops within the watery chyme. The bile salts work like detergents, dispersing the large drops of fat into a fine suspension of smaller droplets. This breaking up, or emulsification, of the fat into droplets produces a greater surface area of fat upon which the lipase enzymes can act, and thus allows the digestion of fat to proceed more rapidly.

After it is produced in the liver, bile is stored and concentrated in the **gallbladder** (the green organ in **figure 25.16**). The arrival of fatty food in the duodenum triggers a neural and endocrine reflex that stimulates the gallbladder to contract, causing bile to be transported through the common bile duct and injected into the duodenum. If the bile duct is blocked by a *gallstone* (formed from a hardened precipitate of cholesterol), contraction of the gallbladder causes pain that is generally felt under the right scapula (shoulder blade).

The digestive system is highly specialized and involves the interactions of many different organs. **Figure 25.17** overviews the different functional areas in the digestive system and the different organs involved. The colored circles indicate the primary areas of digestion and enzyme production: red for protein digestion, orange for carbohydrate digestion, green for fat digestion, and blue for nucleic acid digestion (not really discussed in this chapter as nucleic acids are not a major source of calories in the diet).

Regulatory Functions of the Liver

Because a large vein carries blood from the stomach and intestine directly to the liver, the liver is in a position to chemically modify the substances absorbed in the gastrointestinal tract before they reach the rest of the body. For example, ingested alcohol and other drugs are taken into liver cells and metabolized; this is why the liver is often damaged as a result of alcohol and drug abuse. The liver also removes toxins, pesticides, carcinogens, and other poisons, converting them into less-toxic forms. Also, excess amino acids that may be present in the blood are converted to glucose by liver enzymes. The first step in this conversion is the removal of the amino group ($-NH_2$) from the amino acid, a process called *deamination.* Unlike plants, animals cannot reuse the nitrogen from these amino groups and must excrete it as nitrogenous waste. The product of amino acid deamination, ammonia (NH_3), combines with carbon dioxide to form urea. The urea is released by the liver into the bloodstream, where—as you will learn in chapter 26—the kidneys subsequently remove it.

> **Key Learning Outcome 25.8** The pancreas secretes digestive enzymes and bicarbonate into the pancreatic duct. The liver produces bile, which is stored and concentrated in the gallbladder. The liver and the pancreatic hormones regulate blood glucose concentration.

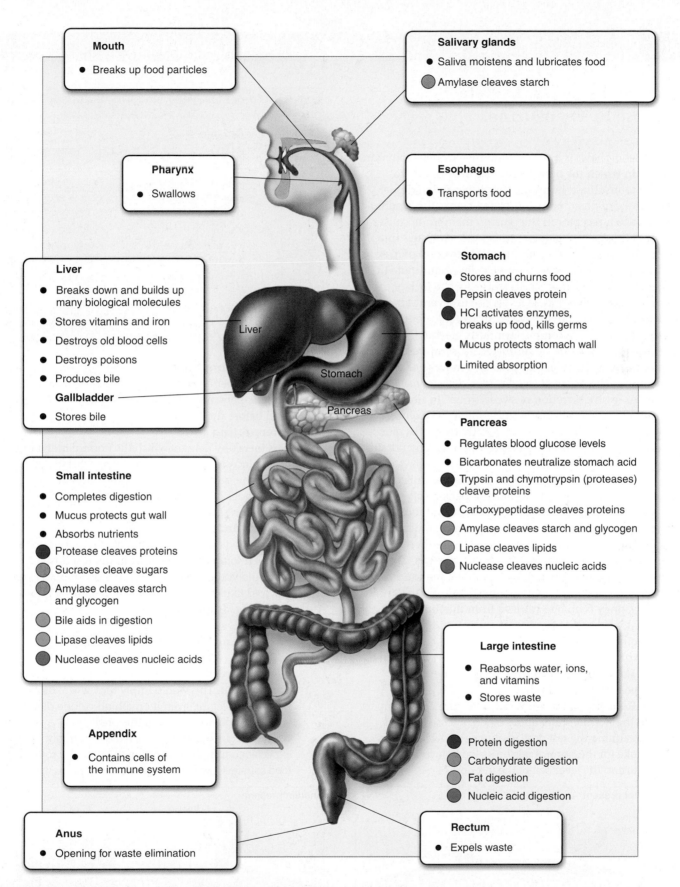

Mouth
- Breaks up food particles

Salivary glands
- Saliva moistens and lubricates food
- Amylase cleaves starch

Pharynx
- Swallows

Esophagus
- Transports food

Liver
- Breaks down and builds up many biological molecules
- Stores vitamins and iron
- Destroys old blood cells
- Destroys poisons
- Produces bile

Gallbladder
- Stores bile

Stomach
- Stores and churns food
- Pepsin cleaves protein
- HCl activates enzymes, breaks up food, kills germs
- Mucus protects stomach wall
- Limited absorption

Pancreas
- Regulates blood glucose levels
- Bicarbonates neutralize stomach acid
- Trypsin and chymotrypsin (proteases) cleave proteins
- Carboxypeptidase cleaves proteins
- Amylase cleaves starch and glycogen
- Lipase cleaves lipids
- Nuclease cleaves nucleic acids

Small intestine
- Completes digestion
- Mucus protects gut wall
- Absorbs nutrients
- Protease cleaves proteins
- Sucrases cleave sugars
- Amylase cleaves starch and glycogen
- Bile aids in digestion
- Lipase cleaves lipids
- Nuclease cleaves nucleic acids

Large intestine
- Reabsorbs water, ions, and vitamins
- Stores waste

Appendix
- Contains cells of the immune system

- Protein digestion
- Carbohydrate digestion
- Fat digestion
- Nucleic acid digestion

Anus
- Opening for waste elimination

Rectum
- Expels waste

Labels on figure: Liver, Stomach, Pancreas

Figure 25.17 The organs of the digestive system and their functions.
The digestive system contains some dozen different organs that act on the food that is consumed, starting with the mouth and ending with the anus. All of these organs must work properly for the body to effectively obtain nutrients.

Why Do Diabetics Excrete Glucose in Their Urine?

Late-onset diabetes is a serious and increasingly common disorder in which the body's cells lose their ability to respond to insulin, a hormone that is needed to trigger their uptake of glucose. As illustrated below, the binding of insulin to a receptor in the plasma membrane causes the rapid insertion of glucose transporter channels into the plasma membrane, allowing the cell to take up glucose. In diabetics, however, glucose molecules accumulate in the blood while the body's cells starve for the lack of them. In mild cases, blood glucose levels rise to several times the normal value of 4 m*M;* in severe, untreated cases, blood glucose levels may become enormously elevated, up to 25 times the normal value. A characteristic symptom of even mild diabetes is the excretion of large amounts of glucose in the urine. The name of the disorder, *diabetes mellitus,* means "excessive secretion of sweet urine." In normal individuals, by contrast, only trace amounts of glucose are excreted. The kidney very efficiently reabsorbs glucose molecules from the fluid passing through it. Why doesn't it do so in diabetic individuals?

The graph on the upper right displays so-called glucose tolerance curves for a normal person (*blue line*) and a diabetic (*red line*). After a night without food, each individual drank a test dose of 100 grams of glucose dissolved in water. Blood glucose levels were then monitored at 30-minute and one-hour intervals. The dotted line indicates the kidney threshold, the maximum concentration of blood glucose molecules (about 10 m*M*) that the kidney is able to retrieve from the fluid passing through it when all of its glucose-transporting channels are being utilized full-bore.

1. **Applying Concepts**
   a. Variable. In the graph, what is the dependent variable?
   b. Reading a Curve. What is the immediate impact on the normal individual's blood glucose levels of consuming the test dose of glucose? How long does it take for the normal person's blood glucose level to return to the level before the test dose?

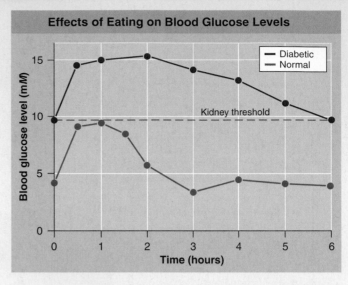

Effects of Eating on Blood Glucose Levels

c. **Comparing Curves.** Is the impact any different for the diabetic person? How long does it take for the diabetic person's blood glucose levels to return to the level before the test dose?

2. **Interpreting Data**
   a. Is there any point at which the normal individual's blood glucose levels exceed the kidney threshold?
   b. Is there any point at which the diabetic individual's blood glucose levels do *not* exceed the kidney threshold?

3. **Making Inferences**
   a. Why do you suppose the diabetic individual took so much longer to recover from the test dose?
   b. Would you expect the normal individual to excrete glucose? Explain. The diabetic individual? Explain.

4. **Drawing Conclusions** Why do diabetic individuals secrete sweet urine?

5. **Further Analysis**
   a. If glucose molecules are being excreted in the urine, then they are not being converted to fatty acids for storage as fat. This would imply that severe diabetics would lose weight, even if on a high-calorie diet. How would you go about testing this prediction?
   b. Denied glucose, the cells of a diabetic might be expected to turn in desperation to the cell's proteins as a food source. How might you test this hypothesis?

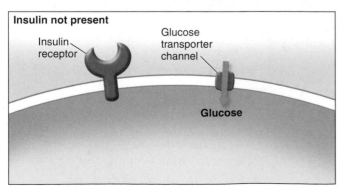

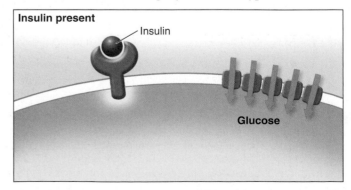

Food Energy and Essential Nutrients

25.1 Food for Energy and Growth

- Animals consume food as a source of energy and as a source of essential molecules and minerals. A balanced diet that is higher in complex carbohydrates, fruits, and vegetables and low in fats and sweets is recommended, as indicated by the pyramid of nutrition shown here from **figure 25.1**.

- The energy from food is either used up through metabolic activity (the BMR and exercise) or is stored as fat in fat cells. A measurement called a body mass index (BMI) is an easy guide to determining whether a person is overweight or obese (**figure 25.2**).

- Many animals must consume fiber in their diets to maintain digestive health. Animals must also consume proteins, fruits, and vegetables to obtain essential amino acids, minerals, and vitamins that the body needs but cannot produce itself.

Digestion

25.2 Types of Digestive Systems

- Single-celled organisms and lower animal phyla, such as sponges, digest food intracellularly. Food is taken up by individual cells and is broken down inside the cells.

- All other animals digest food extracellularly. Digestive enzymes are released into a cavity or tract where they break down food. The products of digestion are then absorbed by cells in the body. The digestive cavity in flatworms and cnidarians, like the *Hydra* shown here from **figure 25.3**, is called the gastrovascular cavity.

- Specialization of the digestive tract, where different regions of the tract are involved in different digestive functions, was possible with the evolution of a one-way digestive tract (**figure 25.4**).

25.3 Vertebrate Digestive Systems

- Vertebrate digestion occurs in a gastrointestinal tract with specialized areas for different digestive functions (**figures 25.5** and **25.6**). The size or structure of specialized digestive areas varies in different animal groups depending on the animal's diet.

25.4 The Mouth and Teeth

- In vertebrates, food is first brought into the mouth. Birds break up food in the gizzard, a compartment of their digestive system (**figure 25.7**). In the gizzard, food is churned and ground up with pebbles that the bird has swallowed. In other vertebrates, teeth are used to chew up food, breaking it into smaller pieces (**figures 25.8** and **25.9**).

- The chewed food mixes with saliva in the mouth. Saliva moistens and lubricates the food and it contains the enzyme salivary amylase that begins the digestion of starches. The moistened food is then swallowed passing from the mouth into the esophagus (**figure 25.10**).

25.5 The Esophagus and Stomach

- Food is moved along the esophagus to the stomach by peristaltic waves of muscle contractions (**figure 25.11**).

- In the stomach, muscle contractions churn up the food with gastric juice, which contains hydrochloric acid and pepsin, a protein-digesting enzyme activated by HCl (**figure 25.12**). The acidic conditions in the stomach help denature proteins so they are easier to digest in the small intestine. The partially digested food and gastric juice that leave the stomach is called chyme.

25.6 The Small and Large Intestines

- The acidic chyme from the stomach passes into the upper portion of the small intestine where it is neutralized and mixed with other digestive enzymes. Some enzymes are secreted by the cells that line the walls of the intestine, but most enzymes and other digestive substances are produced in the pancreas or other accessory organs. The rest of the small intestine is involved in absorption of food molecules and water. The lining of the small intestine is folded into ridges that are covered with fingerlike projections called villi, shown here from **figure 25.13**. The surface of the cells that line the villi are themselves covered with cytoplasmic projections called microvilli. Villi and microvilli increase the surface area for absorption.

- The large intestine collects and compacts solid waste, releasing it through the rectum and anus.

25.7 Variations in Vertebrate Digestive Systems

- In ruminants, cellulose-digesting microorganisms live in a chamber of the stomach called the rumen. Food enters the rumen where microorganisms begin digesting cellulose. The partially digested food is then regurgitated, chewed again, and reswallowed, entering the reticulum. The products of cellulose digestion pass through the other areas of the stomach and then to the small intestine where they are absorbed (**figure 25.14**).

- In other animals, cellulose digestion occurs in the cecum. To gain the nutritional value of cellulose digestion, these animals eat their feces. The structure of digestive systems of animals differs, based on their diets (**figure 25.15**).

25.8 Accessory Digestive Organs

- The pancreas produces the protein-digesting enzymes trypsin and chymotrypsin, the starch-digesting enzyme pancreatic amylase, and the fat-digesting enzyme lipase. The pancreatic fluid also contains bicarbonate, which neutralizes the chyme.

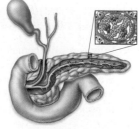

- The liver produces bile (a mixture of bile pigments and bile salts), which breaks down fats. Bile is stored in the gallbladder, shown here from **figure 25.16**, and released into the small intestine. All of the organs of digestion work together (**figure 25.17**).

1. One-way passage of food through the digestive system of many animal groups allows
   a. intracellular digestion.
   b. specialization of different regions of the digestive system.
   c. release of digestive enzymes into the gut.
   d. extracellular digestion.

2. Organisms with longer digestive systems, which help break down difficult to digest food, are usually
   a. herbivores.
   b. carnivores.
   c. omnivores.
   d. detritivores.

3. The purpose of a gizzard, like teeth, is to
   a. hold on to prey.
   b. begin the chemical digestion of food.
   c. release enzymes.
   d. begin the physical digestion of food.

4. When a mammal swallows food, it is prevented from going up into the nasal cavity by the
   a. esophagus.
   b. tongue.
   c. soft palate.
   d. epiglottis.

5. The first site of protein digestion in the digestive system occurs in the
   a. mouth.
   b. esophagus.
   c. stomach.
   d. small intestine.

6. Which of the following statements is *false*?
   a. Carnivores have a reduced or absent cecum.
   b. Only ruminants are able to digest cellulose.
   c. The human digestive system contains bacteria but is not able to gain nutritional value from the digestion of cellulose.
   d. Ruminants are able to regurgitate food.

7. Most of the absorption of food molecules takes place in the
   a. stomach.
   b. liver.
   c. small intestine.
   d. large intestine.

8. The purpose of the villi and microvilli in the small intestine is to
   a. neutralize stomach acid.
   b. produce bile.
   c. produce digestive enzymes.
   d. increase the surface area of the small intestine for absorption of nutrients.

9. The primary function of the large intestine is
   a. the breakdown and absorption of fats.
   b. the absorption of water.
   c. the concentration of solid wastes.
   d. the absorption of vitamin C.

10. The _____ secretes digestive enzymes and bicarbonate solution into the small intestine to aid digestion.
    a. pancreas
    b. liver
    c. gallbladder
    d. All of the above.

Apply Your Understanding

1. **Figure 25.10** "Don't talk with your mouth full" is parental advice that is important for not only social reasons (manners) but for medical reasons. Talking while eating can result in choking. Explain how this could occur.

2. **Figure 25.16** What are the functions of the tan-colored organ and the green-colored organ shown here?

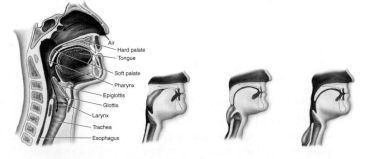

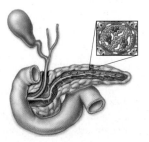

Synthesize What You Have Learned

1. Why is it so important for Popeye, and you, to eat your green leafy vegetables (spinach, chard, turnip and mustard greens, broccoli, cauliflower, cabbage)?

2. You're going on a backpacking trip with three friends. Lisa wants to pack trail snacks of chocolate cupcakes; Chris wants to take beef jerky. Andre insists that you should all pack GORP (*good old raisins and peanuts*—sometimes known as trail mix). They turn to you to decide. Explain which one is best for a quick snack to keep up your energy on a long hike, and why.

26

Maintaining the Internal Environment

The man above is sweating, something each of us has done when our bodies have become overheated from too much exercise or sun. The evaporation of the sweat cools our skin, a clever mechanism to remove heat. Your body, like that of all birds and mammals, tries to maintain a constant body temperature, no matter how hot or cold the surrounding air might be, and sweating is one of the ways it does this. If your body begins to heat above 37°C, you start to sweat and release heat; if you instead begin to cool below 37°C, you shiver and generate heat. Keeping your body temperature constant is only one example of a much broader physiological strategy: Vertebrates maintain relatively constant physiological conditions within their bodies. The pH of your blood, the rate at which you breathe, your blood pressure, the concentrations of water, salt, and glucose in your blood—all are monitored carefully by the brain, which acts continuously to keep each within narrow limits. Your body is constantly making dynamic adjustments in these parameters to counter changes caused by outside factors that would alter the body's internal environment. This steady-state balance of internal conditions, known as homeostasis, is the subject of this chapter. A major objective of your study of biology will be to learn how animals maintain homeostasis.

26.1 How the Animal Body Maintains Homeostasis

As the animal body has evolved, specialization has increased. Each cell is a sophisticated machine, finely tuned to carry out a precise role within the body. Such specialization of cell function is possible only when extracellular conditions are kept within narrow limits. Temperature, pH, the concentration of glucose and oxygen, and many other factors must be held fairly constant for cells to function efficiently and interact properly with one another.

Homeostasis may be defined as the dynamic constancy of the internal environment. The term *dynamic* is used because conditions are never absolutely constant, but fluctuate continuously within narrow limits. Homeostasis is essential for life, and most of the regulatory mechanisms of the vertebrate body are concerned with maintaining homeostasis.

Negative Feedback Loops

To maintain internal constancy, the vertebrate body must have **sensors** that are able to measure each condition of the internal environment (indicated by the green box in **figure 26.1**). These constantly monitor the extracellular conditions and relay this information (usually via nerve signals) to an **integrating center** (the yellow triangle), which contains the *set point,* the proper value for that condition. This set point is analogous to the temperature setting on a house thermostat. In a similar manner, there are set points for body temperature, blood glucose concentration, the tension on a tendon, and so on. The integrating center is often a particular region of the brain or spinal cord, but in some cases it can also be cells of endocrine glands. It receives messages from several sensors, weighs the relative strengths of each sensor input, and then determines whether the value of the condition is deviating from the set point. When a deviation in a condition occurs (the "stimulus" indicated by the red oval), sensors detect it, and the integrating center sends a message to increase or decrease the activity of particular effectors. **Effectors** (the blue box) are generally muscle or glands, and can change the value of the condition in question back toward the set point value, which is "the response" (the purple oval).

To return to the idea of a home thermostat, suppose you set the thermostat at a set point of 70°F. If the temperature of the house rises sufficiently above the set point, the thermostat (equivalent to an integrating center) receives this input from a temperature sensor, like a thermometer within the wall unit. It compares the actual temperature to its set point. When these are different, it sends a signal to an effector. The effector in this case may be an air conditioner, which acts to reverse the deviation and bring it back to the set point.

In humans, if the body temperature exceeds the set point of 37°C, sensors in the brain detect this deviation. Acting via an integrating center (also in the brain), these sensors stimulate effectors (such as sweat glands) that lower the temperature. One can think of the effectors as "defending" the set points of the body against deviations. Because the activity of the effectors is influenced by the effects they produce, and because this regulation is in a negative, or reverse, direction, this type of control system is known as a **negative feedback loop** (figure 26.1).

The nature of the negative feedback loop becomes clear when we again refer to the analogy of the thermostat and air conditioner. After the air conditioner has been on for some time, the room temperature may fall significantly below the set point of the thermostat. When this occurs, the air conditioner will be turned off. The effector (air conditioner) is turned on by high temperature; and when activated, it produces a negative change (lowering of the temperature) that ultimately causes the effector to be turned off. In this way, constancy is maintained.

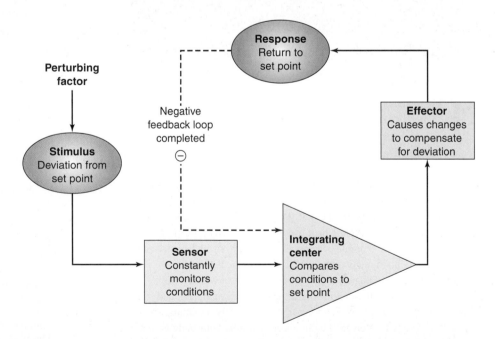

Figure 26.1   A generalized diagram of a negative feedback loop.
Negative feedback loops maintain a state of homeostasis, or dynamic constancy of the internal environment, by correcting deviations from a set point.

Regulating Body Temperature

Humans, together with other mammals and with birds, are endothermic; they can maintain relatively constant body temperatures independent of the environmental temperature. When the temperature of your blood exceeds 37°C (98.6°F), neurons in a part of the brain called the hypothalamus (discussed in chapters 28 and 30) detect the temperature change. Acting through the control of neurons, the hypothalamus responds by promoting the dissipation of heat through sweating, dilation of blood vessels in the skin, and other mechanisms. These responses tend to counteract the rise in body temperature. When body temperature falls, the hypothalamus coordinates a different set of responses, such as shivering and the constriction of blood vessels in the skin, which help to raise body temperature and correct the initial challenge to homeostasis.

Vertebrates other than mammals and birds are ectothermic; their body temperatures are more or less dependent on the environmental temperature. However, to the extent that it is possible, many ectothermic vertebrates attempt to maintain some degree of temperature homeostasis. Certain large fish, including tuna, swordfish, and some sharks, for example, can maintain parts of their body at a significantly higher temperature than that of the water. Reptiles attempt to maintain a constant body temperature through behavioral means—by placing themselves in varying locations of sun and shade. That's why you frequently see lizards basking in the sun. Sick lizards even give themselves a "fever" by seeking warmer locations!

Most invertebrates, like reptiles, modify behaviors to adjust their body temperature. Many butterflies, for example, must reach a certain body temperature before they can fly. In the cool of the morning, they orient their bodies to maximize their absorption of sunlight. Moths and other insects use a shivering reflex to warm their flight muscles.

Regulating Blood Glucose

When you digest a meal containing carbohydrates, you absorb glucose into your blood. This causes a temporary rise in the blood glucose concentration, which is brought back down in a few hours. What counteracts the rise in blood glucose following a meal?

Glucose levels within the blood are constantly monitored by a sensor, cells called the islets of Langerhans in the pancreas (discussed in chapters 25 and 30). When glucose levels increase (the condition of "high blood sugar" in **figure 26.2a**), the islets secrete the hormone *insulin*, which stimulates the uptake of blood glucose into muscles, liver, and adipose tissue. The muscles and liver can convert the glucose into the polysaccharide glycogen. Adipose cells can convert glucose into fat. These actions lower the blood glucose and help to store energy in forms that the body can use later. When enough blood glucose is absorbed, reaching the set point, the release of insulin stops. When blood glucose levels decrease below the set point, as they do between meals, during periods of fasting, and during exercise, the liver secretes glucose into the blood (the center arrow in **figure 26.2b**). This glucose is obtained in part from the breakdown of liver glycogen. The breakdown of liver glycogen is stimulated in two ways: by the hormone *glucagon*, which is also secreted by the islets of Langerhans, and by the hormone adrenaline, which is secreted from the adrenal gland (discussed in much more detail in chapter 30).

> **Key Learning Outcome 26.1** Negative feedback mechanisms correct deviations from a set point for different internal variables. In this way, body temperature and blood glucose, for example, are kept within a normal range.

Figure 26.2 Control of blood glucose levels.

(a) When blood glucose levels are high, cells within the pancreas produce the hormone insulin, which stimulates the liver and muscles to convert blood glucose into glycogen. (b) When blood glucose levels are low, other cells within the pancreas release the hormone glucagon into the bloodstream; in addition, cells within the adrenal gland release the hormone adrenaline into the bloodstream. When they reach the liver, these two hormones act to increase the liver's breakdown of glycogen to glucose.

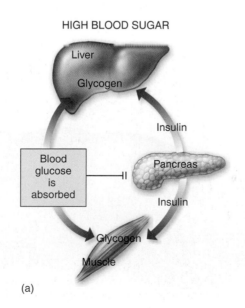

HIGH BLOOD SUGAR

(a)

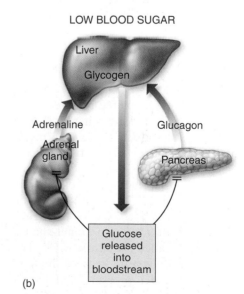

LOW BLOOD SUGAR

(b)

26.2 Regulating the Body's Water Content

Animals must also carefully monitor the water content of their bodies. The first animals evolved in seawater, and the physiology of all animals reflects this origin. Approximately two-thirds of every vertebrate's body is water. If the amount of water in the body of a vertebrate falls much lower than this, the animal dies. Animals use various mechanisms for **osmoregulation,** the regulation of the body's osmotic composition, or how much water and salt it contains. The proper operation of many vertebrate organ systems of the body requires that the osmotic concentration of the blood—the concentration of solutes dissolved within it—be kept within a narrow range.

Animals have evolved a variety of mechanisms to cope with problems of water balance. In many animals and single-celled organisms, the removal of water or salts from the body is coupled with the removal of metabolic wastes through the excretory system. Protists, like the *Paramecium* in **figure 26.3,** employ contractile vacuoles for this purpose, as do the cells of sponges. Water and metabolic wastes are collected by the endoplasmic reticulum and pass through feeder canals to the contractile vacuole. The water and wastes are expelled when the vacuole contracts and releases its contents out a pore. Multicellular animals have a system of excretory tubules (little tubes) that expel fluid and wastes from the body.

In flatworms, these tubules are called *protonephridia* (the green structures you see in **figure 26.4**). They branch throughout the body into bulblike **flame cells,** shown in the enlargement. While these simple excretory structures open to the outside of the body, they do not open to the inside of the body. Rather, the beating action of cilia within the flame cells draw in fluid from the body, which is passed into a collecting tube. Water and metabolites are then reabsorbed, and the substances to be excreted are expelled through excretory pores.

Other invertebrates have a system of tubules that open both to the inside and to the outside of the body. In the earthworm, these tubules are known as *nephridia,* the blue structures you see in **figure 26.5.** The nephridia obtain fluid from the body cavity through a process of filtration into funnel-shaped structures called *nephrostomes.* The term *filtration* is used because the fluid is formed under pressure and passes through small openings, so that molecules larger than a certain size are excluded. This filtered fluid is isotonic

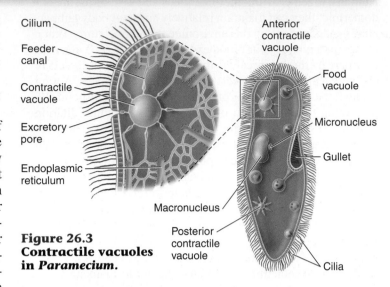

**Figure 26.3
Contractile vacuoles in *Paramecium*.**

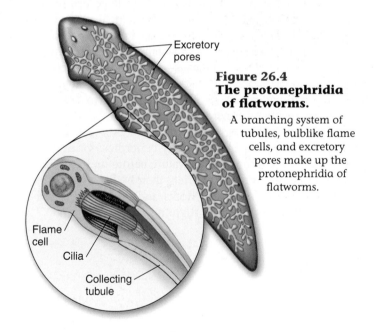

**Figure 26.4
The protonephridia of flatworms.**

A branching system of tubules, bulblike flame cells, and excretory pores make up the protonephridia of flatworms.

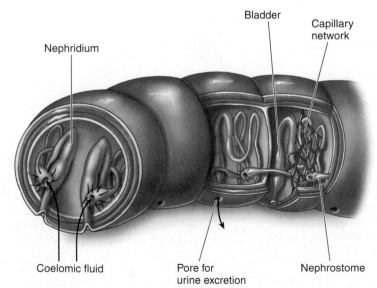

**Figure 26.5
The nephridia of annelids.**

Most invertebrates, such as the annelid shown here, have nephridia. These consist of tubules that receive a filtrate of coelomic fluid, which enters the funnel-like nephrostomes. Salt can be reabsorbed from these tubules, and the fluid that remains, urine, is released from pores into the external environment.

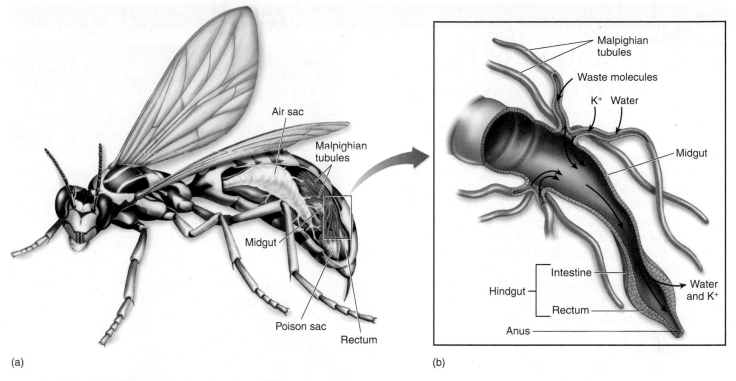

(a)   (b)

Figure 26.6   The Malpighian tubules of insects.
(a) The Malpighian tubules of insects are extensions of the digestive tract that collect water and wastes from the body's circulatory system.
(b) K⁺ is secreted into these tubules, drawing water with it osmotically. Much of this water and K⁺ is reabsorbed across the wall of the hindgut.

(having the same osmotic concentration) to the fluid in the coelom, but as it passes through the tubules of the nephridia, NaCl is removed by active transport processes. A general term for transport out of the tubule and into the surrounding body fluids is *reabsorption.* Because salt is reabsorbed from the filtrate, the urine excreted is more dilute than the body fluids (meaning it is hypotonic).

The excretory organs in insects are called **Malpighian tubules,** the green structures you see in **figure 26.6**. Malpighian tubules are extensions of the digestive tract that branch off anterior to the hindgut. Urine is not formed by filtration in these tubules because there is no pressure difference between the blood in the body cavity and the tubule. Instead, waste molecules and potassium ions (K⁺) are secreted into the tubules by active transport. In *secretion,* ions or molecules are transported from the body fluid into the tubule. The secretion of K⁺ creates an osmotic gradient that causes water to enter the tubules by osmosis from the body's open circulatory system. Most of the water and K⁺ is then reabsorbed into the circulatory system through the epithelium of the hindgut, leaving only small molecules and waste products to be excreted from the rectum along with feces. Malpighian tubules thus provide a very efficient means of water conservation.

Kidneys are the excretory organs in vertebrates and are discussed in more detail in the rest of the chapter. Unlike the Malpighian tubules of insects, kidneys create a tubular fluid by filtration of the blood under pressure. In addition to containing waste products and water, the filtrate contains many small molecules that are of value to the animal, including glucose, amino acids, and vitamins. These molecules and most of the water are reabsorbed from the tubules into the blood, while wastes remain in the filtrate. Additional wastes may be secreted by the tubules and added to the filtrate, and the final waste product, urine, is eliminated from the body.

It may seem odd that the vertebrate kidney should filter out almost everything from blood plasma (except proteins, which are too large to be filtered) and then spend energy to take back or reabsorb what the body needs. But selective reabsorption provides great flexibility; various vertebrate groups have evolved the ability to reabsorb different molecules that are especially valuable in particular habitats. This flexibility is a key factor underlying the successful colonization of many diverse environments by the vertebrates.

Key Learning Outcome 26.2   Many invertebrates filter fluid into a system of tubules and then reabsorb ions and water, leaving waste products for excretion. Insects create an excretory fluid by secreting K⁺ into tubules, which draws water osmotically. The vertebrate kidney produces a filtrate that enters tubules from which water is reabsorbed.

26.3 Evolution of the Vertebrate Kidney

The kidney, first evolving in freshwater fish, is a complex organ consisting of up to a million repeating disposal units called **nephrons.** The nephron pictured below is representative of those found in mammals and birds; the nephron of other vertebrates lacks a looped portion. Blood pressure forces the fluid in blood through a capillary bed, called the *glomerulus,* at the top of each nephron. The glomerulus retains blood cells, proteins, and other useful large molecules in the blood but allows the water, and the small molecules and wastes dissolved in it, to pass into a cuplike structure surrounding the glomerulus and then into the nephron tube. As the filtered fluid passes through the first part of the nephron tube (labeled as the proximal arm in the figure below), useful sugars, amino acids, and ions (such as Ca++) are recovered from it by active transport, leaving the water and metabolic wastes dissolved in a fluid, called urine. Water and salts are reabsorbed later in the nephron.

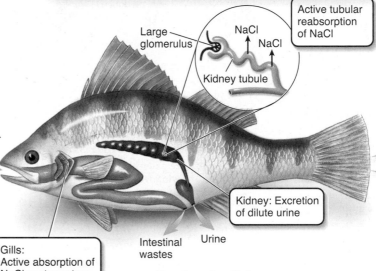

Freshwater fish

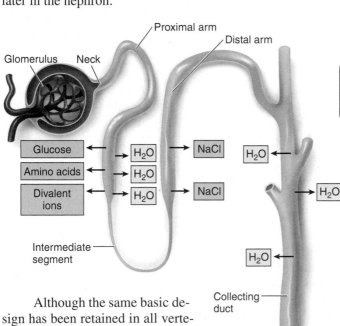

Although the same basic design has been retained in all vertebrate kidneys, there have been a few modifications. Because the original glomerular filtrate is isotonic to blood, all vertebrates can produce a urine that is isotonic to (by reabsorbing ions) or hypotonic to (more dilute than) blood. Only birds and mammals can reabsorb water from their glomerular filtrate to produce a urine that is hypertonic to (more concentrated than) blood.

Freshwater Fish

Kidneys are thought to have evolved first among the freshwater teleosts, or bony fish. Because the body fluids of a freshwater fish have a greater osmotic concentration than the surrounding water, these animals face two serious problems because of osmosis and diffusion: (1) Water tends to enter the body from the environment, and (2) solutes tend to leave the body and enter the environment. Freshwater fish address the first problem by *not* drinking water (water enters the mouth but passes out through the gills—it is not swallowed) and by excreting a large volume of dilute urine, which is hypotonic to their body fluids (as shown in the freshwater fish above). They address the second problem by reabsorbing ions (NaCl) across the nephron tubules, from the glomerular filtrate back into the blood. In addition, they actively transport ions (NaCl) across their gills from the surrounding water into the blood.

Marine Bony Fish

Although most groups of animals seem to have evolved first in the sea, marine bony fish (teleosts) probably evolved from freshwater ancestors. They faced significant new problems in making the transition to the sea because their body fluids are hypotonic to the surrounding seawater. Consequently, water tends to leave their bodies by osmosis across their gills, and they also lose water in their urine. To compensate for this continuous water loss, marine fish drink large amounts of seawater.

Many of the divalent cations in the seawater that a marine fish drinks (principally Ca^{++} and Mg^{++} in the form of $MgSO_4$) remain in the digestive tract and are eliminated through the anus. Some, however, are absorbed into the blood, as are the monovalent ions K^+, Na^+, and Cl^-. Most of the monovalent ions are actively transported out of the blood across the gills, while the divalent ions that enter the blood (represented by $MgSO_4$ in figure) are secreted into the nephron tubules and excreted in the urine. In these two ways, marine bony fish eliminate the ions they get from the seawater they drink. The urine they excrete is isotonic to their body fluids. It is more concentrated than the urine of freshwater fish but not as concentrated as that of birds and mammals.

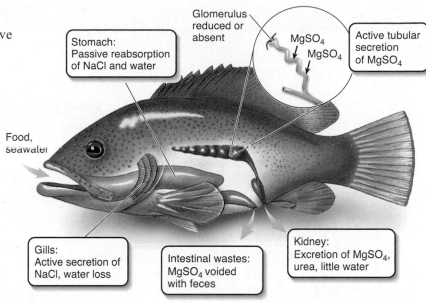

Marine fish

Cartilaginous Fish

The elasmobranchs—sharks, skates, and rays like the one in the photo—are by far the most common subclass in the class Chondrichthyes (cartilaginous fish). Elasmobranchs have solved the osmotic problem posed by their seawater environment in a different way than have the bony fish. Instead of having body fluids that are hypotonic to seawater, so that they have to continuously drink seawater and actively pump out ions, the elasmobranchs reabsorb urea from the nephron tubules and maintain a blood urea concentration that is 100 times higher than that of mammals. This added urea makes their blood approximately isotonic to the surrounding sea. Because there is no net water movement between isotonic solutions, water loss is prevented. Hence, these fish do not need to drink seawater for osmotic balance, and their kidneys and gills do not have to remove large amounts of ions from their bodies. The enzymes and tissues of the cartilaginous fish have evolved to tolerate the high urea concentrations.

Cartilaginous fish

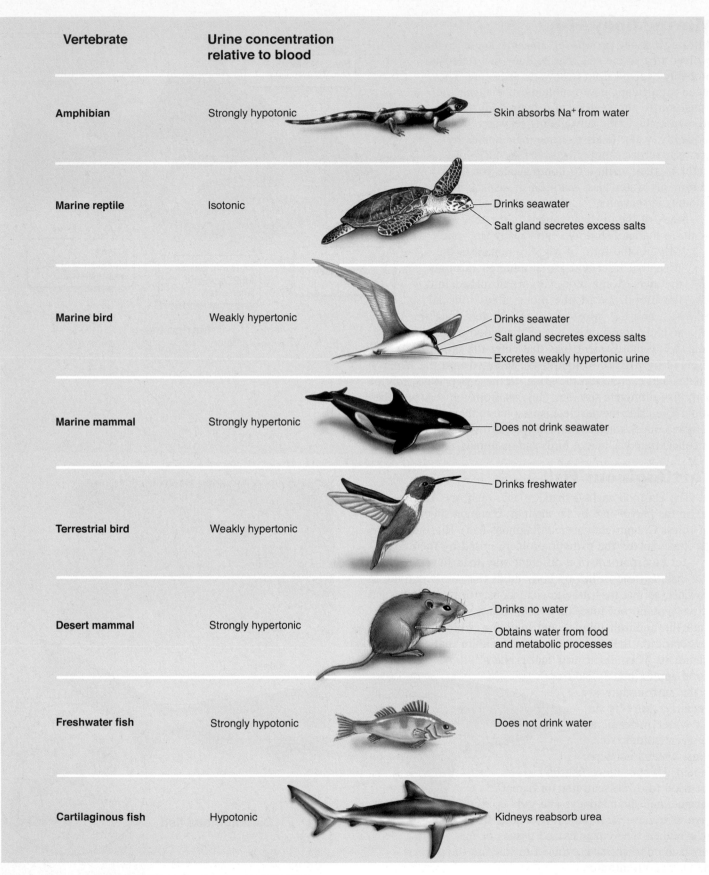

Vertebrate	Urine concentration relative to blood	
Amphibian	Strongly hypotonic	Skin absorbs Na⁺ from water
Marine reptile	Isotonic	Drinks seawater / Salt gland secretes excess salts
Marine bird	Weakly hypertonic	Drinks seawater / Salt gland secretes excess salts / Excretes weakly hypertonic urine
Marine mammal	Strongly hypertonic	Does not drink seawater
Terrestrial bird	Weakly hypertonic	Drinks freshwater
Desert mammal	Strongly hypertonic	Drinks no water / Obtains water from food and metabolic processes
Freshwater fish	Strongly hypotonic	Does not drink water
Cartilaginous fish	Hypotonic	Kidneys reabsorb urea

Figure 26.7 Osmoregulation by some vertebrates.
Only birds and mammals can produce a hypertonic urine and thereby retain water efficiently, but marine reptiles and birds can drink seawater and excrete the excess salt through salt glands.

Amphibians and Reptiles

The first terrestrial vertebrates were the amphibians (pictured at the top of **figure 26.7**), and the amphibian kidney is identical to that of freshwater fish. This is not surprising because amphibians spend a significant portion of their time in freshwater, and when on land, they generally stay in wet places. Like their freshwater ancestors, amphibians produce a very dilute urine and they compensate for their loss of Na^+ by actively transporting Na^+ across their skin from the surrounding water.

Reptiles, on the other hand, live in diverse habitats. Those living mainly in freshwater, like some of the crocodilians, occupy a habitat in many ways similar to that of the freshwater fish and amphibians, and thus have similar kidneys. Marine reptiles, which consist of other crocodilians, turtles (the second entry in **figure 26.7**), sea snakes, and one lizard, possess kidneys similar to those of their freshwater relatives but face opposite problems; they tend to lose water and take in salts. Like marine teleosts (bony fish), they drink the seawater and excrete an isotonic urine. Marine teleosts eliminate the excess salt by transport across their gills, while marine reptiles eliminate excess salt through salt glands near the nose or eye.

The kidneys of terrestrial reptiles also reabsorb much of the salt and water in the nephron tubules, helping somewhat to conserve blood volume in dry environments. Like fish and amphibians, they cannot produce urine that is more concentrated than the blood plasma. However, when their urine enters their cloaca (the common exit of the digestive and urinary tracts), additional water can be reabsorbed.

Mammals and Birds

Mammals and birds are the only vertebrates able to produce urine with a higher osmotic concentration than their body fluids. This allows these vertebrates to excrete their waste products in a small volume of water, so that more water can be retained in the body. Human kidneys can produce urine that is as much as 4.2 times as concentrated as blood plasma, but the kidneys of some other mammals are even more efficient at conserving water. For example, camels, gerbils, and pocket mice, *Perognathus*, can excrete urine 8, 14, and 22 times as concentrated as their blood plasma, respectively. The kidneys of the kangaroo rat shown in **figure 26.8** are so efficient it never has to drink water; it can obtain all the water it needs from its food and from water produced in aerobic cellular respiration!

The production of hypertonic urine is accomplished by the looped portion of the nephron, found only in mammals and birds. A nephron with a long loop, called the *loop of Henle,* extends deeper into the tissue of the kidney and can produce more concentrated urine. Most mammals have some nephrons with short loops and other nephrons with loops that are much longer. Birds, however, have relatively few or no nephrons with long loops, so they cannot produce urine that is as concentrated as that of mammals. At most, they can only reabsorb enough water to produce a urine that is about twice the concentration of their blood. Marine birds solve the problem of water loss by drinking seawater and then excreting the excess salt from salt glands near the eyes, which dribbles down the beak as shown in **figure 26.9**.

Figure 26.8 A desert mammal.

The kangaroo rat *(Dipodomys panamintensis)* has very efficient kidneys that can concentrate urine to a high degree by reabsorbing water, thereby minimizing water loss from the body. This feature is extremely important to the kangaroo rat's survival in dry or desert habitats.

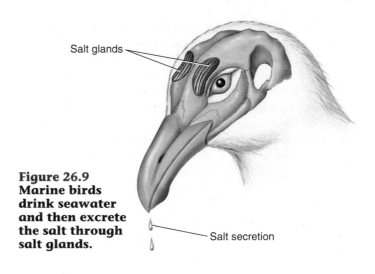

Salt glands

Figure 26.9 Marine birds drink seawater and then excrete the salt through salt glands.

Salt secretion

The moderately hypertonic urine of a bird is delivered to its cloaca, along with the fecal material from its digestive tract. If needed, additional water can be absorbed across the wall of the cloaca to produce a semisolid white paste or pellet, which is excreted.

> **Key Learning Outcome 26.3** The kidneys of freshwater fish must excrete copious amounts of very dilute urine, whereas marine teleosts drink seawater and excrete an isotonic urine. The basic design and function of the nephron of freshwater fish have been retained in the terrestrial vertebrates. Modifications, particularly the loop of Henle, allow mammals and birds to reabsorb water and produce a hypertonic urine.

26.4 | The Mammalian Kidney

In humans, the kidneys are fist-sized organs located in the region of the lower back (**figure 26.10a**). Each kidney receives blood from a renal artery, and it is from this blood that urine is produced. Urine drains from each kidney through a **ureter,** which carries the urine to a **urinary bladder.** Urine passes out of the body through the **urethra.** Within the kidney, the mouth of the ureter flares open to form a funnel-like structure, the *renal pelvis.* The renal pelvis, in turn, has cup-shaped extensions that receive urine from the renal tissue. This tissue is divided into an outer **renal cortex** (containing blood vessels in **figure 26.10b**) and an inner **renal medulla** (containing the cup-shaped structures). Together, these structures perform filtration, reabsorption, secretion, and excretion.

The mammalian kidney is composed of roughly 1 million nephrons (**figure 26.10c**), each of which is composed of four regions:

1. **Filter.** The filtration device at the top of each nephron is called a **Bowman's capsule.** Within each capsule an arteriole enters and splits into a fine network of vessels called a **glomerulus** (labeled ❶ in the figure). The walls of these capillaries act as a filtration device. Blood pressure forces fluid through the capillary walls. These walls withhold proteins and other large molecules in the blood, while passing water, small molecules, ions, and urea, the primary waste product of metabolism.
2. **Proximal tubule.** The Bowman's capsule empties into the proximal tubule, which reclaims most of the water (75%), as well as molecules useful to the body, such as glucose and a variety of ions.
3. **Renal tube.** The proximal tubule is connected to a long, narrow tube called a renal tubule (labeled ❷ through ❹), which is bent back on itself in its center. This long, hairpin loop, called the **loop of Henle,** is a reabsorption device. As the filtrate passes, the renal tubule extracts another 10% of water in the descending loop.
4. **Collecting duct.** The tube empties into a large collection tube called a **collecting duct** ❺ . The collecting duct operates as a water conservation device, reclaiming another 14% of water from the urine so that it is not lost from the body. Human urine is four times as concentrated as blood plasma—that is, the collecting ducts remove much of the water from the filtrate passing through the kidney. Your kidneys achieve this remarkable degree of water conservation by a simple but superbly designed mechanism: The duct bends back alongside the nephron tube, and the duct is permeable to urea. Urea passes out of the collecting duct by diffusion. This greatly increases the local salt (urea) concentration in the tissue surrounding the tube, causing water in urine to pass out of the tube by osmosis. The salty tissue sucks up water from the urine like blotting paper, passing it on to blood vessels that carry it out of the kidneys and back to the bloodstream.

The Kidney at Work

The formation of urine within the mammalian kidney involves the movement of several kinds of molecules between nephrons and the capillaries that surround them. Five steps are involved and indicated in **figure 26.10c**: pressure filtration ❶,

Figure 26.10 The mammalian urinary system contains two kidneys, each of which contain about a million nephrons that lie in the renal cortex and renal medulla.

(a) The urinary system consists of the kidneys, the ureters, which transport urine from the kidneys to the urinary bladder, and the urethra. (b) The kidney is a bean-shaped, reddish brown organ and contains about 1 million nephrons. (c) The glomerulus is enclosed within a filtration device called a Bowman's capsule. Blood pressure forces liquid from blood through the glomerulus and into the proximal tubule of the nephron, where glucose and small proteins are reabsorbed from the filtrate. The filtrate then passes through a loop arrangement consisting of the proximal tubule, the loop of Henle, and the collecting duct, all of which act to remove water from the filtrate. The water is then collected by blood vessels and transported out of the kidney to the systemic (body) circulation.

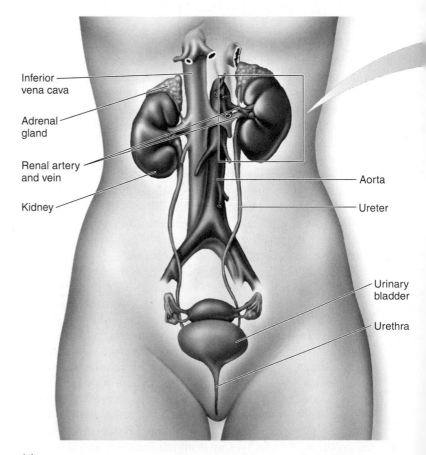

Inferior vena cava
Adrenal gland
Renal artery and vein
Kidney
Aorta
Ureter
Urinary bladder
Urethra

(a)

reabsorption of water ❷, selective reabsorption of ions ❸, tubular secretion ❹, and further reabsorption of water ❺.

Pressure Filtration Driven by the blood pressure, small molecules are pushed across the thin walls of the glomerulus to the inside of the Bowman's capsule ❶. Blood cells and large molecules like proteins cannot pass through, and as a result the blood that enters the glomerulus is divided into two paths: nonfilterable blood components that are retained and leave the glomerulus in the bloodstream, and filterable components that pass across and leave the glomerulus in the urine. This filterable stream is called the **glomerular filtrate.** It contains water, nitrogenous wastes (principally urea), nutrients (principally glucose and amino acids), and a variety of ions.

Reabsorption of Water Filtrate from the glomerulus passes down the proximal tubule into the descending arm of the loop of Henle. The walls of this descending arm are impermeable to either salts or urea but are freely permeable to water. Because the surrounding tissue has a high concentration of urea (for reasons we discuss later), water passes out of the descending arm by osmosis ❷, leaving behind a more concentrated filtrate.

Selective Reabsorption At the turn in the loop, the walls of the tubule become permeable to salts but much less permeable to water. As the concentrated filtrate passes up this ascending arm, these nutrients pass out into the surrounding tissue ❸,

where they are carried away by blood vessels. In the upper region of the ascending arm are active transport channels that pump out salt (NaCl). Left behind in the filtrate is the urea that initially passed through the glomerulus as nitrogenous waste. At this point in the tubule, the urea concentration is becoming very high.

Tubular Secretion In the distal tubule, substances are also added to the urine by a process called tubular secretion ❹. This active transport process secretes into the urine other nitrogenous wastes such as uric acid and ammonia, as well as excess hydrogen ions.

Further Reabsorption of Water The tubule then empties into a collecting duct that passes back through the tissue of the kidney. Unlike the tubule, the lower portions of the collecting duct are permeable to urea, some of which diffuses out into the surrounding tissue (that is why the tissue surrounding the descending arm of the loop of Henle has a high urea concentration indicated by the darker pink color in the figure). The high urea concentration in the tissue causes even more water to pass outward from the filtrate by osmosis ❺. The filtrate that is left in the collecting duct after salts, nutrients, and water have been removed is urine.

Key Learning Outcome 26.4 The mammalian kidney pushes waste molecules through a filter and then reclaims water and useful metabolites and ions from the filtrate before eliminating the residual urine.

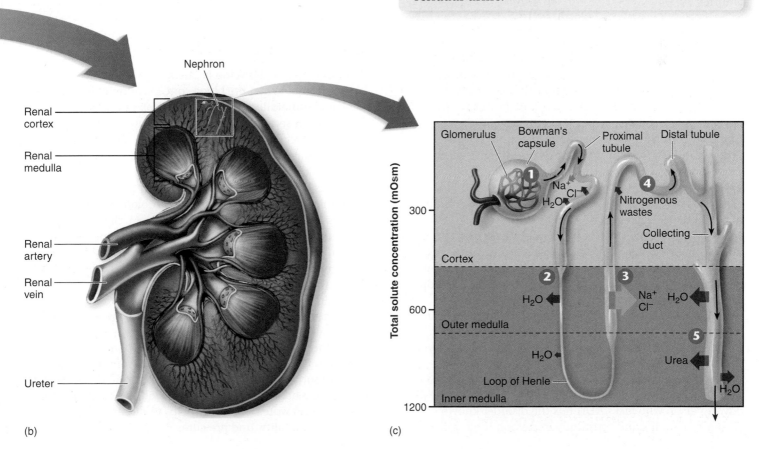

(b)

(c)

Biology and *Staying Healthy*

How Hormones Control Your Kidney's Functions

As in all mammals and birds, the amount of water excreted in your urine varies according to the changing needs of your body. Acting through the mechanisms described in this chapter, your kidneys excrete a hypertonic urine when your body needs to conserve water. If you drink a lot of water, your kidneys will excrete a hypotonic urine. As a result, the volume of your blood, your blood pressure, and the salt levels of your blood plasma are all maintained relatively constant by the kidneys, no matter how much water you drink. The kidneys also regulate the plasma Na^+ and K^+ concentrations and blood pH within narrow limits. These homeostatic functions of the kidneys are coordinated primarily by hormones, which are chemical signals produced in one part of the body that influence events in other parts. Hormones are the subject of chapter 30.

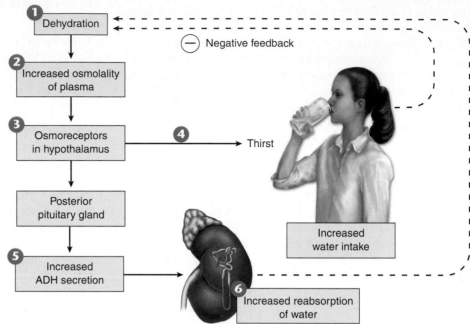

Antidiuretic hormone stimulates the reabsorption of water by the kidneys.

This action completes a negative feedback loop and helps to maintain homeostasis of blood volume and osmolality.

Antidiuretic Hormone. Antidiuretic hormone (ADH) is produced by a portion of your brain called the hypothalamus. The primary stimulus for ADH secretion into your bloodstream is an increase in the *osmolality* (concentration of salt) of the blood plasma. The salt in each milliliter of plasma increases when you are dehydrated or when you eat salty food. Consider the stimulus of dehydration shown in the figure here. Dehydration ❶ causes an increase in the solute concentrations in the blood ❷. Osmoreceptors in the hypothalamus ❸ respond to the elevated blood osmolality by triggering a sensation of thirst ❹ and by an increase in the secretion of ADH ❺.

ADH causes the walls of the distal tubules and collecting ducts in your kidneys (see **figure 26.10c**) to become more permeable to water, thus increasing the amount of water they absorb from the urine as it passes through your kidneys ❻. The increased absorption of water from the kidneys feeds back to the osmoreceptors (the dashed line), causing a reduction in the secretion of ADH. When the secretion of ADH is reduced, the walls become less permeable, and you excrete more water in your urine.

Under conditions of maximal ADH secretion, you excrete only 600 milliliters of highly concentrated urine per day. A person who lacks ADH has the disorder known as diabetes insipidus and constantly excretes a large volume of

dilute urine. Such a person may become severely dehydrated and succumb to dangerously low blood pressure.

Aldosterone. Sodium ion is the major solute in your blood plasma. When the blood concentration of Na^+ falls, the blood osmolality also falls. This drop in osmolality inhibits ADH secretion, causing more water to remain in the collecting duct for excretion in your urine. As a result, your blood volume and blood pressure decrease. A decrease in extracellular Na^+ also causes more water to be drawn into your cells by osmosis, partially offsetting the drop in plasma osmolality but further decreasing your blood volume and blood pressure. If Na^+ deprivation is severe, the blood volume may fall so low that there is insufficient blood pressure to sustain your life. For this reason, salt is necessary for life. Many animals have a "salt hunger" and actively seek salt. This is why a "salt lick" will attract deer.

A drop in blood Na^+ concentration is normally compensated by the kidneys under the influence of another hormone, aldosterone, which is also secreted by your brain. Indeed, under conditions of maximal aldosterone secretion, Na^+ may be completely absent from the urine. The reabsorption of Na^+ is followed by Cl^- and by water, so aldosterone has the net effect of promoting the retention of both salt and water. It thereby helps to maintain your blood volume, osmolality, and pressure.

26.5 Eliminating Nitrogenous Wastes

Amino acids and nucleic acids are nitrogen-containing molecules. When animals catabolize these molecules for energy or convert them into carbohydrates or lipids, they produce nitrogen-containing by-products called **nitrogenous wastes** that must be eliminated from the body.

The first step in the metabolism of amino acids and nucleic acids is the removal of the amino (—NH$_2$) group and its combination with H$^+$ to form **ammonia** (NH$_3$) in the liver, ❶ in figure 26.11. Ammonia is quite toxic to cells and therefore is safe only in very dilute concentrations. The excretion of ammonia is not a problem for the bony fish and tadpoles, which eliminate most of it by diffusion through the gills and the rest by excretion in very dilute urine ❷. In sharks, adult amphibians, and mammals, the nitrogenous wastes are eliminated in the far less toxic form of **urea** ❸. Urea is water-soluble and so can be excreted in large amounts in the urine. It is carried in the bloodstream from its place of synthesis in the liver to the kidneys, where it is excreted in the urine.

Reptiles, birds, and insects excrete nitrogenous wastes in the form of **uric acid** ❹, which is only slightly soluble in water. As a result of its low solubility, uric acid precipitates and thus can be excreted using very little water. Uric acid forms the pasty white material in bird droppings. The ability to synthesize uric acid in these groups of animals is also important because their eggs are encased within shells, and nitrogenous wastes build up as the embryo grows within the egg. The formation of uric acid, while a lengthy process that requires considerable energy, produces a compound that crystallizes and precipitates. As a precipitate, it is unable to affect the embryo's development even though it is still inside the egg.

Mammals also produce some uric acid, but it is a waste product of the degradation of purine nucleotides (see chapter 3), not of amino acids. Most mammals have an enzyme called *uricase*, which converts uric acid into a more soluble derivative, **allantoin.** Only humans, apes, and the dalmatian dog lack this enzyme and so must excrete the uric acid. In humans, excessive accumulation of uric acid in the joints produces a condition known as *gout.*

> **Key Learning Outcome 26.5** The metabolic breakdown of amino acids and nucleic acids produces ammonia as a by-product. Ammonia is excreted by bony fish, but other vertebrates convert nitrogenous wastes into urea and uric acid, which are less toxic nitrogenous wastes.

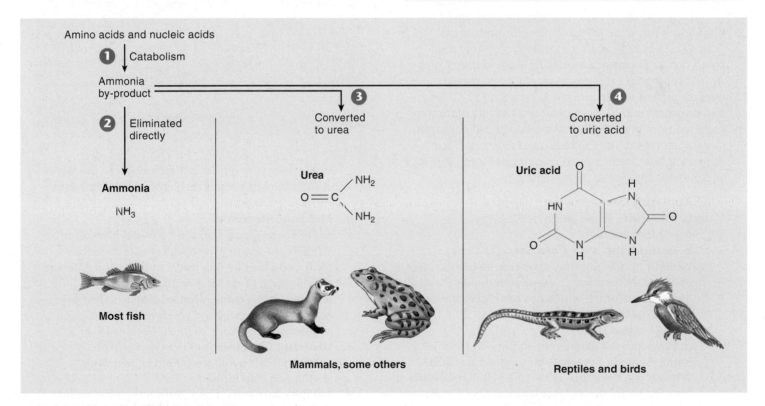

Figure 26.11 Nitrogenous wastes.

When amino acids and nucleic acids are metabolized, the immediate nitrogen by-product is ammonia ❶, which is quite toxic but can be eliminated through the gills of bony fish ❷. Mammals convert ammonia into urea ❸, which is less toxic. Birds and terrestrial reptiles convert it instead into uric acid ❹, which is insoluble in water.

How Do Sleeping Birds Stay Warm?

Mammals and birds are endothermic—they maintain body temperatures regardless of the temperature of their surroundings. This lets them reliably run their metabolism even when external temperatures fall—the rates of most enzyme-catalyzed reactions slow two- to threefold for every 10°C temperature drop. Your body keeps its temperature within narrow bounds at 37°C (98.6°F), and birds maintain even higher temperatures. To stay warm like this, mammals and birds continuously carry out oxidative metabolism, which generates heat. This requires a several-fold increase in metabolic rate, which is expensive, particularly when the animal is not active. The logical solution is to give up the struggle to keep warm and let the body temperature drop during sleep, a condition known as *torpor*. While humans don't adopt this approach, many other mammals and birds do. This raises an interesting question: What prevents a sleeping bird in torpor from freezing? Does its body simply adopt the temperature of its surroundings, or is there a body temperature below which metabolic heating kicks in to avoid freezing?

The graph to the right displays an experiment examining this issue in the tropical hummingbird *Eulampis*. The study examines the effect on metabolic rate (measured as oxygen consumption) of decreasing air temperature. Oxygen consumption was assessed over a range of air temperatures from 3° to 37°C, for two contrasting physiological states: The blue data were collected from birds that were awake, the red data from sleeping, torpid birds. The blue and red lines, called regression lines, were plotted using curve-fitting statistics that provide the best fit to the data.

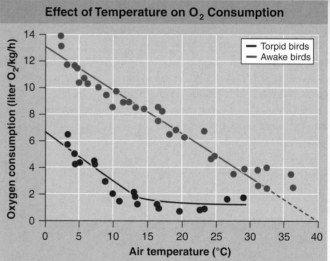

Effect of Temperature on O₂ Consumption

1. **Applying Concepts**
   a. Variable. In the graph, what is the dependent variable?
   b. Comparing Two Data Sets. Do awake hummingbirds maintain the same metabolic rate at all air temperatures? sleeping, torpid ones? At a given temperature, which has the higher metabolic rate, an awake bird or a sleeping one?
2. **Interpreting Data**
   a. How does the oxygen consumption of awake hummingbirds change as air temperature falls? Why do you think this is so? Is the change consistent over the entire range of air temperatures examined?
   b. How does the oxygen consumption of sleeping torpid birds change as air temperature falls? Is this change consistent over the entire range of air temperatures examined? Explain any difference you detect.

c. Are there any significant differences in the slope of the two regression lines below 15°C? What does this suggest to you?
3. **Making Inferences**
   a. For each 5-degree air temperature interval, estimate the average oxygen consumption for awake and for sleeping birds, and plot the difference as a function of air temperature.
   b. Based on this curve, what would you expect to happen to a sleeping bird's body temperature as air temperatures fall from 30° to 20°C? from 15° to 5°C?
4. **Drawing Conclusions** How do *Eulampis* hummingbirds avoid becoming chilled while sleeping on cold nights?
5. **Further Analysis** Flying hummingbirds would be expected to use more metabolic energy than perched, awake ones. How would you expect the level of activity to influence the birds' regulation of body temperature? Explain.

Homeostasis

26.1 How the Animal Body Maintains Homeostasis

- Animals maintain relatively constant internal conditions, a process called homeostasis. Homeostasis refers to a dynamic constancy of the internal environment within narrow ranges.

- The body uses sensors to measure internal conditions. The measurements are sent to an integrating center. The integrating center compares the measurement to a set point for that particular condition. If the measurement is deviating from the set point the body will make adjustments through effectors, muscles or glands, to bring the body back to the set point (**figure 26.1**).

- This type of control system is called a negative feedback loop because the response to correct the deviation is in a reverse or negative direction, bringing the condition back to the set point.

- Examples of homeostasis include body temperature and blood glucose levels (**figure 26.2**).

Osmoregulation

26.2 Regulating the Body's Water Content

- Organisms have evolved various mechanisms to control water balance. Many invertebrates use systems of tubules that collect fluid and reabsorb ions and water. Flatworms use a network of tubules called protonephridia (**figure 26.4**). Nephridia, such as those found in earthworms and shown here from **figure 26.5**, filter fluids and allow the body to reabsorb NaCl. Waste and fluids are excreted as urine through pores. Insects create an osmotic gradient by secreting K$^+$ into Malpighian tubules, causing water to enter the tubules by osmosis. Water and K$^+$ are reabsorbed through the epithelium of the hindgut (**figure 26.6**).

- Kidneys are the excretory organs in vertebrates. They filter fluid under pressure and then reabsorb important molecules.

Osmoregulation in Vertebrates

26.3 Evolution of the Vertebrate Kidney

- The nephron is the basic unit of the kidney (**integrated art, page 552**). Fluid is forced from the blood, and useful molecules, such as sugars, amino acids, ions, water, and salts are reabsorbed.

- The kidneys of freshwater fish excrete dilute urine, whereas marine bony fish drink seawater and excrete isotonic urine (**integrated art, pages 552–553**).

- Cartilaginous fish, in addition to marine bony fish, live in a hypertonic environment. To compensate, their bodies reabsorb urea so that their blood is isotonic to seawater (**integrated art, page 553**).

- The kidneys of amphibians and reptiles are similar to those found in fish. Mammals and birds produce a concentrated urine (**figure 26.7**). They can do this because the kidneys of mammals and birds have a loop of Henle that acts in concentrating the urine to be hypertonic to the blood. Marine birds drink seawater and excrete excess salt from salt glands (**figure 26.9**).

26.4 The Mammalian Kidney

- Each kidney receives blood from a renal artery and produces urine, which then drains from each kidney through a ureter and is carried to the urinary bladder (**figure 26.10a**).

- Each kidney (sectioned here from **figure 26.10b**) is composed of roughly 1 million nephrons. Each nephron contains a Bowman's capsule, a loop of Henle, and a collecting duct. The function of the nephron is to collect the filtrate from the blood, selectively reabsorb ions and water, and expel the waste from the body.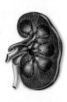

- The blood is first filtered from the glomerulus, a network of blood vessels encapsulated by the Bowman's capsule. Water, small molecules, ions, and urea pass into the capsule. The Bowman's capsule is connected to a long tubule where selected molecules and ions are reabsorbed by the body. The tubule forms a hairpin loop called the loop of Henle. The loop of Henle is longer in mammals, allowing the formation of a hypertonic urine. The tubule empties into a collecting duct, which recovers additional water from the filtrate (**figure 26.10c**).

- Different areas of the loop of Henle are permeable to different substances. In the proximal tubule, much water and important molecules are reclaimed. In the descending arm of the loop of Henle, water passes back into the surrounding tissue by osmosis, but the loop at this point is impermeable to either salts or urea. At the turn of the loop, the walls become permeable to salts and other molecules, which pass out of the tubules and into the surrounding tissue. This results in a higher concentration of solutes in the tissue surrounding the nephron tubule.

- As the filtrate passes into the collecting duct, whose walls are permeable to water, water is absorbed back into the tissues through osmosis. Osmosis occurs because the surrounding tissue is highly concentrated with solutes such as ions and other molecules that were reabsorbed from the ascending loop.

- The fluid that remains in the collecting duct is a highly hypertonic urine that passes to the ureter and is excreted from the body.

26.5 Eliminating Nitrogenous Wastes

- The metabolic breakdown of amino acids and nucleic acids produces ammonia in the liver. Ammonia, a nitrogenous waste product, is toxic and must be eliminated from the body. Different animals use different means of excretion (**figure 26.11**).

- Ammonia is excreted directly in bony fish and tadpoles through gills and in dilute urine.

- In sharks, adult amphibians, and mammals, ammonia is converted to the less toxic urea. Urea is water-soluble and is carried in the bloodstream to the kidneys where it is excreted in the urine.

- Reptiles, birds, and insects excrete nitrogenous wastes as uric acid, which has low toxicity and is only slightly soluble in water and so can be excreted using only small amounts of water. Uric acid is the pasty white substance in bird droppings.

Test Your Understanding

1. The monitoring and adjusting of the body's condition, such as temperature and pH, is known as
   a. exothermy.
   b. homeostasis.
   c. osmoregulation.
   d. ectothermy.
2. Which of the following describes the method of regulation whereby the response of an effector returns conditions to a set point?
   a. inhibitory regulation
   b. homeostasis
   c. positive feedback loop
   d. negative feedback loop
3. If your blood sugar is too low, the hormone glucagon is released by the pancreas. This hormone will cause
   a. the release of insulin.
   b. glycogen to break down.
   c. glycogen to be formed.
   d. fat to be formed.
4. Which of the following is *not* involved in osmoregulation?
   a. pancreas
   b. nephridia
   c. Malpighian tubules
   d. flame cells
5. Which of the following animals use Malpighian tubules for excretion?
   a. birds
   b. ants
   c. kangaroo rats
   d. earthworms
6. To keep the proper concentrations of water and solutes in their blood, freshwater bony fish must drink
   a. lots of water and excrete large volumes of urine that are hypotonic to body fluids.
   b. no water and excrete large volumes of urine that are hypotonic to body fluids.
   c. lots of water and excrete large volumes of urine that are isotonic to body fluids.
   d. no water and excrete large volumes of urine that are isotonic to body fluids.
7. Which of the following animals has the least concentrated urine relative to its blood plasma?
   a. bird
   b. freshwater fish
   c. camel
   d. shark
8. Selective reabsorption of components of the filtrate occurs where?
   a. Bowman's capsule
   b. glomerulus
   c. loop of Henle
   d. ureter
9. Water is removed from kidney filtrate by the process of
   a. diffusion.   c. facilitated diffusion.
   b. active transport.   d. osmosis.
10. Humans excrete their excess nitrogenous wastes as
   a. uric acid crystals.
   b. compounds containing protein.
   c. ammonia.
   d. urea.

Apply Your Understanding

1. **Figure 26.1** Explain the regulation of blood glucose levels using the diagram below. First identify the various components, assuming the stimulus is high blood glucose after eating a meal, then describe what happens to each of the components after the stimulus is detected.

2. **Figure 26.10c** When the body becomes dehydrated it produces a hormone called ADH (see "Biology and Staying Healthy" on page 558). ADH affects the body by making the collecting duct of the kidney more permeable to water. How would this help a dehydrated person? How would the person's urine be affected?

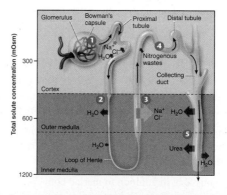

Synthesize What You Have Learned

1. Hummingbirds, pocket mice, and young Chiricahua leopard frogs are all about the same size, and all live in the southwestern deserts. Compare and contrast their nitrogenous wastes, and explain why there is such a difference in their urine.

2. The glomerulus filters out a tremendous amount of water and molecules needed by the body, which must be reabsorbed by the body in the kidney. The Malpighian tubules of insects might seem to function more logically, secreting mainly waste products to be excreted. What advantages might a filtration-reabsorption process provide over a strictly secretion process of elimination?

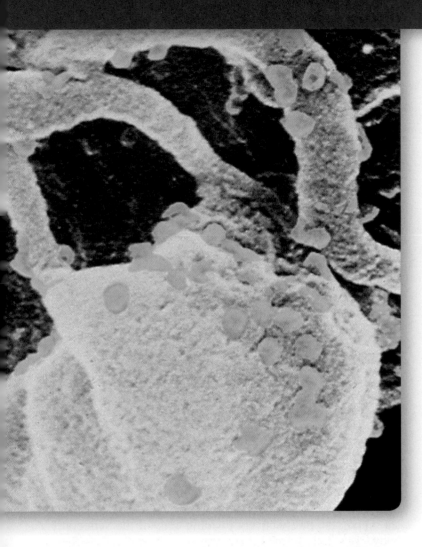

27

How the Animal Body Defends Itself

A ll animals are constantly at war with bacteria and viruses that attempt to use the rich resources of the cellular environment to fuel their own reproduction. Another war is fought on a very different front by animals against their own cells when cells become cancerous and begin to grow without restraint. Both of these wars, against invading microbes and against cancer, are fought with the same defensive weapons: the immune system. This chapter focuses on the vertebrate immune system and how it defends the body in the face of these onslaughts. Sometimes the immune system itself is the target of infection, leading to a loss of the body's ability to defend itself. AIDS results from just this sort of infection by a virus called HIV (human immunodeficiency virus). The cell you see here is a human immune system cell called a macrophage that has been infected by HIV. Progeny HIV viruses are being released from the infected cell, budding out from its surface. These viruses soon spread to neighboring lymphocytes, infecting and killing them. Eventually a large majority of lymphocytes become infected, and the immune defense is destroyed. The individual viruses, colored blue in this scanning electron micrograph, are extremely small; over 200 million would fit on the period at the end of this sentence.

27.1 Skin: The First Line of Defense

Multicellular bodies offer a feast of nutrients for tiny, single-celled creatures, as well as a warm, sheltered environment in which they can grow and reproduce. We live in a world awash with microbes, and no animal can long withstand their onslaught unprotected. Animals survive because they have a variety of very effective defenses against this constant attack.

Overview of the Three Lines of Defense

The vertebrate body is defended from infection the same way knights defended medieval cities. "Walls and moats" make entry difficult; "roaming patrols" attack strangers; and "guards" challenge anyone wandering about and signal for an attack if a proper "ID" is not presented.

1. **Walls and moats.** The outermost layer of the vertebrate body, the skin, is the first barrier to penetration by microbes. Mucous membranes in the respiratory and digestive tracts are also important barriers that protect the body from invasion.
2. **Roaming patrols.** If the first line of defense is penetrated, the response of the body is to mount a *cellular counterattack,* using a battery of cells and chemicals that kill microbes. These defenses act very rapidly after the onset of infection.
3. **Guards.** Lastly, the body is also policed by cells that circulate in the bloodstream and scan the surfaces of every cell they encounter. They are part of the *specific immune response.* One kind of immune cell aggressively attacks and kills any cell identified as foreign, whereas the other type marks the foreign cell or virus for elimination by the roaming patrols.

The Skin

Skin, like the thick, tough skin of the elephants in **figure 27.1,** is the outermost layer of the vertebrate body and provides the first defense against invasion by microbes. Skin is our largest organ, comprising some 15% of our total weight. One square centimeter of skin from your forearm (about the size of a dime) contains 200 nerve endings, 10 hairs and muscles, 100 sweat glands, 15 oil glands, 3 blood vessels, 12 heat-sensing organs, 2 cold-sensing organs, and 25 pressure-sensing organs. The section of skin you see in **figure 27.2** has two distinct layers: an outer **epidermis** and a lower **dermis.** A **subcutaneous layer** lies underneath the dermis. Cells of the outer epidermis are continually being worn away and replaced by cells moving up from below—in one hour your body loses and replaces approximately 1.5 million skin cells!

The epidermis of skin is from 10 to 30 cells thick, about as thick as this page. The outer layer, called the **stratum corneum,** is the one you see when you look at your arm or face.

Figure 27.1 Skin is the body's first line of defense.
This young elephant has tough, leathery skin thicker than a belt, allowing it to follow the herd through dense thickets without injury.

Cells from this layer are continuously subjected to damage. They are abraded, injured, and worn by friction and stress during the body's many activities. They also lose moisture and dry out. The body deals with this damage not by repairing cells but by replacing them. Cells from the stratum corneum are shed continuously, replaced by new cells produced deep within the epidermis (the dark layer of cells at the border of the epidermis and dermis). The cells of this inner **basal layer** are among the most actively dividing cells of the vertebrate body. New cells formed there migrate upward, and as they move they manufacture keratin protein, which makes them tough. Each cell eventually arrives at the outer surface and takes its turn in the stratum corneum, residing there for about a month before it is shed and replaced by a newer cell. Persistent dandruff (psoriasis) is a chronic skin disorder in which new cells reach the epidermal surface every three or four days, about eight times faster than normal.

The dermis of the skin is from 15 to 40 times thicker than the epidermis. It provides structural support for the epidermis, as well as a matrix for many specialized cells residing within the skin. The wrinkling that occurs as we grow older occurs here. The leather used to manufacture belts and shoes is derived from thick animal dermis. The layer of subcutaneous tissue below the dermis is composed of fat-rich cells that act as shock absorbers and provide insulation, which conserves body heat.

The skin not only defends the body by providing a nearly impermeable barrier, but it also reinforces this defense with chemical weapons. For example, the oil glands that occur along the shaft of the hair and sweat glands, which look like yellow coiled spaghetti in **figure 27.2,** make the skin's surface very acidic (pH of between 4 and 5.5), which inhibits the growth of many microbes. Sweat also contains the enzyme lysozyme, which attacks and digests the cell walls of many bacteria.

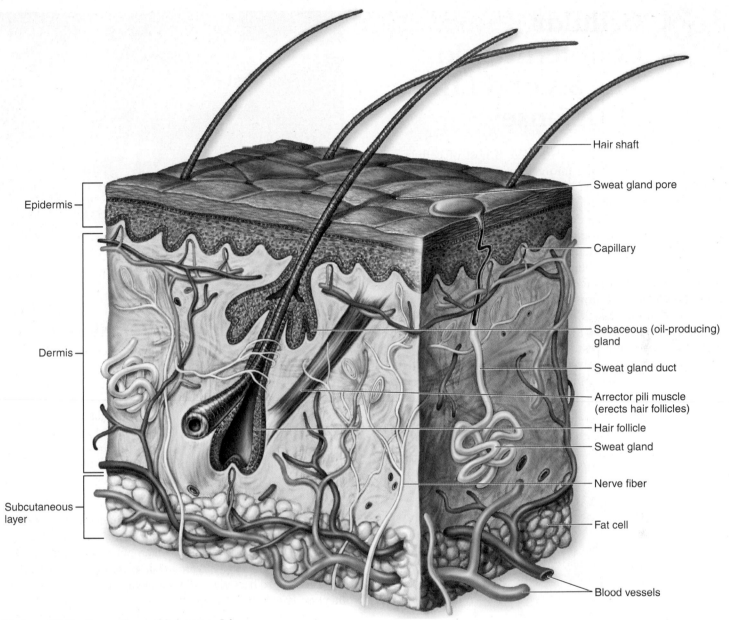

Figure 27.2   A section of human skin.

The skin defends the body by providing a barrier with sweat and oil glands, whose secretions make the skin's surface acidic enough to inhibit the growth of microorganisms.

Other External Surfaces

In addition to the skin, other surfaces, such as the eyes, are exposed to the outside. Like sweat, the tears that bathe the eyes contain lysozyme to fight bacterial infections. Two other potential routes of entry by viruses and microorganisms must be guarded: the *digestive tract* and the *respiratory tract*. Microbes are present in food, but many are killed by saliva (which also contains lysozyme), by the very acidic environment of the stomach (pH of 2), and by digestive enzymes in the intestine. Microorganisms are also present in inhaled air. The cells lining the smaller bronchi and bronchioles secrete a layer of sticky mucus that traps most microorganisms before they can reach the warm, moist lungs, which would provide ideal breeding grounds for them. Other cells lining these passages have cilia that continually sweep the mucus up toward the glottis in the throat, like an escalator. There it can be coughed out or swallowed, carrying potential invaders out of the lungs.

The surface defenses are very effective, but they are occasionally breached. Through breathing, eating, or cuts and nicks, bacteria and viruses now and then enter our bodies. When these invaders reach deeper tissue, a second line of defense comes into play, a cellular counterattack.

Key Learning Outcome 27.1   Skin and the mucous membranes lining the digestive and respiratory tracts are the body's first defenses.

27.2 Cellular Counterattack: The Second Line of Defense

When the body's interior is invaded, a host of cellular and chemical defenses swing into action. Four are of particular importance: (1) cells that kill invading microbes; (2) proteins that kill invading microbes; (3) the inflammatory response, which speeds defending cells to the point of infection; and (4) the temperature response, which elevates body temperature to slow the growth of invading bacteria.

Although these cells and proteins roam throughout the body, there is a central location for their storage and distribution; it is called the **lymphatic system.** The lymphatic system consists of the structures shown in **figure 27.3**: lymph nodes, lymphatic organs, and a network of lymphatic capillaries that drain into lymphatic vessels. Although the lymphatic system has other functions involved with circulation (see chapter 23), it is also involved in the immune response.

Figure 27.4   A macrophage in action.
In this scanning electron micrograph, a macrophage is "fishing" for rod-shaped bacteria.

Cells That Kill Invading Microbes

The most important counterattack to infection is mounted by white blood cells, which attack invading microbes. These cells patrol the bloodstream and await invaders within the tissues. The three basic kinds of killing cells are macrophages and neutrophils, which are phagocytes, and natural killer cells. These three kinds of cells can distinguish the body's cells (self) from foreign cells (nonself) because the body's cells contain self-identifying MHC proteins (discussed in more detail later in this chapter). Each type of killing cell uses a different tactic to kill invading microbes.

Macrophages White blood cells called **macrophages** (Greek, "big eaters") kill bacteria by ingesting them, much as an amoeba ingests a food particle. The macrophage in **figure 27.4** is sending out long, sticky cytoplasmic extensions that catch the sausage-shaped bacteria and draw them back to the cell where they are engulfed. Although some macrophages are anchored within particular organs, particularly the spleen, most patrol the byways of the body, circulating as precursor cells called **monocytes** in the blood, lymph, and fluid between cells. Macrophages are among the most actively mobile cells of the body.

Neutrophils Other white blood cells called **neutrophils** act like kamikazes. In addition to ingesting microbes, they release chemicals (identical to household bleach) to "neutralize" the entire area, killing any bacteria in the neighborhood—and themselves in the process. A neutrophil is like a grenade tossed into an infection. It kills everything in the vicinity. Macrophages, by contrast, kill only one invading cell at a time but live to keep on doing it.

Natural Killer Cells A third kind of white blood cell, called a **natural killer cell,** does not attack invading microbes but rather the body cells that are infected by them. Natural killer cells puncture the membrane of the infected target cell with a molecule called *perforin*. The natural killer cell in **figure 27.5a** is releasing several perforin molecules that insert in the membrane of the infected cell, like boards on a picket fence, forming a pore that allows water to rush in, causing the cell to

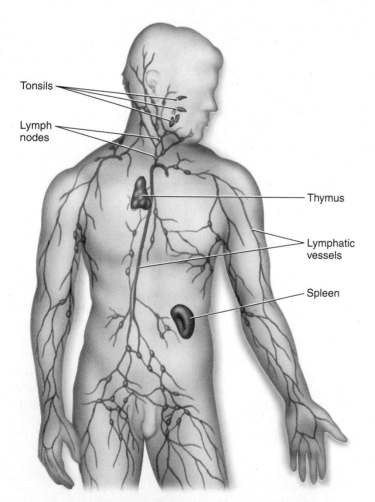

Tonsils

Lymph nodes

Thymus

Lymphatic vessels

Spleen

Figure 27.3   The lymphatic system.
The major lymphatic vessels, organs, and nodes are shown.

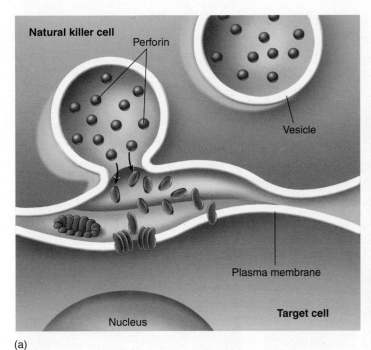

Natural killer cell

Perforin

Vesicle

Plasma membrane

Target cell

Nucleus

(a)

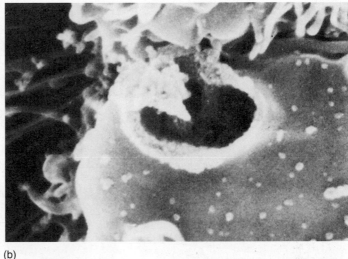

(b)

Figure 27.5 Natural killer cells attack target cells.
(a) The initial event is the tight binding of the natural killer cell to the target cell. Binding initiates a chain of events within the natural killer cell in which vesicles loaded with perforin molecules move to the outer plasma membrane and expel their contents into the intercellular space over the target. The perforin molecules insert into the membrane, forming a pore that admits water and ruptures the cell. (b) A natural killer cell has attacked this cancer cell, punching a hole in its plasma membrane. Water has rushed in, making it balloon out. Soon it will burst.

swell and burst. Natural killer cells are particularly effective at detecting and attacking body cells that have been infected with viruses. They are also one of the body's most potent defenses against cancer. The cancer cell shown in the photo in **figure 27.5b** was killed before it had a chance to develop into a tumor.

Proteins That Kill Invading Microbes

The cellular defenses of vertebrates are complemented by a very effective chemical defense called the **complement system.** This system consists of approximately 20 different proteins that circulate freely in the blood plasma in an inactive state. Their defensive activity is triggered when they encounter the cell walls of bacteria or fungi. The complement proteins then aggregate to form a *membrane attack complex* that inserts itself into the foreign cell's plasma membrane, forming a pore like that produced by natural killer cells. Like the perforin pore, the membrane attack complex in **figure 27.6** allows water to enter the foreign cell causing the cell to swell and burst. The difference between the two is that perforin attacks host cells that have become infected, while complement attacks the foreign cell directly. Aggregation of the complement proteins is also triggered by the binding of antibodies to invading microbes, as we'll see in a later section.

The proteins of the complement system can augment the effects of other body defenses. Some amplify the inflammatory response (discussed next) by stimulating histamine release; others attract phagocytes (monocytes and neutrophils) to the area of infection; and still others coat invading microbes, roughening the microbes' surfaces so that phagocytes may attach to them more readily.

Another class of proteins that play a key role in body defense are interferons. There are three major categories of

interferons: alpha, beta, and gamma. These polypeptides act as messengers that protect other cells in the vicinity from viral infection. The viruses are still able to penetrate the other cells, but the ability of the viruses to replicate and assemble new virus particles is inhibited.

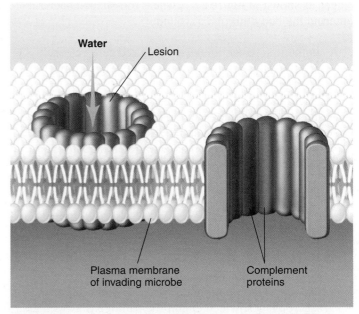

Water

Lesion

Plasma membrane
of invading microbe

Complement
proteins

Figure 27.6 Complement proteins attack invaders.
Complement proteins form a transmembrane channel resembling the perforin-lined lesion made by natural killer cells, but complement proteins are free-floating, and they attach to the invading microbe directly, while perforin molecules insert into infected body cells.

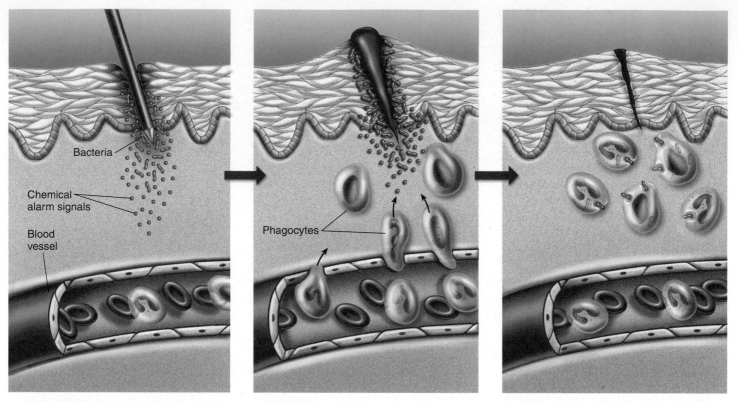

Figure 27.7 The events in a local inflammation.
When an invading microbe has penetrated the skin, chemicals, such as histamine and prostaglandins, act as alarm signals that cause nearby blood vessels to dilate. Increased blood flow brings a wave of phagocytic cells that attack and engulf invading bacteria.

The Inflammatory Response

The aggressive cellular and chemical counterattacks to infection are made more effective by the **inflammatory response.** The inflammatory response can be broken down into three stages as shown in **figure 27.7,** where rod-shaped bacteria are entering the body through a wound:

1. In the first panel, infected or injured cells release chemical alarm signals, most notably histamine and prostaglandins.
2. These chemical alarm signals cause blood vessels to expand, both increasing the flow of blood to the site of infection or injury and, by stretching their thin walls, making the capillaries more permeable. This produces the redness and swelling so often associated with infection.
3. In the second panel, the increased blood flow through larger, leakier capillaries promotes the migration of phagocytes from the blood to the site of infection, squeezing between cells in the capillary walls. Neutrophils arrive first, spilling out chemicals that kill the microbes (as well as tissue cells in the vicinity and themselves). Monocytes follow and become macrophages, which engulf the pathogens as well as the remains of all the dead cells, as shown in the third panel. This counterattack takes a considerable toll; the pus associated with infections is a mixture of dead or dying neutrophils, tissue cells, and pathogens.

The Temperature Response

Human pathogenic bacteria do not grow well at high temperatures. Thus, when macrophages initiate their counterattack, they increase the odds in their favor by sending a message to the brain to raise the body's temperature. The cluster of brain cells that serves as the body's thermostat responds to the chemical signal by boosting the body temperature several degrees above the normal value of 37°C (98.6°F). The higher-than-normal temperature that results is called a **fever.** Although fever is quite effective at inhibiting microbial growth, very high fevers are dangerous because excessive heat can inactivate critical cellular enzymes. In general, temperatures greater than 39.4°C (103°F) are considered dangerous; those greater than 40.6°C (105°F) are often fatal.

The second line of defense, with both chemical and cellular weapons, provides a sophisticated defense against microbial infection. Only occasionally do bacteria or viruses overwhelm this defense. When this happens, they face yet a third line of defense, more difficult to evade than any they have encountered. It is the specific immune response, the most elaborate of the body's defenses.

> **Key Learning Outcome 27.2** Vertebrates respond to infection with a battery of cellular and chemical weapons, including cells and proteins that kill invading microbes, and inflammatory and temperature responses.

27.3 Specific Immunity: The Third Line of Defense

Specific immune defense mechanisms of the body involve the actions of white blood cells, or leukocytes (see also section 23.4). They are very numerous—of the 10 to 100 trillion cells of your body, two in every 100 are white blood cells! Macrophages are white blood cells, as are neutrophils and natural killer cells. In addition, there are T cells, B cells, plasma cells, mast cells, and monocytes (**table 27.1**). T cells and B cells are called **lymphocytes** and are critical to the specific immune response.

After their origin in the bone marrow, **T cells** migrate to the thymus (hence the designation "T"), a gland just above the heart (see **figure 27.3**). There they develop the ability to identify microorganisms and viruses by the antigens exposed on their surfaces. An **antigen** is a molecule that provokes a specific immune response. Antigens are large, complex molecules, such as proteins, and they are generally foreign to the body, usually belonging to bacteria and viruses. Tens of millions of different T cells are made, each specializing in the recognition of one particular antigen. No invader can escape being recognized by at least a few T cells.

Unlike T cells, **B cells** do not travel to the thymus; they complete their maturation in the bone marrow. (B cells are so named because they were originally characterized in a region of chickens called the bursa.) From the bone marrow, B cells are released to circulate in the blood and lymph. Individual B cells, like T cells, are specialized to recognize particular foreign antigens. When a B cell encounters the antigen to which it is targeted, it begins to divide rapidly and its progeny differentiate into *plasma cells* and memory cells. Each plasma cell is a miniature factory producing markers called *antibodies*. These antibodies stick like flags to a particular antigen wherever it occurs in the body, marking any cell bearing that antigen for destruction. So, B cells don't kill foreign invaders directly, but rather they mark these cells so that they are more easily recognized by the other white blood cells that do the dirty work.

B cells and T cells also produce memory cells that provide the body with the ability to recall a previous exposure to an antigen and mount an attack against that antigen very quickly. As described later in this chapter, the initial specific immune response to an antigen encountered for the first time is delayed, which allows the pathogen time to infect the body. A second infection is halted much earlier due to the presence of memory cells, which respond to the pathogen more quickly.

Key Learning Outcome 27.3 T cells develop in the thymus, whereas B cells develop in the bone marrow. T cells can attack cells that carry antigens. When a B cell encounters a specific antigen, it gives rise to plasma cells that produce antibodies. Antibodies tag cells for destruction.

TABLE 27.1	CELLS OF THE IMMUNE SYSTEM
Cell Type	**Function**
Helper T cell	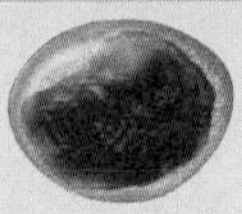 Commander of the immune response; detects infection and sounds the alarm, initiating both T cell and B cell responses
Memory T cell	Provides a quick and effective response to an antigen previously encountered by the body
Cytotoxic T cell	Detects and kills infected body cells; recruited by helper T cells
Suppressor T cell	Dampens the activity of T and B cells, scaling back the defense after the infection has been checked
B cell	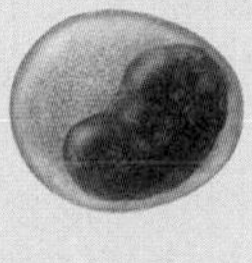 Precursor of plasma and memory cells; specialized to recognize specific foreign antigens
Memory B cell	Provides a quick and effective response to an antigen previously encountered by the body
Neutrophil	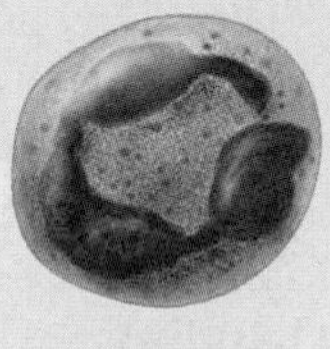 Engulfs invading bacteria and releases chemicals that kill neighboring bacteria
Plasma cell	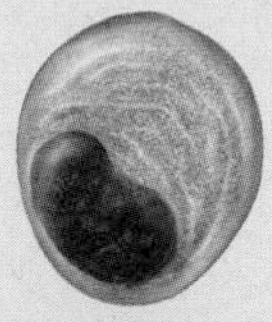 Biochemical factory devoted to the production of antibodies directed against specific foreign antigens
Mast cell	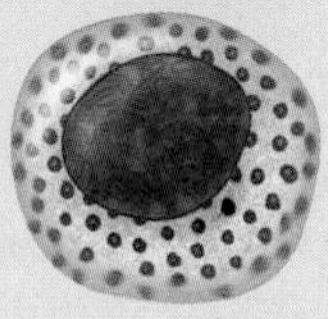 Initiator of the inflammatory response, which aids the arrival of leukocytes at a site of infection; secretes histamine and is important in allergic responses
Monocyte	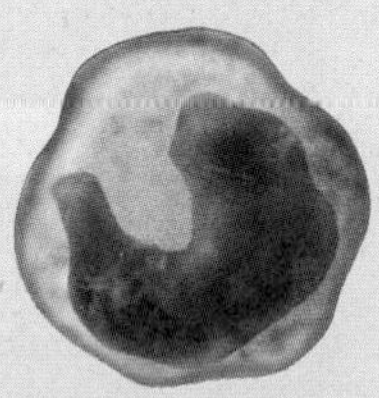 Precursor of macrophage
Macrophage	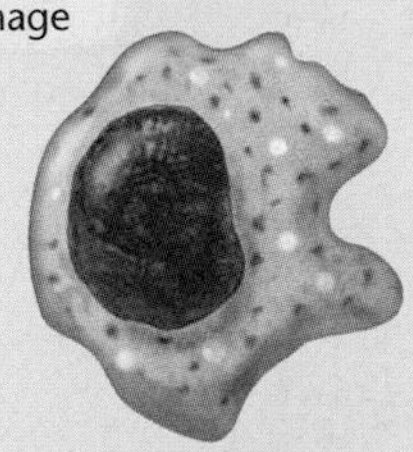 The body's first cellular line of defense; also serves as antigen-presenting cell to B and T cells and engulfs antibody-covered cells
Natural killer cell	 Recognizes and kills infected body cells; natural killer cell detects and kills cells infected by a broad range of invaders

[handwritten annotations: "1st one on scene", "Kill themselves will fighting infection" (near Neutrophil); "B cell can secrete antibodies" (near Plasma cell); "clean up 2nd on scene most abundant" (near Macrophage)]

27.4 Initiating the Immune Response

To understand how this third line of defense works, imagine you have just come down with the flu. Influenza viruses enter your body in small water droplets inhaled into your respiratory system. If they avoid becoming ensnared in the mucus lining the respiratory membranes (first line of defense), and avoid consumption by macrophages (second line of defense), the viruses infect and kill mucous membrane cells.

At this point, macrophages initiate the immune response. Macrophages inspect the surfaces of all cells they encounter. Every cell in the body carries special marker proteins on its surface, called major histocompatibility proteins, or **MHC proteins.** The MHC proteins are different for each individual, much as fingerprints are. The MHC protein on the cell in **figure 27.8a** is exactly the same on all cells in that person's body. As a result, the MHC proteins on the tissue cells serve as "self" markers that enable the individual's immune system to distinguish its cells from foreign cells. For example, the foreign microbe in **figure 27.8b** has different surface proteins that are recognized as antigens.

When a foreign particle infects the body, it is taken in by cells and partially digested. Within the cells, the viral antigens are processed and moved to the surface of the plasma membrane, as shown in **figure 27.8c**. The cells that perform this function are called **antigen-presenting cells** and are usually macrophages. At the membrane, the processed antigens are complexed with the MHC proteins. This process is critical for the function of T cells because T cell receptors can be called into action only when antigens are presented in this way. B cells can interact with free antigens directly.

Macrophages that encounter pathogens—either a foreign cell such as a bacterial cell, which lacks proper MHC proteins, or a virus-infected body cell with telltale viral proteins stuck to its surface—respond by secreting a chemical alarm signal. The alarm signal is a protein called **interleukin-1** (Latin for "between white blood cells"). This protein stimulates **helper T cells.** The helper T cells respond to the interleukin-1 alarm by simultaneously initiating two different parallel lines of immune system defense: the cellular immune response carried out by T cells and the antibody or humoral response carried out by B cells. The immune response carried out by T cells is called the *cellular response* because the T lymphocytes attack the cells that carry antigens. The B cell response is called the *humoral response* because antibodies are secreted into the blood and body fluids (*humor* refers to a body fluid).

> **Key Learning Outcome 27.4** When macrophages encounter cells without the proper MHC proteins, they secrete a chemical alarm that initiates the immune defense.

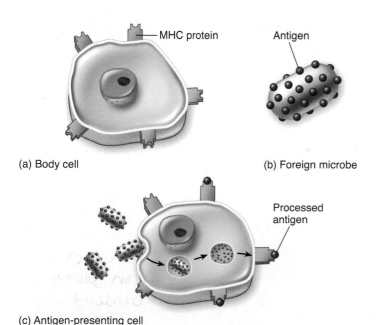

(a) Body cell

(b) Foreign microbe

(c) Antigen-presenting cell

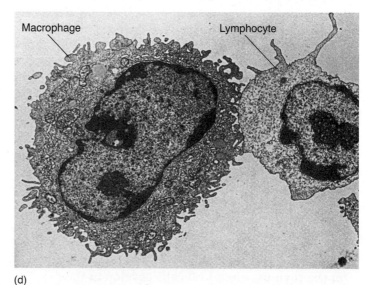

(d)

Figure 27.8 How antigens are presented.

(a) Cells of the body have MHC proteins on their surfaces that identify them as "self" cells. Immune system cells do not attack these cells.
(b) Foreign cells or microbes have antigens on their surfaces. (c) T cells can bind to the antigens to initiate an attack only after the antigens are processed and complexed with MHC proteins on the surface of an antigen-presenting cell. B cells recognize the antigens directly, not requiring an antigen-presenting cell. (d) In this electron micrograph, a lymphocyte (*right*) contacts a macrophage (*left*), an antigen-presenting cell.

T Cells: The Cellular Response

When macrophages process foreign antigens, they trigger the **cellular immune response,** illustrated in figure 27.9. As shown in step ❶, macrophages secrete interleukin-1, which stimulates cell division and proliferation of T cells. Helper T cells become activated when they bind to the complex of MHC proteins and antigens presented to them by the macrophages. The helper T cells then secrete **interleukin-2** ❷, which stimulates the proliferation of **cytotoxic T cells** ❸, which recognize and destroy infected body cells. Cytotoxic T cells destroy infected cells only if those cells display the foreign antigen together with their MHC proteins ❹.

The body makes millions of different versions of T cells. Each version bears a single, unique kind of receptor protein on its membrane, a receptor able to bind to a particular antigen-MHC protein complex on the surface of an antigen-presenting cell. Any cytotoxic T cell whose receptor fits the particular antigen-MHC protein complex present in the body begins to multiply rapidly, soon forming large numbers of T cells ❸ capable of recognizing the complex containing the particular foreign antigen. Large numbers of infected cells can be quickly eliminated, because the single T cell able to recognize the invading virus is amplified in number to form a large clone of identical T cells, all able to carry out the attack. Any of the body's cells that bear traces of viral infection are destroyed. The method used by cytotoxic T cells to kill infected body cells is similar to that used by natural killer cells and complement—they puncture the plasma membrane of the infected cell. Following an infection, some activated T cells give rise to memory T cells ❺ that remain in the body, ready to mount an attack quickly if the antigen is encountered again.

Cytotoxic T cells will also attack any foreign version of the MHC proteins. Thus even though the immune system did not evolve in vertebrates as a defense against tissue transplants, their immune systems will attack transplanted tissue and cause graft rejection. It is for this reason that relatives are often sought for kidney transplants—their MHC proteins are genetically closer to the recipient. The drug cyclosporine is often given to transplant patients because it inactivates cytotoxic T cells.

There is some evidence that cancer cells alter their "self" markers in a way that can be detected by immune cells, creating so-called "cancer-specific antigens," but the potential role of such modified cell surface markers in immunological surveillance against cancer is not well understood.

> **Key Learning Outcome 27.5** The cellular immune response is carried out by T cells, which mount an immediate attack on infected cells, killing any that present unusual surface antigens.

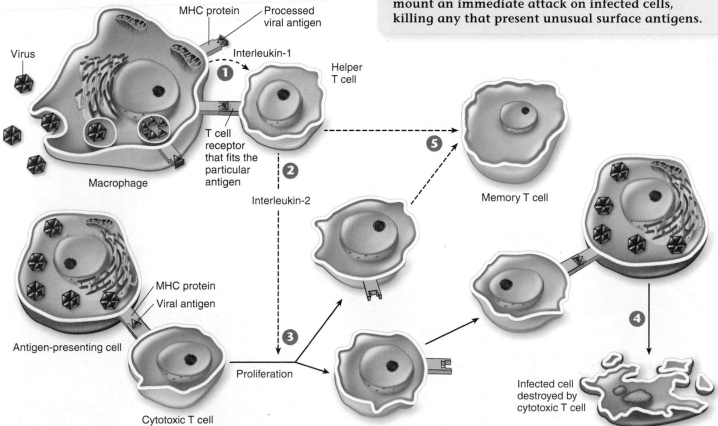

Figure 27.9 The T cell immune defense.

After a macrophage has processed an antigen, it releases interleukin-1, signaling helper T cells to bind to the antigen-MHC protein complex. This triggers the helper T cell to release interleukin-2, which stimulates the multiplication of cytotoxic T cells. In addition, proliferation of cytotoxic T cells is stimulated when a T cell with a receptor that fits the antigen displayed by an antigen-presenting cell binds to the antigen-MHC protein complex. Body cells that have been infected by the antigen are destroyed by the cytotoxic T cells. As the infection subsides, suppressor T cells "turn off" the immune response (not shown here).

27.6 B Cells: The Humoral Response

B cells also respond to helper T cells activated by interleukin-1. Like cytotoxic T cells, B cells have receptor proteins on their surfaces, a different receptor for each version of B cell. B cells recognize invading microbes much as cytotoxic T cells recognize infected cells, but unlike cytotoxic T cells, they do not go on the attack themselves. Rather, they mark the pathogen for destruction by mechanisms that have no "ID check" system of their own. Early in the immune response known as the **humoral immune response,** the markers placed by B cells alert complement proteins to attack the cells carrying them. Later in the response, the markers placed by B cells activate macrophages and natural killer cells.

The way B cells do their marking is simple and foolproof. Unlike the receptors on T cells, which bind only to antigen-MHC protein complexes on antigen-presenting cells, B cell receptors can bind to free, unprocessed antigens, shown in step ❶ of **figure 27.10**. When a B cell encounters an antigen, antigen particles enter the B cell by endocytosis and get processed and placed on the surface complexed with MHC proteins. Helper T cells recognize the specific antigen, bind to the antigen-MHC protein complex on the B cell ❷, and release interleukin-2, which stimulates the B cell to divide. In addition, free, unprocessed antigens stick to antibodies (the green Y-shaped

structures) on the B cell surface. This antigen exposure triggers even more B cell proliferation. B cells divide to produce plasma cells that serve as short-lived antibody factories ❸ and long-lived memory B cells ❹ that remain in the body after the initial infection and are available to mount a quick attack if the antigen enters the body again.

Antibodies are proteins in a class called **immunoglobulins** (abbreviated *Ig*), which is divided into subclasses based on the structures and functions of the antibodies. The five different immunoglobulin subclasses are as follows:

1. **IgM.** This is the first type of antibody to be secreted into the blood when the antigen is first encountered and to serve as a receptor on the B cell surface. These antibodies also promote agglutination reactions (causing antigen-containing particles to stick together, or agglutinate).
2. **IgG.** This is the major form of antibody secreted during a second or subsequent infection and the major one in the blood plasma.
3. **IgD.** These antibodies serve as receptors for antigens on the B cell surface. Their other functions are unknown.
4. **IgA.** This is the form of antibody in external secretions, such as saliva, mucus, and mother's milk.
5. **IgE.** This form of antibody promotes the release of histamine and other agents that produce allergic symptoms, such as those of hay fever.

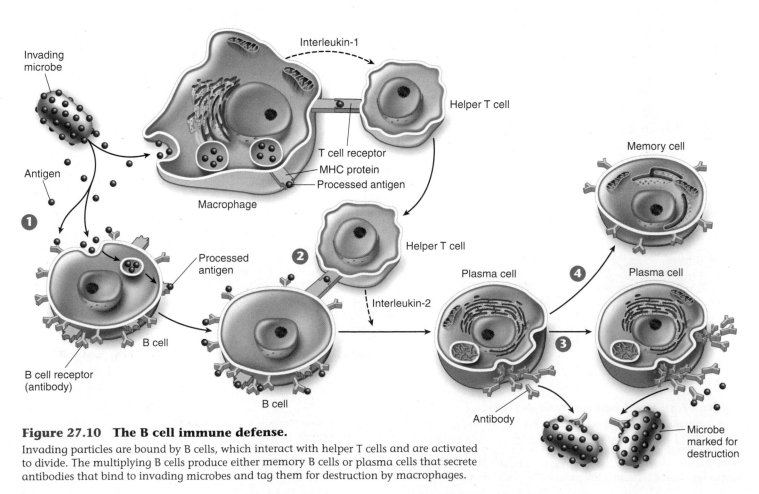

Figure 27.10 The B cell immune defense.
Invading particles are bound by B cells, which interact with helper T cells and are activated to divide. The multiplying B cells produce either memory B cells or plasma cells that secrete antibodies that bind to invading microbes and tag them for destruction by macrophages.

The **plasma cells** that are derived from B cells produce lots of the same particular antibody that was able to bind the antigen in the initial immune response. Flooding through the bloodstream, these antibody proteins (**figure 27.11**) are able to stick to antigens on any cells or microbes that present them, flagging those cells and microbes for destruction. Complement proteins, macrophages, or natural killer cells then destroy the antibody-displaying cells and microbes.

The B cell defense is very powerful because it amplifies the reaction to an initial pathogen encounter a millionfold. It is also a very long-lived defense in that a few of the multiplying B cells do not become antibody producers. Instead they become a line of **memory B cells** that continue to patrol your body's tissues, circulating through your blood and lymph for a long time—sometimes for the rest of your life.

Antibody Diversity

The vertebrate immune system is capable of recognizing as foreign practically any nonself molecule presented to it—literally millions of different antigens. Although vertebrate chromosomes contain only a few hundred receptor-encoding genes, it is estimated that human B cells can make between 10^6 and 10^9 different antibody molecules. How do vertebrates generate millions of different antigen receptors when their chromosomes contain only a few hundred versions of the genes encoding those receptors?

The answer to this question is that the millions of immune receptor genes do not exist as single sequences of nucleotides. Rather, they are assembled by stitching together three or four DNA segments that code for different parts of the receptor molecule. When an antibody is assembled, these different sequences of DNA are brought together to form a composite gene. The antibody molecule in **figure 27.12** was produced by a composite gene: different sections of DNA were used to produce the constant regions (green), the joining regions (red), the diversity regions (blue), and the variable regions (yellow). This process is called **somatic rearrangement.**

Two other processes generate even more sequences. First, the DNA segments are often joined together with one or two nucleotides off-register, shifting the reading frame during gene transcription and so generating a totally different sequence of amino acids in the protein. Second, random mistakes occur during successive DNA replications as the lymphocytes divide during clonal expansion. Both mutational processes produce changes in amino acid sequences, a phenomenon known as **somatic mutation** because it takes place in a somatic cell rather than in a gamete.

Because a cell may end up with any heavy chain gene and any light chain gene during its maturation, the total number of different antibodies possible is staggering: 16,000 heavy chain combinations × 1,200 light chain combinations = 19 million different possible antibodies. If one also takes into account the changes induced by somatic mutation, the total can exceed 200 million! It should be understood that although this discussion has centered on B cells and their receptors, the receptors on T cells are as diverse as those on B cells because they also are subject to similar somatic rearrangements and mutations.

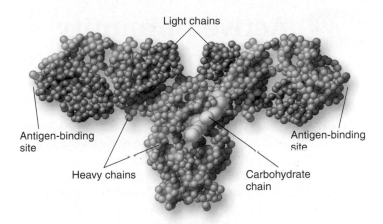

Figure 27.11   An antibody molecule.
In this molecular model of an antibody molecule, each amino acid is represented by a small sphere. Each molecule consists of four protein chains, two "light" (*red*) and two "heavy" (*blue*). The four protein chains wind around one another to form a Y shape. Foreign molecules, called antigens, bind to the arms of the Y.

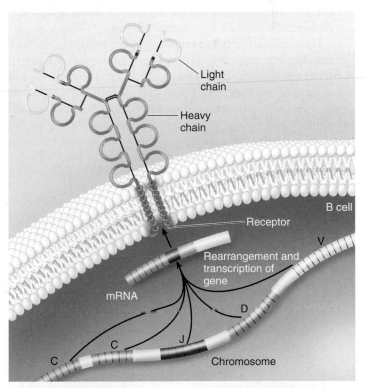

Figure 27.12   The antibody molecule is produced by a composite gene.
Different regions of the DNA code for different regions of the antibody (C, constant regions; J, joining regions; D, diversity regions; and V, variable regions) and are brought together to make a composite gene that codes for the antibody. By combining segments, an enormous number of different antibodies can be produced.

Key Learning Outcome 27.6   In the humoral immune response, B cells label infecting and infected cells with antibodies to tag them for destruction by complement proteins, natural killer cells, and macrophages.

27.7 Active Immunity Through Clonal Selection

As we discussed earlier, B cells and T cells have receptors on their cell surfaces that recognize and bind to specific antigens. When a particular antigen enters the body, it must, by chance, encounter the specific lymphocyte with the appropriate receptor to provoke an immune response. The first time a pathogen invades the body, there are only a few B cells or T cells that may have the receptors that can recognize the invader's antigens. Binding of the antigen to its receptor on the lymphocyte surface, however, stimulates cell division and produces a *clone* (a population of genetically identical cells). This process is known as **clonal selection.** For example, in the first encounter with a chicken pox virus in **figure 27.13** there are only a few cells that can mount an immune response, and the response is relatively weak. This is called a **primary immune response** and is indicated by the first curve, which shows the initial amount of antibody produced upon exposure to the virus.

If the primary immune response involves B cells, some of the cells become plasma cells that secrete antibodies (taking 10 to 14 days to clear the chicken pox virus from the system), and some become memory cells. Some of the T cells involved in the primary response also become memory cells. Because a clone of memory cells specific for that antigen develops after the primary response, the immune response to a second infection by the same pathogen is swifter and stronger, as shown by the second curve. Many memory cells can be produced following the primary response, providing a jump start for the production of antibodies should a second exposure occur. The next time the body is invaded by the same pathogen, the immune system is ready. As a result of the first infection, there is now a large clone of lymphocytes that can recognize that pathogen. This more effective response, elicited by subsequent exposures to an antigen, is called a **secondary immune response.** The "Inquiry and Analysis" feature at the end of this chapter further explores the nature of the secondary immune response.

Memory cells can survive for several decades, which is why people rarely contract chicken pox a second time after they have had it once. Memory cells are also the reason that vaccinations are effective. The viruses causing childhood diseases have surface antigens that change little from year to year, so the same antibody is effective for decades. Other diseases, such as influenza, are caused by viruses whose genes that encode surface proteins mutate rapidly. This rapid genetic change causes new strains to appear every year or so that are not recognized by memory cells from previous infections.

Although the cellular and humoral immune responses were discussed separately, they occur simultaneously in the body. The Key Biological Process illustration on the facing page follows the steps of a viral infection and shows how the cellular and humoral lines of defense work together to produce the body's specific immune response.

When a virus invades the body, viral proteins are displayed on the surfaces of infected cells ❶. Viruses and infected cells are taken up by macrophages through phagocytosis ❷, and viral proteins are displayed on the surface of the macrophage attached to MHC proteins. Stimulated in this way, macrophages release interleukin-1 ❸. Interleukin-1 is an alarm signal that stimulates helper T cells ❹. Activated helper T cells release interleukin-2, which triggers both the cellular (T cell) and humoral (B cell) responses ❺. In the figure, the cellular response follows the green arrows, and the humoral response follows the red arrows.

While some activated T cells become memory T cells ❺ₐ that remain in the body and are able to more quickly fight future infections by the same virus, interleukin-2 also activates cytotoxic T cells. The cytotoxic T cells bind to infected cells that carry the viral antigen and kill them ❻.

Interleukin-2 also activates B cells ❼ that multiply in the cell. Some B cells become memory cells ❽ that remain in the body for future infections by the same virus. Other activated B cells become plasma cells ❾ that produce antibodies directed against the viral surface proteins. The antibodies released into the body will bind to viral proteins that are displayed on the surface of infected body cells ❿ or that are present on the surface of the viruses. Cells or viruses that are tagged with antibodies are destroyed by macrophages that are circulating in the body ⓫. As you can see, both arms of the immune response work together very effectively to rid the body of invaders.

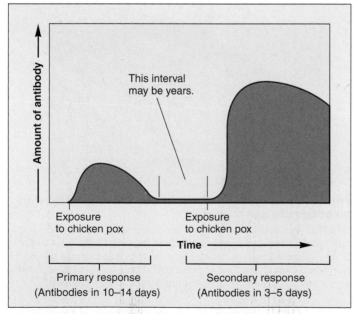

Figure 27.13   The development of active immunity.
Immunity to chicken pox occurs because the first exposure stimulated the development of lymphocyte clones with receptors for the chicken pox virus. As a result of clonal selection, a second exposure stimulates the immune system to produce large amounts of the antibody more rapidly than before, keeping the person from getting sick again.

Key Learning Outcome 27.7   A strong immune response is possible because infecting cells stimulate the few responding B cells and T cells to divide repeatedly, forming clones of responding cells.

KEY BIOLOGICAL PROCESS: The Immune Response

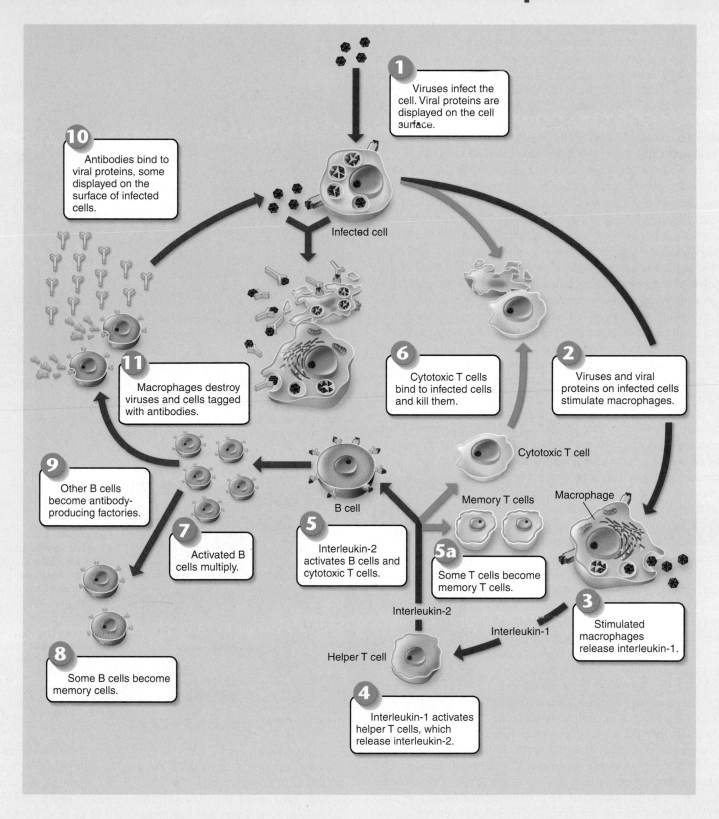

1 Viruses infect the cell. Viral proteins are displayed on the cell surface.

10 Antibodies bind to viral proteins, some displayed on the surface of infected cells.

Infected cell

11 Macrophages destroy viruses and cells tagged with antibodies.

6 Cytotoxic T cells bind to infected cells and kill them.

2 Viruses and viral proteins on infected cells stimulate macrophages.

Cytotoxic T cell

9 Other B cells become antibody-producing factories.

B cell

Memory T cells

Macrophage

7 Activated B cells multiply.

5 Interleukin-2 activates B cells and cytotoxic T cells.

5a Some T cells become memory T cells.

3 Stimulated macrophages release interleukin-1.

Interleukin-2

Interleukin-1

8 Some B cells become memory cells.

Helper T cell

4 Interleukin-1 activates helper T cells, which release interleukin-2.

27.8 Vaccination

In 1796, an English country doctor named Edward Jenner carried out an experiment that marks the beginning of the study of immunology. Smallpox was a common and deadly disease in those days. Jenner observed, however, that milkmaids who had caught a much milder form of "the pox" called cowpox (presumably from cows) rarely caught smallpox. Jenner set out to test the idea that cowpox conferred protection against smallpox. He infected people with mild cowpox (**figure 27.14**), and as he had predicted, many of them became immune to smallpox.

We now know that smallpox and cowpox are caused by two different but similar viruses. Jenner's patients who were injected with the cowpox virus mounted a defense that was also effective against a later infection of the smallpox virus. Jenner's procedure of injecting a harmless microbe to confer resistance to a dangerous one is called vaccination. **Vaccination** is the introduction into a person's body of a dead or disabled pathogen or, more commonly these days, of a harmless microbe with pathogen proteins displayed on its surface. The vaccination triggers an immune response against the pathogen, without an infection ever occurring. Afterward the bloodstream of the vaccinated person contains circulating memory cells (B and T cells) directed against that specific pathogen. The vaccinated person is said to have been "immunized" against the disease.

Through genetic engineering, scientists are now routinely able to produce "piggyback," or subunit, vaccines. These vaccines are made of harmless viruses that contain in their DNA a single gene cut out of a pathogen, a gene encoding a protein normally exposed on the pathogen's surface. By splicing the pathogen gene into the DNA of the harmless host, that host is induced to display the protein on its surface. The harmless virus displaying the pathogen protein is like a sheep in wolf's clothing, unable to hurt you but raising alarm as if it could. Your body responds to its presence by making an antibody directed against the pathogen protein that acts like an alarm to the immune system, should that pathogen ever visit your body.

If the activities of memory cells provide such an effective defense against future infection, why can you catch some diseases like flu more than once? The reason you don't stay immune to flu is that the flu virus has evolved a way to evade the immune system—it changes. The genes encoding the surface proteins of the flu virus mutate very rapidly. Thus, the shapes of these surface proteins alter swiftly. Your memory cells do not recognize viruses with altered surface proteins as being the same viruses they have already successfully defeated or been vaccinated against, because the memory cells' receptors no longer "fit" the new shape of the flu surface proteins. When the new version of flu virus invades your body, you need to mount an entirely new immune defense.

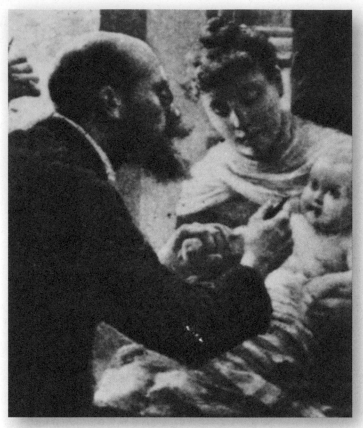

Figure 27.14   The birth of immunology.
This famous painting shows Edward Jenner inoculating patients with cowpox in the 1790s and thus protecting them from smallpox.

**Figure 27.15
The flu epidemic of 1918 killed over 20 million in 18 months.**

With 25 million Americans alone infected during the influenza epidemic, it was hard to provide care for everyone. The Red Cross often worked around the clock.

Sometimes the flu virus surface proteins possess shapes that the immune system does not readily recognize. When mutations arose in a bird flu in 1918 that allowed this flu virus to pass easily from one infected human to another, over 20 million Americans and Europeans died in 18 months (**figure 27.15**). Less-profound changes in flu virus surface proteins occur periodically, resulting in new strains of flu for which we are not immune. The annual flu shots are vaccines against these new strains. Researchers are able to predict the current year's strain of the flu by examining pre-season flu reports from across the globe to prepare a vaccine against the year's dominant strain. However, as you learned in chapter 16, entirely new flu virus strains from birds or pigs can infect humans, and genetic recombination within infected organisms can create even more new combinations of viral surface proteins.

The Search for an AIDS Vaccine

One of the most intensive efforts in the history of medicine is currently under way to develop an effective vaccine against HIV, the virus responsible for AIDS. Researchers are using the "piggyback" method shown in **figure 27.16**. Steps ❶ through ❸ show how the gene that encodes an HIV surface protein is isolated, and then it is inserted into the DNA of a harmless vaccinia virus (steps ❹ and ❺). The genetically engineered vaccinia virus is injected into the body ❻, which triggers the body to begin producing antibodies and memory cells against the HIV surface protein antigen ❼. The HIV virus has nine genes, encoding a variety of proteins. Initial efforts focused on producing a subunit vaccine containing the HIV *env* (envelope) gene, which encodes the protein on the outside of the virus.

Unfortunately, the HIV virus mutates even more rapidly than the flu virus, and vaccines developed from one strain of the HIV virus are not effective against others. The AIDS vaccines are said to solicit **narrowly neutralizing antibodies,** ones that protect against only one or a few of the many hundreds of strains of HIV. This high mutation rate has been the single biggest obstacle to developing a successful AIDS vaccine.

New vaccine approaches look more promising. An intensive effort to identify **broadly neutralizing antibodies** that protect against many or all strains of HIV is yielding very encouraging results. One place where HIV doesn't mutate much is where it attaches to the cell it infects. Researchers created a probe shaped exactly like that key site and used it to fish out antibodies that bind particularly well to it. By 2010, more than a half dozen broadly neutralizing antibodies had been obtained, the strongest of which neutralizes 91% of all known AIDS strains! The next step will be to do a little "reverse engineering," using these broadly neutralizing antibodies to refine the HIV probe. Eventually, the researchers hope that the modified probe can be administered to people as a vaccine that will solicit broadly neutralizing "anti-HIV" antibodies and so protect against AIDS.

> **Key Learning Outcome 27.8** Vaccines introduce antigens similar or identical to those of pathogens, eliciting an immune response that defends against the pathogen too.

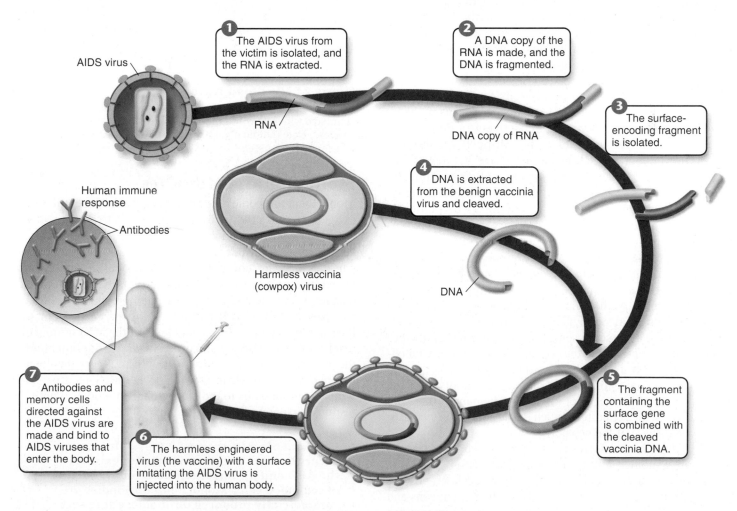

1 The AIDS virus from the victim is isolated, and the RNA is extracted.

2 A DNA copy of the RNA is made, and the DNA is fragmented.

3 The surface-encoding fragment is isolated.

4 DNA is extracted from the benign vaccinia virus and cleaved.

5 The fragment containing the surface gene is combined with the cleaved vaccinia DNA.

6 The harmless engineered virus (the vaccine) with a surface imitating the AIDS virus is injected into the human body.

7 Antibodies and memory cells directed against the AIDS virus are made and bind to AIDS viruses that enter the body.

AIDS virus

RNA

DNA copy of RNA

Human immune response

Antibodies

Harmless vaccinia (cowpox) virus

DNA

Figure 27.16 Researchers are attempting to construct an AIDS vaccine.

27.9 Antibodies in Medical Diagnosis

Blood Typing

A person's blood type indicates the class of antigens found on the red blood cell surface. There are several groups of red blood cell antigens, but the major group is known as the **ABO system.** In terms of the antigens present on the red blood cell surface, a person may be *type A* (with only A antigens), *type B* (with only B antigens), *type AB* (with both A and B antigens), or *type O* (with neither A nor B antigens).

The immune system is tolerant to its own red blood cell antigens. A person who is type A, for example, does not produce anti-A antibodies. However, people with type A blood do make antibodies against the B antigen, and people with blood type B make antibodies against the A antigen. People who are type AB develop tolerance to both antigens and thus do not produce either anti-A or anti-B antibodies. Those who are type O make both anti-A and anti-B antibodies.

If type A blood is mixed on a glass slide with serum from a person with type B blood, the anti-A antibodies in the serum cause the type A blood cells to clump together, or agglutinate (this is shown in the upper right panel of **figure 27.17**). These tests allow the blood types to be matched prior to transfusions, so that agglutination will not occur in the blood vessels, where it could lead to inflammation and organ damage.

Rh Factor Another group of antigens found in most red blood cells is the *Rh factor* (Rh stands for rhesus monkey, in which these antigens were first discovered). People who have these antigens are said to be Rh-positive, whereas those who do not are Rh-negative. There are fewer Rh-negative people because the Rh-positive allele is clinically dominant to the Rh-negative allele and is more common in the human population. The Rh factor is of particular significance when Rh-negative mothers give birth to Rh-positive babies.

Because the fetal and maternal blood are normally kept separate across the placenta (see chapter 31), the Rh-negative mother is not usually exposed to the Rh antigen of the fetus during pregnancy. At the time of birth, however, a varying degree of exposure may occur, and the Rh-negative mother's immune system may become sensitized and produce antibodies against the Rh antigen. If the woman does produce antibodies against the Rh factor, these antibodies can cross the placenta in subsequent pregnancies and cause hemolysis of the Rh-positive red blood cells of the fetus. The baby is therefore born anemic with a condition called erythroblastosis fetalis, or hemolytic disease of the newborn.

Erythroblastosis fetalis can be prevented by injecting the Rh-negative mother with an antibody preparation against the Rh factor within 72 hours after the birth of each Rh-positive baby. The injected antibodies inactivate the Rh antigens and thus prevent the mother from becoming actively immunized to them.

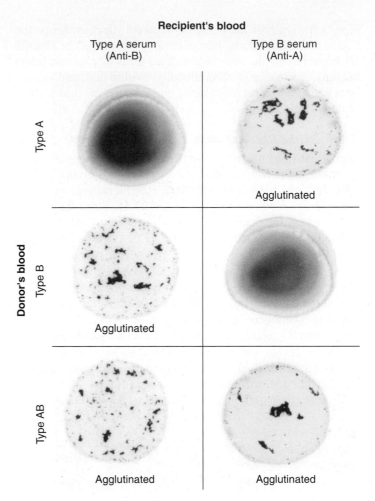

Figure 27.17 Blood typing.
Agglutination of the red blood cells is seen when blood types are mixed with sera containing antibodies against the A and B antigens. Note that no agglutination would be seen if type O blood (not shown) was used.

Monoclonal Antibodies

Monoclonal antibodies are antibodies that are specific to one antigen. Because they provide a very sensitive assay, monoclonals are often commercially prepared for use in clinical laboratory tests. Modern pregnancy tests, for example, use particles that are covered with monoclonal antibodies produced against a pregnancy hormone (abbreviated hCG—see chapter 31) as the antigen. In the blood pregnancy test, these particles are mixed with a sample from a pregnant woman. If the sample contains a significant level of the hCG hormone, it reacts with the antibody and causes a visible agglutination of the particles, indicating a positive test result. Over-the-counter pregnancy tests work in a similar way. hCG in a pregnant woman's urine binds to the monoclonal antibodies within the testing strip and indicates a positive result.

Key Learning Outcome 27.9 Agglutination occurs because different antibodies exist for the ABO and Rh factor antigens on the surface of red blood cells. Monoclonal antibodies are commercially produced antibodies that react against one specific antigen.

27.10 Overactive Immune System

Although the immune system is one of the most sophisticated systems of the vertebrate body, it is still not perfect. Many of the major diseases we face, and some minor irritations as well, reflect an overactive immune system.

Autoimmune Diseases

The ability of T cells and B cells to distinguish cells of your own body—"self" cells—from nonself cells is the key ability of the immune system that makes your body's third line of defense so effective. In certain diseases, this ability breaks down, and the body attacks its own tissues. Such diseases are called **autoimmune diseases.**

Multiple sclerosis is an autoimmune disease that usually strikes people between the ages of 20 and 40. In multiple sclerosis, the immune system attacks and destroys a sheath of fatty material, called myelin (see chapter 28), that insulates motor nerves (like the rubber covering electrical wires). Recall from section 22.6 that nerve impulses travel along the length of the nerve cell, and so degeneration of the myelin sheath interferes with transmission of nerve impulses, until eventually they cannot travel at all. Voluntary functions, such as movement of limbs, and involuntary functions, such as bladder control, are lost, leading finally to paralysis and death. Scientists do not know what stimulates the immune system to attack myelin.

Another autoimmune disease is type I diabetes in which cells are unable to take in glucose because the pancreas fails to produce insulin (recall from chapter 26 that insulin plays a key role in the liver's regulation of levels of glucose in the blood). Type I diabetes is thought to result from an immune attack on the insulin-manufacturing cells of the pancreas. No one knows why the attack occurs. Other autoimmune diseases are rheumatoid arthritis (an immune system attack on the tissues of the joints), lupus (in which the connective tissue and kidneys are attacked), and Graves' disease (in which the thyroid is attacked).

Allergies

Although your immune system provides very effective protection against fungi, parasites, bacteria, and viruses, sometimes it does its job too well, mounting an immune response that is greater than necessary to eliminate an antigen. The antigen in this case is called an **allergen,** and such an immune response is called an **allergy.** Hay fever, sensitivity to even tiny amounts of plant pollen, is a familiar example of an allergy. Many people are also allergic to nuts, eggs, milk, penicillin, and even proteins released from the feces of the house dust mite

Figure 27.18 The house dust mite.

This tiny animal, *Dermatophagoides,* causes an allergic reaction in many people.

(**figure 27.18**). Many people sensitive to feather pillows are in reality allergic to the mites that are residents of the feathers.

What makes an allergic reaction uncomfortable, and sometimes dangerous, is the involvement of antibodies attached to a kind of white blood cell called a **mast cell.** It is the job of the mast cells in an immune response to initiate an inflammatory response. **Figure 27.19** shows what happens when a mast cell encounters something that matches its antibody. Mast cells release histamines and other chemicals that cause capillaries to swell. **Histamines** also increase mucus production by cells of the mucous membranes, resulting in runny noses and nasal congestion (all the symptoms of hay fever). Most allergy medicines relieve these symptoms with antihistamines, chemicals that block the action of histamines.

Asthma is a form of allergic response in which histamines cause the narrowing of air passages in the lungs. People who have asthma have trouble breathing when exposed to substances to which they are allergic.

> **Key Learning Outcome 27.10**   Autoimmune diseases are inappropriate responses to "self" cells, whereas allergies are inappropriate immune responses to harmless antigens.

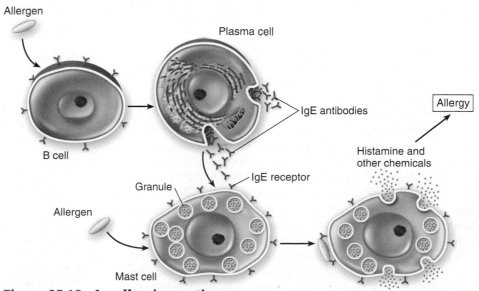

Figure 27.19   An allergic reaction.

In an allergic response, B cells secrete IgE antibodies (see page 572) that attach to the plasma membranes of mast cells, which secrete histamine in response to antigen-antibody binding.

27.11 AIDS: Immune System Collapse

AIDS (acquired immunodeficiency syndrome) was first recognized as a disease in 1981. By the end of 2007 in the United States, 583,298 individuals had died of AIDS, and more than 1 million others were thought to be infected with **HIV** (human immunodeficiency virus), the virus that causes the disease. Worldwide, 33.4 million are infected, over 25 million have died, and 2 million more die each year. HIV apparently evolved from a very similar virus that infects chimpanzees in Africa when a mutation arose that allowed the virus to recognize a human cell surface receptor called CD4. This receptor is present in the human body on certain immune system cells, notably macrophages and helper T cells. It is the identity of these immune system cells that leads to the devastating nature of the disease.

How HIV Attacks the Immune System

HIV attacks and cripples the immune system by inactivating cells that have CD4 receptors (CD4$^+$ cells), including helper T cells. The significance of killing helper T cells is that it leaves the immune system unable to mount a response to *any* foreign antigen. AIDS is a deadly disease for just this reason.

HIV's attack on CD4$^+$ T cells progressively cripples the immune system because HIV-infected cells die after releasing replicated viruses that proceed to infect other CD4$^+$ T cells. Over time, the body's entire population of CD4$^+$ T cells is destroyed. In a normal individual, CD4$^+$ T cells make up 60% to 80% of circulating T cells; in AIDS patients, CD4$^+$ T cells often become too rare to detect, wiping out the human immune defense. With no defense against infection, any of a variety of otherwise commonplace infections proves fatal. Also, with no ability to recognize and destroy cancer cells when they arise, death due to cancer becomes far more likely. Indeed, AIDS was first recognized because of a cluster of cases of a rare cancer, Kaposi's sarcoma. More AIDS victims die of cancer than from any other cause.

The fatality rate of AIDS is 100%; no patient exhibiting the symptoms of AIDS has recovered. The disease is *not* highly contagious because it is only transmitted from one individual to another through the transfer of internal body fluids, typically in semen or vaginal fluid during sexual intercourse and in blood transmitted by needles during drug use. However, symptoms of AIDS do not usually show up for several years after infection with HIV, apparent in the delay of the onset of AIDS in the United States (the red line in **figure 27.20**) after initial infection with HIV (the green line). Because of this symptomless delay, infected individuals can unknowingly spread the virus to others. Awareness campaigns in the United States have helped reduce the numbers of new AIDS cases.

A variety of drugs inhibit HIV in the test tube. These include AZT and its analogs (which inhibit virus nucleic acid replication) and protease inhibitors (which inhibit the production of functional viral proteins). A combination of a protease inhibitor and two AZT analog drugs entirely eliminates the HIV

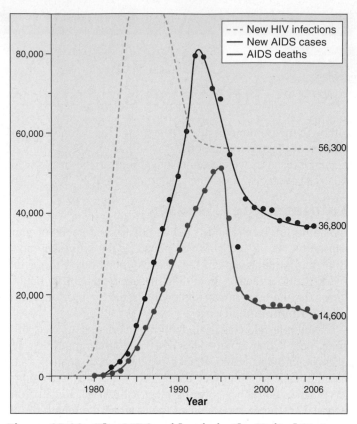

Figure 27.20   The AIDS epidemic in the United States.
The U.S. Centers for Disease Control (CDC) reports that 37,041 new AIDS cases were reported in 2007 in the United States, with a total of over one million cases and nearly 600,000 deaths.

virus from many patients' bloodstreams. Widespread use of this *combination therapy* has cut the U.S. AIDS death rate by almost two-thirds since its introduction in the mid-1990s, from 51,414 AIDS deaths in 1995 to 38,074 in 1996, and ten years later in 2006, deaths remained low, at approximately 14,600.

Unfortunately, this sort of combination therapy does not appear to actually succeed in eliminating HIV from the body. While the virus disappears from the bloodstream, traces of it can still be detected in lymph tissue of the patients. When combination therapy is discontinued, virus levels in the bloodstream once again rise. Because of demanding therapy schedules and many side effects, long-term combination therapy does not seem a promising approach.

Scientists continue to seek a vaccine to protect people from this deadly and incurable disease. But over 30 years and 1 million American AIDS cases later, an effective AIDS vaccine so far still eludes the best efforts of researchers. Because the HIV virus generates mutations at such a prodigious rate, few of those infected have exactly the same virus. For this reason, clinical trials of vaccines targeted against one version of HIV have been ineffective against others. The suggestion that portions of the HIV virus do *not* vary (see page 577) offers encouragement that an effective AIDS vaccine may yet be developed.

Key Learning Outcome 27.11 HIV cripples the vertebrate immune defense by infecting and killing key lymphocytes.

Biology and *Staying Healthy*

AIDS Drugs Target Different Phases of the HIV Infection Cycle

AIDS is so devastating because it dismantles the very process that the body uses to fight infections. While the search for an AIDS vaccine continues, scientists are also trying to find ways to slow down or halt the process whereby the virus spreads throughout the body, infecting other cells. This doesn't cure the person; he or she still has an HIV infection that can be spread to others. But by slowing down HIV's ability to replicate inside a person, these drugs might help to curb the effects of HIV that lead to the development of AIDS.

Many new drugs are being developed that target different phases of the viral infection cycle. To date there are six classes of drugs (described below and keyed to the diagram) that disrupt the ability of HIV to enter the cell, to replicate its DNA, and to form new viral particles. As of yet, there has been little drug development targeting the final phase of the infection cycle, viral exit (**7** where new viruses leave the cell).

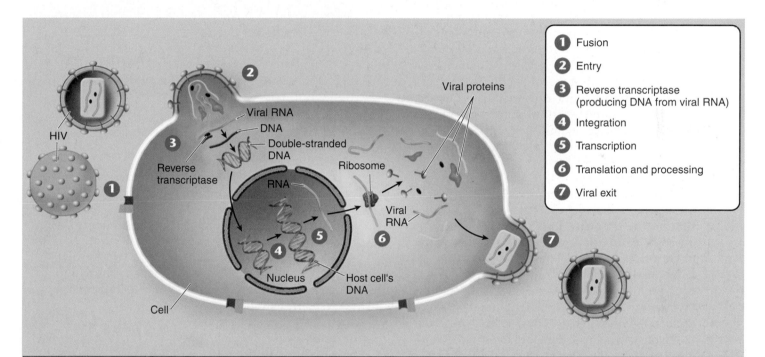

Class	Example	Number of Approved Drugs	Mechanisms of Action	Coming Soon...
❶ Fusion inhibitors	Fuzeon	1	Peptides that bind to a certain receptor protein on the outside of the cell and so prevent fusion of HIV particles to the cell surface	Other experimental drugs that target other receptor proteins
❷ Uptake inhibitors	Maraviroc	1	Bind to a necessary coreceptor protein on the cell surface and so prevent cell uptake of HIV particles	Other experimental drugs that target another coreceptor or the CD4 receptor
❸ Reverse transcriptase inhibitors	Atripla	4	Bind to HIV's "reverse transcriptase" enzyme's active site, gumming it up so that the enzyme cannot function, and so blocking HIV replication.	—
❹ Integrase inhibitors	Raltegravir	2	Bind to HIV "integrase" enzyme that inserts HIV genes into the cell's DNA, and so block HIV replication	A second drug is in the testing phase
❺ Nucleoside analogs	AZT	10	Faulty nucleotides cannot be assembled into functional DNA, and so block HIV replication	Nucleotide analogs that do not require phosphorylation within the cell as nucleosides do
❻ Protease inhibitors	Prezista	10	Bind to HIV "protease" enzyme that cuts primary transcript into functional segments, and so block HIV replication	—

Is Immunity Antigen-Specific?

The immune response provides a valuable protection against infection because it can remember prior experiences. We develop lifelong immunity to many infectious diseases after one childhood exposure. This long-term immunity is why vaccines work. A key question about immune protection is whether or not it is specific. Does exposure to one pathogen confer immunity to only that one, or is the immunity you acquire a more general response, protecting you from a range of infections?

The graph to the right displays the results of an experiment designed to answer this question. A colony of rabbits is immunized once with antigen A, and the level of antibody directed against this antigen monitored in each individual. After 40 days, each rabbit is reinjected, some with antigen A and others with antigen B, and the level of antibody directed against the reinjected antigens is monitored. The red line is typical of results for antigen A, the blue line for antigen B.

1. **Applying Concepts**
   a. Variable. In the graph, what is the dependent variable?
   b. Reading a Continuous Curve. Does each injection of antigen A result in detectable antibody production? antigen B?
2. **Interpreting Data**
   a. The initial response to antigen A is called the "primary" response, and the second response to antigen A administered 40 days later is called the "secondary" response. Compare the speed of the primary and secondary responses—which reaches maximal antibody response quicker?
   b. Compare the magnitude of the primary and secondary immune responses to antigen A—are they similar, or is one response of greater magnitude?

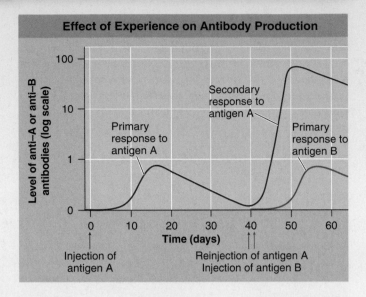

Effect of Experience on Antibody Production

3. **Making Inferences**
   a. Why is the secondary response induced by a second exposure to antigen A different from the primary response?
   b. Is the response to antigen B more similar to the primary or secondary response of antigen A?
4. **Drawing Conclusions**
   a. Does the prior exposure to antigen A have any impact on the speed or magnitude of the primary response to antigen B?
   b. Is the immune response to these antigens antigen-specific?
5. **Further Analysis** If you were to inject both sets of rabbits with antigen B on day 80, what would you expect the results to be? Explain the difference you would expect in the immune responses of the two groups of rabbits to this injection, if any.

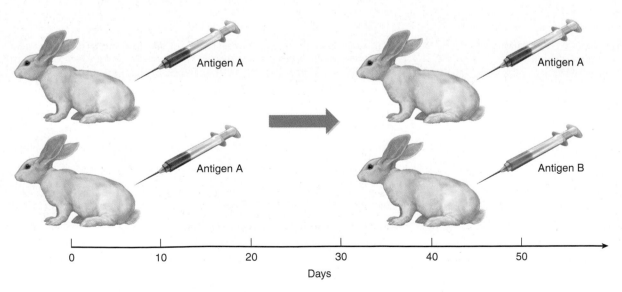

Three Lines of Defense

27.1 Skin: The First Line of Defense

- The body has three lines of defense against infection, the first being skin (**figure 27.2**) and the mucous membranes that line the digestive and respiratory tracts. Skin and mucous membranes form barriers to pathogens.

27.2 Cellular Counterattack: The Second Line of Defense

- The second line of defense is a nonspecific cellular attack. The cells and chemicals used in this line of defense attack all foreign agents that they encounter, and some are stored in the lymphatic system (**figure 27.3**).

- Macrophages and neutrophils attack the invading pathogen, while natural killer cells attack infected cells, killing them before the pathogen can spread to other cells (**figures 27.4** and **27.5**). Free-floating proteins in the blood, called complement, insert into the membranes of foreign cells, killing them (**figure 27.6**).

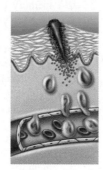

- The inflammatory response, shown here from **figure 27.7**, and the temperature response are also part of the second line of defense.

27.3 Specific Immunity: The Third Line of Defense

- The third line of defense is a specific immune response and involves cells other than macrophages, neutrophils, and natural killer cells (**table 27.1**). Lymphocytes called T cells and B cells are "programmed" by exposure to specific antigens and once programmed, will seek out the antigens or cells carrying the antigens and destroy them.

The Immune Response

27.4 Initiating the Immune Response

- Invading microorganisms display proteins on their surfaces that are different from the person's MHC proteins. Macrophages survey cells for "self" MHC proteins. A cell or virus that exhibits "nonself" proteins is engulfed by a macrophage and is partially digested. Antigens from the microbe are inserted into the surface of the macrophage and it is then called an antigen-presenting cell (**figure 27.8**). This antigen-presenting cell presents these antigens to T cells, activating the T cell response. This cell also secretes interleukin-1. Interleukin-1 stimulates helper T cells that trigger immune responses.

27.5 T Cells: The Cellular Response

- The cellular response involves T cells. Antigen-presenting cells activate helper T cells that release interleukin-2. Interleukin-2 stimulates the cloning of cytotoxic T cells that recognize and kill infected cells that display the specific antigen. Following the infection, memory T cells form and remain in the body to fight subsequent infections (**figure 27.9**).

27.6 B Cells: The Humoral Response

- The humoral response involves B cells. Helper T cells release interleukin-2, which activates the B cells. Activated B cells multiply and "mark" the foreign invaders with specific antibody proteins. The "marked" cells are then attacked by the nonspecific immune response. Activated B cells divide to form plasma cells and memory cells. Plasma cells produce and release antibodies and memory cells circulate in the blood. Memory cells become plasma cells in the event of reinfection by the same antigen (**figure 27.10**).

- Antibody proteins (**figure 27.11**), also called immunoglobulins, that are produced by the plasma cells are divided into five subclasses: IgM, IgG, IgD, IgA, and IgE, each with a different function.

- The vertebrate immune system can produce some 10^9 different antibodies from only a few hundred versions of the genes encoding the antibody receptor proteins. This is accomplished by somatic rearrangement of the genes encoding different parts of the antibody molecule. The various antibodies are produced from composite genes (**figure 27.12**).

27.7 Active Immunity Through Clonal Selection

- The initial immune response triggered by infection is called the primary response. B and T cells are stimulated to begin dividing, producing a clone of cells. This is called clonal selection. The primary immune response is a delayed and somewhat weak response. But through clonal selection, a large population of memory cells is present in the body, so a second infection by the same antigen will trigger a quicker and larger response, called the secondary response (**figure 27.13**).

- Cellular and humoral immune responses occur simultaneously in the body. The activation of helper T cells triggers both immune responses, which also include nonspecific cellular attacks (**Key Biological Process, page 575**).

27.8 Vaccination

- Vaccination takes advantage of the mechanism of the primary and secondary immune responses. Vaccines introduce harmless antigens into the body, triggering the primary response. Later with an actual infection, the body elicits a swift and large secondary immune response (**figure 27.16**).

27.9 Antibodies in Medical Diagnosis

- Antibodies are keen detectors of antigens, and so are used in various medical diagnostic applications such as blood typing and monoclonal antibody assays (**figure 27.17**).

Defeat of the Immune System

27.10 Overactive Immune System

- Sometimes the immune system attacks antigens that are not foreign or pathogenic. In autoimmune responses, the body attacks its own cells. In allergic reactions, the body mounts an attack against a harmless substance (**figure 27.19**).

27.11 AIDS: Immune System Collapse

- AIDS is a fatal disease caused by infection with the HIV virus. HIV attacks macrophages and helper T cells, destroying the cells that protect the body from other infections (**figure 27.20**).

1. Skin is a physical barrier but also provides chemical protection generated with
   a. sweat and oil glands.
   b. mucus.
   c. the skin's basal layer.
   d. the stratum corneum of the skin.
2. The immune system can identify foreign cells in the bloodstream because these foreign cells
   a. are observed destroying other cells by the immune system.
   b. have cell surface proteins that are different from the body's own cell surface proteins.
   c. have CD4 receptors similar to T cells.
   d. All of the above.
3. The purpose of the inflammatory response is to
   a. increase the temperature of an infected area.
   b. reduce pain of an infected area.
   c. increase the number of immune system cells in an infected area.
   d. form a barrier around the infected area.
4. Increasing human body temperature—that is, causing a fever—assists the immune system because
   a. increased temperature speeds up the chemical reactions used by the immune system.
   b. pathogenic bacteria do not grow well at high temperatures.
   c. increased temperature causes body cell proteins to denature.
   d. All of the above.
5. Immune responses tailored to specific antigens involve
   a. T cells.
   b. macrophages.
   c. monocytes.
   d. neutrophils.
6. Antibody production takes place in
   a. T cells.
   b. natural killer cells.
   c. B cells.
   d. mast cells.
7. Cytotoxic T cells
   a. produce antibodies.
   b. destroy pathogens directly.
   c. destroy foreign antigens floating freely in the bloodstream.
   d. destroy cells infected by pathogens.
8. Immunity to future invasion of a specific pathogen is accomplished by production of
   a. plasma cells.
   b. memory T and B cells.
   c. helper T cells.
   d. monocytes.
9. When a body's immune system attacks the body's own cells, this is called
   a. an inflammatory response.
   b. a temperature response.
   c. an autoimmune response.
   d. an allergic response.
10. HIV-infected people with advanced AIDS usually die of an infectious disease or cancer. This is because HIV attacks
    a. helper T cells.
    b. neutrophils.
    c. memory T and B cells.
    d. mast cells.

1. **Figure 27.13** Certain diseases are considered to be primarily "childhood" diseases—measles, mumps, chicken pox—because once individuals have had these illnesses as a child, they don't catch them again when, as parents, they take care of their own children who are sick. Use the graph to help explain the reason.

2. **Figure 27.20** What might account for the large increase in the number of AIDS cases, from the early 1980s to 1992, then the decline, and now the essentially steady state, neither growing nor diminishing by very much, in the past decade?

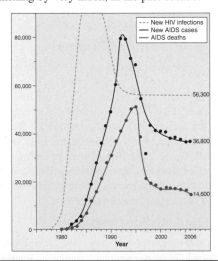

1. Your friend fell yesterday while she was skateboarding and has a deep cut on her forearm from a piece of wire. She shows it to you and complains about how sore it is. You can see the skin is red and swollen around the small puncture wound. Explain to her what is happening.
2. How is the part of an antibody molecule that interacts with an antigen similar to the active site of an enzyme? How is it different?
3. An untreated cut may lead to a surface swelling full of pus. What is pus and why does it accumulate at the cut site?
4. There are two ways to avoid getting chicken pox when you are older, by catching it when you're young or by receiving a vaccination. How do each of these methods work and are there any similarities?

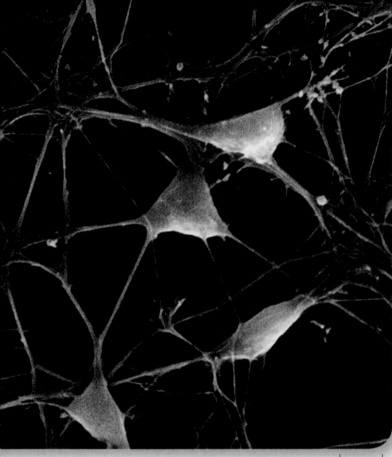

19.94 µm

28

The Nervous System

I n vertebrates, the central nervous system coordinates and regulates the diverse activities of the body, using a network of specialized cells called neurons to direct the voluntary muscles, and a second network not under voluntary control to direct cardiac and smooth muscles. All sensory information is acquired through depolarization of sensory nerve endings. From a knowledge of which neurons are sending signals, and how often they are doing so, the brain builds a picture of the body's internal condition and of the external environment. The network of neurons (nerve cells) seen here, magnified over a thousand times, is transmitting signals within the portion of the brain called the cerebral cortex. The vertebrate brain contains a staggering number of neurons—the human brain contains an estimated 100 billion. The cerebral cortex is a layer of gray matter only a few millimeters thick on the brain's outer surface. Densely packed with neurons and highly convoluted, it is the site of higher mental activities.

Evolution of the Animal Nervous System

An animal must be able to respond to environmental stimuli. To do this, it must have sensory receptors that can detect the stimulus and motor effectors that can respond to it. In most invertebrate phyla and in all vertebrate classes, sensory receptors and motor effectors are linked by way of the **nervous system.** As described in chapter 22, the nervous system consists of neurons (**figure 28.1**) and supporting cells. One type of neuron, called **association neurons** (or **interneurons**), is present in the nervous systems of most invertebrates and all vertebrates. These neurons are located in the brain and spinal cord of vertebrates, together called the **central nervous system (CNS),** represented by the yellow circle in **figure 28.2.**

They help provide more complex reflexes and in the case of the brain, higher associative functions, including learning and memory, which require integration of many sensory inputs.

There are two other types of neurons. **Sensory** (or **afferent**) **neurons** (❶ in **figure 28.1**) carry impulses from sensory receptors to the CNS. **Motor** (or **efferent**) **neurons** ❸ carry impulses away from the CNS to effectors—muscles and glands. The association neurons ❷ link these two types of neurons together in the CNS. Together, motor and sensory neurons constitute the **peripheral nervous system (PNS)** of vertebrates (the bracketed tan circles in **figure 28.2**).

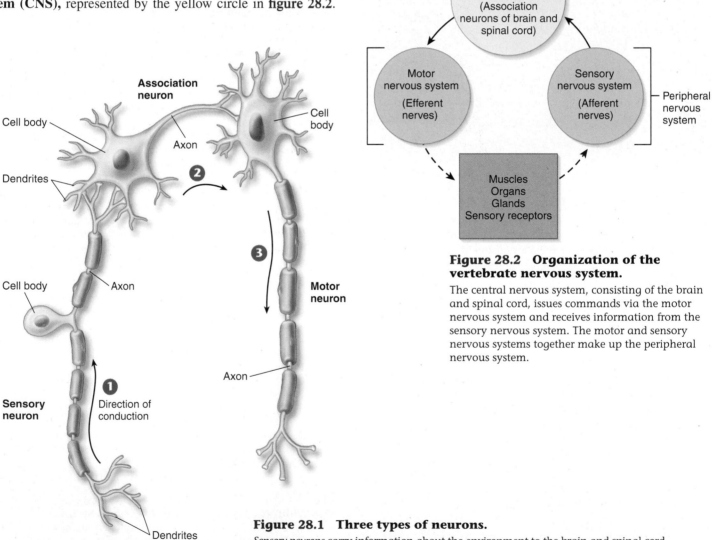

Figure 28.2   Organization of the vertebrate nervous system.

The central nervous system, consisting of the brain and spinal cord, issues commands via the motor nervous system and receives information from the sensory nervous system. The motor and sensory nervous systems together make up the peripheral nervous system.

Figure 28.1   Three types of neurons.

Sensory neurons carry information about the environment to the brain and spinal cord. *Association neurons* are found in the brain and spinal cord and often provide links between sensory and motor neurons. *Motor neurons* carry impulses to muscles and glands (effectors).

Invertebrate Nervous Systems

Sponges are the only major phylum of multicellular animals that lack nerves. If you prick a sponge, the nearby surface contracts slowly. The cytoplasm of each individual cell conducts an impulse that fades within a few millimeters. No messages dart from one part of the sponge body to another, as they do in all other multicellular animals.

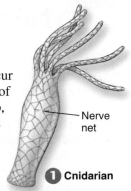

The Simplest Nervous Systems: Reflexes The simplest nervous systems occur among cnidarians, like the *Hydra* ❶. All neurons are similar, each having fibers of approximately equal length. Cnidarian neurons are linked to one another in a web, or *nerve net*, dispersed through the body. Although conduction is slow, a stimulus anywhere can eventually spread through the whole net. There is no associative activity, no control of complex actions, and little coordination. Any motion that results is called a **reflex** because it is an automatic consequence of the nerve stimulation.

Nerve net

❶ Cnidarian

More Complex Nervous Systems: Associative Activities The first associative activity in nervous systems is seen in the free-living flatworms ❷, phylum Platyhelminthes. Running down the bodies of these flatworms are two nerve cords, looking like the uprights of a ladder; peripheral nerves extend outward to the muscles of the body. The two nerve cords converge at the front end of the body, forming an enlarged mass of nervous tissue that also contains associative neurons that connect neurons to one another. This primitive "brain" is a rudimentary central nervous system and permits a far more complex control of muscular responses than is possible in cnidarians.

Associative neurons

Nerve cords

❷ Flatworm

The Evolutionary Path to the Vertebrates All of the subsequent evolutionary changes in nervous systems can be viewed as a series of elaborations on the characteristics already present in flatworms. Five trends can be identified, each becoming progressively more pronounced as nervous systems evolved greater complexity.

1. *More sophisticated sensory mechanisms.* Particularly among the vertebrates, sensory systems become highly complex.
2. *Differentiation into central and peripheral nervous systems.* For example, earthworms ❸ exhibit a central nervous system that is connected to all other parts of the body by peripheral nerves.
3. *Differentiation of sensory and motor nerves.* Neurons operating in particular directions become increasingly specialized (sensory signals traveling to the brain, or motor signals traveling from the brain).
4. *Increased complexity of association.* Central nervous systems with more numerous interneurons evolved, increasing association capabilities dramatically.
5. *Elaboration of the brain.* Coordination of body activities became increasingly localized in mollusks ❹, arthropods ❺, and vertebrates in the front end of the nerve cord, which evolved into the vertebrate brain discussed later in the chapter.

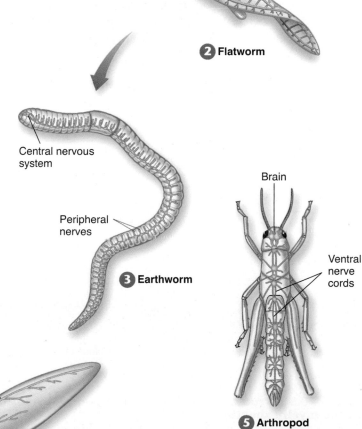

Central nervous system

Peripheral nerves

❸ Earthworm

Brain

Ventral nerve cords

❺ Arthropod

Giant axon

Brain

❹ Mollusk

Key Learning Outcome 28.1 As nervous systems evolved, there was a progressive increase in associative activity, increasingly localized in a brain.

Neurons Generate Nerve Impulses

Neurons

The basic structural unit of the nervous system, whether central, motor, or sensory, is the nerve cell, or **neuron.** All neurons have the same basic structure as you can see by comparing the three cell types in **figure 28.1** and the generalized cell in **figure 28.3a**. The **cell body** in **figure 28.3a** is the flat region of the neuron containing the nucleus. Short, slender branches called **dendrites** extend from one end of a neuron's cell body. Dendrites are input channels. Nerve impulses travel inward along them toward the cell body. Motor and association neurons possess a profusion of highly branched dendrites, enabling those cells to receive information from many different sources simultaneously. Projecting out from the other end of the cell body is a single, long, tubelike extension called an **axon.** Axons are output channels. Nerve impulses travel outward along them, away from the cell body, toward other neurons or to muscles or glands.

Most neurons are unable to survive alone for long; they require the nutritional support provided by companion **neuroglial cells.** More than half the volume of the human nervous system is composed of supporting neuroglial cells. Two of the most important kinds of neuroglial cells are **Schwann cells** and **oligodendrocytes,** which envelop the axons of many neurons with a sheath of fatty material called myelin that acts as an electrical insulator. Schwann cells produce myelin in the PNS, whereas oligodendrocytes produce myelin in the CNS. During development, these cells associate with the axon, as shown at the top in **figure 28.3b**, and begin to wrap themselves around the axon several times to form a **myelin sheath,** an insulating covering consisting of multiple layers of membrane. Axons that have myelin sheaths are said to be myelinated, and those that don't are unmyelinated. The myelin sheath is interrupted at intervals, leaving uninsulated gaps called **nodes of Ranvier** (the nodes are where the yellow underlying axon can be seen). At the node regions, the axon is in direct contact with the surrounding fluid. The nerve impulse jumps from node to node, speeding its travel down the axon. Multiple sclerosis (see chapter 27) and Tay-Sachs (see chapter 10) are debilitating clinical disorders, and, in the case of Tay-Sachs, fatal. They result from the degeneration of the myelin sheath.

The Nerve Impulse

When a neuron is "at rest," not carrying an impulse, active transport channels in the neuron's plasma membrane transport sodium ions (Na^+) out of the cell and potassium ions (K^+) in. This sodium-potassium pump was described in chapter 4. Sodium ions cannot easily move back into the cell once they are pumped out, so the concentration of sodium ions builds up outside the cell. Similarly, potassium ions accumulate inside the cell, although they are not as highly concentrated because many potassium ions are able to diffuse out through open channels. This resting phase is indicated by the yellow coloring in **panel 1** of the Key Biological Process illustration on the facing page. The result is to make the outside of the neuron more positive than the inside, a condition called the *resting membrane potential*. The resting plasma membrane is said to be "polarized."

Neurons are constantly expending energy to pump sodium ions out of the cell, in order to maintain the resting

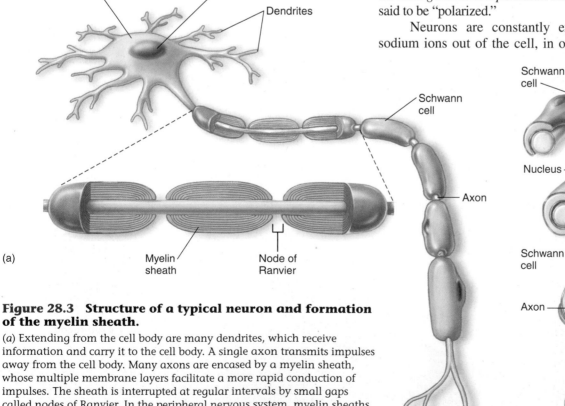

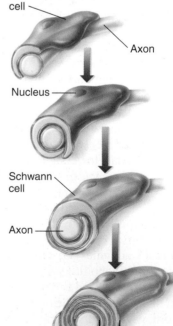

(a)

Cell body
Nucleus
Dendrites
Schwann cell
Axon
Myelin sheath
Node of Ranvier

Schwann cell
Axon
Nucleus
Schwann cell
Axon
(b)
Myelin sheath

Figure 28.3 Structure of a typical neuron and formation of the myelin sheath.

(a) Extending from the cell body are many dendrites, which receive information and carry it to the cell body. A single axon transmits impulses away from the cell body. Many axons are encased by a myelin sheath, whose multiple membrane layers facilitate a more rapid conduction of impulses. The sheath is interrupted at regular intervals by small gaps called nodes of Ranvier. In the peripheral nervous system, myelin sheaths are formed by supporting Schwann cells. (b) The myelin sheath is formed by successive wrappings of Schwann cell membranes around the axon.

membrane potential. The net negative charge of most proteins within the cell also adds to this charge difference. Using sophisticated instruments, scientists have been able to measure the voltage difference between the neuron interior and exterior as –70 millivolts (thousandth of a volt). The resting membrane potential is the starting point for a nerve impulse (**panel 1**).

A nerve impulse travels along the axon and dendrites as an electrical current caused by ions moving in and out of the neuron through **voltage-gated channels** (that is, protein channels in the neuron membrane that open and close in response to an electrical voltage). The impulse starts when pressure or other sensory inputs disturb a neuron's plasma membrane, causing sodium channels on a dendrite to open (the purple channels in **panel 2**). As a result, sodium ions flood into the neuron from outside, down their concentration gradient, and for a brief moment a localized area inside of the membrane is "depolarized," becoming more positive than the outside in that immediate area of the axon (indicated by the pink coloring in **panel 2**).

The sodium channels in the small patch of depolarized membrane remain open for only about a half a millisecond. However, if the change in voltage is large enough, it causes nearby voltage-gated sodium and potassium channels to open (**panel 3**). The sodium channels open first, which starts a wave of depolarization moving down the neuron. The opening of the gated channels causes nearby voltage-gated channels to open, like a chain of falling dominoes. This local reversal of voltage moving along the axon is called an **action potential.** An action potential follows an all-or-none law: A large enough depolarization produces either a full action potential or none at all because the voltage-gated Na^+ channels open completely or not at all. Once they open, an action potential occurs. After a slight delay, potassium voltage-gated channels open and K^+ flows out of the cells down its concentration gradient, making the inside of the cell more negative. The increasingly negative membrane potential (colored green in **panel 4**) causes the voltage-gated sodium channels to snap closed again. This period of time after the action potential has passed and before the resting membrane potential is restored is called the *refractory period.* A second action potential cannot fire during the refractory period, not until the resting potential is restored by the actions of the sodium-potassium pump.

The depolarization and restoration of the resting membrane potential takes only about 5 milliseconds. Fully 100 such cycles could occur, one after another, in the time it takes to say the word *nerve.*

Key Learning Outcome 28.2 Neurons are cells specialized to conduct impulses. Signals typically arrive along any of numerous dendrites, pass over the cell body's surface, and travel outward on a single long axon. Nerve impulses result from ion movements across the neuron plasma membrane through special protein channels that open and close in response to chemical or electrical stimulation.

KEY BIOLOGICAL PROCESS: The Nerve Impulse

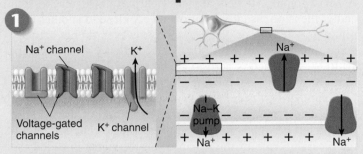

At the resting membrane potential, the inside of the axon is negatively charged because the sodium-potassium pump keeps a higher concentration of Na^+ outside. Voltage-gated ion channels are closed, but there is some leakage of K^+.

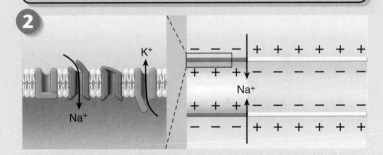

In response to a stimulus, the membrane depolarizes: Voltage-gated Na^+ channels open, Na^+ flows into the cell, and the inside becomes more positive.

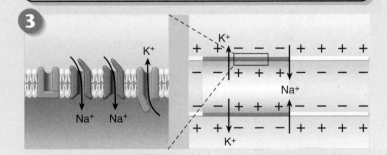

The local change in voltage opens adjacent voltage-gated Na^+ channels, and an action potential is produced.

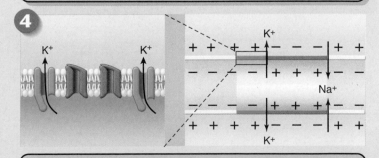

As the action potential travels farther down the axon, voltage-gated Na^+ channels close and K^+ channels open, allowing K^+ to flow out of the cell and restoring the negative charge inside the cell. Ultimately, the sodium-potassium pump restores the resting membrane potential.

28.3 | The Synapse

A nerve impulse travels along a neuron until it reaches the end of the axon, usually positioned very close to another neuron, a muscle cell, or gland. Axons, however, do not actually make direct contact with other cells. Instead, a narrow gap, 10 to 20 nanometers across, called the *synaptic cleft*, separates the axon tip and the target cell. This junction of an axon with another cell is called a **synapse**. A synapse is shown in **figure 28.4**. The membrane on the axon side of the synapse (on the left here) belongs to the **presynaptic cell**; the cell on the receiving side of the synapse (on the right) is called the **postsynaptic cell.**

Neurotransmitters

When a nerve impulse reaches the end of an axon, its message must cross the synapse if it is to continue. Messages do not "jump" across synapses. Instead, they are carried across by chemical messengers called **neurotransmitters.** These chemicals are packaged in tiny sacs, or vesicles, at the tip of the axon. When a nerve impulse arrives at the tip, it causes the sacs to release their contents into the synapse, as shown in **figure 28.5a.** The neurotransmitters diffuse across the synaptic cleft and bind to receptors (the purple structures) in the postsynaptic membrane. The signal passes to the postsynaptic cell when the binding of the neurotransmitter opens special ion channels, allowing ions to enter the postsynaptic cell and cause a change in electrical charge across its membrane. The enlarged view of **figure 28.5a** shows how the channel opens and the ion (the yellow ball) enters the cell. Because these channels open when stimulated by a chemical, they are said to be *chemically gated.*

Why go to all this trouble? Why not just wire the neurons directly together? For the same reason that the wires of your house are not all connected but instead are separated by a host of switches. When you turn on one light switch, you

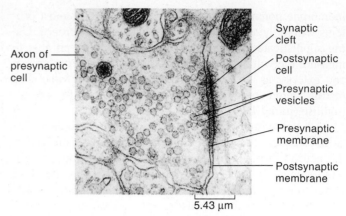

Figure 28.4 A synapse between two neurons.
This micrograph clearly shows the space between the presynaptic and postsynaptic membranes, which is called the synaptic cleft.

don't want every light in the house to go on, the toaster to start heating, and the television to come on! If every neuron in your body were connected to every other neuron, it would be impossible to move your hand without moving every other part of your body at the same time. Synapses are the control switches of the nervous system. However, the control switch must be turned off at some point by getting rid of the neurotransmitter, or the postsynaptic cell would keep firing action potentials. In some cases, the neurotransmitter molecules diffuse away from the synapse. In other cases, the neurotransmitter molecules are either reabsorbed by the presynaptic cell, or are degraded in the synaptic cleft.

Kinds of Synapses

The vertebrate nervous system uses dozens of different kinds of neurotransmitters, each recognized by specific receptors on receiving cells. They fall into two classes, depending on whether they excite or inhibit the postsynaptic cell.

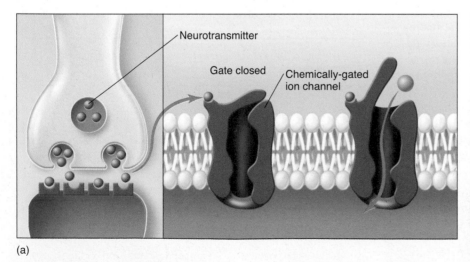

(a)

Figure 28.5   Events at the synapse.
(a) When a nerve impulse reaches the end of an axon, it releases a neurotransmitter into the synaptic cleft. The neurotransmitter molecules diffuse across the synapse and bind to receptors on the postsynaptic cell, opening ion channels. (b) A transmission electron micrograph of the tip of an axon filled with synaptic vesicles.

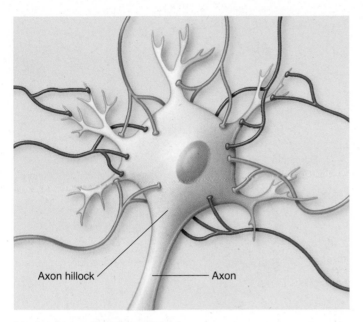

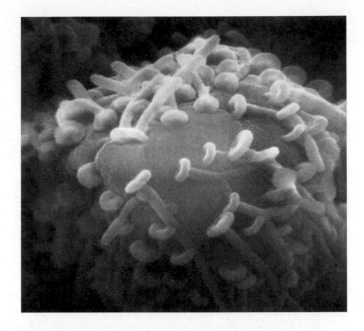

Figure 28.6 Integration.
Many different axons synapse with the cell body and dendrites of the postsynaptic neuron illustrated here. Excitatory synapses are shown in *red* and inhibitory synapses are shown in *blue*. The summed influence of their input at the axon hillock determines whether or not a nerve impulse will be sent down the axon extending below. The electron micrograph shows a neuronal cell body with numerous synapses.

In an *excitatory synapse,* the receptor protein is usually a chemically gated sodium channel, meaning that a sodium channel through the membrane is opened by the neurotransmitter. On binding with a neurotransmitter whose shape fits it, the sodium channel opens, allowing sodium ions to flood inward. If enough sodium ion channels are opened by neurotransmitters, an action potential begins.

In an *inhibitory synapse,* the receptor protein is a chemically gated potassium or chloride channel. Binding with its neurotransmitter opens these channels, leading to the exit of positively charged potassium ions or the influx of negatively charged chloride ions, resulting in a more negative interior in the receiving cell. This inhibits the start of an action potential, because the negative voltage change inside means that even more sodium ion channels must be opened to get a domino effect started among voltage-gated sodium channels, and so start an action potential.

An individual nerve cell, like the neuron in **figure 28.6,** can possess both kinds of synaptic connections to other nerve cells. In the drawing, the excitatory synapses are colored red and the inhibitory synapses are colored blue. When signals from both excitatory and inhibitory synapses reach the cell body of a neuron, the excitatory effects (which cause less internal negative charge) and the inhibitory effects (which cause more internal negative charge) interact with one another. The result is a process of **integration** in which the various excitatory and inhibitory electrical effects tend to cancel or reinforce one another. An area at the base of the axon, called the **axon hillock,** is the site of this integration process. If the result of the integration is a large enough depolarization (that is, the inside of the cell becomes more positive), an action potential will fire. Neurons often receive many inputs. A single

motor neuron in the spinal cord may have as many as 50,000 synapses on it!

Types of Neurotransmitters

Acetylcholine (ACh) is the neurotransmitter released at the neuromuscular junction, the synapse that forms between a neuron and a muscle fiber. ACh forms an excitatory synapse with skeletal muscle but has the opposite effect on cardiac muscle, causing an inhibitory synapse.

Glycine and *GABA* are inhibitory neurotransmitters. This inhibitory effect is very important for neural control of body movements and other brain functions. Interestingly, the drug diazepam (Valium) causes its sedative and other effects by enhancing the binding of GABA to its receptors.

Biogenic amines are a group of neurotransmitters that include *dopamine, norepinephrine, serotonin,* and the hormone *epinephrine.* These neurotransmitters have various effects on the body: Dopamine is important in controlling body movements; norepinephrine and epinephrine are involved in the autonomic nervous system, which will be discussed later; and serotonin is involved in sleep regulation and other emotional states. The drug PCP (angel dust) elicits its actions by blocking the elimination of biogenic amines from the synapse. The symptoms vary depending on the dosage.

Key Learning Outcome 28.3 A synapse is the junction of an axon with another cell. The cells are separated by a gap across which neurotransmitters carry a signal that has either an excitatory or inhibitory effect, depending on which ion channels are opened.

Addictive Drugs Act on Chemical Synapses

Neuromodulators

The body sometimes deliberately prolongs the transmission of a signal across a synapse. It does this by releasing into the synapse special, long-lasting chemicals called **neuromodulators.** Some neuromodulators aid the release of neurotransmitters into the synapse; others inhibit the reabsorption of neurotransmitters so that they remain in the synapse; still others delay the breakdown of neurotransmitters after their reabsorption, leaving them in the tip of the neuron to be released back into the synapse when the next signal arrives.

Mood, pleasure, pain, and other mental states are determined by particular groups of neurons in the brain that use special sets of neurotransmitters and neuromodulators. Mood, for example, is strongly influenced by the neurotransmitter serotonin. Many researchers think that depression results from a shortage of serotonin. Prozac, the world's bestselling antidepressant, inhibits the reabsorption of serotonin, thus increasing the amount in the synapse. The synapse in **figure 28.7** illustrates the effects of Prozac. The red serotonin molecules released into the synapse are usually reabsorbed by the presynaptic cell. As the circled enlargement shows, Prozac inhibits this reabsorption, leaving serotonin in the synapse.

Drug Addiction

When a cell of the body is exposed to a chemical signal for a prolonged period, it tends to lose its ability to respond to the stimulus with its original intensity. (You are familiar with this loss of sensitivity—when you sit in a chair, how long are you aware of the chair?) Nerve cells are particularly prone to this loss of sensitivity. If receptor proteins within synapses are exposed to high levels of neurotransmitter molecules for prolonged periods, that nerve cell often responds by inserting fewer receptor proteins into the membrane. This feedback is a normal part of the functioning of all neurons, a simple mechanism that has evolved to make the cell more efficient by adjusting the number of "tools" (receptor proteins) in the membrane "workshop" to suit the workload.

Cocaine The drug cocaine is a neuromodulator that causes abnormally large amounts of neurotransmitters to remain in the synapses for long periods of time. Cocaine affects nerve cells in the brain's pleasure pathways (the so-called limbic system, an area of the brain discussed in section 28.6). These cells transmit pleasure messages using the neurotransmitter dopamine. **Panel 1** of the Key Biological Process illustration on the facing page shows normal activity at the synapse, where dopamine molecules (the red balls) are reabsorbed by transporters in the presynaptic cell. Using radioactively labeled cocaine molecules, investigators found that cocaine binds tightly to the transporter proteins on presynaptic membranes (**panel 2**). These proteins normally remove

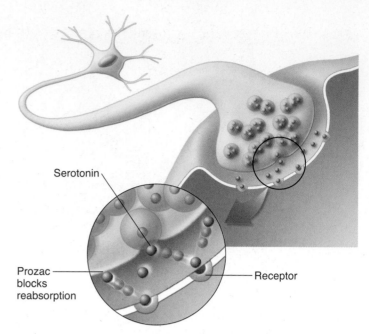

Figure 28.7 Drugs alter transmission of impulses across the synapse.

Depression can result from a shortage of the neurotransmitter serotonin. The antidepressant drug Prozac works by blocking reabsorption of serotonin, keeping serotonin in the synapse longer.

the neurotransmitter dopamine after it has acted. Like a game of musical chairs in which all the chairs become occupied, eventually there are no unoccupied transporter proteins available to the dopamine molecules, so the dopamine stays in the synapse, firing the receptors again and again. As new signals arrive, more and more dopamine is added, firing the pleasure pathway more and more often.

When receptor proteins on limbic system nerve cells are exposed to high levels of dopamine neurotransmitter molecules for prolonged periods of time, the nerve cells "turn down the volume" of the signal by lowering the number of receptor proteins on their surfaces (**panel 3**). They respond to the greater number of neurotransmitter molecules by simply reducing the number of targets available for these molecules to hit. The cocaine user is now addicted. **Addiction** occurs when chronic exposure to a drug induces the nervous system to adapt physiologically. With so few receptors, as in **panel 4**, normal levels of dopamine are not able to trigger an action potential in the postsynaptic cell and so the user needs the drug to maintain even normal levels of limbic activity.

Is Addiction to Smoking Cigarettes Drug Addiction?

Investigators attempting to explore the habit-forming nature of smoking cigarettes used what had been learned about cocaine to carry out what seemed a reasonable experiment—they introduced radioactively labeled nicotine from tobacco into the brain and looked to see what sort of transporter protein it attached itself to. To their great surprise, the nicotine ignored proteins on the presynaptic membrane and instead bound directly to a specific receptor on the postsynaptic cell! This was

KEY BIOLOGICAL PROCESS: Drug Addiction

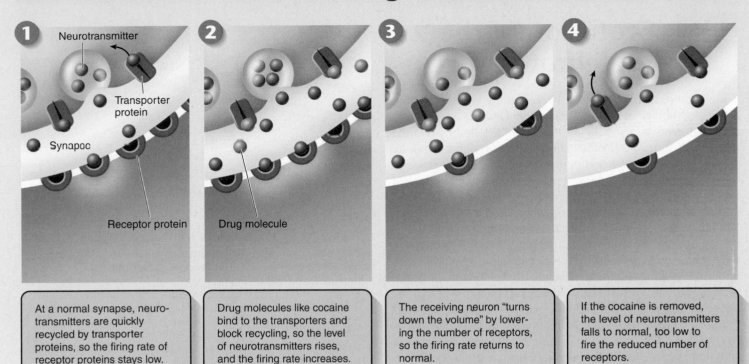

1 At a normal synapse, neurotransmitters are quickly recycled by transporter proteins, so the firing rate of receptor proteins stays low.

2 Drug molecules like cocaine bind to the transporters and block recycling, so the level of neurotransmitters rises, and the firing rate increases.

3 The receiving neuron "turns down the volume" by lowering the number of receptors, so the firing rate returns to normal.

4 If the cocaine is removed, the level of neurotransmitters falls to normal, too low to fire the reduced number of receptors.

totally unexpected, as nicotine does not normally occur in the brain—why should it have a receptor there?

Intensive research followed, and researchers soon learned that the "nicotine receptors" normally served to bind the neurotransmitter acetylcholine. It was just an accident of nature that nicotine, an obscure chemical from a tobacco plant, was also able to bind to them. What, then, is the normal function of these receptors? The target of considerable research, these receptors turned out to be one of the brain's most important tools. The brain uses them to coordinate the activities of many other kinds of receptors, acting to "fine-tune" the sensitivity of a wide variety of behaviors.

When neurobiologists compare the limbic system nerve cells of smokers to those of nonsmokers, they find changes in both the number of nicotine receptors and in the levels of RNA used to make the receptors. They have found that the brain adjusts to prolonged exposure to nicotine by "turning down the volume" in two ways: (1) by making fewer receptor proteins to which nicotine can bind; and (2) by altering the pattern of activation of the nicotine receptors (that is, their sensitivity to neurotransmitters).

It is this second adjustment that is responsible for the profound effect smoking has on the brain's activities. By overriding the normal system the brain uses to coordinate its many activities, nicotine alters the pattern of release of many neurotransmitters into synaptic clefts, including acetylcholine, dopamine, serotonin, and many others. As a result, changes in level of activity occur in a wide variety of nerve pathways within the brain.

Addiction to nicotine occurs because the brain compensates for the many changes nicotine induces by making other changes. Adjustments are made to the numbers and sensitivities of many kinds of receptors within the brain, restoring an appropriate balance of activity. Now what happens if you stop smoking? Everything is out of whack! The newly coordinated system *requires* nicotine to achieve an appropriate balance of nerve pathway activities. This is addiction in any sensible use of the term. The body's physiological response is profound and unavoidable. There is no way to prevent addiction to nicotine with willpower, any more than willpower can stop a bullet when playing Russian roulette with a loaded gun. If you smoke cigarettes for a prolonged period, you will become addicted.

What do you do if you are addicted to smoking cigarettes and you want to stop? When use of an addictive drug like nicotine is stopped, the level of signaling changes to levels far from normal. If the drug is not reintroduced, the altered level of signaling eventually induces the nerve cells to once again make compensatory changes that restore an appropriate balance of activities within the brain. Over time, receptor numbers, their sensitivity, and patterns of release of neurotransmitters all revert to normal, once again producing normal levels of signaling along the pathways.

Key Learning Outcome 28.4 Neuromodulators are long-lasting chemicals that act on synapses to alter nerve function. Many addictive drugs such as cocaine and nicotine act as neuromodulators.

28.5 Evolution of the Vertebrate Brain

The structure and function of the vertebrate brain have long been the subject of scientific inquiry. Despite ongoing research, scientists are still not sure how the brain performs many of its functions. For instance, scientists continue to look for the mechanism the brain employs to store memories, and they do not understand how some memories can be "locked away," only to surface in times of stress. The brain is the most complex vertebrate organ ever to evolve, and it can perform a bewildering variety of complex functions (**figure 28.8**).

Casts of the interior braincases of fossil agnathans, fishes that swam 500 million years ago, have revealed much about the early evolutionary stages of the vertebrate brain. Although small, these brains already had the three divisions, shown in **figure 28.9**, that characterize the brains of all contemporary vertebrates: (1) the hindbrain, or rhombencephalon (colored yellow); (2) the midbrain, or mesencephalon (colored green); and (3) the forebrain, or prosencephalon (colored blue and purple).

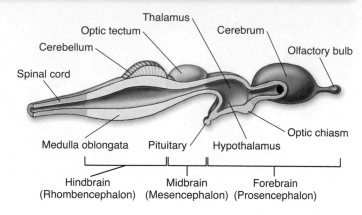

Figure 28.9   The brain of a primitive fish.

The basic organization of the vertebrate brain can be seen in the brains of primitive fishes. The brain is divided into three regions that are found in differing proportions in all vertebrates: the hindbrain, which is the largest portion of the brain in fishes; the midbrain, which in fishes is devoted primarily to processing visual information; and the forebrain, which is concerned mainly with olfaction (the sense of smell) in fishes. In terrestrial vertebrates, the forebrain plays a far more dominant role in neural processing than it does in fishes.

Figure 28.8   Singing well takes practice.

This baby coyote is greeting the approaching evening. His howling is not as impressive as his dad's—a good performance takes practice. His brain is learning by repetition how to control the vocal cords.

The hindbrain was the major component of these early brains, as it still is in fishes today. Composed of the *cerebellum* and *medulla oblongata,* the fish hindbrain may be considered an extension of the spinal cord devoted primarily to coordinating motor reflexes. Tracts containing large numbers of axons run like cables up and down the spinal cord to the hindbrain. The hindbrain integrates the many sensory signals coming from the muscles and coordinates the pattern of motor responses.

Much of this coordination is carried on within a small extension of the hindbrain called the cerebellum ("little cerebrum"). In more advanced vertebrates, the cerebellum plays an increasingly important role as a coordinating center and is correspondingly larger than it is in the fishes. In all vertebrates, the cerebellum processes data on the current position and movement of each limb, the state of relaxation or contraction of the muscles involved, and the general position of the body and its relation to the outside world. These data are gathered in the cerebellum and synthesized, and the resulting commands are issued to efferent pathways.

In fishes, the remainder of the brain is devoted to the reception and processing of sensory information. The midbrain is composed primarily of the **optic lobes** (also called the optic tectum), which receive and process visual information, while the forebrain is devoted to the processing of *olfactory* (smell) information. The brains of fishes continue growing throughout their lives. This continued growth is in marked contrast to the brains of other classes of vertebrates, which generally complete their development by infancy. The human brain continues to develop through early childhood, but no new neurons are produced once development has ceased, except in the hippocampus, involved in long-term memory.

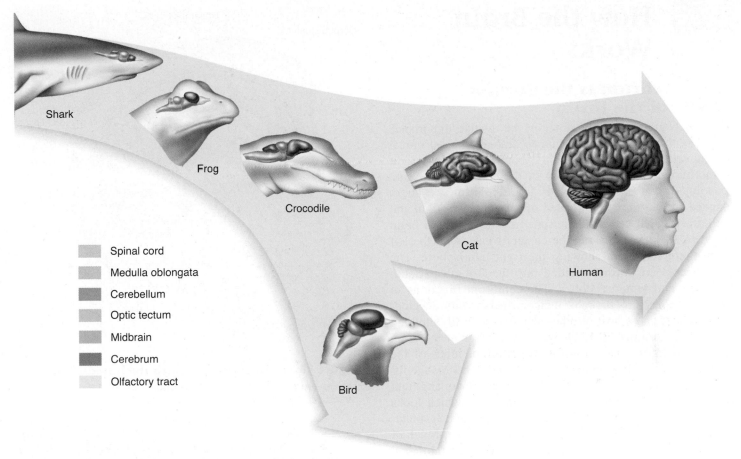

Figure 28.10 The evolution of the vertebrate brain.
In sharks and other fishes, the hindbrain is predominant, and the rest of the brain serves primarily to process sensory information. In amphibians and reptiles, the forebrain is far larger, and it contains a larger cerebrum devoted to associative activity. In birds, which evolved from reptiles, the cerebrum is even more pronounced. In mammals, the cerebrum covers the optic tectum and is the largest portion of the brain. The dominance of the cerebrum is greatest in humans, where it envelops much of the rest of the brain.

The Dominant Forebrain

Starting with the amphibians and continuing more prominently in the reptiles, sensory information is increasingly centered in the forebrain. This pattern was the dominant evolutionary trend in the further development of the vertebrate brain. The areas of the brain are color-coded in **figure 28.10** so that you can follow this trend in brain development, with the cerebrum of the forebrain becoming larger in mammals.

The forebrain in reptiles, amphibians, birds, and mammals is composed of two elements: the diencephalon and the telencephalon. The *diencephalon* consists of the thalamus and hypothalamus (colored blue in **figure 28.9**). The **thalamus** is an integrating and relay center between incoming sensory information and the cerebrum. The **hypothalamus** participates in basic drives and emotions and controls the secretions of the pituitary gland, which in turn regulates many of the other endocrine glands of the body (see chapter 30). Through its connections with the nervous system and endocrine system, the hypothalamus helps coordinate the neural and hormonal responses to many internal stimuli and emotions. The *telencephalon*, or "end brain," is located at the front of the forebrain and is devoted largely to associative activity (colored purple

in **figures 28.9** and **28.10**). In mammals, the telencephalon is called the **cerebrum**.

The Expansion of the Cerebrum

If you examine the relationship between brain size and body size in the animals pictured in **figure 28.10**, you can see a remarkable difference between fishes, amphibians, and reptiles (to the left of the branch point in the figure), and birds and mammals (to the right of the branch point in the figure). Mammals in particular have brains that are particularly large relative to their body size. This is especially true of porpoises and humans. The increase in brain size in mammals largely reflects the great enlargement of the cerebrum, the dominant part of the mammalian brain. The cerebrum is the center for correlation, association, and learning in the mammalian brain. It receives sensory data from throughout the body and issues motor commands to the body.

Key Learning Outcome 28.5 In fishes, the hindbrain forms much of the brain; as terrestrial vertebrates evolved, the forebrain became increasingly more prominent.

How the Brain Works

The Cerebrum Is the Control Center of the Brain

Although vertebrate brains differ in the relative importance of different components, the human brain is a good model of how vertebrate brains function. About 85% of the weight of the human brain is made up of the cerebrum, the tan, convoluted area in **figure 28.11**. The cerebrum is a large, rounded area of the brain divided by a groove into right and left halves called cerebral hemispheres. The sectioned brain in **figure 28.11** is cut along the center groove, with the left hemisphere removed, showing the right hemisphere. The hemispheres are further divided into the frontal, parietal, occipital, and temporal lobes. The cerebrum functions in language, conscious thought, memory, personality development, vision, and a host of other activities we call "thinking and feeling." **Figure 28.12** shows general areas of the brain and the functions they control (the different lobes of the brain are color-coded: yellow for the frontal lobe, orange for the parietal lobe, light green for the occipital lobe, and light purple for the temporal lobe). The cerebrum, which looks like a wrinkled mushroom, is positioned over and surrounding the rest of the brain, like a hand holding a fist. Much of the neural activity of the cerebrum occurs within a thin, gray outer layer only a few millimeters thick called the **cerebral cortex** (*cortex* is Latin for "bark of a tree"). This layer is gray because it is densely packed with neuron cell bodies. The human cerebral cortex contains the cell bodies of more than 10 billion nerve cells, roughly 10% of all the neurons in the brain. The wrinkles in the surface of the cerebral cortex increase its surface area (and number of cell bodies) three-fold. Underneath the cortex is a solid white region of myelinated nerve fibers that shuttle information between the cortex and the rest of the brain.

The right and left cerebral hemispheres are linked by bundles of neurons called **tracts.** These tracts serve as information highways, telling each half of the brain what the other half is doing. Because these tracts cross over, in the area of the brain called the *corpus callosum* (the blue-colored band in **figure 28.11**), each half of the brain controls muscles and glands on the opposite side of the body. Therefore, a touch on the right hand is relayed primarily to the left hemisphere, which may then initiate movement of the right hand in response to the touch.

Researchers have found that the two sides of the cerebrum can operate as two different brains. For instance, in some people the tract between the two hemispheres has been cut by accident or surgery. In laboratory experiments, one eye of an individual with such a "split brain" is covered and a stranger is introduced. If the other eye is then covered instead, the person does not recognize the stranger who was just introduced!

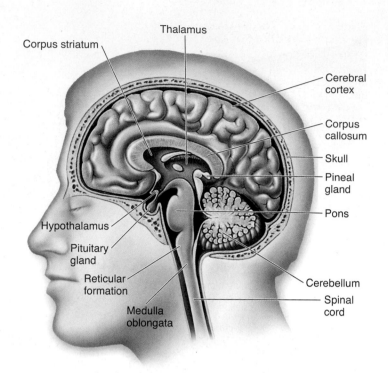

Figure 28.11 A section through the human brain.
The cerebrum occupies most of the brain. Only its outer layer, the cerebral cortex, is visible on the surface.

Sometimes blood vessels in the brain are blocked by blood clots, causing a disorder called a **stroke.** During a stroke, circulation to an area in the brain is blocked and the brain tissue dies. A severe stroke in one side of the cerebrum may cause paralysis of the other side of the body.

The Thalamus and Hypothalamus Process Information

Beneath the cerebrum are the thalamus and hypothalamus, important centers for information processing. The thalamus is the major site of sensory processing in the brain. Auditory (sound), visual, and other information from sensory receptors enter the thalamus and then are passed to the sensory areas of the cerebral cortex. The thalamus also controls balance. Information about posture, derived from the muscles, and information about orientation, derived from sensors within the ear, combine with information from the cerebellum and pass to the thalamus. The thalamus processes the information and channels it to the appropriate motor center on the cerebral cortex.

The hypothalamus integrates all the internal activities of the body. It controls centers in the brain stem that in turn regulate body temperature, blood pressure, respiration, and heartbeat. It also directs the secretions of the brain's major hormone-producing gland, the pituitary gland. The hypothalamus is linked by an extensive network of neurons to some areas of the cerebral cortex. This network, along with parts of the hypothalamus and areas of the brain called the *hippocampus* and *amygdala,* make up the **limbic system.** The areas highlighted

**Figure 28.12 The major
functional regions of the
human brain.**
Specific areas of the cerebral cortex
are associated with different regions
and functions of the body.

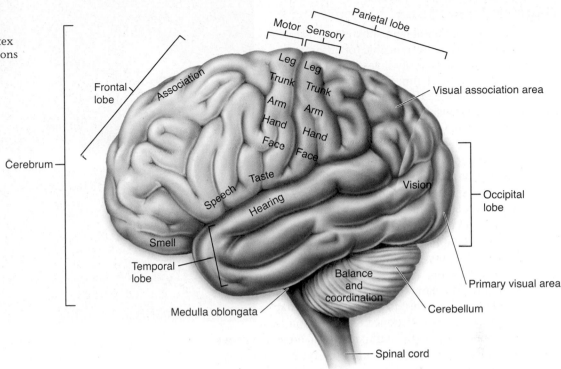

in green in **figure 28.13** indicate the components of the limbic system. The operations of the limbic system are responsible for many of the most deep-seated drives and emotions of vertebrates, including pain, anger, sex, hunger, thirst, and pleasure, centered in the amygdala. You'll recall on page 592 that the limbic system is the area of the brain affected by cocaine. It is also involved in memory, centered in the hippocampus.

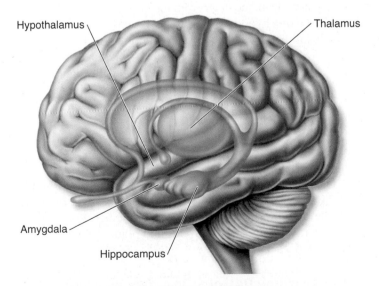

Figure 28.13 The limbic system.
The hippocampus and the amygdala are the major components of the limbic system, which controls our most deep-seated drives and emotions.

The Cerebellum Coordinates Muscle Movements

Extending back from the base of the brain is a structure known as the **cerebellum.** The cerebellum controls balance, posture, and muscular coordination. This small, cauliflower-shaped structure, while well developed in humans and other mammals, is even better developed in birds. Birds perform more complicated feats of balance than we do, because they move through the air in three dimensions. Imagine the kind of balance and coordination needed for a bird to land on a branch, stopping at precisely the right moment without crashing into it.

The Brain Stem Controls Vital Body Processes

The **brain stem,** a term used to collectively refer to the midbrain, pons, and medulla oblongata, connects the rest of the brain to the spinal cord. This stalklike structure contains nerves that control your breathing, swallowing, and digestive processes, as well as the beating of your heart and the diameter of your blood vessels. A network of nerves called the **reticular formation** runs through the brain stem and connects to other parts of the brain. Their widespread connections make these nerves essential to consciousness, awareness, and sleep. One part of the reticular formation filters sensory input, enabling you to sleep through repetitive noises such as traffic yet awaken instantly when a telephone rings.

Language and Other Higher Functions

Although the two cerebral hemispheres seem structurally similar, they are responsible for different activities. The most thoroughly investigated example of this lateralization of function is language. The left hemisphere is the "dominant" hemisphere for language—the hemisphere in which most neural processing related to language is performed—in 90% of right-handed people and nearly two-thirds of left-handed people. There are two language areas in the dominant hemisphere: One is important for language comprehension and the formulation of thoughts into speech, and the other is responsible for the generation of motor output needed for language communication. Different language activities in **figure 28.14** confirm that different parts of the brain are involved.

While the dominant hemisphere for language is adept at sequential reasoning, like that needed to formulate a sentence, the nondominant hemisphere (the right hemisphere in most people) is adept at spatial reasoning, the type of reasoning needed to assemble a puzzle or draw a picture. It is also the hemisphere primarily involved in musical ability— a person with damage to the speech area in the left hemisphere may not be able to speak but may retain the ability to sing! Damage to the nondominant hemisphere may lead to an inability to appreciate spatial relationships and may impair musical activities such as singing. Reading, writing, and oral comprehension remain normal. The nondominant hemisphere is also important for the consolidation of memories of nonverbal experiences.

One of the great mysteries of the brain is the basis of memory and learning. There is no one part of the brain in which all aspects of a memory appear to reside. Although memory is impaired if portions of the brain, particularly the temporal lobes, are removed, it is not lost entirely. Many memories persist in spite of the damage, and the ability to access them gradually recovers with time. Therefore, investigators who have tried to probe the physical mechanisms underlying memory often have felt that they were grasping at a shadow. Although we still do not have complete understanding, we have learned a good deal about the basic processes in which memories are formed.

There appear to be fundamental differences between short-term and long-term memory. Short-term memory is transient, lasting only a few moments. Such memories can readily be erased by the application of an electrical shock, leaving previously stored long-term memories intact. This result suggests that short-term memories are stored electrically in the form of a transient neural excitation. Long-term memory, in contrast, appears to involve structural changes in certain neural connections within the brain. Two parts of the temporal lobes, the hippocampus and the amygdala, are involved in both short-term memory and its consolidation into long-term memory.

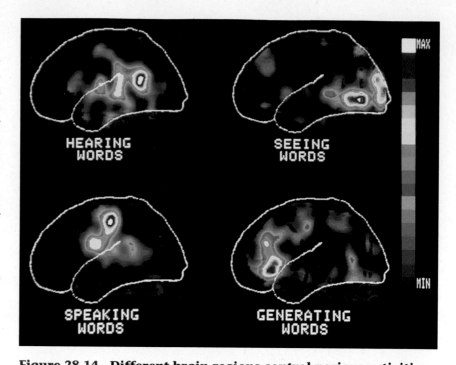

Figure 28.14 Different brain regions control various activities.
This illustration shows how the brain reacts in human subjects asked to listen to a spoken word, to read that same word silently, to repeat the word out loud, and then to speak a word related to the first. Regions of white, red, and yellow show the greatest activity. Compare this to figure 28.12 to see how regions of the brain are mapped.

The Mechanism of Alzheimer Disease Still a Mystery

In the past, little was known about Alzheimer disease, a condition in which the memory and thought processes of the brain become dysfunctional. Scientists disagree about the biological nature of the disease and its cause. Two hypotheses have been proposed: One suggests that nerve cells in the brain are killed from the outside in, the other from the inside out.

In the first hypothesis, external proteins called β-amyloid peptides kill nerve cells. A mistake in protein processing produces an abnormal form of the peptide, which then forms aggregates, or plaques. The plaques begin to fill in the brain, which damages and kills nerve cells. However, recent clinical trails of a drug that inhibits β-amyloid synthesis worsened rather than cured Alzheimer disease.

The second hypothesis maintains that the nerve cells are killed by an abnormal form of an internal protein. This protein, called tau (τ), normally functions to maintain protein transport microtubules. Abnormal forms of τ assemble into helical segments that form tangles, which interfere with the normal functioning of the nerve cells. Researchers continue to study whether tangles and plaques are causes or effects of Alzheimer disease.

> **Key Learning Outcome 28.6** The associative activity of the brain is centered in the cerebral cortex, which lies over the cerebrum. Beneath, the thalamus and hypothalamus process information and integrate body activities. The cerebellum coordinates muscle movements.

The Spinal Cord

The **spinal cord** is a cable of neurons extending from the brain down through the backbone, which is the view in **figure 28.15**. The cross section through the spinal cord in **figure 28.16** shows a darker gray area in the center that consists of neuron cell bodies, which form a column down the length of the cord. This column is surrounded by a sheath of axons and dendrites, which make the outer edges of the cord white because they are coated with myelin. The spinal cord is surrounded and protected by a series of bones called the vertebrae. Spinal nerves pass out to the body from between the vertebrae. Messages between the body and the brain run up and down the spinal cord, like an information highway.

In each segment of the spine, motor nerves extend out of the spinal cord to the muscles. Motor nerves from the spine control most of the muscles below the head. This is why injuries to the spinal cord often paralyze the lower part of the body. A muscle is paralyzed and cannot move if its motor neurons are damaged.

Spinal Cord Regeneration

In the past, scientists have tried to repair severed spinal cords by installing nerves from another part of the body to bridge the gap and act as guides for the spinal cord to regenerate. But most of these experiments have failed because the nerve bridges did not go from white matter to gray matter. Also, there is a factor that inhibits nerve growth in the spinal cord. After discovering that fibroblast growth factor stimulates nerve growth, neurobiologists tried gluing on the nerves, from white to gray matter, with fibrin that had been mixed with the fibroblast growth factor.

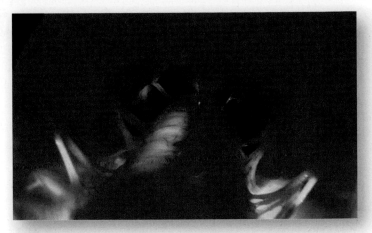

Figure 28.15   A view down the human spinal cord.
Pairs of spinal nerves can be seen extending out from the spinal cord. Along these nerves, the brain and spinal cord communicate with the body.

Three months later, rats with the nerve bridges began to show movement in their lower bodies. In further analyses of the experimental animals, dye tests indicated that the spinal cord nerves had regrown from both sides of the gap. Many scientists are encouraged by the potential to use a similar treatment in human medicine. However, most spinal cord injuries in humans do not involve a completely severed spinal cord; often, nerves are crushed. Also, although the rats with nerve bridges did regain some locomotory ability, tests indicated that they were barely able to walk or stand.

> **Key Learning Outcome 28.7** The spinal cord, protected in vertebrates by a backbone, extends motor nerves to the muscles below the head.

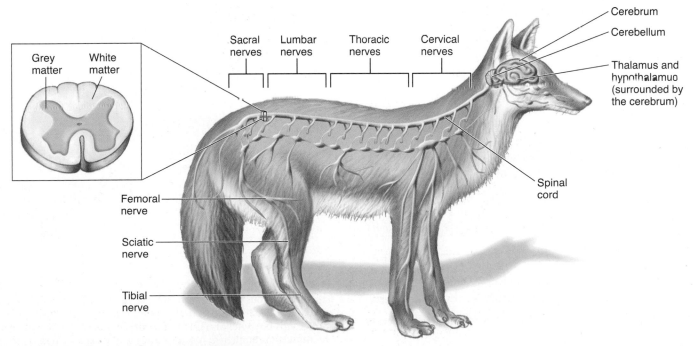

Figure 28.16   The vertebrate nervous system.
The brain is colored *tan* and the spinal cord and nerves are colored *yellow*.

28.8 Voluntary and Autonomic Nervous Systems

As you learned in the opening discussion in section 28.1, the nervous system is divided into two main parts: the central nervous system (the pink boxes in **figure 28.17**, which include the brain and spinal cord) and the peripheral nervous system (the blue boxes, which include the motor and sensory pathways). The motor pathways of the peripheral nervous system of a vertebrate can be further subdivided into the **somatic (voluntary) nervous system,** which relays commands to skeletal muscles, and the **autonomic (involuntary) nervous system,** which stimulates glands and relays commands to the smooth muscles of the body and to cardiac muscle. The voluntary nervous system can be controlled by conscious thought. You can, for example, command your hand to move. The autonomic nervous system, by contrast, cannot be controlled by conscious thought. You cannot, for example, tell the smooth muscles in your digestive tract to speed up their action. The central nervous system issues commands over both voluntary and autonomic systems, but you are conscious of only the voluntary commands.

Voluntary Nervous System

Motor neurons of the voluntary nervous system stimulate skeletal muscles to contract in two ways. First, motor neurons may stimulate the skeletal muscles of the body to contract in response to conscious commands. For example, if you want to bounce a basketball, your CNS sends messages through motor neurons to the muscles in your arms and hands. However, skeletal muscle can also be stimulated as part of reflexes that do not require conscious control.

Reflexes Enable Quick Action The motor neurons of the body have been wired to enable the body to act particularly quickly in time of danger—even before the animal is consciously aware of the threat. These sudden, involuntary movements are called reflexes. A **reflex** produces a rapid motor response to a stimulus because the sensory neuron bringing information about the threat passes the information directly to a motor neuron. The escape reaction of a fly about to be swatted is a reflex. One of the most frequently used reflexes in your body is blinking, a reflex that protects your eyes. If anything approaches your eye, such as an insect or a cloud of dust, the eyelid blinks closed even before you realize what has happened. The reflex occurs before the cerebrum is aware the eye is in danger.

Because they involve passing information between few neurons, reflexes are very fast. Many reflexes never reach the brain. The "danger" nerve impulse travels only as far as the

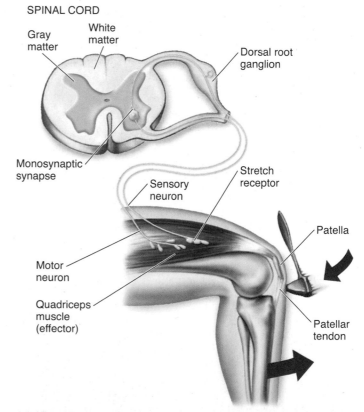

Figure 28.18 The knee-jerk reflex.

The most famous involuntary response, the knee jerk, is produced by activating stretch receptors in the quadriceps muscle. When a rubber mallet taps the patellar tendon, the muscle and stretch receptors in the muscle are stretched. A signal travels up a sensory neuron to the spinal cord, where the sensory neuron stimulates a motor neuron, which sends a signal to the quadriceps muscle to contract.

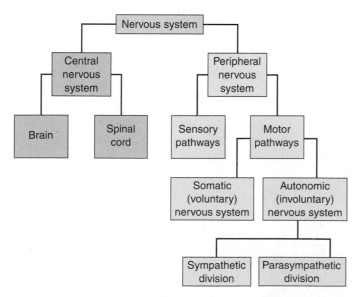

Figure 28.17 The divisions of the vertebrate nervous system.

The motor pathways of the peripheral nervous system are the somatic (voluntary) and autonomic nervous systems.

spinal cord and then comes right back as a motor response. Most reflexes involve a single connecting interneuron between the sensory neuron and the motor neuron. A few, like the knee-jerk reflex in **figure 28.18**, are monosynaptic reflex arcs. You see in the figure that the sensory neuron, a stretch receptor embedded in a muscle, "senses" the stretching of the muscle when the tendon is tapped. This stretching could harm the muscle and so a nerve impulse is sent to the spinal cord where it synapses directly with a motor neuron—there is no interneuron between them. Similarly, if you step on something sharp, your leg jerks away from the danger. The prick causes nerve impulses in sensory neurons, which pass to the spinal cord and then to motor neurons, which cause your leg muscles to contract, jerking your leg up.

Autonomic Nervous System

Some motor neurons are active all the time, even during sleep. These neurons carry messages from the CNS that keep the body going even when it is not active. These neurons make up the autonomic nervous system. The word *autonomic* means involuntary. The autonomic nervous system carries messages to muscles and glands that work without the animal noticing.

The autonomic nervous system is the command network used by the CNS to maintain the body's homeostasis. Using it, the CNS regulates heartbeat and controls muscle contractions in the walls of the blood vessels. It directs the muscles that control blood pressure, breathing, and the movement of food through the digestive system. It also carries messages that help stimulate glands to secrete tears, mucus, and digestive enzymes.

The autonomic nervous system is composed of two divisions that act in opposition to one another. One division, the **sympathetic nervous system,** dominates in times of stress. It controls the "fight-or-flight" reaction, increasing blood pressure, heart rate, breathing rate, and blood flow to the muscles. The sympathetic nervous system is colored pink in **figure 28.19** and consists of a net-work of short motor axons extending out from the spinal cord to clusters of neuron cell bodies, called **ganglia,** indicated by the darker-colored band, located just to the right of the spinal cord in the figure. You can also see this chain of ganglia in **figure 28.16.** Long motor neurons extend from the ganglia directly to each target organ. Another division, the **parasympathetic nervous system,** has the opposite effect. It conserves energy by slowing the heartbeat and breathing rate and by promoting digestion and elimination. The parasympathetic nervous system is colored in blue and consists of a network of long axons extending out from motor neurons within the upper and lower sections of the spinal cord; these axons extend to ganglia in the immediate vicinity of an organ. It also consists of short motor neurons extending from the ganglia to the nearby organ.

Most glands, smooth muscles, and cardiac muscles get constant input from *both* the sympathetic and parasympathetic systems. The CNS controls activity by varying the ratio of the two signals to either stimulate or inhibit the organ.

Key Learning Outcome 28.8 The voluntary nervous system relays commands to skeletal muscles and can be controlled by conscious thought. The autonomic nervous system relays commands to muscles and glands that cannot be controlled by conscious thought.

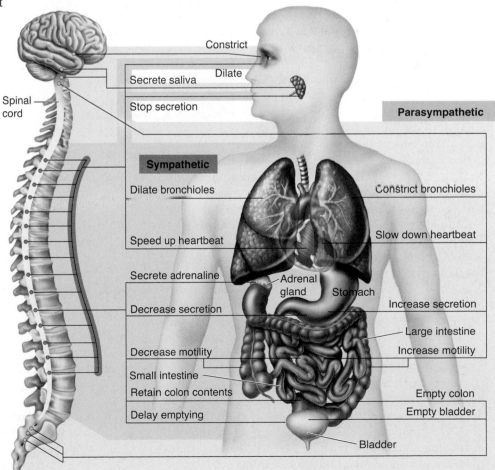

Figure 28.19 How the sympathetic and parasympathetic nervous systems interact.

A nerve path runs from both of the systems to every organ indicated except the adrenal gland, which is only innervated by the sympathetic nervous system.

Spinal cord

Constrict

Dilate

Secrete saliva

Stop secretion

Parasympathetic

Sympathetic

Dilate bronchioles

Constrict bronchioles

Speed up heartbeat

Slow down heartbeat

Secrete adrenaline

Adrenal gland

Stomach

Decrease secretion

Increase secretion

Decrease motility

Large intestine

Increase motility

Small intestine

Retain colon contents

Empty colon

Delay emptying

Empty bladder

Bladder

Are Bigger Nerves Faster?

In this chapter, you learned that the axons of neurons can have an insulating covering called a myelin sheath, but many nerve axons do not have myelin covers. In addition, not all axons are the same. Some are thin, like fine wire, while others are much thicker. A motor axon to a human internal organ might have a diameter of 1 to 5 micrometers, while the giant motor axon to the mantle muscle of a squid is fully 500 micrometers in diameter. Why the great difference in size? Physics tells us that there should be a relationship between the conduction velocity of a nerve axon and its diameter. To be precise, there should be a 10-fold increase in conduction speed for a 100-fold increase in fiber diameter. The squid giant axon is 100 times thicker than the human motor axons to internal organs—is it 10 times faster? Yes. The conduction velocity of the human axon is measured at 2 meters per second, while the conduction velocity of the squid giant axon is fully 25 meters per second—the very high velocity allows squids to contract their mantles fast enough to power their jet propulsion.

Most of the motor nerve axons in a vertebrate body like yours have insulating myelin covers, allowing them to transmit signals much faster, the electrical signal jumping down the axon over the insulated segments. The photo to the right shows a bundle of nerve axons, many of them myelinated and looking somewhat like doughnuts. axons also come in a wide range of sizes, from the 5 micrometer-diameter axons of skin temperature receptors, to the 20 micrometer-diameter fibers travelling to leg muscles. Would you expect fatter myelinated axons to transmit faster? Yes. Why? The transmission speed between axon node regions will depend upon how many ion channels open when an electrical impulse arrives at that node. If you think of the axon as a cylinder or pipe, then the number of ion channels exposed at a node of the axon will be proportional to the exposed surface area of the node, with the larger surface area of bigger axons exposing more ion channels and so transmitting the signal faster to the next node. The surface area considered at any one location of a node is simply the circumference of the axon at that point, which is the diameter times a constant—3.14159265 (called pi, symbolized π). Thus the velocity of a myelinated axon would be expected to be directly proportional to its diameter. Stated simply, doubling diameter should double speed. Is that the case?

With modern electrode technology, it is possible to directly measure the conduction velocity of axons within the body of a vertebrate, and so answer this question.

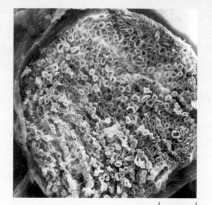

6.25 µm

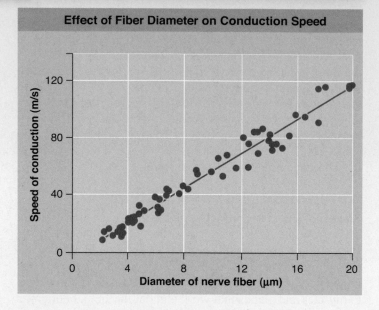

Effect of Fiber Diameter on Conduction Speed

The graph above shows the speed of conduction of myelinated axon fibers of a cat, plotting measured speed of conductance in meters per second against axon fiber diameter measured in micrometers.

1. **Applying Concepts**
   a. What is the dependent variable?
   b. Range. What is the range of nerve fiber diameters examined?
   c. Frequency. Which were more frequently examined, narrow diameters (less than 8 micrometers), or thick diameters (more than 12 micrometers)?

2. **Interpreting Data**
   a. For a nerve fiber diameter of 4 micrometers, what is the speed of conduction?
   b. For a nerve fiber of twice that diameter, 8 micrometers, what is the speed of conduction?
   c. For a nerve fiber twice *that* diameter, 16 micrometers, what is the speed of conduction?

3. **Making Inferences**
   a. Is conduction velocity faster for larger diameter fibers?
   b. When the nerve fiber diameter is doubled, what is the effect on speed of conductance?

4. **Drawing Conclusions** Do these data support the conclusion that conduction velocity is directly proportional to fiber diameter?

5. **Further Analysis** Do you imagine myelinated axons would transmit faster if the distance between nodes were shorter? If it were longer? How might you test this?

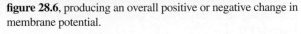
Neurons and How They Work

28.1 Evolution of the Animal Nervous System

- Animals use the nervous system to respond to their environments. In vertebrates, the central nervous system (the brain and spinal cord) receives information from the sensory nervous system and issues commands via the motor nervous system (**figure 28.2**).

- The nervous system consists of neurons and supporting cells. Three types of neurons are found in the nervous system: sensory neurons, motor neurons, and association neurons (**figure 28.1**).

- Nervous systems in animals evolved from nerve nets to more complex systems, with specialized cell types and localization of integration centers in the brain (**integrated art, page 587**).

28.2 Neurons Generate Nerve Impulses

- Neurons have a cell body, dendrites that receive information, and a long axon that conducts impulses from the cell.

- Neurons are supported by neuroglial cells, which make up a large portion of cells in the nervous system. Two of the most important kinds of neuroglial cells are Schwann cells and oligodendrocytes. These cells closely associate with the axons, wrapping them in a fatty material called myelin (**figure 28.3**). The presence of a myelin sheath speeds up the conduction of the nerve impulse.

- When a neuron is at rest, the sodium-potassium pump helps to make the outside of the neuron more positive than the inside, a condition called the resting membrane potential.

- Nerve impulses begin when some kind of sensory input disturbs a neuron's plasma membrane, causing sodium channels to open. This movement of ions in one area of the membrane causes a change in electrical properties, called depolarization. If the change in voltage is large enough, it causes the opening of adjacent ion channels. This can trigger an action potential, a wave of depolarization that will spread down the axon (**Key Biological Process, page 589**).

28.3 The Synapse

- When a nerve impulse reaches the end of the axon, its message must cross a gap between the neuron and other cells. This junction between a neuron and another cell is called a synapse (**figure 28.4**).

- Messages are carried across the synaptic cleft by neurotransmitters, chemicals packaged in sacs at the tip of an axon. The arrival of the nerve impulse at the end of the presynaptic cell triggers the release of neurotransmitters, which diffuse across the synaptic cleft and bind to receptors on the postsynaptic cell. This binding causing chemically gated ion channels to open. Ions flow across the plasma membrane creating electrical responses in the postsynaptic cell (**figure 28.5**).

- Depending on the type of ion that flows into the cell, the synapse is excitatory or inhibitory. All neural inputs are integrated in the postsynaptic cell, shown here from figure 28.6, producing an overall positive or negative change in membrane potential.

- There are many different kinds of neurotransmitters that are active in different types of synapses and elicit different responses.

28.4 Addictive Drugs Act on Chemical Synapses

- Molecules called neuromodulators increase or decrease the effects of neurotransmitters at a synapse. Many mental states are determined by groups of neurons that use certain neurotransmitters and neuromodulators (**figure 28.7**).

- Cocaine and nicotine are addictive because the nervous system makes physical changes to its receptors in response to the drug, so when the drug is removed, the body can't return to normal function (**Key Biological Process, page 593**).

The Central Nervous System

28.5 Evolution of the Vertebrate Brain

- As vertebrates evolved, the forebrain has become increasingly more prominent (**figure 28.10**).

28.6 How the Brain Works

- The cerebrum comprises about 85% of the human brain and functions in language, conscious thought, memory, personality development, vision, and many other higher-level activities. Much of the neural activity of the cerebrum occurs within its outer layer, called the cerebral cortex (**figures 28.11** and **28.12**).

- The thalamus and hypothalamus, which lie underneath the cerebrum, process information and integrate bodily functions. Areas of the hypothalamus, in addition to the hippocampus and amygdala, are also part of the limbic system, which is involved in deep-seated drives and emotions, such as pain, anger, sex, hunger, thirst, and pleasure (**figure 28.13**).

- The cerebellum controls balance, posture, and muscular coordination. The brain stem controls vital functions, such as breathing, swallowing, heart beat, and digestion.

28.7 The Spinal Cord

- The spinal cord is a cable of neurons that extends from the brain down the back and is encased in the bony vertebrae of the backbone (**figure 28.16**). Motor nerves carry impulses from the brain and spinal cord out to the body, and sensory nerves carry impulses from the body to the brain and spinal cord.

The Peripheral Nervous System

28.8 Voluntary and Autonomic Nervous Systems

- The voluntary nervous system relays commands between the CNS and skeletal muscles and can be consciously controlled; however, reflexes, such as the knee-jerk reflex shown here from **figure 28.18**, work without conscious control.

- The autonomic nervous system consists of opposing sympathetic and parasympathetic divisions that unconsciously relay commands between the CNS and muscles and glands (**figure 28.19**).

Test Your Understanding

1. Complexity in animal nervous systems evolved with an increase in
   a. animal body size.
   b. animal body nutritive requirements.
   c. the amount of associative neurons that eventually formed the "brain."
   d. types of animal behavior.
2. Which of the following is *not* found in the peripheral nervous system?
   a. motor neurons   c. sensory neurons
   b. association neurons   d. Schwann cells
3. An action potential is caused by a quick depolarization of the membrane in a nerve cell resulting from the
   a. influx of sodium ions.
   b. actions of the Na^+/K^+ pump.
   c. influx of potassium ions.
   d. All of the above.
4. The sodium-potassium pump
   a. is used to propagate an action potential.
   b. is important for maintenance of the resting membrane potential.
   c. is important only at the synapse.
   d. is a voltage-gated channel.
5. The junction between an axon and another cell is called a(n)
   a. axon hillock.   c. synapse.
   b. axon terminal.   d. action potential.
6. Excitatory neurotransmitters initiate an action potential in a postsynaptic neuron by opening
   a. sodium ion gates in the postsynaptic cell.
   b. potassium ion gates in the postsynaptic cell.
   c. chloride ion gates in the postsynaptic cell.
   d. calcium ion gates in the postsynaptic cell.
7. Most of the neural activity of the cerebrum occurs in the
   a. cerebral cortex.
   b. corpus callosum.
   c. thalamus.
   d. reticular formation.
8. The integration of the internal activities of the body is controlled by the
   a. cerebrum.   c. hypothalamus.
   b. cerebellum.   d. brain stem.
9. The purpose of the limbic system is to
   a. coordinate olfactory information from the nose with visual information from the eyes.
   b. process critical thought and learning.
   c. coordinate optic information with the muscles.
   d. coordinate emotions.
10. The purpose of the autonomic nervous system is to do all of the following *except*
    a. stimulate glands.
    b. relay messages to skeletal muscles.
    c. relay messages to cardiac and smooth muscles.
    d. regulate the body's homeostasis.

Apply Your Understanding

1. **Figure 28.5a** At the synapse, a mutation in the structure of which component would affect only whether or not a signal is transmitted, not how strongly? Explain.

2. **Key Biological Process: Drug Addiction (page 593)** Describe the feelings that might be experienced by someone who goes through the four stages related to drug addiction that are shown in the drawings below.

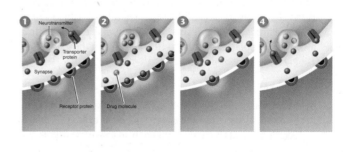

Synthesize What You Have Learned

1. When an investigator stimulates an axon by touching it with an electrode in the middle, action potentials are generated in both directions. If instead the investigator stimulates the axon where it meets the cell body, the action potential only goes outward toward the axon terminals and not inward over the cell body. Can you explain what might be going on at the junction of the axon to the cell body that would prevent further passage of the action potential?

2. Botox, a derivative of the botulinum toxin that can cause fatal food poisoning, acts by inhibiting the release of acetylcholine at neuromuscular junctions. How could treating someone's face with such a toxin produce desired cosmetic effects?

3. Cyanide is a deadly poison that halts cellular respiration by inhibiting the mitochondrial enzyme cytochrome *c* oxidase. Chronic exposure to low levels of cyanide by those who use cassava roots as their primary food source in tropical Africa can eventually lead to permanent paralysis, as cyanide also disables the sodium-potassium pump. Explain how this inhibition might lead to paralysis.

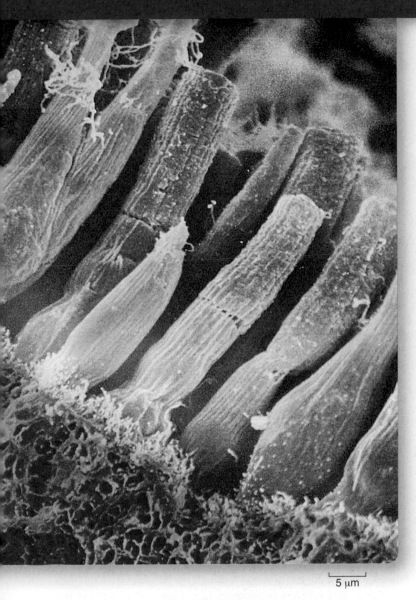

5 μm

29

The Senses

S ensory neurons receive input from different kinds of sensory receptor cells, such as the rods and cones pictured above that are found in the vertebrate eye. All input from sensory neurons to the central nervous system arrives in the same form, as action potentials propagated by afferent (inward-conducting) sensory neurons. Different sensory neurons project to different brain regions, and so are associated with different sensory modalities. The intensity of the sensation depends on the frequency of action potentials conducted by the sensory neuron. A sunset, a symphony, and searing pain are distinguished by the brain only in terms of the identity of the sensory neuron carrying the action potentials and the frequency of these impulses. Thus, if the auditory nerve is artificially stimulated, the brain perceives the stimulation as sound. But if the optic nerve is artificially stimulated in exactly the same manner and degree, the brain perceives a flash of light.

29.1 Processing Sensory Information

Did you ever wonder what it would be like not to know anything about what is going on around you? Imagine if you couldn't hear, or see, or feel, or smell. After a while, a human goes mad if completely deprived of sensory input. The senses are the bridge to experience and perceive the way the body relates to everything around it.

Sensory Receptors

The **sensory nervous system** tells the central nervous system what is happening. Sensory neurons carry impulses to the CNS from more than a dozen different types of sensory cells that detect changes outside and inside the body. Called **sensory receptors,** these specialized sensory cells detect many different things, including changes in blood pressure, strain on ligaments, and smells in the air. Particularly complex sensory receptors, made up of many cell and tissue types, are called **sensory organs.** The eyes and ears (**figure 29.1**) are sensory organs, and so are the taste buds in your mouth.

How does the brain know whether an incoming nerve impulse is light, sound, or pain? This information is built into the "wiring"—into which neurons interact while passing the information to the CNS and into the location in the brain where the information is sent. The brain "knows" it is responding to light because the message from a sensory neuron is wired to light receptor cells. That is why when you press your fingertips gently against the corners of your eyes, you "see stars"—the brain treats any impulse from the eyes as light, even though the eye received no light.

The Path of Sensory Information

The path of sensory information to the CNS is a simple one, composed of three stages:

1. **Stimulation.** A stimulus impinges on a sensory receptor.
2. **Transduction.** Sensory receptors change the stimulus into an electrical potential by initiating the opening or closing of ion channels in a sensory neuron.
3. **Transmission.** The sensory neuron conducts a nerve impulse along an afferent pathway to the CNS.

All sensory receptors are able to initiate nerve impulses by opening or closing **stimulus-gated channels** within sensory neuron membranes. Except for visual photoreceptors, these channels are sodium ion channels that depolarize the membrane and so start an electrical signal. If the stimulus is large enough, the depolarization will trigger an action potential. The greater the sensory stimulus, the greater the depolar-

Figure 29.1 Kangaroo rats have specialized ears.
The ears of kangaroo rats (*Dipodomys*) are adapted to nocturnal life and allow them to hear the low-frequency sounds of their predators, such as an owl's wingbeats or a sidewinder rattlesnake's scales rubbing against the ground. Also, the ears seem to be adapted to the poor sound-carrying quality of dry, desert air.

ization of the sensory receptor and the higher the frequency of action potentials. The channels are opened by chemical or mechanical stimulation, often a disturbance such as touch, heat, or cold. The receptors differ from one another in the nature of the environmental input that triggers the opening of the channel. The body contains many sorts of receptors, each sensitive to a different aspect of the body's condition or to a different quality of the external environment.

Exteroceptors are receptors that sense stimuli that arise in the external environment. Almost all of a vertebrate's exterior senses evolved in water before vertebrates invaded the land. Consequently, many senses of terrestrial vertebrates emphasize stimuli that travel well in water, using receptors that have been retained in the transition from the sea to the land. Hearing, for example, converts an airborne stimulus into a waterborne one, using receptors similar to those that originally evolved in aquatic animals. A few vertebrate sensory systems that function well in the water, such as the electric organs of fish, cannot function well in the air and are not found among terrestrial vertebrates. On the other hand, some land dwellers have sensory systems, such as infrared receptors, that could not function in the sea. Still other organisms seem to be able to detect earth's magnetic field, as we will discuss later on in this chapter.

Interoceptors sense stimuli that arise from within the body. These internal receptors detect stimuli related to muscle length and tension, limb position, pain, blood chemistry, blood volume and pressure, and body temperature. Many of these receptors are simpler than those that monitor the external environment and are believed to bear a closer resemblance to primitive sensory receptors.

Sensing the Internal Environment

Sensory receptors inside the body inform the CNS about the condition of the body. Much of this information passes to a coordinating center in the brain, the hypothalamus. This part of the brain is responsible for maintaining the body's homeostasis—that is, keeping the body's internal environment constant. The vertebrate body uses a variety of different sensory receptors to respond to different aspects of its internal environment.

Temperature change. Two kinds of nerve endings in the skin are sensitive to changes in temperature—one stimulated by cold, the other by warmth. By comparing information from the two, the CNS can learn what the temperature is and if it is changing.

Blood chemistry. Receptors in the walls of arteries sense CO_2 levels in the blood. The brain uses this information to regulate the body's respiration rate, increasing it when CO_2 levels rise above normal.

Pain. Damage to tissue is detected by special nerve endings within tissues, usually near the surface, where damage is most likely to occur. When these nerve endings are physically damaged or deformed, the CNS responds by reflexively withdrawing the body segment and often by changing heartbeat and blood pressure as well.

Muscle contraction. Buried deep within muscles are sensory receptors called stretch receptors (see also pages 600–601). In each, the end of a sensory neuron is wrapped around a muscle fiber, like the receptor shown in **figure 29.2**. When the muscle is stretched, the fiber elongates, stretching the spiral nerve ending (like stretching a spring) and causing repeated nerve impulses to be sent to the brain. From these signals the brain can determine the rate of change of muscle length at any given moment. The CNS uses this information to control movements that require the combined action of several muscles, such as those that carry out breathing or locomotion.

Blood pressure. Blood pressure is sensed by neurons called baroreceptors with highly branched nerve endings within the walls of major arteries. When blood pressure increases, the stretching of the arterial wall, like the expansion of the artery in **figure 29.3**, causes the sensory neuron to increase the rate at which it sends nerve impulses to the CNS. When the wall of the artery is not stretched, the rate of firing of the sensory neuron goes down. Thus, the frequency of impulses provides the CNS with a continuous measure of blood pressure.

Touch. Touch is sensed by pressure receptors buried below the surface of the skin. There are a variety of different types, some specialized to detect rapid changes in pressure, others to measure the duration and extent to which pressure is applied, and still others sensitive to vibrations.

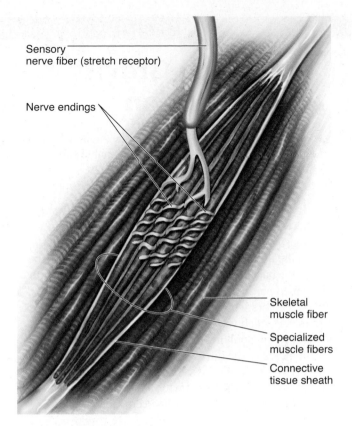

Sensory nerve fiber (stretch receptor)

Nerve endings

Skeletal muscle fiber

Specialized muscle fibers

Connective tissue sheath

Figure 29.2 A stretch receptor embedded within skeletal muscle.

Stretching the muscle elongates the specialized muscle fibers, which deforms the nerve endings, causing them to send a nerve impulse out along the nerve fiber.

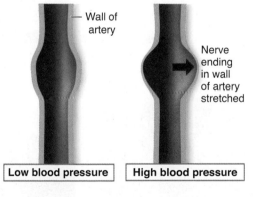

Wall of artery

Nerve ending in wall of artery stretched

Low blood pressure

High blood pressure

Figure 29.3 How a baroreceptor works.

A network of nerve endings covers a region where the wall of the artery is thin. High blood pressure causes the wall to balloon out there, stretching the nerve endings and causing them to fire impulses.

Key Learning Outcome 29.1 Sensory receptors initiate nerve impulses in response to stimulation. All sensory nerve impulses are the same, differing only in the stimulus that fires them and their destination in the brain. A variety of different sensory receptors inform the hypothalamus about different aspects of the body's internal environment, enabling it to maintain the body's homeostasis.

29.2 Sensing Gravity and Motion

Two types of receptors in the ear inform the brain where the body is in three dimensions. This knowledge enables an animal to move freely and maintain its balance. **Figure 29.4** shows the anatomy of the inner ear and the locations of these receptors.

Balance. To keep the body's balance the brain needs a frame of reference, and the reference point it uses is gravity. The sensory receptors that detect gravity are hair cells within the utricle and the saccule of the inner ear ❶. The tips of the hair cells project into a gelatinous matrix with embedded particles called **otoliths** ❷. To illustrate how these receptors work, imagine a pencil standing in a glass. No matter which way you tip the glass, the pencil rolls along the rim, applying pressure to the lip of the glass, which indicates the direction the glass is tipped. Similarly, otoliths in the utricle and saccule will shift in the matrix in response to the pull of gravity, stimulating hair cells, which the brain uses to determine vertical positioning.

Motion. The brain senses motion by employing a receptor in which fluid deflects cilia in a direction opposite that of the motion. Within the inner ear are three fluid-filled **semicircular canals,** each oriented in a different plane at right angles to the other two (❶) so that motion in any direction can be detected. Protruding into the canal are groups of cilia from sensory cells. The cilia from each cell are arranged in a tentlike assembly called a *cupula,* shown in ❸, which is pushed when fluid in the semicircular canals moves in a direction opposite that of the head's movement. Because the three canals are oriented in all three planes, movement in any plane is sensed by at least one of them, and the brain is able to analyze complex movements by comparing the sensory inputs from each canal.

The semicircular canals do not react if the body moves in a straight line because the fluid in the canals does not move. That is why traveling in a car or airplane at a constant speed in one direction gives no sense of motion.

> **Key Learning Outcome 29.2** The body senses gravity and acceleration by the deflection of cilia caused by shifting otoliths or fluid. The body cannot sense motion at a constant velocity and direction.

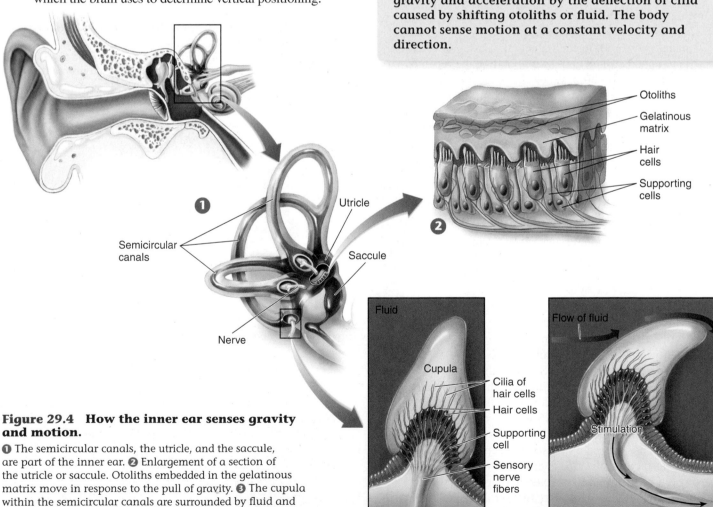

Figure 29.4 How the inner ear senses gravity and motion.
❶ The semicircular canals, the utricle, and the saccule, are part of the inner ear. ❷ Enlargement of a section of the utricle or saccule. Otoliths embedded in the gelatinous matrix move in response to the pull of gravity. ❸ The cupula within the semicircular canals are surrounded by fluid and contain hair cells. Movement in a particular direction causes fluid in the semicircular canal of that plane to move the cupula, stimulating the hair cells.

Sensing Chemicals: Taste and Smell

Vertebrates are able to detect many of the chemicals in air and in food.

Taste

Embedded within the surface of the tongue are *taste buds* located within *papillae,* which are the raised areas on the tongue in **figure 29.5**. Taste buds are onion-shaped structures in the enlarged view of the papillae that contain many taste receptor cells, each of which has fingerlike microvilli that project into an opening called the taste pore. Chemicals from food dissolve in saliva and contact the taste cells through the taste pore. Salty, sour, sweet, bitter, and umami (a "meaty" taste) are perceived because chemicals in food are detected in different ways by taste buds. When the tongue encounters a chemical, information from the taste cells passes to sensory neurons, which transmit the signals to the brain.

Smell

In the nose are chemically sensitive neurons whose cell bodies are embedded within the epithelium of the nasal passage, shown in cross section in **figure 29.6**. When they detect chemicals, these sensory neurons (the red cells in the enlarged view) transmit information to a location in the brain where smell information is processed and analyzed. In many vertebrates (dogs are a familiar example), these neurons are far more sensitive than in humans.

Smell as well as taste is very important in telling an animal about its food. That is why when you have a bad cold and your nose is stuffed up, your food has little taste. Other receptors also play a role. For example, the "hot" sensation of foods such as chili peppers is detected by pain receptors, not chemical receptors.

> **Key Learning Outcome 29.3** Taste and smell are chemical senses. In many vertebrates, the sense of smell is very well developed.

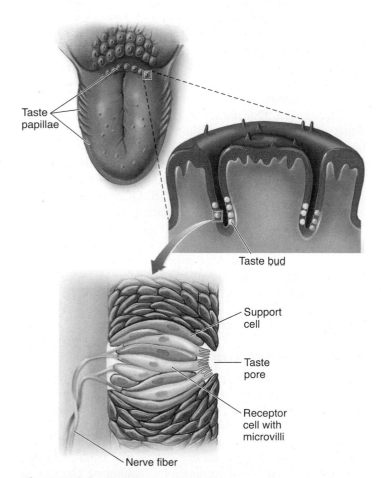

Figure 29.5 Taste.
Taste buds on the human tongue are typically grouped into projections called papillae. Individual taste buds are bulb-shaped collections of taste receptor cells that open out into the mouth through a taste pore.

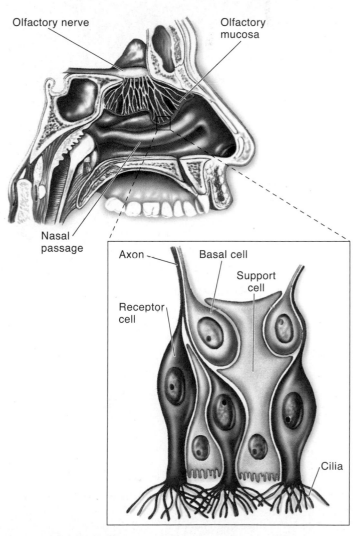

Figure 29.6 Smell.
Humans smell by using receptor cells located in the lining of the nasal passage. The receptor cells are neurons. Axons from these sensory neurons project back through the olfactory nerve directly to the brain.

29.4 Sensing Sounds: Hearing

When you hear a sound, you are detecting the air vibrating—waves of pressure in the air beating against your ear, pushing a membrane called the eardrum in and out. As you can see in **figure 29.7**, on the inner side of the eardrum are three small bones, called ossicles, that act as a lever system to increase the force of the vibration. They transfer the amplified vibration across a second membrane to fluid within the inner ear. The fluid-filled chamber of the inner ear is shaped like a tightly coiled snail shell and is called the cochlea, from the Latin name for "snail." The middle ear, where the ossicles are located, is connected to the throat by the eustachian tube in such a way that there is no difference in air pressure between the middle ear and the outside. That is why your ears sometimes "pop" when you are landing in an airplane—the pressure is equalizing between the two sides of the eardrum. This equalized pressure is necessary for the eardrum to work.

The sound receptors within the cochlea are hair cells that rest on a membrane that runs up and down the middle of the spiraling chamber, separating it into two halves, the upper and lower fluid-filled canals in the enlarged view. The hair cells do not project into the fluid-filled canals of the cochlea; instead, they are covered by a second membrane (the dark blue membrane in the figure). When a sound wave enters the cochlea, it causes the fluid in the chambers to move. The moving fluid causes this membrane "sandwich" to vibrate, bending the hairs pressed against the upper membrane and causing them to send nerve impulses to sensory neurons that travel to the brain.

Sounds of different frequencies cause different parts of the membrane to vibrate, and thus fire different sensory neurons—the identity of the sensory neuron being fired tells the CNS the frequency of the sound. Sound waves of higher frequencies, about 20,000 vibrations (or cycles) per second, also called hertz (Hz), move the membrane in the area closest to the middle ear. Medium-length frequencies, about 2,000 Hz, move the membrane in the area about midway down the length of the cochlea. The lowest-frequency sound waves, about 500 Hz, move the membrane near the tip of the cochlea.

The intensity of the sound is determined by how *often* the neurons fire. Our ability to hear depends upon the flexibility of the membranes within the cochlea. Humans cannot hear low-pitched sounds, below 20 vibrations (or cycles) per second, although some vertebrates can. As children, we can hear high-pitched sounds, up to 20,000 cycles per second, but this ability decreases as we get older. Other vertebrates can hear sounds at far higher frequencies. Dogs readily hear sounds of 40,000 cycles per second and so respond to a high-pitched dog whistle that seems silent to a human.

Frequent or prolonged exposure to loud noises can result in damage to the hair cells and membrane, especially in the high-frequency area to the cochlea. The loss of the ability to detect high-frequency sounds affects a person's ability to hear certain sounds, especially in a noisy setting.

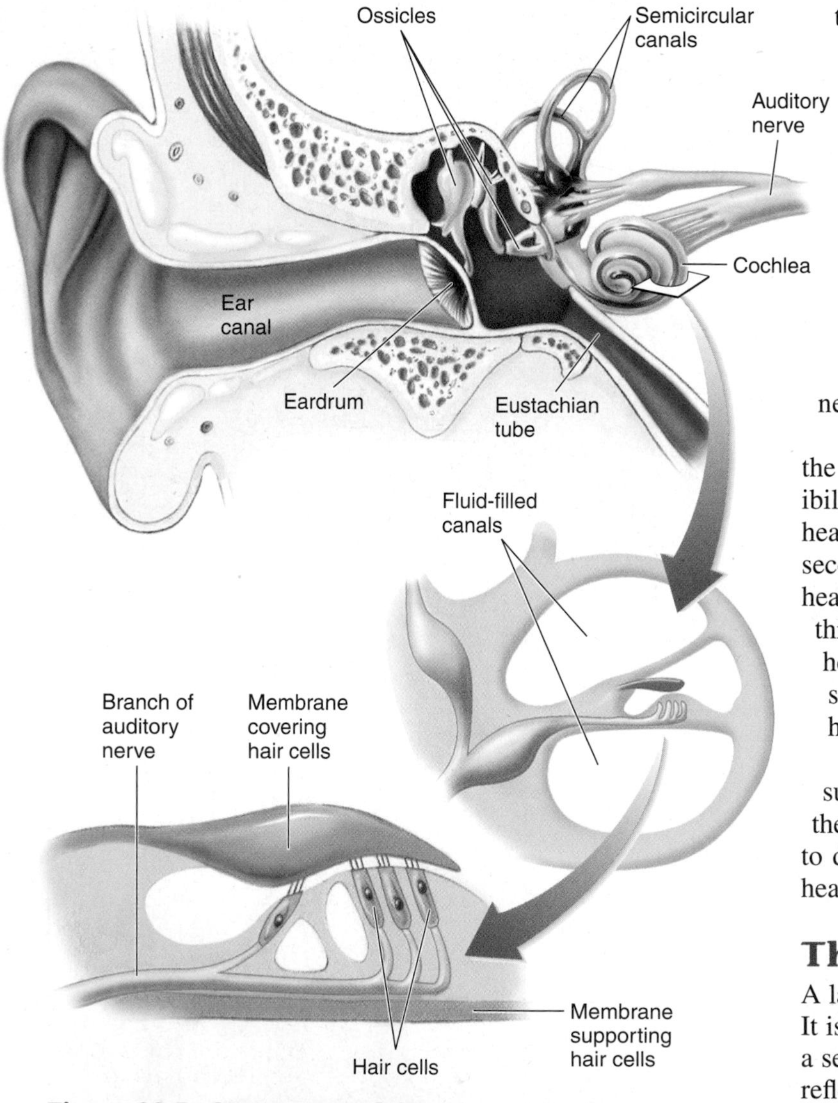

The Lateral Line System

A lateral line system supplements the fish's sense of hearing. It is performed by a different sensory structure, and provides a sense of "distant touch." A fish is able to sense objects that reflect pressure waves and low-frequency vibrations and thus can detect prey, for example, and swim in synchrony with the rest of its school. The lateral line system also enables a blind cave fish to sense its environment by monitoring changes in the patterns of water flow past the lateral line receptors. The same system is found in amphibian larvae, but it is lost at metamorphosis and is not present in any terrestrial vertebrate.

Figure 29.7 Structure and function of the human ear.
Sound waves passing through the ear canal beat on the eardrum, pushing a set of three small bones, or ossicles, against an inner membrane. This sets up a wave motion in the fluid filling the canals within the cochlea. The wave causes the membrane covering the hair cells to move back and forth against the hair cells, which causes associated neurons to fire impulses.

The lateral line system consists of sensory structures within a longitudinal canal in the fish's skin, shown in **figure 29.8**. Canals extend along each side of the body, and there are several canals in the head. As you can see in the enlarged view, openings lead into the canal, which is lined with sensory structures known as hair cells because they have hairlike processes at their surface. The processes of the hair cells project into a gelatinous membrane called a *cupula* (Latin, "little cup"). The hair cells are innervated by sensory neurons that transmit impulses to the brain. Vibrations carried through the fish's environment and down into the canal produce movements of the cupula, which cause the hairs to bend. When the hair cells bend, the associated sensory neurons are stimulated and generate a nerve impulse that is sent to the brain.

Figure 29.9 Using ultrasound to locate a moth.
This bat is emitting high-frequency "chirps" as it flies. It then listens for the sound's reflection against the moth. By timing how long it takes for a sound to return, the bat can "see" the moth even in total darkness.

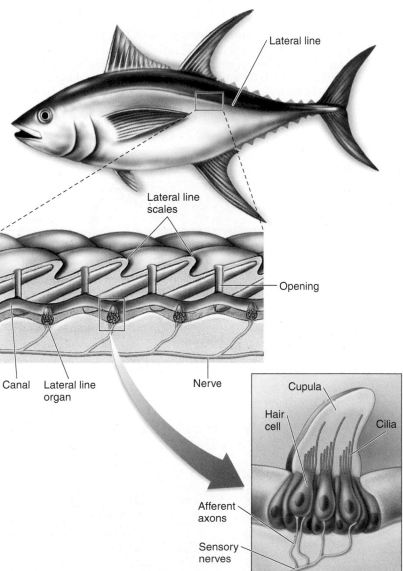

Figure 29.8 The lateral line system.
This system consists of canals running the length of the fish's body beneath the surface of the skin. Within these canals are sensory structures containing hair cells with cilia that project into a gelatinous cupula. Pressure waves traveling through the water in the canals deflect the cilia and depolarize the sensory neurons associated with the hair cells.

Sonar

A few groups of mammals that live and obtain their food in dark environments have circumvented the limitations of darkness. A bat flying in a completely dark room easily avoids objects that are placed in its path—even a wire less than a millimeter in diameter. Shrews use a similar form of "lightless vision" beneath the ground, as do whales and dolphins beneath the sea. All of these mammals perceive distance by means of sonar. They emit sounds and then determine the time it takes these sounds to reach an object and return to the animal. This process is called **echolocation**. The bat in **figure 29.9**, for example, produces clicks that last 2 to 3 milliseconds and are repeated several hundred times per second. The three-dimensional imaging achieved with such an auditory sonar system can be quite sophisticated, and will help this bat find the moth.

> **Key Learning Outcome 29.4** Sound receptors detect air vibrations as waves of pressure pushing against the eardrum. Inside, these waves are amplified and press down hair cells that send signals to the brain. Fish sense pressure waves in water much as an ear senses sound. Many vertebrates sense distant objects by bouncing sounds off of them.

29.5 Sensing Light: Vision

No other stimulus provides as much detailed information about the environment as light. Vision, the perception of light, is carried out by a special sensory apparatus called an eye. All the sensory receptors described to this point have been chemical or mechanical ones. Eyes contain sensory receptors called rods and cones that respond to photons of light. The light energy is absorbed by pigments in the rods and cones, which respond by triggering nerve impulses in sensory neurons.

Evolution of the Eye

Vision begins with the capture of light energy by photoreceptors. Because light travels in a straight line and arrives virtually instantaneously, visual information can be used to determine both the direction and the distance of an object. No other stimulus provides as much detailed information.

Many invertebrates have simple visual systems with photoreceptors clustered in an eyespot. The flatworm in figure 29.10 has an eyespot, consisting of pigment molecules

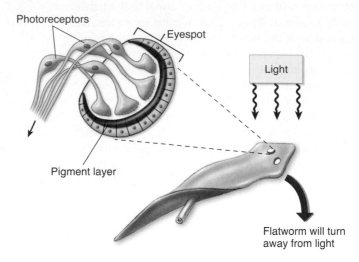

Figure 29.10 Simple eyespots in the flatworm.
Eyespots will detect the direction of light because a pigmented layer on one side of the eyespot screens out light coming from the back of the animal. Light is thus detected more readily coming from the front of the animal; flatworms will respond by turning away from the light.

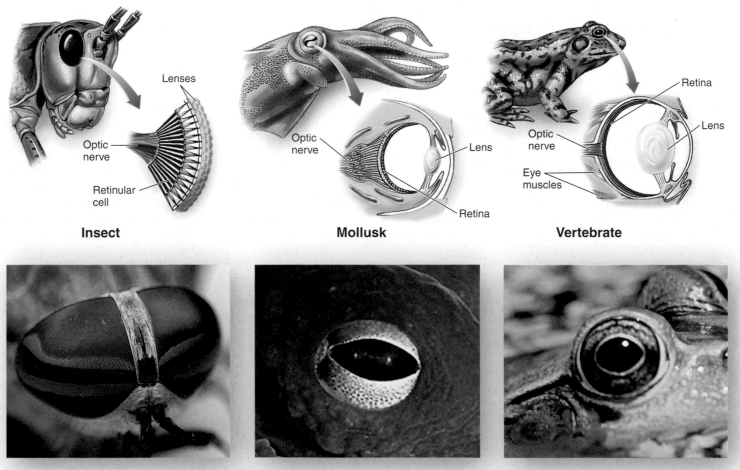

Insect **Mollusk** **Vertebrate**

Figure 29.11 Eyes in three phyla of animals.
Although they are superficially similar, these eyes differ greatly in structure and are not homologous. Each has evolved separately and, despite the apparent structural complexity, has done so from simpler structures.

that are stimulated by light, triggering a nerve impulse in photoreceptor cells. Although an eyespot can perceive the direction of light, it cannot be used to construct a visual image. The members of four phyla—annelids, mollusks, arthropods, and vertebrates—have evolved well-developed, image-forming eyes. True image-forming eyes in these phyla, though they at first seem similar (compare the eyes of an arthropod, mollusk, and vertebrate in **figure 29.11**), are believed to have evolved independently. Interestingly, the photoreceptors in all of them use the same light-capturing molecule, suggesting that not many alternative molecules are able to play this role.

Structure of the Vertebrate Eye

The vertebrate eye works like a lens-focused camera. Light first passes through a transparent protective covering called the **cornea** (the light blue layer in **figure 29.12**), which begins to focus the light onto the rear of the eye. The beam of light then passes through the **lens,** which completes the focusing. The lens is attached by stringlike *suspensory ligaments* to **ciliary muscles.** When these muscles contract and relax, they change the shape of the lens and thus allow the eye to view objects that are near or far. The amount of light entering the eye is controlled by a shutter, called the **iris** (the colored part of your eye), between the cornea and the lens. The transparent zone in the middle of the iris, the **pupil,** gets larger in dim light and smaller in bright light.

The light that passes through the pupil is focused by the lens onto the back of the eye. An array of light-sensitive receptor cells lines the back surface of the eye, called the **retina.** The retina is the light-sensing portion of the eye. The vertebrate retina contains two kinds of photoreceptors, called **rods** and **cones,** which, when stimulated by light, generate nerve impulses that travel to the brain along a short, thick nerve pathway called the optic nerve. Rods, the taller, flat-topped cell in **figure 29.13**, are receptor cells that are extremely sensitive to light, and they can detect various shades of gray even in dim light. However, they cannot distinguish colors, and because they do not detect edges well, they produce poorly defined images. Cones, the pointed-topped cells, are receptor cells that detect color and are sensitive to edges so that they produce sharp images. The center of the vertebrate retina contains a tiny pit, called the **fovea,** densely packed with some 3 million cones. This area produces the sharpest image, which is why we tend to move our eyes so that the image of an object we want to see clearly falls on this area.

The lens of the vertebrate eye is constructed to filter out short-wavelength light. This solves a difficult optical problem: Any uniform lens bends short wavelengths more than it does longer ones, a phenomenon known as chromatic aberration. Consequently, these short wavelengths cannot be brought into focus simultaneously with longer wavelengths. Unable to focus the short wavelengths, the vertebrate eye eliminates them. Insects, whose eyes do not focus light, are able to see these lower, ultraviolet wavelengths quite well and often use them to locate food or mates.

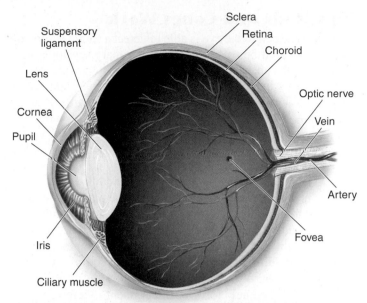

Figure 29.12 The structure of the human eye.
Light passes through the transparent cornea and is focused by the lens on the rear surface of the eye, the retina. The retina is rich in photoreceptors, with a high concentration in an area called the fovea.

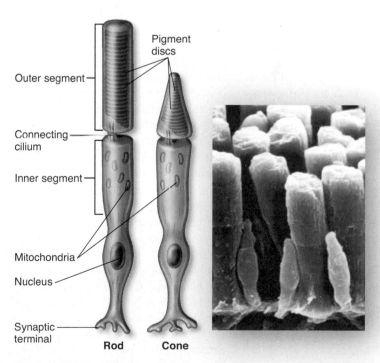

Figure 29.13 Rods and cones.
The broad tubular cell on the *left* is a rod. The shorter, tapered cell next to it is a cone. The electron micrograph of human rods and cones illustrates that rods are typically larger.

How Rods and Cones Work

A rod or cone cell in the eye is able to detect a single photon of light. How can it be so sensitive? The primary sensing event of vision is the absorption of a photon of light by a pigment. The pigments in rods and cones are made from plant pigments called carotenoids. That is why eating carrots is said to be good for night vision—the orange color of carrots is due to the presence of carotenoids called carotenes. The visual pigment in the human eye is a fragment of carotene called **cis-retinal.** The pigment is attached to a protein called **opsin** to form a light-detecting complex called **rhodopsin.**

When it receives a photon of light, the pigment undergoes a change in shape. This change in shape must be large enough to alter the shape of the opsin protein attached to it. When light is absorbed by the *cis*-retinal pigment (the upper molecule in **figure 29.14**), the linear end of the molecule rotates sharply upward, straightening out that end of the molecule. The new form of the pigment is referred to as **trans-retinal** and the dashed outline in the figure shows the shape before it was stimulated by light. This radical change in the pigment's shape induces a change in the shape of the protein opsin to which the pigment is bound, initiating a chain of events that leads to the generation of a nerve impulse.

Each rhodopsin activates several hundred molecules of a protein called transducin. Each of these activates several hundred molecules of an enzyme whose product stimulates sodium channels in the photoreceptor membrane at a rate of about 1,000 per second. This cascade of events allows a single photon to have a large effect on the receptor.

Color Vision

Three kinds of **cone cells** provide us with color vision. Each possesses a different version of the opsin protein (that is, one with a distinctive amino acid sequence and thus a different shape). These differences in shape affect the flexibility of the attached retinal pigment, shifting the wavelength at which it absorbs light. The absorption spectrum in **figure 29.15** shows the wavelength of light that is absorbed by each cone and rod cell. In rods, light is absorbed at 500 nanometers. In cones, the three versions of opsin absorb light at 420 nanometers (blue-absorbing), 530 nanometers (green-absorbing), or 560 nanometers (red-absorbing). By comparing the relative intensities of the signals from the three cones, the brain can calculate the intensity of other colors.

Some people are not able to see all three colors, a condition referred to as *color blindness*. Color blindness is typically due to an inherited lack of one or more types of cones. People with normal vision have all three types of cones. People with only two types of cones lack the ability to detect the third color. For example, people with red-green color blindness lack red cones and have difficulty distinguishing red from green (**figure 29.16**). Color blindness is a sex-linked trait (see chapter 10), and so men are far more likely to be color blind than women.

Most vertebrates, particularly those that are diurnal (active during the day), have color vision, as

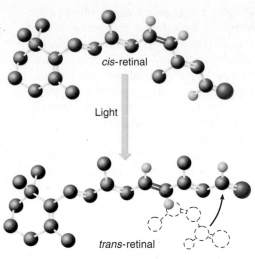

Figure 29.14 Absorption of light.

When light is absorbed by *cis*-retinal, the pigment undergoes a change in shape and becomes *trans*-retinal.

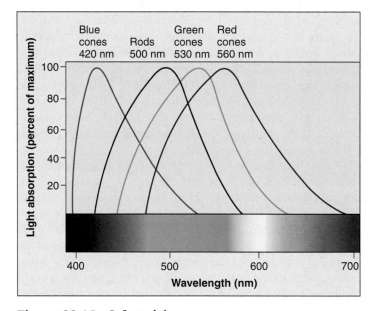

Figure 29.15   Color vision.

The absorption spectrum of *cis*-retinal is shifted in cone cells from the 500 nanometers characteristic of rod cells. The amount of the shift determines what color the cone absorbs: 420 nanometers yields blue absorption; 530 nanometers yields green absorption; and 560 nanometers yields red absorption. Red cones do not peak in the red part of the spectrum, but they are the only cones that absorb red light.

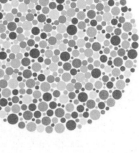

Figure 29.16   Test for color blindness.

People with normal color vision see the number 16, but people that are red-green color blind see just spots and no discernible number.

Source: This image has been reproduced from Ishihara's Tests for Color Deficiency published by KANEHARA TRADING INC., located in Tokyo, Japan. But tests for color deficiency cannot be conducted with this material. For accurate testing, the original plates should be used.

do many insects. Indeed, honeybees can see light in the near-ultraviolet range, which is invisible to the human eye. Color vision requires the presence of more than one photopigment in different receptor cells, but not all animals with color vision have the three-cone system characteristic of humans and other primates. Fish, turtles, and birds, for example, have four or five kinds of cones; the "extra" cones enable these animals to see near-ultraviolet light. Many mammals such as squirrels have only two types of cones.

Conveying the Light Information to the Brain

The path of light through each eye is the reverse of what you might expect. The rods and cones are at the rear of the retina, not the front. If you track the path that light would take in **figure 29.17**, you will see that light passes through several layers of ganglion and bipolar cells before it reaches the rods and cones. Once the photoreceptors are activated, they stimulate bipolar cells, which in turn stimulate ganglion cells. The direction of nerve impulses in the retina is thus opposite to the direction of light.

Action potentials propagated along the axons of ganglion cells are relayed through structures called the *lateral geniculate nuclei* of the thalamus and projected to the occipital lobe of the cerebral cortex. There the brain interprets this information as light in a specific region of the eye's receptive field. The pattern of activity among the ganglion cells across the retina encodes a point-to-point map of the receptive field, allowing the retina and brain to image objects in visual space. In addition, the frequency of impulses in each ganglion cell provides information about the light intensity at each point, while the relative activity of ganglion cells connected (through bipolar cells) with the three types of cones provides color information.

Binocular Vision

Primates (including humans) and most predators have two eyes, one located on each side of the face. When both eyes are trained on the same object, the image that each sees is slightly different because each eye views the object from a different angle. This slight displacement of the images permits **binocular vision,** the ability to perceive three-dimensional images and to sense depth or the distance to an object. Having their eyes facing forward maximizes the field of overlap in which this stereoscopic vision occurs, as seen by the overlapping blue triangles in the human in **figure 29.18**. The triangles are the field of view for each eye.

In contrast, prey animals generally have eyes located to the sides of the head, preventing binocular vision but enlarging the overall receptive field. Depth perception is less important to prey than detection of potential enemies from any angle. The eyes of the American woodcock, for example, are located at exactly opposite sides of its skull so that it has a 360-degree field of view without turning its head! Most birds have laterally placed eyes and, as an adaptation, have two foveas in each retina. One fovea provides sharp frontal vision, like the single fovea in the retina of mammals, and the other fovea provides sharper lateral vision.

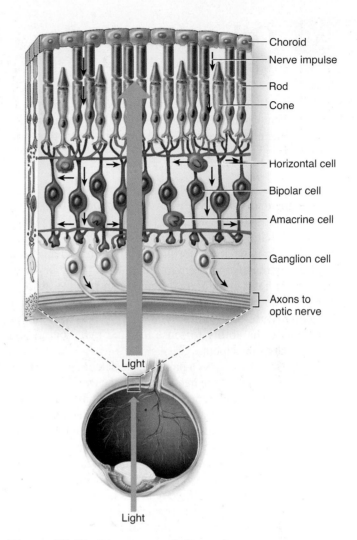

Choroid
Nerve impulse
Rod
Cone
Horizontal cell
Bipolar cell
Amacrine cell
Ganglion cell
Axons to optic nerve
Light
Light

Figure 29.17   Structure of the retina.
The rods and cones are at the rear of the retina. Light passes over four other types of cells in the retina before it reaches the rods and cones. The black arrows indicate how nerve impulses travel through the bipolar cells to the ganglion cells and on to the optic nerve.

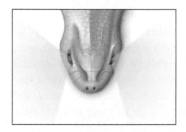

Figure 29.18   Binocular vision.
When the eyes are located on the sides of the head (as on the *left*), the two vision fields do not overlap and binocular vision does not occur. When both eyes are located toward the front of the head (as on the *right*) so that the two fields of vision overlap, depth can be perceived.

Key Learning Outcome 29.5   Vision receptors detect reflected light; binocular vision allows the brain to form three-dimensional images of objects.

29.6 Other Vertebrate Senses

Vision is the primary sense used by all vertebrates that live in a light-filled environment, but visible light is by no means the only part of the electromagnetic spectrum that vertebrates use to sense their environment.

Heat

Electromagnetic radiation with wavelengths longer than those of visible light is too low in energy to be detected by photoreceptors. Radiation from this *infrared* ("below red") portion of the spectrum is what we normally think of as radiant heat. Heat is an extremely poor environmental stimulus in water because water has a high thermal capacity and readily absorbs heat. Air, in contrast, has a low thermal capacity, so heat in air is a potentially useful stimulus. However, the only vertebrates known to have the ability to sense infrared radiation are the snakes known as pit vipers.

The pit vipers possess a pair of heat-detecting pit organs located on either side of the head between the eye and the nostril (**figure 29.19**). The pit organs permit a blindfolded rattlesnake to accurately strike at a warm, dead rat. Each pit organ is composed of two chambers separated by a membrane. The infrared radiation falls on the membrane and warms it. Thermal receptors on the membrane are stimulated. The nature of the pit organ's thermal receptor is not known; it probably consists of temperature-sensitive neurons innervating the two chambers. The two pit organs appear to provide stereoscopic information, in much the same way that two eyes do. Indeed, in snakes the information transmitted from the pit organs is processed in the brain by the homologous structure to the visual center in other vertebrates.

Electricity

While air does not readily conduct an electrical current, water is a good conductor. All aquatic animals generate electrical currents from contractions of their muscles. A number of different groups of fishes can detect these electrical currents. The electrical fish even have the ability to produce electrical discharges from specialized electrical organs. Electrical fish use these weak discharges to locate their prey and mates and to construct a three-dimensional image of their environment even in murky water.

The elasmobranchs (sharks, rays, and skates) have electroreceptors called the ampullae of Lorenzini. The receptor cells are located in sacs that open through jelly-filled canals to pores on the body surface. The jelly is a very good conductor, so a negative charge in the opening of the canal can depolarize the receptor at the base, causing the release of neurotransmitter and increased activity of sensory neurons. This allows sharks, for example, to detect the electrical fields

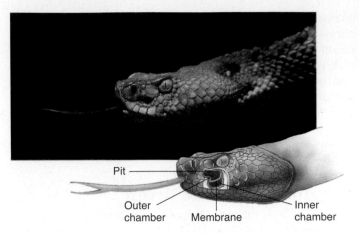

Figure 29.19 "Seeing" heat.
The depression between the nostril and the eye of this rattlesnake opens into the pit organ. In the cutaway portion of the diagram, you can see that the organ is composed of two chambers separated by a membrane. Snakes known as pit vipers have these unique organs which allow them to sense infrared radiation (heat).

generated by the muscle contractions of their prey. Although the ampullae of Lorenzini were lost in the evolution of teleost fish (most of the bony fish), electroreception reappeared in some groups of teleost fish that use sensory structures analogous to the ampullae of Lorenzini. Electroreceptors evolved yet another time, independently, in the duck-billed platypus, an egg-laying mammal. The receptors in its bill can detect the electrical currents created by the contracting muscles of shrimp and fish, enabling the mammal to detect its prey at night and in muddy water.

Magnetism

Eels, sharks, bees, and many birds appear to navigate along the magnetic field lines of the earth. Even some bacteria use such forces to orient themselves. Birds kept in blind cages, with no visual cues to guide them, will peck and attempt to move in the direction in which they would normally migrate at the appropriate time of the year. They will not do so, however, if the cage is shielded from magnetic fields by steel. Indeed, if the magnetic field of a blind cage is deflected 120° clockwise by an artificial magnet, a bird that normally orients to the north will orient toward the east-southeast. There has been much speculation about the nature of the magnetic receptors in these vertebrates, but the mechanism is still very poorly understood.

Key Learning Outcome 29.6 Pit vipers can locate warm prey by infrared radiation (heat), and many aquatic vertebrates can locate prey and ascertain the contours of their environment by means of electroreceptors. Magnetic receptors may aid in bird migration.

How the Platypus Sees With Its Eyes Shut

The duck-billed platypus (*Ornithorhynchus anatinus*) is abundant in freshwater streams of eastern Australia. These mammals have a unique mixture of traits—in 1799, British scientists were convinced that the platypus skin they received from Australia was a hoax. The platypus is covered in soft fur and has mammary glands, but in other ways it seems very reptilian. Females lay eggs as reptiles do, and like reptilian eggs, the yolk of the fertilized egg does not divide. In addition, the platypus has a tail not unlike that of a beaver, a bill not unlike that of a duck, and webbed feet!

It turns out that platypuses also have some very unique behaviors. Until recently, few scientists had studied the platypus in its natural habitat—it is elusive, spending its days in burrows it constructs on the banks of waterways. Also, a platypus is active mostly at night, diving in streams and lagoons to capture bottom-dwelling invertebrates such as shrimps and insect larvae. Interestingly, unlike whales and other marine mammals, a platypus cannot stay under water long. Its dives typically last a minute and a half. (Try holding your breath that long!)

When scientists began to study the platypus' diving behavior, they soon observed a curious fact: The eyes and ears of a platypus are located within a muscular groove, and when a platypus dives, the sides of these grooves close over tightly. Imagine pulling your eyebrows down to your cheeks—effectively blindfolded, you wouldn't be able to see a thing! To complete its isolation, the nostrils at the end of the snout also close. So how in the world does the animal find its prey?

For over a century biologists have known that the soft surface of the platypus bill is pierced by hundreds of tiny openings. In recent years Australian neuroscientists (scientists that study the brain and nervous system) have learned that these pores contain sensitive nerve endings. Nestled in an interior cavity, the nerves are protected from damage by the bill but are linked to the outside

streamwater via the pore. The nerve endings act as sensory receptors, communicating to the brain information about the animal's surroundings. These pores in the platypus bill are its diving "eyes."

Platypuses have two types of sensory cells in these pores. Clustered in the front are so-called mechanoreceptors, which act like tiny pushrods. Anything pushing against them triggers a signal. Your ears work the same way, sound waves pushing against tiny mechanoreceptors within your ears. These pushrods evoke a response over a much larger area of the platypus brain than does stimulation from the eyes and ears—for the diving platypus, the bill is the primary sense organ. What responses do the pushrod receptors evoke? Touching the bill with a fine glass probe reveals the answer—a lightning-fast, snapping movement of its jaws. When the platypus contacts its prey, the pushrod receptors are stimulated, and the jaws rapidly snap and seize the prey.

But how does the platypus locate its prey at a distance, in murky water with its eyes shut? That is where the other sort of sensory receptor comes in. When a platypus feeds, it swims along steadily wagging its bill from side to side, two or three sweeps per second, until it detects and homes in on prey. How does the platypus detect the prey individual and orient itself to it? The platypus does not emit sounds like a bat, which rules out the possibility of sonar as an explanation. Instead, electroreceptors in its bill sense the tiny electrical currents generated by the muscle movements of its prey as the shrimp or insect larva moves to evade the approaching platypus!

It is easy to demonstrate this, once you know what is going on. Just drop a small 1.5-volt battery into the stream. A platypus will immediately orient to it and attack it, from as far away as 30 centimeters. Some sharks and fishes have the same sort of sensory system. In muddy murky waters, sensing the muscle movements of a prey individual is far superior to trying to see its body or hear it move—which is why the platypus that you see in the photo above is hunting with its eyes shut.

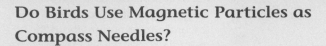

Do Birds Use Magnetic Particles as Compass Needles?

Some migrating birds use infrasound to orient themselves. Others may use visual cues, like the angle of polarizing light or the direction of sunset. Many birds that migrate long distances use the earth's geomagnetic field as a source of compass information. If the magnetic field of a blind "orientation cage" (see photo below) is deflected by 120 degrees clockwise by an artificial magnet, a bird that normally orients to the north will orient toward the southeast.

The sensory system underlying the magnetic compass of these birds is one of the great mysteries of sensory biology. There are two competing hypotheses.

The Magnetite Hypothesis One hypothesis is that crystals of the magnetic mineral magnetite within brain cells of migrating birds act as miniature compass needles. While trace amounts of magnetite are indeed present in some brain cells, intensive research has failed to confirm that information about the orientation of magnetite particles within these cells is transmitted to any other cells of the brain.

The Photoreceptor Hypothesis An alternative hypothesis is that the primary process underlying the compass is instead a magnetically sensitive chemical reaction within the photoreceptors of the bird's eyes. The alignment of photopigment molecules with the earth's magnetic field might alter the visual pattern in a way that could be used to obtain directional information.

Which hypothesis is correct? Experiments have shown that the magnetic detector used by birds in blind cages is light sensitive, as a photoreceptor compass should be—but this would also be true of a light-activated magnetite compass.

In 2004, University of California Irvine researchers devised a clever experiment to distinguish between the two hypotheses. They studied the way in which migrating European robins held in orientation cages use the magnetic field as a source of compass information to hop in the appropriate migratory direction. They found that the robins oriented 16 degrees north during the spring migration, the appropriate direction. To distinguish between the magnetite and photoreceptor hypotheses, the robins in the cages were exposed to oscillating, low-level radio frequencies (7 MHz) that would disrupt the energy state of any light-absorbing photoreceptor molecules involved in sensing the magnetic field, but would not affect the alignment of magnetite particles.

Effect of Radio-Disruption on Orientation

Bird	Mean Heading (degrees)	
	Geomagnetic field only	Radio-disrupted
1	26	110
2	20	126
3	4	86
4	350	17
5	15	162
6	1	330
7	18	297
8	20	220
9	354	58
10	24	261
11	358	278
12	37	3

Mean vector 16°
360° N
270° W   E 90°
S
180°

The chart above presents the results of this study. Each data entry is the mean of three recordings. For each recording, the bird was placed in a 35-inch conical orientation cage lined with coated paper, and the vector (the directional position of first contact with the paper) recorded relative to magnetic north (north = 360 degrees; east = 90 degrees; south = 180 degrees; west = 270 degrees).

1. **Applying Concepts** In the chart, is there a dependent variable? If so, what is it? Discuss.

2. **Interpreting Data** Plot each column on a circle. For birds orienting to the geomagnetic field without radio interference, what is the greatest difference (expressed in degrees) between recorded vectors and the mean vector of 16 degrees? For birds orienting with 7 MHz radio interference?

3. **Making Inferences**
   a. For birds orienting to the geomagnetic field without radio interference, how many of the 12 birds oriented with an accuracy of +/– 30 degrees relative to the mean vector of 16 degrees? For birds orienting with 7 MHz radio interference, how many?
   b. If you were to select a bird at random, what is the probability that it would orient within +/– 30 degrees of the appropriate migration direction (16 degrees north) without radio interference? With radio interference?

4. **Drawing Conclusions** Is the ability of European robins to orient correctly with respect to geomagnetic fields disrupted by 7 MHz radio frequencies? Is it fair to conclude that the birds' compass sense involves a molecule sensitive to radio disruption, such as a photoreceptor? That it does not involve particles not sensitive to radio disruption?

The Sensory Nervous System

29.1 Processing Sensory Information

- Neurons called sensory receptors initiate and carry nerve impulses from throughout the body to the CNS. There are three steps in this information pathway: stimulation, transduction, and transmission.

- Different sensory cells are stimulated by different stimuli. Exteroceptors sense stimuli from the external environment, and interoceptors, such as stretch receptors and baroreceptors, sense stimuli within the body.

- There are different kinds of interoceptors but all function in informing the CNS about the condition of the body. Stretch receptors in muscles allow the CNS to control muscle movements (**figure 29.2**). Baroreceptors in blood vessels (**figure 29.3**) provide the CNS with a continuous measure of blood pressure. These and other receptors communicate with the CNS, and the body can then respond to maintain a stable internal condition.

Sensory Reception

29.2 Sensing Gravity and Motion

- Sensory receptors in the inner ear sense gravity and acceleration. Depending on how these receptors are stimulated, the body can respond to maintain balance.

- Hair cells within the utricle and saccule of the inner ear detect gravity when they are stimulated by the movement of otoliths in a gelatin-like matrix (**figure 29.4**).

- Motion is detected by the deflection of hair cells in the cupula of the semicircular canals (**figure 29.4**). There are three semicircular canals, each oriented in a different plane so that movement in any of the three planes can be detected.

29.3 Sensing Chemicals: Taste and Smell

- Chemicals are detected through the senses of taste, using taste buds on the tongue. Taste receptor cells can detect salty, sour, sweet, bitter, and umami ("meaty") tastes (**figure 29.5**).

- Chemicals are also detected through smell, using olfactory receptors that line the nasal passages (**figure 29.6**).

29.4 Sensing Sound: Hearing

- Sound is detected by the ear when the air vibrates and pushes the eardrum. When the eardrum moves, three small bones called ossicles are pushed, which transfers the vibration across a second membrane to a fluid-filled chamber in the inner ear called the cochlea.

- The tips of the hair cells in the inner chamber of the cochlea are in contact with a membrane that vibrates in response to sound waves. As a sound wave travels the length of the cochlea, different groups of hair cells are stimulated (**figure 29.7**).

- In fish, a lateral line system supplements hearing. Cupula containing hair cells project into a fluid-filled canal. As the water in the canal is displaced by pressure waves, the hair cells are stimulated (**figure 29.8**).

- Some animals use a type of sonar to map out a picture of their environment using sound waves. A sound is emitted by the animal, and sensory receptors can detect the sound as it returns, after bouncing off an object in its environment. This process, called echolocation, is used by both terrestrial and aquatic animals (**figure 29.9**).

29.5 Sensing Light: Vision

- Sensory receptors in the eye called photoreceptors detect light. Simple visual systems that consist of photoreceptors clustered in an eyespot can be seen in some invertebrates. In the flatworm, the eyespot can perceive the direction of light but cannot form a visual image (**figure 29.10**).

- In other invertebrates, and in vertebrates, well-developed, image-forming eyes have evolved independently (**figure 29.11**).

- In the human eye, light passes through the cornea and pupil and is focused by the lens onto the retina, the rear surface of the eye (**figure 29.12**).

- The vertebrate retina contains two kinds of light-sensitive receptor cells (photoreceptors) called rods and cones. Rod cells can detect various shades of gray, even in dim light, while cone cells detect different colors of light and produce sharp images (**figure 29.13**).

- The center of the retina contains an area densely packed with cones called the fovea.

- Rods and cones can detect light because they contain pigment molecules. The visual pigment in humans is *cis*-retinal. When it absorbs a photon of light, the structure of *cis*-retinal changes to *trans*-retinal, which stimulates the sensory receptor (**figure 29.14**).

- Rods absorb light at wavelengths of 500 nanometers (nm), and humans have three kinds of cone cells, absorbing three different wavelengths of light. Blue cones absorb light at 420 nm, green cones absorb light at 530 nm, and red cones absorb light at 560 nm (**figure 29.15**).

- Color-blindness results when a person lacks one or more types of cone cells and so cannot see all the colors of the visual spectrum (**figure 29.16**).

- Light passes through several different layers of cells in the retina before it stimulates the rod and cone cells. The nerve impulse then travels back toward the optic nerve, opposite to the direction of light (**figure 29.17**).

- Eyes positioned on the front of the face, like in humans, allow for binocular vision that aids in depth perception. Other animals have eyes positioned on the sides of the head, allowing for a broader view of their surroundings (**figure 29.18**).

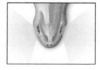

29.6 Other Vertebrate Senses

- Pit vipers can detect infrared radiation with heat-detecting pit organs (**figure 29.19**).

- Electroreceptors have evolved in some types of aquatic animals, and many types of organisms appear to navigate along earth's magnetic field.

Test Your Understanding

1. What is the correct order for the stages of sensory perception?
   a. transduction, stimulation, transmission
   b. transmission, transduction, stimulation
   c. stimulation, transmission, transduction
   d. stimulation, transduction, transmission
2. Which of the following is an example of an exteroceptor?
   a. stretch receptors
   b. rods and cones
   c. baroreceptors
   d. temperature receptors
3. When arm muscles hurt after heavy exercise, the pain is detected by
   a. neurotransmitters.
   b. interoceptors.
   c. associative neurons.
   d. exteroceptors.
4. Which of the following structures of the ear is associated with sensing motion and gravity?
   a. cochlea
   b. ear bones (the ossicles)
   c. semicircular canals
   d. eardrum
5. Which of the following could *not* provide an animal information about its food?
   a. photoreceptors
   b. taste receptors
   c. pain receptors
   d. smell receptors
6. The ear detects sound by the movement of
   a. hair cells within the cupula.
   b. a membrane within the cochlea.
   c. otoliths within a gelatinous matrix.
   d. the eustachian tube.
7. The lateral line system in fish detects
   a. pressure waves.
   b. light.
   c. taste.
   d. magnetism.
8. Which of the following statements is incorrect?
   a. Vertebrates focus the eye by changing the shape of the lens.
   b. The eyes of arthropods and vertebrates use the same light-capturing molecule.
   c. Rod cells detect different colors, and cone cells detect different shades of gray, allowing vision in dim light.
   d. Light changes *cis*-retinal into *trans*-retinal.
9. Binocular vision
   a. is more common in prey animals.
   b. enlarges the overall receptive field.
   c. allows for a better depth of perception.
   d. is only possible when the fields of vision do not overlap.
10. Which of the following can some types of animals detect in their environment?
    a. infrared heat
    b. magnetic fields
    c. electric fields
    d. All of the above.

Apply Your Understanding

1. **Figure 29.3** Your aunt stood up suddenly at Sunday dinner and then fainted. Her fainting from standing up too quickly might involve a problem with what sensory receptor?

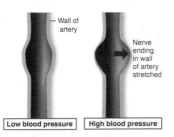

Wall of artery

Nerve ending in wall of artery stretched

Low blood pressure   High blood pressure

2. **Figure 29.11** What do the visual sensory systems of annelids, mollusks, arthropods, and vertebrates have in common, even though they evolved separately?

Synthesize What You Have Learned

1. A friend who knows that you are afraid of snakes told you that if you wrap yourself in insulation when you go into the woods at night, the snakes won't be able to find you. You now realize, after reading this chapter, that your friend was right (sort of). Why?
2. How would you expect the otoliths of an astronaut to respond to zero gravity? Would the astronaut still be able to detect motion? Would the semicircular canals detect angular acceleration in the absence of gravity?
3. If you can taste only five things (salt, bitter, sweet, sour, umami), why do the many foods you eat all seem so distinct?
4. A progressive disease of the eye called macular degeneration causes the death of photoreceptors at the center of the eye. Would you expect the entire field of vision to be affected? Explain.
5. If you enter a dark room, at first you can see nothing but darkness. Within a few minutes, however, you can begin to make out shadowy forms, and after 10 to 30 minutes, you are able to see considerable detail. What adjustment could your eyes be making to cause this effect? [Hint: Focus on rhodopsin.]

30

Chemical Signaling Within the Animal Body

In vertebrates and most other animals, the central nervous system coordinates and regulates the diverse activities of the body by using chemical signals called hormones to effect changes in physiological activities. Many hormones—adrenaline (epinephrine), estrogen, testosterone, insulin, thyroid hormone—are probably familiar to you. Some of these hormones have very different roles in other animals, however. For example, thyroid hormone is needed in amphibians for the metamorphosis of larvae into adults. If the thyroid gland is removed from a tadpole, it will not change into a frog. Conversely, if an immature tadpole is fed pieces of a thyroid gland, it will undergo premature metamorphosis and become a miniature frog! Similarly, melanocyte-stimulating hormone, a peptide hormone, is present in mammals, but we don't know what it does. In reptiles and amphibians, this hormone stimulates color changes. The green anole (*Anolis carolinensis*) shown here in the upper photo has changed to a tan color in the lower photo, in response to an environmental or physiological cue. This color change is the result of dispersal of pigment-containing granules from the center of the anole's skin cells into extensions of the cells, darkening the skin. Triggered by melanocyte-stimulating hormone, the reversible color change takes 5 to 10 minutes. In this chapter you will encounter many other hormones, some familiar, others less so, but all used by vertebrates to regulate their body condition.

30.1 Hormones

A **hormone** is a chemical signal produced in one part of the body that is stable enough to be transported in active form far from where it is produced and that typically acts at a distant site. There are three big advantages to using chemical hormones as messengers rather than speedy electrical signals (like those used in nerves) to control body organs. First, chemical molecules can spread to all tissues via the blood (imagine trying to wire every cell with its own nerve!) and are usually required in only small amounts. Second, chemical signals can persist much longer than electrical ones, a great advantage for hormones controlling slow processes like growth and development. Third, many different kinds of chemicals can act as hormones, so different hormone molecules can be targeted at different tissues. For all these reasons, hormones are excellent messengers for signaling widespread, slow-onset, long duration responses.

Hormones, in general, are produced by glands, most of which are controlled by the central nervous system. Because these glands are completely enclosed in tissue rather than having ducts that empty to the outside, they are called **endocrine glands** (from the Greek, *endon,* within). Hormones are secreted from them directly into the bloodstream (this is in contrast to **exocrine glands,** like sweat glands, that have ducts). Your body has a dozen principal endocrine glands, shown in **figure 30.1,** that together make up the endocrine system.

The *endocrine system* and the *motor nervous system* are the two main routes the central nervous system (CNS) uses to issue commands to the organs of the body. The two are so closely linked that they are often considered a single system—the **neuroendocrine system.** The **hypothalamus** can be considered the main switchboard of the neuroendocrine system. The hypothalamus is continually checking conditions inside the body to maintain a constant internal environment, a condition known as homeostasis. Is the body too hot or too cold? Is it running out of fuel? Is the blood pressure too high? If homeostasis is no longer maintained, the hypothalamus has several ways to set things right again. For example, if the hypothalamus needs to speed up the heart rate, it can send a nerve signal to the medulla oblongata, or it can use a chemical command, causing the adrenal gland to produce the hormone adrenaline, which also speeds up the heart rate. Which command the hypothalamus uses depends on the desired duration of the effect. A chemical message is typically far longer lasting than a nerve signal.

The Chain of Command

The hypothalamus issues commands to a nearby gland, the pituitary, which in turn sends chemical signals to the various hormone-producing glands of the body. The pituitary is suspended from the hypothalamus by a short stalk, across which chemical messages pass from the hypothalamus to the pituitary. The first of these chemical messages to be discovered

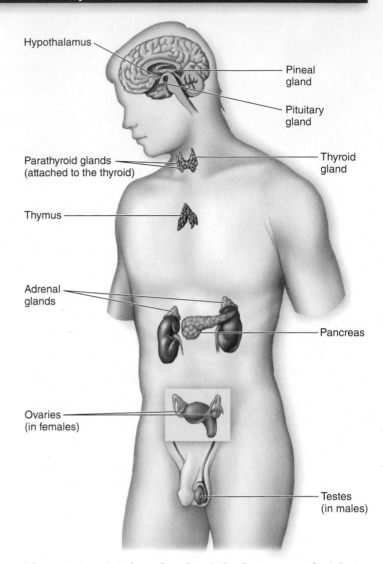

Figure 30.1 Major glands of the human endocrine system.

Hormone-secreting cells are clustered in endocrine glands. The pituitary and adrenal glands are each composed of two glands.

was a short peptide called thyrotropin-releasing hormone (TRH), which was isolated in 1969. The release of TRH from the hypothalamus triggers the pituitary to release a hormone called thyrotropin, or thyroid-stimulating hormone (TSH), which travels to the thyroid and causes the thyroid gland to release thyroid hormones.

Several other hypothalamic hormones have since been isolated, which together govern the pituitary. Thus, the CNS regulates the body's hormones through a chain of command. The "releasing" hormones made by the hypothalamus cause the pituitary to synthesize a corresponding pituitary hormone, which travels to a distant endocrine gland and causes that gland to begin producing its particular endocrine hormone. The hypothalamus also secretes inhibiting hormones that keep the pituitary from secreting specific pituitary hormones.

KEY BIOLOGICAL PROCESS: Hormonal Communication

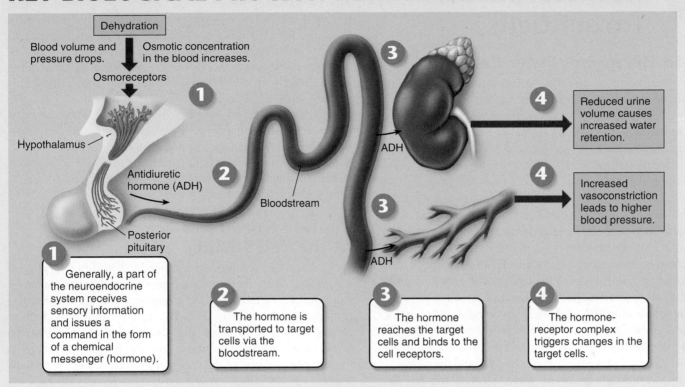

Dehydration

Blood volume and pressure drops.

Osmotic concentration in the blood increases.

Osmoreceptors

1

Hypothalamus

Antidiuretic hormone (ADH)

2

Bloodstream

Posterior pituitary

3

ADH

3

ADH

4 Reduced urine volume causes increased water retention.

4 Increased vasoconstriction leads to higher blood pressure.

1 Generally, a part of the neuroendocrine system receives sensory information and issues a command in the form of a chemical messenger (hormone).

2 The hormone is transported to target cells via the bloodstream.

3 The hormone reaches the target cells and binds to the cell receptors.

4 The hormone-receptor complex triggers changes in the target cells.

How Hormones Work

The key reason why hormones are effective messengers within the body is because a particular hormone can influence a specific target cell. How does the target cell recognize that hormone, ignoring all others? Embedded in the plasma membrane or within the target cell are receptor proteins that match the shape of the potential signal hormone like a hand fits a glove. As you recall from chapter 28, nerve cells have highly specific receptors within their synapses, each receptor shaped to "respond" to a different neurotransmitter molecule. Similarly, cells that the body has targeted to respond to a particular hormone have receptor proteins shaped to fit that hormone and no other. Thus, chemical communication within the body involves *two* elements: a molecular signal (the hormone) and a protein receptor on or in target cells. The system is highly specific because each protein receptor has a shape that only a particular hormone fits.

Hormones secreted by endocrine glands belong to four different chemical categories:

1. **Polypeptides** are composed of chains of amino acids that are shorter than about 100 amino acids. Some important examples include insulin and antidiuretic hormone (ADH).
2. **Glycoproteins** are composed of polypeptides significantly longer than 100 amino acids to which are attached carbohydrates. Examples include follicle-stimulating hormone (FSH) and luteinizing hormone (LH).

3. **Amines,** derived from the amino acids tyrosine and tryptophan, include hormones secreted by the adrenal medulla, thyroid, and pineal glands.
4. **Steroids** are lipids derived from cholesterol, and include the hormones testosterone, estrogen, progesterone, aldosterone, and cortisol.

The path of communication taken by a hormonal signal can be visualized as the series of simple steps shown in the Key Biological Process illustration above:

1. **Issuing the command.** The hypothalamus of the CNS controls the release of many hormones. Some hormones produced in cells in the hypothalamus are stored in the posterior pituitary and are released into the bloodstream in response to a signal from the brain.
2. **Transporting the signal.** While hormones can act on an adjacent cell, most are transported throughout the body by the bloodstream.
3. **Hitting the target.** When a hormone encounters a cell with a matching receptor, called a target cell, the hormone binds to that receptor.
4. **Having an effect.** When the hormone binds to the receptor protein, the protein responds by changing shape, which triggers a change in cell activity.

> **Key Learning Outcome 30.1** Hormones are effective because they are recognized by specific receptors. Thus, only cells possessing the appropriate receptor will respond to a particular hormone.

30.2 How Hormones Target Cells

Steroid Hormones Enter Cells

Some protein receptors designed to recognize hormones are located in the cytoplasm or nucleus of the target cell. The hormones in these cases are lipid-soluble molecules, typically **steroid hormones.** The chemical shapes of these molecules are multi-ring structures resembling chicken wire. All steroid hormones are manufactured from cholesterol, a complex molecule composed of four rings. The hormones that promote the development of secondary sexual characteristics are steroids. They include testosterone, as well as estrogen and progesterone, which control the female reproductive system and are discussed in chapter 31. Cortisol is also a steroid hormone.

Steroid hormones, like estrogen, "E" in **figure 30.2,** can pass across the lipid bilayer of the cell plasma membrane ❶, and bind to receptors within the cell and often, as in the case with estrogen, within the nucleus. This complex of receptor and hormone then binds to the DNA in the nucleus ❷, and activates the gene for a progesterone receptor protein, which is transcribed ❸. The protein is synthesized ❹, and the receptor is available to bind progesterone when it enters the cell ❺, which itself activates another set of genes.

Anabolic steroids, which are used by some weight lifters and other athletes, are synthetic compounds that resemble the male sex hormone testosterone. Their injection into muscles activates growth genes and causes the muscle cells to produce more protein, resulting in bigger muscles and increased strength. However, anabolic steroids have many dangerous side effects in both men and women, including liver damage, heart disease, high blood pressure, acne, balding, and psychological disorders. Men can also experience the suppression of testicular function and feminization, and women can undergo masculinization. Use in adolescents can also result in stunted growth and accelerated puberty changes. Anabolic steroids are illegal, and for many sports, athletes are tested for their presence.

Peptide Hormones Act at the Cell Surface

Other hormone receptors are embedded within the plasma membrane, with their recognition region directed outward from the cell surface. Peptide hormones, like the one binding to the receptor in **figure 30.3** ❶, are typically short peptide chains (although some are full-sized proteins). The binding of the **peptide hormone** to the receptor triggers a change in the cytoplasmic end of the receptor protein. This change then triggers events within the cell cytoplasm, usually through intermediate within-cell signals called **second messengers** ❷,

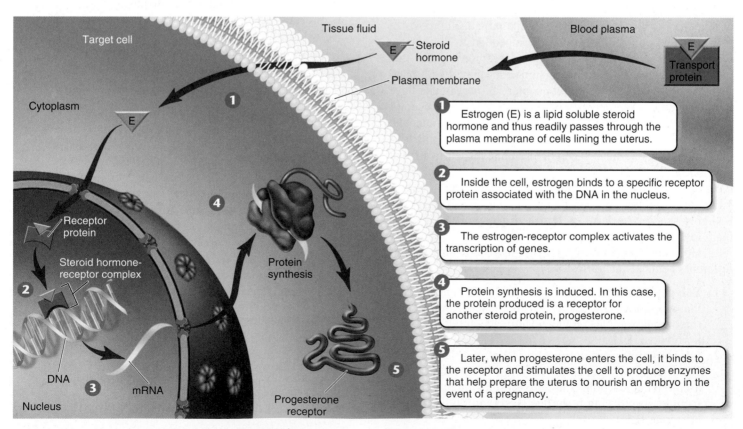

❶ Estrogen (E) is a lipid soluble steroid hormone and thus readily passes through the plasma membrane of cells lining the uterus.

❷ Inside the cell, estrogen binds to a specific receptor protein associated with the DNA in the nucleus.

❸ The estrogen-receptor complex activates the transcription of genes.

❹ Protein synthesis is induced. In this case, the protein produced is a receptor for another steroid protein, progesterone.

❺ Later, when progesterone enters the cell, it binds to the receptor and stimulates the cell to produce enzymes that help prepare the uterus to nourish an embryo in the event of a pregnancy.

Figure 30.2 How steroid hormones work.

KEY BIOLOGICAL PROCESS: Second Messengers

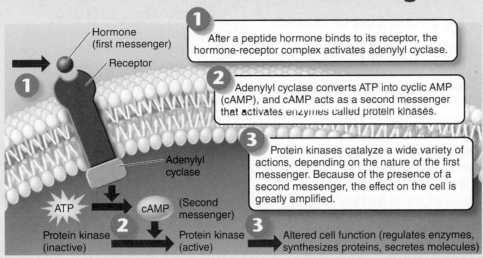

1. After a peptide hormone binds to its receptor, the hormone-receptor complex activates adenylyl cyclase.

2. Adenylyl cyclase converts ATP into cyclic AMP (cAMP), and cAMP acts as a second messenger that activates enzymes called protein kinases.

3. Protein kinases catalyze a wide variety of actions, depending on the nature of the first messenger. Because of the presence of a second messenger, the effect on the cell is greatly amplified.

Hormone (first messenger)
Receptor
Adenylyl cyclase
ATP
cAMP (Second messenger)
Protein kinase (inactive)
Protein kinase (active)
Altered cell function (regulates enzymes, synthesizes proteins, secretes molecules)

which greatly amplify the original signal and result in changes in the cell ❸.

How does a second messenger amplify a hormone's signal? Second messengers activate enzymes. One of the most common second messengers, cyclic AMP (cAMP), is shown in the Key Biological Process illustration above. Cyclic AMP is made from ATP by an enzyme that removes two phosphate units, forming AMP; the ends of the AMP join, forming a circle. A single hormone molecule binding to a receptor in the plasma membrane can result in the formation of many second messengers in the cytoplasm. Each second messenger can activate many molecules of a certain enzyme, and sometimes each of these enzymes can in turn activate many other enzymes. Thus, second messengers enable each hormone molecule to have a tremendous effect inside the cell, far greater than if the hormone had simply entered the cell and sought out a single target.

Insulin is one of many hormones that acts through second messenger systems, and it provides a well-studied example of how peptide hormones achieve their effects within target cells. Most human cells have insulin receptors in their membranes—typically only a few hundred but far more in tissues involved in glucose metabolism. A single liver cell, for example, may have 100,000 of them. When insulin binds to one of these insulin receptors, the receptor protein changes its shape, prodding an adjacent signal-modulating protein on the cell interior to activate the release of Ca^{++} ions. The Ca^{++} acts as a second messenger, activating a variety of cellular enzymes in a cascading series of events that greatly amplifies the strength of the original signal.

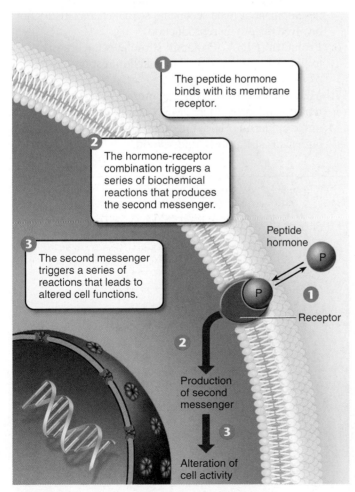

1. The peptide hormone binds with its membrane receptor.

2. The hormone-receptor combination triggers a series of biochemical reactions that produces the second messenger.

3. The second messenger triggers a series of reactions that leads to altered cell functions.

Peptide hormone
Receptor
Production of second messenger
Alteration of cell activity

Figure 30.3 How peptide hormones work.

Key Learning Outcome 30.2 Steroid hormones pass through the cell's plasma membrane and bind to receptors in the cell, forming a complex that alters the transcription of specific genes. Peptide hormones do not enter cells. Instead, they bind to receptors on the target cell surface, triggering a cascade of enzymic activations within the cell.

30.3 The Hypothalamus and the Pituitary

The hypothalamus, the "control center" of the neuroendocrine system, exerts its control by releasing hormones that influence the nearby **pituitary gland,** located in a bony recess in the brain just below the hypothalamus. The pituitary in turn produces hormones that influence the body's other endocrine glands. Hormones produced by the back portion of the pituitary, or *posterior lobe,* regulate water conservation, milk letdown and uterine contraction in women; hormones produced by the front portion, or *anterior lobe,* regulate the other endocrine glands.

The Posterior Pituitary

The posterior pituitary contains axons that originate in cells within the hypothalamus (see **figure 30.5**). The hormones released from the posterior pituitary are actually produced by neuron cell bodies located in the hypothalamus. The hormones are transported to the posterior pituitary through axon tracts and are stored and released from the posterior pituitary.

The role of the posterior pituitary first became evident in 1912, when a remarkable medical case was reported: A man who had been shot in the head developed a surprising disorder—he began to urinate every 30 minutes, unceasingly. The bullet had lodged in his pituitary gland, and subsequent research demonstrated that surgical removal of the pituitary also produces these unusual symptoms. Pituitary extracts were shown to contain a substance that makes the kidneys conserve water, and in the early 1950s the peptide hormone **vasopressin** (also called **antidiuretic hormone, ADH**) was isolated. As you learned in chapter 26, ADH regulates the kidney's retention of water. When ADH is missing, the kidneys cannot retain water, which is why the bullet caused excessive urination. Excessive alcohol and caffeine consumption, which inhibit ADH secretion, have a similar effect.

The posterior pituitary also releases a second hormone, **oxytocin,** of very similar structure—both are short peptides composed of nine amino acids—but very different function. Oxytocin initiates uterine contraction during childbirth and milk release in mothers. Here is how milk release works: Sensory receptors in the mother's nipples, when stimulated by sucking, send messages to the hypothalamus, causing the hypothalamus to stimulate the release of oxytocin from the posterior pituitary. The oxytocin travels in the bloodstream to the breasts where it stimulates contraction of the muscles around the ducts into which the mammary glands secrete milk. Both oxytocin and ADH are produced in the cell bodies of the hypothalamus but stored and released from the posterior pituitary.

The Anterior Pituitary

The anterior pituitary gland produces seven major peptide hormones (blue in **figure 30.4**), each controlled by a particular releasing signal secreted from the hypothalamus:

1. **Thyroid-stimulating hormone (TSH).** TSH stimulates the thyroid gland to produce the thyroid hormone thyroxine, which in turn stimulates oxidative respiration.

2. **Adrenocorticotropic hormone (ACTH).** ACTH stimulates the adrenal gland to produce a variety of steroid hormones. Some regulate the production of glucose from fat; others regulate the balance of sodium and potassium ions in the blood.

3. **Growth hormone (GH).** GH stimulates the growth of muscle and bone throughout the body.

4. **Follicle-stimulating hormone (FSH).** FSH is significant in the female menstrual cycle by triggering the maturation of egg cells and stimulating the release of estrogen. In males, it stimulates cells in the testes, regulating development of the sperm.

5. **Luteinizing hormone (LH).** LH plays an important role in the female menstrual cycle by triggering ovulation, the release of a mature egg. It also stimulates the male gonads to produce testosterone, which initiates and maintains the development of male secondary sexual characteristics not involved directly in reproduction.

6. **Prolactin (PRL).** Prolactin stimulates the breasts to produce milk, which is released in response to oxytocin.

7. **Melanocyte-stimulating hormone (MSH).** In reptiles and amphibians, MSH stimulates color changes in the epidermis. The function of this hormone in humans is still poorly understood.

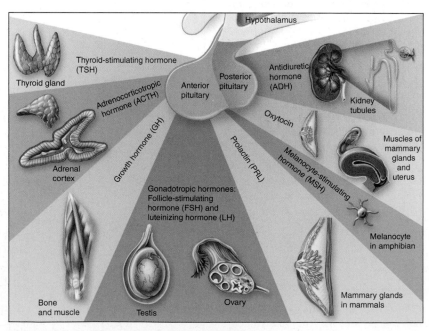

Figure 30.4 The role of the pituitary.

How the Hypothalamus Controls the Anterior Pituitary

As noted earlier, the hypothalamus controls production and secretion of the anterior pituitary hormones by means of a family of special hormones. Neurons in the hypothalamus secrete these releasing and inhibiting hormones into blood capillaries at the base of the hypothalamus. **Figure 30.5** shows the relationship of the two groups of neurons in the hypothalamus. As discussed earlier, some neurons (colored blue in the figure) extend into the posterior pituitary where axons deliver the hormones for storage and release. Other neurons of the hypothalamus (colored yellow in the figure) produce releasing and inhibiting hormones and release them into capillaries. These capillaries drain into small veins that run within the stalk of the pituitary to a second bed of capillaries in the anterior pituitary. This unusual system of vessels is known as the *hypothalamohypophyseal portal system*. It is called a portal system because it has a second capillary bed downstream from the first; the only other body location with a similar system is the liver.

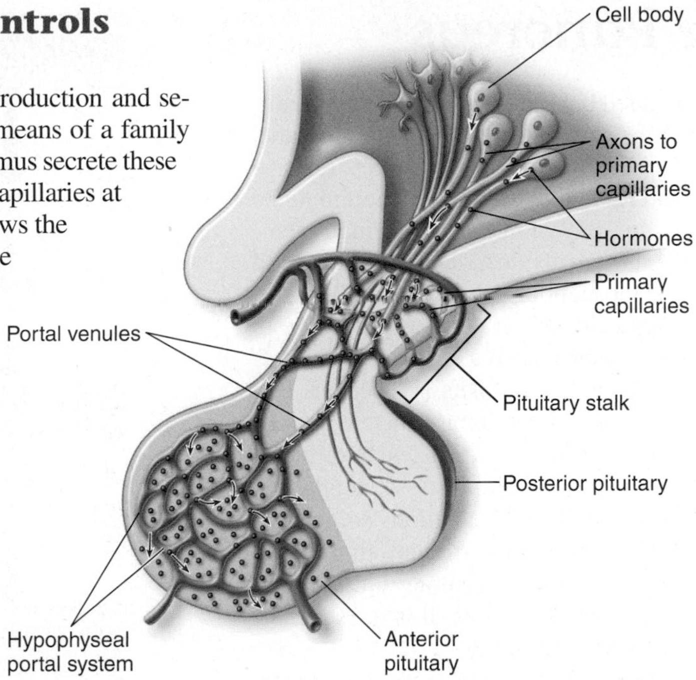

Each releasing hormone delivered to the anterior pituitary by this portal system regulates the secretion of a specific anterior pituitary hormone. For example, thyrotropin-releasing hormone (TRH) stimulates the release of TSH; corticotropin-releasing hormone (CRH) stimulates the release of ACTH; gonadotropin-releasing hormone (GnRH) stimulates the release of FSH and LH; growth-hormone-releasing hormone (GHRH) stimulates the release of GH; and prolactin-releasing factor (PRF) stimulates the release of prolactin—however, this factor has not yet been chemically characterized and may actually be a chemical similar to thyrotropin-releasing hormone.

The hypothalamus also secretes hormones that inhibit the release of certain anterior pituitary hormones. To date, three such hormones have been discovered: Somatostatin inhibits the secretion of GH; prolactin-inhibiting hormone (PIH), possibly dopamine, inhibits the secretion of prolactin; and melanotropin-inhibiting hormone (MIH) inhibits the secretion of MSH.

Because hypothalamic hormones control the secretions of the anterior pituitary gland, and the anterior pituitary hormones control the secretions of some other endocrine glands, it may seem that the hypothalamus functions as a "master gland," in charge of hormonal secretion in the body. This idea is not generally valid, however, for two reasons. First, a number of endocrine organs, such as the adrenal medulla and the pancreas, are not directly regulated by this control system. Second, the hypothalamus and the anterior pituitary gland are themselves controlled by the very hormones whose secretion they stimulate! In most cases this is an inhibitory control. **Figure 30.6** shows how the target gland hormones inhibit the hypothalamus and anterior pituitary. When enough of the target gland hormone has been produced, the hormone then feeds back and inhibits the release of the stimulating hormones from the hypothalamus and anterior pituitary, indicated by the

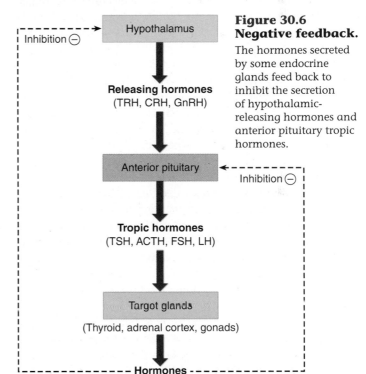

Figure 30.5 Hormonal control of the anterior pituitary gland by the hypothalamus.

Neurons in the hypothalamus secrete hormones that are carried by short blood vessels to the anterior pituitary gland, where they either stimulate or inhibit the secretion of anterior pituitary hormones.

Labels: Cell body; Axons to primary capillaries; Hormones; Primary capillaries; Pituitary stalk; Posterior pituitary; Anterior pituitary; Hypophyseal portal system; Portal venules

Figure 30.6 Negative feedback.

The hormones secreted by some endocrine glands feed back to inhibit the secretion of hypothalamic-releasing hormones and anterior pituitary tropic hormones.

Hypothalamus
Inhibition ⊖
Releasing hormones (TRH, CRH, GnRH)
Anterior pituitary
Inhibition ⊖
Tropic hormones (TSH, ACTH, FSH, LH)
Target glands (Thyroid, adrenal cortex, gonads)
Hormones

dashed lines. This type of control system is an example of **negative feedback (or feedback inhibition),** which was also discussed in chapter 26.

Key Learning Outcome 30.3 The posterior pituitary gland contains axons originating from neurons in the hypothalamus that produce hormones. The anterior pituitary responds to hormonal signals from the hypothalamus and produces a family of pituitary hormones that are carried to distant glands that produce specific hormones.

30.4 The Pancreas

The **pancreas** gland is located behind the stomach and is connected to the front end of the small intestine by a narrow tube. It secretes a variety of digestive enzymes into the digestive tract through this tube, and for a long time it was thought to be solely an exocrine gland. In 1869, however, a German medical student named Paul Langerhans described some unusual clusters of cells scattered throughout the pancreas. In 1893, doctors concluded that these clusters of cells, which came to be called islets of Langerhans, produced a substance that prevented diabetes mellitus. **Diabetes mellitus** is a serious disorder in which affected individuals' cells are unable to take up glucose from the blood, even though their levels of blood glucose become very high. Some individuals lose weight and literally starve; others develop poor circulation, sometimes resulting in amputation of limbs with restricted circulation. Diabetes is the leading cause of blindness among adults, and it accounts for one-third of all kidney failures. It is the seventh-leading cause of death in the United States.

The substance produced by the islets of Langerhans, which we now know to be the peptide hormone *insulin,* was not isolated until 1922. Two young doctors working in a Toronto hospital injected an extract purified from beef pancreas glands into a 13-year-old boy, a diabetic whose weight had fallen to 29 kilograms (65 pounds) and who was not expected to survive. The hospital record gives no indication of the historic importance of the trial, only stating, "15 cc of MacLeod's serum. 7-1/2 cc into each buttock." With this single injection, the glucose level in the boy's blood fell 25%—his cells were taking up glucose. A more potent extract soon brought levels down to near normal.

This was the first instance of successful insulin therapy. The islets of Langerhans in the pancreas produce two hormones that interact to govern the levels of glucose in the blood. These hormones are *insulin* and *glucagon.* Insulin is a storage hormone, designed to put away nutrients for leaner times. It promotes the accumulation of glycogen in the liver and triglycerides in fat cells. When food is consumed (left side of **figure 30.7**), beta cells in the islets of Langerhans secrete insulin, causing the cells of the body to take up and store glucose as glycogen and triglycerides to be used later. When body activity causes the level of glucose in the blood to fall as it is used as fuel (right side of **figure 30.7**), other cells in the islets of Langerhans, called alpha cells, secrete glucagon, which causes liver cells to release stored glucose and fat cells to break down triglycerides for energy use. The two hormones work together to keep glucose levels in the blood within a narrow range.

Over 23 million people in the United States, and over 246 million people worldwide, have **diabetes.** There are two kinds of diabetes mellitus. About 5% to 10% of affected individuals suffer from type I diabetes, an autoimmune disease in which the immune system attacks the islets of Langerhans, resulting in abnormally low insulin secretion. Called juvenile-onset diabetes, this type usually develops before age 20. Affected individuals can be treated by daily injections of insulin. Active research on the possibility of transplanting islets

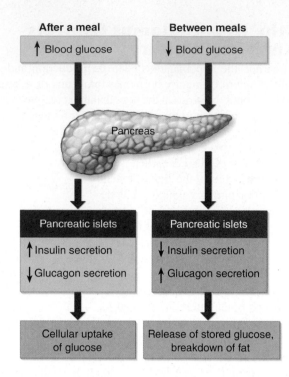

Figure 30.7   Insulin and glucagon secreted by the pancreas regulate blood glucose levels.

After a meal, an increased secretion of insulin by the beta cells of the pancreatic islets of Langerhans promotes the movement of glucose from blood into tissue cells. Between meals, an increased secretion of glucagon by the alpha cells of the pancreatic islets and decreased secretion of insulin cause the release of stored glucose and the breakdown of fat.

of Langerhans holds much promise as a lasting treatment for this form of diabetes.

In type II diabetes, the level of insulin in the blood is often higher than normal, but the cells don't respond to insulin. This form of diabetes usually develops in people over 40 years of age. It is almost always a consequence of excessive weight; in the United States, 80% of those who develop type II diabetes are obese. The cells of some type II diabetics, overwhelmed with food, adjust their appetite for glucose downward by reducing their sensitivity to insulin. Similar to the way a drug addict's neurons reduce their number of neurotransmitter receptors after continued exposure to a drug, the obese individual's cells reduce their number of insulin receptors. To compensate, the pancreas pumps out ever-more insulin, and, in some people, the insulin-producing cells are unable to keep up with the ever-heavier workload and stop functioning. Type II diabetes is usually treatable with diet and exercise, and most affected individuals do not need daily injections of insulin.

Key Learning Outcome 30.4   Clusters of cells within the pancreas secrete the hormones insulin and glucagon. Insulin stimulates the storage of glucose as glycogen, while glucagon stimulates glycogen breakdown to glucose. Working together, these hormones keep glucose levels within a narrow range.

The Type II Diabetes Epidemic

We Americans love to eat, but recently the Centers for Disease Control and Prevention released a report warning we are eating ourselves into a diabetes epidemic. Diabetes affected 7 million Americans in 1991. By the end of 2007, the number was over 23 million, more than 7% of all Americans—an alarming increase in just 16 years!

The same explosion of diabetes is being seen worldwide. Diabetes now affects 246 million people, and kills 3.8 million each year. Every 10 seconds one person dies of diabetes. In the same 10 seconds two more people develop the disease.

Diabetes is a disorder in which the body's cells fail to take up glucose from the blood. Tissues waste away as glucose-starved cells are forced to consume their own proteins. Diabetes is the leading cause of kidney failure, blindness, and amputation in adults. Almost all the increase in diabetes in the last decade is in the 90% of diabetics who suffer from type II, or "adult-onset," diabetes. These individuals lack the ability to use the hormone insulin.

Your body manufactures insulin after a meal as a way to alert cells that higher levels of glucose are coming soon. The insulin signal attaches to special receptors on the cell surfaces, which respond by causing the cell to turn on its glucose-transporting machinery. Some individuals who suffer from type II diabetes have normal or even elevated levels of insulin in their blood, and normal insulin receptors, but for some reason the binding of insulin to their cell receptors does not turn on the glucose-transporting machinery like it is supposed to do. For 30 years researchers have been trying to figure out why not.

How does insulin act to turn on a normal cell's glucose transporting machinery? Proteins called IRS proteins (the names refer not to tax collectors, but to *insulin receptor substrate*) snuggle up against the insulin receptor inside the cell. When insulin attaches to the receptor protein, the receptor responds by adding a phosphate group onto the IRS molecules. Like being touched by a red-hot poker, this galvanizes the IRS molecules into action. Dashing about, they activate a variety of processes, including an enzyme that turns on the glucose-transporting machinery.

When the IRS genes are deliberately taken out of action in so-called "knockout" mice, type II diabetes results. Are defects in the genes for IRS proteins responsible for type II diabetes? Probably not. When researchers look for IRS gene mutations in inherited type II diabetes, they don't find them. The IRS genes are normal.

This suggests that in type II diabetes something is interfering with the action of the IRS proteins. What might it be? An estimated 80% of those who develop type II diabetes are obese, a tantalizing clue. Look at the graph. Over the same 16 years that diabetes has undergone its explosive increase, the obesity rate increased from 12% of the U.S. population to 34%.

What is the link between diabetes and obesity? Recent research suggests an answer to this key question. A team

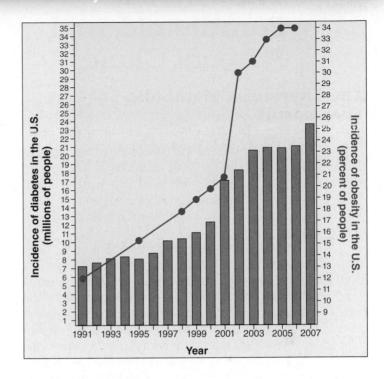

of scientists at the University of Pennsylvania School of Medicine had been investigating why a class of drugs called thiazolidinediones (TZDs) helped combat diabetes. They found that TZDs cause the body's cells to use insulin more effectively, and this suggested to them that the TZD drug might be targeting a hormone.

The researchers then set out to see if they could find such a hormone in mice. In search of a clue, they started by looking to see which mouse genes were activated or deactivated by TZD. Several were. Examining them, they were able to zero in on the hormone they sought. Dubbed *resistin,* the hormone is produced by fat cells and prompts tissues to resist insulin. The same *resistin* gene is present in humans too. The researchers speculate that resistin may have evolved to help the body deal with periods of famine.

Mice given resistin by the researchers lost much of their ability to take up blood sugar. When given a drug that lowers resistin levels, these mice recovered the lost glucose-transporting ability.

Researchers don't yet know how resistin acts to lower insulin sensitivity, although blocking the action of IRS proteins seems a likely possibility.

Importantly, dramatically high levels of the hormone were found in mice obese from overeating. Finding this sort of result is like ringing a dinner bell to diabetes researchers. If obesity is causing high resistin levels in humans, leading to type II diabetes, then resistin-lowering drugs might offer a diabetes cure!

On the scent of something important, resistin researchers are now shifting their efforts from mice to humans. Much needs to be checked, as there are no guarantees that what works in a mouse will do so in the same way in a human. Still, the excitement is tangible.

The Thyroid, Parathyroid, and Adrenal Glands

The Thyroid: A Metabolic Thermostat

The **thyroid gland** is shaped like a shield (its name comes from *thyros,* the Greek word for "shield") and lies just below the Adam's apple in the front of the neck. The thyroid makes several hormones, the two most important of which are **thyroxine,** which increases metabolic rate and promotes growth, and **calcitonin,** which inhibits the release of calcium from bones.

Thyroxine regulates the level of metabolism in the body in several important ways. Without adequate thyroxine, growth is retarded. For example, children with underactive thyroid glands are not able to carry out carbohydrate breakdown and protein synthesis at normal rates, a condition called cretinism, which results in stunted growth. Mental retardation can also result because thyroxine is needed for normal development of the central nervous system. The thyroid is stimulated to produce thyroxine by the hypothalamus, which is inhibited by thyroxine via negative feedback. Recall from chapter 26 that in negative feedback, the target gland's hormone inhibits the stimulation of the gland. The dashed lines in **figure 30.8a** indicate that thyroxine inhibits the release of TRH and TSH from the hypothalamus and anterior pituitary, respectively. Thyroxine contains iodine, and if the amount of iodine in the diet is too low, the thyroid cannot make adequate amounts of thyroxine to keep the hypothalamus inhibited. The hypothalamus will then continue to stimulate the thyroid, which will grow larger in a futile attempt to manufacture more thyroxine. The greatly enlarged thyroid gland that results is called a goiter (**figure 30.8b**). This need for iodine in the diet is why iodine is added to table salt. You probably have had your thyroid examined. A doctor will probe the area around your throat, feeling for lumps or enlarged areas. An overactive thyroid, which produces too much thyroxine, can cause hyperthyroidism. The overproduction of thyroxine results in elevated metabolism, which causes increased heart rate, weight loss, elevated body temperature—all symptoms of elevated metabolism.

Calcitonin, which will be discussed later, plays a key role in maintaining proper calcium levels in the body.

The Parathyroids: Regulating Calcium

The **parathyroid glands** are four small glands attached to the thyroid. Small and unobtrusive, they were ignored by researchers until well into the last century. The first suggestion that the parathyroids produce a hormone came from experiments in which they were removed from dogs: The concentration of calcium in the dogs' blood plummeted to less than half the normal level. However, if an extract of the parathyroid gland was administered, calcium levels returned to normal. If

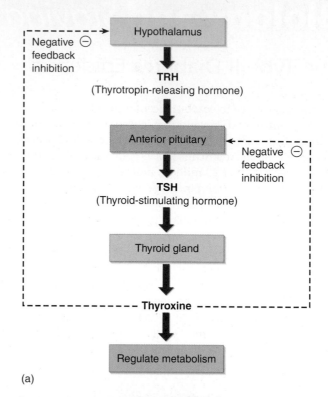

(a)

(b)

Figure 30.8   The thyroid gland secretes thyroxine.
(*a*) Thyroxine exerts negative feedback control of the hypothalamus and anterior pituitary. (*b*) A goiter is caused by a lack of iodine in the diet, which causes thyroxine secretion to decrease. As a result, there is less negative feedback, TSH is not inhibited from stimulating the thyroid gland, and the thyroid gland becomes enlarged.

an excess was administered, calcium levels in the blood became *too* high, and the bones of the dogs were literally dismantled by the extract. It was clear that the parathyroid glands were producing a hormone that acted on calcium, both its uptake into and release from bones.

The hormone produced by the parathyroids is **parathyroid hormone (PTH).** It is one of only two hormones in the body that is absolutely essential for survival (the other is aldosterone, a hormone produced by the adrenal glands, discussed on the next page). PTH regulates the level of calcium in blood. Recall that calcium ions are the key actors in muscle

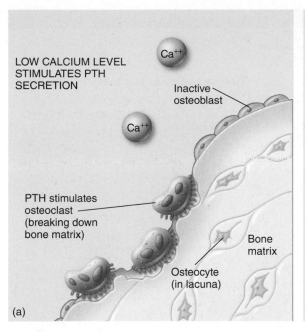

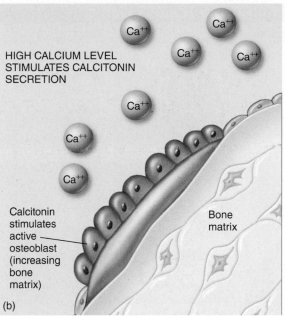

LOW CALCIUM LEVEL STIMULATES PTH SECRETION

Inactive osteoblast

PTH stimulates osteoclast (breaking down bone matrix)

Bone matrix

Osteocyte (in lacuna)

(a)

HIGH CALCIUM LEVEL STIMULATES CALCITONIN SECRETION

Calcitonin stimulates active osteoblast (increasing bone matrix)

Bone matrix

(b)

Figure 30.9 Maintenance of proper calcium levels in the blood. (a) When calcium levels in the blood become too low, the parathyroid gland produces additional amounts of PTH, which stimulates the breakdown of bone, releasing calcium. (b) Conversely, abnormally high levels of calcium in the blood trigger the thyroid gland to secrete calcitonin, which inhibits the release of calcium from bone, and promotes the activity of osteoblasts to remove calcium from the blood and deposit it in bone.

contraction—by initiating calcium release, nerve impulses cause muscles to contract. A vertebrate cannot live without the muscles that pump the heart and drive the body, and these muscles cannot function if calcium levels are not kept within narrow limits.

PTH acts as a fail-safe to make sure calcium levels never fall too low. If they do (**figure 30.9a**), PTH is released into the bloodstream, travels to the bones, and acts on the osteoclast cells (the blue cells) within bones, stimulating them to dismantle bone tissue and release calcium into the bloodstream. PTH also acts on the kidneys to reabsorb calcium ions from the filtrate and leads to activation of vitamin D, necessary for calcium absorption by the intestine. A diet deficient in vitamin D leads to poor bone formation, a condition called rickets. When PTH is synthesized by the parathyroids in response to falling levels of calcium in the blood, the body is essentially sacrificing bone to keep calcium levels within the narrow limits necessary for proper functioning of muscle and nerve tissues. Calcitonin, a hormone referred to earlier, is released from the thyroid gland and acts in reverse of PTH. When calcium levels in the blood rise (**figure 30.9b**), calcitonin activates osteoblast cells (the orange cells) to take up calcium, and rebuild bone.

The Adrenals: Two Glands in One

Mammals have two **adrenal glands,** one located just above each kidney (see **table 30.1** on the next page). Each adrenal gland is composed of two parts: (1) an inner core, the **medulla,** which produces the hormones adrenaline (also called epinephrine) and norepinephrine; and (2) an outer shell, the **cortex,** which produces the steroid hormones cortisol and aldosterone.

The Adrenal Medulla: Emergency Warning Siren The medulla releases **epinephrine** (adrenaline) and **norepinephrine** in times of stress. These hormones act as emergency signals that stimulate rapid deployment of body fuel. The "alarm" response these hormones produce throughout the body is

identical to the individual effects achieved by the sympathetic nervous system, but it is much longer lasting. Among the effects of these hormones are an accelerated heartbeat, increased blood pressure, higher levels of blood sugar, and increased blood flow to the heart and lungs.

The Adrenal Cortex: Maintaining the Proper Amount of Salt The adrenal cortex produces the steroid hormone **cortisol.** Cortisol (also called hydrocortisone) acts on many different cells in the body to maintain nutritional well-being. It stimulates carbohydrate metabolism and reduces inflammation. Synthetic derivatives of this hormone, such as prednisone, have widespread medical use as anti-inflammatory agents. Cortisol is also called the *stress hormone,* released in times of stress to help the body deal with acute stress. Problems arise when the body experiences chronic stress and cortisol levels remain high in the body. This can lead to problems with high blood pressure, reduced immune function, fat accumulation, and maintaining blood sugar, among others. These chronic effects of cortisol are unhealthy.

The adrenal cortex also produces **aldosterone.** Aldosterone acts primarily in the kidney to promote the uptake of sodium and other salts from the urine, which also increases the reabsorption of water. Sodium ions play crucial roles in nerve conduction and many other body functions. Water is needed to maintain blood volume and blood pressure. Aldosterone is, with PTH, one of the two endocrine hormones essential for survival. Removal of the adrenal glands is invariably fatal.

> **Key Learning Outcome 30.5** The thyroid acts as a metabolic thermostat, secreting hormones that adjust metabolic rate. Parathyroid hormone regulates calcium levels in the blood. The adrenal medulla releases epinephrine and norepinephrine. The adrenal hormone aldosterone promotes the uptake of sodium and other salts in the kidney.

TABLE 30.1 THE PRINCIPAL ENDOCRINE GLANDS

Endocrine Gland and Hormone	Target	Principal Actions
Adrenal Cortex		
Aldosterone	Kidney tubules	Maintains proper balance of sodium and potassium ions
Cortisol	General	Adaptation to long-term stress; raises blood glucose level; mobilizes fat
Adrenal Medulla		
Epinephrine (adrenaline) and norepinephrine (noradrenaline)	Smooth muscle, cardiac muscle, blood vessels, skeletal muscle	Initiate stress responses; increase heart rate, blood pressure, metabolic rate; dilate blood vessels; mobilize fat; raise blood glucose level
Hypothalamus		
Thyrotropin-releasing hormone (TRH)	Anterior pituitary	Stimulates TSH release from anterior pituitary
Corticotropin-releasing hormone (CRH)	Anterior pituitary	Stimulates ACTH release from anterior pituitary
Gonadotropin-releasing hormone (GnRH)	Anterior pituitary	Stimulates FSH and LH release from anterior pituitary
Prolactin-releasing factor (PRF)	Anterior pituitary	Stimulates PRL release from anterior pituitary
Growth-hormone-releasing hormone (GHRH)	Anterior pituitary	Stimulates GH release from anterior pituitary
Prolactin-inhibiting hormone (PIH)	Anterior pituitary	Inhibits PRL release from anterior pituitary
Growth-hormone-inhibiting hormone (somatostatin)	Anterior pituitary	Inhibits GH release from anterior pituitary
Melanotropin-inhibiting hormone (MIH)	Anterior pituitary	Inhibits MSH release from anterior pituitary
Ovary		
Estrogen	General; female reproductive structures	Stimulates development of secondary sex characteristics in females and growth of sex organs at puberty; prompts monthly preparation of uterus for pregnancy
Progesterone	Uterus, breasts	Completes preparation of uterus for pregnancy; stimulates development of breasts
Pancreas		
Insulin	General	Lowers blood glucose level; increases storage of glycogen in liver
Glucagon	Liver, adipose tissue	Raises blood glucose level; stimulates breakdown of glycogen in liver
Parathyroid Glands		
Parathyroid hormone (PTH)	Bone, kidneys, digestive tract	Increases blood calcium level by stimulating bone breakdown; stimulates calcium reabsorption in kidneys; activates vitamin D

TABLE 30.1 *(continued)*

Endocrine Gland and Hormone	Target	Principal Actions
Pineal Gland		
Melatonin	Hypothalamus	Function not well understood; may help control onset of puberty in humans and help regulate sleep cycle
Posterior Lobe of Pituitary		
Oxytocin (OT)	Uterus	Stimulates contraction of uterus
	Mammary glands	Stimulates ejection of milk
Vasopressin (antidiuretic hormone, ADH)	Kidneys	Conserves water; increases blood pressure
Anterior Lobe of Pituitary		
Growth hormone (GH)	General	Stimulates growth by promoting protein synthesis and breakdown of fatty acids
Prolactin (PRL)	Mammary glands	Sustains milk production after birth
Thyroid-stimulating hormone (TSH)	Thyroid gland	Stimulates secretion of thyroid hormones
Adrenocorticotropic hormone (ACTH)	Adrenal cortex	Stimulates secretion of adrenal cortical hormones
Follicle-stimulating hormone (FSH)	Gonads	Stimulates ovarian follicle growth and secretion of estrogen in females; stimulates production of sperm cells in males
Luteinizing hormone (LH)	Ovaries and testes	Stimulates ovulation and corpus luteum formation in females; stimulates secretion of testosterone in males
Melanocyte-stimulating hormone (MSH)	Skin	Stimulates color change in reptiles and amphibians; unknown function in mammals
Testes		
Testosterone	General; male reproductive structures	Stimulates development of secondary sex characteristics in males and growth spurt at puberty; stimulates development of sex organs; stimulates sperm production
Thyroid Gland		
Thyroid hormone (thyroxine, T4, and others)	General	Stimulates metabolic rate; essential to normal growth and development
Calcitonin	Bone	Lowers blood calcium level by inhibiting release of calcium from bone
Thymus		
Thymosin	White blood cells	Promotes production and maturation of white blood cells

How Strong Is the Association Between Smoking and Lung Cancer?

About a third of all cases of cancer in the United States are directly attributable to cigarette smoking. The association between smoking and cancer is particularly striking for lung cancer. The lung you see in the photograph below, riddled with cancer, is that of a smoker. A cancerous tumor has almost completely taken over the top half, and the black discoloration is due to tars. Cancer cells can migrate from the lungs into the lymph and blood vessels and spread through the body. Often, victims of lung cancer die of secondary tumors that form in other parts of the body, such as the brain. Over half a million people died of cancer in the United States in 2009; about 28% of them died of lung cancer.

All Americans die. The tragedy of this statistic is that so many die unnecessarily soon—fully 87% of the lung cancer deaths were cigarette smokers. Smoking is a popular pastime among Americans. In the United States, 21% of adults and 23% of teens smoke, and U.S. smokers consumed 389 billion cigarettes in 2005. The smoke emitted from these cigarettes contains some 3,000 chemical components, including vinyl chloride, benzo[a] pyrenes, and N-nitrosonornicotine, all potent mutagens. Smoking places these mutagens into direct contact with the tissues of the lungs, with cancer the result.

How strong is the correlation between the number of cigarettes smoked per day and the incidence of lung cancer? To find out, a detailed study was made of the incidence of lung cancer among American men, and of the cigarettes smoked per day. The results are presented in the graph above.

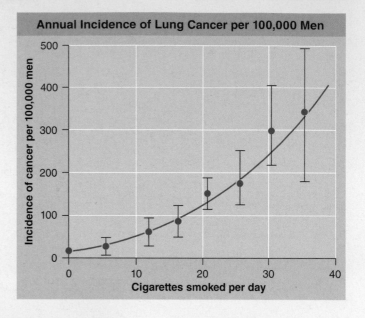

Annual Incidence of Lung Cancer per 100,000 Men

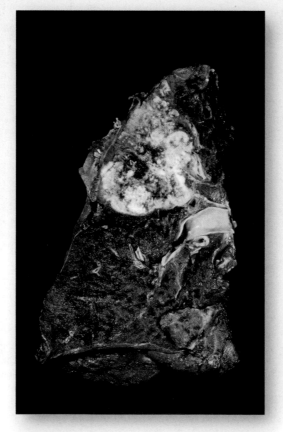

1. **Applying Concepts**
   a. In the graph, what is the dependent variable?
   b. The vertical lines drawn through the points on the graph are "error bars." How much estimation error is associated with the estimate of cancer incidence among men smoking 20 cigarettes a day? 30 cigarettes?

2. **Interpreting Data**
   a. If the incidence of cancer per 100,000 men is 100, what is the percent of men with cancer?
   b. Do you see a trend in the magnitude of the error bars in the graph? What might account for this?

3. **Making Inferences**
   a. There are 20 cigarettes in a pack. What is the incidence of cancer among "a-pack-a-day" smokers?
   b. What is the incidence of lung cancer among nonsmokers?
   c. Compare the risk of contracting lung cancer among individuals who smoke one pack a day to the risk among nonsmokers.
   d. Is the relationship between cigarettes smoked per day and incidence of lung cancer linear? Why do you think the relationship is this way? What do you think this says about the risks of heavier smoking?

4. **Drawing Conclusions** Do these results support the hypothesis that cigarette smoking causes cancer? Do they prove it? Explain.

The Neuroendocrine System

30.1 Hormones

- Hormones are chemical signals produced in glands or other endocrine tissues and transported to distant sites in the body. Endocrine glands produce hormones and release the hormones into the bloodstream (**figure 30.1**).

- Endocrine glands and tissues are under the control of the central nervous system, primarily the hypothalamus. A command from the hypothalamus often causes the release of a hormone from an endocrine gland. Only cells that have receptors for the hormone respond and are called "target cells." The hormone binds to the receptor and elicits a response in the cell, often a change in cellular activity or genetic expression (**Key Biological Process, page 623**).

30.2 How Hormones Target Cells

- Steroid hormones are lipid-soluble molecules. They pass through the plasma membrane of the target cell and bind to receptors in the cytoplasm or nucleus. The hormone-receptor complex binds to DNA, causing a change in gene expression that alters cell function (**figure 30.2**).

- Peptide hormones are unable to pass through the plasma membrane. Instead, they bind to transmembrane protein receptors (**figure 30.3**). The binding of the hormone causes a change in the internal side of the receptor which activates a second messenger (**Key Biological Process, page 625**). Second messengers, such as cyclic AMP, activate enzymes in the cell. The enzymes then trigger a change in cellular activity. The second messenger system is a cascade of reactions that amplifies the signal and facilitates the change in cellular activity.

The Major Endocrine Glands

30.3 The Hypothalamus and the Pituitary

- The pituitary gland is actually two glands: the posterior and anterior pituitary glands. The posterior pituitary develops as an extension of the hypothalamus and contains axons that extend from cell bodies in the hypothalamus.

- The hormones released from the posterior pituitary are actually produced in the hypothalamus and transported by the axons to the posterior pituitary for storage and release. The hormones of the posterior pituitary include antidiuretic hormone (ADH), which regulates water retention in the kidneys, and oxytocin, which initiates uterine contractions during childbirth and milk release in the mother.

- The anterior pituitary originates from epithelial tissue and produces the hormones it releases. Seven hormones are produced in the anterior pituitary (**figure 30.4**). They are thyroid-stimulating hormone (TSH), adrenocorticotropic hormone (ACTH), growth hormone (GH), follicle-stimulating hormone (FSH), luteinizing hormone (LH), prolactin (PRL), and melanocyte-stimulating hormone (MSH). The hypothalamus controls the anterior pituitary. The hypothalamus produces hormones that are released into blood capillaries that surround the pituitary stalk, called the hypothalamohypophyseal portal system. They travel a short distance to the anterior pituitary, as shown here from **figure 30.5**.

- Many hormones, including those released by the hypothalamus and the pituitary are controlled by negative feedback. When enough of the hormone has been released, it feeds back to inhibit the hormone-production process (**figure 30.6**).

30.4 The Pancreas

- The pancreas secretes two hormones—insulin and glucagon—into the blood. These hormones interact to maintain stable blood glucose levels. Insulin stimulates cell uptake of glucose from the blood. Glucagon stimulates the breakdown of glycogen to glucose. Two different types of cells in the islets of Langerhans produce insulin and glucagon.

- These hormones work opposite to each other (**figure 30.7**). An increase in blood glucose levels triggers the release of insulin and a decrease in blood glucose levels triggers the release of glucagon. When insulin is not available or cells fail to respond to insulin, diabetes mellitus can result.

30.5 The Thyroid, Parathyroid, and Adrenal Glands

- The thyroid gland is a shield-shaped organ that lies just beneath the Adam's apple in the front of the neck. The thyroid makes several hormones, but the two most important hormones produced by the thyroid are thyroxine, which increases metabolism and growth, and calcitonin, which stimulates calcium uptake by bones. Thyroxine is controlled by negative feedback, and the over- or underproduction of thyroxine can lead to serious health problems (**figure 30.8**).

- The parathyroid glands are four small glands that are attached to the thyroid. The parathyroids produce parathyroid hormone. PTH regulates the levels of calcium in the blood. Low calcium ion concentrations stimulate the release of PTH from the parathyroid glands. PTH acts on the bones to dismantle bone tissue, releasing Ca^{++} into the blood, shown here from **figure 30.9**. When Ca^{++} levels again increase, calcitonin is released from the thyroid and stimulates the uptake of Ca^{++} by bone cells and the synthesis of new bone tissue.

- The adrenal gland is actually two glands: The adrenal medulla is the inner core, and the adrenal cortex is the outer shell. The adrenal medulla secretes epinephrine and norepinephrine, and the adrenal cortex secretes aldosterone, which, like PTH, is necessary for survival. It promotes the uptake of sodium and water from urine.

Test Your Understanding

1. One advantage chemical signaling has over electrical signaling is that
   a. reaction to stimuli can happen very quickly.
   b. although it takes large amounts of chemicals, the chemical signals are efficient.
   c. chemical signals stick around longer than electrical signals and can be used for slow processes.
   d. chemical signals are used in response to external and internal stimuli.

2. A coordination center for some of the endocrine system is the
   a. hypothalamus.   c. thyroid gland.
   b. adrenal gland.   d. pancreas.

3. Hormones and neurotransmitters are similar because they
   a. fit into receptors specifically shaped for them.
   b. are proteins.
   c. are released into the bloodstream.
   d. All of the above.

4. The action of steroid hormones is different from peptide hormones because
   a. peptide hormones must enter the cell to begin action, whereas steroid hormones must begin action on the external surface of the cell membrane.
   b. steroid hormones must enter the cell to begin action, whereas peptide hormones must begin action on the external surface of the cell membrane.
   c. peptide hormones produce a hormone receptor complex that works directly on the DNA, whereas steroid hormones cause the release of a secondary messenger that triggers enzymes.
   d. No answer is correct.

5. The hormone that regulates water concentration in the urine is released from the
   a. thyroid gland.
   b. thymus.
   c. anterior pituitary gland.
   d. posterior pituitary gland.

6. _____ is the hormone that stimulates the adrenal gland to produce a number of steroid hormones.
   a. ACTH
   b. LH
   c. TSH
   d. MSH

7. Type I diabetes is caused by an abnormality in endocrine cells of the
   a. pancreas.   c. adrenal glands.
   b. thymus.   d. hypothalamus.

8. The release of the thyroid hormone calcitonin is triggered by
   a. too much glucose in the blood.
   b. too much sodium in the blood.
   c. too much calcium in the blood.
   d. too much iodine in the blood.

9. Epinephrine mimics the effects of the
   a. somatic nervous system.
   b. central nervous system.
   c. parasympathetic nervous system.
   d. sympathetic nervous system.

10. A person afflicted with a goiter suffers from
    a. lack of iodine.   c. excess thyroxine.
    b. low calcium levels.   d. an underactive hypothalamus.

Apply Your Understanding

1. **Figure 30.1** Some of the body systems are located primarily in one area of the body, or are obviously connected. The respiratory system, for instance, is located in the head and upper portion of the body. The skeletal system is articulated, almost every bone connected to others. The endocrine system, however, is spread out, a batch of glands that do not appear connected with one another. Speculate on why this is so.

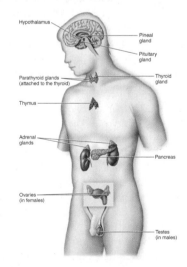

2. **Figure 30.4** A hypothetical patient has a disorder of the hypothalamus in which it can no longer secrete its "inhibiting" hormones. Which of the anterior pituitary hormones will be affected?

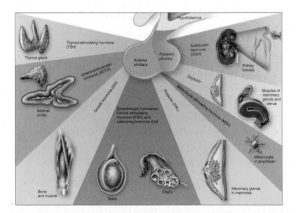

Synthesize What You Have Learned

1. Tad, the younger brother of your friend Sofia, wants to be a sports star in high school. Although he is only in the eighth grade, he brags that he is taking steroids he gets from a friend in order to "bulk up" for next year. He asks if you have ever heard of any problems for kids his age—he only wants to take them "a couple of years" to get a football scholarship so his family can afford for him to go to college. How would you advise him?

2. Since your bones are so very important, why would your body ever need a system that includes osteoclasts, which literally break down your bone tissue?

31

Reproduction and Development

Few subjects pervade our everyday thinking more than sex; few urges are more insistent. They are no accident, these strong feelings. They are a natural part of being human. All animals share them. The cry of a cat in heat, insects chirping outside the windows, frogs croaking in swamps, wolves howling in a frozen northern scene—all these are the sounds of the living world's essential act, an urgent desire to reproduce that has been patterned by a long history of evolution. It is a pattern that all of us share. The reproduction of our families spontaneously elicits in us a sense of rightness and fulfillment. It is difficult not to return the smile of a new infant, not to feel warmed by it and by the look of wonder and delight to be seen on the faces of parents like this nursing mother. This chapter deals with sex and reproduction among the vertebrates, of which we human beings are one kind. Few subjects are of more direct concern to students than sex. Because many students must make important decisions about sex, the subject is of far more than academic interest, and is one about which all students need to be well informed.

31.1 Asexual and Sexual Reproduction

Not all reproduction involves two parents. Asexual reproduction, in which the offspring are genetically identical to one parent, is the primary means of reproduction among protists, cnidarians, and tunicates, and also occurs in some more complex animals.

Through mitosis, genetically identical cells are produced from a single parent cell. This permits asexual reproduction to occur in the *Euglena* in **figure 31.1** by division of the organism, or **fission.** The DNA replicates and cell structures, such as the flagellum, duplicate. The nucleus divides with identical nuclei going to each daughter cell. Cnidaria commonly reproduce by **budding,** where a part of the parent's body becomes separated from the rest and differentiates into a new individual. The new individual may become an independent animal or may remain attached to the parent, forming a colony.

Unlike asexual reproduction, sexual reproduction occurs when a new individual is formed by the union of *two* cells. These cells are called **gametes,** and the two kinds that combine are generally called *sperm* and *eggs* (or *ova*). The union of a sperm and an egg produces a fertilized egg, or **zygote,** that develops by mitotic division into a new multicellular organism. The zygote and the cells it forms by mitosis are diploid; they contain both members of each homologous pair of chromosomes. The gametes, formed by meiosis in the sex organs, or **gonads**—the *testes* and *ovaries*—are haploid (see chapter 9). The processes of spermatogenesis (sperm formation) and oogenesis (egg formation) are described in later sections.

Different Approaches to Sex

Parthenogenesis, a type of reproduction in which offspring are produced from unfertilized eggs, is common in many species of arthropods. Some species are exclusively parthenogenic, whereas others switch between sexual reproduction and parthenogenesis in different generations. In honeybees, for example, a queen bee mates only once and stores the sperm. She then can control the release of sperm. If no sperm are released, the eggs develop parthenogenetically into drones, which are males; if sperm are allowed to fertilize the eggs, the fertilized eggs develop into other queens or worker bees, which are female.

The Russian biologist Ilya Darevsky reported in 1958 one of the first cases of unusual modes of reproduction among

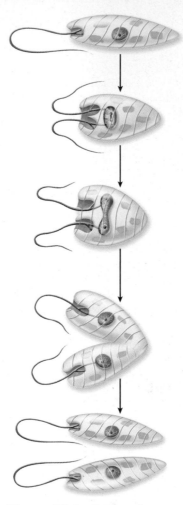

Figure 31.1   Asexual reproduction in protists.

The protist *Euglena* reproduces asexually: A mature individual divides by fission, and two complete individuals result.

vertebrates. He observed that some populations of small lizards of the genus *Lacerta* were exclusively female, and he suggested that these lizards could lay eggs that were viable even if they were not fertilized. In other words, they were capable of asexual reproduction in the absence of sperm, a type of parthenogenesis. Further work has shown that parthenogenesis occurs among populations of other lizard genera.

Hermaphroditism, another variation in reproductive strategy, is when one individual has both testes and ovaries and so can produce both sperm and eggs. The hamlet bass in **figure 31.2a** are hermaphroditic, producing both eggs and sperm. During mating each fish switches from producing eggs that are fertilized by its partner, to producing sperm that fertilizes its partner's eggs. A tapeworm is hermaphroditic and can fertilize itself as well as cross fertilize, a useful strategy because it is unlikely to encounter another tapeworm living inside its host. Most hermaphroditic animals, however, require another individual to reproduce. Two earthworms, for example, are required for reproduction—like the hamlet bass, each functions as both male and female. Each leaves the encounter with fertilized eggs.

Sequential hermaphroditism, in which individuals can change their sex, occurs in numerous fish genera. Among coral reef fish, for example, both *protogyny* ("first female," a change from female to male) and *protandry* ("first male," a change from male to female) occur. In the protogynous bluehead wrasse in **figure 31.2b**, the sex change appears to be under social control. These fish commonly live in large groups, or schools, where successful reproduction is typically limited to one or a few large, dominant males. If those males are removed, the largest female rapidly changes sex and becomes a dominant male (the blue-headed fish in the photo).

Sex Determination

Among the fish just described, and in some species of reptiles, environmental changes can cause changes in the sex of the animal. In mammals, the sex is determined early in embryonic development. The reproductive systems of human males and females appear similar for the first 40 days after conception. During this time, the cells that will give rise to ova or sperm migrate to the embryonic gonads, which have the potential to become either ovaries in females or testes in males. If the embryo is XY, it is a male and will carry a gene on the Y chromosome whose product converts the gonads into testes (as on the left in **figure 31.3**). In females, who are XX, this

(a)

(b)

Figure 31.2 Hermaphroditism and protogyny.

(*a*) The hamlet bass (genus *Hypoplectrus*) is a deep-sea fish that is a hermaphrodite. In the course of a single pair-mating, one fish may switch sexual roles as many as four times. Here, the fish acting as a male curves around its motionless partner, fertilizing the upward-floating eggs. (*b*) The bluehead wrasse *Thalassoma bifasciatium* is protogynous. Here a large male, or sex-changed female, is seen among females, which are typically much smaller.

Figure 31.3 Sex determination.

Sex determination in mammals is made by a gene on the Y chromosome designated *SRY*. Testes are formed when the Y chromosome and *SRY* are present; ovaries are formed when they are absent.

Y chromosome gene and the protein it encodes are absent, and the gonads become ovaries (as on the right). Recent evidence suggests that the sex-determining gene may be one known as **SRY** (for "*sex-determining region of the Y* chromosome"). The *SRY* gene appears to have been highly conserved during the evolution of different vertebrate groups.

Once testes form in the embryo, they secrete testosterone and other hormones that promote the development of the male external genitalia and accessory reproductive organs (indicated in the blue box). If testes do not form, the embryo develops female external genitalia and accessory reproductive organs. The ovaries do not promote this development of female organs because the ovaries are

nonfunctional at this stage. In other words, all mammalian embryos will develop female sex accessory organs and external genitalia by default unless they are masculinized by the secretions of the testes.

Key Learning Outcome 31.1 Sexual reproduction is most common among animals, but many reproduce asexually by fission, budding, or parthenogenesis. Sexual reproduction generally involves the fusion of gametes derived from different individuals of a species, but some species are hermaphroditic.

Evolution of Vertebrate Sexual Reproduction

Vertebrate sexual reproduction evolved in the ocean before vertebrates colonized the land. The females of most species of marine bony fish produce eggs, or ova, in batches and release them into the water. The males generally release their sperm into the water containing the eggs where the union of the free gametes occurs. This process is known as **external fertilization.**

Although seawater is not a hostile environment for gametes, it does cause the gametes to disperse rapidly, so their release by females and males must be almost simultaneous. Thus, most marine fish restrict the release of their eggs and sperm to a few brief and well-defined periods. Some reproduce just once a year, while others do so more frequently. There are few seasonal cues in the ocean that organisms can use as signals for synchronizing reproduction, but one all-pervasive signal is the cycle of the moon. Once each month, the moon approaches closer to the earth than usual, and when it does, its increased gravitational attraction causes somewhat higher tides. Many marine organisms sense the tidal changes and entrain the production and release of their gametes to the lunar cycle.

Fertilization is external in most fish but internal in most other vertebrates. The invasion of land posed the new danger of desiccation (drying out), a problem that was especially severe for the small and vulnerable gametes. On land, the gametes could not simply be released near each other, because they would soon dry up and perish. Consequently, there was intense selective pressure for terrestrial vertebrates (as well as some groups of fish) to evolve **internal fertilization**—that is, the introduction of male gametes into the female reproductive tract. By this means, fertilization still occurs in a nondesiccating environment, even when the adult animals are fully terrestrial. The vertebrates that practice internal fertilization have three strategies for embryonic and fetal development. Depending upon the relationship of the developing embryo to the mother and egg, those vertebrates with internal fertilization may be classified as oviparous, ovoviviparous, or viviparous.

Oviparity. In oviparity, the eggs, after being fertilized internally, are deposited outside the mother's body to complete their development. This is found in some bony fish, most reptiles, some cartilaginous fish, some amphibians, a few mammals, and all birds.

Ovoviviparity. In ovoviviparity, the fertilized eggs are retained within the mother to complete their development, but the embryos still obtain all of their nourishment from the egg yolk. The young are fully developed when they are hatched and released from the mother. This is found in some bony fish (including mollies, guppies, and mosquito fish), some cartilaginous fish, and many reptiles.

Viviparity. In viviparity, the young develop within the mother and obtain nourishment directly from their mother's blood, rather than from the egg yolk. This is found in most cartilaginous fish (**figure 31.4**), some amphibians, a few reptiles, and almost all mammals.

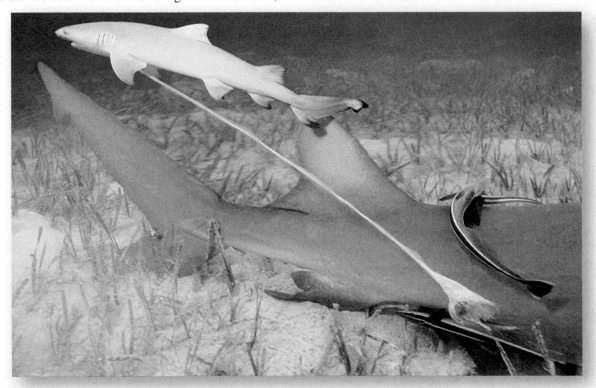

Figure 31.4 Viviparous vertebrates carry live, mobile young within their bodies.
The young complete their development within the body of the mother and are then released as small but competent adults. Here a lemon shark has just given birth to a young shark, which is still attached by the umbilical cord.

Fish and Amphibians

Most fish and amphibians, unlike other vertebrates, reproduce by means of external fertilization.

Fish. Fertilization in most species of bony fish (teleosts) is external, and the eggs contain only enough yolk to sustain the developing embryo for a short time. After the initial supply of yolk has been exhausted, the young fish must seek its food from the waters around it. Development is speedy, and the young that survive mature rapidly. Although thousands of eggs are fertilized in a single mating, many of the resulting individuals succumb to microbial infection or predation, and few grow to maturity.

   In marked contrast to the bony fish, fertilization in most cartilaginous fish is internal. The male introduces sperm into the female through a modified pelvic fin. Development of the young in these vertebrates is generally viviparous.

Amphibians. The amphibians invaded the land without fully adapting to the terrestrial environment, and their life cycle is still tied to the water. Amphibians, like the red-spotted newt in **figure 31.5**, reproduce in the water and have aquatic larval stages before moving to the land. Fertilization is external in many amphibians, just as it is in most species of bony fish. Gametes from both males and females are released through the cloaca, a common opening used by the digestive, reproductive, and urinary systems. Among the frogs and toads, the male grasps the female and discharges fluid containing the sperm onto the eggs as they are released into the water (**figure 31.6**). Although the eggs of most amphibians develop in the water, there are some interesting exceptions shown in **figure 31.7**. In two species of frogs (one being the Darwin's frog in **figure 31.7d**), the eggs develop in the vocal sacs and the stomach, and the young frogs leave through their parent's mouth!

Figure 31.6   The eggs of frogs are fertilized externally.

When frogs mate, as these two are doing, the clasp of the male induces the female to release a large mass of mature eggs, over which the male discharges his sperm.

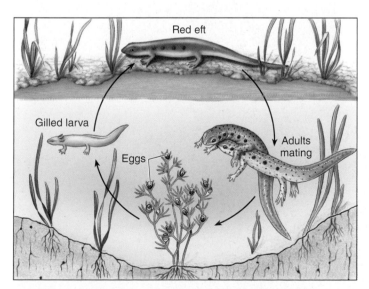

Figure 31.5   Life cycle of the red-spotted newt.

Many salamanders have both aquatic and terrestrial stages in their life cycle. In the red-spotted newt (*Notophthalmus viridescens*), eggs are laid in water and hatch into aquatic larvae with external gills and a finlike tail. After a period of growth, the larvae can metamorphose into a terrestrial "red eft" stage that later metamorphoses again to produce aquatic, breeding adults.

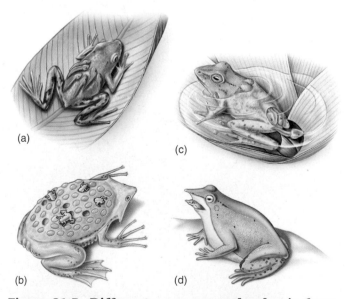

Figure 31.7   Different ways young develop in frogs.

(*a*) In the poison arrow frog, the male carries the tadpoles on his back. (*b*) In the female Surinam frog, froglets develop from eggs in special brooding pouches on the back. (*c*) In the South American pygmy marsupial frog, the female carries the developing larvae in a pouch on her back. (*d*) Tadpoles of the Darwin's frog develop into froglets in the vocal pouch of the male and emerge from the mouth.

The time required for development of most amphibians is much longer than that for fish, but amphibian eggs do not include a significantly greater amount of yolk. Instead, the process of development in most amphibians is divided into embryonic, larval, and adult stages, in a way reminiscent of the life cycles found in some insects. The embryo develops within the egg, obtaining nutrients from the yolk. After hatching from the egg, the aquatic larva then functions as a free-swimming, food-gathering machine, often for a considerable period of time. The larvae may increase in size rapidly; some tadpoles, which are the larvae of frogs and toads, grow in a matter of weeks from creatures no bigger than the tip of a pencil into individuals as big as a goldfish. When the larva has grown large enough, it undergoes a developmental transition, or metamorphosis, into the terrestrial adult form.

Reptiles and Birds

Most reptiles and all birds are oviparous, laying amniotic eggs that are protected by watertight membranes from desiccation. After the eggs are fertilized internally, they are deposited outside of the mother's body to complete their development. Like most vertebrates that fertilize internally, most male reptiles use a cylindrical organ, the penis, to inject sperm into the female, a process called copulation (**figure 31.8**). The penis, containing erectile tissue, can become quite rigid and penetrate far into the female reproductive tract. Reptiles exhibit all three types of internal fertilization. Most are oviparous, laying eggs and then abandoning them. These eggs are surrounded by a leathery shell that is deposited as the egg passes through the oviduct, the part of the female reproductive tract leading from the ovary. Other species of reptiles are ovoviviparous (forming eggs that develop into embryos and hatch within the body of the mother) or viviparous (developing inside the mother while being nourished by her rather than by the yolk of an egg).

All birds practice internal fertilization, though most male birds lack a penis. In some of the larger birds (including swans, geese, and ostriches), however, the male cloaca extends to form a false penis. **Figure 31.9a** shows the formation of a bird's egg. As the fertilized egg (ovum) passes along the oviduct (from top to bottom here), glands secrete albumin proteins (the egg white) and the hard, calcareous shell that distinguishes bird eggs from reptilian eggs. While modern reptiles are poikilotherms (animals whose body temperature varies with the temperature of their environment), birds are homeotherms (animals that maintain a relatively constant body temperature independent of environmental temperatures). Hence, most birds incubate their eggs after laying them to keep them warm (**figure 31.9b**).

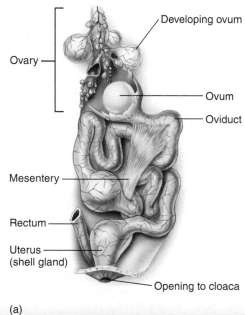

(a)

(b)

Figure 31.9   Egg formation and incubation in birds.
(*a*) In birds, fertilization of the egg (ovum) takes place within the female, in the upper portion of the oviduct. As the fertilized egg passes down the oviduct, albumin (egg white), shell membranes, and the shell is secreted around the egg. (*b*) Bird eggs must be kept warm after they are laid. This blue-footed booby is incubating its two eggs.

Figure 31.8   The introduction of sperm by the male into the female's body is called copulation.

The shelled eggs of reptiles and birds constitute one of the most important adaptations of these vertebrates to life on land, because shelled eggs can be laid in dry places. Such eggs are known as amniotic eggs because the embryo develops within a fluid-filled cavity surrounded by a membrane called the amnion. The amnion is an extraembryonic membrane—that is, a membrane formed from embryonic cells but located outside the body of the embryo. Amniotic eggs contain three other extraembryonic membranes, one of which is a yolk sac. In contrast, the eggs of fish and amphibians contain only one extraembryonic membrane, the yolk sac.

Mammals

Some mammals are seasonal breeders, reproducing only once a year, such as dogs, foxes, and bears, whereas others, such as horses and sheep, have multiple, short reproductive cycles throughout a given time of the year. Among the latter, the females generally undergo the reproductive cycles, whereas the males are more constant in their reproductive activity. Cycling in females involves the periodic release of a mature ovum from the ovary in a process known as ovulation. Most female mammals are "in heat," or sexually receptive to males, only around the time of ovulation. This period of sexual receptivity is called *estrus,* and the reproductive cycle is therefore called an *estrous cycle.* Females continue to cycle until they become pregnant.

In the estrous cycle of most mammals, changes in the secretion of follicle-stimulating hormone (FSH) and luteinizing hormone (LH) by the anterior pituitary gland cause changes in egg cell development and hormone secretion in the ovaries. Humans and apes have menstrual cycles that are similar to the estrous cycles of other mammals in their cyclic pattern of hormone secretion and ovulation. Unlike mammals with estrous cycles, however, human and some ape females bleed when they shed the inner lining of their uterus, a process called menstruation, and may engage in copulation at any time during the cycle.

Rabbits and cats differ from most other mammals in that they are *induced ovulators*. Instead of ovulating in a cyclic fashion regardless of sexual activity, the females ovulate only after copulation as a result of a reflex stimulation of LH secretion (described later). This makes them extremely fertile.

Monotremes (the most primitive mammals, consisting solely of the duck-billed platypus and the echidna), are oviparous, like the reptiles from which they evolved. They incubate their eggs in a nest (**figure 31.10a**) or specialized pouch, and the young hatchlings obtain milk from their mother's mammary glands by licking her skin, because monotremes lack nipples. All other mammals are viviparous, divided into two subcategories based on how they nourish their young.

Figure 31.10 Reproduction in mammals.
(*a*) Monotremes, like the duck-billed platypus shown here, lay eggs in a nest. (*b*) Marsupials, such as this kangaroo, give birth to small fetuses that complete their development in a pouch. (*c*) In placental mammals, such as this doe nursing her fawn, the young remain inside the mother's uterus for a longer period of time and are born relatively more developed.

(a) Monotremes

(b) Marsupials

(c) Placentals

Marsupials, a group that includes opossums and kangaroos, give birth to fetuses that are incompletely developed. They complete their development in a pouch of their mother's skin, where they obtain nourishment from nipples of her mammary glands (**figure 31.10b**).

Placental mammals (**figure 31.10c**) retain their young for a much longer period of development within the mother's uterus. The fetuses are nourished by a structure known as the placenta, which is derived from both an extraembryonic membrane and from the mother's uterine lining. Because the fetal and maternal blood vessels are in very close proximity in the placenta, the fetus can obtain nutrients by diffusion from the mother's blood. The placenta is discussed in more detail later in this chapter.

> **Key Learning Outcome 31.2** Fertilization is external in frogs and most bony fish and internal in other vertebrates. Birds and most reptiles lay watertight eggs, as do monotreme mammals. All other mammals are viviparous.

31.3 | Males

The human male gamete, or **sperm,** is highly specialized for its role as a carrier of genetic information. Produced after meiosis, sperm cells have 23 chromosomes instead of the 46 found in other cells of the male body. Sperm do not successfully complete their development at 37°C (98.6°F), the normal human body temperature. The sperm-producing organs, the **testes** (singular, **testis**), move during the course of fetal development into a sac called the **scrotum (figure 31.11)**, which hangs between the legs of the male, maintaining the two testes at a temperature about 3°C cooler than the rest of the body. The testes contain cells that secrete the male sex hormone **testosterone.**

Male Gametes Are Formed in the Testes

An internal view of the testes in **figure 31.12 ❶** shows that they are composed of several hundred compartments, each packed with large numbers of tightly coiled tubes called **seminiferous tubules** (seen in cross section in ❷). Sperm production, *spermatogenesis,* takes place inside the tubules. The process of spermatogenesis begins in germinal cells toward the outside of the tubule (shown in the enlarged view in ❸). As the cells undergo

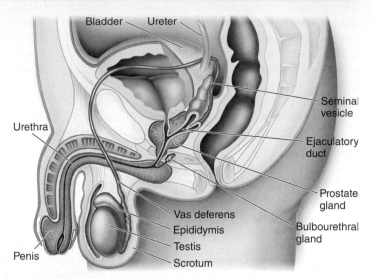

Figure 31.11 The male reproductive organs.

The testis is where sperm are formed. Cupped above the testis is the epididymis, a highly coiled passageway within which sperm complete their maturation. Extending away from the epididymis is a long tube, the vas deferens.

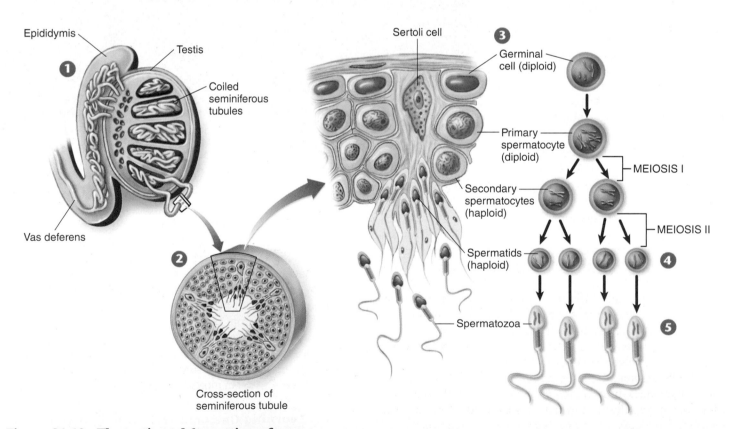

Figure 31.12 The testis and formation of sperm.

Inside the testis ❶, the seminiferous tubules ❷ are the sites of sperm formation. Germinal cells in the seminiferous tubules ❸ give rise to primary spermatocytes (diploid), which undergo meiosis to form haploid spermatids ❹. Spermatids develop into mobile spermatozoa, or sperm ❺. Sertoli cells are nongerminal cells within the walls of the seminiferous tubules. They assist spermatogenesis in several ways, such as helping to convert spermatids into spermatozoa.

meiosis, they move toward the lumen of the tubule **4**, with spermatozoa being released into the tubule **5**. The number of sperm produced is truly incredible. A typical adult male produces several hundred million sperm each day of his life! Those that are not ejaculated from the body are broken down, and their materials are reabsorbed and recycled.

After a sperm cell is manufactured within the testes through intermediate stages of meiosis, it is delivered to a long, coiled tube called the **epididymis** (see **figure 31.12**), where it matures. A sperm cell is not motile when it arrives in the epididymis, and it must remain there for at least 18 hours before its motility develops. Mature sperm are relatively simple cells, consisting of a head, body, and tail (**figure 31.13**). The head encloses a compact nucleus and is capped by a vesicle called an *acrosome*. The acrosome contains enzymes that aid in the penetration of the protective layers surrounding the egg. The body and tail provide a propulsive mechanism: Within the tail is a flagellum, and inside the body are centrioles, which act as a basal body for the flagellum, and mitochondria, which generate the energy needed for flagellar movement.

From the epididymis, the sperm is delivered to another long tube, the **vas deferens.** When sperm are released during intercourse, they travel through a tube from the vas deferens to the **urethra,** where the reproductive and urinary tracts join, emptying through the penis. Sperm is released in a fluid called **semen,** which also contains secretions mostly from the **seminal vesicles** and the **prostate gland** that provide metabolic energy sources for the sperm. Benign enlargement of the prostate occurs in 90% of men by age 70, but it can be cancerous. Prostate cancer is the second most common cancer in men and can be treated effectively if detected early during physical examinations before it spreads.

Male Gametes Are Delivered by the Penis

In the case of humans and some other mammals, the **penis** is an external tube containing two long cylinders of spongy tissue side by side (seen in cross section in **figure 31.14**). Below and between them runs a third cylinder of spongy tissue that contains in its center the urethra, through which both semen (during ejaculation) and urine (during urination) pass. Why this unusual design? The penis is designed to inflate. The spongy tissues that make up the three cylinders are riddled with small spaces between the cells, and when nerve impulses from the CNS cause the arterioles leading into this tissue to expand, blood collects within these spaces. Like blowing up a balloon, this causes the penis to become erect and rigid. Continued stimulation by the CNS is required for erection to continue.

Erection can be achieved without any physical stimulation of the penis, but physical stimulation is required for semen to be delivered. Stimulation of the penis, as by repeated thrusts into the vagina of a female, leads first to the mobilization of the sperm. In this process, muscles encircling the vas deferens contract, moving the sperm along the vas deferens into the urethra. The *bulbourethral glands* also secrete a clear, slippery fluid that neutralizes

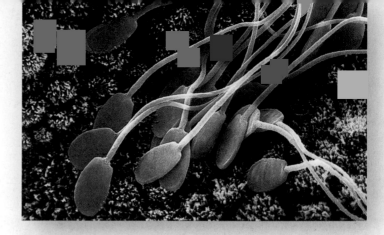

Figure 31.13 Human sperm cells.

Each sperm possesses a long tail that propels the sperm and a head that contains the nucleus. The tip, or acrosome, contains enzymes to help the sperm cell digest a passageway into the egg for fertilization.

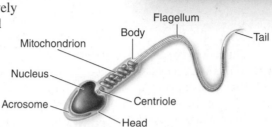

Flagellum
Body
Mitochondrion
Nucleus
Acrosome
Tail
Centriole
Head

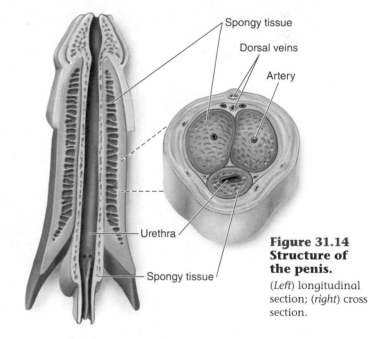

Spongy tissue
Dorsal veins
Artery
Urethra
Spongy tissue

Figure 31.14 Structure of the penis.

(*Left*) longitudinal section; (*right*) cross section.

the acidity of any residual urine and lubricates the head of the penis. Further penis stimulation then leads to the strong contraction of the muscles at the base of the penis. The result is **ejaculation,** the forceful ejection of 2 to 5 milliliters of semen. Within this small 5-milliliter volume are several hundred million sperm. Because the odds against any one individual sperm cell successfully completing the long journey to the egg and fertilizing it are extraordinarily high, successful fertilization requires a high sperm count. Males with fewer than 20 million sperm per milliliter are generally considered sterile.

Key Learning Outcome 31.3   Male testes continuously produce large numbers of male gametes, sperm, which mature in the epididymis, are stored in the vas deferens, and are delivered through the penis into the female.

31.4 Females

In females, eggs develop from cells called **oocytes,** located in the outer layer of compact masses of cells called **ovaries** within the abdominal cavity (**figure 31.15**). Recall that in males the gamete-producing cells are constantly dividing. In females all of the oocytes needed for a lifetime are already present at birth. During each reproductive cycle, one or a few of these oocytes are initiated to continue their development in a process called **ovulation;** the others remain in developmental holding patterns.

Only One Female Gamete Matures Each Month

At birth, a female's ovaries contain some 2 million oocytes, all of which have begun the first meiotic division. At this stage they are called *primary oocytes* (❶ in **figure 31.16**). Each primary oocyte waits to receive the proper developmental "go" signal before continuing on with meiosis. Until then, its meiosis remains arrested in prophase of the first meiotic division. Very few ever receive the awaited signal, which turns out to be the pituitary hormones FSH and LH, which you recall were discussed in chapter 30.

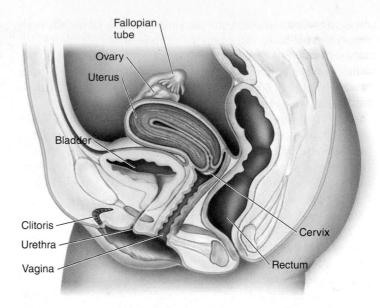

Figure 31.15   The female reproductive system.
The organs of the female reproductive system are specialized to produce gametes and to provide a site for embryonic development if the gamete is fertilized.

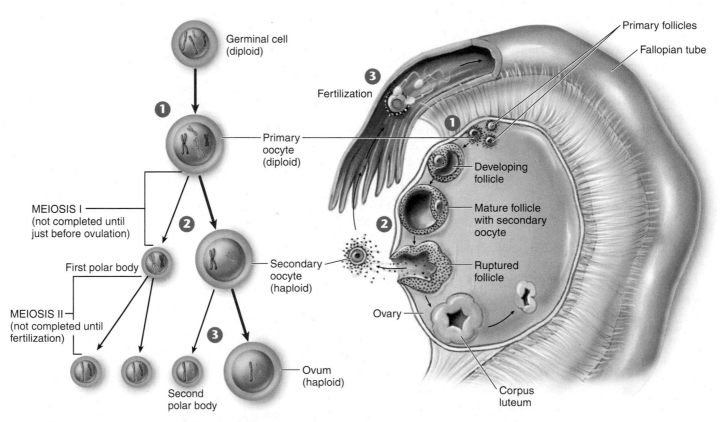

Figure 31.16   The ovary and formation of an ovum.
In this figure, the maturation of the ovum through meiosis is shown on the left, and the developmental journey of the ovum is on the right, with corresponding stages numbered on each. At birth, a human female's ovaries contain about 2 million egg-forming cells called oocytes, which have begun the first meiotic division and stopped. At this stage, they are called primary oocytes ❶, and their further development is halted until they receive the proper developmental signals, which is the hormones FSH and LH. At puberty, an adult monthly cycle of hormone secretion is established. When the hormones FSH and LH are released, meiosis resumes in a few oocytes, but only one oocyte usually continues to mature while the others regress. The primary oocyte (diploid) completes the first meiotic division, and one division product becomes a nonfunctional polar body. The other product, the secondary oocyte, is released during ovulation ❷, along with the polar body. The secondary oocyte does not complete the second meiotic division until fertilization ❸; that division yields two more nonfunctional polar bodies and a single haploid egg, or ovum. Fusion of the haploid egg with a haploid sperm during fertilization produces a diploid zygote.

With the onset of puberty, females mature sexually. At this time, the release of FSH and LH initiates the resumption of the first meiotic division in a few oocytes. The first meiotic division produces the *secondary oocyte* and a nonfunctional *polar body* ❷. In humans, usually only a single oocyte is ovulated and the others regress. In some instances more than one oocyte develops; if both are fertilized, they become fraternal twins. Approximately every 28 days after that, another oocyte matures and is ovulated, although the exact timing may vary from month to month. Only about 400 of the approximately 2 million oocytes a woman is born with mature and are ovulated during her lifetime.

Fertilization Occurs in the Oviducts

The **oviducts** (also called **fallopian tubes** or uterine tubes) transport eggs from the ovaries to the **uterus.** In humans, the uterus is a muscular, pear-shaped organ about the size of a fist that narrows to a muscular ring called the **cervix,** which leads to the vagina (**figure 31.17a**). Mammals other than primates have more complex female reproductive tracts, where part of the uterus divides to form uterine "horns."

The uterus is lined with a stratified epithelial membrane, the **endometrium.** In humans, the surface of the endometrium is shed approximately once a month during menstruation, while the underlying portion remains to generate a new surface during the next cycle. After ovulation, smooth muscles lining the fallopian tubes contract rhythmically, moving the egg down the tube to the uterus in much the same way that food is moved down through your intestines, pushing it along by squeezing the tube behind it. The journey of the egg through the fallopian tube is a slow one, taking from five to seven days to complete. However, if the egg is not fertilized within 24 hours of ovulation, it loses its capacity to develop.

During sexual intercourse, sperm are deposited within the vagina, a thin-walled muscular tube about 7 centimeters long that leads to the mouth of the uterus. Using their flagella, sperm entering the uterus swim up to and enter the fallopian tubes. Sperm can remain viable within the female reproductive tract for up to 6 days. If sexual intercourse takes places 5 days before ovulation or 1 day after, a viable egg will be present high up in the fallopian tubes. Of the several hundred million sperm that are ejaculated, only a few dozen make it to the egg. Once they reach the egg, the sperm must penetrate through two protective layers that surround the secondary oocyte (**figure 31.17b ❶**): a layer of granulosa cells and a protein layer called the zona pellucida. Enzymes within the acrosome cap of the sperm (see **figure 31.13**) help digest the second of these layers ❷. The first sperm to make it through the second layer stimulates the oocyte to block the entry of other sperm ❸ and to complete meiosis II. Meiosis II produces the **ovum** (plural, **ova**) and two more nonfunctional polar bodies (see **figure 31.16 ❸**). When the female haploid nucleus within the ovum meets the male haploid nucleus, the egg is fertilized and becomes a zygote. The zygote then begins a series of cell divisions while traveling down the fallopian tube. After about 6 days, it reaches the uterus, attaches itself to the endometrial lining, and continues the long developmental journey that eventually leads to the birth of a child.

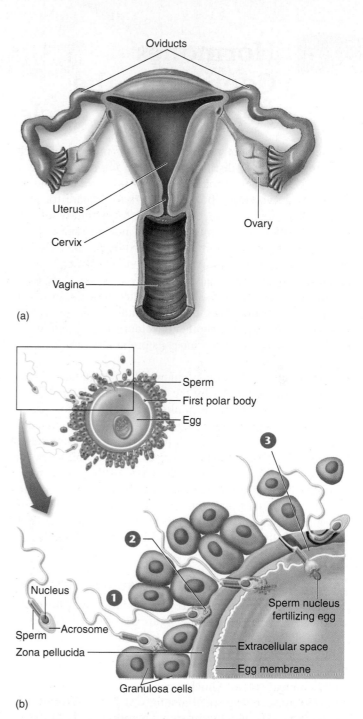

Figure 31.17 Fertilization occurs in the oviducts.
(*a*) The oviducts extend out from the uterus. Sperm are deposited in the vagina and travel to the oviducts. (*b*) Fertilization occurs in the oviduct when a sperm penetrates the outer layers of the egg cell.

Key Learning Outcome 31.4 In human females, hormones trigger the development of one or a few oocytes each 28 days. After ovulation, an egg cell travels down the fallopian tube and, if fertilized during its journey, implants in the wall of the uterus.

31.5 Hormones Coordinate the Reproductive Cycle

The female reproductive cycle, called a **menstrual cycle,** is composed of two distinct phases: the *follicular phase,* in which an egg reaches maturation and is ovulated, and the *luteal phase,* where the body continues to prepare for pregnancy. These phases are coordinated by a family of hormones. Hormones play many roles in human reproduction. Sexual development is initiated by hormones, released from the anterior pituitary and ovary, that coordinate simultaneous sexual development in many kinds of tissues. The production of gametes is another closely orchestrated process, involving a series of carefully timed developmental events. Successful fertilization initiates yet another developmental "program," in which the female body continues its preparation for the many changes of pregnancy.

Production of the sex hormones that coordinate all these processes is coordinated by the hypothalamus, which sends releasing hormones to the pituitary, directing it to produce particular sex hormones. Negative feedback, discussed in chapter 30, plays a key role in regulating these activities of the hypothalamus. When target organs receive a pituitary hormone, they begin to produce a hormone of their own, which circulates back to the hypothalamus, shutting down production of the pituitary hormone. In addition, *positive feedback* mechanisms play a role too. In these cases, a hormone circulates back to the hypothalamus and increases the production of a pituitary hormone.

Triggering the Maturation of an Egg

The first phase of the menstrual cycle, the **follicular phase,** corresponds to days 0 through 14 in **figure 31.18.** During this time, several follicles (an oocyte and its surrounding tissue is called a **follicle**) are stimulated to develop. This development is carefully regulated by hormones. The anterior pituitary, after receiving a chemical signal (GnRH) from the hypothalamus, starts the cycle by secreting small amounts of **follicle-stimulating hormone (FSH)** and **luteinizing hormone (LH)** ❶. These hormones stimulate follicular growth ❷ and the secretion of the female sex hormone **estrogen** ❸, more technically known as *estradiol,* from the developing follicles. Several follicles are stimulated to grow under FSH stimulation.

Initially, the relatively low levels of estrogen have a negative-feedback effect on FSH and LH secretion. The low but rising levels of estrogen in the bloodstream feed back to the hypothalamus, which responds to the rising estrogen by commanding the anterior pituitary to decrease production of FSH and LH. As FSH levels fall, usually only one follicle achieves maturity. Late in the follicular phase, estrogen levels in the blood have increased drastically, and these higher levels of estrogen begin to have a positive-feedback effect on

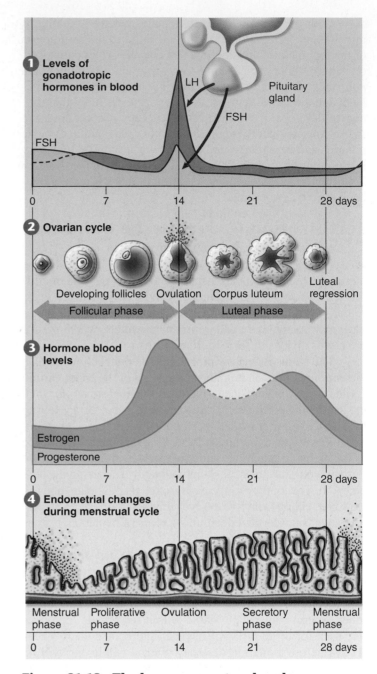

Figure 31.18   The human menstrual cycle.
Ovulation and the preparation of the uterine lining for implantation is controlled by a group of four hormones during the menstrual cycle.

FSH and LH secretion. The rise in estrogen levels signals the completion of the follicular phase of the menstrual cycle.

Preparing the Body for Fertilization

The second phase of the cycle, the **luteal phase** (days 14 through 28), follows smoothly from the first. In a positive-feedback response to high levels of estrogen, the hypothalamus causes the anterior pituitary to rapidly secrete large amounts of LH and FSH (see day 14 in ❶). The surge of LH

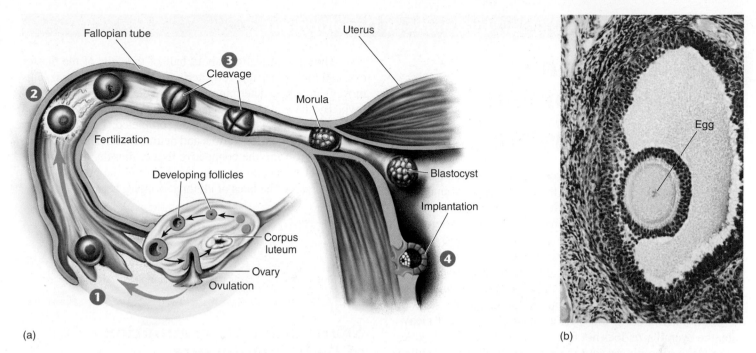

Figure 31.19 The journey of an ovum.

(a) Produced within a follicle and released at ovulation, an ovum is swept up into a fallopian tube ❶ and carried down by waves of contraction of the tube walls. Fertilization occurs within the tube ❷, by sperm journeying upward. Several mitotic divisions occur while the fertilized ovum undergoes cleavage and continues its journey down the fallopian tube ❸, becoming first a morula then a blastocyst. The blastocyst implants itself within the wall of the uterus ❹, where it continues its development. (b) A mature egg within an ovarian follicle. In each menstrual cycle, a few follicles are stimulated to grow under the influence of FSH and LH, but usually only one achieves full maturity and ovulation.

is larger than the surge of FSH and can last up to 24 hours. The peak in LH secretion triggers ovulation: LH causes the wall of the follicle to burst, and the egg within the follicle is released into one of the fallopian tubes extending from the ovary to the uterus (see ❶ in **figure 31.19**).

After the egg's release and departure, estrogen levels decrease, and LH directs the repair of the ruptured follicle, which fills in and becomes yellowish. In this condition, it is called the **corpus luteum,** which is simply the Latin phrase for "yellow body." The corpus luteum soon begins to secrete the hormone **progesterone** (the light green curve in **figure 31.18** ❸), in addition to small levels of estrogen. Increased levels of progesterone and estrogen have a negative-feedback effect on the secretion of FSH and LH, preventing further ovulations. Progesterone completes the body's preparation of the uterus for fertilization including the thickening of the endometrium (**figure 31.18** ❹). If fertilization does *not* occur soon after ovulation, however, production of progesterone slows and eventually ceases, marking the end of the luteal phase. The decreasing levels of progesterone cause the thickened layer of blood-rich tissue to be sloughed off, a process that results in the bleeding associated with menstruation. **Menstruation,** or "having a period," usually occurs about midway between successive ovulations (shown in **figure 31.18** at 28 days), although its timing varies widely for individual females.

At the end of the luteal phase, neither estrogen nor progesterone is being produced. In their absence, the anterior pituitary can again initiate production of FSH and LH, thus starting another reproductive cycle. Each cycle begins immediately after the preceding one ends. A cycle usually occurs every 28 days, or a little more frequently than once a month, although this varies in individual cases. The Latin word for "month" is *mens,* which is why the reproductive cycle is called the menstrual cycle, or monthly cycle.

If fertilization does occur high in the fallopian tube (❷ in **figure 31.19**), the zygote undergoes a series of cell divisions called cleavage ❸, while traveling toward the uterus. At the blastocyst stage, it implants in the lining of the uterus ❹. The tiny embryo secretes human chorionic gonadotropin (hCG), an LH-like hormone, which maintains the corpus luteum. By maintaining the corpus luteum, hCG keeps the levels of estrogen and progesterone high, thereby preventing menstruation, which would terminate the pregnancy. Because hCG comes from the embryo and not from the mother, it is hCG that is tested in all pregnancy tests.

> **Key Learning Outcome 31.5** Humans and apes have menstrual cycles driven by cyclic patterns of hormone secretion and ovulation. The cycle is composed of two distinct phases, follicular and luteal, coordinated by a family of hormones.

31.6 Embryonic Development

Cleavage: Setting the Stage for Development

Fertilization begins a carefully orchestrated series of developmental events. **Table 31.1** traces the major stages of mammalian development, beginning with fertilization. Follow down the table as the stages of development are discussed here.

The first major event in human embryonic development is the rapid division of the zygote into a larger and larger number of smaller and smaller cells, becoming first 2 cells, then 4, then 8, and so on. The first of these divisions occurs about 30 hours after union of the egg and the sperm, and the second, 30 hours later. During this period of division, called **cleavage,** the overall size does not increase from that of the zygote. The resulting tightly packed mass of about 32 cells is called a **morula,** and each individual cell in the morula is referred to as a **blastomere.** The cells of the morula continue to divide, each cell secreting a fluid into the center of the cell mass. Eventually, a hollow ball of 500 to 2,000 cells is formed. This is the **blastocyst,** which contains a fluid-filled cavity called the **blastocoel.** Within the ball is an *inner cell mass* concentrated at one pole that goes on to form the developing embryo. The outer sphere of cells, called the *trophoblast,* releases the hCG hormone, discussed earlier.

During cleavage, the morula journeys down the mother's fallopian tube. On about the sixth day, the blastocyst has formed and reaches the uterus; it attaches to the uterine lining and penetrates into the tissue of the lining. The blastocyst now begins to grow rapidly, initiating the formation of the membranes that will later surround, protect, and nourish it. One of these membranes, the **amnion,** will enclose the developing embryo, whereas another, the **chorion,** which forms from the trophoblast, will interact with uterine tissue to form the **placenta,** which will nourish the growing embryo (see **figures 20.23** and **31.21**). The placenta connects the developing embryo to the blood supply of the mother. Fully 61 of the cells at the 64-celled stage develop into the trophoblast and only 3 into the embryo proper.

Gastrulation: The Onset of Developmental Change

Ten to eleven days after fertilization, certain groups of cells move inward from the surface of the cell mass in a carefully orchestrated migration called **gastrulation.** First, the lower cell layer of the blastocyst cell mass differentiates into **endoderm,** one of the three primary embryonic tissues, and the upper layer into **ectoderm.** Just after this differentiation, much of the **mesoderm** arises by the invagination of cells that move from the upper layer of the cell mass *inward,* along the edges of a furrow that appears at the embryo midline, the **primitive streak.**

During gastrulation, about half of the cells of the blastocyst cell mass move into the interior of the human embryo. This movement largely determines the future development of the embryo. By the end of gastrulation, distribution of cells into the three primary germ layers has been completed. The ectoderm is destined to form the epidermis and neural tissue. The mesoderm is destined to form the connective tissue, muscle, and vascular elements. The endoderm forms the lining of the gut and its derivative organs. The fates of all three primary germ layers are:

Ectoderm	Epidermis, central nervous system, sense organs, neural crest
Mesoderm	Skeleton, muscles, blood vessels, heart, gonads
Endoderm	Lining of digestive and respiratory tracts; liver, pancreas

Neurulation: Determination of Body Architecture

In the third week of embryonic development, the three primary cell types begin their development into the tissues and organs of the body. This stage in development is called **neurulation.**

The first characteristic vertebrate feature to form is the **notochord,** a flexible rod. Soon after gastrulation is complete, it forms along the midline of the embryo, below its dorsal surface. Then the second characteristic vertebrate feature, the **neural tube,** forms above the notochord and later differentiates into the spinal cord and brain. Just before the neural tube closes, two strips of cells break away and form the **neural crest.** These neural crest cells give rise to neural structures found in the vertebrate body.

While the neural tube is forming from ectoderm, the rest of the basic architecture of the human body is being rapidly determined by changes in the mesoderm. On either side of the developing notochord, segmented blocks of tissue form. Ultimately, these blocks, or **somites,** give rise to the muscles, vertebrae, and connective tissues. As development continues, more and more somites are formed. Within another strip of mesoderm that runs alongside the somites, many of the significant glands of the body, including the kidneys, adrenal glands, and gonads, develop. The remainder of the mesoderm layer moves out and around the inner endoderm layer of cells and eventually surrounds it entirely. As a result, the mesoderm forms two layers. The outer layer is associated with the body wall and the inner layer is associated with the gut. Between these two layers of mesoderm is the **coelom,** which becomes the body cavity of the adult.

By the end of the third week, over a dozen somites are evident, and the blood vessels and gut have begun to develop. At this point the embryo is about 2 millimeters (less than a tenth of an inch) long.

Key Learning Outcome 31.6 The vertebrate embryo develops in three stages: cleavage, a hollow ball of cells forms; gastrulation, cells move into the interior, forming the primary tissues; neurulation, organs begin to form.

	Stage (age)	Description
TABLE 31.1 STAGES OF MAMMALIAN DEVELOPMENT		
Sperm / Ovum	**Fertilization** (day 1)	The haploid male and female gametes fuse to form a diploid zygote.
Blastocyst — Inner cell mass — Trophoblast — Blastocoel	**Cleavage** (days 2–10)	The zygote rapidly divides into many cells, with no overall increase in size. These divisions affect future development because different cells receive different portions of the egg cytoplasm and, hence, different regulatory signals.
Amniotic cavity — Ectoderm — Endoderm	**Gastrulation** (days 11–15)	The cells of the embryo move, forming three primary germ layers: Ectoderm and endoderm form first, followed by the formation of mesoderm.
Primitive streak — Ectoderm — Mesoderm — Endoderm — Formation of extraembryonic membranes		
Neural groove — Notochord	**Neurulation** (days 16–25)	In all chordates, the first organ to form is the notochord; the second is the neural tube.
Neural crest — Neural tube — Notochord		During neurulation, the neural crest is produced as the neural tube is formed. The neural crest gives rise to several uniquely vertebrate structures such as sensory neurons, sympathetic neurons, Schwann cells, and other cell types.
	Organogenesis (days 26+)	Cells from the three primary cell layers combine in various ways to produce the organs of the body.

31.7 Fetal Development

The Fourth Week: Organogenesis

In the fourth week of pregnancy, the body organs begin to form, a process called **organogenesis** (figure 31.20a; the drawing helps identify the structures in the photo). The eyes form, and the heart begins a rhythmic beating and develops four chambers. At 70 beats per minute, the little heart is destined to beat more than 2.5 billion times during a lifetime of about 70 years. More than 30 pairs of somites are visible by

the end of the fourth week, and the arm and leg buds have begun to form. The embryo more than doubles in length during this week, reaching about 5 millimeters.

By the end of the fourth week, the developmental scenario is far advanced, although most women are not yet aware that they are pregnant. This is a crucial time in development because the proper course of events can be interrupted easily. For example, alcohol use by pregnant women during the first months of pregnancy is one of the leading causes of birth defects, producing **fetal alcohol syndrome,** in which the baby is born with a deformed face and often severe mental retarda-

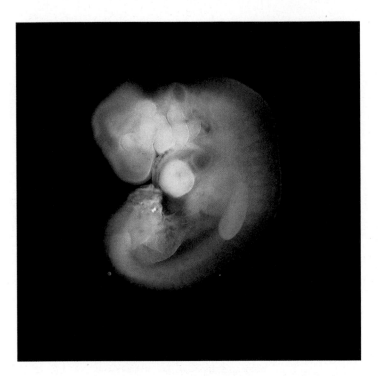

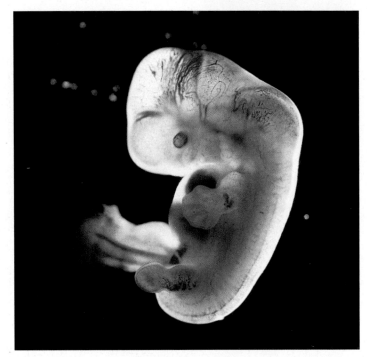

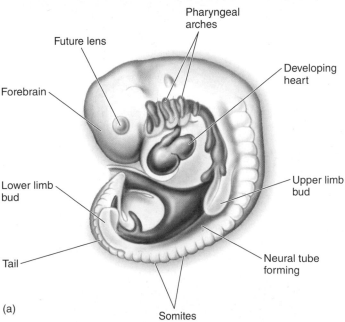

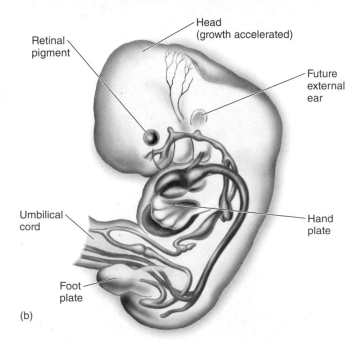

(a)

(b)

Figure 31.20   The developing human.

(a) Four weeks; (b) seven weeks; (c) three months; and (d) four months.

tion. One in 250 newborns in the United States is affected by fetal alcohol syndrome. Also, most spontaneous abortions (miscarriages) occur during this period.

The Second Month: The Embryo Takes Shape

During the second month of pregnancy, great changes in morphology occur as the embryo takes shape (**figure 31.20b**). The miniature limbs of the embryo assume their adult shapes. The arms, legs, knees, elbows, fingers, and toes can all be seen as well as a short, bony tail. The bones of the embryonic tail, an evolutionary reminder of our past, later fuse to form the coccyx, or tailbone. Within the body cavity, the major internal organs are evident, including the liver and pancreas. By the end of the second month, the embryo has grown to about 25 millimeters in length—it is 1 inch long. It weighs perhaps a gram and is beginning to look distinctly human.

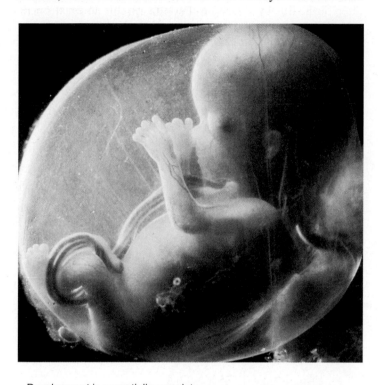

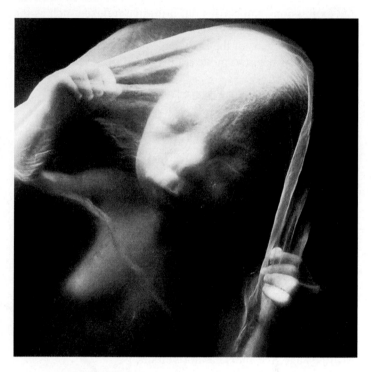

Development is essentially complete.

The developing human is now referred to as a fetus.

Facial expressions and primitive reflexes are carried out.

All of the major body organs have been established.

Arms and legs begin to move.

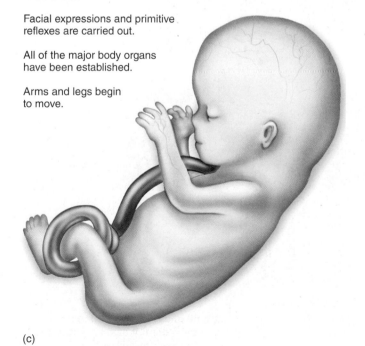

(c)

Figure 31.20 (continued)

Bones actively enlarge.

Mother can feel baby kicking.

Following a period of rapid growth, the fetus is born.

Neurological growth continues after birth.

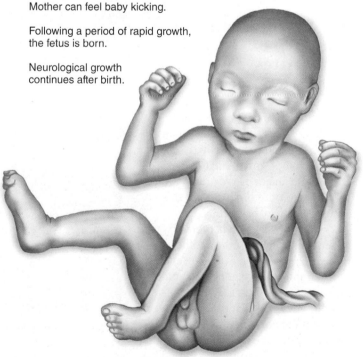

(d)

The Third Month: Completion of Development

Development of the embryo is essentially complete except for the lungs and brain. The lungs don't complete development until the third trimester, and the brain continues to develop even after birth. From this point on, the developing human is referred to as a **fetus** rather than an embryo. What remains is essentially growth. The nervous system and sense organs develop during the third month. The fetus begins to show facial expressions and carries out primitive reflexes such as the startle reflex and sucking. By the end of the third month, all of the major organs of the body have been established and the arms and legs begin to move (**figure 31.20c**).

The Second Trimester: The Fetus Grows in Earnest

The second trimester is a time of growth. In the fourth (**figure 31.20d**) and fifth months of pregnancy, the fetus grows to about 175 millimeters in length (almost 7 in long), with a body weight of about 225 grams. Bone formation occurs actively during the fourth month. During the fifth month, the head and body become covered with fine hair. This downy body hair, called **lanugo,** is another evolutionary relic and is lost later in development. By the end of the fourth month, the mother can feel the baby kicking; by the end of the fifth month, she can hear its rapid heartbeat with a stethoscope.

In the sixth month, growth accelerates. By the end of the sixth month, the baby is over 0.3 meters (1 ft) long and weighs 0.6 kilograms (about 1.5 lb)—and most of its prebirth growth is still to come. At this stage, the fetus cannot yet survive outside the uterus without special medical intervention.

The Third Trimester: The Pace of Growth Accelerates

The third trimester is a period of rapid growth. In the seventh, eighth, and ninth months of pregnancy, the weight of the fetus more than doubles. This increase in bulk is not the only kind of growth that occurs. Most of the major nerve tracts are formed within the brain during this period, as are new brain cells.

All of this growth is fueled by nutrients provided by the mother's bloodstream, passing into the fetal blood supply within the placenta. The placenta in **figure 31.21** contains blood vessels that extend from the **umbilical cord** into tissues that line the uterus (the *decidua basalis* in the figure). The mother's blood bathes this tissue so that nutrients

can pass from the mother's blood into the blood vessels that carry it back to the fetus without the two blood systems ever mixing blood. The undernourishment of the fetus by a malnourished mother can adversely affect this growth and result in severe retardation of the infant. Retardation resulting from fetal malnourishment is a severe problem in many underdeveloped countries, where poverty and hunger are common.

By the end of the third trimester, the neurological growth of the fetus is far from complete and, in fact, continues long after birth. But by this time the fetus is able to exist on its own. Why doesn't the fetus continue to develop within the uterus until its neurological development is complete? What's the rush to get out and be born? Because physical growth is continuing as well, and the fetus is about as large as it can get and still be delivered through the pelvis without damage to mother or child. As any woman who has had a baby can testify, it is a tight fit. Birth takes place as soon as the probability of survival is high.

Birth

At approximately 40 weeks from the last menstrual cycle, the process of birth begins. Hormonal changes in the mother, in addition to physical and possibly hormonal signals from the fetus, initiate the onset of **labor.** During labor (and delivery), the cervix gradually dilates (the opening becomes larger), the

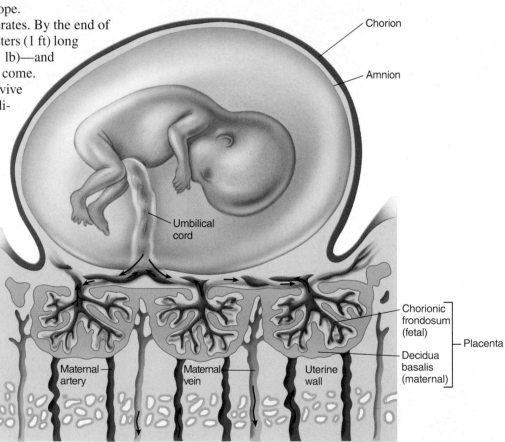

Figure 31.21 Structure of the placenta.
The placenta contains a fetal component, the chorionic frondosum, and a maternal component, the decidua basalis. Oxygen and nutrients enter the fetal blood from the maternal blood by diffusion. Waste substances enter the maternal blood from the fetal blood, also by diffusion.

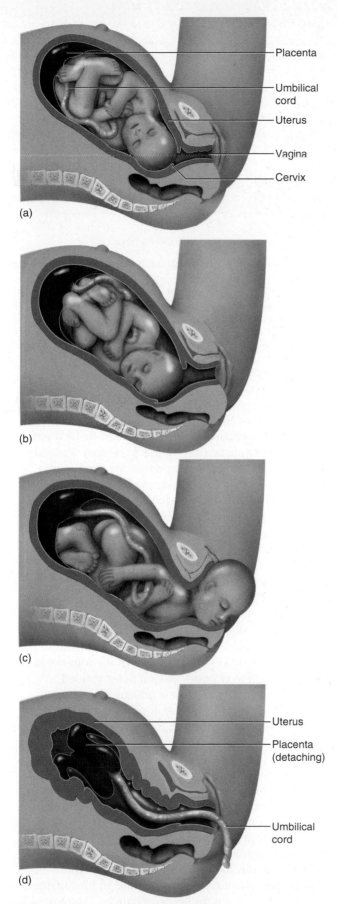

(a)

- Placenta
- Umbilical cord
- Uterus
- Vagina
- Cervix

(b)

(c)

(d)

- Uterus
- Placenta (detaching)
- Umbilical cord

Figure 31.22 Stages of childbirth.

amnion ruptures causing amniotic fluid to flow out through the vagina (sometimes referred to as the "water breaking"), and uterine contractions become strong and regular, usually resulting in the expulsion of the fetus from the uterus. Hormones called **oxytocin** and **prostaglandins** work in a positive feedback mechanism to stimulate and increase uterine contractions. The fetus is usually in a head-down position near the end of pregnancy. In a vaginal birth, the fetus is pushed down through the cervix and out through the vagina (**figure 31.22**). The umbilical cord is still attached to the baby, and a doctor, nurse, or parent clamps and cuts the cord. The baby transitions from living in a fluid environment to a gaseous one, and many of its organ systems undergo major changes. After the birth of the fetus, continuing uterine contractions expel the placenta and associated membranes, collectively called the "afterbirth." In some cases, such as when a vaginal delivery would cause harm to the fetus or the mother, the fetus and placenta are surgically removed from the uterus in a procedure called a *caesarian section (C-section).*

In the mother, hormones during late pregnancy prepare the *mammary glands* (see section 20.8) for nourishing the baby after birth. For the first couple of days after childbirth, the mammary glands produce a fluid called *colostrum,* which contains protein and lactose but little fat. Then milk production is stimulated by the anterior pituitary hormone **prolactin,** usually by the third day after delivery. When the infant suckles at the breast, the posterior pituitary hormone oxytocin is released, initiating milk release, or milk "letdown."

Postnatal Development

Growth continues rapidly after birth. Babies typically double their birth weight within a few months. Different organs grow at different rates, however, and the body proportions of infants are different than those of adults. The head, for example, is disproportionately large in newborns, but after birth it grows more slowly than the rest of the body. Such a pattern of growth, in which different components grow at different rates, is referred to as *allometric growth.*

At birth, the developing nervous system of humans is generating new nerve cells at an average rate of more than 250,000 per minute. Then, about six months after birth, this production of new neurons essentially ceases (recent research has shown new neurons continue to be produced into adulthood in a few small regions of the brain). The fact that the human brain continues to grow significantly for the first few years of postnatal life means that adequate nutrition and a safe environment are particularly crucial during this period.

Key Learning Outcome 31.7 Most of the key events in fetal development occur early. Organs begin to form in the fourth week, and by the end of the second month the developing body looks distinctly human. The development of the embryo is essentially complete before the woman may even know she is pregnant. What remains in the second and third trimester is primarily growth.

Why Don't Men Get Breast Cancer?

In 2007, an estimated 240,510 new cases of breast cancer were expected among women in the United States and 2,030 new cases among men—over 99% of the new breast cancer victims were women. Similarly, of 40,460 breast cancer deaths anticipated that year, all but 450 of them were women. It is impossible not to wonder why so few men? An obvious answer would be that men don't have breasts, but men do have breast tissue, it just isn't as developed as a woman's. So why so few men?

While 5% of breast cancers are due to inherited genetic mutations (*BRCA1* and *BRCA2*), the cause of 95%—the overwhelming majority—remains a mystery. Hormone differences seem the most promising place to start, as the female sex hormone estrogen controls breast development in women; men, by contrast, lack physiologically significant amounts of estrogen. A logical suggestion is that in breast cancer patients, the effect of estrogen on breast cells is being altered by exposure to a so-called endocrine disrupter. Endocrine disrupters are man-made chemicals that mimic hormones. By sheer chance, their molecules are perfectly shaped to fit particular hormone receptors. In this case, the culprit would be a chemical mimic of estrogen that promotes cancerous growth in breast cells.

One candidate is bisphenol A (BPA), a molecule that is structurally similar to estrogen, with carbon rings at each end tipped with OH groups. Used to form the plastic packaging of many foods and drinks, as well as the clear plastic liners of metal food and beverage cans, BPA is a chemical to which all of us are exposed daily. Six billion pounds are produced worldwide each year.

It has been known since 1938 that BPA promotes excess estrogen production in rats. Alarmingly, in 1993 BPA was shown to have the same effects on human breast cancer cells growing in culture. What was alarming was that the effect could be measured at concentrations as low as 2 parts per billion, not much above the levels to which we humans are routinely exposed.

Does BPA in fact induce breast cancer? To test this possibility, Dr. Ana Soto of Tufts University School of Medicine and her research team in 2006 tested its effects on rats, which get breast cancer in much the same way as humans do. They exposed pregnant female rats to a range of BPA concentrations, and after 50 days sacrificed them for examination of their breast tissue. The researchers looked in particular for aberrant cell growth patterns in breast tissue called ductal hyperplasias, which in both rats and people are considered to be the precursors of

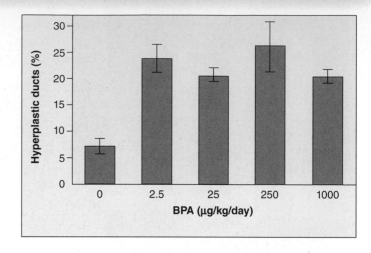

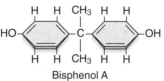

Bisphenol A

Estrogen

breast cancer. BPA was administered to four groups of rats. Some received low doses not unlike what humans are exposed to, while others received much higher doses. In a fifth group, which served as a control, no BPA was administered.

The histogram above shows what the researchers found. Amounts of BPA administered to the rats are reported in micrograms of BPA per kilogram of body weight per day. The incidence of breast cancer detected is presented as the percent of examined breast tissue ducts which were hyperplastic (that is, precancerous). Vertical black lines (called error bars) represent 5% confidence intervals (a measure of scatter among the data—95% of replicate experiments would be expected to fall within the ranges of the error bars).

What are we to make of the data the researchers obtained? Did any of the four doses of BPA they administered result in a percent of hyperplastic ducts significantly higher than that seen in the BPA control group? (Hint: If their error bars overlap with the control's, the difference is not significant.) The answer—yes, all four doses are significant. Do higher doses of BPA increase the incidence of hyperplastic ducts? Do the error bars of the four doses fail to overlap? No, all doses yielded similar results. It is difficult to avoid the conclusion that exposure to even low levels of bisphenol A induces ductal hyperplasias in laboratory rats.

These results suggest the rather alarming conclusion that BPA, a chemical to which we are all exposed every day, may cause breast cancer. The study is small, and a rat is not a human, but the possibility that BPA is a carcinogen certainly merits further investigation. Most government-funded breast cancer research has focused on the search for more effective breast cancer treatments; far less money is spent on searching for the causes of breast cancer. The most encouraging aspect of this study is its suggestion that such a search, if it became a priority, might prove fruitful.

31.8 Contraception and Sexually Transmitted Diseases

Contraception

Not all couples want to initiate a pregnancy every time they have sex, yet sexual intercourse may be a necessary and important part of their emotional lives together. The solution to this dilemma is to find a way to avoid reproduction without avoiding sexual intercourse, an approach that is commonly called **birth control,** or contraception.

Abstinence The simplest and most reliable way to avoid pregnancy is not to have sex at all. Of all methods of birth control, this is the most certain—and the most limiting, because it denies a couple the emotional support of a sexual relationship. A variant of this approach is to avoid sex only on the days when successful fertilization is likely to occur. The rest of the sexual cycle is considered relatively "safe" for intercourse. This approach, called the rhythm method, or **natural family planning,** when other indicators are also monitored, is satisfactory in principle but difficult in application because ovulation is not easy to predict and may occur unexpectedly. Failure rate can be as high as 20% to 30%.

Prevention of Egg Maturation A widespread form of birth control in the United States has been the daily ingestion of hormones, or **birth control pills.** These pills contain estrogen and progesterone, which shut down production of the pituitary hormones FSH and LH. The ovarian follicles do not ripen in the absence of FSH, and ovulation does not occur in the absence of LH. Other methods of hormone delivery include medroxy progesterone (Depo-Provera), which is injected every one to three months, the seven-day birth control patch, which releases the hormones through the skin, and surgically implanted capsules that release hormones. Failure rates of these methods are less than 2%.

Emergency contraception, called **Plan B,** is a high-dose progesterone pill that can block ovulation if taken soon after unprotected sex. Its failure rate varies and should not be used as a primary method of birth control.

Prevention of Embryo Implantation The insertion of a coil or other irregularly shaped object into the uterus is an effective means of birth control. The irritation in the uterus prevents the implantation of the descending embryo within the uterine wall. Such **intrauterine devices (IUDs)** are very effective because once they are inserted, they can be forgotten. They have a failure rate of less than 2%.

A chemical means of preventing embryo implantation or ending an early pregnancy is **RU-486.** This pill blocks the action of progesterone, causing the endometrium to slough off. RU-486 must be administered under a doctor's care because of potentially serious side effects.

Sperm Blockage Fertilization cannot occur without sperm. One way to prevent the delivery of sperm is to encase the penis within a thin rubber bag, or **condom.** In principle, this method is easy to apply and foolproof, but in practice, it proves to be less effective than you might expect, with a failure rate of up to 15%. A second way to prevent the entry of sperm is to cover the cervix with a rubber dome called a **diaphragm,** inserted immediately before intercourse. Because the dimensions of individual cervices vary, diaphragms must be fitted by a physician. Failure rates average 20%.

Sperm Destruction A third general approach to birth control is to destroy the sperm within the vagina. Sperm can be destroyed with **spermicidal jellies, suppositories,** and **foams** applied immediately before intercourse. The failure rate varies widely, from 3% to 22%.

Sexually Transmitted Diseases

Sexually transmitted diseases (STDs) are diseases that spread from one person to another through sexual contact. AIDS, discussed in chapter 27, is a deadly viral STD. Other significant sexually transmitted diseases include:

Gonorrhea. The primary symptom of this disease, which is caused by the bacterium *Neisseria gonorrhoeae*, is discharge from the penis or vagina. It can be treated with antibiotics. If left untreated in women, gonorrhea can cause pelvic inflammatory disease (PID), a condition in which the fallopian tubes become scarred and blocked. PID can eventually lead to sterility.

Chlamydia. Caused by the bacterium *Chlamydia trachomatis,* this disease is sometimes called the "silent STD" because women usually experience no symptoms until after the infection has become established. Like gonorrhea, chlamydia can cause PID in women if left untreated.

Syphilis. Caused by the bacterium *Treponema pallidum,* this disease is one of the most potentially devastating STDs. Left untreated, the disease progresses to heart disease, mental deficiency, and nerve damage that may include loss of motor function or blindness.

Genital herpes. Caused by the herpes simplex virus type 2 (HSV-2), this disease is the most common STD in the United States. The virus causes red blisters on the penis or on the labia, vagina, or cervix that scab over.

Cervical cancer. About 70% of cervical cancer is caused by HPV (human papillomavirus), a sexually transmitted virus. Gardasil, a newly developed vaccine that blocks HPV in women not yet exposed to the virus, could cut worldwide deaths (about 290,000 women each year) by two-thirds.

> **Key Learning Outcome 31.8** A variety of birth control methods are available, many of them quite effective. Sexually transmitted diseases are spread through sexual contact.

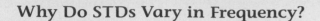
Why Do STDs Vary in Frequency?

As a general rule, the incidence of a sexually transmitted disease is expected to increase with increasing frequencies of unprotected sexual contact. With the emergence of AIDS, intense publicity and education has lessened such dangerous behavior. Both the number of sexual partners and the frequency of unprotected sex have fallen significantly in the United States in the last decade. It would follow, then, that the frequencies of sexually transmitted diseases (STDs) like syphilis, gonorrhea, and chlamydia should also be falling.

However, the level of one STD sometimes rises while another falls. What are we to make of this? The simplest explanation of such a difference is that the two STDs are occurring in different populations, and one population has rising levels of sexual activity, while the other has falling levels. However, nationwide statistics encompass all population subgroups, and there is no reason to expect subgroups to contain different STDs. Certainly each major subgroup contains all three major STDs mentioned above. So this would seem an unlikely explanation for the frequency of one STD to be rising while another falls.

A second possible explanation would be a change in the infectivity of one of the STDs. A less-infective STD would tend to fall in frequency in the population, for the simple reason that fewer sexual contacts result in infection. To assess this possibility, we must examine the individual STDs more closely.

Syphilis is most infective in its initial stage, but this stage lasts only about a month. Most transmissions occur during the much longer second stage, marked by a pink rash and sores in the mouth. The bacteria can be transmitted at this stage by kissing or shared liquids. Any drop in infectivity of this STD would be expected to shorten this stage—but no such shortening has been observed.

Gonorrhea can be transmitted by various forms of sexual contact with an infected individual at any time during the infection. There has been no drop in infectivity per sexual contact reported.

Chlamydia offers the most interesting possibility of changes in infectivity, because of its unusual nature. *Chlamydia trachomatis* is genetically a bacterium but is an obligate intracellular parasite, much like a virus in this respect—it can reproduce only inside human

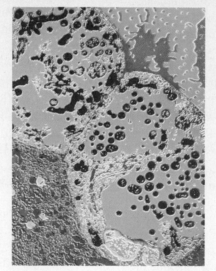

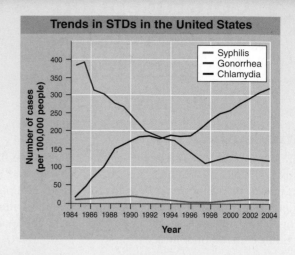

Trends in STDs in the United States

cells. The red structures in the photo are chlamydia bacteria inside human cells. Like gonorrhea, chlamydia is transmitted through vaginal, anal, or oral intercourse with an infected person. With chlamydia, the person may show no symptoms. Because the disease agent lives inside cells, its infectivity would not be expected to change unless the number of cells of an infected individual to which his or her sex partner would be exposed during intercourse were to change, a very unlikely possibility. So a drop in infectivity doesn't seem very likely. There is, however, a third possible explanation for why the frequency of one STD in a population might rise while the frequency of another STD in that same population falls. To grasp this third possible explanation, we will need to examine carefully the trends in the incidence in the United States of the gonorrhea, chlamydia, and syphilis. Detailed yearly statistics are reported in the graph above.

1. **Applying Concepts** What is the dependent variable?

2. **Making Inferences**
   a. Gonorrhea: What is the incidence in 1985? in 1995? Has the frequency declined or increased? In general, are individuals aware they are infected when they transmit the STD?
   b. Chlamydia: What is the incidence in 1985? in 1995? Has the frequency declined or increased? In general, are individuals aware they are infected when they transmit the STD?

3. **Drawing Conclusions** How might heightened public awareness explain why the trend in levels of gonorrhea differs from that of chlamydia?

Vertebrate Reproduction

31.1 Asexual and Sexual Reproduction

- Asexual reproduction through fission (**figure 31.1**) or budding is the primary means of reproduction among protists and some animals, but most animals reproduce sexually.

- Sexual reproduction is most common in animals, but parthenogenesis and hermaphroditism are two variations. In parthenogenesis, offspring are produced from unfertilized eggs. In hermaphroditism, an individual has both testes and ovaries, producing both sperm and eggs.

- In mammals, sex is genetically determined and appears during embryonic development. An embryo that is XY develops into a male and an XX embryo develops into a female (**figure 31.3**).

31.2 Evolution of Vertebrate Sexual Reproduction

- Fertilization in most fish and many amphibians occurs externally, as shown here from **figure 31.6**, but internal fertilization occurs in most other vertebrates, through a process called copulation. Even with internal fertilization, development may occur inside or outside of the mother.

- Birds and most reptiles lay watertight eggs that are protected from desiccation (**figure 31.9**).

The Human Reproductive System

31.3 Males

- Sperm is produced in the testes (**figure 31.11**). The testes contain a large number of tightly coiled tubes called seminiferous tubules, where sperm develop in a process called spermatogenesis (**figure 31.12**). As sperm develop and undergo meiosis, they move toward the lumen of the tubules, and from there they pass into the epididymis. Once matured, they are stored in the vas deferens. During sexual intercourse, sperm are delivered through the penis (**figure 31.14**) into the female.

31.4 Females

- Female gametes, eggs or ova, develop in the ovary from oocytes. The ovaries are located in the lower abdomen (**figure 31.15**). A female is born with some 2 million oocytes, all arrested during the first meiotic division. The hormones FSH and LH initiate the resumption of meiosis I in a few oocytes, but usually only one oocyte completes development.

- The egg ruptures from the ovary, called ovulation, and enters the fallopian tube (**figure 31.16**). Sperm deposited in the vagina travel up through the cervix and uterus, shown here from **figure 31.17**, and into the fallopian tube, or oviduct, to reach the egg. Usually only one sperm cell penetrates the egg's protective layers. At this point, the oocyte completes meiosis II, and fertilization occurs. The fertilized egg, called a zygote, is transported to the uterus through the oviduct. The zygote attaches to the endometrial lining of the uterus where it completes development.

31.5 Hormones Coordinate the Reproductive Cycle

- The human reproductive cycle, called a menstrual cycle, is divided into two phases: the follicular and luteal phases. The follicular phase begins with the secretion of FSH and LH, which stimulates the resumption of oocyte development and the secretion of estrogen. The estrogen acts as a negative feedback signal to stop the secretion of FSH from the anterior pituitary. The luteal phase begins with a surge of LH, which causes ovulation and the formation of the corpus luteum (**figure 31.18**). The corpus luteum begins secreting progesterone, which acts to prepare the uterus for implantation of the zygote (**figure 31.19**).

- If a zygote implants in the lining of the uterus, estrogen and progesterone levels remain high due to the release of human chorionic gonadotropin (hCG) from the embryo. The uterus is maintained and no further egg maturation occurs.

- If fertilization does not occur, estrogen and progesterone levels drop and the endometrial lining of the uterus sloughs off, a process called menstruation, and a new cycle begins.

The Course of Development

31.6 Embryonic Development

- The vertebrate embryo develops in three stages. The first stage, called cleavage, involves hundreds of cell divisions that eventually produce a hollow ball of cells called a blastocyst (**table 31.1**). The second stage, called gastrulation, involves the orchestrated movement of cells, forming the three germ layers: endoderm, ectoderm, and mesoderm. In the third stage, neurulation, the notochord and neural tube form.

31.7 Fetal Development

- Organs begin forming by the fourth week, and by the end of the second month the embryo looks distinctly human. By the end of the third month, all major organs except the brain and lungs are developed (**figure 31.20**).

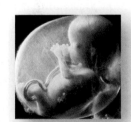

- The second and third trimesters are periods of considerable growth, with the fetus receiving nourishment from the placenta (**figure 31.21**).

- During labor and delivery, the fetus and placenta are expelled from the uterus (**figure 31.22**). Hormones coordinate the production of milk in the mother for nourishing the newborn. Brain development continues in the baby after birth.

Birth Control and Sexually Transmitted Diseases

31.8 Contraception and Sexually Transmitted Diseases

- Various birth control methods are available and work by preventing egg maturation, preventing embryo implantation, and blocking or killing sperm.

- Sexually transmitted diseases are spread through sexual contact. AIDS is a deadly STD. Other STDs may not be as fatal as AIDS but are quite destructive, especially if left untreated.

1. If the offspring are not genetically identical to each other or the parent, then the organism reproduces through
   a. fission.
   b. sexual reproduction.
   c. budding.
   d. All of the above.
2. In mammals, the embryonic gonads will develop into ovaries
   a. if the *SRY* gene is expressed.
   b. if both sex chromosomes are X.
   c. if the sex chromosomes are X and Y.
   d. within the first 40 days.
3. Embryonic development in dogs is an example of
   a. viviparity.
   b. ovoviviparity.
   c. oviparity.
   d. parthenogenesis.
4. Temperature regulation of spermatogenesis in human males is controlled by the location of the
   a. seminiferous tubules.
   b. epididymis.
   c. vas deferens.
   d. scrotum.
5. Oocyte development in human females requires the hormones
   a. estrogen and testosterone.
   b. FSH and LH.
   c. progesterone and testosterone.
   d. oxytocin and prolactin.

6. When pregnancy occurs, the endometrium is maintained by the
   a. embryo releasing hCG.
   b. decrease in levels of progesterone.
   c. hypothalamus releasing GnRH.
   d. increasing levels of FSH.
7. A human embryo has formed the three germ layers from which all tissues form by the time
   a. the blastula forms.
   b. neurulation is complete.
   c. the blastocyst forms.
   d. gastrulation is complete.
8. In a developing human, the first tissues to begin forming are the
   a. skeletal.
   b. muscular.
   c. neural.
   d. digestive.
9. Contractions of the uterus during labor are stimulated by the hormone
   a. estrogen.
   b. prolactin.
   c. oxytocin.
   d. progesterone.
10. Which of the following is *not* a method of contraception?
    a. destruction of the egg
    b. prevention of egg maturation
    c. sperm blockage
    d. prevention of embryo implantation

1. **Figure 31.6** Why do you think many amphibians and many fish have external fertilization, whereas no lizards, birds, and mammals do?

2. **Figure 31.17a** Sometimes a zygote implants in the fallopian tube rather than in the uterus. This is called an ectopic pregnancy (ectopic meaning "out of place"). It is necessary to terminate an ectopic pregnancy because it endangers the mother and the fetus cannot survive. Explain why the mother is in danger and why the fetus can't survive.

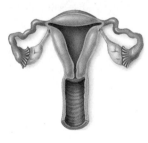

1. Why are all the parents of parthenogenic offspring female?
2. Males produce their sperm throughout their lives, while females produce all of their eggs at one time during sexual development. Would you expect most gene mutations in newborns to have originated from the father, or would you expect both parents to contribute a similar risk of mutation? Explain.
3. My daughter tells me that the most reliable method of contraception is baby sitting. Explain what she means.
4. Oviparity is the rule in the animal kingdom. It is common in insects, fishes, amphibians, and reptiles, and universal in birds. Among mammals, the most primitive (the echidna and the platypus) are also oviparous. However, all mammals that evolved later (marsupials and placentals) are viviparous. What do you think was responsible for this major shift in reproductive strategy?

32
Evolution of Plants

Learning Objectives

Plants

32.1 Adapting to Terrestrial Living
1. List three environmental challenges overcome by plants as they adapted to life on land.
2. Diagram a stoma and explain its functioning.
3. Contrast a plant's gametophyte and sporophyte generations.

32.2 Plant Evolution
1. List four key innovations in the evolution of plants.

Seedless Plants

32.3 Nonvascular Plants
1. Compare the water-conducting abilities of mosses and hornworts.

32.4 The Evolution of Vascular Tissue
1. Differentiate between xylem and phloem and between primary and secondary growth.

32.5 Seedless Vascular Plants
1. Compare the gametophyte and sporophyte generations of mosses with those of ferns.

The Advent of Seeds

32.6 Evolution of Seed Plants
1. Discriminate between gymnosperms and angiosperms and between a microspore and a megaspore.
2. Diagram the structure of a seed, labeling the three visible parts.
3. List the four ways in which seeds are adapted to life on land.

32.7 Gymnosperms
1. Describe the life cycle of the most widespread phylum of gymnosperms.

The Evolution of Flowers

32.8 Rise of the Angiosperms
1. Discuss two reasons why angiosperms have been so much more evolutionarily successful than gymnosperms.
2. Name, locate, and describe the four whorls of a flower.

32.9 Why Are There Different Kinds of Flowers?
1. Explain why there are different-colored flowers.

32.10 Double Fertilization
1. Outline the process of double fertilization and explain the advantages double fertilization confers.
2. Differentiate between monocots and dicots.

32.11 Fruits
1. Explain why angiosperms have fruits, whereas gymnosperms do not.
2. Discuss the evolutionary advantages conferred by fruits.

Inquiry & Analysis: How Does Arrowgrass Tolerate Salt?

P lants are thought to be descendants of green algae. Plants first invaded the land over 455 million years ago in partnership with fungi—the most ancient surviving plants have mycorrhizal associations. One of the many challenges posed by the terrestrial environment was the difficulty of finding a mate when anchored to one spot. The solution adopted by most early plants was for male individuals to cast their gametes—pollen—into the wind, and let air currents carry pollen grains to nearby female plants. This strategy works particularly well in dense stands, in which there are many nearby individuals of the same species. The massive redwood tree seen here, one of the largest individual organisms living on land, grows in dense stands and is wind pollinated. An alternative solution proved even better, however. Plants evolved flowers, devices to attract insects. When insects visit the flower to obtain nectar, they become coated with pollen. When they then visit another flower seeking more nectar, they deposit some of this pollen, pollinating that flower. It doesn't matter how far apart the two plants are from each other—the insect will seek them out. In this chapter, you will explore this and other evolutionary challenges met by plants as they colonized the land.

32.1 Adapting to Terrestrial Living

Plants are complex multicellular organisms that are terrestrial **autotrophs**—that is, they occur almost exclusively on land and feed themselves by photosynthesis. The name *autotroph* comes from the Greek, *autos,* self, and *trophos,* feeder. Today, plants are the dominant organisms on the surface of the earth (**figure 32.1**). Nearly 300,000 species are now in existence, covering almost every part of the terrestrial landscape. In this chapter, we see how plants adapted to life on land.

Figure 32.1 Plants dominate life on land.
Plants are astonishingly diverse and are key components in the biosphere. For example, photosynthesis conducted by plants creates oxygen in the atmosphere, plants are the main source of food for humans and many other organisms (directly and indirectly), and plants provide us with wood, cloth, paper, and many other nonfood products.

The green algae that were probably the ancestors of today's plants are aquatic organisms that are not well adapted to living on land. Before their descendants could live on land, they had to overcome many environmental challenges. For example, they had to absorb minerals from the rocky surface. They had to find a means of conserving water. They had to develop a way to reproduce on land.

Absorbing Minerals

Plants require relatively large amounts of six inorganic minerals: nitrogen, potassium, calcium, phosphorus, magnesium, and sulfur. Each of these minerals constitutes 1% or more of a plant's dry weight. Algae absorb these minerals from water, but where is a plant on land to get them? From the soil. Soil is the weathered outer layer of the earth's crust. It is composed of a mixture of ingredients, which may include sand, rocks, clay, silt, humus (partly decayed organic material), and various other forms of mineral and organic matter. The soil is also rich in microorganisms that break down and recycle organic debris. Plants absorb these materials, along with water, through their *roots* (described in chapter 33). Most roots are found in topsoil, which is a mixture of mineral particles, living organisms, and humus. When topsoil is lost due to erosion, the soil loses its ability to hold water and nutrients.

The first plants seem to have developed a special relationship with fungi, which was a key factor in their ability to absorb minerals in terrestrial habitats. Within the roots or underground stems of many early fossil plants like *Cooksonia* (see **figure 32.6**) and *Rhynia,* fungi can be seen living intimately within and among the plant's cells. As you may recall from chapter 18, these kinds of symbiotic associations are called **mycorrhizae.** In plants with mycorrhizae, the fungi enable the plant to take up phosphorus and other nutrients from rocky soil, while the plant supplies organic molecules to the fungus.

Conserving Water

One of the key challenges to living on land is to avoid drying out. To solve this problem, plants have a watertight outer covering called a **cuticle.** The covering is formed from a waxy substance that is impermeable to water. Like the wax on a shiny car, the cuticle prevents water from entering or leaving the stem or leaves. Water enters the plant only through the roots, while the cuticle prevents water loss to the air. Passages do exist through the cuticle, in the form of specialized pores called **stomata** (singular, **stoma**) in the leaves and sometimes the green portions of the stems. **Figure 32.2** shows a stoma on the underside of a leaf. The cutaway view allows you to see the placement of the stoma in relation to other cells in the

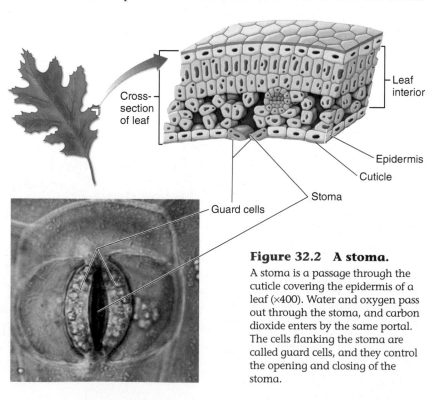

Cross-section of leaf

Leaf interior

Epidermis

Cuticle

Stoma

Guard cells

Figure 32.2 A stoma.
A stoma is a passage through the cuticle covering the epidermis of a leaf (×400). Water and oxygen pass out through the stoma, and carbon dioxide enters by the same portal. The cells flanking the stoma are called guard cells, and they control the opening and closing of the stoma.

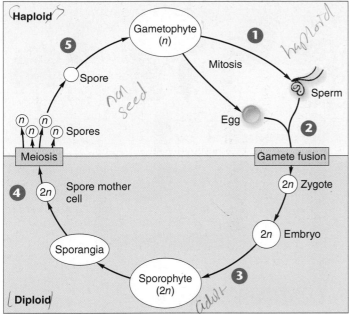

handwritten annotations: alternation Generation, Haploid, non seed, adult, Diploid

Figure 32.3 Generalized plant life cycle.
In a plant life cycle, a diploid generation alternates with a haploid one. Gametophytes, which are haploid (*n*), alternate with sporophytes, which are diploid (2*n*). ❶ Gametophytes give rise by mitosis to sperm and eggs. ❷ The sperm and egg ultimately come together to produce the first diploid cell of the sporophyte generation, the zygote. ❸ The zygote undergoes cell division, ultimately forming the sporophyte. ❹ Meiosis takes place within the sporangia, the spore-producing organs of the sporophyte, resulting in the production of the spores, which are haploid and are the first cells ❺ of gametophyte generations.

leaf. Stomata, which occur on at least some portions of all plants except liverworts, allow carbon dioxide to pass into the plant bodies (by diffusion) for photosynthesis and allow water and oxygen gas to pass out of them. The two *guard cells* that border the stoma swell (opening the stoma) and shrink (closing the stoma) when water moves into and out of them by osmosis. The opening and closing of stomata controls the loss of water from the leaf while allowing the entrance of carbon dioxide (see also **figure 33.13**). In most plants, water enters through the roots as a liquid and exits through the underside of the leaves as water vapor. Water movement within a plant is discussed in more detail in section 33.6.

Reproducing on Land

To reproduce sexually on land, it is necessary to pass gametes from one individual to another, which is a challenge because plants cannot move about. In the first plants, the eggs were surrounded by a jacket of cells, and a film of water was required for the sperm to swim to the egg and fertilize it. In later plants, pollen evolved, providing a means of transferring gametes without drying out. The pollen grain is protected from drying and allows plants to transfer gametes by wind or insects.

Changing the Life Cycle Among many algae, haploid cells occupy the major portion of the life cycle (see section 17.2). The zygote formed by the fusion of gametes is the only diploid

handwritten annotation: diplod

cell, and it immediately undergoes meiosis to form haploid cells again. In early plants, by contrast, meiosis was delayed and the cells of the zygote divided to produce a multicellular diploid structure. Thus for a significant portion of the life cycle, the cells were diploid. This change resulted in an **alternation of generations,** in which a diploid generation alternates with a haploid one. **Figure 32.3** shows a life cycle that exhibits an alternation of generations. Botanists call the haploid generation (the yellow boxed area) the **gametophyte** because it forms haploid gametes by mitosis ❶. The diploid generation (the blue boxed area) is called the **sporophyte** because it forms haploid spores by meiosis ❹.

When you look at primitive plants such as liverworts and mosses, you see mostly gametophyte tissue, the green leafy structures in the photo below—the sporophytes are smaller brown structures (not visible in the photo) attached to or enclosed within the tissues of the larger gametophyte.

Free-living gametophyte of a moss

When you look at plants that evolved later, such as a pine tree, you see mostly sporophyte tissue. The gametophytes of these plants are always much smaller than the sporophytes and are often enclosed within sporophyte tissues. The photo below shows the male gametophyte (pollen grains) of a pine tree. They are not photosynthetic cells and are dependent upon the sporophyte tissue in which they are usually enclosed.

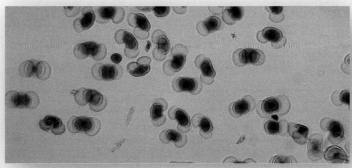

Male gametophytes of a pine

Key Learning Outcome 32.1 Plants are multicellular terrestrial photosynthesizers that evolved from green algae. Plants adapted to life on land by developing ways to absorb minerals in partnership with fungi, to conserve water with watertight coverings, and to reproduce on land.

32.2 Plant Evolution

Once plants became established on land, they gradually developed many other features that aided their evolutionary success in this new, demanding habitat. For example, among the first plants there was no fundamental difference between the aboveground and the belowground parts. Later, roots and shoots with specialized structures evolved, each suited to its particular below- or aboveground environment. The evolution of specialized *vascular tissue* allowed plants to grow larger. For example, compare the size of a moss, a more primitive type of plant, to the more recently evolved tree on which it grows.

As we explore plant diversity, we will examine several key evolutionary innovations that have given rise to the wide variety of plants we see today. **Table 32.1** provides you with an overview of the plant phyla and their characteristics. While other interesting and important changes also arose, four key innovations discussed in this chapter serve to highlight the evolutionary trends exhibited by the plant kingdom. These innovations lead to the evolution of the major plant groups, as shown in **figure 32.4**.

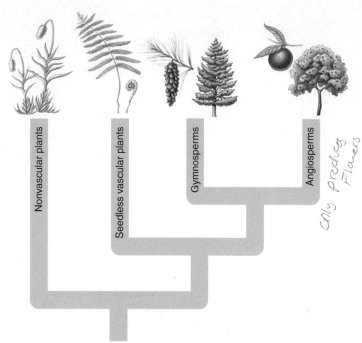

Figure 32.4 The evolution of plants.

TABLE 32.1	**PLANT PHYLA**			
Phylum	**Typical Examples**		**Key Characteristics**	**Approximate Number of Living Species**
Nonvascular Plants				
Hepaticophyta (liverworts)	*Marchantia*		Without true vascular tissues; lack true roots and leaves; live in moist habitats and obtain water and nutrients by osmosis and diffusion; require water for fertilization; gametophyte is dominant structure in the life cycle; the three phyla were once grouped together	15,600
Anthocerophyta (hornworts)	*Anthoceros*			
Bryophyta (mosses)	*Polytrichum, Sphagnum* (hairy cap and peat moss)			
Seedless Vascular Plants				
Lycophyta (lycopods)	*Lycopodium* (club mosses)		Seedless vascular plants, some are similar in appearance to mosses but diploid; require water for fertilization; sporophyte is dominant structure in life cycle; found in moist woodland habitats	1,150
Pterophyta (ferns)	*Azolla, Sphaeropteris* (water and tree ferns)		Seedless vascular plants; require water for fertilization; sporophytes diverse in form and dominate the life cycle	11,000
	Equisetum (horsetails)			
	Psilotum (whisk ferns)			

Four key evolutionary innovations serve to trace the evolution of the plant kingdom:

1. **Alternation of generations.** Although algae exhibit a haploid and a diploid phase, the diploid phase is not a significant portion of their life cycle. By contrast, even in early plants (the nonvascular plants indicated by the first vertical bar in **figure 32.4**), the diploid sporophyte is a larger structure and offers protection for the egg and developing embryo. The dominance of the sporophyte, both in size and the proportion of time devoted to it in the life cycle, becomes greater throughout the evolutionary history of plants.

2. **Vascular tissue.** A second key innovation was the emergence of vascular tissue. Vascular tissue transports water and nutrients throughout the plant body and provides structural support. With the evolution of vascular tissue, plants were able to supply the upper portions of their bodies with water absorbed from the soil and had some rigidity, allowing the plants to grow larger and in drier conditions. The first vascular plants were the seedless vascular plants, the second vertical bar in **figure 32.4**.

3. **Seeds.** The evolution of *seeds* (see section 32.6) was a key innovation that allowed plants to dominate their terrestrial environments. Seeds provide nutrients and a tough, durable cover that protects the embryo until it encounters favorable growing conditions. The first plants with seeds were the gymnosperms, the third vertical bar in **figure 32.4**.

4. **Flowers and fruits.** The evolution of flowers and fruits were key innovations that improved the chances of successful mating in sedentary organisms and facilitated the dispersal of their seeds. Flowers both protected the egg and improved the odds of its fertilization, allowing plants that were located at considerable distances to mate successfully. Fruit, which surrounds the seed and aids in its dispersal, allows plant species to better invade new and possibly more favorable environments. The angiosperms, the fourth vertical bar in **figure 32.4**, are the only plants to produce flowers and fruits.

Key Learning Outcome 32.2 Plants evolved from freshwater green algae and eventually developed more dominant diploid phases of the life cycle, conducting systems of vascular tissue, seeds that protected the embryo, and flowers and fruits that aided in fertilization and distribution of the seeds.

TABLE 32.1 (continued)			
Phylum	**Typical Examples**	**Key Characteristics**	**Approximate Number of Living Species**
Seed Plants			
Coniferophyta (conifers)	Pines, spruce, fir, redwood, cedar	Gymnosperms; wind pollinated; ovules partially exposed at time of pollination; flowerless; seeds are dispersed by the wind; sperm lack flagella; sporophyte is dominant structure in life cycle; leaves are needlelike or scalelike; most species are evergreens and live in dense stands; among the most common trees on earth	601
Cycadophyta (cycads)	Cycads, sago palms	Gymnosperms; wind pollination or possibly insect pollination; very slow growing, palmlike trees; sperm have flagella; trees are either male or female; sporophyte dominant in the life cycle	206
Gnetophyta (shrub teas)	Mormon tea, *Welwitschia*	Gymnosperms; nonmotile sperm; shrubs and vines; wind pollination and possibly insect pollination; plants are either male or female; sporophyte is dominant in the life cycle	65
Ginkgophyta (ginkgo)	Ginkgo trees	Gymnosperms; fanlike leaves that are dropped in winter (deciduous); seeds fleshy and ill-scented; motile sperm; trees are either male or female; sporophyte is dominant in the life cycle	1
Anthophyta (flowering plants, also called angiosperms)	Oak trees, corn, wheat, roses	Flowering; pollination by wind, animal, and water; characterized by ovules that are fully enclosed by the carpel; fertilization involves two sperm nuclei: one forms the embryo, the other fuses with the polar body to form endosperm for the seed; after fertilization, carpels and the fertilized ovules (now seeds) mature to become fruit; sporophyte is dominant in life cycle	250,000

32.3 Nonvascular Plants

Liverworts and Hornworts

The first successful land plants had no vascular system—no tubes or pipes to transport water and nutrients throughout the plant. This greatly limited the maximum size of the plant body because all materials had to be transported by osmosis and diffusion (see chapter 4). However, these simple plants are highly adapted to a diversity of terrestrial environments. Only two phyla of living plants, the **liverworts** (phylum Hepaticophyta) and the **hornworts** (phylum Anthocerophyta), completely lack a vascular system. The word *wort* meant *herb* in medieval Anglo-Saxon when these plants were named. Liverworts are the simplest of all living plants. About 6,000 species of liverworts and 100 species of hornworts survive today, usually growing in moist and shady places.

Primitive Conducting Systems: Mosses

Another phylum of plants, the **mosses** (phylum Bryophyta), were the first plants to evolve strands of specialized cells that conduct water and carbohydrates up the stem of the gametophyte. The conducting cells do not have specialized wall thickenings; instead they are like nonrigid pipes and cannot carry water very high. Because these conducting cells could at the most be considered a primitive vascular system, mosses are usually grouped by botanists with the liverworts and hornworts as "nonvascular" plants. Today about 9,500 species of mosses grow in moist places all over the world. In the Arctic and Antarctic, mosses are the most abundant plants. "Peat moss" (genus *Sphagnum*) can be used as a fuel or a soil conditioner. The moss *Physcomitrella patens,* whose genome was sequenced in 2006, has been the subject of many genetic studies. In **figure 32.5,** you can see that the majority of the moss life cycle consists of the haploid gametophyte generation (the green part of the plant), which exists as male or female plants. The diploid sporophyte (the brown stalk with the swollen head) grows out of the gametophyte of the female plant ❸, after the egg cell has been fertilized. Cells within the sporophyte undergo meiosis to produce haploid spores ❹ that grow into gametophytes ❺.

> **Key Learning Outcome 32.3**
> While liverworts and hornworts totally lack a vascular system, mosses have simple soft strands of conducting cells.

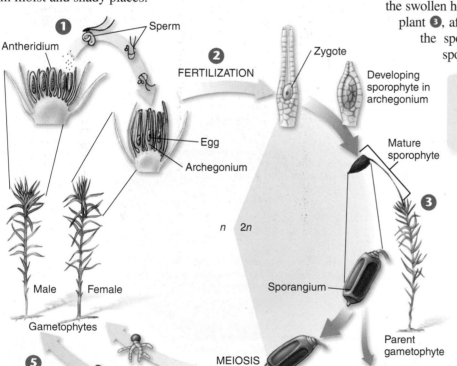

Figure 32.5 The life cycle of a moss.
On the haploid gametophytes, sperm are released from each antheridium (sperm-producing structure) ❶. They then swim through free water to an archegonium (egg-producing structure) and down its neck to the egg. Fertilization takes place there ❷; the resulting zygote develops into a diploid sporophyte. The sporophyte grows out of the archegonium, forming the sporangium at its apex ❸. The sporophyte grows on the gametophyte, as shown in the photo, and eventually produces spores as a result of meiosis. The spores are shed from the sporangium ❹. The spores germinate, giving rise to gametophytes ❺.

A hair-cup moss, *Polytrichum*

32.4 The Evolution of Vascular Tissue

The remaining seven phyla of plants, which have efficient vascular systems made of highly specialized cells, are called **vascular plants.** The first vascular plant appeared approximately 430 million years ago, but only incomplete fossils have been found. The first vascular plants for which we have relatively complete fossils, the extinct phylum Rhyniophyta, lived 410 million years ago. Among them is the oldest known vascular plant, *Cooksonia.* The fossil in **figure 32.6** clearly shows that the plant had branched, leafless shoots that formed spores at the tips in structures called *sporangia.*

Cooksonia and the other early plants that followed became successful colonizers of the land through the development of efficient water- and food-conducting systems known as **vascular tissues** (Latin, *vasculum,* vessel or duct). These tissues, discussed in detail in chapter 33, consist of strands of specialized cylindrical or elongated cells that form a network of tubelike structures throughout a plant, extending from near the tips of the roots (when present), through the stems, and into the leaves (when present; **figure 32.7**). One type of vascular tissue, *xylem,* conducts water and dissolved minerals upward from the roots; another type of vascular tissue, *phloem,* conducts carbohydrates throughout the plant. The presence of a cuticle and stomata are also characteristic of vascular plants.

Most early vascular plants seem to have grown by cell division at the tips of the stem and roots. Imagine stacking dishes—the stack can get taller but not wider! This sort of growth is called **primary growth** and was quite successful. During the so-called Coal Age (between 350 and 290 million years ago), when much of the world's fossil fuel was formed, the lowland swamps that covered Europe and North America were dominated by an early form of seedless tree called a lycophyte. Lycophyte trees grew to heights of 10 to 35 meters (33 to 115 ft), and their trunks did not branch until they attained most of their total height. The pace of evolution was rapid during this period, for the world's climate was changing, growing dryer and colder. As the world's swamplands began to dry up, the lycophyte trees vanished, disappearing abruptly from the fossil record. They were replaced by tree-sized ferns, a form of vascular plant that will be described in more detail in section 32.5. Tree ferns grew to heights of more than 20 meters (66 ft) with trunks 30 centimeters (12 in) thick. Like the lycophytes, the trunks of tree ferns were formed entirely by primary growth.

About 380 million years ago, vascular plants developed a new pattern of growth, in which a cylinder of cells beneath the bark divides, producing new cells in regions around the plant's periphery. This growth is called **secondary growth.** Secondary growth makes it possible for a plant stem to increase in diameter. Only after the evolution of secondary growth could vascular plants become thick-trunked and, therefore, tall. Redwood trees today reach heights of up to 117 meters (384 ft) and trunk diameters in excess of 11 meters (36 ft). This evolutionary advance made possible the dominance of the tall forests that today cover northern North

America. You are familiar with the product of plant secondary growth as **wood.** The growth rings that are visible in cross sections of trees are zones of secondary growth (spring–summer) spaced by zones of little growth (fall–winter).

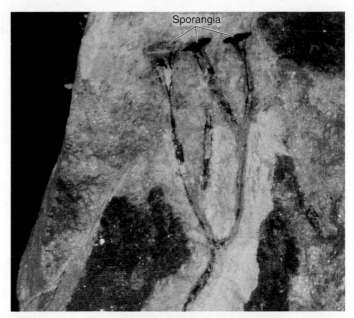

Figure 32.6   The earliest vascular plant.
The earliest vascular plant of which we have complete fossils is *Cooksonia.* This fossil shows a plant that lived some 410 million years ago; its upright branched stems terminated in spore-producing sporangia at the tips.

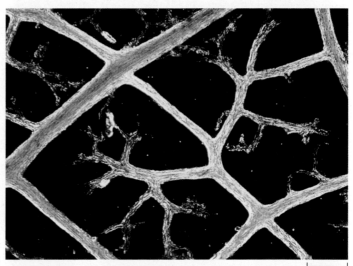

260 µm

Figure 32.7   The vascular system of a leaf.
The veins of a vascular plant contain strands of specialized cells for conducting food and water.

> **Key Learning Outcome 32.4   Vascular plants have specialized vascular tissue composed of hollow tubes that conduct water from the roots to the leaves. Another type of vascular tissue forms cylinders that conduct food from the leaves to the rest of the plant.**

32.5 Seedless Vascular Plants

The earliest vascular plants lacked seeds, and two of the seven phyla of modern-day vascular plants do not have them. One phyla of living seedless vascular plants is the ferns, phylum Pterophyta. This phylum includes the typical ferns seen growing on forest floors, like those shown in **figure 32.8a,b**. It also includes the whisk ferns (**figure 32.8c**) and the horsetails (**figure 32.8d**). The other phylum, Lycophyta, contains the club mosses (**figure 32.8e**). These phyla have free-swimming sperm that require the presence of free water for fertilization.

By far the most abundant of seedless vascular plants are the **ferns,** with about 11,000 living species. Ferns are found throughout the world,

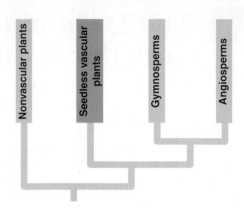

(a)

(b)

(c)

(d)

(e)

Figure 32.8   Seedless vascular plants.

(a) A tree fern in the forests of Malaysia (phylum Pterophyta). The ferns are by far the largest group of spore-producing vascular plants. (b) Ferns on the floor of a redwood forest. (c) A whisk fern. Whisk ferns have no roots or leaves. (d) A horsetail, *Equisetum telmateia*. This species forms two kinds of erect stems; one is green and photosynthetic, and the other, which terminates in a spore-producing "cone," is mostly light brown. (e) The club moss *Lycopodium lucidulum,* recently renamed *Huperzia lucidula.* Although superficially similar to the gametophytes of mosses, the conspicuous club moss plants shown here are sporophytes.

although they are much more abundant in the tropics than elsewhere. Many are small, only a few centimeters in diameter, but some of the largest plants that live today are also ferns. Descendants of ancient tree ferns, they can have trunks more than 24 meters (79 ft) tall and leaves up to 5 meters (16 ft) long!

The Life of a Fern

In ferns, the life cycle of plants begins a revolutionary change that culminates later with seed plants. Nonvascular plants like mosses are made largely of gametophyte (haploid) tissue. Vascular seedless plants like ferns have both gametophyte and sporophyte individuals, each independent and self-sufficient. The gametophyte (the heart-shaped plant at the top of **figure 32.9**) produces eggs and sperm. After sperm swim through water and fertilize the egg, the zygote grows into a sporophyte. The sporophyte bears haploid spores on the underside of its leaves, in

brown clusters called sori (singular, sorus). The spores are released from the sorus and float to the ground where they germinate, growing into haploid gametophytes. The fern gametophytes are small, thin, heart-shaped photosynthetic plants that live in moist places. The fern sporophytes are much larger and more complex, with long, vertical leaves called **fronds.** When you see a fern, you are almost always looking at the sporophyte.

> **Key Learning Outcome 32.5** Ferns are among the vascular plants that lack seeds, reproducing with spores as nonvascular plants do.

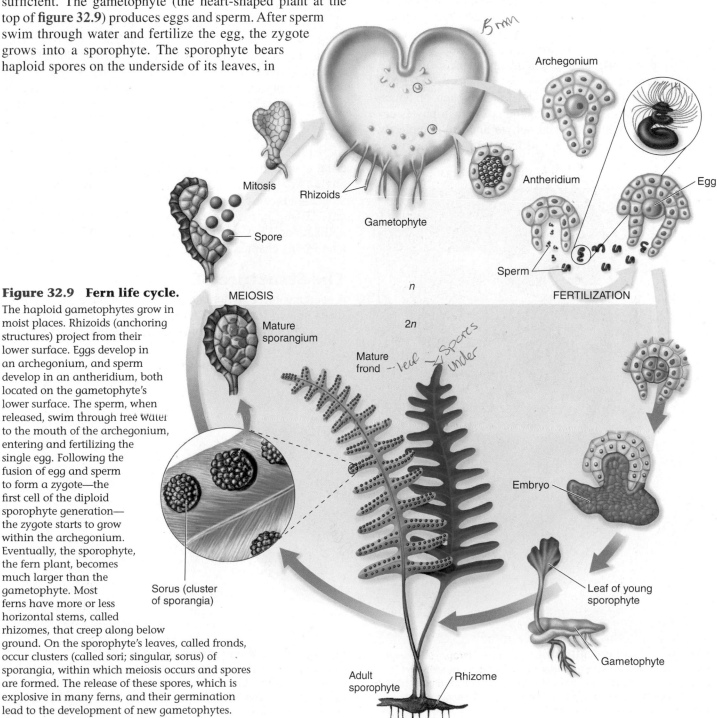

Figure 32.9   Fern life cycle.
The haploid gametophytes grow in moist places. Rhizoids (anchoring structures) project from their lower surface. Eggs develop in an archegonium, and sperm develop in an antheridium, both located on the gametophyte's lower surface. The sperm, when released, swim through free water to the mouth of the archegonium, entering and fertilizing the single egg. Following the fusion of egg and sperm to form a zygote—the first cell of the diploid sporophyte generation—the zygote starts to grow within the archegonium. Eventually, the sporophyte, the fern plant, becomes much larger than the gametophyte. Most ferns have more or less horizontal stems, called rhizomes, that creep along below ground. On the sporophyte's leaves, called fronds, occur clusters (called sori; singular, sorus) of sporangia, within which meiosis occurs and spores are formed. The release of these spores, which is explosive in many ferns, and their germination lead to the development of new gametophytes.

32.6 Evolution of Seed Plants

A key evolutionary advance among the vascular plants was the development of a protective cover for the embryo called a **seed.** The seed is a crucial adaptation to life on land because it protects the embryonic plant when it is at its most vulnerable stage. The plant in **figure 32.10** is a cycad and its seeds (the green balls in the photo) develop on the edges of the scales of the cone. The embryonic plants are inside the seeds where they are protected. The evolution of the seed was a critical step in allowing plants to dominate life on land.

The dominance of the sporophyte (diploid) generation in the life cycle of vascular plants reaches its full force with the advent of the seed plants. Seed plants produce two kinds of gametophytes—male and female, each of which consists of just a few cells. Both kinds of gametophytes develop separately within the sporophyte and are completely dependent on it for their nutrition. Male gametophytes, commonly referred to as **pollen grains,** arise from **microspores.** The pollen grains become mature when sperm are produced. The sperm are carried to the egg in the female gametophyte without the need for free water in the environment. A female gametophyte contains the egg and develops from a **megaspore** produced within an **ovule.** The transfer of pollen to an ovule by insects, wind, or other agents is referred to as **pollination.** The pollen grain then cracks open and sprouts, or germinates, and the pollen tube, containing the sperm cells, grows out, transporting the sperm directly to the egg. Thus free water is not required in the pollination and fertilization process.

Botanists generally agree that all seed plants are derived from a single common ancestor. There are five living phyla. In four of them, collectively called the **gymnosperms** (Greek, *gymnos,* naked, and *sperma,* seed), the ovules are not completely enclosed by sporophyte tissue at the time of pollination. Gymnosperms were the first seed plants. From gymnosperms evolved the fifth group of seed plants, called **angiosperms** (Greek, *angion,* vessel, and *sperma,* seed), phylum Anthophyta. Angiosperms, or flowering plants, are the most recently evolved of all the plant phyla. Angiosperms differ from all gymnosperms in that their ovules are completely enclosed by a vessel of sporophyte tissue in the flower called the **carpel** at the time of pollination. We will discuss gymnosperms and angiosperms later in this chapter.

The Structure of a Seed

A seed has three parts that are visible in the corn and bean seeds shown in **figure 32.11**: (1) a sporophyte plant embryo, (2) a source of food for the developing embryo called

Figure 32.10   A seed plant.

The seeds of this cycad, like all seeds, consist of a plant embryo and a protective covering. A cycad is a gymnosperm (naked-seeded plant), and its seeds develop out in the open on the edges of the cone scales.

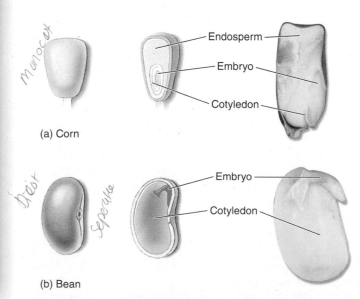

(a) Corn

(b) Bean

Figure 32.11   Basic structure of seeds.

A seed contains a sporophyte (diploid) embryo and a source of food, either endosperm (*a*) or food stored in the cotyledons (*b*). A seed coat, formed of sporophytic tissue from the parent, surrounds the seed and protects the embryo.

Figure 32.12   Seeds allow plants to bypass the dry season.
Seeds remain dormant until conditions are favorable for growth. When it does rain, seeds can germinate, and plants can grow rapidly to take advantage of the relatively short periods when water is available. This palo verde desert tree (*Cercidium floridum*) has tough seeds (*inset*) that germinate only after they are cracked. Rains leach out the chemicals in the seed coats that inhibit germination, and the hard coats of the seeds may be cracked when they are washed down gullies in temporary floods.

endosperm in flowering plants (the endosperm makes up most of the seed in corn and is the white part of popcorn), and (3) a drought-resistant protective cover. In some seeds, the endosperm is used up during the development of the embryo and is stored as food by the embryo in thick "leaf-like" structures called **cotyledons.** The endosperm is replaced by the cotyledon, as in the bean seed. Seeds are one way in which plants, anchored by their roots to one place in the ground, are able to disperse their progeny to new locations. The hard cover of the seed (formed from the tissue of the parent plant) protects the seed while it travels to a new location. The seed travels by many means such as by air, water, and animals. Many airborne seeds have devices to aid in carrying them farther. Most species of pine, for example, have seeds with thin flat wings attached. These wings help catch air currents, which carry the seeds to new areas. Seed dispersal will be discussed further later in this chapter and in chapter 34.

Once a seed has fallen to the ground, it may lie there dormant for many years. When conditions are favorable, however, and particularly when moisture is present, the seed germinates and begins to grow into a young plant (**figure 32.12**). Most seeds have abundant food stored in them to provide a ready source of energy for the new plant as it starts its growth.

The advent of seeds had an enormous influence on the evolution of plants. Seeds are particularly adapted to life on land in at least four respects:

1. **Dispersal.** Most importantly, seeds facilitate the migration and dispersal of plant offspring into new habitats.
2. **Dormancy.** Seeds permit plants to postpone development when conditions are unfavorable, as during a drought, and to remain dormant until conditions improve.
3. **Germination.** By making the reinitiation of development dependent upon environmental factors such as temperature, seeds permit the course of embryonic development to be synchronized with critical aspects of the plant's habitat, such as the season of the year.
4. **Nourishment.** The seed offers the young plant nourishment during the critical period just after germination, when the seedling must establish itself.

Key Learning Outcome 32.6   A seed is a dormant diploid embryo encased with food reserves in a hard protective coat. Seeds play critical roles in improving a plant's chances of successfully reproducing in a varied environment.

32.7 Gymnosperms

Four phyla constitute the gymnosperms (**figure 32.13**): the conifers (Coniferophyta), the cycads (Cycadophyta), the gnetophytes (Gnetophyta), and the ginkgo (Ginkgophyta). The conifers are the most familiar of the four phyla of gymnosperms and include pine, spruce, hemlock, cedar, redwood, yew, cypress, and fir trees, such as the Douglas firs in **figure 32.13a**. Conifers are trees that produce their seeds in cones. The seeds (ovules) of conifers develop on scales within the cones and are exposed at the time of pollination. Most of the conifers have needlelike leaves, an evolutionary adaptation for retarding water loss. Conifers are often found growing in moderately dry regions of the world, including the vast taiga forests of the northern latitudes. Many are very important as sources of timber and pulp.

There are about 600 living species of conifers. The tallest living vascular plant, the coastal sequoia (*Sequoia sempervirens*), found in coastal California and Oregon, is a conifer and reaches over 100 meters (328 ft). The biggest redwood, however, is the mountain sequoia redwood species (*Sequoiadendron gigantea*) of the Sierra Nevadas. The largest individual tree is nicknamed after General Sherman of the Civil War, and it stands more than 83 meters (274 ft) tall while measuring 31 meters (102 ft) around its base. Another much smaller type of conifer, the bristlecone pines in Nevada, may be the oldest trees in the world—about 5,000 years old.

The other three gymnosperm phyla are much less widespread. Cycads (**figure 32.13b**), the predominant land

(a)

(b)

Figure 32.13 Gymnosperms.
(*a*) These Douglas fir trees, a type of conifer, often occur in vast forests. (*b*) An African cycad, *Encephalartos transvenosus*, phylum Cycadophyta. The cycads have fernlike leaves and seed-forming cones, like the ones shown here. (*c*) *Welwitschia mirabilis*, phylum Gnetophyta, is found in the extremely dry deserts of southwestern Africa. In *Welwitschia*, two enormous, beltlike leaves grow from a circular zone of cell division that surrounds the apex of the carrot-shaped root. (*d*) Maidenhair tree, *Ginkgo biloba*, the only living representative of the phylum Ginkgophyta, a group of plants that was abundant 200 million years ago. Among living seed plants, only the cycads and ginkgo have swimming sperm.

(c)

(d)

plant in the golden age of dinosaurs, the Jurassic period (213–144 million years ago), have short stems and palmlike leaves. They are still widespread throughout the tropics. The gnetophytes, phylum Gnetophyta, consist of only three kinds of plants, all unusual. One of them is perhaps the most bizarre of all plants, *Welwitschia,* shown in **figure 32.13c**, which grows on the exposed sands of the harsh Namibian Desert of southwestern Africa. *Welwitschia* acts like a plant standing on its head! Its two beltlike, leathery leaves are generated continuously from their base, splitting as they grow out over the desert sands. There is only one living species of ginkgo, the maidenhair tree, which has fan-shaped leaves (shown in **figure 32.13d**) that are shed in the autumn. Because ginkgos are resistant to air pollution, they are commonly planted along city streets.

The fossil record indicates that members of the ginkgo phylum were once widely distributed, particularly in the Northern Hemisphere; today, only one living species, the maidenhair tree (*Ginkgo biloba*), remains. The reproductive structures of ginkgos are produced on separate trees. The fleshy outer coverings of the seeds of female ginkgo plants exude the foul smell of rancid butter caused by butyric and isobutyric acids. In many Asian countries, however, the seeds are considered a delicacy. In Western countries, because of the seed odor, male plants that are vegetatively propagated are preferred for cultivation.

The Life of a Gymnosperm

We will examine conifers as typical gymnosperms. The conifer life cycle is illustrated in **figure 32.14**. Conifer trees form two kinds of cones. Seed cones ❸ contain the female gametophytes, with their egg cells; pollen cones ❶ contain pollen grains. Conifer pollen grains ❷ are small and light and are carried by the wind to seed cones. Because it is very unlikely that any particular pollen grain will succeed in being carried to a seed cone (the wind can take it anywhere), a great many pollen grains are produced to be sure that at least a few succeed in pollinating seed cones. For this reason, pollen grains are shed from their cones in huge quantities, often appearing as a sticky yellow layer on the surfaces of ponds and lakes—and even on windshields.

When a grain of pollen settles down on a scale of a female cone, a slender tube grows out of the pollen cell up into the scale, delivering the male gamete to the female gametophyte containing the egg, or ovum. Fertilization occurs when the sperm cell fuses with the egg ❺, forming a zygote that develops into an embryo. This zygote is the beginning of the sporophyte generation. What happens next is the essential improvement in reproduction achieved by seed plants.

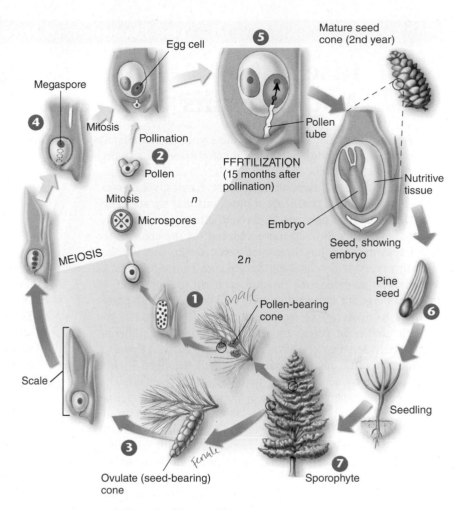

Figure 32.14 Life cycle of a conifer.

In all seed plants, the gametophyte generation is greatly reduced. In conifers such as pines, the relatively delicate pollen-bearing cones ❶ contain microspores, which give rise to pollen grains ❷, the male gametophytes. The familiar seed-bearing cones of pines ❸ are much heavier and more substantial structures than the pollen-bearing cones. Two ovules, and ultimately two seeds, are borne on the upper surface of each scale, which contains the megaspores that give rise to the female gametophytes. After a pollen grain reaches a scale, it germinates, and a slender pollen tube grows toward the egg. When the pollen tube grows to the vicinity of the female gametophyte ❹, sperm are released ❺, fertilizing the egg and producing a zygote. The development of the zygote into an embryo takes place within the ovule, which matures into a seed ❻. Eventually, the seed falls from the cone and germinates, the embryo resuming growth and becoming a new pine tree ❼.

Instead of the zygote growing immediately into an adult sporophyte—just as you grew directly into an adult from a fertilized zygote—the fertilized ovule forms a seed ❻. The pine seed contains a sail-like structure that helps the seed to be carried by the wind. The seed can then be dispersed into new habitats. If conditions are favorable where the seed lands, it will germinate and begin to grow, forming a new sporophyte plant ❼.

Key Learning Outcome 32.7 Gymnosperms are seed plants in which the ovules are not completely enclosed by diploid tissue at pollination. Gymnosperms do not have flowers.

32.8 Rise of the Angiosperms

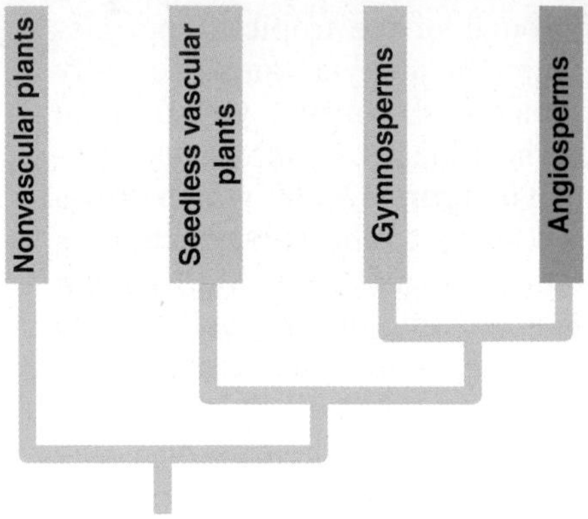

Angiosperms are plants in which the ovule is completely enclosed by sporophyte tissue when it is fertilized. Ninety percent of all living plants are angiosperms, about 250,000 species, including many trees, shrubs, herbs, grasses, vegetables, and grains—in short, nearly all of the plants that we see every day. Virtually all of our food is derived, directly or indirectly, from angiosperms. In fact, more than half of the calories we consume come from just three species: rice, corn, and wheat.

In a very real sense, the remarkable evolutionary success of the angiosperms is the apparent culmination of the plant kingdom's adaptation to life on land, although plants continue to evolve. Angiosperms successfully meet the last difficult challenge posed by terrestrial living: the inherent conflict between the need to obtain nutrients (solved by roots, which anchor the plant to one place) and the need to find mates (solved by accessing plants of the same species). This challenge has never really been overcome by gymnosperms, whose pollen grains are carried passively by the wind on the chance that they might by luck encounter a female cone. Think about how inefficient this is! Angiosperms are able to deliver their pollen directly, as if in an addressed envelope, from one individual of a species to another. How? *By inducing insects and other animals to carry it for them!* The tool that makes this animal-dictated pollination possible, the great advance of the angiosperms, is the flower. While some very successful later-evolving angiosperms like grasses have reverted to wind pollination, the directed pollination of flowering plants has led to phenomenal evolutionary success.

The Flower

Flowers are the reproductive organs of angiosperm plants. A flower is a sophisticated pollination machine. It employs bright colors to attract the attention of insects (or birds or small mammals), nectar to induce the insect to enter the flower, and structures that coat the insect with pollen grains while it is visiting. Then, when the insect visits another flower, it carries the pollen with it into that flower.

The basic structure of a flower consists of four concentric circles, or **whorls,** connected to a base called the **receptacle:**

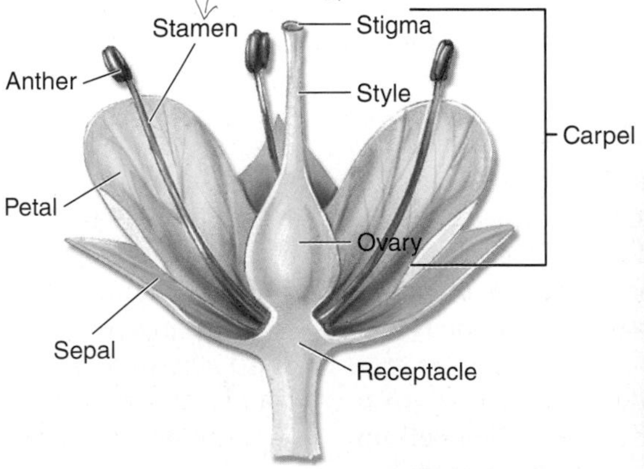

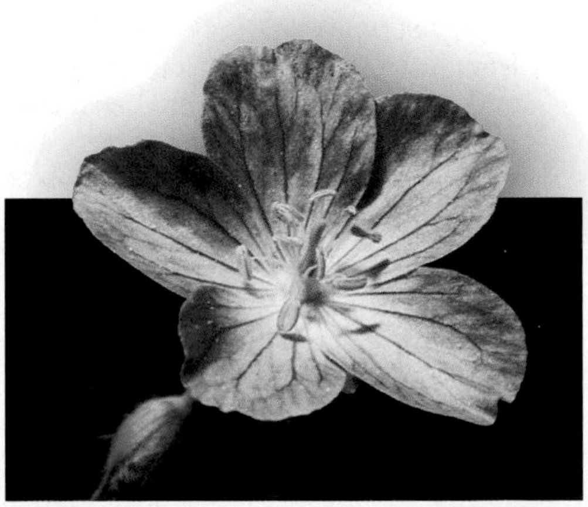

1. The outermost whorl, called the **sepals** of the flower, typically serves to protect the flower from physical damage. These, the green leaflike structures in the diagram above, are in effect modified leaves that protect the flower while it is a bud. The sepals can be seen surrounding the flower bud to the lower left of the flower in the photo below.
2. The second whorl, called the **petals** of the flower, serves to attract specific pollinators. Petals have particular pigments, often vividly colored like the light purple color in the flower diagramed below.
3. The third whorl, called the **stamens** of the flower, contains the "male" parts that produce the pollen grains. Stamens are the slender, threadlike filaments in the diagram with a swollen **anther** at the tip containing pollen.
4. The fourth and innermost whorl, called the **carpel** of the flower, contains the "female" parts that produce eggs. The carpel is the vase-shaped structure in the diagramed flower. The carpel is sporophyte tissue that completely encases the ovules within which the egg cell develops. The ovules occur in the bulging lower portion of the carpel, called the **ovary;** usually there is a slender stalk rising from the ovary called the **style,** with a sticky tip called a **stigma,** which receives pollen. When the flower is pollinated, a pollen tube grows down from the pollen grain on the stigma through the style to the ovary to fertilize the egg.

Key Learning Outcome 32.8 Angiosperms are seed plants in which the ovule is completely enclosed by diploid tissue at pollination and that use flowers to attract pollinators.

32.9 | Why Are There Different Kinds of Flowers?

If you were to watch insects visiting flowers, you would quickly discover that the visits are not random. Instead, certain insects are attracted by particular flowers. Insects recognize a particular color pattern and odor and search for flowers that are similar. Insects and plants have coevolved (see chapters 34 and 35) so that certain insects specialize in visiting particular kinds of flowers. As a result, a particular insect carries pollen from one individual flower to another *of the same species*. It is this keying to particular species that makes insect pollination so effective.

Of all insect pollinators, the most numerous are bees. Bees evolved soon after flowering plants, some 125 million years ago. Today there are over 20,000 species. Bees locate sources of nectar largely by odor at first (that is why flowers smell sweet) and then focus in on the flower's color and shape. Bee-pollinated flowers are usually yellow or blue, like the yellow flower in **figure 32.15a**. They frequently have guiding stripes or lines of dots to indicate the position in the flower of the nectar, usually in the throat of the flower, but these markings may not always be visible to the human eye. For example, the yellow flower in **figure 32.15a** looks very different through an ultraviolet filter in **figure 32.15b**. The UV rays show a dark area in the middle of the flower, where the nectar is located. Why have hidden signals? Because they are not hidden from bees that can detect UV rays. While inside the flower, the bee becomes coated with pollen, as in **figure 32.15c**. When the bee leaves this flower and visits another, it takes the pollen along for the ride, pollinating a neighboring flower.

Many other insects pollinate flowers. Butterflies tend to visit flowers of plants like phlox that have "landing platforms" on which they can perch. These flowers typically have long, slender floral tubes filled with nectar that a butterfly can reach by uncoiling its long proboscis (a hoselike tube extending out from the mouth). Moths, which visit flowers at night, are attracted to white or very pale-colored flowers, often heavily scented, that are easy to locate in dim light. Flowers pollinated by flies, such as members of the milkweed family, are usually brownish in color and foul smelling.

Red flowers, interestingly, are not typically visited by insects, most of which cannot "see" red as a distinct color. Who pollinates these flowers? Hummingbirds and sunbirds (**figure 32.16**)! To these birds, red is a very conspicuous color, just as it is to us. Birds do not have a well-developed sense of smell, and do not orient to odor, which is why red flowers often do not have a strong smell.

Some angiosperms have reverted to the wind pollination practiced by their ancestors, notably oaks, birches, and, most important, the grasses. The flowers of these plants are small, greenish, and odorless. Other angiosperm species are aquatic, and while some have developed specialized pollination systems where the pollen is transported underwater or

(a)   (b)

(c)

Figure 32.15 How a bee sees a flower.
(a) A yellow flower photographed in normal light and (b) with a filter that selectively transmits ultraviolet light. To a bee, this flower appears as if it has a conspicuous central bull's-eye. (c) When inside the flower, the bee becomes covered in pollen, which it takes to a neighboring flower.

Figure 32.16 Red flowers are pollinated by hummingbirds.
This long-tailed hermit hummingbird is extracting nectar from the red flowers of *Heliconia imbricata* in the forests of Costa Rica. Note the pollen on the bird's beak.

floats from one plant to another, most aquatic angiosperms are either wind pollinated or insect pollinated like their terrestrial ancestors. Their flowers extend up above the surface of the water.

Key Learning Outcome 32.9 Flowers can be viewed as pollinator-attracting devices, with different kinds of pollinators attracted to different kinds of flowers.

Double Fertilization

The seeds of gymnosperms often contain food to nourish the developing plant in the critical time immediately after germination, but the seeds of angiosperms have greatly improved on this aspect of seed function. Angiosperms produce a special, highly nutritious tissue called **endosperm** within their seeds. Here is how it happens. The angiosperm life cycle is presented in **figure 32.17**, but there are actually two parts to the cycle, a male and a female part, indicated by the two sets of arrows at the top of the cycle. We'll begin with the flower of the sporophyte on the left side of the cycle ❶. The development of the male gametophyte (the pollen grain) occurs in the anthers and is indicated in the upper set of arrows. The anthers are shown in cross section in ❷ so you can see the microspores mother cells that develop into pollen grains. The pollen grain contains two haploid sperm. Upon adhering to the stigma at the top of the carpel (the female organ in which the egg cell is produced), the pollen begins to form a pollen tube ❹. The yellow pollen tube grows down into the carpel until it reaches the ovule in the ovary ❺. The two sperm (the small purplish cells) travel down the pollen tube and into the ovary. The first sperm fuses with the egg (the green cell at the base of the ovary), as in all sexually reproducing organisms, forming the zygote that develops into the embryo. The other sperm cell fuses with two other products of meiosis, called polar nuclei, to form a triploid (three copies of the chromosomes, $3n$) endosperm cell. This cell divides much more rapidly than the zygote, giving rise to the nutritive endosperm tissue within the seed (the tan material surrounding the embryo ❻). This process of fertilization with two sperm to produce both a zygote and endosperm is called **double fertilization.** Double fertilization forming endosperm is exclusive to angiosperms.

In some angiosperms, such as the common pea or bean, the endosperm is fully used up by the time the seed is mature. Food reserves are stored by the embryo in swollen, fleshy leaves called cotyledons, or seed leaves. In other angiosperms, such as corn, the mature seed contains abundant endosperm, which is used after germination. It also contains a cotyledon, but its seed leaf is used to protect the plant during germination and not as a food source.

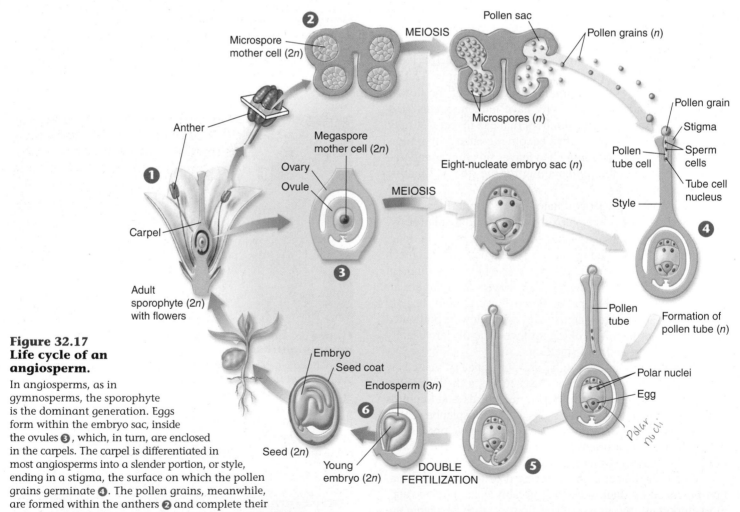

Figure 32.17 Life cycle of an angiosperm.

In angiosperms, as in gymnosperms, the sporophyte is the dominant generation. Eggs form within the embryo sac, inside the ovules ❸, which, in turn, are enclosed in the carpels. The carpel is differentiated in most angiosperms into a slender portion, or style, ending in a stigma, the surface on which the pollen grains germinate ❹. The pollen grains, meanwhile, are formed within the anthers ❷ and complete their differentiation to the mature, three-celled stage either before or after grains are shed. Fertilization is distinctive in angiosperms, being a double process ❺. A sperm and an egg come together, producing a zygote; at the same time, another sperm fuses with the two polar nuclei, producing the primary endosperm nucleus, which is triploid. The zygote and the primary endosperm nucleus divide mitotically, giving rise, respectively, to the embryo and the endosperm ❻. The endosperm is the tissue, almost exclusive to angiosperms, that nourishes the embryo and young plant.

Some angiosperm embryos have two cotyledons, and are called *dicotyledons,* or **dicots.** The first angiosperms were like this. Dicots typically have leaves with netlike branching of veins and flowers with four to five parts per whorl (**figure 32.18,** *top*). Oak and maple trees are dicots, as are many shrubs.

The embryos of other angiosperms, which evolved somewhat later, have a single cotyledon and are called *monocotyledons,* or **monocots.** Monocots typically have leaves with parallel veins and flowers with three parts per whorl (**figure 32.18,** *bottom*). Grasses, one of the most abundant of all plants, are wind-pollinated monocots. There are also differences in the organization of vascular tissue in the stems of monocots and dicots, which will be compared in more detail in chapter 33.

Key Learning Outcome 32.10 Two sperm fertilize each angiosperm ovule. One fuses with the egg to form the zygote, the other with two polar nuclei to form triploid ($3n$) nutritious endosperm.

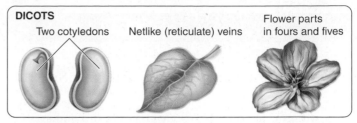

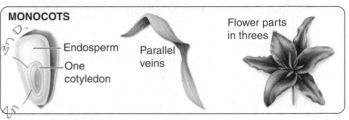

Figure 32.18   Dicots and monocots.

Dicots have two cotyledons and netlike (reticulate) veins. Their flower parts occur in fours and fives. Monocots are characterized by one cotyledon, parallel veins, and the occurrence of flower parts in threes (or multiples of three).

32.11 Fruits

Just as a mature ovule becomes a seed, so a mature ovary that surrounds the ovule becomes all or a part of the **fruit.** This is why fruit forms in angiosperms and not in gymnosperms. Both types of plants have tissue that surrounds the egg, called the ovule, which becomes the seed in both. But, in angiosperms, the ovule is surrounded by another layer of tissue, the ovary, which develops into the fruit. A fruit is a mature ripened ovary containing fertilized seeds. Fruits provide angiosperms with a second way of dispersing their progeny than simply sending their seeds off on the wind. Just as in pollination, they employ animals. By making fruits fleshy and tasty to animals, like the berries in **figure 32.19a,** angiosperms encourage animals to eat them. The seeds within the fruit are resistant to chewing and digestion. They pass out of the animal with the feces, undamaged and ready to germinate at a new location far from the parent plant.

Although many fruits are dispersed by animals, some fruits are dispersed by water, like the coconut in **figure 32.19b,** and many plant fruits are specialized for wind dispersal. The small, nonfleshy fruits of the dandelion, for example, have a plumelike structure that allows them to be carried long distances on wind currents. The fruits of many grasses are small particles, so light wind bears them easily. Maples have long wings attached to the fruit, as in **figure 32.19c,** that allow them to be carried by the wind before reaching the ground. In tumbleweeds, the whole plant breaks off and is blown across open country by the wind, scattering seeds as it rolls. Fruits will be discussed in more detail in chapter 34.

Key Learning Outcome 32.11 A fruit is a mature ovary containing fertilized seeds, often specialized to aid in seed dispersal.

(a)   (b)   (c)

Figure 32.19   Different ways of dispersing fruits.

(a) Berries are fruits that are dispersed by animals. (b) Coconuts are fruits dispersed by water, where they are carried off to new island habitats. (c) The fruits of maples are dry and winged, carried by the wind like small helicopters floating through the air.

How Does Arrowgrass Tolerate Salt?

Plants grow almost everywhere on earth, thriving in many places where exposure, drought, and other severe conditions challenge their survival. In deserts, a common stress is the presence of high levels of salt in the soils. Soil salinity is also a problem for millions of acres of abandoned farmland because the accumulation of salt from irrigation water restricts growth. Why does excess salt in the soil present a problem for a plant? For one thing, high levels of sodium ions that are taken up by the roots are toxic. For another, a plant's roots cannot obtain water when growing in salty soil. Osmosis (the movement of water molecules to areas of higher solute concentrations, see page 96) causes water to move in the opposite direction, drawn out of the roots by the soil's high levels of salt. And yet plants do grow in these soils. How do they manage?

To investigate this, researchers have studied seaside arrowgrass (*Triglochin maritima*), the plant you see below. Arrowgrass plants are able to grow in very salty seashore soils, where few other plants survive. How are they able to survive? Researchers found that their roots do not take up salt, and so do not accumulate toxic levels of salt.

However, this still leaves the arrowgrass plant the challenge of preventing its root cells from losing water to the surrounding salty soil. How then do the roots achieve osmotic balance? In an attempt to find out, researchers grew arrowgrass plants in nonsalty soil for two weeks, then transferred them to one of several soils which differed in salt level. After ten days, shoots were harvested and analyzed for amino acids, because accumulating amino acids could be one way that the cells maintain osmotic balance. Results are presented in the graph.

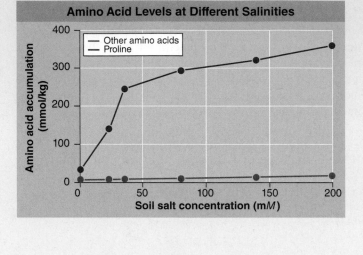

Amino Acid Levels at Different Salinities

Legend:
- Other amino acids
- Proline

x-axis: Soil salt concentration (mM)
y-axis: Amino acid accumulation (mmol/kg)

1. **Applying Concepts**
   a. Variable. What is the dependent variable?
   b. Concentration. What do the abbreviations "mM" and "mmol/Kg" mean?
2. **Interpreting Data**
   a. In salt-free soil (that is, the mM soil salt concentration = 0), how much proline has accumulated in the roots after ten days? How much of other amino acids?
   b. In salty beach soils with salt levels of 35 mM, how much proline has accumulated in the roots after ten days? How much of other amino acids?
3. **Making Inferences**
   a. In general, what is the effect of soil salt concentration on arrowgrass plant's accumulation of the amino acid proline? of other amino acids?
   b. Is the effect of salt on proline accumulation the same at low salt (below 50 mM) as at high salt (above 50 mM)?
4. **Drawing Conclusions** Are these results consistent with the hypothesis that arrowgrass accumulates proline to achieve osmotic balance with salty soils?
5. **Further Analysis** What do you think might account for the different rates of proline accumulation in low-salt and high-salt soils? Can you think of a way to test this hypothesis?

Plants

32.1 Adapting to Terrestrial Living

- Plants are complex multicellular autotrophs, producing their own food through photosynthesis.

- Plants acquire water and minerals from the soil. They control water loss through a watertight layer called the cuticle. Openings, called stomata, allow for gas exchange with the air (**figure 32.2**).

- The adaptation of pollen allowed land plants to transfer gametes on land under dry conditions. The pollen grain protects the gamete from drying out. Plant life cycles involve an alternation of generations where a haploid gametophyte alternates with a diploid sporophyte, with the sporophyte dominating the life cycle more and more as plants evolved (**figure 32.3**).

32.2 Plant Evolution

- Plants evolved from green algae, helped by four innovations: an alternation of generations life cycle, vascular tissue, seeds, and flower and fruits (**figure 32.4** and **table 32.1**).

- In an alternation of generations, the plant spends a portion of its life in a multicellular haploid phase that produces gametes and another portion in a multicellular diploid phase. As plants evolved, the diploid sporophyte became a more dominant structure, both in size and in its duration of time in the life cycle.

- The evolution of vascular tissue allowed plants to transport water and minerals from the soil and up through their bodies, distributing food throughout the plant and allowing them to grow taller.

- Seeds provide protection and nutrients for the developing embryo. Flowers and fruits improve the chances of successful mating and the distribution of seeds.

Seedless Plants

32.3 Nonvascular Plants

- Liverworts and hornworts don't have vascular tissue. Mosses have specialized cells for conducting water and carbohydrates in the plant, but not rigid vascular tissue. These plants are grouped together as nonvascular plants. Their life cycles are dominated by the haploid gametophyte (**figure 32.5**).

32.4 The Evolution of Vascular Tissue

- Vascular tissue consists of specialized cylindrical or elongated cells that form a network throughout the plant. These cells conduct water from the roots and carbohydrates manufactured in the leaves throughout the plant (**figure 32.7**).

32.5 Seedless Vascular Plants

- Among the most primitive vascular plants are the Pterophyta (ferns) and Lycophyta (**figure 32.8**). These plants can grow very tall due to vascularization, but they lack seeds and require water for fertilization. The sporophyte, shown here from **figure 32.9**, is larger than its gametophyte.

The Advent of Seeds

32.6 Evolution of Seed Plants

- In seed plants, the sporophyte becomes the dominant structure in the life cycle. The plants contain separate male and female gametophytes. The male gametophyte, called a pollen grain, produces sperm that are carried to the egg, which is contained in the small female gametophyte. Pollination in seed plants doesn't require water.

- The seed was an evolutionary innovation that provided protection for the plant embryo. The seed contains the embryo and food inside a drought-resistant cover (**figure 32.11**). Seeds are a key adaptation to land-dwelling plants because they improve the dispersal of embryos, allow dormancy when needed, germinate when conditions are favorable, and provide nourishment during germination.

32.7 Gymnosperms

- Gymnosperms are nonflowering seed plants and include conifers, cycads, gnetophytes, and the ginkgo. The seeds are produced in cones where pollination occurs. Pollen, produced in smaller cones, travels by wind to the seed-bearing cones, where fertilization of the egg takes place. The ovules are not completely enclosed by diploid tissue, as is the case with flowering plants. The seed, like the one shown here from **figure 32.14**, disperses the embryo.

The Evolution of Flowers

32.8 Rise of the Angiosperms

- Angiosperms, flowering plants, are plants in which the ovule is completely enclosed by sporophyte tissue.

- The flowers are the reproductive structures. The basic structure of a flower includes four concentric whorls (**integrated art, page 674**). The outermost green sepals protect the flower. Next, the colorful petals attract pollinators. The third whorl contains the stalklike stamen with the pollen-bearing anthers. The innermost whorl is a vaselike structure called the carpel, which contains the egg within the ovary. Flowers improved the efficiency of mating.

32.9 Why Are There Different Kinds of Flowers?

- Flowers vary greatly in size, shape, and color so that they are identifiable by a particular pollinator. This improves the likelihood that the pollen will be carried to the appropriate mate (**figures 32.15** and **32.16**).

32.10 Double Fertilization

- Double fertilization provides food for the germinating plant. The pollen grain produces two sperm cells that travel through the pollen tube, as shown here from **figure 32.17**. One fuses with the egg, and the other fuses with two polar nuclei, producing triploid endosperm. The angiosperms are divided into two groups: monocots and dicots (**figure 32.18**).

32.11 Fruits

- A further advancement in angiosperms was the development of the ovary into fruit tissue. Fruits aid in the dispersal of seeds to new habitats (**figure 32.19**).

1. A major consideration for the evolution of terrestrial plants is the problem of
   a. too much sunlight.
   b. predators.
   c. dehydration.
   d. not enough carbon.

2. Which of the following structures or systems is not unique to plant evolution?
   a. chloroplasts
   b. vascular tissue
   c. seeds
   d. flowers and fruits

3. Mosses, liverworts, and hornworts do not reach a large size because
   a. they lack chlorophyll.
   b. they do not have specialized vascular tissue to transport water very high.
   c. photosynthesis does not take place at a very fast rate.
   d. alternation of generations does not allow the plants to grow very tall before reproduction.

4. One characteristic that separates ferns from complex vascular plants is that ferns do not have
   a. a vascular system.
   b. chloroplasts.
   c. alternation of generations in their life cycle.
   d. seeds.

5. As seed plants evolved, the _____ form became more dominant in the life cycle.
   a. gametophyte
   b. gymnosperm
   c. sporophyte
   d. angiosperm

6. In seeds, the endosperm helps with
   a. fertilization.
   b. photosynthesis.
   c. nourishment.
   d. dispersal.

7. What separates the gymnosperms from other seed plants?
   a. a vascular system
   b. wind dispersion of pollen
   c. ovules not completely covered by the sporophyte
   d. fruits and flowers

8. What separates the angiosperms from other seed plants?
   a. a vascular system
   b. wind dispersion of pollen
   c. ovules not completely covered by the sporophyte
   d. fruits and flowers

9. Flower shape and color can be linked to the process of
   a. pollination.
   b. photosynthesis.
   c. germination.
   d. secondary growth.

10. Monocots and dicots differ in all of the following ways except
    a. vein patterns in the leaves.
    b. the process of double fertilization.
    c. the number of flower parts.
    d. the number of cotyledons.

Apply Your Understanding

1. **Figure 32.4** Identify which key innovation leads to the evolution of each group and explain why the innovation was significant.

2. **Figure 32.11** When farmers harvest corn, soybeans, wheat, and rice, they collect the seeds and get rid of the rest of the plant. Why do we eat the seeds rather than the stems or roots of these plants?

3. **Figure 32.12** The seeds that fall from this tree are scattered about on the ground nearby. What are the benefits or drawbacks to having: (a) chemicals in the seed to inhibit sprouting, and (b) hard seed coats?

Synthesize What You Have Learned

1. In New Zealand, large gymnosperms are far more common than large angiosperms. Under what conditions might gymnosperms have an evolutionary advantage over angiosperms, and why?

2. Gymnosperms and angiosperms are thought to have arisen at the same time in the late Carboniferous. However, while there are plenty of gymnosperm fossils from that time, the earliest angiosperm fossils are not seen until the late Jurassic, 150 million years later. What might account for the gap?

3. Why does a pine tree, which might live more than 100 years, put so much energy and resources into producing hundreds of pine cones each year, each with 30 to 50 seeds?

33

Plant Form and Function

O f all the many kinds of plants, trees are the largest, rising high above the surrounding vegetation to capture the sun's rays. This hardwood forest, green with the leaves of spring, can be thought of as an enormous photosynthesis machine, each of its trees competing with its neighbors for light and soil nutrients, and putting the raw materials it captures to work producing the organic molecules necessary for growth and reproduction. Typical of plants, a tree captures light with the green pigment chlorophyll, which gives its leaves their characteristic color. A tree captures soil nutrients with its roots, which spread out in a fine network through the surrounding soil. Connecting the leaves of a tree with its roots is the stem, the massive, tall woody cylinder that makes up most of the mass of the tree. The stem of a tall tree is an engineering marvel, piping water and dissolved soil nutrients to the leaves many meters higher up in the air, and sending back down to the roots the carbohydrates produced by photosynthesis in the leaves. In this chapter, we will journey through the plant body, one of nature's most interesting creations.

33.1 Organization of a Vascular Plant

Most plants possess the same fundamental architecture and the same three major groups of organs—roots, stems, and leaves. Vascular plants have within their stems vascular tissue, which conducts water, minerals, and food throughout the plant. The cells and tissues of vascular plants, and how the plant body carries out the functions of living, are the focus of this chapter. We discuss the fundamental differences between the roots, which are usually belowground, and the shoot, which is typically aboveground. We also examine the structural and functional relationships between them.

A vascular plant is organized along a vertical axis (**figure 33.1**). The part belowground is called the **root,** and the part aboveground is called the **shoot** (although in some instances roots may extend above the ground, and some shoots can extend below it). Although roots and shoots differ in their basic structure, growth at the tips throughout the life of the individual is characteristic of both. The root penetrates the soil and absorbs water and various minerals, which are crucial for plant nutrition. It also anchors the plant. The shoot consists of stem and leaves. The **stem** serves as a framework for the positioning of the **leaves,** where most photosynthesis takes place. The arrangement, size, and other characteristics of the leaves are critically important in the plant's production of food. Flowers, and ultimately fruits and seeds, are also formed on the shoot.

Meristems

When an animal grows taller, all parts of its body lengthen—when you grew taller as a child, your arms and legs lengthened and so did your torso. A plant doesn't grow this way. Instead, it adds tissues to the tips of its roots and shoots. If you grew like this, your legs would get longer, and your head taller, while the central portion of your body would not change.

Why do plants grow in this way? The plant body contains growth zones of unspecialized cells called **meristems.** Meristems are areas with actively dividing cells that result in plant growth but also continually replenish themselves. That is, one cell divides to give rise to two cells. One remains meristematic, while the other is free to differentiate and contribute to the plant body, resulting in plant growth. In this way, meristem cells function much like "stem cells" in animals (see chapter 13). Molecular genetic evidence supports the hypothesis that animal stem cells and plant meristem cells may also share some common pathways of gene expression.

In plants, **primary growth** is initiated at the tips by the **apical meristems,** regions of active cell division that occur at the tips of roots and shoots, colored lime green in **figure 33.1.** The growth of these meristems results primarily in the extension of the plant body. As the body tip elongates, it forms what is known as the primary plant body, which is made up of the primary tissues.

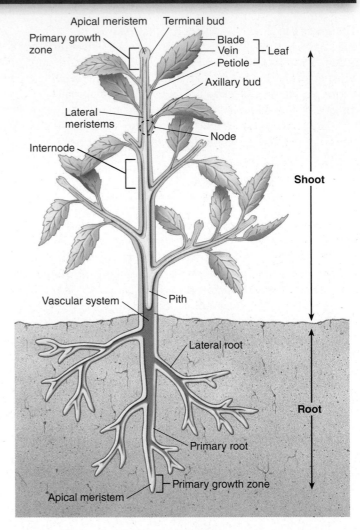

Figure 33.1 The body of a plant.
The body of this dicot plant consists of an aboveground portion called the shoot (stems and leaves) and a belowground portion called the root. Elongation of the plant, so-called primary growth, takes place when clusters of cells called the apical meristems *(lime green areas)* divide at the ends of the roots and the stems. Thickening of the plant, so-called secondary growth, takes place in the lateral meristems *(yellow areas)* of the stem, allowing the plant to increase in girth like letting out a belt.

Growth in thickness, **secondary growth,** involves the activity of the **lateral meristems,** which are cylinders of meristematic tissue, colored yellow in **figure 33.1.** The continued division of their cells results primarily in the thickening of the plant body. There are two kinds of lateral meristems: the *vascular cambium,* which gives rise to ultimately thick accumulations of secondary xylem and phloem, and the *cork cambium,* from which arise the outer layers of bark on both roots and shoots.

> **Key Learning Outcome 33.1** The body of a vascular plant is a continuous structure, a grouping of tubes connecting roots to leaves, with growth zones called meristems.

33.2 Plant Tissue Types

The organs of a plant—the roots, stem, leaves, and in some cases, flowers and fruits—are composed of different combinations of tissues, just as your legs are composed of bone, muscle, and connective tissue. A tissue is a group of similar cells—cells that are specialized in the same way—organized into a structural and functional unit. Most plants have three major tissue types: (1) *ground tissue*, in which the vascular tissue is embedded; (2) *dermal tissue*, the outer protective covering of the plant; and (3) *vascular tissue*, which conducts water and dissolved minerals up the plant and conducts the products of photosynthesis throughout.

Each major tissue type is composed of distinctive kinds of cells, whose structures are related to the functions of the tissues in which they occur. For example, vascular tissue is composed of *xylem*, which conducts water and dissolved minerals, and *phloem*, which conducts carbohydrates (mostly sugars) that the plant uses as food.

Ground Tissue

Parenchyma cells are the least specialized and the most common of all plant cell types; they form masses in leaves, stems, and roots. Parenchyma cells, unlike some other cell types, are characteristically alive at maturity, with fully functional cytoplasm and a nucleus. They are the cells that carry out the basic functions of living, including photosynthesis, cellular respiration, and food and water storage. The edible parts of most fruits and vegetables are composed of parenchyma cells. They are capable of cell division and are important in cell regeneration and wound healing. As you can see here, parenchyma cells have only thin cell walls, called **primary cell walls,** which are mostly cellulose that is laid down while the cells are still growing.

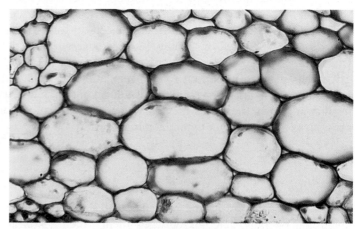
Parenchyma

Collenchyma cells, like parenchyma cells, are living at maturity and may live for many years. The cells, which are usually a little longer than they are wide, have walls that vary in thickness (as you can see in this photograph). Collenchyma cells provide support for plant organs. They are relatively flexible, allowing the organs to bend without breaking. They often form strands or continuous cylinders beneath the epidermis of stems or leaf stalks (petioles) and along veins in leaves. Strands of collenchyma provide much of the support for stems in which secondary growth has not taken place. The parts of celery that we eat (petioles) have "strings" that consist mainly of collenchyma and vascular bundles (conducting tissues).

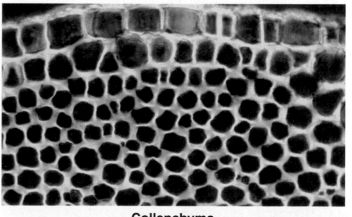
Collenchyma

Sclerenchyma cells, in contrast to parenchyma and collenchyma cells, have tough, thick cell walls called **secondary cell walls;** they usually do not contain living cytoplasm when mature. The secondary cell wall is laid down inside of the primary cell wall after the cell has stopped growing and expanding in size. The secondary cell wall provides cells with strength and rigidity. There are two types of sclerenchyma: **fibers,** which are long, slender cells that usually form strands, and **sclereids,** which are variable in shape but often branched. Sclereids, the reddish-colored cells in the photograph, are sometimes called "stone cells." Clusters of sclereids form the gritty texture you feel in the flesh of pears. Both fibers and sclereids are thick-walled and strengthen the tissues in which they occur. Compare the thickness of the cell walls in the sclereid cells with the green-stained parenchyma cells that surround them.

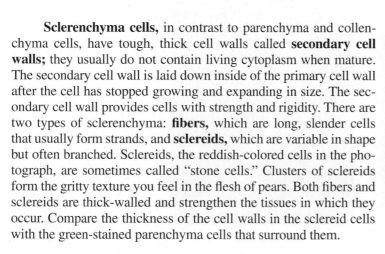
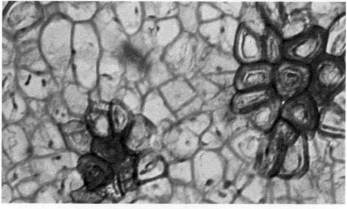
Sclerenchyma

Dermal Tissue

All parts of the outer layer of a primary plant body are covered by flattened epidermal cells. These are the most abundant cells in the plant *epidermis,* or outer covering. The epidermis is one cell layer thick and is a protective layer that provides an effective barrier against water loss. The epidermis is often covered with a thick, waxy layer called the **cuticle,** which protects against ultraviolet light damage and water loss. In some cases, the dermal tissue is more extensive and forms the bark of trees.

Trichomes are outgrowths of the epidermis that occur on the shoot, on the surfaces of stems and leaves. Trichomes vary greatly in form in different kinds of plants, from rounded-tip forms to globular-tipped ones. A "fuzzy" or "woolly" leaf is covered with trichomes, which when viewed under the microscope look like a thicket of fibers. Trichomes play an important role in regulating the heat and water balance of the leaf, much as the hairs of an animal's coat provide insulation. Other trichomes are glandular, secreting sticky or toxic substances that may deter potential herbivores.

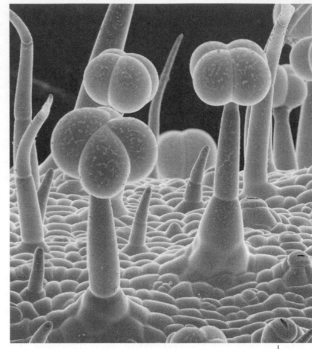

Trichomes   186 µm

Guard cells are paired cells flanking a mouth-shaped epidermal opening that lies between them called a **stoma** (plural, **stomata**). Guard cells, unlike other epidermal cells, contain chloroplasts. Guard cells and stomata occur frequently in the epidermis of leaves and occasionally on other parts of the shoot, such as on stems or fruits. Oxygen, carbon dioxide, and water vapor pass across the epidermis almost exclusively through the stomata, which open and shut in response to external factors such as supply of moisture and light. There are from 1,000 to more than 1 million stomata per square centimeter of leaf surface. In many plants, stomata are more numerous on the lower epidermis of the leaf than on the upper epidermis, a design that minimizes water loss.

Trichome   Epidermal cells   Stomatal opening   Guard cells

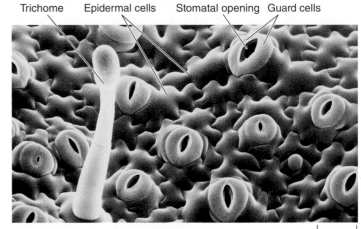

Stomata   137 µm

Root hairs are tubular outgrowths of individual epidermal cells near the tips of young growing roots. Because a root hair is simply an extension of a cell's cytoplasm and not a separate cell, there is no cross-wall isolating it from the epidermal cell. Root hairs keep the root in intimate contact with the surrounding particles of soil. Because root hairs greatly increase the surface area of the root, they profoundly increase the efficiency of absorption from the soil by the root. Indeed, most of the absorption of water and minerals occurs through root hairs, especially in herbaceous plants. As the root grows, root hairs at the older end slough off while new ones are produced behind the growing root tip.

Root hairs

Root hairs

Vascular Tissue

Vascular plants contain two kinds of conducting, or vascular, tissue: xylem and phloem.

Xylem is the plant's principal water-conducting tissue, forming a continuous system that runs throughout the plant body. Within this system, water (and minerals dissolved in it) passes from the roots up through the shoot in an unbroken stream. When water reaches the leaves, much of it passes into the air as water vapor, through the stomata.

The two principal types of conducting cells in the xylem are **tracheids** and **vessel elements,** both of which have thick secondary walls that are laid down inside the primary cell wall. These cells are elongated, and they have no living cytoplasm (they are dead) at maturity. Tracheids are elongated cells that overlap at the ends. In conducting elements composed of tracheids, water flows from tracheid to tracheid through openings called *pits* in the secondary walls. In contrast, vessel elements are elongated cells that line up end-on-end. The end walls of vessel elements may be almost completely open or may have bars or strips of wall material perforated by pores through which water flows. A linked row of vessel elements forms a vessel (in the upper right, a scanning electron micrograph of the red maple *Acer rubrum* shows tracheids and vessels). Primitive angiosperms and other vascular plants have only tracheids, but the majority of angiosperms have vessels. Vessels conduct water much more efficiently than do strands of tracheids.

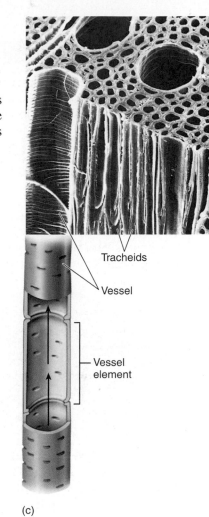

Tracheids
Vessel

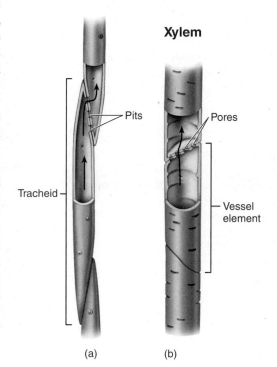

Xylem

Tracheid

Pits

Pores

Vessel element

Vessel element

(a)   (b)   (c)

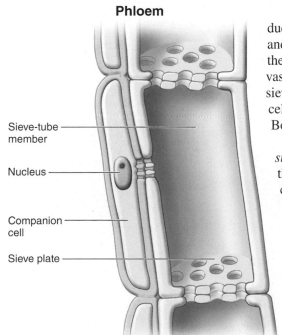

Phloem

Sieve-tube member

Nucleus

Companion cell

Sieve plate

Phloem is the principal food-conducting tissue in vascular plants. Food conduction in phloem is carried out through two kinds of elongated cells: **sieve cells** and **sieve-tube members**. The cells differ in the extent of the perforations between the cells, with the sieve cells having smaller perforations between cells. Seedless vascular plants and gymnosperms have only sieve cells; most angiosperms have sieve-tube members. Clusters of pores known as *sieve areas* occur on both kinds of cells and connect the cytoplasms of adjoining sieve cells and sieve-tube members. Both cell types are living, but their nuclei are lost during maturation.

In sieve-tube members, some sieve areas have larger pores and are called *sieve plates*. Sieve-tube members occur end to end, as shown in the figure to the left, forming longitudinal series called **sieve tubes.** Specialized parenchyma cells known as **companion cells** occur regularly in association with sieve-tube members. The companion cells can be seen in the figure associated with the left side of the sieve-tube members. Companion cells apparently carry out some of the metabolic functions that are needed to maintain the associated sieve-tube member; their cytoplasms are connected to the sieve-tube members through openings called *plasmodesmata.*

In addition to conducting cells, xylem and phloem also contain fibers (sclerenchyma cells) and parenchyma cells. The parenchyma cells function in storage, and the fibers provide support and some storage. Xylem fibers are used in the production of paper.

> **Key Learning Outcome 33.2 Plants contain a variety of ground tissue, dermal tissue (outer coverings), and vascular tissues (conducting tissues).**

We now consider the three kinds of vegetative organs that form the body of a plant: roots, stems, and leaves. While we will examine their basic structures, it is important to understand that these organs are often modified for different functions. For example, roots and stems can be modified for water and food storage, and leaves can be modified for defenses, such as cactus spines.

33.3 Roots

Root Structure

Roots have a simpler pattern of organization and development than do stems, and we will examine them first. Although different patterns exist, the kind of root shown here is found in many dicots.

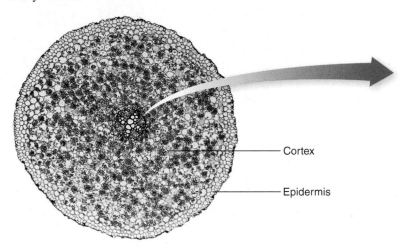

Cortex

Epidermis

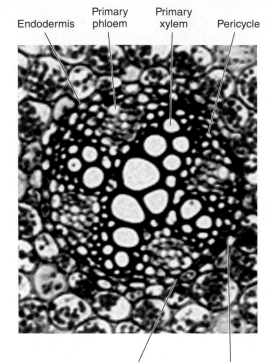

Endodermis | Primary phloem | Primary xylem | Pericycle

Casparian strip

Sectioned endodermal cells

H_2O H_2O

The outer layer of the root is the epidermis. The mass of parenchyma in which the root's vascular tissue is located is the *cortex*. In the photographs above, you can also see that the dicot root contains both xylem and phloem. Focusing on the core, you can see a central column of xylem with radiating arms. Alternating with the radiating arms of xylem are strands of primary phloem. Surrounding the column of vascular tissue, and forming its outer boundary, is a cylinder of cells one or more cell layers thick called the **pericycle.** Branch, or lateral, roots are formed from cells of the pericycle. Just outside the pericycle is the **endodermis,** a single layer of specialized cells that regulate the flow of water between the vascular tissues and the root's outer portion.

Endodermal cells are encircled by a thickened, waxy band called the **Casparian strip.** Here you see a drawing of endodermal cells showing how the wax substance that makes up the Casparian strip surrounds each cell. As the black arrows indicate, the Casparian strip blocks the movement of water *between* cells and instead directs the movement of water *through* the plasma membrane of the endodermal cells. In this way, the Casparian strip controls the passage of minerals into the xylem because transport through the endodermal cells is regulated by special channels embedded in the plasma membrane.

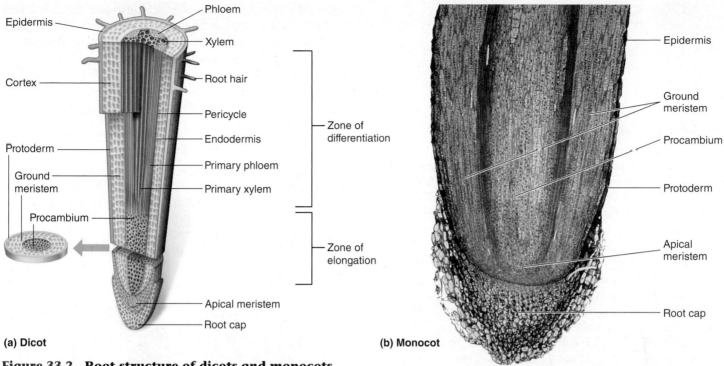

(a) Dicot

Epidermis — Phloem — Xylem — Root hair — Pericycle — Endodermis — Primary phloem — Primary xylem — Apical meristem — Root cap

Cortex — Protoderm — Ground meristem — Procambium

Zone of differentiation — Zone of elongation

(b) Monocot

Epidermis — Ground meristem — Procambium — Protoderm — Apical meristem — Root cap

Figure 33.2 Root structure of dicots and monocots.
(*a*) Diagram of primary meristems in a dicot root, showing their relation to the apical meristem. The three primary meristems are the protoderm, which differentiates further into epidermis; the procambium, which differentiates further into primary vascular strands; and the ground meristem, which differentiates further into ground tissue. (*b*) Median longitudinal section of a monocot root tip in corn, *Zea mays*, showing the differentiation of protoderm, procambium, and ground meristem.

Meristems

The apical meristem of the root (shown as a group of cells at the base of the root in **figure 33.2a,b**) divides and produces cells both inwardly, back toward the body of the plant, and outwardly. The three primary meristems, shown in the "cutaway" portion of the dicot root in **figure 33.2a**, are the **protoderm,** which becomes the epidermis; the **procambium,** which produces primary vascular tissues (primary xylem and primary phloem); and the **ground meristem,** which differentiates further into ground tissue that is composed of parenchyma cells. Outward cell division results in the formation of a thimblelike mass of relatively unorganized cells, called the **root cap,** which you can clearly see in the photo above. The root cap covers and protects the root's apical meristem as it grows through the soil.

The root elongates relatively rapidly just behind its tip in the area called the *zone of elongation.* Abundant root hairs (see **figure 33.14**), extensions of single epidermal cells, form above that zone, in the area called the *zone of differentiation.* Virtually all water and minerals are absorbed from the soil through the root hairs, which greatly increase the root's surface area and absorptive powers.

Lateral Roots

One of the fundamental differences between roots and shoots has to do with the nature of their branching. In stems, branching occurs from buds on the stem surface; in roots, branching is initiated well back of the root tip as a result of cell divisions in the pericycle. The developing lateral roots (the red-stained mass of cells in **figure 33.3**) grow out through the cortex to-

ward the surface of the root, eventually breaking through and becoming established as lateral roots. In some plants, roots may arise along a stem, or in some place other than the root of the plants. These roots are called *adventitious roots.* Adventitious roots occur in ivy, bulb plants such as onions, perennial grasses, and other plants that produce rhizomes, which are horizontal stems that grow underground.

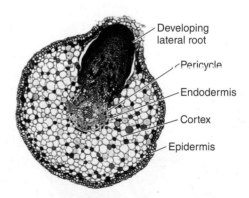

Developing lateral root — Pericycle — Endodermis — Cortex — Epidermis

Figure 33.3 Lateral roots.
A lateral root growing out through the cortex of the black willow, *Salix nigra.* Lateral roots originate beneath the surface of the main root, whereas lateral stems originate at the surface.

> **Key Learning Outcome 33.3** Roots, the belowground portion of the plant body, are adapted to absorb water and minerals from the soil.

33.4 Stems

Stems serve as the main structural support of the plant and the framework for the positioning of the leaves. Often experiencing both primary and secondary growth, stems are the source of an economically important product—wood.

Primary Growth

In the primary growth of a shoot, leaves first appear as leaf primordia (singular, primordium), rudimentary young leaves that cluster around the apical meristem, unfolding and growing as the stem itself elongates. The places on the stem at which leaves form are called nodes (indicated by the small bracket in **figure 33.4**). The portions of the stem between these attachment points are called the internodes (the larger bracket). As the leaves expand to maturity, a bud—a tiny undeveloped side shoot—develops in the **axil** of each leaf, the angle between a leaf and the stem from which it arises. These buds, which have their own immature leaves (shown in **figure 33.4**), may elongate and form lateral branches, or they may remain small and dormant. A hormone moving downward from the terminal bud of the shoot continually suppresses the expansion of the lateral buds in the upper portions of the stem. Lateral buds begin forming lower down when the stem lengthens to the point where the amount of hormone reaching the lower portion of the stem is reduced, or if the terminal bud is removed, such as when you prune a plant.

Within the soft, young stems, the strands of vascular tissue, xylem and phloem, are arranged differently in dicots versus monocots. In dicots, the vascular bundles contain xylem and phloem and are arranged around the outside of the stem as a cylinder (**figure 33.5a**). In monocots, the vascular bundles are scattered throughout the stem (**figure 33.5b**). This difference in vascular tissue organization, in addition to other char-acteristics discussed in chapter 32, illustrates the differences between these two major groups of angiosperms. The vascular bundles contain both primary xylem and primary phloem. At the stage when only primary growth has occurred, the inner portion of the ground tissue of a dicot stem is called the **pith** (the center pink-stained cells in **figure 33.5a**), and the outer portion is the **cortex** (the light green-stained cells located toward the outside).

Secondary Growth

In stems, secondary growth (the thickening, but not the lengthening, of the stem) is initiated by the differentiation of a lateral meristem called the **vascular cambium,** a thin cylinder of actively dividing cells located between the bark and the main stem in woody plants. The vascular cambium develops from cells within the vascular bundles of the stem,

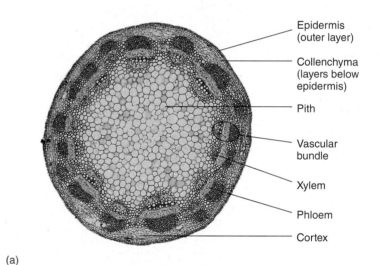

(a)

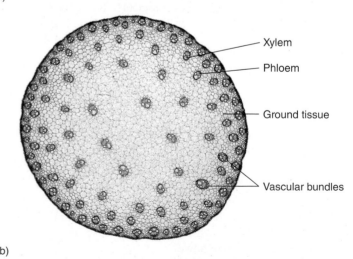

(b)

Figure 33.5 A comparison of dicot and monocot stems.

(a) Transection of a young stem of a dicot, the common sunflower, *Helianthus annuus,* in which the vascular bundles are arranged around the outside of the stem. (b) Transection of a monocot stem, corn, *Zea mays,* with the scattered vascular bundles characteristic of the group.

Figure 33.4 A woody twig.

This twig shows key stem structures, including the node and internode areas, the axillary bud in the axil, and leaves.

between the xylem (the purple-colored areas in **figure 33.6**) and the phloem (the light green area). The cylindrical form of the vascular cambium is completed by the differentiation of some of the parenchyma cells that lie between the bundles. Once established, the vascular cambium consists of elongated and somewhat flattened cells with large vacuoles. The cells that divide from the vascular cambium outwardly, toward the bark, become secondary phloem; those that divide from it inwardly become secondary xylem.

While the vascular cambium is becoming established, a second kind of lateral meristem, the **cork cambium,** develops in the stem's outer layers. The cork cambium usually consists of plates of dividing cells that move deeper and deeper into the stem as they divide. Outwardly, the cork cambium splits off densely packed **cork cells;** they contain a fatty substance and are nearly impermeable to water. Cork cells are dead at maturity. Inwardly, the cork cambium divides to produce a layer of parenchyma cells. The cork, the cork cambium that produces it, and this layer of parenchyma cells make up a layer called the **periderm** (see **figure 33.6**), which is the plant's outer protective covering.

Cork covers the surfaces of mature stems or roots. The term **bark** refers to all of the tissues of a mature stem or root outside of the vascular cambium. Because the vascular cambium has the thinnest-walled cells that occur anywhere in a secondary plant body, it is the layer at which bark breaks away from the accumulated secondary xylem.

Wood is one of the most useful, economically important, and beautiful products obtained from plants. Anatomically, wood is accumulated secondary xylem (the light purple pie-shaped areas in **figure 33.6**). As the secondary xylem ages, its cells become infiltrated with gums and resins, and the wood may become darker. For this reason, the wood located nearer the central regions of a given trunk, called heartwood, can be darker and denser than the wood nearer the vascular cambium, called sapwood, which is still actively involved in water transport within the plant.

Because of the way it is accumulated, wood often displays rings. The rings that you see in the section of pine in **figure 33.7** reflect the fact that the vascular cambium of trees divides more actively in the spring and summer, when water is plentiful and temperatures are suitable for growth, than in the fall and winter, when water is scarce and the weather is cold. As a result, layers of larger, thinner-walled cells formed during the growing season (the lighter rings) alternate with the smaller, darker layers of thick-walled cells formed during the rest of the year. New rings are laid down each year toward the outer edge of the stem. A count of such annual rings in a tree trunk can be used to calculate the tree's age, and the width of rings can reveal information about environmental

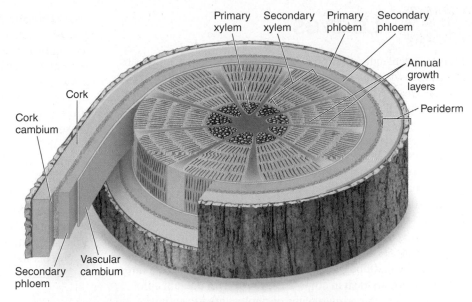

Figure 33.6   Vascular cambium and secondary growth.
The vascular cambium and cork cambium (lateral meristems) produce secondary tissues, causing the stem's girth to increase. Each year, a new layer of secondary tissue is laid down, forming rings in the wood.

Figure 33.7   Annual rings in a section of pine.
To test your understanding of how they form, answer this question: Are the broad inner rings older or younger than the narrow outer rings?

factors. For example, the region of thinner rings could indicate a period of prolonged drought conditions that was followed by wetter years. Can you estimate the age of the tree shown in **figure 33.7**?

> **Key Learning Outcome 33.4**   Stems, the aboveground framework of the plant body, grow both at their tips and in circumference.

33.5 Leaves

Leaves are usually the most prominent shoot organs and are structurally diverse (**figure 33.8**). As outgrowths of the shoot apex, leaves are the major light-capturing organs of most plants. Most of the chloroplast-containing cells of a plant are within its leaves, and it is there where the bulk of photosynthesis occurs (see chapter 6). Exceptions to this are found in some plants, such as cacti, whose green stems have largely taken over the function of photosynthesis for the plant. Photosynthesis is conducted mainly by the "greener" parts of plants because they contain more chlorophyll, the most efficient photosynthetic pigment. In some plants, other pigments may also be present, giving the leaves a color other than green. Recall in chapter 6, we described accessory pigments that absorb light of other wavelengths. Thus, although coleus plants and red maple trees have leaves that are reddish in color, these leaves still contain chlorophyll and are the primary sites of photosynthetic activity in the plant.

The apical meristems of stems and roots are capable of growing indefinitely under appropriate conditions. Leaves, in contrast, grow by means of **marginal meristems,** which flank their thick central portions. These marginal meristems grow outward and ultimately form the **blade** (flattened portion) of the leaf, while the central portion becomes the midrib. Once a leaf is fully expanded, its marginal meristems cease to grow.

In addition to the flattened blade, most leaves have a slender stalk, the **petiole.** Two leaflike organs, the **stipules,** may flank the base of the petiole where it joins the stem. Veins, consisting of both xylem and phloem, run through the leaves. As mentioned in chapter 32, in most dicots the pattern is net, or reticulate, venation—as you can see in **figure 33.9a.** In many monocots, the veins are parallel, like the parallel veins that pass vertically up through the monocot leaf in **figure 33.9b.**

Leaf blades come in a variety of forms from oval to deeply lobed to having separate leaflets (the blade being divided but attached to a

(a)　(b)　(c)　(d)　(e)　(f)

Figure 33.8　Leaves.
Leaves are stunningly variable. (*a*) *Simple leaves* from a gray birch, in which there is a single blade. (*b*) A simple leaf, its margin lobed, from the vine maple. (*c*) A *pinnately compound* leaf of a black walnut tree, where leaflets occur in pairs along the central axis of the main vein. (*d*) *Palmately compound* leaves of a horse chestnut tree, in which the leaflets radiate out from a single point. (*e*) The leaves of pine trees are tough and needlelike. (*f*) Many unusual types of modified leaves occur in different kinds of plants. For example, some plants produce floral leaves or bracts; the most conspicuous parts of this poinsettia flower are the red bracts, which are modified leaves that surround the small yellowish true flowers in the center.

single petiole like the black walnut leaf in **figure 33.8c**). In **simple leaves** (see **figure 33.8a,b**), such as those of birch or maple trees, there is a single blade, undivided, but some simple leaves may have teeth, indentations, or lobes, such as the leaves of maples and oaks. In **compound leaves,** such as those of ashes, box elders, and walnuts, the blade is divided into leaflets. If the leaflets are arranged in pairs along a common axis—the equivalent of the main central vein, or *midrib,* in simple leaves—the leaf is **pinnately compound,** such as in the black walnut (see **figure 33.8c**). If, however, the leaflets radiate out from a common point at the blade end of the petiole, the leaf is **palmately compound,** such as in horse chestnuts (**figure 33.8d**) and Virginia creepers.

The positioning of leaves can also vary. Leaves may be **alternately** arranged (alternate leaves usually spiral around a shoot, like the ivy shown below) or they may be in **opposite** pairs (like the periwinkle). Less often, three or more leaves may be in a **whorl,** a circle of leaves at the same level at a node (like the sweet woodruff).

(a)   (b)

Figure 33.9 Dicot and monocot leaves.

(*a*) The leaves of dicots have netted, or reticulate, veins; (*b*) those of monocots have parallel veins.

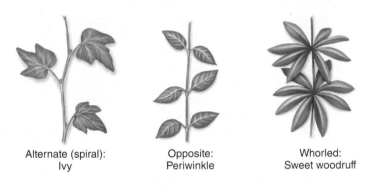

Alternate (spiral):   Opposite: Whorled:
Ivy   Periwinkle   Sweet woodruff

A typical leaf contains masses of parenchyma, called **mesophyll** ("middle leaf"), through which the vascular bundles, or veins, run. Beneath the upper epidermis of a leaf are one or more layers of closely packed, columnlike parenchyma cells called **palisade mesophyll** (the red-stained cells in the

photo in **figure 33.10**). These cells contain more chloroplasts than other cells in the leaf and so are more capable of carrying out photosynthesis. This makes sense when you consider that the cells on the surface receive more sun. The rest of the leaf interior, except for the veins, consists of a tissue called **spongy mesophyll.** Between the spongy mesophyll cells are large intercellular spaces that function in gas exchange and particularly in the passage of carbon dioxide from the atmosphere to the mesophyll cells. You can see the spongy mesophyll in the photo of **figure 33.10**, but the air spaces that are the basis of this tissue's function might be easier to see in the drawing. These intercellular spaces are connected, directly or indirectly, with the stomata in the lower epidermis.

> **Key Learning Outcome 33.5** Leaves, the photosynthetic organs of the plant body, are varied in shape and arrangement.

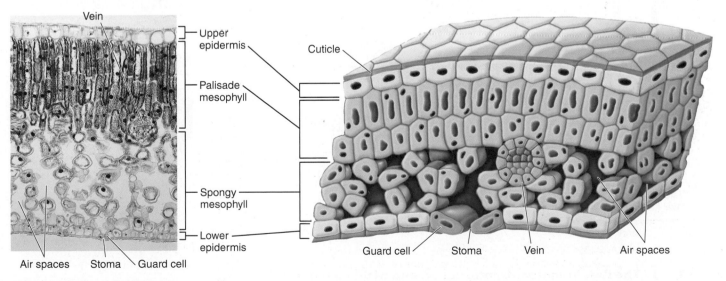

Vein
Upper epidermis
Cuticle
Palisade mesophyll
Spongy mesophyll
Lower epidermis
Air spaces   Stoma   Guard cell
Guard cell   Stoma   Vein   Air spaces

Figure 33.10 A leaf in cross section.

Cross section of a leaf, showing the arrangement of palisade and spongy mesophyll, a vascular bundle or vein, and the epidermis, with paired guard cells flanking the stoma.

33.6 Water Movement

Vascular plants have a conducting system, as humans do, for transporting fluids and nutrients from one part to another. Functionally, a plant is essentially a bundle of tubes with its base embedded in the ground. At the base of the tubes are roots, and at their tops are leaves. For a plant to function, two kinds of transport processes must occur: First, the carbohydrate molecules produced in the leaves by photosynthesis must be carried to all of the other living plant cells. To accomplish this, liquid, with these carbohydrate molecules dissolved in it, must move both up and down the tubes. Second, minerals and water in the soil must be taken up by the roots and ferried to the leaves and other plant cells. In this process, liquid moves up the tubes. Plants accomplish these two processes by using chains of specialized cells. Cells of the phloem transport photosynthetically produced carbohydrates up and down the plant (red arrows in **figure 33.11**), and those of the xylem carry water and minerals upward (blue arrows in **figure 33.11**).

Cohesion-Adhesion-Tension Theory

Many of the leaves of a large tree may be more than 10 stories off the ground. How does a tree manage to raise water so high? Several factors are at work to move water up the height of a plant. The initial movement of water into the roots of a plant involves osmosis. Water moves into the cells of the root because the fluid in the xylem contains more solutes than the surroundings—recall from chapter 4 that water will move across a membrane from an area of lower solute concentration to an area of higher solute concentration. However, this force, called *root pressure,* is not by itself strong enough to "push" water up a plant's stem.

Capillary action adds a "pull." *Capillary action* results from the tiny electrical attractions of polar water molecules to surfaces that carry an electrical charge, a process called *adhesion.* In the laboratory, a column of water rises up a tube of glass because the attraction of the water molecules to the charged molecules on the interior surface of the glass tube "pulls" the water up in the tube. In **figure 33.12**, which illustrates this process, why does the water travel higher up in the narrower tube? The water molecules are attracted to the glass molecules, and the water travels up farther in the narrower tube because the amount of surface area available for adhesion is greater than in the larger-diameter tube.

However, although capillary action can produce enough force to raise water a meter or two, it cannot account for the movement of water to the tops of tall trees. A second very strong "pull" accomplishes this, provided by transpiration, to be discussed later. Opening up the tube and blowing air across its upper end demonstrates how transpiration draws water up a plant stem. The stream of relatively dry air causes water molecules at the water column's exposed top surface to evaporate

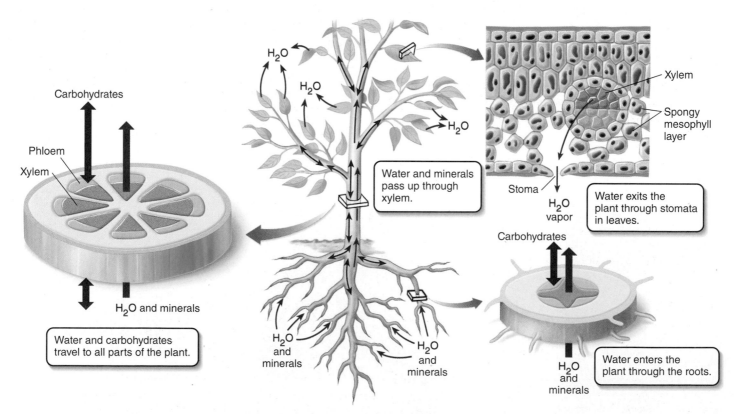

Figure 33.11 The flow of materials into, out of, and within a plant.
Water and minerals enter through the roots of a plant and are transported through the xylem to all parts of the plant body (*blue arrows*). Water leaves the plant through the stomata in the leaves. Carbohydrates synthesized in the leaves are circulated throughout the plant by the phloem (*red arrows*).

Figure 33.12   Capillary action.
Capillary action causes the water within a narrow tube to rise above the surrounding water; the attraction of the water molecules to the glass surface, which draws water upwards, is stronger than the force of gravity, which tends to draw it down. The narrower the tube, the greater the surface area available for adhesion for a given volume of water, and the higher the water rises in the tube.

from the tube. The water level in the tube does not fall, because as water molecules are drawn from the top, they are replenished by new water molecules pulled up from the bottom. This, in essence, is what happens in plants. The passage of air across leaf surfaces results in the loss of water by evaporation, creating a "pull" at the open upper end of the plant. New water molecules that enter the roots are pulled up the plant. Adhesion of water molecules to the walls of the narrow vessels in plants also helps to maintain water flow to the tops of plants.

A column of water in a tall tree does not collapse simply due to its weight because water molecules have an inherent strength that arises from their tendency to form hydrogen bonds with one another. These hydrogen bonds cause *cohesion* of the water molecules (see chapter 2); in other words, a column of water resists separation. The beading of water droplets illustrates the property of cohesion. This resistance, called *tensile strength,* varies inversely with the diameter of the column; that is, the smaller the diameter of the column, the greater the tensile strength. Therefore, plants must have very narrow transporting vessels to take advantage of tensile strength.

How the combination of gravity, adhesion, and tensile strength due to cohesion affects water movement in plants is called the **cohesion-adhesion-tension theory.** It is important to note that the movement of water up through a plant is a passive process and requires no expenditure of energy on the part of the plant.

Transpiration

The process by which water leaves a plant is called **transpiration.** More than 90% of the water taken in by plant roots is ultimately lost to the atmosphere, almost all of it from the leaves. It passes out primarily through the stomata in the evaporation of water vapor, as you can see in **panel 1** of the Key Biological Process illustration below. On its journey from the plant's interior to the outside, a molecule of water first diffuses from the xylem into the spongy mesophyll cells of the leaf. Then, water passes into the pockets of air within the leaf by evaporating from the walls of the spongy mesophyll that line the intercellular spaces. These intercellular spaces open to the outside of the leaf by way of the stomata. The water that evaporates from these surfaces of the spongy mesophyll cells is continuously replenished from the tips of the veinlets in the leaves. Molecules of water diffusing from the xylem replace evaporating water molecules. Because the strands of xylem conduct water within the plant in an unbroken stream all the way from the roots to the leaves, when a portion of the water vapor in the intercellular spaces passes out through the stomata, the supply of water vapor in these spaces is continually renewed from lower down in the column (**panel 2**) and ultimately from the roots (**panel 3**). Because the process of transpiration is dependent upon evaporation, factors that affect evaporation also affect transpiration. In addition to the movement of air across the stomata, mentioned earlier, humidity levels

KEY BIOLOGICAL PROCESS: Transpiration

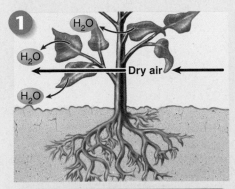

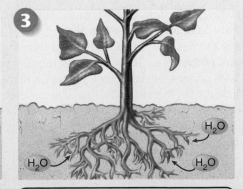

Dry air passes across the leaves and causes water vapor to evaporate out of the stomata.

The loss of water from the leaves creates a type of "suction" that draws water up the stem through the xylem.

New water enters the plant through the roots to replace the water moving up the stem.

in the air will affect the rate of evaporation—high humidity reducing it and low humidity increasing it. Temperature will also affect the rate of evaporation—high temperatures increase it and lower temperatures reduce it. This temperature effect is especially important because evaporation also acts to cool plant tissues.

Structural features such as the stomata, the cuticle, and the intercellular spaces in leaves have evolved in response to two contradictory requirements: minimizing the loss of water to the atmosphere, on the one hand, and admitting carbon dioxide, which is essential for photosynthesis, on the other. How plants resolve this problem is discussed next.

Regulation of Transpiration: Open and Closed Stomata

The only way plants can control water loss on a short-term basis is to close their stomata. Many plants can do this when subjected to water stress. But the stomata must be open at least part of the time so that carbon dioxide, which is necessary for photosynthesis, can enter the plant. In its pattern of opening or closing its stomata, a plant must respond to both the need to conserve water and the need to admit carbon dioxide.

The stomata open and close because of changes in the water pressure of their guard cells. Stomatal guard cells are long, sausage-shaped cells attached at their ends. These are the green cells in **figure 33.13**. The cellulose microfibrils of their cell wall wrap around the cell such that when the guard cells are **turgid** (plump and swollen with water), they expand in length, causing the cells to bow, thus opening the stomata as wide as possible, shown on the left side of the figure. Turgor in guard cells results from the active uptake of ions, causing water to enter osmotically as a consequence.

A number of environmental factors affect the opening and closing of stomata. The most important is water loss. The stomata of plants that are wilted because of a lack of water tend to close. An increase in carbon dioxide concentration also causes the stomata of most species to close. In most plant species, stomata open in the light and close in the dark.

Water Absorption by Roots

Most of the water absorbed by plants comes in through the root hairs, extensions of epidermal cells. These give a root the feathery appearance shown in **figure 33.14**. These root hairs greatly increase the surface area and therefore the absorptive powers of the roots. Root hairs are turgid—plump and swollen with water—because they contain a higher concentration of dissolved minerals and other solutes than does the water in the soil solution; water, therefore, tends to move into them steadily. Once inside the roots, water passes inward to the conducting elements of the xylem.

Water is not the only substance that enters the roots by passing into the cells of root hairs. Minerals also enter the root. Membranes of root hair cells contain a variety of ion transport channels that actively pump specific ions into the plant, even against large concentration gradients. These ions, many of which are plant nutrients, are then transported throughout the plant as a component of the water flowing through the xylem.

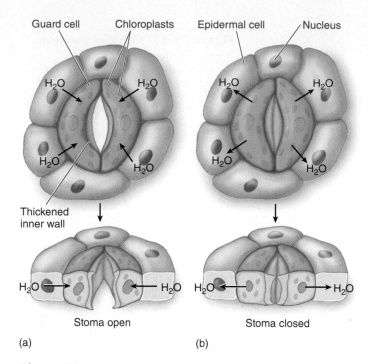

Figure 33.13 How guard cells regulate the opening and closing of stomata.

(*a*) When guard cells contain a high level of solutes, water enters the guard cells by osmosis, causing them to swell and bow outward. This bowing opens the stoma. (*b*) When guard cells contain a low level of solutes, water leaves the guard cells, causing them to become flaccid. This flaccidity closes the stoma.

Figure 33.14 Root hairs.

Abundant fine root hairs can be seen in the back of the root apex of this germinating seedling of radish, *Raphanus sativus*.

Key Learning Outcome 33.6 Water is drawn up the plant stem from the roots by transpiration from the leaves.

33.7 | Carbohydrate Transport

Most of the carbohydrates manufactured in plant leaves and other green parts are moved through the phloem to other parts of the plant. This process, known as **translocation,** makes suitable carbohydrate building blocks available at the plant's actively growing regions. The carbohydrates are converted into transportable molecules, such as sucrose, and moved through the plant.

The pathway that sugars and other substances travel within the plant has been demonstrated precisely by using radioactive isotopes and aphids, a group of insects that suck the sap of plants. Aphids thrust their piercing mouthparts into the phloem cells of leaves and stems to obtain the abundant sugars there. When the aphids are cut off of the leaf, the liquid continues to flow from the detached mouthparts protruding from the plant tissue and is thus available in pure form for analysis. The liquid in the phloem contains 10% to 25% dissolved solid matter, almost all of which is usually sucrose. The harvesting of sap from maple trees uses a similar process. The starch, stored over the winter, is converted into sap that is carried up throughout the plant. A hole is drilled in the tree and the sugar-rich fluid is drained from the tree using tubing and collected in buckets. The sap is then processed into maple syrup.

Using aphids to obtain the critical samples and radioactive tracers to mark them, researchers have learned that movement of substances in the phloem can be remarkably fast—rates of 50 to 100 centimeters per hour have been measured. This translocation movement is a passive process that does not require the expenditure of energy by the plant. The **mass flow** of materials transported in the phloem occurs because of water pressure, which develops as a result of osmosis. The Key Biological Process illustration on the right walks you through the process of translocation. Sucrose produced as a result of photosynthesis is actively "loaded" into the sieve tubes (or sieve cells) of the vascular bundles (**panel 1**). This loading increases the solute concentration of the sieve tubes, so water passes into them by osmosis (**panel 2**). An area where the sucrose is made is called a *source;* an area where sucrose is delivered from the sieve tubes is called a *sink.* Sinks include the roots and other regions of the plant that are not photosynthetic, such as young leaves and fruits. Water flowing into the phloem forces the sugary substance in the phloem to flow down the plant (**panel 3**). The sucrose is unloaded and stored in sink areas (**panel 4**). There the solute concentration of the sieve tubes is decreased as the sucrose is removed. As a result of these processes, water moves through the sieve tubes from the areas where sucrose is being added into those areas where it is being withdrawn, and the sucrose moves passively with the water. This is called the *pressure-flow hypothesis.*

Key Learning Outcome 33.7 Carbohydrates move through the plant by the passive osmotic process of translocation.

KEY BIOLOGICAL PROCESS: Translocation

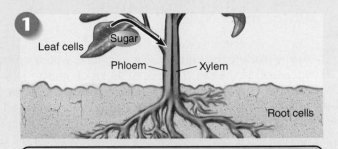

Sugar created in the leaves by photosynthesis ("source") enters the phloem by active transport.

When the sugar concentration in the phloem increases, water is drawn into phloem cells from the xylem by osmosis.

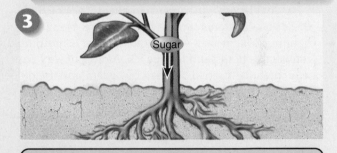

The addition of water from the xylem causes pressure to build up inside the phloem and pushes the sugar down.

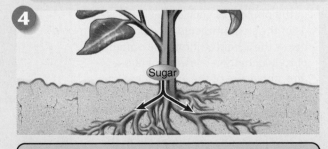

Sugar from the phloem enters the root cells ("sink") by active transport.

Does Water Move Up a Tree Through Phloem or Xylem?

Before reading this chapter, you may have wondered how water gets to the top of a tree, 10 stories above its roots. A column of water that tall weighs an awful lot. If you were to make a tube of drinking straws that tall and fill it with water, you would not be able to lift it. The answer to this puzzle was first proposed by biologist Otto Renner in Germany in 1911. He suggested that dry air moving across the tree's leaves captured water molecules by evaporation, and that this water was replaced with other water molecules coming in from the roots. Renner's idea, which was essentially correct, forms the core of the cohesion-adhesion-tension theory described in this chapter. Essential to the theory is that there is an unbroken water column from leaves to roots, a "pipe" from top to bottom through which the water can move freely.

There are two candidates for the role of water pipe, each a long series of narrow vessels that runs the length of the stem of a tree. As you have learned earlier, these two vessel systems are called xylem and phloem. In principle, either xylem or phloem could provide the plumbing through which water moves up a tree trunk or other stem. Which is it?

An elegant experiment demonstrates which of these vessel systems carries water up a tree stem. The bottom end of a stem was placed in water containing the radioactive potassium isotope ^{42}K. A piece of wax paper was carefully inserted between the xylem and the phloem in a 23-cm section of the stem to prevent any lateral transport of water between xylem and phloem.

After enough time had elapsed to allow water movement up the stem, the 23-cm section of the stem was removed, cut into six segments, and the amounts of ^{42}K measured both in the xylem and in the phloem of each segment, as well as in the stem immediately above and below the 23-cm section. The amount of radioactivity recorded provides a direct measure of the amount of water that has moved up from the bottom of the stem through either the xylem or phloem.

The results are presented in the graph above.

1. **Applying Concepts** What is the dependent variable?

2. **Interpreting Data**
   a. In the portion of the stem below where the 23-cm section

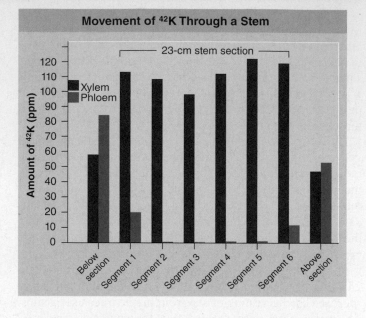

Movement of ^{42}K Through a Stem

was removed, do xylem and phloem both contain radioactivity? How about in the portion above where the 23-cm section was removed?
   b. In the central portion of the 23-cm segment of the stem (segments 2, 3, 4, 5), do xylem and phloem both contain radioactivity?

3. **Making Inferences**
   a. In the 23-cm section, is more ^{42}K found in xylem or phloem? What might you conclude from this?
   b. Above and below the 23-cm section, is more ^{42}K found in xylem or phloem? How would you account for this? [Hint: These sections did not contain the wax paper barrier that prevents lateral transport between xylem and phloem.]
   c. Within the 23-cm section, the phloem in segments 1 and 6 contains more ^{42}K than interior segments. What best accounts for this?
   d. Is it fair to infer that water could move through either xylem or phloem vessel systems?

4. **Drawing Conclusions** Does water move up a stem through phloem or xylem? Explain.

5. **Further Analysis** Devise an experiment along similar lines to test which vessel system is responsible for transporting sugars produced by photosynthesis in a plant's leaves to the cells in its roots.

Structure and Function of Plant Tissues

33.1 Organization of a Vascular Plant

- Most plants possess roots, stems, and leaves, although they may not always look the same. Vascular tissue extends throughout the plant, connecting roots, stems, and leaves (**figure 33.1**).

- Growth occurs in regions called meristems. The tips of the roots and shoots contain apical meristems, which are the sites of primary growth. Primary growth extends the plant body lengthwise. Extending the thickness or girth of the plant, called secondary growth, occurs at the lateral meristems, which are cylinders of meristematic tissue.

33.2 Plant Tissue Types

- Ground tissue makes up the main body of the plant and contains several different cell types (**integrated art, page 683**). Parenchyma cells are the most common type of cell in plants. They carry out functions such as photosynthesis, and food and water storage. The edible parts of fruits and vegetables are primarily parenchyma cells.

- Collenchyma cells form strands that provide support, especially for plants that do not have secondary growth. They are usually elongated cells with unevenly thickened primary cell walls.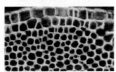

- Sclerenchyma cells have thick secondary cell walls that provide strength and rigidity. The secondary cell wall is laid down after the cell has stopped growing. They form long fibers or branched structures called sclereids or "stone cells."

- Dermal tissue makes up the outer layer of the plant body (**integrated art, page 684**), which consists of epidermal cells covered with a waxy layer called the cuticle (root cells lack a cuticle). This outer layer protects the plant and the cuticle provides a barrier to water loss. Paired guard cells are specialized cells in the epidermis. The space between the two guard cells, called the stoma, opens (allowing for gas exchange) and closes in response to external factors.

- Trichomes and root hairs are extensions of epidermal cells. Trichomes help regulate heat and water balance in the leaves, and some secrete toxic substances as a defense mechanism. Root hairs increase the surface area of the roots, allowing them to take up more water and minerals from the soil.

- Vascular tissue is composed of xylem and phloem (**integrated art, page 685**). Xylem contains water-conducting cells, like the tracheids and vessel elements shown here. They have thick secondary cell walls that provide structural support. They form long strands that are connected by pits and pores through which water passes.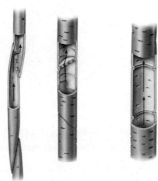

- Phloem contains food-conduction cells, sieve cells, and sieve-tube members. The cells fit end-to-end, and food is conducted through pores between the cells. Companion cells associated with the phloem cells carry out metabolic functions needed to maintain the sieve-tube members.

The Plant Body

33.3 Roots

- Roots are organs adapted to absorb water and minerals from the soil. Vascular tissue extends into the root core.

- A single layer of cells called the endodermis surrounds the vascular tissue. A waxy Casparian strip encircles the endodermal cells and blocks the passage of water between the cells (**integrated art, page 686**). Water is forced through the cells into the xylem.

- The root grows at the tip in the area called the zone of elongation (**figure 33.2**). New cells are added by the apical meristem. Symbiotic relationships with fungi and bacteria can increase the absorption of water and minerals.

33.4 Stems

- The stem serves as a framework for positioning the leaves. Primary growth occurs at the apical meristem. Leaves grow out of the stems at node areas, as shown here from **figure 33.4**. Secondary growth occurs at the lateral meristems, with the differentiation of vascular cambium into xylem and phloem, and the cork cambium into the layers of cork inside bark (**figures 33.6**). Wood forms from the accumulation of secondary xylem, which is thicker in spring and summer, resulting in rings in the wood.

33.5 Leaves

- Leaves are the primary site for photosynthesis. They grow out from the stem by means of marginal meristems that form the blade. They vary in size, shape, and arrangement (**figure 33.8**). Photosynthetic palisade mesophyll cells lie toward the surface. An underlying spongy mesophyll cell layer has large intercellular spaces that function in gas exchange (**figure 33.10**).

Plant Transport and Nutrition

33.6 Water Movement

- Carbohydrates and water are transported by phloem and xylem, respectively (**figure 33.11**).

- Water enters the roots by osmosis. Root pressure and capillary action (**figure 33.12**) cause the water to pass up into the tissues. However, for water to travel up the length of the stem, it requires a stronger force—the combination of cohesion and adhesion, known as the cohesion-adhesion-tension theory. Transpiration, the evaporation of water vapor from the leaves, creates the "pull" that raises water through the xylem (**Key Biological Process, page 693**).

- The guard cells flanking stomata will swell up when water is plentiful (**figure 33.13**). This turgid pressure opens stomata, letting water vapor out. Under conditions of water stress, water leaves the guard cells and stomata close, reducing water loss.

- Minerals enter the plant along with water through ion channels in the cells of root hairs (**figure 33.14**).

33.7 Carbohydrate Transport

- Carbohydrates produced in the leaves travel throughout the plant in phloem tissue. Translocation involves osmotic movement of water into the phloem cells, forcing the sugars to "sinks" for carbohydrate storage until needed (**Key Biological Process, page 695**).

Test Your Understanding

1. Growth in vascular plants originates in
   a. photosynthetic tissue.
   b. root tissue.
   c. meristematic tissue.
   d. leaf epidermal tissue.
2. In vascular plants, phloem tissue primarily
   a. transports water.
   b. transports carbohydrates.
   c. transports minerals.
   d. supports the plant.
3. The ground tissue that carries out most of the metabolic and storage functions is
   a. parenchyma cells.
   b. collenchyma cells.
   c. sclerenchyma cells
   d. sclereid cells.
4. In roots, growth of lateral branches begins
   a. on the root epidermis.
   b. on the root hairs.
   c. at the ground meristem.
   d. at the pericycle.
5. In stems, the tissue responsible for secondary growth is the
   a. collenchyma.
   b. pith.
   c. cambium.
   d. cortex.
6. One difference between monocot and dicot plant stems is the
   a. absence of buds in monocots.
   b. organization of vascular tissue.
   c. presence of guard cells.
   d. absence of stomata.
7. In vascular plant leaves, gases enter and leave the plant through pores called
   a. stomata.
   b. meristems.
   c. chloroplasts.
   d. trichomes.
8. Which of the following is *not* a process that directly assists in water movement from the roots to the leaves?
   a. photosynthesis
   b. root pressure
   c. capillary action
   d. transpiration
9. The passive process of moving carbohydrates throughout a plant is called
   a. transpiration.
   b. translocation.
   c. translation.
   d. evaporation.
10. The movement of carbohydrates through the plant is driven by
    a. facilitated diffusion.
    b. active transport.
    c. evaporation.
    d. osmosis.

Apply Your Understanding

1. **Figure 33.2** What is the purpose of the root cap covering the apical meristem of the root?

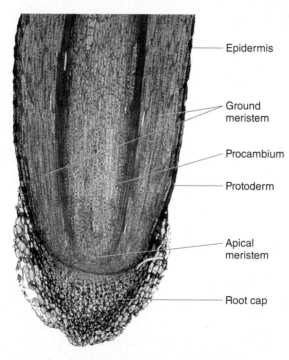

- Epidermis
- Ground meristem
- Procambium
- Protoderm
- Apical meristem
- Root cap

2. **Figure 33.7** Explain how the change in seasons produces tree rings. How might wet and dry years affect this?

3. **Figure 33.13** If you went on vacation for several days and left your houseplants in a warm, stuffy apartment, would the stomata look like the one on the left or the one on the right? Explain.

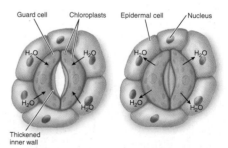

Guard cell   Chloroplasts   Epidermal cell   Nucleus
H_2O
Thickened inner wall

Synthesize What You Have Learned

1. Why do land plants need sclerenchyma cells?
2. Some plants in cold climates lose their leaves in the winter. They have bark, and their stems are not green. In desert climates, plants such as palo verde and ocotillo lose some or all of their leaves in the hot, dry summer to minimize water loss. These plants do have green stems. Why the difference?
3. A friend just returned from a family trip to northern Michigan, where he visited a maple tree farm where they made maple syrup. On the maple trees, they make just one relatively small cut all the way through the bark (or two cuts on larger trees) and hang a bucket beneath to catch the sap. Why, he asks you, don't they just make a cut completely around the tree and collect much more sap, much faster? How would you answer him?

34

Plant Reproduction and Growth

S eeds are one of the cleverest adaptations of plants. Carried by wind or animals, they are capable of transporting the next generation to distant locations and so ensuring the plant an opportunity to occupy any available habitats. A seed is a protected package of genetic information, an embryonic individual kept in a dormant state by a variety of mechanisms, such as a watertight covering that keeps the seed's interior free of water. When the seed is deposited in suitable soil, the watertight covering splits open and the embryo begins to grow, a process called germination. These seeds of a soybean plant are germinating, with leaves thrusting upward and roots downward toward the soil. For many seeds, moisture and moderate temperatures are sufficient to trigger germination. For some seeds, however, more extreme cues are required. The seeds of many species of pine tree, for example, will not germinate unless exposed to extreme temperatures of the sort experienced in forest fires; periodic forest fires provide openings for sunlight to reach the seedlings, and abundant nutrients enter the soil from the tissues of fire-killed trees.

34.1 Angiosperm Reproduction

Although reproduction varies greatly among the members of the plant kingdom, we focus in this chapter on reproduction among flowering plants. While the evolution of their unique sexual reproductive features, flowers and fruits, have contributed to their success, angiosperms also reproduce asexually.

Asexual Reproduction

In **asexual reproduction,** an individual inherits all of its chromosomes from a single parent and is, therefore, genetically identical to that parent. Asexual reproduction produces a "clone" of the parent.

In a stable environment, asexual reproduction may prove more advantageous than sexual reproduction because it allows individuals to reproduce with a lower investment of energy and maintain successful traits. A common type of asexual reproduction, called *vegetative reproduction,* results when new individuals are simply cloned from parts of the parent.

Vegetative reproduction in plants varies:

Runners. Some plants reproduce by means of runners—long, slender stems that grow along the surface of the soil. The strawberry plant shown here reproduces by runners. Notice that

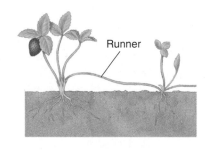

Runner

at node regions on the stem, adventitious roots form, extending into the soil. Leaves and flowers form, and a new shoot is sent out, continuing the runner.

Rhizomes. Rhizomes are underground horizontal stems that create a network underground. As in runners, nodes give rise to new flowering shoots. The noxious character of many weeds results from this type of growth pattern,

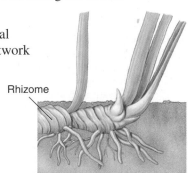

Rhizome

but so do grasses and many garden plants such as irises. Other specialized stems, called tubers, function in food storage and reproduction. White potatoes are specialized underground stems that store food, and the "eyes" give rise to new plants.

Suckers. The roots of some plants produce "suckers," or sprouts, that give rise to new plants, such as found in cherry, apple, raspberry, and blackberry plants. When the root of a dandelion is broken, which may occur if one tries to pull it from the ground, each root fragment may give rise to a new plant.

Adventitious Plantlets. In a few species, even the leaves are reproductive. One example is the house plant *Kalanchoë daigremontiana,* familiar to many people as the "maternity plant," or "mother of thousands." The common names of this plant are based on the fact that numerous plantlets arise from meristematic tissue located in notches along the leaves, as you can see in the photo shown here. The maternity plant is ordinarily propagated by means of these small plants, which drop to the soil and take root when they mature.

Sexual Reproduction

Plant sexual life cycles are characterized by an alternation of generations, in which a diploid *sporophyte generation* gives rise to a haploid *gametophyte generation,* as described in chapter 32. In angiosperms, the developing gametophyte generation is completely enclosed within the tissues of the parent sporophyte. The male gametophytes are **pollen grains,** and they develop from *microspores.* The female gametophyte is the **embryo sac,** which develops from a *megaspore.* Microspores and megaspores are discussed further on the following pages. Pollen grains and the embryo sac both are produced in separate, specialized structures of the angiosperm flower. The pollen grains form on filaments that extend out from the flower, and the embryo sac sits at the base of the flower, as shown below.

Female gametophyte (embryo sac) is inside base of flower

Parent sporophyte tissue (main plant body and outer whorls of flowers)

Male gametophytes (pollen grains) are on these filaments

Key Learning Outcome 34.1 Reproduction in angiosperms involves asexual and sexual reproduction.

34.2 Structure of the Flower

Flowers contain the organs for sexual reproduction in angiosperms. Like animals, angiosperms have separate structures for producing male and female gametes (eggs and sperm), but the reproductive organs of angiosperms are different from those of animals in two ways. First, in angiosperms, both male and female structures usually, but not always, occur together in the same individual flower. Second, angiosperm reproductive structures are not permanent parts of the adult individual. Angiosperm flowers and reproductive organs develop seasonally; these flowering seasons correspond to times of the year most favorable for pollination, the transfer of pollen to the female parts of a flower.

Flowers often contain male and female parts. The male parts, called *stamens,* are the long filamentous structures you see in the cut-away diagram of a flower in **figure 34.1a**. At the tip of each filament is a swollen portion, called the *anther,* that contains pollen. The female part, called the *carpel,* is the vase-shaped structure in the figure. The carpel consists of a lower bulging portion called the *ovary,* a slender stalk called the *style,* and a sticky tip called the *stigma* that receives pollen. The ovary contains the egg cell. Often flowers contain both stamens and carpels, as in **figure 34.1a**, but there are some exceptions. Various species of flowering plants, such as willows and some mulberries, have *imperfect* flowers, flowers containing only male or only female parts that occur on separate plants. Plants that contain imperfect flowers with only ovules or only pollen are called *dioecious,* from the Greek words for "two houses." These plants cannot self-pollinate and must rely on outcrossing. In other plants, there are separate male and female flowers, but they occur on the same plant. These plants are called *monoecious,* meaning "one house." In monoecious plants, the male and female flowers may mature at different times, increasing the chances of outcrossing.

Even in plants that have functional stamens and carpels present in each flower, these organs may reach maturity at different times, which keeps the flower from pollinating itself. For example, the fireweed flowers shown in **figure 34.1b,c** contain both stamens and carpels, but these organs mature at different times. First, the stamens mature, as shown in **figure 34.1b**, and the anthers shed pollen. Then after about two days, the style elongates above the stamen, as shown in **figure 34.1c**, and the four lobes of the stigma are ready to receive pollen. However, all the flowers aren't always on the same schedule. In the case of the fireweed, the lower flowers open first and so they are in their female phase when the upper flowers are opening and shedding pollen. This encourages pollen to be transferred to the stigma of another flower, promoting outcrossing.

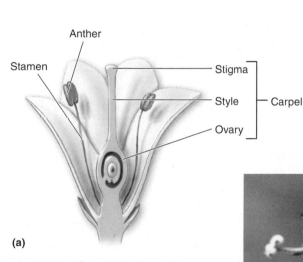

Figure 34.1 The structure of a flower.

(*a*) Most flowers contain both male and female parts. The flowers of the fireweed (*Epilobium angustifolium*) contain both male and female structures but they mature at different times. The male flower (*b*) opens first. The female phase (*c*) follows with the elongation of the style above the stamen.

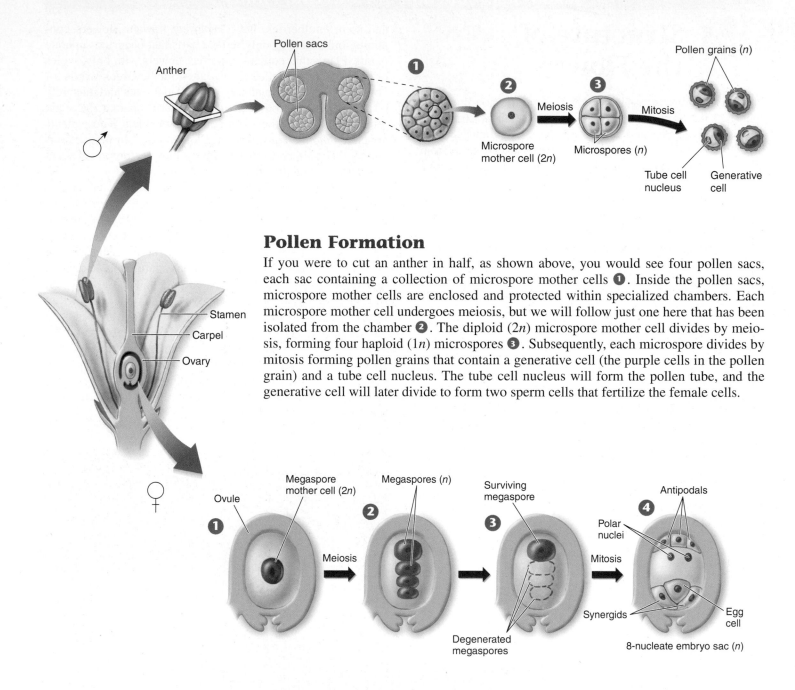

Pollen sacs

Anther

① ② ③

Pollen grains (n)

Meiosis

Mitosis

Microspore
mother cell (2n)

Microspores (n)

Tube cell
nucleus

Generative
cell

Stamen

Carpel

Ovary

Pollen Formation

If you were to cut an anther in half, as shown above, you would see four pollen sacs, each sac containing a collection of microspore mother cells ❶. Inside the pollen sacs, microspore mother cells are enclosed and protected within specialized chambers. Each microspore mother cell undergoes meiosis, but we will follow just one here that has been isolated from the chamber ❷. The diploid (2n) microspore mother cell divides by meiosis, forming four haploid (1n) microspores ❸. Subsequently, each microspore divides by mitosis forming pollen grains that contain a generative cell (the purple cells in the pollen grain) and a tube cell nucleus. The tube cell nucleus will form the pollen tube, and the generative cell will later divide to form two sperm cells that fertilize the female cells.

Ovule

Megaspore
mother cell (2n)

Megaspores (n)

Surviving
megaspore

Antipodals

Polar
nuclei

① ② ③ ④

Meiosis

Mitosis

Synergids

Egg
cell

Degenerated
megaspores

8-nucleate embryo sac (n)

Egg Formation

Eggs develop in the **ovule** of the angiosperm flower, which is contained within the ovary at the base of the carpel. Within each ovule is a megaspore mother cell ❶. Each diploid megaspore mother cell undergoes meiosis to produce four haploid megaspores ❷. In most plants, only one of these megaspores, however, survives; the rest are absorbed by the ovule. The lone remaining megaspore ❸ undergoes repeated mitotic divisions to produce eight haploid nuclei, which are enclosed within a structure called an *embryo sac* ❹. Within the embryo sac, the eight nuclei are arranged in precise positions. One nucleus (the green egg cell) is located near the opening of the embryo sac. Two nuclei are located in a single cell above the egg cell and are called polar nuclei. Two nuclei reside in cells that flank the egg cell and are called synergids; and the other three nuclei are located in cells at the top of the embryo sac, opposite the egg cell, and are called antipodals.

Figure 34.2 Different kinds of flowers.

The various flowers shown here all function in plant reproduction using the same structures, but they look very different. The birch tree flowers, in the upper left, are imperfect flowers. The male flowers are dangling down on the left and the female flowers are the clustered conelike structures on the right. The birch is a wind-pollinated plant and so the flowers are drab, lacking the brilliant colors seen in the other flowers that are insect-pollinated. These other plants use colorful flowers to attract their pollinators.

Although all flowers contain the same basic structures, and serve the same function, they don't all look the same. There is an amazing amount of variety in colors, shapes, sizes, and features of flowers. **Figure 34.2** shows a mere sampling of the wide variety of flowers.

Key Learning Outcome 34.2 Sexual reproduction in angiosperms involves flowers, which contain the male and female reproductive organs. Pollen grains develop in the anther, and an embryo sac develops within the ovule.

34.3 Gametes Combine Within the Flower

Pollination

Pollination is the process by which pollen is transferred from the anther to the stigma (**figure 34.3 ❶**). The pollen may be carried to the flower by wind or by animals, or it may originate within the individual flower itself. When pollen from a flower's anther pollinates the same flower's stigma, the process is called *self-pollination,* which can lead to *self-fertilization.* For some plants, self-pollination and self-fertilization occur because self-pollination eliminates the need for animal pollinators and maintains beneficial phenotypes in stable environments. However, other plants are adapted to *outcrossing,* the crossing of two different plants of the same species. The presence of only male or female flowers on a plant (a dioecious plant) requires outcrossing, as does the different timing of appearance of the male and female parts on a flower. Even when a flower's stamen and stigma mature at the same time, some plants exhibit **self-incompatibility.** Self-incompatibility results when the pollen and stigma recognize each other as being genetically related, and pollination of the flower is blocked, as shown on the left in **figure 34.4a**.

In many angiosperms, the pollen grains are carried from flower to flower by insects and other animals that visit the flowers for food or other rewards or are deceived into doing so because the flower's characteristics suggest such rewards. A liquid called **nectar,** which is rich in sugar as well as amino acids and other substances, is often the reward sought by animals. Successful pollination depends on the plants attracting insects and other animals regularly enough that the pollen is carried from one flower of that particular species to another (**figure 34.4b**).

The relationship between such animals, known as *pollinators,* and flowering plants has been important to the evolution of both groups, a process called **coevolution.** By using insects to transfer pollen, the flowering plants can disperse

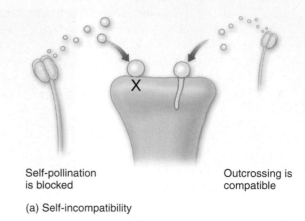

Self-pollination is blocked   Outcrossing is compatible

(a) Self-incompatibility

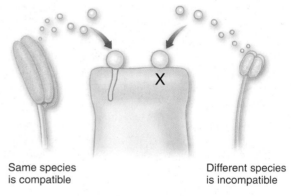

Same species is compatible   Different species is incompatible

(b) Controlled-pollination by insects

Figure 34.4 Incompatibility in pollination.
(*a*) Self-pollination is not permitted in some plant species. If a pollen grain from the same plant lands on the stigma, pollination is blocked, but in outcrossing, pollination occurs. (*b*) Controlled-pollination by insects or animals increases the likelihood that a plant will be pollinated by its same species and not by a different species.

Figure 34.3 Pollination and fertilization.

When pollen lands on the stigma of a flower, the pollen tube cell grows toward the embryo sac, forming a pollen tube. While the pollen tube is growing, the generative cell divides to form two sperm cells. When the pollen tube reaches the embryo sac, it bursts through one of the synergids and releases the sperm cells. In a process called double fertilization, one sperm cell nucleus fuses with the egg cell to form the diploid (2*n*) zygote, and the other sperm cell nucleus fuses with the two polar nuclei to form the triploid (3*n*) endosperm nucleus.

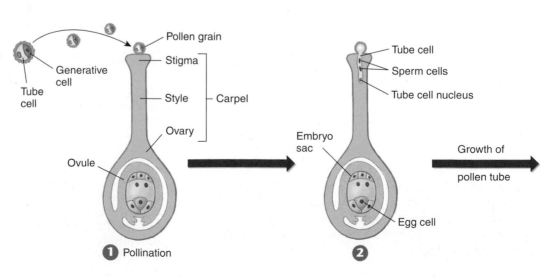

Pollen grain
Generative cell
Tube cell
Stigma
Style — Carpel
Ovary
Ovule
❶ Pollination

Tube cell
Sperm cells
Tube cell nucleus
Embryo sac
Egg cell
❷

Growth of pollen tube

their gametes on a regular and more or less controlled basis, despite their being anchored to the ground. The more attractive the plant is to the pollinator, the more frequently the plant will be visited. Therefore, any changes in the phenotype of the plant that result in more visits by pollinators offer a selective advantage. This has resulted in the evolution of a wide variety of angiosperm species that have very different flowers, as you saw on page 703.

For pollination by animals to be effective, a particular insect or other animal must visit plant individuals of the same species. A flower's color and form have been shaped by evolution to promote such specialization. Yellow and blue flowers are particularly attractive to bees (**figure 34.5a**), whereas red flowers attract birds but are not particularly noticed by most insects. Some flowers have very long floral tubes with the nectar produced deep within them; only the long, slender beaks of hummingbirds or the long, coiled proboscis of moths or butterflies can reach such nectar supplies. You can see the long proboscis in the butterfly reaching down inside a flower in **figure 34.5b**.

In certain angiosperms and all gymnosperms, pollen is blown about by the wind and reaches the stigmas passively. For such a system to operate efficiently, the individuals of a given plant species must grow relatively close together because wind does not carry pollen very far or very precisely, compared to transport by insects or other animals. Because gymnosperms, such as spruces or pines, grow in dense stands, wind pollination is very effective. Wind-pollinated angiosperms, such as birches, grasses, and ragweed, also tend to grow in dense stands. The flowers of wind-pollinated angiosperms are usually small, greenish, and odorless, and their petals are either reduced in size or absent altogether. They typically produce large quantities of pollen.

Double Fertilization

Once a pollen grain has been spread by wind, an animal, or self-pollination, it adheres to the sticky, sugary substance that covers the stigma and begins to grow a **pollen tube,** which pierces the style. The pollen tube, nourished by the sugary substance, grows, and the generative cell within the pollen

Figure 34.5 Insect pollination.
(a) Bees are usually attracted to yellow flowers. (b) This alfalfa butterfly (*Colias eurytheme*) has a long proboscis that allows it to feed on nectar deep in the flower.

grain tube cell divides to form two sperm cells, shown in **figure 34.3 ❷**. Growth of the pollen tube continues until it reaches the ovule in the ovary ❸.

When the pollen tube reaches the entry to the embryo sac in the ovule, one of the nuclei flanking the egg cell degenerates, and the pollen tube enters that cell. The tip of the pollen tube bursts and releases the two sperm cells ❹. One of the sperm cells fertilizes the egg cell, forming a zygote. The other sperm cell fuses with the two polar nuclei located at the center of the embryo sac, forming the triploid (3n) primary endosperm nucleus ❺. The primary endosperm nucleus eventually develops into the endosperm, which will nourish the developing embryo (see chapter 32). This process of fertilization in angiosperms in which two sperm cells are used is called **double fertilization.**

Key Learning Outcome 34.3 In pollination, pollen is transferred to the female stigma. After growth of the pollen tube, double fertilization leads to the development of an embryo and endosperm.

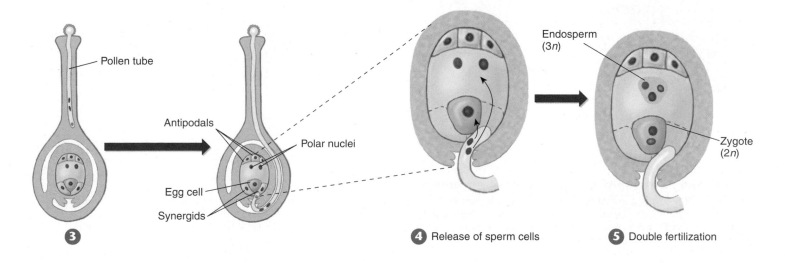

3 Pollen tube / Antipodals / Polar nuclei / Egg cell / Synergids

4 Release of sperm cells

5 Double fertilization / Endosperm (3n) / Zygote (2n)

34.4 Seeds

The entire series of events that occurs between fertilization and maturity is called *development*. During development, cells become progressively more specialized, or differentiated. Look at the lower series of panels of **figure 34.6** and you can see that the first stage in the development of a plant is active cell division to form an organized mass of cells, the embryo shown in **⑥**. In angiosperms, the differentiation of cell types within the embryo begins almost immediately after fertilization. By the fifth day, the principal tissue systems can be detected within the embryo mass, and within another day, the root and shoot apical meristems can be detected, as shown in **⑧**. The developing embryo is first nourished by the endosperm, and then later in some plants by the seed leaves, thick leaflike food storage structures called **cotyledons.**

Early in the development of an angiosperm embryo, a profoundly significant event occurs: The embryo simply stops developing and becomes dormant as a result of drying. In many plants, embryo development is arrested at the point shown in **⑧**, soon after apical meristems and the cotyledons are differentiated. The ovule of the plant has now matured into a **seed,** which includes the dormant embryo and a source of stored food, both surrounded by a protective and relatively impermeable seed coat that develops from the outermost coverings of the ovule.

Once the seed coat fully develops around the embryo, most of the embryo's metabolic activities cease; a mature seed contains only about 10% water. Under these conditions, the seed and the young plant within it are very stable.

Germination, or the resumption of metabolic activities that leads to the growth of a mature plant, cannot take place until water and oxygen reach the embryo, a process that sometimes involves cracking the seed. Seeds of some plants have been known to remain viable for hundreds, and in some cases thousands, of years. The seed will germinate when conditions are favorable for the plant's survival.

> **Key Learning Outcome 34.4** A seed contains a dormant embryo and substantial food reserves, encased within a tough drought-resistant coat.

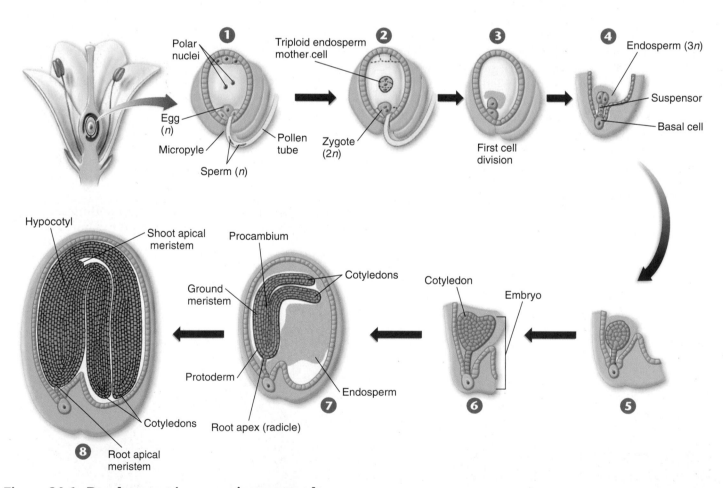

Figure 34.6 Development in an angiosperm embryo.
After the zygote forms, the first cell division is asymmetric ❸. After another division, the basal cell, the one nearest the opening through which the pollen tube entered, undergoes a series of divisions and forms a narrow column of cells called the suspensor ❹. The other three cells continue to divide and form a mass of cells arranged in layers ❺. By about the fifth day of cell division, the principal tissue systems of the developing plant can be detected within this mass ❼.

34.5 Fruit

During seed formation, the flower ovary begins to develop into fruit. The evolution of flowers was key to the success and diversification of the angiosperms. But of equal importance to angiosperm success has been the evolution of fruits, which aid in seed dispersal. Fruits form in many ways and exhibit a wide array of modes of specialization.

Three layers of the ovary wall can have distinct fates and account for the diversity of fruit types, from fleshy to dry and hard. There are three main kinds of fleshy fruits: berries, drupes, and pomes. In *berries*—such as grapes, tomatoes (**figure 34.7a**), and peppers—which are typically many-seeded, the inner layers of the ovary wall are fleshy. In *drupes*—such as peaches (**figure 34.7b**), olives, plums, and cherries—the inner layer of the fruit is stony and adheres tightly to a single seed. In *pomes*—such as apples (**figure 34.7c**) and pears—the fleshy portion of the fruit forms from the portion of the flower that is embedded in the receptacle (the swollen end of the flower stem that holds the petals and sepals). The inner layer of the ovary is a tough, leathery membrane that encloses the seeds.

Fruits that have fleshy coverings, often black, bright blue, or red (as in **figure 34.7d**), are normally dispersed by birds and other vertebrates. By feeding on these fruits, the animals carry seeds from place to place before excreting the seeds as solid waste. The seeds, not harmed by the animal digestive system, thus are transferred from one suitable habitat to another. Other fruits that are dispersed by wind, or by attaching themselves to the fur of mammals or the feathers of birds, are called dry fruits because they lack the fleshy tissue of edible fruits, and their ovaries form hard layers rather than fleshy tissue. Dry fruits can have structures that aid in their dispersion, as seen in the plumed dandelion in **figure 34.7e** or the spiny cocklebur in **figure 34.7f**, which catches onto fur (or socks or pants!) and is carried to new habitats. Still other fruits, such as those of mangroves, coconuts, and certain other plants that characteristically occur on or near beaches or swamps, are spread from place to place by water.

> **Key Learning Outcome 34.5** Fruits are specialized to achieve widespread dispersal by wind, by water, by attachment to animals, or, in the case of fleshy fruits, by being eaten.

(a) Berries   (b) Drupes   (c) Pomes

(d) Eaten by animals   (e) Dispersed by wind   (f) Dispersed by attaching to animals

Figure 34.7 Types of fruits and common modes of dispersion.
(*a*) Tomatoes are a type of fleshy fruit called berries, which have multiple seeds. (*b*) Peaches are a type of fleshy fruit called drupes; they contain a single large seed. (*c*) Apples are a type of fleshy fruit called pomes, which contain multiple seeds. (*d*) The bright red berries of this honeysuckle, *Lonicera,* are highly attractive to birds. Birds may carry the berry seeds either internally or stuck to their feet for great distances. (*e*) The seeds of this dandelion, *Taraxacum officinale,* are enclosed in a dry fruit with a "parachute" structure that aids its dispersal by wind. (*f*) The spiny fruits of this cocklebur, *Xanthium strumarium,* adhere readily to any passing animal.

34.6 Germination

What happens to a seed when it encounters conditions suitable for its germination? First, it absorbs water. Seed tissues are so dry at the start of germination that the seed takes up water with great force, after which metabolism resumes. Initially, the metabolism may be anaerobic, but when the seed coat ruptures, aerobic metabolism takes over. At this point, oxygen must be available to the developing embryo because plants, which drown for the same reason people do, require oxygen for active growth (see chapter 7). Few plants produce seeds that germinate successfully underwater, although some, such as rice, have evolved a tolerance of anaerobic conditions and can initially respire anaerobically. **Figure 34.8** shows the development of a dicot (on the left) and monocot (on the right) from germination through early stages. The first stage in both cases is the emergence of the roots. Following that, in dicots, the cotyledons emerge from underground along with the stem. The cotyledons eventually wither and the first leaves begin the process of photosynthesis. In monocots, the cotyledon doesn't emerge from underground; instead a structure called the *coleoptile* (a sheath wrapped around the emerging shoot) pushes through to the surface where the first leaves emerge and begin photosynthesis.

Key Learning Outcome 34.6 Germination is the resumption of a seed's growth and reproduction, triggered by water.

Figure 34.8 Development of angiosperms.

Dicot development in a soybean. The first structure to emerge is the embryonic root followed by the two cotyledons of the dicot. The cotyledons are pulled up through the soil along with the hypocotyl (the stem below the cotyledons). The cotyledons are the seed leaves that provide nutrients to the growing plant. As other leaves develop, they provide nutrients through photosynthesis, and the cotyledons shrivel and fall off the stem. Flowers develop in buds at the nodes.
Monocot development in corn. The first structure to emerge is the radicle or primary root. Monocots have one cotyledon, which does not emerge from underground. The coleoptile is a tubular sheath; it encloses and protects the shoot and leaves as they push their way up through the soil.

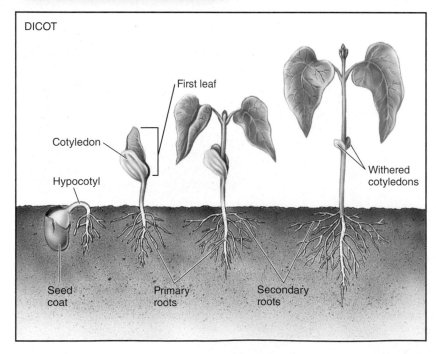

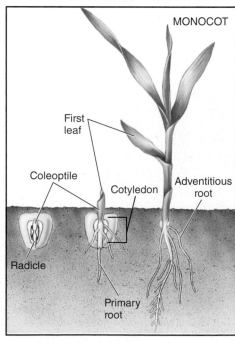

Growth and Nutrition

Just as human beings need certain nutrients, such as carbohydrates, amino acids, and vitamins, to survive, plants also need various nutrients to grow and remain healthy. Lack of an important nutrient may slow a plant's growth or make the plant more susceptible to disease or even death.

Macronutrients

Plants require a number of nutrients. Some of these are macronutrients, which plants need in relatively large amounts, and others are micronutrients, which are required in trace amounts. These nutrients are called *essential* because the plant cannot manufacture them; they have to be brought into the plant. For example, a plant can manufacture amino acids used to build proteins, but it can't manufacture the carbon or nitrogen atoms that make up the amino acids; therefore, these are essential nutrients. There are nine macronutrients: carbon, hydrogen, and oxygen—the three elements found in all organic compounds—as well as nitrogen (essential for amino acids), potassium, calcium, phosphorus, magnesium (the center of the chlorophyll molecule), and sulfur. Each of these nutrients approaches, or as in the case with carbon, may greatly exceed, 1% of the dry weight of a healthy plant.

Macronutrients are involved in plant metabolism in many ways. *Nitrogen* (N), acquired from the soil with the help of nitrogen-fixing bacteria, is an essential part of proteins and nucleic acids. *Potassium* (K) ions regulate the **turgor pressure** (the pressure within a cell that results from water moving into the cell) of guard cells and therefore the rate at which the plant loses water and takes in carbon dioxide. *Calcium* (Ca) is an essential component of the middle lamellae, the structural elements laid down between plant cell walls, and it also helps to maintain the physical integrity of membranes. *Magnesium* (Mg) is a part of the chlorophyll molecule. The presence of *phosphorus* (P) in many key biological molecules such as nucleic acids and ATP has been explored in detail in earlier chapters. *Sulfur* (S) is a key component of an amino acid (cysteine), essential in building proteins.

Micronutrients

The seven micronutrient elements—iron, chlorine, copper, manganese, zinc, molybdenum, and boron—constitute from less than one to several hundred parts per million in most plants. While the macronutrients were generally discovered in the last century, most micronutrients have been detected much more recently when technology developed that allowed investigators to identify and work with very small quantities.

Identifying Essential Nutrients

Nutritional requirements are assessed in hydroponic cultures as shown in **figure 34.9**; a plant's roots are suspended in aerated water containing nutrients. The solutions contain all the necessary nutrients in the correct proportions but with certain known or suspected nutrients left out. The plants are then allowed to grow and are studied for the presence of abnormal symptoms that might indicate a need for the missing element. To give an idea of how small the quantities of micronutrients may be, the standard dose of molybdenum added to seriously deficient soils in Australia amounts to about 34 grams (about one handful) per hectare, once every 10 years! Most plants grow satisfactorily in hydroponic culture, and the method, although expensive, is occasionally practical for commercial purposes. Analytical chemistry has made it much easier to take plant material and test for levels of different molecules. One application has been the investigation of elevated levels of carbon (a result of global warming) on plant growth. With increasing levels of CO_2, the leaves of some plants increase in size, but the amount of nitrogen decreases relative to carbon. This decreases the nutritional value of the leaves to herbivores.

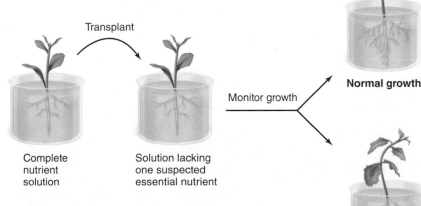

Figure 34.9 Identifying nutritional requirements of plants.

A seedling is first grown in a complete nutrient solution. The seedling is then transplanted to a solution that lacks one suspected essential nutrient. The growth of the seedling is then studied for the presence of abnormal symptoms, such as discolored leaves and stunted growth. If the seedling's growth is normal, the nutrient that was left out may not be essential; if the seedling's growth is abnormal, the lacking nutrient is essential for growth.

> **Key Learning Outcome 34.7** All plants require significant amounts of nine macronutrients to survive. They also require trace amounts of seven other elements.

34.8 Plant Hormones

After a seed germinates, the pattern of growth and differentiation that was established in the embryo is repeated indefinitely until the plant dies. But differentiation in plants, unlike that in animals, is largely reversible. Botanists first demonstrated in the 1950s that individual differentiated cells isolated from mature individuals could give rise to entire individuals. In an experiment, shown in **figure 34.10**, F. C. Steward was able to induce isolated bits of phloem tissue taken from carrots to form new plants, plants that were normal in appearance and fully fertile. Regeneration of entire plants from differentiated tissue has since been carried out in many plants, including cotton, tomatoes, and cherries. These experiments clearly demonstrate that the original differentiated phloem tissue still contains cells that retain all of the genetic potential needed for the differentiation of entire plants. No information is lost during plant tissue differentiation in these cells, and no irreversible steps are taken.

Once a seed has germinated, the plant's further development depends on the activities of the meristematic tissues, which interact with the environment through hormones (discussed next). The shoot and root apical meristems give rise to all of the other cells of the adult plant. Differentiation, or the formation of specialized tissues, occurs in five stages in plants and is shown in **figure 34.11**. The establishment of the shoot and root apical meristems occurs at stage 2; after that point, the tissues become more and more differentiated.

The tissue regeneration experiments of Steward and many others have led to the general conclusion that some nucleated cells in differentiated plant tissue are capable of expressing their hidden genetic information when provided with suitable environmental signals. What halts the expression of genetic potential when the same kinds of cells are incorporated into normal, growing plants? As we will see, the expression of some of these genes is controlled by plant hormones.

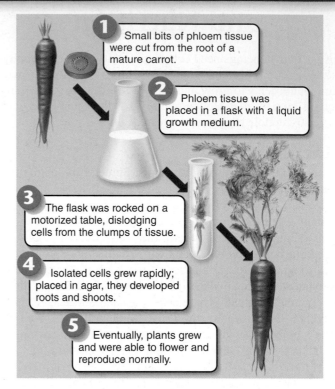

1. Small bits of phloem tissue were cut from the root of a mature carrot.

2. Phloem tissue was placed in a flask with a liquid growth medium.

3. The flask was rocked on a motorized table, dislodging cells from the clumps of tissue.

4. Isolated cells grew rapidly; placed in agar, they developed roots and shoots.

5. Eventually, plants grew and were able to flower and reproduce normally.

Figure 34.10   How Steward regenerated a plant from differentiated tissue.

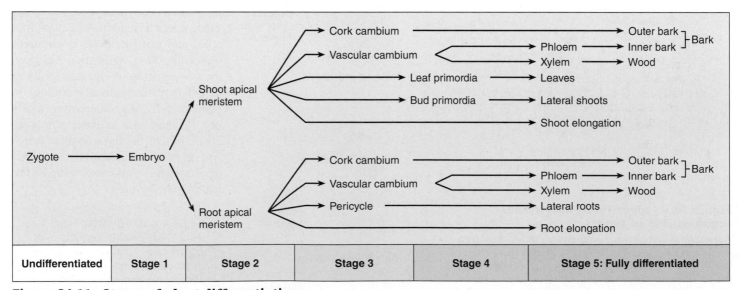

Figure 34.11   Stages of plant differentiation.

As this diagram shows, the different cells and tissues in a plant all originate from the shoot and root apical meristems. It is important to remember, however, that this is showing the origin of the tissue, not the location of the tissue in the plant. For example, the vascular tissues of xylem and phloem arise from the vascular cambium, but these tissues are present throughout the plant, in the leaves, shoots, and roots.

Hormones are chemical substances produced in small (often minute) quantities in one part of an organism and then transported to another part of the organism, where they stimulate certain physiological processes and inhibit others. How they act in a particular instance is influenced both by which hormones are present and by how they affect the particular tissue that receives their message.

In animals, there are several organs, called endocrine glands, that are solely involved with hormone production (hormones are produced in other organs as well). In plants, on the other hand, all hormones are produced in tissues that are not specialized for that purpose and carry out many other functions.

At least five major kinds of hormones are found in plants: auxin, gibberellins, cytokinins, ethylene, and abscisic acid. Their chemical structures and descriptions are provided in **table 34.1**. Other kinds of plant hormones certainly exist but are less well understood. Hormones have multiple functions in the plant; the same hormone may work differently in different parts of the plant, at different times, and interact with other hormones in different ways. The study of plant hormones, especially how hormones produce their effects, is today an active and important field of research.

> **Key Learning Outcome 34.8** The development of plant tissues is controlled by the actions of hormones. They act on the plant by regulating the expression of key genes.

TABLE 34.1   FUNCTIONS OF THE MAJOR PLANT HORMONES

Hormone	Major Functions	Where Produced or Found in Plant	Practical Applications
Auxin (IAA)	Promotes stem elongation and growth; forms adventitious roots; inhibits leaf abscission; promotes cell division (with cytokinins); induces ethylene production; promotes lateral bud dormancy	Apical meristems; other immature parts of plants	Seedless fruit production; synthetic auxins act as herbicides
Gibberellins (GA_1, GA_2, GA_3, etc.)	Promotes stem elongation; stimulates enzyme production in germinating seeds	Root and shoot tips; young leaves; seeds	Uniform seed germination for production of barley malt used in brewing; early seed production of biennial plants; increasing size of grapes by allowing more space for growth
Cytokinins	Stimulates cell division, but only in the presence of auxin; promotes chloroplast development; delays leaf aging; promotes bud formation	Root apical meristems; immature fruits	Tissue culture and biotechnology; pruning trees and shrubs, which causes them to "fill out"
Ethylene	Controls leaf, flower, and fruit abscission; promotes fruit ripening	Roots, shoot apical meristems; leaf nodes; aging flowers; ripening fruits	Fruit ripening of agricultural products that are picked early to retain freshness
Abscisic acid (ABA)	Controls stomatal closure; some control of seed dormancy; inhibits effects of other hormones	Leaves, fruits, root caps, seeds	Research on stress tolerance in plants, specifically drought tolerance

34.9 Auxin

In his later years, the great evolutionist Charles Darwin became increasingly devoted to the study of plants. In 1881, he and his son Francis published a book called *The Power of Movement in Plants,* in which they reported their systematic experiments concerning the way in which growing plants bend toward light, a phenomenon known as **phototropism.**

After conducting a series of experiments shown in **figure 34.12,** they observed that plants grew toward light (**panel 1**). If the tip of the seedling was covered, the plant didn't bend toward the light (**panel 2**). A control experiment showed that the cap was not interfering with the directional growth pattern (**panel 3**). Another control showed that covering the lower portions of the plant did not block the directional growth (**panel 4**). The Darwins hypothesized that when plant shoots were illuminated from one side, an "influence" that arose in the uppermost part of the shoot was then transmitted downward, causing the shoot to bend. Later, several botanists conducted a series of experiments that demonstrated that the substance causing the shoots to bend was a chemical we call **auxin.**

How auxin controls plant growth was discovered in 1926 by Frits Went, a Dutch plant physiologist, in the course of studies for his doctoral dissertation. From his experiments, described in **figure 34.13,** Went was able to show that the substance that flowed into agar from the tips of the light-grown grass seedlings (**steps 1** and **2**) enhanced cell elongation (shown in **step 3**). This chemical messenger caused the tissues on the side of the seedling into which it flowed to grow more than those on the opposite side (**step 4**). Control experiments indicated that the effects were not due to properties of the agar (**steps 1***a* and **2***a*). He named the substance that he had discovered auxin, from the Greek word *auxin,* meaning "to increase."

Went's experiments provided a basis for understanding the responses the Darwins had obtained some 45 years earlier: Grass seedlings bend toward the light because the side of the shoot that is in the shade has more auxin; therefore, its cells elongate more than those on the lighted side, bending the plant toward the light as in the enlargement in **figure 34.14.** Later experiments showed that auxin in normal plants migrates away from the illuminated side toward the dark side in response to light and thus causes the plant to bend toward the light.

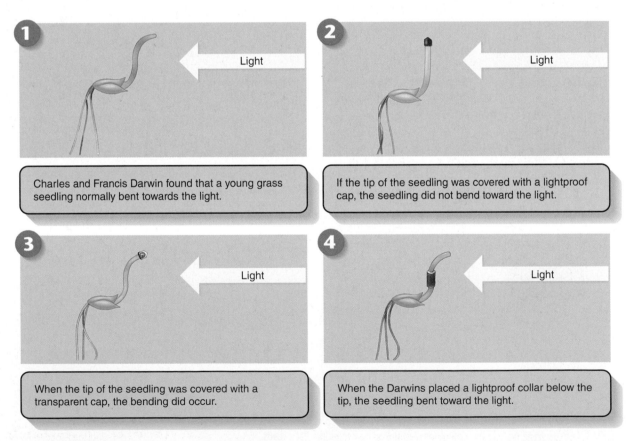

1 Light

Charles and Francis Darwin found that a young grass seedling normally bent towards the light.

2 Light

If the tip of the seedling was covered with a lightproof cap, the seedling did not bend toward the light.

3 Light

When the tip of the seedling was covered with a transparent cap, the bending did occur.

4 Light

When the Darwins placed a lightproof collar below the tip, the seedling bent toward the light.

Figure 34.12   The Darwins' experiment with phototropism.
From these experiments, the Darwins concluded that, in response to light, an "influence" that causes bending was transmitted from the tip of the seedling to the area below the tip, where bending usually occurs.

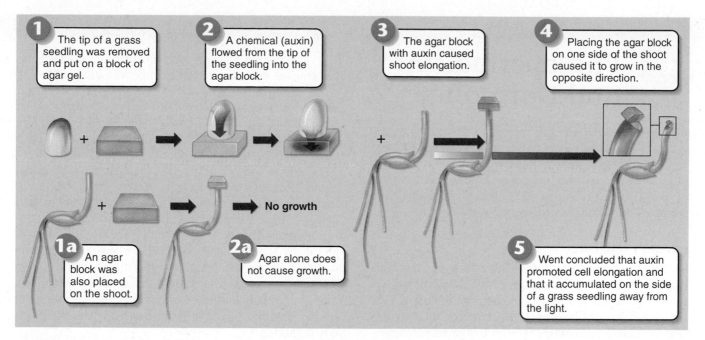

1. The tip of a grass seedling was removed and put on a block of agar gel.

2. A chemical (auxin) flowed from the tip of the seedling into the agar block.

3. The agar block with auxin caused shoot elongation.

4. Placing the agar block on one side of the shoot caused it to grow in the opposite direction.

No growth

1a. An agar block was also placed on the shoot.

2a. Agar alone does not cause growth.

5. Went concluded that auxin promoted cell elongation and that it accumulated on the side of a grass seedling away from the light.

Figure 34.13   How Went demonstrated the effects of auxin on plant growth.

The experiment showed how a chemical at the tip of the seedling caused the shoot to elongate and to bend. Steps 1a and 2a show the control experiment.

Auxin appears to act by increasing the stretchability of the plant cell wall within minutes of its application. Researchers speculate that the covalent bonds linking the polysaccharides of the cell wall to one another change extensively in response to auxin, allowing the cells to take up water and thus enlarge.

Synthetic auxins are routinely used to control weeds. When applied as herbicides, they are used in higher concentrations than those at which auxin normally occurs in plants. One of the most important of the synthetic auxins used in this way is 2,4-dichlorophenoxyacetic acid, usually known as 2,4-D. It kills weeds in lawns without harming the grass because 2,4-D affects only broad-leaved dicots. When treated, the weeds literally "grow to death," rapidly reducing ATP production so that no energy remains for transport or other essential functions.

Closely related to 2,4-D is the herbicide 2,4,5-trichlorophenoxyacetic acid (2,4,5-T), which is widely used to kill woody seedlings and weeds. Notorious as the Agent Orange of the Vietnam War, 2,4,5-T is easily contaminated with a by-product of its manufacture, dioxin. Dioxin is harmful to people because it is an **endocrine disrupter,** a chemical that interferes with the course of human development. The growing release of endocrine disrupters as by-products of modern chemical manufacturing is a subject of great environmental concern.

> **Key Learning Outcome 34.9**   The primary growth-promoting hormone of plants is auxin, which increases the plasticity of plant cell walls, allowing growth in specific directions.

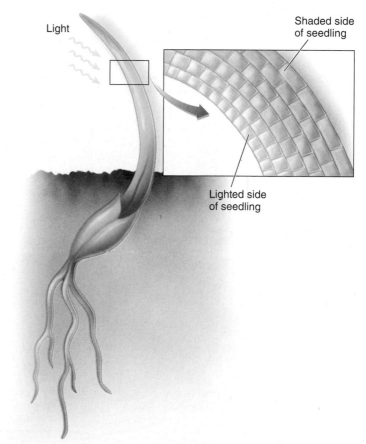

Light

Shaded side of seedling

Lighted side of seedling

Figure 34.14   Auxin causes cells to elongate.

Plant cells that are on the shaded side have more auxin and grow faster, elongating more, than cells on the lighted side, causing the plant to bend toward light.

34.10 Photoperiodism and Dormancy

Plants respond to different environmental stimuli in a variety of ways. As discussed earlier in the chapter, plants bend toward light as they grow in response to this environmental stimulus. A host of other plant responses, including flowering, dropping of leaves, and yellowing of leaves due to loss of chlorophyll, are also prompted by various environmental stimuli.

Photoperiodism

Essentially all eukaryotic organisms are affected by the cycle of night and day, and many features of plant growth and development are keyed to changes in the proportions of light and dark in the daily 24-hour cycle. Such responses constitute **photoperiodism,** a mechanism by which organisms measure seasonal changes in relative day and night length. One of the most obvious of these photoperiodic reactions concerns angiosperm flower production.

Day length changes with the seasons; the farther from the equator you are, the greater the variation. Plants' flowering responses fall into three basic categories in relation to day length: long-day plants, short-day plants, and day-neutral plants. Long-day plants, like the iris in **panel 1** of the Key Biological Process illustration below, initiate flowers in the summer, when nights become shorter than a certain length (and days become longer). Short-day plants, on the other hand, begin to form flowers when nights become longer than a critical length (and days become shorter); the goldenrod in **panel 2** doesn't flower in summer, but instead flowers in fall. Thus, many spring and early summer flowers are long-day plants, and many fall flowers are short-day plants. The "interrupted night" experiment in **panel 3** makes it clear that it is actually the length of uninterrupted dark that is the flowering trigger. The flash of light during a long night triggers flowering in the iris and inhibits flowering in the goldenrod, even though the day is shorter.

In addition to long-day and short-day plants, a number of plants are described as day-neutral. Day-neutral plants produce flowers without regard to day length.

KEY BIOLOGICAL PROCESS: Photoperiodism

1

Long-day plants   Short-day plants

Early summer. Short periods of darkness induce flowering in long-day plants, such as iris, but not in short-day plants, such as goldenrod.

2

Long-day plants   Short-day plants

Late fall. Long periods of darkness induce flowering in short-day plants, such as goldenrod, but not in long-day plants, such as iris.

3

Long-day plants   Short-day plants

Interrupted night. If the long night of winter is artificially interrupted by a flash of light, the goldenrod will not bloom and the iris will.

The Chemical Basis of Photoperiodism

Flowering responses to daylight and darkness are controlled by several chemicals that interact in complex ways. Although the nature of some of these chemicals has been deduced, how the various chemicals work together to promote or inhibit flowering responses is still being debated.

Plants contain a pigment, **phytochrome,** that exists in two interconvertible forms, P_r (inactive) and P_{fr} (active). When P_{fr} is present, biological reactions like flowering are influenced. When P_r absorbs red light (660 nanometers—orangish red), it is instantly converted into P_{fr}. Conversely, when P_{fr} is left in the dark or absorbs far-red light (730 nm—deep red), it is instantly converted to P_r and the biological response ceases.

In short-day plants, the presence of P_{fr} leads to a biological reaction that suppresses flowering. The amount of P_{fr} steadily declines in darkness, the molecules converting to P_r. When the period of darkness is long enough, the suppression reaction ceases and the flowering response is triggered. However, a single flash of red light at a wavelength of about 660 nm converts most of the molecules of P_r to P_{fr}, and the flowering reaction is blocked.

Dormancy

Plants respond to their external environment largely by changes in growth rate. Plants' ability to stop growing altogether when conditions are not favorable—to become dormant—is critical to their survival.

In temperate regions, dormancy is generally associated with winter, when low temperatures and the unavailability of water because of freezing make it impossible for plants to grow. During this season, the buds of deciduous trees and shrubs remain dormant, and the apical meristems remain well protected inside enfolding scales. Perennial herbs spend the winter underground as stout stems or roots packed with stored food. Many other kinds of plants, including most annuals, pass the winter as seeds.

> **Key Learning Outcome 34.10** Plant growth and reproduction are sensitive to photoperiod, using chemicals to link flowering to season.

34.11 Tropisms

Tropisms are directional and irreversible growth responses of plants to external stimuli. They control patterns of plant growth and thus plant appearance. Three major classes of plant tropisms include phototropism (**figure 34.15***a,* as discussed earlier), gravitropism, and thigmotropism.

Gravitropism

Gravitropism causes stems to grow upward and roots downward in response to gravity. Both of these responses clearly have adaptive significance. Stems, like the one growing from a tipped over flower pot in **figure 34.15***b,* grow upward and are apt to receive more light than those that do not; roots that grow downward are more apt to encounter a more favorable environment than those that do not.

Thigmotropism

Still another commonly observed response of plants is **thigmotropism,** a name derived from the Greek root *thigma,* meaning "touch." Thigmotropism is defined as the response of plants to touch. Examples include plant tendrils, which rapidly curl around and cling to stems or other objects, and twining plants, such as bindweed, which also coil around objects (**figure 34.15***c*). These behaviors result from rapid growth responses to touch. Specialized groups of cells in the plant epidermis appear to be concerned with thigmotropic reactions, but again, their exact mode of action is not well understood.

> **Key Learning Outcome 34.11** Growth of the plant body is often sensitive to light, gravity, or touch.

(a)

(b)

(c)

Figure 34.15 Tropism guides plant growth.

(*a*) Phototropism is exhibited by this *Coriandrum* plant growing toward light. (*b*) Gravitropism is exhibited by this plant, *Phaseolus vulgaris.* Note the negative gravitational response of the shoot. (*c*) The thigmotropic response of these twining stems causes them to coil around the object with which they have come in contact.

Are Pollinators Responsible for the Evolution of Flower Color?

Evolution results from many types of interactions among organisms, including predator-prey relationships, competition, and mate selection. An important type of coevolution among plants and animals involves flowering plants and their pollinators. Pollinators need flowers for food, and plants need pollinators for reproduction. It is logical then, to hypothesize that evolutionary changes in flower shape, size, odor, and color are driven, to a large extent, by pollinators such as bees. We know that insects respond to variation in flower traits by visiting flowers with certain features, but few studies have been carried out to predict and then evaluate the response of plant populations to selection by pollinators. In wild radish populations, honeybees preferentially visit yellow and white flowers, while syrphid flies prefer pink flowers. Rebecca Irwin and Sharon Strauss at the University of California, Davis studied the response of wild radish flower color to selection by pollinators in natural populations. They compared the frequency of four flower colors (yellow, pink, white, and bronze) in two populations of wild radishes. Bees created the first population by visiting flowers based on their color preferences. The scientists produced the second population by hand pollinating wild radish flowers with no discrimination due to flower color. However, the pollen used for artificial pollinations contained varying proportions of each type of plant, based on the frequencies of each in the wild. In the graph shown here, you can see the distribution of flower colors in these two populations. The blue bars indicate the number of plants with yellow, white, pink, and bronze flowers in the population generated by the bee pollination. The red bars show the number of each type of flower in the population generated by researcher pollination, with no selection for flower color.

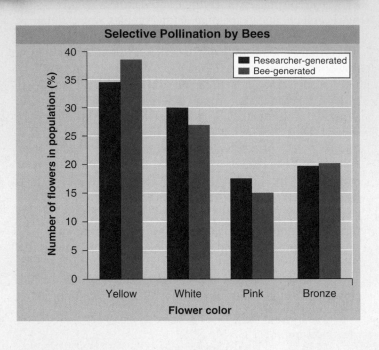

Selective Pollination by Bees

Legend: Researcher-generated, Bee-generated

y-axis: Number of flowers in population (%)
x-axis: Flower color — Yellow, White, Pink, Bronze

An uncommon visitor: here a syrphid fly is visiting a yellow wild radish flower, usually preferred by bees.

1. **Applying Concepts**
   a. Variable. In the graph, what is the dependent variable?
   b. Reading a bar graph. Does this graph reflect data on flower colors from syrphid fly pollinations?

2. **Interpreting Data**
   a. Which flower color was most common?
   b. Which flower color(s) did the bees preferentially appear to visit?

3. **Making Inferences**
   a. Which type of insect do you suppose was most abundant in the region where this study was carried out, and why?
   b. Because there were pink flowers in these populations, can you say that syrphid flies *had* to be present in the area?

4. **Drawing Conclusions**
   a. Does it appear that insects are influencing the evolution of flower color in wild radish populations?
   b. Why did the researcher-pollinated population of flowers exhibit a similar pattern of flower colors as the bee-pollinated population? Were the researchers showing some experimental bias in their color selections?

5. **Further Analysis** Assume the wild radish population in this study is visited again in 10 years and although a slight increase in yellow-flowered plants is observed, the proportion is not as high as would be predicted by this study. Provide some explanation for a slower than expected increase in the yellow-flowered plants.

Flowering Plant Reproduction

34.1 Angiosperm Reproduction

- Angiosperms reproduce sexually and asexually. In asexual reproduction, offspring are genetically identical to the parent; this often involves vegetative reproduction. Plants use many forms of vegetative reproduction including runners, rhizomes, suckers, and adventitious plantlets (**integrated art, page 700**).

- Sexual reproduction in plants involves an alternation of generations. A diploid sporophyte gives rise to a haploid gametophyte that produces gametes—egg and pollen (**integrated art, page 700**).

Sexual Reproduction in Flowering Plants

34.2 Structure of the Flower

- Flowers have male and female parts (**figure 34.1**). The male parts include the stamen and the pollen-producing anthers. Pollen grains, the male gametophyte, are produced in the anthers. The female parts include the stigma, style, and ovary, which make up the carpel. The embryo sac is the female gametophyte, and it forms in the ovule, within the ovary (**integrated art, page 702**).

34.3 Gametes Combine Within the Flower

- Pollen grains are carried to flowers by wind or animals. A pollen grain lands on the stigma and extends a pollen tube through the style to the base of the ovule (**figure 34.3**). Two sperm cells travel down the pollen tube. One sperm fertilizes the egg and the other fuses with the two polar nuclei. This process is called double fertilization. Pollination and fertilization bring the gametes together.

34.4 Seeds

- The fertilized egg begins dividing, forming the embryo (**figure 34.6**). After the shoot and root apical meristems form, the embryo stops growing and becomes dormant in a structure called a seed.

- The endosperm provides a source of food for the plant embryo. The outer layer of the ovule becomes the seed coat.

34.5 Fruit

- During seed formation, the flower's ovary begins to develop into fruit that surrounds the seed. The walls of the ovary can develop differently which accounts for the variety of fruits. Fruits are dispersed in different ways. Fleshy fruits are eaten by animals and then dispersed  through their feces. Dry fruits are usually dispersed by wind, water, or animals. Dry fruits have structures that aid in their dispersal (**figure 34.7**).

34.6 Germination

- A seed germinates under favorable conditions. The seed absorbs water and uses the endosperm or cotyledons as a food source. The seed coat cracks open and the plant begins to grow. The overall process is similar in monocots and dicots, but there are differences in the structures involved (**figure 34.8**).

34.7 Growth and Nutrition

- All plants need key elements to survive and thrive. There are nine macronutrients and seven micronutrients required by plants. Nutritional requirements of plants can be assessed using hydroponic cultures (**figure 34.9**).

Regulating Plant Growth

34.8 Plant Hormones

- Differentiation in plants is largely reversible. New plants can be grown from parts of adult plants (**figure 34.10**). The shoot and root meristems differentiate early in development and give rise to all of the other cell types (**figure 34.11**).

- Hormones are chemicals produced in very small quantities that affect growth and development. There are at least five major kinds of plant hormones: auxin, gibberellins, cytokinins, ethylene, and abscisic acid (**table 34.1**).

34.9 Auxin

- Early researchers, including Darwin and his son, described a process, now called phototropism, where plants grow toward light. In a series of experiments, they showed that the tip of a plant will grow toward light (**figure 34.12**). Went identified a chemical he called auxin as the hormone involved in phototropism (**figure 34.13**). When a plant is exposed to light on one side, auxin is released from the tip and causes cells on the shady side of the plant to elongate. This causes the plant to grow toward the light (**figure 34.14**).

Plant Responses to Environmental Stimuli

34.10 Photoperiodism and Dormancy

- The length of daylight affects flowering, a process called photoperiodism. Some plants flower in response to short days, some to long days, and some are day-neutral (**Key Biological Process, page 714**). A plant pigment, phytochrome, exists in two forms converted by darkness. The active form of phytochrome, P_{fr}, inhibits flowering. Darkness converts P_{fr} to P_r, which allows flowering to occur. Plants survive unfavorable conditions by entering a phase of dormancy, when they stop growing.

34.11 Tropisms

- Tropisms are irreversible growth patterns in response to external stimuli. Phototropism is a growth response toward light. Gravitropism is growth in response to the pull of gravity; this causes stems to grow upward and roots to grow downward. Thigmotropism is growth in response to touch (**figure 34.15**).

Test Your Understanding

1. Sexual reproduction in angiosperms requires
   a. pollen.
   b. runners.
   c. rhizomes.
   d. suckers.

2. The male parts of a flower include all of the following except
   a. anthers.
   b. stigma.
   c. stamen.
   d. microspores.

3. Angiosperm plants that contain both male and female flowers are called
   a. dioecious.
   b. monoecious.
   c. imperfect.
   d. incomplete.

4. The functions of flower shape, scent, color, and presence of nectar in the flowers of some angiosperms are to
   a. deter predators.
   b. attract animal pollinators.
   c. deter insect pests.
   d. enhance symbiosis.

5. For a seed to germinate, the dormant plant embryo must get
   a. carbon dioxide and water.
   b. nitrogen and water.
   c. oxygen and nitrogen.
   d. oxygen and water.

6. Fruit forms from a flower's
   a. ovary.
   b. sepals.
   c. carpels.
   d. stigma.

7. Meristematic tissue regulates plant growth and development through the use of
   a. carbohydrates.
   b. amount of available water.
   c. hormones.
   d. phototropism.

8. The hormone auxin causes
   a. cells in plant stems to shorten by releasing water.
   b. fruit to ripen.
   c. cells in plant stems to elongate.
   d. growth of more lateral branches.

9. Angiosperm flowering is controlled by
   a. temperature.
   b. gibberellins.
   c. photoperiod.
   d. magnesium.

10. Sensitivity of plants to touch is known as
   a. thigmotropism.
   b. photoperiodism.
   c. phototropism.
   d. gravitropism.

Apply Your Understanding

1. **Figure 34.5** Bees tend to pollinate yellow flowers, and so would not be attracted to the light pink flowers shown below that the butterfly is visiting. Give another reason why a bee probably wouldn't pollinate these pink flowers.

2. **Figure 34.7** From a culinary standpoint, plant products that are sweet and juicy are considered fruits, while other edible plant parts are often considered vegetables. However, a fruit isn't always sweet and juicy. Consider the following plant products and indicate if you eat the fruit, the seed, or other parts of the plant.

   (1) corn   (2) tomato   (3) cucumber   (4) peanut
   (5) carrot   (6) potato   (7) peas   (8) green peppers
   (9) lettuce

(a) Berries   (b) Drupes   (c) Pomes
(d) Eaten by animals   (e) Dispersed by wind   (f) Dispersed by attaching to animals

Synthesize What You Have Learned

1. On January 12, 2010, a magnitude 7.0 earthquake struck Haiti. Many thousands fled from the devastated cities to rural farms, too many people for Haiti's farms to feed. To aid the hungry, an agricultural company offered free seed to Haitian farmers. Not genetically engineered, the seed grows highly productive hybrid plants, far more productive than local crops and so able to feed many more people. However, a Haitian farmers' organization refused the donated seed, pointing out that the local farmers cannot save seeds from the hybrid crops to plant the following year, as the offspring of hybrid crops are puny. If Haiti's poor farmers accept the short-term gain of hybrid crops to feed their relatives, they ask, where would they get seed for *next* year's crop? How would you advise the Haitian farmers?

2. If ivy is planted next to a building, its stems attach to the side of the building and, in time, the mat of ivy will grow upward and eventually cover the building. How is ivy able to do this?

35

Populations and Communities

Often the most significant ecological events that occur in a particular ecosystem involve the organisms that inhabit it. The swarming insects you see here are migratory locusts, *Locusta migratoria,* moving across farmland in North Africa in 1988. In most years, the locusts are not plentiful and do not swarm. In particularly favorable years, however, when food is plentiful and the weather mild, the abundance of resources leads to greater-than-usual growth of locust populations. When high population densities are reached, the locusts exhibit different hormonal and physical characteristics and take off as a swarm. Moving over the landscape, the swarm eats every available plant, denuding the landscape. Swarming locusts, although not common in North America, are a legendary plague of large areas of Africa and Eurasia. In this chapter, we examine how natural populations grow, and what factors limit this growth. The organisms of the living world have evolved many accommodations to facilitate living together, creating complex evolutionary arrangements. When these arrangements are disturbed by unusual weather—or human intervention—the consequences can be catastrophic.

35.1 What Is Ecology?

Ecology is the study of how organisms interact with each other and with their environment. Ecology also encompasses the study of the distribution and abundance of organisms, which includes population growth and the limits and influences on population growth. The word *ecology* was coined in 1866 by the great German biologist Ernst Haeckel and comes from the Greek words *oikos* (house, place where one lives) and *logos* (study of). Our study of ecology, then, is a study of the house in which we live. Do not forget this simple analogy built into the word ecology—most of our environmental problems could be avoided if we treated the world in which we live the same way we treat our own homes. Would you pollute your own house?

Levels of Ecological Organization

Ecologists consider groups of organisms at six progressively more encompassing levels of organization. As mentioned in chapter 1, new characteristics called *emergent properties* arise at each higher level, resulting from the way components of each level interact.

1. **Populations.** Individuals of the same species that live together are members of a population. They potentially interbreed with one another, share the same habitat, and use the same pool of resources the habitat provides.
2. **Species.** All populations of a particular kind of organism form a species. Populations of the species can interact and affect the ecological characteristics of the species as a whole.
3. **Communities.** Populations of different species that live together in the same place are called communities. Different species typically use different resources within the habitat they share (**figure 35.1**).
4. **Ecosystems.** A community and the nonliving factors with which it interacts is called an **ecosystem**. An ecosystem is affected by the flow of energy, ultimately derived from the sun, and the cycling of the essential elements on which the lives of its constituent organisms depend. The redwood forest community pictured in **figure 35.1** is part of an ecosystem, where the giant trees and other organisms interact with each other and with their physical surroundings.
5. **Biomes.** Biomes are major terrestrial assemblages of plants, animals, and microorganisms that occur over wide geographical areas that have distinct physical characteristics. Examples include deserts, tropical forests, and grasslands. Similar types of groupings occur in marine and freshwater habitats.
6. **The biosphere.** All the world's biomes, along with its marine and freshwater assemblages, together constitute an interactive system we call the biosphere. Changes in one biome can have profound consequences for others.

Figure 35.1 The redwood community.

(*a*) The redwood forest of coastal California and southwestern Oregon is dominated by the population of redwoods *(Sequoia sempervirens)*. Other species in the redwood community include (*b*) redwood sorrel *(Oxalis oregana)*, (*c*) sword ferns *(Polystichum munitum)*, and (*d*) ground beetles *(Scaphinotus velutinus)*, this one feeding on a slug on a sword fern leaf.

Although we include biomes and the biosphere as higher levels of ecological organization in this list of organizational levels, the *ecosystem* is viewed as the "basic functional unit," in much the same way the cell rather than tissues or organs is considered the basic unit of living organisms.

Some ecologists, called *population ecologists,* focus on a particular species and how its populations grow. Other ecologists, called *community ecologists,* study how the different species living in a place interact with one another. Still other ecologists, called *systems ecologists,* are interested in how biological communities interact with their physical environment.

We will begin our study of ecology at the basic levels, by examining populations and communities. We will then work our way up the hierarchy by examining ecosystems, biomes, and ending with a critical look at the conditions of the biosphere. Although we break these topics into separate chapters, we should not overlook the fact that an organism does not live in a vacuum. Individuals interact with each other and with their physical environment and these interactions introduce challenges and obstacles to survival.

The Environmental Challenge

The nature of the physical environment determines to a great extent which organisms live in a particular climate or region. Key elements of the environment include:

Temperature. Most organisms are adapted to live within a relatively narrow range of temperatures and will not thrive if temperatures are colder or warmer. The growing season of plants, for example, is strongly influenced by temperature.

Water. All organisms require water. On land, water is often scarce, so patterns of rainfall have a major influence on life.

Sunlight. Almost all ecosystems rely on energy captured by photosynthesis, and so the availability of sunlight influences the amount of life an ecosystem can support, particularly below the surface in marine environments.

Soil. The physical consistency, pH, and the availability of minerals in the soil often severely limit plant growth, particularly the amount of nitrogen and phosphorus present in the soil.

During the course of a day, a season, or a lifetime, an individual organism must cope with a range of living conditions. Many organisms are able to adapt to environmental change by making physiological, morphological, or behavioral adjustments. For example, you sweat when it is hot, increasing heat loss through evaporation and thus preventing overheating. Morphological adaptations in some mammals may include growing a thicker coat of fur in winter (**figure 35.2**). And, many animals deal with variations in the environment through behavior, such as moving from one place to another, thereby avoiding areas that are unsuitable. For example, a tropical lizard manages to maintain a fairly uniform body temperature by

Figure 35.2   Wolf in winter.
This gray wolf grows a thicker coat of fur in the winter to insulate its body. Escaping body heat is trapped in the air surrounding the hairs of its coat, holding in heat and thus helping to maintain the wolf's body temperature in the cold winter.

Figure 35.3   Costa Rican lizard.
This green iguana escapes to the shade in the heat of the day, helping to keep its body cooler as the temperature outside rises.

basking in the sunlight but then retreating to the shade when it becomes too hot (**figure 35.3**). These physiological, morphological, or behavioral abilities are a product of natural selection acting in a particular environmental setting over time, which explains why an individual organism that is moved to a different environment may not survive.

> **Key Learning Outcome 35.1   Ecology is the study of how the organisms that live in a place interact with each other and with their physical environment. An ecosystem is a dynamic ecological system that challenges organisms to adjust to its changing physical conditions.**

35.2 Population Range

Organisms live as members of **populations,** groups of individuals that occur together at one place and time. Whether the population is a group of birds, insects, plants, or humans, ecologists can study several key elements of populations and learn more about them.

The term "population" can be defined narrowly or broadly. This flexibility allows us to speak in similar terms of the world's human population, the population of protists in the gut of an individual termite, or the population of deer that inhabit a forest. Sometimes the boundaries defining a population are sharp, such as the edge of an isolated mountain lake for trout, and sometimes they are fuzzier, such as when individual deer readily move back and forth between two forests separated by a cornfield.

Five aspects of populations are particularly important: *population range,* which is the area throughout which a population occurs; *population distribution,* which is the pattern of spacing of individuals within that range; *population size,* which is the number of individuals a population contains; *population density,* which is how many individuals share an area; and *population growth,* which describes whether a population is growing or shrinking, and at what rate. We will consider each in turn.

Population Ranges

No population, not even one composed of humans, occurs in all habitats throughout the world. Most species, in fact, have relatively limited geographic ranges, and the range of some species is minuscule. The Devil's Hole pupfish, for example, lives in a single hot water spring in southern Nevada, and the Socorro isopod is known from a single spring system in New Mexico. **Figure 35.4** shows a collection of other species that are found in a single population in an isolated habitat. At the other extreme, some species are widely distributed. The common dolphin (*Delphinus delphis*), for example, is found throughout all of the world's oceans.

Organisms must be adapted for the environment in which they occur. Polar bears are exquisitely adapted to survive the cold of the Arctic, but you won't find them in the tropical rain forest. Certain prokaryotes can live in the near-boiling waters of Yellowstone's geysers, but they do not occur in cooler streams nearby. Each population has its own requirements—temperature, humidity, certain types of food, and a host of other factors—that determine where it can live and reproduce and where it can't. In addition, in places that are otherwise suitable, the presence of predators, competitors, or parasites may prevent a population from occupying an area, a topic we will take up later in this chapter.

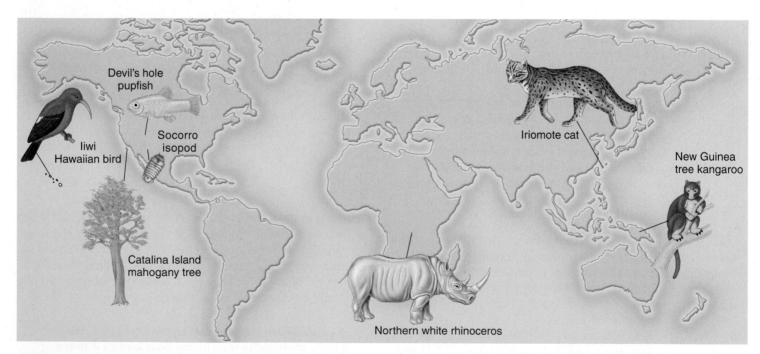

Figure 35.4 Species that occur in only one place.
These species, and many others, are only found in a single population. All are endangered species, and should anything happen to their single habitat, the population, and the species, would go extinct.

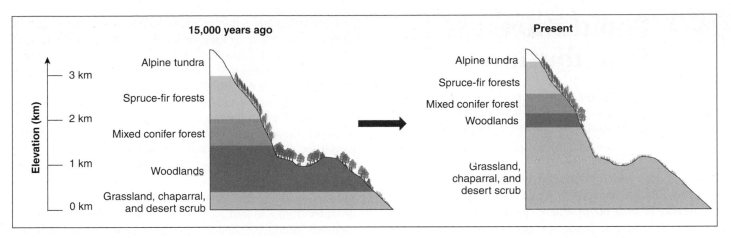

Figure 35.5 Altitudinal shifts in population ranges in the mountains of southwestern North America.
During the glacial period 15,000 years ago, conditions were cooler than they are now. As the climate has warmed, tree species that require colder temperatures have shifted their range upward in altitude so that they live in the climatic conditions to which they are adapted.

Range Expansions and Contractions

Population ranges are not static; rather, they change through time. These changes occur for two reasons. In some cases, the environment changes. For example, as the glaciers retreated at the end of the last Ice Age, approximately 10,000 years ago, many North American plant and animal populations expanded northward. At the same time, as climates warmed, species experienced shifts in the elevation at which they could live. The temperatures at higher elevations are cooler than at lower elevations. For example, the range for trees that survive better in colder temperatures shifts farther up a mountain when temperatures increase in an area as shown in **figure 35.5**.

In addition, populations can expand their ranges when they are able to move from inhospitable habitats to suitable, previously unoccupied areas. For example, cattle egrets native to Africa appeared in northern South America some time in the late 1800s. These birds made the nearly 2,000-mile transatlantic crossing, perhaps aided by strong winds. Since then, they have steadily expanded their range and now can be found throughout most of the United States (**figure 35.6**).

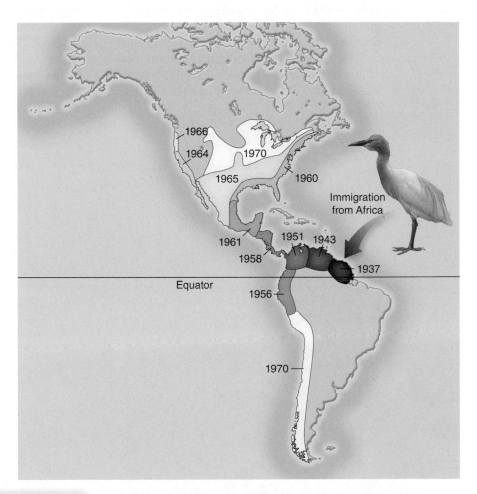

Figure 35.6 Range expansion of the cattle egret.
The cattle egret—so-named because it follows cattle and other hoofed animals, catching any insects or small vertebrates that they disturb—first arrived in South America in the late 1800s. Since the 1930s, the range expansion of this species has been well-documented, as it has moved westward and up into much of North America, as well as down the western side of the Andes to near the southern tip of South America.

> **Key Learning Outcome 35.2**   A population is a group of individuals of the same species existing together in an area. Its range, the area a population occupies, changes over time.

35.3 Population Distribution

A key characteristic affecting a species' range is the way in which individuals of its populations are distributed. They may be randomly spaced, uniformly spaced, or clumped (**figure 35.7**).

Randomly Spaced

Individuals are randomly spaced within populations when they do not interact strongly with one another or with nonuniform aspects of their environment. Random distributions are not common in nature. Some species of trees, however, appear to exhibit random distributions in Panamanian rainforests (**figure 35.7b**).

Uniformly Spaced

Uniform spacing within a population may often, but not always, result from competition for resources. The means by which it is accomplished, however, varies.

In animals, uniform spacing often results from behavioral interactions, such as those seen in **figure 35.8**. In many species, individuals of one or both sexes defend a territory from which other individuals are excluded. These territories provide the owner with exclusive access to resources such as food, water, hiding refuges, or mates, and individuals tend to be evenly spaced across the habitat. Even in nonterritorial species, individuals often maintain a defended space into which other animals are not allowed to intrude.

Among plants, uniform spacing also is a common result of competition for resources (**figure 35.7b**). In this case, however, the spacing results from direct competition for the resources. Closely spaced individual plants will compete for available sunlight, nutrients, or water. These contests can be direct, as when one plant casts a shadow over another, or indirect, as when two plants compete by extracting nutrients or water from a shared area. In addition, some plants, such as creosote, produce chemicals in the surrounding soil that are toxic to other members of their species. In all of these cases, only plants that are spaced an adequate distance from each other will be able to coexist, leading to uniform spacing.

Clumped Spacing

Individuals clump into groups or clusters in response to uneven distribution of resources in their immediate environments (see **figure 35.7b**). Clumped distributions are common in nature because individual animals, plants, and microorganisms tend to prefer microhabitats defined by soil type, moisture, or other aspects of the environment to which they are best adapted.

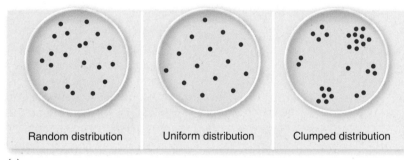

Random distribution   Uniform distribution   Clumped distribution

(a)

Figure 35.7 Population distribution.
The different patterns of spacing are exhibited by
(a) different arrangements of bacterial colonies and
(b) three different species of trees from the same locality in Panama.

Source: Data from Elizabeth Losos, Center for Tropical Forest Science, Smithsonian Tropical Research Institute.

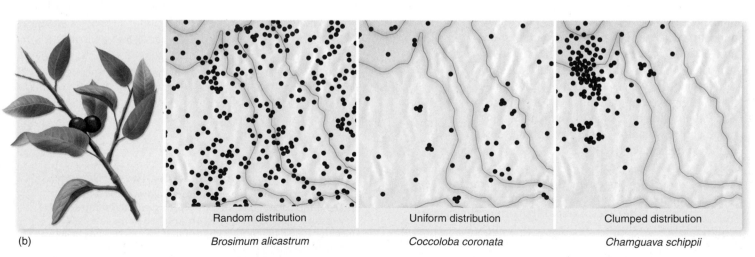

(b)   Random distribution   Uniform distribution   Clumped distribution

    Brosimum alicastrum   *Coccoloba coronata*   *Chamguava schippii*

Figure 35.8 **Uniform distribution in a population of gannets in New Zealand.**

Social interactions also can lead to clumped distributions. Many species live and move around in large groups, which go by a variety of names (for example, flock, herd, pride). Such groupings can provide many advantages, including increased awareness of and defense against predators, decreased energetic cost of moving through air and water, and access to the knowledge of all group members.

At a broader scale, populations are often most densely populated in the interior of their range and less densely distributed toward the edges. Such patterns usually result from the manner in which the environment changes in different areas. Populations are often best adapted to the conditions in the interior of their distribution. As environmental conditions change, individuals are less well-adapted, and thus densities decrease.

The Human Effect

By altering the environment, humans have allowed some species, such as coyotes, to expand their ranges and move into areas they previously did not occupy. Moreover, humans have served as an agent of dispersal for many species. Some of these transplants have been widely successful, as discussed in more detail in chapter 38. For example, 100 starlings were introduced into New York City in 1896 in a misguided attempt to establish every species of bird mentioned by Shakespeare. Their population steadily spread such that by 1980, they occurred throughout the United States. Similar stories could be told for countless numbers of plants and animals, and the list increases every year. Unfortunately, the success of these invaders often comes at the expense of native species, as discussed further in chapter 38.

Dispersal Mechanisms

Dispersal to new areas can occur in many ways. Lizards, for example, have colonized many distant islands, probably due to individuals or their eggs floating or drifting on vegetation. Bats are often the only mammals on distant islands because they can fly to them. Seeds of many plants are designed to disperse in many ways. Some seeds are aerodynamically designed to be blown long distances by the wind. Others have structures that stick to the fur or feathers of animals, so that they are carried long distances before falling to the ground. Still others are enclosed in fleshy fruits. These seeds can pass through the digestive systems of mammals or birds and then germinate at the spot upon which they are defecated. Finally, seeds of *Arceuthobium* are violently propelled from the base of the fruit in an explosive discharge. Although the probability of long-distance dispersal events leading to successful establishment of new populations is slim, over millions of years, many such dispersals have occurred.

> **Key Learning Outcome 35.3** The distribution of individuals within a population can be random, uniform, or clumped and is determined in part by the availability of resources.

35.4 Population Growth

Within its range, a species typically is found living in local populations, separated to at least some extent from other populations of that species. In this section, we will focus on the factors that influence whether a population will grow or shrink, and at what rate, and these factors are also important when considering our own population. Although we humans picture ourselves as different from populations of animals living in the wild, factors that affect wild populations affect human populations in similar ways.

One of the critical properties of any population is its **population size**—the number of individuals in the population. For example, if an entire species consists of only one or a few small populations, that species is likely to become extinct, especially if it occurs in areas that have been or are being radically changed. In addition to population size, **population density**—the number of individuals that occur in a unit area, such as per square kilometer—is often an important characteristic. The density of a population, how closely individuals associate with each other, is an indication of how they live. Animals that live in small family groups, like the Siberian tiger seen in **figure 35.9a**, often have few predators, while animals that live in large groups, such as the herd of wildebeests seen in **figure 35.9b**, may find safety in numbers.

In addition to size and density, another key characteristic of any population is its capacity to grow. To understand populations we must consider this, and what factors in nature limit **population growth.**

The Exponential Growth Model

The simplest model of population growth assumes a population growing without limits at its maximal rate. This rate, symbolized r and called the **biotic potential,** is the rate at which a population of a given species will increase when no limits are placed on its rate of growth. In mathematical terms, this is defined by the following formula:

$$growth\ rate = G = r_i N$$

where N is the number of individuals in the population, G is the change in its numbers over time (growth rate), and r_i is the *intrinsic* rate of natural increase for that population—its innate capacity for growth.

The *actual* rate of population increase r is defined as the difference between the birthrate b and the death rate d corrected for any movement of individuals in or out of the population, whether emigration (movement out of the area e) or immigration (movement into the area i). Thus,

$$r = (b - d) + (i - e)$$

Movements of individuals can have a major impact on population growth rates. For example, the increase in human population in the United States during the closing decades of the twentieth century was mostly due to immigrants. Less than half of the increase came from the reproduction of the people already living there.

The innate capacity for growth of any population is exponential, and is called *exponential growth.* Even when the *rate* of increase remains constant, the actual increase in the *number* of individuals accelerates rapidly as the size of the population grows. Rapid exponential growth is indicated by the red line in **figure 35.10**. This sort of growth pattern is similar to that obtained by compounding interest on an investment. In practice, such patterns prevail only for short periods, usually when an organism reaches a new habitat with abundant resources. Natural examples include dandelions reaching the fields, lawns, and meadows of North America from Europe for the first time; algae colonizing a newly formed pond; or the first plants arriving on an island recently thrust up from the sea.

(a)

(b)

Figure 35.9 Population density.
(a) Siberian tigers occupy enormous territories (typically 60–100 km² for an adult male) because of the relative lack of prey in the dense Siberian forests, especially in winter. (b) This Serengeti wildebeest herd numbers over 1 million animals.

Carrying Capacity

No matter how rapidly populations grow, they eventually reach a limit imposed by shortages of important environmental factors such as space, light, water, or nutrients. A population usually ultimately stabilizes at a certain size, called the **carrying capacity** of the particular place where it lives, and the size of the population levels off, like the blue line in **figure 35.10**. The carrying capacity, symbolized by *K,* is the maximum number of individuals that an area can support.

The Logistic Growth Model

As a population approaches its carrying capacity, its rate of growth slows greatly, because fewer resources remain for each new individual to use. The growth curve of such a population, which is always limited by one or more factors in the environment, can be approximated by the following **logistic growth equation** that adjusts the growth rate to account for the lessening availability of limiting factors:

$$G = rN \left(\frac{K - N}{K} \right)$$

In this logistic model of population growth, the growth rate of the population (*G*) equals its rate of increase (*r* multiplied by *N*, the number of individuals present at any one time), adjusted for the amount of resources available. The adjustment is made by multiplying *rN* by the fraction of *K* still unused (*K* minus *N*, divided by *K*). As *N* increases (the population grows in size), the fraction by which *r* is multiplied (the remaining resources) becomes smaller and smaller, and the rate of increase of the population declines.

In mathematical terms, as *N* approaches *K*, the rate of population growth (*G*) begins to slow, until it reaches 0 when *N* = *K* (the blue line in **figure 35.10**). In practical terms, factors such as increasing competition among more individuals for a given set of resources, the buildup of waste, or an increased rate of predation causes the decline in the rate of population growth.

Graphically, if you plot *N* versus *t* (time), you obtain an S-shaped **sigmoid growth curve** characteristic of most biological populations. The curve is called "sigmoid" because its shape has a double curve like the letter *S*. As the size of a population stabilizes at the carrying capacity, its rate of growth slows down, eventually coming to a halt. The fur seal population in **figure 35.11** has a carrying capacity of about 10,000 breeding male seals.

Processes such as competition for resources, emigration, and the accumulation of toxic wastes all tend to increase as a population approaches its carrying capacity for a particular habitat. The resources for which the members of the population are competing may be food, shelter, light, mating sites, mates, or any other factor needed to survive and reproduce.

> **Key Learning Outcome 35.4** The size at which a population stabilizes in a particular place is defined as the carrying capacity of that place for that species. Populations increase in size to the carrying capacity of their environment.

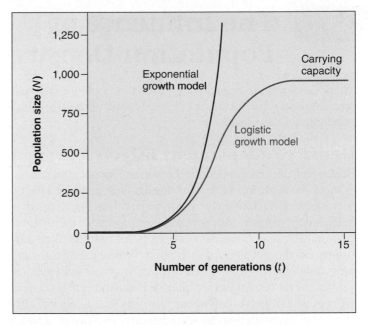

Figure 35.10 Two models of population growth.

The *red line* illustrates the exponential growth model for a population with an *r* of 1.0. The *blue line* illustrates the logistic growth model in a population with *r* = 1.0 and *K* = 1,000 individuals. At first, logistic growth accelerates exponentially, and then, as resources become limiting, the birthrate decreases or the death rate increases, and growth slows. Growth ceases when the death rate equals the birthrate. The carrying capacity (*K*) ultimately depends on the resources available in the environment.

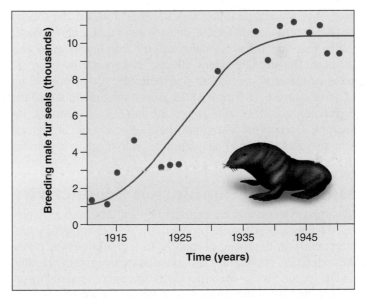

Figure 35.11 Most natural populations exhibit logistic growth.

These data present the history of a fur seal (*Callorhinus ursinus*) population on St. Paul Island, Alaska. Driven almost to extinction by hunting in the late 1800s, the fur seal made a comeback after hunting was banned in 1911. Today the number of breeding males with "harems" oscillates around 10,000 individuals, presumably the carrying capacity of the island.

35.5 The Influence of Population Density

Many factors act to regulate the growth of populations in nature. Some of these factors act independently of the size of the population; others do not.

Density-Independent Effects

Effects that are independent of the size of a population and act to regulate its growth are called **density-independent effects.** A variety of factors may affect populations in a density-independent manner. Most of these are aspects of the external environment, such as weather (extremely cold winters, droughts, storms, floods) and physical disruptions (volcanic eruptions and fire). Individuals often will be affected by these activities regardless of the size of the population. Populations that occur in areas in which such events occur relatively frequently will display erratic population growth patterns, increasing rapidly when conditions are relatively good, but suffering extreme reductions whenever the environment turns hostile.

Density-Dependent Effects

Effects that are dependent on the size of the population and act to regulate its growth are called **density-dependent effects.** Among animals, these effects may be accompanied by hormonal changes that can alter behavior that will directly affect the ultimate size of the population. One striking example occurs in migratory locusts (which you encountered at the beginning of this chapter). When they become crowded, the locusts produce hormones that cause them to enter a migratory phase; the locusts take off as a swarm and fly long distances to new habitats. Density-dependent effects, in general, have an increasing effect as population size increases. As the population of song sparrows in **figure 35.12** grows, the individuals in the population compete with increasing intensity for limited resources. Darwin proposed that these effects result in natural selection and improved adaptation as individuals compete for the limiting factors.

Maximizing Population Productivity

In natural systems that are exploited by humans, such as fisheries, the aim is to maximize productivity by exploiting the population early in the rising portion of its sigmoid growth curve. At such times, populations and individuals are growing rapidly, and net productivity—in terms of the amount of material incorporated into the bodies of these organisms—is highest.

Commercial fisheries attempt to operate so that they are always harvesting populations in the steep, rapidly growing parts of the curve. The point of *maximal sustainable yield* (the red line in **figure 35.13**) lies partway up the sigmoid curve. Harvesting the population of an economically desirable species near this point will result in the best sustained yields. Overharvesting a population that is smaller than this critical size can destroy its productivity for many years or even drive it to extinction. This evidently happened in the Peruvian anchovy fishery after the populations had been depressed by

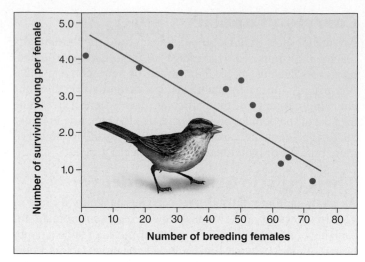

Figure 35.12   Density-dependent effects.
Reproductive success of the song sparrow (*Melospiza melodia*) decreases as population size increases.

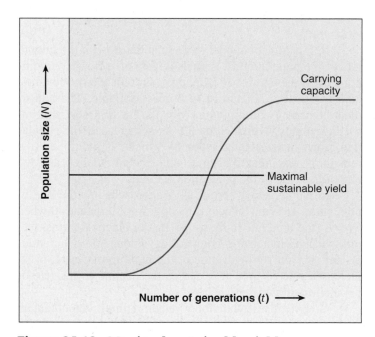

Figure 35.13   Maximal sustainable yield.
The goal of harvesting organisms for commercial purposes is to harvest just enough organisms to maximize current yields but also to sustain the population for future yields. Harvesting the organisms when the population is in the rapid growth phase of the sigmoidal curve, but not overharvesting, will result in sustained yields.

the 1972 El Niño. It is often difficult to determine population levels of commercially valuable species, and without this information it is equally difficult to determine the yield most suitable for long-term, productive harvesting.

> **Key Learning Outcome 35.5** Density-independent effects are controlled by factors that operate regardless of population size; density-dependent effects are caused by factors that come into play particularly when the population size is larger.

35.6 Life History Adaptations

Populations of many species, including annual plants, some insects, and most bacteria, can have very fast rates of growth when not limited by dwindling environmental resources. Habitats with more available resources than the population requires favor very rapid reproduction rates, which often approximate the exponential growth model discussed earlier.

Populations of most animals have much slower rates of growth with numbers limited by available resources. Growth slows as available resources become limiting, producing a sigmoid growth curve approximating the logistic growth model discussed earlier. Habitats with limited resources lead to more intense competition for resources, and favor individuals that can survive and successfully reproduce more efficiently. The number of individuals that can survive at this limit is the carrying capacity of the environment, or K.

The complete life cycle of an organism constitutes its *life history*. Life histories are very diverse, with different organisms having different adaptations in response to their environments. Some life history adaptations of a population favor very rapid growth in a habitat with abundant resources, or in unpredictable or volatile environments in which organisms have to take advantage of the resources when they are available. In these situations, reproducing early, producing many small offspring that mature quickly, and engaging in other aspects of "big bang" reproduction are favored. Using the terms of the exponential model, these adaptations, all favoring a high rate of increase r, are called **r-selected adaptations.** Examples of organisms displaying r-selected life history adaptations include dandelions, aphids, mice, and cockroaches (**figure 35.14**).

Other life history adaptations favor survival in an environment in which individuals are competing for limited resources. These features include reproducing late, having small numbers of large offspring that mature slowly and receive intensive parental care, and other aspects of "carrying capacity" reproduction. In terms of the logistic model, these adaptations, all favoring reproduction near the carrying capacity of the environment K, are called **K-selected adaptations.** Examples of organisms displaying K-selected life history adaptations include coconut palms, whooping cranes, and whales.

The r/K concept of life histories often provides a powerful way to examine more closely related organisms living in different types of habitats. In general, populations living in rapidly changing habitats tend to exhibit r-selected adaptations, whereas populations of closely related organisms living in more stable and competitive habitats exhibit more K-selected adaptations. Most natural populations show life history adaptations that exist along a continuum, ranging from completely r-selected traits to completely K-selected traits. **Table 35.1** outlines the adaptations at the extreme ends of the continuum.

Key Learning Outcome 35.6 Some life history adaptations favor near-exponential growth, while others favor the more competitive logistic growth.

Figure 35.14 The consequences of exponential growth.

All organisms have the potential to produce populations larger than those that actually occur in nature. The German cockroach (*Blatella germanica*), a major household pest, produces 80 young every six months. If every cockroach that hatched survived for three generations, kitchens might look like this theoretical culinary nightmare concocted by the Smithsonian Museum of Natural History.

TABLE 35.1		
	\multicolumn{2}{}{r-SELECTED AND K-SELECTED LIFE HISTORY ADAPTATIONS}	
Adaptation	**r-Selected Populations**	**K-Selected Populations**
Age at first reproduction	Early	Late
Homeostatic capability	Limited	Often extensive
Life span	Short	Long
Maturation time	Short	Long
Mortality rate	Often high	Usually low
Number of offspring produced per reproductive episode	Many	Few
Number of reproductions per lifetime	Usually one	Often several
Parental care	None	Often extensive
Size of offspring or eggs	Small	Large

Population Demography

Demography is the statistical study of populations. The term comes from two Greek words: *demos,* "the people" (the same root we see in the word *democracy*), and *graphos,* "measurement." Demography therefore means measurement of people, or, by extension, of the characteristics of populations. Demography is the science that helps predict how population sizes will change in the future. Populations grow if births outnumber deaths and shrink if deaths outnumber births. Because birthrates and death rates also depend on age and sex, the future size of a population depends on its present age structure and sex ratio.

Age Structure

Many annual plants and insects time their reproduction to particular seasons of the year and then die. All members of these populations are the same age. Perennial plants and longer-lived animals contain individuals of more than one generation, so that in any given year individuals of different ages are reproducing within the population. A group of individuals of the same age is referred to as a **cohort.**

Within a population, every cohort has a characteristic birthrate, or **fecundity,** defined as the number of offspring produced in a standard time (for example, per year), and a characteristic death rate, or **mortality,** the number of individuals that die in that period. The rate of a population's growth depends on the difference between these two rates.

The relative number of individuals in each cohort defines a population's age structure. Because individuals of different ages have different fecundity and death rates, age structure has a critical impact on a population's growth rate. A population with a large proportion of young individuals, for example, tends to grow rapidly because an increasing proportion of its individuals are reproductive.

Sex Ratio

The proportion of males and females in a population is its **sex ratio.** The number of births is usually directly related to the number of females, but it may not be as closely related to the number of males in species where a single male can mate with several females. In deer, elk, lions, and many other animals, a reproductive male guards a "harem" of females with which he mates, while preventing other males from mating with them. In such species, a reduction in the number of males simply changes the identities of the reproductive males without reducing the number of births. Among monogamous species like many birds, by contrast, where pairs form long-lasting reproductive relationships, a reduction in the number of males can directly reduce the number of births.

Mortality and Survivorship Curves

A population's intrinsic rate of increase depends on the ages of the organisms in it and the reproductive perfor-

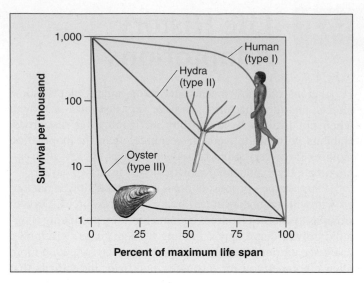

Figure 35.15 Survivorship curves.
By convention, survival (the *vertical axis*) is plotted on a log scale. Humans have a type I life cycle, the hydra (an animal related to jellyfish) type II, and oysters type III.

mance of the individuals in the various age groups. When a population lives in a constant environment for a few generations, its **age distribution**—the proportion of individuals in different age categories—tends to stabilize. This distribution differs greatly from species to species and even, to some extent, from population to population within a given species. Depending on the mating system of the species, sex ratio and generation time can also have a significant effect on population growth. A population whose size remains fairly constant through time is called a stable population. In such a population, births plus immigration must balance deaths plus emigration.

One way to express the age distribution characteristics of populations is through a **survivorship curve.** Survivorship is defined as the percentage of an original population that survives to a given age. Examples of different kinds of survivorship curves are shown in **figure 35.15**. In hydra, animals related to jellyfish, individuals are equally likely to die at any age, as indicated by the straight survivorship curve (the **blue line,** type II). Oysters, like plants, produce vast numbers of offspring, only a few of which live to reproduce. However, once they become established and grow into reproductive individuals, their mortality is extremely low (**red line,** type III survivorship curve). Finally, even though human babies are susceptible to death at relatively high rates, mortality in humans, as in many animals and protists, rises in the postreproductive years (**green line,** type I survivorship curve).

> **Key Learning Outcome 35.7** The growth rate of a population is a sensitive function of its age structure. In some species, mortality is higher among the young, and in others, among the old; in only a few is mortality independent of age.

35.8 Communities

Almost any place on earth is occupied by species, sometimes by many of them, as in the rain forests of the Amazon, and sometimes by only a few, as in the near-boiling waters of Yellowstone's geysers (where a number of microbial species live). The term **community** refers to the species that occur at any particular locality, like the array of plants and animals that you see in the savanna photo in **figure 35.16a** as well as the ones you can't see (like fungi, protists, and microbes). Communities can be characterized either by their constituent species, a list of all species present in the community, or by their properties, such as species richness (the number of different species present) or primary productivity.

Interactions among community members govern many ecological and evolutionary processes. These interactions, such as predation (**figure 35.16b**), competition (**figure 35.16c**), and mutualism, affect the population biology of particular species—whether a population increases or decreases in abundance for example—as well as the ways in which energy and nutrients cycle through the ecosystem. As explored in more detail in the next chapter, an *ecosystem* includes a community of living organisms and the nonliving components that surround them.

Scientists study biological communities in many ways, ranging from detailed observations to elaborate, large-scale experiments. In some cases, such studies focus on the entire community, whereas in other cases only a subset of species that are likely to interact with each other are studied. Regardless of how they are studied, two views exist on the makeup and functioning of communities.

The *individualistic concept* of communities, first championed by H. A. Gleason of the University of Chicago early in the twentieth century, holds that a community is nothing more than an aggregation of species that happen to co-occur at one place. By contrast, the *holistic concept* of communities, which can be traced to the work of F. E. Clements, also about a century ago, views communities as an integrated unit. In this sense, the community could be viewed as a superorganism whose constituent species have coevolved to the extent that they function as a part of a greater whole, just as the kidneys, heart, and lungs all function together within an animal's body. In this view, then, a community would amount to more than the sum of its parts.

Most ecologists today favor the individualistic concept. For the most part, species seem to respond independently to changing environmental conditions. As a result, community composition changes gradually across landscapes as some species appear and become more abundant, while others decrease in abundance and eventually disappear. Competition is an important factor that affects individuals and in so doing affects the community.

(a)

(b)

(c)

**Figure 35.16
A Tanzanian savanna community.**

A community consists of all the species—plants, animals, fungi, protists, and prokaryotes—that occur at a locality. (a) A savanna community in Lake Manyara National Park in Tanzania. Species within a community interact with each other, such as through predation in (b) or competition for a resource in (c).

Key Learning Outcome 35.8 A community consists of all species that occur at a site. Their interactions shape ecological and evolutionary patterns.

35.9 The Niche and Competition

Within a community, each organism occupies a particular biological role, or **niche.** The niche that an organism occupies is the sum total of all the ways it utilizes the resources of its environment. A niche may be described in terms of space utilization, food consumption, temperature range, appropriate conditions for mating, requirements for moisture, and other factors. *Niche* is not synonymous with **habitat,** the place where an organism lives. *Habitat* is a place, and *niche* is a pattern of living. Many species can share a habitat, but as we shall see, no two species can long occupy exactly the same niche.

Sometimes species are not able to occupy their entire niche because of the presence or absence of other species. Species can interact with each other in a number of ways, and these interactions can either have positive or negative effects. **Competition** describes the interaction when two organisms attempt to use the same resource when there is not enough of the resource to satisfy both.

Competition between individuals of different species is called **interspecific competition.** Interspecific competition is often greatest between organisms that obtain their food in similar ways and between organisms that are more similar. Another type of competition, called **intraspecific competition,** is competition between individuals of the same species.

The Realized Niche

Because of competition, organisms may not be able to occupy the entire niche they are theoretically capable of using, called the **fundamental niche** (or theoretical niche). The actual niche the organism is able to occupy in the presence of competitors is called its **realized niche.**

In a classic study, J. H. Connell of the University of California, Santa Barbara, investigated competitive interactions between two species of barnacles that grow together on rocks along the coast of Scotland. Barnacles are marine animals (crustaceans) that have free-swimming larvae. The larvae eventually settle down, cementing themselves to rocks and remaining attached for the rest of their lives. Of the two species Connell studied, *Chthamalus stellatus* (the smaller barnacle in **figure 35.17**) lives in shallower water, where tidal action often exposes it to air, and *Semibalanus balanoides* (the larger barnacle) lives at lower depths, where it is rarely exposed to the atmosphere. In the deeper zone, *Semibalanus* could always outcompete *Chthamalus* by crowding it off the rocks, undercutting it, and replacing it even where it had begun to grow. When Connell removed *Semibalanus* from the area, however, *Chthamalus* was easily able to occupy the deeper zone, indicating that no physiological or other general obstacles prevented it from becoming established there. In contrast, *Semibalanus* could not survive in

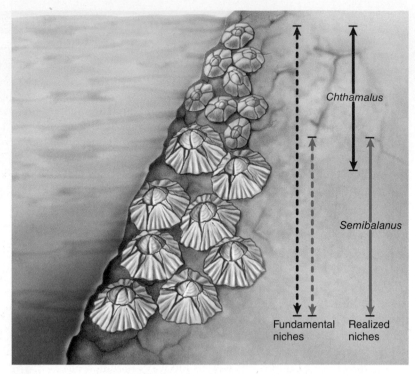

Figure 35.17 Competition among two species of barnacles limits niche use.

Chthamalus can live in both deep and shallow zones (its fundamental niche), but *Semibalanus* forces *Chthamalus* out of the part of its fundamental niche that overlaps the realized niche of *Semibalanus.*

the shallow-water habitats where *Chthamalus* normally occurs; it evidently does not have the special physiological and morphological adaptations that allow *Chthamalus* to occupy this zone. Thus, the fundamental niche of the barnacle *Chthamalus* in Connell's experiments in Scotland included that of *Semibalanus* (the red dashed arrow), but its realized niche was much narrower (the red solid arrow) because *Chthamalus* was outcompeted by *Semibalanus* in its fundamental niche.

Predators, as well as competitors, can limit the realized niche of a species. In the previous example, *Chthamalus* was able to fully occupy its fundamental niche when there was no competition. This is often the case when a species first enters a new, very favorable habitat that presents it with adequate resources, little or no competition, and no predators. However, once resources are limited, the population approaches carrying capacity, and other species begin to compete for the same resources. Also, predators may begin to more frequently recognize the species, and the population will be forced into its realized niche. For example, a plant called the St. John's wort was introduced and became widespread in open rangeland habitats in California. It occupied all of its fundamental niche until a species of beetle that feeds on the plant was introduced into the habitat. Populations of the plant then quickly decreased, and it is now only found in shady sites where the beetle cannot thrive.

Competitive Exclusion

In classic experiments carried out between 1934 and 1935, Russian ecologist G. F. Gause studied competition among three species of *Paramecium,* a tiny protist. All three species grew well alone in culture tubes (**figure 35.18a**), preying on bacteria and yeasts that fed on oatmeal suspended in the culture fluid. However, when Gause grew *P. aurelia* together with *P. caudatum* in the same culture tube (**figure 35.18b**), the numbers of *P. caudatum* (the green line) always declined to extinction, leaving *P. aurelia* the only survivor. Why? Gause found *P. aurelia* was able to grow six times faster than its competitor, *P. caudatum,* because it was able to better use the limited available resources.

From experiments such as this, Gause formulated what is now called the *principle of competitive exclusion*. This principle states that if two species are competing for a resource, the species that uses the resource more efficiently will eventually eliminate the other locally—no two species with the same niche can coexist.

Niche Overlap

In a revealing experiment, Gause challenged *P. caudatum*—the defeated species in his earlier experiments—with a third species, *P. bursaria.* Because he expected these two species to also compete for the limited bacterial food supply, Gause thought one would win out, as had happened in his previous experiments. But that's not what happened. Instead, both species survived in the culture tubes (**figure 35.18c**); the paramecia found a way to divide the food resources. How did they do it? In the upper part of the culture tubes, where the oxygen concentration and bacterial density were high, *P. caudatum* dominated because it was better able to feed on bacteria. However, in the lower part of the tubes, the lower oxygen concentration favored

the growth of a different potential food, yeast, and *P. bursaria* was better able to eat this food. The fundamental niche of each species was the whole culture tube, but the realized niche of each species was only a portion of the tube. This graph also demonstrates the negative effect competition had on the participants: Competition was always detrimental to both species involved. Both species more than doubled their densities when grown without a competitor as when grown together.

Gause's principle of competitive exclusion can be restated to say that no two species can occupy the same niche indefinitely when resources are limiting. Certainly species can and do coexist while competing for the same resources; we have seen many examples of such relationships. Nevertheless, Gause's theory predicts that when two species are able to coexist on a long-term basis, either resources must not be limited or their niches will always differ in one or more features; otherwise, one species outcompetes the other and the extinction of the second species inevitably results through competitive exclusion.

Niche is a complex concept, involving all facets of the environment that are important to individual species. In recent years, a vigorous debate has arisen concerning the role of competitive exclusion, not only in determining the structure of communities but also in setting the course of evolution. When resources are abundant, species may overlap substantially in their use. However, when one or more resources suddenly become sharply limiting, as in periods of drought, the role of competition becomes much more obvious. When niches overlap, two outcomes are possible: competitive exclusion (winner takes all), or *resource partitioning* (dividing up resources to create two realized niches). It is only through resource partitioning, the topic of the next section, that the species can continue to coexist over long periods.

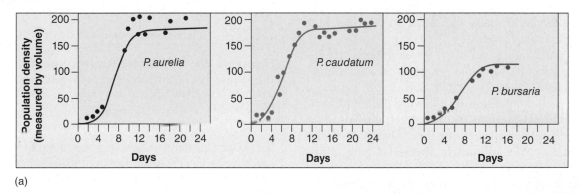

(a)

Figure 35.18 Competitive exclusion among three species of *Paramecium*.

In the microscopic world, *Paramecium* is a ferocious predator. Paramecia eat by ingesting their prey; their plasma membranes surround bacterial or yeast cells, forming a food vacuole containing the prey cell. In his experiments, (*a*) Gause found that three species of *Paramecium* grew well alone in culture tubes. (*b*) However, *P. caudatum* declined to extinction when grown with *P. aurelia* because they shared the same realized niche, and *P. aurelia* outcompeted *P. caudatum* for food resources. (*c*) *P. caudatum* and *P. bursaria* were able to coexist, although in smaller populations, because the two have different realized niches and thus avoid competition.

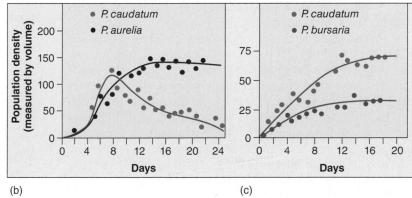

(b)   (c)

Figure 35.19 Resource partitioning among lizard species.

Species of *Anolis* lizards in the Caribbean partition their tree habitats in a variety of ways. Some species of anoles occupy the canopy of trees (*a*), others use twigs on the periphery (*b*), and still others are found at the base of the trunk (*c*). In addition, some use grassy areas in the open (*d*). This same pattern of resource partitioning has evolved independently on different Caribbean islands.

Resource Partitioning

Gause's exclusion principle has a very important consequence: Persistent and intense competition between two species is rare in natural communities. Either one species drives the other to extinction, or natural selection reduces the competition between them, such as through **resource partitioning**. In resource partitioning, species that live in the same geographical area avoid competition by living in different portions of the habitat or by using different food or other resources. A clear example of this is seen in *Anolis* lizards (**figure 35.19**), where species may live in different parts of a tree habitat to avoid competition for food and space with other species that may live on the twigs, trunks, or grass.

Resource partitioning can often be seen in closely related species that occupy the same geographical area. Called **sympatric species** (Greek, *syn,* same, and *patria,* country), these species avoid competition by evolving different adaptations to use different portions of the habitat, food or other resources. Closely related species that do not live in the same geographical area, called **allopatric species** (Greek, *allos,* other, and *patria,* country), often use the same habitat locations and food resources—because they are not in competition, natural selection does not favor evolutionary changes that subdivide their niche.

When a pair of closely related species occur in the same place, they tend to exhibit greater differences in morphology and behavior than the same two species do when living in different areas. Called **character displacement,** the differences evident between sympatric species are thought to have been favored by natural selection as a mechanism to facilitate resource partitioning and thus reduce competition. Character displacement can be seen clearly among Darwin's finches. The two Galápagos finches in **figure 35.20** have beaks of similar size when each is living on an island where the other does

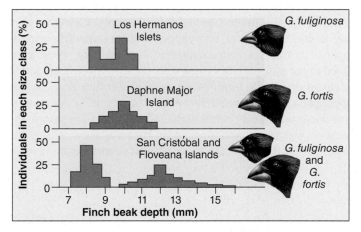

Figure 35.20 Character displacement.

These two species of Galápagos finches (genus *Geospiza*) have beaks of similar sizes when living apart, but different sizes when living together.

not occur. On islands where they are found living together, the two species have evolved beaks of different sizes, one adapted to larger seeds, the other to smaller ones. In essence, the two finches have subdivided the food niche, creating two new smaller niches. By partitioning the available food resources, the two species have avoided direct competition with each other, and so are able to live together in the same habitat.

> **Key Learning Outcome 35.9** A niche may be defined as the way in which an organism uses its environment. No two species can occupy the same niche indefinitely without competition driving one to extinction if resources are limiting. Sympatric species partition available resources, reducing competition between them.

Character Displacement Among Darwin's Finches

While evidence of character displacement can be seen in many laboratory populations, it has been difficult to observe the process in action in natural populations. Recently a particularly clear view of character displacement in action has been reported among finches on one of the smaller Galápagos islands, Daphne Major. For 33 years, researchers led by the husband-and-wife team of Peter and Rosemary Grant from Princeton University have marked each bird on the island, recording births and deaths, beak sizes, eating habits, and a wealth of other information. The seed-eating medium ground finch (*Geospiza fortis*) has lived on this island with essentially no competition for food from other finches. It shares the island only with the cactus finch, which uses its pointed beak to eat cactus fruit and pollen, and does not eat seeds. The absence of other seed-eating finches means that *G. fortis* could take maximum advantage of all seeds available on the island. While it prefers smaller, more tender seeds, when those are not available, *G. fortis* will eat larger, tougher seeds. During such difficult times, the *G. fortis* individuals with larger beaks are better able to crack open the large, tough seeds than smaller-beaked birds are, and so are more likely to survive and leave offspring.

As described in chapter 14, a severe drought in 1977 acted to weed out the *G. fortis* population on Daphne Major. The only food left for the medium ground finches was larger seeds. Most finches died that year on Daphne Major, but the birds with smaller beaks died in higher numbers.

The selection for larger beaks in a drought—beaks better able to break open the hard case of large, dry seeds—was precisely what evolutionary theory would predict, a clear example of directional selection.

However, a severe drought in 2003 and 2004 produced a dramatically different result! In the year immediately following that drought, the beak sizes in the population of medium ground finches showed a shift to smaller beaks, not larger ones. What had happened during the 2003/2004 drought that was different?

The answer, scientists found, was a new competitor. During the previous severe drought of 1977, the only competition that *G. fortis* experienced was from within its own species, between larger-beaked and smaller-beaked individuals. But in the drought of 2003 a new species of seed-

Geospiza magnirostris

eating finch was present on the island, vying for the same limited food resources as *G. fortis*.

How did this happen? Prior to 1982, the large ground finch *G. magnirostris* (seen in this photo) was sometimes a visitor to Daphne Major but never stayed and bred on the island. Heavy rains in 1982 changed all that. The abundant food on the island after the rains induced two females and three males to stay and start a breeding population. Their numbers grew substantially over the next 10 years. By 2004, the large ground finch—weighing 30 grams, almost twice the size of *G. fortis*—made up nearly 40% of the ground finch population on the island.

Then the hammer fell. The drought that year was devastating. No small, new seeds were produced by the island's drought-stricken plants, and the two species of ground finches soon exhausted the supply of larger, dry seeds left from previous years. The shortage nearly wiped out both populations. But, unlike in previous droughts, small-beaked *G. fortis* individuals fared better than large-beak birds of that species—only 13% of large-billed *G. fortis* birds survived.

Within a year, the average beak size of the *G. fortis* finch population on Daphne Major was considerably smaller than in the years prior to the drought, which was exactly opposite to the results following earlier droughts. Two droughts, two different results. Why? In the drought of 1977, *G. fortis* experienced selective pressure favoring large beak size, as this helped them eat the only food available—large, dry seeds. In the drought of 2003/2004, *G. fortis* experienced this same selective pressure, but also experienced a new and even stronger selective pressure exerted by a competitor that it had never had to face on Daphne Major in the past. The large ground finch, with its stout, large beak, was better adapted than *G. fortis* at cracking open the bigger seeds and ate most of the remaining large seeds. As a result, although pickings were slim, this time more small seeds were available than large seeds for *G. fortis*, and individuals with smaller beaks were more likely to survive. When the drought ended, *G. fortis* finches used these smaller beaks to eat smaller seeds, leaving the large seeds to *G. magnirostris*.

This competition-driven shift in beak size is a clear example of character displacement. Often observed in laboratory experiments, this study is the first time it has been documented in the wild.

35.10 Coevolution and Symbiosis

The previous section described the "winner take all" results of competition between two species whose niches overlap. Other relationships in nature are less competitive and more cooperative.

Coevolution

The plants, animals, protists, fungi, and prokaryotes that live together in communities have changed and adjusted to one another continually over millions of years. For example, many features of flowering plants have evolved in relation to the dispersal of the plant's gametes by animals (**figure 35.21**). These animals, in turn, have evolved a number of special traits that enable them to obtain food or other resources efficiently from the plants they visit, often from their flowers. In addition, the

seeds of many flowering plants have features that make them more likely to be dispersed to new areas of favorable habitat.

Such interactions, which involve the long-term, mutual evolutionary adjustment of the characteristics of the members of biological communities, are examples of **coevolution**. Coevolution is the adaptation of two or more species to each other. In this section, we consider the many ways species interact, some of which involve coevolution.

Symbiosis Is Widespread

In symbiotic relationships, two or more kinds of organisms live together in often elaborate and more or less permanent relationships. All symbiotic relationships carry the potential for coevolution between the organisms involved, and in many instances the results of this coevolution are fascinating. Examples of symbiosis include lichens, which are associations of certain fungi with green algae or cyanobacteria (see chapter 18). Other important examples are mycorrhizae, the associations between fungi and the roots of most kinds of plants. The fungi expedite the plant's absorption of certain nutrients, and the plants in turn provide the fungi with carbohydrates. Similarly, root nodules that occur in legumes and certain other kinds of plants contain bacteria that fix atmospheric nitrogen and make it available to their host plants.

The major kinds of symbiotic relationships include: (1) **mutualism**, in which both participating species benefit; (2) **parasitism**, in which one species benefits but the other is harmed; and (3) **commensalism**, in which one species benefits while the other neither benefits nor is harmed. Parasitism can also be viewed as a form of predation (discussed later), although the organism that is preyed upon does not necessarily die.

Mutualism

Mutualism is a symbiotic relationship among organisms in which both species benefit. Examples of mutualism are of fundamental importance in determining the structure of biological communities. Some of the most spectacular examples of mutualism occur among flowering plants and their animal visitors, including insects, birds, and bats. As we discussed in chapter 32, during the course of their evolution, the characteristics of flowers have evolved in large part in relation to the characteristics of the animals that visit them for food and, in doing so, spread their pollen from individual to individual. At the same time, characteristics of the animals have changed, increasing their specialization for obtaining food or other substances from particular kinds of flowers.

Another example of mutualism involves ants and aphids. Aphids, also called greenflies, are small insects that suck fluids with their piercing mouthparts from the phloem of living plants. They extract a certain amount of the sucrose and other nutrients from this fluid, but they excrete much of it in an altered form through their anus. Certain ants have taken advantage of this—in effect, domesticating the aphids

Figure 35.21 Pollination by bat.
Many flowering plants have coevolved with other species to facilitate pollen transfer. Insects are widely known as pollinators, but they're not the only ones. Notice the cargo of pollen on the bat's snout.

Figure 35.22 Mutualism: ants and aphids.
These ants are tending to aphids (small green organisms), feeding on the "honeydew" that the aphids excrete continuously, moving the aphids from place to place, and protecting them from potential predators.

(figure 35.22). The ants carry the aphids to new plants, where they come into contact with new sources of food, and then consume as food the "honeydew" that the aphids excrete.

Parasitism

Parasitism is a symbiotic relationship that may be regarded as a special form of predator/prey relationship, discussed later. In this symbiotic relationship the predator, or parasite, is much smaller than the prey, or host, and remains closely associated with it. Parasitism is harmful to the host organism and beneficial to the parasite, but unlike a predator/prey relationship, a parasite often does not kill its host. The concept of parasitism seems obvious, but individual instances are often surprisingly difficult to distinguish from predation and from other kinds of symbiosis.

External Parasites Parasites that feed on the exterior surface of an organism are external parasites, or **ectoparasites.** Lice, which live their entire lives on the bodies of vertebrates—mainly birds and mammals—are normally considered parasites. Mosquitoes are not considered parasites, even though they draw food from birds and mammals in a similar manner to lice, because their interaction with their host is so brief.

Internal Parasites Vertebrates are parasitized internally by **endoparasites,** members of many different phyla of animals and protists. Invertebrates also have many kinds of parasites that live within their bodies. Bacteria and viruses are not usually considered parasites, even though they fit our definition precisely.

Internal parasitism is generally marked by much more extreme specialization than external parasitism, as shown by the many protist and invertebrate parasites that infect humans. The structure of an internal parasite is often simplified, and unnecessary armaments and structures are lost as it evolves.

Commensalism

Commensalism is a symbiotic relationship that benefits one species and neither hurts nor helps the other. In nature, individuals of one species are often physically attached to members of another. For example, epiphytes are plants that grow on the branches of other plants. In general, the host plant is unharmed, and the epiphyte that grows on it benefits. Similarly, various marine animals, such as barnacles, grow on other, often actively moving, sea animals like whales and thus are carried passively from place to place without harming their hosts. These "passengers" presumably gain more protection from predation than they would if they were fixed in one place, and they also reach new sources of food. The increased water circulation that such animals receive as their host moves around may be of great importance, particularly if the passengers are filter-feeders.

Examples of Commensalism The best-known examples of commensalism involve the relationships between certain small tropical fishes and sea anemones, which are marine animals that have stinging tentacles (see chapter 19). Certain species of tropical fish have evolved the ability to live among the tentacles of sea anemones, even though these tentacles would quickly paralyze other fishes that touched them. The anemone fishes feed on the detritus left from the meals of the host anemone, remaining uninjured under remarkable circumstances.

On land, an analogous relationship exists between birds called oxpeckers and grazing animals such as cattle or rhinoceroses. The oxpecker birds spend most of their time clinging to the animals, picking off parasites and other insects, carrying out their entire life cycles in close association with the host animals.

When Is Commensalism Commensalism? In each of these instances, it is difficult to be certain whether the second partner receives a benefit or not; there is no clear-cut boundary between commensalism and mutualism. For instance, it may be advantageous to the sea anemone to have particles of food removed from its tentacles; it may then be better able to catch other prey. Similarly, the grazing animals may benefit from the relationship with the oxpeckers or cattle egrets if the bugs that are picked off of them are harmful, such as ticks or fleas. If so, then the relationship is mutualism. However, if the birds also pick at scabs, causing bleeding and possibly infections, the relationship may be parasitic. In true commensalism, only one of the partners benefits and the other neither benefits nor is harmed. If the grazing animals are not harmed by either the ticks that are eaten off of their bodies or by the oxpeckers that feed off of them, then it is an example of commensalism.

> **Key Learning Outcome 35.10** Coevolution is a term that describes the long-term evolutionary adjustments of species to one another. In symbiosis, two or more species live together. Mutualism involves cooperation between species, to the mutual benefit of both. In parasitism, one organism serves as a host to another organism, usually to the host's disadvantage. Commensalism is the benign use of one organism by another.

35.11 Predator-Prey Interactions

In the previous section, we considered parasitism, a symbiotic relationship, a specialized form of **predator-prey interaction** in which the predator is much smaller than its prey and does not generally kill it. **Predation** is the consuming of one organism by another, usually of a similar or larger size. In this sense, predation includes everything from a leopard capturing and eating an antelope to a whale grazing on microscopic ocean plankton.

In nature, predators often have large effects on prey populations. Some of the most dramatic examples involve situations in which humans have either added or eliminated predators from an area. For example, the elimination of large carnivores from much of the eastern United States has led to population explosions of white-tailed deer, which strip the habitat of all edible plant life within their reach. Similarly, when sea otters were hunted to near extinction on the western coast of the United States, populations of sea urchins, a principal prey item of the otters, exploded. Appearances, however, sometimes can be deceiving. On Isle Royale in Lake Superior, moose reached the island by crossing over ice in an unusually cold winter and multiplied freely there in isolation. When wolves later reached the island by crossing over the ice, naturalists widely assumed that the wolves were playing a key role in controlling the moose population. More careful studies have demonstrated that this is not in fact the case. The moose that the wolves eat are, for the most part, old or diseased animals that would not survive long anyway. In general, the moose are controlled by food availability, disease, and other factors rather than by the wolves (figure 35.23).

Predator-Prey Cycles

Population cycles are characteristic of some species of small mammals, such as lemmings, and they appear to be stimulated, at least in some situations, by their predators. Ecologists have studied cycles in hare populations since the 1920s (figure 35.24). They have found that the North American snowshoe hare, *Lepus americanus,* follows a "10-year cycle" (in reality, it varies from 8 to 11 years). Its numbers fall 10-fold to 30-fold in a typical cycle, and 100-fold changes can occur. Two factors appear to be generating the cycle: food plants and predators.

> **Key Learning Outcome 35.11** Prey populations can be affected by their predators. Some populations of predators and their prey oscillate in a cyclic manner.

Figure 35.23 Wolves chasing a moose—what will the outcome be?

On Isle Royale, Michigan, a large pack of wolves pursue a moose. They chased this moose for almost 2 kilometers; it then turned and faced the wolves, who by that time were exhausted from running through chest-deep snow. The wolves lay down, and the moose walked away.

(a)

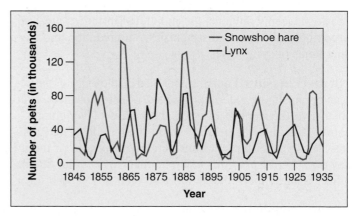

(b)

Figure 35.24 A predator-prey cycle.

(a) A snowshoe hare being chased by a lynx. (b) The numbers of lynxes and snowshoe hares oscillate in tune with each other in northern Canada. The data are based on numbers of animal pelts from 1845 to 1935. As the number of hares grows, so does the number of lynxes, with the cycle repeating about every 10 years. Both predators (lynxes) and available food resources control the number of hares. The number of lynxes is controlled by the availability of prey (snowshoe hares).

35.12 Mimicry

Different strategies have evolved among prey to deter predation. Some species use physical or chemical defenses. Organisms that contain toxins may advertise this fact with *warning (aposematic) coloration.* Interestingly, during the course of their evolution, many nonpoisonous animals have come to resemble distasteful or dangerous ones that exhibit aposematic coloration. Also protected species can mimic each other.

Batesian Mimicry

Batesian mimicry is named for Henry Bates, the nineteenth-century British naturalist who first brought this type of mimicry to general attention in 1857. In his journeys to the Amazon region of South America, Bates discovered many instances of

(a) Model

(b) Batesian mimic

Figure 35.25   A Batesian mimic.

(a) The model. Monarch butterflies (*Danaus plexippus*) are protected from birds and other predators by the cardiac glycosides they incorporate from the milkweeds and dogbanes they feed on as larvae. Adult monarch butterflies advertise their poisonous nature with warning coloration. (b) The mimic. Viceroy butterflies, *Limenitis archippus,* are Batesian mimics of the poisonous monarch. Although the viceroy is not related to the monarch, it looks a lot like it, so predators that have learned not to eat distasteful monarchs avoid viceroys, too.

palatable insects that resembled brightly colored, distasteful species. He reasoned that the mimics are avoided by predators, who are fooled by the disguise into thinking the mimic actually is the distasteful model.

Many of the best-known examples of Batesian mimicry occur among butterflies and moths. Obviously, predators in systems of this kind must use visual cues to hunt for their prey; otherwise, similar color patterns would not matter to potential predators. There is also increasing evidence indicating that Batesian mimicry can also involve nonvisual cues, such as olfaction, although such examples are less obvious to humans.

The kinds of butterflies that provide the models in Batesian mimicry are members of groups whose caterpillars feed on one or a few closely related plant families that are strongly protected by toxic chemicals. The model butterflies incorporate the poisonous molecules from these plants into their bodies. The mimic butterflies, in contrast, belong to groups in which the feeding habits of the caterpillars are not so restricted. As caterpillars, these butterflies feed on a number of different plant families unprotected by toxic chemicals.

One often-studied mimic among North American butterflies is the viceroy, *Limenitis archippus* (**figure 35.25b**). This butterfly, which resembles the poisonous monarch (in **figure 35.25a**), ranges from central Canada through much of the United States and into Mexico. The caterpillars feed on willows and cottonwoods, and neither caterpillars nor adults were thought to be distasteful to birds, although recent findings may dispute this. Interestingly, the Batesian mimicry seen in the adult viceroy butterfly does not extend to the caterpillars: Viceroy caterpillars are camouflaged on leaves, resembling bird droppings, whereas the monarch's distasteful caterpillars are very conspicuous.

Batesian mimicry also occurs in vertebrates. Probably the most famous case is the scarlet king snake, whose red, black, and yellow bands mimic those of the venomous coral snake.

Müllerian Mimicry

Another kind of mimicry, **Müllerian mimicry,** was named for German biologist Fritz Müller, who first described it in 1878. In Müllerian mimicry, several unrelated but protected animal species come to resemble one another. Thus, different kinds of stinging wasps have yellow-and-black-striped abdomens, but they may not all be descended from a common yellow-and-black-striped ancestor. In general, yellow-and-black and bright red tend to be common color patterns that warn predators relying on vision. If animals that are all poisonous or dangerous resemble one another, they gain an advantage because a predator learns more quickly to avoid them.

> **Key Learning Outcome 35.12**   In Batesian mimicry, unprotected species resemble others that are distasteful. Both species exhibit aposematic coloration. In Müllerian mimicry, two or more unrelated but protected species resemble one another, thus achieving a kind of group defense.

Today's Biology

Invasion of the Killer Bees

One of the harshest lessons of environmental biology is that the unexpected does happen. Precisely because science operates at the edge of what we know, scientists sometimes stumble over the unexpected. This lesson has been brought clearly to mind in recent years, with reports of killer bees being discovered east of the Mississippi River.

In the continental United States, we are used to the mild-mannered European honeybee, a subspecies called *Apis mellifera mellifera.* The African variety, subspecies *A. m. scutelar,* looks very much like them, but they are hardly mild mannered. They are in fact very aggressive critters with a chip on their shoulder, and when swarming, they do not have to be provoked to start trouble.

A few individuals may see you at a distance, "lose it," and lead thousands of bees in a concerted attempt to do you in. The only thing you can do is run—fast. They will keep after you for up to a mile. It doesn't do any good to duck under water, as they just wait for you.

They are nicknamed "killer" bees not because any one sting is worse than the European kind, but rather because so many of the bees try to sting you. An average human can survive no more than 300 bee stings. A horrible total of more than 8,000 bee stings is not unusual for a killer bee attack. An American graduate student attacked and killed in a Costa Rican jungle had 10,000 stings.

Killer bees remind us of the "law of unintended consequences" because their invasion of this continent is the direct result of scientists stumbling over the unexpected.

Killer bees were brought from Africa to Brazil 47 years ago by a prominent Brazilian scientist, Warwick Estevan Kerr. A famous geneticist, Kerr is the only Brazilian to be a member of the U.S. National Academy of Sciences. What he was doing in 1956, at the request of the Brazilian government, was attempting to establish tropical bees in Brazil to expand the commercial bee industry (bees pollinate crops and make commercial honey in the process). African bees seemed ideal candidates, better adapted to the tropics than European bees and more prolific honey producers.

Kerr established a quarantine colony of African bees at a remote field station outside the city of Rio Claro, several hundred miles from his university at São Paulo.

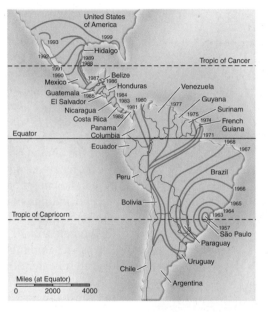

Although he had brought back many queens from South Africa, the colony came to be dominated by the offspring of a single very productive queen from Tanzania. Kerr noted at the time that she appeared unusually aggressive.

In the fall of 1957, the field station was visited by a beekeeper. As no one else was around that day, the visitor performed the routine courtesy of tending the hives. A hive with a queen in it has a set of bars across the door so the queen can't get out. Called a "queen excluder," the bars are far enough apart that the smaller worker bees can squeeze through. Once a queen starts to lay eggs, she never leaves the hive, so there is little point in slowing down entry of workers to the hive, and the queen excluder is routinely removed. On that day, the visitor saw the Tanzanian queen laying eggs in the African colony hive, and so removed the queen excluder.

One and a half days later, when a staff member inspected the African colony, the Tanzanian queen and 26 of her daughter queens had decamped. Out into the neighboring forest they went. And that's what was unexpected. European queen bees never leave the hive after they have started to lay eggs. No one could have guessed that the Tanzanian queen would behave differently. But she did.

By 1970 the superaggressive African bees had blanketed Brazil, totally replacing local colonies. They reached Central America by 1980, Mexico by 1986, and Texas by 1990, having conquered 5 million square miles in 33 years. In the process they killed an estimated 1,000 people and over 100,000 cows. The first American to be killed, a rancher named Lino Lopez, died of multiple stings in Texas in 1993.

All during the following decades, the bees have continued their invasion of the United States. All of Arizona and much of Texas has been occupied. A few years ago, Los Angeles county was officially declared colonized, and it looks like African bees will eventually move at least halfway up the state of California.

Soon, however, the invasion is predicted to cease, on a line roughly from San Francisco, California, to Richmond, Virginia. Winter cold is expected to limit any further northward advance of the invading hoards.

For the states below this line, sure as the sun rises, the bees are coming, not caring one bit that they were unanticipated. The deep lesson—that unexpected things do happen—is being driven home by millions of tiny, aggressive teachers.

35.13 Ecological Succession

Competition, predation, and cooperation often produce dramatic changes in communities. This results in changes in ecosystems, by the orderly replacement of one community with another, from simple to complex in a process known as **succession.** This process is familiar to anyone who has seen a vacant lot or cleared woods slowly become occupied by an increasing number of plants, or a pond become dry land as it is filled with vegetation encroaching from the sides.

Secondary Succession

If a wooded area is cleared and left alone, plants slowly reclaim the area. Eventually, traces of the clearing disappear and the area is again woods. Similarly, intense flooding may clear a stream bed of many organisms, leaving mostly sand and rock; afterward, the bed is progressively reinhabited by protists, invertebrates, and other aquatic organisms. This kind of succession, which occurs in areas where an existing community has been disturbed, is called **secondary succession.**

Primary Succession

In contrast, **primary succession** occurs on bare, lifeless substrate, such as rocks. Primary succession occurs in lakes left behind after the retreat of glaciers, on volcanic islands that rise above the sea, and on land exposed by retreating glaciers. Primary succession on glacial moraines provides an example. The graph in **figure 35.26** shows how the concentration of nitrogen in the soil changes as primary succession occurs. On bare, mineral-poor soil, lichens grow first, forming small pockets of soil. Acidic secretions from the lichens help to break down the substrate and add to the accumulation of soil. Mosses then colonize these pockets of soil (**figure 35.26a**), eventually building up enough nutrients in the soil for alder shrubs to take hold (**figure 35.26b**). These first plants to appear form a **pioneering community.** Over 100 years, the alders (**figure 35.26c** and *inset photo*) build up the soil nitrogen levels until spruce are able to thrive, eventually crowding out the alder and forming a dense spruce forest (**figure 35.26d**).

Primary successions end with a community called a **climax community,** whose populations remain relatively stable and are characteristic of the region as a whole. However, because local climate keeps changing, the process of succession is often very slow, and human activities have a major impact; many successions do not reach climax.

Why Succession Happens

Succession happens because species alter the habitat and the resources available in it, often in ways that favor other species. Three dynamic concepts are of critical importance in the process: tolerance, facilitation, and inhibition.

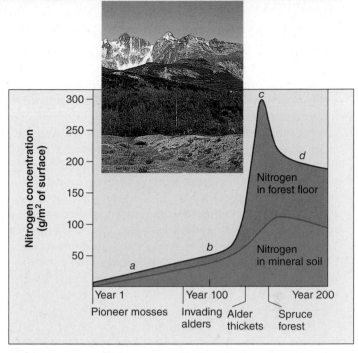

Figure 35.26 Plant succession produces progressive changes in the soil.

Initially the glacial moraine at Glacier Bay, Alaska, had little soil nitrogen, but nitrogen-fixing alders (photo above) led to a buildup of nitrogen in the soil, encouraging the subsequent growth of the conifer forest.

1. **Tolerance.** Early successional stages are characterized by weedy *r*-selected species that do not compete well in established communities but are tolerant of the harsh, abiotic conditions in barren areas.
2. **Facilitation.** The weedy early successional stages introduce local changes in the habitat that favor other, less weedy species. Thus, the mosses in the Glacier Bay succession of **figure 35.26** fix nitrogen, which allows alders to invade. The alders in turn lower soil pH as their fallen leaves decompose, allowing spruce and hemlock, which require acidic soil, to invade.
3. **Inhibition.** Sometimes the changes in the habitat caused by one species, while favoring other species, inhibit the growth of the species that caused them. Alders, for example, do not grow as well in acidic soil as the spruce and hemlock that replace them.

As ecosystems mature, and more *K*-selected species replace *r*-selected ones, species richness and total biomass increase but net productivity decreases. Because earlier successional stages are more productive than later ones, agricultural systems are intentionally maintained in early successional stages to keep net productivity high.

> **Key Learning Outcome 35.13** In succession, communities change through time, often in a predictable sequence.

Are Island Populations of Song Sparrows Density Dependent?

When island populations are isolated, receiving no visitors from other populations, they provide an attractive opportunity to test the degree to which a population's growth rate is affected by its size. A population's size can influence the rate at which it grows because increased numbers of individuals within a population tend to deplete available resources, leading to an increased risk of death by deprivation. Also, predators tend to focus their attention on common prey, resulting in increasing rates of mortality as populations grow. However, simply knowing that a population is decreasing in numbers does not tell you that the decrease has been caused by the size of the population. Many factors, such as severe weather, volcanic eruption, and human disturbance, can influence island population sizes too.

The graph to the right displays data collected from 13 song sparrow populations on Mandarte Island (see map below). In an attempt to gauge the impact of population size on the evolutionary success of these populations, each population was censused, and its juvenile mortality rate estimated. On the graph, these juvenile mortality rates have been plotted against the number of breeding adults in each population. Although the data appear scattered, the "best-fit" regression line is statistically significant (**statistically significant** means that there is a less than 5% chance that there is, in fact, no correlation between dependent and independent variables).

1. **Applying Concepts**
   a. Variable. In the graph, what is the dependent variable?
   b. Analyzing Scattered Data. What is the size of the song sparrow population (based on breeding adults) with the least juvenile mortality? with the most?

2. **Interpreting Data**
   a. What is the average juvenile mortality of all 13 populations, estimated from the 13 points on the graph?

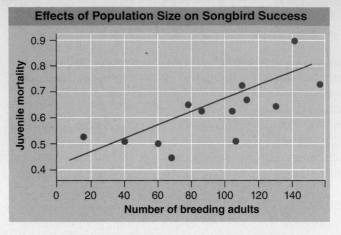

Effects of Population Size on Songbird Success

y-axis: Juvenile mortality; x-axis: Number of breeding adults

b. How many populations were observed to have juvenile mortality rates *below* this average value? What is the average size of these populations?
c. How many populations were observed to have juvenile mortality rates *above* this average value? What is the average size of these populations?

3. **Making Inferences** Are the populations with lower juvenile mortality bigger or smaller than the populations with higher juvenile mortality?

4. **Drawing Conclusions** Do the population sizes of these song sparrows appear to exhibit density dependence?

5. **Further Analysis**
   a. The fact that the song sparrow populations with lower juvenile mortality are a different size than those with higher juvenile mortality does not, in itself, establish that the difference is statistically significant. How would you go about testing these data to see if the relationship between juvenile mortality and population size is real?
   b. What would you expect to happen if the researchers supplemented the food available to the birds? Explain.
   c. What would you expect to happen if the researchers removed individuals from populations with more than 100 breeding adults, reducing each to 100?

Ecology

35.1 What Is Ecology?

- Ecology is the study of how organisms interact with each other and with their physical environment. There are six levels of ecological organization: populations, species, communities, ecosystems, biomes, and the biosphere (**figure 35.1**).

- Individual organisms must cope with a range of environmental conditions (**figures 35.2** and **35.3**).

Populations

35.2 Population Range

- A population is a group of individuals of the same species that live together and influence each other's survival. The area a population occupies, the population range, can change in response to environmental changes or as a result of migration to previously unavailable habitats (**figures 35.5** and **35.6**).

35.3 Population Distribution

- Availability of resources largely determines how individuals are distributed within populations (**figure 35.7**).

35.4 Population Growth

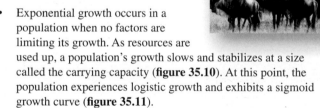

- Population size, density, and growth are other key characteristics of populations (**figure 35.9**).

- Exponential growth occurs in a population when no factors are limiting its growth. As resources are used up, a population's growth slows and stabilizes at a size called the carrying capacity (**figure 35.10**). At this point, the population experiences logistic growth and exhibits a sigmoid growth curve (**figure 35.11**).

35.5 The Influence of Population Density

- Factors such as weather and physical disruptions are density-independent effects and act on population growth regardless of population size.

- Density-dependent effects are factors, such as resources, that are affected by increases in population size. As resources are used up, individuals die off, reducing the size of the population (**figure 35.12**). A population is less affected by losses during the rising portion of the sigmoid growth curve, a point called the maximal sustainable yield (**figure 35.13**).

35.6 Life History Adaptations

- Populations whose resources are abundant experience little competition and reproduce rapidly; these organisms exhibit *r*-selected adaptations. Populations that experience competition over limited resources tend to be more reproductively efficient and exhibit *K*-selected adaptations (**table 35.1**).

35.7 Population Demography

- Population demography is the statistical study of populations, and often predictions can be made about population sizes in the future. Survivorship curves illustrate the impact of mortality rates among different age groups in a population (**figure 35.15**).

How Competition Shapes Communities

35.8 Communities

- The array of organisms that live together in an area is called a community. These individuals compete and cooperate with each other to make the community stable (**figure 35.16**).

35.9 The Niche and Competition

- A niche is the way an organism uses all available resources in its environment. Competition limits an organism from using its entire niche (**figure 35.17**). Two species cannot use the same niche; one will either out-compete the other, driving it to extinction, called competitive exclusion (**figure 35.18a,b**), or they will divide the niche into two smaller niches, called resource partitioning (**figures 35.18c** and **35.19**).

- As each species adapts to its portion of the niche, resource partitioning can affect morphological characteristics, called character displacement (**figure 35.20**).

Species Interactions

35.10 Coevolution and Symbiosis

- Coevolution is the adaptation of two or more species to each other. Symbiotic relationships involve two or more organisms of different species that live together and form a somewhat permanent relationship. Symbiotic relationships, such as lichens and mycorrhizae, can lead to coevolution. The major kinds of symbioses include mutualism (**figure 35.22**), parasitism, and commensalism.

35.11 Predator-Prey Interactions

- In predator-prey relationships, the predator kills and consumes the prey. Sometimes, in the absence of a predator, a prey population can grow rapidly. Other times, relationships between predators and prey are more complicated (**figure 35.23**).

- Predator and prey populations often exhibit cycles, the prey population being hunted to a low number, which begins to negatively affect predator population size (**figure 35.24**).

35.12 Mimicry

- Mimicry occurs when one organism takes advantage of the warning coloration of another organism. Batesian mimicry occurs when a harmless species has come to resemble a harmful species and so is avoided (**figure 35.25**). In Müllerian mimicry, a group of harmful species has a similar warning coloration pattern.

Community Stability

35.13 Ecological Succession

- Succession is the replacement of one community with another. Secondary succession occurs following the disturbance of an existing community, and primary succession is the emergence of a pioneering community where no life existed before (**figure 35.26**).

1. In the levels of ecological organization, the lowest level, composed of individuals of a single species who live near each other, share the same resources, and can potentially mate, is a(n)
   a. population.
   b. community
   c. ecosystem.
   d. biome.

2. When the number of organisms in a population remains more or less the same over time in the specific place where these organisms live, it is said that this population of organisms has reached its
   a. dispersion.
   b. biotic potential.
   c. carrying capacity.
   d. population density.

3. Which of the following is a density-dependent effect on a population?
   a. earthquake
   b. increased competition for food
   c. habitat destruction by humans
   d. seasonal flooding

4. Which of the following traits is *not* a characteristic of an organism that has *K*-selected adaptations?
   a. short life span
   b. few offspring per breeding season
   c. extensive parental care of offspring
   d. low mortality rate

5. If the age structure of a population shows more older organisms than younger organisms, then the fecundity
   a. will increase, and the mortality will decrease.
   b. will decrease, and the mortality will increase.
   c. and the mortality will be equal.
   d. and the mortality will not change.

6. All the organisms that live in the same location make up a(n)
   a. biome.
   b. population.
   c. ecosystem.
   d. community.

7. For similar species to occupy the same space, their niches must be different in some way. One way for these species to both survive is through
   a. competitive exclusion.
   b. interspecific competition.
   c. resource partitioning.
   d. intraspecific competition.

8. A relationship between two species where one species benefits and the other is neither hurt nor helped is known as
   a. parasitism.
   b. commensalism.
   c. mutualism.
   d. competition.

9. The yellow-and-black-striped patterns of many kinds of stinging wasps are examples of
   a. parasitism.
   b. Müllerian mimicry.
   c. commensalism
   d. Batesian mimicry.

10. Succession that occurs on abandoned agricultural fields is best described as
    a. coevolution.
    b. primary succession.
    c. secondary succession.
    d. prairie succession.

1. **Figure 35.10** What factors in an environment could cause the carrying capacity of a population to decrease? Explain.

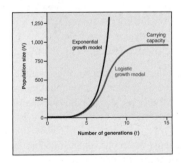

2. **Figure 35.17** How would the realized niche change for *Chthamalus* if *Semibalanus* was removed from the community? How would the realized niche change for *Semibalanus* if *Chthamalus* was removed from the community?

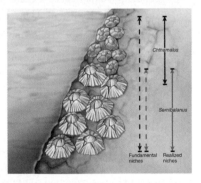

1. Give at least two examples, from your own area, of limiting factors on a population that are density-dependent factors, and two that are density-independent factors on the same population.

2. There are many examples of coevolution between flowers and their pollinators. A group of flowers, often referred to as carrion flowers, smell like a dead, rotting animal. What insects do you think pollinate these flowers? Explain why you think so.

3. Many U.S. communities struggle with issues of deer overpopulation. Even in dense housing developments, people complain that deer eat all the flowers in their gardens; also traffic accidents involving hitting deer are increasingly common on suburb streets. Explain the cause of this problem. What sort of remedy would you propose?

4. Why are climax communities rare?

36

Ecosystems

The earth provides living organisms with much more than a place to stand or swim. Many chemicals cycle between our bodies and the physical environment around us. We live in a delicate balance with our physical surroundings, one easily disturbed by human activities. The collection of organisms that live in a place, and all the physical aspects of the environment that affect how they live, operate as a fundamental biological unit, the ecological system or ecosystem. This mountain meadow, in the High Sierras of California, is an ecosystem, and so is the desert of Death Valley. All of the earth's surface—mountains and deserts and deep-sea floors—is teeming with life, although it may not always appear so. The same ecological principles apply to the organization of all the earth's communities, both on land and in the sea, although the details may differ greatly. Ecology is the study of ecosystems and the organisms that live within them. In this chapter, we focus on principles that govern the functioning of communities of organisms living together, and on the physical and biological factors that determine why particular kinds of organisms can be found living in one place and not in another. A proper understanding of how ecosystems function will be critical to preserving the living world in this new century.

36.1 Energy Flows Through Ecosystems

What Is an Ecosystem?

The ecosystem is the most complex level of biological organization. Collectively, the organisms in ecosystems regulate the capture and expenditure of energy and the cycling of chemicals. All organisms depend on the ability of other organisms—plants, algae, and some bacteria—to recycle the basic components of life.

Ecologists, the scientists who study ecology, view the world as a patchwork quilt of different environments, all bordering on and interacting with one another. Consider for a moment a patch of forest, the sort of place a deer might live. Ecologists call the collection of creatures that live in a particular place a **community**—all the animals, plants, fungi, and microorganisms that live together in a forest, for example, are the forest community. Ecologists call the place where a community lives its **habitat**—the soil and the water flowing through it are key components of a forest habitat. The sum of these two, community and habitat, is an ecological system, or **ecosystem.** An ecosystem is a largely self-sustaining collection of organisms and their physical environment. An ecosystem can be as large as a forest or as small as a tidepool.

The Path of Energy: Who Eats Whom in Ecosystems

Energy flows into the biological world from the sun, which shines a constant beam of light on our earth. Life exists on earth because some of that continual flow of light energy can be captured and transformed into chemical energy through the process of photosynthesis and used to make organic molecules such as carbohydrates, nucleic acids, proteins, and fats. These organic molecules are what we call food. Living organisms use the energy in food to make new materials for growth, to repair damaged tissues, to reproduce, and to do myriad other things that require energy, like turning the pages of this text.

You can think of all the organisms in an ecosystem as chemical machines fueled by energy captured in photosynthesis. The organisms that first capture the energy, the **producers,** are plants, algae, and some bacteria, which produce their own energy-storing molecules by carrying out photosynthesis. They are also referred to as *autotrophs.* All other organisms in an ecosystem are **consumers,** obtaining their energy-storing molecules by consuming plants or other animals, and are referred to as *heterotrophs.* Ecologists

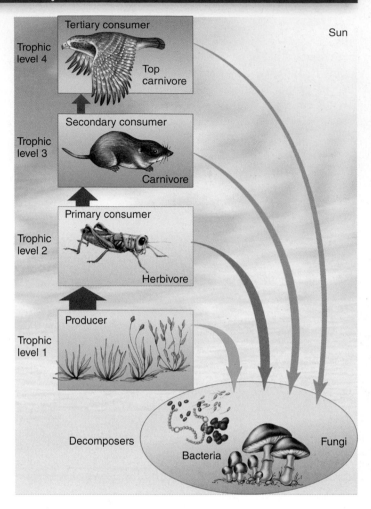

Figure 36.1 Trophic levels within an ecosystem.
Ecologists assign all the members of a community to various trophic levels based on feeding relationships.

assign every organism in an ecosystem to a trophic (or feeding) level, depending on the source of its energy. A **trophic level** is composed of those organisms within an ecosystem whose source of energy is the same number of consumption "steps" away from the sun. Thus, as shown in **figure 36.1**, a plant's trophic level is 1, while *herbivores* (animals that graze on plants) are in trophic level 2, and *carnivores* (animals that eat these grazers) are in trophic level 3. Higher trophic levels exist for animals that eat higher on the food chain (like the top carnivore in **figure 36.1**). Food energy passes through an ecosystem from one trophic level to another. When the path is a simple linear progression, like the links of a chain, it is called a **food chain.** The chain ends with *decomposers,* who break down dead organisms, or their excretions, and return the organic matter to the soil.

Producers

The lowest trophic level of any ecosystem is occupied by the producers—green plants in most land ecosystems (and, usually, algae in freshwater). Plants use the energy of the sun to build energy-rich sugar molecules. They also absorb carbon dioxide from the air, and nitrogen and other key substances from the soil, and use them to build biological molecules. It is important to realize that plants consume as well as produce. The roots of a plant, for example, do not carry out photosynthesis—there is no sunlight underground. Roots obtain their energy the same way you do, by using energy-storing molecules produced elsewhere (in this case, in the leaves of the plant).

Producers and herbivores

Herbivores

At the second trophic level are **herbivores,** animals that eat plants. They are the *primary consumers* of ecosystems. Deer and zebras are herbivores, and so are rhinoceroses, chickens (primarily herbivores), and caterpillars. Most herbivores rely on "helpers" to aid in the digestion of cellulose, a structural material found in plants. A cow, for instance, has a thriving colony of bacteria in its gut that digests cellulose for it. So does a termite. Humans cannot digest cellulose because we lack these bacteria—that is why a cow can live on a diet of grass and you cannot.

Carnivores

Carnivores

At the third trophic level are animals that eat herbivores, called **carnivores** (meat-eaters). They are the *secondary consumers* of ecosystems. Tigers and wolves are carnivores, and so are mosquitoes and blue jays. Some animals, like bears and humans, eat both plants and animals and are called **omnivores.** They use the simple sugars and starches stored in plants as food and not the cellulose. Many complex ecosystems contain a fourth trophic level, composed of animals that consume other carnivores. They are called *tertiary consumers,* or *top carnivores.* A weasel that eats a blue jay is a tertiary consumer. Only rarely do ecosystems contain more than four trophic levels, for reasons we will discuss later.

Omnivore

Detritivores and Decomposers

Detritivore

In every ecosystem there is a special class of consumers that include **detritivores,** organisms that eat dead organisms (also referred to as scavengers) and **decomposers,** organisms that break down organic substances making the nutrients available to other organisms. They obtain their energy from all trophic levels. Worms, crabs, and vultures are examples of detritivores. Bacteria and fungi are the principal decomposers in land ecosystems.

Decomposer

Energy Flows Through Trophic Levels

How much energy passes through an ecosystem? **Primary productivity** is the total amount of light energy converted by photosynthetic organisms into organic compounds in a given area per unit of time. An ecosystem's **net primary productivity** is the total amount of energy fixed by photosynthesis per unit of time, minus that which is expended by photosynthetic organisms to fuel metabolic activities. In short, it is the energy stored in organic compounds that is available to heterotrophs. The total weight of all ecosystem organisms, called the ecosystem's **biomass,** increases as a result of the ecosystem's net productivity. Some ecosystems, such as cattail swamps, which are wetlands, have a high net primary productivity. Others, such as tropical rain forests, also have a relatively high net primary productivity, but a rain forest has a much larger biomass than a wetlands area. Consequently, a rain forest's net primary productivity is much lower in relation to its biomass.

When a plant uses the energy from sunlight to make structural molecules such as cellulose, it loses a lot of the energy as heat. In fact, only about half of the energy captured by the plant ends up stored in its molecules. The other half of the energy is lost. This is the first of many such losses as the energy passes through the ecosystem. When the energy flow through an ecosystem is measured at each trophic level, we find that 80% to 95% of the energy available at one trophic level is not transferred to the next. In other words, only 5% to 20% of the available energy passes from one trophic level to the next. For example, the amount of energy that ends up in the beetle's body in **figure 36.2** is approximately only 17% of the energy present in the plant molecules it eats. Similarly, when a carnivore eats the herbivore, a comparable amount of energy is lost from the amount of energy present in the herbivore's molecules. This is why food chains generally consist of only three or four steps. So much energy is lost at each step that very little usable energy remains in the system after

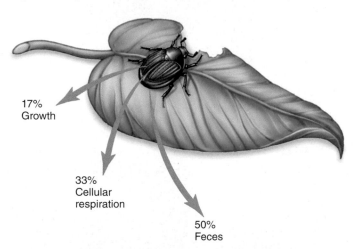

Figure 36.2 How heterotrophs use food energy.

A heterotroph assimilates only a fraction of the energy it consumes. For example, if a "bite" is composed of 500 Joules of energy (1 Joule = 0.239 calories), about 50%, 250 J, is lost in feces; about 33%, 165 J, is used to fuel cellular respiration; and about 17%, 85 J, is converted into consumer biomass. Only this 85 J is available to the next trophic level.

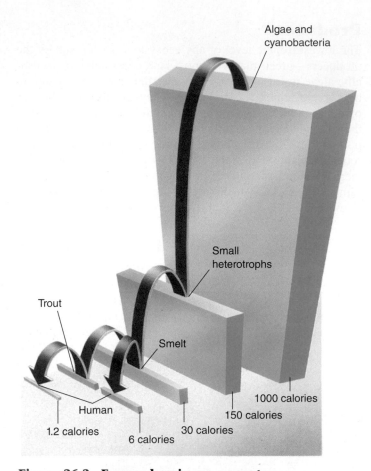

Figure 36.3 Energy loss in an ecosystem.

In a classic study of Cayuga Lake in New York, the path of energy was measured precisely at all points in the food web.

it has been incorporated into the bodies of organisms at four successive trophic levels.

Lamont Cole of Cornell University studied the flow of energy in a freshwater ecosystem in Cayuga Lake in upstate New York. In **figure 36.3**, each block represents the energy obtained by a different trophic level, with the producers, the algae and cyanobacteria, being the largest block. He calculated that about 150 of each 1,000 calories of potential energy fixed by algae and cyanobacteria are transferred into the bodies of animal plankton (small heterotrophs). Of these, about 30 calories are incorporated into the bodies of a type of small fish called a smelt, the principal secondary consumers of the system. If humans eat the smelt, they gain about 6 of the 1,000 calories that originally entered the system. If trout eat the smelt and humans eat the trout, humans gain only about 1.2 of the 1,000 calories. Thus, in most ecosystems, the path of energy is not a simple linear one because individual animals often feed at several trophic levels. This creates a more complicated path of energy flow called a **food web,** as shown in **figure 36.4.**

Key Learning Outcome 36.1 Energy moves through ecosystems from producers, to herbivores, to carnivores, and finally to detritivores and decomposers, which consume the dead bodies of all the others. Much energy is lost at each stage of a food chain.

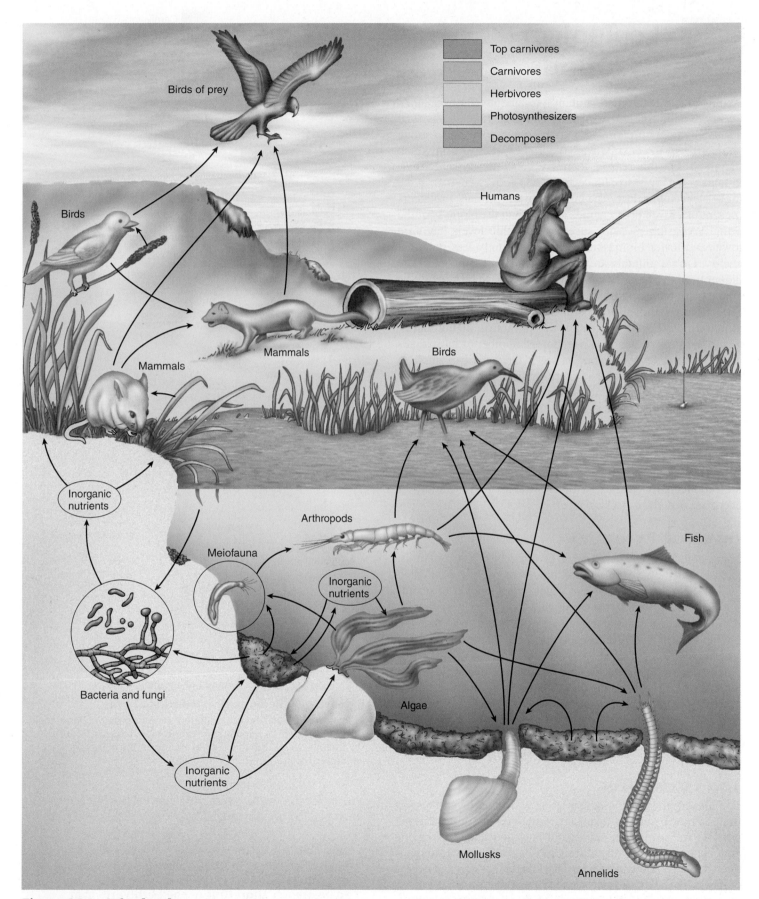

Figure 36.4 A food web.

A food web is much more complicated than a linear food chain. The path of energy passes from one trophic level to another and back again in complex ways.

Legend:
- Top carnivores
- Carnivores
- Herbivores
- Photosynthesizers
- Decomposers

Labels in figure: Birds of prey, Birds, Mammals, Humans, Birds, Mammals, Inorganic nutrients, Arthropods, Meiofauna, Inorganic nutrients, Fish, Bacteria and fungi, Algae, Inorganic nutrients, Mollusks, Annelids

36.2 Ecological Pyramids

As previously mentioned, a plant fixes about 1% of the sun's energy that falls on its green parts. The successive members of a food chain, in turn, process into their own bodies on average about 10% of the energy available in the organisms on which they feed. For this reason, there are generally far more individuals at the lower trophic levels of any ecosystem than at the higher levels. Similarly, the biomass of the primary producers present in a given ecosystem is greater than the biomass of the primary consumers, with successive trophic levels having a lower and lower biomass and correspondingly less potential energy. Larger animals are characteristically members of the higher trophic levels. To some extent, they must be larger to be able to capture enough prey in the lower trophic levels.

These relationships, if shown diagrammatically, appear as pyramids. Ecologists speak of "pyramids of numbers," where the sizes of the blocks reflect the number of individuals at each trophic level. As you can see in the aquatic example in **figure 36.5a,** the producers in the green box represent the largest number of individuals. Similarly, the producers (plankton) represent the largest group in "pyramids of biomass," shown in **figure 36.5b** where the sizes of the blocks reflect the total weight of all the organisms at each trophic level. The inverted pyramid in **figure 36.5c** is an exception and is discussed below. The "pyramid of energy" in **figure 36.5d** shows the producers as the largest block, reflecting the amount of energy stored at this trophic level.

Inverted Pyramids

Some aquatic ecosystems have inverted biomass pyramids, like in **figure 36.5c.** In a planktonic ecosystem—dominated by small organisms floating in water—the turnover of photosynthetic phytoplankton at the lowest level is very rapid, with zooplankton consuming phytoplankton so quickly that the phytoplankton (the producers at the base of the food chain) can never develop a large population size. Because the phytoplankton reproduce very rapidly, the community can support a population of heterotrophs that is larger in biomass and more numerous than the phytoplankton. However, don't confuse the sizes of these bars with the energy present at each level. The zooplankton that eat the phytoplankton are present in greater numbers but they contain about only 10% of the energy.

Top Carnivores

The loss of energy that occurs at each trophic level places a limit on how many top-level carnivores a community can support. As we have seen, only about one-thousandth of the energy captured by photosynthesis passes all the way through a three-stage food chain to a tertiary consumer such as a snake or hawk. This explains why there are no predators that subsist on lions or eagles—the biomass of these animals is simply insufficient to support another trophic level.

In the pyramid of numbers, top-level predators tend to be fairly large animals. Thus, the small residual biomass available at the top of the pyramid is concentrated in a relatively small number of individuals.

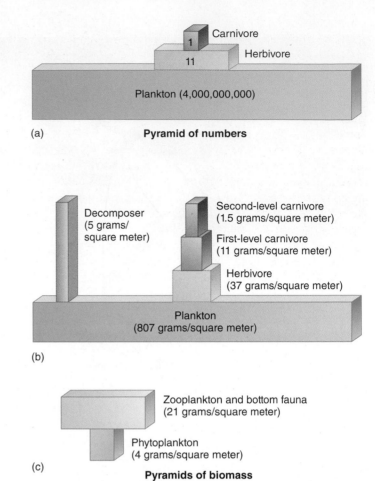

(a) **Pyramid of numbers**

Carnivore: 1
Herbivore: 11
Plankton (4,000,000,000)

(b)

Decomposer (5 grams/square meter)
Second-level carnivore (1.5 grams/square meter)
First-level carnivore (11 grams/square meter)
Herbivore (37 grams/square meter)
Plankton (807 grams/square meter)

(c) **Pyramids of biomass**

Zooplankton and bottom fauna (21 grams/square meter)
Phytoplankton (4 grams/square meter)

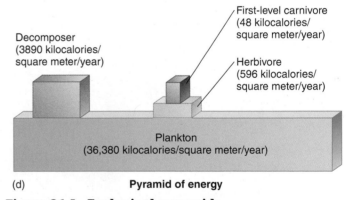

(d) **Pyramid of energy**

First-level carnivore (48 kilocalories/square meter/year)
Decomposer (3890 kilocalories/square meter/year)
Herbivore (596 kilocalories/square meter/year)
Plankton (36,380 kilocalories/square meter/year)

Figure 36.5 Ecological pyramids.

Ecological pyramids measure different characteristics of each trophic level. (*a*) Pyramid of numbers. Pyramids of biomass, both normal (*b*) and inverted (*c*). (*d*) Pyramid of energy. In the above aquatic examples, the producers are plankton.

> **Key Learning Outcome 36.2** Because energy is lost at every step of a food chain, the biomass of primary producers (photosynthesizers) tends to be greater than that of the herbivores that consume them, and herbivore biomass tends to be greater than the biomass of the carnivores that consume them.

36.3 The Water Cycle

Unlike energy, which flows through the earth's ecosystems in one direction (from the sun to producers to consumers), the physical components of ecosystems are passed around and reused within ecosystems. Ecologists speak of such constant reuse as recycling or, more commonly, **cycling.** Materials that are constantly recycled include all the chemicals that make up the soil, water, and air. While many are important and will be considered later, the proper cycling of four materials is particularly critical to the health of any ecosystem: water, carbon, and the soil nutrients nitrogen and phosphorus.

The paths of water, carbon, and soil nutrients as they pass from the environment to living organisms and back form closed circles, or cycles. In each cycle, the chemical resides for a time in an organism and then returns to the nonliving environment, often referred to as a *biogeochemical cycle.*

Of all the nonliving components of an ecosystem, water has the greatest influence on the living portion. The availability of water and the way in which it cycles in an ecosystem in large measure determines the biological richness of that ecosystem—how many different kinds of creatures live there and how many of each.

Water cycles within an ecosystem in two ways: the environmental water cycle and the organismic water cycle. Both cycles are shown in **figure 36.6**.

The Environmental Water Cycle

In the environmental water cycle, water vapor in the atmosphere condenses and falls to the earth's surface as rain or snow (called precipitation in **figure 36.6**). Heated there by the sun, it reenters the atmosphere by **evaporation** from lakes, rivers, and oceans, where it condenses and falls to the earth again.

The Organismic Water Cycle

In the organismic water cycle, surface water does not return directly to the atmosphere. Instead, it is taken up by the roots of plants. After passing through the plant, the water reenters the atmosphere through tiny openings (stomata) in the leaves, evaporating from their surface. This evaporation from leaf surfaces is called **transpiration.** Transpiration is also driven by the sun: The sun's heat creates wind currents that draw moisture from the plant by passing air over the leaves.

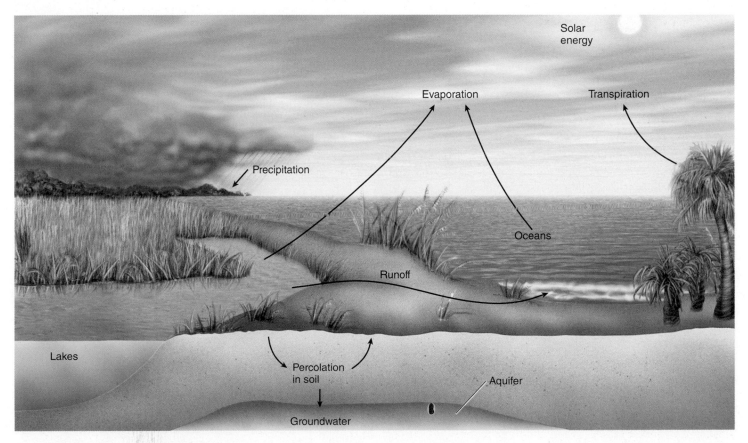

Figure 36.6   The water cycle.

Precipitation on land eventually makes its way to the ocean via groundwater, lakes, and finally, rivers. Solar energy causes evaporation, adding water to the atmosphere. Plants give off excess water through transpiration, also adding water to the atmosphere. Atmospheric water falls as rain or snow over land and oceans, completing the water cycle.

Breaking the Cycle

In very dense forest ecosystems, such as tropical rain forests, more than 90% of the moisture in the ecosystem is taken up by plants and then transpired back into the air. Because so many plants in a rain forest are doing this, the vegetation is the primary source of local rainfall. In a very real sense, these plants create their own rain: The moisture that travels up from the plants into the atmosphere falls back to earth as rain.

Where forests are cut down, the organismic water cycle is broken, and moisture is not returned to the atmosphere. Water drains off to the sea instead of rising to the clouds and falling again on the forest. During his expeditions from 1799 to 1805, the great German explorer Alexander von Humboldt reported that stripping the trees from a tropical rain forest in Colombia prevented water from returning to the atmosphere and created a semiarid desert. It is a tragedy of our time that just such a transformation is occurring in many tropical areas, as tropical and temperate rain forests are being clear-cut or burned in the name of "development" (**figure 36.7**).

Groundwater

Much less obvious than the surface waters seen in streams, lakes, and ponds is the groundwater, which occurs in permeable, saturated, underground layers of rock, sand, and gravel called *aquifers*. In many areas, groundwater is the most important water reservoir; for example, in the United States, more than 96% of all freshwater is groundwater. Groundwater flows much more slowly than surface water, anywhere from a few millimeters to as much as a meter or so per day. In the United States, groundwater provides about 25% of the water used for all purposes and provides about 50% of the population with drinking water. Rural areas tend to depend on groundwater almost exclusively, and its use is growing at about twice the rate of surface water use.

Because of the greater rate at which groundwater is being used, the increasing chemical pollution of groundwater is a very serious problem. Pesticides, herbicides, and fertilizers are key sources of groundwater pollution. Because of the large volume of water, its slow rate of turnover, and its inaccessibility, removing pollutants from aquifers is virtually impossible.

> **Key Learning Outcome 36.3** Water cycles through ecosystems in the atmosphere via precipitation and evaporation, some of it passing through plants on the way.

Figure 36.7 Burning or clear-cutting forests breaks the water cycle.
The high density and large size of plants in a forest translate into great quantities of water being transpired to the atmosphere, creating rain over the forests. In this way, rain forests perpetuate the wet climate that supports them. Tropical deforestation permanently alters the climate in these areas, creating arid zones.

36.4 | The Carbon Cycle

The earth's atmosphere contains plentiful carbon, present as carbon dioxide (CO_2) gas. This carbon cycles between the atmosphere and living organisms, often being locked up for long periods of time in organisms or deep underground. The cycle is begun by plants that use CO_2 in photosynthesis to build organic molecules—in effect, they trap the carbon atoms of CO_2 within the living world. The carbon atoms are returned to the atmosphere's pool of CO_2 through respiration, combustion, and erosion. The carbon cycle, seemingly more complex than the water cycle, is shown in **figure 36.8**.

Respiration

Most of the organisms in ecosystems respire—that is, they extract energy from organic food molecules by stripping away the carbon atoms and combining them with oxygen to form CO_2. At times, plants respire, as do the herbivores, which eat the plants, and the carnivores, which eat the herbivores. All of these organisms use oxygen to extract energy from food, and CO_2 is what is left when they are done. This by-product of respiration is released into the atmosphere.

Combustion

A lot of carbon is tied up in wood, and it may stay trapped there for many years, only returning to the atmosphere when the wood is burned or after it decomposes. Sometimes the duration of the carbon's visit to the organic world is long indeed. Plants that become buried in sediment, for example, may be gradually transformed by pressure into coal or oil. The carbon originally trapped by these plants is only released back into the atmosphere when the coal or oil (called **fossil fuels**) is burned.

Erosion

Very large amounts of carbon are present in seawater as dissolved CO_2. Substantial amounts of this carbon are extracted from the water by marine organisms, which use it to build their calcium carbonate shells. When these marine organisms die, their shells sink to the ocean floor, become covered with sediments, and form limestone. Eventually, the ocean recedes and the limestone becomes exposed to weather and erodes; as a result the carbon washes back and is dissolved in oceans where it is returned to the cycle through diffusion.

> **Key Learning Outcome 36.4** Carbon captured from the atmosphere by photosynthesis is returned to it through respiration, combustion, and erosion.

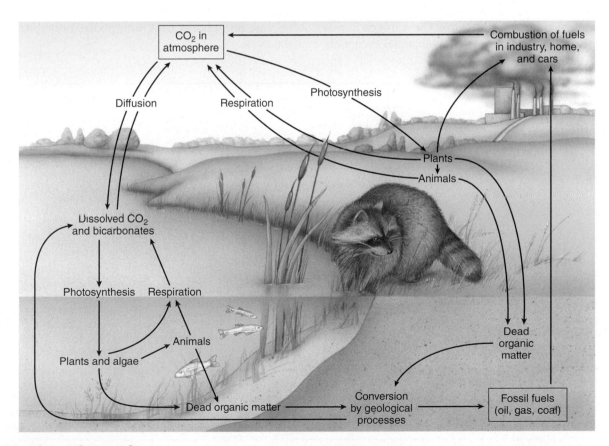

Figure 36.8 The carbon cycle.
Carbon from the atmosphere and from water is fixed by photosynthetic organisms and returned through respiration, combustion, and erosion.

36.5 Soil Nutrients and Other Chemical Cycles

The Nitrogen Cycle

Organisms contain a lot of nitrogen (a principal component of protein) and so does the atmosphere, which is 78.08% nitrogen gas (N_2). However, the chemical connection between these two reservoirs is very delicate because most living organisms are unable to use the N_2 so plentifully available in the air surrounding them. The two nitrogen atoms of N_2 are bound together by a particularly strong "triple" covalent bond that is very difficult to break. Luckily, a few kinds of bacteria can break the nitrogen triple bond and bind its nitrogen atoms to hydrogen (forming "fixed" nitrogen, ammonia [NH_3], which becomes ammonium ion [NH_4^+]) in a process called **nitrogen fixation.**

Bacteria evolved the ability to fix nitrogen early in the history of life, before photosynthesis had introduced oxygen gas into the earth's atmosphere, and that is still the only way the bacteria are able to do it—even a trace of oxygen poisons the process. In today's world, awash with oxygen, these bacteria live encased within bubbles called cysts that admit no oxygen or within special airtight cells in nodules of tissue on the roots of beans, aspen trees, and a few other plants. **Figure 36.9** shows the workings of the nitrogen cycle. Bacteria make needed nitrogen available to other organisms. The nitrogen moves up the food chain as one organism eats another and eventually returns following their deaths or through their excretions. Decomposing bacteria and then ammonifying bacteria return the nitrogen to ammonia and ammonium ion forms. Continuing the cycle, nitrifying bacteria can convert ammonium ion into nitrate (NO_3^-), and denitrifying bacteria are able to convert nitrate back into atmospheric nitrogen (N_2).

The growth of plants in ecosystems is often severely limited by the availability of "fixed" nitrogen in the soil, which is why farmers fertilize fields. This agricultural practice is a very old one, known even to primitive societies—the American Indians instructed the pilgrims to bury fish, a rich source of fixed nitrogen, with their corn seeds. Today most fixed nitrogen added to soils by farmers is not organic but instead is produced in factories by industrial rather than bacterial nitrogen fixation, a process that accounts for a prodigious 30% of the entire nitrogen cycle.

The Phosphorus Cycle

Phosphorus is an essential element in all living organisms, a key part of both ATP and DNA. Phosphorus is often in very limited supply in the soil of particular ecosystems, and because phosphorus does not form a gas, none is available in the atmosphere. Most phosphorus exists in soil and rock as the mineral calcium phosphate, which, as shown in **figure 36.10**, dissolves in water to form phosphate ions (Coca-Cola is a sweetened solution of phosphate ions). These phosphate ions are absorbed by the roots of plants and used by them to build organic molecules like ATP and DNA. When the plants and animals die and decay, bacteria in the soil convert the organic phosphorus back into phosphorus ions, completing the cycle.

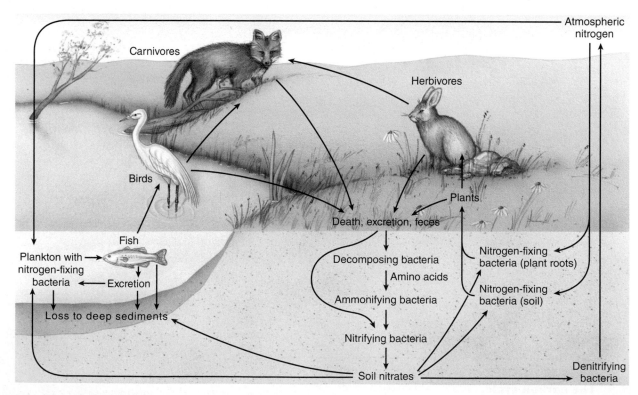

Figure 36.9 The nitrogen cycle.
Relatively few kinds of organisms—all of them bacteria—can convert atmospheric nitrogen into forms that can be used for biological processes.

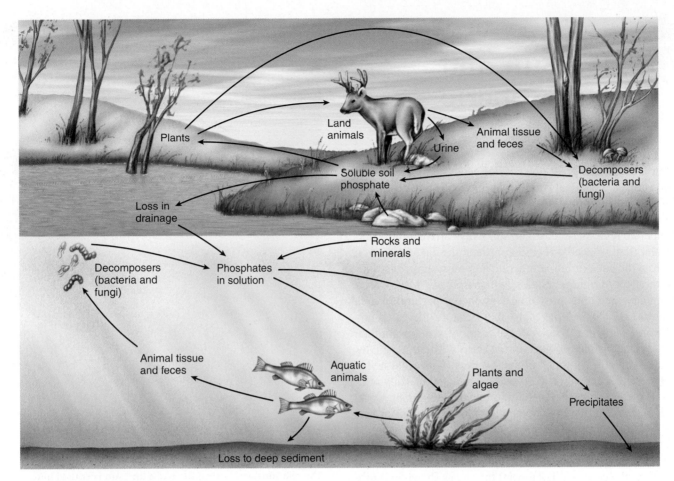

Figure 36.10 The phosphorus cycle.
Phosphorus plays a critical role in plant nutrition; next to nitrogen, phosphorus is the element most likely to be so scarce that it limits plant growth.

The phosphorus level in freshwater lake ecosystems is often quite low, preventing much growth of photosynthetic algae in these systems. Such ecosystems are particularly vulnerable to the inadvertent addition of phosphorus by human activity. For example, agricultural fertilizers and many commercial detergents are rich in phosphorus. Pollution of a lake by the addition of phosphorus to its waters first produces a green scum of algal growth on the surface of the lake, which then, if the pollution continues, proceeds to "kill" the lake. After the initial bloom of rapid algal growth, aging algae die, and bacteria feeding on the dead algae cells use up so much of the lake's dissolved oxygen that fish and invertebrate animals suffocate. Such rapid, uncontrolled growth caused by excessive nutrients in an aquatic ecosystem is called **eutrophication.**

The Cycling of Other Chemicals

Many other chemicals cycle through an ecosystem and must be maintained in a balanced state for the ecosystem to be healthy. Proper balance is important. Some chemicals can become harmful when their concentrations exceed normal levels for cycling, as we saw with phosphorus. Other chemicals, when in excess of normal cycling levels, can have similar devastating effects on an ecosystem.

Sulfur, a chemical that cycles through the atmosphere, can harm an ecosystem when large amounts of it are pumped into the atmosphere through coal-burning power plants. The excess sulfur combines with water vapor and oxygen, producing sulfuric acid. This acid then reenters the ecosystem as precipitation. This "acid rain" is discussed further in chapter 38.

Heavy metals, which include mercury, cadmium, and lead, are particularly damaging as they cycle through biological food chains, as they tend to progressively accumulate in organisms of higher trophic levels. This process, called *biological magnification,* is discussed further in chapter 38.

> **Key Learning Outcome 36.5** Most of the earth's atmosphere is diatomic nitrogen gas that cannot be used by most organisms. Certain bacteria are able to convert this nitrogen gas into ammonia through nitrogen fixation. These nitrogen atoms then cycle through the earth's ecosystem. Phosphorus, critical to organisms, is available in soil and dissolved in water. It cycles between organisms and the environment and is often the limiting factor in determining what organisms are able to live in an ecosystem.

36.6 The Sun and Atmospheric Circulation

The world contains a great diversity of ecosystems because its climate varies a great deal from place to place. On a given day, Miami and Boston often have very different types of weather. There is no mystery about this. The tropics are warmer than the temperate regions because the sun's rays arrive almost perpendicular (that is, dead on) at regions near the equator. As you move from the equator into temperate latitudes, sunlight strikes the earth at more oblique angles, which spreads it out over a much greater area, thus providing less energy per unit of area (**figure 36.11**). This simple fact—that because the earth is a sphere some parts of it receive more energy from the sun than others—is responsible for much of the earth's different climates and thus, indirectly, for much of the diversity of its ecosystems.

The earth's annual orbit around the sun and its daily rotation on its own axis are also both important in determining world climate. Because of the daily cycle, the climate at a given latitude is relatively constant. Because of the annual cycle and the inclination of the earth's axis, all parts away from the equator experience a progression of seasons. In summer in the Southern Hemisphere, the earth is tilted toward the sun as shown in **figure 36.11**, and rays hit more directly, leading to higher temperatures; as the earth reaches the opposite position in its annual orbit, the Northern Hemisphere receives more direct rays from the sun and experiences summer.

The major atmospheric circulation patterns result from the interactions of six large air masses. These great air masses (shown as circulating arrows below) occur in pairs, with one air mass of the pair occurring in the northern latitudes and the other occurring in the southern latitudes. These air masses affect

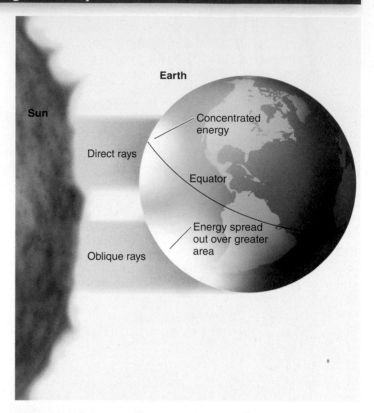

Figure 36.11 Latitude affects climate.
The relationship between the earth and sun is critical in determining the nature and distribution of life on earth. The tropics are warmer than the temperate regions because the sun's rays strike at a direct angle, providing more energy per unit of area.

climate because the rising and falling of an air mass influence its temperature, which, in turn, influences its moisture-holding capacity.

Near the equator, warm air rises and flows toward the poles (indicated by arrows at the equator that rise and circle toward the poles). As it rises and cools, this air loses most of its moisture because cool air holds less water vapor than warm air. (This explains why it rains so much in the tropics where the air is warm.) When this air has traveled to about 30 degrees north and south latitudes, the cool, dry air sinks and becomes reheated, soaking up water like a sponge as it warms, producing a broad zone of low rainfall. It is no accident that all of the great deserts of the world lie near 30 degrees north or 30 degrees south latitude. Air at these latitudes is still warmer than it is in the polar regions, and thus it continues to flow toward the poles. At about 60 degrees north and south latitudes, air rises and cools and sheds its moisture, and such are the locations of the great temperate forests of the world. Finally, this rising air descends near the poles, producing zones of very low precipitation.

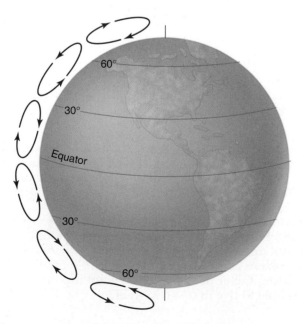

Key Learning Outcome 36.6 The sun drives circulation of the atmosphere, causing rain in the tropics and a band of deserts at 30 degrees latitude.

36.7 Latitude and Elevation

Temperatures are higher in tropical ecosystems for a simple reason: More sunlight per unit area falls on tropical latitudes (see **figure 36.11**). Solar radiation is most intense when the sun is directly overhead, and this occurs only in the tropics, where sunlight strikes the equator perpendicularly. Temperature also varies with elevation, with higher altitudes becoming progressively colder. At any given latitude, air temperature falls about 6°C for every 1,000-meter increase in elevation. The ecological consequences of temperature varying with elevation are the same as temperature varying with latitude. **Figure 36.12** illustrates this principle comparing changes in ecosystems that occur with increasing latitudes in North America with the ecosystem changes that occur with increasing elevation at the tropics. A 1,000-meter increase in elevation on a mountain in southern Mexico (**figure 36.12b**) results in a temperature drop equal to that of an 880-kilometer increase in latitude on the North American continent (**figure 36.12a**). This is why the "timberline" (the elevation above which trees do not grow) occurs at progressively lower elevations as one moves farther from the equator.

Rain Shadows

When a moving body of air encounters a mountain, it is forced upward, and as it is cooled at higher elevations, the air's moisture-holding capacity decreases, producing the rain you see on the windward side of the mountains shown below—the side from which the wind is blowing. Thus, moisture-laden winds from the Pacific Ocean rise and are cooled when they encounter the Sierra Nevada mountains. As the winds cool, their moisture-holding capacity decreases and precipitation occurs.

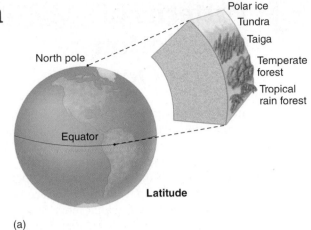

(a)

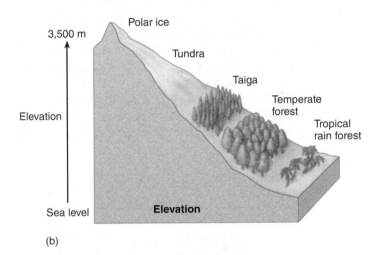

(b)

Figure 36.12   How elevation affects ecosystems.
The same land ecosystems that normally occur as latitude increases north and south of the equator at sea level (*a*) can occur in the tropics as elevation increases (*b*).

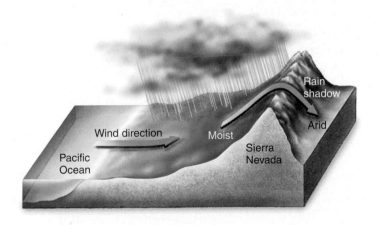

The effect on the other side of the mountain—the leeward side—is quite different. As the air passes the peak and descends on the far side of the mountains, it is warmed, so its moisture-holding capacity increases. Sucking up all available moisture, the air dries the surrounding landscape, often producing a desert. This effect, called a **rain shadow,** is responsible for deserts such as Death Valley, which is in the rain shadow of Mount Whitney, the tallest mountain in the Sierra Nevada.

Similar effects can occur on a larger scale. Regional climates are areas that are located on different parts of the globe but share similar climates because of similar geography. A so-called Mediterranean climate results when winds blow from a cool ocean onto warm land during the summer. As a result, the air's moisture-holding capacity is increased and precipitation is blocked, similar to what occurs on the leeward side of mountains. This effect accounts for dry, hot summers and cool, moist winters in areas with a Mediterranean climate such as portions of southern California or Oregon, central Chile, southwestern Australia, and the Cape region of South Africa. Such a climate is unusual on a world scale. In the regions where it occurs, many unusual kinds of endemic (local in distribution) plants and animals have evolved.

Key Learning Outcome 36.7 Temperatures fall with increasing latitude and also with increasing elevation. Rainfall is higher on the windward side of mountains, with air losing its moisture as it rises up the mountain; descending on the far side, the dry air warms and sucks up moisture, creating deserts.

Patterns of Circulation in the Ocean

Patterns of oceanic circulation are determined by the patterns of atmospheric circulation, which means that indirectly, the currents are driven by solar energy. The radiant input of heat from the sun sets the atmosphere in motion as already described, and then the winds set the ocean in motion. Oceanic circulation is dominated by the movement of surface waters in huge spiral patterns called gyres, which move around the subtropical zones of high pressure between approximately 30 degrees north and south latitudes. These gyres, indicated by the red and blue arrows in **figure 36.13**, move clockwise in the Northern Hemisphere and counterclockwise in the Southern Hemisphere. The way they redistribute heat profoundly affects life not only in the oceans but also on coastal lands. For example, the Gulf Stream, in the North Atlantic (the red-colored arrow, meaning it carries warm waters), swings away from North America near Cape Hatteras, North Carolina, and reaches Europe near the southern British Isles. Because of the Gulf Stream, western Europe is much warmer and more temperate than eastern North America at similar latitudes. As a general principle, western sides of continents in temperate zones of the Northern Hemisphere are warmer than their eastern sides; the opposite is true of the Southern Hemisphere.

Off the western coast of South America, the Humboldt Current carries phosphorus-rich cold water northward up the west coast. Phosphorus is brought up from the ocean depths by the upwelling of cool water that occurs as offshore winds blow from the mountainous slopes that border the Pacific Ocean. This nutrient-rich current helps make possible the abundance of marine life that supports the fisheries of Peru and northern Chile. Marine birds, which feed on these organisms, are responsible for the commercially important, phosphorus-rich guano deposits on the seacoasts of these countries.

El Niño Southern Oscillations and Ocean Ecology

Every Christmas a warm current sweeps down the coast of Peru and Ecuador from the tropics, reducing the fish population slightly and giving local fishers some time off. The local fishers named this Christmas current *El Niño* (literally, "the child," after the Christ Child). Scientists have adopted the term

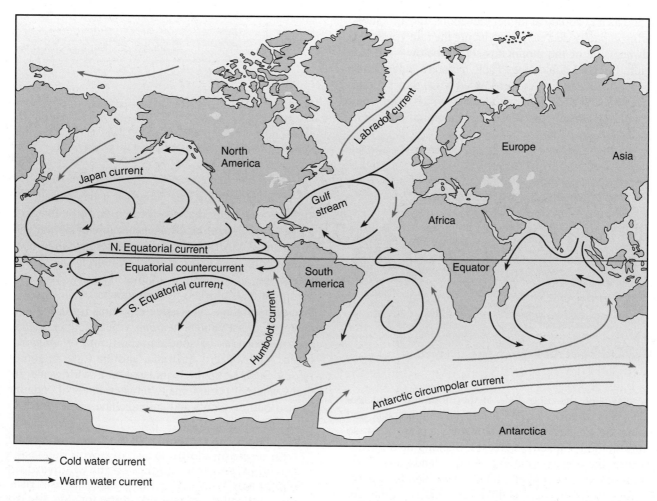

Figure 36.13 Oceanic circulation.

The circulation in the oceans moves in great surface spiral patterns called gyres; oceanic circulation affects the climate on adjacent lands.

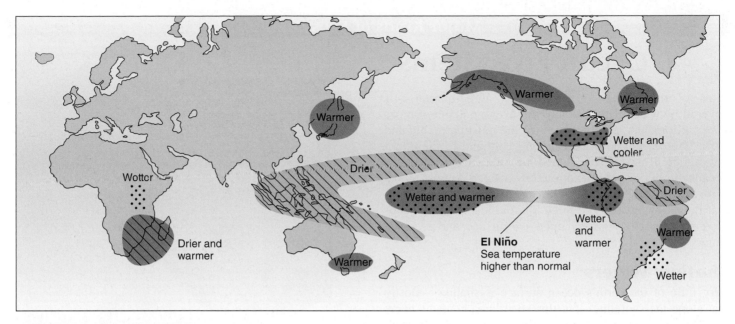

Figure 36.14 Effects of El Niño.

El Niño currents produce unusual weather patterns all over the world as warm waters from the western Pacific move eastward.

El Niño Southern Oscillation (ENSO) to refer to the dramatic version of the same phenomenon, one that occurs every two to seven years and is not only felt locally but on a global scale.

Scientists now have a pretty good idea of what goes on in an El Niño. Normally the Pacific Ocean is fanned by constantly blowing, east-to-west trade winds. These winds push warm surface water away from the ocean's eastern side (Peru, Ecuador, and Chile) and allow cold water to well up from the depths in its place, carrying nutrients that feed plankton and hence fish. This warm surface water piles up in the west, around Australia and the Philippines, making it several degrees warmer and a meter or so higher than the eastern side of the ocean. But if the winds slacken briefly, warm water begins to slosh back across the ocean (indicated by the darker red band stretching between Australia and the northern coast of South America in **figure 36.14**), causing an El Niño.

We now know that a slight weakening of the trade winds is actually part of a change in wind circulation patterns that recurs irregularly every 2 to 7 years. Once the winds weaken a bit, the eastern ocean warms, and the air above it becomes warmer and lighter. The eastern air becomes more similar to the air on the western side. This reduces the difference in pressure across the ocean. Because a pressure difference is what makes winds blow, the easterly trades weaken further, letting the warm water continue its eastward advance.

The end result is to shift the weather systems of the western Pacific Ocean 6,000 kilometers eastward. The tropical rainstorms that usually drench Indonesia and the Philippines are caused when warm seawater abutting these islands causes the air above it to rise, cool, and condense its moisture into clouds. When the warm water moves east, so do the clouds, leaving the previously rainy areas in drought (indicated by the light pink, hatched areas in **figure 36.16**). Conversely, the western edge of South America, its coastal

waters usually too cold to trigger much rain, gets a soaking (the dotted red areas), while the upwelling slows down because of the warm water. During an El Niño, commercial fish stocks virtually disappear from the waters of Peru and northern Chile, and plankton drops to a twentieth of its normal abundance. The commercially valuable anchovy fisheries of Peru were essentially destroyed by the 1972 and 1997 El Niños.

That is just the beginning. El Niño's effects are propagated across the world's weather systems. Violent winter storms lash the coast of California, accompanied by flooding, and El Niño produces colder and wetter winters than normal in Florida and along the Gulf Coast. The U.S. Midwest experiences heavier than normal rains (the blue dotted area).

La Niña. El Niño is an extreme phase of a naturally occurring climatic cycle, but as in all cycles, there is an opposite side to it. While El Niño is characterized by unusually warm ocean temperatures in the eastern Pacific, *La Niña* is characterized by unusually cold ocean temperatures in the eastern Pacific. The strengthening of the east-to-west trade winds intensifies the cold upwelling along the eastern Pacific, with coastal water temperatures along the South American coast falling as much as 7°F below normal. Although not as well known as El Niño, La Niña causes equally extreme effects that are nearly opposite to those of El Niño. In the United States, the effects of La Niña are most apparent during the winter months.

> **Key Learning Outcome 36.8** The world's oceans circulate in huge gyres deflected by continental landmasses. Disturbances in ocean currents like an El Niño and La Niña can have profound influences on world climate.

36.9 Ocean Ecosystems

Most of the earth's surface—nearly three-quarters—is covered by water. The seas have an average depth of more than 3 kilometers, and they are, for the most part, cold and dark. Photosynthetic organisms are confined to the upper few hundred meters (the light blue area in **figure 36.15**) because light does not penetrate any deeper. Almost all organisms that live below this level feed on organic debris that rains downward. The three main kinds of marine ecosystems are shallow waters, open-sea surface, and deep-sea waters (**figure 36.15**).

Shallow Waters

Very little of the earth's ocean surface is shallow—mostly that along the shoreline—but this small area contains many more species than other parts of the ocean (**figure 36.16a**). The world's great commercial fisheries occur on banks in the coastal zones, where nutrients derived from the land are more abundant than in the open ocean. Part of this zone consists of the **intertidal region,** which is exposed to the air whenever the tides recede. Partly enclosed bodies of water, such as those that often form at river mouths and in coastal bays, where the salinity is intermediate between that of seawater and freshwater, are called **estuaries.** Estuaries are among the most naturally fertile areas in the world, often containing rich stands of submerged and emergent plants, algae, and microscopic organisms. They provide the breeding grounds for most of the coastal fish and shellfish that are harvested both in the estuaries and in open water.

Open-Sea Surface

Drifting freely in the upper, better-illuminated waters of the ocean is a diverse biological community of microscopic organisms. Most of the plankton occurs in the top 100 meters of the sea. Many fish swim in these waters as well, feeding on the plankton and one another (**figure 36.16b**). Some members of the plankton, including algae and some bacteria, are photosynthetic and are called phytoplankton. Collectively, these organisms are responsible for about 40% of all photosynthesis that takes place on earth. Over half of this is carried out by organisms less than 10 micrometers in diameter—at the lower limits of size for organisms—and almost all of it near the surface of the sea, in the zone into which light from the surface penetrates freely.

Deep-Sea Waters

In the deep waters of the sea, below the top 300 meters, little light penetrates. Very few organisms live there, compared to the rest of the ocean, but those that do include some of the most bizarre organisms found anywhere on earth. Many deep-sea inhabitants have bioluminescent (light-producing) body parts that they use to communicate or to attract prey (**figure 36.17a**).

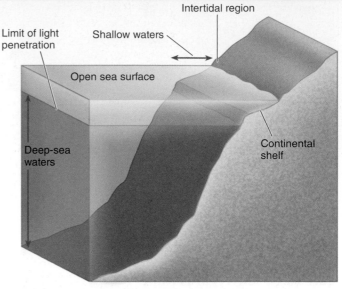

Figure 36.15 Ocean ecosystems.
There are three primary ecosystems found in the earth's oceans. Shallow water ecosystems occur along the shoreline and at areas of coral reefs. Open sea surface ecosystems occur in the upper 100–200 meters where light can penetrate. Finally, deep-sea water ecosystems are areas below 300 meters.

(a)

(b)

Figure 36.16 Shallow waters and open sea surface.
(a) Fish and many other kinds of animals find food and shelter among the coral in the coastal waters of some regions. (b) The upper layers of the open ocean contain plankton and large schools of fish, like these bigeye snapper.

The supply of oxygen can often be critical in the deep ocean, and as water temperatures become warmer, the water holds less oxygen. For this reason, the amount of available oxygen becomes an important limiting factor for deep-sea organisms in warmer marine regions of the globe. Carbon dioxide, in contrast, is almost never limited in the deep ocean. The distribution of minerals is much more uniform in the ocean than it is on land, where individual soils reflect the composition of the parent rocks from which they have weathered.

Frigid and bare, the floors of the deep sea have long been considered a biological desert. Recent close-up looks taken by marine biologists, however, paint a different picture (**figure 36.17c**). The ocean floor is teeming with life. Often kilometers deep, thriving in pitch darkness under enormous pressure, crowds of marine invertebrates have been found in hundreds of deep samples from the Atlantic and Pacific. Rough estimates of deep-sea diversity have soared to hundreds of thousands of species. Many appear endemic (local). The diversity of species is so high it may rival that of tropical rain forests! This profusion is unexpected. New species usually require some kind of barrier to diverge (see chapter 14), and the ocean floor seems boringly uniform. However, little migration occurs among deep populations, and this lack of movement may encourage local specialization and species formation. A patchy environment may also contribute to species formation there; deep-sea ecologists find evidence that fine but nonetheless formidable resource barriers arise in the deep sea.

No light falls in the deep ocean. From where do deep-sea organisms obtain their energy? While some utilize energy falling to the ocean floor as debris from above, other deep-sea organisms are autotrophic, gaining their energy from **hydrothermal vent systems,** areas in which seawater circulates through porous rock surrounding fissures where molten material from beneath the earth's crust comes close to the surface. Hydrothermal vent systems, also called deep-sea vents, support a broad array of heterotrophic life (**figure 36.17b**). Water in the area of these hydrothermal vents is heated to temperatures in excess of 350°C, and contains high concentrations of hydrogen sulfide. Prokaryotes that live by these deep-sea vents obtain energy and produce carbohydrates through chemosynthesis instead of photosynthesis. Like plants, they are autotrophs; they extract energy from hydrogen sulfide to manufacture food, much as a plant extracts energy from the sun to manufacture its food. These prokaryotes live symbiotically within the tissues of heterotrophs that live around the deep-sea vents. The animals provide a place for the prokaryotes to live and obtain nutrients, and in turn the prokaryotes supply the animal with organic compounds to use as food.

Despite the many new forms of small invertebrates now being discovered on the seafloor, and the huge biomass that occurs in the sea, more than 90% of all *described* species of organisms occur on land. Each of the largest groups of organisms, including insects, mites, nematodes, fungi, and plants, has marine representatives, but they constitute only a very small fraction of the total number of described species.

Figure 36.17 Deep-sea waters.

(*a*) The luminous spot below the eye of this deep-sea fish results from the presence of a symbiotic colony of luminous bacteria. (*b*) These giant beardworms live along vents where water jets from fissures at 350°C and then cools to the 2°C of the surrounding water. (*c*) Looking for all the world like some undersea sunflower, these two sea anemones (actually animals) use a glass-sponge stalk to catch "marine snow," food particles raining down on the ocean floor from the ocean surface several kilometers above.

Key Learning Outcome 36.9 The three principal types of ocean ecosystems are shallow waters, open-sea surface, and deep-sea waters. Both intertidal shallows and deep-sea communities are very diverse.

36.10 Freshwater Ecosystems

Freshwater ecosystems (lakes, ponds, rivers, and wetlands) are distinct from both ocean and land ecosystems, and they are very limited in area. Inland lakes cover about 1.8% of the earth's surface and rivers, streams, and wetlands about 0.4%. All freshwater habitats are strongly connected to land habitats, with marshes and swamps (wetlands) constituting intermediate habitats. In addition, a large amount of organic and inorganic material continually enters bodies of freshwater from communities growing on the land nearby (**figure 36.18a**). Many kinds of organisms are restricted to freshwater habitats (**figure 36.18b,c**). When they occur in rivers and streams, they must be able to attach themselves in such a way as to resist or avoid the effects of current or risk being swept away.

Like the ocean, ponds and lakes have three zones in which organisms live (**figure 36.19a**): a shallow "edge" zone (the littoral zone), an open-water surface zone (the limnetic zone), and a deep-water zone where light does not penetrate (the profundal zone). Also, lakes can be divided into two categories, based on their

(a)

(c)

(b)

**Figure 36.18
Freshwater ecosystems.**

(*a*) In this stream in the northern coastal mountains of California, as in all streams, much organic material falls or seeps into the water from communities along the edges. This input is responsible for much of the stream's biological productivity. Organisms such as this speckled darter (*b*) and this giant waterbug with eggs on its back (*c*) can only live in freshwater habitats.

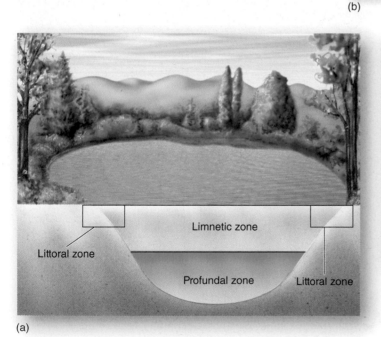

Limnetic zone

Littoral zone

Profundal zone

Littoral zone

(a)

(b) Oligotrophic lake

(c) Eutrophic lake

Figure 36.19   Characteristics of ponds and lakes.

(*a*) Ponds and lakes can be divided into three zones based on the types of organisms that live in each. A shallow "edge" (littoral) zone lines the periphery of the lake where attached algae and their insect herbivores live. An open-water surface (limnetic) zone lies across the entire lake and is inhabited by floating algae, zooplankton, and fish. A dark, deep-water (profundal) zone overlies the sediments at the bottom of the lake. The profundal zone contains numerous bacteria and wormlike organisms that consume dead debris settling at the bottom of the lake. Lakes can be oligotrophic (*b*), containing scarce amounts of organic material, or eutrophic (*c*), containing abundant amounts of organic material.

production of organic material. In **oligotrophic lakes** (figure **36.19b**), organic matter and nutrients are relatively scarce. Such lakes are often deep, and their deep waters are always rich in oxygen. Oligotrophic lakes are highly susceptible to pollution from excess phosphorus from such sources as fertilizer runoff, sewage, and detergents. **Eutrophic lakes,** on the other hand, have an abundant supply of minerals and organic matter (**figure 36.19c**). Oxygen is depleted at the lower depths in the summer because of the abundant organic material and high rate at which aerobic decomposers in the lower layer use oxygen. These stagnant waters circulate to the surface in the fall (during the fall overturn, as discussed below) and are then infused with more oxygen.

Thermal stratification, characteristic of the larger lakes in temperate regions, is the process whereby water at a temperature of 4°C (which is when water is most dense) sinks beneath water that is either warmer or cooler. Follow through the changes in a large lake in **figure 36.20** beginning in winter ❶, where water at 4°C sinks beneath cooler water that freezes at the surface at 0°C. Below the ice, the water remains between 0° and 4°C, and plants and animals survive there. In spring ❷, as the ice melts, the surface water is warmed to 4°C and sinks below the cooler water, bringing the cooler water

to the top with nutrients from the lake's lower regions. This process is known as the *spring overturn.*

In summer ❸, warmer water forms a layer over the cooler water that lies below. In the area between these two layers, called the *thermocline,* temperature changes abruptly. You may have experienced the existence of these layers if you have dived into a pond in temperate regions in the summer. Depending on the climate of the particular area, the warm upper layer may become as much as 20 meters thick during the summer. In autumn ❹, its surface temperature drops until it reaches that of the cooler layer underneath—4°C. When this occurs, the upper and lower layers mix—a process called the *fall overturn.* Therefore, colder waters reach the surfaces of lakes in the spring and fall, bringing up fresh supplies of dissolved nutrients.

> **Key Learning Outcome 36.10** Freshwater ecosystems cover only about 2% of the earth's surface; all are strongly tied to adjacent terrestrial ecosystems. In some, organic materials are common, and in others, scarce. The temperature zones in lakes overturn twice a year, in spring and fall.

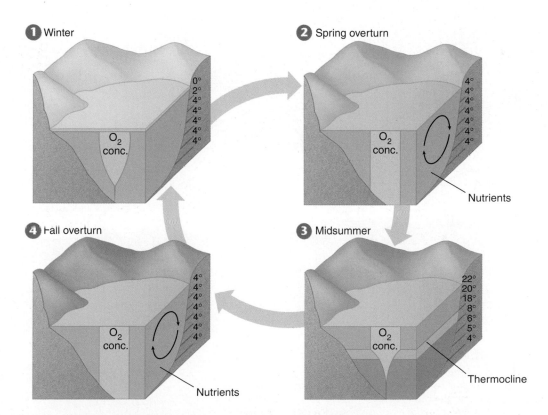

Figure 36.20 Spring and fall overturns in freshwater ponds or lakes.

The pattern of stratification in a large pond or lake in temperate regions is upset in the spring and fall overturns. Of the three layers of water shown in midsummer (*lower right*), the densest water occurs at 4°C. The warmer water at the surface is less dense. The thermocline is the zone of abrupt change in temperature that lies between them. In summer and winter, oxygen concentrations are lower at greater depths, whereas in the spring and fall, they are more similar at all depths.

Land Ecosystems

Living on land ourselves, we humans tend to focus much of our attention on terrestrial ecosystems. A **biome** is a terrestrial ecosystem that occurs over a broad area. Each biome is characterized by a particular climate and a defined group of organisms.

While biomes can be classified in a number of ways, the seven most widely occurring biomes (color-coded in **figure 36.21**) are (1) tropical rain forest (dark green), (2) savanna (pink), (3) desert (pale yellow), (4) temperate grassland (tan), (5) temperate deciduous forest (brown), (6) taiga (purple), and (7) tundra (light blue). The reason that there are seven primary biomes, and not one or 80, is that they have evolved to suit the climate of the region, and the earth has seven principal climates. The seven biomes differ remarkably from one another but show many consistencies within; a particular biome often looks similar, with many of the same types of creatures living there, wherever it occurs on earth.

There are seven other less widespread biomes also shown in **figure 36.21**: chaparral; polar ice; mountain zone; temperate evergreen forest; warm, moist evergreen forest; tropical monsoon forest; and semidesert.

If there were no mountains and no climatic effects caused by the irregular outlines of the continents and by different sea temperatures, each biome would form an even belt around the globe. In fact, their distribution is greatly affected by these factors, especially by elevation. Thus, the summits of the Rocky Mountains are covered with a vegetation type that resembles tundra, whereas other forest types that resemble taiga occur farther down. It is for reasons such as these that the distributions of the biomes are so irregular. One trend that is apparent is that those biomes that normally occur at high latitudes also follow an altitudinal gradient along mountains. That is, biomes found far north and far south of the equator at sea level also occur in the tropics but at high mountain elevations (see **figure 36.12**).

Distinctive features of the seven major biomes—tropical rain forest, savanna, desert, temperate grassland, temperate deciduous forest, taiga, and tundra—along with several of the less widespread biomes are now discussed in more detail.

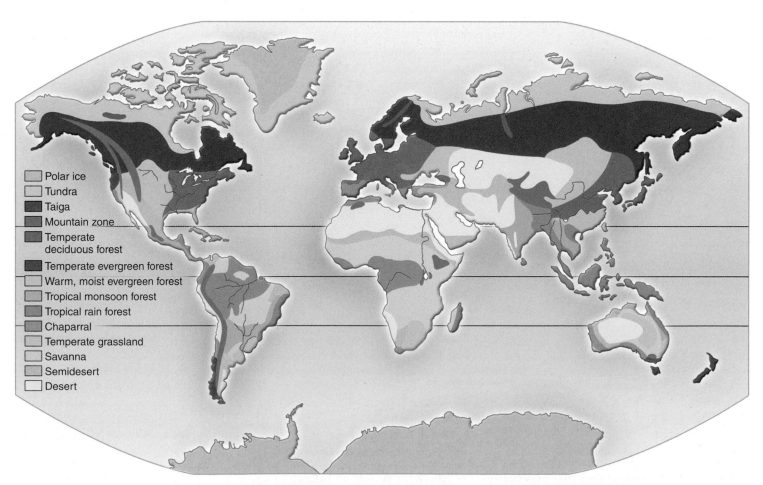

Figure 36.21 Distribution of the earth's biomes.
The seven primary types of biomes are tropical rain forest, savanna, desert, temperate grassland, temperate deciduous forest, taiga, and tundra. In addition, seven less widespread biomes are shown.

Legend:
- Polar ice
- Tundra
- Taiga
- Mountain zone
- Temperate deciduous forest
- Temperate evergreen forest
- Warm, moist evergreen forest
- Tropical monsoon forest
- Tropical rain forest
- Chaparral
- Temperate grassland
- Savanna
- Semidesert
- Desert

Lush Tropical Rain Forests

Rain forests, which experience over 250 centimeters of rain a year, are the richest ecosystems on earth. They contain at least half of the earth's species of terrestrial plants and animals—more than 2 million species! In a single square mile of tropical forest in Rondonia, Brazil, there are 1,200 species of butterflies—twice the total number found in the United States and Canada combined. The communities that make up tropical rain forests are diverse in that each kind of animal, plant, or microorganism is often represented in a given area by very few individuals. There are extensive tropical rain forests in South America, Africa, and Southeast Asia. But the world's tropical rain forests are being destroyed, and with them, countless species, many of them never seen by humans. Perhaps a quarter of the world's species will disappear with the rain forests during the lifetime of many of us.

Tropical rain forest

Savannas: Dry Tropical Grasslands

In the dry climates that border the tropics are found the world's great grasslands, called **savannas.** Landscapes are open, often with widely spaced trees, and rainfall (75 to 125 cm annually) is seasonal. Many of the animals and plants are active only during the rainy season. The huge herds of grazing animals that inhabit the African savanna are familiar to all of us. Such animal communities occurred in the temperate grasslands of North America during the Pleistocene epoch but have persisted mainly in Africa. On a global scale, the savanna biome is transitional between tropical rain forest and desert. As these savannas are increasingly converted to agricultural use to feed rapidly expanding human populations in subtropical areas, their inhabitants are finding it difficult to survive. The elephant, rhino, and cheetah are now endangered species; the lion and giraffe will soon follow them.

Savanna

Deserts: Burning Hot Sands

In the interior of continents are found the world's great deserts, especially in Africa (the Sahara), Asia (the Gobi), and Australia (the Great Sandy Desert). **Deserts** are dry places where less than 25 centimeters of rain falls in a year—an amount so low that vegetation is sparse and survival depends on water conservation. One quarter of the world's land surface is desert. The plants and animals that live in deserts may restrict their activity to favorable times of the year, when water is present. To avoid high temperatures, most desert vertebrates live in deep, cool, and sometimes even somewhat moist burrows. Those that are active over a greater portion of the year emerge only at night, when temperatures are relatively cool. Some, such as camels, can drink large quantities of water when it is available and then survive long, dry periods. Many animals simply migrate to or through the desert, where they exploit food that may be abundant seasonally.

Desert

Grasslands: Seas of Grass

Halfway between the equator and the poles are temperate regions where rich **grasslands** grow. These grasslands once covered much of the interior of North America, and they were widespread in Eurasia and South America as well. Such grasslands are often highly productive when converted to agriculture. Many of the rich agricultural lands in the United States and southern Canada were originally occupied by **prairies,** another name for temperate grasslands. The roots of perennial grasses characteristically penetrate far into the soil, and grassland soils tend to be deep and fertile. Temperate grasslands are often populated by herds of grazing mammals. In North America, the prairies were once inhabited by huge herds of bison and pronghorns. The herds are almost all gone now, with most of the prairies having been converted to the richest agricultural region on earth.

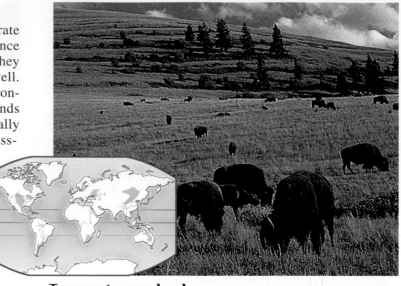

Temperate grassland

Deciduous Forests: Rich Hardwood Forests

Mild climates (warm summers and cool winters) and plentiful rains promote the growth of **deciduous** ("hardwood") **forests** in Eurasia, the northeastern United States, and eastern Canada. A deciduous tree is one that drops its leaves in the winter. Deer, bears, beavers, and raccoons are the familiar animals of the temperate regions. Because the temperate deciduous forests represent the remnants of more extensive forests that stretched across North America and Eurasia several million years ago, these remaining areas—especially those in eastern Asia and eastern North America—share animals and plants that were once more widespread. Alligators, for example, are found only in China and in the southeastern United States. The deciduous forest in eastern Asia is rich in species because climatic conditions have remained constant.

Temperate deciduous forest

Taiga: Trackless Conifer Forests

A great ring of northern forests of coniferous trees (spruce, hemlock, larch, and fir) extends across vast areas of Asia and North America. Coniferous trees are ones with leaves like needles that are kept all year long. This ecosystem, called **taiga,** is one of the largest on earth. Here, the winters are long and cold. Rain, often as little as in hot deserts, falls in the summer. Because it has too short a growing season for farming, few people live there. Many large mammals, including elk, moose, deer, and such carnivores as wolves, bears, lynx, and wolverines, live in the taiga. Traditionally, fur trapping has been extensive in this region. Lumber production is also important. Marshes, lakes, and ponds are common and are often fringed by willows or birches. Most of the trees occur in dense stands of one or a few species.

Taiga

Tundra: Cold Boggy Plains

In the far north, above the great coniferous forests and below the polar ice, there are few trees. There the grassland, called **tundra,** is open, windswept, and often boggy. Enormous in extent, this ecosystem covers one-fifth of the earth's land surface. Very little rain or snow falls. When rain does fall during the brief arctic summer, it sits on frozen ground, creating a sea of boggy ground. **Permafrost,** or permanent ice, usually exists within a meter of the surface. Trees are small and are mostly confined to the margins of streams and lakes. Large grazing mammals, including musk-oxen, caribou, reindeer, and carnivores such as wolves, foxes, and lynx, live in the tundra. Lemming populations rise and fall on a long-term cycle, with important effects on the animals that prey on them.

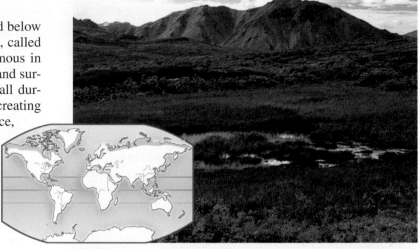
Tundra

Chaparral

Chaparral consists of evergreen, often spiny shrubs and low trees that form communities in regions with what is called a "Mediterranean," dry summer climate. These regions include California, central Chile, the Cape region of South Africa, southwestern Australia, and the Mediterranean area itself. Many plant species found in chaparral can germinate only when they have been exposed to the hot temperatures generated during a fire. The chaparral of California and adjacent regions is historically derived from deciduous forests.

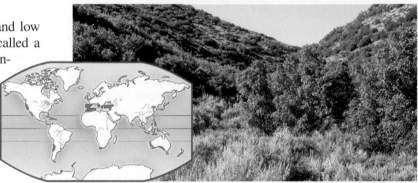
Chaparral

Polar Ice Caps

Polar ice caps lie over the Arctic Ocean in the north and Antarctica in the south. The poles receive almost no precipitation, so although ice is abundant, freshwater is scarce. The sun barely rises in the winter months. Life in Antarctica is largely limited to the coasts. Because the Antarctic ice cap lies over a landmass, it is not warmed by the latent heat of circulating ocean water and becomes very cold. As a result, only prokaryotes, algae, and some small insects inhabit the vast Antarctic interior.

Polar ice

Tropical Monsoon Forest

Tropical upland forests occur in the tropics and semitropics at slightly higher latitudes than rain forests or where local climates are drier. Most trees in these forests are deciduous, losing many of their leaves during the dry season. This loss of leaves allows sunlight to penetrate to the understory and ground levels of the forest, where a dense layer of shrubs and small trees grow rapidly. Rainfall is typically very seasonal, measuring several inches daily in the monsoon season and approaching drought conditions in the dry season, particularly in locations far from oceans, such as in central India.

Tropical monsoon forest

Key Learning Outcome 36.11 Biomes are major terrestrial communities defined largely by temperature and rainfall patterns.

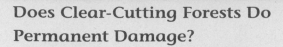
Does Clear-Cutting Forests Do Permanent Damage?

The lumber industry practice called "clear-cutting" has been common in many states. Loggers find it more efficient to simply remove all trees from a watershed, and sort the logs out later, than to selectively cut only the most desirable mature trees. While the open cuts seem a desolation to the casual observer, the loggers claim that new forests can become established more readily in the open cut as sunlight now more easily reaches seedlings at ground level. Ecologists counter that clear-cutting permanently changes the forest in ways that cannot be reversed.

Who is right? The most direct way to find out is to clear-cut an area and watch it very carefully. Just this sort of massive field test was carried out in a now-classic experiment at the Hubbard Brook Experimental Forest in New Hampshire. Hubbard Brook is the central stream of a large watershed that drains a region of temperate deciduous forest in northern New Hampshire. The research team, led by then-Dartmouth College professors Herbert Bormann and Gene Likens, first gathered a great deal of information about the forest watershed. Starting in 1963, they censused the trees, measured the flow of water through the watershed, and carefully documented the levels of minerals and other nutrients in the water leaving the ecosystem via Hubbard Brook. To keep track, they constructed concrete dams across each of the six streams that drain the forest and monitored the runoff, chemically analyzing samples. The undisturbed forest proved very efficient at retaining nitrogen and other nutrients. The small amounts of nutrients that entered the ecosystem in rain and snow were approximately equal to the amounts of nutrients that ran out of the valleys into Hubbard Brook.

Now came the test. In the winter of 1965 the investigators felled all the trees and shrubs in 48 acres drained by one stream (as shown in the photo), and examined the water running off. The immediate effect was dramatic: The amount of water running out of the valley increased by 40%. Water that otherwise would have been taken up by vegetation and released into the atmosphere through evaporation was now simply running off.

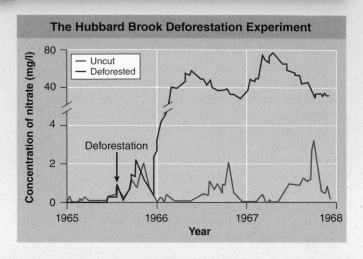

The Hubbard Brook Deforestation Experiment

It was clear that the forest was not retaining water as well, but what about the soil nutrients, the key to future forest fertility?

The red line in the graph above shows nitrogen minerals leaving the ecosystem in the runoff water of the stream draining the clear-cut area; the blue line shows the nitrogen runoff in a neighboring stream draining an adjacent uncut portion of the forest.

1. **Applying Concepts**
   a. Variable. In the graph, what is the dependent variable?
   b. Scale. What is the significance of the break in the vertical axis between 4 and 40?

2. **Interpreting Data**
   a. What is the approximate concentration of nitrogen in the runoff of the uncut valley before cutting? of the cut valley before cutting?
   b. What is the approximate concentration of nitrogen in the runoff of the uncut valley one year after cutting? of the clear-cut valley one year after cutting?

3. **Making Inferences**
   a. Is there any yearly pattern to the nitrogen runoff in the uncut forest? Can you explain it?
   b. How does the loss of nitrogen from the ecosystem in the clear-cut forest compare with nitrogen loss from the uncut forest?

4. **Drawing Conclusions**
   a. What is the impact of this forest's trees upon its ability to retain nitrogen?
   b. Has clear-cutting harmed this ecosystem? Explain.

The Energy in Ecosystems

36.1 Energy Flows Through Ecosystems

- An ecosystem includes the community and the habitat present in a particular area. Energy constantly flows into an ecosystem from the sun, and is passed among organisms in a food chain. Energy from the sun is captured by photosynthetic producers, which are eaten by herbivores, which are in turn eaten by carnivores, as shown here from **figure 36.1**. Organisms at all trophic levels die and are consumed by detritivores and decomposers (**integrated art, page 747**).

- An ecosystem's net primary productivity is the total amount of energy that is captured by producers. Energy is lost at every level in a food chain, such that only 5% to 20% of available energy is passed on to the next trophic level (**figures 36.2** and **36.3**).

- A food chain is organized linearly, but in nature the flow of energy is more complex, and is called a food web (**figure 36.4**).

36.2 Ecological Pyramids

- Because energy is lost as it passes up through the trophic levels of the food chain, there tends to be more individuals at the lower trophic levels. Similarly, the amount of biomass is also less at the higher trophic levels, as is the amount of energy. Ecological pyramids illustrate this distribution of number of individuals, biomass, and energy (**figure 36.5**).

Materials Cycle Within Ecosystems

36.3 The Water Cycle

- Physical components of the ecosystem cycle through the ecosystem, being used then recycled and reused. This cycling of materials often involves living organisms and is referred to as a biogeochemical cycle.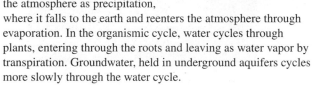

- Water availability can limit the number of organisms in an ecosystem. There are two water cycles: the environmental cycle and the organismic cycle. In the environmental cycle, shown here from **figure 36.6**, water cycles from the atmosphere as precipitation, where it falls to the earth and reenters the atmosphere through evaporation. In the organismic cycle, water cycles through plants, entering through the roots and leaving as water vapor by transpiration. Groundwater, held in underground aquifers cycles more slowly through the water cycle.

36.4 The Carbon Cycle

- Carbon cycles from the atmosphere through plants, via carbon fixation of CO_2 in photosynthesis. Carbon then returns to the atmosphere as CO_2 via cellular respiration, but some carbon is also stored in the tissues of the organisms. Eventually, this carbon reenters the atmosphere by the burning of fossil fuels and by diffusion after erosion (**figure 36.8**).

36.5 Soil Nutrients and Other Chemical Cycles

- Nitrogen gas in the atmosphere cannot be readily used by organisms and needs to be fixed by certain types of bacteria.

Animals eat plants that have taken up fixed nitrogen. Nitrogen reenters the ecosystem through animal excretion and decomposition (**figure 36.9**).

- Phosphorus also cycles through the ecosystem and may limit growth when not available (**figure 36.10**) or it can cause problems in aquatic ecosystems when found in excess amounts.

- Other chemicals, such as sulfur and heavy metals, also cycle through the ecosystem. Excess amounts of these chemicals can cause problems in the ecosystem.

How Weather Shapes Ecosystems

36.6 The Sun and Atmospheric Circulation

- The heating power of the sun and air currents affect evaporation, causing certain parts of the globe, such as the tropics, to have larger amounts of precipitation (**figure 36.11**).

36.7 Latitude and Elevation

- Temperature and precipitation are similarly affected by elevation and latitude. Changes in ecosystems from the equator to the poles are similarly reflected in the changes in ecosystems from sea level to mountaintops (**figure 36.12**). Changes in temperature cause the rain shadow effect, where precipitation is deposited on the windward side of mountains, causing deserts on the leeward side.

36.8 Patterns of Circulation in the Ocean

- The earth's oceans circulate in patterns that distribute warmer and cooler waters to different areas of the world. These ocean patterns affect climates across the globe (**figures 36.13** and **36.14**).

Major Kinds of Ecosystems

36.9 Ocean Ecosystems

- There are three primary ocean ecosystems: shallow waters, open-sea surfaces, and deep-sea bottoms (**figure 36.15**). Each is affected by light and temperature.

36.10 Freshwater Ecosystems

- Freshwater ecosystems are closely tied to the terrestrial environments that surround them. Freshwater ecosystems are affected by light, temperature, and nutrients. The penetration of light divides a lake into three zones with varying amounts of light (**figure 36.19**). Temperature variations in a lake, called thermal stratification, also bring about an overturning of the lake that distributes nutrients (**figure 36.20**).

36.11 Land Ecosystems

- Biomes are terrestrial communities found throughout the world. Each biome contains its own group of organisms based on temperature and rainfall patterns (**figure 36.21**).

1. Energy from the sun is captured and converted into chemical energy by
   a. herbivores.
   b. carnivores.
   c. producers.
   d. detritivores.

2. As energy is transferred from one trophic level to the next, substantial amounts of energy are lost to/as
   a. undigestible biomass.
   b. heat.
   c. metabolism.
   d. All answers are correct.

3. The number of carnivores found at the top of an ecological pyramid is limited by the
   a. number of organisms below the top carnivores.
   b. number of trophic levels below the top carnivores.
   c. amount of biomass below the top carnivores.
   d. amount of energy transferred to the top carnivores.

4. Hydrologists, scientists who study the movements and cycles of water, refer to the return of water from the ground to the air as evapotranspiration. The first part of the word refers to evaporation. The second part of the word refers to transpiration, which is evaporation of water
   a. from plants.
   b. through animal perspiration.
   c. off the ground shaded by plants.
   d. from the surface of rivers.

5. The carbon cycle includes a store of carbon as fossil fuels that is released through
   a. respiration.
   b. combustion.
   c. erosion.
   d. All answers are correct.

6. The element phosphorus is needed in organisms to build
   a. proteins.
   b. carbohydrates.
   c. ATP.
   d. steroids.

7. A rain shadow results in
   a. extremely wet conditions due to the lack of wind over a mountain range.
   b. dry air moving toward the poles that cools and sinks in regions 15 to 30 degrees north/south latitude.
   c. global polar regions that rarely receive moisture from the warmer, tropical regions, and are therefore dryer.
   d. desert conditions on the downwind side of a mountain due to increased moisture-holding capacity of the winds as the air heats up.

8. As one travels from northern Canada south to the United States, the timberline increases in elevation. This is because as latitude
   a. increases, temperature increases.
   b. decreases, temperature increases.
   c. increases, humidity decreases.
   d. decreases, humidity increases.

9. In freshwater lakes during the summer, layers of sudden temperature change called the _____ form.
   a. eutrophy
   b. profundal zone
   c. oligotrophy
   d. thermocline

10. Which of the following biomes is *not* found south of the equator?
    a. polar ice cap
    b. savanna
    c. tundra
    d. tropical monsoon forest

1. **Figure 36.3**
   a. One thousand calories of energy is harvested from sunlight by the algae in a stream. Explain the probable efficiency if you were to eat, respectively, (1) the algae itself, (2) the heterotrophs, (3) the smelt, or (4) the trout.
   b. How many more people could be fed if they all ate algae?
   c. Assuming there were enough algae, why can't we survive by eating just algae?

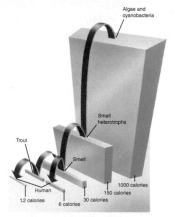

2. **Figure 36.6** Imagine that you are a molecule of water. Describe the journey of your existence, starting with falling as part of a drop of rain onto the earth, and ending in a cloud ready to fall once again. Be sure to take a trip through a plant along the way.

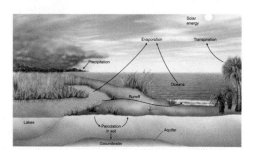

1. Given the amount of sunlight that hits the plants on our planet, and the ability of plants for rapid growth and reproduction, explain why we aren't all hip deep in plants?

2. Many experiments by ecologists have shown that species-rich communities are more productive than species-poor ones. If this is so, how is it that American farming, based almost entirely on monocultures, is so productive?

3. Why do increasing latitude and increasing elevation each affect which plant species grow in a place, and why do the two factors do so in the same way?

4. Pick two different biomes that you personally have encountered. Compare and contrast the sunshine, rainfall, major temperature features, and the plants and animals found there.

37

Behavior and the Environment

T his dog's bow is its way of saying to another dog, "Let's play." Its rear shoved high in the air, its forelegs flat on the ground, the pup looks up, hopeful that its partner will agree. Every dog invites play in exactly this way, a terrier like this, a retriever—even a wolf. The bow is an innate behavior that they all share, part of what it means to be a canine. In other ways, this dog may behave like no other dog. It may learn to "sit" or "roll over"—or to catch Frisbees, herd sheep, or bring in the morning paper. It can solve surprisingly complex problems. Wolves and other wild canines are social animals, living in packs that cooperate in hunting and raising their young. Of course, there are some things that a dog cannot learn, no matter how much effort is expended in trying. Although a dog can bark, howl, or whine, it will never be able to speak. Behavioral biologists investigate how animals behave, and why they behave in one way and not another. They have learned that there are surprisingly few differences other than language between humans and apes, for example. Even you and this puppy share more behavior than you might expect.

37.1 Approaches to the Study of Behavior

Animals respond to their environment in many ways. Beavers build dams in the fall that create lakes, and birds sing in the spring. Bees search for honey, and when they find it, they fly back to the hive and spread the good news. To understand these behaviors, we need to appreciate both the internal factors that shape the way an animal behaves, as well as the aspects of the external environment that trigger an individual behavior.

Behavior can be defined as the way an organism responds to stimuli in its environment. These stimuli might be as simple as the odor of food. In this sense, a bacterial cell "behaves" by moving toward higher concentrations of sugar. In animals, behaviors are far more complex than this simple bacterial one. This is particularly true for animals with nervous systems. Using eyes, ears, and a variety of other sense organs, they are able to perceive environmental stimuli, process the information, and dictate appropriate body responses, which can be both complex and subtle.

Explaining Behavior

When we observe animal behavior, we can examine it in two different ways. First, we might ask *how* it all works. How do the animal's senses, nerve networks, and internal state act together physiologically to produce the behavior? Like a mechanic studying the behavior of a car, we are asking how the machine works. A psychologist would say we would be asking a question of **proximate causation.** To analyze the proximate cause of behavior, we might measure hormone levels, or record the impulse activity of particular neurons in the brain. The field of psychology often focuses on proximate causes.

We might also ask *why* it all works the way it does. Why did the behavior evolve in this way? What is the adaptive value of this particular response to the environment? This is a question of **ultimate causation.** To study the ultimate cause of a behavior, we would attempt to find how it influenced the animal's survival or reproductive success. The field of animal behavior typically focuses on ultimate causes.

Any behavior can be looked at either way. For example, a male songbird may sing during the breeding season (**figure 37.1**). Why? One explanation is that longer spring days have induced in his body elevated levels of the steroid sex hormone testosterone, which at these high levels bind to hormone receptors in the songbird's brain and trigger the production of song. Elevated testosterone would be the proximate cause of the male songbird's song.

Another explanation is that the male is exhibiting a pattern of behavior produced by natural selection to better adapt it to its environment. Seen in this light, the male songbird sings to defend a territory from other males, and to attract a female to reproduce. These reproductive motives are the

Figure 37.1 Two ways to look at a behavior.
This male songbird can be said to be singing because his levels of testosterone are elevated, triggering innate "song" programs in his brain. Viewed in a different light, he is singing to defend his territory and attract a mate, behaviors that have evolved to increase his reproductive fitness.

ultimate, or evolutionary, explanation of the male songbird's behavior.

A Controversial Field of Biology

The study of behavior has had a long history of controversy. One source of controversy has been the question of whether an animal's behavior is determined more by an individual's genes, or by its learning and experience. In other words, is behavior the result of nature (instinct) or nurture (learning)? In the past, this question has been considered an "either-or" proposition, but we now know that instinct and learning both play significant roles, often interacting in complex ways to produce the final behavior. We will begin our study of animal behavior by examining more closely the scientific study of instinct and learning, and the ways in which they interact to determine behavior, both proximately and ultimately.

> **Key Learning Outcome 37.1** Animal behavior is the way an animal responds to stimuli in its environment. Some biologists study the physiological mechanisms producing the behavior, others the evolutionary forces responsible for its development.

37.2 Instinctive Behavioral Patterns

Early research in the field of animal behavior focused on behavioral patterns that appeared to be instinctive (or "innate"). Because behavior in animals is often stereotyped (that is, appearing in the same way in different individuals of a species), behavioral scientists argued that the behaviors must be based on preset paths in the nervous system. In their view, these neural paths are structured from genetic blueprints and cause animals to show essentially the same behavior from the first time it is produced throughout their lives.

This study of the instinctive nature of animal behavior was typically carried out in the field rather than on laboratory animals. The study of animal behavior in natural conditions is called **ethology.** The three scientists most responsible for founding the field of ethology were Karl von Frisch, Konrad Lorenz, and Niko Tinbergen. These scientists were awarded the Nobel Prize in Physiology or Medicine in 1973 for their path-making contributions.

An Example of Innate Behavior

The study by Konrad Lorenz of the egg retrieval behavior of geese provides a clear example of what ethologists mean by innate behavior. The drawings in **figure 37.2a** illustrate how when a goose is incubating its eggs in a nest and it notices that an egg has been knocked out of the nest, it will extend its neck toward the egg, get up, and roll the egg back into the nest with a side-to-side motion of its neck while the egg is tucked beneath its bill. Even if the egg is removed during retrieval, the goose completes the behavior, as if driven by a program released by the initial sight of the egg outside the nest.

According to ethologists like Lorenz, egg retrieval behavior is triggered by a **sign stimulus** (also called a key stimulus), which in this case is the appearance of an egg outside the nest. The way in which nerves are connected in the goose's brain, the **innate releasing mechanism,** responds to the sign stimulus by providing the neural instructions for the motor program, or **fixed action pattern,** which causes the goose to carry out the intricate egg retrieval behavior.

More generally, the sign stimulus is a "signal" in the environment that triggers a behavior. The innate releasing mechanism is the hard-wired element of the brain, and the fixed action pattern is the stereotyped act.

Studying fixed action patterns in birds and other animals, ethologists have discovered that in some situations, a wide variety of objects will trigger a fixed action pattern. For example, geese will attempt to roll baseballs, and even beer cans, back into their nests!

A clear example of the general nature of some sign stimuli is seen in the mating behavior of male stickleback fish studied by Niko Tinbergen. During the breeding season, males develop bright red coloration on their undersides. The males are very territorial, reacting aggressively to the approach of other males. They first perform an aggressive display (shown on the right in

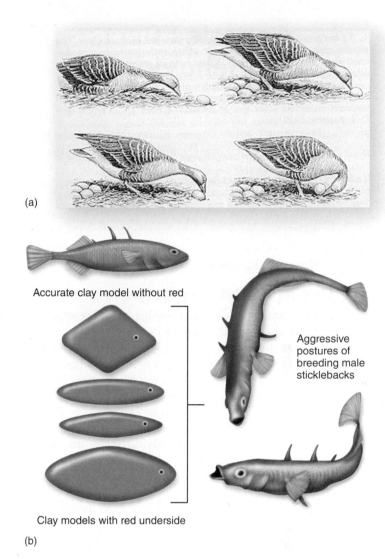

(a)

Accurate clay model without red

Clay models with red underside

(b)

Aggressive postures of breeding male sticklebacks

Figure 37.2 Sign stimulus and fixed action pattern.

(a) The series of movements used by a goose to retrieve an egg is a fixed action pattern. Once it detects the sign stimulus (in this case, an egg outside the nest), the goose goes through the entire set of movements: It will extend its neck toward the egg, get up, and roll the egg back into the nest with a side-to-side motion of its neck while the egg is tucked underneath its bill. (b) In stickleback fish, a red color acts as a sign stimulus to trigger a fixed action pattern in males: aggressive threat displays or postures. When the clay models above are presented to a male stickleback, he will display less often to the first model, which looks more like a male stickleback but lacks the red belly characteristic of breeding males.

figure 37.2b), and if the invading male is not deterred, they attack it. However, when Niko Tinbergen observed a male stickleback in a laboratory aquarium display aggressively when a red firetruck passed by the window, he realized that the red color was the sign stimulus. In experiments using models shown on the left in **figure 37.2b**, he was able to produce the aggressive display in males by challenging them with the many unfishlike models, so long as the models had a red strip.

> **Key Learning Outcome 37.2** The ethological approach to studying animal behavior has emphasized innate, instinctive behaviors that are the result of preset pathways in the nervous system.

37.3 Genetic Effects on Behavior

Although most animal behaviors are not "hard-wired" instincts, such as those studied by the early ethologists, animal behaviorists who followed these pioneer researchers have clearly demonstrated that many animal behaviors are strongly influenced by genes passed from parent to offspring. In other words, "nature" plays a key role in determining patterns of behavior.

If genes determine behavior, then it should be possible to study their inheritance, much as Mendel studied the inheritance of flower color in garden peas. This sort of investigation is called **behavioral genetics.**

Studies of Genetic Hybrids

Behavioral genetic studies have revealed many examples of behaviors that seem to be inherited in a Mendelian manner. William Dilger of Cornell University examined two species of lovebirds that differ in the way they carry twigs, paper, and other materials used to build a nest. One species, Fisher's lovebird, holds nest materials in its beak, while another, the peachfaced lovebird, carries material tucked beneath its flank (tail) feathers. When Dilger crossed the two species to produce hybrids, he found that the hybrids carry nest material in a way that is intermediate between that of the parents: They repeatedly shift material between the bill and the flank feathers. Other studies conducted on courtship songs in crickets and tree frogs also demonstrate the intermediate nature of hybrid behavior; hybrids, possessing alleles from both parental species, produce songs that are a combination of the songs of their parents.

Studies of Twins

The influence of genes on behavior can also be seen in humans by comparing the behavior of identical twins. Identical twins are, as their name implies, genetically identical. Because most sets of identical twins are raised together, any similarities in their behavior might result either from identical genes, or from shared experiences as they grow up. However, in some instances twins have been separated at birth and raised apart in different families. A recent study of 50 such sets of twins revealed many similarities in personality, temperament, and even leisure-time activities, even though the twins were often raised in very different circumstances. These results show that genes play a key role in determining human behavior, although the relative importance of genes versus environment is still hotly debated.

A Detailed Look at How One Gene Affects a Behavior

One well-studied gene mutation in mice provides a clear look at how a particular gene influences a behavior. In 1996 behavioral geneticists discovered a new gene, *fosB*, that seems to determine whether or not female mice will nurture their young. Females with both *fosB* alleles knocked

(a)

(b)

Figure 37.3   A gene alters maternal care.
In mice, normal mothers (*a*) take very good care of their offspring, retrieving them if they move away and crouching over them. Mothers with the mutant *fosB* allele (*b*) perform neither of these behaviors, leaving their pups exposed.

out (experimentally removed) will initially investigate their newborn babies, but then ignore them, in stark contrast to the caring and protective maternal care provided by normal females (figure 37.3).

This inattentiveness appears to result from a chain reaction. When mothers of new babies initially inspect them, information from their auditory, olfactory, and tactile senses is transmitted to the hypothalamus. There, *fosB* alleles are activated, producing a particular protein, which in turn activates both enzymes and other genes that affect the neural circuitry within the hypothalamus. These modifications within the brain cause the female to react maternally toward her offspring. In a general way, the information gained from inspecting the newborn babies can be viewed as acting like a sign stimulus, the *fosB* gene as an innate releasing mechanism, and the maternal behavior as the resulting action pattern.

In mothers lacking the *fosB* allele, this innate behavioral pattern is stopped midway. No protein is activated, the brain's neural circuitry is not rewired, and maternal behavior does not result.

> **Key Learning Outcome 37.3**   The conclusion that genes play a key role in many behaviors is supported by a broad range of studies in many animals, including humans.

37.4 How Animals Learn

Many of the behavioral patterns displayed by animals are not the result solely of instinct. In many cases, animals alter their behavior as a result of previous experiences, a process termed **learning.** The simplest type of learning, **nonassociative learning,** does not require an animal to form an association between two stimuli, or between a stimulus and a response. One form of nonassociative learning is *sensitization,* in which repeating a stimulus produces a greater response. Another form of nonassociative learning is *habituation,* a decrease in response to a repeated stimulus. In many cases, the stimulus evokes a strong response when it is first encountered, but the magnitude of the response gradually declines with repeated exposure. As an everyday example, are you still conscious of the chair you are sitting in? Habituation can be thought of as learning not to respond to a stimulus. Being able to ignore unimportant stimuli is critical when facing a barrage of stimuli in a complex environment.

A change in behavior that involves an association between two stimuli, or between a stimulus and a response, is called **associative learning.** The behavior is modified, or *conditioned,* through the association. This form of learning is more complex than habituation. The two major types of associative learning are called classical conditioning and operant conditioning. They differ in the way the association is established.

Classical Conditioning

In **classical conditioning,** the paired presentation of two kinds of stimuli causes the animal to form an association between the stimuli. When the Russian psychologist Ivan Pavlov presented meat powder, an *unconditioned stimulus,* to a dog, the dog responded by salivating. If an unrelated stimulus, such as the ringing of a bell, was present at the same time as the meat powder, then over repeated trials the dog would come to salivate in response to the sound of the bell alone. The dog had learned to associate the unrelated sound stimulus with the meat powder stimulus. Its response to the sound stimulus had become conditioned; the sound of the bell was now a *conditioned stimulus.*

Operant Conditioning

In **operant conditioning,** an animal learns to associate its behavioral response with a reward or punishment. Psychologist B. F. Skinner studied operant conditioning in rats by placing them in an experimental cage nicknamed a "Skinner box." As the rat explored the box, it would occasionally press a lever by accident, causing a pellet of food to appear. At first, the rat would ignore the lever, eat the food pellet, and continue to move about. Soon, however, it would learn to associate pressing the lever (the behavioral response) with food (the reward). When a conditioned rat was hungry, it would spend all its time pressing the lever. This sort of trial-and-error learning is of major importance to most vertebrates.

Figure 37.4 An unlikely parent.
The eager goslings follow ethologist Konrad Lorenz as if he were their mother. He is the first object they saw when they hatched, and they used him as a model for imprinting.

Imprinting

As an animal matures, it may form preferences or social attachments to other individuals that will profoundly influence behavior later in life. This process, called **imprinting,** is sometimes considered a type of learning. In *filial imprinting,* social attachments form between parents and offspring. For example, young birds of some species begin to follow their mother within a few hours after hatching, forming a strong bond between mother and young. This is a form of associative learning—it is the association the young bird forms during a critical window of time (roughly 13 to 16 hours in geese, for example) that determines how the imprint will be established. Birds will follow the first object they see after hatching, and direct their social behavior toward that object as their mother. Ethologist Konrad Lorenz raised geese from eggs, and when he offered himself as a model for imprinting, the goslings treated him as if he were their parent, following him dutifully (**figure 37.4**).

> **Key Learning Outcome 37.4** Habituation and sensitization are simple forms of learning in which there is no association between stimulus and response. In contrast, associative learning (conditioning and imprinting) involves the formation of an association between two stimuli or between a stimulus and a response.

37.5 Instinct and Learning Interact

Some animals have innate predispositions toward forming certain associations. Certain pairs of stimuli can be linked by operant conditioning, others not. For example, pigeons can learn to associate food with colors but not with sounds; on the other hand, they can associate danger with sounds, but not with colors. This sort of *learning preparedness* demonstrates that what an animal can learn is biologically influenced; that is, learning is possible only within the boundaries set by instinct.

The innate programs that make up an animal's instincts have evolved because each of them reinforces an adaptive response. The seed a pigeon eats may have a distinctive color that the pigeon can see, but it makes no sound the pigeon can hear. The approach of a predator a pigeon fears may generate noise, but involves no distinctive color.

Behavior Often Reflects Ecological Factors

Knowledge of an animal's ecology is key to understanding its behavior, as the genetic component of behaviors has evolved to match animals to their habitats. For example, some species of birds, like Clark's nutcracker, feed on seeds. These birds bury seeds in the ground when seeds are abundant so they will have food during the winter. Thousands of seeds may be buried by a bird and then later recovered, sometimes as much as nine months later. One would expect these birds to have an extraordinary spatial memory, and this is indeed what researchers have found. One Clark's nutcracker can remember the locations of up to 2,000 seeds, using features of the landscape and other surrounding objects as spatial references to memorize their locations. When examined, Clark's nutcrackers turn out to have an unusually large hippocampus, the center for memory storage in the brain.

The Interaction Between Instinct and Learning

The way in which white-crowned sparrows first acquire their courtship songs provides an excellent example of the interaction between instinct and learning in the development of behavior. Courtship songs are sung by mature birds and are species specific. By rearing male birds in soundproof incubators provided with speakers and microphones, animal behaviorist Peter Marler could control what a bird heard as it matured, and then record the song it produced as an adult, a recording called a sonogram. When compared to a normal sonogram, in **figure 37.5a**, he found that white-crowned sparrows that heard no song at all during development had a poorly developed song as adults, shown in the sonogram in **figure 37.5b**. The same thing happened if they heard only the song of a different species, the song sparrow. But birds that heard the song of their own species sang a fully developed white-crowned

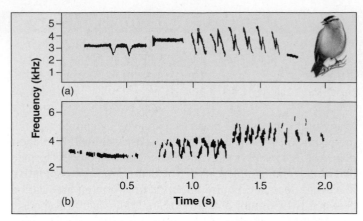

Figure 37.5 Song development in birds involves both instinct and learning.

The sonograms of songs produced by male white-crowned sparrows that had been exposed to their own species' song during development (*a*) are different from those of male sparrows that heard no song during rearing (*b*). This difference indicates that the genetic program itself is insufficient to produce a normal song.

sparrow song as adults. This was true even if the young birds also heard the song sparrow song along with their own.

Marler's results suggest that these birds have a genetic template, or instinctive program, that guides them to learn the appropriate song. During a critical period in development, the template will accept the correct song as a model. Thus, song acquisition depends on learning, but only the song of the correct species can be learned.

Although the song template is genetically determined, Marler found that learning also plays a prominent role in song development. If a young white-crowned sparrow becomes deaf *after* it hears its species' song during the critical period, it will sing a poorly developed song as an adult. The bird must "practice" listening to himself sing, matching what he hears to the model his template has accepted.

The males of some bird species have no opportunity to hear the song of their own species. In such cases, it appears that the males instinctively "know" their own species song. For example, cuckoos are brood parasites; females lay their eggs in the nest of another species of bird, and the young that hatch are reared by the foster parents. When the cuckoos become adults, they sing the song of their own species rather than that of their foster parents. Because male brood parasites would hear the song of their host species during development, it is adaptive for them to ignore such "incorrect" stimuli. They hear no adult males of their own species singing, so no correct song models are available. In this species, natural selection has provided the male with a genetically programmed song totally guided by instinct.

> **Key Learning Outcome 37.5** Behavior is both instinct (influenced by genes) and learned through experience. Genes are thought to limit the extent to which behavior can be modified and the types of associations that can be made.

Animal Cognition

For many decades, students of animal behavior flatly rejected the notion that nonhuman animals can think. Instead, the prevailing approach was to treat animals as though they responded to the environment through instinct and simple innately programmed learning.

In recent years, serious attention has been given by researchers to the topic of animal awareness. The central question is whether animals other than humans show **cognitive behavior**—that is, do they process information and respond in a manner that suggests thinking?

Evidence of Conscious Planning

What kinds of behavior would demonstrate cognition? Some birds in urban areas remove the foil caps from nonhomogenized milk bottles to get at the cream beneath. Japanese macaques (a kind of monkey) learn to float grain on water to separate it from sand, and teach other macaques to do it. A chimpanzee pulls the leaves off of a tree branch and uses the stick to probe the entrance to a termite nest and gather termites, suggesting that the ape is consciously planning ahead, with full knowledge of what it intends to do. A sea otter will use a rock as an "anvil," against which it bashes a clam to break it open, often keeping a favorite rock for a long time, as though it had a clear idea of its future use of the rock.

Problem Solving

Some instances of problem solving by animals are hard to explain in any other way than as a result of some sort of cognitive process—of what, if we were doing it, we would call reasoning. For example, in a series of classic experiments conducted in the 1920s, a chimpanzee was left in a room with a banana hanging from the ceiling out of reach. Also in the room were several boxes, each lying on the floor. After unsuccessful attempts to jump up and grab the banana, the chimpanzee suddenly looked at the boxes and immediately proceeded to move them underneath the banana, placing one on top of another, and climbed up the boxes to claim its prize. Many humans would not have solved the problem so quickly.

It is not surprising to find obvious intelligence in animals so closely related to us as chimpanzees. Perhaps more surprising, however, are recent studies finding that other animals also show evidence of cognition. Ravens have always been considered among the most intelligent of birds. Bernd Heinrich of the University of Vermont conducted an experiment using a group of hand-reared ravens that lived in an outdoor aviary. Heinrich placed a piece of meat on the end of a string and hung it from a branch in the aviary. The birds like to eat meat, but had never seen string before and were unable to get at the meat. After several hours, during which time the birds periodically looked at the meat but did nothing else, one bird flew to the branch, reached down, grabbed the string with its beak, pulled it up, and placed it under his foot. He then reached down and pulled up another length of the

Figure 37.6 Problem solving by a raven.
Confronted with a problem it had never previously encountered, the raven figures out how to get the meat at the end of the string by repeatedly pulling up a bit of string and stepping on it.

string, repeating this action over and over, each time bringing the meat closer (**figure 37.6**). Eventually the meat was within reach, and was grasped and eaten by the bird. The raven, presented with a completely novel problem, had devised a solution. Eventually, three of the other five ravens also figured out how to get the meat. This result can leave little doubt that ravens have advanced cognitive abilities.

Key Learning Outcome 37.6 Research on the cognitive abilities of animals is in its infancy, but some examples argue compellingly that animals can reason.

37.7 Behavioral Ecology

The investigation of animal behavior can be conveniently divided into three sorts of studies: (1) *The study of its development.* Lorenz's study of imprinting in geese was a study of this sort. (2) *The study of its physiological basis.* Analysis of the impact of the *fosB* gene on maternal behavior in mice was a study of this sort. (3) *The study of its function* (that is, its evolutionary significance). This third sort of study is addressed by biologists working in the field of **behavioral ecology.** Behavioral ecology is the study of how natural selection shapes behavior.

Behavioral ecology examines the survival value of behavior. How does an animal's behavior allow it to stay alive and reproduce, or keep its offspring alive to reproduce? Research in behavioral ecology thus focuses on a behavior's adaptive significance—that is, on the contribution a behavior makes to an animal's reproductive success, or fitness.

It is important to remember that all genetic differences in behavior need not have survival value. Many genetic differences in natural populations are the result of random mutations that accidentally become common, a process called genetic drift. It is only by experiment that we can learn if a particular behavior has been favored by natural selection.

Nobel laureate Niko Tinbergen's pioneering study of seagull nesting provides an excellent example of how a behavioral ecologist investigates the potential evolutionary significance of a behavior. Tinbergen observed that after gull nestlings hatched from their eggs, the parent birds quickly remove the eggshells from the nest. Why? What possible evolutionary advantage would this behavior confer on the birds?

To investigate this, Tinbergen camouflaged chicken eggs by painting them to resemble gulls' eggs that blend in with the natural background where the gull nests were located (**figure 37.7**), and distributed them on the ground throughout the nesting area. He placed broken eggshells next to some of the eggs, and, as a control, he left other camouflaged eggs alone without eggshells. He then watched to see which eggs were found more easily by crows. Because the crows could use the white interior of a broken eggshell as a cue, they repeatedly ate the camouflaged eggs that were near broken eggshells, and tended to ignore solitary camouflaged eggs that sat on the ground in plain sight. Tinbergen concluded that eggshell removal behavior is adaptive, that it does confer an evolutionary advantage on birds. Removing broken eggshells from the nest reduces predation of unhatched eggs (and probably of newborn chicks) and thus increases the chance that offspring will survive.

It is not always so easy to learn how an adaptive trait confers its evolutionary advantage. Some behaviors, like eggshell removal, reduce predation. Other behaviors

Figure 37.7 The adaptive value of eggshell removal.

Niko Tinbergen painted chicken eggs to resemble the mottled brown camouflage of gull eggs. Mottled eggs look like the rocky ground around a gull nest. The eggs were used to test the hypothesis that camouflaged eggs are more difficult for a predator to find and thus increase the young's chance of survival. He then placed broken eggshells near the camouflaged ones to test the hypothesis that the white interior of broken eggshells attracts predators.

enhance energy intake, allowing an increased number of offspring to be supported. Still others reduce exposure or increase resistance to disease, enhance the ability to acquire a mate, or in some other way increase an individual's fitness, its ability to contribute offspring to the next generation.

Key Learning Outcome 37.7 Behavioral ecology is the study of how natural selection shapes behavior.

A Cost-Benefit Analysis of Behavior

One important way in which behavioral ecologists examine the evolutionary advantage of a behavior is to ask if it provides an evolutionary benefit greater than its cost. Thus, for example, a behavior may be favored by natural selection if it increases the intake of food by parents. This is a clear adaptive benefit when it increases survival of offspring, but it comes at a cost. Searching for food or defending food supplies can expose a parent to predation, decreasing the probability that the parent will survive to raise its offspring. To understand these sorts of behaviors, it is necessary to carefully evaluate their costs and benefits.

Foraging Behavior

For many animals, the food that they eat can be found in many sizes, and in many places. An animal must choose what food to select, and how far to go in search of it. These choices are called the animal's **foraging behavior.** Each choice involves benefits and associated costs. Thus, although a larger food may contain more energy, it may be harder to capture and less abundant. In addition, more desirable foods may be farther away than other types. Hence, an animal's foraging behavior involves a trade-off between a food's energy content and the cost of obtaining it.

The net energy (in calories) gained by feeding on each kind of food available to a foraging animal is simply the energy content of the food minus the energy costs of pursuing and handling it. At first glance, one might expect that evolution would favor foraging behaviors that are as energetically efficient as possible. This sort of reasoning has led to what is known as the **optimal foraging theory,** which predicts that animals will select food items that maximize their net energy intake per unit of foraging time.

Is the optimal foraging theory correct? Many foragers do preferentially use food items that maximize the energy return per unit time. Shore crabs, for example, tend to feed primarily on intermediate-sized mussels, which provide the greatest energy return (as discussed in the Inquiry & Analysis at the end of the chapter). Larger mussels provide more energy, but also take considerably more energy to crack open. Many other animals also behave to maximize energy acquisition.

The key question, however, is whether increased energy resources acquired by optimal foraging leads to increased reproductive success. In many cases, it does. In a diverse group of animals that includes ground squirrels (**figure 37.8**), zebra finches, and orb-weaving spiders, the number of offspring raised successfully increases when parents have access to more food energy.

Figure 37.8 Optimal foraging.

Optimal foraging in this golden mantled ground squirrel pays off with increased reproductive success.

In other cases, however, the costs of foraging seem to outweigh the benefits. An animal in danger of being eaten itself is often better off to minimize the amount of time it spends foraging. Many animals alter their foraging behavior when predators are present, reflecting this trade-off between food and risk.

Territorial Behavior

Animals often move over a large area, or *home range*. In many species, the home ranges of several individuals overlap, but each individual defends only a portion of its home range and uses it exclusively. This behavior is called **territoriality.**

Territories are defended by displays that advertise that the territories are occupied, and by overt aggression. A bird sings from its perch within its territory to prevent invasion of its territory by a neighboring bird. If the intruding bird is not deterred by the song, the territory owner may attack and attempt to drive the invader away.

Figure 37.9 The benefit of territoriality.

Sunbirds, found in Africa and ecologically similar to hummingbirds, increase nectar availability by defending flowers. A sunbird will expend 3,000 calories per hour chasing intruders away.

Why aren't all animals territorial? The answer involves a cost-benefit analysis. The actual adaptive value of an animal's territorial behavior depends on the trade-off between the behavior's benefits and its costs. Territoriality offers clear benefits, including increased food intake from nearby resources (**figure 37.9**), access to refuges from predators, and exclusive access to mates. The costs of territorial behavior, however, may also be significant. The singing of a bird, for example, is energetically expensive, and attacks from competitors can lead to injury. In addition, advertisement through song or visual display can reveal one's location to a predator. In many instances, particularly when food sources are abundant, defending easily obtained resources is simply not worth the cost.

> **Key Learning Outcome 37.8** Natural selection tends to favor the evolution of foraging and territorial behaviors that maximize energy gain, although other considerations such as avoiding predators are also important.

Migratory Behavior

Many animals breed in one part of the world, and spend the rest of the year in another. Long-range, two-way annual movements like this are called **migrations** (see figure 37.11). Migratory behavior is particularly common in birds. Ducks and geese migrate southward along flyways from northern Canada across the United States each fall, overwinter, and then return northward each spring to nest. Warblers and many other insect-eating songbirds winter in the tropics and breed in the United States and Canada in spring and summer, when insects are plentiful. Monarch butterflies migrate each fall from central and eastern North America to overwinter in several small, geographically isolated areas of coniferous forests in the mountains of central Mexico. Gray whales feed in summer in the Arctic Ocean, then swim 10,000 kilometers to the warm waters off Baja, California, where they breed during winter months.

Biologists have studied migration with great interest. In attempting to understand how animals are able to navigate accurately over such long distances, it is important to understand the difference between *compass sense* (an innate ability to move in a particular direction, called "following a bearing") and *map sense* (a learned ability to adjust a bearing depending on the animal's location). Experiments on starlings shown in **figure 37.10** indicate that inexperienced birds migrate using a compass sense, and older birds that have migrated previously also employ a map sense to help them navigate—in essence, they learn the route. Migrating birds were captured in Holland, the halfway point of their migration, and were taken to Switzerland where they were released. Inexperienced birds (the red arrows) kept flying in their original direction, while experienced birds (the blue arrow) were able to adjust course and reached their normal wintering grounds.

The Compass Sense

In birds we now have a good understanding of how the compass sense is achieved. Many migrating birds have the ability to detect the earth's magnetic field and to orient themselves with respect to it. In a closed indoor cage, they will attempt to move in the correct geographical direction, even though there are no visible external clues. However, the placement of a powerful magnet near the cage can alter the direction in which the birds attempt to move.

The first migration of young birds appears to be innately guided by the earth's magnetic field. Inexperienced birds also use the sun and particularly the stars to orient themselves (migrating birds fly mainly at night).

The indigo bunting, which flies during the day and uses the sun to set its bearing, compensates for the movement of the sun in the sky as the day progresses by reference to the North Star, which does not move in the sky. Starlings compensate for the sun's apparent movement in the sky by using an internal clock. If captive starlings are shown an experimental sun in a fixed position, they will change their orientation to it at a constant rate of about 15 degrees per hour—the same rate the sun moves across the sky.

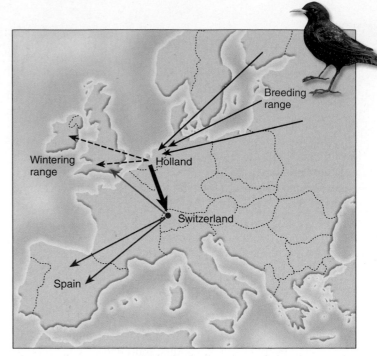

Figure 37.10 Starlings learn how to navigate.

The navigational abilities of inexperienced birds differ from those of adults that have made the migratory journey before. Starlings were captured in Holland, halfway along their full migratory route from Baltic breeding grounds to wintering grounds in the British Isles. These captured birds were transported to Switzerland and released. Experienced older birds compensated for the displacement and flew toward the normal wintering grounds (*blue* arrow). Inexperienced young birds kept flying in the same direction as before, on a course that took them toward Spain (*red* arrows).

The Map Sense

Much less is known about how migrating birds and other animals acquire their map sense. During their first migration, young birds move with a flock of experienced older birds that know the route, and during the course of the journey they appear to learn to recognize certain cues, such as the position of mountains and coastline.

Animals that migrate through featureless terrain present more of a puzzle. Consider the green sea turtle (figure 37.11). Every year great numbers of these large 400-pound turtles migrate with incredible precision from Brazil halfway across the Atlantic to Ascension Island, through 1,400 miles of open ocean, where females lay their eggs. Plowing head down through the waves, how do they find this tiny rocky island, over the horizon, more than a thousand miles away? No one knows for sure, although recent studies suggest that the direction of wave action provides an important navigational clue.

> **Key Learning Outcome 37.9** Many animals migrate in predictable ways, navigating by looking at the sun and stars, and in some cases by detecting magnetic fields. In many instances, young individuals learn the route by following experienced ones.

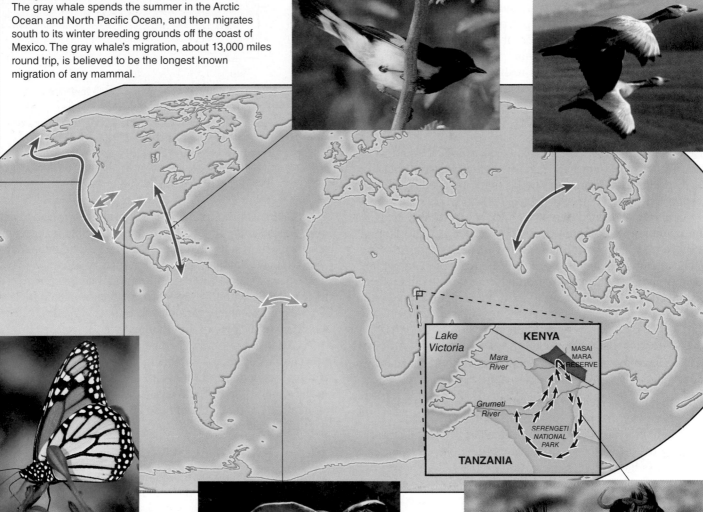

American Redstart

The American redstart is a "neotropical migrant," a species of bird that spends the spring and summer in North America and then flies south to Central America and South America for the winter. The populations of many neotropical migrants are in decline due to fragmentation and degradation of both the winter and summer habitat.

Bar-headed Goose

The bar-headed goose nests in Tibet and then flies south over the Himalayas to winter in India. During this 1,000-mile trip, the geese may fly directly over Mt. Everest, making them one of the world's highest-altitude migrants.

Gray Whale

The gray whale spends the summer in the Arctic Ocean and North Pacific Ocean, and then migrates south to its winter breeding grounds off the coast of Mexico. The gray whale's migration, about 13,000 miles round trip, is believed to be the longest known migration of any mammal.

Monarch

In the fall, monarch butterflies undergo an incredible migration of up to 2,500 miles to spend the winter roosting in trees. Monarchs west of the Rocky Mountains migrate to southern California, and monarchs east of the Rockies migrate all the way to Mexico, and amazingly, two to five generations may be produced during the migration.

Green Sea Turtle

Some populations of green sea turtles will migrate 1,300 miles across the Atlantic Ocean between their Ascension Island nesting grounds and their Brazilian coast feeding grounds. Usually, individuals will return to nest on the same beach where they were born.

Wildebeests

Every year, over 1.4 million wildebeests, along with over 200,000 zebras and gazelles, migrate in a clockwise circular pattern that follows the rains and the rivers. The herd moves from Kenya's Masai Mara in the north to the southern Serengeti of Tanzania, where the calves are born.

Figure 37.11 Examples of migration.

37.10 Reproductive Behaviors

Animals exhibit many different types of behaviors (**table 37.1**), but reproductive behavior is particularly complex, encompassing a variety of behaviors, including courtship and parental care. The reproductive success of an animal is directly affected by its reproductive behaviors because these behaviors influence how long the individual lives, how frequently it mates, and how many offspring it produces per mating. The second of these factors, competition for mating opportunities, has been termed **sexual selection.**

Sexual selection involves both *intrasexual selection,* or interactions between members of one sex ("the power to conquer other males in battle," as Darwin put it), and *intersexual selection* ("the power to charm").

Intrasexual selection leads to the evolution of structures used in combat with other males (such as a deer's antlers or a ram's horns). Combat for mates, whether ritual or real, is a form of *agonistic behavior,* a confrontation waged by threats, displays, or actual combat.

Intersexual selection, also called **mate choice,** leads to the evolution of complex courtship behaviors, and of ornaments used to "persuade" members of the opposite sex to mate, such as long tail feathers or bright plumage. For example, the male peacock will parade in front of females with tail feathers displayed. **Figure 37.12** shows that the more eyespots on a male's tail feathers, the more mates he will attract.

The Benefits of Mate Choice

Why did mating preferences evolve? What is their adaptive value? Biologists have proposed several reasons:

1. In many species of birds and mammals, males help raise the offspring. In these cases, females would benefit by choosing the male that can provide the best care—the better the male parent, the more offspring she is likely to rear successfully.
2. In other species, males provide no care, but maintain territories that provide food, nesting sites, and predator refuges. In such species, females that choose males with the best territories will maximize reproductive success.
3. In some species, males provide no direct benefits of any kind to the female. If a female selects a more vigorous male, probably at least to some degree the result of a good genetic makeup, the female will be ensuring that her offspring receive good genes from their father.

Mating Systems

Reproductive behavior in animals varies substantially from one species to the next. Some animals mate with many partners during the breeding season, others with only one. The typical number of mates an animal has during its breeding season is called the **mating system.** Among animals, there are three principal mating systems: *monogamy* (one male mates with one female), *polygyny* (one male mates with more than one female), and *polyandry* (one female mates with more than one male).

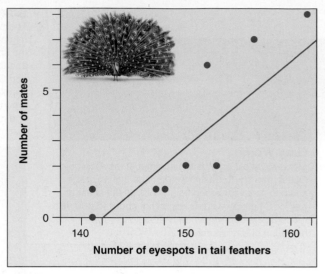

Figure 37.12 The male peacock's tail feathers are a product of sexual selection.
Experiments show that female peahens, whose tail feathers are drab, prefer to mate with males with greater numbers of colorful "eyespots" in their tail feathers.

Like mate choice, mating systems have evolved to maximize reproductive fitness. For instance, a male may defend a territory with resources sufficient for more than one female.

A male holding such a high-quality territory may already have a mate, but it is still more advantageous for a female to breed with that male than with an unmated male that defends a low-quality territory. In this way, natural selection would favor the evolution of polygyny.

Although polygyny is much more common in animals, polyandrous systems—in which one female mates with several males—are known in a variety of animals. For example, in the spotted sandpiper, a seashore bird, males take care of all incubation and parenting, and females mate and leave eggs with two or more males.

Reproductive Strategies

During the breeding season, animals make several important "decisions" concerning their choice of mates (mate choice), how many mates to have (mating system), and how much time and energy to devote to rearing offspring (parenting). These decisions are all aspects of an animal's reproductive strategy, a set of behaviors that has evolved to maximize that species' reproductive success. The two sexes of a species often have different reproductive strategies, reflecting the different parental investment each sex makes in producing and rearing offspring. In most animal species, females exercise far more mate choice, and in these species female parental investment is much greater than that of males. In species with biparental care, male and female parental investments are roughly equal, and both sexes tend to exercise mate choice.

> **Key Learning Outcome 37.10** Natural selection has favored the evolution of mate choice, mating system, and parenting behaviors that maximize reproductive success.

TABLE 37.1   ANIMAL BEHAVIORS

Foraging Behavior

Select, obtain, and consume food

Oystercatchers forage for food by stabbing the ground or rocks in search of arthropods or mollusks.

Territorial Behavior

Defend portion of home range and use it exclusively

Male elephant seals fight with each other for possession of territories. Only the largest males can hold territories, which contain many females.

Migratory Behavior

Move to a new location for part of the year

Wildebeests undergo an annual migration in search of new pastures and sources of water. Migratory herds can contain up to a million individuals and can stretch thousands of miles.

Courtship

Attract and communicate with potential mates

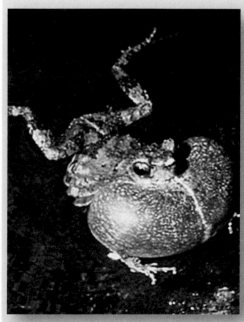

This male frog is vocalizing, producing a call that attracts females.

Parental Care

Produce and rear offspring

Female lions share the responsibility of raising the pride's young, increasing the probability that the young will survive into adulthood.

Social Behavior

Communicate information and interact with members of a social group

These leaf-cutter ants are members of different castes (or worker classes) in an insect society. The large ant is a worker, carrying leaves to the nest, and the smaller ants are protecting the worker from attack.

37.11 Communication Within Social Groups

Many insects, fish, birds, and mammals live in social groups in which information is communicated between group members. For example, some individuals in mammalian societies serve as "guards." When a predator appears, the guards give an **alarm call,** and group members respond by seeking shelter. Social insects, such as ants and honeybees, secrete chemicals called **alarm pheromones** that trigger attack behavior. Ants also deposit **trail pheromones** between the nest and a food source to lead other colony members to food (**figure 37.13**). Honeybees have an extremely complex **dance language** that directs nestmates to rich nectar sources.

Figure 37.13 Ants following a pheromone trail.
Trail pheromones organize cooperative foraging. The trails taken by the first ants to travel to a food source are soon followed by most of the other ants due to the release of pheromones by the first ants.

The Dance Language of the Honeybee

The European honeybee, *Apis mellifera,* lives in hives consisting of 30,000 to 40,000 individuals whose behaviors are integrated into a complex colony. Worker bees may forage miles from the hive, collecting nectar and pollen from a variety of plants on the basis of how energetically rewarding their food is. The food sources used by bees tend to occur in patches, and each patch offers much more food than a single bee can transport to the hive. A colony is able to exploit the resources of a patch because of the behavior of scout bees, which locate patches and communicate their location to hivemates through a *dance language.* Over many years, Nobel laureate Karl von Frisch was able to unravel the details of this communication system.

After a successful scout bee returns to the hive, she performs a remarkable behavior pattern called a *waggle dance* on a vertical comb (**figure 37.14**). The path of the bee during the dance resembles a figure-eight. On the straight part of the path, the bee vibrates or waggles her abdomen while producing bursts of sound. She may stop periodically to give her hivemates a sample of the nectar she has carried back to the hive in her crop. As she dances, she is followed closely by other bees, which soon appear as foragers at the new food source.

Von Frisch and his colleagues claimed that the other bees use information in the waggle dance to locate the food source. According to their explanation, the scout bee indicates the *direction* of the food source by representing the angle between the food source, the hive, and the sun as the deviation from vertical of the straight part of the dance performed on the hive wall (that is, if the bee moved straight up, then the food source would be in the direction of the sun, but if the food was at a 30-degree angle relative to the sun's position, then the bee would move upward at a 30-degree angle

from vertical). The *distance* to the food source is indicated by the tempo, or degree of vigor, of the dance.

Adrian Wenner, a scientist at the University of California, did not believe that the dance language communicated anything about the location of food, and he challenged von Frisch's explanation. Wenner maintained that flower odor was the most important cue allowing recruited bees to arrive at a new food source. A heated controversy ensued as the two groups of researchers published articles supporting their positions.

Such controversies can be very beneficial, because they often generate innovative experiments. In this case, the "dance language controversy" was resolved (in the minds of

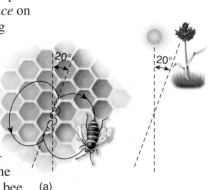

(a)

(b)

Figure 37.14 The waggle dance of honeybees.
(*a*) The angle between the food source, the nest, and the sun is represented by a dancing bee as the angle between the straight part of the dance and vertical. Here the food is 20 degrees to the right of the sun, and the straight part of the bee's dance on the hive is 20 degrees to the right of vertical. (*b*) A scout bee dancing in the hive.

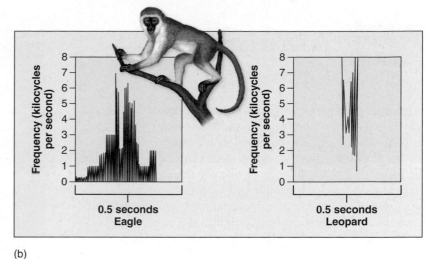

(a)
(b)

Figure 37.15   Primate semantics.

(a) This leopard, an efficient predator of primates, has attacked and will eat a vervet monkey. (b) Escaping a leopard attack presents a very different challenge to a vervet monkey than avoiding an eagle, another key predator of monkeys. Vervet monkeys give different alarm calls when leopards are sighted by troop members than they do when a member sees an eagle. Each distinctive call elicits a different and adaptive escape behavior.

most scientists) in the mid-1970s by the creative research of James L. Gould. Gould devised an experiment in which hive members were tricked into misinterpreting the directions given by the scout bee's dance. As a result, Gould was able to manipulate where the hive members would go if they were using visual signals. If odor was the cue they were using, hive members would have appeared at the food source, but instead they appeared exactly where Gould predicted. This confirmed von Frisch's ideas.

Recently, researchers have extended the study of the honeybee dance language by building robot bees whose dances can be completely controlled. Their dances are programmed by a computer and perfectly reproduce the natural honeybee dance; the robots even stop to give food samples! The use of robot bees has allowed scientists to determine precisely which cues direct hivemates to food sources.

Primate Language

Some primates have a "vocabulary" that allows individuals to communicate the identity of specific predators. The vocalizations of African vervet monkeys, for example, distinguish eagles, leopards, and snakes. The two sonograms in **figure 37.15b** show the alarm calls for an eagle and a leopard, each eliciting different responses in other members of the troop. Chimpanzees and gorillas can learn to recognize a large number of symbols and use them to communicate abstract concepts.

The complexity of human language would at first appear to defy biological explanation, but closer examination suggests that the differences are in fact superficial—all languages share many basic structural similarities. All of the roughly 3,000 human languages draw from the same set of 40 consonant sounds (English uses two dozen of them), and any human can learn them. Researchers believe these similarities reflect

the way our brains handle abstract information, a genetically determined characteristic of all humans.

Language develops at an early age in humans. Human infants are capable of recognizing the 40 consonant sounds characteristic of speech, including those not present in the particular language they will learn, while they ignore other sounds. In contrast, individuals who have not heard certain consonant sounds as infants can only rarely distinguish or produce them as adults. That is why English speakers have difficulty mastering the throaty French "r," French speakers typically replace the English "th" with "z," and native Japanese often substitute "r" for the unfamiliar English "l." Children go through a "babbling" phase, in which they learn by trial and error how to make the sounds of language. Even deaf children go through a babbling phase using sign language. Next, children quickly and easily learn a vocabulary of thousands of words. Like babbling, this phase of rapid learning seems to be genetically programmed. It is followed by a stage in which children form simple sentences that, though they may be grammatically incorrect, can convey information. Learning the rules of grammar constitutes the final step in language acquisition.

While language is the primary channel of human communication, odor and other nonverbal signals (such as "body language") may also convey information. However, it is difficult to determine the relative importance of these other communication channels in humans.

Key Learning Outcome 37.11   The study of animal communication involves analysis of the specificity of signals, their information content, and the methods used to produce and receive them.

37.12 | Altruism and Group Living

Altruism—the performance of an action that benefits another individual at a cost to the actor—occurs in many guises in the animal world. In many bird species, for example, parents are assisted in raising their young by other birds, which are called *helpers at the nest.* In species of both mammals and birds, individuals that spy a predator will give an alarm call, alerting other members of their group, even though such an act would seem to call the predator's attention to the caller. Finally, lionesses with cubs will allow all cubs in the pride to nurse, including cubs of other females.

The existence of altruism has long perplexed evolutionary biologists. If altruism imposes a cost to an individual, how could an allele for altruism be favored by natural selection? One would expect such alleles to be at a disadvantage, and thus their frequency in the gene pool should decrease through time.

A number of explanations have been put forward to explain the evolution of altruism. One suggestion often heard on television documentaries is that such traits evolve for the good of the species. The problem with such explanations is that natural selection operates on individuals within species, not on species themselves. Thus, natural selection will not favor alleles that lead an individual to act in ways that benefit others at a cost to itself; it is even possible for traits to evolve that are detrimental to the species as a whole, as long as they benefit the individual.

In some cases, selection can operate on groups of individuals, but this is rare. For example, if an allele for supercannibalism evolved within a population, individuals with that allele would be favored, as they would have more to eat; however, the group might eventually eat itself to extinction, and the allele would be removed from the species. In certain circumstances, such **group selection** can occur, but the conditions for it to occur are rarely met in nature. In most cases, consequently, the "good of the species" cannot explain the evolution of altruistic traits.

Another possibility is that seemingly altruistic acts aren't altruistic after all. For example, helpers at the nest are often young and gain valuable parenting experience by assisting established breeders. Moreover, by hanging around an area, such individuals may inherit the territory when the established breeders die. Similarly, alarm callers (**figure 37.16**) may actually benefit by causing other animals to panic. In the ensuing confusion, the caller may be able to slip off undetected. Detailed field studies in recent years have demonstrated that some acts truly are altruistic, but others are not as they seemed.

Reciprocity

Robert Trivers, now of Rutgers University, proposed that individuals may form "partnerships" in which mutual exchanges of altruistic acts occur because it benefits both participants to do so. In the evolution of such reciprocal altruism, "cheaters"

Figure 37.16 An altruistic act—or is it?
A meerkat sentinel on duty. Meerkats, *Suricata suricata,* are a species of highly social mongoose living in the semiarid sands of the Kalahari Desert in southern Africa. This meerkat is taking its turn to act as a lookout for predators. Under the security of its vigilance, the other members of the group can focus their attention on foraging. The sentinel puts his own life at risk when he gives an alarm, an apparent example of altruistic behavior.

(nonreciprocators) are discriminated against and are cut off from receiving future aid. According to Trivers, if the altruistic act is relatively inexpensive, the small benefit a cheater receives by not reciprocating is far outweighed by the potential cost of not receiving future aid. Under these conditions, cheating should not occur.

For example, vampire bats roost in hollow trees in groups of 8 to 12 individuals. Because these bats have a high metabolic rate, individuals that have not fed recently may die. Bats that have found a host imbibe a great deal of blood; giving up a small amount presents no great energy cost to the donor, and it can keep a roostmate from starvation. Vampire bats tend to share blood with past reciprocators. If an individual fails to give blood to a bat from which it had received blood in the past, it will be excluded from future bloodsharing.

Kin Selection

The most influential explanation for the origin of altruism was presented by William D. Hamilton in 1964. It is perhaps best introduced by quoting a passing remark made in a pub in 1932 by the great population geneticist J. B. S. Haldane. Haldane said that he would willingly lay down his life for two brothers or eight first cousins. Evolutionarily speaking, Haldane's statement makes sense, because for each allele Haldane received from his parents, his brothers each had a 50% chance of receiving the same allele. Consequently, it is statistically expected that two of his brothers would pass on as many of Haldane's particular combination of alleles to the next generation as Haldane himself would. Similarly, Haldane and a first cousin would share an eighth of their alleles. Their parents, which are siblings, would each share half their alleles, and each of their children would receive half of these, of which half on the average would be in common: $1/2 \times 1/2 \times 1/2 = 1/8$. Eight first cousins would therefore pass on as many of those alleles to the next generation as Haldane himself would. Hamilton saw Haldane's point clearly: Natural selection will favor any strategy that increases the net flow of an individual's alleles to the next generation.

Hamilton showed that by directing aid toward kin, or close genetic relatives, an altruist may increase the reproductive success of its relatives enough to compensate for the reduction in its own fitness. Because the altruist's behavior increases the propagation of its own alleles in relatives, it will be favored by natural selection. Selection that favors altruism directed toward relatives is called **kin selection.** Although the behaviors being favored are cooperative, the genes are actually "behaving selfishly," because they encourage the organisms to support copies of themselves in other individuals. In other words, if an individual has a dominant allele that causes altruism, any action that increases the fitness of relatives and thus increases the frequency of this allele in future generations will be favored, even if that action is detrimental to the particular individual taking the action.

Examples of Kin Selection

Hamilton's kin selection model predicts that altruism is likely to be directed toward close relatives. The more closely related two individuals are, the greater the potential genetic payoff.

Many examples of kin selection are known from the animal world. Belding's ground squirrels give alarm calls when they spot a predator such as a coyote or a badger. Such predators may attack a calling squirrel, so giving a signal places the caller at risk. The social unit of a ground squirrel colony consists of a female and her daughters, sisters, aunts, and nieces. When they mature, males disperse long distances from where they are born; thus, adult males in the colony are not genetically related to the females. By marking all squirrels in a colony with an individual dye pattern on their fur and by recording which individuals gave calls and the social circumstances of their calling, researchers found that females

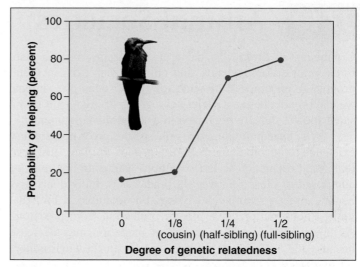

Figure 37.17 Kin selection is common among vertebrates.

In the white-fronted bee-eaters (*Merops bullockoides*), nonbreeding individuals will help raise the offspring of others. Most helpers are close relatives, and the probability that a bird will help another increases with genetic relatedness.

who have relatives living nearby are more likely to give alarm calls than females with no kin nearby. Males tend to call much less frequently, as would be expected because they are not related to most colony members.

Another example of kin selection comes from a bird called the white-fronted bee-eater, which lives along rivers in Africa in colonies of 100 to 200 birds. In contrast to the ground squirrels, it is the males that usually remain in the colony in which they were born, and the females that disperse to join new colonies. Many bee-eaters do not raise their own offspring, but rather help others. Many of these birds are relatively young, but helpers also include older birds whose nesting attempts have failed. The presence of a single helper, on average, doubles the number of offspring that survive. Two lines of evidence support the idea that kin selection is important in determining helping behavior in this species. First, helpers are normally males, which are usually related to other birds in the colony, and not females, which are not related. Second, when birds have the choice of helping different parents, they almost invariably choose the parents to which they are most closely related. The graph in **figure 37.17** compares the probability of a bird helping in a nest (on the *y* axis) and the relationship of the helper (on the *x* axis). The more closely related the helper (toward the right side of the graph), the higher the probability that it will be a helper in the nest.

> **Key Learning Outcome 37.12** Many factors could be responsible for the evolution of altruistic behaviors. Individuals may benefit directly if altruistic acts are reciprocated; also kin selection explains how alleles for altruism can increase in frequency if altruistic acts are directed toward relatives.

Animal Societies

Organisms as diverse as bacteria, cnidarians, insects, fish, birds, prairie dogs, whales, and chimpanzees exist in social groups. To encompass the wide variety of social phenomena, we can broadly define a **society** as a group of organisms of the same species that are organized in a cooperative manner.

Why have individuals in some species given up a solitary existence to become members of a group? We have just seen that one explanation is kin selection, where groups may be composed of close relatives. In other cases, individuals may benefit directly from social living. For example, a bird that joins a flock may receive greater protection from predators. As flock size increases, the risk of predation may decrease because there are more individuals to scan the environment for predators.

Figure 37.18   Savanna-dwelling African weaver birds form colonial nests.

Insect Societies

In insects, sociality has chiefly evolved in two orders, the Hymenoptera (ants, bees, and wasps) and the Isoptera (termites), although a few other insect groups include social species. These social insect colonies are composed of different **castes,** groups of individuals that differ in size and morphology and that perform different tasks, such as workers and soldiers.

Honeybees   In honeybees, the queen maintains her dominance in the hive by secreting a pheromone, called "queen substance," that suppresses development of the ovaries in other females, turning them into sterile workers. Drones (male bees) are produced only for purposes of mating. When the colony grows larger in the spring, some members do not receive a sufficient quantity of queen substance, and the colony begins preparations for swarming. Workers make several new queen chambers, in which new queens begin to develop. Scout workers look for a new nest site and communicate its location to the colony. The old queen and a swarm of female workers then move to the new site. Left behind, a new queen emerges, kills the other potential queens, flies out to mate, and returns to assume "rule" of the hive.

Leaf-Cutter Ants   The leaf-cutter ants provide another fascinating example of the remarkable lifestyles of social insects. Leaf-cutters live in colonies of up to several million individuals, growing crops of fungi beneath the ground. Their mound-like nests are underground "cities" covering more than 100 square meters, with hundreds of entrances and chambers as deep as 5 meters beneath the ground. The division of labor among the worker ants is related to their size. Every day, workers travel along trails from the nest to a tree or a bush, cut its leaves into small pieces, and carry the pieces back to the nest. Smaller workers chew the leaf fragments into a mulch, which they spread like a carpet in the underground fungus chambers. Even smaller workers implant fungal hyphae in the mulch (recent molecular studies suggest that ants have been cultivating these fungi for more than 50 million years!). Soon a luxuriant garden of fungi is growing. While other workers weed out undesirable kinds of fungi, nurse ants carry the larvae of the nest to choice spots in the garden, where the larvae graze. Some of these larvae grow into reproductive queens that will disperse from the parent nest and start new colonies, repeating the cycle.

Vertebrate Societies

In contrast to the highly structured and integrated insect societies and their remarkable forms of altruism, vertebrate social groups are usually less rigidly organized and cohesive. Each social group of vertebrates has a certain size, stability of members, number of breeding males and females, and type of mating system. Behavioral ecologists have learned that the way a vertebrate group is organized is influenced most often by ecological factors such as food type and predation.

African weaver birds, which construct nests from vegetation, provide an excellent example to illustrate the relationship between ecology and social organization. Their roughly 90 species can be divided according to the type of social group they form. One set of species lives in the forest and builds camouflaged, solitary nests. Males and females are monogamous; they forage for insects to feed their young. The second group of species nests in colonies in trees on the savanna (**figure 37.18**). They are polygynous and feed in flocks on seeds. The feeding and nesting habits of these two sets of species are correlated with their mating systems. In the forest, insects are hard to find, and both parents must cooperate in feeding the young. The camouflaged nests do not call the attention of predators to their brood. On the open savanna, building a hidden nest is not an option. Rather, savanna-dwelling weaver birds protect their young from predators by nesting in trees, which are not very abundant. This shortage of safe nest sites means that birds must nest together in colonies. Because seeds occur abundantly, a female can acquire all the food needed to rear young without a male's help. The male, free from the duties of parenting, spends his time courting many females—a polygynous mating system.

Key Learning Outcome 37.13   Insect societies are very structured and are composed of different castes. The extent of vertebrate sociality is affected by environmental conditions.

37.14 Human Social Behavior

One of the most profound lessons of biology is that we human beings are animals, quite close relatives of the chimpanzee, and not some special form of life set apart from the earth's other creatures. This biological view of human life raises an important issue as we conclude this chapter. To what degree are the animal behaviors we have described in this chapter characteristic of the social behavior of humans?

Genes and Human Behavior

As we have seen repeatedly, genes play a key role in determining many aspects of animal behavior. From maternal behavior in mice to migratory behavior in songbirds, changes in genes have been shown to have a profound impact. This leads to an important prediction. If behaviors have a genetic basis, and if behaviors affect an animal's ability to survive and raise young, then surely behaviors must be subject to natural selection, as are any other gene-mediated traits that impact survival. Ethology, the study of animal behavior in natural conditions, has provided ample evidence that behaviors do indeed evolve.

Figure 37.19   A New York City street scene.

To the degree that social behaviors are affected by genes (and both theory and evidence strongly suggest they are), the complex social behaviors of animals, including humans, should also evolve. The study of this facet of animal behavior, often called **sociobiology,** was pioneered by E. O. Wilson of Harvard University. It has proven highly controversial.

Genes certainly affect human behavior in important ways. Human facial expressions are similar the world over, no matter the culture or language, arguing that they have a deep genetic basis. Infants born blind still smile and frown, even though they have never seen these expressions on another face. Similarly, studies of identical twins demonstrate that personality and intelligence are highly heritable—well over 50% of the variance (the "scatter" among individuals) in these traits is due to the contribution of genes.

It is also true, however, that learning has a huge impact on how humans behave. Of the 40 different consonant sounds that humans can make, babies in the United States learn some 24, and then most quickly lose the ability to make the others. Just as birds learn songs and migration patterns from experienced adults, so human babies learn to speak from listening to the adults around them.

Diversity Is the Hallmark of Human Culture

When a group of biologists who studied social behavior in chimpanzees among seven widely separated areas compared notes, they identified 39 behaviors involving such things as social behavior, courtship, and tool use that were common in some groups but absent in others. Each population seemed to have a distinct repertoire of behaviors. It seems that each population has developed its own collection of customary behaviors, which each generation teaches its children. In a word, chimpanzees have culture.

No animal, not even our close relative the chimpanzee, exhibits cultural differences to the degree seen in human populations (**figure 37.19**). There are polyandrous, polygynous, and monogamous societies. There are human groups in which warfare is common, others in which it never occurs. In some cultures, marriage to first cousins is forbidden; in others, it is encouraged. The variation in social behavior of other species is vanishingly small compared to the enormous diversity of human cultures.

To what degree human culture is the product of evolution acting on behavior-determining genes is a question that is hotly debated by behavioral biologists. However one chooses to answer this question, it is perfectly clear that much of our behavior is molded by experience. Human cultures, and the behaviors that produce them, change very rapidly, far too rapidly to reflect genetic evolution. Human cultural diversity, a hallmark of our species, is certainly strongly influenced, if not largely determined, by learning and experience.

> **Key Learning Outcome 37.14   In humans, as in other animals, behavior is molded by experience within limits set by our genes. Both heredity and learning play key roles in determining how we behave.**

Do Crabs Eat Sensibly?

Many behavioral ecologists claim that animals exhibit so-called optimal foraging behavior. The idea is that because an animal's choice in seeking food involves a trade-off between the food's energy content and the cost of obtaining it, evolution should favor foraging behaviors that optimize the trade-off.

While this all makes sense, it is not at all clear that this is what animals would actually do. This optimal foraging approach makes a key assumption—that maximizing the amount of energy acquired will lead to increased reproductive success. In some cases this is clearly true. As discussed earlier in this chapter, in ground squirrels, zebra finches, and orb-weaving spiders, researchers have found a direct relationship between net energy intake and the number of offspring raised successfully.

However, animals have other needs besides energy, and sometimes these needs conflict. One obvious "other need," important to many animals, is to avoid predators. It makes little sense for you to eat a little more food if doing so greatly increases the probability that you yourself will be eaten. Often the behavior that maximizes energy intake increases predation risk. A shore crab foraging for mussels on a beach exposes itself to predatory gulls and other shore birds with each foray. Thus the behavior that maximizes fitness may reflect a trade-off, obtaining the most energy with the least risk of being eaten. Not surprisingly, a wide variety of animals use a more cautious foraging behavior when predators are present—becoming less active and staying nearer to cover.

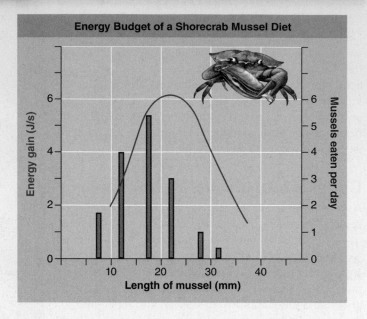

Energy Budget of a Shorecrab Mussel Diet

So what does a shore crab do? To find out, an investigator looked to see if shore crabs in fact feed on those mussels that provide the most energy, as the theory predicts. He found that the mussels on the beach he studied come in a range of sizes, from small ones less than 10 mm in length that are easy for a crab to open but yield the least amount of energy, to large mussels over 30 mm in length that yield the most energy but also take considerably more energy to pry open. To obtain the most net energy, the optimal approach, described by the blue curve in the graph above, would be for shore crabs to feed primarily on intermediate-sized mussels about 22 mm in length. Is this in fact what shore crabs do? To find out, the researcher carefully monitored the size of the mussels eaten each day by the beach's population of shore crabs. The results he obtained—the numbers of mussels of each size actually eaten—are presented in the red histogram above.

1. **Applying Concepts** What is the dependent variable in the curve? in the histogram?
2. **Making Inferences**
   a. What is the most energetically optimal mussel size for the crabs to eat, in mm?
   b. What size mussel is most frequently eaten by crabs, in mm?
3. **Drawing Conclusions** Do shore crabs tend to feed on those mussels that provide the most energy?
4. **Further Analysis** What factors might be responsible for the slight difference in peak prey length relative to the length optimal for maximal energy gain?

Some Behavior Is Genetically Determined

37.1 Approaches to the Study of Behavior

- The study of animal behavior, how animals respond to stimuli in their environment, includes examining how and why the behavior occurs. Both instinct (nature) and learning (nurture) play significant roles in behavior (**figure 37.1**).

37.2 Instinctive Behavioral Patterns

- Instinctive, or innate, behaviors are those that are the same in all individuals of a species and appear to be controlled by preset pathways in the nervous system. A sign stimulus triggers the behavior, called a fixed action pattern, such as egg retrieval in geese shown here from **figure 37.2**.

37.3 Genetic Effects on Behavior

- Most behaviors are not "hard-wired" instincts. Instead they are strongly influenced by genes and so they can be studied as inherited traits. Hybrids, twins, and genetically altered mice (**figure 37.3**) have been used to study genetically influenced behaviors.

Behavior Can Also Be Influenced by Learning

37.4 How Animals Learn

- Many behaviors are learned, having been formed or altered based on previous experiences. Classical conditioning results when two stimuli are paired such that the animal learns to associate the two stimuli. Operant conditioning results when an animal associates a behavior with a reward or punishment. Imprinting is when an animal forms social attachments, usually during a critical window of time (**figure 37.4**).

37.5 Instinct and Learning Interact

- Behavior is often both genetically determined (instinctive) and modified by learning. Genes can limit the extent to which a behavior can be modified through learning (**figure 37.5**). Ecology has a lot to do with behavior, and knowing an animal's ecological niche can reveal much about its behavior.

37.6 Animal Cognition

- While humans have evolved a great capacity for cognitive thought, studies show that other animals possess varying degrees of cognitive abilities. Some behaviors in animals show conscious planning ahead. Still other animals show problem-solving abilities. When presented with a novel situation, such as a piece of meat dangling out of reach of a raven (shown here from **figure 37.6**), these animals respond to the situation in a problem-solving capacity.

Evolutionary Forces Shape Behavior

37.7 Behavioral Ecology

- Behavioral ecology is the study of how natural selection shapes behavior. Only behaviors that have a genetic basis and offer some advantage for survival or reproduction can be acted upon through natural selection (**figure 37.7**).

37.8 A Cost Benefit Analysis of Behavior

- For every behavior that offers an individual an advantage for survival, there is usually an associated cost. For example, foraging and territorial behaviors, exhibited by this sunbird from **figure 37.9**, offer a benefit by providing food and shelter for the individual and their offspring but may endanger the parents through predation or expenditure of energy. The benefits have to outweigh the costs in order for the behaviors to be favored by natural selection.

37.9 Migratory Behavior

- Migration is a behavior that changes throughout the life of an animal. Inexperienced animals seem to rely on compass sense (following a direction), while experienced animals may rely more on map sense (learned ability to alter the path based on location) (**figure 37.10**).

37.10 Reproductive Behaviors

- Behaviors that maximize reproduction are favored by natural selection. Often these behaviors involve mate choice, mating systems, and parenting behaviors. Mate choice has lead to the evolution of complex courtship behaviors and ornate physical characteristics (**figure 37.12**).

Social Behavior

37.11 Communication Within Social Groups

- Communication is a behavior found in animals that live in groups or societies. Some animals secrete chemical pheromones to communicate information to others (**figure 37.13**). Others use movements, like the waggle dance of the honeybee shown here from **figure 37.14**. Although not as complex as human language, other animals are able to communicate a great deal of information through auditory signals (**figure 37.15**).

37.12 Altruism and Group Living

- Altruistic behaviors evolved in animals that live in groups. The reason why may involve the reciprocation of altruistic acts or benefits for relatives, called kin selection (**figure 37.17**).

37.13 Animal Societies

- Many types of animals live in social groups, or societies (**figure 37.18**). Some insect societies are highly structured.

37.14 Human Social Behavior

- Both genetics and learning play key roles in human behaviors, but the extent of each is hotly debated.

1. Innate behavior patterns
   a. can be modified if the stimulus changes.
   b. cannot be modified, as these behaviors seem built into the brain and nervous system.
   c. can be modified if environmental conditions begin to vary over a long period, a year or more.
   d. cannot be modified, as these behaviors are learned while very young.
2. The study of mouse maternal behavior shows a clearer genetic effect on behavior than lovebird or human twin studies because
   a. there is a clear link between presence or absence of a specific gene, a specific metabolic pathway, and a specific behavior.
   b. the behavior of mice was less complex and easier to study than that of lovebirds or human twins.
   c. parental care has a larger influence on mice than on lovebirds and humans.
   d. None of the above.
3. Training a dog to perform tricks using verbal commands and treats is an example of
   a. nonassociative learning.
   b. operant conditioning.
   c. classical conditioning.
   d. imprinting.
4. Behavior can be tied to ecology and evolution by considering what the behavior does to increase
   a. body size.
   b. number of breeding sites.
   c. reproductive fitness.
   d. territory size.
5. The selection of foods and the journey to seek those foods is called
   a. territoriality.
   b. imprinting.
   c. migratory behavior.
   d. foraging behavior.
6. Courtship rituals are thought to have come about through
   a. intrasexual selection.
   b. agonistic behavior.
   c. intersexual selection.
   d. kin selection.
7. A mating system where the female mates with more than one male is
   a. protandry.
   b. polyandry.
   c. polygyny.
   d. monogamy.
8. Sharing of blood meals between fed and hungry vampire bats and the shunning of a bat who takes but doesn't share are examples of
   a. helpers.
   b. kin selection.
   c. reciprocity.
   d. group selection.
9. Bird offspring who help their parents care for younger offspring are showing
   a. brood parasitism.
   b. kin selection.
   c. reciprocity.
   d. group selection.
10. While vertebrate animal societies are more loosely structured than insect societies, the organization of these societies is most influenced by
    a. which females are receptive to reproduction.
    b. migration patterns.
    c. how large the territory is compared to neighboring societies.
    d. ecological factors such as food type and predation.

Apply Your Understanding

1. **Figure 37.2b** What does the classic experiment illustrated here reveal about innate behaviors? How might you modify the experiment to attempt to identify innate behaviors in humans?

   Accurate clay model without red

   Aggressive postures of breeding male sticklebacks

   Clay models with red underside

   (b)

2. **Figure 37.15b** To what degree do the data in these graphs support the hypothesis that nonhuman primates communicate?

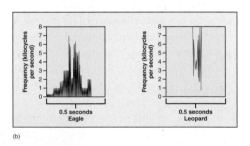

   (b)

Synthesize What You Have Learned

1. The American cowbird is a brood parasite that lays its eggs in the nests of other birds, leaving the chicks to be raised by the other birds. The cowbird chicks are aggressive, taking far more than their share of the nourishment provided by the nest parents, and often killing their chicks. What sort of natural selection has favored the evolution of this situation?
2. Some behaviors require both genetic (nature) and environmental (nurture) input to be carried out. Explain how these two forces interact to affect the song of the white-crowned sparrow.
3. Optimal foraging theory balances costs and benefits of the various methods of finding and hoarding food items, and the danger of being exposed to predators. Discuss the probable differences in foraging behavior of a skunk, a porcupine, and a field mouse.

38

Human Influences on the Living World

The girl gazing from this now-classic *National Geographic* photo faces an uncertain future. An Afghani refugee, the whims of war destroyed her home, her family, and all that was familiar to her. Her expression carries a message about our own future: The problems humanity faces on an increasingly unstable, overcrowded, and polluted earth are no longer hypothetical. They are with us today and demand solutions. This chapter provides an overview of the problems and then focuses on solutions—on what can be done to address very real problems. As a concerned citizen, your first task must be to clearly understand the nature of the problem. You cannot hope to preserve what you do not understand. The world's environmental problems are acute, and a knowledge of biology is an essential tool you will need to contribute to the effort to solve them. It has been said that we do not inherit the earth from our parents—we borrow it from our children. We must preserve for them a world in which they can live. That is our challenge for the future, and it is a challenge that must be met soon. In many parts of the world, the future is happening right now.

38.1   Pollution

Our world is one ecological continent, one highly interactive biosphere, and damage done to any one ecosystem can have ill effects on many others. Burning high-sulfur coal in Illinois kills trees in Vermont, while dumping refrigerator coolants in New York destroys atmospheric ozone over Antarctica and leads to increased skin cancer in Madrid. Biologists call such widespread effects on the worldwide ecosystem **global change.** The pattern of global change that has become evident within recent years, including chemical pollution, acid precipitation, the ozone hole, the greenhouse effect, and the loss of biodiversity, is one of the most serious problems facing humanity's future.

Chemical Pollution

The problem posed by chemical pollution has grown very serious in recent years, both because of the growth of heavy industry and because of an overly casual attitude in industrialized countries. In one example, a poorly piloted oil tanker named the *Exxon Valdez* ran aground in Alaska in 1989 and spilled oil over many kilometers of North American coastline, killing many of the organisms that live there and coating the land with a thick layer of sludge. If the tanker had been loaded no higher than the waterline, little oil would have been lost, but it was loaded far higher than that, and the weight of the above-waterline oil forced thousands of tons of oil out the hole in the ship's hull. Why do policies permit overloading like this?

Chemicals are released into both the air and into water; therefore, their effects are far reaching.

Air Pollution   Air pollution is a major problem in the world's cities. In Mexico City, oxygen is sold routinely on corners for patrons to inhale. Cities such as New York, Boston, and Philadelphia are known as gray-air cities because the pollutants in the air are usually sulfur oxides emitted by industry. Cities such as Los Angeles, however, are called brown-air cities because the pollutants in the air undergo chemical reactions in the sunlight to form smog.

Water Pollution   Water pollution is a very serious consequence of our casual attitude about pollution. "Flushing it down the sink" doesn't work in today's crowded world. There is simply not enough water available to dilute the many substances that the enormous human population produces continuously. Despite improved methods of sewage treatment, lakes and rivers throughout the world are becoming increasingly polluted with sewage. In addition, fertilizers and insecticides also get washed from the land to the water in great quantities.

Agricultural Chemicals

The spread of "modern" agriculture, and particularly the Green Revolution, which brought high-intensity farming to developing countries, has caused very large amounts of many kinds of new chemicals to be introduced into the global ecosystem,

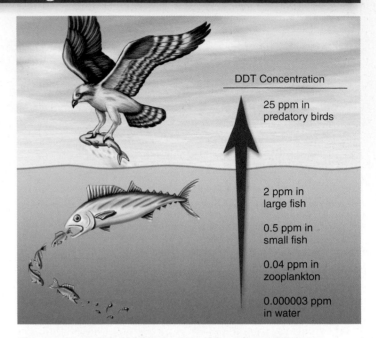

Figure 38.1   Biological magnification of DDT.
Because DDT accumulates in animal fat, the compound becomes increasingly concentrated in higher levels of the food chain.

particularly pesticides, herbicides, and fertilizers. Industrialized countries like the United States now attempt to carefully monitor side effects of these chemicals. Unfortunately, large quantities of many toxic chemicals, although no longer manufactured, still circulate in the ecosystem.

For example, the chlorinated hydrocarbons, compounds that include DDT, chlordane, lindane, and dieldrin, have been banned in the United States, where they were once widely used. They are still manufactured in the United States and exported to other countries, where their use continues. Chlorinated hydrocarbon molecules break down slowly and accumulate in animal fat tissue. Furthermore, as they pass through a food chain, they become increasingly concentrated in a process called **biological magnification. Figure 38.1** shows how a minute concentration of DDT in plankton increases to significant levels as it is passed up through this aquatic food chain. In the United States and elsewhere, DDT caused serious ecological problems by leading to the production of thin, fragile eggshells in many predatory bird species, such as peregrine falcons, bald eagles, osprey, and brown pelicans. In the late 1960s, DDT was banned in time to save the birds from extinction. Chlorinated compounds have other undesirable side effects and exhibit hormonelike activities in the bodies of animals.

Key Learning Outcome 38.1   All over the globe, increasing industrialization is leading to higher levels of pollution.

38.2 Acid Precipitation

The smokestacks you see in **figure 38.2** are those of the Four Corners power plant in New Mexico. This facility burns coal, sending the smoke high into the atmosphere through these tall stacks. The smoke contains high concentrations of sulfur dioxide and other sulfates, which produce acid when they combine with water vapor in the air. The first tall stacks were introduced in Britain in the mid-1950s, and the design rapidly spread through Europe and the United States. The intent of having tall smokestacks was to release the sulfur-rich smoke high in the atmosphere, where winds would disperse and dilute it, carrying the acids far away.

However, in the 1970s, scientists began noticing that the acids from the sulfur-rich smoke were having devastating effects. Throughout northern Europe, lakes were reported to have suffered drastic drops in biodiversity, some even becoming devoid of life. The trees of the great Black Forest of Germany seemed to be dying—and the damage was not limited to Europe. In the eastern United States and Canada, many of the forests and lakes have been seriously damaged.

It turns out that when the sulfur introduced into the upper atmosphere combined with water vapor to produce sulfuric acid, the acid was taken far from its source, but it later fell along with water as acidic rain and snow. This pollution-acidified precipitation is called **acid rain** (but the term acid precipitation is actually more correct). Natural rainwater rarely has a pH lower than 5.6; however, rain and snow in many areas of the United States have pH values less than 5.3, and in the northeastern U.S., pHs of 4.2 or below have been recorded, with occasional storms as low as 3.0.

Acid precipitation destroys life. Many of the forests of the northeastern United States and Canada have been seriously damaged. In fact, it is now estimated that at least 1.4 million acres of forests in the Northern Hemisphere have been adversely affected by acid precipitation (**figure 38.3**). In addition, thousands of lakes in Sweden and Norway no longer support fish—these lakes are now eerily clear. In the northeastern United States and Canada, tens of thousands of lakes are dying biologically as their pH levels fall to below 5.0. At pH levels below 5.0, many fish species and other aquatic animals die, unable to reproduce.

The solution seems like it would be easy—clean up the sulfur emissions. But there have been serious problems with implementing this solution. First, it is expensive. Estimates of the cost of installing and maintaining the necessary "scrubbers" in the United States are on the order of $5 billion a year. An additional difficulty is that the polluter and the recipient of the pollution are far from one another, and neither wants to pay so much for what they view as someone else's problem. Clean Air legislation has begun to address this problem by mandating some cleaning of emissions in the United States, although much still remains to be done worldwide.

Figure 38.2 Tall stacks export pollution.
Tall stacks like those of the Four Corners coal-burning power plant in New Mexico send pollution far up into the atmosphere.

Figure 38.3 Acid precipitation.
Acid precipitation is killing many of the trees in North American and European forests. Much of the damage is done to the mycorrhizae, fungi growing within the cells of the tree roots. Trees need mycorrhizae in order to extract nutrients from the soil.

> **Key Learning Outcome 38.2** Pollution-acidified precipitation—loosely called acid rain—is destroying forest and lake ecosystems in Europe and North America. The solution is to clean up the emissions.

Global Warming

For over 150 years, the growth of our industrial society has been fueled by cheap energy, much of it obtained by burning fossil fuels—coal, oil, and gas. Coal, oil, and gas are the remains of ancient plants, transformed by pressure and time into carbon-rich "fossil fuels." When such fossil fuels are burned, this carbon is combined with oxygen atoms, producing carbon dioxide (CO_2). Industrial society's burning of fossil fuels has released huge amounts of carbon dioxide into the atmosphere. No one paid any attention to this because the carbon dioxide was thought to be harmless and because the atmosphere was thought to be a limitless reservoir, able to absorb and disperse any amount. It turns out neither assumption is true, and in recent decades, the levels of carbon dioxide in the atmosphere have risen sharply and are continuing to rise.

What is alarming is that the carbon dioxide doesn't just sit in the air doing nothing. The chemical bonds in carbon dioxide molecules transmit radiant energy from the sun but trap the longer wavelengths of infrared light, or heat, that are reflected off the earth's surface and prevent them from radiating back into space. This creates what is known as the **greenhouse effect.** Planets that lack this type of "trapping" atmosphere are much colder than those that possess one. If the earth did not have a "trapping" atmosphere, the average earth temperature would be about –20°C, instead of the actual +15°C.

Global Warming Due to Greenhouse Gases

The rise in average global temperatures during recent decades, a profound change in the earth's atmosphere referred to as **global warming** (shown as the red line in **figure 38.4**), is correlated with increased carbon dioxide concentrations in the atmosphere (the blue line). The suggestion that global warming might in fact be caused by the accumulation of greenhouse gases (carbon dioxide, CFCs, nitrogen oxides, and methane) in the atmosphere has been controversial, and is examined in detail in the Inquiry & Analysis feature at the end of this chapter. After serious examination of the evidence, the overwhelming consensus among scientists is that indeed greenhouse gases are causing global warming.

Increases in the amounts of greenhouse gases increase average global temperatures from 1° to 4°C, which could have serious impact on rain patterns, prime agricultural lands, and changes in sea levels.

Effects on Rain Patterns Global warming is predicted to have a major effect on rainfall patterns. Areas that have already been experiencing droughts may see even less rain, contributing to even greater water shortages. Recent increases in the frequency of El Niño events (see chapter 36) and catastrophic hurricanes may indicate that climatic changes due to global warming are already beginning to occur.

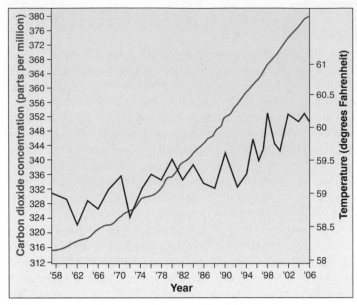

Figure 38.4   The greenhouse effect.

The concentration of carbon dioxide in the atmosphere has shown a steady increase for many years *(blue line)*. The *red line* shows the average global temperature for the same period of time. Note the general increase in temperature since the 1950s and, specifically, the sharp rise beginning in the 1980s. Data from the National Center for Atmospheric Research and other sources.

Effects on Agriculture Both positive and negative effects of global warming on agriculture are predicted. Warmer temperatures and increased levels of carbon dioxide in the atmosphere would be expected to increase the yields of some crops but have a negative impact on others. Droughts that may result from global warming will also negatively affect crops. Plants in the tropics are growing at near their maximal temperature limits; any further increases in temperature will probably begin to have a negative impact on agricultural yields of tropical farms.

Rising Sea Levels Much of the water on earth is locked into ice in glaciers and polar ice caps. As global temperatures increase, these large stores of ice have begun to melt. Most of the water from the melted glaciers ends up in the oceans, causing water levels to rise (but because the Arctic ice cap floats, its melting will not raise sea levels, any more than melting ice raises the level of water in a glass). Higher water levels can be expected to cause increased flooding of low-lying lands.

There is considerable disagreement among governments about what ought to be done about global warming. The Clean Air Act of 1990 and the Kyoto Treaty have established goals for reducing the emission of greenhouse gases. Countries across the globe are making progress toward reducing emissions, but much more needs to be done.

> **Key Learning Outcome 38.3** Humanity's burning of fossil fuels has greatly increased atmospheric levels of CO_2 leading to global warming.

The Ozone Hole

For 2 billion years, life was trapped in the oceans because radiation from the sun seared the earth's surface unchecked. Nothing could survive that bath of destructive energy. Living things were able to leave the oceans and colonize the surface of the earth only after a protective shield of ozone had been added to the atmosphere by photosynthesis. Imagine if that shield were taken away. Alarmingly, it appears that we are destroying it ourselves. Starting in 1975, the earth's ozone shield began to disintegrate. Over the South Pole in September of that year, satellite photos revealed that the ozone concentration was unexpectedly less than elsewhere in the earth's atmosphere. It was as if some "ozone eater" were chewing it up in the Antarctic sky, leaving a mysterious zone of lower-than-normal ozone concentration, an **ozone hole.** For many years after that, more of the ozone was depleted, and the hole grew bigger and deeper. The satellite image in **figure 38.5** shows lower levels of ozone as light purple (Antarctica is also colored purple, indicating that the ozone hole completely covers it). The graph indicates the size of the ozone hole over a 10-year period, with the largest hole appearing in September of 2000 (the blue line).

What is eating the ozone? Scientists soon discovered that the culprit was a class of chemicals that everyone had thought to be harmless: **chlorofluorocarbons (CFCs).** CFCs were invented in the 1920s, a miracle chemical that was stable, harmless, and a near-ideal heat exchanger. Throughout the world, CFCs are used in large amounts as coolants in refrigerators and air conditioners, as the gas in aerosol dispensers, and as the foaming agent in Styrofoam containers. All of these CFCs eventually escape into the atmosphere, but no one worried about this until recently, both because CFCs were thought to be chemically inert and because everyone tends to think of the atmosphere as limitless. But CFCs are very stable chemicals, and have continually accumulated in the atmosphere.

It turned out that the CFCs were causing mischief the chemists had not imagined. High over the South and North Poles, nearly 50 kilometers up, where it is very, very cold, the CFCs stick to frozen water vapor and act as catalysts of a chemical reaction. Just as an enzyme carries out a reaction in your cells without being changed itself, so the CFCs catalyze the conversion of ozone (O_3) into oxygen (O_2) without being used up themselves. Very stable, the CFCs in the atmosphere just keep at it—little machines that never stop. They are still there, still doing it, today. The drop in ozone worldwide is now over 3%.

Ultraviolet radiation is a serious human health concern. Every 1% drop in the atmospheric ozone content is estimated to lead to a 6% increase in the incidence of skin cancers. At middle latitudes, the drop of approximately 3% that has occurred worldwide is estimated to have led to an increase of perhaps as much as 20% in lethal melanoma skin cancers.

Experts generally agree that levels of ozone-killing chemicals in the upper atmosphere are leveling off since more than 180 countries in the 1980s signed an international agreement, which phases out the manufacture of most CFCs. The 2005 ozone hole peaked at about 25 million square kilometers (the size of North America), below the 2000 record size of about 28.4 million square kilometers. Current computer models suggest the Antarctic ozone hole should recover by 2065, and the lesser-damaged ozone layer over the Arctic by about 2023.

> **Key Learning Outcome 38.4**   CFCs and other chemicals are catalytically destroying the ozone in the upper atmosphere, exposing the earth's surface to dangerous radiation. International attempts to solve the problem appear to be succeeding.

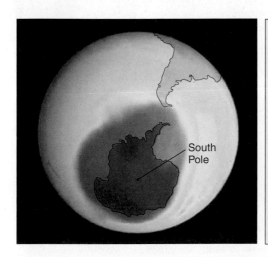
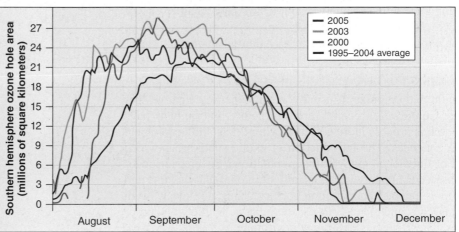

Figure 38.5 The ozone hole over Antarctica.

For decades NASA satellites have tracked the extent of ozone depletion over Antarctica. Every year since 1975 an ozone "hole" has appeared in August when sunlight triggers chemical reactions in cold air trapped over the South Pole during Antarctic winter. The hole intensifies during September before tailing off as temperatures rise in November–December. In 2000, the 28.4-million-square-kilometer hole (*purple* in the satellite image) covered an area larger than the United States, Canada, and Mexico combined, the largest hole ever recorded. In September 2000, the hole extended over Punta Arenas, a city of about 120,000 people in southern Chile, exposing residents to very high levels of UV radiation.

38.5 Loss of Biodiversity

Just as death is as necessary to a normal life cycle as reproduction, so extinction is as normal and necessary to a stable world ecosystem as species formation. Most species, probably all, go extinct eventually. More than 99% of species known to science (most from the fossil record) are now extinct. However, current rates of extinctions are alarmingly high. The extinction rate for birds and mammals was about one species every decade from 1600 to 1700, but it rose to one species every year during the period from 1850 to 1950, and four species per year between 1986 and 1990. It is this increase in the rate of extinction that is the heart of the **biodiversity** crisis.

Factors Responsible for Extinction

What factors are responsible for extinction? Studying a wide array of recorded extinctions, and many species currently threatened with extinction, biologists have identified three factors that seem to play a key role in many extinctions: habitat loss, species overexploitation, and introduced species (**figure 38.6**).

Habitat Loss Habitat loss is the single most important cause of extinction. Given the tremendous amounts of ongoing destruction of all types of habitat, from rain forest to ocean floor, this should come as no surprise. Natural habitats may be adversely affected by human influences in four ways: (1) destruction, (2) pollution, (3) human disruption, and (4) habitat fragmentation (dividing up the habitat into small isolated areas). As you can see in **figure 38.7**, destruction of rain forest habitat is occurring rapidly in Madagascar, which is endangering species.

Species Overexploitation Species that are hunted or harvested by humans have historically been at grave risk of extinction, even when the species populations are initially very abundant. There are many examples in our recent history of overexploitation: passenger pigeons, bison, many species of whales, commercial fish such as Atlantic bluefin tuna, and mahogany trees in the West Indies are but a few.

Introduced Species Occasionally, a new species will enter a habitat and colonize it, usually at the expense of native species. Colonization occurs rarely in nature, but humans have made this process more common, with devastating ecological consequences. The introduction of exotic species has wiped out or threatened many native populations. Species introductions occur in many ways, usually unintentionally. Plants and animals can be transported in nursery plants, as stowaways in boats, cars, and planes, and as beetle larvae within wood products. These species enter new environments where they have no native predators to keep their population sizes in check, and they crowd out native species.

> **Key Learning Outcome 38.5** Loss of biodiversity can be attributed to one of a few main causes, including habitat loss, overexploitation, and introduced species.

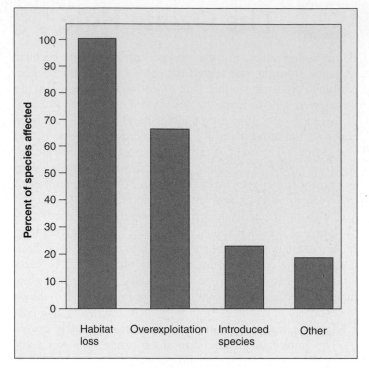

Figure 38.6 Factors responsible for animal extinction.

These data represent known extinctions of mammals in Australia, Asia, and the Americas. Some extinctions have more than one cause.

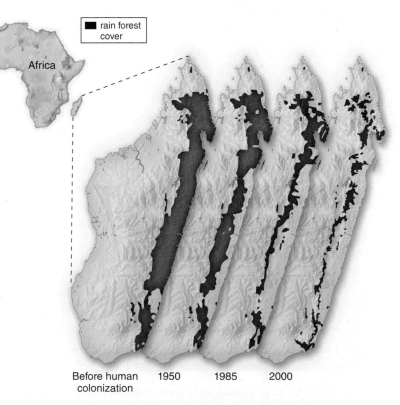

Figure 38.7 Extinction and habitat destruction.

The rain forest covering the eastern coast of Madagascar, an island off the coast of East Africa, has been progressively destroyed as the island's human population has grown. Ninety percent of the original forest cover is now gone. Many species have become extinct, and many others are threatened, including 16 of Madagascar's 31 primate species.

Today's *Biology*

The Global Decline in Amphibians

Sometimes important things happen, right before our eyes, without anyone noticing. That thought occurred to David Bradford as he stood looking at a quiet lake high in the Sierra Nevada Mountains of California in the summer of 1988. Bradford, a biologist, had hiked all day to get to the lake, and when he got there his worst fears were confirmed. The lake was on a list of mountain lakes that Bradford had been visiting that summer in Sequoia-Kings Canyon National Parks while looking for a little frog with yellow legs. The frog's scientific name was *Rana muscosa,* and it had lived in the lakes of the parks for as long as anyone had kept records. But this silent summer evening, the little frog was gone. The last major census of the frog's populations within the parks had been taken in the mid-1970s, and *R. muscosa* had been everywhere, a common inhabitant of the many freshwater ponds and lakes within the parks. Now, for some reason Bradford did not understand, the frogs had disappeared from 98% of the ponds that had been their homes.

After Bradford reported this puzzling disappearance to other biologists, an alarming pattern soon became evident. Throughout the world, local populations of amphibians (frogs, toads, and salamanders) were becoming extinct. Waves of extinction have swept through high-elevation amphibian populations in the western United States, and have also cut through the frog populations of Central America and coastal Australia.

Amphibians have been around for 350 million years, since long before the dinosaurs. Their sudden disappearance from so many of their natural homes sounded an alarm among biologists. What are we doing to our world? If amphibians cannot survive the world we are making, can we?

In 1998 the U.S. National Research Council brought scientists together from many disciplines in a serious attempt to address the problem. After years of intensive investigation, they have begun to sort out the reasons for the global decline in amphibians. Like many important questions in science, this one does not have a simple answer.

Five factors seem to be contributing in a major way to the worldwide amphibian decline: (1) habitat deterioration and destruction, particularly clear-cutting of forests, which drastically lowers the humidity (water in the air) that amphibians require; (2) the introduction of exotic species that outcompete local amphibian populations; (3) chemical pollutants that are toxic to amphibians; (4) fatal infections by pathogens; and (5) global warming, which is making some habitats unsuitable.

Infection by parasites appears to have played a particularly important role in the western United States and coastal Australia. Amphibian ecology expert James Collins of Arizona State University has reported one clear instance of infection leading to amphibian decline. When Collins examined populations of salamanders living on the Kaibab Plateau along the Grand Canyon rim, he found many sick salamanders. Their skin was covered with white pustules, and most infected ones died, their hearts and spleens collapsed. The infectious agent proved to be a virus common in fish called a ranavirus. Ranavirus isolated by Collins from one sick salamander would cause the disease in a healthy salamander, so there was no doubt that ranavirus was the culprit responsible for the salamander decline on the Kaibab Plateau.

Ranavirus outbreaks eliminate small populations, but in larger ones a few individuals survive infection, sloughing off their pustule-laden skin. These populations slowly recover.

A second kind of infection, very common in Australia but also seen in the United States, is having more widespread effects. Populations infected with this microbe, a kind of fungus called a chytrid (pronounced "kit-rid," see chapter 18), do not recover. Usually a harmless soil fungus that decomposes plant material, this particular chytrid (with the Latin name of *Batrachochytrium dendrobatidis*) is far from harmless to amphibians. It dissolves and absorbs the keratinous mouthparts of amphibian larvae, killing them.

This killer chytrid appeared in Australia near Melbourne in the early 1980s. Now almost all Australia is affected. How did the disease spread so rapidly? Apparently it traveled by truck. Infected frogs moved all across Australia in wooden boxes with bunches of bananas. In one year, 5,000 frogs were collected from banana crates in one Melbourne market alone.

In other parts of the world, infection does not seem to play as important a role as acid precipitation, habitat loss, and introduction of exotic species. This complex pattern of cause and effect only serves to emphasize the take-home lesson: Worldwide amphibian decline has no one culprit. Instead, all five factors play important roles. It is their total impact that has shifted the worldwide balance toward extinction.

To reverse the trend toward extinction, we must work to lessen the impact of all of these factors. It is important that we not get discouraged at the size of the job, however. Any progress we make on any one factor will help shift the balance back toward survival. Extinction is only inevitable if we let it be.

38.6 Reducing Pollution

The pattern of global change that is overtaking our world is very disturbing. Human activities are placing a severe stress on the biosphere, and we must quickly find ways to reduce the harmful impact. There are four key areas in which it will be particularly important to meet the challenge successfully: reducing pollution, finding other sources of energy, preserving nonreplaceable resources, and curbing population growth.

To solve the problem of industrial pollution, it is first necessary to understand the cause of the problem. In essence, it is a failure of our economy to set a proper price on environmental health. To understand how this happens, we must think for a moment about money. The economy of the United States (and much of the rest of the industrial world) is based on a simple feedback system of supply and demand. As a commodity gets scarce, its price goes up, and this added profit acts as an incentive for more of the item to be produced; if too much is produced, the price falls and less of it is made because it is no longer so profitable to produce it.

This system works very well and is responsible for the economic strength of our nation, but it has one great weakness. If demand is set by price, then it is very important that all the costs be included in the price. Imagine that the person selling the item were able to pass off part of the production cost to a third person. The seller would then be able to set a lower price and sell more of the item! Driven by the lower price, the buyer would purchase more than if all the costs had been added into the price.

Unfortunately, that sort of pricing error is what has driven the pollution of the environment by industry. The true costs of energy and of the many products of industry are composed of direct production costs, such as materials and wages, and of indirect costs, such as pollution of the ecosystem. Economists have identified an "optimum" amount of pollution based on how much it costs to reduce pollution versus the social and environmental cost of allowing pollution. The economically optimum amount of pollution is indicated by the blue dot in figure 38.8. If more pollution than the optimum is allowed, the social cost is too high, but if less than the optimum is allowed, the economic cost is too high.

The indirect costs of pollution are usually not taken into account. However, the indirect costs do not disappear because we ignore them. They are simply passed on to future generations, which must pay the bill in terms of damage to the ecosystems on which we all depend. Increasingly, the future is now. Our world, unable to support more damage, is demanding that something be done—that we finally pay up.

Antipollution Laws

Two effective approaches have been devised to curb pollution in this country. The first is to pass laws forbidding it. In the last 20 years, laws have begun to significantly curb the spread of pollution by setting stiff standards for what

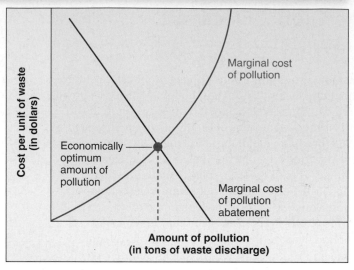

Figure 38.8 Is there an optimum amount of pollution?
Economists identify the "optimum" amount of pollution as the point at which eliminating the next unit of pollution (the marginal cost of pollution abatement) equals the cost in damages caused by that unit of pollution (the marginal cost of pollution).

can be released into the environment. For example, all cars are required to have effective catalytic converters to eliminate automobile smog. Similarly, the Clean Air Act of 1990 requires that power plants eliminate sulfur emissions. They can accomplish this by either installing scrubbers on their smokestacks or by burning low-sulfur coal (clean-coal technology), which is more expensive. The effect is that the consumer pays to avoid polluting the environment. The cost of the converters makes cars more expensive, and the cost of the scrubbers increases the price of the energy. The new, higher costs are closer to the true costs, lowering consumption to more appropriate levels.

Pollution Taxes

A second approach to curbing pollution has been to increase the consumer costs directly by placing a tax on the pollution, in effect an artificial price hike imposed by the government as a tax added to the price of production. This added cost lowers consumption too, but by adjusting the tax, the government can attempt to balance the conflicting demands of environmental safety and economic growth. Such taxes, often imposed as "cap-and-trade pollution permits," are becoming an increasingly important part of antipollution laws.

Key Learning Outcome 38.6 Free market economies often foster pollution when prices do not include environmental costs. Laws and taxes are being designed in an attempt to compensate.

38.7 Preserving Nonreplaceable Resources

Among the many ways ecosystems are being damaged, one problem stands out as more serious than all the rest: consuming or destroying resources that we all share in common but cannot replace in the future (figure 38.9). Although a polluted stream can be cleaned, no one can restore an extinct species. In the United States, three sorts of nonreplaceable resources are being reduced at alarming rates: topsoil, groundwater, and biodiversity.

Topsoil

Soil is composed of a mixture of rocks and minerals with partially decayed organic matter called humus. Plant growth is strongly affected by soil composition. Minerals like nitrogen and phosphorus are critical to plant growth, and are abundant in humus-rich soils.

The United States is one of the most productive agricultural countries on earth, largely because much of it is covered with particularly fertile soils. Our Midwestern farm belt sits astride what was once a great prairie. The soil of that ecosystem accumulated bit by bit from countless generations of animals and plants until, by the time humans came to plow, the humus-rich soil extended down several feet.

We cannot replace this rich **topsoil**, the capital upon which our country's greatness is built, yet we are allowing it to be lost at a rate of centimeters every decade. Our country has lost one-quarter of its topsoil since 1950! By repeatedly tilling (turning the soil over) to eliminate weeds, we permit rain to wash more and more of the topsoil away, into rivers and eventually out to sea. New approaches are desperately needed to lessen the reliance on intensive cultivation. Some possible solutions include using genetic engineering to make crops resistant to weed-killing herbicides and terracing to recapture lost topsoil.

Groundwater

A second resource that we cannot replace is **groundwater,** water trapped beneath the soil within porous rock reservoirs called aquifers. This water seeped into its underground reservoir very slowly during the last ice age over 12,000 years ago. We should not waste this treasure, for we cannot replace it.

In most areas of the United States, local governments exert relatively little control over the use of groundwater. As a result, a very large portion is wasted watering lawns, washing cars, and running fountains. A great deal more is inadvertently being polluted by poor disposal of chemical wastes—and once pollution enters the groundwater, there is no effective means of removing it. Some cities, like Phoenix and Las Vegas, may completely deplete their groundwater within several decades.

"Freedom in a Commons Brings Ruin to All"

The essence of Hardin's original essay:

Picture a pasture open to all. It is expected that each herdsman will try to keep as many cattle as possible on [this] commons....What is the utility...of adding one more animal?...Since the herdsman receives all the proceeds from the sale of the additional animal, the positive utility [to the herdsman] is nearly +1.... Since, however, the effects of overgrazing are shared by all the herdsmen, the negative utility for any particular decision-making herdsman is only a fraction of -1. Adding together the...partial utilities, the rational herdsman concludes that the only sensible course for him to pursue is to add another animal to [the] herd. And another; and another.... Therein is the tragedy. Each man is locked into a system that [causes] him to increase his herd without limit—in a world that is limited....Freedom in a commons brings ruin to all.

—G. Hardin, "The Tragedy of the Commons,"
Science **162,** 1243 (1968), p. 1244

Figure 38.9 The tragedy of the commons.
In a now-famous essay, ecologist Garrett Hardin argues that destruction of the environment is driven by freedom without responsibility.
Reprinted with permission from "The Tragedy of the Commons," by G. Hardin, Science, *162, p. 1244. Copyright 1968 AAAS.*

(a)

(b)

(c)

Figure 38.10 Tropical rain forest destruction.
(a) These fires are destroying the rain forest in Brazil, which is being cleared for cattle pasture. (b) The flames are so widespread and so high that their smoke can be viewed from space. (c) The consequences of deforestation can be seen on these middle-elevation slopes in Ecuador. The slopes now support only low-grade pastures where they used to support highly productive forest.

Biodiversity

The number of species in danger of extinction during your lifetime is far greater than the number that became extinct with the dinosaurs. This disastrous loss of biodiversity is important to every one of us, because as these species disappear, so does our chance to learn about them and their possible benefits for ourselves. The fact that our entire supply of food is based on 20 kinds of plants, out of the 250,000 available, should give us pause. Like burning a library without reading the books, we don't know what it is we waste. All we can be sure of is that we cannot retrieve it. Extinct is forever.

Over the last 20 years, about half of the world's tropical rain forests have been either burned to make pasture land or cut for timber (**figure 38.10**). Over 6 million square kilometers have been destroyed. Every year the rate of loss increases as the human population of the tropics grows. About 160,000 square kilometers were cut each year in the 1990s, a rate greater than 0.6 hectares (1.5 acres) per second! At this rate, all the rain forests of the world will be gone in your lifetime. In the process, it is estimated that one-fifth or more of the world's species of animals and plants will become extinct—more than a million species. This would be an extinction event unparalleled for at least 65 million years, since the Age of Dinosaurs.

You should not be lulled into thinking that loss of biodiversity is a problem limited to the tropics. The ancient forests of the Pacific Northwest are being cut at a ferocious rate today, largely to supply jobs (the lumber is exported), with much of the cost of cutting it down subsidized by our government (the Forest Service builds the necessary access roads, for example). At the current rate, very little will remain in a decade. Nor is the problem restricted to one area. Throughout our country, natural forests are being "clear-cut," replaced by pure stands of lumber trees planted in rows like so many lines of corn. It is difficult to scold those living in the tropics when we ourselves do such a poor job of preserving our own country's biodiversity.

But what is so bad about losing species? What is the value of biodiversity? Loss of a species entails three costs: (1) the direct economic value of the products we might have obtained from species; (2) the indirect economic value of benefits produced by species without our consuming them, such as nutrient recycling in ecosystems; and (3) their ethical and aesthetic value. It is not difficult to see the value in protecting species that we use to obtain food, medicine, clothing, energy, and shelter, but other species are vitally important to maintaining healthy ecosystems; by destroying biodiversity, we are creating conditions of instability and lessened productivity. Other species add beauty to the living world, no less crucial because it is hard to set a price upon.

Key Learning Outcome 38.7 Nonreplaceable resources are being consumed at an alarming rate all over the world; key among them are topsoil, groundwater, and biodiversity.

38.8 Curbing Population Growth

If we were to solve all the problems mentioned in this chapter, we would merely buy time to address the fundamental problem: There are getting to be too many of us.

Humans first reached North America at least 12,000 to 13,000 years ago, crossing the narrow straits between Siberia and Alaska and moving swiftly to the southern tip of South America. By 10,000 years ago, when the continental ice sheets withdrew and agriculture first developed, about 5 million people lived on earth, distributed over all the continents except Antarctica. With the new and much more dependable sources of food that became available through agriculture, the human population began to grow more rapidly. By the time of Christ, 2,000 years ago, an estimated 130 million people lived on earth. By the year 1650, the world's population had doubled, and doubled again, reaching 500 million. Starting in the early 1700s, changes in technology have given humans more control over their food supply, have led to the development of cures for many diseases, and have produced improvements in shelter and storage capabilities that make humans less vulnerable to climatic uncertainties. Recall from chapter 35 that populations grow exponentially until they reach the limits of their environment, called the carrying capacity. These changes allowed humans to expand the carrying capacity of the habitats in which they lived and thus to escape the confines of logistic growth and reenter the exponential phase of the sigmoidal growth curve, shown by the explosive growth in figure 38.11.

Although the human population has grown explosively for the last 300 years, the average human birthrate has stabilized at about 21 births per year per 1,000 people worldwide. However, with the spread of better sanitation and improved medical techniques, the death rate has fallen steadily, to its present level of 9 per 1,000 per year. The difference between birth and death rates amounts to a population growth rate of 1.2% per year, which seems like a small number, but it is not, given the large population size.

The world population reached 7 billion people in 2011, and the annual increase now amounts to about 78 million people, which leads to a doubling of the world population in about 58 years. Put another way, more than 214,000 people are added to the world population each day, or almost 150 every minute. At this rate, the world's population will continue to grow and perhaps stabilize at a figure around 10 billion. Such growth cannot continue, because our world cannot support it. Just as a cancer cannot grow unabated in your body without eventually killing you, so humanity cannot continue to grow unchecked in the biosphere without killing it.

Most countries are devoting considerable attention to slowing the growth rate of their populations and there are genuine signs of progress, but the world population may still gain another 1 to 4 billion people before it stabilizes. No one knows whether the world can support so many people.

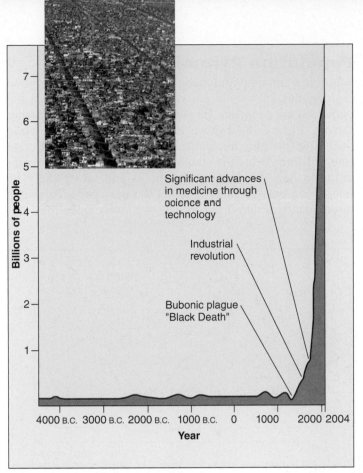

Figure 38.11 Growth of the human population.
Over the past 300 years, the world population has been growing steadily. Currently, there are over 6 billion people on the earth. Mexico City (inset photo), one of the world's largest cities, has about 26 million inhabitants.

Population Growth Rate Declining

The world population growth rate has been declining, from a high of 2.0% in the period 1965–70 to 1.2% in 2004. Nonetheless, because of the larger population, this amounts to an increase of 78 million people per year to the world population, compared to 53 million per year in the 1960s.

The United Nations attributes the decline to increased family planning efforts and the increased economic power and social status of women. As family size decreases in developing countries, education programs improve, leading to increased education levels for women, which in turn tends to result in further decreases in family size.

No one knows whether the world can sustain today's population of 7 billion people, much less the far greater numbers expected in the future. We cannot reasonably expect to expand the world's carrying capacity indefinitely. The population will begin scaling back in size, as predicted by logistic growth models; indeed it is already happening. In the sub-Saharan area of Africa, population projections for the year 2025 have been scaled back from 1.33 billion to 1.05 billion because of the impact of AIDS. If we are to avoid catastrophic increases in death rates, such as the tragedy we are seeing in sub-Saharan Africa, the birthrates must continue to fall dramatically.

Population Pyramids

While the human population as a whole continues to grow rapidly, this growth is not occurring uniformly over the planet. Some countries, like Mexico, are currently growing rapidly. **Figure 38.12** shows how Mexico's birthrate, while declining (the blue line), still greatly exceeds its death rate (the red line), which is stabilizing. There is often a correlation in how developed a country is and how rapidly its population grows. **Table 38.1**, on the next page, compares three countries that differ in how well developed they are.

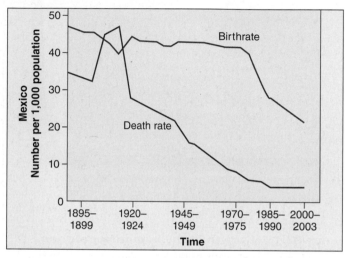

Figure 38.12   Why Mexico's population is growing.

The death rate (*red line*) in Mexico has been falling, while the birthrate (*blue line*) remained fairly steady until 1970. The difference between birth and death rates has fueled a high growth rate. Efforts begun in 1970 to reduce the birthrate have been quite successful. Although the growth rate remains rapid, it is expected to begin leveling off in the near future as the birthrate continues to drop.

Ethiopia, a developing country, has a higher fertility rate, which results in a higher birthrate than either Brazil or the United States. But Ethiopia also has a much higher infant mortality rate and a lower life expectancy. Overall, the population in Ethiopia will double much more quickly than the population of Brazil or the United States. The rate at which a population can be expected to grow in the future can be assessed graphically by means of a population pyramid—a bar graph displaying the numbers of people in each age category (some examples are shown in **figure 38.13**). Males are conventionally shown to the left of the vertical age axis (colored blue here) and females to the right (colored red). In most human population pyramids, the number of older females is disproportionately large compared with the number of older males, because females in most regions have a longer life expectancy than males. This is apparent in the upper portion of the 2005 U.S. pyramid.

Viewing such a pyramid, one can predict demographic trends in births and deaths. In general, rectangular pyramids are characteristic of countries whose populations are stable; their numbers are neither growing nor shrinking. A triangular pyramid, like the 2005 Kenya pyramid, is characteristic of a country that will exhibit rapid future growth, as most of its population has not yet entered the child-bearing years. Inverted triangles are characteristic of populations that are shrinking.

Compare the differences in the population pyramids for the United States and Kenya in **figure 38.13**. In the somewhat more rectangular population pyramid for the United States in 2005, the cohort (group of individuals) 40 to 59 years old represents the "baby boom," the large number of babies born following World War II. When the media refers to the "graying of America," they are referring to the aging of this disproportionately large cohort that

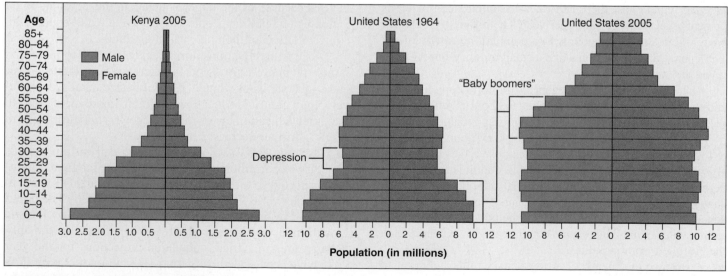

Figure 38.13   Population pyramids.

Population pyramids are graphed according to a population's age distribution. Kenya's pyramid has a broad base because of the great number of individuals below child-bearing age. When all of the young people begin to bear children, the population will experience rapid growth. The 2005 U.S. pyramid demonstrates a larger number of individuals in the "baby boom" cohort—the pyramid bulges because of an increase in births between 1945 and 1964, as shown at the base of the 1964 pyramid. The 25 to 34 cohort in the 1964 pyramid represents people born during the Depression and is smaller in size than the cohorts in the preceding and following years.

TABLE 38.1	A COMPARISON OF 2006 POPULATION DATA IN DEVELOPED AND DEVELOPING COUNTRIES		
	United States (highly developed)	Brazil (moderately developed)	Ethiopia (developing)
Fertility rate	2.1	2.3	5.4
Doubling time at current rate (yr)	72.2	55.5	27.9
Infant mortality rate (infant deaths/1,000 births)	6.5	27	77
Life expectancy (yr)	78	72	49
Per capita income (U.S. dollar equivalent)	$44,260	$8,800	$1,190

will impact the health-care system and other age-related systems in the future. The very triangular pyramid of Kenya, by contrast, predicts explosive future growth. The population of Kenya is predicted to double in less than 20 years. However, it is important to note that these estimates do not take into account the huge impact that natural disasters and epidemics such as AIDS will have on population sizes. In sub-Saharan Africa, the AIDS epidemic has reduced the life expectancy at birth by 20 years. **Figure 38.14** shows two population pyramid projections for Botswana, Africa, where over 36% of the population is living with HIV or AIDS. The uncolored portions of the bars indicate the population projections in 2025 without the effect of the AIDS epidemic, and the colored bars reflect actual projections with AIDS.

The Level of Consumption in the Developed World Is Also a Problem

The world population is expected to stabilize sometime in this century at about 10 billion. We in the developed countries of the world need to pay more attention to lessening the impact of our resource consumption. Indeed, the wealthiest 20% of the world's population accounts for 86% of the world's consumption of resources and produces 53% of the world's carbon dioxide emissions, whereas the poorest 20% of the world is responsible for only 1.3% of consumption and 3% of CO_2 emissions.

One way of quantifying this disparity is by calculating what has been termed the **ecological footprint,** which is the amount of productive land required to support an individual at the standard of living of a particular population through the course of his or her life. As **figure 38.15** illustrates, the ecological footprint of an individual in the United States is more than 10 times greater than that of someone in India. Based on these measurements, researchers have calculated that resource use by humans is now one-third greater than the amount that nature can sustainably replace; if all humans lived at the standard of living in the developed world, two additional planet earths would be needed.

> **Key Learning Outcome 38.8** The problem at the core of all other environmental concerns is the rapid growth of the world's human population. Serious efforts are being made to slow its growth.

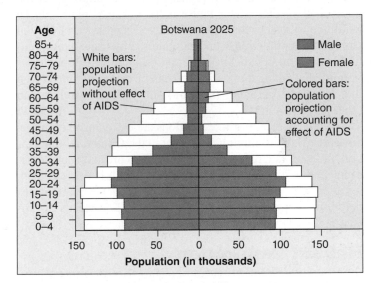

Figure 38.14 Projected AIDS effect on Botswana population (year 2025).

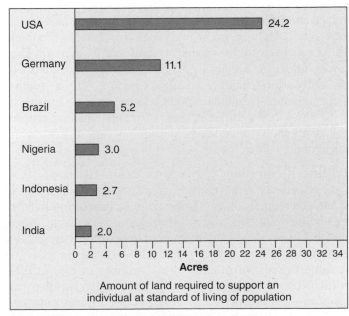

Figure 38.15 Ecological footprint of individuals in different countries in 2003.

The ecological footprint calculates how much land is required to support a person through his or her life, including the acreage used for production of food, forest products, and housing, in addition to the forest required to absorb the carbon dioxide produced by the combustion of fossil fuels.

38.9 Preserving Endangered Species

Once you understand the reasons why a particular species is endangered, it becomes possible to think of designing a recovery plan. If the cause is commercial overharvesting, regulations can be designed to lessen the impact and protect the threatened species. If the cause is habitat loss, plans can be instituted to restore lost habitat. Loss of genetic variability in isolated subpopulations can be countered by transplanting individuals from genetically different populations. Populations in immediate danger of extinction can be captured, introduced into a captive breeding program, and later reintroduced to other suitable habitats.

Of course, all of these solutions are extremely expensive. As Bruce Babbitt, Interior Secretary in the Clinton administration, noted, it is much more economical to prevent such "environmental trainwrecks" from occurring than it is to clean them up afterward. Preserving ecosystems and monitoring species before they are threatened is the most effective means of protecting the environment and preventing extinctions.

Habitat Restoration

Conservation biology typically concerns itself with preserving populations and species in danger of decline or extinction. Conservation, however, requires that there be something left to preserve, while in many situations, conservation is no longer an option. Species, and in some cases whole communities, have disappeared or have been irretrievably modified. The clear-cutting of the temperate forests of Washington State leaves little behind to conserve, nor does converting a piece of land into a wheat field or an asphalt parking lot. Redeeming these situations requires restoration rather than conservation.

Three quite different sorts of habitat restoration programs might be undertaken, depending very much on the cause of the habitat loss.

Pristine Restoration In situations where all species have been effectively removed, one might attempt to restore the plants and animals that are believed to be the natural inhabitants of the area, when such information is available. When abandoned farmland is to be restored to prairie (**figure 38.16**), how do you know what to plant? Although it is in principle possible to reestablish each of the original species in their original proportions, rebuilding a community requires that you know the identity of all of the original inhabitants, and the ecologies of each of the species. We rarely ever have this much information, so no restoration is truly pristine.

Removing Introduced Species Sometimes the habitat of a species has been destroyed by a single introduced species. In such a case, habitat restoration involves removal of the intro-

(a)

(b)

Figure 38.16 Habitat restoration.
The University of Wisconsin–Madison Arboretum has pioneered restoration ecology. (a) The restoration of the prairie was at an early stage in November, 1935. (b) The prairie as it looks today. This picture was taken at approximately the same location as the 1935 photograph.

duced species. For example, Lake Victoria, Africa, was home to over 300 species of cichlid fishes, small perchlike fishes that display incredible diversity. However, in 1954, the Nile perch, a commercial fish with a voracious appetite, was introduced into Lake Victoria. For decades, these perch did not seem to have a significant impact, and then something happened to cause the Nile perch to explode and spread rapidly through the lake, eating their way through the cichlids. By 1986, over 70% of cichlid species had disappeared, including all open-water species.

The situation has been compounded by a second factor, the introduction into Lake Victoria of a floating water weed from South America, the water hyacinth *Eichornia crassipes*. Extremely prolific under eutrophic conditions, thick mats of water hyacinth soon covered entire bays and inlets, choking off the coastal habitats of non-open-water cichlids.

Restoration of the once-diverse cichlid fishes to Lake Victoria will require more than breeding and restocking the endangered species. Eutrophication will have to be reversed, and the introduced water hyacinth and Nile perch populations brought under control or removed.

Cleanup and Rehabilitation Habitats seriously degraded by chemical pollution cannot be restored until the pollution is cleaned up. The successful restoration of the Nashua River in

New England, discussed later in this chapter, is one example of how a concerted effort can succeed in restoring a heavily polluted habitat to a relatively pristine condition.

Captive Propagation

Recovery programs, particularly those focused on one or a few species, often must involve direct intervention in natural populations to avoid an immediate threat of extinction. Introducing wild-caught individuals into captive breeding programs is being used in an attempt to save the black-footed ferret and California condor populations in immediate danger of disappearing. Several other such captive propagation programs have had success.

Case History: The Peregrine Falcon U.S. populations of birds of prey such as the peregrine falcon (*Falco peregrinus*) began an abrupt decline shortly after World War II. Of the approximately 350 breeding pairs east of the Mississippi River in 1942, all had disappeared by 1960. The culprit proved to be the chemical pesticide DDT and related organochlorine pesticides. Birds of prey are particularly vulnerable to DDT because they feed at the top of the food chain, where DDT becomes concentrated. DDT interferes with the deposition of calcium in the bird's eggshells, causing most of the eggs to break before they hatch.

The use of DDT was banned by federal law in 1972, causing levels in the eastern United States to fall quickly. There were no peregrine falcons left in the eastern United States to reestablish a natural population, however. Falcons from other parts of the country were used to establish a captive breeding program at Cornell University in 1970, with the intent of reestablishing the peregrine falcon in the eastern United States by releasing offspring of these birds. By the end of 1986, over 850 birds had been released in 13 eastern states, producing an astonishingly strong recovery.

Sustaining Genetic Diversity

One of the chief obstacles to a successful species recovery program is that a species is generally in serious trouble by the time a recovery program is instituted. When populations become very small, much of their genetic diversity is lost. If a program is to have any chance of success, every effort must be made to sustain as much genetic diversity as possible.

Case History: The Black Rhino All five species of rhinoceros are critically endangered. The three Asian species live in a forest habitat that is rapidly being destroyed, while the two African species are illegally killed for their horns. Fewer than 11,000 individuals of all five species survive today. The problem is intensified by the fact that many of the remaining animals live in very small, isolated populations. The 2,400 wild-living individuals of the black rhino, *Diceros bicornis* (figure 38.17), live in approximately 75 small, widely separated groups that are adapted to local conditions throughout the species' range. All six subspecies appear to have low genetic variability; in three of the subspecies, only a few dozen animals remain. Analysis of mitochondrial DNA suggests that in these populations most individuals are genetically very similar.

This lack of genetic variability represents one of the greatest challenges to the future of the species. Much of the range of the black rhino is still open and not yet subject to human encroachment. To have any significant chance of success, a species recovery program will have to find a way to sustain the genetic diversity that remains in this species. Heterozygosity could be best maintained by bringing all black rhinos together in a single breeding population, but this is not a practical possibility. A more feasible solution would be to move individuals between populations. Managing the black rhino populations for genetic diversity could prevent the loss of genetic variation, which might prove fatal to this species.

Placing black rhinos from a number of different locations together in a sanctuary to increase genetic diversity raises a potential problem: Local subspecies may be adapted in different ways to their immediate habitats—what if these local adaptations are crucial to their survival? Homogenizing the black rhino populations by pooling their genes risks destroying such local adaptations, perhaps at great cost to survival.

Figure 38.17 Sustaining genetic diversity.

The black rhino is highly endangered, living in 75 small, widely separated populations. Only about 2,400 individuals survive in the wild. Conservation biologists have the difficult job of finding ways to preserve genetic diversity in small, isolated populations.

Preserving Keystone Species

Keystone species are species that exert a particularly strong influence on the structure and functioning of their ecosystem. Their removal can have disastrous consequences.

Case History: Flying Foxes The severe decline of many species of pteropodid bats, or "flying foxes," in the Old World tropics is an example of how the loss of a keystone species can have dramatic effects on the other species living within an ecosystem, sometimes even leading to a cascade of further extinctions (figure 38.18). These bats have very close relationships with important plant species on the islands of the Pacific and Indian Oceans. The family Pteropodidae contains nearly 200 species, approximately a quarter of them in the genus *Pteropus,* and is widespread on the islands of the South Pacific, where they are the most important—and often the only—pollinators and seed dispersers. A study in Samoa found that 80% to 100% of the seeds landing on the ground during the dry season were deposited by flying foxes. Many species are entirely dependent on these bats for pollination.

In Guam, where the two local species of flying fox have recently been driven extinct or nearly so, the impact on the ecosystem appears to be substantial. Many plant species are not fruiting, or are doing so only marginally, with fewer fruits than normal. Fruits are not being dispersed away from parent plants, so offspring shoots are being crowded out by the adults.

Flying foxes are being driven to extinction by human hunting. They are hunted for food, for sport, and by orchard farmers, who consider them pests. Flying foxes are particularly vulnerable because they live in large, easily seen groups of up to a million individuals. Because they move in regular and predictable patterns and can be tracked to their home roost, hunters can easily bag thousands at a time.

Species preservation programs aimed at preserving particular species of flying foxes are only just beginning. One particularly successful example is the program to save the Rodrigues fruit bat, *Pteropus rodricensis,* which occurs only on Rodrigues Island in the Indian Ocean near Madagascar. The population dropped from about 1,000 individuals in 1955 to fewer than 100 by 1974, the drop reflecting largely the loss of the fruit bat's forest habitat to farming. Since 1974 the species has been legally protected, and the forest area of the island is being increased through a tree-planting program. Eleven captive breeding colonies have been established, and the bat population is now increasing rapidly. The combination of legal protection, habitat restoration, and captive breeding has in this instance produced a very effective preservation program.

Figure 38.18 Preserving keystone species.

The flying fox is a keystone species in many Old World tropical islands. It pollinates many of the plants, and is a key disperser of seeds. Its elimination by hunting and habitat loss is having a devastating effect on the ecosystems of many South Pacific islands.

Conservation of Ecosystems

Habitat fragmentation is one of the most pervasive enemies of biodiversity conservation efforts. Some species simply require large patches of habitat to thrive, and conservation efforts that cannot provide suitable habitat of such a size are doomed to failure. As it has become clear that isolated patches of habitat lose species far more rapidly than large preserves do, conservation biologists have promoted the creation, particularly in the tropics, of so-called mega-reserves—large areas of land containing a core of one or more undisturbed habitats.

In addition to this focus on maintaining large enough reserves, in recent years, conservation biologists also have recognized that the best way to preserve biodiversity is to focus on preserving intact ecosystems, rather than focusing on particular species. For this reason, attention in many cases is turning to identifying those ecosystems most in need of preservation and devising the means to protect not only the species within the ecosystem, but the functioning of the ecosystem itself.

> **Key Learning Outcome 38.9** Recovery programs at the species level must deal with habitat loss and fragmentation, and often with a marked reduction in genetic diversity.

38.10 Finding Cleaner Sources of Energy

Modern society's propensity to burn fossil fuels has recycled an enormous amount of CO_2 back into the earth's atmosphere. To gain some idea, focus for a moment on your own personal contribution to the carbon cycle. Every mile you drive your car releases about a pound of CO_2 into the air. How many miles do you drive in a year? You see the point? Think about the natural gas that heats your home, the electricity that lights it (mostly generated by the burning of fossil fuels). Your life is having a significant impact on the earth's carbon balance. And you are not alone. Three hundred million other Americans are having a similar impact. In 2005 alone, the United States released 6 billion metric tons of carbon dioxide into earth's atmosphere from the burning of fossil fuels (5,973,000,000,000,000 pounds!).

This massive flow of carbon dioxide into the atmosphere is having an unintended and very grave consequence—the earth is getting warmer. What can we do about it? Switch to *alternative energy sources*.

Many countries are turning to nuclear power for their growing energy needs. In 2007, 436 nuclear reactors were producing power worldwide, generating 14% of the world's electricity. Seventy-eight percent of France's electricity is now produced by nuclear power plants.

Nuclear power plants have not been as popular in this country as in the rest of the world because we have ample access to cheap coal and because the public fears the consequences of an accident. A reactor partial meltdown at the Three Mile Island nuclear plant in Pennsylvania in 1979 released little radiation into the environment but galvanized these fears. There has been little nuclear power development in this country since then (**figure 38.19**).

In theory, nuclear power can provide plentiful, cheap energy, but the reality is less encouraging. Nuclear power presents several problems—safety, waste disposal, security—that must be overcome if it is to provide a significant portion of the energy that will fuel our future world.

Alternative Energy Sources

A variety of other sources of cleaner energy can help reduce our use of fossil fuels. Many of these are *renewable energy*—sources of energy, such as solar power, that are naturally replenished. The solar panels in **figure 38.20a** capture the energy of sunlight to heat water or other fluids to make steam that turns a turbine, generating electricity. Smaller applications use solar panels connected to photovoltaic cells, which convert solar energy directly into electricity. Other sources of renewable energy include wind (see **figure 38.20b**) and, in particular, biomass—plants like corn and sugar cane that can be used to produce ethanol, replacing gasoline in cars.

Figure 38.19   Three Mile Island nuclear power plant.
Since a nuclear accident here in 1979, the building of nuclear power stations in the United States has slowed dramatically.

(a)

(b)

Figure 38.20   Alternate energy sources.
(*a*) Solar energy uses large mirrors to collect energy from the sun. These solar panels absorb heat from the sun that is used to boil water (or other fluids), which creates steam. The steam turns large turbines (not pictured), generating electricity. (*b*) Wind-powered energy is an old technology modernized for large-scale use. Large wind fields harness the kinetic energy in wind, converting it into electricity.

Looking More Closely at Ethanol

Ethanol is a simple two-carbon alcohol, CH_3CH_2OH—the same alcohol found in beer and wine. Rich in energy-storing C-H bonds, ethanol makes a good fuel. Burning a gallon of ethanol in your car releases about 80% as much car-powering energy as burning a gallon of gasoline.

How can we burn ethanol and not add more CO_2 to the atmosphere? Focus on the word "more." Ethanol is produced through yeast fermentation of sugars found in plants, the same fermentation process that is used to make beer and wine. If our automobiles burn carbon molecules recently produced via photosynthesis by living plants, then they are simply returning to the atmosphere the CO_2 recently extracted from it! No net increase in atmospheric CO_2 occurs.

To understand this, focus on where the carbon atoms come from. Burning a fossil fuel like gasoline releases into the atmosphere stores of carbon that had been trapped for thousands of years as oil buried deep in the earth. Burning ethanol also releases carbon, but in this case the carbon dioxide released into the atmosphere has just been taken from it. Think of the atmosphere as a fountain. A fountain recycles the water it shoots into the air. The level of water in the pool stays the same because the water added to the pool by the falling spray is recycled back to the pump to shoot up again. That is how ethanol works with regard to carbon dioxide emissions. Carbon dioxide is taken from the atmosphere and used by the plant to build plant tissue; that tissue is then used to make ethanol. Now imagine if there is a nearby tank of water (representing fossil fuels), and water from the tank was pumped and sprayed into the fountain's pool. Not only would the tank get depleted, but the pool would overflow (too much carbon dioxide). That is what happens when fossil fuels are burned.

When used as fuel, ethanol is typically added to gasoline rather than burned by itself. New cars called Flexible Fuel Vehicles (FFVs) have a redesigned engine that can burn 100% gasoline, but can also burn a gasoline blend called E85 that is 85% ethanol and 15% gasoline. This adoption of ethanol for automobile fuel is good news for combating global warming.

In the United States, commercial ethanol fuel is traditionally produced by fermenting sugars obtained from starch stored in corn kernels. How might we get ethanol out of the rest of the corn plant?—of what sort of molecules are the stalk, leaves, and cob composed? They are made largely of three kinds of organic molecules: 40% cellulose, 40% hemicellulose, and 10% lignins.

Focus first on the cellulose. Fully one-half of all the organic carbon in the living world is cellulose. Like starch, cellulose consists of chains of glucose sugars linked together. Why not use the sugars in cellulose to make ethanol? Be-cause there is a subtle but very important chemical difference between starch and cellulose. Each of the glucose sugars in a starch molecule has six carbon atoms arranged in a ring like a group of children holding hands in a circle. In a cellulose molecule, the rings of the glucose molecules are inside-out, as if the children are facing with their backs to the center of the circle. Yeast enzymes do not attack links between these kinds of sugars. To use cellulose to produce ethanol, bioengineers must find a way to teach yeasts to break these links.

There are microbes with enzymes that can do this, otherwise cows could not survive by eating grass, nor termites wood. Using the sort of genetic engineering technology discussed in chapter 13, it is possible to "bioengineer" a yeast to be able to ferment cellulose. Researchers in Spain have succeeded in commercially producing ethanol from cellulose biomass by adding DNA to yeast, DNA taken from plant-digesting bacteria in the gut of termites—DNA containing the genes these bacteria use to break down cellulose and free sugars.

Nor should we forget hemicellulose, which contains one-fifth of the corn plant's carbon. Hemicellulose is like cellulose, but with five-carbon sugars. Genetically engineered bacteria containing the enzymes necessary to break down hemicellulose into free five-carbon sugars have already been constructed (The United States Patent Office Patent Number 5,000,000). So, this approach seems rich with promise.

Nor is corn the only plant that can be grown to produce biomass. There are other fast-growing plants that could be dedicated to fuel production, including switchgrass and trees such as hybrid poplar and willows. These plants can be grown on lands that are not suitable for row crops like corn, especially erosion-prone soils, leaving fertile soil for consumable crops such as corn, soybeans, and wheat. Other sources of cellulose from industrial and commercial waste, such as sawdust and paper pulp, could also be used to produce ethanol, and so could leaves and yard waste and the paper and cardboard that make up the bulk of municipal dumps.

It is difficult to imagine a more attractive long-term investment in our country's future than developing strains of ethanol-producing yeasts that ferment cellulose and hemicellulose as well as starch. Our country should massively increase its funding of research in this area. With a serious national commitment to developing biomass-to-ethanol technology, we might all, in the not-too-distant future, be filling up our tanks with farm-grown fuels rather than fossil fuels, to the world's great benefit.

> **Key Learning Outcome 38.10** Renewable energy alternatives, particularly auto fuels from biomass, are becoming increasingly important as cleaner energy sources.

Individuals Can Make the Difference

The development of appropriate solutions to the world's environmental problems must rest partly on the shoulders of politicians, economists, bankers, engineers—many kinds of public and commercial activity will be required. However, it is important not to lose sight of the key role often played by informed individuals in solving environmental problems. Often one person has made the difference; two examples serve to illustrate the point.

The Nashua River

Running through the heart of New England, the Nashua River was severely polluted by mills established in Massachusetts in the early 1900s. By the 1960s, the river was clogged with pollution and declared ecologically dead. When Marion Stoddart moved to a town along the river in 1962, she was appalled. She approached the state about setting aside a "greenway" (trees running the length of the river on both sides), but the state wasn't interested in buying land along a filthy river. So Stoddart organized the Nashua River Cleanup Committee and began a campaign to ban the dumping of chemicals and wastes into the river. The committee presented bottles of dirty river water to politicians, spoke at town meetings, recruited businesspeople to help finance a waste treatment plant, and began to clean garbage from the Nashua's banks. This citizen's campaign, coordinated by Stoddart, greatly aided passage of the Massachusetts Clean Water Act of 1966. Industrial dumping into the river is now banned, and the river has largely recovered (**figure 38.21**).

Lake Washington

A large, 86-square-kilometer freshwater lake east of Seattle, Lake Washington became surrounded by Seattle suburbs in the building boom following the Second World War. Between 1940 and 1953, a ring of 10 municipal sewage plants discharged their treated effluent into the lake. Safe enough to drink, the effluent was believed "harmless." By the mid-1950s a great deal of effluent had been dumped into the lake (try multiplying 80 million liters/day × 365 days/year × 10 years). In 1954, an ecology professor at the University of Washington in Seattle, W. T. Edmondson, noted that his research students were reporting filamentous blue-green algae growing in the lake. Such algae require plentiful nutrients, which deep freshwater lakes usually lack—the sewage had been fertilizing the lake! Edmondson, alarmed, began a campaign in 1956 to educate public officials to the danger: Bacteria decomposing dead algae would soon so deplete the lake's oxygen that the lake would die. After five years, joint municipal taxes financed the building of a sewer to carry the effluent out to sea. The lake is now clean (**figure 38.22**).

Solving Environmental Problems

It is easy to become discouraged when considering the world's many environmental problems, but do not lose track of the single most important conclusion that emerges from our examination of these problems—the fact that each is solvable. A polluted lake can be cleaned; a dirty smokestack can be altered to remove noxious gas; waste of key resources can be stopped. What is required is a clear understanding of the problem and a commitment to doing something about it. The extent to which U.S. families **recycle** aluminum cans and newspapers is evidence of the degree to which people want to become part of the solution, rather than part of the problem.

> **Key Learning Outcome 38.11** In solving environmental problems, the commitment of one person can make a critical difference.

Figure 38.21 Cleaning up the Nashua River.
The Nashua River, seen on the left in the 1960s, was severely polluted by factories along its banks dumping their wastes directly into the river. Seen on the right today, the river is mostly clean.

Figure 38.22 Lake Washington, Seattle.
Lake Washington in Seattle is surrounded by residences, businesses, and industries. By the 1950s, the dumping of sewage and the runoff of fertilizers had caused an algal bloom in the lake, which would eventually deplete the lake's oxygen. Efforts to clean up the lake began in 1956. The lake is now clean.

How Real Is Global Warming?

The controversy over global warming has two aspects. The first contentious issue is the claim that global temperatures are rising significantly, a profound change in the earth's atmosphere and oceans referred to as "global warming." The second contentious issue is the assertion that global warming is the consequence of elevated concentrations of carbon dioxide in the atmosphere as a consequence of the widespread burning of fossil fuels.

  Resolution of the second issue requires detailed science and is only now reaching consensus acceptance. Resolution of the first issue is a simpler proposition, because it is, in essence, a data statement. The graph to the right displays the data in question—global air temperatures for the last century and a half. Temperature data is collected from measuring stations across the globe, as shown in the image below, and averaged. The bars of the histogram represent mean yearly global air temperatures for each year since 1850. In order to dampen the effects of random year-to-year variations and so better reveal accumulating influences, the data are presented as an anomaly histogram (in an **anomaly histogram**, each bar presents the deviation of the value during that period from the average value determined for some standard period). In this instance, the anomaly histogram shows the deviation of each year's global mean air temperature from the mean of these values observed over a standard 30-year period between 1961 and 1990.

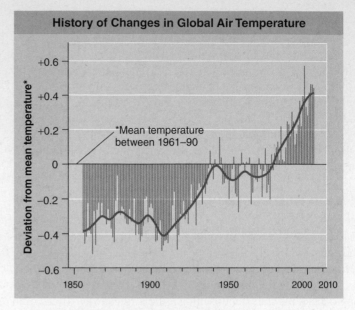

History of Changes in Global Air Temperature

*Mean temperature between 1961–90

1. **Applying Concepts**
   a. Variable. In the plot, is there a dependent variable? If so, what is it?
   b. Anomaly Histograms. What fraction of the 155 years do not deviate from the 1961–1990 mean value? What fraction deviates more than +0.2°C? more than −0.2°C? more than +0.4°C? more than −0.4°C?
2. **Interpreting Data**
   a. Of the years that deviate more than +0.2°C, how many are before 1940? between 1940 and 1980? after 1980? What fraction occurs after 1980?
   b. Of the years that deviate more than +0.4°C, how many are before 1980? after 2000? What fraction occurs after 2000?

c. Of the years that deviate more than −0.2 °C, how many are before 1940? between 1940 and 1980? after 1980? What fraction occurs before 1940?
d. Of the years that deviate more than −0.4°C, how many are before 1940? 1900? What fraction occurs before 1900?
3. **Making Inferences** If you were to pick a year at random between 1850 and 1900, would it be most likely to deviate +0.2, +0.4, 0, −0.2, or −0.4? a year between 1900 and 1940? a year between 1940 and 1980? a year after 1980? a year after 2000?
4. **Drawing Conclusions** Has the global air temperature been warming progressively over the last century and a half?

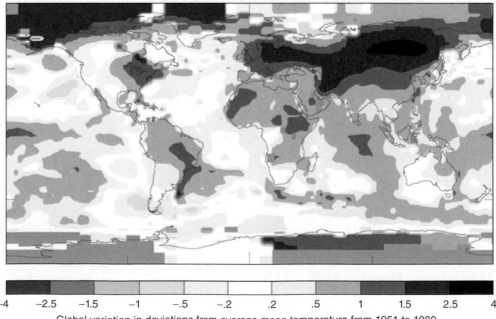

Global variation in deviations from average mean temperature from 1951 to 1980

Global Change

38.1 Pollution

- Pollution leads to global change because its effects can spread far from the source. Air and water become polluted when chemicals that are harmful to organisms are released into the ecosystem.

- The use of agricultural chemicals, such as pesticides, herbicides, and fertilizers, has been widespread with devastating effects on animals. Biological magnification occurs when harmful chemicals become more concentrated as they pass up through the food chain, as shown here from **figure 38.1**.

38.2 Acid Precipitation

- The burning of coal releases sulfur into the atmosphere, where it combines with water vapor, forming sulfuric acid. This acid falls back to earth in rain and snow, commonly called acid rain, far from the source of the pollution, killing animals and vegetation (**figures 38.2** and **38.3**).

38.3 Global Warming

- The burning of fossil fuels releases carbon dioxide into the atmosphere, more carbon dioxide than can be cycled back through the ecosystem. It remains in the atmosphere, where it traps infrared light (heat) from the sun, a phenomenon known as the greenhouse effect.

- The average global temperatures have been steadily increasing, a process known as global warming (**figure 38.4**). Global warming is predicted to have major impacts on global rain patterns, agriculture, and rising sea levels.

38.4 The Ozone Hole

- Ozone (O_3) forms a protective shield in the earth's upper atmosphere that blocks out harmful UV rays from the sun. In the mid-1970s, scientists determined that ozone was being depleted.

- Chlorofluorocarbons (CFCs), used in refrigeration systems, react with ozone, converting it to oxygen gas (O_2), which doesn't block UV rays. Termed the ozone hole, this reduction in ozone over and extending from the South Pole, as shown here from **figure 38.5**, is resulting in dangerously high levels of radiation reaching the earth.

38.5 Loss of Biodiversity

- Extinction is a fact of life, but the current rate of species loss is alarmingly high. Three factors mostly responsible for present-day extinctions include loss of habitat, overexploitation, and the introduction of exotic species (**figure 38.6**). Loss of habitat is the most devastating (**figure 38.7**).

Saving Our Environment

38.6 Reducing Pollution

- Human activities are placing severe stress on the biosphere. Reducing pollution requires examining the costs associated with pollution (**figure 38.8**). Antipollution laws and pollution taxes are ways to begin factoring in the costs of pollution.

38.7 Preserving Nonreplaceable Resources

- The consumption or destruction of nonreplaceable resources is perhaps the most serious problem humans face. Topsoil, necessary for agriculture, is being depleted rapidly. Groundwater, which percolates through the soil to underground reservoirs, is our  primary source of drinking water, but it is being wasted and polluted. Biodiversity is being reduced through extinctions, due primarily to loss of habitat, such as rain forests (**figure 38.10**).

38.8 Curbing Population Growth

- The root of all environmental problems is the rapid growth of the human population. More people means more resources are depleted, more land is developed, and more pollution is created.

- Technology has allowed the human population to grow exponentially for the last 300 years to a current population of over 6.5 billion (**figure 38.11**).

- Human populations grow at different rates, with developing countries' populations growing more rapidly than developed countries' populations. However, it takes more resources to support populations in developed countries (**figures 38.12–38.15**).

Solving Environmental Problems

38.9 Preserving Endangered Species

- In an attempt to slow the loss of biodiversity, recovery programs are under way, designed to save endangered species. These programs include habitat restoration, breeding in captivity, and conservation of ecosystems (**figures 38.16–38.18**).

38.10 Finding Cleaner Sources of Energy

- The burning of fossil fuels leads to pollution, depletion of valuable resources, and global warming. Alternative sources of energy are needed, and some countries have looked to nuclear power, but nuclear power has its own drawbacks. Renewable energy sources such as solar and wind power and ethanol are promising alternatives (**figure 38.20**).

38.11 Individuals Can Make the Difference

- There are environmental success stories, where one or a few individuals made a difference and reversed an ecological disaster (**figures 38.21** and **38.22**).

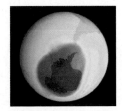

1. "Gray-air cities" are the result of
   a. biological magnification of air pollutants.
   b. chlorinated hydrocarbons as a major air pollutant.
   c. pesticides as a major air pollutant.
   d. sulfur oxides as a major air pollutant.

2. The main cause of acid rain is
   a. car and truck exhaust.
   b. coal-powered industry.
   c. chlorofluorocarbons.
   d. chlorinated hydrocarbons.

3. Global warming affects all of the following except
   a. rain patterns.
   b. rising sea levels.
   c. ozone levels.
   d. agriculture.

4. Destruction of the ozone layer is due to
   a. car and truck exhaust.
   b. coal-powered industry.
   c. chlorofluorocarbons.
   d. chlorinated hydrocarbons.

5. The factor most responsible for present-day extinctions is
   a. habitat loss.
   b. overexploitation of species.
   c. introduction of new species.
   d. All of these are equally responsible.

6. Free market economies often promote pollution. This is because
   a. environmental costs are hardly ever recognized as part of the economy.
   b. supply never keeps up with demand, so industry must increase output to address the demand.

   c. the costs of energy and raw materials are so variable.
   d. laws about pollution are unenforceable.

7. Preserving biodiversity is
   a. needed to preserve possible direct value from species, such as new medicines.
   b. not needed as extinction is a "natural" cycle and should not be disturbed.
   c. needed to make sure all niches are filled.
   d. not needed as it interferes with industrial development.

8. Which factor is *not* responsible for the large increase in the human population over the last 300 or so years?
   a. larger and more reliable food reserves from the modernization of farming techniques
   b. decreasing mortality rate due to improvements in medicine
   c. increasing amounts of open space as countries develop
   d. increased sanitation practices

9. Removal of the endangered black-footed ferret and California condor populations from the wild for breeding programs in zoos and field laboratories are examples of preservation through
   a. pristine restoration.
   b. habitat restoration.
   c. habitat rehabilitation.
   d. captive propagation.

10. If the removal of a species causes an ecosystem to collapse, that species is known as a(n)
    a. keystone species.
    b. endangered species.
    c. threatened species.
    d. None of the above.

Apply Your Understanding

1. **Figure 38.5** Your friend tells you that his father complains that environmentalists are trying to save "every confounded bug and weed on the planet." He says, "Things have always gone extinct, why should now be any different?" Use the graph to frame a response to your friend's father.

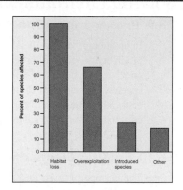

2. **Figures 38.11** and **35.10** Discuss the human growth curve on the left in terms of the "carrying capacity" graph on the right.

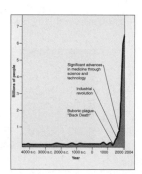

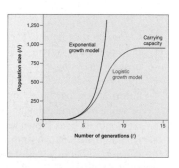

Synthesize What You Have Learned

1. Explain why exposure to even very tiny amounts of chemical pollutants like BPA that mimic estrogen can, over time, be hazardous to your health. Should one sex be more concerned than the other? Explain.

2. Economists tend to look at the world in terms of costs. They say, "If you increase pollution, then you increase the social costs to people's health, but if you try to decrease pollution, then you increase the economic costs of cleaning it up." Discuss whether you think these two costs are paid equally by the same set of people.

3. Consider the stories of Marion Stoddart and W. T. Edmondson, and the following statement: "Never doubt that a small group of thoughtful, committed citizens can change the world, indeed it's the only thing that ever has."— Margaret Mead, Anthropologist. What *could* you do to make your neighborhood/area/community a better, healthier place? What *will* you do?

Appendix A

Answers to Test Your Understanding Questions

Chapter 1 The Science of Biology

1. a. kingdoms 2. c. cellular organization 3. b. atom, molecule, organelle, cell, tissue, organ, organ system, organism, population, species, community, ecosystem 4. d. emergent properties 5. b. evolution, energy flow, cooperation, structure determines function, and homeostasis 6. a. inductive reasoning 7. b. test each hypothesis, using appropriate controls, to rule out as many as possible 8. c. After sufficient testing, you can accept it as probable, being aware that it may be revised or rejected in the future 9. d. all living organisms consist of cells, and all cells come from other cells 10. c. is contained in a long molecule called DNA

Chapter 2 The Chemistry of Life

1. b. an atom 2. c. an ion 3. d. ionic, covalent, and hydrogen 4. b. it can form four single covalent bonds 5. d. All of the above 6. a. hydrogen bonds between the individual water molecules 7. c. heat storage and heat of vaporization 8. a. cohesion 9. a. (1) acids and (2) bases 10. c. A buffer stops water from ionizing

Chapter 3 Molecules of Life

1. b. proteins, carbohydrates, lipids, and nucleic acids 2. c. polypeptides 3. c. structure, function 4. d. All of the above 5. d. are information storage devices found in body cells 6. a. Adenine forms hydrogen bonds with thymine 7. a. structure and energy 8. a. glycogen 9. d. All of these are characteristics of fats. 10. c. energy storage and for some hormones

Chapter 4 Cells

1. a. cells are the smallest living things. Nothing smaller than a cell is considered alive 2. b. a double lipid layer with proteins inserted in it, which surrounds every cell individually 3. d. prokaryotes, eukaryotes 4. a. a nucleolus 5. c. endoplasmic reticulum and the Golgi complex

6. d. mitochondria and the chloroplasts 7. b. Eukaryotic cells in plants and fungi, and all prokaryotes, have a cell wall 8. d. slowly disperse throughout the water; this is because of diffusion 9. b. endocytosis and phagocytosis 10. c. energy and specialized pumps or channels

Chapter 5 Energy and Life

1. c. energy 2. d. says that energy can change forms, but cannot be made or destroyed 3. b. says that entropy, or disorder, continually increases in a closed system 4. a. exergonic and release energy 5. b. enzymes 6. c. temperature and pH 7. d. All of the above 8. c. an inhibitor molecule competes with the substrate for the active site on the enzyme 9. a. active site 10. a. the breaking of phosphate bonds in ATP

Chapter 6 Photosynthesis: Acquiring Energy from the Sun

1. c. photosynthesis 2. b. with molecules called pigments that absorb photons and use their energy 3. c. a small portion in the middle of the spectrum 4. a. red and blue 5. d. All of the above 6. c. only go through the system once; they are obtained by splitting a water molecule 7. b. chemiosmosis 8. a. electron transport system of photosystem I, Calvin cycle 9. c. build sugar molecules 10. b. use C_4 photosynthesis or CAM

Chapter 7 How Cells Harvest Energy from Food

1. a. breaking down the organic molecules that were consumed 2. c. substrate-level phosphorylation 3. b. glycolysis 4. b. makes ATP by splitting glucose and capturing the energy 5. b. NAD$^+$, the electron transport chain 6. c. mitochondria of the cell and are broken down in the presence of O_2 to make more ATP 7. d. during the electron transport chain 8. b. fermentation 9. a. pyruvate 10. c. each type of macromolecule is broken down into its subunits, which enter the oxidative respiration pathway

Chapter 8 Mitosis

1. a. copying DNA then undergoing binary fission 2. c. in the production of genetically identical daughter cells 3. c. and most eukaryotes have between 10 and 50 pairs of chromosomes 4. c. carry information about the same traits located in the same places on the chromosome 5. d. All of the above 6. b. metaphase 7. c. cytokinesis 8. a. a series of checkpoints 9. b. cancer 10. d. All of the above

Chapter 9 Meiosis

1. d. the egg and sperm only have half the number of chromosomes found in the parents due to meiosis 2. d. 23 3. a. 1n gametes (haploid), followed by 2n zygotes (diploid) 4. a. Homologous chromosomes randomly orient themselves on the metaphase plate, called independent assortment 5. c. The duplicated sister chromatids separate 6. a. prophase I 7. c. homologous chromosomes become closely associated along their lengths 8. b. homologous chromosomes exchange chromosomal material 9. a. cells that are genetically identical to the parent cells/ haploid cells 10. c. has a lot of genetic reasortment due to processes in meiosis I

Chapter 10 Foundations of Genetics

1. d. All of the above 2. a. all purple flowers 3. c. 3/4 purple and 1/4 white flowers 4. b. some factor, or information, about traits to their offspring and it may or may not be expressed 5. a. dihybrid cross 6. d. multiple genes 7. c. codominant traits 8. b. sex-linked eye color in fruit flies 9. d. All of the above 10. b. chromosomal karyotyping

Chapter 11 DNA: The Genetic Material

1. b. hereditary information can be added to cells from other cells 2. d. DNA 3. a. structure of DNA 4. c. the type of nitrogen base 5. d. adenine and guanine 6. b. TAACGTA 7. b. splits down the

middle into two single strands, and each one then acts as a template to build its complement 8. a. a primer 9. c. by mutation or by recombination 10. a. germ-line tissues and are passed on to future generations

Chapter 12 How Genes Work

1. a. nRNA (nuclear RNA) 2. d. a codon 3.c. 64 4. c. transcription 5. a. promoter 6. d. translation 7. c. AUG 8. d. some genes are always off as long as a repressor is bound 9. d. all of the above 10. c. translating a gene as if it is being transcribed

Chapter 13 The New Biology

1. c. genome 2. b. the exons used to make a specific mRNA can be rearranged to form different proteins 3. c. restriction enzyme 4. b. exposing the mRNA of the desired eukaryotic gene to reverse transcriptase 5. c. DNA fingerprinting becomes more and more reliable as more probes are used. 6. a. the drug to be produced in far larger amounts than in the past 7. a. increased yield 8. d. harm to the crop itself from mutations 9. a. immunological rejection of the tissue by the patient 10. c. viruses

Chapter 14 Evolution and Natural Selection

1. c. populations are capable of geometric increase, yet remain at constant levels 2. a. natural selection 3. b. seems to agree with Darwin's original ideas 4. a. homologous structures 5. c. no mutation within the population 6. a. 0.20 7. d. genetic drift 8. c. directional selection 9. d. reproductive isolation 10. c. hybrid infertility

Chapter 15 How We Name Living Things

1. a. the red fox is in the same family, but different genus than dogs and wolves 2. b. phylogeny 3. a. physical, behavioral, and molecular characteristics 4. c. share a more recent common ancestor than those organisms that are farther apart 5. d. cell structure and DNA sequence 6. b. Archaea 7. d. are prokaryotes 8. c. Protista 9. b. multicellular 10. d. endosymbiosis of bacteria

Chapter 16 Prokaryotes: The First Single-Celled Creatures

1. b. The primordial atmosphere of the earth contained the same amount of oxygen as today 2. c. RNA 3. d. All of the above 4. b. chemoautotrophs 5. d. All of the above 6. d. producing the oxygen that is in the atmosphere 7. a. protein coats that contain DNA or RNA 8. c. lytic cycle 9. b. matching a marker on the surface of the virus to a complementary marker on the surface of a cell 10. c. use an enzyme called reverse transcriptase to convert the viral RNA to DNA, which is then integrated into the cell's chromosome

Chapter 17 Protists: Advent of the Eukaryotes

1. b. mitochondria and chloroplasts have their own DNA 2. c. cysts 3. d. mushrooms 4. d. All of the above 5. b. brown algae 6. a. have both a sexual and an asexual phase 7. c. green algae 8. b. choanoflagellates 9. a. cytoplasm 10. c. slime molds

Chapter 18 Fungi Invade the Land

1. c. cell specialization 2. b. cellulose cell walls 3. d. mycelium 4. a. both sexually and asexually 5. b. Ascomycota 6. c. a heterokaryon 7. a. a special sac called the ascus 8. b. basidia 9. b. algae and funi 10. d. minerals

Chapter 19 Evolution of the Animal Phyla

1. c. conserved anatomical characteristics 2. b. choanocytes 3. c. extracellular digestion 4. c. specialization of digestive tract 5. b. endoderm; mesoderm 6. a. segmentation 7. b. specialization of segments to carry out different functions 8. c. weight of the thick exoskeleton needed to support very large insects 9. a. an animal with a sessile lifestyle, rather than one that moves through the environment 10. b. sea cucumbers

Chapter 20 History of the Vertebrates

1. b. era, period, epoch, age 2. d. arthropods and the chordates 3. a. amphibians 4. d. after the Cretaceous extinction 5. b. jaws 6. c. an internal skeleton made of cartilage 7. c. middle ear bones 8. d. thin, hollow bones in the skeleton 9. b. They have high metabolic rates 10. c. hair

Chapter 21 How Humans Evolved

1. b. opposable digits on hands 2. a. Africa 3. c. lemurs 4. d. All of the above 5. d. bipedalism 6. a. brain size 7. b. *Australopithecus anamensis* 8. d. *Homo habilis* 9. d. *Homo erectus* 10. b. Africa

Chapter 22 The Animal Body and How It Moves

1. b. more flexible movement as individual segments can move independently of each other 2. a. cells, tissues, organs, organ systems, organism 3. c. move the body 4. d. red blood cells 5. c. osteoblasts; osteoclasts 6. b. neurotransmitters 7. a. skeletal 8. b. axial skeleton 9. a. a single muscle can only pull and not push 10. d. expose myosin attachment sites on actin

Chapter 23 Circulation

1. d. All of the above are functions of the circulatory system 2. c. by passing warm blood near cold blood in the extremities to warm the blood 3. a. capillaries 4. d. carry fluids 5. d. erythrocyte 6. c. A heart with separate chambers is first seen in fishes 7. b. better separation of oxygenated and deoxygenated blood 8. a. Only arteries carry oxygenated blood 9. d. endothermy 10. c. sphygmomanometer

Chapter 24 Respiration

1. c. the animal's blood to be continually exposed to water of higher oxygen concentration 2. b. increase the surface area available for gas exchange 3. a. energy need for different vertebrate classes increases 4. c. fish 5. c. It takes three cycles of breathing for air to pass through the bird's respiratory system 6. c. contracting the muscles around the thoracic cavity pulls your chest out 7. a. hemoglobin in red blood cells 8. b. as bicarbonate in the blood plasma 9. a. bicarbonate 10. d. All of the above

Chapter 25 The Path of Food Through the Animal Body

1. b. specialization of different regions of the digestive system 2. a. herbivores 3. d. begin the physical digestion of food 4. c. soft palate 5. c. stomach 6. b. Only ruminants are able to digest cellulose 7. c. small intestine 8. d. increase the surface area of the small intestine for absorption of nutrients 9. c. the concentration of solid wastes 10. a. pancreas

Chapter 26 Maintaining the Internal Environment

1. b. homeostasis 2. d. negative feedback loop 3. b. glycogen to break down 4. a. pancreas 5. b. ants 6. b. no water and excrete large volumes of urine that are hypotonic to body fluids 7. b. freshwater fish 8. c. loop of Henle 9. d. osmosis 10. d. urea

Chapter 27 How the Animal Body Defends Itself

1. a. sweat and oil glands 2. b. have cell surface proteins that are different from the body's own cell surface proteins 3. c. increase the number of immune system cells in an infected area 4. b. pathogenic bacteria do not grow well at high temperatures 5. a. T cells 6. c. B cells 7. d. destroy cells infected by pathogens 8. b. memory T and B cells 9. c. an autoimmune response 10. a. helper T cells

Chapter 28 The Nervous System

1. c. the amount of associative neurons that eventually formed the "brain" 2. b. association neurons 3. a. influx of sodium ions 4. b. is important for maintenance of the resting membrane potential 5. c. synapse 6. a. sodium ion gates in the postsynaptic cell 7. a. cerebral cortex 8. c. hypothalamus 9. d. coordinate emotions 10. b. relay messages to skeletal muscles.

Chapter 29 The Senses

1. d. stimulation, transduction, transmission 2. b. rods and cones 3. b. interoceptors 4. c. semicircular canals 5. a. photoreceptors 6. b. a membrane within the cochlea 7. a. pressure waves 8. c. Rod cells detect different colors, and cone cells detect different shades of gray, allowing vision in dim light 9. c. allows for a better depth of perception 10. d. all of the above

Chapter 30 Chemical Signaling Within the Animal Body

1. c. chemical signals stick around longer than electrical signals and can be used for slow processes 2. a. hypothalamus 3. a. fit into receptors specifically shaped for them 4. b. steroid hormones must enter the cell to begin action, whereas peptide hormones must begin action on the external surface of the cell membrane 5. d. posterior pituitary gland 6. a. ACTH 7. a. pancreas 8. c. too much calcium in the blood 9. d. sympathetic nervous system 10. a. lack of iodine

Chapter 31 Reproduction and Development

1. b. sexual reproduction 2. b. if both sex chromosomes are X 3. a. viviparity 4. d. scrotum 5. b. FSH and LH 6. a. embryo releasing hCG 7. d. gastrulation is complete 8. c. neural 9. c. oxytocin 10. a. destruction of the egg

Chapter 32 Evolution of Plants

1. c. dehydration 2. a. chloroplasts 3. b. they do not have specialized vascular tissue to transport water very high 4. d. seeds 5. c. sporophyte 6. c. nourishment 7. c. ovules not completely covered by the sporophyte 8. d. fruits and flowers 9. a. pollination 10. b. the process of double fertilization

Chapter 33 Plant Form and Function

1. c. meristematic tissue 2. b. transports carbohydrates 3. a. parenchyma cells 4. d. at the pericycle 5. c. cambium 6. b. organization of vascular tissue 7. a. stomata 8. a. photosynthesis 9. b. translocation 10. d. osmosis

Chapter 34 Plant Reproduction and Growth

1. a. pollen 2. b. stigma 3. b. monoecious 4. b. attract animal pollinators 5. d. oxygen and water 6. a. ovary 7. c. hormones 8. c. cells in plant stems to elongate 9. c. photoperiod 10. a. thigmotropism

Chapter 35 Populations and Communities

1. a. population 2. c. carrying capacity 3. b. increased competition for food 4. a. short life span 5. b. will decrease, and the mortality will increase 6. d. community 7. c. resource partitioning 8. b. commensalism 9. b. Müllerian mimicry 10. c. secondary succession

Chapter 36 Ecosystems

1. c. producers 2. d. All answers are correct 3. d. amount of energy transferred to the top carnivores 4. a. from plants 5. b. combustion 6. c. ATP 7. d. desert conditions on the downwind side of a mountain due to increased moisture-holding capacity of the winds as the air heats up 8. b. decreases, temperature increases 9. d. thermocline 10. c. tundra

Chapter 37 Behavior and the Environment

1. b. cannot be modified, as these behaviors seem built into the brain and nervous system 2. a. there is a clear link between presence or absence of a specific gene, a specific metabolic pathway, and a specific behavior 3. b. operant conditioning 4. c. reproductive fitness 5. d. foraging behavior 6. c. intersexual selection 7. b. polyandry 8. c. reciprocity 9. b. kin selection 10. d. ecological factors such as food type and predation

Chapter 38 Human Influences on the Living World

1. d. sulfur oxides as a major air pollutant 2. b. coal-powered industry 3. c. ozone levels 4. c. chlorofluorocarbons 5. a. habitat loss 6. a. environmental costs are hardly ever recognized as part of the economy 7. a. needed to preserve possible direct value from species, such as new medicines 8. c. increasing amounts of open space as countries develop 9. d. captive propagation 10. a. keystone species

Glossary

Terms & Concepts

A

absorption (L. *absorbere*, to swallow down) The movement of water and substances dissolved in water into a cell, tissue, or organism.

acid Any substance that dissociates to form H^+ ions when dissolved in water. Having a pH value less than 7.

acoelomate (Gr. *a*, not + *koiloma*, cavity) A bilaterally symmetrical animal not possessing a body cavity, such as a flatworm.

actin (Gr. *actis*, ray) One of the two major proteins that make up myofilaments (the other is myosin). It provides the cell with mechanical support and plays major roles in determining cell shape and cell movement.

action potential A single nerve impulse. A transient all-or-none reversal of the electrical potential across a neuron membrane. Because it can activate nearby voltage-sensitive channels, an action potential propagates along a nerve cell.

activation energy The energy a molecule must acquire to undergo a specific chemical reaction.

activator A regulatory protein that binds to the DNA and makes it more accessible for transcription.

active transport The transport of a solute across a membrane by protein carrier molecules to a region of higher concentration by the expenditure of chemical energy. One of the most important functions of any cell.

adaptation (L. *adaptare*, to fit) Any peculiarity of structure, physiology, or behavior that promotes the likelihood of an organism's survival and reproduction in a particular environment.

adenosine triphosphate (ATP) A molecule composed of ribose, adenine, and a triphosphate group. ATP is the chief energy currency of all cells. Cells focus all of their energy resources on the manufacture of ATP from ADP and phosphate, which requires the cell to supply 7 kilocalories of energy obtained from photosynthesis or from electrons stripped from foodstuffs to form 1 mole of ATP. Cells then use this ATP to drive endergonic reactions.

adhesion (L. *adhaerere*, to stick to) The molecular attraction exerted between the surfaces of unlike bodies in contact, as water molecules to the walls of the narrow tubes that occur in plants.

aerobic (Gr. *aer*, air + *bios*, life) Oxygen-requiring.

allele (Gr. *allelon*, of one another) One of two or more alternative forms of a gene.

allele frequency The relative proportion of a particular allele among individuals of a population. Not equivalent to gene frequency, although the two terms are sometimes confused.

allosteric interaction (Gr. *allos*, other + *stereos*, shape) The change in shape that occurs when an activator or repressor binds to an enzyme. These changes result when specific, small molecules bind to the enzyme, molecules that are not substrates of that enzyme.

alternation of generations A reproductive life cycle in which the multicellular diploid phase produces spores that give rise to the multicellular haploid phase and the multicellular haploid phase produces gametes that fuse to give rise to the zygote. The zygote is the first cell of the multicellular diploid phase.

alveolus, *pl.* alveoli (L. *alveus*, a small cavity) One of the many small, thin-walled air sacs within the lungs in which the bronchioles terminate.

amniotic egg An egg that is isolated and protected from the environment by a more or less impervious shell. The shell protects the embryo from drying out, nourishes it, and enables it to develop outside of water.

anaerobic (Gr. *an*, without + *aer*, air + *bios*, life) Any process that can occur without oxygen. Includes glycolysis and fermentation. Anaerobic organisms can live without free oxygen.

anaphase In mitosis and meiosis II, the stage initiated by the separation of sister chromatids, during which the daughter chromosomes move to opposite poles of the cell; in meiosis I, marked by separation of replicated homologous chromosomes.

angiosperms The flowering plants, one of five phyla of seed plants. In angiosperms, the ovules at the time of pollination are completely enclosed by tissues.

anterior (L. *ante*, before) Located before or toward the front. In animals, the head end of an organism.

anther (Gr. *anthos*, flower) The part of the stamen of a flower that bears the pollen.

antibody (Gr. *anti*, against) A protein substance produced by a B cell lymphocyte in response to a foreign substance (antigen) and released into the bloodstream. Binding to the antigen, antibodies mark them for destruction by other elements of the immune system.

anticodon The three-nucleotide sequence of a tRNA molecule that is complementary to, and base pairs with, an amino acid-specifying codon in mRNA.

antigen (Gr. *anti*, against + *genos*, origin) A foreign substance, usually a protein, that stimulates lymphocytes to proliferate and secrete specific antibodies that bind to the foreign substance, labeling it as foreign and destined for destruction.

apical meristem (L. *apex*, top + Gr. *meristos*, divided) In vascular plants, the growing point at the tip of the root or stem.

aposematic coloration An ecological strategy of some organisms that "advertise" their poisonous nature by the use of bright colors.

appendicular skeleton (L. *appendicula*, a small appendage) The skeleton of the limbs of the human body containing 126 bones.

archaea A group of prokaryotes that are among the most primitive still in existence, characterized by the absence of peptidoglycan in their cell walls, a feature that distinguishes them from bacteria.

asexual Reproducing without forming gametes. Asexual reproduction does not involve sex. Its outstanding characteristic is that an individual offspring is genetically identical to its parent.

association neuron A nerve cell found only in the CNS that acts as a functional link between sensory neurons and motor neurons. Also called interneuron.

atom (Gr. *atomos*, indivisible) A core (nucleus) of protons and neutrons surrounded by an orbiting cloud of electrons. The chemical behavior of an atom is largely determined by the distribution of its electrons, particularly the number of electrons in its outermost level.

atomic number The number of protons in the nucleus of an atom. In an atom that does not bear an electric charge (that is, one that is not an ion), the atomic number is also equal to the number of electrons.

autonomic nervous system (Gr. *autos*, self + *nomos*, law) The motor pathways that carry commands from the central nervous system to regulate the glands and nonskeletal muscles of the body. Also called the involuntary nervous system.

autosome (Gr. *autos*, self + *soma*, body) Any of the 22 pairs of human chromosomes that are similar in size and morphology in both males and females.

autotroph (Gr. *autos*, self + *trophos*, feeder) An organism that can harvest light energy from the sun or from the oxidation of inorganic compounds to make organic molecules.

axial skeleton The skeleton of the head and trunk of the human body containing 80 bones.

axon (Gr., axle) A process extending out from a neuron that conducts impulses away from the cell body.

B

bacterium, *pl.* **bacteria (Gr.** *bakterion,* **dim. of** *baktron,* **a staff)** The simplest cellular organism. Its cell is smaller and prokaryotic in structure, and it lacks internal organization.

basal body In eukaryotic cells that contain flagella or cilia, a form of centriole that anchors each flagellum.

base Any substance that combines with H^+ ions thereby reducing the H^+ ion concentration of a solution. Having a pH value above 7.

Batesian mimicry After Henry W. Bates, English naturalist. A situation in which a palatable or nontoxic organism resembles another kind of organism that is distasteful or toxic. Both species exhibit warning coloration.

B cell A lymphocyte that recognizes invading pathogens much as T cells do, but instead of attacking the pathogens directly, it marks them with antibodies for destruction by the nonspecific body defenses.

bilateral symmetry (L. *bi,* **two +** *lateris,* **side; Gr.** *symmetria,* **symmetry)** A body form in which the right and left halves of an organism are approximate mirror images of each other.

binary fission (L. *binarius,* **consisting of two things or parts +** *fissus,* **split)** Asexual reproduction of a cell by division into two equal, or nearly equal, parts. Bacteria divide by binary fission.

binomial system (L. *bi,* **twice, two + Gr.** *nomos,* **usage, law)** A system of nomenclature that uses two words. The first names the genus, and the second designates the species.

biomass (Gr. *bios,* **life +** *maza,* **lump or mass)** The total weight of all of the organisms living in an ecosystem.

biome (Gr. *bios,* **life +** *-oma,* **mass, group)** A major terrestrial assemblage of plants, animals, and microorganisms that occur over wide geographical areas and have distinct characteristics. The largest ecological unit.

buffer A substance that takes up or releases hydrogen ions (H^+) to maintain the pH within a certain range.

C

calorie (L. *calor,* **heat)** The amount of energy in the form of heat required to raise the temperature of 1 gram of water 1 degree Celsius.

calyx (Gr. *kalyx,* **a husk, cup)** The sepals collectively. The outermost flower whorl.

cancer Unrestrained invasive cell growth. A tumor or cell mass resulting from uncontrollable cell division.

capillary (L. *capillaris,* **hairlike)** A blood vessel with a very small diameter. Blood exchanges gases and metabolites across capillary walls. Capillaries join the end of an arteriole to the beginning of a venule.

carbohydrate (L. *carbo,* **charcoal +** *hydro,* **water)** An organic compound consisting of a chain or ring of carbon atoms to which hydrogen and oxygen atoms are attached in a ratio of approximately 1:2:1. A compound of carbon, hydrogen, and oxygen having the generalized formula $(CH_2O)_n$, where n is the number of carbon atoms.

carcinogen (Gr. *karkinos,* **cancer + -gen)** Any cancer-causing agent.

cardiovascular system (Gr. *kardia,* **heart + L.** *vasculum,* **vessel)** The blood circulatory system and the heart that pumps it. Collectively, the blood, heart, and blood vessels.

carpel (Gr. *karpos,* **fruit)** A leaflike organ in angiosperms that encloses one or more ovules.

carrying capacity The maximum population size that a habitat can support.

catabolism (Gr. *katabole,* **throwing down)** A process in which complex molecules are broken down into simpler ones.

catalysis (Gr. *katalysis,* **dissolution +** *lyein,* **to loosen)** The enzyme-mediated process in which the subunits of polymers are positioned so that their bonds undergo chemical reactions.

catalyst (Gr. *kata,* **down +** *lysis,* **a loosening)** A general term for a substance that speeds up a specific chemical reaction by lowering the energy required to activate or start the reaction. An enzyme is a biological catalyst.

cell (L. *cella,* **a chamber or small room)** The smallest unit of life. The basic organizational unit of all organisms. Composed of a nuclear region containing the hereditary apparatus within a larger volume called the cytoplasm bounded by a lipid membrane.

cell cycle The repeating sequence of growth and division through which cells pass each generation.

cellular respiration The process in which the energy stored in a glucose molecule is released by oxidation. Hydrogen atoms are lost by glucose and gained by oxygen.

central nervous system The brain and spinal cord, the site of information processing and control within the nervous system.

centromere (Gr. *kentron,* **center +** *meros,* **a part)** A constricted region of the chromosome joining two sister chromatids, to which the kinetochore is attached.

chemical bond The force holding two atoms together. The force can result from the attraction of opposite charges (ionic bond) or from the sharing of one or more pairs of electrons (a covalent bond).

chemiosmosis The cellular process responsible for almost all of the adenosine triphosphate (ATP) harvested from food and for all the ATP produced by photosynthesis.

chemoautotroph An autotrophic bacterium that uses chemical energy released by specific inorganic reactions to power its life processes, including the synthesis of organic molecules.

chiasma, *pl.* **chiasmata (Gr. a cross)** In meiosis, the points of crossing over where portions of chromosomes have been exchanged during synapsis. A chiasma appears as an X-shaped structure under a light microscope.

chloroplast (Gr. *chloros,* **green +** *plastos,* **molded)** A cell-like organelle present in algae and plants that contains chlorophyll (and usually other pigments) and is the site of photosynthesis.

chromatid (Gr. *chroma,* **color + L. -id, daughters of)** One of two daughter strands of a duplicated chromosome that is joined by a single centromere.

chromatin (Gr. *chroma,* **color)** The complex of DNA and proteins of which eukaryotic chromosomes are composed.

chromosome (Gr. *chroma,* **color +** *soma,* **body)** The vehicle by which hereditary information is physically transmitted from one generation to the next. In a eukaryotic cell, long threads of DNA that are associated with protein and that contain hereditary information.

cilium, *pl.* **cilia (L. eyelash)** Refers to flagella, which are numerous and organized in dense rows. Cilia propel cells through water. In human tissue, they move water or mucus over the tissue surface.

cladistics A taxonomic technique used for creating hierarchies of organisms based on derived characters that represent true phylogenetic relationship and descent.

class A taxonomic category ranking below a phylum (division) and above an order.

clone (Gr. *klon,* **twig)** A line of cells, all of which have arisen from the same single cell by mitotic division. One of a population of individuals derived by asexual reproduction from a single ancestor. One of a population of genetically identical individuals.

codominance In genetics, a situation in which the effects of both alleles at a particular locus are apparent in the phenotype of the heterozygote.

codon (L. code) The basic unit of the genetic code. A sequence of three adjacent nucleotides in DNA or mRNA that codes for one amino acid or for polypeptide termination.

coelom (Gr. *koilos,* **a hollow)** A body cavity formed between layers of mesoderm and in which the digestive tract and other internal organs are suspended.

coenzyme A cofactor of an enzyme that is a nonprotein organic molecule.

coevolution (L. *co-,* **together +** *e-,* **out +** *volvere,* **to fill)** A term that describes the long-term evolutionary adjustment of one group of organisms to another.

commensalism (L. *cum,* **together with +** *mensa,* **table)** A symbiotic relationship in which one species benefits while the other neither benefits nor is harmed.

community (L. *communitas,* **community, fellowship)** The populations of different species that live together and interact in a particular place.

competition Interaction between individuals for the same scarce resources. Intraspecific competition is competition between individuals of a single species. Interspecific competition is competition between individuals of different species.

competitive exclusion The hypothesis that if two species are competing with one another for the same limited resource in the same place, one will be able to use that resource more efficiently than the other and eventually will drive that second species to extinction locally.

complement system The chemical defense of a vertebrate body that consists of a battery of proteins that insert in bacterial and fungal cells, causing holes that destroy the cells.

concentration gradient The concentration difference of a substance as a function of distance. In a cell, a greater concentration of its molecules in one region than in another.

condensation The coiling of the chromosomes into more and more tightly compacted bodies begun during the G_2 phase of the cell cycle.

conjugation (L. *conjugare*, to yoke together) An unusual mode of reproduction in unicellular organisms in which genetic material is exchanged between individuals through tubes connecting them during conjugation.

consumer In ecology, a heterotroph that derives its energy from living or freshly killed organisms or parts thereof. Primary consumers are herbivores; secondary consumers are carnivores or parasites.

cortex (L. bark) In vascular plants, the primary ground tissue of a stem or root, bounded externally by the epidermis and internally by the central cylinder of vascular tissue. In animals, the outer, as opposed to the inner, part of an organ, as in the adrenal, kidney, and cerebral cortexes.

cotyledon (Gr. *kotyledon*, a cup-shaped hollow) Seed leaf. Monocot embryos have one cotyledon, and dicots have two.

countercurrent flow In organisms, the passage of heat or of molecules (such as oxygen, water, or sodium ions) from one circulation path to another moving in the opposite direction. Because the flow of the two paths is in opposite directions, a concentration difference always exists between the two channels, facilitating transfer.

covalent bond (L. *co-*, together + *valare*, to be strong) A chemical bond formed by the sharing of one or more pairs of electrons.

crossing over An essential element of meiosis occurring during prophase when nonsister chromatids exchange portions of DNA strands.

cuticle (L. *cutis*, skin) A very thin film covering the outer skin of many plants.

cytokinesis (Gr. *kytos*, hollow vessel + *kinesis*, movement) The C phase of cell division in which the cell itself divides, creating two daughter cells.

cytoplasm (Gr. *kytos*, hollow vessel + *plasma*, anything molded) A semifluid matrix that occupies the volume between the nuclear region and the cell membrane. It contains the sugars, amino acids, proteins, and organelles (in eukaryotes) with which the cell carries out its everyday activities of growth and reproduction.

cytoskeleton (Gr. *kytos*, hollow vessel + *skeleton*, a dried body) In the cytoplasm of all eukaryotic cells, a network of protein fibers that supports the shape of the cell and anchors organelles, such as the nucleus, to fixed locations.

D

deciduous (L. *decidere*, to fall off) In vascular plants, shedding all the leaves at a certain season.

dehydration reaction Water-losing. The process in which a hydroxyl (OH) group is removed from one subunit of a polymer and a hydrogen (H) group is removed from the other subunit, linking the subunits together and forming a water molecule as a by-product.

demography (Gr. *demos*, people + *graphein*, to draw) The statistical study of population. The measurement of people or, by extension, of the characteristics of people.

density The number of individuals in a population in a given area.

deoxyribonucleic acid (DNA) The basic storage vehicle or central plan of heredity information. It is stored as a sequence of nucleotides in a linear nucleotide polymer. Two of the polymers wind around each other like the outside and inside rails of a circular staircase.

depolarization The movement of ions across a cell membrane that wipes out locally an electrical potential difference.

deuterostome (Gr. *deuteros*, second + *stoma*, mouth) An animal in whose embryonic development the anus forms from or near the blastopore, and the mouth forms later on another part of the blastula. Also characterized by radial cleavage.

dicot Short for dicotyledon; a class of flowering plants generally characterized by having two cotyledons, netlike veins, and flower parts in fours or fives.

diffusion (L. *diffundere*, to pour out) The net movement of molecules to regions of lower concentration as a result of random, spontaneous molecular motions. The process tends to distribute molecules uniformly.

dihybrid (Gr. *dis*, twice + L. *hibrida*, mixed offspring) An individual heterozygous for two genes.

dioecious (Gr. *di*, two + *eikos*, house) Having male and female flowers on separate plants of the same species.

diploid (Gr. *diploos*, double + *eidos*, form) A cell, tissue, or individual with a double set of chromosomes.

directional selection A form of selection in which selection acts to eliminate one extreme from an array of phenotypes. Thus, the genes promoting this extreme become less frequent in the population.

disaccharide (Gr. *dis*, twice + *sakcharon*, sugar) A sugar formed by linking two monosaccharide molecules together. Sucrose (table sugar) is a disaccharide formed by linking a molecule of glucose to a molecule of fructose.

disruptive selection A form of selection in which selection acts to eliminate rather than favor the intermediate type.

diurnal (L. *diurnalis*, day) Active during the day.

division Traditionally, a major taxonomic group of the plant kingdom comparable to a phylum of the animal kingdom. Today divisions are called phyla.

dominant allele An allele that dictates the appearance of heterozygotes. One allele is said to be dominant over another if an individual heterozygous for that allele has the same appearance as an individual homozygous for it.

dorsal (L. *dorsum*, the back) Toward the back, or upper surface. Opposite of ventral.

double fertilization A process unique to the angiosperms, in which one sperm nucleus fertilizes the egg and the second one fuses with the polar nuclei. These two events result in the formation of the zygote and the primary endosperm nucleus, respectively.

E

ecdysis (Gr. *ekdysis*, stripping off) The shedding of the outer covering or skin of certain animals. Especially the shedding of the exoskeleton by arthropods.

ecology (Gr. *oikos*, house + *logos*, word) The study of the relationships of organisms with one another and with their environment.

ecosystem (Gr. *oikos*, house + *systema*, that which is put together) A community, together with the nonliving factors with which it interacts.

ectoderm (Gr. *ecto*, outside + *derma*, skin) One of three embryonic germ layers that forms in the gastrula; giving rise to the outer epithelium and to nerve tissue.

ectothermic Referring to animals whose body temperature is regulated by their behavior or their surroundings.

electron A subatomic particle with a negative electrical charge. The negative charge of one electron exactly balances the positive charge of one proton. Electrons orbit the atom's positively charged nucleus and determine its chemical properties.

electron transport chain A collective term describing the series of membrane-associated electron carriers embedded in the inner mitochondrial membrane. It puts the electrons harvested from the oxidation of glucose to work driving proton-pumping channels.

electron transport system A collective term describing the series of membrane-associated electron carriers embedded in the thylakoid membrane of the chloroplast. It puts the electrons harvested from water molecules and energized by photons of light to work driving proton-pumping channels.

element A substance that cannot be separated into different substances by ordinary chemical methods.

emergent properties Novel properties in the hierarchy of life that were not present at the simpler levels of organization.

endergonic (Gr. *endon*, within + *ergon*, work) Reactions in which the products contain more energy than the reactants and require an input of usable energy from an outside source before they can proceed. These reactions are not spontaneous.

endocrine gland (Gr. *endon*, within + *krinein*, to separate) A ductless gland producing hormonal secretions that pass directly into the bloodstream or lymph.

endocrine system The dozen or so major endocrine glands of a vertebrate.

endocytosis (Gr. *endon*, within + *kytos*, cell) The process by which the edges of plasma membranes fuse together and form an enclosed chamber called a vesicle. It involves the incorporation of a portion of an exterior medium into the cytoplasm of the cell by capturing it within the vesicle.

endoderm (Gr. *endon*, outside + *derma*, skin) One of three embryonic germ layers that forms in the gastrula; giving rise to the epithelium that lines internal organs and most of the digestive and respiratory tracts.

endoplasmic reticulum (ER) (L. *endoplasmic*, within the cytoplasm + *reticulum*, little net) An extensive network of membrane compartments within a eukaryotic cell; attached ribosomes synthesize proteins to be exported.

endoskeleton (Gr. *endon*, within + *skeletos*, hard) In vertebrates, an internal scaffold of bone or cartilage to which muscles are attached.

endosperm (Gr. *endon*, within + *sperma*, seed) A nutritive tissue characteristic of the seeds of angiosperms that develops from the union of a male nucleus and the polar nuclei of the embryo sac. The endosperm is either digested by the growing embryo or retained in the mature seed to nourish the germinating seedling.

endosymbiotic (Gr. *endon*, within + *bios*, life) theory A theory that proposes how eukaryotic cells arose from large prokaryotic cells that engulfed smaller ones of a different species. The smaller cells were not consumed but continued to live and function within the larger host cell. Organelles that are believed to have entered larger cells in this way are mitochondria and chloroplasts.

endothermic The ability of animals to maintain an elevated body temperature using their metabolism.

energy The capacity to bring about change, to do work.

enhancer A site of regulatory protein binding on the DNA molecule distant from the promoter and start site for a gene's transcription.

entropy (Gr. *en*, in + *tropos*, change in manner) A measure of the disorder of a system. A measure of energy that has become so randomized and uniform in a system that the energy is no longer available to do work.

enzyme (Gr. *enzymos*, leavened; from *en*, in + *zyme*, leaven) A protein capable of speeding up specific chemical reactions by lowering the energy required to activate or start the reaction but that remains unaltered in the process.

epidermis (Gr. *epi*, on or over + *derma*, skin) The outermost layer of cells. In vertebrates, the nonvascular external layer of skin of ectodermal origin; in invertebrates, a single layer of ectodermal epithelium; in plants, the flattened, skinlike outer layer of cells.

epistasis (Gr. *epistasis*, a standing still) An interaction between the products of two genes in which one modifies the phenotypic expression produced by the other.

epithelium (Gr. *epi*, on + *thele*, nipple) A thin layer of cells forming a tissue that covers the internal and external surfaces of the body. Simple epithelium consists of the membranes that line the lungs and major body cavities and that are a single cell layer thick. Stratified epithelium (the skin or epidermis) is composed of more complex epithelial cells that are several cell layers thick.

erythrocyte (Gr. *erythros*, red + *kytos*, hollow vessel) A red blood cell, the carrier of hemoglobin. Erythrocytes act as the transporters of oxygen in the vertebrate body. During the process of their maturation in mammals, they lose their nuclei and mitochondria, and their endoplasmic reticulum is reabsorbed.

estrus (L. *oestrus*, frenzy) The period of maximum female sexual receptivity. Associated with ovulation of the egg. Being "in heat."

estuary (L. *aestus*, tide) A partly enclosed body of water, such as those that often form at river mouths and in coastal bays, where the salinity is intermediate between that of saltwater and freshwater.

ethology (Gr. *ethos*, habit or custom + *logos*, discourse) The study of patterns of animal behavior in nature.

euchromatin (Gr. *eu*, true + *chroma*, color) Chromatin that is extended except during cell division, from which RNA is transcribed.

eukaryote (Gr. *eu*, true + *karyon*, kernel) A cell that possesses membrane-bounded organelles, most notably a cell nucleus, and chromosomes whose DNA is associated with proteins; an organism composed of such cells. The appearance of eukaryotes marks a major event in the evolution of life, as all organisms on earth other than bacteria and archaea are eukaryotes.

eumetazoan (Gr. *eu*, true + *meta*, with + *zoion*, animal) A "true animal." An animal with a definite shape and symmetry and nearly always distinct tissues.

eutrophic (Gr. *eutrophos*, thriving) Refers to a lake in which an abundant supply of minerals and organic matter exists.

evaporation The escape of water molecules from the liquid to the gas phase at the surface of a body of water.

evolution (L. *evolvere*, to unfold) Genetic change in a population of organisms over time (generations). Darwin proposed that natural selection was the mechanism of evolution.

exergonic (L. *ex*, out + Gr. *ergon*, work) Any reaction that produces products that contain less free energy than that possessed by the original reactants and that tends to proceed spontaneously.

exocytosis (Gr. *ex*, out of + *kytos*, cell) The extrusion of material from a cell by discharging it from vesicles at the cell surface. The reverse of endocytosis.

exoskeleton (Gr. *exo*, outside + *skeletos*, hard) An external hard shell that encases a body. In arthropods, comprised mainly of chitin.

experiment The test of a hypothesis. An experiment that tests one or more alternative hypotheses and those that are demonstrated to be inconsistent with experimental observation are rejected.

F

facilitated diffusion The transport of molecules across a membrane by a carrier protein in the direction of lowest concentration.

family A taxonomic group ranking below an order and above a genus.

feedback inhibition A regulatory mechanism in which a biochemical pathway is regulated by the amount of the product that the pathway produces.

fermentation (L. *fermentum*, ferment) A catabolic process in which the final electron acceptor is an organic molecule.

fertilization (L. *ferre*, to bear) The union of male and female gametes to form a zygote.

fitness The genetic contribution of an individual to succeeding generations, relative to the contributions of other individuals in the population.

flagellum, *pl.* flagella (L. *flagellum*, whip) A fine, long, threadlike organelle protruding from the surface of a cell. In bacteria, a single protein fiber capable of rotary motion that propels the cell through the water. In eukaryotes, an array of microtubules with a characteristic internal 9 + 2 microtubule structure that is capable of vibratory but not rotary motion. Used in locomotion and feeding. Common in protists and motile gametes. A cilium is a short flagellum.

food web The food relationships within a community. A diagram of who eats whom.

founder effect The effect by which rare alleles and combinations of alleles may be enhanced in new populations.

frequency In statistics, defined as the proportion of individuals in a certain category, relative to the total number of individuals being considered.

fruit In angiosperms, a mature, ripened ovary (or group of ovaries) containing the seeds.

G

gamete (Gr. wife) A haploid reproductive cell. Upon fertilization, its nucleus fuses with that of another gamete of the opposite sex. The resulting diploid cell (zygote) may develop into a new diploid individual, or in some protists and fungi, may undergo meiosis to form haploid somatic cells.

gametophyte (Gr. *gamete*, wife + *phyton*, plant) In plants, the haploid (*n*), gamete-producing generation, which alternates with the diploid (2*n*) sporophyte.

ganglion, *pl.* **ganglia (Gr. a swelling)** A group of nerve cells forming a nerve center in the peripheral nervous system.

gastrulation The inward movement of certain cell groups from the surface of the blastula.

gene (Gr. *genos*, birth, race) The basic unit of heredity. A sequence of DNA nucleotides on a chromosome that encodes a polypeptide or RNA molecule and so determines the nature of an individual's inherited traits.

gene expression The process in which an RNA copy of each active gene is made, and the RNA copy directs the sequential assembly of a chain of amino acids at a ribosome.

gene frequency The frequency with which individuals in a population possess a particular gene. Often confused with allele frequency.

genetic code The "language" of the genes. The mRNA codons specific for the 20 common amino acids constitute the genetic code.

genetic drift Random fluctuations in allele frequencies in a small population over time.

genetic map A diagram showing the relative positions of genes.

genetics (Gr. *genos*, birth, race) The study of the way in which an individual's traits are transmitted from one generation to the next.

genome (Gr. *genos*, offspring + L. *oma*, abstract group) The genetic information of an organism.

genomics The study of genomes as opposed to individual genes.

genotype (Gr. *genos*, offspring + *typos*, form) The total set of genes present in the cells of an organism. Also used to refer to the set of alleles at a single gene locus.

genus, *pl.* **genera (L. race)** A taxonomic group that ranks below a family and above a species.

germination (L. *germinare*, to sprout) The resumption of growth and development by a spore or seed.

gland (L. *glandis*, acorn) Any of several organs in the body, such as exocrine or endocrine, that secrete substances for use in the body. Glands are composed of epithelial tissue.

glomerulus (L. a little ball) A network of capillaries in a vertebrate kidney, whose walls act as a filtration device.

glycolysis (Gr. *glykys*, sweet + *lyein*, to loosen) The anaerobic breakdown of glucose; this enzyme-catalyzed process yields two molecules of pyruvate with a net of two molecules of ATP.

golgi complex Flattened stacks of membrane compartments that collect, package, and distribute molecules made in the endoplasmic reticulum.

gravitropism (L. *gravis*, heavy + *tropes*, turning) The response of a plant to gravity, which generally causes shoots to grow up and roots to grow down.

greenhouse effect The process in which carbon dioxide and certain other gases, such as methane, that occur in the earth's atmosphere transmit radiant energy from the sun but trap the longer wavelengths of infrared light, or heat, and prevent them from radiating into space.

guard cells Pairs of specialized epidermal cells that surround a stoma. When the guard cells are turgid, the stoma is open; when they are flaccid, it is closed.

gymnosperm (Gr. *gymnos*, naked + *sperma*, seed) A seed plant with seeds not enclosed in an ovary. The conifers are the most familiar group.

H

habitat (L. *habitare*, to inhabit) The place where individuals of a species live.

half-life The length of time it takes for half of a radioactive substance to decay.

haploid (Gr. *haploos*, single + *eidos*, form) The gametes of a cell or an individual with only one set of chromosomes.

Hardy-Weinberg equilibrium After G. H. Hardy, English mathematician, and G. Weinberg, German physician. A mathematical description of the fact that the relative frequencies of two or more alleles in a population do not change because of Mendelian segregation. Allele and genotype frequencies remain constant in a random-mating population in the absence of inbreeding, selection, or other evolutionary forces. Usually stated as: If the frequency of allele A is p and the frequency of allele a is q, then the genotype frequencies after one generation of random mating will always be $(p + q)^2 = p^2 + 2pq + q^2$.

Haversian canal After Clopton Havers, English anatomist. Narrow channels that run parallel to the length of a bone and contain blood vessels and nerve cells.

helper T cell A class of white blood cells that initiates both the cell-mediated immune response and the humoral immune response; helper T cells are the targets of the AIDS virus (HIV).

hemoglobin (Gr. *haima*, blood + L. *globus*, a ball) A globular protein in vertebrate red blood cells and in the plasma of many invertebrates that carries oxygen and carbon dioxide.

herbivore (L. *herba*, grass + *vorare*, to devour) Any organism that eats only plants.

heredity (L. *heredis*, heir) The transmission of characteristics from parent to offspring.

heterochromatin (Gr. *heteros*, different + *chroma*, color) That portion of a eukaryotic chromosome that remains permanently condensed and therefore is not transcribed into RNA. Most centromere regions are heterochromatic.

heterokaryon (Gr. *heteros*, other + *karyon*, kernel) A fungal hypha that has two or more genetically distinct types of nuclei.

heterotroph (Gr. *heteros*, other + *trophos*, feeder) An organism that does not have the ability to produce its own food. *See also* autotroph.

heterozygote (Gr. *heteros*, other + *zygotos*, a pair) A diploid individual carrying two different alleles of a gene on its two homologous chromosomes.

hierarchical (Gr. *hieros*, sacred + *archos*, leader) Refers to a system of classification in which successively smaller units of classification are included within one another.

histone (Gr. *histos*, tissue) A complex of small, very basic polypeptides rich in the amino acids arginine and lysine. A basic part of chromosomes, histones form the core around which DNA is wrapped.

homeostasis (Gr. *homeos*, similar + *stasis*, standing) The maintaining of a relatively stable internal physiological environment in an organism or steady-state equilibrium in a population or ecosystem.

homeotherm (Gr. *homeo*, similar + *therme*, heat) An organism, such as a bird or mammal, capable of maintaining a stable body temperature.

hominid (L. *homo*, man) Human beings and their direct ancestors. A member of the family Hominidae. *Homo sapiens* is the only living member.

homologous chromosome (Gr. *homologia*, agreement) One of the two nearly identical versions of each chromosome. Chromosomes that associate in pairs in the first stage of meiosis. In diploid cells, one chromosome of a pair that carries equivalent genes.

homology (Gr. *homologia*, agreement) A condition in which the similarity between two structures or functions is indicative of a common evolutionary origin.

homozygote (Gr. *homos*, same or similar + *zygotos*, a pair) A diploid individual whose two copies of a gene are the same. An individual carrying identical alleles on both homologous chromosomes is said to be homozygous for that gene.

hormone (Gr. *hormaein*, to excite) A chemical messenger, often a steroid or peptide, produced in a small quantity in one part of an organism and then transported to another part of the organism, where it brings about a physiological response.

hybrid (L. *hybrida*, the offspring of a tame sow and a wild boar) A plant or animal that results from the crossing of dissimilar parents.

hybridization The mating of unlike parents of different taxa.

hydrogen bond A molecular force formed by the attraction of the partial positive charge of one hydrogen atom of a water molecule with the partial negative charge of the oxygen atom of another.

hydrolysis reaction (Gr. *hydro*, water + *lyse*, break) The process of tearing down a polymer by adding a molecule of water. A hydrogen is attached to one subunit and a hydroxyl to the other, which breaks the covalent bond. Essentially the reverse of a dehydration reaction.

hydrophilic (Gr. *hydro*, water + *philic*, loving) Describes polar molecules, which form hydrogen bonds with water and therefore are soluble in water.

hydrophobic (Gr. *hydro*, water + *phobos*, hating) Describes nonpolar molecules, which do not form hydrogen bonds with water and therefore are not soluble in water.

hydroskeleton (Gr. *hydro*, water + *skeletos*, hard) The skeleton of most soft-bodied invertebrates that have neither an internal nor an external skeleton. They use the relative incompressibility of the water within their bodies as a kind of skeleton.

hypertonic (Gr. *hyper*, above + *tonos*, tension) A cell that contains a higher concentration of solutes than its surrounding solution.

hypha, *pl.* hyphae (Gr. *hyphe*, web) A filament of a fungus. A mass of hyphae comprises a mycelium.

hypothalamus (Gr. *hypo*, under + *thalamos*, inner room) The region of the brain under the thalamus that controls temperature, hunger, and thirst and that produces hormones that influence the pituitary gland.

hypothesis (Gr. *hypo*, under + *tithenai*, to put) A proposal that might be true. No hypothesis is ever proven correct. All hypotheses are provisional—proposals that are retained for the time being as useful but that may be rejected in the future if found to be inconsistent with new information. A hypothesis that stands the test of time—often tested and never rejected—is called a theory.

hypotonic (Gr. *hypo*, under + *tonos*, tension) A solution surrounding a cell that has a lower concentration of solutes than does the cell.

I

inbreeding The breeding of genetically related plants or animals. In plants, inbreeding results from self-pollination. In animals, inbreeding results from matings between relatives. Inbreeding tends to increase homozygosity.

incomplete dominance The ability of two alleles to produce a heterozygous phenotype that is different from either homozygous phenotype.

independent assortment Mendel's second law: The principle that segregation of alternative alleles at one locus into gametes is independent of the segregation of alleles at other loci. Only true for gene loci located on different chromosomes or those so far apart on one chromosome that crossing over is very frequent between the loci.

industrial melanism (Gr. *melas*, black) The evolutionary process in which a population of initially light-colored organisms becomes a population of dark organisms as a result of natural selection.

inflammatory response (L. *inflammare*, to flame) A generalized nonspecific response to infection that acts to clear an infected area of infecting microbes and dead tissue cells so that tissue repair can begin.

integument (L. *integumentum*, covering) The natural outer covering layers of an animal. Develops from the ectoderm.

interneuron A nerve cell found only in the CNS that acts as a functional link between sensory neurons and motor neurons. Also called association neuron.

internode The region of a plant stem between nodes where stems and leaves attach.

interoception (L. *interus*, inner + Eng. [re]ceptive) The sensing of information that relates to the body itself, its internal condition, and its position.

interphase That portion of the cell cycle preceding mitosis. It includes the G_1 phase, when cells grow, the S phase, when a replica of the genome is synthesized, and a G_2 phase, when preparations are made for genomic separation.

intron (L. *intra*, within) A segment of DNA transcribed into mRNA but removed before translation. These untranslated regions make up the bulk of most eukaryotic genes.

ion An atom in which the number of electrons does not equal the number of protons. An ion carries an electrical charge.

ionic bond A chemical bond formed between ions as a result of the attraction of opposite electrical charges.

ionizing radiation High-energy radiation, such as X rays and gamma rays.

isolating mechanisms Mechanisms that prevent genetic exchange between individuals of different populations or species.

isotonic (Gr. *isos*, equal + *tonos*, tension) A cell with the same concentration of solutes as its environment.

isotope (Gr. *isos*, equal + *topos*, place) An atom that has the same number of protons but different numbers of neutrons.

J

joint The part of a vertebrate where one bone meets and moves on another.

K

karyotype (Gr. *karyon*, kernel + *typos*, stamp or print) The particular array of chromosomes that an individual possesses.

kinetic energy The energy of motion.

kinetochore (Gr. *kinetikos*, putting in motion + *choros*, chorus) A disk of protein bound to the centromere to which microtubules attach during cell division, linking chromatids to the spindle.

kingdom The chief taxonomic category. This book recognizes six kingdoms: Archaea, Bacteria, Protista, Fungi, Animalia, and Plantae.

L

lamella, *pl.* lamellae (L. a little plate) A thin, platelike structure. In chloroplasts, a layer of chlorophyll-containing membranes. In bivalve mollusks, one of the two plates forming a gill. In vertebrates, one of the thin layers of bone laid concentrically around the Haversian canals.

ligament (L. *ligare*, to bind) A band or sheet of connective tissue that links bone to bone.

linkage The patterns of assortment of genes that are located on the same chromosome. Important because if the genes are located relatively far apart, crossing over is more likely to occur between them than if they are close together.

lipid (Gr. *lipos*, fat) A loosely defined group of molecules that are insoluble in water but soluble in oil. Oils such as olive, corn, and coconut are lipids, as well as waxes, such as beeswax and earwax.

lipid bilayer The basic foundation of all biological membranes. In such a layer, the nonpolar tails of phospholipid molecules point inward, forming a nonpolar zone in the interior of the bilayers. Lipid bilayers are selectively permeable and do not permit the diffusion of water-soluble molecules into the cell.

littoral (L. *litus*, shore) Referring to the shoreline zone of a lake or pond or the ocean that is exposed to the air whenever water recedes.

locus, *pl.* loci (L. place) The position on a chromosome where a gene is located.

loop of Henle After F. G. J. Henle, German anatomist. A hairpin loop formed by a urine-conveying tubule when it enters the inner layer of the kidney and then turns around to pass up again into the outer layer of the kidney.

lymph (L. *lympha*, clear water) In animals, a colorless fluid derived from blood by filtration through capillary walls in the tissues.

lymphatic system An open circulatory system composed of a network of vessels that function to collect the water within blood plasma forced out during passage through the capillaries and to return it to the bloodstream. The lymphatic system also returns proteins to the circulation, transports fats absorbed from the intestine, and carries bacteria and dead blood cells to the lymph nodes and spleen for destruction.

lymphocyte (Gr. *lympha*, water + Gr. *kytos*, hollow vessel) A white blood cell. A cell of the immune system that either synthesizes antibodies (B cells) or attacks virus-infected cells (T cells).

lyse (Gr. *lysis*, loosening) To disintegrate a cell by rupturing its plasma membrane.

M

macromolecule (Gr. *makros*, large + L. *moliculus*, a little mass) An extremely large molecule. Refers specifically to carbohydrates, lipids, proteins, and nucleic acids.

macrophage (Gr. *makros*, large + -*phage*, eat) A phagocytic cell of the immune system able to engulf and digest invading bacteria, fungi, and other microorganisms, as well as cellular debris.

marrow The soft tissue that fills the cavities of most bones and is the source of red blood cells.

mass flow The overall process by which materials move in the phloem of plants.

mass number The mass number of an atom consists of the combined mass of all of its protons and neutrons.

meiosis (Gr. *meioun*, to make smaller) A special form of nuclear division that precedes gamete formation in sexually reproducing eukaryotes. It results in four haploid daughter cells.

Mendelian ratio After Gregor Mendel, Austrian monk. Refers to the characteristic 3:1 segregation ratio that Mendel observed, in which pairs of alternative traits are expressed in the F_2 generation in the ratio of three-fourths dominant to one-fourth recessive.

menstruation (L. *mens*, month) Periodic sloughing off of the blood-enriched lining of the uterus when pregnancy does not occur.

meristem (Gr. *merizein*, to divide) In plants, a zone of unspecialized cells whose only function is to divide.

mesoderm (Gr. *mesos*, middle + *derma*, skin) One of the three embryonic germ layers that form in the gastrula. Gives rise to muscle, bone, and other connective tissue; the peritoneum; the circulatory system; and most of the excretory and reproductive systems.

mesophyll (Gr. *mesos*, middle + *phyllon*, leaf) The photosynthetic parenchyma of a leaf, located within the epidermis. The vascular strands (veins) run through the mesophyll.

metabolism (Gr. *metabole*, change) The process by which all living things assimilate energy and use it to grow.

metamorphosis (Gr. *meta*, after + *morphe*, form + *osis*, state of) Process in which form changes markedly during postembryonic development—for example, tadpole to frog or larval insect to adult.

metaphase (Gr. *meta*, middle + *phasis*, form) The stage of mitosis characterized by the alignment of the chromosomes on a plane in the center of the cell.

metastasis, *pl.* metastases (Gr. to place in another way) The spread of cancerous cells to other parts of the body, forming new tumors at distant sites.

microevolution (Gr. *mikros*, small + L. *evolvere*, to unfold) Refers to the evolutionary process itself. Evolution within a species. Also called adaptation.

microtubule (Gr. *mikros*, small + L. *tubulus*, little pipe) In eukaryotic cells, a long, hollow cylinder about 25 nanometers in diameter and composed of the protein tubulin. Microtubules influence cell shape, move the chromosomes in cell division, and provide the functional internal structure of cilia and flagella.

mimicry (Gr. *mimos*, mime) The resemblance in form, color, or behavior of certain organisms (mimics) to other more powerful or more protected ones (models), which results in the mimics being protected in some way.

mitochondrion, *pl.* mitochondria (Gr. *mitos*, thread + *chondrion*, small grain) A tubular or sausage-shaped organelle 1 to 3 micrometers long. Bounded by two membranes, mitochondria closely resemble the aerobic bacteria from which they were originally derived. As chemical furnaces of the cell, they carry out its oxidative metabolism.

mitosis (Gr. *mitos*, thread) The M phase of cell division in which the microtubular apparatus is assembled, binds to the chromosomes, and moves them apart. This phase is the essential step in the separation of the two daughter cell genomes.

mole (L. *moles*, mass) The atomic weight of a substance, expressed in grams. One mole is defined as the mass of 6.0222×10^{23} atoms.

molecule (L. *moliculus*, a small mass) The smallest unit of a compound that displays the properties of that compound.

monocot Short for monocotyledon; flowering plant in which the embryos have only one cotyledon, the flower parts are often in threes, and the leaves typically are parallel-veined.

monomers (Gr. *mono*, single + *meris*, part) Simple molecules that can join together to form polymers.

monosaccharide (Gr. *monos*, one + *sakcharon*, sugar) A simple sugar.

morphogenesis (Gr. *morphe*, form + *genesis*, origin) The formation of shape. The growth and differentiation of cells and tissues during development.

motor endplate The point where a neuron attaches to a muscle. A neuromuscular synapse.

multicellularity A condition in which the activities of the individual cells are coordinated and the cells themselves are in contact. A property of eukaryotes alone and one of their major characteristics.

muscle (L. *musculus*, mouse) The tissue in the body of humans and animals that can be contracted and relaxed to make the body move.

muscle cell A long, cylindrical, multinucleated cell that contains numerous myofibrils and is capable of contraction when stimulated.

muscle spindle A sensory organ that is attached to a muscle and sensitive to stretching.

mutagen (L. *mutare*, to change) A chemical capable of damaging DNA.

mutation (L. *mutare*, to change) A change in a cell's genetic message.

mutualism (L. *mutuus*, lent, borrowed) A symbiotic relationship in which both participating species benefit.

mycelium, *pl.* mycelia (Gr. *mykes*, fungus) In fungi, a mass of hyphae.

mycology (Gr. *mykes*, fungus) The study of fungi. A person who studies fungi is called a mycologist.

mycorrhiza, *pl.* mycorrhizae (Gr. *mykes*, fungus + *rhiza*, root) A symbiotic association between fungi and plant roots.

myofibril (Gr. *myos*, muscle + L. *fibrilla*, little fiber) An elongated structure in a muscle fiber, composed of myosin and actin.

myosin (Gr. *myos*, muscle + *in*, belonging to) One of two protein components of myofilaments. (The other is actin.)

N

natural selection The differential reproduction of genotypes caused by factors in the environment. Leads to evolutionary change.

nematocyst (Gr. *nema*, thread + *kystos*, bladder) A coiled, threadlike stinging structure of cnidarians that is discharged to capture prey and for defense.

nephron (Gr. *nephros*, kidney) The functional unit of the vertebrate kidney. A human kidney has more than 1 million nephrons that filter waste matter from the blood. Each nephron consists of a Bowman's capsule, glomerulus, and tubule.

nerve A bundle of axons with accompanying supportive cells, held together by connective tissue.

nerve impulse A rapid, transient, self-propagating reversal in electrical potential that travels along the membrane of a neuron.

neuromodulator A chemical transmitter that mediates effects that are slow and longer lasting and that typically involve second messengers within the cell.

neuromuscular junction The structure formed when the tips of axons contact (innervate) a muscle fiber.

neuron (Gr. nerve) A nerve cell specialized for signal transmission.

neurotransmitter (Gr. *neuron*, nerve + L. *trans*, across + *mitere*, to send) A chemical released at an axon tip that travels across the synapse and binds a specific receptor protein in the membrane on the far side.

neurulation (Gr. *neuron*, nerve) The elaboration of a notochord and a dorsal nerve cord that marks the evolution of the chordates.

neutron (L. *neuter*, neither) A subatomic particle located within the nucleus of an atom. Similar to a proton in mass, but as its name implies, a neutron is neutral and possesses no charge.

neutrophil An abundant type of white blood cell capable of engulfing microorganisms and other foreign particles.

niche (L. *nidus*, nest) The role an organism plays in the environment; realized niche is the niche that an organism occupies under natural circumstances; fundamental niche is the niche an organism would occupy if competitors were not present.

nitrogen fixation The incorporation of atmospheric nitrogen into nitrogen compounds, a process that can be carried out only by certain microorganisms.

nocturnal (L. *nocturnus*, night) Active primarily at night.

node (L. *nodus*, knot) The place on the stem where a leaf is formed.

node of Ranvier After L. A. Ranvier, French histologist. A gap formed at the point where two Schwann cells meet and where the axon is in direct contact with the surrounding intercellular fluid.

nondisjunction The failure of homologous chromosomes to separate in meiosis I. The cause of Down syndrome.

nonrandom mating A phenomenon in which individuals with certain genotypes sometimes mate with one another more commonly than would be expected on a random basis.

notochord (Gr. *noto*, back + L. *chorda*, cord) In chordates, a dorsal rod of cartilage that forms between the nerve cord and the developing gut in the early embryo.

nucleic acid A nucleotide polymer. A long chain of nucleotides. Chief types are deoxyribonucleic acid (DNA), which is double-stranded, and ribonucleic acid (RNA), which is typically single-stranded.

nucleosome (L. *nucleus*, kernel + *soma*, body) The basic packaging unit of eukaryotic chromosomes, in which the DNA molecule is wound around a ball of histone proteins. Chromatin is composed of long strings of nucleosomes, like beads on a string.

nucleotide A single unit of nucleic acid, composed of a phosphate, a five-carbon sugar (either ribose or deoxyribose), and a purine or a pyrimidine.

nucleolus A region inside the nucleus where rRNA and ribosomes are produced.

nucleus (L. *a kernel*, dim. Fr. *nux*, nut) A spherical organelle (structure) characteristic of eukaryotic cells. The repository of the genetic information that directs all activities of a living cell. In atoms, the central core, containing positively charged protons and (in all but hydrogen) electrically neutral neutrons.

O

oocyte (Gr. *oion*, egg + *kytos*, vessel) A cell in the outer layer of the ovary that gives rise to an ovum. A primary oocyte is any of the 2 million oocytes a female is born with, all of which have begun the first meiotic division.

operon (L. *operis*, work) A cluster of functionally related genes transcribed onto a single mRNA molecule. A common mode of gene regulation in prokaryotes; it is rare in eukaryotes other than fungi.

order A taxonomic category ranking below a class and above a family.

organ (L. *organon*, tool) A complex body structure composed of several different kinds of tissue grouped together in a structural and functional unit.

organelle (Gr. *organella*, little tool) A specialized compartment of a cell. Mitochondria are organelles.

organism Any individual living creature, either unicellular or multicellular.

organ system A group of organs that function together to carry out the principal activities of the body.

osmoconformer An animal that maintains the osmotic concentration of its body fluids at about the same level as that of the medium in which it is living.

osmoregulation The maintenance of a constant internal solute concentration by an organism, regardless of the environment in which it lives.

osmosis (Gr. *osmos*, act of pushing, thrust) The diffusion of water across a membrane that permits the free passage of water but not that of one or more solutes. Water moves from an area of low solute concentration to an area with higher solute concentration.

osmotic pressure The increase of hydrostatic water pressure within a cell as a result of water molecules that continue to diffuse inward toward the area of lower water concentration (the water concentration is lower inside than outside the cell because of the dissolved solutes in the cell).

osteoblast (Gr. *osteon*, bone + *blastos*, bud) A bone-forming cell.

osteocyte (Gr. *osteon*, bone + *kytos*, hollow vessel) A mature osteoblast.

outcross A term used to describe species that interbreed with individuals other than those like themselves.

oviparous (L. *ovum*, egg + *parere*, to bring forth) Refers to reproduction in which the eggs are developed after leaving the body of the mother, as in reptiles.

ovulation The successful development and release of an egg by the ovary.

ovule (L. *ovulum*, a little egg) A structure in a seed plant that becomes a seed when mature.

ovum, *pl.* ova (L. egg) A mature egg cell. A female gamete.

oxidation (Fr. *oxider*, to oxidize) The loss of an electron during a chemical reaction from one atom to another. Occurs simultaneously with reduction. Is the second stage of the 10 reactions of glycolysis.

oxidative metabolism A collective term for metabolic reactions requiring oxygen.

oxidative respiration Respiration in which the final electron acceptor is molecular oxygen.

P

parasitism (Gr. *para*, beside + *sitos*, food) A symbiotic relationship in which one organism benefits and the other is harmed.

parthenogenesis (Gr. *parthenos*, virgin + Eng. *genesis*, beginning) The development of an adult from an unfertilized egg. A common form of reproduction in insects.

partial pressures (P) The components of each individual gas—such as nitrogen, oxygen, and carbon dioxide—that together constitute the total air pressure.

pathogen (Gr. *pathos*, suffering + Eng. *genesis*, beginning) A disease-causing organism.

pedigree (L. *pes*, foot + *grus*, crane) A family tree. The patterns of inheritance observed in family histories. Used to determine the mode of inheritance of a particular trait.

peptide (Gr. *peptein*, to soften, digest) Two or more amino acids linked by peptide bonds.

peptide bond A covalent bond linking two amino acids. Formed when the positive (amino, or NH_2) group at one end and a negative (carboxyl, or COOH) group at the other end undergo a chemical reaction and lose a molecule of water.

peristalsis (Gr. *peri*, around + *stellein*, to wrap) The rhythmic sequences of waves of muscular contraction in the walls of a tube.

pH Refers to the concentration of H^+ ions in a solution. The numerical value of the pH is the negative of the exponent of the molar concentration. Low pH values indicate high concentrations of H^+ ions (acids), and high pH values indicate low concentrations (bases).

phagocyte (Gr. *phagein*, to eat + *kytos*, hollow vessel) A cell that kills invading cells by engulfing them. Includes neutrophils and macrophages.

phagocytosis (Gr. *phagein*, to eat + *kytos*, hollow vessel) A form of endocytosis in which cells engulf organisms or fragments of organisms.

phenotype (Gr. *phainein*, to show + *typos*, stamp or print) The realized expression of the genotype. The observable expression of a trait (affecting an individual's structure, physiology, or behavior) that results from the biological activity of proteins or RNA molecules transcribed from the DNA.

pheromone (Gr. *pherein*, to carry + [hor]mone) A chemical signal emitted by certain animals as a means of communication.

phloem (Gr. *phloos*, bark) In vascular plants, a food-conducting tissue basically composed of sieve elements, various kinds of parenchyma cells, fibers, and sclereids.

phosphodiester bond The bond that results from the formation of a nucleic acid chain in which individual sugars are linked together in a line by the phosphate groups. The phosphate group of one sugar binds to the hydroxyl group of another, forming an—O—P—O bond.

photon (Gr. *photos*, light) The unit of light energy.

photoperiodism (Gr. *photos*, light + *periodos*, a period) A mechanism that organisms use to measure seasonal changes in relative day and night length.

photorespiration A process in which carbon dioxide is released without the production of ATP or NADPH. Because it produces neither ATP nor NADPH, photorespiration acts to undo the work of photosynthesis.

photosynthesis (Gr. *photos*, light + *-syn*, together + *tithenai*, to place) The process by which plants, algae, and some bacteria use the energy of sunlight to create from carbon dioxide (CO_2) and water (H_2O) the more complicated molecules that make up living organisms.

phototropism (Gr. *photos*, light + *trope*, turning to light) A plant's growth response to a unidirectional light source.

phylogeny (Gr. *phylon*, race, tribe) The evolutionary relationships among any group of organisms.

phylum, *pl.* phyla (Gr. *phylon*, race, tribe) A major taxonomic category, ranking above a class.

physiology (Gr. *physis*, nature + *logos*, a discourse) The study of the function of cells, tissues, and organs.

pigment (L. *pigmentum*, paint) A molecule that absorbs light.

pili (pilus) Short flagella that occur on the cell surface of some prokaryotes.

pinocytosis (Gr. *pinein*, to drink + *kytos*, cell) A form of endocytosis in which the material brought into the cell is a liquid containing dissolved molecules.

pistil (L. *pistillum*, pestle) Central organ of flowers, typically consisting of ovary, style, and stigma; a pistil may consist of one or more fused carpels and is more technically and better known as the gynoecium.

plankton (Gr. *planktos*, wandering) The small organisms that float or drift in water, especially at or near the surface.

plasma (Gr. form) The fluid of vertebrate blood. Contains dissolved salts, metabolic wastes, hormones, and a variety of proteins, including antibodies and albumin. Blood minus the blood cells.

plasma membrane A lipid bilayer with embedded proteins that control the cell's permeability to water and dissolved substances.

plasmid (Gr. *plasma*, a form or something molded) A small fragment of DNA that replicates independently of the bacterial chromosome.

platelet (Gr. dim of *plattus*, flat) In mammals, a fragment of a white blood cell that circulates in the blood and functions in the formation of blood clots at sites of injury.

pleiotropy (Gr. *pleros*, more + *trope*, a turning) A gene that produces more than one phenotypic effect.

polarization The charge difference of a neuron so that the interior of the cell is negative with respect to the exterior.

polar molecule A molecule with positively and negatively charged ends. One portion of a polar molecule attracts electrons more strongly than another portion, with the result that the molecule has electron-rich (−) and electron-poor (+) regions, giving it magnetlike positive and negative poles. Water is one of the most polar molecules known.

pollen (L. fine dust) A fine, yellowish powder consisting of grains or microspores, each of which contains a mature or immature male gametophyte. In flowering plants, pollen is released from the anthers of flowers and fertilizes the pistils.

pollen tube A tube that grows from a pollen grain. Male reproductive cells move through the pollen tube into the ovule.

pollination The transfer of pollen from the anthers to the stigmas of flowers for fertilization, as by insects or the wind.

polygyny (Gr. *poly*, many + *gyne*, woman, wife) A mating choice in which a male mates with more than one female.

polymer (Gr. *polus*, many + *meris*, part) A large molecule formed of long chains of similar molecules called subunits.

polymerase chain reaction (PCR) A process by which DNA polymerase is used to copy a sequence of DNA repeatedly, making millions of copies of the same DNA.

polymorphism (Gr. *polys*, many + *morphe*, form) The presence in a population of more than one allele of a gene at a frequency greater than that of newly arising mutations.

polynomial system (Gr. *polys*, many + [bi]nomial) Before Linnaeus, naming a genus by use of a cumbersome string of Latin words and phrases.

polyp A cylindrical, pipe-shaped cnidarian usually attached to a rock with the mouth facing away from the rock on which it is growing. Coral is made up of polyps.

polypeptide (Gr. *polys*, many + *peptein*, to digest) A general term for a long chain of amino acids linked end to end by peptide bonds. A protein is a long, complex polypeptide.

polysaccharide (Gr. *polys*, many + *sakcharon*, sugar) A sugar polymer. A carbohydrate composed of many monosaccharide sugar subunits linked together in a long chain.

population (L. *populus*, the people) Any group of individuals of a single species, occupying a given area at the same time.

posterior (L. *post*, after) Situated behind or farther back.

potential difference A difference in electrical charge on two sides of a membrane caused by an unequal distribution of ions.

potential energy Energy with the potential to do work. Stored energy.

predation (L. *praeda*, prey) The eating of other organisms. The one doing the eating is called a predator, and the one being consumed is called the prey.

primary growth In vascular plants, growth originating in the apical meristems of shoots and roots, as contrasted with secondary growth; results in an increase in length.

primary plant body The part of a plant that arises from the apical meristems.

primary producers Photosynthetic organisms, including plants, algae, and photosynthetic bacteria.

primary structure of a protein The sequence of amino acids that makes up a particular polypeptide chain.

primordium, *pl.* primordia (L. *primus*, first + *ordiri*, begin) The first cells in the earliest stages of the development of an organ or structure.

productivity The total amount of energy of an ecosystem fixed by photosynthesis per unit of time. Net productivity is productivity minus that which is expended by the metabolic activity of the organisms in the community.

prokaryote (Gr. *pro*, before + *karyon*, kernel) A simple organism that is small, single-celled, and has little evidence of internal structure.

promoter An RNA polymerase binding site. The nucleotide sequence at the end of a gene to which RNA polymerase attaches to initiate transcription of mRNA.

prophase (Gr. *pro*, before + *phasis*, form) The first stage of mitosis during which the chromosomes become more condensed, the nuclear envelope is reabsorbed, and a network of microtubules (called the spindle) forms between opposite poles of the cell.

protein (Gr. *proteios*, primary) A long chain of amino acids linked end to end by peptide bonds. Because the 20 amino acids that occur in proteins have side groups with very different chemical properties, the function and shape of a protein is critically affected by its particular sequence of amino acids.

protist (Gr. *protos*, first) A member of the kingdom Protista, which includes unicellular eukaryotic organisms and some multicellular lines derived from them.

proton A subatomic particle in the nucleus of an atom that carries a positive charge. The number of protons determines the chemical character of the atom because it dictates the number of electrons orbiting the nucleus and available for chemical activity.

protostome (Gr. *protos*, first + *stoma*, mouth) An animal in whose embryonic development the mouth forms at or near the blastopore. Also characterized by spiral cleavage.

protozoa (Gr. *protos*, first + *zoion*, animal) The traditional name given to heterotrophic protists.

pseudocoel (Gr. *pseudos*, false + *koiloma*, cavity) A body cavity similar to the coelom except that it forms between the mesoderm and endoderm.

punctuated equilibrium A hypothesis of the mechanism of evolutionary change that proposes that long periods of little or no change are punctuated by periods of rapid evolution.

Q

quaternary structure of a protein A term to describe the way multiple protein subunits are assembled into a whole.

R

radial symmetry (L. *radius*, a spoke of a wheel + Gr. *summetros*, symmetry) The regular arrangement of parts around a central axis so that any plane passing through the central axis divides the organism into halves that are approximate mirror images.

radioactivity The emission of nuclear particles and rays by unstable atoms as they decay into more stable forms. Measured in curies, with 1 curie equal to 37 billion disintegrations a second.

radula (L. scraper) A rasping, tonguelike organ characteristic of most mollusks.

recessive allele An allele whose phenotype effects are masked in heterozygotes by the presence of a dominant allele.

recombination The formation of new gene combinations. In bacteria, it is accomplished by the transfer of genes into cells, often in association with viruses. In eukaryotes, it is accomplished by reassortment of chromosomes during meiosis and by crossing over.

reducing power The use of light energy to extract hydrogen atoms from water.

reduction (L. *reductio*, a bringing back; originally, "bringing back" a metal from its oxide) The gain of an electron during a chemical reaction from one atom to another. Occurs simultaneously with oxidation.

reflex (**L.** *reflectere,* **to bend back**) An automatic consequence of a nerve stimulation. The motion that results from a nerve impulse passing through the system of neurons, eventually reaching the body muscles and causing them to contract.

refractory period The recovery period after membrane depolarization during which the membrane is unable to respond to additional stimulation.

renal (**L.** *renes,* **kidneys**) Pertaining to the kidney.

repression (**L.** *reprimere,* **to press back, keep back**) The process of blocking transcription by the placement of the regulatory protein between the polymerase and the gene, thus blocking movement of the polymerase to the gene.

repressor (**L.** *reprimere,* **to press back, keep back**) A protein that regulates transcription of mRNA from DNA by binding to the operator and so preventing RNA polymerase from attaching to the promoter.

resolving power The ability of a microscope to distinguish two points as separate.

respiration (**L.** *respirare,* **to breathe**) The utilization of oxygen. In terrestrial vertebrates, the inhalation of oxygen and the exhalation of carbon dioxide.

resting membrane potential The charge difference that exists across a neuron's membrane at rest (about 70 millivolts).

restriction endonuclease A special kind of enzyme that can recognize and cleave DNA molecules into fragments. One of the basic tools of genetic engineering.

restriction fragment-length polymorphism (RFLP) An associated genetic mutation marker detected because the mutation alters the length of DNA segments.

retrovirus (**L.** *retro,* **turning back**) A virus whose genetic material is RNA rather than DNA. When a retrovirus infects a cell, it makes a DNA copy of itself, which it can then insert into the cellular DNA as if it were a cellular gene.

ribonucleic acid (RNA) A nucleic acid that contains the sugar ribose and the pyrimidine uracil and that is used in protein production; includes mRNA, tRNA, rRNA, and siRNA.

ribose A five-carbon sugar.

ribosome A cell structure composed of protein and RNA that translates RNA copies of genes into protein.

RNA interference A type of gene silencing in which mRNA is prevented from being translated; small interfering RNAs (siRNAs) have been found to bind to mRNA and target its degradation or block its translation

RNA polymerase The enzyme that transcribes RNA from DNA.

S

saltatory conduction A very fast form of nerve impulse conduction in which the impulses leap from node to node over insulated portions.

sarcoma (**Gr.** *sarx,* **flesh**) A cancerous tumor that involves connective or hard tissue, such as muscle.

sarcomere (**Gr.** *sarx,* **flesh** + *meris,* **part of**) The fundamental unit of contraction in skeletal muscle. The repeating bands of actin and myosin that appear between two Z lines.

sarcoplasmic reticulum (**Gr.** *sarx,* **flesh** + *plassein,* **to form, mold; L.** *reticulum,* **network**) The endoplasmic reticulum of a muscle cell. A sleeve of membrane that wraps around each myofilament.

scientific creationism A view that the biblical account of the origin of the earth is literally true, that the earth is much younger than most scientists believe, and that all species of organisms were individually created just as they are today.

secondary growth In vascular plants, growth that results from the division of a cylinder of cells around the plant's periphery. Secondary growth causes a plant to grow in diameter.

secondary structure of a protein The folding and bending of a polypeptide chain, which is held in place by hydrogen bonds.

second messenger An intermediary compound that couples extracellular signals to intracellular processes and also amplifies a hormonal signal.

seed A structure that develops from the mature ovule of a seed plant. Contains an embryo and a food source surrounded by a protective coat.

selection The process by which some organisms leave more offspring than competing ones and their genetic traits tend to appear in greater proportions among members of succeeding generations than the traits of those individuals that leave fewer offspring.

self-fertilization The transfer of pollen from an anther to a stigma in the same flower or to another flower of the same plant.

sepal (**L.** *sepalum,* **a covering**) A member of the outermost whorl of a flowering plant. Collectively, the sepals constitute the calyx.

septum, *pl.* **septa** (**L.** *saeptum,* **a fence**) A partition or cross-wall, such as those that divide fungal hyphae into cells.

sex chromosomes In humans, the X and Y chromosomes, which are different in the two sexes and are involved in sex determination.

sex-linked characteristic A genetic characteristic that is determined by genes located on the sex chromosomes.

sexual reproduction Reproduction that involves the regular alternation between syngamy and meiosis. Its outstanding characteristic is that an individual offspring inherits genes from two parent individuals.

shoot In vascular plants, the aboveground parts, such as the stem and leaves.

sieve cell In the phloem (food-conducting tissue) of vascular plants, a long, slender sieve element with relatively unspecialized sieve areas and with tapering end walls that lack sieve plates. Found in all vascular plants except angiosperms, which have sieve-tube members.

soluble Refers to polar molecules that dissolve in water and are surrounded by a hydration shell.

solute The molecules dissolved in a solution. *See also* solution, solvent.

solution A mixture of molecules, such as sugars, amino acids, and ions, dissolved in water.

solvent The most common of the molecules in a solution. Usually a liquid, commonly water.

somatic cells (**Gr.** *soma,* **body**) All the diploid body cells of an animal that are not involved in gamete formation.

somite A segmented block of tissue on either side of a developing notochord.

species, *pl.* **species** (**L.** **kind, sort**) A level of taxonomic hierarchy; a species ranks next below a genus.

sperm (**Gr.** *sperma,* **sperm, seed**) A sperm cell. The male gamete.

spindle The mitotic assembly that carries out the separation of chromosomes during cell division. Composed of microtubules and assembled during prophase at the centrioles of the dividing cell.

spore (**Gr.** *spora,* **seed**) A haploid reproductive cell, usually unicellular, that is capable of developing into an adult without fusion with another cell. Spores result from meiosis, as do gametes, but gametes fuse immediately to produce a new diploid cell.

sporophyte (**Gr.** *spora,* **seed** + *phyton,* **plant**) The spore-producing, diploid ($2n$) phase in the life cycle of a plant having alternation of generations.

stabilizing selection A form of selection in which selection acts to eliminate both extremes from a range of phenotypes.

stamen (**L.** **thread**) The part of the flower that contains the pollen. Consists of a slender filament that supports the anther. A flower that produces only pollen is called staminate and is functionally male.

steroid (**Gr.** *stereos,* **solid** + **L.** *ol,* **from oleum, oil**) A kind of lipid. Many of the molecules that function as messengers and pass across cell membranes are steroids, such as the male and female sex hormones and cholesterol.

steroid hormone A hormone derived from cholesterol. Those that promote the development of the secondary sexual characteristics are steroids.

stigma (**Gr. mark**) A specialized area of the carpel of a flowering plant that receives the pollen.

stoma, *pl.* **stomata** (**Gr. mouth**) A specialized opening in the leaves of some plants that allows carbon dioxide to pass into the plant body and allows water vapor and oxygen to pass out of them.

stratum corneum The outer layer of the epidermis of the skin of the vertebrate body.

substrate (**L.** *substratus,* **strewn under**) A molecule on which an enzyme acts.

substrate-level phosphorylation The generation of ATP by coupling its synthesis to a strongly exergonic (energy-yielding) reaction.

succession In ecology, the slow, orderly progression of changes in community composition that takes place through time. Primary succession occurs in nature on bare substrates, over long periods of time. Secondary succession occurs when a climax community has been disturbed.

sugar Any monosaccharide or disaccharide.

surface tension A tautness of the surface of a liquid, caused by the cohesion of the liquid molecules. Water has an extremely high surface tension.

surface-to-volume ratio Describes cell size increases. Cell volume grows much more rapidly than surface area.

symbiosis (Gr. *syn*, together with + *bios*, life) The condition in which two or more dissimilar organisms live together in close association; includes parasitism, commensalism, and mutualism.

synapse (Gr. *synapsis*, a union) A junction between a neuron and another neuron or muscle cell. The two cells do not touch. Instead, neurotransmitters cross the narrow space between them.

synapsis (Gr. *synapsis*, contact, union) The close pairing of homologous chromosomes that occurs early in prophase I of meiosis. With the genes of the chromosomes thus aligned, a DNA strand of one homologue can pair with the complementary DNA strand of the other.

syngamy (Gr. *syn*, together with + *gamos*, marriage) Fertilization. The union of male and female gametes.

T

taxonomy (Gr. *taxis*, arrangement + *nomos*, law) The science of the classification of organisms.

T cell A type of lymphocyte involved in cell-mediated immune responses and interactions with B cells. Also called a T lymphocyte.

tendon (Gr. *tenon*, stretch) A strap of connective tissue that attaches muscle to bone.

tertiary structure of a protein The three-dimensional shape of a protein. Primarily the result of hydrophobic interactions of amino acid side groups and, to a lesser extent, of hydrogen bonds between them. Forms spontaneously.

test cross A cross between a heterozygote and a recessive homozygote. A procedure Mendel used to further test his hypotheses.

theory (Gr. *theorein*, to look at) A well-tested hypothesis supported by a great deal of evidence.

thigmotropism (Gr. *thigma*, touch + *trope*, a turning) The growth response of a plant to touch.

thorax (Gr. a breastplate) The part of the body between the head and the abdomen.

thylakoid (Gr. *thylakos*, sac + *-oides*, like) A flattened, saclike membrane in the chloroplast of a eukaryote. Thylakoids are stacked on top of one another in arrangements called grana and are the sites of photosystem reactions.

tissue (L. *texere*, to weave) A group of similar cells organized into a structural and functional unit.

trachea, *pl*. tracheae (L. windpipe) In vertebrates, the windpipe.

tracheid (Gr. *tracheia*, rough) An elongated cell with thick, perforated walls that carries water and dissolved minerals through a plant and provides support. Tracheids form an essential element of the xylem of vascular plants.

transcription (L. *trans*, across + *scribere*, to write) The first stage of gene expression in which the RNA polymerase enzyme synthesizes an mRNA molecule whose sequence is complementary to the DNA.

translation (L. *trans*, across + *latus*, that which is carried) The second stage of gene expression in which a ribosome assembles a polypeptide, using the mRNA to specify the amino acids.

translocation (L. *trans*, across + *locare*, to put or place) In plants, the process in which most of the carbohydrates manufactured in the leaves and other green parts of the plant are moved through the phloem to other parts of the plant.

transpiration (L. *trans*, across + *spirare*, to breathe) The loss of water vapor by plant parts, primarily through the stomata.

transposon (L. *transponere*, to change the position of) A DNA sequence carrying one or more genes and flanked by insertion sequences that confer the ability to move from one DNA molecule to another. An element capable of transposition (the changing of chromosomal location).

trophic level (Gr. *trophos*, feeder) A step in the movement of energy through an ecosystem.

tropism (Gr. *trop*, turning) A plant's response to external stimuli. A positive tropism is one in which the movement or reaction is in the direction of the source of the stimulus. A negative tropism is one in which the movement or growth is in the opposite direction.

turgor pressure (L. *turgor*, a swelling) The pressure within a cell that results from the movement of water into the cell. A cell with high turgor pressure is said to be turgid.

U

unicellular Composed of a single cell.

urea (Gr. *ouron*, urine) An organic molecule formed in the vertebrate liver. The principal form of disposal of nitrogenous wastes by mammals.

urine (Gr. *ouron*, urine) The liquid waste filtered from the blood by the kidneys.

V

vaccination The injection of a harmless microbe into a person or animal to confer resistance to a dangerous microbe.

vacuole (L. *vacuus*, empty) A cavity in the cytoplasm of a cell that is bound by a single membrane and contains water and waste products of cell metabolism. Typically found in plant cells.

van der Waals forces Weak chemical attractions between atoms that can occur when atoms are very close to each other.

variable Any factor that influences a process. In evaluating alternative hypotheses about one variable, all other variables are held constant so that the investigator is not misled or confused by other influences.

vascular bundle In vascular plants, a strand of tissue containing primary xylem and primary phloem. These bundles of elongated cells conduct water with dissolved minerals and carbohydrates throughout the plant body.

vascular cambium In vascular plants, the meristematic layer of cells that gives rise to secondary phloem and secondary xylem. The activity of the vascular cambium increases stem or root diameter.

ventral (L. *venter*, belly) Refers to the bottom portion of an animal. Opposite of dorsal.

vertebrate An animal having a backbone made of bony segments called vertebrae.

vesicle (L. *vesicula*, a little (ladder) Membrane-enclosed sacs within eukaryotic cells.

vessel element In vascular plants, a typically elongated cell, dead at maturity, that conducts water and solutes in the xylem.

villus, *pl*. villi (L. a tuft of hair) In vertebrates, fine, microscopic, fingerlike projections on epithelial cells lining the small intestine that serve to increase the absorptive surface area of the intestine.

vitamin (L. *vita*, life + *amine*, of chemical origin) An organic substance that the organism cannot synthesize, but is required in minute quantities by an organism for growth and activity.

viviparous (L. *vivus*, alive + *parere*, to bring forth) Refers to reproduction in which eggs develop within the mother's body and young are born free-living.

voltage-gated channel A transmembrane pathway for an ion that is opened or closed by a change in the voltage, or charge difference, across the cell membrane.

W

water vascular system The system of water-filled canals connecting the tube feet of echinoderms.

whorl A circle of leaves or of flower parts present at a single level along an axis.

wood Accumulated secondary xylem. Heartwood is the central, nonliving wood in the trunk of a tree. Hardwood is the wood of dicots, regardless of how hard or soft it actually is. Softwood is the wood of conifers.

X

xylem (Gr. *xylon*, wood) In vascular plants, a specialized tissue, composed primarily of elongate, thick-walled conducting cells, that transports water and solutes through the plant body.

Y

yolk (O.E. *geolu*, yellow) The stored substance in egg cells that provides the embryo's primary food supply.

Z

zygote (Gr. *zygotos*, paired together) The diploid ($2n$) cell resulting from the fusion of male and female gametes (fertilization).

Credits

Photographs

Chapter 0

Opener: © Anup Shah/Riser/Getty Images; p. 3: © BananaStock/JupiterImages RF; p. 4, 5, 8(corn, sunbather): © Corbis RF; p. 8(duck): © Index Stock RF; p. 8(cells): © University of Wisconsin-Madison News & Public Affairs; p. 8(Darwin): © Huntington Library/Superstock; p. 8(runner): © Pete Saloutos/Corbis; p. 8(smokers): © Brand X Pictures/PunchStock RF; p. 9(top): © Corbis RF; p. 9(middle): The McGraw-Hill Companies, Inc./Lars A. Niki, photographer; p. 9(bottom): Ozone Processing Team, Goddard Space Flight Center, NASA.

Chapter 1

Opener: © Brand X Picture/Getty RF; 1.1(Achaea): © R. Robinson/Visuals Unlimited; 1.1(bacteria): © Alfred Pasieka/Science Photo Library/Photo Researchers; 1.1(protista, plantae): © Corbis RF; 1.1(fungi, animalia): © Getty RF; 1.2: © Tom E. Adams/Visuals Unlimited; 1.3: © Corbis RF; 1.4(goose & crane): © Jim Bailey; 1.4(geese on lake): © Raymond Gehman/Corbis; 1.4(cranes on lake): © Corbis RF; 1.4(ecosystem): © Winfried Wisniewski/zefa/Corbis; p. 19(rock pigeon): © blickwinkel/Alamy; p. 19(red fantail pigeon): © Kenneth Fink/Photo Researchers; p. 19(fairy swallow pigeon): © Tom McHugh/Photo Researchers; p. 19(leaves, eagle, hippo): © Corbis RF; p. 19(moth): © George Grall/National Geographic/Getty Images; 1.5 (both): Courtesy of Bill Ober; 1.8: Ozone Processing Team, Goddard Space Flight Center, NASA; 1.9: © Laurel Hungerford; 1.10: © Dennis Kunkel/Phototake; 1.13: © Leonard Lessin/Peter Arnold; p. 30: © Michael & Patricia Fogden/Corbis.

Chapter 2

Opener: © Stan Kujawa/Alamy; 2.1: © Thinkstock Images/Jupiter RF; 2.7(both): Courtesy of Todd M. Blodgett, MD, University of Pittsburgh Medical Center; p. 38: The McGraw-Hill Companies, Inc.; p. 39: © Bettmann/Corbis; p. 40, 2.10a: © Corbis RF; 2.10b: © Hermann Eisenbeiss/National Audubon Society Collection/Photo Researchers; 2.12(top): © The McGraw-Hill Companies, Inc./Bob Coyle, photographer; 2.12(middle): © The McGraw-Hill Companies, Inc./Jacques Cornell photographer; 2.12(bottom): © The McGraw-Hill Companies, Inc./Jill Braaten photographer; p. 44: © Gilbert S. Grant/National Audubon Society Collection/Photo Researchers; p. 46(both): © Photo Archives South Tyrol Museum of Archaeology - www.iceman.it.

Chapter 3

Opener: © Michael Viard/Peter Arnold; 3.1(label, popcorn): © The McGraw-Hill Companies, Inc.; 3.4a: © Corbis; 3.4b: © Phototake/Alamy; 3.4c–d: © Getty RF; 3.4e: © Dennis Kunkel Ph.D.; 3.4f: © Photodisc/Getty RF; 3.7: © Dr. Gopal Murti/Visuals Unlimited/

Getty Images; p. 59: © A. Barrington Brown/Photo Researchers; 3.14: © J.D. Litvay/Visuals Unlimited; p. 61(cow, potatoes, leaves): © Corbis RF; p. 61(sugarcane): © PhotoLink/Getty RF; p. 61(arm): © Getty RF; p. 61(lobster): © image100/PunchStock RF; 3.15b: © Getty RF; 3.15c: © C Squared Studios/Getty RF; p. 63: © Otto Greule Jr/Getty; p. 64: © Burke/Triolo/Brand X Pictures RF; 3.16c–d: © Brand X Pictures/PunchStock RF; p. 66: © Dr. David M. Phillips/Visuals Unlimited.

Chapter 4

Opener: © Manfred Kage/Peter Arnold; p. 73 (brightfield): © David M. Phillips/Visuals Unlimited; p. 73(differential interference & darkfield): © Mike Abbey/Visuals Unlimited; p. 73(flourescence): © K.G. Murti/Visuals Unlimited; p. 73(phase contrast): © David M. Phillips/Visuals Unlimited; p. 73(confocal): © David Becker/Science Photo Library/Photo Researchers; p. 73(TEM): © Microworks/Phototake; p. 73(SEM): © Stanley Flegler/Visuals Unlimited; 4.5a: © SPL/Photo Researchers; 4.5b: © Andrew Syred/Photo Researchers; 4.5c: © Alfred Pasieka/Photo Researchers; 4.5d: © Microfield Scientific Ltd/Photo Researchers; p. 80: © SIU/Visuals Unlimited; p. 83: From "MicrobodyLike Organelles in Leaf Cells," Sue Ellen Frederick and Eldon H. Newcomb, SCIENCE, Vol. 163: 1353–1355 © 21 March 1969. Reprinted with permission from AAAS; 4.9: © Don W. Fawcett/Visuals Unlimited; 4.10: Courtesy of Dr. Kenneth Miller, Brown University; 4.13: © Biophoto Associates/Photo Researchers; 4.14: © Lester V. Bergman/Corbis; 4.15b: © Alamy RF; 4.16: © Stanley Flegler/Visuals Unlimited; 4.18: © Biophoto Associates/Photo Researchers; 4.21(all): © David M. Phillips/Visuals Unlimited; 4.23b: Courtesy of Dr. Birgit Satir, Albert Einstein College of Medicine; 4.24 (all): Courtesy of M.M. Perry & A.B. Gilbert, Cell Science, 39: 257–272, 1979.

Chapter 5

Opener: © Jane Buron/Bruce Coleman/Photoshot; 5.3(both): © Keith Eng, 2008.

Chapter 6

Opener: © Skip Moody/Dembinsky Photo Associates; p. 118: © Beateworks/Corbis RF; p. 121: © Corbis RF; 6.4(both): © Eric Soder; p. 131: © Dr. Rob Stepney/Science Photo Library/Photo Researchers; p. 132: © Philip Sze/Visuals Unlimited.

Chapter 7

Opener: © John Gerlach/Animals Animals-Earth Scenes; 7.1: © Robert A. Caputo/Aurora & Quanta Productions; p. 141(tree): © Corbis RF; p. 141(zebra & lion): © Getty RF; p. 141 (scavengers): © Anup Shah/Photodisc/Getty RF; p. 141 (butterflies): © Edward S. Ross; p. 147: © ColorBlind Images/Corbis; p. 149: © Corbis RF; p. 150: U.S. Fish and Wildlife Service.

Chapter 8

Opener: © Andrew S. Bajer; 8.1a: © Lee D. Simon/Photo Researchers; 8.3: © SPL/Photo Researchers; 8.5(all): © Andrew S. Bajer; 8.6a: © David M. Phillips/Visuals Unlimited; 8.7: © Cabisco/Visuals Unlimited; 8.12: © Moredun Animal Health LTD/Science Photo Library/Photo Researchers; p. 168: Courtesy of Priv. Doz. Dr. Roland Zell.

Chapter 9

Opener: © Adrian T Sumner/SPL/Photo Researchers; 9.2: © MIXA/Getty; 9.7(all): © C.A. Hasenkampf/Biological Photo Service; p. 181: Sir John Tenniel; Inquiry & Analysis: Reprinted with permission from Science Vol. 311, no. 5759 20 January 2006. Image: Khodjakov. Copyright 2006 AAAS.

Chapter 10

Opener: © Richard Gross/Biological Photography; 10.1: Image # 238141 American Museum of Natural History; 10.4: Courtesy R. W. Van Norman; 10.12a: From Albert & Blakeslee "Corn and Man," *Journal of Heredity* V. 5, pg. 511, 1914, Oxford University Press; 10.15a–b: © Fred Bruemmer; p. 199: © RubberBall Productions/Getty RF; 10.17(left): © Richard Hutchings/Photo Researchers; 10.17(middle left): © Cheryl A. Ertelt/Visuals Unlimited; 10.17(middle right): © William H. Mullins/Photo Researchers; 10.17(right): © Gerard Lacz/Peter Arnold; 10.18: © Corbis; 10.20(both): © CBS/Phototake; 10.23: © SPL/Photo Researchers; 10.25a: Courtesy of Loris McGavran, Denver Children's Hospital; 10.25b: © Stockbyte/Veer RF; 10.28: The Field Museum #CSA118, Chicago; 10.29: Reproduced from Ishihara's Tests for Colour Deficiency published by Kanehara Trading, Tokyo, Japan Tests for color deficiency cannot be conducted with this material. For accurate testing, the original plates should be used; p. 209: © Bettmann/Corbis; 10.31(both): © Stanley Flegler/Visuals Unlimited; 10.36: © Yoav Levy/Phototake; 10.37: © Pascal Goetgheluck/Photo Researchers; Inquiry & Analysis: *Journal of Heredity* 1932, v.23, pp.345–352, author O.L. Mohr. Used with permission.

Chapter 11

Opener: © Michael Dunning/Getty; 11.4a(top): © Science Source/Photo Researchers; 11.4a(bottom): From J.D. Watson, *The Double Helix Atheneum*, New York, 1968. Cold Spring Harbor Lab; 11.4b: © A.C. Barrington Brown/Photo Researchers; 11.8: © Don W. Fawcett/Visuals Unlimited; 11.9: © David Scharf; p. 227(top): Courtesy of Lifecodes Corp, Stamford, CT; p. 227(bottom): © AP Wide World; p. 230: © Brand X Pictures/PunchStock RF; p. 231: © Corbis RF.

Chapter 12

Opener: N. Ban, P. Nissen, J. Hansen, P.B. Moore & T.A. Steitz, "The Complete Atomic Structure of the Large Ribosomal Subunit at 2.4 A Resolution," reprinted with permission from *Science* v. 289 #5481, p. 917, © American Association for the Advancement of Science; 12.1: R C Williams, *PNAS* 74 (1977); 2313; 12.10: Courtesy of Dr. Oscar L. Miller; 12.15: Courtesy of Dr. Victoria Foe; 12.18(both): Courtesy of Dr. Harold Echols.

Chapter 13

Opener: © OSF/Bromhall, Derek/Animals Animals-Earth Scenes; 13.4(top): Courtesy of Dr. Ken Culver, Photo by John Crawford, National Institutes of Health; 13.4(middle): Courtesy of Robert H. Devlin/ Fisheries & Oceans Canada; 13.4(bottom left): Courtesy of Monsanto Corporation; 13.4(bottom right): © Dr. Gopal Murti/Science Photo Library/ Photo Researchers; p. 262: Courtesy of Lifecodes Corp., Stamford, CT; p. 263: © Getty RF; 13.8: R.L. Brinster, U. of Pennsylvania Sch. of Vet. Med.; 13.11: Monsanto Co.; 13.12: Courtesy of Ingo Potrykus & Peter Beyer, photo by Peter Beyer; p. 269(top): © AFP/Getty Images; p. 269(bottom left): © Digital Vision/Getty Images RF; p. 269(bottom right): Courtesy of Bio-Rad Laboratories; 13.13: © Getty; 13.14: © AP Photo/Paul Clements; 13.15(left): © AP Photo/Hwang Woo-suk; 13.15(right): © Seoul National University/Handout/Reuters/Corbis; 13.16: © University of Wisconsin-Madison News & Public Affairs; 13.17: © Alex Wong/Getty; 13.19: © SUW-Madison News & Public Affairs University of Wisconsin-Madison, Photo by Jeff Miller; 13.21: C. Garon & J. Rose/NIAID/NIH; p. 278: © Getty RF.

Chapter 14

Openers(top left & bottom right): Dr. Robert H. Rothman, Department of Biological Sciences, Rochester Institute of Technology; (top right): © Tui De Roy/Minden Pictures; (bottom left): © D. Parera & E. Parera-Cook/Auscape/Minden Pictures; 14.1: © Huntington Library/Superstock; 14.2: From Darwin, the Life of a Tormented Evolutionist, by Adrian Desmond; 14.7: © Digital Vision/Getty RF; 14.8: © Mary Evans Picture Library/Photo Researchers; 14.10(both): Courtesy of Dr. Arkhat Abzhanov, Harvard School of Dental Medicine, Photographer: Dr. Peter Grant of Princeton University; 14.13: © Andy Crawford/ Getty; 14.15: Courtesy of Michael Richardson and Ronan O'Rahilly; 14.21: © Ted Daeschler/ VIREO; 14.24: Courtesy of Dr. Victor A. McKusick, Johns Hopkins University; 14.32(both): © Breck P. Kent/Animals Animals-Earth Scenes; p. 309(all): Michael W. Nachman, Hopi E. Hoekstra, and Susan L. D'Agostino, "The genetic basis of adaptive melanism in pocket mice," *PNAS* April 29, 2003 vol. 100 no. 9 5268–5273. © 2003 National Academy of Sciences, U.S.A.; 14.35: Courtesy of H. Rodd; p. 312: © Peter E. Smith/The Natural Sciences Image Library; 14.36a–b: © Corbis RF; 14.36c: © Porterfield/Chickering/Photo Researchers; 14.37–1: © John Shaw/Tom Stack & Associates; 14.37–2: © Rob & Ann Simpson/Visuals Unlimited; 14.37–3: © Suzanne L. Collins & Joseph T. Collins/ National Audubon society Collection/Photo Researchers; 14.37–4: © Phil A. Dotson/National Audubon Society Collection/Photo Researchers.

Chapter 15

Opener: .© Dave Watts/Tom Stack & Associates; 15.2a(top): © Corbis RF; 15.2a(bottom): © Henry Ausloos/Animals Animals-Earth Scenes; 15.2b(top): © PunchStock/Getty Images (RF); 15.2b(bottom): © Corbis RF; 15.2c(top): © U. S. Fish and Wildlife Service/Lee Karney photographer; 15.2c(bottom): © Ingram Publishing/SuperStock RF; 15.4(top left): © Juniors Bildarchiv/Alamy; 15.4(top right): © Photodisc/Getty RF; 15.4(bottom): © Alamy RF; 15.7(cat, cheetah, lynx, tiger, lion, leopard): © Corbis RF; 15.7(puma, bobcat, ocelot, caracal, snow leopard, jaguar): © Getty RF; 15.8: © PhotoLink/Getty RF; p. 329(left): © B. Steele/ VIREO; p. 329(right): © S. Greer/VIREO; p. 329(all caterpillars): © Daniel H. Janzen; 15.11b: © OSF/ Animals Animals-Earth Scenes; p. 334(all): U.S. Fish and Wildlife Service.

Chapter 16

Opener: © AP Photo/Jean-Marc Bouju; 16.1: © Getty RF; p. 341(top): NASA; p. 341(bottom): NASA/JPL/Cornell; 16.4(E. coli): © Getty RF; 16.4(Streptococcus): © S. Lowry/University Ulster/Getty Images; 16.4(Spirillum): © John D. Cunningham/Visuals Unlimited; 16.4(Helicobacter): © CAMR/A. Barry Dowsett/Photo Researchers, Inc.; 16.4(Cyanobacteria): © Dr. James Richardson/ Visuals Unlimited, Inc.; 16.4(Gliding): © Phototake, Inc.; 16.5: Courtesy of Dr. Charles Brinton; p. 344(top): © Abraham & Beachey/BPS/Tom Stack & Associates; p. 344(bottom): © F. Widell/ Visuals Unlimited; 16.6: © Science VU/EPA/ Visuals Unlimited; 16.7: © Leonard Lessin/Peter Arnold; 16.8: © Dwight R. Kuhn; 16.10a: © Dept. of Microbiology, Biozentrum/Science Photo Library/ Photo Researchers; 16.12: © Scott Camazine/Photo Researchers; p. 354(duck): © Index Stock RF; p. 354(chimp): © Getty RF; p. 354(Ebola): © Corbis RF; p. 354(mouse): © Gene Ott; p. 354(bat): © Eric and David Hosking/Corbis; p. 354(crow): © Tim Zurowski/Corbis; p. 356: © CDC/Science Source/ Photo Researchers.

Chapter 17

Opener: © Linda Sims/Visuals Unlimited; 17.3a: © Brian Parker/Tom Stack & Associates; 17.3b: © Manfred Kage/Peter Arnold; 17.5: © Dennis Kunkel Ph.D.; 17.6: © Stephen Durr; 17.7: © Andrew H. Knoll, Harvard University; p. 368(middle): © Dr. David M. Phillips/Visuals Unlimited; 17.9: © Michael Abbey/Visuals Unlimited; 17.10a: © Edward S. Ross; 17.10b: © Eye of Science/ Photo Researchers; 17.11a: © Randy Morse/Earth Scenes/Animals Animals-Earth Scenes; 17.11b: © Jan Hinsch/Photo Researchers; 17.14: © Kevin Schafer/Peter Arnold; 17.16: © Rick Harbo; 17.17a: © Andrew Syred/Photo Researchers; 17.17b: © Wim van Egmond/Visuals Unlimited; 17.18: © Bob Barnes/Alamy; 17.19: Image by D.J. Patterson, provided by micro*scope (http: //microsccope. mbl.edu); 17.20a: © John D. Cunningham/Visuals Unlimited; 17.20b: © Peter Parks/Oxford Scientific Films/Photolibrary; 17.21: © Manfred Kage/Peter Arnold; 17.22: © Ric Ergenbright/Corbis; 17.23: © Phil A. Harrington/Peter Arnold; 17.25: © Edward S. Ross; p. 376: © Sue Ford/Photo Researchers.

Chapter 18

Opener: © Corbis RF; 18.1: © Kjell Sandved/Visuals Unlimited; 18.2: © imagebroker/Alamy RF; 18.3: © Garry T. Cole/Biological Photo Service; 18.4: © Bill Keogh/Visuals Unlimited; 18.5: © L. West/ Photo Researchers; 18.6b: © Cabisco/Phototake; 18.7b, 18.8b: © Corbis RF; 18.9: Courtesy of Dr. Bruce Waldman, School of Biological Sciences, © University of Canterbury; 18.10: © H.C. Huang/ Visuals Unlimited; 18.11: © David M. Phillips/ Visuals Unlimited; 18.12: Ralph Williams, USDA, Forest Service; 18.13a: © Image Source/Getty RF; 18.13b: © Luca DiCecco/Alamy RF; 18.13c: © Gallo Images-Nigel Dennis/Digital Vision/Getty RF; 18.14: © D.H. Marx/Visuals Unlimited; 18.15: © Corbis RF; p. 390(both): *Diseases of Aquatic Organisms*, Vol. 68: 51–63, 2005; "Life cycle stages of the amphibian chytrid Batrachochytrium dendrobatidis," Lee Berger, Alex D. Hyatt, Rick Speare, Joyce E. Longcore.

Chapter 19

Opener: © James H. Robinson/Animals Animals-Earth Scenes; p. 394(bear, brittle star, butterfly, jellyfish): © Corbis RF; p. 394(cancer cells): © David M. Phillips/Visuals Unlimited; p. 395(millipede): © Alan Morgan RF; p. 395(turtles): © Cleveland P. Hickman; p. 395(frog egg): © Cabisco/Phototake; p. 395(neuromuscular junctions): © Ed Reschke; 19.4a: © ImageState Alamy RF; 19.4b: © Corbis RF; 19.5a: © Gwen Fidler/Tom Stack & Associates; 19.5b–d: © Corbis RF; 19.10a: © T.E. Adams/Visuals Unlimited; 19.10b: © Stan Elems/Visuals Unlimited; 19.14a: © Larry Jensen/Visuals Unlimited; 19.14b: © T.E. Adams/Visuals Unlimited; 19.15a: © Image Ideas/PictureQuest RF; 19.15b: © Comstock Images/ PictureQuest RF; 19.15c: © Fred Bavendam/Minden Pictures; 19.16a: © David M. Dennis; 19.16b: © Darlyne A. Murawski/National Geographic/Getty; 19.20: © NGS/Getty; 19.21a: © George D. Dodge & Kathleen M. Dodge; 19.21b: © Anne Moreton/Tom Stack & Associates; 19.22a: © Comstock Images/ PictureQuest RF; 19.22b: © Edward S. Ross; 19.22c: © MJ Photography/Alamy RF; 19.24a: © Alex Kerstitch/Visuals Unlimited; 19.24b: © Edward S. Ross; 19.26a: © IT Stock/ Alamy RF; 19.26b: © Edward S. Ross; 19.26c: © IT Stock/ age fotostock; 19.26d: © John Gerlach/Visuals Unlimited; 19.26e: © Edward S. Ross; 19.26f: © J.A. Alcock/Visuals Unlimited; 19.26g: © Cleveland P. Hickman; 19.28a: © Randy Morse; 19.28b: © Carl Roessler; 19.28c: © William C. Ober; 19.28d: © Andrew J. Martinez/ Photo Researchers; 19.28e: © Daniel W. Gotshall; 19.28f: © Alex Kerstitch/Visuals Unlimited; 19.29a: © Corbis RF; 19.29b: © Heather Angel; 19.30: © Eric N. Olsen/The University of Texas MD Anderson Cancer Center; p. 430(all): © The Smithsonian Institution.

Chapter 20

Opener: © Getty RF; 20.2a: © Laurie O'Keefe/ Photo Researchers; 20.2b: © Alex Kerstitch/Visuals Unlimited; 20.3: © Life File/Photodisc/Getty RF; 20.4: © Michael Long/Natural History Museum, London; 20.5: © Louie Psihoyos/Corbis; 20.6a–c: © Karen Carr; 20.7: © E.R. Degginger/Animals Animals-Earth Scenes; 20.10: © Tom McHugh/Photo Researchers; 20.12: © Stephen Frink Collection/ Alamy; 20.13: © Alex Kerstitch/Visuals Unlimited; 20.16: © Natural History Museum/J. Sibbick; 20.17:

Chapter 36

Opener: © Bill Ross/Corbis; p. 747(zebras): © Dave G. Houser/Corbis; p. 747(wolves, bear): © Corbis RF; p. 747(crab, decomposer): © Edward S. Ross; 36.7: © Gerry Ellis/Minden Pictures; 36.16a: © Digital Vision/PictureQuest RF; 36.16b: © Getty RF; 36.17a: © David Fleetham/Alamy; 36.17b: Courtesy of J. Frederick Grassel, Woods Hole Oceanographic Institution; 36.17c: © Kenneth L. Smith; 36.18a: © Edward S.Ross; 36.18b: © Fred Rhode/Visuals Unlimited; 36.18c: © Dwight Kuhn; 36.19b–c: © Corbis RF; p. 765(top): © Michael Graybill & Jan Hodder/Biological Photo Service; p. 765(middle): © E.R. Degginger/ Photo Researchers; p. 765(bottom): © Corbis RF; p. 766(top): © J. Weber/Visuals Unlimited; p. 766(middle): © IFA/Peter Arnold; p. 766(bottom): © Charlie Ott/The National Audubon Society Collection/Photo Researchers; p. 767(top): © John Shaw/Tom Stack & Associates; p. 767(second): © Tom McHugh/Photo Researchers; p. 767(third): © Dave Watts/Tom Stack & Associates; p. 767 (bottom): © E.R. Degginger/Animals Animals-Earth Scenes; p. 768: U.S. Forest Service.

Chapter 37

Opener: © Renee Lynn/Photo Researchers; 37.1: © Stone/Getty; 37.3a–b: From J.R. Brown et al, "A defect in nurturing in mice lacking the immediate early gene fosB" *Cell* v. 86, 1996 pp 297–308, © Cell Press; 37.4: © Thomas McAvoy, Life Magazine/ Time/Getty; 37.6: Courtesy of Bernd Heinrich; 37.7: © Nina Leen/Time & Life Pictures/Getty; 37.8: © Photolibrary/Alamy RF; 37.9: © Bios(C. Thouvenin)/Peter Arnold; 37.11(whale): © Francois Gohier/Photo Researchers; 37.11(redstart): © Rolf Nussbaumer Photography/Alamy; 37.11(goose): © John Downer/Getty Images; 37.11(butterfly): © Corbis RF; 37.11(turtle): © Purestock/PunchStock; 37.11(wildebeest): © Image Source/PunchStock RF; p. 783(bird): © Eric & David Hosling/Corbis; p. 783(seal): © Marc Moritsch/National Geographic Image Collection; p. 783(wildebeest): © Alamy RF; p. 783(frog): Courtesy James Traniello; p. 783 (lion): © K. Ammann/Bruce Coleman/Photoshot; p. 783(ants): © Mark Moffett/Minden Pictures; 37.13: © Dr. Don W. Fawcett/Visuals Unlimited; 37.14b: © Scott Camazine/Photo Researchers; 37.15a: © S. Osolinski/OSF/Animals Animals-Earth Scenes; 37.16: © Nigel Dennis/National Audubon Society Collection/Photo Researchers; 37.18: © David Hosking/Photo Researchers; 37.19: © Jim Pickerell/Alamy; p. 790: © Alamy RF.

Chapter 38

Opener: © Steve McCurry/Magnum Photos; 38.2: © Grant Heilman/Grant Heilman Photography; 38.3: © Rob & Ann Simpson/Visuals Unlimited; 38.5: NASA; p. 799: © 1999 Ed Ely/Biological Photo Service; 38.9: © Gary Griffen/Animals Animals-Earth Scenes; 38.10a: © Digital Vision/PunchStock; 38.10b: NASA; 38.10c: © Frans Lanting; 38.11: © Stephanie Maze/Woodfin Camp & Associates; 38.16a–b: University of Wisconsin-Madison Arboretum; 38.17: © Kennan Ward/Corbis; 38.18: © Merlin D. Tuttle, Bat Conservation International; 38.19: © Byron Augustine; 38.20a: © Corbis RF; 38.20b: © Creatas/PunchStock RF; p. 810: © Corbis RF; 38.21(both): Courtesy of Nashua River Watershed Association; 38.22: © T. Henneghan/ ImageState; p. 812: NASA.

Line Art and Text

Chapter 0

Fig 0.9(inset): Data from http://jwocly.gsfc.nasa. gov/mulit/oz_hole_area.jpg

Chapter 1

Fig 1.8: Data from http://jwocly.gsfc.nasa.gov/ mulit/oz_hole_area.jpg

Chapter 15

Fig 15.3: From Niles Eldredge, *Life in the Balance*, in Natural History, June 1998. Reprinted by permission of Patricia J. Wynne.

Chapter 22

Page 494: *Schmidt Nielson Animal Physiology*, 3/e Cambridge Univ. Press 1983, Fig. 6.19 page 110.

Chapter 23

Page 512: *Schmidt Nielson Animal Physiology*, 3/e Cambridge Univ. Press 1983, Fig. 4.11 page 110.

Chapter 24

Page 526: *Schmidt Nielson Animal Physiology*, 3/e Cambridge Univ. Press 1983, Fig. 3.7 page 79.

Chapter 25

Fig 25.1: Data from www.mypyramid.gov US Department of Agriculture, US Department of Health and Human Services.; Fig 25.2 Data from "*Shape Up America*," National Institutes of Health.

Chapter 35

Fig 35.20: Data from E.J. Heske, et al, *Ecology*, 1994.

Index

Drug addiction, 592 *fig.,* 592–593
Drupes, 707, 707 *fig.*
Dubois, Eugene, 465, 465 *fig.*
Duck(s), behavioral isolation of, 314
Duck-billed platypus, 319, 454, 617, 643, 643 *fig.*
Dugesia, 410, 410 *fig.*
Duodenum, 538
Duplicated sequences, 259 *fig.*
Duplications, segmental, 259
Dynein, 91

E

Eagle, 19 *table*
Earthworm, 416 *fig.,* 416–417
    circulatory system of, 498
    digestive system of, 532, 532 *fig.*
    hydraulic skeleton of, 488, 488 *fig.*
    nervous systems of, 587
Ebola hemorrhagic fever, 354, 355 *table*
Ebola virus, 337
Ecdysozoans, 397, 397 *fig.*
ECG (electrocardiogram), 510, 510 *fig.*
Echidna, 454, 454 *fig.*
Echinarachnius parma, 426 *fig.*
Echinoderms (Echinodermata), 401 *table,* 426,
    426 *fig.,* 436–437
    nervous systems of, 587
Echolocation, 611, 611 *fig.*
ECM (extracellular matrix) of animal cells, 92, 92 *fig.*
Ecological footprint, 805, 805 *fig.*
Ecological isolation, 313 *table,* 314, 314 *fig.*
Ecological pyramids, 750, 750 *fig.*
Ecological succession, 741, 741 *fig.*
Ecology, 720–730. *See also* Communities;
    Ecosystems; Niches
    behavioral, 778, 778 *fig.*
    levels of ecological organization and, 720 *fig.,*
    720–721
Ecosystems, 17, 17 *fig.,* 720, 745–769. *See also*
    Communities; Niches
    conservation of, 808
    energy in, 746–750
    freshwater, 762 *fig.,* 762–763, 763 *fig.*
    land, 764–767
    materials cycles in, 751–755
    ocean, 760 *fig.,* 760–761, 761 *fig.*
    weather and, 756–759
Ectoderm, 396, 650, 651 *table*
    of vertebrates, 475 *table,* 478
Ectoparasites, 737
Ectoprocta, 401 *table*
Edema, 503
Edentata, 455 *table*
Edessa rufomarginata, 423 *fig.*
Edmondson, W. T., 811
Education, evolution and, 295
Edward VII, King of England, 209
Effectors, 548, 548 *fig.*
EGF (epidermal growth factor), 162
Egg(s). *See also* Ovum(a)
    formation in flowers, 702
    of reptiles, 446, 446 *fig.*
Egg-laying mammals, 454, 454 *fig.,* 643
Egret, cattle, 723, 723 *fig.*
Eichornia crassipes, 806
Ejaculation, 645, 645 *fig.*
EKG (electrocardiogram), 510, 510 *fig.*
Elasmobranchs, 616
Elastic connective tissue, 480, 481 *table*
Electrocardiogram (ECG or EKG), 510, 510 *fig.*
Electromagnetic spectrum, 122, 122 *fig.*
Electron(s), 34, 34 *fig.,* 35
    energy carried by, 35, 35 *fig.*
    harvest from chemical bonds, 140, 140 *fig.,* 142,
    143 *fig.*

Electronegativity, 39
Electron shells, 35, 35 *fig.*
Electron transport chain, 144–145
Electron transport system (ETS), 124, 126
Electroreceptors of vertebrates, 616, 617
Elephant, 433
Elevation, weather and, 757, 757 *fig.*
Embryonic development
    of animals, 395 *table,* 424–429
    mammalian, 650, 651 *table*
Embryonic stem cells, 272 *fig.,* 272–273, 273 *fig.,*
    274 *fig.*
    ethics of research using, 274
Embryo sacs of flowers, 700, 702
Emergent properties, 17, 720
Emerging viruses, 354
Emerson, R. A., 198
Emigration, 303
Encephalartos transvenosus, 672 *fig.*
Endangered species, preserving, 806–808,
    806–808 *fig.*
Endergonic reactions, 108, 108 *fig.*
Endler, John, 311
Endocrine disrupters, 713
Endocrine glands, 622, 622 *fig.. See also specific*
    glands and hormones
Endocrine system of vertebrates, 477
Endocytosis, 98, 98 *fig.,* 101 *table*
    receptor-mediated, 99, 99 *fig.,* 101 *table*
Endoderm, 396, 650, 651 *table*
    of cnidarians, 404
    of vertebrates, 475 *table,* 478
Endodermis of root, 686
Endomembrane system, eukaryotic, 78, 78 *fig.*
Endometrium, 647
Endomycorrhizae, 389
Endonucleases, restriction, 260–261, 261 *fig.*
Endoparasites, 737
Endoplasmic reticulum (ER), eukaryotic, 82, 88 *table*
Endoskeletons, 488, 488 *fig.*
    of echinoderms, 426
    human, 488–489, 489 *fig.*
Endosperm of seeds, 670 *fig.,* 670–671, 676, 676 *fig.*
Endospores, 343
Endosymbiosis, 85, 85 *fig.,* 333, 333 *fig.*
Endosymbiotic theory, 360–363, 361 *fig.*
Endothelium
    pseudostratified, 478, 479 *table*
    stratified, 478, 479 *table*
Endothermy of mammals, 453
Energy, 105–115
    absorption by chlorophyll, 121
    activation, 108, 108 *fig.*
    ATP and. *See* Adenosine triphosphate (ATP)
    carried by electrons, 35, 35 *fig.*
    cell's recycling of, 102
    chemical, conversion of light energy to. *See*
    Photosynthesis
    chemical reactions and, 106, 108, 108 *fig.*
    definition of, 106
    in ecosystems, 746–750
    entropy and, 107
    enzymes and, 109–111
    finding cleaner sources of, 809 *fig.,* 809–810
    flows of. *See* Energy flows
    food for, 530 *fig.,* 530–531, 531 *fig.*
    harvest from food, 135–150. *See also* Cellular
    respiration; Fermentation; Glycolysis;
    Krebs cycle
    heat, 107
    kinetic, 106, 106 *fig.*
    laws of thermodynamics and, 106, 107, 107 *fig.*
    light. *See* Light; Sunlight
    plants' capture from sunlight, 122 *fig.,* 122–123,
    123 *fig.*

    potential, 35, 106, 106 *fig.*
    renewable, 809, 809 *fig.*
Energy flows, 18, 19 *table*
    through ecosystems, 746–750
    in living things, 106, 106 *fig.*
Entropy, 107, 107 *fig.*
Environment
    adaptation challenges and, 721, 721 *fig.*
    effects on alleles, 197 *fig.,* 197–198, 199
    genetically modified crops and, 268, 278
    prokaryotes and, 345, 345 *fig.*
Environmental water cycle, 751
Enzyme(s), 51, 52 *fig.,* 53, 109–111
    actions of, 109 *fig.,* 109–110
    activating, 239
    allosteric, 111
    attachment to substrates, 114
    catalysis by, 114
    factors affecting activity of, 110, 110 *fig.*
    function of, 55 *fig.*
    regulation by cells, 111, 111 *fig.*
    restriction, 260–261, 261 *fig.*
Enzyme polymorphism, 316
Eosinophils, 505 *fig.*
Epidermal growth factor (EGF), 162
Epidermis, 564, 565 *fig.*
    of plants, 684
    of roots, 686, 687 *fig.*
Epididymis, 644 *fig.,* 645
Epiglottis, 535, 535 *fig.*
Epinephrine, 591, 632 *table*
Epistasis, 198, 198 *fig.,* 200, 200 *fig.*
Epithelial tissues, of vertebrates, 476, 476 *fig.,* 478,
    478 *fig.,* 479 *table*
Epithelium, simple, 478, 479 *table*
Epochs, 434, 434 *fig.*
Epperson, Susan, 295
Equilibrium, 94
    punctuated, 430
Equisetum, 664 *table*
Equisetum talmeteia, 668 *fig.*
ER (endoplasmic reticulum), eukaryotic, 82,
    88 *table*
Eras, 434, 434 *fig.*
Erosion in carbon cycle, 753
Erythrocytes, 52 *fig.,* 480, 481 *table,* 504–505,
    505 *fig.*
Erythropoietin, genetically engineered, 264 *table*
Escherichia coli, 244 *fig.,* 244–245, 349
    genetically engineered, 260 *fig.*
Esophagus, 543 *fig.*
    structure and function of, 536, 536 *fig.*
Essay on the Principle of Population (Malthus), 285
Essential amino acids, 531
Essential nutrients for plants, 709, 709 *fig.*
Estradiol, 648
Estrogen, 632 *table,* 648
Estrous cycle, 643
Estrus, 643
Estuaries, 760
Ethanol
    as energy source, 810
    fermentation producing, 146, 146 *fig.*
Ethics of stem cell research, 274
Ethology, 773
Ethylene, 711, 711 *table*
ETS (electron transport system), 124, 126
Euglena, 367 *table,* 368, 638, 638 *fig.*
Euglenoids, 366 *fig.,* 367 *table,* 368 *fig.,* 368–369,
    369 *fig.*
Euglenozoa, 366 *fig.,* 367 *table*
Eukaryotes (Eukarya), 28 *fig.,* 331 *table. See also*
    Protists (Protista)
    cells of. *See* Eukaryotic cells
    complex cell cycle of, 155

digestive system of, 533, 533 *fig.*
  environmental conservation and, 800–805
  global change and, 794–799
  recent evolution of, 468, 468 *fig.*
  social behavior of, 789, 789 *fig.*
  solving environmental problems and, 806–811
Human activity, mass extinction due to, 436
Human culture, diversity and, 789, 789 *fig.*
Human genome, 194, 257 *table,* 258–259
  geography of, 259 *fig.*
  noncoding DNA and, 259, 259 *fig.*
Human growth hormone (HGH), genetically
  engineered, 264 *fig.,* 264 *table*
Human herpesvirus 3 (HHV-3), 355 *table*
Human immunodeficiency virus. *See* AIDS; HIV
Human reproductive system, 644–649
  female, 646 *fig.,* 646–647, 647 *fig.*
  hormonal coordination of, 648 *fig.,* 648–649,
    649 *fig.*
  male, 644 *fig.,* 644–645, 645 *fig.*
  sexually transmitted diseases and, 658
Hummingbird, 560
Humoral immune response, 570, 572–573, 573 *fig.*
Hunter, John, 319
Huntington's disease, 211, 211 *fig.*
Huperzia lucidula, 668 *fig.*
Huxley, Thomas, 295
Hydra, 404, 404 *fig.,* 406
  circulatory system of, 498
  gastrovascular cavity of, 532, 532 *fig.*
Hydration shells, 42
Hydraulic skeletons, 488, 488 *fig.*
Hydrogen, atomic number and mass number of,
    34 *table*
Hydrogenated fats, 62
Hydrogen bonds, 40
  in water, 40, 41 *fig.,* 41 *table,* 41–42, 42 *fig.*
Hydrogen ions, 43, 53
  buffers and, 45, 45 *fig.*
  pH and, 43–45
Hydrolysis, 51, 51 *fig.*
Hydrophilic molecules, 42
Hydrophobic molecules, 42
Hydrothermal vent systems, 761
Hydroxide ions, 43
Hydroxyapatite, 480
Hydroxyl groups, 50 *fig.*
Hylobates, 461
Hypercholesterolemia, 64
Hypertonic solutions, 96–97, 97 *fig.*
Hyphae of fungi, 380, 380 *fig.,* 381
Hypogonadism, 63
Hypothalamohypophyseal portal system, 627
Hypothalamus, 596 *fig.,* 596–597, 597 *fig.,* 622,
    632 *table*
  evolution of, 594 *fig.,* 595
  pituitary gland and, 626 *fig.,* 626–627, 627 *fig.*
Hypotheses, 20, 22, 23, 25
  alternative, 23
  null, 301
  testing of. *See* Experiments
Hypotonic solutions, 96–97, 97 *fig.*

I

Ice, formation of, 41, 41 *fig.*
Icthyosaurs (Ichthyosauria), 446, 447 *table*
ID (intelligent design), 295, 338
  critical analysis of, 298–299
Ig(s) (immunoglobulins), 572
Iguana, temperature adaptation and, 721, 721 *fig.*
Iguanodon, 449
Ileum, 538
Illustrations, learning from, 7, 7 *fig.*
Immigration, 303
Immovable joints, 489

Immune connective tissue, 480, 481 *table*
Immune response, 569–582
  active immunity through clonal selection and,
    574 *fig.,* 574–575
  antigen specificity of, 582
  cellular, 570, 571, 571 *fig.*
  humoral, 570, 572–573, 573 *fig.*
  initiation of, 570, 570 *fig.*
  primary, 574 *fig.,* 574–575
  secondary, 574, 574 *fig.*
  specific immunity and, 564, 569, 569 *table*
  vaccination and, 576 *fig.,* 576–577, 577 *fig.*
Immune system
  defense as circulatory system function and,
    499
  HIV attack on, 580, 580 *fig.*
  of vertebrates, 477
Immunoglobulins (Igs), 572
Imperfect flowers, 701, 701 *fig.*
Imperfect fungi, 383 *table,* 387, 387 *fig.*
Imprinting, 775, 775 *fig.*
Inclusion-cell disease, 83
Incomplete dominance, 197, 197 *fig.*
Independent assortment, 174, 174 *fig.,* 180, 180 *fig.*
  law of, 193, 193 *fig.*
Independent variables, 10
Indexes, 132
Indigo bunting, 780
Individualistic concept of communities, 731
Induced ovulators, 643
Inductive reasoning, 20, 20 *fig.*
Industrial melanism, 308 *fig.,* 308–309, 309 *fig.*
Inferior vena cava, 508 *fig.,* 509
Inflammatory response, 568, 568 *fig.*
Influenza, 351, 354, 355 *table,* 576
Influenza vaccine, 576–577
Ingram, Vernon, 269
Ingroups, 324
Inhalation, 521
Inheritance. *See also* Gene(s); Genetics; Heredity
  chromosomal theory of, 28, 28 *fig.*
  molecular basis of, 26, 26 *fig.,* 27 *fig.*
  polygenic, 196, 196 *fig.*
Inhibition, succession and, 741
Inhibitory synapses, 591, 591 *fig.*
Initiation complexes, 247, 247 *fig.,* 248 *fig.*
Innate behavior, 773, 773 *fig.*
Innate releasing mechanism, 773
Inner cell mass, 650
Innocence Project, 227, 263
Inquiry & Analysis feature, 12
Insect(s) (Insecta), 422, 423 *fig. See also specific
    insects*
  eyes of, 612 *fig.,* 613
  flower color and, 675, 675 *fig.*
  Malpighian tubules of, 551, 551 *fig.*
  pollination by, 675, 675 *fig.,* 705, 705 *fig.*
  segmentation in, 418, 418 *fig.*
  societies of, 788
Insectivores (Insectivora), 455 *table,* 540, 541 *fig.*
Insertion of muscles, 490, 490 *fig.*
Insertions (in DNA), 229, 229 *table*
Instinct, interaction with learning, 776, 776 *fig.*
Insulin, 549, 549 *fig.,* 628, 629, 632 *table*
  genetically engineered production of, 260 *fig.,*
    264, 264 *table,* 345
Insulin receptor substrate, 629
Insulin therapy, 628
Integrase inhibitors, 581
Integrating center, 548, 548 *fig.*
Integration of inhibitory and excitatory effects, 591,
    591 *fig.*
Integrins, 92, 92 *fig.*
Integumentary system. *See also* Hair; Skin
  of vertebrates, 477

Intelligent design (ID), 295, 338
  critical analysis of, 298–299
Intercellular coordination, 380
Interferons, 567
  genetically engineered, 264 *table*
Interleukin(s)
  genetically engineered, 264 *table*
  interleukin-1, 570
  interleukin-2, 571, 571 *fig.*
Intermediate filaments, 86–87
Intermembrane space, 144–145
Internal fertilization, 640
Interneurons, 487, 487 *table,* 586, 586 *fig.*
Interoceptors, 606
Interphase, 155, 158, 158 *fig.*
  microtubule formation in, 182
Intersexual selection, 782
Interspecific competition, 732
Interstitial fluid, 499
Intertidal region, 760, 760 *fig.*
Intestines
  large, 538, 543 *fig.*
  small, 538, 539 *fig.,* 543 *fig.*
Intrasexual selection, 782
Intraspecific competition, 732
Intrauterine devices (IUDs), 657
Introns, 241–242, 242 *fig.,* 259 *fig.,* 261
Invertebrates, 395 *table*
  genomes of, 257 *table*
  nervous systems of, 586–587
Involuntary nervous system, 600, 601, 601 *fig.*
Ion(s), 36, 36 *fig.*
  in blood plasma, 504
Ionic bonds, 38, 412
Ionization, 43, 43 *fig.*
I.Q. scores, environmental effects on, 199
Iris, 613, 613 *fig.*
Irish elk, 439 *table*
Irish kings, tracing DNA of, 228
Iron
  atomic number and mass number of, 34 *table*
  phytoplankton growth and, 132
Islets of Langerhans, 542, 542 *fig.*
Isosmotic solutions, 97
Isotonic solutions, 96–97, 97 *fig.*
Isotopes, 36 *fig.,* 36–37
  half-life of, 37
  radioactive, dating of fossils using, 37, 37 *fig.,* 46
  radioactive, medical uses of, 36, 37 *fig.*
IUDs (intrauterine devices), 657
Ivy, 691

J

Janzen, Daniel, 329
Java man, 465, 465 *fig.*
Jaws
  evolution in fishes, 441, 441 *fig.*
  of theropods, 449
Jeffreys, Alec, 227, 269
Jejunum, 538
Jellyfish, 404, 404 *fig.,* 406
Jenner, Edward, 265, 576, 576 *fig.*
Joint(s), human, 489, 489 *fig.,* 490
Jointed appendages of arthropods, 418–422
Jones, Marion, 63
Jumping spider, 420
Jurassic period, 434 *fig.,* 436, 437 *fig.,* 448

K

Kalanchoâ daigremontiana, 700
Kangaroo, 454 *fig.,* 462 *fig.*
Kangaroo rats, 19 *table,* 30, 555, 555 *fig.*
Karyokinesis. *See* Mitosis
Karyotype, 156 *fig.,* 156–157
kcals (kilocalories), 494

van Leeuwenhoek, Anton, 26, 72
Van Valen, Lee, 456
Vaporization, heat of, of water, 41
Variables, 10
Variance, 199
Varicella-zoster virus, 355 *table*
Variola virus, 355 *table*
Vascular cambium, 682, 688–689, 689 *fig.*
Vascular plants, 667–673, 668 *fig.,* 668–669, 669 *fig.*
  carbohydrate transport in, 695
  dermal tissue of, 683, 684
  evolution of vascular tissue and, 667, 667 *fig.*
  ground tissue of, 683
  leaves of, 682, 682 *fig.,* 690 *fig.,* 690–691, 691 *fig.*
  organization of, 682, 682 *fig.*
  roots of, 682, 682 *fig.,* 686–687, 687 *fig.,* 692, 694, 694 *fig.*
  seed, 670 *fig.,* 670–671. *See also* Angiosperms; Gymnosperms; Seed(s)
  seedless, 668 *fig.,* 668–669, 669 *fig.*
  stems of, 682, 682 *fig.,* 688 *fig.,* 688–689, 689 *fig.,* 692–694
  vascular tissue of. *See* Vascular tissue of plants
  water movement in, 692–694, 692–694 *fig.,* 696
Vascular system of echinoderms, 426
Vascular tissue of plants, 664, 665, 683, 685
  evolution of, 667, 667 *fig.*
Vas deferens, 644 *fig.,* 645
Vasopressin, 626, 633 *table*
Vectors for gene therapy, 276–277
Vegetative reproduction, 700
Veins, 498, 501 *fig.,* 502, 502 *fig.*
Velociraptor, 449, 451
Vena cava, 508 *fig.,* 509
Venter, Craig, 269
Ventral, definition of, 407, 407 *fig.*
Ventricles
  of fish heart, 506, 506 *fig.*
  left, of mammalian and bird heart, 508, 508 *fig.*
Vertebral column, 428
  of fishes, 440
  human, 489, 489 *fig.*
  of theropods, 449
Vertebrates, 395 *table,* 428 *fig.,* 428–429, 433–457. *See also specific animals*
  aquatic, respiratory system of, 517, 517 *fig.*
  digestive systems of, 533 *fig.,* 533–541
  electroreceptors of, 616
  evolution of, 434–439
  evolution of brain in, 594 *fig.,* 594–595, 595 *fig.*
  eyes of, 612 *fig.,* 613
  family tree of, 440 *fig.*
  genomes of, 257 *table*
  heat sense of, 616, 616 *fig.*
  invasion of land by, 435
  kidneys of, evolution of, 552–553, 555, 555 *fig.*
  magnetic receptors in, 616, 618
  nervous system of, 477, 585, 600, 600 *fig.*
  organs of, 476
  organ systems of, 477, 477 *fig.*
  osmoregulation by, 554 *fig.,* 555
  reproduction in, evolution of, 640 *fig.,* 640–643, 642 *fig.*
  societies of, 788
  teeth of, 534 *fig.,* 534–535, 535 *fig.*
  terrestrial, respiratory system of, 518 *fig.,* 518–519, 519 *fig.*
  tissues of, 476, 476 *fig.,* 478–487

Vesicles, 82
  eukaryotic, 78, 78 *fig.*
Vessel elements, 685
Vestigial organs, 292
Vibrio cholerae, 347 *table*
  conversion of, 350
Viceroy butterfly, 739, 739 *fig.*
Victoria, Queen of England, 209, 209 *fig.*
Villus(i), small intestinal, 538, 539 *fig.*
Viral diseases, 354, 355 *table. See also specific viruses and diseases*
Viruses, 337, 348–356
  animal, entry into cells, 352 *fig.,* 352–353, 353 *fig.*
  bacteriophage entry into prokaryotic cells and, 349 *fig.,* 349–350, 350 *fig.*
  emerging, 354
  replication by, 352–353
  structure of, 348, 348 *fig.*
Visible light, 122, 122 *fig.*
Vision, 612–615
  binocular, 460, 615, 615 *fig.*
  color, 614 *fig.,* 614–615
  transmission of light information to brain and, 615, 615 *fig.*
Vital capacity, 521
Vitamin(s), 531
Vitamin K, synthesis of, 540
Viviparity, 640
Voltage-gated channels, 589
Voluntary nervous system, 600 *fig.,* 600–601
Volvox, 365, 365 *fig.,* 372
von Frisch, Karl, 773, 784
von Humboldt, Alexander, 752
Vorticella, 364, 390 *fig.*

W

Waggle dance, 784 *fig.,* 784–785
Wallace, Alfred Russel, 286, 467
Wallace Line, 467
Warney, Douglas, 263
Washington, Lake, cleanup of, 811, 811 *fig.*
Wastes
  in blood plasma, 504
  nitrogenous, elimination of, 559, 559 *fig.*
Water
  absorption by roots, 694, 694 *fig.*
  groundwater, 752, 801
  hydrogen bonds in, 40, 41 *fig.,* 41 *table,* 41–42, 42 *fig.*
  ionization of, 43, 43 *fig.*
  movement across plasma membrane, 76
  movement in vascular plants, 692–694, 692–694 *fig.,* 696
  plant conservation of, 662 *fig.,* 662–663
  properties of, 41 *table,* 41–42
Water cycle, 751 *fig.,* 751–752, 752 *fig.*
Water ferns, 664 *table*
Water molds, 367 *table,* 369
Water pollution, 794
Water supply, adaptations and, 721, 721 *fig.*
Water vascular system of echinoderms, 426
Watson, James, 26, 59, 220, 221 *fig.,* 269
Wax as food source, 540
Weather, 756–759
  latitude and elevation and, 757, 757 *fig.*
  sun and atmospheric circulation and, 756, 756 *fig.*
Weight, 34
Weight loss, 149

Welwitschia, 665 *table*
Welwitschia mirabilis, 672 *fig.,* 673
Wenner, Adrian, 784
Went, Frits, 712
West Nile virus, 354
Whales, evolution of, 291
Wheat, 665 *table*
  genetically modified, 268 *table*
Wheel animals (Rotifera), 401 *table,* 412, 412 *fig.*
Whisk fern, 664 *table,* 668 *fig.*
White blood cells, 52 *fig.,* 90, 505, 505 *fig.*
  HIV infection of, 356
White Cliffs of Dover, 374, 374 *fig.*
Whorls
  of flowers, 674, 674 *fig.*
  leaf arrangement in, 691
Wildebeests, 781 *fig.*
Wild lettuce, temporal isolation of, 314
Wilkins, Maurice, 59, 269
Willow oak, 320, 320 *fig.*
Wilmut, Ian, 269, 271
Wilson, E. O., 789
Wind
  pollination by, 675
  seed dispersal by, 707, 707 *fig.*
Winder, Ernst, 230
Wine making, 147
Wolf, temperature adaptation and, 721, 721 *fig.*
Wolves as carnivores, 747
Wood, 667, 689, 689 *fig.*
Woolly hair, 214
Worms. *See also* Earthworm; Flatworms (Platyhelminthes)
  round. *See* Nematodes (Nematoda); Roundworms
  segmented, 400 *table,* 416 *fig.,* 416–417, 550, 550 *fig.*
  solid, 407 *fig.,* 407–411, 408 *fig.*

X

Xanthium strumarium, 707 *fig.*
x axis, 66
X chromosome
  chromosomal theory of inheritance and, 202–203, 203 *fig.*
  nondisjunction involving, 206, 206 *fig.*
Xylem, 667, 683, 685, 687 *fig.,* 688, 688 *fig.*
  water transport in, 696

Y

Yamagiwa, Katsusaburo, 230
Yamanaka, Shinya, 269
Y chromosome, 638–639, 639 *fig.*
  chromosomal theory of inheritance and, 202–203, 203 *fig.*
  nondisjunction involving, 206
Yeasts, 147, 383 *table,* 385, 385 *fig.,* 387, 387 *fig.*
Yellow fever, 355 *table*
Yersinia pestis, 347 *table*

Z

Zamecnik, Paul, 252
Zea mays. See Corn
Zebra as herbivore, 747
Zone diet, 149
Zone of differentiation of root, 687, 687 *fig.*
Zone of elongation of root, 687, 687 *fig.*
Zygomycetes, 383 *table,* 384, 384 *fig.*
Zygosporangium, 384, 384 *fig.*
Zygote, 172, 638
Zygotic meiosis, 363, 363 *fig.,* 365